PRENTICE HALL
mygeologyplace ™
Where geology comes to you.

Registration Instructions for MyGeologyPlace

1. Go to *www.prenhall.com/tarbuck*

2. Click the cover for *Earth: An Introduction to Physical Geology*, Ninth Edition, by Tarbuck/Lutgens.

3. Cllck "Register."

4. Using a coin (not a knife) scratch off the metallic coating below to reveal your Access Code.

5. Complete the online registration form, choosing your own personal Login Name and Password.

6. Enter your pre-assigned Access Code exactly as it appears below.

7. Complete the online registration form by entering your School Location information.

8. After your personal Login Name and Password are confirmed by e-mail, go back to *www.prenhall.com/tarbuck*, click your book's cover, enter your new Login Name and Password, and click "Log In".

Your Access Code is:

If there is no metallic coating covering the access code above, the code may no longer be valid and you will need to purchase online access using a major credit card to use the Website. To do so, go to *www.prenhall.com/tarbuck*, click the cover for Tarbuck/Lutgens, *Earth: An Introduction to Physical Geology,* Ninth Edition, then click the "Get Access" button, and follow the instructions under "Students".

Important: Please read the Subscription and End-User License Agreement located on the "Log In" screen before using the Tarbuck/Lutgens, *Earth: An Introduction to Physical Geology,* Ninth Edition MyGeologyPlace. By using the Website, you indicate that you have read, understood, and accepted the terms of the agreement.

Minimum System Requirements

PC Operating Systems:

Windows 2000, or Windows XP

Pentium II 233 MHz processor. 64 MB RAM In addition to the minimum memory required by your OS.

Internet Explorer 6.0 or Internet Explorer 7.0

Macintosh Operating Systems:

Macintosh Power PC with OS X (10.4)

In addition to the RAM required by your OS, this application requires 64 MB RAM, with 40 MB Free RAM, with Virtual Memory enabled

Safari 2.0

Macromedia Shockwave (TM)

8.50 release 326 plugin

Macromedia Flash Player 6.0.79 & 7.0

Acrobat Reader 6.0.1

800 x 600 pixel screen resolution

Technical Support Call 1-800-677-6337.
Phone support is available Monday–Friday, 8am to 8pm and Sunday 5pm to 12am, Eastern time.
Visit our support site at *http://247.prenhall.com*.
E-mail support is available 24/7.

GEODe: Earth CD-ROM
Edward J. Tarbuck and Frederick K. Lutgens
Illustrated by Dennis Tasa
ISBN-13: 978-0-13-156692-7 ISBN-10: 0-13-156692-X
CD License Agreement
© 2005 Pearson Education, Inc.
Pearson Prentice Hall
Pearson Education, Inc.
Upper Saddle River, NJ 07458

YOU SHOULD CAREFULLY READ THE TERMS AND CONDITIONS BEFORE USING THE CD-ROM PACKAGE. USING THIS CD-ROM PACKAGE INDICATES YOUR ACCEPTANCE OF THESE TERMS AND CONDITIONS.

Pearson Education, Inc. provides this program and licenses its use. You assume responsibility for the selection of the program to achieve your intended results, and for the installation, use, and results obtained from the program. This license extends only to use of the program in the United States or countries in which the program is marketed by authorized distributors.

LICENSE GRANT

You hereby accept a nonexclusive, nontransferable, permanent license to install and use the program ON A SINGLE COMPUTER at any given time. You may copy the program solely for backup or archival purposes in support of your use of the program on the single computer. You may not modify, translate, disassemble, decompile, or reverse engineer the program, in whole or in part.

TERM

The License is effective until terminated. Pearson Education, Inc. reserves the right to terminate this License automatically if any provision of the License is violated. You may terminate the License at any time. To terminate this License, you must return the program, including documentation, along with a written warranty stating that all copies in your possession have been returned or destroyed.

LIMITED WARRANTY

THE PROGRAM IS PROVIDED "AS IS" WITHOUT WARRANTY OF ANY KIND, EITHER EXPRESSED OR IMPLIED, INCLUDING, BUT NOT LIMITED TO, THE IMPLIED WARRANTIES OR MERCHANTABILITY AND FITNESS FOR A PARTICULAR PURPOSE. THE ENTIRE RISK AS TO THE QUALITY AND PERFORMANCE OF THE PROGRAM IS WITH YOU. SHOULD THE PROGRAM PROVE DEFECTIVE, YOU (AND NOT PEARSON EDUCATION, INC. OR ANY AUTHORIZED DEALER) ASSUME THE ENTIRE COST OF ALL NECESSARY SERVICING, REPAIR, OR CORRECTION. NO ORAL OR WRITTEN INFORMATION OR ADVICE GIVEN BY PEARSON EDUCATION, INC., ITS DEALERS, DISTRIBUTORS, OR AGENTS SHALL CREATE A WARRANTY OR INCREASE THE SCOPE OF THIS WARRANTY.

SOME STATES DO NOT ALLOW THE EXCLUSION OF IMPLIED WARRANTIES, SO THE ABOVE EXCLUSION MAY NOT APPLY TO YOU. THIS WARRANTY GIVES YOU SPECIFIC LEGAL RIGHTS AND YOU MAY ALSO HAVE OTHER LEGAL RIGHTS THAT VARY FROM STATE TO STATE.

Pearson Education, Inc. does not warrant that the functions contained in the program will meet your requirements or that the operation of the program will be uninterrupted or error-free.

However, Pearson Education, Inc. warrants the CD-ROM(s) on which the program is furnished to be free from defects in material and workmanship under normal use for a period of ninety (90) days from the date of delivery to you as evidenced by a copy of your receipt.

The program should not be relied on as the sole basis to solve a problem whose incorrect solution could result in injury to person or property. If the program is employed in such a manner, it is at the user's own risk and Pearson Education, Inc. explicitly disclaims all liability for such misuse.

LIMITATION OF REMEDIES

Pearson Education, Inc.'s entire liability and your exclusive remedy shall be: 1. the replacement of any CD-ROM not meeting Pearson Education, Inc.'s "LIMITED WARRANTY" and that is returned to Pearson Education, or 2. if Pearson Education is unable to deliver a replacement CD-ROM that is free of defects in materials or workmanship, you may terminate this agreement by returning the program.

IN NO EVENT WILL PEARSON EDUCATION, INC. BE LIABLE TO YOU FOR ANY DAMAGES, INCLUDING ANY LOST PROFITS, LOST SAVINGS, OR OTHER INCIDENTAL OR CONSEQUENTIAL DAMAGES ARISING OUT OF THE USE OR INABILITY TO USE SUCH PROGRAM EVEN IF PEARSON EDUCATION, INC. OR AN AUTHORIZED DISTRIBUTOR HAS BEEN ADVISED OF THE POSSIBILITY OF SUCH DAMAGES, OR FOR ANY CLAIM BY ANY OTHER PARTY.

SOME STATES DO NOT ALLOW FOR THE LIMITATION OR EXCLUSION OF LIABILITY FOR INCIDENTAL OR CONSEQUENTIAL DAMAGES, SO THE ABOVE LIMITATION OR EXCLUSION MAY NOT APPLY TO YOU.

GENERAL

You may not sublicense, assign, or transfer the license of the program. Any attempt to sublicense, assign or transfer any of the rights, duties, or obligations hereunder is void.

This Agreement will be governed by the laws of the State of New York.

Should you have any questions concerning this Agreement, you may contact Pearson Education, Inc. by writing to:

ESM Media Development

Higher Education Division

Pearson Education, Inc.

1 Lake Street

Upper Saddle River, NJ 07458

Should you have any questions concerning technical support, you may write to:
New Media Production
Higher Education Division
Pearson Education, Inc.
1 Lake Street
Upper Saddle River, NJ 07458

YOU ACKNOWLEDGE THAT YOU HAVE READ THIS AGREEMENT, UNDERSTAND IT, AND AGREE TO BE BOUND BY ITS TERMS AND CONDITIONS. YOU FURTHER AGREE THAT IT IS THE COMPLETE AND EXCLUSIVE STATEMENT OF THE AGREEMENT BETWEEN US THAT SUPERSEDES ANY PROPOSAL OR PRIOR AGREEMENT, ORAL OR WRITTEN, AND ANY OTHER COMMUNICATIONS BETWEEN US RELATING TO THE SUBJECT MATTER OF THIS AGREEMENT.

Installation & Use Instructions

Windows

GEODe: Earth has no installation or setup program.

To start the GEODe: Earth program simply insert "GEODe Earth" CD-ROM into your CD-ROM drive. If you have tray notification enabled, the program will automatically start.

If your display is larger than 800 x 600 pixels, there will be a black border around the 800 x 600 display window.

Note for NT users: If you cannot hear the narration audio, verify that you have installed the latest version of your computer's sound drivers.

Mac

GEODe: Earth has no installation or setup program.

To start the GEODe: Earth program simply insert "GEODe Earth" CD-ROM into your CD-ROM drive and double-click the icon for your operating system of choice ("GEODe Earth for OS X" or "GEODe Earth for OS 8.6 to 9.2").

If your display is larger than 800 x 600 pixels, there will be a black border around the 800 x 600 display window.

System Requirements:

Windows

Operating System: Windows 98, ME, 2000, NT version 4 or later, XP

Processor: In addition to the processor required by your OS, this application requires at least a Pentium II 200 MHz processor.

RAM: In addition to the RAM required by your OS, this application requires at least an additional 32 Meg RAM.

Required hardware: CD drive, speakers, and a printer is required to print student scores.

Monitor resolution: 800 x 600 pixels; millions of colors

Required third-party software: QuickTime(TM) (version 5.0.2 or later)

A newer version of QuickTime may be available from the QuickTime website: http://www.apple.com/quicktime/download

This CD-ROM is intended for standalone use only. It is not for use on a network.

Mac

Operating System: MAC OS X (10.1 or later) and MAC OS 8.6 to 9.2

Processor: In addition to the processor required by your OS, this application requires at least a PowerPC running at 180 Mhz (a G3 is recommended for OS 8.6 to 9.2 and a G3 is required for OS X).

RAM: In addition to the RAM required by your OS, this application requires at least an additional 32 Meg RAM (128 Megs RAM for OS X).

Required hardware: CD drive, speakers, and a printer is required to print student scores.

Monitor resolution: 800 x 600 pixels; millions of colors

Required third-party software: QuickTime™ (version 5.0.2 or later)

A newer version of QuickTime may be available from the QuickTime website: http://www.apple.com/quicktime/download

This CD-ROM is intended for standalone use only. It is not for use on a network

Technical Support

If you are having problems with this software, call (800) 677-6337 between 8:00 a.m. and 5:00 p.m. CST, Monday through Friday. You can also get support by filling out the web form located at:

http://247.prenhall.com/mediaform

Our technical staff will need to know certain things about your system in order to help us solve your problems more quickly and efficiently. If possible, please be at your computer when you call for support. You should have the following information ready:

- Textbook ISBN
- CD-ROM ISBN
- corresponding product and title
- computer make and model
- Operating System (Windows or Macintosh) and Version
- RAM available
- hard disk space available
- sound card? Yes or No
- printer make and model
- network connection
- detailed description of the problem, including the exact wording of any error messages.

NOTE: Pearson does not support and/or assist with the following:

- third-party software (i.e. Apple's QuickTime)
- homework assistance
- Textbooks and CD-ROM's purchased used are not supported and are non-replaceable.

To purchase a new CD-ROM contact Pearson Individual Order Copies at 1-800-282-0693.

EARTH

NINTH EDITION

EARTH

An Introduction to Physical Geology

Edward J. Tarbuck

Frederick K. Lutgens

Illustrated by
Dennis Tasa

PEARSON

Prentice
Hall

Upper Saddle River, NJ 07458

Library of Congress Cataloging-in-Publication Data

Tarbuck, Edward J.
  Earth : an introduction to physical geology / Edward J. Tarbuck, Frederick K. Lutgens ;
illustrated by Dennis Tasa.—9th ed.
    p. cm.
    Includes index.
    ISBN 0-13-156684-9
    1. Physical geology—Textbooks.   I. Lutgens, Frederick K. II Title.
QE28.2.T37 2008
551—dc22

Publisher, Geosciences: *Daniel Kaveney*
Editor in Chief, Science: *Nichole Folchetti*
Executive Managing Editor: *Kathleen Schiaparelli*
Production Editor: *Edward Thomas*
Media Production Editor: *Richard Barnes*
Media Editor: *Andrew Sobel*
Assistant Editor: *Sean Hale*
Editorial Assistant: *John DeSantis*
Marketing Director, Science: *Patrick Lynch*
Marketing Manager: *Amy Porubsky*
Marketing Assistant: *Jessica Muraviov*
Manufacturing Manager: *Alexis Heydt-Long*
Manufacturing Buyer: *Alan Fischer*
Director of Creative Services: *Paul Belfanti*
Creative Director: *Juan Lopez*
Art Director: *Maureen Eide*
AV Production Editor: *Daniel Missildine*
Interior and Cover Design: *Suzanne Behnke*
Senior Managing Editor, Art Production and Management: *Patty Burns*

Manager, Production Technologies: *Matthew Haas*
Managing Editor, Art Management: *Abigail Bass*
Art Production Editors: *Dan Missildine, Eric Day*
Director, Image Resource Center: *Melinda Reo*
Manager, Rights and Permissions: *Zina Arabia*
Interior Image Specialist: *Beth Boyd-Brenzel*
Cover Image Specialist: *Karen Sanatar*
Photo Researcher: *Yvonne Gerin*
Copy Editor: *Marcia Youngman*
Proofreader: *Alison Lorber*
Image Permission Coordinator: *Debbie Hewitson*
Color Scanning Supervisor: *Joseph Conti*
Production Assistant: *Nancy Bauer*
Composition: *Pine Tree Composition*
Cover Photo: *Hiker in Antelope Canyon, near Paige, Arizona
    (Galen Rowell/Mountain Light)*
Title Page Photo: *Delicate Arch in Arches National Park, Utah
    (David Muench/Muench Photography, Inc.)*

© 2008, 2005, 2002, 1999, 1996 Pearson Education, Inc.
Pearson Prentice Hall
Pearson Education, Inc.
Upper Saddle River, NJ 07458

Earlier editions © 1993 by Macmillan Publishing Company;
© 1990, 1987, 1984 by Merrill Publishing Company.

Pearson Prentice Hall™ is a trademark of Pearson Education, Inc.

Printed in the United States of America

10 9 8 7 6 5 4 3 2 1

ISBN 0-13-156684-9

Pearson Education Ltd., *London*
Pearson Education Australia Pty., Limited, *Sydney*
Pearson Education *Singapore,* Pte. Ltd
Pearson Education North Asia Ltd., *Hong Kong*
Pearson Education Canada, Ltd., *Toronto*
Pearson Educación de Mexico, S.A. de C.V.
Pearson Education—Japan, *Tokyo*
Pearson Education Malaysia, Pte. Ltd.

To our wives, Joanne and Nancy,
for their support and patience.

Brief Contents

GEODe: EARTH Contents

Contents

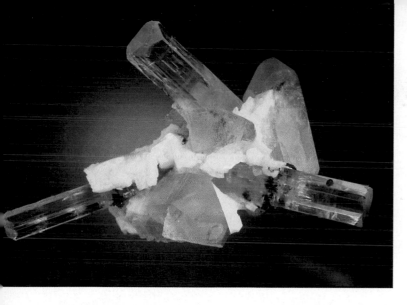

CHAPTER 6
Weathering and Soil   165

CHAPTER 7
Sedimentary Rocks   193

CHAPTER 8
Metamorphism and Metamorphic
Rocks   221

CHAPTER 12
Earth's Interior   325

CHAPTER 13
Divergent Boundaries: Origin and
Evolution of the Ocean Floor   349

CHAPTER 14
Convergent Boundaries: Origin
of Mountains   377

CHAPTER 21
Global Climate Change   567

CHAPTER 22
Earth's Evolution through Geologic Time   595

Preface

Earth is a very small part of a vast universe, but it is our home. It provides the resources that support our modern society and the ingredients necessary to maintain life. Therefore, knowledge of our planet is critical to our well-being and, indeed, vital to our survival. The science of geology contributes greatly to our understanding of planet Earth.

Media reports frequently remind us of the geological forces at work on our planet. News stories graphically portray the violent force of a volcanic eruption, the devastation generated by a strong earthquake, and the large numbers left homeless by landslides and flooding. Such events, and many others as well, are destructive to life and property, and we must learn to deal with them. Moreover, we also face many basic environmental issues that have a significant geological component. Examples include groundwater contamination, soil erosion, and the many impacts of extracting energy and mineral resources. To understand these issues and try to find solutions to the problems related to them requires an awareness of how science is done and the scientific principles that influence our planet, its rocks, mountains, atmosphere, and oceans.

The ninth edition of *Earth: An Introduction to Physical Geology*, like its predecessors, is a college-level text that is intended to be a meaningful non-technical survey for students taking their first course in geology. In addition to being informative and up-to-date, a major goal of *Earth* is to meet the need of students for a readable and user-friendly text, a book that is a highly usable "tool" for learning the basic principles and concepts of geology.

Basic Organization

Earth is organized to reflect the unifying role that the theory of plate tectonics plays in our understanding of planet Earth. Since the late 1960s scientists have come to realize that Earth's outer shell is broken into segments called *plates*. Driven by heat from Earth's interior, these huge slabs gradually move in relation to one another. Where landmasses split apart, new ocean basins are created between diverging continents. Meanwhile, older portions of the seafloor plunge back into Earth's interior. These movements generate earthquakes, cause volcanoes to form, and lead to the creation of Earth's major mountain belts.

Chapter 1 presents an introduction to the science of geology, followed by a look at the nature of scientific inquiry, and a discussion of the birth and early evolution of planet Earth. Next, Chapter 2 traces the historical development of the theory of plate tectonics as a way of providing insight into how science and scientists work. This is immediately followed by an overview of the plate tectonics theory. Developing a basic understanding of this model of how Earth

works will aid students as they explore many of the phenomena discussed in the chapters that follow.

With the basic framework of plate tectonics firmly established, we turn to discussions of Earth materials and the related processes of volcanism, metamorphism, and weathering. Along the way, students will clearly see the relationships between these phenomena and the theory of plate tectonics. Next, the fundamental concepts of geologic time are presented in detail. This is followed by an examination of earthquakes, Earth's internal structure, and the processes that deform rocks.

We revisit plate tectonics again in Chapters 13 and 14. These chapters expand on earlier discussions as we consider the nature of Earth's ocean basins and mountains. Chapter 13 explores the origin and structure of the ocean floor. Students are asked to examine how the seafloor is generated, why it is continually being destroyed, and what clues it can provide about events that occurred earlier in Earth's history. Then Chapter 14 considers the role of plate tectonics in the formation of Earth's major mountain belts. After this examination of Earth's large-scale features, we examine the geological work of gravity, water, wind, and ice. It is these processes that modify and sculpt Earth's surface, creating many of its varied landforms.

Following the multichapter examination of Earth's major surface processes in Chapters 15 through 20, two new chapters focus on global climate change and Earth history. These new additions are described in more detail in the next section of the preface. Finally, *Earth, Ninth Edition* concludes with a chapter on energy and mineral resources followed by a significantly revised chapter on planetary geology.

As in previous editions of this text, we have designed each chapter as a self-contained unit so that material may be taught in a different sequence according to the preference of the instructor or the dictates of the laboratory. Thus, the instructor who wishes to discuss erosional processes prior to earthquakes, plate tectonics, and mountain building may do so without difficulty.

Two New Chapters

Two all-new chapters have been added—Chapter 21, "Global Climate Change" and Chapter 22, "Earth's Evolution through Geologic Time." This was done in response to *extensive* pre-revision reviews and surveys that indicated a strong desire among professors to see these topics added. The intent is not to expand the amount of material presented in introductory geology, but to allow for greater instructor flexibility and variety when determining course content.

Chapter 21, "Global Climate Change" provides an excellent opportunity to explore many interrelationships in the Earth system. The crucial questions of climate change are those of process and response, and cause and effect. What factors cause Earth's climate to change? How does the Earth system respond and on what time scales? This chapter presents an overview that examines both natural variability and climate change related to human activities. Today, climate change is more than just a subject that is of "academic interest." Rather it is a topic that is making headlines. This chapter presents an up-to-date look at a major global environmental issue.

Chapter 22, "Earth's Evolution through Geologic Time" presents a clear, concise summary of Earth history beginning with an engaging introduction entitled "Is Earth Unique?" The chapter includes easy-to-follow discussions on the birth and early evolution of the planet and on the origin of continents, the atmosphere, and oceans. To allow for maximum instructor flexibility, there are separate discussions of Earth's physical history and the evolution of life through geologic time.

Distinguishing Features

Readability

The language of this book is straightforward and written to be understood. Clear, readable discussions with a minimum of technical language are the rule. The frequent headings and subheadings help students follow discussions and identify the important ideas presented in each chapter. In the ninth edition, improved readability was achieved by examining chapter organization and flow, and writing in a more personal style. Large portions of the text were substantially rewritten in an effort to make the material more understandable.

Illustrations and Photographs

Geology is highly visual. Therefore, photographs and artwork are a critical part of an introductory book. *Earth, Ninth Edition,* contains dozens of new high-quality photographs that were carefully selected to aid understanding, add realism, and heighten the interest of the reader.

There has been substantial revision and improvement of the art program. Clearer, easier-to-understand line drawings show greater color and shading contrasts. More figures combine the use of diagrams and/or maps and photos together. Moreover, many new art pieces have additional labels that "narrate" the process being illustrated. The result is an art program that illustrates ideas and concepts more clearly than ever before. As in the eight previous editions, we are grateful to Dennis Tasa, a gifted artist and respected geological illustrator, for his outstanding work.

Focus on Learning

When a chapter has been completed, several useful devices help students review. First, the Summary recaps all of the major points. This is followed by Review Questions that help students examine their knowledge of significant facts and ideas. Next is a checklist of Key Terms with page references. Learning the language of geology helps students learn the material. This is followed by a reminder to visit MyGeologyPlace, the online study guide for *Earth, Ninth Edition* (**http://www.prenhall.com/tarbuck**). It contains many excellent opportunities for review and exploration. Finally, each chapter closes with two frames from the *GEODe: Earth* CD-ROM to remind the student about this unique and effective learning aid.

Earth as a System

An important occurrence in modern science has been the realization that Earth is a giant multidimensional system. Our planet consists of many separate but interacting parts. A change in any one part can produce changes in any or all of the other parts—often in ways that are neither obvious nor immediately apparent. Although it is not possible to study the entire system at once, it is possible to develop an awareness and appreciation for the concept and for many of the system's important interrelationships. Therefore, beginning with an expanded discussion in Chapter One, the theme of "Earth as a System" recurs at appropriate places throughout the book. It is a thread that "weaves" through the chapters and helps tie them together. Several new and revised special-interest boxes relate to "Earth as a System."

People and the Environment

Because knowledge about our planet and how it works is necessary to our survival and well-being, the treatment of environmental issues has always been an important part of *Earth.* Such discussions serve to illustrate the relevance and application of geological knowledge. With each new edition, this focus has been given greater emphasis. This is certainly the case with the ninth edition. The text integrates a great deal of information about the relationship between people and the natural environment and explores the application of geology to understanding and solving problems that arise from these interactions. In addition to many basic text discussions, more than 20 of the text's special-interest boxes involve the "People and the Environment" theme.

Understanding Earth

As members of a modern society, we are constantly reminded of the benefits derived from science. But what exactly is the nature of scientific inquiry? Developing an understanding of how science is done and how scientists work is another important theme that appears throughout this book, beginning with the section on "The Nature of Scientific Inquiry" in Chapter 1. Students will examine some of the difficulties encountered by scientists as they attempt to acquire reliable data about our planet and some of the ingenious methods that have been developed to overcome these difficulties. Students will also explore many examples of how hypotheses are formulated and tested as well as learn about the evolution and development of some major scientific theories. Many basic text discussions as well as a number of the special-interest boxes on "Understanding Earth" provide the reader with a sense of the observational techniques and reasoning processes involved in developing scientific knowledge. The emphasis is not just on what scientists know, but how they figured it out.

More About the Ninth Edition

The ninth edition of *Earth* represents a thorough revision. Every part of the book was examined carefully with the dual goals of keeping topics current and improving the clarity of text discussions. In addition to the two new chapters that have already been described, those familiar with previous editions of *Earth* will find many other changes. Here are some examples:

- Chapter 12 "Earth's Interior" has been *completely* revised and rewritten by Michael Wysession, Earth and Planetary Sciences Department, Washington University. Professor Wysession is not only an expert on geophysics and Earth's inner structure, but he is also an accomplished geoscience educator. Basic coverage has been expanded to include heat flow and temperature distribution in Earth's interior, Earth's gravity, the use of seismic tomography to study Earth's interior, and a more in-depth treatment of Earth's magnetic field. The chapter not only has broader, more up-to-date coverage, but explanations of these sometimes complex ideas are written in a more conversational, easier to understand style. The revised chapter is supported with a strong new art program and many new images.

- Chapter 24, "Planetary Geology," has received a major revision with the assistance of professors Teresa Tarbuck and Mark Watry of Rocky Mountain College. There are new discussions on solar system evolution, lunar history, and the dynamic geologic evolution of Mars. Also included is a new section on dwarf planets (Pluto now has this status) and the latest on the Kuiper Belt and Oort cloud.

- Chapter 5, "Volcanoes and Other Igneous Activity," includes a revised discussion on the nature of volcanic eruptions, an expanded look at Yellowstone-type calderas and a new section on "Living with Volcanoes."

- Chapter 7, "Sedimentary Rocks," has a new introduction that is followed by an all new section on the "Origins of Sedimentary Rocks."

- Chapter 8, "Metamorphism and Metamorphic Rocks," has new and revised discussions of "Contact/Thermal Metamorphism" and "Burial Metamorphism" and now concludes with a section on "Interpreting Metamorphic Environments."

- Chapter 15, "Mass Wasting," begins with a new section on "Landslides as Natural Disasters" and also has a new case study (Box 15.1) of a deadly event at La Conchita, California.

- Chapter 16, "Running Water," has been reorganized and almost entirely rewritten so that the discussion of streams progresses in a manner that is clearer and more logical for the beginning student.

- Chapter 18, "Glaciers and Glaciation," includes a new discussion and art on "Ice Dams and Proglacial Lakes" and a related new box on "Glacial Lake Missoula, Megafloods, and the Channeled Scablands." An updated and expanded section on "Causes of Glaciation" wraps up the chapter.

- Chapter 20, "Shorelines," has a new section on "Hurricanes—The Ultimate Coastal Hazard," that includes a look at the impact of Hurricane Katrina on the Gulf Coast. A new special interest box (Box 20.2) "Examining Hurricane Katrina from Space" accompanies the discussion.

- Chapter 23, "Energy and Mineral Resources," has a revised and rewritten discussion on "Wind Energy," and a new box (Box 23.2) on "Maintaining the Flow of Geothermal Energy at The Geysers."

The Teaching and Learning Package

For the Instructor:

Instructor's Resource Center (IRC) on DVD The IRC puts all of your lecture resources in one easy-to-reach place:

- *All* of the line art, tables, and photos from the text in .jpg files (Are illustrations central to your lecture? Check out the Student Lecture Notebook.)
- Animations of dozens of key geological processes
- *Images of Earth* photo gallery
- PowerPoint™ presentations
- *Instructor's Manual* in Microsoft Word
- *Test Item File* in Microsoft Word
- TestGen EQ test generation and management software

Animations The Prentice Hall Geoscience Animation Library includes over 100 animations illuminating the most difficult-to-visualize topics of physical geology. Created through a unique collaboration among five of Prentice Hall's leading geoscience authors, these animations represent a most significant leap forward in lecture presentation aids. They are provided both as Flash files and, for your convenience, pre-loaded into PowerPoint™ slides.

PowerPoint™ Presentations Found on the IRC are *three* PowerPoint files for each chapter. Cut down on your preparation time, no matter what your lecture needs:

1. Exclusively Art — All of the photos, art, and tables from the text, in order, loaded into PowerPoint slides.
2. Lecture Outline — Authored by Stan Hatfield of Southwestern Illinois College, this set averages 35

slides per chapter and includes customizable lecture outlines with supporting art.

3. Animations—Each animation preloaded into slides for easy cut-and-paste into your presentation.

All art is modified for projection—labels are enlarged, colors brightened, and contrasts sharpened. The slides are designed for *clear viewing,* even in the largest lecture halls, and *brightness* so you need not dim the lights as much.

Images of Earth Photo Gallery   Supplement your personal and text-specific slides with this amazing collection of over 300 geologic photos contributed by Marli Miller (University of Oregon) and other professionals in the field. Photos are grouped by geologic concept and are available on the Instructor's Resource Center.

Transparencies   Every table and most of Dennis Tasa's illustrations in *Earth, Ninth Edition* are available on full-color, projection-enhanced transparencies. (Are illustrations central to your lecture? Check out the *Student Lecture Notebook* on the Student Resources Page).

Instructor's Manual with Tests   Authored by Stanley Hatfield (Southwestern Illinois College), the *Instructor's Manual* contains: learning objectives, chapter outlines, answers to end-of-chapter questions and suggested, short demonstrations to spice up your lecture. The *Test Item File* incorporates art and averages 75 multiple-choice, true/false, short-answer, and critical-thinking questions per chapter.

TestGen EQ   Use this electronic version of the *Test Item File* to build and customize your tests. Create multiple versions, add or edit questions, add illustrations—your customization needs are easily addressed by this powerful software.

Blackboard and WebCT   Already have your own website set up? We will provide a *Test Item File* in Blackboard or WebCT formats for importation. Additional course resources are available on the *IRC* and are available for use, with permission.

For the Student:

GEODe: Earth   Somewhere between a text and a tutor, *GEODe: Earth* reinforces key concepts using animations, video, narration, interactive exercises, and practice quizzes. The quizzes are randomly generated and the results can be printed, making *GEODe: Earth* a great, easy-to-grade homework assignment. A copy of *GEODe: Earth* is automatically included in every copy of the text purchased from Prentice Hall.

Student Lecture Notebook   All of the line art from the text and transparency set are reproduced in this full color notebook, with space for notes. Students can now fully focus on the lecture and not be distracted by attempting to replicate figures. Each page is three-hole punched for easy integration with other course materials.

MyGeologyPlace   www.prenhall.com/tarbuck   This website has been designed to provide students with all the tools needed for online study and review. A self-quiz is available for each key concept within a chapter, and hints and feedback are provided for guidance. Links to other resources are also included for further study. An access code for MyGeologyPlace is bound into the front of every new copy of this textbook.

Acknowledgments

Writing a college textbook requires the talents and cooperation of many individuals. Working with Dennis Tasa, who is responsible for all of the text's outstanding illustrations and much of the developmental work of *GEODe: Earth,* is always special for us. We not only value his outstanding artistic talents and imagination but his friendship.

We were privileged to have Michael Wysession of Washington University collaborate on the revision of Chapter 12, "Earth's Interior." His expertise, insights, and writing skills greatly improved this chapter. We also appreciate the assistance of Teresa Tarbuck and Mark Watry of Rocky Mountain College in updating and revising Chapter 24, "Planetary Geology."

Our sincere appreciation goes to those colleagues who prepared in-depth reviews. Their critical comments and thoughtful input helped guide our work and clearly strengthened the text. Special thanks to

Jessica Barone, *Monroe Community College*

Sam Boggs, Jr., *University of Oregon*

John H. Burris, *San Juan College*

Beth Christensen, *Georgia State University*

Michael Clark, *University of Tennessee*

Christopher J. Crow, *Indiana University-Purdue University Fort Wayne*

Joachim Dorsch, *St. Louis Community College – Meramec*

Anne Gardulski, *Tufts University*

Cyrena Goodrich, *Kingsborough Community College*

Paul D. Howell, *University of Kentucky*

James A. Hyatt, *Eastern Connecticut State University*

Aaron W. Johnson, *University of Virginia's College at Wise*

Brennan Jordan, *Whitman College*

Daniel Karner, *Sonoma State University*

David Kirschner, *Saint Louis University*

Emily M. Klein, *Duke University*

Richard Lambert, *Skyline College*

Jennifer T. McGuire, *Texas A&M University*

Giuseppina Kysar Mattietti, *George Mason University*

Leslie A. Melim, *Western Illinois University*

Philip M. Novack-Gottshall, *University of West Georgia*

Edward J. Perantoni, *Lindenwood University*

Robert W. Pinker, *Johnson County Community College*

Nicholas Pinter, *Southern Illinois University*

Emma C. Rainforth, *Ramapo College of New Jersey*

Bethany D. Rinard, *Tarleton State University*

Laura Sanders, *Northeastern Illinois University*

Edward L. Simpson, *Kutztown University*

Roger N. Weller, *Cochise College*

Thomas C. Wynn, *Lock Haven University of Pennsylvania*

We also want to acknowledge the team of professionals at Prentice Hall. We sincerely appreciate the company's continuing strong commitment to excellence and innovation. Thanks to our former Executive Editor, Patrick Lynch, who guided this project through the critical planning and review stages and to Dan Kaveney, Publisher for Geosciences, who saw *Earth, Ninth Edition* through the remaining stages of development and design. We truly value the leadership of both men and appreciate their genuine interest in what we do.

The production team, led by Ed Thomas, has once again done an outstanding job. The strong visual impact of *Earth, Ninth Edition* benefited greatly from the work of photo researcher Yvonne Gerin and image permission coordinator, Debbie Hewitson. Thanks also to Marcia Youngman for her excellent copyediting skills and to proofreader Alison Lorber. All are true professionals with whom we are very fortunate to be associated.

Ed Tarbuck

Fred Lutgens

An Introduction to Geology

Pakistan's Charakusa Valley with the bold peaks of the Karakoram Range in the background. (Photo by Jimmie Chin/National Geographic/Getty)

The spectacular eruption of a volcano, the terror brought by an earthquake, the magnificent scenery of a mountain valley, and the destruction created by a landslide are all subjects for the geologist (Figure 1.1). The study of geology deals with many fascinating and practical questions about our physical environment. What forces produce mountains? Will there soon be another great earthquake in California? What was the Ice Age like? Will there be another? How were these ore deposits formed? Should we look for water here? Is strip mining practical in this area? Will oil be found if a well is drilled at that location?

The Science of Geology

The subject of this text is **geology,** from the Greek *geo,* "Earth" and *logos,* "discourse." It is the science that pursues an understanding of planet Earth. Geology is traditionally divided into two broad areas—physical and historical. **Physical geology,** which is the primary focus of this book, examines the materials composing Earth and seeks to understand the many processes that operate beneath and upon its surface. The aim of **historical geology,** on the other hand, is to understand the origin of Earth and its development through time. Thus, it strives to establish an orderly chronological arrangement of the multitude of physical and biological changes that have occurred in the geologic past. The study of physical geology logically precedes the study of Earth history because we must first understand how Earth works before we attempt to unravel its past. It should also be pointed out that physical and historical geology are divided into many areas of specialization. Table 1.1 provides a partial list. Every chapter of this book represents one or more areas of specialization in geology.

To understand Earth is challenging because our planet is a dynamic body with many interacting parts and a complex

FIGURE 1.1 Reflection Lake in Washington's Mount Rainier National Park. Glaciers are still sculpting this large volcanic mountain. (Photo by Art Wolfe)

TABLE 1.1	Different Areas of Geologic Study*
Archaeological Geology	Paleoclimatology
Biogeosciences	Paleontology
Engineering Geology	Petrology
Geochemistry	Planetary Geology
Geomorphology	Sedimentary Geology
Geophysics	Seismology
History of Geology	Structural Geology
Hydrogeology	Tectonics
Mineralogy	Volcanology
Ocean Sciences	

*This is a partial list of interest sections and specialties of associated societies affiliated with the Geological Society of America (www .geosociety.org) and the American Geophysical Union (www.agu.org), two professional societies to which many geologists belong.

history. Throughout its long existence, Earth has been changing. In fact, it is changing as you read this page and will continue to do so into the foreseeable future. Sometimes the changes are rapid and violent, as when landslides or volcanic eruptions occur. Just as often, change takes place so slowly that it goes unnoticed during a lifetime. Scales of size and space also vary greatly among the phenomena that geologists study. Sometimes they must focus on phenomena that are submicroscopic, and at other times they must deal with features that are continental or global in scale.

Geology is perceived as a science that is done in the out of doors, and rightly so. A great deal of geology is based on observations and experiments conducted in the field. But geology is also done in the laboratory where, for example, the study of various Earth materials provides insights into many basic processes. Frequently geology requires an understanding and application of knowledge and principles from physics, chemistry, and biology. Geology is a science that seeks to expand our knowledge of the natural world and our place in it.

Geology, People, and the Environment

The primary focus of this book is to develop an understanding of basic geological principles, but along the way, we will explore numerous important relationships between people and the natural environment. Many of the problems and issues addressed by geology are of practical value to people.

Natural hazards are a part of living on Earth. Every day they adversely affect literally millions of people worldwide and are responsible for staggering damages (Figure 1.2). Among the hazardous Earth processes studied by geologists are volcanoes, floods, tsunami, earthquakes, and landslides. Of course, geologic hazards are simply *natural* processes. They become hazards only when people try to live where these processes occur (Figure 1.3).

Resources represent another important focus of geology that is of great practical value to people. They include water and soil, a great variety of metallic and nonmetallic minerals, and energy. Together they form the very foundation of modern civilization. Geology deals not only with the formation and occurrence of these vital resources but also with maintaining supplies and with the environmental impact of their extraction and use.

Complicating all environmental issues is rapid world population growth and everyone's aspiration to a better standard of living. The population of our planet is about 6.5 billion people and is gaining about 100 million more people each year. This means a ballooning demand for resources and a growing pressure for people to dwell in environments having significant geologic hazards.

Not only do geologic processes have an impact on people but we humans can dramatically influence geologic processes

FIGURE 1.2   Natural hazards are part of living on Earth. Aerial view of earthquake destruction at Muzaffarabad, Pakistan, January 31, 2006. (Photo by Danny Kemp/AFP/Getty Images)

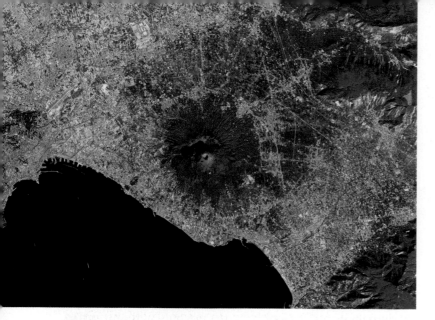

FIGURE 1.3 This is an image of Italy's Mt. Vesuvius in September 2000. This major volcano is surrounded by the city of Naples and the Bay of Naples. In 79 A.D. Vesuvius explosively erupted, burying the towns of Pompeii and Herculanaeum in volcanic ash. Will it happen again? Geologic hazards are *natural* processes. They only become hazards when people try to live where these processes occur. (Image courtesy of NASA)

as well. For example, river flooding is natural, but the magnitude and frequency of flooding can be changed significantly by human activities such as clearing forests, building cities, and constructing dams. Unfortunately, natural systems do not always adjust to artificial changes in ways that we can anticipate. Thus, an alteration to the environment that was intended to benefit society often has the opposite effect.

At appropriate places throughout this book, you will have the opportunity to examine different aspects of our relationship with the physical environment. It will be rare to find a chapter that does not address some aspect of natural hazards, environmental issues, or resources. Significant parts of some chapters provide the basic geologic knowledge and principles needed to understand environmental problems. Moreover, a number of the book's special-interest boxes focus on geology, people, and the environment by providing case studies or highlighting a topical issue.

Some Historical Notes about Geology

The nature of our Earth—its materials and processes—has been a focus of study for centuries. Writings about such topics as fossils, gems, earthquakes, and volcanoes date back to the early Greeks, more than 2300 years ago.

Certainly the most influential Greek philosopher was Aristotle. Unfortunately, Aristotle's explanations about the natural world were not based on keen observations and experiments. Instead, they were arbitrary pronouncements. He believed that rocks were created under the "influence" of the stars and that earthquakes occurred when air crowded into the ground, was heated by central fires, and escaped explosively. When confronted with a fossil fish, he explained that "a great many fishes live in the earth motionless and are found when excavations are made."

Although Aristotle's explanations may have been adequate for his day, they unfortunately continued to be expounded for many centuries, thus thwarting the acceptance of more up-to-date accounts. Frank D. Adams states in *The Birth and Development of the Geological Sciences* (New York: Dover, 1938) that "throughout the Middle Ages Aristotle was regarded as the head and chief of all philosophers; one whose opinion on any subject was authoritative and final."

Catastrophism

In the mid-1600s, James Ussher, Anglican Archbishop of Armagh, Primate of all Ireland, published a major work that had immediate and profound influences. A respected scholar of the Bible, Ussher constructed a chronology of human and Earth history in which he determined that Earth was only a few thousand years old, having been created in 4004 B.C. Ussher's treatise earned widespread acceptance among Europe's scientific and religious leaders, and his chronology was soon printed in the margins of the Bible itself.

During the 17th and 18th centuries the doctrine of **catastrophism** strongly influenced people's thinking about Earth. Briefly stated, catastrophists believed that Earth's landscapes had been shaped primarily by great catastrophes. Features such as mountains and canyons, which today we know take great periods of time to form, were explained as having been produced by sudden and often worldwide disasters produced by unknowable causes that no longer operate. This philosophy was an attempt to fit the rates of Earth processes to the then-current ideas on the age of Earth.

The relationship between catastrophism and the age of Earth has been summarized as follows:

> That the earth had been through tremendous adventures and had seen mighty changes during its obscure past was plainly evident to every inquiring eye; but to concentrate these changes into a few brief millenniums required a tailor-made philosophy, a philosophy whose basis was sudden and violent change.*

The Birth of Modern Geology

Modern geology began in the late 1700s when James Hutton, a Scottish physician and gentleman farmer, published his *Theory of the Earth* (Figure 1.4). In this work, Hutton put forth a fundamental principle that is a pillar of geology today: **uniformitarianism.** It simply states that the *physical, chemical, and biological laws that operate today have also operated in the geologic past.* This means that the forces and processes that we observe presently shaping our planet have been at work for a very long time. Thus, to understand ancient rocks, we must first understand present-day processes and their results. This idea is commonly expressed by saying, "The present is the key to the past."

---

*H. E. Brown, V. E. Monnett, and J. W. Stovall, *Introduction to Geology* (New York: Blaisdell, 1958).

FIGURE 1.4 James Hutton (1726–1797), a founder of modern geology. (Photo courtesy of The Natural History Museum, London)

the present gives us insight into the past and that the physical, chemical, and biological laws that govern geological processes remain unchanging through time. However, we also understand that the doctrine should not be taken too literally. To say that geological processes in the past were the same as those occurring today is not to suggest that they always had the same relative importance or that they operated at precisely the same rate. Moreover, some important geologic processes are not currently observable, but evidence that they occur is well established. For example, we know that Earth has experienced impacts from large meteorites even though we have no human witnesses. Such events altered Earth's crust, modified its climate, and strongly influenced life on the planet.

The acceptance of uniformitarianism meant the acceptance of a very long history for Earth. Although Earth processes vary in intensity, they still take a very long time to create or destroy major landscape features (Figure 1.5).

For example, geologists have established that mountains once existed in portions of present-day Minnesota, Wisconsin, and Michigan. Today the region consists of low hills and plains. Erosion (processes that wear land away) gradually destroyed these peaks. Estimates indicate that the North American continent is being lowered at a rate of about 3 centimeters per 1000 years. At this rate it would take 100 million years for water, wind, and ice to lower mountains that were 3000 meters (10,000 feet) high.

But even this time span is relatively short on the time scale of Earth history, for the rock record contains evidence that shows Earth has experienced many cycles of mountain building and erosion. Concerning the ever-changing nature of

Prior to Hutton's *Theory of the Earth*, no one had effectively demonstrated that geological processes occur over extremely long periods of time. However, Hutton persuasively argued that forces that appear small could, over long spans of time, produce effects that were just as great as those resulting from sudden catastrophic events. Unlike his predecessors, Hutton carefully cited verifiable observations to support his ideas.

For example, when he argued that mountains are sculpted and ultimately destroyed by weathering and the work of running water, and that their wastes are carried to the oceans by processes that can be observed, Hutton said, "We have a chain of facts which clearly demonstrate . . . that the materials of the wasted mountains have traveled through the rivers"; and further, "There is not one step in all this progress . . . that is not to be actually perceived." He then went on to summarize this thought by asking a question and immediately providing the answer: "What more can we require? Nothing but time."

Today the basic tenets of uniformitarianism are just as viable as in Hutton's day. Indeed, we realize more strongly than ever that

FIGURE 1.5 Weathering and erosion have gradually sculpted these striking rock formations in Arizona's Monument Valley. Geologic processes often act so slowly that changes may not be visible during an entire human lifetime. (Photo by David Muench Photography, Inc.)

Earth through great expanses of geologic time, Hutton made a statement that was to become his most famous. In concluding his classic 1788 paper published in the *Transactions of the Royal Society of Edinburgh,* he stated, "The results, therefore, of our present enquiry is, that we find no vestige of a beginning—no prospect of an end." A quote from William L. Stokes sums up the significance of Hutton's basic concept:

> In the sense that uniformitarianism implies the operation of timeless, changeless laws or principles, we can say that nothing in our incomplete but extensive knowledge disagrees with it.*

In the chapters that follow, we will be examining the materials that compose our planet and the processes that modify it. It is important to remember that, although many features of our physical landscape may seem to be unchanging over the decades we observe them, they are nevertheless changing, but on time scales of hundreds, thousands, or even many millions of years.

Geologic Time

Although Hutton and others recognized that geologic time is exceedingly long, they had no methods to accurately determine the age of Earth. However, in 1896 radioactivity was discovered. Using radioactivity for dating was first attempted in 1905 and has been refined ever since. Geologists are now able to assign fairly accurate dates to events in Earth history.† For example, we know the dinosaurs became extinct about 65 million years ago. Today the age of Earth is put at about 4.5 billion years.

Relative Dating and the Geologic Time Scale

During the 19th century, long before the advent of radiometric dating, a geologic time scale was developed using principles of relative dating. **Relative dating** means that events are placed in their proper sequence or order without knowing their age in years. This is done by applying principles such as the **law of superposition** (*super* = over, *positum* = to place), which states that in layers of sedimentary rocks or lava flows, the youngest layer is on top and the oldest is on the bottom (assuming that nothing has turned the layers upside down, which sometimes happens). Arizona's Grand Canyon provides a fine example in which the oldest rocks are located in the inner gorge and the youngest rocks are found on the rim (see Chapter 9 opening photo, p. 246, and Figure 15.2, p. 403). So the law of superposition establishes the *sequence* of rock layers—but not, of course, their numerical ages (Figure 1.6). Today such a proposal appears to be elementary, but 300 years ago it amounted to a major breakthrough in scientific reasoning by establishing a rational basis for relative time measurements.

Fossils, the remains or traces of prehistoric life, were also essential to the development of the geologic time scale (Figure 1.7). Fossils are the basis for the **principle of fossil succession,** which states that *fossil organisms succeed one another in a definite and determinable order, and therefore any time period can be recognized by its fossil content.* This principle was laboriously worked out over decades by collecting fossils from countless rock layers around the world. Once established, it allowed geologists to identify rocks of the same age in widely separated places and to build the geologic time scale shown in Figure 1.8.

Notice that units having the same designations do not necessarily extend for the same number of years. For example, the Cambrian period lasted about 50 million years, whereas the Silurian period spanned only about 26 million years. As we will emphasize again in Chapter 9, this situation exists because the basis for establishing the time scale was not the regular rhythm of a clock but the changing character of life forms through time. Specific dates were added long after the time scale was established. A glance at Figure 1.8 also reveals that the Phanerozoic eon is divided into many more units than earlier eons, even though it encompasses only about 12 percent of Earth history. The meager fossil record for these earlier eons is the primary reason for the lack of detail on this portion of the time scale. Without abundant fossils, geologists lose their primary tool for subdividing geologic time.

The Magnitude of Geologic Time

The concept of geologic time is new to many nongeologists. People are accustomed to dealing with increments of time that are measured in hours, days, weeks, and years. Our history books often examine events over spans of centuries, but even a century is difficult to appreciate fully. For most of us, someone or something that is 90 years old is *very old,* and a 1000-year-old artifact is *ancient.*

By contrast, those who study geology must routinely deal with vast time periods—millions or billions (thousands of millions) of years. When viewed in the context of Earth's 4.5-billion-year history, a geologic event that occurred 100 million years ago may be characterized as "recent" by a geologist, and a rock sample that has been dated at 10 million years may be called "young."

An appreciation for the magnitude of geologic time is important in the study of geology because many processes are so gradual that vast spans of time are needed before significant changes occur.

How long is 4.5 billion years? If you were to begin counting at the rate of one number per second and continued 24 hours a day, 7 days a week and never stopped, it would take about two lifetimes (150 years) to reach 4.5 billion! Another interesting basis for comparison is as follows:

> Compress, for example, the entire 4.5 billion years of geologic time into a single year. On that scale, the oldest rocks we know date from about mid-March. Living things first appeared in the sea in May. Land plants and animals emerged in late November and the widespread swamps that formed

---

Essentials of Earth History (Englewood Cliffs, New Jersey: Prentice Hall, 1966), p. 34.

†Chapter 9 is devoted to a much more complete discussion of geologic time.

FIGURE 1.6   These rock layers are exposed in Minnewaska State Park, New York. Their relative ages can be determined by applying the law of superposition. The youngest rocks are on top, and the oldest are at the bottom. (Photo by Carr Clifton)

the Pennsylvanian coal deposits flourished for about four days in early December. Dinosaurs became dominant in mid-December, but disappeared on the 26th, at about the time the Rocky Mountains were first uplifted. Manlike creatures appeared sometime during the evening of December 31st, and the most recent continental ice sheets began to recede from the Great Lakes area and from northern Europe about 1 minute and 15 seconds before midnight on the 31st. Rome ruled the Western world for 5 seconds from 11:59:45 to 11:59:50. Columbus discovered America 3 seconds before midnight, and the science of geology was born with the writings of James Hutton just slightly more than one second before the end of our eventful year of years.*

---

*Don L. Eicher, *Geologic Time*, 2nd ed. (Englewood Cliffs, New Jersey: Prentice Hall, 1978), pp. 18–19. Reprinted by permission.

The foregoing is just one of many analogies that have been conceived in an attempt to convey the magnitude of geologic time. Although helpful, all of them, no matter how clever, only begin to help us comprehend the vast expanse of Earth history.

The Nature of Scientific Inquiry

All science is based on the assumption that the natural world behaves in a consistent and predictable manner that is comprehensible through careful, systematic study. The overall goal of science is to discover the underlying patterns in nature and then to use this knowledge to make predictions about what should or should not be expected, given certain facts or circumstances. For example, by knowing how oil deposits form, geologists are able to predict the most favorable sites for exploration and, perhaps as important, how to avoid regions having little or no potential.

The development of new scientific knowledge involves some basic logical processes that are universally accepted. To determine what is occurring in the natural world, scientists collect scientific "*facts*" through observation and measurement. Because some error is inevitable, the accuracy of a particular measurement or observation is always open to question. Nevertheless, these data are essential to science and serve as the springboard for the development of scientific theories (see Box 1.1).

Hypothesis

Once facts have been gathered and principles have been formulated to describe a natural phenomenon, investigators try to explain how or why things happen in the manner observed. They often do this by constructing a tentative (or

FIGURE 1.7   Fossils are important tools for the geologist. In addition to being very important in relative dating, fossils can be useful environmental indicators. **A.** A fossil fish of Eocene age from the Green River Formation in Wyoming. (Photo by John Cancalosi/DRK Photo) **B.** Fossil ferns from the coal-forming Pennsylvanian Period, St. Clair, Pennsylvania. (Photo by Breck p. Kent)

A.

B.

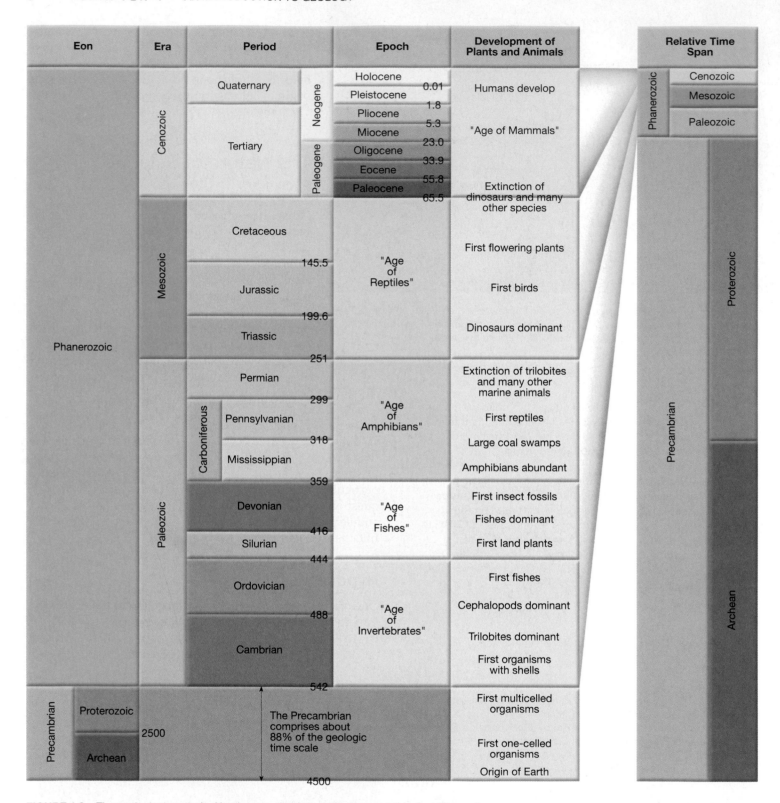

FIGURE 1.8 The geologic time scale. Numbers on the time scale represent time in millions of years before the present. These dates were added long after the time scale had been established using relative dating techniques. The Precambrian accounts for more than 88 percent of geologic time. (Data from Geological Society of America)

BOX 1.1 ▶ UNDERSTANDING EARTH

Studying Earth from Space

Scientific facts are gathered in many ways, including laboratory studies and field observations and measurements. Satellite images like the one in Figure 1.A are another useful source of data. Such images provide perspectives that are difficult to gain from more traditional sources. Moreover, the high-tech instruments aboard many satellites enable scientists to gather information from remote regions where data are otherwise scarce.

The image in Figure 1.A makes use of the Advanced Spaceborne Thermal Emission and Reflection Radiometer (ASTER). Because different materials reflect and emit energy in different ways, ASTER can provide detailed information about the composition of Earth's surface. Figure 1.A is a three-dimensional view looking north over Death Valley, California. The data have been computer enhanced to exaggerate the color variations that highlight differences in types of surface materials.

Salt deposits on the floor of Death Valley appear in shades of yellow, green, purple, and pink, indicating the presence of carbonate, sulfate, and chloride minerals. The Panamint Mountains to the west (left) and the Black Mountains to the east are made up of sedimentary limestones, sandstones, shales, and metamorphic rocks. The bright red areas are dominated by the mineral quartz, found in sandstone; green areas are limestone. In the lower center of the image is Badwater, the lowest point in North America.

The image in Figure 1.B is from NASA's *Tropical Rainfall Measuring Mission (TRMM)*. Rainfall patterns over land have been studied for many years using ground-based radar and other instruments. Now the instruments aboard the *TRMM* satellite have greatly expanded our ability to collect precipitation data. In addition to data for land areas, this satellite provides extremely precise measurements of rainfall over the oceans where conventional land-based instruments cannot see. This is especially important because much of Earth's rain falls in ocean-covered tropical areas, and a great deal of the globe's weather-producing energy comes from heat exchanges involved in the rainfall process. Until the *TRMM*, information on the intensity and amount of rainfall over the tropics was scanty. Such data are crucial to understanding and predicting global climate change.

FIGURE 1.A   This satellite image shows detailed information about the composition of surface materials in Death Valley, California. It was produced by superimposing nighttime thermal infrared data, acquired on April 7, 2000, over topographic data from the U.S. Geological Survey. (Image courtesy of NASA)

FIGURE 1.B   This map of rainfall for December 7–13, 2004, in Malaysia was constructed using TRMM data. Over 800 millimeters (32 inches) of rain fell along the east coast of the peninsula (darkest red area). The extraordinary rains caused extensive flooding and triggered many mudflows. (NASA/TRMM image)

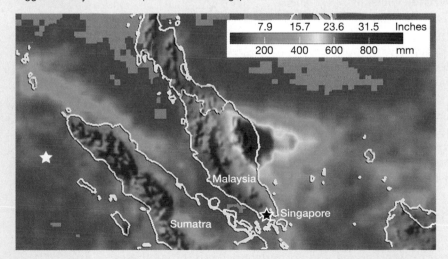

untested) explanation, which is called a scientific **hypothesis** or **model.** (The term *model,* although often used synonymously with hypothesis, is a less precise term because it is sometimes used to describe a scientific theory as well.) It is best if an investigator can formulate more than one hypothesis to explain a given set of observations. If an individual scientist is unable to devise multiple models, others in the scientific community will almost always develop alternative explanations. A spirited debate frequently ensues. As a result, extensive research is conducted by proponents of opposing models, and the results are made available to the wider scientific community in scientific journals.

Before a hypothesis can become an accepted part of scientific knowledge, it must pass objective testing and analysis. (If a hypothesis cannot be tested, it is not scientifically useful, no matter how interesting it might seem.) The verification process requires that *predictions* be made based on the model being considered and that the predictions be tested by comparing them against objective observations of nature. Put another way, hypotheses must fit observations other than those used to formulate them in the first place. Those hypotheses that fail rigorous testing are ultimately discarded. The history of science is littered with discarded hypotheses. One of the best known is the Earth-centered model of the universe—a proposal that was supported by the apparent daily motion of the Sun, Moon, and stars around Earth. As the mathematician Jacob Bronowski so ably stated, "Science is a great many things, but in the end they all return to this: Science is the acceptance of what works and the rejection of what does not."

Theory

When a hypothesis has survived extensive scrutiny and when competing models have been eliminated, a hypothesis may be elevated to the status of a scientific **theory.** In everyday language we may say, "That's only a theory." But a scientific theory is a well-tested and widely accepted view that the scientific community agrees best explains certain observable facts.

Theories that are extensively documented are held with a very high degree of confidence. Theories of this stature that are comprehensive in scope have a special status. They are called **paradigms** because they explain a large number of interrelated aspects of the natural world. For example, the theory of plate tectonics is a paradigm of the geological sciences that provides the framework for understanding the origin of mountains, earthquakes, and volcanic activity. In addition, plate tectonics explains the evolution of the continents and the ocean basins through time—a topic we will consider later in this chapter.

Scientific Methods

The process just described, in which researchers gather facts through observations and formulate scientific hypotheses and theories, is called the *scientific method.* Contrary to popular belief, the scientific method is not a standard recipe that scientists apply in a routine manner to unravel the secrets of our natural world. Rather, it is an endeavor that involves creativity and insight. Rutherford and Ahlgren put it this way: "Inventing hypotheses or theories to imagine how the world works and then figuring out how they can be put to the test of reality is as creative as writing poetry, composing music, or designing skyscrapers."*

There is not a fixed path that scientists always follow that leads unerringly to scientific knowledge. Nevertheless, many scientific investigations involve the following steps: (1) the collection of scientific facts through observation and measurement (Figure 1.9); (2) the development of one or more working hypotheses or models to explain these facts; (3) development of observations and experiments to test the hypotheses; and (4) the acceptance, modification, or rejection of the model based on extensive testing (see Box 1.2).

Other scientific discoveries may result from purely theoretical ideas, which stand up to extensive examination. Some researchers use high-speed computers to simulate

---

*F. James Rutherford and Andrew Ahlgren, *Science for All Americans* (New York: Oxford University Press, 1990), p. 7.

FIGURE 1.9   This field geologist is checking a seismograph. (Photo by Andrew Rafkind/Getty Images Inc.—Stone Allstock)

BOX 1.2 ▶ UNDERSTANDING EARTH

Do Glaciers Move? An Application of the Scientific Method

The study of glaciers provides an early application of the scientific method. High in the Alps of Switzerland and France, small glaciers exist in the upper portions of some valleys. In the late 18th and early 19th centuries, people who farmed and herded animals in these valleys suggested that glaciers in the upper reaches of the valleys had previously been much larger and had occupied downvalley areas. They based their explanation on the fact that the valley floors were littered with angular boulders and other rock debris that seemed identical to the materials that they could see in and near the glaciers at the heads of the valleys.

Although the explanation of these observations seemed logical, others did not accept the notion that masses of ice hundreds of meters thick were capable of movement. The disagreement was settled after a simple experiment was designed and carried out to test the hypothesis that glacial ice can move.

Markers were placed in a straight line completely across an alpine glacier. The position of the line was marked on the valley walls so that if the ice moved, the change in position could be detected. After a year or two the results were clear. The markers on the glacier had advanced down the valley, proving that glacial ice indeed moves. In addition, the experiment demonstrated that ice within a glacier does not move at a uniform rate, because the markers in the center advanced farther than did those along the margins. Although most glaciers move too slowly for direct visual detection, the experiment succeeded in demonstrating that movement nevertheless occurs. In the years that followed, this experiment was repeated many times with greater accuracy using more modern surveying techniques. Each time, the basic relationships established by earlier attempts were verified.

The experiment illustrated in Figure 1.C was carried out at Switzerland's Rhône Glacier later in the 19th century. It not only traced the movement of markers within the ice but also mapped the position of the glacier's terminus. Notice that even though the ice within the glacier was advancing, the ice front was retreating. As often occurs in science, experiments and observations designed to test one hypothesis yield new information that requires further analysis and explanation.

FIGURE 1.C   Ice movement and changes in the terminus at Rhône Glacier, Switzerland. In this classic study of a valley glacier, the movement of stakes clearly show that glacial ice moves and that movement along the sides of the glacier is slower than movement in the center. Also notice that even though the ice front was retreating, the ice within the glacier was advancing.

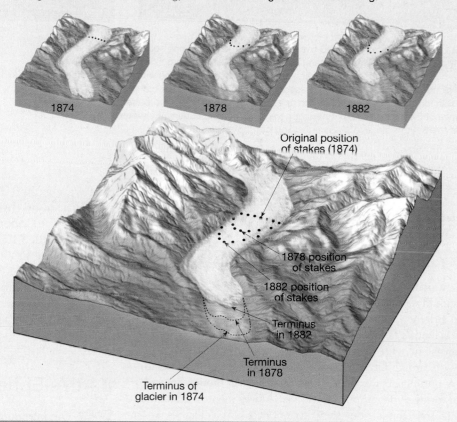

1874   1878   1882

Original position of stakes (1874)

1878 position of stakes

1882 position of stakes

Terminus in 1882

Terminus in 1878

Terminus of glacier in 1874

---

what is happening in the "real" world. These models are useful when dealing with natural processes that occur on very long time scales or take place in extreme or inaccessible locations. Still other scientific advancements are made when a totally unexpected happening occurs during an experiment. These serendipitous discoveries are more than pure luck, for as Louis Pasteur said, "In the field of observation, chance favors only the prepared mind."

Scientific knowledge is acquired through several avenues, so it might be best to describe the nature of scientific inquiry as the methods of science rather than the scientific method. In addition, it should always be remembered that even the most compelling scientific theories are still simplified explanations of the natural world.

Plate Tectonics and Scientific Inquiry

There are many opportunities in the pages of this book to develop and reinforce your understanding of how science works and, in particular, how the science of geology works. You will learn about the methods involved in gathering data and develop a sense of the observational techniques and reasoning processes used by geologists. Chapter 2, "Plate Tectonics: A Scientific Revolution Unfolds," is an excellent example.

A.

B.

FIGURE 1.10 A. View that greeted the *Apollo 8* astronauts as their spacecraft emerged from behind the Moon. (NASA Headquarters) **B.** Africa and Arabia are prominent in this image of Earth taken from *Apollo 17.* The tan cloud-free zones over the land coincide with major desert regions. The band of clouds across central Africa is associated with a much wetter climate that in places sustains tropical rain forests. The dark blue of the oceans and the swirling cloud patterns remind us of the importance of the oceans and the atmosphere. Antarctica, a continent covered by glacial ice, is visible at the South Pole. (NASA/Science Source/Photo Researchers, Inc.)

Students Sometimes Ask . . .

In class you compared a hypothesis to a theory. How is each one different from a scientific law?

A scientific *law* is a basic principle that describes a particular behavior of nature that is generally narrow in scope and can be stated briefly—often as a simple mathematical equation. Because scientific laws have been shown time and time again to be consistent with observations and measurements, they are rarely discarded. Laws may, however, require modifications to fit new findings. For example, Newton's laws of motion are still useful for everyday applications (NASA uses them to calculate satellite trajectories), but they do not work at velocities approaching the speed of light. For these circumstances, they have been supplanted by Einstein's theory of relativity.

Within the past several decades, a great deal has been learned about the workings of our dynamic planet. This period has seen an unequaled revolution in our understanding of Earth. The revolution began in the early part of the 20th century with the radical proposal of *continental drift*—the idea that the continents moved about the face of the planet. This hypothesis contradicted the established view that the continents and ocean basins are permanent and stationary features on the face of Earth. For that reason, the notion of drifting continents was received with great skepticism and even ridicule. More than 50 years passed before enough

data were gathered to transform this controversial hypothesis into a sound theory that wove together the basic processes known to operate on Earth. The theory that finally emerged, called the *theory of plate tectonics*, provided geologists with the first comprehensive model of Earth's internal workings.

As you read Chapter 2, you will not only gain insights into the workings of our planet, you will also see an excellent example of the way geological "truths" are uncovered and reworked.

Earth's Spheres

 GEODe An Introduction to Geology
   ▶ A View of Earth

The classic view of Earth, shown in Figure 1.10A, provided the *Apollo 8* astronauts and the rest of humanity with a unique perspective of our home. Seen from space, Earth is breathtaking in its beauty and startling in its solitude. Such an image reminds us that our home is, after all, a planet—small, self-contained, and in some ways even fragile.

As we look more closely at our planet from space, it becomes apparent that Earth is much more than rock and soil (Figure 1.10B). In fact, the most conspicuous features are not the continents but swirling clouds suspended above the surface and the vast global ocean. These features emphasize the importance of air and water to our planet.

The closer view of Earth from space, shown in Figure 1.10B, helps us appreciate why the physical environment is traditionally divided into three major parts: the water portion of our planet, the hydrosphere; Earth's gaseous envelope, the atmosphere; and, of course, the solid Earth, or geosphere. It needs to be emphasized that our environment is highly integrated and not dominated by rock, water, or air alone. Rather, it is characterized by continuous interactions as air comes in contact with rock, rock with water, and water with air. Moreover, the biosphere, which is the totality of all plant and animal life on our planet, interacts with each of the three physical realms and is an equally integral part of the planet. Thus, Earth can be thought of as consisting of four major spheres: the hydrosphere, atmosphere, geosphere, and biosphere.

The interactions among Earth's four spheres are incalculable. Figure 1.11 provides us with one easy-to-visualize ex-

FIGURE 1.11   The shoreline is one obvious meeting place for rock, water, and air. In this scene, ocean waves that were created by the force of moving air break against the rocky shore. The force of the water can be powerful, and the erosional work that is accomplished can be great. (Photo by Galen Rowell)

ample. The shoreline is an obvious meeting place for rock, water, and air. In this scene, ocean waves that were created by the drag of air moving across the water are breaking against the rocky shore. The force of the water can be powerful, and the erosional work that is accomplished can be great.

Hydrosphere

Earth is sometimes called the *blue* planet. Water more than anything else makes Earth unique. The **hydrosphere** is a dynamic mass of water that is continually on the move, evaporating from the oceans to the atmosphere, precipitating to the land, and running back to the ocean again. The global ocean is certainly the most prominent feature of the hydrosphere, blanketing nearly 71 percent of Earth's surface to an average depth of about 3800 meters (12,500 feet). It accounts for about 97 percent of Earth's water. However, the hydrosphere also includes the fresh water found underground and in streams, lakes, and glaciers. Moreover, water is an important component of all living things.

Although these latter sources constitute just a tiny fraction of the total, they are much more important than their meager percentage indicates. In addition to providing the fresh water that is so vital to life on land, streams, glaciers, and groundwater are responsible for sculpting and creating many of our planet's varied landforms.

Atmosphere

Earth is surrounded by a life-giving gaseous envelope called the **atmosphere.** Compared with the solid Earth, the atmosphere is thin and tenuous. One half lies below an altitude of 5.6 kilometers (3.5 miles), and 90 percent occurs within just 16 kilometers (10 miles) of Earth's surface. By comparison, the radius of the solid Earth (distance from the surface to the center) is about 6400 kilometers (nearly 4000 miles)! Despite its modest dimensions, this thin blanket of air is an integral part of the planet. It not only provides the air that we breathe but also protects us from the Sun's intense heat and dangerous ultraviolet radiation. The energy exchanges that continually occur between the atmosphere and Earth's surface and between the atmosphere and space produce the effects we call weather and climate.

If, like the Moon, Earth had no atmosphere, our planet would not only be lifeless but many of the processes and interactions that make the surface such a dynamic place could not operate. Without weathering and erosion, the face of our planet might more closely resemble the lunar surface, which has not changed appreciably in nearly 3 billion years.

Biosphere

The **biosphere** includes all life on Earth. Ocean life is concentrated in the sunlit surface waters of the sea. Most life on land is also concentrated near the surface, with tree roots and burrowing animals reaching a few meters underground and flying insects and birds reaching a kilometer or so into the atmosphere. A surprising variety of life forms are also adapted to extreme environments. For example, on the ocean floor where pressures are extreme and no light penetrates, there are places where vents spew hot, mineral-rich fluids that support communities of exotic life forms. On land, some bacteria thrive in rocks as deep as 4 kilometers (2.5 miles) and in boiling hot springs. Moreover, air currents can carry microorganisms many kilometers into the atmosphere. But even when we consider these extremes, life still must be thought of as being confined to a narrow band very near Earth's surface.

Plants and animals depend on the physical environment for the basics of life. However, organisms do not just respond to their physical environment. Indeed, the biosphere powerfully influences the other three spheres. Without life, the makeup and nature of the geosphere, hydrosphere, and atmosphere would be very different.

Geosphere

Lying beneath the atmosphere and the oceans is the solid Earth, or **geosphere.** The geosphere extends from the surface to the center of the planet, a depth of nearly 6400 kilometers, making it by far the largest of Earth's four spheres. Much of our study of the solid Earth focuses on the more accessible surface features. Fortunately, many of these features represent the outward expressions of the dynamic behavior of Earth's interior. By examining the most prominent surface features and their global extent, we can obtain clues to the dynamic processes that have shaped our planet. A first look at the structure of Earth's interior and at the major surface features of the geosphere will come later in the chapter.

Soil, the thin veneer of material at Earth's surface that supports the growth of plants, may be thought of as part of all four spheres. The solid portion is a mixture of weathered rock debris (geosphere) and organic matter from decayed plant and animal life (biosphere). The decomposed and disintegrated rock debris is the product of weathering processes that require air (atmosphere) and water (hydrosphere). Air and water also occupy the open spaces between the solid particles.

Earth As a System

Anyone who studies Earth soon learns that our planet is a dynamic body with many separate but interacting parts or *spheres.* The hydrosphere, atmosphere, biosphere, and geosphere and all of their components can be studied separately. However, the parts are not isolated. Each is related in some way to the others to produce a complex and continuously interacting whole that we call the *Earth system.*

Earth System Science

A simple example of the interactions among different parts of the Earth system occurs every winter as moisture evaporates from the Pacific Ocean and subsequently falls as rain in the hills and mountains of southern California, triggering destructive landslides. The processes that move water from the hydrosphere to the atmosphere and then to the solid Earth have a profound impact on the plants and animals (including humans) that inhabit the affected regions (Figure 1.12).

Scientists have recognized that in order to more fully understand our planet, they must learn how its individual components (land, water, air, and life forms) are interconnected. This endeavor, called **Earth system science,** aims to study Earth as a *system* composed of numerous interacting parts, or *subsystems.* Using an interdisciplinary approach, those who practice Earth system science attempt to achieve the level of understanding necessary to comprehend and solve many of our global environmental problems.

What Is a System? Most of us hear and use the term *system* frequently. We may service our car's cooling *system,* make use of the city's transportation *system,* and be a participant in the political *system.* A news report might inform us of an approaching weather *system.* Further, we know that Earth is just a small part of a larger system known as the *solar system,* which, in turn, is a subsystem of an even larger system called the Milky Way Galaxy.

Loosely defined, a **system** can be any size group of interacting parts that form a complex whole. Most natural systems are driven by sources of energy that move matter and/or energy from one place to another. A simple analogy is a car's cooling system, which contains a liquid (usually water and antifreeze) that is driven from the engine to the radiator

FIGURE 1.12 This deadly landslide on Leyte Island in the Philippines occurred in February 2006, and was triggered by heavy rains that saturated materials on the mountainside. This is a simple example of interactions among different parts of the Earth system. (Photo by AP Photo/Wally Santana)

and back again. The role of this system is to transfer heat generated by combustion in the engine to the radiator, where moving air removes it from the system. Hence, the term cooling system.

Systems like a car's cooling system are self-contained with regard to matter and are called **closed systems**. Although energy moves freely in and out of a closed system, no matter (liquid in the case of our auto's cooling system) enters or leaves the system. (This assumes you don't get a leak in your radiator). By contrast, most natural systems are **open systems** and are far more complicated than the foregoing example. In an open system both energy and matter flow into and out of the system. In a weather system such as a hurricane, factors such as the quantity of water vapor available for cloud formation, the amount of heat released by condensing water vapor, and the flow of air into and out of the storm can fluctuate a great deal. At times the storm may strengthen; at other times it may remain stable or weaken.

Feedback Mechanisms   Most natural systems have mechanisms that tend to enhance change, as well as other mechanisms that tend to resist change and thus stabilize the system. For example, when we get too hot, we perspire to cool down. This cooling phenomenon works to stabilize our body temperature and is referred to as a **negative feedback mechanism.** Negative feedback mechanisms work to maintain the system as it is or, in other words, to maintain the status quo. By contrast, mechanisms that enhance or drive change are called **positive feedback mechanisms.**

Most of Earth's systems, particularly the climate system, contain a wide variety of negative and positive feedback mechanisms. For example, substantial scientific evidence indicates that Earth has entered a period of global warming. One consequence of global warming is that some of the world's glaciers and ice caps have begun to melt. Highly reflective snow- and ice-covered surfaces are gradually being replaced by brown soils, green trees, or blue oceans, all of which are darker, so they absorb more sunlight. The result is a positive feedback that contributes to the warming.

On the other hand, an increase in global temperature also causes greater evaporation of water from Earth's land-sea surface. One result of having more water vapor in the air is an increase in cloud cover. Because cloud tops are white and highly reflective, more sunlight is reflected back to space, which diminishes the amount of sunshine reaching Earth's surface and thus reduces global temperatures. Further, warmer temperatures tend to promote the growth of vegetation. Plants in turn remove carbon dioxide (CO_2) from the air. Since carbon dioxide is one of the atmosphere's *greenhouse gases,* its removal has a negative impact on global warming.*

In addition to natural processes, we must also consider the human element. Extensive cutting and clearing of the

tropical rain forests and the burning of fossil fuels (oil, natural gas, and coal) result in an increase in atmospheric CO_2. Such activity is contributing to the increase in global temperature that our planet is experiencing. One of the daunting tasks of Earth system scientists is to predict what the climate will be like in the future by taking into account many variables, including technological changes, population trends, and the overall impact of the numerous competing positive and negative feedback mechanisms. Chapter 21 on "Global Climate Change" explores this topic in some detail.

The Earth System

The Earth system has a nearly endless array of subsystems in which matter is recycled over and over again (Figure 1.13). One example that you will learn about in Chapter 7 traces the movements of carbon among Earth's four spheres. It shows us, for example, that the carbon dioxide in the air and the carbon in living things and in certain sedimentary rocks is all part of a subsystem described by the *carbon cycle.*

Cycles in the Earth System   A more familiar loop or subsystem is the *hydrologic cycle.* It represents the unending circulation of Earth's water among the hydrosphere, atmosphere, biosphere, and geosphere. Water enters the atmosphere by evaporation from Earth's surface and by transpiration from plants. Water vapor condenses in the atmosphere to form clouds, which in turn produce precipitation that falls back to Earth's surface. Some of the rain that falls onto the land sinks in to be taken up by plants or become groundwater, and some flows across the surface toward the ocean.

Viewed over long time spans, the rocks of the geosphere are constantly forming, changing, and reforming (Figure 1.13). The loop that involves the processes by which one rock changes to another is called the *rock cycle* and will be discussed at some length later in the chapter. The cycles of the Earth system, such as the hydrologic and rock cycles, are not independent of one another. To the contrary, there are many places where they interface. An **interface** is a common boundary where different parts of a system come in contact and interact. For example, in Figure 1.13, weathering at the surface gradually disintegrates and decomposes solid rock. The work of gravity and running water may eventually move this material to another place and deposit it. Later, groundwater percolating through the debris may leave behind mineral matter that cements the grains together into solid rock (a rock that is often very different from the rock we started with). This changing of one rock into another could not have occurred without the movement of water through the hydrologic cycle. There are many places where one cycle or loop in the Earth system interfaces with and is a basic part of another.

Energy for the Earth System   The Earth system is powered by energy from two sources. The Sun drives external processes that occur in the atmosphere and hydrosphere and at Earth's surface. Weather and climate, ocean circulation,

---

*Greenhouse gases absorb heat energy emitted by Earth and thus help keep the atmosphere warm.

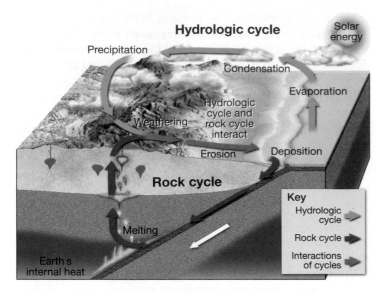

Hydrologic cycle

Precipitation

Solar energy

Condensation

Evaporation

Hydrologic cycle and rock cycle interact

Weathering

Erosion

Deposition

Rock cycle

Melting

Earth's internal heat

Key

Hydrologic cycle

Rock cycle

Interactions of cycles

FIGURE 1.13 Each part of the Earth system is related to every other part to produce a complex interacting whole. The Earth system involves many cycles, including the hydrologic cycle and the rock cycle. Such cycles are not independent of each other. There are many places where they interface.

and erosional processes are driven by energy from the Sun. Earth's interior is the second source of energy. Heat remaining from when our planet formed, and heat that is continuously generated by decay of radioactive elements powers the internal processes that produce volcanoes, earthquakes, and mountains.

The Parts Are Linked The parts of the Earth system are linked so that a change in one part can produce changes in any or all of the other parts. For example, when a volcano erupts, lava from Earth's interior may flow out at the sur-

face and block a nearby valley. This new obstruction influences the region's drainage system by creating a lake or causing streams to change course. The large quantities of volcanic ash and gases that can be emitted during an eruption might be blown high into the atmosphere and influence the amount of solar energy that can reach Earth's surface. The result could be a drop in air temperatures over the entire hemisphere.

Where the surface is covered by lava flows or a thick layer of volcanic ash, existing soils are buried. This causes the soil-forming processes to begin anew to transform the new surface material into soil (Figure 1.14). The soil that eventually forms will reflect the interactions among many parts of the Earth system—the volcanic parent material, the climate, and the impact of biological activity. Of course, there would also be significant changes in the biosphere. Some organisms and their habitats would be eliminated by the lava and ash, whereas new settings for life, such as the lake, would be created. The potential climate change could also impact sensitive life forms.

The Earth system is characterized by processes that vary on spatial scales from fractions of millimeters to thousands of kilometers. Time scales for Earth's processes range from milliseconds to billions of years. As we learn about Earth, it becomes increasingly clear that despite significant separations in distance or time, many processes are connected, and a change in one component can influence the entire system.

Humans are *part of* the Earth system, a system in which the living and nonliving components are entwined and interconnected. Therefore, our actions produce changes in all of the other parts. When we burn gasoline and coal, dispose of our wastes, and clear the land, we cause other parts of the system to respond, often in unforeseen ways. Throughout this book you will learn about many of Earth's subsystems, including the hydrologic system, the tectonic (mountain-building) system, and the rock cycle, to name a few. Remember that these components *and we humans* are all part of the complex interacting whole we call the Earth system.

FIGURE 1.14 When Mount St. Helens erupted in May 1980, the area shown here was buried by a volcanic mudflow. Now plants are reestablished and new soil is forming. (Jack W. Dykinga Associates)

Early Evolution of Earth

Recent earthquakes caused by displacements of Earth's crust, along with lavas erupted from active volcanoes, represent only the latest in a long line of events by which our planet has attained its present form and structure. The geologic processes operating in Earth's interior can be best understood when viewed in the context of much earlier events in Earth history.

Origin of Planet Earth

The following scenario describes the most widely accepted views of the origin of our solar system. Although this model is presented as fact, keep in mind that like all scientific hypotheses, this one is subject to revision and even outright rejection. Nevertheless, it remains the most consistent set of ideas to explain what we observe today.

Our scenario begins about 14 billion years ago with the *Big Bang*, an incomprehensibly large explosion that sent all matter of the universe flying outward at incredible speeds. In time, the debris from this explosion, which was almost entirely hydrogen and helium, began to cool and condense into the first stars and galaxies. It was in one of these galaxies, the Milky Way, that our solar system and planet Earth took form.

Earth is one of eight planets that, along with several dozen moons and numerous smaller bodies, revolve around the Sun. The orderly nature of our solar system leads most researchers to conclude that Earth and the other planets formed at essentially the same time and from the same primordial material as the Sun. The **nebular hypothesis** proposes that the bodies of our solar system evolved from an enormous rotating cloud called the **solar nebula** (Figure 1.15). Besides the hydrogen and helium atoms generated during the Big Bang, the solar nebula consisted of microscopic dust grains and the ejected matter of long-dead stars. (Nuclear fusion in stars converts hydrogen and helium into the other elements found in the universe.)

Nearly 5 billion years ago this huge cloud of gases and minute grains of heavier elements began to slowly contract due to the gravitational interactions among its particles. Some external influence, such as a shock wave traveling from a catastrophic explosion (*supernova*), may have triggered the collapse. As this slowly spiraling nebula contracted, it rotated faster and faster for the same reason ice skaters do when they draw their arms toward their bodies. Eventually the inward pull of gravity came into balance with the outward force caused by the rotational motion of the nebula (Figure 1.15). By this time the once vast cloud had assumed a flat disk shape with a large concentration of material at its center called the *protosun* (pre-Sun). (Astronomers are fairly confident that the nebular cloud formed a disk because similar structures have been detected around other stars.)

During the collapse, gravitational energy was converted to thermal energy (heat), causing the temperature of the inner portion of the nebula to dramatically rise. At these high temperatures, the dust grains broke up into molecules and excited atomic particles. However, at distances beyond the orbit of Mars, the temperatures probably remained quite low. At $-200°C$, the tiny particles in the outer portion of the nebula were likely covered with a thick layer of ices made of frozen water, carbon dioxide, ammonia, and methane. The disk-shaped cloud also contained appreciable amounts of the lighter gases hydrogen and helium.

The formation of the Sun marked the end of the period of contraction and thus the end of gravitational heating. Temperatures in the region where the inner planets now reside began to decline. The decrease in temperature caused those substances with high melting points to condense into tiny particles that began to coalesce (join together). Materials such as iron and nickel and the elements of which the rock-forming minerals are composed—silicon, calcium, sodium, and so forth—formed metallic and rocky clumps that orbited the Sun (Figure 1.15). Repeated collisions caused these masses to coalesce into larger asteroid-size bodies, called *planetesimals*, which in a few tens of millions of years accreted into the four inner planets we call Mercury, Venus, Earth, and Mars. Not all of these clumps of matter were incorporated into the protoplanets. Those rocky and metallic pieces that remained in orbit are called *meteorites* when they survive an impact with Earth.

As more and more material was swept up by the planets, the high-velocity impact of nebular debris caused the temperature of these bodies to rise. Because of their relatively high temperatures and weak gravitational fields, the inner planets were unable to accumulate much of the lighter components of the nebular cloud. The lightest of these, hydrogen and helium, were eventually whisked from the inner solar system by the solar wind.

At the same time that the inner planets were forming, the larger, outer planets (Jupiter, Saturn, Uranus, and Neptune), along with their extensive satellite systems, were also developing. Because of low temperatures far from the Sun, the material from which these planets formed contained a high percentage of ices—water, carbon dioxide, ammonia, and methane—as well as rocky and metallic debris. The accumulation of ices accounts in part for the large size and low density of the outer planets. The two most massive planets, Jupiter and Saturn, had a surface gravity sufficient to attract and hold large quantities of even the lightest elements—hydrogen and helium.

Formation of Earth's Layered Structure

As material accumulated to form Earth (and for a short period afterward), the high-velocity impact of nebular debris and the decay of radioactive elements caused the temperature of our planet to steadily increase. During this time of intense heating, Earth became hot enough that iron and nickel began to melt. Melting produced liquid blobs of heavy metal that sank toward the center of the planet. This process occurred rapidly on the scale of geologic time and produced Earth's dense iron-rich core.

The early period of heating resulted in another process of chemical differentiation, whereby melting formed buoyant masses of molten rock that rose toward the surface, where they solidified to produce a primitive crust. These rocky materials were enriched in oxygen and "oxygen-seeking" elements, particularly silicon and aluminum, along with lesser amounts of calcium, sodium, potassium, iron, and magnesium. In addition, some heavy metals such as gold, lead, and uranium, which have low melting points or were highly soluble in the ascending molten masses, were scavenged from Earth's interior and concentrated in the developing crust. This early period of chemical segregation established the three basic divisions of Earth's interior—the iron-rich *core*; the thin *primitive crust*; and Earth's largest layer, called the *mantle*, which is located between the core and crust.

FIGURE 1.15   Formation of the solar system according to the nebular hypothesis. **A.** The birth of our solar system began as dust and gases (nebula) started to gravitationally collapse. **B.** The nebula contracted into a rotating disk that was heated by the conversion of gravitational energy into thermal energy. **C.** Cooling of the nebular cloud caused rocky and metallic material to condense into tiny solid particles. **D.** Repeated collisions caused the dust-size particles to gradually coalesce into asteroid-size bodies. **E.** Within a few million years these bodies accreted into the planets.

An important consequence of this early period of chemical differentiation is that large quantities of gaseous materials were allowed to escape from Earth's interior, as happens today during volcanic eruptions. By this process a primitive atmosphere gradually evolved. It is on this planet, with this atmosphere, that life as we know it came into existence.

Following the events that established Earth's basic structure, the primitive crust was lost to erosion and other geologic processes, so we have no direct record of its makeup. When and exactly how the continental crust—and thus Earth's first landmasses—came into existence is a matter of ongoing research. Nevertheless, there is general agreement that the continental crust formed gradually over the last 4 billion years. (The oldest rocks yet discovered are isolated fragments found in the Northwest Territories of Canada that have radiometric dates of about 4 billion years.) In addition, as you will see in Chapter 2, Earth is an evolving planet whose continents (and ocean basins) have continually changed shape and even location during much of this period.

Earth's Internal Structure

GEODe An Introduction to Geology
▶ Earth's Layered Structure

In the preceding section, you learned that the segregation of material that began early in Earth's history resulted in the formation of three major layers defined by their chemical composition—the crust, mantle, and core. In addition to these compositionally distinct layers, Earth can be divided into layers based on physical properties. The physical properties used to define such zones include whether the layer is solid or liquid and how weak or strong it is. Knowledge of both types of layers is essential to our understanding of basic geologic processes, such as volcanism, earthquakes, and mountain building (Figure 1.16).

Earth's Crust

The **crust,** Earth's relatively thin, rocky outer skin, is of two different types—continental crust and oceanic crust. Both share the word "crust," but the similarity ends there. The oceanic crust is roughly 7 kilometers (5 miles) thick and composed of the dark igneous rock *basalt*. By contrast, the continental crust averages 35 to 40 kilometers (25 miles) thick but may exceed 70 kilometers (40 miles) in some mountainous regions such as the Rockies and Himalayas. Unlike the oceanic crust, which has a relatively homogeneous chemical composition, the continental crust consists of many rock types. Although the upper crust has an average composition of a *granitic rock* called *granodiorite*, it varies considerably from place to place.

Continental rocks have an average density of about 2.7 g/cm^3, and some have been discovered that are 4 billion years old. The rocks of the oceanic crust are younger (180 million years or less) and denser (about 3.0 g/cm^3) than continental rocks.*

Earth's Mantle

More than 82 percent of Earth's volume is contained in the **mantle,** a solid, rocky shell that extends to a depth of about 2900 kilometers (1800 miles). The boundary between the crust and mantle represents a marked change in chemical composition. The dominant rock type in the uppermost mantle is *peridotite,* which is richer in the metals magnesium and iron than the minerals found in either the continental or oceanic crust.

The Upper Mantle   The upper mantle extends from the crust–mantle boundary down to a depth of about 660 kilometers (410 miles). The upper mantle can be divided into three different parts. The top portion of the upper mantle is part of the stiff *lithosphere,* and beneath that is the weaker **asthenosphere.** The bottom part of the upper mantle is called the *transition zone.*

The **lithosphere** (sphere of rock) consists of the entire crust and uppermost mantle and forms Earth's relatively cool, rigid outer shell. Averaging about 100 kilometers in thickness, the lithosphere is more than 250 kilometers thick below the oldest portions of the continents (Figure 1.16). Beneath this stiff layer to a depth of about 350 kilometers lies a soft, comparatively weak layer known as the *asthenosphere* ("weak sphere"). The top portion of the asthenosphere has a temperature/pressure regime that results in a small amount of melting. Within this very weak zone the lithosphere is mechanically detached from the layer below. The result is that the lithosphere is able to move independently of the asthenosphere, a fact we will consider in the next chapter.

It is important to emphasize that the strength of various Earth materials is a function of both their composition and of the temperature and pressure of their environment. You should not get the idea that the entire lithosphere behaves like a brittle solid similar to rocks found on the surface. Rather, the rocks of the lithosphere get progressively hotter and weaker (more easily deformed) with increasing depth. At the depth of the uppermost asthenosphere, the rocks are close enough to their melting temperature (some melting may actually occur) that they are very easily deformed. Thus, the uppermost asthenosphere is weak because it is near its melting point, just as hot wax is weaker than cold wax.

From about 410 kilometers to about 660 kilometers in depth is the part of the upper mantle called the **transition zone.** The top of the transition zone is identified by a sudden increase in density from about 3.5 to 3.7 gm/cm^3. This change occurs because minerals in the rock peridotite respond to the

---

*Liquid water has a density of 1 g/cm^3; therefore, the density of basalt is three times that of water.

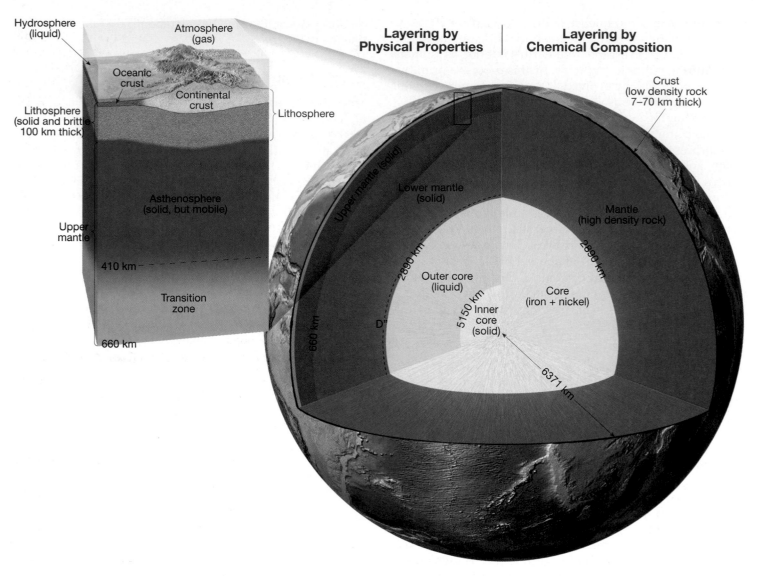

FIGURE 1.16 Views of Earth's layered structure. The right side of the large cross section shows that Earth's interior is divided into three different layers based on compositional differences—the crust, mantle, and core. The left side of the large cross section depicts layers of Earth's interior based on physical properties—the lithospere, asthenosphere, transition zone, D″ layer, outer core, and inner core. The block diagram to the left of the large cross section shows an enlarged view of the upper portion of Earth's interior.

increase in pressure by forming new minerals with closely packed atomic structures.

The Lower Mantle From a depth of 660 kilometers to the top of the core, at a depth of 2900 kilometers (1800 miles), is the **lower mantle.** Because of an increase in pressure (caused by the weight of the rock above) the mantle gradually strengthens with depth. Despite their strength however, the rocks within the lower mantle are very hot and capable of very gradual flow.

In the bottom few hundred kilometers of the mantle is a highly variable and unusual layer called the **D″ layer** (pronounced "dee double-prime"). The nature of this boundary layer between the rocky mantle and the hot liquid iron outer core will be examined in Chapter 12.

Earth's Core

The composition of the **core** is thought to be an iron-nickel alloy with minor amounts of oxygen, silicon, and sulfur—elements that readily form compounds with iron. At the extreme pressure found in the core, this iron-rich material has an average density of nearly 11 g/cm^3 and approaches 14 times the density of water at Earth's center.

The core is divided into two regions that exhibit very different mechanical strengths. The **outer core** is a *liquid layer* 2270 kilometers (1410 miles) thick. It is the movement of metallic iron within this zone that generates Earth's magnetic field. The **inner core** is a sphere having a radius of 1216 kilometers (754 miles). Despite its higher temperature, the iron in the inner core is *solid* due to the immense pressures that exist in the center of the planet.

How Do We Know What We Know?

At this point you may be asking yourself, "How did we learn about the composition and structure of Earth's interior?" You might suspect that the internal structure of Earth has been sampled directly. However, the deepest mine in the world (the Western Deep Levels mine in South Africa) is only about 4 kilometers (2.5 miles) deep, and the deepest drilled hole in the world (completed in the Kola Peninsula of Russia in 1992) goes down only about 12 kilometers (7.5 miles). In essence, humans have never drilled a hole into the mantle (and will never drill into the core) in order to directly sample these materials.

Despite these limitations, theories that describe the nature of Earth's interior have been developed that closely fit most observational data. Thus, our model of Earth's interior represents the best inferences we can make based on the available data. For example, the layered structure of Earth has been established using indirect observations. Every time there is an earthquake, waves of energy (called *seismic waves*) penetrate Earth's interior, much like X rays penetrate the human body. Seismic waves change their speed and are bent and reflected as they move through zones having different properties. An extensive series of monitoring stations around the world detect and record this energy. With the aid of computers, these data are analyzed and used to determine the structure of Earth's interior. For more about how this is done, see Chapter 12, "Earth's Interior."

What evidence do we have to support the alleged composition of our planet's interior? It may surprise you to learn that rocks that originated in the mantle have been collected at Earth's surface. These include diamond-bearing samples, which laboratory studies indicate can form only in very high-pressure environments. Since these rocks must have crystallized at depths exceeding 200 kilometers (120 miles), they are inferred to be samples of the mantle that underwent very little alteration during their ascent to the surface. In addition, we have been able to examine slivers of the uppermost mantle and overlying oceanic crust that have been thrust high above sea level in locations such as Cyprus, Newfoundland, and Oman.

Establishing the composition of the core is another matter altogether. Because of its great depth and high density, not a single sample of the core has reached the surface. Nevertheless, we have significant evidence to suggest that this layer consists primarily of iron.

Surprisingly, meteorites provide important clues to the composition of the core and mantle. (Meteorites are solid, extraterrestrial objects that strike Earth's surface.) Most meteorites are fragments derived from the collisions of larger bodies, principally from the asteroid belt between the orbits of Mars and Jupiter. They are important because they represent samples of the material (*planetesimals*) from which the inner planets, including Earth, formed. Meteorites are composed mainly of an iron-nickel alloy (*irons*), silicate minerals (*stones*), or a combination of both (*stony-irons*) of these materials. The stones have an average composition that is close to that calculated for the mantle. Irons, on the other hand, contain a much higher percentage of this metallic material than is presently found in either Earth's crust or mantle. If Earth did indeed form from the same material in the solar nebula that generated the meteorites and the other terrestrial planets, it must contain a much higher percentage of iron than is found in crustal rock. Consequently, we can conclude that the core is greatly enriched in this heavy metal.

This view is also supported by studies of the Sun's composition, which indicate that iron is the most abundant substance found in the solar system that possesses the density that has been calculated for the core. Furthermore, Earth's magnetic field requires that the core be made of a material that conducts electricity, such as iron. Because all of the available evidence points to the core as being composed largely of iron, we take this to be a fact, at least until new evidence tells us differently.

The Face of Earth

 GEODe

An Introduction to Geology
▶ Features of the Continents and Floor of the Ocean

The two principal divisions of Earth's surface are the continents and the ocean basins (Figure 1.17). A significant difference between these two areas is their relative levels. The continents are remarkably flat features that have the appearance of plateaus protruding above sea level. With an average elevation of about 0.8 kilometer (0.5 mile), continental blocks lie close to sea level, except for limited areas of mountainous terrain. By contrast, the average depth of the ocean floor is about 3.8 kilometers (2.4 miles) below sea level, or about 4.5 kilometers (2.8 miles) lower than the average elevation of the continents.

The elevation difference between the continents and ocean basins is primarily the result of differences in their respective densities and thicknesses. Recall that the continents average 35–40 kilometers in thickness and are composed of granitic rocks having a density of about 2.7 g/cm^3. The basaltic rocks that comprise the oceanic crust average only 7 kilometers thick and have an average density of about 3.0 g/cm^3. Thus, the thicker and less dense continental crust is more buoyant than the oceanic crust. As a result, continental crust floats on top of the deformable rocks of the mantle at a higher level than oceanic crust for the same reason that a large, empty (less dense) cargo ship rides higher than a small, loaded (more dense) one.

Major Features of the Continents

The largest features of the continents can be grouped into two distinct categories: extensive, flat stable areas that have been eroded nearly to sea level, and uplifted regions of deformed rocks that make up present-day mountain belts. Notice in Figure 1.18 that the young mountain belts tend to be long, narrow topographic features at the margins of continents, and the flat, stable areas are typically located in the interior of the continents.

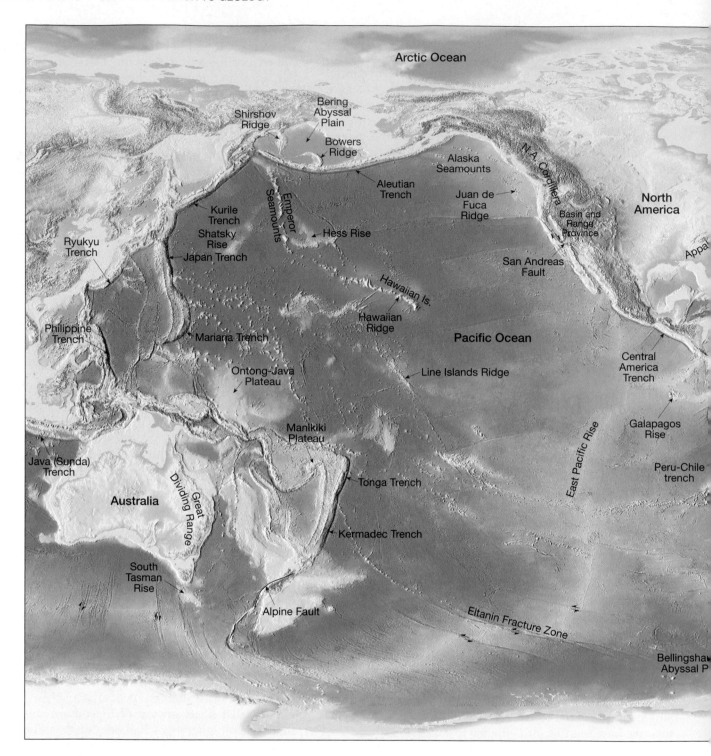

FIGURE 1.17   The topography of Earth's solid surface is shown on these two pages.

Mountain Belts   The most prominent topographic features of the continents are linear mountain belts. Although the distribution of mountains appears to be random, this is not the case. When the youngest mountains are considered (those less than 100 million years old), we find that they are located principally in two major zones. The circum-Pacific belt (the region surrounding the Pacific Ocean) includes the mountains of the western Americas and continues into the western Pacific in the form of volcanic island arcs (Figure

1.17). Island arcs are active mountainous regions composed largely of volcanic rocks and deformed sedimentary rocks. Examples include the Aleutian Islands, Japan, the Philippines, and New Guinea.

The other major mountainous belt extends eastward from the Alps through Iran and the Himalayas and then dips southward into Indonesia. Careful examination of mountainous terrains reveals that most are places where thick sequences of rocks have been squeezed and highly deformed,

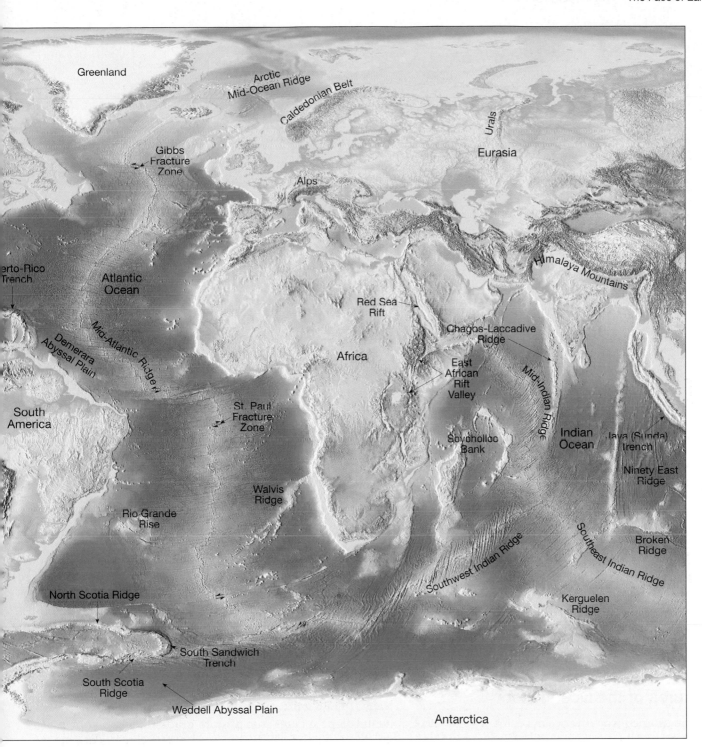

FIGURE 1.17 (continued)

as if placed in a gigantic vise. Older mountains are also found on the continents. Examples include the Appalachians in the eastern United States and the Urals in Russia. Their once lofty peaks are now worn low, the result of millions of years of erosion.

The Stable Interior Unlike the young mountain belts, which have formed within the last 100 million years, the interiors of the continents, called **cratons,** have been relatively

stable (undisturbed) for the last 600 million years, or even longer. Typically these crustal blocks were involved in a mountain-building episode much earlier in Earth's history.

Within the stable interiors are areas known as **shields,** which are expansive, flat regions composed of deformed crystalline rock. Notice in Figure 1.18 that the Canadian Shield is exposed in much of the northeastern part of North America. Radiometric dating of various shields has revealed that they are truly ancient regions. All contain Precambrian-age rocks

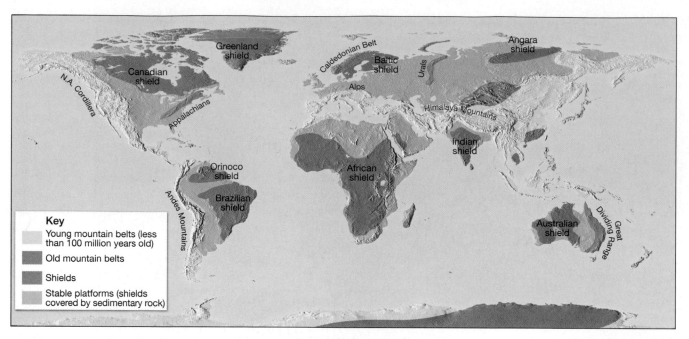

FIGURE 1.18 This map shows the general distribution of Earth's mountain belts, stable platforms, and shields.

that are more than 1 billion years old, with some samples approaching 4 billion years in age. Even these oldest-known rocks exhibit evidence of enormous forces that have folded, faulted, and metamorphosed them. Thus, we conclude that these rocks were once part of an ancient mountain system that has since been eroded away to produce these expansive, flat regions.

Other flat areas of the craton exist in which highly deformed rocks, like those found in the shields, are covered by a relatively thin veneer of sedimentary rocks. These areas are called **stable platforms.** The sedimentary rocks in stable platforms are nearly horizontal except where they have been warped to form large basins or domes. In North America a major portion of the stable platform is located between the Canadian Shield and the Rocky Mountains.

Major Features of the Ocean Floor

If all water were drained from the ocean basins, a great variety of features would be seen, including linear chains of volcanoes, deep canyons, plateaus, and large expanses of monotonously flat plains. In fact, the scenery would be nearly as diverse as that on the continents (see Figure 1.17).

During the past 50 years, oceanographers using modern depth-sounding equipment have slowly mapped significant portions of the ocean floor. From these studies they have delineated three major topographically distinct units: *continental margins, deep-ocean basins,* and *oceanic (mid-ocean) ridges.*

Continental Margins The **continental margin** is that portion of the seafloor adjacent to major landmasses. It may include the *continental shelf,* the *continental slope,* and the *continental rise.*

Although land and sea meet at the shoreline, this is not the boundary between the continents and the ocean basins. Rather, along most coasts a gently sloping platform of material, called the **continental shelf,** extends seaward from the shore. Because it is underlain by continental crust, it is clearly a flooded extension of the continents. A glance at Figure 1.17 shows that the width of the continental shelf is variable. For example, it is broad along the East and Gulf coasts of the United States but relatively narrow along the Pacific margin of the continent.

The boundary between the continents and the deep-ocean basins lies along the **continental slope,** which is a relatively steep dropoff that extends from the outer edge of the continental shelf to the floor of the deep ocean (Figure 1.17). Using this as the dividing line, we find that about 60 percent of Earth's surface is represented by ocean basins and the remaining 40 percent by continents.

In regions where trenches do not exist, the steep continental slope merges into a more gradual incline known as the **continental rise.** The continental rise consists of a thick accumulation of sediments that moved downslope from the continental shelf to the deep-ocean floor.

Deep-Ocean Basins Between the continental margins and oceanic ridges lie the **deep-ocean basins.** Parts of these regions consist of incredibly flat features called **abyssal plains.** The ocean floor also contains extremely deep depressions that are occasionally more than 11,000 meters (36,000 feet) deep. Although these **deep-ocean trenches** are relatively narrow and represent only a small fraction of the ocean floor, they are nevertheless very significant features. Some trenches are located adjacent to young mountains that flank the continents. For example, in Figure 1.17 the

Peru–Chile trench off the west coast of South America parallels the Andes Mountains. Other trenches parallel linear island chains called *volcanic island arcs.*

Dotting the ocean floor are submerged volcanic structures called **seamounts,** which sometimes form long narrow chains. Volcanic activity has also produced several large *lava plateaus,* such as the Ontong Java Plateau located northeast of New Guinea. In addition, some submerged plateaus are composed of continental-type crust. Examples include the Campbell Plateau southeast of New Zealand and the Seychelles plateau northeast of Madagascar.

Oceanic Ridges　The most prominent feature on the ocean floor is the **oceanic** or **mid-ocean ridge.** As shown in Figure 1.17, the Mid-Atlantic Ridge and the East Pacific Rise are parts of this system. This broad elevated feature forms a continuous belt that winds for more than 70,000 kilometers (43,000 miles) around the globe in a manner similar to the seam of a baseball. Rather than consisting of highly deformed rock, such as most of the mountains on the continents, the oceanic ridge system consists of layer upon layer of igneous rock that has been fractured and uplifted.

Understanding the topographic features that comprise the face of Earth is critical to our understanding of the mechanisms that have shaped our planet. What is the significance of the enormous ridge system that extends through all the world's oceans? What is the connection, if any, between young, active mountain belts and oceanic trenches? What forces crumple rocks to produce majestic mountain ranges? These are questions that will be addressed in the next chapter as we begin to investigate the dynamic processes that shaped our planet in the geologic past and will continue to shape it in the future.

Rocks and the Rock Cycle

GEODe　An Introduction to Geology
▸ Rock Cycle

Rock is the most common and abundant material on Earth. To a curious traveler, the variety seems nearly endless. When a rock is examined closely, we find that it consists of smaller crystals or grains called minerals. *Minerals* are chemical compounds (or sometimes single elements), each with its own composition and physical properties. The grains or crystals may be microscopically small or easily seen with the unaided eye.

The nature and appearance of a rock is strongly influenced by the minerals that compose it. In addition, a rock's *texture*—the size, shape, and/or arrangement of its constituent minerals—also has a significant effect on its appearance. A rock's mineral composition and texture, in turn, are a reflection of the geologic processes that created it.

The characteristics of the rocks in Figure 1.19 provided geologists with the clues they needed to determine the processes that formed them. This is true of all rocks. Such analyses are critical to an understanding of our planet. This understanding has many practical applications, as in the search for basic mineral and energy resources and the solution of environmental problems.

Basic Rock Types

Geologists divide rocks into three major groups: igneous, sedimentary, and metamorphic. In Figure 1.19, the lava flow in northern Arizona is classified as igneous, the sandstone in Utah's Zion National Park is sedimentary, and the schist exposed at the bottom of the Grand Canyon is metamorphic. What follows is a brief look at these three basic rock groups. As you will learn, each group is linked to the others by the processes that act upon and within the planet.

Igneous Rocks　**Igneous rocks** (ignis = fire) form when molten rock, called *magma,* cools and solidifies. Magma is melted rock that can form at various levels deep within Earth's crust and upper mantle. As magma cools, crystals of various minerals form and grow. When magma remains deep within the crust, it cools slowly over thousands of years. This gradual loss of heat allows relatively large crystals to develop before the entire mass is completely solidified. Coarse-grained igneous rocks that form far below the surface are called *intrusive.* The cores of many mountains consist of igneous rock that formed in this way. Only subsequent uplift and erosion expose this rock at the surface. A common and important example is *granite* (Figure 1.20). This coarse-grained intrusive rock is rich in the light-colored silicate minerals quartz and feldspar. Granite and related rocks are major constituents of the continental crust.

Sometimes magma breaks through at Earth's surface, as during a volcanic eruption. Because it cools quickly in a surface environment, the molten rock solidifies rapidly, and there is not sufficient time for large crystals to grow. Rather, there is the simultaneous formation of many tiny crystals. Igneous rocks that form at Earth's surface are described as *extrusive* and are usually fine-grained. An abundant and important example is *basalt* (Figure 1.19A). This black to dark-green rock is rich in silicate minerals that contain significant iron and magnesium. Because of its higher iron content, basalt is denser than granite. Basalt and related rocks make up the oceanic crust as well as many volcanoes both in the ocean and on the continents.

Sedimentary Rocks　*Sediments,* the raw materials for **sedimentary rocks,** accumulate in layers at Earth's surface. They are materials derived from preexisting rocks by the processes of *weathering.* Some of these processes physically break rock into smaller pieces with no change in composition. Other weathering processes decompose rock—that is, chemically change minerals into new minerals and into substances that readily dissolve in water.

The products of weathering are usually transported by water, wind, or glacial ice to sites of deposition where they form relatively flat layers called *beds.* Sediments are commonly turned into rock or *lithified* by one of two processes. *Compaction* takes place as the weight of overlying materials

A.

B.

C.

FIGURE 1.19 **A.** This fine-grained black rock, called *basalt,* is part of a lava flow from Sunset Crater in northern Arizona. It formed when molten rock erupted from the volcano hundreds of years ago and solidified. (Photo by David Muench) **B.** This rock is exposed in the walls of southern Utah's Zion National Park. This layer, known as the Navajo Sandstone, consists of durable grains of the glassy mineral quartz that once covered this region with mile after mile of drifting sand dunes. (Photo by Tom Bean/DRK Photo) **C.** This rock unit, known as the Vishnu Schist, is exposed in the inner gorge of the Grand Canyon. Its formation is associated with environments far below Earth's surface where temperatures and pressures are high and with ancient mountain-building processes that occurred in Precambrian time. (Photo by Tom Bean/DRK Photo)

squeezes sediments into denser masses. *Cementation* occurs as water containing dissolved substances percolates through the open spaces between sediment grains. Over time the material dissolved in water is chemically precipitated onto the grains and cements them into a solid mass.

Sediments that originate and are transported as solid particles are called *detrital sediments,* and the rocks they form are called *detrital sedimentary rocks.* Particle size is the primary basis for naming the members in this category. Two common examples are *shale* and *sandstone.* Shale is a fine-grained rock consisting of clay-size (less than 1/256 mm) and silt-size (1/256 to 1/16 mm) particles. The deposition of these tiny grains is associated with "quiet" environments such as swamps, river floodplains, and portions of deep-ocean basins. *Sandstone* is the name given sedimentary rocks in which sand-size (1/16 to 2 mm) grains predominate. Sandstones are associated with a variety of environments, including beaches and dunes (Figure 1.19B).

Chemical sedimentary rocks form when material dissolved in water precipitates. Unlike detrital sedimentary rocks that are subdivided on the basis of particle size, the primary basis for distinguishing among chemical sedimentary rocks is their mineral composition. Limestone, the most common chemical sedimentary rock, is composed chiefly of the mineral calcite (calcium carbonate, $CaCO_3$). There are many varieties of limestone (Figure 1.21). The most abundant types have a biochemical origin, meaning that water-dwelling organisms extract dissolved mineral matter and create hard parts such as shells. Later these hard parts accumulate as sediment.

Geologists estimate that sedimentary rocks account for only about 5 percent (by volume) of Earth's outer 16 kilometers (10 miles). However, the importance of this group of rocks is far greater than this percentage implies. If you sampled the rocks exposed at Earth's surface, you would find that the great majority are sedimentary. In fact, about 75 percent of all rock outcrops on the continents are sedimentary. Therefore, we can think of sedimentary rocks as comprising a relatively thin and somewhat discontinuous layer in the uppermost portion of the crust. This makes sense because sediment accumulates at the surface.

It is from sedimentary rocks that geologists reconstruct many details of Earth's history. Because sediments are deposited in a variety of different settings at the surface, the rock layers that they eventually form hold many clues to past surface environments. They may also exhibit characteristics that allow geologists to decipher information about the method and distance of sediment transport. Furthermore, it is sedimentary rocks that contain fossils, which are vital sources of data in the study of the geologic past.

A.

B.

FIGURE 1.20 Granite is an intrusive igneous rock that is especially abundant in Earth's continental crust. **A.** Erosion has uncovered this mass of granite in California's Yosemite National Park. **B.** This sample of granite shows its coarse-grained texture. (Photo by E. J. Tarbuck)

Metamorphic Rocks **Metamorphic rocks** are produced from preexisting igneous, sedimentary, or even other metamorphic rocks. Thus, every metamorphic rock has a parent rock—the rock from which it formed. *Metamorphic* is an appropriate name because it literally means to "change form." Most changes occur at the elevated temperatures and pressures that exist deep in Earth's crust and upper mantle.

The processes that create metamorphic rocks often progress incrementally, from slight changes (low-grade metamorphism) to substantial changes (high-grade metamorphism). For example, during low-grade metamorphism, the common sedimentary rock shale becomes the more compact metamorphic rock *slate.* By contrast, high-grade metamorphism causes a transformation that is so complete that the identity of the parent rock cannot be determined. Furthermore, when rocks at depth (where temperatures are high)

FIGURE 1.21 Limestone is a chemical sedimentary rock in which the mineral calcite predominates. There are many varieties. The topmost layer in Arizona's Grand Canyon, known as the Kaibab Formation, is limestone of Permian age that had a marine origin. (Photo by E. J. Tarbuck)

are subjected to directed pressure, they gradually deform to generate intricate folds. In the most extreme metamorphic environments, the temperatures approach those at which rocks melt. However, *during metamorphism the rock must remain essentially solid,* for if complete melting occurs, we have entered the realm of igneous activity.

Most metamorphism occurs in one of three settings:

1. When rock is intruded by a magma body, *contact* or *thermal metamorphism* may take place. Here, change is driven by a rise in temperature within the host rock surrounding an igneous intrusion.
2. *Hydrothermal metamorphism* involves chemical alterations that occur as hot, ion-rich water circulates through fractures in rock. This type of metamorphism is usually associated with igneous activity that provides the heat required to drive chemical reactions and circulate these fluids through rock.
3. During mountain building, great quantities of deeply buried rock are subjected to the directed pressures and high temperatures associated with large-scale deformation called *regional metamorphism.*

The degree of metamorphism is reflected in the rock's texture and mineral makeup. During regional metamorphism the crystals of some minerals will recrystallize with an orientation that is perpendicular to the direction of the compressional force. The resulting mineral alignment often gives the rock a layered or banded appearance termed a *foliated texture. Schist* and *gneiss* are two examples (Figure 1.22A).

Not all metamorphic rocks have a foliated texture. Such rocks are said to exhibit a *nonfoliated texture.* Metamorphic rocks composed of only one mineral that forms equidimensional crystals are, as a rule, not visibly foliated. For example, limestone, if pure, is composed of only a single mineral, calcite. When a fine-grained limestone is metamorphosed, the small calcite crystals combine to form larger, interlocking crystals. The resulting rock resembles a coarse-grained igneous rock. This nonfoliated metamorphic equivalent of limestone is called *marble* (Figure 1.22B).

Extensive areas of metamorphic rocks are exposed on every continent. Metamorphic rocks are an important component of many mountain belts, where they make up a large portion of a mountain's crystalline core. Even the stable continental interiors, which are generally covered by sedimentary rocks, are underlain by metamorphic basement rocks. In all of these settings, the metamorphic rocks are usually highly deformed and intruded by igneous masses. Indeed, significant parts of Earth's continental crust are composed of metamorphic and associated igneous rocks.

The Rock Cycle: One of Earth's Subsystems

Earth is a system. This means that our planet consists of many interacting parts that form a complex whole. Nowhere is this idea better illustrated than when we examine the rock cycle (Figure 1.23). The **rock cycle** allows us to view many of the interrelationships among different parts of the Earth system. It helps us understand the origin of igneous, sedimentary, and metamorphic rocks and to see that each type is linked to the others by processes that act upon and within the planet. Learn the rock cycle well; you will be examining its interrelationships in greater detail throughout this book.

The Basic Cycle We begin at the bottom of Figure 1.23 with magma, molten rock that forms deep beneath Earth's surface. Over time, magma cools and solidifies. This process,

FIGURE 1.22 Common metamorphic rocks. **A.** The foliated rock gneiss often has a banded appearance and frequently has a mineral composition similar to the igneous rock granite. **B.** Marble is a coarse, crystalline, nonfoliated rock whose parent rock is limestone. (Photos by E. J. Tarbuck)

A.

B.

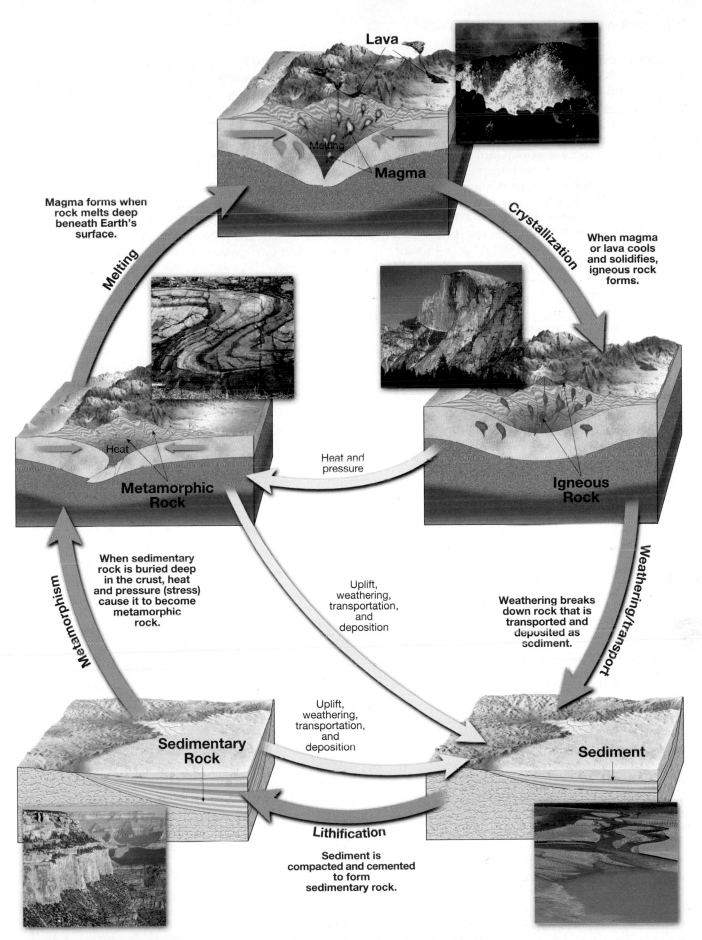

Lava

Melting

Magma

Magma forms when rock melts deep beneath Earth's surface.

Melting

Crystallization

When magma or lava cools and solidifies, igneous rock forms.

Heat and pressure

Metamorphic Rock

Heat

Igneous Rock

When sedimentary rock is buried deep in the crust, heat and pressure (stress) cause it to become metamorphic rock.

Metamorphism

Uplift, weathering, transportation, and deposition

Weathering/transport

Weathering breaks down rock that is transported and deposited as sediment.

Sedimentary Rock

Uplift, weathering, transportation, and deposition

Sediment

Lithification

Sediment is compacted and cemented to form sedimentary rock.

FIGURE 1.23 Viewed over long spans, rocks are constantly forming, changing, and reforming. The rock cycle helps us understand the origin of the three basic rock groups. Arrows represent processes that link each group to the others.

called *crystallization*, may occur either beneath the surface or, following a volcanic eruption, at the surface. In either situation, the resulting rocks are called *igneous rocks*.

If igneous rocks are exposed at the surface, they will undergo *weathering*, in which the day-in and day-out influences of the atmosphere slowly disintegrate and decompose rocks. The materials that result are often moved downslope by gravity before being picked up and transported by any of a number of erosional agents, such as running water, glaciers, wind, or waves. Eventually these particles and dissolved substances, called *sediment*, are deposited. Although most sediment ultimately comes to rest in the ocean, other sites of deposition include river floodplains, desert basins, swamps, and dunes.

Next the sediments undergo *lithification*, a term meaning "conversion into rock." Sediment is usually lithified into *sedimentary rock* when compacted by the weight of overlying layers or when cemented as percolating groundwater fills the pores with mineral matter.

If the resulting sedimentary rock is buried deep within Earth and involved in the dynamics of mountain building or intruded by a mass of magma, it will be subjected to great pressures and/or intense heat. The sedimentary rock will react to the changing environment and turn into the third rock type, *metamorphic rock*. When metamorphic rock is subjected to additional pressure changes or to still higher temperatures, it will melt, creating magma, which will eventually crystallize into igneous rock.

Processes driven by heat from Earth's interior are responsible for creating igneous and metamorphic rocks. Weathering and erosion, external processes powered by energy from the Sun, produce the sediment from which sedimentary rocks form.

Alternative Paths   The paths shown in the basic cycle are not the only ones that are possible. To the contrary, other paths are just as likely to be followed as those described in the preceding section. These alternatives are indicated by the blue arrows in Figure 1.23.

Igneous rocks, rather than being exposed to weathering and erosion at Earth's surface, may remain deeply buried. Eventually these masses may be subjected to the strong compressional forces and high temperatures associated with mountain building. When this occurs, they are transformed directly into metamorphic rocks.

Metamorphic and sedimentary rocks, as well as sediment, do not always remain buried. Rather, overlying layers may be stripped away, exposing the once buried rock. When this happens, the material is attacked by weathering processes and turned into new raw materials for sedimentary rocks.

Although rocks may seem to be unchanging masses, the rock cycle shows that they are not. The changes, however, take time—great amounts of time.

Summary

- *Geology* means "the study of Earth." The two broad areas of the science of geology are (1) *physical geology*, which examines the materials composing Earth and the processes that operate beneath and upon its surface; and (2) *historical geology*, which seeks to understand the origin of Earth and its development through time.

- The relationship between people and the natural environment is an important focus of geology. This includes natural hazards, resources, and human influences on geologic processes.

- During the 17th and 18th centuries, *catastrophism* influenced the formulation of explanations about Earth. Catastrophism states that Earth's landscapes have been developed primarily by great catastrophes. By contrast, *uniformitarianism*, one of the fundamental principles of modern geology advanced by *James Hutton* in the late 1700s, states that the physical, chemical, and biological laws that operate today have also operated in the geologic past. The idea is often summarized as "The present is the key to the past." Hutton argued that processes that appear to be slow-acting could, over long spans of time, produce effects that were just as great as those resulting from sudden catastrophic events.

- Using the principles of *relative dating*, the placing of events in their proper sequence or order without knowing their age in years, scientists developed a geologic time scale during the 19th century. Relative dates can be established by applying such principles as the *law of superposition* and the *principle of fossil succession*.

- All science is based on the assumption that the natural world behaves in a consistent and predictable manner. The process by which scientists gather facts and formulate scientific *hypotheses* and *theories* is called the *scientific method*. To determine what is occurring in the natural world, scientists often (1) collect facts, (2) develop a scientific hypothesis, (3) construct experiments to test the hypothesis, and (4) accept, modify, or reject the hypothesis on the basis of extensive testing. Other discoveries represent purely theoretical ideas that have stood up to extensive examination. Still other scientific advancements have been made when a totally unexpected happening occurred during an experiment.

- Earth's physical environment is traditionally divided into three major parts: the solid Earth or *geosphere*; the water portion of our planet, the *hydrosphere*; and Earth's gaseous envelope, the *atmosphere*. In addition, the *biosphere,* the totality of life on Earth, interacts with each of the three physical realms and is an equally integral part of Earth.

- Although each of Earth's four spheres can be studied separately, they are all related in a complex and continuously interacting whole that we call the *Earth system. Earth system science* uses an interdisciplinary approach to integrate the knowledge of several academic fields in the study of our planet and its global environmental problems.

- A *system* is a group of interacting parts that form a complex whole. *Closed systems* are those in which energy moves freely in and out, but matter does not enter or leave the system. In an open system, both energy and matter flow into and out of the system.

- Most natural systems have mechanisms that tend to enhance change, called *positive feedback mechanisms,* and other mechanisms, called *negative feedback mechanisms,* that tend to resist change and thus stabilize the system.

- The two sources of energy that power the Earth system are (1) the Sun, which drives the external processes that occur in the atmosphere, hydrosphere, and at Earth's surface; and (2) heat from Earth's interior, which powers the internal processes that produce volcanoes, earthquakes, and mountains.

- The *nebular hypothesis* describes the formation of the solar system. The planets and Sun began forming about 5 billion years ago from a large cloud of dust and gases. As the cloud contracted, it began to rotate and assume a disk shape. Material that was gravitationally pulled toward the center became the *protosun.* Within the rotating disk, small centers, called *protoplanets,* swept up more and more of the cloud's debris. Because of the high temperatures near the Sun, the inner planets were unable to accumulate many of the elements that vaporize at low temperatures. Because of the very cold temperatures existing far from the Sun, the large outer planets consist of huge amounts of ices and lighter materials. These substances account for the comparatively large sizes and low densities of the outer planets.

- Earth's internal structure is divided into layers based on differences in chemical composition and on the basis of changes in physical properties. Compositionally, Earth is divided into a thin outer *crust,* a solid rocky *mantle,* and a dense *core.* Other layers, based on physical properties, include the *lithosphere, asthenosphere, transition zone, lower mantle, D" layer, outer core,* and *inner core.*

- Two principal divisions of Earth's surface are the *continents* and *ocean basins.* A significant difference is their relative levels. The elevation differences between continents and ocean basins is primarily the result of differences in their respective densities and thicknesses.

- The largest features of the continents can be divided into two categories: *mountain belts* and the *stable interior.* The ocean floor is divided into three major topographic units: *continental margins, deep-ocean basins,* and *oceanic ridges.*

- The *rock cycle* is one of the many cycles or loops of the Earth system in which matter is recycled. The rock cycle is a means of viewing many of the interrelationships of geology. It illustrates the origin of the three basic rock types and the role of various geologic processes in transforming one rock type into another.

Review Questions

1. Geology is traditionally divided into two broad areas. Name and describe these two subdivisions.

2. Briefly describe Aristotle's influence on the science of geology.

3. How did the proponents of catastrophism perceive the age of Earth?

4. Describe the doctrine of uniformitarianism. How did the advocates of this idea view the age of Earth?

5. About how old is Earth?

6. The geologic time scale was established without the aid of radiometric dating. What principles were used to develop the time scale?

7. How is a scientific hypothesis different from a scientific theory?

8. List and briefly describe the four "spheres" that constitute our environment.

9. How is an open system different from a closed system?

10. Contrast positive feedback mechanisms and negative feedback mechanisms.

11. What are the two sources of energy for the Earth system?

12. Briefly describe the events that led to the formation of the solar system.

13. List and briefly describe Earth's compositional divisions.

14. Contrast the asthenosphere and the lithosphere.

15. Describe the general distribution of Earth's youngest mountains.

16. Distinguish between shields and stable platforms.

17. List the three major topographic units of the ocean floor.

18. Name each of the rocks described below:
    - light-colored, coarse-grained intrusive rock
    - detrital rock rich in clay-size particles
    - a fine-grained black rock that makes up the oceanic crust
    - nonfoliated rock, for which limestone is its parent rock

19. For each characteristic below, indicate whether it is associated with igneous, sedimentary, or metamorphic rocks:
    - may be intrusive or extrusive
    - lithified by compaction and cementation
    - sandstone is an example
    - some members of this group are foliated
    - this group is divided into detrital and chemical categories
    - gneiss is a member of this group

20. Using the rock cycle, explain the statement "One rock is the raw material for another."

Key Terms

abyssal plain (p. 24)
asthenosphere (p. 19)
atmosphere (p. 13)
biosphere (p. 13)
catastrophism (p. 4)
closed system (p. 15)
continental margin (p. 24)
continental rise (p. 24)
continental shelf (p. 24)
continental slope (p. 24)
core (p. 20)
craton (p. 23)
crust (p. 19)
D" layer (p. 20)
deep-ocean basin (p. 24)

deep-ocean trench (p. 24)
Earth system science (p. 14)
fossil succession, principle of (p. 6)
geology (p. 2)
geosphere (p. 14)
historical geology (p. 2)
hydrosphere (p. 13)
hypothesis (p. 10)
igneous rock (p. 25)
inner core (p. 20)
interface (p. 15)
lithosphere (p. 19)
lower mantle (p. 20)
mantle (p. 19)

metamorphic rock (p. 27)
model (p. 10)
nebular hypothesis (p. 17)
negative feedback mechanism (p. 15)
oceanic (mid-ocean) ridge (p. 25)
open system (p. 15)
outer core (p. 20)
paradigm (p. 10)
physical geology (p. 2)
positive feedback mechanism (p. 15)
relative dating (p. 6)
rock cycle (p. 28)

seamount (p. 25)
sedimentary rock (p. 25)
shield (p. 23)
solar nebula (p. 17)
stable platform (p. 24)
superposition, law of (p. 6)
system (p. 14)
theory (p. 10)
transition zone (p. 19)
uniformitarianism (p. 4)

Web Resources

The *Earth* Website uses the resources and flexibility of the Internet to aid in your study of the topics in this chapter. Written and developed by geology instructors, this site will help improve your understanding of geology. Visit **http://www. prenhall.com/tarbuck** and click on the cover of *Earth 9e* to find:

- Online review quizzes.
- Critical thinking exercises.
- Links to chapter-specific Web resources.
- Internet-wide key-term searches.

http://www.prenhall.com/tarbuck

GEODe: Earth

GEODe: Earth makes studying faster and more effective by reinforcing key concepts using animation, video, narration, interactive exercises and practice quizzes. A copy is included with every copy of *Earth*.

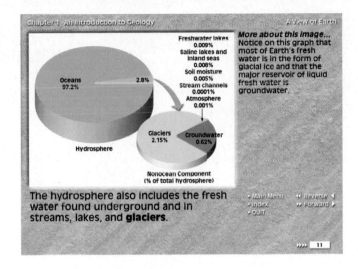

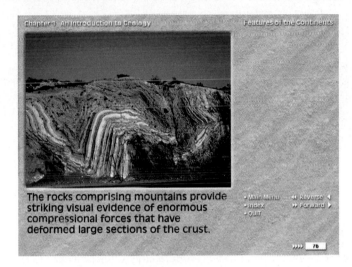

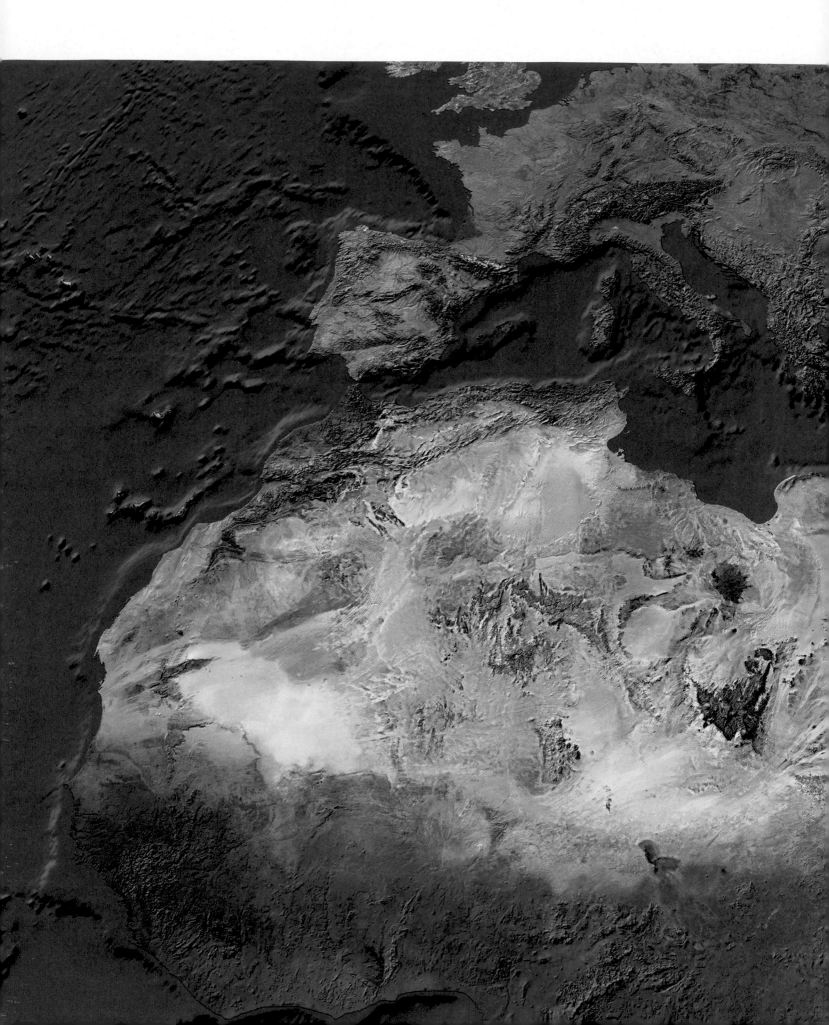

Plate Tectonics: A Scientific Revolution Unfolds

Composite satellite image of Europe, North Africa, and the Arabian Peninsula. (Image © by Worldsat International, Inc., 2001, www.worldsat.ca. All Rights Reserved)

The notion that continents gradually drift about the face of Earth was introduced at the beginning of the 20th century. This proposal was in stark contrast to the established view that the ocean basins and continents are permanent features of great antiquity. This view was reinforced by evidence gathered from the study of earthquake waves that revealed a solid, rocky mantle extending halfway to the center of Earth. The concept of a solid mantle led most investigators to conclude that Earth's outer shell was incapable of movement.

During this period, the conventional wisdom of the scientific community was that mountains form due to compressional forces created as Earth gradually cooled from a once molten state. Simply, the explanation went like this—as the interior cooled and contracted, Earth's solid outer skin was deformed by folding and faulting to fit the shrinking planet. Mountains were viewed as being somewhat analogous to the wrinkles on a dried-out piece of fruit. This model of Earth's tectonic* processes, however inadequate, was firmly entrenched in the geologic thought of the time.

Since the 1960s, our understanding of the nature and workings of our planet has dramatically improved. Scientists have come to realize that Earth's outer shell is mobile and that continents gradually migrate across the globe. Further, landmasses occasionally split apart to create new ocean basins between diverging continental blocks. Meanwhile, older portions of the seafloor plunge back into the mantle in the vicinity of deep-ocean trenches. Because of these movements, blocks of continental material collide and generate Earth's great mountain ranges (Figure 2.1). In short, a revolutionary new model of Earth's tectonic processes has emerged.

This profound reversal of scientific understanding has been appropriately described as a *scientific revolution.* The revolution began as a relatively straightforward proposal by Alfred Wegener, called *continental drift.* After many years of heated debate, Wegener's hypothesis of drifting continents was rejected by the vast majority of the scientific community. The concept of a mobile Earth was particularly distasteful to North American geologists, perhaps because much of the supporting evidence had been gathered from the southern continents, with which most North American geologists were unfamiliar.

During the 1950s and 1960s, new kinds of evidence began to rekindle interest in this nearly abandoned proposal. By 1968 these new developments led to the unfolding of a far more encompassing explanation that incorporated aspects of continental drift and seafloor spreading—a theory known as *plate tectonics.*

In this chapter, we will examine the events that led to this dramatic reversal of scientific opinion in an attempt to provide some insight into how science works. We will also briefly trace the developments of the concept of continental drift, examine why it was first rejected, and consider the evidence that finally led to the acceptance of the theory of plate tectonics.

Tectonics refers to the study of the processes that deform Earth's crust and the major structural features produced by that deformation, such as mountains, continents, and ocean basins.

FIGURE 2.1 Climbers camping on a shear rock face of a mountain known as K7 in Pakistan's Karakoram, a part of the Himalayas. These mountains formed as India collided with Eurasia. (Photo by Jimmy Chin/National Geographic/Getty)

Continental Drift: An Idea Before Its Time

The idea that continents, particularly South America and Africa, fit together like pieces of a jigsaw puzzle originated with the development of reasonably accurate world maps. However, little significance was given this notion until 1915, when Alfred Wegener, a German meteorologist and geophysicist, published *The Origin of Continents and Oceans*. In his book, which was published in several editions, Wegener set forth the basic outline of his radical hypothesis of **continental drift.**

Wegener suggested that a single *supercontinent* called **Pangaea** (*pan* = all, *gaea* = Earth) had once existed (Figure 2.2). He further hypothesized that in the Mesozoic era, about 200 million years ago, this supercontinent began to fragment into smaller continents, which then gradually "drifted" to their present positions (see Box 2.1). It is believed that Wegener's notion that continents might break apart may have been triggered when he observed the breakup of sea ice during an expedition to Greenland between 1906 and 1908.

Wegener and others who advocated the continental drift hypothesis collected substantial evidence to support their point of view. The fit of South America and Africa and the geographic distribution of fossils and ancient climates all seemed to buttress the idea that these now separate landmasses were once joined. Let us examine this evidence.

Fit of the Continents

Like a few others before him, Wegener first suspected that the continents might once have been joined when he noticed the remarkable similarity between the coastlines on opposite sides of the Atlantic Ocean. However, Wegener's use of present-day shorelines to make a fit of the continents was challenged immediately by other Earth scientists. These opponents correctly argued that shorelines are continually modified by erosional and depositional processes. Even if continental displacement had taken place, such a fit today would be unlikely. Wegener was apparently aware of this fact since his original jigsaw fit of the continents was only very crude (Figure 2.2B).

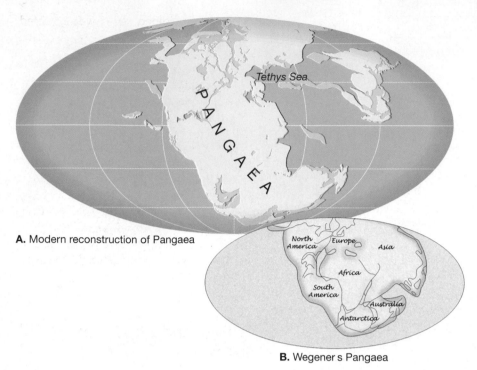

A. Modern reconstruction of Pangaea

B. Wegener s Pangaea

FIGURE 2.2   Reconstructions of Pangaea as it is thought to have appeared 200 million years ago. **A.** Modern reconstruction. **B.** Reconstruction done by Wegener in 1915.

Scientists have determined that a much better approximation of the true outer boundary of a continent is the seaward edge of its continental shelf, which lies submerged a few hundred meters below sea level. In the early 1960s, Sir Edward Bullard and two associates produced a map that attempted to fit the edges of the continental shelves of South America and Africa at depths of 900 meters. The remarkable fit that was obtained is shown in Figure 2.3. Although the continents overlap in a few places, these are regions where streams have deposited large quantities of sediment, thus enlarging the continental shelves. The overall fit was even better than these researchers suspected it would be.

Fossil Evidence

Although the seed for Wegener's hypothesis came from the remarkable similarities of the continental margins on opposite sides of the Atlantic, he thought the idea of a mobile Earth was improbable. Not until he learned that identical fossil organisms were known from rocks in both South America and Africa did he begin to seriously pursue this idea. Through a review of the literature, Wegener learned that most paleontologists (scientists who study the fossilized remains of organisms) were in agreement that some type of land connection was needed to explain the existence of identical fossils of Mesozoic life forms on widely separated landmasses. (Just as modern life forms native to North America are quite different from those of Africa, one would expect that during the Mesozoic era, organisms on widely separated continents would be quite distinct.)

Mesosaurus   To add credibility to his argument for the existence of a supercontinent, Wegener cited documented cases of several fossil organisms that were found on different landmasses despite the unlikely possibility that their living forms could have crossed the vast ocean presently separating these continents. The classic example is *Mesosaurus,* an aquatic fish-catching reptile whose fossil remains are found

FIGURE 2.3   This shows the best fit of South America and Africa along the continental slope at a depth of 500 fathoms (about 900 meters). The areas where continental blocks overlap appear in brown. (After A. G. Smith, "Continental Drift," in *Understanding the Earth*, edited by I. G. Gass.)

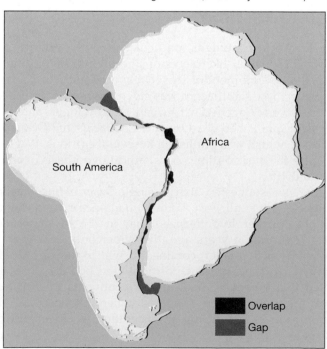

South America

Africa

Overlap

Gap

Students Sometimes Ask . . .

If all the continents were joined during the time of Pangaea, what did the rest of Earth look like?

When all the continents were together, there must also have been one huge ocean surrounding them. This ocean is called *Panthalassa* (*pan* = all, *thalassa* = sea). Panthalassa included several smaller seas, one of which was the centrally located and shallow *Tethys Sea* (see Figure 2.2). About 180 million years ago the supercontinent of Pangaea began to split apart, and the various continental masses we know today started to drift toward their present geographic positions. Today all that remains of Panthalassa is the Pacific Ocean, which has been decreasing in size since the breakup of Pangaea.

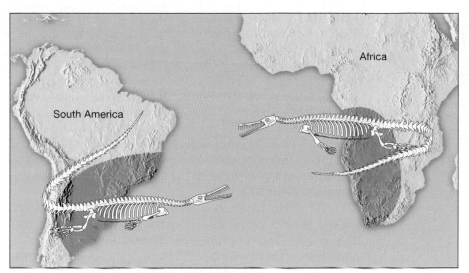

FIGURE 2.4 Fossils of *Mesosaurus* have been found on both sides of the South Atlantic and nowhere else in the world. Fossil remains of this and other organisms on the continents of Africa and South America appear to link these landmasses during the late Paleozoic and early Mesozoic eras.

only in black shales of Permian age (about 260 million years ago) in eastern South America and southern Africa (Figure 2.4). If *Mesosaurus* had been able to make the long journey across the vast South Atlantic Ocean, its remains should be more widely distributed. As this is not the case, Wegener argued that South America and Africa must have been joined during that period of Earth history.

How did scientists during Wegener's era explain the existance of identical fossil organisms in places separated by thousands of kilometers of open ocean? Transoceanic land bridges were the most widely accepted explanations of such migrations (Figure 2.5). We know, for example, that during the recent Ice Age the lowering of sea level allowed animals to cross the narrow Bering Strait between Asia and North America. Was it possible that land bridges once connected Africa and South America but later subsided below sea level? Modern maps of the seafloor substantiate Wegener's contention that land bridges of this magnitude had not existed. If they had, remnants would still lie below sea level.

Glossopteris Wegener also cited the distribution of the fossil fern *Glossopteris* as evidence for the existence of Pangaea. This plant, identified by its large seeds that were not easily distributed, was known to be widely dispersed among Africa, Australia, India, and South America during the late Paleozoic era. Later, fossil remains of *Glossopteris* were discovered in Antarctica as well. Wegener also learned that these seed ferns and associated flora grew only in a subpolar climate. Therefore, he concluded that when these

landmasses were joined, they were located much closer to the South Pole.

Present-day Organisms In a later edition of his book, Wegener also cited the distribution of present-day organisms as evidence to support the drifting of continents. For example, modern organisms with similar ancestries clearly had to evolve in isolation during the last few tens of millions of years. Most obvious of these are the Australian marsupials (such as kangaroos), which have a direct fossil link to the marsupial opossums found in the Americas. After the breakup of Pangaea, the Australian marsupials followed a different evolutionary path than related life forms in the Americas.

Rock Type and Structural Similarities

Anyone who has worked a picture puzzle knows that in addition to the pieces fitting together, the picture must be continuous as well. The "picture" that must match in the "continental drift puzzle" is one of rock types and mountain belts found on the continents. If the continents were once together, the rocks found in a particular region on one continent should closely match in age and type those found in adjacent positions on the adjoining continent. Wegener found evidence of 2.2-billion-year-old igneous rocks in Brazil that closely resembled similarly aged rocks in Africa.

FIGURE 2.5 These sketches by John Holden illustrate various explanations for the occurrence of similar species on landmasses that are presently separated by vast oceans. (Reprinted with permission of John Holden)

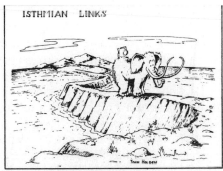

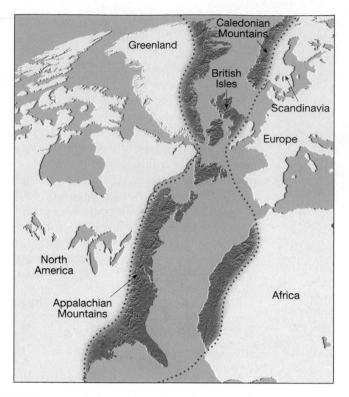

FIGURE 2.6 Matching mountain ranges across the North Atlantic. The Appalachian Mountains trend along the eastern flank of North America and disappear off the coast of Newfoundland. Mountains of comparable age and structure are found in Greenland, the British Isles, and Scandinavia. When these landmasses are placed in their predrift locations, these ancient mountain chains form a nearly continuous belt. These folded mountain belts formed roughly 300 million years ago as the landmasses collided during the formation of the supercontinent of Pangaea.

Similar evidence exists in the form of mountain belts that terminate at one coastline, only to reappear on landmasses across the ocean. For instance, the mountain belt that includes the Appalachians trends northeastward through the eastern United States and disappears off the coast of Newfoundland. Mountains of comparable age and structure are found in Greenland, the British Isles, and Scandinavia. When these landmasses are reassembled, as in Figure 2.6, the mountain chains form a nearly continuous belt.

Wegener must have been convinced that the similarities in rock structure on both sides of the Atlantic linked these landmasses when he said, "It is just as if we were to refit the torn pieces of a newspaper by matching their edges and then check whether the lines of print run smoothly across. If they do, there is nothing left but to conclude that the pieces were in fact joined in this way."*

Paleoclimatic Evidence

Because Alfred Wegener was a meteorologist by profession, he was keenly interested in obtaining paleoclimatic (*paleo* = ancient, *climatic* = climate) data to support conti-

nental drift. His efforts were rewarded when he found evidence for apparently dramatic global climatic changes during the geologic past. In particular, he learned of ancient glacial deposits that indicated that near the end of the Paleozoic era (about 300 million years ago), ice sheets covered extensive areas of the Southern Hemisphere and India (Figure 2.7A). Layers of glacially transported sediments of the same age were found in southern Africa and South America, as well as in India and Australia. Much of the land area containing evidence of this late Paleozoic glaciation presently lies within 30 degrees of the equator in subtropical or tropical climates.

Could Earth have gone through a period of sufficient cooling to have generated extensive ice sheets in areas that are presently tropical? Wegener rejected this explanation because during the late Paleozoic, large tropical swamps existed in the Northern Hemisphere. These swamps, with their lush vegetation, eventually became the major coal fields of the eastern United States, Europe, and Siberia.

Fossils from these coal fields indicate that the tree ferns that produced the coal deposits had large fronds, which are indicative of tropical settings. Furthermore, unlike trees in colder climates, these trees lacked growth rings, a characteristic of tropical plants that grow in regions having minimal fluctuations in temperature.

Wegener suggested that a more plausible explanation for the late Paleozoic glaciation was provided by the

FIGURE 2.7 Paleoclimatic evidence for continental drift. **A.** Near the end of the Paleozoic era (about 300 million years ago) ice sheets covered extensive areas of the Southern Hemisphere and India. Arrows show the direction of ice movement that can be inferred from glacial grooves in the bedrock. **B.** Shown are the continents restored to their former position with the South Pole located roughly between Antarctica and Africa. This configuration accounts for the conditions necessary to generate a vast glacial ice sheet and also explains the directions of ice movement that radiated away from the South Pole.

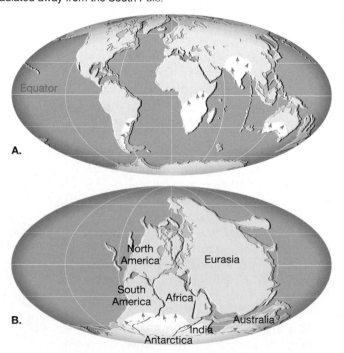

BOX 2.1 ▶ UNDERSTANDING EARTH

The Breakup of Pangaea

Wegener used evidence from fossils, rock types, and ancient climates to create a jig-saw-puzzle fit of the continents—thereby creating his supercontinent of Pangaea. In a similar manner, but employing modern tools not available to Wegener, geologists have recreated the steps in the breakup of this supercontinent, an event that began nearly 180 million years ago. From this work, the dates when individual crustal fragments separated from one another and their relative motions have been well established (Figure 2.A).

An important consequence of the breakup of Pangaea was the creation of a "new" ocean basin: the Atlantic. As you can see in part B of Figure 2.A, splitting of the supercontinent did not occur simultaneously along the margins of the Atlantic. The first split developed between North America and Africa.

Here, the continental crust was highly fractured, providing pathways for huge quantities of fluid lavas to reach the surface. Today these lavas are represented by weathered igneous rocks found along the Eastern Seaboard of the United States—primarily buried beneath the sedimentary rocks that form the continental shelf. Radiometric dating of these solidified lavas indicate that rifting began in various stages between 180 million and 165 million years ago. This time span can be used as the "birth date" for this section of the North Atlantic.

By 130 million years ago, the South Atlantic began to open near the tip of what is now South Africa. As this zone of rifting migrated northward, it gradually opened the South Atlantic (compare Figure 2.A, parts B and C). Continued breakup of the southern landmass led to the separation of Africa and Antarctica and sent India on a

northward journey. By the early Cenozoic, about 50 million years ago, Australia had separated from Antarctica, and the South Atlantic had emerged as a full-fledged ocean (Figure 2.A, part D).

A modern map (Figure 2.A, part F) shows that India eventually collided with Asia, an event that began about 45 million years ago and created the Himalayas as well as the Tibetan Highlands. About the same time, the separation of Greenland from Eurasia completed the breakup of the northern landmass. During the last 20 million years or so of Earth history, Arabia rifted from Africa to form the Red Sea, and Baja California separated from Mexico to form the Gulf of California (Figure 2.A, part E). Meanwhile, the Panama Arc joined North America and South America to produce our globe's familiar, modern appearance.

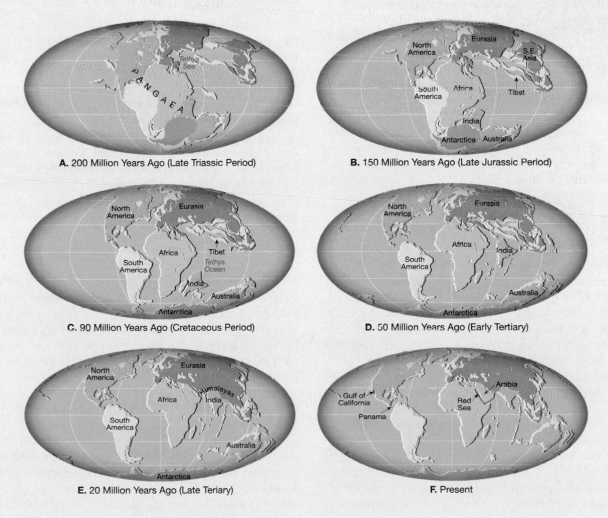

A. 200 Million Years Ago (Late Triassic Period)

B. 150 Million Years Ago (Late Jurassic Period)

C. 90 Million Years Ago (Cretaceous Period)

D. 50 Million Years Ago (Early Tertiary)

E. 20 Million Years Ago (Late Teriary)

F. Present

FIGURE 2.A   Several views of the breakup of Pangaea over a period of 200 million years.

supercontinent of Pangaea. In this configuration the southern continents are joined together and located near the South Pole (Figure 2.7B). This would account for the conditions necessary to generate extensive expanses of glacial ice over much of the Southern Hemisphere. At the same time, this geography would place today's northern landmasses nearer the equator and account for their vast coal deposits. Wegener was so convinced that his explanation was correct that he wrote, "This evidence is so compelling that by comparison all other criteria must take a back seat."

How does a glacier develop in hot, arid central Australia? How do land animals migrate across wide expanses of open water? As compelling as this evidence may have been, 50 years passed before most of the scientific community accepted the concept of continental drift and the logical conclusions to which it led.

The Great Debate

Wegener's proposal did not attract much open criticism until 1924, when his book was translated into English, French, Spanish, and Russian. From that point until his death, in 1930, his drift hypothesis encountered a great deal of hostile criticism. The respected American geologist R. T. Chamberlain stated, "Wegener's hypothesis in general is of the foot-loose type, in that it takes considerable liberty with our globe, and is less bound by restrictions or tied down by awkward, ugly facts than most of its rival theories. Its appeal seems to lie in the fact that it plays a game in which there are few restrictive rules and no sharply drawn code of conduct."

W. B. Scott, former president of the American Philosophical Society, expressed the prevalent American view of continental drift in fewer words when he described the hypothesis as "utter damned rot!"

Rejection of the Continental Drift Hypothesis

One of the main objections to Wegener's hypothesis appears to have stemmed from his inability to identify a mechanism that was capable of moving the continents across the globe. Wegener suggested two possible mechanisms for continental drift. One of these was the gravitational force that the Moon and Sun exert on Earth to produce the tides. Wegener argued that tidal forces would chiefly affect Earth's outermost layer, which would slide as detached continental fragments over the interior. However, the prominent physicist Harold Jeffreys correctly countered that tidal forces of the magnitude needed to displace the continents would bring Earth's rotation to a halt in a matter of a few years.

Wegener also incorrectly suggested that the larger and sturdier continents broke through the oceanic crust, much like ice breakers cut through ice. However, no evidence existed to suggest that the ocean floor was weak enough to permit passage of the continents without the continents being appreciably deformed in the process.

By 1929 strong opposition to Wegener's ideas was pouring in from all areas of the scientific community. Despite these affronts, Wegener wrote the fourth edition of his book, maintaining his basic hypothesis and adding supporting evidence.

In 1930 Wegener made his fourth and final trip to the Greenland Ice Sheet. Although the primary focus of this expedition was to study the harsh winter weather on the ice-covered island, Wegener continued to test his continental drift hypothesis. Wegener felt strongly that repeated determinations of longitude on exactly the same spot would verify the westward drift of Greenland with respect to Europe. Although early efforts using astronomical methods seemed promising, Danish workers who took measurements in 1927, 1936, 1938, and 1948 found no evidence of drift. Thus, Wegener's critical test proved to be a failure and his hypothesis was discredited. Today modern techniques allow scientists to measure the gradual displacement of the continents that Wegener had hoped to detect.

In November 1930, while returning from Eismitte (an experimental station located in the center of Greenland), Wegener perished along with a companion (see Box 2.2). His intriguing idea, however, did not die with him.

Continental Drift and the Scientific Method

What went wrong? Why was Wegener unable to overturn the established scientific views of his day? First, although the core of his drift hypothesis was correct, it contained many incorrect details. For example, the continents do not break through the ocean floor, and tidal energy is much too weak to cause continents to move. Moreover, in order for any comprehensive scientific theory to gain wide acceptance, it must stand up to critical testing from all areas of science. This same idea was stated very well by Wegener himself in response to his critics when he said, "Scientists still do not appear to understand sufficiently that all Earth sciences must contribute evidence toward unveiling the state of our planet in earlier times, and the truth of the matter can only be reached by combining all this evidence." Wegener's great contribution to our understanding of Earth notwithstanding, not *all* of the evidence supported the continental drift hypothesis as he had formulated it. Therefore, Wegener himself answered the very question he must have asked many times: "Why do they reject my proposal?"

Although many of Wegener's contemporaries opposed his views, even to the point of open ridicule, some considered his ideas plausible. Among the most notable of this latter group were the eminent South African geologist Alexander du Toit and the well-known Scottish geologist Arthur Holmes. In 1937 du Toit published *Our Wandering Continents*, in which he eliminated some of Wegener's weaker arguments and added a wealth of new evidence in support of this revolutionary idea. In 1928 Arthur Holmes proposed the first plausible driving mechanism for continental drift. In Holmes's book *Physical Geology*, he elaborated on this idea by suggesting that convection currents operating within the mantle were responsible for propelling the continents across the globe.

BOX 2.2 ▶ UNDERSTANDING EARTH

Alfred Wegener (1880–1930): Polar Explorer and Visionary

Alfred Wegener, polar explorer and visionary, was born in Berlin in 1880. His undergraduate and graduate studies were completed in Heidelberg and Innsbruck. Although he received his doctorate in astronomy (1905), he also developed a strong interest in meteorology. In 1906 he and his brother Kurt set an endurance record for balloon flight by staying aloft for 52 hours, breaking the previous record by 17 hours. Later that year, he joined a Danish expedition to northeastern Greenland where he may have first entertained the possibility of continental drift. That journey marked the beginning of his lifelong dedication to exploring this ice-covered island where he would perish nearly 25 years later.

Following his first expedition to Greenland, Wegener returned to Germany in 1908 and took an academic position as a lecturer in meteorology and astronomy. During that time, he authored a paper on continental drift as well as a book on meteorology. Wegener returned to Greenland from 1912–1913 with colleague J. P. Koch for an expedition that distinguished Wegener as the first person to make a scientific crossing of the 1200-kilometer (700-mile) glacial core of the island (Figure 2.B).

Shortly after returning from Greenland, Wegener married Else Köppen, daughter of Wladimir Köppen, an eminent climatologist who developed a classification of world climates that is still used today. Shortly after his marriage, Wegener was a combatant in the First World War, during which he was wounded twice, but remained in the army until the war was over. During his convalescence, Wegener wrote his controversial book on continental drift, entitled *The Origin of Continents and Oceans*. Wegener authored revised editions in 1920, 1922, and 1929.

In addition to his passion for finding supporting evidence for continental drift, Wegener also wrote numerous scientific papers on meteorology and geophysics. In 1924 he coauthored a book on ancient climatic changes (paleoclimates) with his father-in-law, Köppen.

In the spring of 1930, Wegener departed on his fourth and final expedition to his beloved Greenland. One of the goals of the journey was to establish a mid-ice camp (Eismitte station) located 400 kilometers (250 miles) from Greenland's west coast at an elevation of nearly 3000 meters (10,000 feet). Because unusually bad weather hampered attempts to establish this outpost, only a fraction of the supplies that were needed for the two scientists stationed there made it to the camp.

As the expedition leader, Wegener led a relief party consisting of fellow meteorologist Fritz Lowe and 13 Greenlanders to resupply Eismitte Station. Heavy snow and temperatures that fell below −50° C (−58° F) caused all but one Greenlander to return to base camp. Wegener, Lowe, and Rasmus Villumsen trudged on.

Forty days later, on October 30, 1930, Wegener and his two companions reached Eismitte Station. Unable to communicate with base camp, the researchers who were believed to be in dire need of supplies had managed to dig an ice cave for shelter and intended to stretch their supplies through the winter. The heroic supply run had been unnecessary.

Lowe decided to winter at Eismitte due to exhaustion and frostbitten limbs. Wegener, however, was reported to have "looked as fresh, happy and fit as if he had just been on a walk." Two days later, on November 1, 1930, they celebrated Wegener's 50th birthday, and he and his Greenlander companion, Rasmus Villumsen, set off on the downhill trek back to the coast. They never arrived.

Because of the inability to maintain contact between stations during the winter months, the two were thought to have wintered over at Eismitte. While the exact date and cause of Wegener's death remain unknown, a search team located his body beneath the snow, approximately halfway between Eismitte and the coast. Because Wegener was known to be physically fit and his body showed no signs of trauma, starvation, or exposure, it is thought that he may have suffered a fatal heart attack. Villumsen, Wegener's companion, is assumed to have died during the journey as well, although his remains were never located.

The search team buried Wegener in the position in which he was found and respectfully constructed a snow-block monument. Later, a 20-foot iron cross was erected at the site. All have long since vanished beneath the snow to eventually become part of this great ice sheet.

FIGURE 2.B Alfred Wegener shown waiting out the 1912–1913 Arctic winter during an expedition to Greenland, where he made a 1200-kilometer traverse across the widest part of the island's ice sheet. (Photo courtesy of Bildarchiv Preussischer Kulturbesitz, Berlin)

For those geologists who continued the search, the exciting concept of continents in motion held their interest. Others viewed continental drift as a solution to previously unexplainable observations. Nevertheless, most of the scientific community, particularly in North America, either rejected continental drift outrightly or at least treated it with considerable scepticism.

Continental Drift and Paleomagnetism

Little additional light was shed on the continental drift hypothesis during the two decades following Wegener's death in 1930. However, by the mid-1950s two new lines of evidence began to emerge, which seriously questioned science's basic understanding of how Earth works. One line of evidence came from explorations of the seafloor and will be considered later. The other line of evidence came from a comparatively new field—*paleomagnetism.*

Earth's Magnetic Field and Fossil Magnetism

Anyone who has used a compass to find direction knows that Earth's magnetic field has a north and south magnetic pole. Today these magnetic poles align closely, but not exactly, with the geographic poles. (The geographic poles, or true north and south poles, are where Earth's rotational axis intersects the surface.) Earth's magnetic field is similar to that produced by a simple bar magnet. Invisible lines of force pass through the planet and extend from one magnetic pole to the other, as shown in Figure 2.8. A compass needle, itself a small magnet free to rotate on an axis, becomes aligned with the magnetic lines of force and points to the magnetic poles.

Unlike the pull of gravity, we cannot feel Earth's magnetic field, yet its presence is revealed because it deflects a compass needle. In a similar manner, certain rocks contain minerals that serve as "fossil compasses." These iron-rich minerals, such as *magnetite*, are abundant in lava flows of basaltic composition.* When heated above a temperature known as the **Curie point,** these magnetic minerals lose their magnetism. However, when these iron-rich grains cool below their Curie point (about 585°C for magnetite), they gradually become magnetized in the direction of the existing magnetic lines of force. Once the minerals solidify, the magnetism they possess will usually remain "frozen" in this position. In this regard, they behave much like a compass needle; they "point" toward the position of the magnetic poles at the time of their formation. Then, if the rock is moved, the rock magnetism will retain its original alignment. Rocks that formed thousands or millions of years ago and contain a "record" of the direction of the magnetic poles at the time of their formation are said to possess **fossil magnetism,** or **paleomagnetism.**

Another important aspect of rock magnetism is that the magnetized minerals not only indicate the direction to the

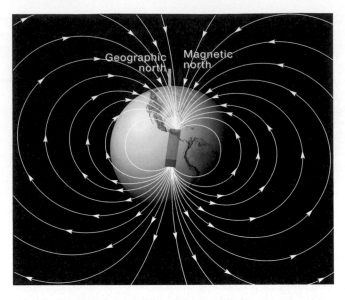

FIGURE 2.8 Earth's magnetic field consists of lines of force much like those a giant bar magnet would produce if placed at the center of Earth.

poles (like a compass) but also provide a means of determining the latitude of their origin. To envision how latitude can be established from paleomagnetism, imagine a compass needle mounted in a vertical plane rather than horizontally, like an ordinary compass. As shown in Figure 2.9, when this modified compass (*dip needle*) is situated over the north magnetic pole, it aligns with the magnetic lines of force and points straight down. However, as this dip needle

FIGURE 2.9 Earth's magnetic field causes a dip needle (compass oriented in a vertical plane) to align with the lines of magnetic force. The dip angle decreases uniformly from 90 degrees at the magnetic poles to 0 degrees at the magnetic equator. Consequently, the distance to the magnetic poles can be determined from the dip angle.

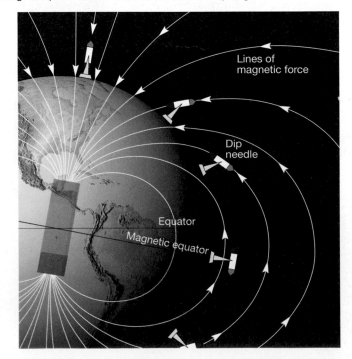

---

*Some sediments and sedimentary rocks contain enough iron-bearing mineral grains to acquire a measurable amount of magnetization.

is moved closer to the equator, the angle of inclination is reduced until the needle becomes horizontal at the equator. Thus, from the dip needle's angle of inclination, one can determine the latitude.

In a similar manner, the inclination of the paleomagnetism in rocks indicates the latitude of the rock *at the time it became magnetized*. Figure 2.10 shows the relationship between the magnetic inclination determined for a rock sample and the latitude where it formed. By knowing the latitude where a rock sample was magnetized, its distance to the magnetic poles can also be determined. For example, lavas that are forming today in Hawaii (about 20° N latitude) are roughly 70 degrees from the north magnetic pole. (This assumes that the average position of the north magnetic pole is the same as the geographic north pole, which is 90° N latitude.) Therefore, rocks in the distant past with magnetization that indicates they formed at 40° N latitude would have been 50° from the north magnetic pole at the time of their formation. If these same rocks were found today at the equator, we could measure their magnetism and determine that they moved 40 degrees in a southerly direction since their formation.

In summary, rock magnetism provides a record of the direction and distance to the magnetic poles at the time a rock unit was magnetized.

Apparent Polar Wandering

A study of rock magnetism conducted during the 1950s in Europe by S. K. Runcorn and his associates led to an interesting discovery. The magnetic alignment in the iron-rich minerals in lava flows of different ages indicated that many different peleomagnetic poles once existed. A plot of the apparent positions of the magnetic north pole with respect to Europe revealed that during the past 500 million years, the location of the pole had gradually wandered from a location near Hawaii northward through eastern Siberia and finally to its present location (Figure 2.11A). This was strong evidence that either the magnetic poles had migrated through

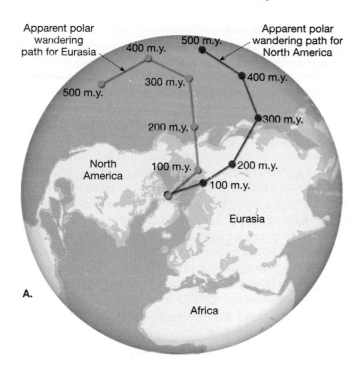

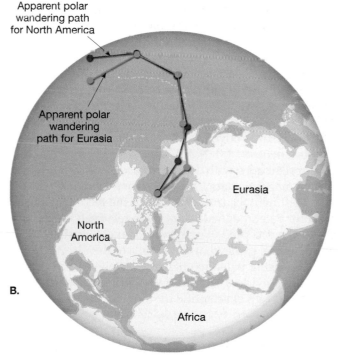

A.

B.

FIGURE 2.11 Simplified apparent polar-wandering paths as established from North American and Eurasian paleomagnetic data. **A.** The more westerly path determined from North American data was caused by the westward movement of North America by about 30 degrees from Eurasia. **B.** The positions of the wandering paths when the landmasses are reassembled.

FIGURE 2.10 Magnetic inclination and the corresponding latitude.

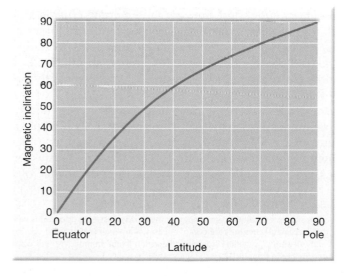

time, an idea known as *polar wandering*, or that the lava flows moved—in other words, Europe had drifted in relation to the poles.

Although the magnetic poles are known to move in an erratic path around the geographic poles, studies of paleomagnetism from numerous locations show that the positions of the magnetic poles, averaged over thousands of years, correspond closely to the positions of the geographic poles. Therefore, a more acceptable explanation for the apparent polar wandering paths was provided by Wegener's hypothesis. If the magnetic poles remain stationary, their *apparent movement* is produced by continental drift.

The latter idea was further supported by comparing the latitude of Europe as determined from fossil magnetism with evidence obtained from paleoclimatic studies. Recall that during the Pennsylvanian period (about 300 million years ago) coal-producing swamps covered much of Europe. During this same time period, paleomagnetic evidence places Europe near the equator—a fact consistent with the tropical environment indicated by these coal deposits.

Further evidence for continental drift came a few years later when a polar-wandering path was constructed for North America (Figure 2.11A). It turned out that paths for North America and Europe had similar shapes but were separated by about 30° of longitude. At the time these rocks crystallized, could there have been two magnetic north poles that migrated parallel to each other? Investigators found no evidence to support this possibility. The differences in these migration paths, however, can be reconciled if the two presently separated continents are placed next to one another, as we now believe they were prior to opening of the Atlantic Ocean. Notice in Figure 2.11B that these apparent wandering paths nearly coincided during the period from about 400 to 160 million years ago. This is evidence that North America and Europe were joined during this time period and moved relative to the poles as part of the same continent.

For researchers with knowledge of, and confidence in, the paleomagnetic data, this was strong evidence that continental drift had occurred. However, the techniques used in extracting paleomagnetic data were relatively new and not universally accepted. Furthermore, most geologists were unfamiliar and somewhat suspicious of the results provided by studies using paleomagnetism. Despite these issues, paleomagnetic evidence reinstated continental drift to a respectable subject of scientific inquiry. A new era had begun!

A Scientific Revolution Begins

Following World War II, oceanographers equipped with new marine tools and ample funding from the U.S. Office of Naval Research embarked on an unprecedented period of oceanographic exploration. Over the next two decades a much better picture of large expanses of the seafloor slowly and painstakingly began to emerge. From this work came the discovery of a global **oceanic ridge system** that winds through all of the major oceans in a manner similar to the seams on a baseball. One of the segments of this interconnected feature extends down the middle of the Atlantic Ocean and hence was named the *Mid-Atlantic Ridge.* Also of importance was the discovery of a central rift valley extending the length of the Mid-Atlantic Ridge. This structure is evidence that tensional forces

are actively pulling the ocean crust apart at the ridge crest. In addition, high heat flow and volcanism were found to characterize the oceanic ridge system.

In other parts of the ocean, new discoveries were also being made. Earthquake studies conducted in the western Pacific demonstrated that tectonic activity was occuring at great depths beneath deep-ocean trenches. Flat-topped seamounts, named *guyots,* were discovered hundreds of meters below sea level. These structures were thought to be former volcanic islands whose tops had been eroded prior to subsiding below sea level. Of equal importance was the fact that dredging of the seafloor did not bring up any oceanic crust that was older than 180 million years. Further, sediment accumulations in the deep-ocean basins were found to be thin, not the thousands of meters that were predicted.

Many of these discoveries were unexpected and difficult to fit into the existing model of Earth's tectonic processes. Recall that geologists thought cooling and contraction in Earth's interior produced the compressional forces that deformed the crust by folding and fracturing. Evidence from the Mid-Atlantic ridge demonstrated that here, at least, the crust was actually being pulled apart. Furthermore, the thin sediment cover blanketing the seafloor requires that the rate of sedimentation in the geologic past must have been much less than the current rate, or that the ocean floor was indeed much younger than previously thought.

The Seafloor-Spreading Hypothesis

In the early 1960s, Harry Hess of Princeton University incorporated these newly discovered facts into a hypothesis that was later named **seafloor spreading.** In Hess's now classic paper, he proposed that oceanic ridges are located above zones of convective upwelling in the mantle (Figure 2.12). As rising material from the mantle spreads laterally, seafloor is carried in a conveyor-belt fashion away from the ridge crest. Here, tensional forces fracture the crust and provide pathways for magma to intrude and generate new slivers of oceanic crust. Thus, as the seafloor moves away from the ridge crest, newly formed crust replaces it. Hess further proposed that the descending limb of a convection current in the mantle occurs in the vicinity of deep-ocean trenches.* These, Hess suggested, are sites where ocean crust is drawn back into Earth's interior. As a consequence, the older portions of the seafloor are gradually consumed as they descend into the mantle. As one researcher summarized, "No wonder the ocean floor was young—it was constantly being renewed!"

One of Hess's central ideas was that "convective flow in the mantle caused the Earth's entire outer shell to move." Thus, unlike Wegener's hypothesis that had continents plowing through the seafloor, Hess proposed that the continents were carried passively by the horizontal part of the

---

*Although Hess proposed that convection in the Earth consists of upcurrents coming from the deep mantle beneath the ocean ridges, it is now clear that these upcurrents are shallow features not related to deep convection in the mantle. We will consider this topic further in Chapter 13.

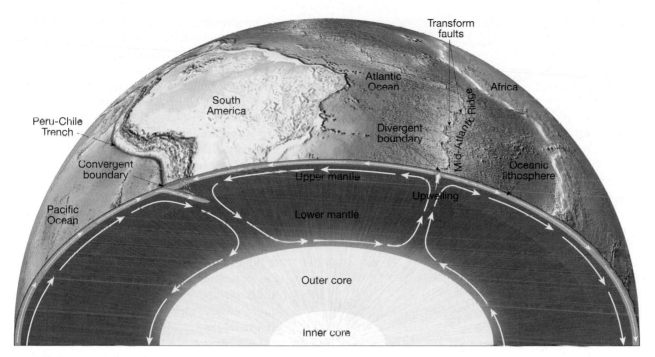

FIGURE 2.12   Seafloor spreading. Harry Hess proposed that upwelling of mantle material along the mid-ocean ridge system created new seafloor. The convective motion of mantle material carries the seafloor in a convoyor-bell fashion to the deep-ocean trenches, where the seafloor descends into the mantle.

convective flow in the mantle. Moreover, the youth of the seafloor and the thinness of sediments were accounted for in Hess's proposal. Despite its logical appeal, seafloor spreading remained a highly controversial topic for the next few years.

Hess presented his paper as an "essay in geopoetry," which may have reflected the speculative nature of this idea. Or, as others have suggested, he may have wanted to deflect criticism from those who were still hostile to continental drift. In either case, his hypothesis provided specific testable ideas, which is one of the hallmarks of good science.

With the seafloor-spreading hypothesis in place, Harry Hess had initiated another phase of this scientific revolution. The conclusive evidence to support his ideas came a few years later from the work of a young Cambridge University graduate student, Fred Vine, and his supervisor, D. H. Matthews. The significance of the Vine-Matthews hypothesis was that it connected two ideas previously thought to be unrelated: Hess's seafloor-spreading hypothesis and the newly discovered geomagnetic reversals (see Box 2.3).

Geomagnetic Reversals: Evidence for Seafloor Spreading

About the time Hess formulated the concept of seafloor spreading, geophysicists were beginning to accept the fact that over periods of hundreds of thousands of years, Earth's magnetic field periodically reverses polarity. During a **geomagnetic reversal** the north magnetic pole becomes the south magnetic pole, and vice versa. Lava solidifying dur-

ing a period of reverse polarity will be magnetized with the polarity opposite that of rocks being formed today. When rocks exhibit the same magnetism as the present magnetic field, they are said to possess **normal polarity**, whereas rocks exhibiting the opposite magnetism are said to have **reverse polarity.**

Evidence for magnetic reversals was obtained when investigators measured the magnetism of lavas and sediments of various ages around the world. They found that normally and reversely magnetized rocks of a given age in one location matched the magnetism of rocks of the same age found in all other locations. This was convincing evidence that Earth's magnetic field had indeed reversed.

Once the concept of magnetic reversals was confirmed, researchers set out to establish a time scale for magnetic reversals. The task was to measure the magnetic polarity of hundreds of lava flows and use radiometric dating techniques to establish their ages (Figure 2.13). Figure 2.14 shows the **magnetic time scale** established for the last few million years. The major divisions of the magnetic time scale are called *chrons* and last for roughly 1 million years. As more measurements became available, researchers realized that several, short-lived reversals (less than 200,000 years long) occur during any one chron.

Meanwhile, oceanographers had begun to do magnetic surveys of the ocean floor in conjunction with their efforts to construct detailed maps of seafloor topography. These magnetic surveys were accomplished by towing very sensitive instruments called **magnetometers** behind research vessels. The goal of these geophysical surveys was to map variations

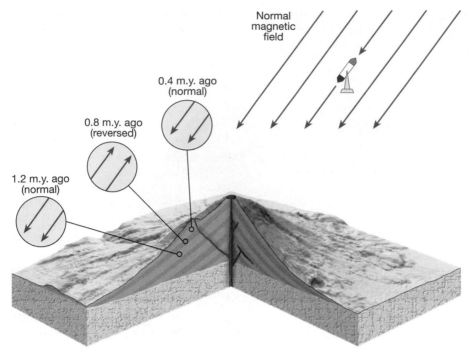

FIGURE 2.13 Schematic illustration of paleomagnetism preserved in lava flows of various ages. Data such as these from various locales were used to establish the time scale of polarity reversals shown in Figure 2.14.

in the strength of Earth's magnetic field that arise from differences in the magnetic properties of the underlying crustal rocks.

The first comprehensive study of this type was carried out off the Pacific Coast of North America and had an unexpected outcome. Researchers discovered alternating strips of high- and low-intensity magnetism as shown in Figure 2.15.

This relatively simple pattern of magnetic variation defied explanation until 1963, when Fred Vine and D. H. Matthews demonstrated that the high- and low-intensity strips supported Hess's concept of seafloor spreading. Vine and Matthews suggested that the strips of high-intensity magnetism are regions where the paleomagnetism of the ocean crust exhibits normal polarity (Figure 2.16). Consequently, these rocks *enhance* (reinforce) Earth's magnetic field. Conversely, the low-intensity strips are regions where the ocean crust is polarized in the reverse direction and therefore *weakens* the existing magnetic field. But how do parallel strips of normally and reversely magnetized rock become distributed across the ocean floor?

FIGURE 2.14 Time scale of Earth's magnetic field in the recent past. This time scale was developed by establishing the magnetic polarity for lava flows of known age. (Data from Allen Cox and G. B. Dalrymple)

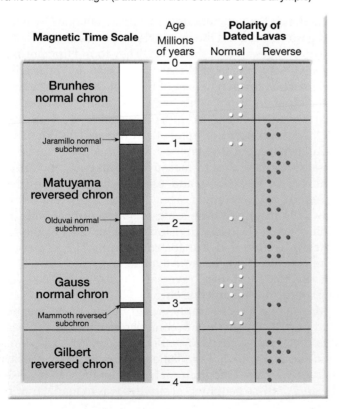

FIGURE 2.15 Pattern of alternating strips of high- and low-intensity magnetism discovered off the Pacific Coast of North America.

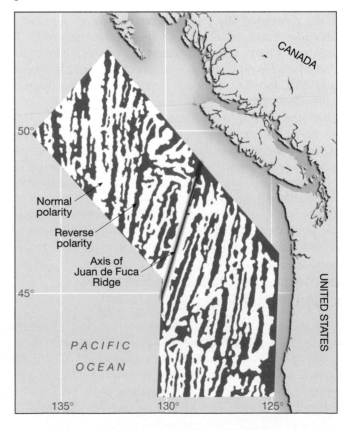

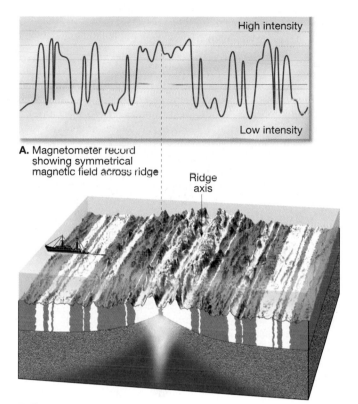

A. Magnetometer record showing symmetrical magnetic field across ridge

Ridge axis

B. Research vessel towing magnetometer across ridge crest

FIGURE 2.16 The ocean floor as a magnetic tape recorder. **A.** Schematic representation of magnetic intensities recorded as a magnetometer is towed across a segment of the oceanic ridge. **B.** Notice the symmetrical strips of low- and high-intensity magnetism that parallel the ridge crest. Vine and Matthews suggested that the strips of high-intensity magnetism occur where normally magnetized oceanic basalts enhance the existing magnetic field. Conversely, the low-intensity strips are regions where the crust is polarized in the reverse direction, which weakens the existing magnetic field.

Ridge just south of Iceland revealed a pattern of magnetic strips exhibiting a remarkable degree of symmetry to the ridge axis.

The Last Piece of the Puzzle

The 1960s have been characterized as a period of chaos in regard to the tectonics debate. Some geologists believed in seafloor spreading and continental drift, whereas others held that an expanding Earth could better account for the displacement occurring at the crests of oceanic ridges. According to the latter view, landmasses once covered the entire surface of Earth, as shown in Figure 2.18. As Earth expanded, the continents split into their current configurations, while new seafloor "filled in" the space between them as they drifted apart (Figure 2.18).

Against this background entered J. Tuzo Wilson, a Canadian physicist turned geologist. In a paper published in 1965, Wilson provided the missing piece needed to formulate the theory of plate tectonics. Wilson suggested that large faults connected the global mobile belts into a continuous network that divides Earth's outer shell into several "rigid plates." In addition, Wilson described the three types of plate margins and how the solid blocks of Earth's outer shell moved relative to one another. At oceanic ridges, plates are moving apart, whereas along deep-ocean trenches, plates move

Vine and Matthews reasoned that as magma solidifies along narrow rifts at the crest of an oceanic ridge, it is magnetized with the polarity of the existing magnetic field (Figure 2.17). Because of seafloor spreading, this strip of magnetized crust would gradually increase in width. When Earth's magnetic field reverses polarity, any newly formed seafloor (having the opposite polarity) would form in the middle of the old strip. Gradually, the two parts of the old strip are carried in opposite directions away from the ridge crest. Subsequent reversals would build a pattern of normal and reverse strips as shown in Figure 2.17. Because new rock is added in equal amounts to both trailing edges of the spreading ocean floor, we should expect that the pattern of strips (size and polarity) found on one side of an oceanic ridge to be a mirror image of the other side. A few years later a survey across the Mid-Atlantic

FIGURE 2.17 As new basalt is added to the ocean floor at mid-ocean ridges, it is magnetized according to Earth's existing magnetic field. Hence, it behaves much like a tape recorder as it records each reversal of the planet's magnetic field.

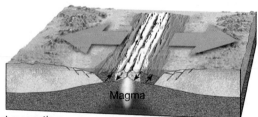

A. Period of normal magnetism

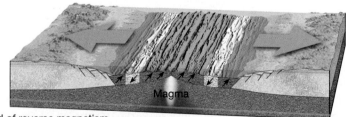

B. Period of reverse magnetism

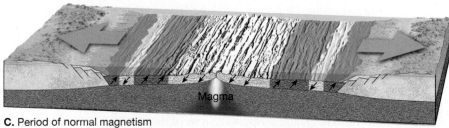

C. Period of normal magnetism

BOX 2.3 ▶ UNDERSTANDING EARTH

Priority in the Sciences

The priority, or credit, for a scientific idea or discovery is *usually* given to the researcher, or group of researchers, who first publish their findings in a scientific journal. However, it is not uncommon for two, or even more researchers, to reach similar conclusions almost simultaneously. Two well-known examples are the independent discoveries of organic evolution by Charles Darwin and Alfred Wallace and development of the calculus by Isaac Newton and Gottfried W. Leibniz. Similarly, some of the main ideas that led to the tectonics revolution in the Earth sciences were also discovered independently by more than one group of investigators.

Although the continental drift hypothesis is rightfully associated with the name Alfred Wegener, he was not the first to suggest continental mobility. In fact, Francis Bacon, in 1620, pointed out the similarities of the outlines of Africa and South America; however, he did not develop this idea further. Nearly three centuries later, in 1910, two years before Wegener made a formal presentation of his ideas, American geologist F. B. Taylor published the first paper to outline the concept we now call continental drift. Why, then, is Wegener credited with this idea?

Because the papers authored by Taylor had relatively little impact among the scientific community, Wegener was not aware of Taylor's work. Hence, it is believed that Wegener independently and simultaneously reached the same conclusion. More important, however, Wegener made great efforts throughout his professional life to provide a wide range of evidence to support his hy-

pothesis. By contrast, Taylor appeared content to simply state, "There are many bonds of union which show that Africa and South America were once joined." Further, whereas Taylor viewed continental drift as a somewhat speculative idea, Wegener was *certain* that the continents had drifted. According to H. W. Menard in his book, *The Ocean of Truth*, Taylor was uncomfortable having his ideas coupled with Wegener's hypothesis. Menard quotes Taylor as writing, "Wegener was a young professor of meteorology. Some of his ideas are very different from mine and he went much further in his speculation."

Another controversy concerning priority came with the development of the seafloor spreading hypothesis. In 1960, Harry Hess of Princeton University wrote a paper that outlined his ideas on seafloor spreading. Rather than rushing it to publication, Hess mailed copies of the manuscript to numerous colleagues, a common practice among researchers. In the meantime, and apparently independently, Robert Dietz of Scripps Institution of Oceanography published a similar paper in the respected journal *Nature* (1961), titled "Continents and Ocean Basin Evolution by Spreading of the Sea Floor." When Dietz became aware of Hess's earlier, although unpublished paper, he acknowledged priority for the idea of seafloor spreading to Hess. It is interesting to note that the basic ideas in Hess's paper actually appeared in a textbook authored by Arthur Holmes in 1944. Therefore, priority for seafloor spreading may rightfully belong to Holmes. Nevertheless, Dietz and Hess both presented

new ideas that were influential to the development of the theory of plate tectonics. Thus, historians associate the names Hess and Dietz with the discovery of seafloor spreading with occasional mention of contributions by Holmes.

Perhaps the most controversial issue of scientific priority came in 1963 when Fred Vine and D. H. Matthews published their paper that linked the seafloor spreading hypothesis with the newly discovered data on magnetic reversals. However, nine months earlier a similar paper by Canadian geophysicist, L. W. Morley was not accepted for publication. One reviewer of Morley's paper commented, "Such speculation makes interesting talk at cocktail parties, but is not the sort of thing that ought to be published under serious, scientific aegis." Morley's paper was eventually published in 1964, but priority had already been established, and the idea became known as the Vine-Matthews hypothesis. In 1971, N. D. Watkins wrote of Morley's paper, "The manuscript certainly had substantial historical interest, ranking as probably the most significant paper in the earth sciences to ever be denied publication."

With the development of the theory of plate tectonics came many other races for priority by researchers from various competing institutions. Some of the new ideas that unfolded from this body of work will be presented in this and later chapters. Because priority for scientific ideas is complicated by the frequency of independent and nearly simultaneous discoveries, it became prudent for investigators to publish their ideas as quickly as possible.

---

together. Furthermore, along great faults, which he named *transform faults*, plates slide past one another. In a broad sense, Wilson had presented what would later be called the *theory of plate tectonics*, a topic we will consider next.

Once the key concepts of plate tectonics had been set forth, the hypothesis-testing phase moved forward very quickly. Some of the evidence that these researchers uncovered to support the plate tectonics model will be presented in this and other chapters. Much of the supporting evidence for the plate tectonics model already existed. What this theory provided was a unified explanation for what seemed to be numerous unrelated observations from the fields of geology, paleontology, geophysics, and oceanography among others.

By the end of the 1960s the tide of scientific opinion had indeed turned! However, some opposition to plate tectonics

continued for at least a decade. Nevertheless, Wegener had been vindicated and the revolution in geology was nearing an end.

Plate Tectonics: The New Paradigm

GEODe Plate Tectonics
EARTH ▶ Introduction

By 1968 the concepts of continental drift and seafloor spreading were united into a much more encompassing theory known as **plate tectonics** (*tekton* = to build). Plate tectonics can be defined as the composite of a great variety of ideas that explain the observed motion of Earth's outer shell through

FIGURE 2.18 An alternate hypothesis to continental drift was an expanding Earth. According to this model Earth was once only half its current diameter and covered by a layer of continents. As Earth expanded the continents split into their current configurations, while new seafloor "filled in" the spaces as they drifted apart.

the mechanisms of subduction and seafloor spreading, which, in turn, generate Earth's major features, including continents, mountains, and ocean basins. The implications of plate tectonics are so far-reaching that this theory has become the basis for viewing most geologic processes.

Earth's Major Plates

According to the plate tectonics model, the uppermost mantle, along with the overlying crust, behave as a strong, rigid layer, known as the **lithosphere** (*lithos* = stone, *sphere* = a ball), which is broken into pieces called **plates** (Figure 2.19). Lithospheric plates are thinnest in the oceans where their thickness may vary from as little as a few kilometers at the oceanic ridges to 100 kilometers in the deep-ocean basins. By contrast, continental lithosphere is generally about 100 kilometers thick but may be more than 250 kilometers thick below older portions of landmasses. The lithosphere overlies a weaker region in the mantle known as the

asthenosphere (*asthenos* = weak, *sphere* = a ball). The temperature/pressure regime in the upper asthenosphere is such that the rocks there are very near their melting temperatures. This results in a very weak zone that permits the lithosphere to be effectively detached from the layers below. Thus, the weak rock within the upper asthenosphere allows Earth's rigid outer shell to move.

The lithosphere is broken into numerous segments, called **lithospheric** or **tectonic plates,** that are in motion with respect to one another and are continually changing in shape and size. As shown in Figure 2.20, seven major lithospheric plates are recognized. They are the *North American, South American, Pacific, African, Eurasian, Australian-Indian,* and *Antarctic plates.* The largest is the Pacific plate, which encompasses a significant portion of the Pacific Ocean basin. Notice from Figure 2.20 that most of the large plates include an entire continent plus a large area of ocean floor (for example, the South American plate). This is a major departure from Wegener's continental drift hypothesis, which proposed that the continents moved through the ocean floor, not with it. Note also that none of the plates are defined entirely by the margins of a continent.

Intermediate-sized plates include the *Caribbean, Nazca, Philippine, Arabian, Cocos, Scotia,* and *Juan de Fuca plates.* In addition, there are over a dozen smaller plates that have been identified but are not shown in Figure 2.20.

One of the main tenets of the plate tectonic theory is that plates move as coherent units relative to all other plates. As plates move, the distance between two locations on the same plate—New York and Denver, for example—remains relatively constant, whereas the distance between sites on different plates, such as New York and London, gradually changes. (Recently it has been shown that plates can suffer *some* internal deformation, particularly oceanic lithosphere.)

FIGURE 2.19 Illustration of some of Earth's lithospheric plates.

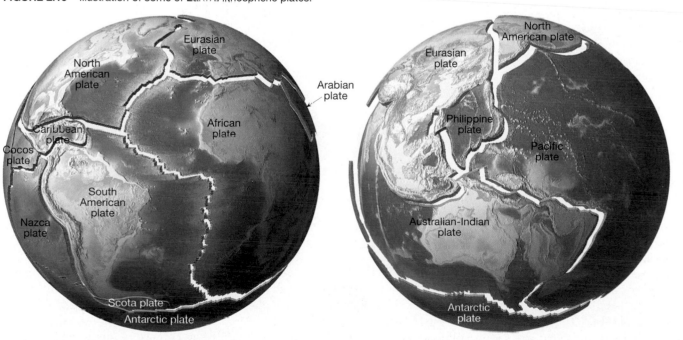

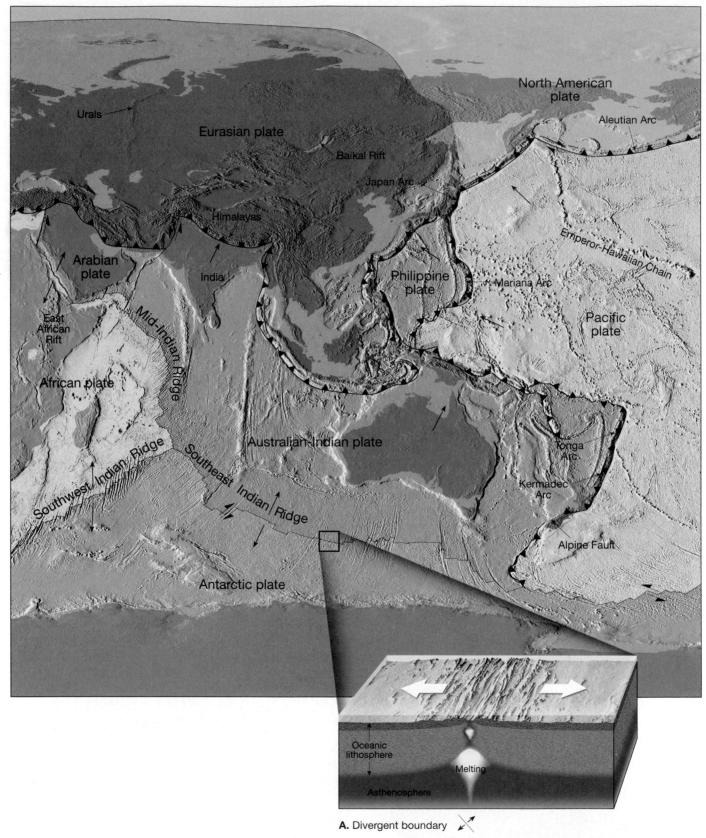

A. Divergent boundary

FIGURE 2.20 A mosaic of rigid plates constitutes Earth's outer shell. (After W. B. Hamilton, U.S. Geological Survey)

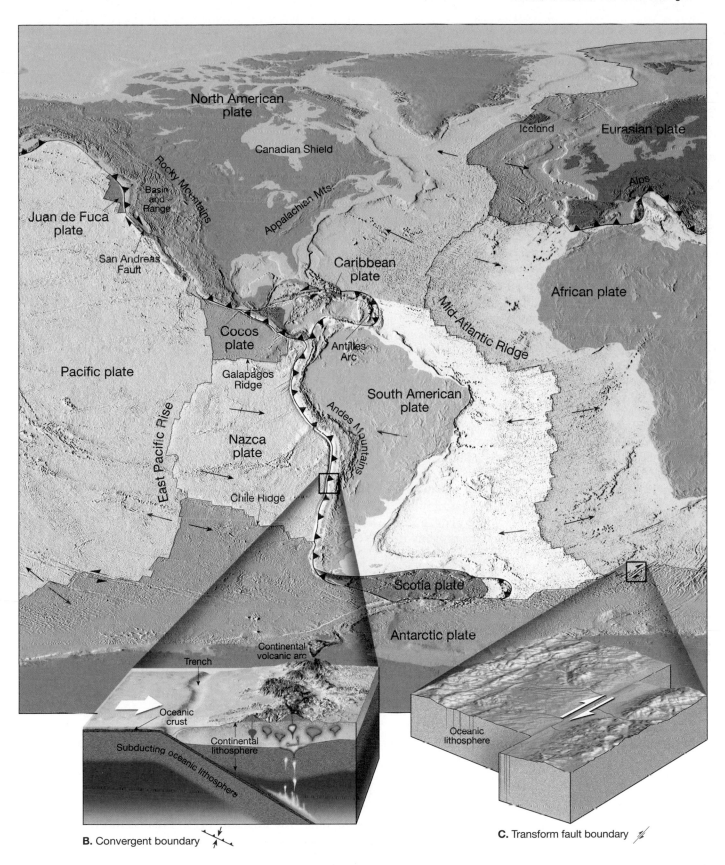

North American
plate

Canadian Shield

Iceland   Eurasian plate

Rocky Mountains

Basin
and
Range

Appalachian Mts.

Alps

Juan de Fuca
plate

San Andreas
Fault

Cocos
plate

Caribbean
plate

African plate

Antilles
Arc

Mid-Atlantic Ridge

Pacific plate

Galapagos
Ridge

South American
plate

East Pacific Rise

Nazca
plate

Andes Mountains

Chile Ridge

Scotia plate

Antarctic plate

B. Convergent boundary

Continental
volcanic arc

Trench

Oceanic
crust

Continental
lithosphere

Subducting oceanic lithosphere

C. Transform fault boundary

Oceanic
lithosphere

Lithospheric plates move relative to each other at a very slow but continuous rate that averages about 5 centimeters (2 inches) per year. This movement is ultimately driven by the unequal distribution of heat within Earth. Hot material found deep in the mantle moves slowly upward and serves as one part of our planet's internal convection system. Concurrently, cooler, denser slabs of oceanic lithosphere descend into the mantle, setting Earth's rigid outer shell into motion. Ultimately, the titanic, grinding movements of Earth's lithospheric plates generate earthquakes, create volcanoes, and deform large masses of rock into mountains.

Plate Boundaries

Tectonic plates move as coherent units relative to all other plates. Although the interiors of plates may experience some deformation, all major interactions among individual plates (and therefore most deformation) occur along their *boundaries*. In fact, plate boundaries were first established by plotting the locations of earthquakes. Moreover, plates are bounded by three distinct types of boundaries, which are differentiated by the type of movement they exhibit. These boundaries are depicted at the bottom of Figure 2.20 and are briefly described here:

1. **Divergent boundaries** (*constructive margins*)—where two plates move apart, resulting in upwelling of material from the mantle to create new seafloor (Figure 2.20A).
2. **Convergent boundaries** (*destructive margins*)—where two plates move together, resulting in oceanic lithosphere descending beneath an overriding plate, eventually to be reabsorbed into the mantle, or possibly in the collision of two continental blocks to create a mountain system (Figure 2.20B).
3. **Transform fault boundaries** (*conservative margins*)—where two plates grind past each other without the production or destruction of lithosphere (Figure 2.20C).

Each plate is bounded by a combination of these three types of plate margins. For example, the Juan de Fuca plate has a divergent zone on the west, a convergent boundary on the east, and numerous transform faults, which offset segments of the oceanic ridge (Figure 2.20). Although the total surface area of Earth does not change, individual plates may diminish or grow in area depending on any imbalance between the growth rate at divergent boundaries and the rate at which lithosphere is destroyed at convergent boundaries. The Antarctic and African plates are almost entirely bounded by divergent boundaries and hence are growing larger by adding new lithosphere to their margins. By contrast, the Pacific plate is being consumed into the mantle along its northern and western flanks faster than it is being replaced, and therefore is diminishing in size.

It is also important to note that plate boundaries are not fixed but move about. For example, the westward drift of the South American plate is causing it to override the Nazca plate. As a result, the boundary that separates these plates is gradually being displaced as well. Moreover, since the Antarctic plate is surrounded by constructive margins and is growing larger, the divergent boundaries are migrating away from the continent of Antarctica.

New plate boundaries can be created in response to changes in the forces acting on these rigid slabs. For example, a relatively new divergent boundary is located in the Red Sea. Less than 20 million years ago the Arabian Peninsula began to rift away from Africa. At other locations, plates carrying continental crust are presently moving toward one another. Eventually these continents may collide and be sutured together. In this case, the boundary that once separated two plates disappears as the plates become one. The result of such a continental collision is a majestic mountain range such as the Himalayas.

In the following sections we will briefly summarize the nature of the three types of plate boundaries.

Divergent Boundaries

 GEODe Plate Tectonics
▸ Divergent Boundaries

Most **divergent** (*di* = apart, *vergere* = to move) **boundaries** are located along the crests of oceanic ridges and can be thought of as *constructive plate margins* since this is where new oceanic lithosphere is generated (Figure 2.21). Divergent boundaries are also called **spreading centers,** because seafloor spreading occurs at these boundaries. Here, as the plates move away from the ridge axis, the fractures that form are filled with molten rock that wells up from the hot mantle below. Gradually, this magma cools to produce new slivers of seafloor. In a continuous manner, adjacent plates spread apart and new oceanic lithosphere forms between them. As we shall see later, divergent boundaries are not confined to the ocean floor but can also form on the continents.

Oceanic Ridges and Seafloor Spreading

Along well-developed divergent plate boundaries, the seafloor is elevated, forming the *oceanic ridge.* The interconnected oceanic ridge system is the longest topographic feature on Earth's surface, exceeding 70,000 kilometers (43,000 miles) in length. Representing 20 percent of Earth's surface, the oceanic ridge system winds through all major ocean basins like the seam on a baseball. Although the crest of the oceanic ridge is commonly 2 to 3 kilometers higher than the adjacent ocean basins, the term "ridge" may be misleading because this feature is not narrow but has widths from 1000 to 4000 kilometers. Further, along the axis of some ridge segments is a deep down-faulted structure called a **rift valley.**

The mechanism that operates along the oceanic ridge system to create new seafloor is appropriately called *seafloor spreading.* Typical rates of spreading average around

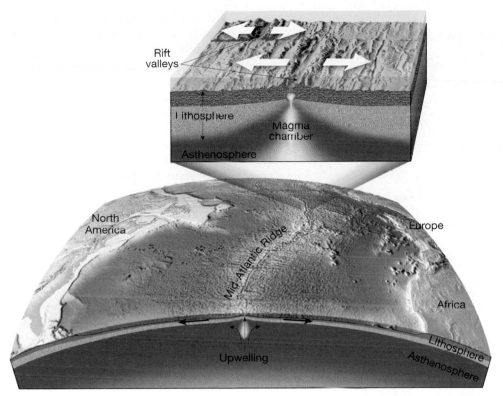

FIGURE 2.21   Most divergent plate boundaries are situated along the crests of oceanic ridges.

Continental Rifting

Divergent plate boundaries can also develop within a continent, in which case the landmass may split into two or more smaller segments, as Alfred Wegener had proposed for the breakup of Pangaea. The splitting of a continent is thought to begin with the formation of an elongated depression called a *continental rift*. A modern example of a continental rift is the East African Rift. Whether this rift will develop into a full-fledged spreading center and eventually split the continent of Africa is a matter of much speculation.

Nevertheless, the East African Rift represents the initial stage in the breakup of a continent (see Figure 13.20, page 366). Here tensional forces have stretched and thinned the continental crust. As a result, molten rock ascends from the asthensophere and initiates volcanic activity at the surface (Figure 2.22A). Large volcanic mountains such as Kilimanjaro and Mount Kenya exemplify the extensive volcanic activity that accompanies continental rifting. Research suggests that if tensional forces are maintained, the rift valley will lengthen and deepen, eventually extending out to the margin of the plate, splitting it in two (Figure 2.22C). At this point, the rift becomes a narrow sea with an outlet to the ocean, similar to the Red Sea. The Red Sea formed when the Arabian Peninsula rifted from Africa, an event that began about 20 million years ago. Consequently, the Red Sea provides oceanographers with a view of how the Atlantic Ocean may have looked in its infancy.

5 centimeters (2 inches) per year. This is roughly the same rate at which human fingernails grow. Comparatively slow spreading rates of 2 centimeters per year are found along the Mid-Atlantic Ridge, whereas spreading rates exceeding 15 centimeters (6 inches) have been measured along sections of the East Pacific Rise. Although these rates of lithospheric production are slow on a human time scale, they are nevertheless rapid enough to have generated all of Earth's ocean basins within the last 200 million years. In fact, none of the ocean floor that has been dated exceeds 180 million years in age.

The primary reason for the elevated position of the oceanic ridge is that newly created oceanic crust is hot, and occupies more volume, which makes it less dense than cooler rocks. As new lithosphere is formed along the oceanic ridge, it is slowly yet continually displaced away from the zone of upwelling along the ridge axis. Thus, it begins to cool and contract, thereby increasing in density. This thermal contraction accounts for the greater ocean depths that exist away from the ridge crest. It takes about 80 million years before the cooling and contracting cease completely. By this time, rock that was once part of the elevated oceanic ridge system is located in the deep-ocean basin, where it may be buried by substantial accumulations of sediment. In addition, cooling causes the mantle rocks below the oceanic crust to strengthen, thereby adding to the plate's thickness. Stated another way, the thickness of oceanic lithosphere is age-dependent. The older (cooler) it is, the greater its thickness.

Convergent Boundaries

GEODe   Plate Tectonics
▸ Convergent Boundaries

Although new lithosphere is constantly being produced at the oceanic ridges, our planet is not growing larger—its total surface area remains constant. To balance the addition of newly created lithosphere, older, denser portions of oceanic lithosphere descend into the mantle along **convergent** (*con* = together, *vergere* = to move) **boundaries.** Because lithosphere is "destroyed" at convergent boundaries, they are also called *destructive plate margins* (Figure 2.23A).

Convergent plate margins occur where two plates move toward each other and the leading edge of one is bent downward, allowing it to slide beneath the other. The surface expression produced by the descending plate is a **deep-ocean trench,** such as the Peru–Chile trench (see Figure 13.9, page 357). Trenches formed in this manner may be thousands of

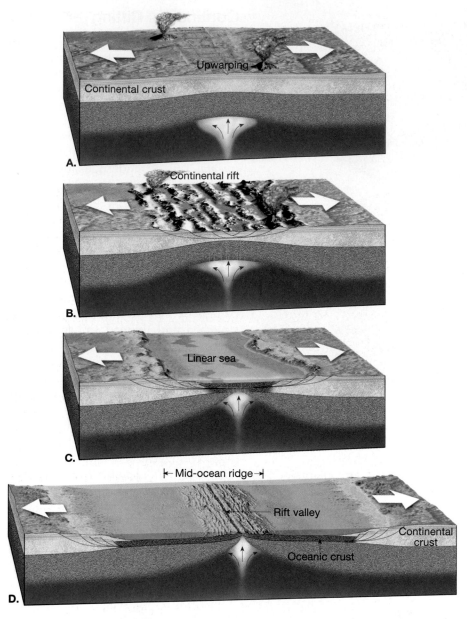

FIGURE 2.22 Continental rifting and the formation of a new ocean basin. **A.** Continental rifting is thought to occur where tensional forces stretch and thin the crust. As a result, molten rock ascends from the asthenosphere and initiates volcanic activity at the surface. **B.** As the crust is pulled apart, large slabs of rock sink, generating a rift valley. **C.** Further spreading generates a narrow sea. **D.** Eventually, an expansive ocean basin and ridge system are created.

kilometers long, 8 to 12 kilometers deep, and between 50 and 100 kilometers wide.

Convergent boundaries are also called **subduction zones,** because they are sites where lithosphere is descending (being subducted) into the mantle. Subduction occurs because the density of the descending tectonic plate is greater than the density of the underlying asthenosphere. In general, oceanic lithosphere is more dense than the asthenosphere whereas continental lithosphere is less dense and resists subduction. As a consequence, it is always oceanic lithosphere that is subducted.

Slabs of oceanic lithosphere descend into the mantle at angles that vary from a few degrees to nearly vertical (90 de-

grees), but average about 45 degrees. The angle at which oceanic lithosphere descends depends largely on its density. For example, when a spreading center is located near a subduction zone, the lithosphere is young and, therefore, warm and buoyant. Hence, the angle of descent is small. This is the situation along parts of the Peru–Chile trench. Low dip angles usually result in considerable interaction between the descending slab and the overriding plate. Consequently, these regions experience great earthquakes.

As oceanic lithosphere ages (gets farther from the spreading center), it gradually cools, which causes it to thicken and increase in density. Once oceanic lithosphere is about 15 million years old, it becomes more dense than the supporting asthenosphere and will sink when given the opportunity. In parts of the western Pacific, some oceanic lithosphere is more than 180 million years old. This is the thickest and most dense in today's oceans. The subducting slabs in this region typically plunge into the mantle at angles approaching 90 degrees.

Although all convergent zones have the same basic characteristics, they are highly variable features. Each is controlled by the type of crustal material involved and the tectonic setting. Convergent boundaries can form between two oceanic plates, one oceanic and one continental plate, or two continental plates. All three situations are illustrated in Figure 2.23.

Oceanic–Continental Convergence

Whenever the leading edge of a plate capped with continental crust converges with a slab of oceanic lithosphere, the buoyant continental block remains "floating," while the denser oceanic slab sinks into the mantle (Figure 2.23A). When a descending oceanic slab reaches a depth of about 100 kilometers, melting is triggered within the wedge of hot asthenosphere that lies above it. But how does the subduction of a cool slab of oceanic lithosphere cause mantle rock to melt? The answer lies in the fact that volatiles (mainly water) act like salt does to melt ice. That is, "wet" rock, in a high-pressure environment, melts at substantially lower temperatures than "dry" rock of the same composition.

Sediments and oceanic crust contain a large amount of water which is carried to great depths by a subducting plate. As the plate plunges downward, water is "squeezed" from the pore spaces as confining pressure increases. At even greater depths, heat and pressure drive water from hydrated (water-rich) minerals such as the *amphiboles*. At a depth of roughly 100 kilometers, the mantle is sufficiently hot that

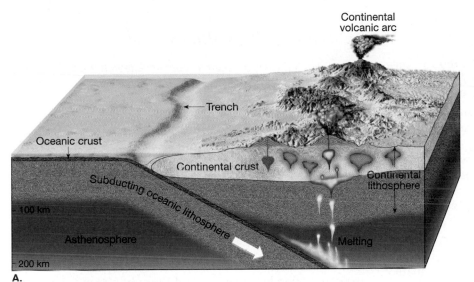

A.

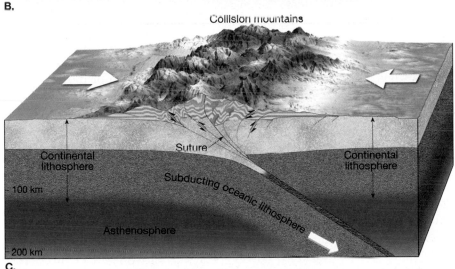

B.

C.

FIGURE 2.23   Zones of plate convergence. **A.** Oceanic–continental **B.** Oceanic–oceanic **C.** Continental–continental.

ment, these mantle-derived magmas may ascend through the crust and give rise to a volcanic eruption. However, much of this molten rock never reaches the surface; rather, it solidifies at depth where it acts to thicken the crust.

Partial melting of mantle rock generates molten rock that has a *basaltic composition*, similar to what erupts on the Island of Hawaii. In a continental setting, however, basaltic magma typically melts and assimilates some of the crustal rocks through which it ascends. The result is the formation of a silica-rich (SiO_2) magma. On occasions when silica-rich magmas reach the surface, they often erupt explosively, generating large columns of volcanic arc and gases. A classic example of such an eruption was the 1980 eruption of Mount St. Helens. You will learn more about the formation of magma and its influence on the explosiveness of volcanic eruptions in Chapters 4 and 5.

The volcanoes of the towering Andes are the product of magma generated by the subduction of the Nazca plate beneath the South American continent (see Figure 2.20). Mountains such as the Andes, which are produced in part by volcanic activity associated with the subduction of oceanic lithosphere, are called **continental volcanic arcs.** The Cascade Range of Washington, Oregon, and California is another volcanic arc which consists of several well-known volcanic mountains, including Mount Rainier, Mount Shasta, and Mount St. Helens (see Figure 5.19, p. 142). (This active volcanic arc also extends into Canada, where it includes Mount Garibaldi, Mount Silverthrone, and others.)

Oceanic–Oceanic Convergence

An oceanic–oceanic convergent boundary has many features in common with oceanic–continental plate margins. The differences are mainly attributable to the nature of the crust capping the overriding plate. Where two oceanic slabs converge, one descends beneath the other, initiating volcanic activity by the same mechanism that operates at oceanic–continental plate boundaries. Water "squeezed" from the subducting slab of oceanic lithosphere triggers melting in the hot wedge of

the introduction of water leads to some melting. This process, called **partial melting,** generates as little as 10 percent molten material, which is intermixed with unmelted mantle rock. Being less dense than the surrounding mantle, this hot mobile material gradually rises toward the surface as a teardrop-shaped structure. Depending on the environ-

mantle rock that lies above. In this setting, volcanoes grow up from the ocean floor, rather than upon a continental platform. When subduction is sustained, it will eventually build a chain of volcanic structures that emerge as islands. The volcanic islands are spaced about 80 kilometers apart and are built upon submerged ridges of volcanic material a few hundred kilometers wide. This newly formed land consisting of an arc-shaped chain of small volcanic islands is called a **volcanic island arc,** or simply an **island arc** (Figure 2.23B).

The Aleutian, Mariana, and Tonga islands are examples of volcanic island arcs. Island arcs such as these are generally located 100 to 300 kilometers (60 to 200 miles) from a deep-ocean trench. Located adjacent to the island arcs just mentioned are the Aleutian trench, the Mariana trench, and the Tonga trench (see Figure 1.17).

Most volcanic island arcs are located in the western Pacific. Only two volcanic island arcs are located in the Atlantic—the Lesser Antilles arc adjacent to the Caribbean Sea and the Sandwich Islands in the South Atlantic. The Lesser Antilles are a product of the subduction of the Atlantic beneath the Caribbean plate. Located within this arc is the island of Martinique, where Mount Pelée erupted in 1902, destroying the town of St. Pierre and killing an estimated 28,000 people; and the island of Montserrat, where volcanic activity has occurred very recently.*

Relatively young island arcs are fairly simple structures that are underlain by deformed oceanic crust that is generally less than 20 kilometers (12 miles) thick. Examples include the arcs of Tonga, the Aleutians, and the Lesser Antilles. By contrast, older island arcs are more complex and are underlain by crust that ranges in thickness from 20 to 35 kilometers. Examples include the Japanese and Indonesian arcs, which are built upon material generated by earlier episodes of subduction or sometimes on a small piece of continental crust.

Continental–Continental Convergence

As you saw earlier, when an oceanic plate is subducted beneath continental lithosphere, an Andean-type volcanic arc develops along the margin of the continent. However, if the subducting plate also contains continental lithosphere, continued subduction eventually brings the two continental blocks together (Figure 2.23C). Whereas oceanic lithosphere is relatively dense and sinks into the asthenosphere, continental lithosphere is buoyant, which prevents it from being subducted to any great depth. The result is a collision between the two continental fragments (Figure 2.23C).

Such a collision occurred when the subcontinent of India "rammed" into Asia, producing the Himalayas—the most spectacular mountain range on Earth (Figure 2.24). During this collision, the continental crust buckled, fractured, and was generally shortened and thickened. In addition to the Himalayas, several other major mountain systems, including the Alps, Appalachians, and Urals, formed during continental collisions.

Prior to a continental collision, the landmasses involved were separated by an ocean basin (Figure 2.24A). As the continental blocks converge, the intervening seafloor is subducted beneath one of the plates. Subduction initiates partial melting in the overlying mantle, which in turn results in the growth of a volcanic arc. Depending on the location of the subduction zone, the volcanic arc could develop on either of

*More on these volcanic events is found in Chapter 5.

the converging landmasses, or if the subduction zone developed several hundred kilometers seaward from the coast, a volcanic island arc would form. Eventually, as the intervening seafloor is consumed, these continental masses collide (Figure 2.24B). This folds and deforms the accumulation of sediments and sedimentary rocks along the continental margin as if they had been placed in a gigantic vise. The result is the formation of a new mountain range composed of deformed and metamorphosed sedimentary rocks, fragments of the volcanic arc, and often slivers of oceanic crust.

Transform Fault Boundaries

 GEODe Plate Tectonics
▶ Transform Fault Boundaries

The third type of plate boundary is the **transform** (*trans* = across, *forma* = form) **fault,** where plates slide horizontally past one another without the production or destruction of lithosphere (*conservative plate margins*). The nature of transform faults was discovered in 1965 by J. Tuzo Wilson of the University of Toronto. Wilson suggested that these large faults connect the global active belts (convergent boundaries, divergent boundaries, and other transform faults) into a continuous network that divides Earth's outer shell into several rigid plates. Thus, Wilson became the first to suggest that Earth was made of individual plates, while at the same time identifying the faults along which relative motion between the plates is made possible.

Most transform faults join two segments of an oceanic ridge (Figure 2.25). Here, they are part of prominent linear breaks in the oceanic crust known as **fracture zones,** which include both the active transform faults as well as their inactive extentions into the plate interior. These fracture zones are present approximately every 100 kilometers along the trend of a ridge axis. As shown in Figure 2.25, active transform faults lie *only between* the two offset ridge segments. Here seafloor produced at one ridge axis moves in the opposite direction as seafloor produced at an opposing ridge segment. Thus, between the ridge segments these adjacent slabs of oceanic crust are grinding past each other along the fault. Beyond the ridge

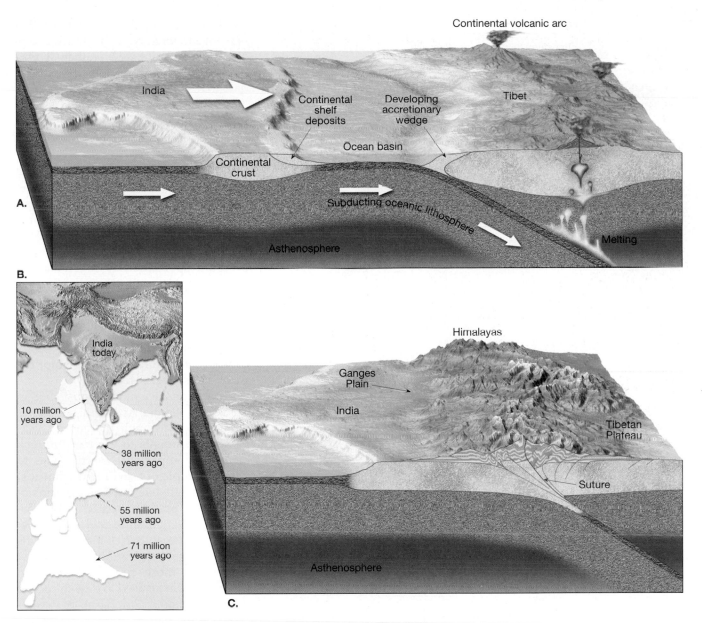

FIGURE 2.24   The ongoing collision of India and Asia, starting about 45 million years ago, produced the majestic Himalayas. **A.** Converging plates generated a subduction zone, while partial melting triggered by the subducting oceanic slab produced a continental volcanic arc. Sediments scraped from the subducting plate were added to the accretionary wedge. **B.** Position of India in relation to Eurasia at various times. (Modified after Peter Molnar) **C.** Eventually the two landmasses collided, deforming and elevating the sediments that had been deposited along the continental margins. In addition, slices of the Indian crust were thrust up onto the Indian plate.

crests are the inactive zones, where the fractures are preserved as linear topographic scars. The trend of these fracture zones roughly parallels the direction of plate motion at the time of their formation. Thus, these structures can be used to map the direction of plate motion in the geologic past.

In another role, transform faults provide the means by which the oceanic crust created at ridge crests can be transported to a site of destruction: the deep-ocean trenches. Figure 2.26 illustrates this situation. Notice that the Juan de Fuca plate moves in a southeasterly direction, eventually being subducted under the west coast of the United States. The southern end of this plate is bounded by the Mendocino

fault. This transform fault boundary connects the Juan de Fuca ridge to the Cascadia subduction zone (Figure 2.26). Therefore, it facilitates the movement of the crustal material created at the ridge crest to its destination beneath the North American continent (Figure 2.26).

Although most transform faults are located within the ocean basins, a few cut through continental crust. Two examples are the earthquake-prone San Andreas Fault of California and the Alpine Fault of New Zealand. Notice in Figure 2.26 that the San Andreas Fault connects a spreading center located in the Gulf of California to the Cascadia subduction zone and the Mendocino fault located along the northwest

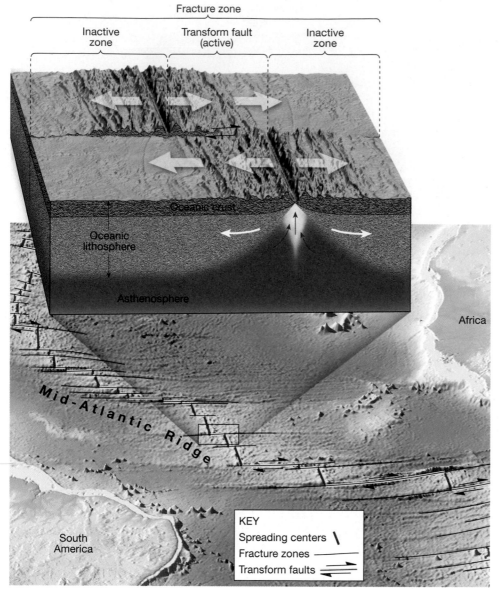

FIGURE 2.25   Diagram illustrating a transform fault boundary offsetting segments of the Mid-Atlantic Ridge.

tal in solidifying the support for this new idea follows. Note that much of this evidence was not new; rather, it was new interpretations of already existing data that swayed the tide of opinion.

Evidence from Ocean Drilling

Some of the most convincing evidence confirming seafloor spreading has come from drilling directly into the ocean floor. From 1968 until 1983, the source of these important data was the Deep Sea Drilling Project, an international program sponsored by several major oceanographic institutions and the National Science Foundation. The primary goal was to gather firsthand information about the age of the ocean basins and processes that formed them. To accomplish this, a new drilling ship, the *Glomar Challenger,* was built.

Operations began in August 1968 in the South Atlantic. At several sites holes were drilled through the entire thickness of sediments to the basaltic rock below. An important objective was to gather samples of sediment from just above the igneous crust as a means of dating the seafloor at each site.* Because sedimentation begins immediately after the oceanic crust forms, remains of microorganisms found in the oldest sediments—those resting directly on the crust—can be used to date the ocean floor at that site.

When the oldest sediment from each drill site was plotted against its distance from the ridge crest, the plot demonstrated that the age of the sediment increased with increasing distance from the ridge. This finding supported the seafloor-spreading hypothesis, which predicted the youngest oceanic crust would be found at the ridge crest and the oldest oceanic crust would be at the continental margins.

The data from the Deep Sea Drilling Project also reinforced the idea that the ocean basins are geologically youthful because no seafloor with an age in excess of 180 million years was found. By comparison, continental crust that exceeds 4 billion years in age has been dated.

The thickness of ocean-floor sediments provided additional verification of seafloor spreading. Drill cores from the *Glomar Challenger* revealed that sediments are almost entirely absent on the ridge crest and the sediment thickness with increasing distance from the ridge. Because the ridge crest is

coast of the United States. Along the San Andreas Fault, the Pacific plate is moving toward the northwest, past the North American plate. If this movement continues, that part of California west of the fault zone, including the Baja Peninsula, will eventually become an island off the West Coast of the United States and Canada. It could eventually reach Alaska. However, a more immediate concern is the earthquake activity triggered by movements along this fault system.

Testing the Plate Tectonics Model

With the development of the theory of plate tectonics, researchers from all of the Earth sciences began testing this new model of how Earth works. Some of the evidence supporting continental drift and seafloor spreading has already been presented. In addition, some of the evidence that was instrumen-

---

*Radiometric dates of the ocean crust itself are unreliable because of the alteration of basalt by seawater.

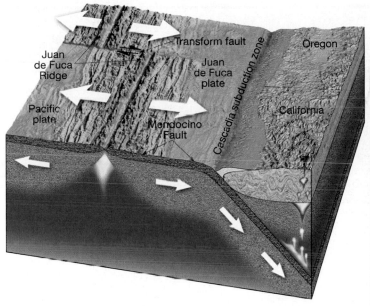

FIGURE 2.26 The Mendocino transform fault permits seafloor generated at the Juan de Fuca ridge to move southeastward past the Pacific plate and beneath the North American plate. Thus, this transform fault connects a divergent boundary to a subduction zone. Furthermore, the San Andreas Fault, also a transform fault, connects two spreading centers: the Juan de Fuca ridge and a divergent zone located in the Gulf of California.

younger than the areas farther away from it, this pattern of sediment distribution should be expected if the seafloor-spreading hypothesis is correct.

The Ocean Drilling Program succeeded the Deep Sea Drilling Project and, like its predecessor, was a major international program. The more technologically advanced drilling ship, the *JOIDES Resolution,* continued the work of the *Glomar Challenger* (see Box 2.4).* The *JOIDES Resolution* can drill in water depths as great as 8200 meters (27,000 feet) and contains onboard laboratories equipped with a large and varied array of seagoing scientific research equipment (Figure 2.27).

In October 2003, the *JOIDES Resolution* became part of a new program, the Integrated Ocean Drilling Program (IODP). This new international effort does not rely on just one drilling ship but uses multiple vessels for exploration. One of the new additions is the massive 210-meter-long *Chikyu,* which began operations in 2006.

Hot Spots and Mantle Plumes

Mapping of seamounts (submarine volcanoes) in the Pacific Ocean revealed several linear chains of volcanic structures. One of the most studied chains consists of at least 129 volcanoes and extends from the Hawaiian Islands to Midway Island and continues northward toward the Aleutian trench a

distance of nearly 6000 kilometers (Figure 2.28). This nearly continuous string of volcanic islands and seamounts is called the Hawaiian Island–Emperor Seamount chain. Radiometric dating of these structures showed that the volcanoes increase in age with increasing distance from Hawaii. Hawaii, the youngest volcanic island in the chain, rose from the seafloor less than a million years ago, whereas Midway Island is 27 million years old, and Suiko Seamount, near the Aleutian trench, is more than 60 million years old (Figure 2.28).

Taking a closer look at the Hawaiian Islands, we see a similar increase in age from the volcanically active island of Hawaii, at the southeastern end of the chain, to the inactive volcanoes that make up the island of Kauai in the northwest (Figure 2.28).

Researchers are in agreement that a rising plume of mantle material is located beneath the island of Hawaii. As the ascending **mantle plume** enters the low-pressure environment at the base of the lithosphere, melting occurs. The surface manifestation of this activity is a **hot spot,** an area of volcanism, high heat flow, and crustal uplifting that is a few hundred kilometers across. As the Pacific plate moved over this hot spot, successive volcanic structures were built. As shown in Figure 2.28, the age of each volcano indicates the time when it was situated over the relatively stationary mantle plume. These chains of volcanic structures, known as **hot spot tracks,** trace the direction of plate motion.

Kauai is the oldest of the large islands in the Hawaiian chain. Five million years ago, when it was positioned over

---

*JOIDES stands for Joint Oceanographic Institutions for Deep Earth Sampling.

BOX 2.4 ▶ UNDERSTANDING EARTH

Sampling the Ocean Floor

A fundamental aspect of scientific inquiry is the gathering of basic facts through observation and measurement. Formulating and testing hypotheses requires reliable data. The acquisition of such information is no easy task when it comes to sampling the vast storehouse of data contained in seafloor sediments and oceanic crust. Acquiring samples is technically challenging and very expensive.

The ship in Figure 2.C, the *JOIDES Resolution,* is a research vessel that is capable of drilling into the seafloor and collecting long cylinders (cores) of sediment and rock. The "JOIDES" in the ship's name stands for Joint Oceanographic Institutions for Deep Earth Sampling. The "Resolution" honors the ship HMS *Resolution,* commanded more than 200 years ago by the prolific English explorer Captain James Cook.

The *JOIDES Resolution* has a tall metal derrick that is used to conduct *rotary drilling,* while the ship's thrusters hold it in a fixed position at sea (Figure 2.C). Individual sections of drill pipe are fitted together to make a single string of pipe up to 8200 meters (27,000 feet) long. The drill bit, located at the end of the pipe string, rotates as it is pressed against the ocean bottom and can drill up to 2100 meters (6900 feet) into the seafloor. Like twirling a soda straw into a layer cake, the drilling operation cuts through sediments and rock and retains a cylinder of material (a core sample) on the inside of the hollow pipe, which can then be raised on board the ship and analyzed in state-of-the-art laboratory facilities.

Since 1985, the ship has drilled more than 1700 holes worldwide. The result has been the recovery of more than 210,000 meters (more than 130 miles!) of core samples that represent millions of years of Earth history and are used by scientists to study many aspects of the Earth sciences including changes in global climate. Although the number of holes drilled into the ocean floor is impressive, it represents just one hole per area about the size of Colorado.

In September 2003, the *JOIDES Resolution* completed its 110th and final expedition as part of the highly successful Ocean Drilling Program (ODP). In October 2003, it became part of a new program, the Integrated Ocean Drilling Program (IOPD). This new international effort does not rely on just one drilling ship, but uses multiple vessels for exploration. One of the new additions is the massive 210-meter- (nearly 700-foot-) long *Chikyu* (meaning "planet Earth" in Japanese) which began operations in 2006. The primary objective of the new program is to collect cores that will allow scientists to better understand Earth history and Earth system processes, including properties of the deep crust, climate change patterns, earthquake mechanisms, and the microbiology of the deep-ocean floor.

FIGURE 2.C   The *JOIDES Resolution* drills into the ocean floor and collects cores of sediment and rock for analysis. The ship's dynamic positioning system consists of powerful thrusters (small propellers) that allow it to remain stationary above a drill site. Previous drill sites can be reused years later and are located by bouncing sound waves between the ship's hydrophones and sonar beacons. A remote television camera aids in positioning the drill pipe into the reentry cone.

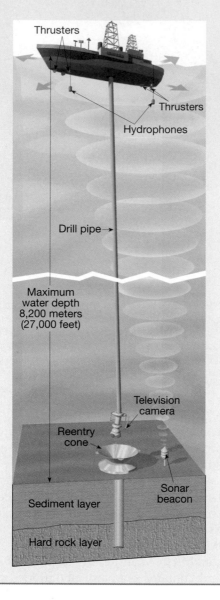

the hot spot, Kauai was the only Hawaiian Island (Figure 2.28). Visible evidence of the age of Kauai can be seen by examining its extinct volcanoes, which have been eroded into jagged peaks and vast canyons. By contrast, the relatively young island of Hawaii exhibits young lava flows, and one of its volcanoes, Kilauea, remains active.

Notice in Figure 2.28 that the Hawaiian Island–Emperor Seamount chain bends. This particular bend in the track occurred about 50 million years ago when the motion of the Pacific plate changed from nearly due north to a northwesterly path. Similarly, hot spots found on the floor of the Atlantic have increased our understanding of the migration of landmasses following the breakup of Pangaea.

Research suggests that at least some mantle plumes originate at great depth, perhaps at the mantle-core boundary. Others, however, may have a much shallower origin. Of the 40 or so hot spots that have been identified, over a dozen are located near spreading centers. For example, the mantle plume located beneath Iceland is responsible for the large accumulation of volcanic rocks found along the northern section of the Mid-Atlantic Ridge.

The existence of mantle plumes and their association with hot spots is well documented. Most mantle plumes are long-lived features that appear to maintain relatively fixed positions within the mantle. However, recent evidence has shown that some hot spots may slowly migrate. If this is the case, models of past plate motion that are based on a fixed hot spot frame of reference will need to be reevaluated.

Measuring Plate Motion

A number of methods have been employed to establish the direction and rate of plate motion. As noted earlier, hot spot tracks like those of the Hawaiian Island–Emperor Seamount

FIGURE 2.27 The *JOIDES Resolution*, one of the drilling ships of the Integrated Ocean Drilling Program. (Photo courtesy of Ocean Drilling Program)

moving relative to Europe at the rate of roughly 2 centimeters per year. Recall that the direction of seafloor spreading can be established from the fracture zones found on the seafloor.

Measuring Plate Velocities from Space

It is currently possible, with the use of space-age technology, to directly measure the relative motion between plates. This is accomplished by periodically establishing the exact locations, and hence the distance, between two observing stations situated on opposite sides of a plate boundary. Two of the methods used for this calculation are *Very Long Baseline Interferometry* (VLBI) and a satellite positioning technique that employs the *Global Positioning System* (GPS). The Very Long Baseline Interferometry system utilizes large radio telescopes to record signals from very distant quasars (quasi-stellar objects) (Figure 2.29). Quasars lie billions of light-years from Earth, so they act as stationary reference points. The millisecond differences in the arrival times of the same signal at different Earth-bound observatories provide a means of establishing the precise distance between receivers. A typical survey may take a day to perform and involves two widely spaced radio telescopes ob-

chain trace the direction of movement of the Pacific Plate relative to the mantle below. Further, by measuring the length of this volcanic chain and the time interval between the formation of the oldest structure (Suiko Seamount) and youngest structure (Hawaii), an average rate of plate motion can be calculated. In this case the volcanic chain is roughly 3000 kilometers long and has formed over the past 65 million years—making the average rate of movement about 9 centimeters (4 inches) per year. The accuracy of this calculation hinges on a fixed position of the hot spot in the mantle.

Paleomagnetism and Plate Motions

The paleomagnetism stored in rocks on the floor of the ocean also provides a method for measuring rates of plate motion—at least as averaged over millions of years. Recall that a symmetrical pattern of magnetic strips occurs on both sides of the oceanic ridge. Soon after this discovery, investigators began to assign ages to magnetic strips using the magnetic time scale produced from lava flows on land. Once the age of the magnetic strip and its distance from the ridge crest is known, a calculation of the average rate of plate motion can be made.

For example, the boundary between the Gauss and Matuyama epochs occurred about 2.5 million years ago. Along a section of the Mid-Atlantic Ridge, the distance from the ridge axis to that boundary is about 25 kilometers in both directions, for a total distance of 50 kilometers. The rate of seafloor spreading along this section of the Mid-Atlantic Ridge is 50 kilometers for every 2.5 million years, or 2 centimeters per year. Thus, North America is

FIGURE 2.28 The chain of islands and seamounts that extends from Hawaii to the Aleutian trench results from the movement of the Pacific plate over an apparently stationary hot spot. Radiometric dating of the Hawaiian Islands shows that the volcanic activity decreases in age toward the island of Hawaii.

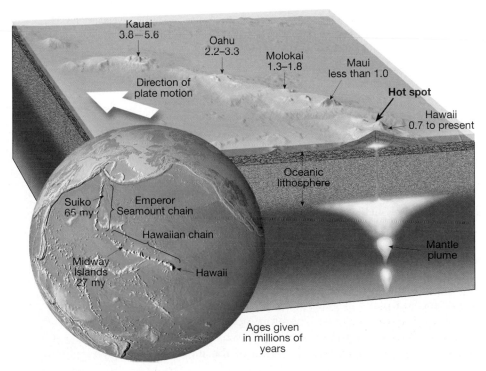

Kauai
3.8–5.6

Oahu
2.2–3.3

Molokai
1.3–1.8

Maui
less than 1.0

Direction of plate motion

Hot spot

Hawaii
0.7 to present

Oceanic lithosphere

Suiko
65 my

Emperor Seamount chain

Hawaiian chain

Midway Islands
27 my

Hawaii

Mantle plume

Ages given in millions of years

If the continents move, do other features, like segments of the mid-ocean ridge, also move?

That's a good observation, and yes, they do! It is interesting to note that very little is really fixed in place on Earth's surface. When we talk about movement of features on Earth, we must consider the question, "Moving relative to what?" Certainly, the mid-ocean ridge moves relative to the continents (which sometimes causes segments of the mid-ocean ridges to be subducted beneath the continents). In addition, the mid-ocean ridge is moving relative to a fixed location outside Earth. This means that an observer orbiting above Earth would notice, after only a few million years, that all continental and seafloor features—as well as plate boundaries—are indeed moving.

serving perhaps a dozen quasars, 5 to 10 times each. This scheme provides an estimate of the distance between these observatories, which is accurate to about 2 centimeters. By repeating this experiment at a later date, researchers can establish the relative motion of these sites. This method has been particularly useful in establishing large-scale plate motions, such as the separation that is occurring between the United States and Europe.

You may be familiar with the Global Positioning System, which is part of the navigation system used in automobiles to locate one's position and to provide directions to some other location. The Global Positioning System employs numerous satellites rather than an extragalactic source to precisely measure the location of a particular site on Earth's surface. By using two widely spaced GPS receivers, signals obtained by these instruments can be used to calculate their relative positions with considerable accuracy. Techniques using GPS receivers have been shown to be useful in establishing small-scale crustal movements such as those that occur along faults in regions known to be tectonically active.

Data obtained from these and other techniques confirm the fact that real plate motion has been detected. Calculations show that Hawaii is moving in a northwesterly direction and approaching Japan at 8.3 centimeters per year (Figure 2.30). A site located in Maryland is retreating from one in England at a rate of about 1.7 centimeters per year—a rate that is close to the 2.3-centimeters-per-year spreading rate that was established from paleomagnetic evidence.

What Drives Plate Motions?

The plate tectonics theory *describes* plate motion and the role that this motion plays in generating and/or modifying the major features of Earth's crust. Therefore, acceptance of plate tectonics does not rely on knowing exactly what drives plate motion. This is fortunate, because none of the models yet proposed can account for all major facets of plate tectonics. Nevertheless, researchers generally agree on the following:

1. Convective flow in the rocky 2900-kilometer-thick mantle—in which warm, buoyant rock rises and cooler, denser material sinks under its own weight—is the underlying driving force for plate movement.
2. Mantle convection and plate tectonics are part of the same system. Subducting oceanic plates drive the cold downward-moving portion of convective flow while shallow upwelling of hot rock along the oceanic ridge and buoyant mantle plumes are the upward-flowing arm of the convective mechanism.
3. The slow movements of Earth's plates and mantle are ultimately driven by the unequal distribution of heat within Earth's interior. Furthermore, this flow is the mechanism that transports heat away from Earth's core and up through the mantle.

What is not known with any high degree of certainty is the precise nature of this convective flow.

Some researchers have argued that the mantle is like a giant layer cake, divided at a depth of 660 kilometers.

Will plate tectonics eventually "turn off" and cease to operate on Earth?

Because plate tectonic processes are powered by heat from within Earth (which is of a finite amount), the forces will slow sometime in the distant future to the point that the plates will cease to move. The work of external processes, however, will continue to erode Earth's surface features, most of which will eventually become relatively flat. What a different world it will be—an Earth with no earthquakes, no volcanoes, and no mountains. Flatness will prevail!

FIGURE 2.29   Radio telescopes like this one located at Green Bank, West Virginia, are used to accurately determine the distance between two distant sites. Data collected by repeated measurements have detected relative plate motions of 1 to 15 centimeters per year between various sites worldwide. (Courtesy of National Radio Astronomy Observatory)

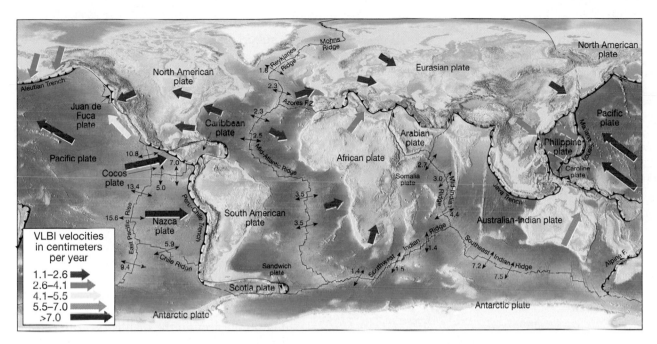

FIGURE 2.30 This map illustrates directions and rates of plate motion in centimeters per year. Seafloor-spreading velocities (as shown with black arrows and labels) are based on the spacing of dated magnetic strips (anomalies). The colored arrows show Very Long Baseline Interferometry (VLBI) data of plate motion at selected locations. The data obtained by these methods are typically consistent. (Seafloor data from DeMets and others, VLBI data from Ryan and others)

Convection operates in both layers, but mixing between layers is minimal. At the other end of the spectrum is a model that loosely resembles a pot of just barely boiling soup, churning ever so slowly from top to bottom over eons of geologic time. Neither model fits all of the available data. We will first look at some of the mechanisms that are thought to contribute to plate motion and then examine a few of the models that have been proposed to describe plate-mantle convection.

Forces that Drive Plate Motion

There is general agreement that the subduction of cold, dense slabs of oceanic lithosphere is the main driving force of plate motion (Figure 2.31). As these slabs sink into the asthenosphere, they "pull" the trailing plate along. This phenomenon, called **slab pull,** occurs because old slabs of oceanic lithosphere are more dense than the underlying asthenosphere and hence "sink like a rock."

Another important driving force is called **ridge push** (Figure 2.31). This gravity-driven mechanism results from the elevated position of the oceanic ridge, which causes slabs of lithosphere to "slide" down the flanks of the ridge. Ridge push appears to contribute far less to plate motions than slab pull. The primary evidence for this comes from known spreading rates on ridge seg-

ments having different elevations. For example, despite its greater average height above the seafloor, spreading rates along the Mid-Atlantic Ridge are considerably less than spreading rates along the less steep East Pacific Rise (Figure 2.30). The fact that fast-moving plates are being subducted along a large percentage of their margins also supports the idea that slab pull has more significant impact than ridge push. Examples of fast-moving plates include the Pacific, Nazca, and Cocos plates, all of which have spreading rates that exceed 10 centimeters per year.

FIGURE 2.31 Illustration of some of the forces that act on tectonic plates.

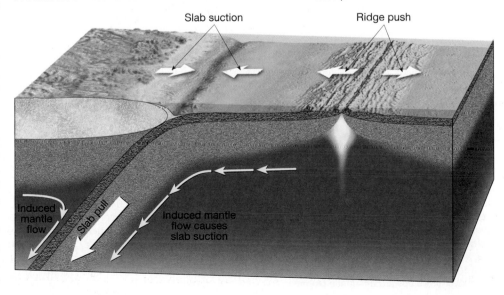

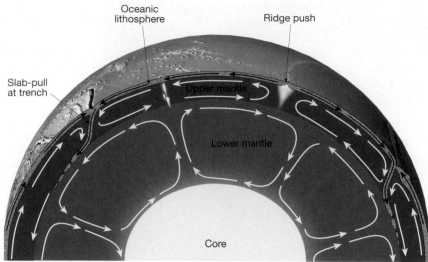

A. Layering at 660 kilometers

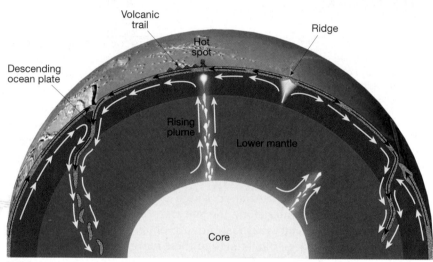

B. Whole-mantle convection

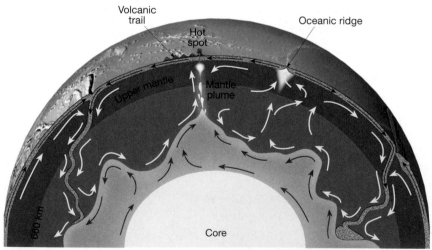

C. Deep-layer model

FIGURE 2.32 Proposed models for mantle convection. **A.** The model shown in this illustration consists of two convection layers—a thin, convective layer above 660 kilometers and a thick one below. **B.** In this whole-mantle convection model, cold oceanic lithosphere descends into the lowermost mantle while hot mantle plumes transport heat toward the surface. **C.** This deep-layer model suggests that the mantle operates similar to a lava lamp on a low setting. Earth's heat causes these layers of convection to slowly swell and shrink in complex patterns without substantial mixing. Some material from the lower layer flows upward as mantle plumes.

Yet another driving force arises from the drag of a subducting slab on the adjacent mantle. The result is an induced mantle circulation that pulls both the subducting and overriding plates toward the trench. Because this mantle flow tends to "suck" in nearby plates (similar to pulling the plug on a bathtub), it is called **slab suction** (Figure 2.31). Even if a subducting slab becomes detached from the overlying plate, its descent will continue to create flow in the mantle and hence will continue to drive plate motion.

Models of Plate-Mantle Convection

Any model of mantle-plate convection must be consistent with observed physical and chemical properties of the mantle. When seafloor spreading was first proposed, geologists suggested that convection in the mantle consisted of upcurrents coming from deep in the mantle beneath oceanic ridges. Upon reaching the base of the lithosphere, these currents were thought to spread laterally and drag the plates along. Thus, plates were viewed as being carried passively by the flow in the mantle. However, based on physical evidence, it became clear that upwelling beneath oceanic ridges is shallow and not related to deep convection in the lower mantle. It is the horizontal movement of lithospheric plates away from the ridge that causes mantle upwelling, not the other way around. We have also learned that plate motion is the dominant source of convective flow in the mantle. As the plates move, they drag the adjacent material along, thereby inducing flow in the mantle. Thus, modern models have plates being an integral part of mantle convection and perhaps even its most active component.

In addition, any acceptable model must explain compositional variations known to exist in the mantle. For example, both the basaltic lavas that erupt along oceanic ridges and those that are generated by hot-spot volcanism, such as Hawaii, have mantle sources. Yet ocean ridge basalts are fairly homogeneous in composition and depleted in certain trace elements, while high concentrations of these same elements are associated with hot-spot eruptions. Because basaltic lavas from different tectonic settings have different concentrations of certain elements, they are assumed to be derived from chemically distinct mantle reservoirs.

Layering at 660 Kilometers Earlier we referred to the "layer cake" version of mantle convection. As shown in Figure 2.32A, one layered model has two zones of convection—a thin convective layer in the upper mantle (above 660 kilometers) and a thick one located below. This model successfully explains why basaltic lavas that erupt along the oceanic ridges have a somewhat different composition than those that erupt in Hawaii as a result of hot-spot activity. The mid-ocean ridge basalts come from the upper convective layer, which is well mixed, whereas the mantle plume that feeds the Hawaiian volcanoes taps a deeper, more primitive magma source that resides in the lower convective layer.

Despite evidence that supports this model, data gathered from the study of earthquake waves has shown that at least some subducting slabs of cold oceanic lithosphere penetrate the 660-kilometer boundary and descend deep into the mantle. The subducting lithosphere should serve to mix the upper and lower layers. As a result, the layered mantle structure proposed in this model would be destroyed.

Whole-Mantle Convection   Because of problems with the layered model, researchers began to favor the whole-mantle convection model. In this model, slabs of dense, cold oceanic lithosphere sink into the lower mantle, while hot, buoyant plumes originating near the mantle-core boundary transport heat toward the surface (Figure 2.32B).

However, recent work has shown that whole-mantle mixing would cause the entire mantle to completely mix in a matter of a few hundred million years. This, in turn, would eliminate chemically distinct magma sources—those that are observed in hot-spot volcanism and those associated with volcanic activity along oceanic ridges.

Deep-Layer Model   Another hypothesis favors layering deep within the mantle. One deep-layer model has been described as analogous to a lava lamp on a low setting. As shown in Figure 2.32C, the lower one-third of the mantle is somewhat like the colored fluid in the bottom of a lava lamp. Like a lava lamp, heat from Earth's interior causes the lower layer to slowly swell and shrink in complex patterns without substantial mixing with the layer above. In this model, small amounts of material from the lower layer form mantle plumes which generate hot-spot volcanism at the surface. Thus, this model provides the two chemically distinct mantle sources, one in the upper mantle and a deeper source near the core-mantle boundary. Furthermore, the

deep-layer model is compatible with evidence that shows lithospheric plates penetrating the 660-kilometer layer. Despite its plausibility, there is little evidence to suggest that a deep layer of this nature exists, except for the very thin D" layer located just above the mantle-core boundary.

Although there is still much to be learned about the mechanisms that cause plate tectonics to operate, some facts are clear. The unequal distribution of heat in Earth's interior generates some type of thermal convection that ultimately drives plate-mantle motion. The primary driving force for this flow is provided by dense, sinking lithospheric plates that serve to transport cold material into the mantle. Moreover, mantle plumes that are generated near the core-mantle boundary carry heat upward from the core into the mantle.

The Importance of the Plate Tectonics Theory

Plate tectonics is the first theory to provide a comprehensive view of the processes that produced Earth's major surface features, including the continents and ocean basins. As such, it has linked many aspects of geology that previously had been considered unrelated. Various branches of geology have joined to provide a better understanding of the workings of our dynamic planet. Within the framework of plate tectonics, geologists have found explanations for the geologic distribution of earthquakes, volcanoes, and mountain belts. Further, we are now better able to explain the distributions of plants and animals in the geologic past, as well as the distribution of economically significant mineral deposits.

Despite its usefulness in explaining many of the large-scale geologic processes operating on Earth, plate tectonics is not completely understood. The model that was set forth in 1968 was simply a basic framework that left much of the detail to future research. Through critical testing, this initial model has been modified and expanded to become the theory we know today. The current theory will certainly be refined as additional data and observations are obtained. The theory of plate tectonics, although a powerful tool, is nonetheless an evolving model of Earth's dynamic processes.

Summary

- In the early 1900s *Alfred Wegener* set forth the *continental drift* hypothesis. One of its major tenets was that a supercontinent called *Pangaea* began breaking apart into smaller continents about 200 million years ago. The smaller continental fragments then "drifted" to their present positions. To support the claim that the now separate continents were once joined, Wegener and others used the *fit of South America and Africa, fossil evidence, rock types and structures,* and *ancient climates.* One of the main objections to the continental drift hypothesis was its inability to provide an acceptable mechanism for the movement of continents.

- From the study of *paleomagnetism,* researchers learned that the continents had wandered as Wegener proposed. In 1962, Harry Hess formulated the idea of *seafloor spreading,* which states that new seafloor is continually being generated at mid-ocean ridges and old, dense seafloor is being consumed at the deep-ocean trenches. Support for seafloor spreading followed, with the discovery of alternating strips of high- and low-intensity magnetism that parallel the ridge crests.

- By 1968 continental drift and seafloor spreading were united into a far more encompassing theory known as *plate tectonics.* According to plate tectonics, Earth's rigid outer layer (*lithosphere*) overlies a weaker region called the *asthenosphere*. Further, the lithosphere is broken into seven large and numerous smaller segments, called *plates*, that are in motion and continually changing in shape and size. Plates move as relatively coherent units and are deformed mainly along their boundaries.

- *Divergent plate boundaries* occur where plates move apart, resulting in upwelling of material from the mantle to create new seafloor. Most divergent boundaries occur along the axis of the oceanic ridge system and are associated with seafloor spreading, which occurs at rates of 2 to 15 centimeters per year. New divergent boundaries may form within a continent (for example, the East African rift valleys), where they may fragment a landmass and develop a new ocean basin.

- *Convergent plate boundaries* occur where plates move together, resulting in the subduction of oceanic lithosphere into the mantle along a deep oceanic trench. Convergence between an oceanic and continental block results in subduction of the oceanic slab and the formation of a *continental volcanic arc* such as the Andes of South America. Oceanic–oceanic convergence results in an arc-shaped chain of volcanic islands called a *volcanic island arc*. When two plates carrying continental crust converge, both plates are too buoyant to be subducted. The result is a "collision" resulting in the formation of a mountain belt such as the Himalayas.

- *Transform fault boundaries* occur where plates grind past each other without the production or destruction of lithosphere. Most transform faults join two segments of a mid-ocean ridge. Others connect spreading centers to subduction zones and thus facilitate the transport of oceanic crust created at a ridge crest to its site of destruction, at a deep-ocean trench. Still others, like the San Andreas Fault, cut through continental crust.

- The theory of plate tectonics is supported by the ages and thickness of *sediments* from the floors of the deep-ocean basins and the existence of island chains that formed over *hot spots* and provide a frame of reference for tracing the direction of plate motion.

- Three basic models for mantle convection are currently being evaluated. Mechanisms that contribute to this convective flow are slab pull, ridge push, and mantle plumes. *Slab pull* occurs where cold, dense oceanic lithosphere is subducted and pulls the trailing lithosphere along. *Ridge push* results when gravity sets the elevated slabs astride oceanic ridges in motion. Hot, buoyant *mantle plumes* are considered the upward flowing arms of mantle convection. One model suggests that mantle convection occurs in two layers separated at a depth of 660 kilometers. Another model proposes whole-mantle convection that stirs the entire 2900-kilometer-thick rocky mantle. Yet another model suggests that the bottom third of the mantle gradually bulges upward in some areas and sinks in others without appreciable mixing.

Review Questions

1. Who is credited with developing the continental drift hypothesis?

2. What was probably the first evidence that led some to suspect that the continents were once connected?

3. What was Pangaea?

4. List the evidence that Wegener and his supporters gathered to substantiate the continental drift hypothesis.

5. Explain why the discovery of the fossil remains of *Mesosaurus* in both South America and Africa, but nowhere else, supports the continental drift hypothesis.

6. Early in the 20th century what was the prevailing view of how land animals migrated across vast expanses of ocean?

7. How did Wegener account for the existence of glaciers in the southern landmasses, while at the same time areas in North America, Europe, and Siberia supported lush tropical swamps?

8. Explain how paleomagnetism can be used to establish the latitude of a specific place at some distant time.

9. What is meant by seafloor spreading? Who is credited with formulating this important concept? Where is active seafloor spreading occurring today?

10. Describe how Fred Vine and D. H. Matthews related the seafloor-spreading hypothesis to magnetic reversals.

11. Where is lithosphere being formed? Consumed? Why must the production and destruction of the lithosphere be going on at about the same rate?

12. Why is oceanic lithosphere subducted while continental lithosphere is not?

13. Briefly describe how the Himalaya Mountains formed.

14. Differentiate between transform faults and the two other types of plate boundaries.

15. Some people predict that California will sink into the ocean. Is this idea consistent with the theory of plate tectonics?

16. What is the age of the oldest sediments recovered by deep-ocean drilling? How do the ages of these sediments compare to the ages of the oldest continental rocks?

17. Applying the idea that hot spots remain fixed, in what direction was the Pacific plate moving while the Hawaiian Islands were forming? (See Figure 2.28, p. 63)

18. With which of the three types of plate boundaries are the following places or features associated: Himalayas, Aleutian Islands, Red Sea, Andes Mountains, San Andreas Fault, Iceland, Japan, Mount St. Helens?

19. Briefly describe the three models proposed for mantle-plate convection. What is lacking in each of these models?

Key Terms

asthenosphere (p. 51)
continental drift (p. 37)
continental volcanic arc (p. 57)
convergent boundary (p. 55)
Curie point (p. 44)
deep-ocean trench (p. 55)
divergent boundary (p. 54)
fossil magnetism (p. 44)
fracture zone (p. 58)

geomagnetic reversal (p. 47)
hot spot (p. 61)
hot-spot tracks (p. 61)
island arc (p. 58)
lithosphere (p. 51)
lithospheric plate (p. 51)
magnetometer (p. 47)
magnetic time scale (p. 47)
mantle plume (p. 61)
normal polarity (p. 47)

oceanic ridge system (p. 46)
paleomagnetism (p. 44)
Pangaea (p. 37)
partial melting (p. 57)
plate (p. 51)
plate tectonics (p. 50)
reverse polarity (p. 47)
ridge push (p. 65)
rift, or rift valley (p. 54)
seafloor spreading (p. 46)

slab pull (p. 65)
slab suction (p. 66)
spreading center (p. 54)
subduction zone (p. 56)
tectonic plate (p. 51)
transform fault (p. 58)
transform fault boundary (p. 54)
volcanic island arc (p. 58)

Web Resources

The *Earth* Website uses the resources and flexibility of the Internet to aid in your study of the topics in this chapter. Written and developed by geology instructors, this site will help improve your understanding of geology. Visit **http://www.prenhall .com/tarbuck** and click on the cover of *Earth 9e* to find:

- Online review quizzes.
- Critical thinking exercises.
- Links to chapter-specific Web resources.
- Internet-wide key-term searches.

http://www.prenhall.com/tarbuck

GEODe: Earth

GEODe: Earth makes studying faster and more effective by reinforcing key concepts using animation, video, narration, interactive exercises and practice quizzes. A copy is included with every copy of *Earth*.

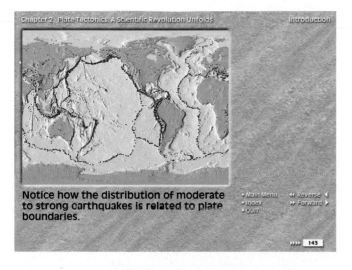

Notice how the distribution of moderate to strong earthquakes is related to plate boundaries.

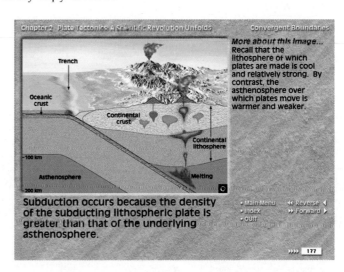

Subduction occurs because the density of the subducting lithospheric plate is greater than that of the underlying asthenosphere.

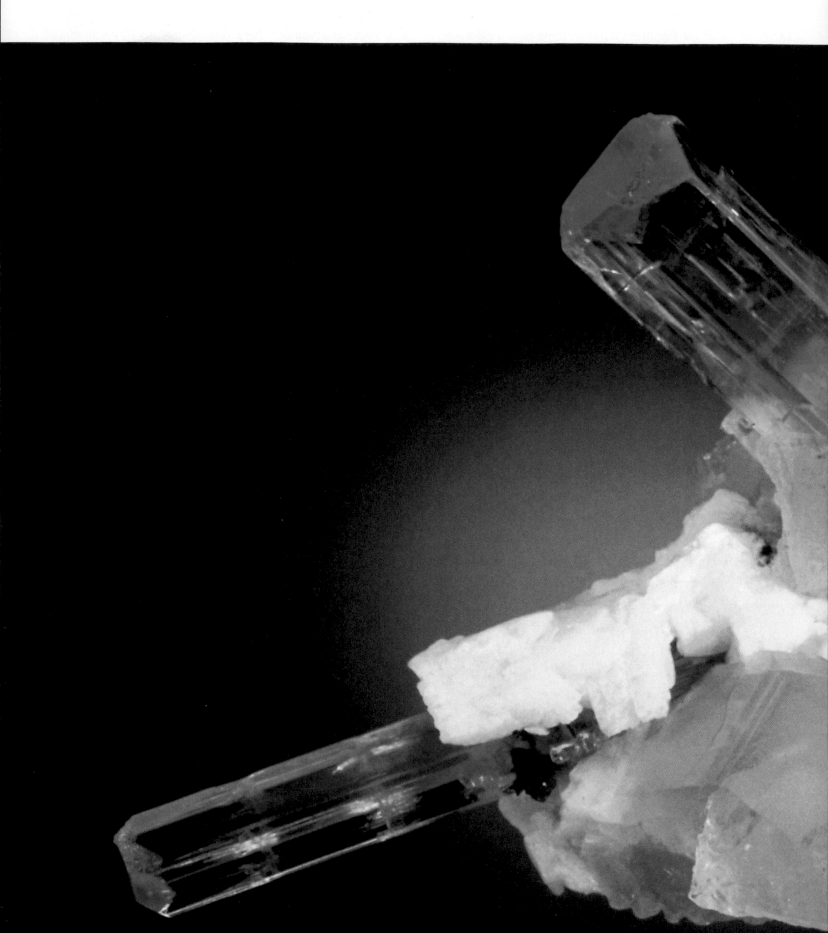

Matter and Minerals

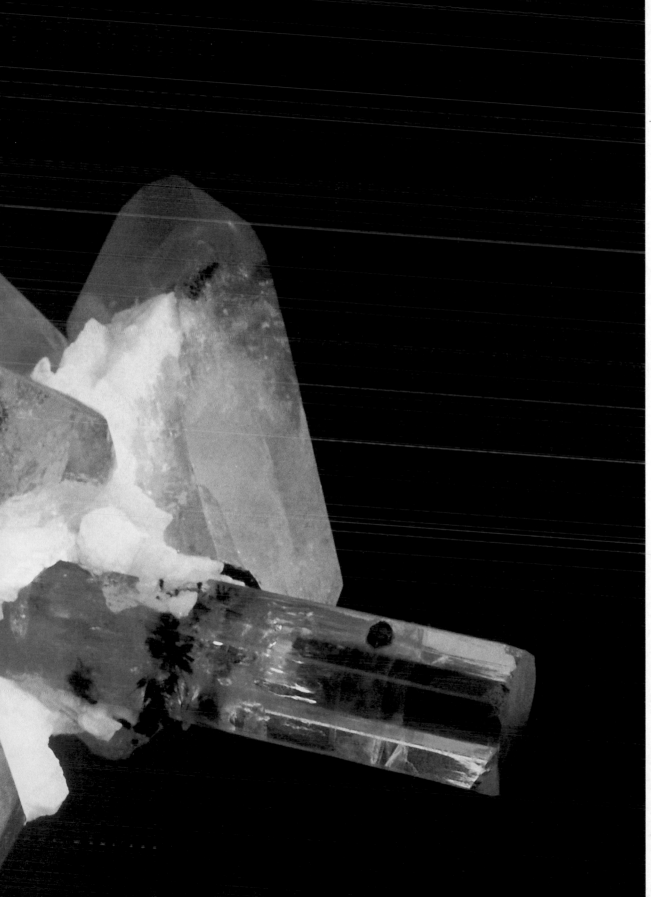

Crystals of Beryl (blueish-green), Feldspar (white), and Quartz (clear). (Photo by Jeff Scovil)

Earth's crust and oceans are the source of a wide variety of useful and essential minerals. Most people are familiar with the common uses of many basic metals, including aluminum in beverage cans, copper in electrical wiring, and gold and silver in jewelry. But some people are not aware that pencil lead contains the greasy-feeling mineral graphite and that bath powders and many cosmetics contain the mineral talc. Moreover, many do not know that drill bits impregnated with diamonds are employed by dentists to drill through tooth enamel, or that the common mineral quartz is the source of silicon for computer chips. In fact, practically every manufactured product contains materials obtained from minerals. As the mineral requirements of modern society grow, the need to locate the additional supplies of useful minerals also grows, becoming more challenging as well.

In addition to the economic uses of rocks and minerals, all of the processes studied by geologists are in some way dependent on the properties of these basic Earth materials. Events such as volcanic eruptions, mountain building, weathering and erosion, and even earthquakes involve rocks and minerals. Consequently, a basic knowledge of Earth materials is essential to the understanding of all geologic phenomena.

FIGURE 3.1 The Coronation Crown of England, made of solid gold and set with 444 gemstones. Known as the St. Edward's Crown, the stones were reset in 1911. (Photo by Tim Graham/Getty Images)

Minerals: Building Blocks of Rocks

 GEODe
EARTH

Minerals: Building Blocks of Rocks
▶ Introduction

We begin our discussion of Earth materials with an overview of **mineralogy** (*mineral* = mineral, *ology* = the study of), because minerals are the building blocks of rocks. In addition, minerals have been employed by humans for both useful and decorative purposes for thousands of years (Figure 3.1). The first minerals mined were flint and chert, which people fashioned into weapons and cutting tools. As early as 3700 B.C., Egyptians began mining gold, silver, and copper; and by 2200 B.C. humans discovered how to combine copper with tin to make bronze, a tough, hard alloy. The Bronze Age began its decline when the ability to extract iron from minerals such as hematite was discovered. By about 800 B.C., iron-working technology had advanced to the point that weapons and many everyday objects were made of iron rather than copper, bronze, or wood. During the Middle Ages, mining of a variety of minerals was common throughout Europe and the impetus for the formal study of minerals was in place.

The term *mineral* is used in several different ways. For example, those concerned with health and fitness extol the benefits of vitamins and minerals. The mining industry typically uses the word when referring to anything taken out of the ground, such as coal, iron ore, or sand and gravel. The guessing game known as *Twenty Questions* usually begins with the question, *Is it animal, vegetable, or mineral?* What criteria do geologists use to determine whether something is a mineral?

Geologists define a **mineral** as *any naturally occurring inorganic solid that possesses an orderly crystalline structure and a well-defined chemical composition.* Thus, those Earth materials that are classified as minerals exhibit the following characteristics:

1. **Naturally occurring.** Minerals form by natural, geologic processes. Consequently, synthetic diamonds and rubies, as well as a variety of other useful materials produced in a laboratory, are not considered minerals.

2. **Solid substance.** Minerals are solids within the temperature ranges normally experienced at Earth's surface. Thus, ice (frozen water) is considered a mineral, whereas liquid water and water vapor are not.

3. **Orderly crystalline structure.** Minerals are crystalline substances, which means their atoms are arranged in an orderly, repetitive manner. This orderly packing of atoms is reflected in the regularly shaped objects we call crystals (Figure 3.2). Some naturally occurring solids, such as volcanic glass (obsidian), lack a repetitive atomic structure and are not considered minerals.

4. **Well-defined chemical composition.** Most minerals are chemical compounds having compositions that are given by their chemical formulas. The mineral pyrite (FeS_2), for example, consists of one iron (Fe) atom, for every two sulfur (S) atoms. In nature, however, it is not uncommon for some atoms within a crystal structure to be replaced by others of similar size without changing the internal structure or properties of that mineral. Therefore, the chemical compositions of minerals may vary, but they *vary within specific, well-defined limits.*

5. **Generally inorganic.** Inorganic crystalline solids, as exemplified by ordinary table salt (halite), that are found naturally in the ground are considered minerals. Organic compounds, on the other hand, are generally not. Sugar, a crystalline solid like salt but which comes from sugarcane or sugar beets, is a common example of such an organic compound. However, many marine animals secrete inorganic compounds, such as calcium carbonate (calcite), in the form of shells and coral reefs. These materials are considered minerals by most geologists.

In contrast to minerals, rocks are more loosely defined. Simply, a **rock** is any solid mass of mineral, or mineral-like, matter that occurs naturally as part of our planet. A few rocks are composed almost entirely of one mineral. A common example is the sedimentary rock *limestone*, which consists of impure masses of the mineral calcite. However, most rocks, like the common rock granite shown in Figure 3.3, occur as aggregates of several kinds of minerals. Here, the term *aggregate* implies that the minerals are joined in such a way that the properties of each mineral are retained. Note that you can easily identify the mineral constituents of the granite in Figure 3.3.

A few rocks are composed of nonmineral matter. These include the volcanic rocks *obsidian* and *pumice*, which are noncrystalline glassy substances, and *coal*, which consists of solid organic debris (see Box 3.1).

FIGURE 3.2 Collection of well-developed quartz crystals found near Hot Springs, Arkansas. (Photo by Jeff Scovil)

Granite
(Rock)

Quartz
(Mineral)

Hornblende
(Mineral)

Feldspar
(Mineral)

FIGURE 3.3 Most rocks are aggregates of two or more minerals.

BOX 3.1 ▶ PEOPLE AND THE ENVIRONMENT

Making Glass from Minerals

Many everyday objects are made of glass, including windowpanes, jars and bottles, and the lenses of some eyeglasses. People have been making glass for at least 2000 years. Glass is manufactured by melting naturally occurring materials and cooling the liquid quickly before the atoms have time to arrange themselves into an orderly crystalline form. (This is the same way that natural glass, called *obsidian,* is generated from lava.)

It is possible to produce glass from a variety of materials, but the primary ingredient (75 percent) of most commercially produced glass is the mineral quartz (SiO_2). Lesser amounts of the minerals calcite (calcium carbonate) and trona (sodium carbonate) are added to the mix. These materials lower the melting temperature and improve the workability of the molten glass.

In the United States, high-quality quartz (usually quartz sandstone) and calcite (limestone) are readily available in many areas. Trona, on the other hand, is mined almost exclusively in the Green River area of southwestern Wyoming. In addition to its use in making glass, trona is used in making detergents, paper, and even baking soda.

Manufacturers can change the properties of glass by adding minor amounts of several other ingredients (Figure 3.A). Coloring agents include iron sulfide (amber), selenium (pink), cobalt oxide (blue), and iron oxides (green, yellow, brown). The addition of lead imparts clarity and brilliance to glass and is therefore used in the manufacture of fine crystal tableware. Ovenware, such as Pyrex®, owes its heat resistance to boron, whereas aluminum makes glass resistant to weathering.

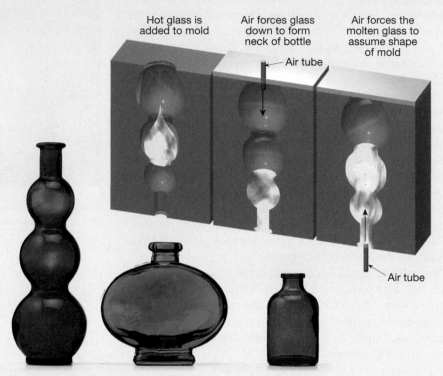

FIGURE 3.A   Glass bottles are made by first pouring molten glass into a mold. Then air forces the molten glass to assume the shape of the mold. Metallic compounds are mixed with the raw ingredients to color the glass. (Photo by Guy Ryecart)

Students Sometimes Ask . . .

Are the minerals you talked about in class the same as those found in dietary supplements?

Not ordinarily. From a geologic perspective, a mineral must be a *naturally occurring* crystalline solid. Minerals found in dietary supplements are man-made inorganic compounds that contain *elements* needed to sustain life. These dietary minerals typically contain elements that are metals—calcium, potassium, phosphorus, magnesium, and iron—as well as trace amounts of a dozen other elements, such as copper, nickel, and vanadium. Although these two types of "minerals" are different, they are related. The source of the elements used to make dietary supplements is in fact the naturally occurring minerals of Earth's crust. It should also be noted that vitamins are *organic compounds* produced by living organisms, not *inorganic compounds,* such as minerals.

Although this chapter deals primarily with the nature of minerals, keep in mind that most rocks are simply aggregates of minerals. Because the properties of rocks are determined largely by the chemical composition and crystalline structure of the minerals contained within them, we will first consider these Earth materials. Then, in Chapters 4, 7, and 8 we will take a closer look at Earth's major rock groups.

Elements: Building Blocks of Minerals

You are probably familiar with the names of many elements, including copper, iron, oxygen, and carbon. An **element** is a substance that cannot be broken down into simpler substances by chemical or physical means. There are about 90 different elements found in nature, and consequently about 90 different kinds of atoms. In addition, scientists have succeeded in creating about 23 synthetic elements.

The elements can be organized into rows so that those with similar properties are in the same column. This arrangement,

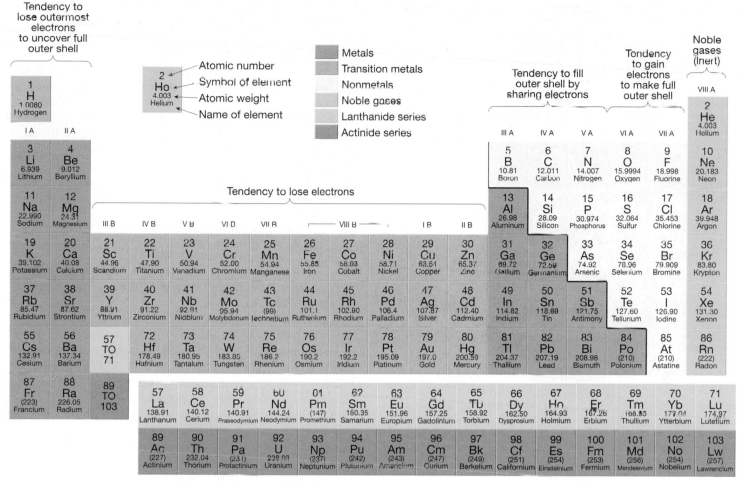

FIGURE 3.4 Periodic table of the elements.

called the **periodic table** is shown in Figure 3.4. Notice that symbols are used to provide a shorthand way of representing an element. Each element is also known by its atomic number, which is shown above each symbol on the periodic table.

Some elements, such as copper (number 29) and gold (number 79), exist in nature with single atoms as the basic unit. Thus, native copper and gold are minerals made entirely from one element. However, many elements are quite reactive and join together with atoms of one or more other elements to form **chemical compounds.** As a result, most minerals are chemical compounds consisting of two or more different elements.

To understand how elements combine to form compounds, we must first consider the atom. The **atom** (*a* = not, *tomos* = cut) is the smallest particle of matter that retains the essential characteristics of an element. It is this extremely small particle that does the combining.

The central region of an atom is called the **nucleus.** The nucleus contains protons and neutrons. **Protons** are very dense particles with positive electrical charges. **Neutrons** have the same mass as a proton but lack an electrical charge.

Orbiting the nucleus are **electrons,** which have negative electrical charges. For convenience, we sometimes diagram

atoms to show the electrons orbiting the nucleus, like the orderly orbiting of the planets around the Sun (Figure 3.5A). However, electrons move in a less predictable way than planets. As a result, they create a sphere-shaped negative zone around the nucleus. A more realistic picture of the positions of electrons can be obtained by envisioning a cloud of negatively charged electrons surrounding the nucleus (Figure 3.5B).

Studies of electron configurations predict that individual electrons move within regions around the nucleus called **principal shells,** or **energy levels.** Furthermore, each of these shells can hold a specific number of electrons, with the outermost principal shell containing the **valence electrons.** Valence electrons are important because, as you will see later, they are involved in chemical bonding.

Some atoms have only a single proton in their nuclei, while others contain more than 100 protons. The number of protons in the nucleus of an atom is called the **atomic number.** For example, all atoms with six protons are carbon atoms; hence, the atomic number of carbon is 6. Likewise, every atom with eight protons is an oxygen atom.

Atoms in their natural state also have the same number of electrons as protons, so the atomic number also equals the

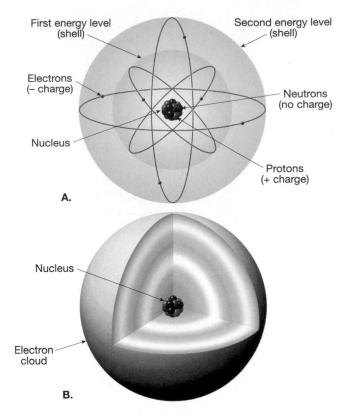

First energy level (shell)

Second energy level (shell)

Electrons (– charge)

Neutrons (no charge)

Nucleus

Protons (+ charge)

A.

Nucleus

Electron cloud

B.

FIGURE 3.5   Two models of the atom. **A.** A very simplified view of the atom, which consists of a central nucleus, consisting of protons and neutrons, encircled by high-speed electrons. **B.** Another model of the atoms showing spherically shaped electron clouds (energy level shells). Note that these models are not drawn to scale. Electrons are minuscule in size compared to protons and neutrons, and the relative space between the nucleus and electron shells is much greater than illustrated.

number of electrons surrounding the nucleus. Therefore, carbon has six electrons to match its six protons, and oxygen has eight electrons to match its eight protons. Neutrons have no charge, so the positive charge of the protons is exactly balanced by the negative charge of the electrons. Consequently, atoms in the natural state are neutral electrically and have no overall electrical charge.

Why Atoms Bond

Why do elements join together to form compounds? From experimentation it has been learned that the forces holding the atoms together are electrical. Further, it is known that chemical bonding results in a change in the electron configuration of the bonded atoms. As we noted earlier, it is the valence electrons (outer-shell electrons) that are generally involved in chemical bonding. Figure 3.6 shows a shorthand way of representing the number of electrons in the outer principal shell (valence electrons). Notice that the elements in group I have one valence electron, those in group II have two, and so forth up to group VIII, which has eight valence electrons.

Other than the first shell, which can hold a maximum of two electrons, *a stable configuration occurs when the valence*

shell contains eight electrons. Only the noble gases, such as neon and argon, have a complete outermost principal shell. Hence, the noble gases are the least chemically reactive and are designated as "inert."

When an atom's outermost shell does not contain the maximum number of electrons (eight), the atom is likely to chemically bond with one or more other atoms. A *chemical bond* is the sharing or transfer of electrons to attain a stable electron configuration among the bonding atoms. If the electrons are transferred, the bond is an *ionic bond.* If the electrons are shared, the bond is called a *covalent bond.* In either case, the bonding atoms get stable electron configurations, which usually consist of eight electrons in the outer shell.

Ionic Bonds: Electrons Transferred

Perhaps the easiest type of bond to visualize is an **ionic bond.** In ionic bonding, one or more valence electrons are transferred from one atom to another. Simply, one atom gives up its valence electrons, and the other uses them to complete its outer shell. A common example of ionic bonding is sodium (Na) and chlorine (Cl) joining to produce sodium chloride (common table salt). This is shown in Figure 3.7A. Notice that sodium gives up its single valence electron to chlorine. As a result, sodium achieves a stable configuration having eight electrons in its outermost shell. By acquiring the electron that sodium loses, chlorine—which has seven valence electrons—gains the eighth electron needed to complete its outermost shell. Thus, through the transfer of a single electron, both the sodium and chlorine atoms have acquired a stable electron configuration.

Once electron transfer takes place, atoms are no longer electrically neutral. By giving up one electron, a neutral sodium atom becomes *positively charged* (11 protons/10 electrons). Similarly, by acquiring one electron, the neutral chlorine atom becomes *negatively charged* (17 protons/18 electrons). Atoms which have an electrical charge because of the unequal numbers of electrons and protons, are called **ions.** Atoms that pick up extra electrons and become negatively charged are called *anions*, whereas atoms that lose electrons and become positively charged are called *cations.*

Electron Dot Diagrams for Some Representative Elements							
I	II	III	IV	V	VI	VII	VIII
H ·							He:
Li ·	·Be·	·B·	·C·	·N·	:O·	:F·	:Ne:
Na·	·Mg·	·Al·	·Si·	·P·	:S·	:Cl·	:Ar:
K ·	·Ca·	·Ga·	·Ge·	·As·	:Se·	:Br·	:Kr:

FIGURE 3.6   Dot diagrams for some representative elements. Each dot represents a valence electron found in the outermost principal shell.

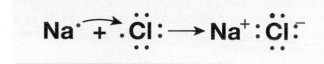

A.

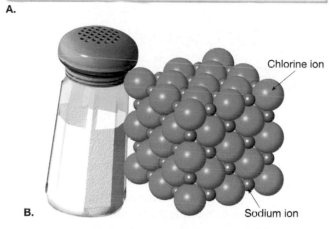

B.

FIGURE 3.7 Chemical bonding of sodium chloride (table salt). **A.** Through the transfer of one electron in the outer shell of a sodium atom to the outer shell of a chlorine atom, the sodium becomes a positive ion (cation) and chlorine a negative ion (anion). **B.** Diagram illustrating the arrangement of sodium and chlorine ions in table salt.

We know that ions with like charges repel, and those with unlike charges attract. Thus, an *ionic bond* is the attraction of oppositely charged ions to one another, producing an electrically neutral compound. Figure 3.7B illustrates the arrangement of sodium and chlorine ions in ordinary table salt. Notice that salt consists of alternating sodium and chlorine ions, positioned in such a manner that each positive ion is attracted to and surrounded on all sides by negative ions, and vice versa. This arrangement maximizes the attraction between ions with unlike charges while minimizing the repulsion between ions with like charges. Thus, *ionic compounds consist of an orderly arrangement of oppositely charged ions assembled in a definite ratio that provides overall electrical neutrality.*

The properties of a chemical compound are *dramatically different* from the properties of the elements comprising it. For example, sodium, a soft, silvery metal, is extremely reactive and poisonous. If you were to consume even a small amount of elemental sodium, you would need immediate medical attention. Chlorine, a green poisonous gas, is so toxic it was used as a chemical weapon during World War I. Together, however, these elements produce sodium chloride, a harmless flavor enhancer that we call table salt. When elements combine to form compounds, their properties change dramatically.

Covalent Bonds: Electrons Shared

Not all atoms combine by transferring electrons to form ions. Some atoms *share* electrons. For example, the gaseous elements oxygen (O_2), hydrogen (H_2), and chlorine (Cl_2) exist as stable molecules consisting of two atoms bonded together, without a complete transfer of electrons.

Figure 3.8 illustrates the sharing of a pair of electrons between two chlorine atoms to form a molecule of chlorine gas

(Cl_2). By overlapping their outer shells, these chlorine atoms share a pair of electrons. Thus, each chlorine atom has acquired, through cooperative action, the needed eight electrons to complete its outer shell. The bond produced by the sharing of electrons is called a **covalent bond.**

A common analogy may help you visualize a covalent bond. Imagine two people at opposite ends of a dimly lit room, each reading under a separate lamp. By moving the lamps to the center of the room, they are able to combine their light sources so each can see better. Just as the overlapping light beams meld, the shared electrons that provide the "electrical glue" in covalent bonds are indistinguishable from each other. The most common mineral group, the silicates, contains the element silicon, which readily forms covalent bonds with oxygen.

Other Bonds

As you might suspect, many chemical bonds are actually hybrids. They consist to some degree of electron sharing, as in covalent bonding, and to some degree of electron transfer,

---

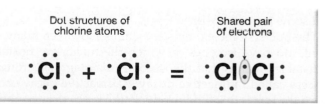

Dot structures of chlorine atoms   Shared pair of electrons

FIGURE 3.8 Dot diagrams used to illustrate the sharing of electrons between two chlorine atoms to form a chlorine molecule. Notice that by sharing a pair of electrons, both chlorine atoms achieve a full outer shell (8 electrons).

as in ionic bonding. For example, silicate minerals are composed of silicon and oxygen atoms that are joined together by bonds that display characteristics of both ionic and covalent bonds. These basic building blocks are, in turn, ionically bonded to metallic ions, producing a great variety of electrically neutral chemical compounds.

Another chemical bond exists in which valence electrons are free to migrate from one ion to another. The mobile valence electrons serve as the electrical glue. This type of electron sharing is found in metals such as copper, gold, aluminum, and silver and is called **metallic bonding.** Metallic bonding accounts for the high electrical conductivity of metals, the ease with which metals are shaped, and numerous other special properties of metals.

Isotopes and Radioactive Decay

Subatomic particles are so incredibly small that a special unit, called an *atomic mass unit,* was devised to express their mass. A proton or a neutron has a mass just slightly more than one atomic mass unit, whereas an electron is only about one two-thousandth of an atomic mass unit. Thus, although electrons play an active role in chemical reactions, they do not contribute significantly to the mass of an atom.

The **mass number** of an atom is simply the total of its neutrons and protons. Atoms of the same element always have the same number of protons. But the number of neutrons for atoms of the same element can vary. Atoms with the same number of protons but different numbers of neutrons are **isotopes** of that element. Isotopes of the same element are labeled by placing the mass number after the element's name or symbol.

For example, carbon has three well-known isotopes. One has a mass number of 12 (carbon-12), another has a mass number of 13 (carbon-13), and the third, carbon-14, has a mass number of 14. Since all atoms of the same element have the same number of protons, and carbon has six, carbon-12 also has *six neutrons* to give it a mass number of 12. Likewise, carbon-14 has six protons plus *eight neutrons* to give it a mass number of 14.

In chemical behavior, all isotopes of the same element are nearly identical. To distinguish among them is like trying to differentiate identical twins, with one being slightly heavier. Because isotopes of an element react the same chemically, different isotopes can become parts of the same mineral. For example, when the mineral calcite forms from calcium, carbon, and oxygen, some of its carbon atoms are carbon-12 and some are carbon-14.

The nuclei of most atoms are stable. However, many elements do have isotopes in which the nuclei are unstable. "Unstable" means that the isotopes disintegrate through a process called **radioactive decay.** Radioactive decay occurs when the forces that bind the nucleus are not strong enough.

During radioactive decay, unstable atoms radiate energy and emit particles. Some of this energy powers the movements of Earth's crust and upper mantle. The rates at which unstable atoms decay are measurable. Therefore, certain radioactive atoms can be used to determine the ages of fossils, rocks, and minerals. A discussion of radioactive decay and its applications in dating past geologic events is found in Chapter 9.

Crystals and Crystallization

Many people associate the word crystal with delicate water or wine goblets or, perhaps glassy objects with smooth sides and gem-like shapes. Mineralogists, on the other hand, use the term **crystal** or **crystalline** in reference to *any natural solid with an ordered, repetitive, atomic structure.*

By this definition, crystals may or may not have smooth-sided faces. The specimen shown in Figure 3.2, for example, exhibits the characteristic crystal form we associate with the mineral quartz—a six-sided prismatic shape with pyramidal ends. However, the quartz crystals in the sample of granite shown in Figure 3.3 do not display well-defined faces. Crystals that exhibit perfect geometric shapes with smooth faces are called *euhedral,* as opposed to crystals with irregular surfaces, which are known as *anhedral.* All minerals are crystalline, but those with well-developed faces are relatively uncommon.

How Do Minerals Form?

Minerals form through the process of **crystallization,** in which the constituent molecules, or ions, chemically bond to form an orderly internal structure. Perhaps the most familiar type of crystallization results when an aqueous (water) solution containing dissolved ions evaporates. As you might expect, the concentration of dissolved material increases as water is removed during evaporation. Once saturation occurs, the atoms bond together to form solid crystalline substances that precipitate from (settle out of) the solution. Inland bodies of salty water such as the Dead Sea or the Great Salt Lake often precipitate the minerals halite, sylvite, gypsum, and other soluble salts. Worldwide, extensive sedimentary rock layers, some exceeding 300 meters (1000 feet) in thickness, provide evidence of ancient seas that have long since evaporated (see Figure 3.35, p. 97).

Minerals can also precipitate from slowly moving groundwater where they fill fractures and voids in rocks. One interesting example, called a *geode,* is a somewhat spherical shaped body with inward-projecting crystals (Figure 3.9). Geodes often contain spectacular crystals of quartz, calcite, or other minerals.

A decrease in temperature may also initiate crystallization, as in the growth of ice from liquid water. Likewise, the crystallization of igneous minerals from molten magma, although more complicated, is similar to water freezing. When magma is very hot, the atoms are very mobile; but when the molten material cools, the atoms slow and begin to combine. Crystallization of a magma body generates igneous rocks that consist of a mosaic of intergrown crystals that lack well-developed faces (see Figure 3.3, p. 73).

According to the textbook, thick beds of halite and gypsum formed when ancient seas evaporated. Has this happened in the recent past?

Yes. During the past 6 million years, the Mediterranean Sea may have dried up and then refilled several times. When 65 percent of seawater evaporates, the mineral gypsum begins to precipitate, meaning it comes out of solution and settles to the bottom. When 90 percent of the water is gone, halite crystals form, followed by salts of potassium and magnesium. Deep-sea drilling in the Mediterranean has encountered the presence of thick beds of gypsum and salt deposits (mostly halite) sitting one atop the other to a maximum depth of 2 kilometers (1.2 miles). These deposits are inferred to have resulted from tectonic events that periodically closed and reopened the connection between the Atlantic Ocean and the Mediterranean Sea (the modern-day Straits of Gibraltar) over the past several million years. During periods when the Mediterranean was cut off from the Atlantic, the warm and dry climate in this region caused the Mediterranean to nearly evaporate. Then, when the connection to the Atlantic was opened, the Mediterranean basin would refill with seawater of normal salinity. This cycle was repeated over and over again, producing the layers of gypsum and salt found on the Mediterranean seafloor.

Metamorphic processes, which typically occur where temperatures and pressures are high, can also generate new minerals (see Box 3.2). Under these extreme conditions, weathered mineral matter or preexisting minerals are recrystallized into new minerals that tend to exhibit larger grain sizes than the parent material. For example, the metamorphism of mudstones, which are made of minute clay

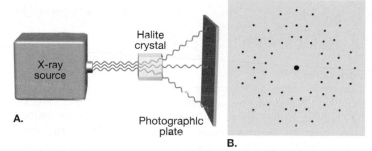

A.

B.

FIGURE 3.10 The principle of X-ray diffraction. **A.** When a parallel beam of X rays pass through a crystal, they are diffracted (scattered) in such a way as to produce a pattern of dark spots on a photographic plate. **B.** The diffraction pattern shown is for the mineral halite (NaCl).

minerals, generates rocks that contain minerals such as mica, quartz, and garnet (see Figure 3.34, p. 94).

The crystallization of minerals occurs in other ways as well. Sulfur deposits are often found near volcanic vents where crystals are deposited directly from hot, sulfur-rich vapors. This mechanism is similar to how snowflakes form—water vapor is deposited as ice crystals without ever entering the liquid phase. Organisms, such as microscopic bacteria, are also responsible for initiating the crystallization of substantial quantities of mineral matter. For example, the mineral pyrite (FeS_2), when found in shale and coal beds, is typically generated by the action of sulfate-reducing bacteria. In addition, mollusks (clams) and other marine invertebrates secrete shells composed of the carbonate minerals calcite and aragonite. The remains of these shells are major components of many limestone beds.

Crystal Structures

The smooth faces and symmetry possessed by well-developed crystals are manifestations of the orderly packing of atoms or molecules that constitute a mineral's internal structure. Although this relationship between crystallization and atomic structure had been proposed much earlier, experimental verification did not occur until 1912 when it was discovered that X rays are diffracted (scattered) by crystals. One X-ray diffraction technique emits a parallel beam of X rays that pass through a crystal and expose a sheet of photographic film on the far side (Figure 3.10A). This is similar to how medical and dental X-ray images are produced. Here, the beam interacts with atoms and is diffracted in a manner that produces a pattern of dark spots on the film (Figure 3.10B). The diffraction pattern produced is a function of the spacing and distribution of atoms in the crystal, and is unique to each mineral. Hence, X-ray diffraction is an important tool for mineral identification.

With the development of X-ray diffraction techniques, mineralogists were able to directly investigate the internal structures of minerals. What followed was the discovery that minerals have highly ordered atomic arrangements that can be described as spherical atoms, or ions, held together by ionic, covalent, or metallic bonds. Native metals that are

FIGURE 3.9 Geode partially filled with amethyst, a purple variety of quartz, Brazil. (Photo by Jeff Scovil)

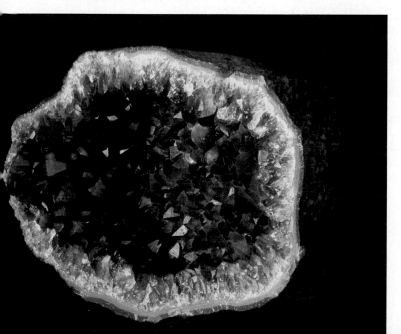

BOX 3.2 ▶ PEOPLE AND THE ENVIRONMENT

Asbestos: What Are the Risks?

Once considered safe enough to use in toothpaste, asbestos became one of the most feared contaminants on Earth. The asbestos panic in the United States began in 1986 when the Environmental Protection Agency (EPA) instituted the Asbestos Hazard Emergency Response Act. It required inspection of all public and private schools for asbestos. This brought asbestos to public attention and raised parental fears that children could contract asbestos-related cancers because of high levels of airborne fibers in schools.

What is Asbestos?

Asbestos is a commercial term applied to a variety of silicate minerals that readily separate into thin, strong fibers that are highly flexible, heat resistant, and relatively inert (Figure 3.B). These properties make asbestos a desirable material for the manufacture of a wide variety of products, including insulation, fireproof fabrics, cement, floor tiles, and car-brake linings. In addition, wall coatings rich in asbestos fibers were used extensively during the U.S. building boom of the 1950s and early 1960s.

The mineral *chrysotile*, marketed as "white asbestos," belongs to the serpentine mineral group and accounts for the vast majority of asbestos sold commercially. All other forms of asbestos are amphiboles and constitute less than 10 percent of asbestos used commercially. The two most common amphibole asbestos minerals are informally called "brown" and "blue" asbestos, respectively.

Exposure and Risk

Health concerns about asbestos stem largely from claims of high death rates among asbestos miners attributed to asbestosis (lung scarring from asbestos fiber inhalation), mesothelioma (cancer of chest and abdominal cavity), and lung cancer. The degree of concern created by these claims is demonstrated by the growth of an entire industry built around asbestos removal from buildings.

The stiff, straight fibers of brown and blue (amphibole) asbestos are known to readily pierce, and remain lodged in, the linings of human lungs. The fibers are physically and chemically stable and are not broken down in the human body. These forms of asbestos are therefore a genuine cause for concern. White asbestos, however, being a different mineral, has different properties. The curly fibers of white asbestos are readily expelled from the lungs and, if they are not expelled, can dissolve within a year.

The U.S. Geological Survey has taken the position that the risks from the most widely used form of asbestos (chrysotile or "white asbestos") are minimal to non-existent. They cite studies of miners of white asbestos in northern Italy that show mortality rates from mesothelioma and lung cancer differ very little from the general public. Another study was conducted on people living in the area of Thetford Mines, Quebec, once the largest chrysotile mine in the world. For many years there were no dust controls, so these people were exposed to extremely high levels of airborne asbestos. Nevertheless, they exhibited normal levels of the diseases thought to be associated with asbestos exposure.

Despite the fact that more than 90 percent of all asbestos used commercially is white asbestos, a number of countries have moved to ban the use of asbestos in many applications, because the different mineral forms of asbestos are not distinguished. Very little of this once exalted mineral is presently used in the United States. Perhaps future studies will determine whether the asbestos panic, in which billions of dollars have been spent on testing and removal, was warranted or not.

◀— 2 cm —▶

FIGURE 3.B Chrysotile asbestos. This sample is a fibrous form of the mineral serpentine. (Photo by E. J. Tarbuck)

composed of only one element, such as gold and silver, consist of atoms that are packed together in a rather simple three-dimensional network that minimizes voids. Imagine a group of same-sized cannon balls in which the layers are stacked so that the spheres in one layer nestle in the hollows between spheres in the adjacent layers.

The atomic arrangement in most minerals is more complicated than those of native metals, because they consist of various sized positive and negative ions. Figure 3.11 illustrates the relative sizes of some of the most common ions found in minerals. Notice that anions, which are atoms that gain electrons, tend to be larger than cations, which lose electrons.

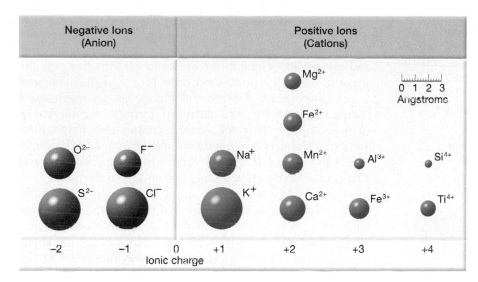

The chart shows relative sizes and ionic charges.

Negative Ions (Anion)	Positive Ions (Cations)				

Angstroms scale: 0 1 2 3

Negative Ions: O²⁻, F⁻, S²⁻, Cl⁻

Positive Ions: Mg²⁺, Fe²⁺, Na⁺, Mn²⁺, Al³⁺, Si⁴⁺, K⁺, Ca²⁺, Fe³⁺, Ti⁴⁺

Ionic charge: −2, −1, 0, +1, +2, +3, +4

FIGURE 3.11 The relative sizes and ionic charges of various cations and anions commonly found in minerals. Ionic radii are usually expressed in angstroms (1 angstrom equals 10^{-8} cm).

Other things being equal, minerals containing large cations are typically less dense than those in which the cations are small. For example, the density of sylvite (KCl), which contains the relatively large potassium cation, is less than half that of pyrrhotite (FeS), even though their molecular weights are similar. This does not hold true for large cations that have very high atomic weights.

Most crystal structures can be considered three-dimensional arrays of large spheres (anions) with small spheres (cations) located in the spaces between them, so the positive and negative charges cancel each other out. Consider the mineral halite (NaCl), which has a relatively simple framework composed of an equal number of positively charged sodium ions and negatively charged chlorine ions. Because anions repel anions, and cations repel cations, ions of similar charge are spaced as far apart from each other as possible. Consequently, in the mineral halite, each sodium ion (Na^{1+}) is surrounded on all sides by chlorine ions (Cl^{1-}) and vice versa (Figure 3.12). This particular packing arrangement results in basic building blocks, called *unit cells*, that have cubic shapes. As shown in Figure 3.12C, these cubic unit cells combine to form cubic-shaped halite crystals. Salt crystals, including the ones that come out of a salt shaker, are often perfect cubes.

The basic building blocks of halite stack together in a regular manner, creating the whole crystal. In addition, the shape and symmetry of these building blocks relate to the shape and symmetry of the entire crystal.

Thus, despite the fact that natural crystals are rarely perfect, the angles between equivalent crystal faces of the same mineral are remarkably consistent. This observation was first made by Nicolas Steno in 1669. Steno found that the angles between adjacent prism faces of quartz crystals are 120 degrees, regardless of sample size, the size of the crystal faces, or where the crystals were collected (Figure 3.13). This observation is commonly called **Steno's Law,** or the **Law of Constancy of Interfacial Angles,** because it applies to all minerals. Steno's Law states that *angles between equivalent faces of crystals of the same mineral are always the same.* For this reason, **crystal shape** is frequently a valuable tool in mineral identification.

It is important to note, however, that minerals can be constructed of geometrically similar building blocks yet exhibit different external forms. For example, the minerals fluorite, magnetite, and garnet are all constructed of cubic unit cells. However, cubic cells can stick together to produce crystals of many shapes. Typically, fluorite crystals are cubes, while magnetite crystals are octahedrons, and garnets form dodecahedrons built up of many small cubes as shown in Figure 3.14. Because the building blocks are so small, the resulting crystal faces are smooth, flat surfaces.

FIGURE 3.12 This diagram illustrates the orderly arrangement of sodium and chloride ions in the mineral halite. The arrangement of atoms into basic building blocks having a cubic shape results in regularly shaped cubic crystals. (Photo by M. Claye/Jacana Scientific Control/Photo Researchers, Inc.)

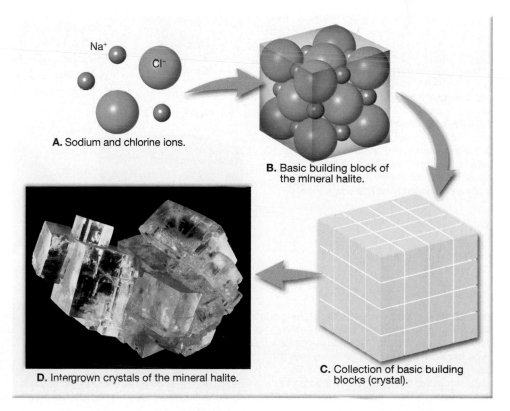

Na⁺ Cl⁻

A. Sodium and chlorine ions.

B. Basic building block of the mineral halite.

C. Collection of basic building blocks (crystal).

D. Intergrown crystals of the mineral halite.

Compositional Variations in Minerals

Using sophisticated analytical techniques, researchers discovered that the chemical composition of some minerals varies substantially from sample to sample. These compositional variations are possible because ions of similar size can substitute readily for one another without disrupting a mineral's internal framework. An analogy can be made with a bricklayer who builds a wall using bricks of different colors and materials. As long as the bricks are all the same size, the shape of the wall is unaffected; only its composition changes.

Let us consider the mineral olivine as an example of chemical variability. The chemical formula for olivine is $(Mg, Fe)_2SiO_4$, where the variable components magnesium and iron are shown in parentheses and separated by a comma. Notice in Figure 3.11 that the cations magnesium (Mg^{2+}) and iron (Fe^{2+}) are nearly the same size and have the same electrical charge. At one extreme, olivine may contain magnesium without any iron, or vice-versa—but most samples of olivine have some of both cations in their structure. Consequently, olivine can be thought of as having a range of combinations from fosterite (Mg_2SiO_4) at one end to fayalite (Fe_2SiO_4) at the other. Nevertheless, all specimens of olivine have the same internal structure and exhibit similar, but not identical, properties. For example, iron-rich olivines have a higher density than magnesium-rich specimens, a reflection of the greater atomic weight of iron as compared to magnesium.

In contrast to olivine, minerals such as quartz (SiO_2) and fluorite (CaF_2) tend to have chemical compositions that vary little from their chemical formulas. However, even these minerals contain extremely small amounts of other less common elements, referred to as *trace elements*. Although trace elements have little effect on most mineral properties, they can significantly influence a mineral's color.

Structural Variations in Minerals

It is also common for two minerals with exactly the same chemical composition to have different internal structures and, hence. different external forms. Minerals of this type are called **polymorphs** (*poly* = many, *morph* = form). Graphite and diamonds are particularly good examples of polymorphism because they both consist exclusively of carbon atoms yet have drastically different properties. Graphite is the soft gray material of which pencil "lead" is made, whereas

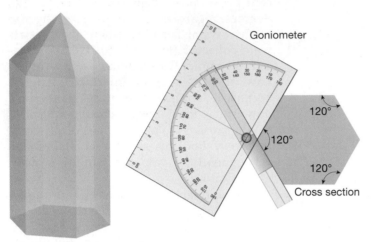

A. Quartz crystal

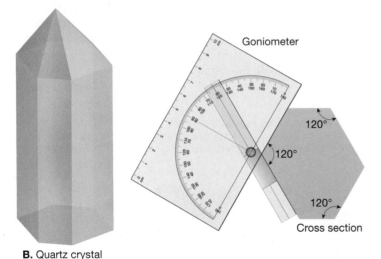

B. Quartz crystal

FIGURE 3.13 Illustration of Steno's Law. Because some faces of a crystal may grow larger than others, two crystals of the same mineral may *not* have identical shapes. Nevertheless, the angles between equivalent faces are remarkably consistent.

FIGURE 3.14 Cubic unit cells stack together in different ways to produce crystals that exhibit different shapes. Fluorite (**A**) tends to display cubic crystals, whereas magnetite crystals (**B**) are typically octahedrons, and garnets (**C**) usually occur as dodecahedrons. (Photos by Dennis Tasa)

A. Cube (fluorite)   **B.** Octahedron (magnetite)   **C.** Dodecahedron (garnet)

FIGURE 3.15 Comparing the structures of diamond and graphite. Both are natural substances with the same chemical composition— carbon atoms. Nevertheless, their internal structure and physical properties reflect the fact that each formed in a very different environment. **A.** All carbon atoms in diamond are covalently bonded into a compact, three-dimensional framework, which accounts for the extreme hardness of the mineral. **B.** In graphite the carbon atoms are bonded into sheets that are joined in a layered fashion by very weak electrical forces. These weak bonds allow the sheets of carbon to readily slide past each other, making graphite soft and slippery, and thus useful as a dry lubricant. (**A.** Photographer Dane Pendland, courtesy of Smithsonian Institution; **B.** E. J. Tarbuck)

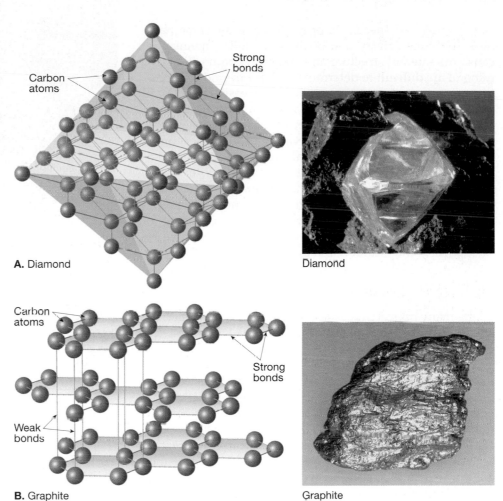

A. Diamond

Diamond

B. Graphite

Graphite

diamond is the hardest known mineral. The differences between these minerals can be attributed to the condition under which they were formed. Diamonds form at depths approaching 200 kilometers (nearly 125 miles), where extreme pressures and temperatures produce the compact structure shown in Figure 3.15A. Graphite, on the other hand, forms under comparatively low pressures and consists of sheets of carbon atoms that are widely spaced and weakly bonded (Figure 3.15B). Because these carbon sheets will easily slide past one another, graphite has a greasy feel and makes an excellent lubricant.

Scientists have learned that by heating graphite under high pressure they can produce diamonds. Because synthetic diamonds often contain flaws, they are generally not gem quality. However, due to their hardness, they have many industrial uses. Further, because all diamonds form in environments of extreme pressures and temperatures, they are somewhat unstable at Earth's surface. Fortunately for jewelers, "diamonds are forever" because the rate at which they change to their more stable form, graphite, is infinitesimally slow.

Students Sometimes Ask ...

Are there any artificial materials harder than diamonds?

Yes, but you won't be seeing them anytime soon. A hard form of carbon nitride (C_3N_4), described in 1989 and synthesized in a laboratory shortly thereafter, may be harder than diamond but hasn't been produced in large enough amounts for a proper test. In 1999, researchers discovered that a form of carbon made from fused spheres of 20 and 28 carbon atoms—relatives of the famous "buckyballs"—could also be as hard as a diamond. These materials are expensive to produce, so diamonds continue to be used as abrasives and in certain kinds of cutting tools. Synthetic diamonds, produced since 1955, are now widely used in these industrial applications.

Two other minerals with identical chemical compositions, calcium carbonate ($CaCO_3$), but different internal structures are calcite and aragonite. Calcite forms mainly through biochemical processes and is the major constituent of the sedimentary rock limestone. Aragonite, a less common polymorph of $CaCO_3$, is often deposited by hot springs and is an important constituent of pearls and the shells of some marine organisms. Because aragonite gradually changes to the more stable crystalline structure of calcite, it is rare in rocks older than 50 million years.

The transformation of one polymorph to another is called a *phase change*. In nature, certain minerals go through phase changes as they move from one environment to another. For example, when a slab of ocean crust composed of olivine-rich basalt is carried to great depths by a subducting plate, olivine changes to a more compact polymorph called spinel.

Physical Properties of Minerals

 GEODe **Matter and Minerals**
▶ **Physical Properties of Minerals**

Each mineral has a definite crystalline structure and chemical composition that give it a unique set of physical and chemical properties shared by all samples of that mineral.

For example, all specimens of halite have the same hardness, the same density, and break in a similar manner. Because the internal structure and chemical composition of a mineral are difficult to determine without the aid of sophisticated tests and equipment, the more easily recognized physical properties are frequently used in identification.

The diagnostic physical properties of minerals are those that can be determined by observation or by performing a simple test. The primary physical properties that are commonly used to identify hand samples are luster, color, streak, crystal shape (habit), tenacity, hardness, cleavage, fracture, and density or specific gravity. Secondary (or "special") properties that are exhibited by a limited number of minerals include magnetism, taste, feel, smell, double refraction, and chemical reaction to hydrochloric acid.

Optical Properties

Of the many optical properties of minerals, four—luster, the ability to transmit light, color, and streak—are most frequently used for mineral identification.

Luster   The appearance or quality of light reflected from the surface of a mineral is known as **luster**. Minerals that have the appearance of metals, regardless of color, are said to have a *metallic luster* (Figure 3.16). Some metallic minerals, such as native copper and galena, develop a dull coating or tarnish when exposed to the atmosphere. Because they are not as shiny as samples with freshly broken surfaces, these samples are often said to exhibit a *submetallic luster.*

Most minerals have a *nonmetallic luster* and are described using various adjectives such as *vitreous* or *glassy*. Other nonmetallic minerals are described as having a *dull* or *earthy luster* (a dull appearance like soil), or a *pearly luster* (such as a pearl, or the inside of a clamshell). Still others exhibit a *silky luster* (like satin cloth), or a *greasy luster* (as though coated in oil).

The Ability to Transmit Light   Another optical property used in the identification of minerals is the ability to transmit light. When no light is transmitted, the mineral is described as *opaque;* when light but not an image is transmitted through a mineral it is said to be *translucent.* When light and an image are visible through the sample, the mineral is described as *transparent.*

Color   Although **color** is generally the most conspicuous characteristic of any mineral, it is considered a diagnostic property of only a few minerals. Slight impurities in the common mineral quartz, for example, give it a variety of tints including pink, purple, yellow, white, gray, and even black (see Figure 3.32, p. 93). Other minerals, such as tourmaline, also exhibit a variety of hues, with multiple colors sometimes occurring in the same sample. Thus, the use of color as a means of identification is often ambiguous or even misleading.

Streak   Although the color of a sample is not always helpful in identification, **streak**—the color of the powdered min-

FIGURE 3.16   The freshly broken sample of galena (right) displays a metallic luster, while the sample on the left is tarnished and has a submetallic luster. (Photo courtesy of E. J. Tarbuck)

eral—is often diagnostic. The streak is obtained by rubbing the mineral across a piece of unglazed porcelain, termed a *streak plate* and observing the color of the mark it leaves (Figure 3.17). Although the color of a mineral may vary from sample to sample, the streak usually does not.

Streak can also help distinguish between minerals with metallic luster and those having nonmetallic luster. Metallic minerals generally have a dense, dark streak, whereas minerals with nonmetallic luster have a streak that is typically light colored.

It should be noted that not all minerals produce a streak when using a streak plate. For example, if the mineral is harder than the streak plate, no streak is observed.

Crystal Shape or Habit

Mineralogists use the term **habit** to refer to the common or characteristic shape of a crystal or aggregate of crystals. A few minerals exhibit somewhat regular polygons that are

FIGURE 3.17   Although the color of a mineral is not always helpful in identification, the streak, which is the color of the powdered mineral, can be very useful.

helpful in their identification. Recall that magnetite crystals sometimes occur as octahedrons, garnets often form dodecahedrons, and halite and fluorite crystals tend to grow as cubes or near cubes (see Figure 3.14). While most minerals have only one common habit, a few such as pyrite have two or more characteristic crystal shapes (Figure 3.18).

By contrast, some minerals rarely develop perfect geometric forms. Many of these do, however, develop a characteristic shape that is useful for identification. Some minerals tend to grow equally in all three dimensions, whereas others tend to be elongated in one direction, or flattened if growth in one dimension is suppressed. Commonly used terms to describe these and other crystal habits include *equant* (equidimensional), *bladed, fibrous, tabular, prismatic, platy, blocky,* and *botryoidal* (Figure 3.19).

Mineral Strength

How easily minerals break or deform under stress relates to the type and strength of the chemical bonds that hold the crystals together. Mineralogists use terms including tenacity, hardness, cleavage, and fracture to describe mineral strength and how minerals break when stress is applied.

Tenacity   The term **tenacity** describes a mineral's toughness, or its resistance to breaking or deforming. Minerals which are ionically bonded, such as fluorite and halite, tend to be *brittle* and shatter into small pieces when struck. By contrast, minerals with metallic bonds, such as native copper, are *malleable,* or easily hammered into different shapes. Minerals, including gypsum and talc, that can be cut into thin shavings are described as *sectile*. Still others, notably the micas, are *elastic* and will bend and snap back to their original shape after the stress is released.

FIGURE 3.18   Although most minerals exhibit only one common crystal shape, some, such as pyrite, have two or more characteristic habits. (Photos by Dennis Tasa)

A. Bladed   **B.** Prismatic

C. Banded   **D.** Botryoidal

FIGURE 3.19   Some common crystal habits. **A.** *Bladed.* Elongated crystals that are flattened in one direction. **B.** *Prismatic.* Elongated crystals with faces that are parallel to a common direction. **C.** *Banded.* Minerals that have stripes or bands of different color or texture. **D.** *Botryoidal.* Groups of intergrown crystals resembling a bunch of grapes.

Hardness   One of the most useful diagnostic properties is **hardness,** a measure of the resistance of a mineral to abrasion or scratching. This property is determined by rubbing a mineral of unknown hardness against one of known hardness, or vice versa. A numerical value of hardness can by obtained by using the **Mohs scale** of hardness, which consists of 10 minerals arranged in order from 1 (softest) to 10 (hardest), as shown in Figure 3.20A. It should be noted that the Mohs scale is a relative ranking, and it does not imply that mineral number 2, gypsum, is twice as hard as mineral 1, talc. In fact, gypsum is only slightly harder than talc as shown in Figure 3.20B.

In the laboratory, other common objects can be used to determine the hardness of a mineral. These include a human fingernail, which has a hardness of about 2.5, a copper penny 3.5, and a piece of glass 5.5. The mineral gypsum, which has a hardness of 2, can be easily scratched with a fingernail. On the other hand, the mineral calcite, which has a hardness of 3, will scratch a fingernail but will not scratch glass. Quartz, one of the hardest common minerals, will easily scratch glass. Diamonds, hardest of all, scratch anything.

Cleavage   Some minerals have atomic structures that are not the same in every direction and chemical bonds that vary in strength. As a result, these minerals tend to break, so that the broken fragments are bounded by more or less flat, planar surfaces, a property called **cleavage** (*kleiben* = carve). Cleavage can be recognized by rotating a sample and looking for smooth, even surfaces that reflect light like a mirror. Note, however, that cleavage surfaces can occur in small, flat segments arranged in stair-step fashion.

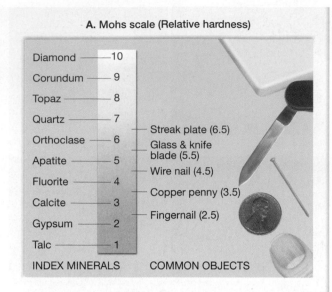

A. Mohs scale (Relative hardness)

INDEX MINERALS		COMMON OBJECTS
Diamond	10	
Corundum	9	
Topaz	8	
Quartz	7	
Orthoclase	6	Streak plate (6.5)
Apatite	5	Glass & knife blade (5.5)
Fluorite	4	Wire nail (4.5)
Calcite	3	Copper penny (3.5)
Gypsum	2	Fingernail (2.5)
Talc	1	

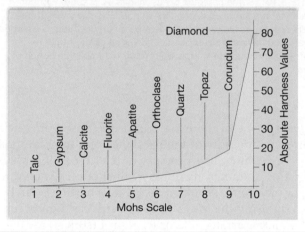

B. Comparison of Mohs scale and an absolute scale

Diamond

Absolute Hardness Values: 10, 20, 30, 40, 50, 60, 70, 80

Talc, Gypsum, Calcite, Fluorite, Apatite, Orthoclase, Quartz, Topaz, Corundum

Mohs Scale: 1 2 3 4 5 6 7 8 9 10

FIGURE 3.20 Hardness scales. **A.** Mohs scale of hardness, with the hardness of some common objects. **B.** Relationship between Mohs relative hardness scale and an absolute hardness scale.

The simplest type of cleavage is exhibited by the micas (Figure 3.21). Because the micas have much weaker bonds in one direction than in the others, they cleave to form thin, flat sheets. Some minerals have excellent cleavage in several directions, whereas others exhibit fair or poor cleavage, and still others have no cleavage at all. When minerals break evenly in more than one direction, cleavage is described by the *number of cleavage planes and the angle(s) at which they meet* (Figure 3.22).

Do not confuse cleavage with crystal shape. When a mineral exhibits cleavage, it will break into pieces that have the same geometry as each other. By contrast, the smooth-sided quartz crystals shown in Figure 3.2 do not have cleavage. If broken, they fracture into shapes that do not resemble each other or the original crystals.

Fracture Minerals that have structures that are equally, or nearly equally, strong in all directions **fracture** to form irreg-

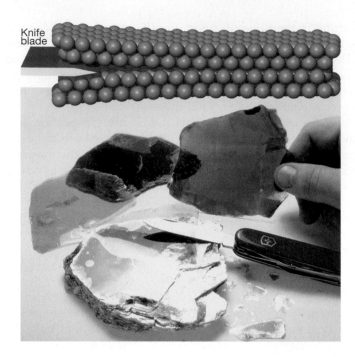

FIGURE 3.21 The thin sheets shown here were produced by splitting a mica (muscovite) crystal parallel to its perfect cleavage. (Photo by Breck P. Kent)

ular surfaces. Those, such as quartz, that break into smooth curved surfaces resembling broken glass exhibit a *conchoidal fracture* (Figure 3.23). Others break into splinters or fibers, but most minerals display an irregular fracture.

Density and Specific Gravity

Density is an important property of matter defined as mass per unit volume usually expressed as grams per cubic centimeter. Mineralogists often use a related measure called *specific gravity* to describe the density of minerals. **Specific gravity** is a unitless number representing the ratio of a mineral's weight to the weight of an equal volume of water.

Most common rock-forming minerals have a specific gravity of between 2 and 3. For example, quartz has a specific gravity of 2.65. By contrast, some metallic minerals such as pyrite, native copper and magnetite are more than twice as dense as quartz. Galena, which is an ore of lead, has a specific gravity of roughly 7.5, whereas the specific gravity of 24-karat gold is approximately 20.

With a little practice, you can estimate the specific gravity of a mineral by hefting it in your hand. Ask yourself, does this mineral feel about as "heavy" as similar sized rocks you have handled? If the answer is "yes," the specific gravity of the sample will likely be between 2.5 and 3.

Other Properties of Minerals

In addition to the properties already discussed, some minerals can be recognized by other distinctive properties. For example, halite is ordinary salt, so it can be quickly identified through taste. Talc and graphite both have distinctive feels;

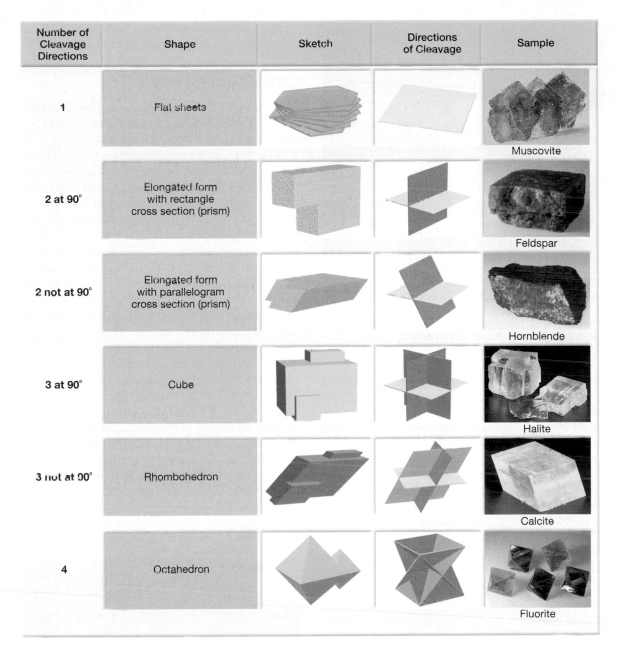

Number of Cleavage Directions	Shape	Sketch	Directions of Cleavage	Sample
1	Flat sheets			Muscovite
2 at 90°	Elongated form with rectangle cross section (prism)			Feldspar
2 not at 90°	Elongated form with parallelogram cross section (prism)			Hornblende
3 at 90°	Cube			Halite
3 not at 90°	Rhombohedron			Calcite
4	Octahedron			Fluorite

FIGURE 3.22 Common cleavage directions exhibited by minerals. (Photos by E. J. Tarbuck and Dennis Tasa)

talc feels soapy, and graphite feels greasy. Further, the streak of many sulfur-bearing minerals smells like rotten eggs. A few minerals, such as magnetite, have a high iron content and can be picked up with a magnet, while some varieties (lodestone) are natural magnets and will pick up small iron-based objects such as pins and paper clips (see Figure 3.36, p. 97).

Moreover, some minerals exhibit special optical properties. For example, when a transparent piece of calcite is placed over printed material, the letters appear twice. This optical property is known as *double refraction* (Figure 3.24).

One very simple chemical test involves placing a drop of dilute hydrochloric acid from a dropper bottle onto a freshly broken mineral surface. Certain minerals, called carbonates, will effervesce (fizz) as carbon dioxide gas is released

(Figure 3.25). This test is especially useful in identifying the common carbonate mineral calcite.

How Are Minerals Named and Classified?

Nearly 4000 minerals have been named, and 30 to 50 new ones are identified each year. Fortunately, for students beginning their study of minerals, no more than a few dozen are abundant. Collectively, these few make up most of the rocks of Earth's crust and, as such, are often referred to as the *rock-forming minerals*.

FIGURE 3.23 Conchoidal fracture. The smooth curved surfaces result when minerals break in a glasslike manner. (Photo by E. J. Tarbuck)

Although less abundant, many other minerals are used extensively in the manufacture of products that drive our modern society and are called *economic minerals*. It should be noted that rock-forming minerals and economic minerals are not mutually exclusive groups. When found in large deposits, some rock-forming minerals are economically significant. One example is the mineral calcite, which is the primary component of the sedimentary rock limestone and has many uses including the production of cement.

Determining what is or is not a mineral is in the hands of the International Mineralogical Association. Individuals who believe they have discovered a new mineral must apply to this organization with a detailed description of their findings. If confirmed, the discoverer is allowed to choose a name for the new mineral. Minerals have been named for people, locations where the mineral was discovered, appearance, and chemical composition. Although

FIGURE 3.24 Double refraction illustrated by the mineral calcite. (Photo by Chip Clark)

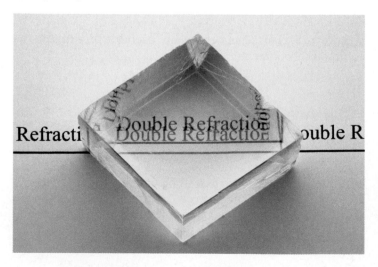

nearly 50 percent are named in honor of a person, rules prohibit naming a mineral after oneself.

Classifying Minerals

Minerals are placed into categories in much the same way as plants and animals are classified by biologists. Mineralogists use the term *mineral species* for a collection of specimens that exhibit similar internal structures and chemical compositions. Some common mineral species include quartz, calcite, galena, and pyrite. However, just as individual plants and animals within a species differ somewhat from one another, so do most specimens within a mineral species.

Mineral species are usually assigned to a *mineral class* based on their anions, or anion complexes as shown in Table 3.1. Some of the important mineral classes include the silicates, (SiO_4^{4-}), carbonates (CO_3^{2-}), halides, (Cl^{1-}, F^{1-}, Br^{1-}), and sulfates (SO_4^{2-}). Minerals within each class tend to have similar internal structures and, hence, similar properties. For example, minerals that belong to the carbonate class react chemically with acid—albeit to varying degrees—and many exhibit rhombic cleavage. Furthermore, minerals within the same class are often found together in the same rock. The halides, for example, commonly occur together in evaporite deposits.

Mineral classes are further divided into *groups* (or, in some cases subclasses) based on similarities in atomic structures or compositions. For example, mineralogists group the feldspar minerals (orthoclase, albite, and anorthite) together because they all have similar internal structures. The minerals kyanite,

FIGURE 3.25 Calcite reacting with a weak acid. (Photo by Chip Clark)

TABLE 3.1 Major Mineral Classes

Class	Anion, Anion Complex, or Elements	Example (Mineral species)	Chemical Formula
silicates	$(SiO_4)^{4-}$	quartz	SiO_2
halides	Cl^-, F^-, Br^-, I^-	halite	$NaCl$
oxides	O^{2-}	corundum	Al_2O_3
hydroxides	$(OH)^-$	gibbsite	$Al(OH)_3$
carbonates	$(CO_3)^{2-}$	calcite	$CaCO_3$
nitrates	$(NO_3)^-$	nitratite	$NaNO_3$
sulfates	$(SO_4)^{2-}$	gypsum	$CaSO_4 \cdot 2H_2O$
phosphates	$(PO_4)^{3-}$	apatite	$Ca_5(PO_4)_3(OH, F, Cl)$
native elements	Cu, Ag, S	copper	Cu
sulfides	S^{2-}	pyrite	FeS_2

andalusite, and sillimanite are grouped together because they have the same chemical composition (Al_2SiO_5). Recall that minerals having the same composition, but different crystal structures are called *polymorphs*. Some common mineral groups include the feldspars, pyroxenes, amphiboles, micas, and olivine. Two members (species) of the mica group are biotite and muscovite. Although they have similar atomic structures, biotite contains iron and magnesium, which give it a greater density and darker appearance than muscovite.

Some mineral species are further subdivided into *mineral varieties*. For example, pure quartz (SiO_2) is colorless and transparent. However, when small amounts of aluminum are incorporated into its atomic structure, quartz appears quite dark, a variety called *smoky quartz*. *Amethyst*, another variety of quartz, owes its violet color to the presence of trace amounts of iron.

Major Mineral Classes

It is interesting to point out that only eight elements make up the bulk of rock-forming minerals and represent over 98 percent (by weight) of the continental crust (Figure 3.26). These elements, in order of abundance are oxygen (O), silicon (Si), aluminum (Al), iron (Fe), calcium (Ca), sodium (Na), potassium (K), and magnesium (Mg). As shown in Figure 3.26, silicon and oxygen are by far the most common elements in Earth's crust. Furthermore, these two elements readily combine to form the framework for the most dominant mineral class, the **silicates**, which account for more than 90 percent of Earth's crust. More than 800 species of silicate minerals are known.

Because other mineral classes are far less abundant than the silicates, they are often grouped together under the heading **nonsilicates**. Although not as common as the silicates, some nonsilicate minerals are very important economically. They provide us with the iron and aluminum to build our automobiles, gypsum for plaster and drywall to construct our homes, and copper for wire to carry electricity and connect us to the Internet. In addition to their economic importance, these mineral groups include members that are major constituents in sediments and sedimentary rocks.

We will first discuss the most common mineral class, the silicates, and then consider some of the prominent nonsilicate mineral groups.

The Silicates

Every silicate mineral contains the elements oxygen and silicon. Further, most contain one or more of the other common elements of Earth's crust. Together, these elements give rise to hundreds of minerals with a wide variety of properties, including hard quartz, soft talc, sheet-like mica, fibrous asbestos, green olivine and blood-red garnet.

The Silicon–Oxygen Tetrahedron

All silicate minerals have the same fundamental building block, the **silicon–oxygen tetrahedron**. This structure consists of four oxygen anions surrounding one comparatively small silicon cation, forming a tetrahedron—a pyramid shape with four identical faces (Figure 3.27). These tetrahedra are not chemical compounds, but rather complex anions (SiO_4^{4-}) having a net charge of –4. Because minerals must have a balanced charge, silicon-oxygen tetrahedra bond to other positively charged ions.

Specifically, each O^{2-} has one of its valence electrons balance by bonding with the Si^{4+} located at the center of the tetrahedron. The remaining –1 charge on each oxygen anion is available to bond with another cation, or with the silicon cation in an adjacent tetrahedron.

Independent Tetrahedra In nature, the simplest way for *independent tetrahedra* to become neutral compounds is through the addition of positively charged ions. For example, in the mineral olivine, magnesium (Mg^{2+}) and/or iron (Fe^{2+}) cations pack between larger independent SiO_4^{4-} units, forming a dense three-dimensional structure. Garnet, another

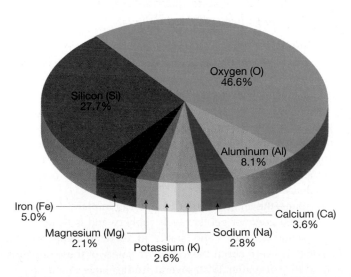

FIGURE 3.26 Relative abundance of the eight most abundant elements in the continental crust.

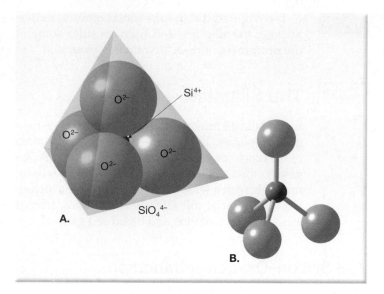

FIGURE 3.27 Two representations of the silicon–oxygen tetrahedron. **A.** The four large spheres represent oxygen ions, and the blue sphere represents a silicon ion. The spheres are drawn in proportion to the radii of the ions. **B.** An expanded view of the tetrahedron using rods to depict the bonds that connect the ions.

common silicate, is also composed of independent tetrahedra ionically bonded by cations. Both olivine and garnet form dense, hard, equidimensional crystals that lack cleavage.

Other Silicate Structures   One reason for the great variety of silicate minerals is the ability of the silicate anion (SiO_4^{4-}) to link together in a variety of configurations. Vast numbers of tetrahedra connect to form *single chains, double chains,* or *sheet structures* as shown in Figure 3.28. This phenomenon, called *polymerization,* is achieved by the sharing of oxygen atoms between adjacent tetrahedra.

To understand how this sharing occurs, select one of the silicon ions (small blue spheres) near the middle of the single-chain structure shown in Figure 3.28. Notice that this silicon ion is completely surrounded by four larger oxygen ions (you are looking through one of the four to see the blue silicon ion). Also notice that, of the four oxygen ions, half are bonded to two silicon ions, whereas the other two are not shared in this manner. *It is the linkage across the shared oxygen ions that join the tetrahedra into a chain structure.* Now examine a silicon ion near the middle of the sheet structure and count the number of shared and unshared oxygen ions surrounding it. The sheet structure is the result of three of the four oxygen atoms being shared by adjacent tetrahedra.

Other silicate structures exist. In the most common of these, all four oxygen ions are shared and produce a complex three-dimensional framework. Each silicate framework forms the skeleton for a particular group of silicate minerals.

By now you can see that the ratio of oxygen ions to silicon ions differs in each of the silicate structures. In the independent tetrahedron, there are four oxygen ions for every silicon ion. In single chains, the oxygen-to-silicon ratio is 3:1, and in three-dimensional frameworks this ratio is 2:1. As more of the oxygen ions are shared, the percentage of silicon in the structure increases. Silicate minerals are, therefore, described as having a high or low silicon content based on their ratio of oxygen to silicon. This difference in silicon content is important, as we shall see in the chapter on igneous rocks.

Joining Silicate Structures

Except for quartz (SiO_2), the framework (chains, sheets, or three-dimensional networks) of most other silicate minerals has a net negative charge. Therefore, cations are required to bring the overall charge into balance and to serve as the "mortar" that holds these structures together. The cations that most often link silicate structures are iron (Fe^{2+}), magnesium (Mg^{2+}), potassium (K^{1+}), sodium (Na^{1+}), aluminum (Al^{3+}), and calcium (Ca^{2+}). These cations generally fit into spaces between unshared oxygen atoms that occupy the corners of the tetrahedra.

As a general rule, the partly covalent bonds between silicon and oxygen are stronger than the ionic bonds that hold one silicate framework to the next. Consequently, properties such as cleavage and, to some extent, the hardness of these minerals are controlled by the nature of the silicate framework. Quartz, which has only silicon–oxygen bonds, has great hardness and lacks cleavage, mainly because of equally strong bonds in all directions. By contrast, the sheet silicates, such as mica, have perfect cleavage in one direction and low hardness due to weak bonding between the sheet structures.

In addition, there tends to be a gradual decrease in density as the number of shared oxygens increases. For example, olivine and garnet, which are composed of independent tetrahedra (no sharing of oxygen atoms), tend to form quite compact structures having a specific gravity of about 3.2 to 4.4. By contrast, quartz and feldspar, which have a relatively open

Students Sometimes Ask . . .

Are these silicates the same materials used in silicon computer chips and silicone breast implants?

Not really, but all three contain the element silicon (Si). Further, the source of silicon for numerous products, including computer chips and breast implants, comes from silicate minerals. Pure silicon (without the oxygen that silicates have) is used to make computer chips, giving rise to the term "Silicon Valley" for the high-tech region of San Francisco, California's south bay area, where many of these devices are designed. Manufacturers of computer chips engrave silicon wafers with incredibly narrow conductive lines, squeezing millions of circuits into every fingernail-size chip.

Silicone—the material used in breast implants—is a silicon–oxygen polymer gel that feels rubbery and is water repellent, chemically inert, and stable at extreme temperatures. Although concern about the long-term safety of these implants limited their use after 1992, several studies have found no evidence linking them to various diseases.

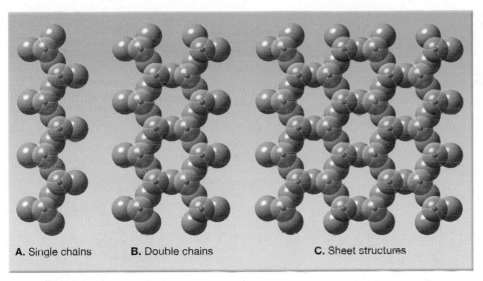

A. Single chains   B. Double chains   C. Sheet structures

FIGURE 3.28 Three types of silicate structures. **A.** Single chains. **B.** Double chains. **C.** Sheet structures.

whereas quartz crystallizes at much lower temperatures.

In addition, some silicate minerals form at Earth's surface from the weathered products of other silicate minerals. Still other silicate minerals are formed under the extreme pressures associated with mountain building. Each silicate mineral, therefore, has a structure and a chemical composition that *indicate the conditions under which it formed*. Thus, by carefully examining the mineral constituents of rocks, geologists can often determine the circumstances under which the rocks formed.

We will now examine some of the most common silicate minerals, which we divide into two major groups on the basis of their chemical makeup.

three-dimensional framework (complete sharing of oxygen atoms), have a specific gravity that is in the range of 2.6 to 2.8.

Recall that cations of approximately the same size are able to substitute freely for one another. For instance, the mineral olivine contains iron (Fe^{2+}) and magnesium (Mg^{2+}) ions that substitute for each other without altering that mineral's structure. This also holds true for some feldspars in which calcium and sodium cations occupy the same site in a crystal structure. In addition, aluminum (Al) often substitutes for silicon (Si) in silicon–oxygen tetrahedra.

Because most silicate structures will readily accommodate different cations at a given bonding site, individual specimens of a particular mineral may contain varying amounts of certain elements. As a result, many silicate minerals form a *mineral group* that exhibits a range of compositions between two end members. Some of the common mineral groups are given in Figure 3.29, and include the olivines, pyroxenes, amphiboles, micas, and feldspars.

Common Silicate Minerals

GEODe
Minerals: Building Blocks of Rocks
▶ Mineral Groups

The major silicate groups and common examples are given in Figure 3.29. The feldspars are by far the most plentiful silicate group, comprising more than 50 percent of Earth's crust. Quartz, the second most abundant mineral in the continental crust, is the only common mineral made completely of silicon and oxygen.

Most silicate minerals form when molten rock cools and crystallizes. Cooling can occur at or near Earth's surface (low temperature and pressure) or at great depths (high temperature and pressure). The environment during crystallization and the chemical composition of the molten rock determine to a large degree which minerals are produced. For example, the silicate mineral olivine crystallizes at high temperatures,

The Light Silicates

The **light** (or **nonferromagnesian**) **silicates** are generally light in color and have a specific gravity of about 2.7, which is considerably less than the dark (ferromagnesian) silicates. These differences are mainly attributable to the presence or absence of iron and magnesium. The light silicates contain varying amounts of aluminum, potassium, calcium, and sodium rather than iron and magnesium.

Feldspar Group Feldspar, the most common mineral group, can form under a wide range of temperatures and pressures, a fact that partially accounts for its abundance (Figure 3.30). All of the feldspars have similar physical properties. They have two planes of cleavage meeting at or near 90-degree angles, are relatively hard (6 on the Mohs scale), and have a luster that ranges from glassy to pearly. As one component in a rock, feldspar crystals can be identified by their rectangular shape and rather smooth shiny faces (Figure 3.29).

Two different feldspar structures exist. One group of feldspar minerals contains potassium ions in its structure and is therefore referred to as *potassium feldspar*. (*Orthoclase* and *microcline* are common members of the potassium feldspar group.) The other group, called *plagioclase feldspar*, contains both sodium and calcium ions that freely substitute for one another depending on the environment during crystalization.

Potassium feldspar is usually light cream, salmon pink, or occasionally bluish-green in color. The plagioclase feldspars, on the other hand, range in color from white to medium gray. However, color should not be used to distinguish these groups. The only sure way to distinguish the feldspars physically is to look for a multitude of fine parallel lines, called *striations*. Striations are found on some cleavage planes of plagioclase feldspar but are not present on potassium feldspar (Figure 3.31).

Quartz Quartz is the only common silicate mineral consisting entirely of silicon and oxygen. As such, the term *silica* is

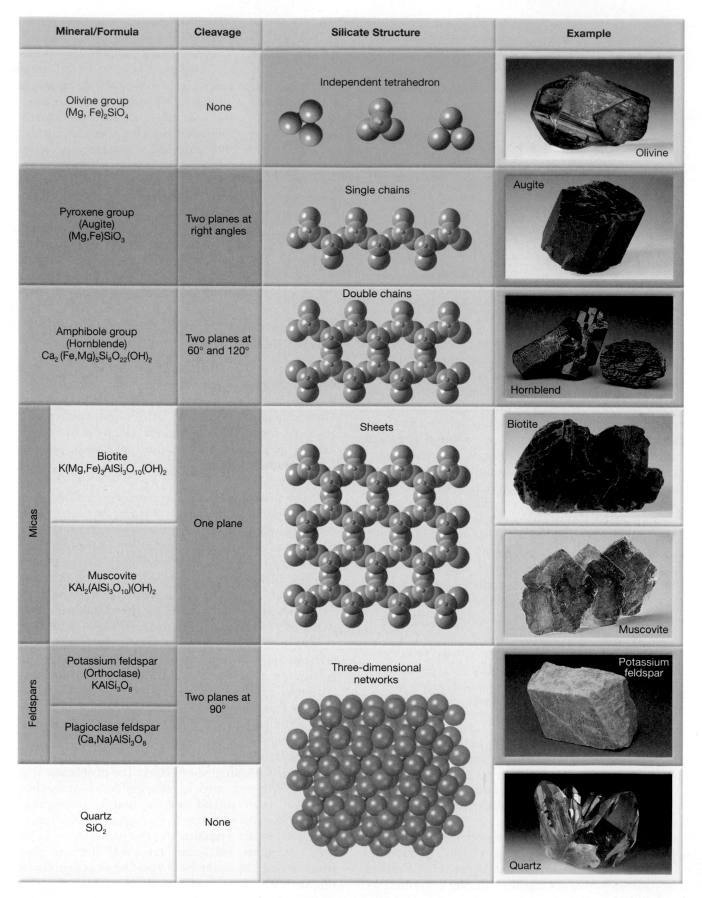

Mineral/Formula	Cleavage	Silicate Structure	Example
Olivine group $(Mg, Fe)_2SiO_4$	None	Independent tetrahedron	Olivine
Pyroxene group (Augite) $(Mg,Fe)SiO_3$	Two planes at right angles	Single chains	Augite
Amphibole group (Hornblende) $Ca_2(Fe,Mg)_5Si_8O_{22}(OH)_2$	Two planes at 60° and 120°	Double chains	Hornblend
Micas — Biotite $K(Mg,Fe)_3AlSi_3O_{10}(OH)_2$	One plane	Sheets	Biotite
Micas — Muscovite $KAl_2(AlSi_3O_{10})(OH)_2$	One plane	Sheets	Muscovite
Feldspars — Potassium feldspar (Orthoclase) $KAlSi_3O_8$	Two planes at 90°	Three-dimensional networks	Potassium feldspar
Feldspars — Plagioclase feldspar $(Ca,Na)AlSi_3O_8$	Two planes at 90°	Three-dimensional networks	
Quartz SiO_2	None	Three-dimensional networks	Quartz

FIGURE 3.29 Common silicate minerals. Note that the complexity of the silicate structure increases down the chart. (Photos by Dennis Tasa and E. J. Tarbuck)

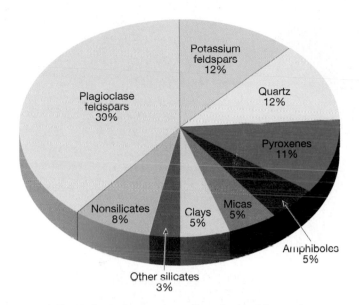

FIGURE 3.30 Estimated percentages (by volume) of the most common minerals in Earth's crust.

applied to quartz, which has the chemical formula SiO_2. Because the structure of quartz contains a ratio of two oxygen ions (O^{7-}) for every silicon ion (Si^{4+}), no other positive ions are needed to attain neutrality.

In quartz, a three-dimensional framework is developed through the complete sharing of oxygen by adjacent silicon atoms. Thus, all of the bonds in quartz are of the strong silicon–oxygen type. Consequently, quartz is hard, resistant to weathering, and does not have cleavage. When broken, quartz generally exhibits conchoidal fracture (See Figure 2.3). In a pure form, quartz is clear and, if allowed to grow without interference, will form hexagonal crystals that develop pyramid-shaped ends. However, like most other clear minerals, quartz is often colored by inclusions of various ions (impurities) and forms without developing good crystal faces. The most common varieties of quartz are milky (white), smoky (gray), rose (pink), amethyst (purple), and rock crystal (clear) (Figure 3.32).

Muscovite Muscovite is a common member of the mica family. It is light in color and has a pearly luster. Like other micas, muscovite has excellent cleavage in one direction. In thin sheets, muscovite is clear, a property that accounts for its use as window "glass" during the Middle Ages. Because muscovite is very shiny, it can often be identified by the sparkle it gives a rock. If you have ever looked closely at beach sand, you may have seen the glimmering brilliance of the mica flakes scattered among the other sand grains.

Clay Minerals Clay is a term used to describe a variety of complex minerals that, like the micas, have a sheet structure. Unlike other common silicates, such as quartz and feldspar, clays do not form in igneous environments. Rather, most clay minerals originate as products of the chemical weathering of other silicate minerals. Thus, clay minerals make up a large percentage of the surface material we call soil. Because of the importance of soil in agriculture, and because of its role as a supporting material for buildings, clay minerals are extremely important to humans. In addition, clays account for nearly half the volume of sedimentary rocks.

Clay minerals are generally very fine grained, which makes identification difficult, unless studied microscopically. Their layered structure and weak bonding between layers give them a characteristic feel when wet. Clays are common in shales, mudstones, and other sedimentary rocks. Although clays are fine grained, they can form very thick beds or layers.

One of the most common clay minerals is *kaolinite,* which is used in the manufacture of fine chinaware and as a coating for high-gloss paper, such as that used in this textbook. Further, some clay minerals absorb large amounts of water, which allows them to swell to several times their normal size. These clays have been used commercially in a variety of ingenious ways, including as an additive to thicken milkshakes in fast-food restaurants.

FIGURE 3.32 Quartz. Some minerals, such as quartz, occur in a variety of colors. These samples include crystal quartz (colorless), amethyst (purple quartz), citrine (yellow quartz), and smoky quartz (gray to black). (Photo courtesy of E. J. Tarbuck)

FIGURE 3.31 These parallel lines, called striations, are a distinguishing characteristic of the plagioclase feldspars. (Photo by E. J. Tarbuck)

The Dark Silicates

The **dark** (or **ferromagnesian**) **silicates** are those minerals containing ions of iron (iron = *ferro*) and/or magnesium in their structure. Because of their iron content, ferromagnesian silicates are dark in color and have a greater specific gravity, between 3.2 and 3.6, than nonferromagnesian silicates. The most common dark silicate minerals are olivine, the pyroxenes, the amphiboles, dark mica (biotite), and garnet.

Olivine Group Olivine is a family of high-temperature silicate minerals that are black to olive green in color and have a glassy luster and a conchoidal fracture (Figure 3.29). Rather than developing large crystals, olivine commonly forms small, rounded crystals that give rocks consisting largely of olivine a granular appearance. Olivine is composed of individual tetrahedra, which are bonded together by a mixture of iron and magnesium ions positioned so as to link the oxygen atoms and magnesium atoms together. Because the three-dimensional network generated in this fashion does not have its weak bonds aligned, olivine does not possess cleavage.

Pyroxene Group The *pyroxenes* are a group of complex minerals that are important components of Earth's mantle. The most common member, *augite,* is a black, opaque mineral with two directions of cleavage that meet at nearly a 90-degree angle (Figure 3.29). Its crystalline structure consists of single chains of tetrahedra bonded together by ions of iron and magnesium. Because the silicon–oxygen bonds are stronger than the bonds joining the silicate structures, augite cleaves parallel to the silicate chains, as shown in Figure 3.33. Augite is one of the dominant minerals in basalt, a common igneous rock of the oceanic crust and volcanic areas on the continents.

Amphibole Group Hornblende is the most common member of a chemically complex group of minerals called *amphiboles* (Figure 3.29). Hornblende is usually dark green to black in color, and except for its cleavage angles, which are about 60 degrees and 120 degrees, it is very similar in appearance to augite. The double chains of tetrahedra in the hornblende structure account for its particular cleavage (Figure 3.33). In a rock, hornblende often forms elongated crystals. This helps distinguish it from pyroxene, which forms rather blocky crystals. Hornblende is found in igneous rocks, where it often makes up the dark portion of an otherwise light-colored rock.

Biotite Biotite is the dark iron-rich member of the mica family (Figure 3.29). Like other micas, biotite possesses a sheet struc-

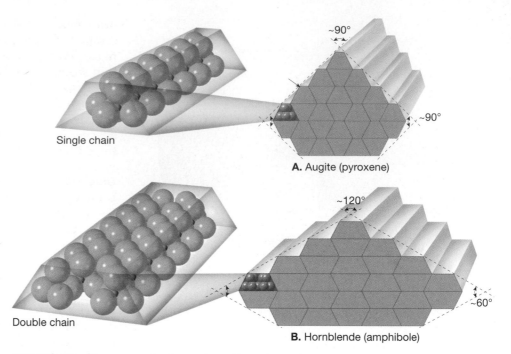

FIGURE 3.33 Cleavage angles for augite and hornblende. Because hornblende's double chains are more weakly bonded than augite's single chains, cleavage is better developed in hornblende.

ture that gives it excellent cleavage in one direction. Biotite also has a shiny black appearance that helps distinguish it from the other dark ferromagnesian minerals. Like hornblende, biotite is a common constituent of igneous rocks, including the rock granite.

Garnet Garnet is similar to olivine in that its structure is composed of individual tetrahedra linked by metallic ions. Also like olivine, garnet has a glassy luster, lacks cleavage, and exhibits conchoidal fracture. Although the colors of garnet are varied, this mineral is most often brown to deep red. Garnet readily forms equidimensional crystals that are most commonly found in metamorphic rocks (Figure 3.34). When garnets are transparent, they may be used as gemstones.

FIGURE 3.34 A deep-red garnet crystal embedded in a light-colored, mica-rich metamorphic rock. (Photo by E. J. Tarbuck)

Important Nonsilicate Minerals

Minerals: Building Blocks of Rocks
▶ Mineral Groups

Nonsilicate minerals are typically subdivided into *classes*, based on the anion (negatively charged ion) or complex anion that the members have in common (Table 3.2). For example, the *oxides* contain the negative oxygen ion (O^{2-}), which is bonded to one or more kinds of positive ions. Thus, within each mineral class, the basic structure and type of bonding is similar. As a result, the minerals in each group have similar physical properties that are useful in mineral identification.

Although the nonsilicates make up only about 8 percent of Earth's crust, some minerals, such as gypsum, calcite, and

halite, occur as constituents in sedimentary rocks in significant amounts. Furthermore, many others are important economically. Table 3.2 lists some of the nonsilicate mineral classes and a few examples of each. A brief discussion of a few of the more common nonsilicate minerals follows.

Some of the most common nonsilicate minerals belong to one of three classes of minerals—the carbonates (CO_3^{2-}), the sulfates (SO_4^{2-}), and the halides (Cl^{1-}, F^{1-}, Br^{1-}). The carbonate minerals are much simpler structurally than the silicates. This mineral group is composed of the carbonate ion (CO_3^{2-}) and one or more kinds of anions. The two most common carbonate minerals are *calcite*, $CaCO_3$ (calcium carbonate), and *dolomite*, $CaMg(CO_3)_2$ (calcium/magnesium carbonate). Because these minerals are similar both physically and chemically, they are difficult to distinguish from each other. Both have a vitreous luster, a hardness between 3 and 4, and nearly perfect rhombic cleavage. They can, however, be distinguished by using dilute hydrochloric acid. Calcite reacts vigorously with this acid, whereas dolomite reacts much more slowly. Calcite and dolomite are usually found together as the primary constituents in the sedimentary rocks limestone and dolostone. When calcite is the dominant mineral, the rock is called *limestone*, whereas *dolostone* results from a predominance of dolomite. Limestone has many uses, including as road aggregate, as building stone, and as the main ingredient in Portland cement.

Two other nonsilicate minerals frequently found in sedimentary rocks are *halite* and *gypsum*. Both minerals are commonly found in thick layers that are the last vestiges of ancient seas that have long since evaporated (Figure 3.35). Like limestone, both are important nonmetallic resources. Halite is the mineral name for common table salt (NaCl)

TABLE 3.2   Common Nonsilicate Mineral Classes

Mineral Groups (Key ions or elements)	Mineral Name	Chemical Formula	Economic Use
Carbonates (CO_3^{2-})	Calcite	$CaCO_3$	Portland cement, lime
	Dolomite	$CaMg(CO_3)_2$	Portland cement, lime
Halides (Cl^{1-}, F^{1-}, Br^{1-})	Halite	NaCl	Common salt
	Fluorite	CaF_2	Used in steelmaking
	Sylvite	KCl	Fertilizer
Oxides (O^{2-})	Hematite	Fe_2O_3	Ore of iron, pigment
	Magnetite	Fe_3O_4	Ore of iron
	Corundum	Al_2O_3	Gemstone, abrasive
	Ice	H_2O	Solid form of water
Sulfides (S^{2-})	Galena	PbS	Ore of lead
	Sphalerite	ZnS	Ore of zinc
	Pyrite	FeS_2	Sulfuric acid production
	Chalcopyrite	$CuFeS_2$	Ore of copper
	Cinnabar	HgS	Ore of mercury
Sulfates (SO_4^{2-})	Gypsum	$CaSO_4 \cdot 2\,H_2O$	Plaster
	Anhydrite	$CaSO_4$	Plaster
	Barite	$BaSO_4$	Drilling mud
Native elements (single elements)	Gold	Au	Trade, jewelry
	Copper	Cu	Electrical conductor
	Diamond	C	Gemstone, abrasive
	Sulfur	S	Sulfa drugs, chemicals
	Graphite	C	Pencil lead, dry lubricant
	Silver	Ag	Jewelry, photography
	Platinum	Pt	Catalyst

BOX 3.3 ▶ UNDERSTANDING EARTH

Gemstones

Precious stones have been prized since antiquity. But misinformation abounds regarding gems and their mineral makeup. This stems partly from the ancient practice of grouping precious stones by color rather than mineral makeup. For example, *rubies* and red *spinels* are very similar in color, but they are completely different minerals. Classifying by color led to the more common spinels being passed off to royalty as rubies. Even today, with modern identification techniques, common *yellow quartz* is sometimes sold as the more valuable gemstone *topaz*.

Naming Gemstones

Most precious stones are given names that differ from their parent mineral. For example, *sapphire* is one of two gems that are varieties of the same mineral, *corundum*. Trace elements can produce vivid sapphires of nearly every color (Figure 3.C). Tiny amounts of titanium and iron in corundum produce the most prized blue sapphires. When the mineral corundum contains a sufficient quantity of chromium, it exhibits a brilliant red color, and the gem is called *ruby*. Further, if a specimen is not suitable as a gem, it simply goes by the mineral name *corundum*. Because of its hardness, corundum that is not of gem quality is often crushed and sold as an abrasive.

To summarize, when corundum exhibits a red hue, it is called *ruby*, but if it exhibits any other color, the gem is called *sapphire*. Whereas corundum is the base mineral for two gems, quartz is the parent of more than a dozen gems. Table 3.A lists some well-known gemstones and their parent minerals.

What Constitutes a Gemstone?

When found in their natural state, most gemstones are dull and would be passed over by most people as "just another rock." Gems must be cut and polished by experienced professionals before their true beauty is displayed (Figure 3.C). (One of the methods used to shape a gemstone is *cleaving*, the act of splitting the mineral along one of its planes of weakness, or cleavage.) Only those mineral specimens that are of such quality that they can command a price in excess of the cost of processing are considered gemstones.

Gemstones can be divided into two categories: precious and semiprecious. A *precious* gem has beauty, durability, and rarity, whereas a *semiprecious* gem generally has only one or two of these qualities. The gems traditionally in highest esteem are diamonds, rubies, sapphires, emeralds, and some varieties of opal (Table 3.A). All other gemstones are classified as semiprecious. However, large, high-quality specimens of semiprecious stones often command a very high price.

Today translucent stones with evenly tinted colors are preferred. The most favored hues are red, blue, green, purple, rose, and yellow. The most prized stones are pigeon-blood rubies, blue sapphires, grass-green emeralds, and canary-yellow diamonds. Colorless gems are generally less than desirable except for diamonds that display "flashes of color" known as *brilliance*.

The durability of a gem depends on its hardness; that is, its resistance to abrasion by objects normally encountered in everyday living. For good durability, gems should be as hard or harder than quartz as defined by the Mohs scale of hardness. One notable exception is opal, which is comparatively soft (hardness 5 to 6.5) and brittle. Opal's esteem comes from its "fire," which is a display of a variety of brilliant colors, including greens, blues, and reds.

It seems to be human nature to treasure that which is rare. In the case of gemstones, large, high-quality specimens are much rarer than smaller stones. Thus, large rubies, diamonds, and emeralds, which are rare in addition to being beautiful and durable, command the very highest prices.

FIGURE 3.C Australian sapphires showing variation in cuts and colors. (Photo by Fred Ward, Black Star)

Table 3.A	Important Gemstones	
Gem	**Mineral Name**	**Prized Hues**
Precious		
Diamond	Diamond	Colorless, yellows
Emerald	Beryl	Greens
Opal	Opal	Brilliant hues
Ruby	Corundum	Reds
Sapphire	Corundum	Blues
Semiprecious		
Alexandrite	Chrysoberyl	Variable
Amethyst	Quartz	Purples
Cat's-eye	Chrysoberyl	Yellows
Chalcedony	Quartz (agate)	Banded
Citrine	Quartz	Yellows
Garnet	Garnet	Reds, greens
Jade	Jadeite or nephrite	Greens
Moonstone	Feldspar	Transparent blues
Peridot	Olivine	Olive greens
Smoky quartz	Quartz	Browns
Spinel	Spinel	Reds
Topaz	Topaz	Purples, reds
Tourmaline	Tourmaline	Reds, blue-greens
Turquoise	Turquoise	Blues
Zircon	Zircon	Reds

FIGURE 3.35 Thick bed of halite (salt) at an underground mine in Grand Saline, Texas. (Photo by Tom Bochsler)

A.

B.

FIGURE 3.36 Magnetite (**A**) and hematite (**B**) are both oxides and are both important ores of iron. (Photos by E. J. Tarbuck)

Students Sometimes Ask . . .

What is the largest diamond ever found?

Weighing in at 3106 carats (1.37 pounds), and mined in 1905 in South Africa, the Cullian diamond is the largest ever discovered. During the process of removing its flaws through cleaving, this enormous diamond was broken into nine fragments that were fashioned into gems. The largest of these is a 550.2 carat pear-shaped gem that was mounted in the British royal scepter. By comparison, the Hope diamond, perhaps the most-recognized gem in the world, was only about 112 carats when discovered. Having been cut at least twice, this fabulous 45.52 carat, blue diamond now resides in the National Museum of Natural History of the Smithsonian Institution.

Gypsum ($CaSO_4 \cdot 2H_2O$), which is calcium sulfate with water bound into the structure, is the mineral of which plaster and other similar building materials are composed.

Most nonsilicate mineral classes contain members that are prized for their economic value. This includes the oxides, whose members hematite and magnetite are important ores of iron (Figure 3.36). Also significant are the sulfides, which are basically compounds of sulfur (S) and one or more metals. Examples of important sulfide minerals include galena (lead), sphalerite (zinc), and chalcopyrite (copper). In addition, native elements, including gold, silver, and carbon (diamonds), plus a host of other nonsilicate minerals—fluorite (flux in making steel), corundum (gemstone, abrasive), and uraninite (a uranium source)—are important economically.

Summary

- A *mineral* is any naturally occurring inorganic solid that possesses an orderly crystalline structure and a well-defined chemical composition. Most *rocks* are aggregates composed of two or more minerals.
- The building blocks of minerals are *elements*. An *atom* is the smallest particle of matter that still retains the characteristics of an element. Each atom has a *nucleus*, which contains *protons* (particles with positive electrical charges) and *neutrons* (particles with neutral electrical charges). Orbiting the nucleus of an atom in regions called *energy levels*, or *principal shells*, are *electrons*, which have negative electrical charges. The number of protons in an atom's nucleus determines its *atomic number* and the name of the element. An element is a large collection of electrically neutral atoms, all having the same atomic number.
- Atoms combine with each other to form more complex substances called *compounds*. Atoms bond together by either gaining, losing, or sharing electrons with other atoms. In *ionic bonding*, one or more electrons are transferred from one atom to another, giving the atoms a net positive or negative charge. The resulting electrically charged atoms are called *ions*. Ionic compounds consist of oppositely charged ions assembled in a regular, crystalline structure that allows for the maximum attraction of ions, given their sizes. Another type of bond, the *covalent bond*, is produced when atoms share electrons.

- *Isotopes* are variants of the same element but with a different *mass number* (the total number of neutrons plus protons found in an atom's nucleus). Some isotopes are unstable and disintegrate naturally through a process called *radioactivity*.
- Mineralogists use the term *crystal* or *crystalline* in reference to *any natural solid with an ordered, repetitive, atomic structure*. Minerals form through the process of *crystallization*, which occurs when material precipitates out of a solution, as magma cools and crystallizes, or in high-temperature and high-pressure metamorphic environments.
- The basic building blocks of minerals, called *unit cells*, consist of an array of cations and anions arranged so that the positive and negative charges cancel each other out. These unit cells stack together in a regular manner that relates to the shape and symmetry of a crystal. Thus, the *angles between equivalent faces of crystals of the same mineral are always the same*, an observation known as *Steno's Law*.
- The chemical composition of some minerals varies from sample to sample, because ions of similar size can substitute for one another. It is also common for two minerals with exactly the same chemical composition to have different internal structures, and hence, different external forms. Minerals of this type are called *polymorphs*.
- The properties of minerals include *crystal shape (habit), luster, color, streak, hardness, cleavage, fracture,* and *density* or *specific gravity*. In addition, a number of special physical and chemical properties (*taste, smell, elasticity, feel, magnetism, double refraction,* and *chemical reaction to hydrochloric acid*) are useful in identifying certain minerals. Each mineral has a unique set of properties that can be used for identification.
- Of the nearly 4000 minerals, no more than a few dozen make up most of the rocks of Earth's crust and, as such, are classified as rock-forming minerals. Eight elements (oxygen, silicon, aluminum, iron, calcium, sodium, potassium, and magnesium) make up the bulk of these minerals and represent over 98 percent (by weight) of Earth's continental crust.
- The most common mineral group is the *silicates*. All silicate minerals have the negatively charged *silicon–oxygen tetrahedron* as their fundamental building block. In some silicate minerals the tetrahedra are joined in chains (the pyroxene and amphibole groups); in others, the tetrahedra are arranged into sheets (the micas—biotite and muscovite), or three-dimensional networks (the feldspars and quartz). The tetrahedra and various silicate structures are often bonded together by the positive ions of iron, magnesium, potassium, sodium, aluminum, and calcium. Each silicate mineral has a structure and a chemical composition that indicates the conditions under which it formed.
- The *nonsilicate* mineral groups, which contain several economically important minerals, include the *oxides* (e.g., the mineral hematite, mined for iron), *sulfides* (e.g., the mineral sphalerite, mined for zinc, and the mineral galena, mined for lead), *sulfates, halides,* and *native elements* (e.g., gold and silver). The more common nonsilicate rock-forming minerals include the *carbonate minerals,* calcite and dolomite. Two other nonsilicate minerals frequently found in sedimentary rocks are halite and gypsum.

Review Questions

1. List five characteristics an Earth material should have in order to be considered a mineral.
2. Define the term *rock*.
3. List the three main particles of an atom and explain how they differ from one another.
4. If the number of electrons in a neutral atom is 35 and its mass number is 80, calculate the following:
   a. the number of protons
   b. the atomic number
   c. the number of neutrons
5. What is the significance of valence electrons?
6. Briefly distinguish between ionic and covalent bonding.
7. What occurs in an atom to produce a cation? An anion?
8. What is an isotope?
9. What do minerologists mean when they use the word *crystal*?
10. Describe three ways a mineral might form (crystallize).
11. What is Steno's Law?
12. What are polymorphs? How are they similar?
13. Why might it be difficult to identify a mineral by its color?
14. If you found a glassy-appearing mineral while rock hunting and had hopes that it was a diamond, what simple test might help you make a determination?
15. Explain the use of corundum as given in Table 3.2 (p. 95) in terms of the Mohs hardness scale.
16. Gold has a specific gravity of almost 20. If a 25-liter pail of water weighs 25 kilograms, how much would a 25-liter pail of gold weigh?
17. Explain the difference between the terms *silicon* and *silicate*.
18. What is meant when we refer to a mineral's tenacity? List three terms that describe tenacity.
19. On what basis are minerals placed in mineral classes?
20. Describe the silicon–oxygen tetrahedron.
21. What do ferromagnesian minerals have in common? List examples of ferromagnesian minerals.
22. What do muscovite and biotite have in common? How do they differ?
23. Should color be used to distinguish between orthoclase and plagioclase feldspar? What is the best means of distinguishing between these two types of feldspar?
24. Each of the following statements describes a silicate mineral or mineral group. In each case, provide the appropriate name:

a. the most common member of the amphibole group

b. the most common nonferromagnesian member of the mica family

c. the only silicate mineral made entirely of silicon and oxygen

d. a high-temperature silicate with a name that is based on its color

e. one that is characterized by striations

f. one that originates as a product of chemical weathering

25. What simple test can be used to distinguish calcite from dolomite?

Key Terms

atom (p. 75)
atomic number (p. 75)
chemical compound (p. 75)
cleavage (p. 85)
color (p. 84)
Constancy of Interfacial Angles, Law of (p. 81)
covalent bond (p. 77)
crystal (p. 78)
crystalline (p. 78)
crystallization (p. 78)
crystal shape (p. 81)
dark silicates (p. 94)

density (p. 86)
electron (p. 75)
element (p. 75)
energy levels, or shells (p. 75)
ferromagnesian silicates (p. 94)
fracture (p. 86)
habit (p. 84)
hardness (p. 85)
ion (p. 76)
ionic bond (p. 76)
isotope (p. 78)

light silicates (p. 91)
luster (p. 84)
mass number (p. 78)
metallic bond (p. 78)
mineral (p. 72)
mineralogy (p. 72)
Mohs scale (p. 85)
neutron (p. 75)
nonferromagnesian silicates (p. 91)
nonsilicates (p. 89)
nucleus (p. 75)
periodic table (p. 75)

polymorph (p. 82)
principal shells (p. 75)
proton (p. 75)
radioactive decay (p. 78)
rock (p. 73)
silicate (p. 89)
silicon–oxygen tetrahedron (p. 89)
specific gravity (p. 86)
Steno's law (p. 81)
streak (p. 84)
tenacity (p. 85)
valence electron (p. 75)

Web Resources

The *Earth* Website uses the resources and flexibility of the Internet to aid in your study of the topics in this chapter. Written and developed by geology instructors, this site will help improve your understanding of geology. Visit **http://www.prenhall .com/tarbuck** and click on the cover of *Earth 9e* to find:

- Online review quizzes.
- Critical thinking exercises.
- Links to chapter-specific Web resources.
- Internet-wide key-term searches.

http://www.prenhall.com/tarbuck

GEODe: Earth

GEODe: Earth makes studying faster and more effective by reinforcing key concepts using animation, video, narration, interactive exercises and practice quizzes. A copy is included with every copy of *Earth*.

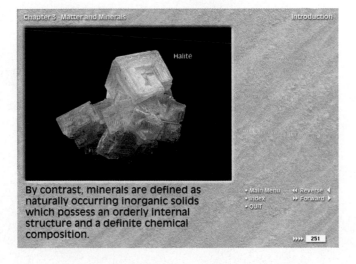

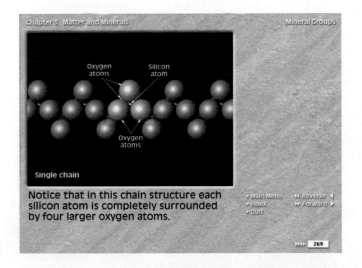

Igneous Rocks

Two climbers on the summit of a spire in the Sierra Nevada. (Photo by Brian Bailey/Getty Images)

gneous rocks and metamorphic rocks derived from igneous "parents" make up about 95 percent of Earth's crust. Furthermore, the mantle, which accounts for more than 82 percent of Earth's volume, is composed entirely of igneous rock. Thus, Earth can be described as a huge mass of igneous rock covered with a thin veneer of sedimentary rock and having a relatively small iron-rich core.

Magma: The Parent Material of Igneous Rock

 GEODe Igneous Rocks
▸ Introduction

In our discussion of the rock cycle, it was pointed out that **igneous rocks** (*ignis* = fire) form as molten rock cools and solidifies. Abundant evidence supports the fact that the parent material for igneous rocks, called **magma,** is formed by a process called *partial melting*. Partial melting occurs at various levels within Earth's crust and upper mantle to depths of perhaps 250 kilometers (about 150 miles). We will explore the origin of magma later in this chapter.

Once formed, a magma body buoyantly rises toward the surface because it is less dense than the surrounding rocks. Occasionally molten rock breaks through, producing a spec-tacular volcanic eruption. Magma that reaches Earth's surface is called **lava.** Sometimes lava is emitted as fountains that are produced when escaping gasses propel molten rock from a magma chamber. On other occasions, magma is explosively ejected from a vent, producing a catastrophic eruption. However, not all eruptions are violent; many volcanoes emit quiet outpourings of very fluid lava (Figure 4.1).

Igneous rocks that form when molten rock solidifies *at the surface* are classified as **extrusive** (*ex* = out, *trudere* = thrust) or **volcanic** (after the fire god Vulcan). Extrusive igneous rocks are abundant in western portions of the Americas, including the volcanic cones of the Cascade Range and the extensive lava flows of the Columbia Plateau. In addition, many oceanic islands, typified by the Hawaiian chain, are composed almost entirely of volcanic igneous rocks.

FIGURE 4.1 Fluid basaltic lava emitted from Hawaii's Kilauea Volcano. (Photo by Philip Rosenberg/Pacific Stock)

Magma that loses its mobility before reaching the surface eventually crystallizes at depth. Igneous rocks that *form at depth* are termed **intrusive** (*in* = into, *trudere* = thrust) or **plutonic** (after Pluto, the god of the lower world in classical mythology). Intrusive igneous rocks would never outcrop at the surface if portions of the crust were not uplifted and the overlying rocks stripped away by erosion. (When a mass of crustal rock is exposed—not covered with soil—it is called an *outcrop.*) Exposures of intrusive igneous rocks occur in many places, including Mount Washington, New Hampshire; Stone Mountain, Georgia; the Black Hills of South Dakota; and Yosemite National Park, California (Figure 4.2).

The Nature of Magma

Magma is completely or partly molten material, which on cooling solidifies to form an igneous rock. Most magmas consist of three distinct parts—a liquid component, a solid component, and a gaseous phase.

The liquid portion, called **melt,** is composed of mobile ions of those elements commonly found in Earth's crust. Melt is made up mostly of ions of silicon and oxygen, along with lesser amounts of aluminum, potassium, calcium, sodium, iron, and magnesium.

The solid components (if any) in magma are silicate minerals that have crystallized from the melt. As a magma body cools, the size and number of crystals increases. During the last stage of cooling, a magma body is mostly a crystalline solid with only minor amounts of melt.

The gaseous components of magma, called **volatiles,** are materials that will vaporize (form a gas) at surface pressures. The most common volatiles found in magma are water vapor (H_2O), carbon dioxide (CO_2), and sulfur dioxide (SO_2), which are confined by the immense pressure exerted by the overlying rocks. These gases tend to separate from magma as it moves toward the surface (low-pressure environment), where they may generate a steam eruption. Further, when deeply buried magma bodies crystallize, the remaining volatiles form hot, water-rich fluids that migrate through the surrounding rocks. These hot fluids play an im-

Students Sometimes Ask . . .

Are lava and magma the same thing?

No, but their *composition* might be similar. Both are terms that describe molten or liquid rock: Magma exists beneath Earth's surface, and lava is molten rock that has reached the surface. That's the reason why they can be similar in composition. Lava is produced from magma, but it generally has lost materials that escape as a gas, such as water vapor.

portant role in metamorphism and will be considered in Chapter 8.

From Magma to Crystalline Rock

When magma is at its hottest, ions and groups of ions join together and break apart constantly. Then, as magma cools, the ions begin to move more slowly and eventually join together into orderly crystalline structures. This process, called **crystallization,** generates various silicate minerals that reside within the remaining melt.

Before we examine how magma crystallizes, let us first examine how a simple crystalline solid melts. In any crystalline solid, the ions are arranged in a closely packed regular pattern. However, they are not without some motion. They exhibit a sort of restricted vibration about fixed points. As the temperature rises, the ions vibrate more rapidly and consequently collide with ever-increasing vigor with their neighbors. Thus, heating causes the ions to occupy more space, which in turn causes the solid to expand. When the ions are vibrating rapidly enough to overcome the force of the chemical bonds, the solid begins to melt. At this stage the ions are able to slide past one another, and their orderly crystalline structure disintegrates. Thus, melting converts a solid consisting of tight, uniformly packed ions into a liquid composed of unordered ions moving randomly about.

In the process of crystallization, cooling reverses the events of melting. As the temperature of the liquid drops, the ions pack closer and closer together as they slow their rate of movement. When cooled sufficiently, the forces of the chemical bonds will again confine the ions to an orderly crystalline arrangement.

When magma cools, it is generally the silicon and oxygen atoms that link together first to form silicon–oxygen tetrahedra, the basic building blocks of the silicate minerals. As a magma continues to lose heat to its surroundings, the tetrahedra join with each other and with other ions to form embryonic crystal nuclei. Slowly each nucleus grows as ions lose their mobility and join the crystalline network.

The earliest formed minerals have space to grow and tend to have better-developed crystal faces than do the later ones that fill the remaining space. Eventually all of the melt is transformed into a solid mass of interlocking silicate minerals that we call an *igneous rock* (Figure 4.3).

FIGURE 4.2 Mount Rushmore National Memorial, located in the Black Hills of South Dakota, is carved from intrusive igneous rocks. (Photo by Marc Muench)

FIGURE 4.3 **A.** Close-up of interlocking crystals in a coarse-grained igneous rock. The largest crystals are about 2 centimeters in length. **B.** Photomicrograph of interlocking crystals in a coarse-grained igneous rock. (Photos by E. J. Tarbuck)

As you will see later, the crystallization of magma is much more complex than just described. Whereas a single compound, such as water, crystallizes at a specific temperature, solidification of magma with its diverse chemistry spans a temperature range of 200°C, or more. During crystallization, the composition of the melt continually changes as ions are selectively removed and incorporated into the earliest formed minerals. If the melt should separate from the earliest formed minerals, its composition will be different from that of the original magma. Thus, a single magma may generate rocks with widely differing compositions. As a consequence, a great variety of igneous rocks exist. We will return to this important idea later in the chapter.

Although the crystallization of magma is complex, it is nevertheless possible to classify igneous rocks based on their mineral composition and the conditions under which they formed. Their environment during crystallization can be roughly inferred from the size and arrangement of the mineral grains, a property called *texture*. Consequently, *igneous rocks are most often classified by their texture and mineral composition.* We will consider these two rock characteristics in the following sections.

Igneous Textures

 Igneous Rocks
▶ **Igneous Textures**

The term **texture,** when applied to an igneous rock, is used to describe the overall appearance of the rock based on the size, shape, and arrangement of its interlocking crystals (Figure 4.4). Texture is an important characteristic because it reveals a great deal about the environment in which the rock formed. This fact allows geologists to make inferences about a rock's origin while working in the field where sophisticated equipment is not available.

Factors Affecting Crystal Size

Three factors contribute to the textures of igneous rocks: (1) *the rate at which magma cools;* (2) *the amount of silica present;* and (3) *the amount of dissolved gases in the magma.* Of these, the rate of cooling is the dominant factor, but like all generalizations, this one has exceptions.

As a magma body loses heat to its surroundings, the mobility of its ions decreases. A very large magma body located at great depth will cool over a period of perhaps tens or hundreds of thousands of years. Initially, relatively few crystal nuclei form. Slow cooling permits ions to migrate freely until they eventually join one of the existing crystalline structures. Consequently, slow cooling promotes the growth of fewer but larger crystals.

On the other hand, when cooling occurs more rapidly—for example, in a thin lava flow—the ions quickly lose their mobility and readily combine to form crystals. This results in the development of numerous embryonic nuclei, all of which compete for the available ions. The result is a solid mass of small intergrown crystals.

When molten material is quenched quickly, there may not be sufficient time for the ions to arrange into an ordered crystalline network. Rocks that consist of unordered ions are referred to as **glass.**

Types of Igneous Textures

As you saw, the effect of cooling on rock textures is fairly straightforward. Slow cooling promotes the growth of large crystals, whereas rapid cooling tends to generate smaller crystals. We will consider the other two factors affecting crystal growth as we examine the major textural types.

Aphanitic (fine-grained) Texture Igneous rocks that form at the surface or as small masses within the upper crust where cooling is relatively rapid possess a very fine-grained texture termed **aphanitic** (*a* = not, *phaner* = visible). By definition, the crystals that make up aphanitic rocks are so small that individual minerals can only be distinguished with the aid of a microscope (Figure 4.4A). Because mineral identification is not possible, we commonly characterize fine-grained rocks as being light, intermediate, or dark in color. Using this system of grouping, light-colored aphanitic rocks are those containing primarily light-colored nonferromagnesian silicate materials and so forth (see the section titled "Common Silicate Minerals" in Chapter 3).

Commonly seen in many aphanitic rocks are the voids left by gas bubbles that escape as lava solidifies. These spherical or elongated openings are called *vesicles,* and the rocks that contain them are said to have a **vesicular texture.** Rocks that exhibit a vesicular texture usually form in the upper zone of a lava flow, where cooling occurs rapidly enough to "freeze" the lava, thereby preserving the openings produced by the expanding gas bubbles (Figure 4.5).

Phaneritic (Coarse-Grained) Texture When large masses of magma slowly solidify far below the surface, they form igneous rocks that exhibit a coarse-grained texture described

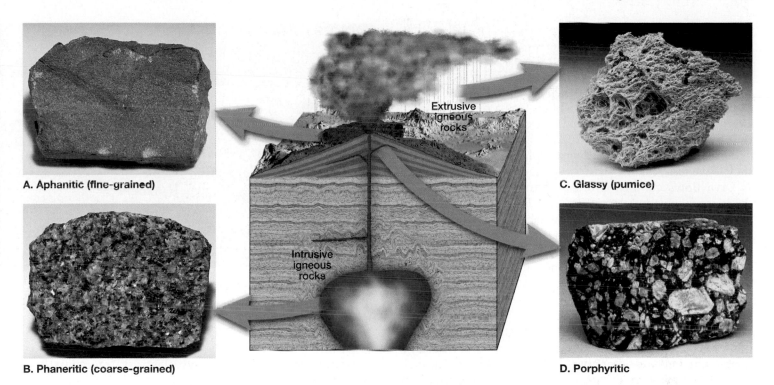

A. Aphanitic (fine-grained)

B. Phaneritic (coarse-grained)

Extrusive
igneous
rocks

Intrusive
igneous
rocks

C. Glassy (pumice)

D. Porphyritic

FIGURE 4.4   Igneous rock textures. **A.** Igneous rocks that form at or near Earth's surface cool quickly and often exhibit a fine-grained (aphanitic) texture. **B.** Coarse-grained (phaneritic) igneous rocks form when magma slowly crystallizes at depth. **C.** During a volcanic eruption in which silica-rich lava is ejected into the atmosphere, a frothy glass called pumice may form. **D.** A porphyritic texture results when magma that already contains some large crystals migrates to a new location where the rate of cooling increases. The resulting rock consists of larger crystals (phenocrysts) embedded within a matrix of smaller crystals (groundmass).
(Photos courtesy of E. J. Tarbuck)

as **phaneritic** (*phaner* = visible). These coarse-grained rocks consist of a mass of intergrown crystals, which are roughly equal in size and large enough so that the individual minerals can be identified without the aid of a microscope (Figure 4.4B). (Geologists often use a small magnifying lens to aid in identifying coarse-grained minerals.) Because

phaneritic rocks form deep within Earth's crust, their exposure at Earth's surface results only after erosion removes the overlying rocks that once surrounded the magma chamber.

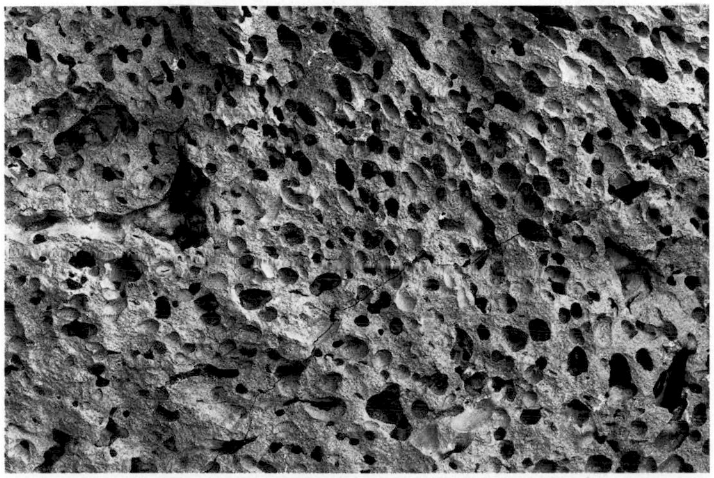

Porphyritic Texture   A large mass of magma located at depth may require tens to hundreds of thousands of years to solidify. Because different minerals crystallize at different temperatures (as well as at differing rates), it is possible for some crystals to become quite large before others even begin to form. If magma containing some large crystals should change environments—for example, by erupting at the surface—the remaining liquid portion of the lava would cool relatively quickly. The resulting rock, which has large crystals embedded in a matrix of smaller crystals, is said to have a **porphyritic texture** (Figure 4.4D). The large crystals in such a rock are referred to as **phenocrysts** (*pheno* = show, *cryst* = crystal) whereas the matrix of smaller crystals is called **groundmass.** A rock with such a texture is termed a **porphyry.**

Glassy Texture   During some volcanic eruptions, molten rock is ejected into the atmosphere, where it is quenched quickly. Rapid cooling of this type may generate rocks having a **glassy texture** (Figure 4.4C). As we indicated earlier,

FIGURE 4.5   Vesicular texture displayed on a freshly broken surface of the volcanic rock scoria. Vesicles are small holes left by escaping gas bubbles. (Photo by Michael Collier)

glass results when unordered ions are "frozen" before they are able to unite into an orderly crystalline structure. *Obsidian,* a common type of natural glass, is similar in appearance to a dark chunk of manufactured glass (Figure 4.6). Because of its excellent conchodial fracture and ability to hold a sharp, hard edge, obsidian was a prized material from which Native Americans chipped arrowheads and cutting tools (Figure 4.6 inset). Today, scalpels made from obsidian are being used for delicate plastic surgery because they leave less scarring than those made of steel.

Lava flows composed of obsidian a few hundred feet thick occur in some places (Figure 4.7). Thus, rapid cooling is not the only mechanism by which a glassy texture can form. As a general rule, magmas with a high silica content tend to form long, chainlike structures before crystallization is complete. These structures in turn impede ionic transport and increase the magma's viscosity. (*Viscosity* is a measure of a fluid's resistance to flow.)

Granitic magma, which is rich in silica, may be extruded as an extremely viscous mass that eventually solidifies to form obsidian. By contrast, basaltic magma, which is low in silica, forms very fluid lavas that upon cooling usually generate fine-grained crystalline rocks. However, the surface of basaltic lava may be quenched rapidly enough to form a thin, glassy skin. Moreover, Hawaiian volcanoes sometimes generate lava fountains, which spray basaltic lava tens of meters into the air. Such activity can produce strands of volcanic glass called *Pele's hair,* after the Hawaiian goddess of volcanoes.

Pyroclastic (Fragmental) Texture Some igneous rocks are formed from the consolidation of individual rock fragments that are ejected during a violent volcanic eruption (Figure 4.8). The ejected particles might be very fine ash, molten blobs, or large angular blocks torn from the walls of the vent during the eruption. Igneous rocks composed of these rock

FIGURE 4.7 This obsidian flow was extruded from a vent along the south wall of Newberry Caldera, Oregon. Note the road for scale. (Photo by Marli Miller)

fragments are said to have a **pyroclastic** or **fragmental texture** (Figure 4.8 inset).

A common type of pyroclastic rock called *welded tuff* is composed of fine fragments of glass that remained hot enough during their flight to fuse together upon impact. Other pyroclastic rocks are composed of fragments that

FIGURE 4.8 Rocks that exhibit a pyroclastic texture are a result of the consolidation of rock fragments that were ejected during a violent volcanic eruption. (Photo by Steve Kaufman / DRK)

FIGURE 4.6 Obsidian, a natural glass, was used by Native Americans for making arrowheads and cutting tools. (Inset photo by Jeffrey Scovil)

FIGURE 4.9 A granite pegmatite composed mainly of quartz and feldspar (salmon color). The elongated, dark quartz crystal on the right is about the size of a person's index finger. (Photo by Colin Keates)

solidified before impact and became cemented together at some later time. Because pyroclastic rocks are made of individual particles or fragments rather than interlocking crystals, their textures often appear to be more similar to sedimentary rocks than to other igneous rocks.

Pegmatitic Texture   Under special conditions, exceptionally coarse-grained igneous rocks, called **pegmatites,** may form. These rocks, which are composed of interlocking crystals all larger than a centimeter in diameter, are said to have a **pegmatitic texture** (Figure 4.9). Most pegmatites are found around the margins of large plutons as small masses or thin veins that commonly extend into the adjacent host rock.

Pegmatites form in the late stages of crystallization, when water and other volatiles, such as chlorine, fluorine, and sulfur, make up an unusually high percentage of the melt. Because ion migration is enhanced in these fluid-rich environments, the crystals that form are abnormally large. Thus, the large crystals in pegmatites are not the result of inordinately long cooling histories; rather, they are the consequence of the fluid-rich environment that enhances crystallization.

The composition of most pegmatites is similar to that of granite. Thus, pegmatites contain large crystals of quartz, feldspar, and muscovite. However, some contain significant quantities of comparatively rare and hence valuable minerals.

Igneous Compositions

 GEODe Igneous Rocks
▸ Igneous Compositions

Igneous rocks are mainly composed of silicate minerals (see Box 4.1). Furthermore, the mineral makeup of a particular igneous rock is ultimately determined by the chemical composition of the magma from which it crystallizes. Recall that magma is composed largely of the eight elements that are the major constituents of the silicate minerals. Chemical analysis shows that silicon and oxygen (usually expressed as the silica [SiO_2] content of a magma) are by far the most abundant constituents of igneous rocks. These two elements, plus ions of aluminum (Al), calcium (Ca), sodium (Na), potassium (K), magnesium (Mg), and iron (Fe), make up roughly 98 percent by weight of most magmas. In addition, magma contains small amounts of many other elements, including titanium and manganese, and trace amounts of much rarer elements such as gold, silver, and uranium.

As magma cools and solidifies, these elements combine to form two major groups of silicate minerals. The *dark* (or *ferromagnesian*) *silicates* are rich in iron and/or magnesium and comparatively low in silica. *Olivine, pyroxene, amphibole,* and *biotite mica* are the common dark silicate minerals of Earth's crust. By contrast, the *light* (or *nonferromagnesian*) *silicates* contain greater amounts of potassium, sodium, and calcium rather than iron and magnesium. As a group, these minerals are richer in silica than the dark silicates. The light silicates include *quartz, muscovite mica,* and the most abundant mineral group, the *feldspars*. The feldspars make up at least 40 percent of most igneous rocks. Thus, in addition to feldspar, igneous rocks contain some combination of the other light and/or dark silicates listed above.

Granitic (Felsic) versus Basaltic (Mafic) Compositions

Despite their great compositional diversity, igneous rocks (and the magmas from which they form) can be divided into broad groups according to their proportions of light and dark minerals (Figure 4.10). Near one end of the continuum are rocks composed almost entirely of light-colored silicates—quartz and feldspar. Igneous rocks in which these are the dominant minerals have a **granitic composition.** Geologists also refer to granitic rocks as being **felsic,** a term derived from *fel*dspar and *si*lica (quartz). In addition to quartz and feldspar, most granitic rocks contain about 10 percent dark silicate minerals, usually biotite mica and amphibole. Granitic rocks are rich in silica (about 70 percent) and are major constituents of the continental crust.

Rocks that contain substantial dark silicate minerals and calcium-rich plagioclase feldspar (but no quartz) are said to have a **basaltic composition** (Figure 4.10). Because basaltic rocks contain a high percentage of ferromagnesian minerals, geologists also refer to them as **mafic** (from *ma*gnesium and *f*errum, the Latin name for iron). Because of their iron content, mafic rocks are typically darker and denser than granitic rocks. Basaltic rocks make up the ocean floor as well as many of the volcanic islands located within the ocean basins. Basalt also forms extensive lava flows on the continents.

Other Compositional Groups

As you can see in Figure 4.10, rocks with a composition between granitic and basaltic rocks are said to have an **intermediate,** or **andesitic composition** after the common volcanic

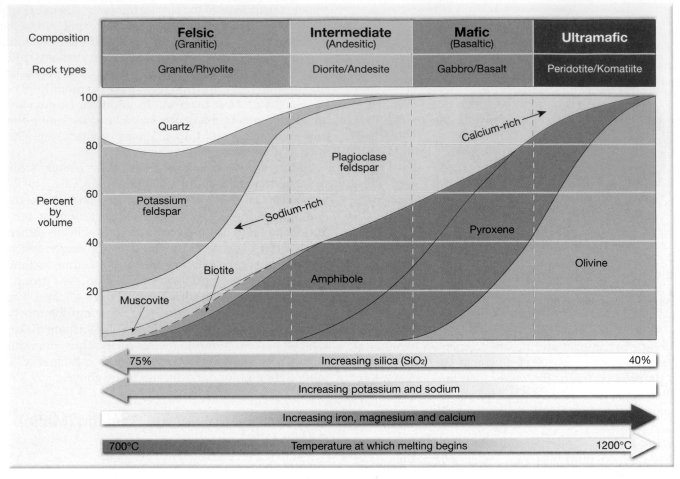

FIGURE 4.10 Mineralogy of common igneous rocks and the magmas from which they form. (After Dietrich, Daily, and Larsen)

rock *andesite.* Intermediate rocks contain at least 25 percent dark silicate minerals, mainly amphibole, pyroxene, and biotite mica with the other dominant mineral being plagioclase feldspar. This important category of igneous rocks is associated with volcanic activity that is typically confined to the margins of the continents.

Another important igneous rock, *peridotite,* contains mostly olivine and pyroxene and thus falls on the opposite side of the compositional spectrum from granitic rocks (Figure 4.10). Because peridotite is composed almost entirely of ferromagnesian minerals, its chemical composition is referred to as **ultramafic.** Although ultramafic rocks are rare at Earth's surface, peridotite is believed to be the main constituent of the upper mantle.

Silica Content As an Indicator of Composition

An important aspect of the chemical composition of igneous rocks is their silica (SiO_2) content. Recall that silicon and oxygen are the two most abundant elements in igneous rocks. Typically, the silica content of crustal rocks ranges from a low of about 45 percent in ultramafic rocks to a high

of over 70 percent in granitic rocks (Figure 4.10). The percentage of silica in igneous rocks actually varies in a systematic manner that parallels the abundance of other elements. For example, rocks comparatively low in silica contain large amounts of iron, magnesium, and calcium. By contrast, rocks high in silica contain very small amounts of those elements but are enriched instead in sodium and potassium. Consequently, the chemical makeup of an igneous rock can be inferred directly from its silica content.

Further, the amount of silica present in magma strongly influences its behavior. Granitic magma, which has a high silica content, is quite viscous (thick) and exists as a liquid at temperatures as low as 700°C. On the other hand, basaltic magmas are low in silica and are generally more fluid. Further, basaltic magmas crystallize at higher temperatures than granitic magmas and are completely solid when cooled to about 1000°C.

In summary, igneous rocks can be divided into broad groups according to the proportions of light and dark minerals they contain. Granitic (felsic) rocks, which are almost entirely composed of the lightcolored minerals quartz and feldspar, are at one end of the compositional spectrum (Figure 4.10). Basaltic (mafic) rocks, which contain abundant dark silicate minerals in addition to plagioclase feldspar,

BOX 4.1 ▶ UNDERSTANDING EARTH

Thin Sections and Rock Identification

Igneous rocks are classified on the basis of their mineral composition and texture. When analyzing specimens, geologists examine them closely to identify the minerals present and to determine the size and arrangement of the interlocking crystals. When out in the field, geologists use megascopic techniques to study rocks. The *megascopic* characteristics of rocks are those features that can be determined with the unaided eye or by using a low-magnification (10X) hand lens. When practical to do so, geologists collect hand samples that can be taken back to the laboratory where *microscopic*, or high-magnification, methods can be employed. Microscopic examination is important to identify trace minerals, as well as those textural features that are too small to be visible to the unaided eye.

Because most rocks are not transparent, microscopic work requires the preparation of a very thin slice of rock known as a *thin section* (Figure 4.A, part B). First, a saw containing diamonds embedded in its blade is used to cut a narrow slab from the sample. Next, one side of the slab is polished using grinding powder and then cemented to a microscope slide. Once the mounted sample is firmly in place, the other side of it is ground to a thickness of about 0.03 millimeter. When a slice of rock is that thin, it is usually transparent. Nevertheless, some metallic minerals, such as pyrite and magnetite, remain opaque.

Once produced, thin sections are examined under a specially designed microscope called a *polarizing microscope*. Such an instrument has a light source beneath the stage so that light can be transmitted upward through the thin section. Because minerals have crystalline structures that influence polarized light in a measurable way, this procedure allows for the identification of even the smallest components of a rock. Part C of Figure 4.A is a photomicrograph (photo taken through a microscope) of a thin section of granite shown under polarized light. The mineral constituents are identified by their unique optical properties. In addition to assisting in the study of igneous rocks, microscopic techniques are used with great success in analyzing sedimentary and metamorphic rocks as well.

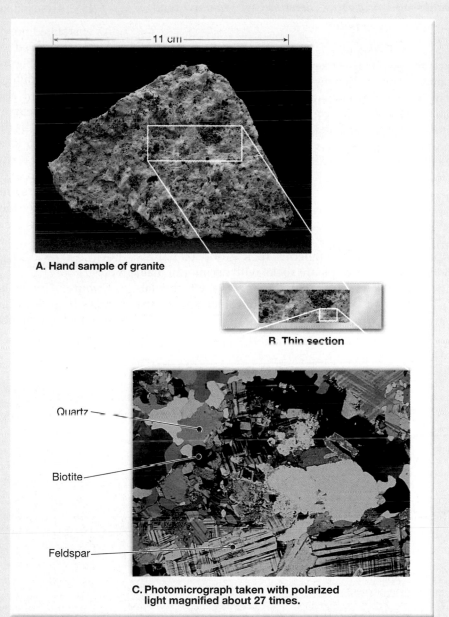

11 cm

A. Hand sample of granite

B. Thin section

Quartz

Biotite

Feldspar

C. Photomicrograph taken with polarized light magnified about 27 times.

FIGURE 4.A Thin sections are very useful in identifying the mineral constituents in rocks. **A.** A slice of rock is cut from a hand sample using a diamond saw. **B.** This slice is cemented to a microscope slide and ground until it is transparent to light (about 0.03 millimeter thick) This very thin slice of rock is called a *thin section.* **C.** A thin section of granite viewed under polarized light. (Photos by E. J. Tarbuck)

Students Sometimes Ask ...

I've heard some igneous rocks described as "granitic." Is all granitic rock really granite?

Technically, no. True granite is a coarse-grained intrusive rock with a certain percentage of key minerals, mostly light-colored quartz and feldspar, with other accessory dark minerals. However, it has become common practice among geologists to apply the term *granite* to any coarse-grained intrusive rock composed predominantly of light-colored silicate minerals. Further, some rocks are polished and sold as granites for use as countertops or floor tile that, in addition to not being granite, are not even igneous rocks!

make up the other major igneous rock group of Earth's crust. Between these groups are rocks with an intermediate (andesitic) composition, while ultramafic rocks, which totally lack light-colored minerals, lie at the other end of the compositional spectrum from granitic rocks.

Naming Igneous Rocks

 GEODe Igneous Rocks
▸ Naming Igneous Rocks

As was stated previously, igneous rocks are most often classified, or grouped, on the basis of their texture and mineral composition (Figure 4.11). The various igneous textures result mainly from different cooling histories, whereas the mineral composition of an igneous rock is the consequence of the chemical makeup of its parent magma. Because igneous rocks are classified on the basis of their mineral composition and texture, two rocks may have similar mineral constituents but have different textures and hence different names. For example, *granite,* a coarse-grained plutonic rock, has a fine-grained volcanic equivalent called *rhyolite.* Although these rocks are mineralogically the same, they have different cooling histories and do not look at all alike (Figure 4.12).

Felsic (Granitic) Igneous Rocks

Granite Granite is perhaps the best known of all igneous rocks (Figure 4.12). This is partly because of its natural beauty, which is enhanced when it is polished, and partly because

Chemical Composition			**Felsic** (Granitic)	**Intermediate** (Andesitic)	**Mafic** (Basaltic)	**Ultramafic**
Dominant Minerals			Quartz Potassium feldspar Sodium-rich plagioclase feldspar	Amphibole Sodium- and calcium-rich plagioclase feldspar	Pyroxene Calcium-rich plagioclase feldspar	Olivine Pyroxene
Accessory Minerals			Amphibole Muscovite Biotite	Pyroxene Biotite	Amphibole Olivine	Calcium-rich plagioclase feldspar
T E X T U R E	Phaneritic (coarse-grained)		**Granite**	**Diorite**	**Gabbro**	**Peridotite**
	Aphanitic (fine-grained)		**Rhyolite**	**Andesite**	**Basalt**	**Komatiite** (rare)
	Porphyritic		"Porphyritic" precedes any of the above names whenever there are appreciable phenocrysts			
	Glassy		**Obsidian** (compact glass) **Pumice** (frothy glass)			Uncommon
	Pyroclastic (fragmental)		**Tuff** (fragments less than 2 mm) **Volcanic Breccia** (fragments greater than 2 mm)			
Rock Color (based on % of dark minerals)			0% to 25%	25% to 45%	45% to 85%	85% to 100%

FIGURE 4.11 Classification of the major igneous rock groups based on mineral composition and texture. Coarse-grained rocks are plutonic, solidifying deep underground. Fine-grained rocks are volcanic, or solidify as shallow, thin plutons. Ultramafic rocks are dark, dense rocks, composed almost entirely of minerals containing iron and magnesium. Although relatively rare on Earth's surface, these rocks are major constituents of the upper mantle.

Texture	Composition		
	Felsic (Granitic)	**Intermediate** (Andesitic)	**Mafic** (Basaltic)
Phaneritic (course-grained)	Granite	Diorito	Gabbro
Aphanitic (fine-grained)	Rhyolite	Andocito	Basalt
Porphyritic	Granite porphyry	Andesite porphyry	Basalt porphyry

FIGURE 4.12   Common igneous rocks. (Photos by E.J. Tarbuck)

of its abundance in the continental crust. Slabs of polished granite are commonly used for tombstones and monuments and as building stones. Well-known areas in the United States where granite is quarried include Barre, Vermont; Mount Airy, North Carolina; and St. Cloud, Minnesota.

Granite is a phaneritic rock composed of about 25 percent quartz and roughly 65 percent feldspar, mostly potassium- and sodium-rich varieties. Quartz crystals, which are roughly spherical in shape, are often glassy and clear to light gray in color. By contrast, feldspar crystals are not as glassy, are generally white to gray or salmon pink in color, and exhibit a rectangular rather than spherical shape (see Figure 4.3A). Other minor constituents of granite include muscovite and some dark silicates, particularly biotite and amphibole. Although the dark components generally make up less than 10 percent of most granites, dark minerals appear to be more prominent than their percentage would indicate.

When potassium feldspar is dominant and dark pink in color, granite appears reddish (Figure 4.12). This variety of granite is popular for monuments and building stone. However, the feldspar grains are more often white to gray, so when they are mixed with lesser amounts of dark silicates, granite appears light gray in color (Figure 4.13). In addition, some granites have a porphyritic texture. These specimens contain elongated feldspar crystals a few centimeters in length that are scattered among smaller crystals of quartz and amphibole.

Granite and other related crystalline rocks are often the by-products of mountain building. Because granite is very resistant to weathering, it frequently forms the core of eroded mountains. For example, Pikes Peak in the Rockies, Mount Rushmore in the Black Hills, the White Mountains of New Hampshire, Stone Mountain in Georgia, and Yosemite National Park in the Sierra Nevada are all areas where large quantities of granite are exposed at the surface (Figure 4.13)

FIGURE 4.13 Rocks contain information about the processes that produce them. This massive granitic monolith (El Capitan) located in Yosemite National Park, California, was once a molten mass found deep within Earth. (Photo by Tim Fitzharris/Minden Pictures)

Granite is a very abundant rock. However, it has become common practice among geologists to apply the term *granite* to any coarse-grained intrusive rock composed predominantly of light silicate materials that contain quartz. We will follow this practice for the sake of simplicity. You should keep in mind that this use of the term *granite* covers rocks having a wide range of mineral compositions.

Rhyolite Rhyolite is the extrusive equivalent of granite and, like granite, is composed essentially of the light-colored silicates (Figure 4.12). This fact accounts for its color, which is usually buff to pink or occasionally very light gray. Rhyolite is aphanitic and frequently contains glass fragments and voids, indicating rapid cooling in a surface environment. When rhyolite contains phenocrysts, they are

A. Large block of obsidian

B. Hand sample of obsidian

FIGURE 4.14 Obsidian is a dark-colored, glassy rock formed from silica-rich lava. The scene in part A shows an obsidian lava flow at Newberry Caldera, Oregon. (Photos courtesy of E. J. Tarbuck)

112

small and composed of either quartz or potassium feldspar. In contrast to granite, which is widely distributed as large plutonic masses, rhyolite deposits are less common and generally less voluminous. Yellowstone Park is one well-known exception. Here, rhyolite lava flows and thick ash deposits of similar composition are extensive.

Obsidian   *Obsidian* is a dark-colored glassy rock that usually forms when silica-rich lava is quenched quickly (Figure 4.14). In contrast to the orderly arrangement of ions characteristic of minerals, *the ions in glass are unordered.* Consequently, glassy rocks such as obsidian are not composed of minerals in the same sense as most other rocks.

Although usually black or reddish-brown in color, obsidian has a composition that is more akin to light-colored igneous rocks such as granite, rather than to dark rocks such as basalt. Obsidian's dark color results from small amounts of metallic ions in an otherwise relatively clear, glassy substance. If you examine a thin edge, obsidian will appear nearly transparent (see Figure 4.6).

Pumice   *Pumice* is a volcanic rock with a glassy texture that forms when large amounts of gas escape through silica-rich lava to generate a gray, frothy mass (Figure 4.15). In some samples, the voids are quite noticeable, whereas in others the pumice resembles fine shards of intertwined glass. Because of the large percentage of voids, many samples of pumice will float when placed in water. Oftentimes flow lines are visible in pumice, indicating that some movement occurred before solidification was complete. Moreover, pumice and obsidian can often be found in the same rock mass, where they exist in alternating layers.

FIGURE 4.15   Pumice, a glassy rock, is very lightweight because it contains numerous vesicles. (Inset photo by Chip Clark)

Intermediate (Andesitic) Igneous Rocks

Andesite   Andesite is a medium-gray, fine-grained rock of volcanic origin. Its name comes from South America's Andes Mountains, where numerous volcanoes are composed of this rock type. In addition to the volcanoes of the Andes, many of the volcanic structures occupying the continental margins that surround the Pacific Ocean are of andesitic composition. Andesite commonly exhibits a porphyritic texture (Figure 4.12). When this is the case, the phenocrysts are often light, rectangular crystals of plagioclase feldspar or black, elongated amphibole crystals. Andesite often resembles rhyolite, so their identification usually requires microscopic examination to verify mineral makeup.

Diorite   Diorite is the plutonic equivalent of andesite. It is a coarse-grained intrusive rock that looks somewhat similar to gray granite. However, it can be distinguished from granite by the absence of visible quartz crystals and because it contains a higher percentage of dark silicate minerals. The mineral makeup of diorite is primarily sodium-rich plagioclase feldspar and amphibole, with lesser amounts of biotite. Because the light-colored feldspar grains and dark amphibole crystals appear to be roughly equal in abundance, diorite has a salt-and-pepper appearance (Figure 4.12).

Mafic (Basaltic) Igneous Rocks

Basalt   Basalt is a very dark green to black fine-grained volcanic rock composed primarily of pyroxene and calcium-rich plagioclase feldspar, with lesser amounts of olivine and amphibole present (Figure 4.12). When porphyritic, basalt commonly contains small light-colored feldspar phenocrysts or green, glassy-appearing olivine phenocrysts embedded in a dark groundmass.

Basalt is the most common extrusive igneous rock. Many volcanic islands, such as the Hawaiian Islands and Iceland, are composed mainly of basalt. Further, the upper layers of the oceanic crust consist of basalt. In the United States, large portions of central Oregon and Washington were the sites of

FIGURE 4.16   Basalt flows along the Columbia River above The Dalles, Oregon. Note the train for scale. (Photo by Michael Collier)

extensive basaltic outpourings (Figure 4.16). At some locations these once fluid basaltic flows have accumulated to thicknesses approaching 3 kilometers.

Gabbro   Gabbro is the intrusive equivalent of basalt (Figure 4.12). Like basalt, it tends to be dark green to black in color and composed primarily of pyroxene and calcium-rich plagioclase feldspar. Although gabbro is not a common constituent of the continental crust, it undoubtedly makes up a significant percentage of the oceanic crust.

Pyroclastic Rocks

Pyroclastic rocks are composed of fragments ejected during a volcanic eruption. One of the most common pyroclastic rocks, called *tuff,* is composed mainly of tiny ash-size fragments that were later cemented together (Figure 4.17). In situations where the ash particles remained hot enough to fuse, the rock is called *welded tuff.* Although welded tuff consists mostly of tiny glass shards, it may contain walnut-size pieces of pumice and other rock fragments.

Welded tuffs blanket vast portions of once volcanically active areas of the western United States. Some of these tuff deposits are hundreds of feet thick and extend for tens of miles from their source. Most formed millions of years ago as volcanic ash spewed from large volcanic structures (calderas) in an avalanche style, spreading laterally at speeds approaching 100 kilometers per hour. Early investigators of these deposits incorrectly classified them as rhyolite lava flows. Today we know that silica-rich lava is too viscous (thick) to flow more than a few miles from a vent.

Pyroclastic rocks composed mainly of particles larger than ash are called *volcanic breccia.* The particles in volcanic breccia can consist of streamlined fragments that solidified in air, blocks broken from the walls of the vent, crystals, and glass fragments.

Unlike most igneous rock names, such as granite and basalt, the terms *tuff* and *volcanic breccia* do not imply mineral composition. Thus, they are frequently used with a modifier, as, for example, rhyolite tuff.

Origin of Magma

Considerable evidence indicates that most magma originates in the uppermost mantle. It is also clear that plate tectonics plays a major role in generating most magma. The greatest quantities are produced at divergent plate boundaries in association with seafloor spreading, whereas

FIGURE 4.17   Outcrop of welded tuff (tan) interbedded with obsidian (black) near Shoshone, California. Tuff is composed mainly of ash-sized particles and may contain larger fragments of pumice or other volcanic rocks. (Photo by Breck P. Kent)

Students Sometimes Ask . . .

At a store, I saw a barbecue grill with material the clerk called "lava rock." Is this really a volcanic rock?

Not only is "lava rock" at your hardware store, it's also at home-improvement stores for use as a building and landscaping material and is frequently found in aquarium-supply stores. Geologists call this material *scoria,* which is a red or dark mafic rock characterized by a vesicular texture (full of holes). It's also called *volcanic cinder.* In gas barbecue grills, lava rock is used to absorb and reradiate heat to ensure even cooking.

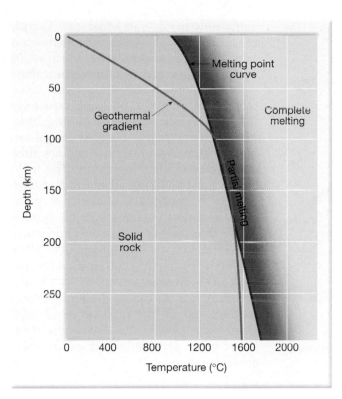

FIGURE 4.18 A highly schematic diagram illustrating a typical geothermal gradient (increase of temperature with depth) for the continental crust and upper mantle. Also illustrated is an idealized curve that depicts the melting point temperatures for mantle rocks. Notice that the geothermal gradient just crosses the partial melting curve for mantle material between depths of about 100 and 200 kilometers. In this zone the mantle is very near or at its melting temperature and in some tectonic settings partial melting occurs. At greater and shallower depths, however, the mantle should be completely solid. Keep in mind that the geothermal gradient varies somewhat from one tectonic setting to another.

Generating Magma from Solid Rock

Based on available scientific evidence, *Earth's crust and mantle are composed primarily of solid, not molten, rock.* Although the outer core is a fluid, this iron-rich material is very dense and remains deep within Earth. So what is the source of magma that produces igneous activity?

Geologists conclude that most magma originates when essentially solid rock, located in the crust and upper mantle, melts. The most obvious way to generate magma from solid rock is to raise the temperature above the rock's melting point.

Role of Heat What source of heat is sufficient to melt rock? Workers in underground mines know that temperatures get higher as they go deeper. Although the rate of temperature change varies from place to place, it *averages* between 20°C and 30°C per kilometer in the *upper* crust. This increase in temperature with depth is known as the **geothermal gradient** (Figure 4.18). Estimates indicate that the temperature at a depth of

100 kilometers ranges between 1200°C and 1400°C*. At these high temperatures, rocks in the upper mantle are near their melting points and in some tectonic settings partial melting may occur (Figure 4.18).

Role of Pressure If temperature were the only factor that determined whether or not rock melts, our planet would be a molten ball covered with a thin, solid outer shell. This, of course, is not the case. The reason is that pressure also increases with depth.

Melting, which is accompanied by an increase in volume, *occurs at higher temperatures at depth* because of greater confining pressure (Figure 4.19). Consequently, an increase in confining pressure causes an increase in the rock's melting temperature. Conversely, reducing confining pressure lowers a rock's melting temperature. When confining pressure drops enough, **decompression melting** is triggered. Such melting occurs where mantle rock ascends in zones of convective upwelling, thereby moving into regions of lower pressure. Decompression melting is responsible for generating magma along divergent plate boundaries (oceanic ridges) where plates are rifting apart (Figure 4.20). Below the ridge crest, hot mantle rock rises to replace the material that shifted horizontally away from the ridge axis. Decompression melting also occurs within ascending mantle plumes, such as the one responsible for the volcanic activity that created the Hawaiian Islands.

Role of Volatiles Another important factor affecting the melting temperature of rock is its water content. Water and other volatiles act as salt does to melt ice. That is, volatiles cause rock to melt at lower temperatures. Further, the effect of volatiles is magnified by increased pressure. Consequently, "wet" rock buried at depth has a much lower melting

---

*We will consider the heat sources for the geothermal gradient in Chapter 12.

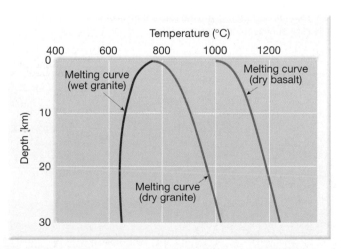

FIGURE 4.19 Idealized melting temperature curves. These curves portray the minimum temperatures required to melt rock within Earth's crust. Notice that dry granite and dry basalt melt at higher temperatures with increasing depth. By contrast, the melting temperature of wet granite actually decreases as the confining pressure increases.

lesser amounts form at subduction zones, where oceanic lithosphere descends into the mantle. Some igneous activity occurs far from plate boundaries, indicating that not all magma is produced in these relatively narrow zones.

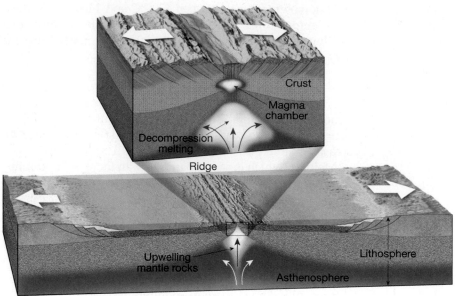

FIGURE 4.20 As hot mantle rock ascends, it continually moves into zones of lower pressure. This drop in confining pressure can trigger melting, even without additional heat.

sufficiently to generate some melt. Laboratory studies have shown that the melting point of basalt can be lowered by as much as 100°C by the addition of only 0.1 percent water.

When enough mantle-derived basaltic magma forms, it will buoyantly rise toward the surface. In a continental setting, basaltic magma may "pond" beneath crustal rocks, which have a lower density and are already near their melting temperature. This may result in some melting of the crust and the formation of a secondary, silica-rich magma.

In summary, magma can be generated under three sets of conditions: (1) *Heat* may be added; for example, a magma body from a deeper source intrudes and melts crustal rock; (2) *a decrease in pressure* (without the addition of heat) can result in *decompression melting*; and (3) the *introduction of volatiles* (principally water) can lower the melting temperature of mantle rock sufficiently to generate magma.

temperature than does "dry" rock of the same composition and under the same confining pressure (Figure 4.19). Therefore, in addition to a rock's composition, its temperature, depth (confining pressure), and water content determine whether it exists as a solid or liquid.

Volatiles play an important role in generating magma at convergent plate boundaries where cool slabs of oceanic lithosphere descend into the mantle (Figure 4.21). As an oceanic plate sinks, both heat and pressure drive water from the subducting crustal rocks. These volatiles, which are very mobile, migrate into the wedge of hot mantle that lies directly above. This process lowers the melting temperature of mantle rock

How Magmas Evolve

Because a large variety of igneous rocks exists, it is logical to assume that an equally large variety of magmas must also exist. However, geologists have observed that a single volcano may extrude lavas exhibiting quite different compositions (Figure 4.22). Data of this type led them to examine the possibility that magma might change (evolve) and thus become the parent to a variety of igneous rocks. To explore this idea, a pioneering investigation into the crystallization of magma was carried out by N. L. Bowen in the first quarter of the 20th century.

FIGURE 4.21 As an oceanic plate descends into the mantle, water and other volatiles are driven from the subducting crustal rocks. These volatiles lower the melting temperature of mantle rock sufficiently to trigger melting.

Bowen's Reaction Series and the Composition of Igneous Rocks

Recall that ice freezes at a single temperature, whereas magma crystallizes through at least 200°C of cooling. In a laboratory setting Bowen and his coworkers demonstrated that as a basaltic magma cools, minerals tend to crystallize in a systematic fashion based on their melting points. As shown in Figure 4.23, the first mineral to crystallize from a basaltic magma is the ferromagnesian mineral olivine. Further cooling generates calcium-rich plagioclase feldspar as well as pyroxene, and so forth down the diagram.

During the crystallization process, the composition of the liquid portion of the magma continually changes. For example, at the stage when about a third of the magma

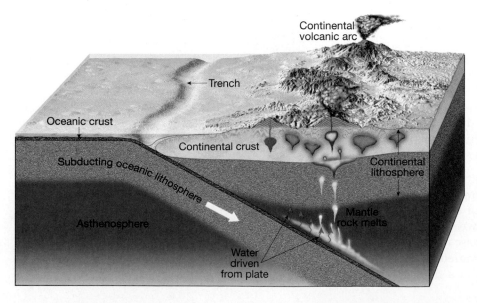

FIGURE 4.22 Ash and pumice ejected during a large eruption of Mount Mazama (Crater Lake). Notice the gradation from light-colored, silica-rich ash near the base to dark-colored rocks at the top. It is likely that prior to this eruption the magma began to segregate as the less dense, silica-rich magma migrated toward the top of the magma chamber. The zonation seen in the rocks resulted because a sustained eruption tapped deeper and deeper levels of the magma chamber. Thus, this rock sequence is an inverted representation of the compositional zonation in the magma body; that is, the magma from the top of the chamber erupted first and is found at the base of these ash deposits and vice versa. (Photo by E. J. Tarbuck)

has solidified, the melt will be nearly depleted of iron, magnesium, and calcium because these elements are constituents of the earliest-formed minerals. The removal of these elements from the melt will cause it to become enriched in sodium and potassium. Further, because the original basaltic magma contained about 50 percent silica (SiO_2), the crystallization of the earliest-formed mineral, olivine, which is only about 40 percent silica, leaves the remaining melt richer in (SiO_2). Thus, the silica component of the melt becomes enriched as the magma evolves.

Bowen also demonstrated that if the solid components of a magma remain in contact with the remaining melt, they will chemically react and evolve into the next mineral in the sequence shown in Figure 4.23. For this reason, this arrangement of minerals became known as **Bowen's reaction series** (Box 4.2). As you will see, in some natural settings the earliest-formed minerals can be separated from the melt, thus halting any further chemical reaction.

FIGURE 4.23 Bowen's reaction series shows the sequence in which minerals crystallize from a magma. Compare this figure to the mineral composition of the rock groups in Figure 4.11. Note that each rock group consists of minerals that crystallize in the same temperature range.

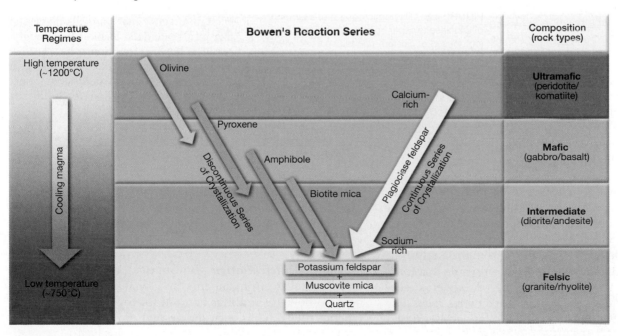

BOX 4.2 ▶ UNDERSTANDING EARTH

A Closer Look at Bowen's Reaction Series

Although it is highly idealized, Bowen's reaction series provides us with a visual representation of the order in which minerals crystallize from a magma of average composition (see Figure 4.23). This model assumes that the magma cools slowly at depth in an otherwise unchanging environment. Notice that Bowen's reaction series is divided into two branches—a discontinuous series and a continuous series.

Discontinuous Reaction Series.

The upper left branch of Bowen's reaction series indicates that as a magma cools, olivine is the first mineral to crystallize. Once formed, olivine will chemically react with the remaining melt to form the mineral pyroxene (see Figure 4.23). In this reaction, olivine, which is composed of individual silicon–oxygen tetrahedra, incorporates more silica into its structure, thereby linking its tetrahedra into single-chain structures of the mineral pyroxene. (Note: pyroxene has a lower crystallization temperature than olivine and is more stable at lower temperatures.) As the magma body cools further, the pyroxene crystals will in turn react with the melt to generate the double-chain structure of amphibole. This reaction will continue until the last mineral in this series, biotite mica, crystallizes. In nature, these reactions do not usually run to completion, so that various amounts of each of the minerals in the series may exist at any given time, and some minerals such as biotite may never form.

This branch of Bowen's reaction series is called a *discontinuous reaction series* because at each step a different silicate structure emerges. Olivine, the first mineral in the sequence, is composed of isolated tetrahedra, whereas pyroxene is composed of single chains, amphibole of double chains, and biotite of sheet structures.

Continuous Reaction Series.

The right branch of the reaction series, called the *continuous reaction series*, illustrates that calcium-rich plagioclase feldspar crystals react with the sodium ions in the melt to become progressively more sodium-rich (see Figure 4.23). Here the sodium ions diffuse into the feldspar crystals and displace the calcium ions in the crystal lattice. Often, the rate of cooling occurs rapidly enough to prohibit a complete replacement of the calcium ions by sodium ions. In these instances, the feldspar crystals will have calcium-rich interiors surrounded by zones that are progressively richer in sodium (Figure 4.B).

During the last stage of crystallization, after much of the magma has solidified,

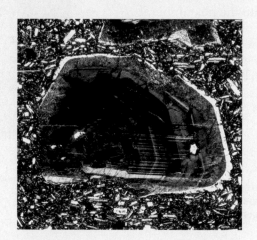

FIGURE 4.B Photomicrograph of a zoned plagioclase feldspar crystal. After this crystal (composed of calcium-rich feldspar) solidified, further cooling resulted in sodium ions displacing calcium ions. Because replacement was not complete, this feldspar crystal has a calcium-rich interior surrounded by zones that are progressively richer in sodium. (Photo courtesy of E. J. Tarbuck)

potassium feldspar forms. (Muscovite will form in pegmatites and other plutonic igneous rocks that crystallize at considerable depth.) Finally, if the remaining melt has excess silica, the mineral quartz will form.

Testing Bowen's Reaction Series.

During an eruption of Hawaii's Kilauea volcano in 1965, basaltic lava poured into a pit crater, forming a lava lake that became a natural laboratory for testing Bowen's reaction series. When the surface of the lava lake cooled enough to form a crust, geologists drilled into the magma and periodically removed samples that were quenched to preserve the melt, and minerals that were growing within it. By sampling the lava at successive stages of cooling, a history of crystallization was recorded.

As Bowen's reaction series predicts, olivine crystallized early but later ceased to form, and was partly reabsorbed into the cooling melt. (In a larger magma body that cooled more slowly, we would expect most, if not all, of the olivine to react with the melt and change to pyroxene.) Most important, the melt changed composition throughout the course of crystallization. In contrast to the original basaltic lava, which contained about 50 percent silica (SiO_2), the final melt contained more than 75 percent silica and had a composition similar to granite.

Although the lava in this setting cooled rapidly compared to rates experienced in deep magma chambers, it was slow enough to verify that minerals do crystallize in a systematic fashion that roughly parallels Bowen's reaction series. Further, had the melt been separated at any stage in the cooling process, it would have formed a rock with a composition much different from the original lava.

The diagram of Bowen's reaction series in Figure 4.23 depicts the sequence that minerals crystallize from a magma of average composition under laboratory conditions. Evidence that this highly idealized crystallization model approximates what can happen in nature comes from the analysis of igneous rocks. In particular, we find that minerals that form in the same general temperature regime on Bowen's reaction series are found together in the same igneous rocks. For example, notice in Figure 4.23 that the minerals quartz, potassium feldspar, and muscovite, which are located in the same region of Bowen's diagram, are typically found together as major constituents of the plutonic igneous rock *granite.*

Magmatic Differentiation Bowen demonstrated that minerals crystallize from magma in a systematic fashion. But how do Bowen's findings account for the great diversity of igneous rocks? It has been shown that, at one or more stages during crystallization, a separation of the solid and liquid components of a magma can occur. One example is called **crystal settling.** This process occurs when the earlier-formed minerals are denser (heavier) than the liquid portion and sink toward the bottom of the magma chamber, as shown in Figure 4.24. When the remaining melt solidifies—either in place or in another location if it migrates into fractures in the surrounding rocks—it will form a rock with a chemical composition much different from the parent magma (Figure 4.24). The formation of one or more secondary magmas from a single parent magma is called **magmatic differentiation.**

A classic example of magmatic differentiation is found in the Palisades Sill, which is a 300-meter-thick tabular mass of dark igneous rock exposed along the west bank of the lower Hudson River. Because of its great thickness and subsequent slow rate of solidification, crystals of olivine (the first mineral to form) sank and make up about 25 percent of the lower portion of the Palisades Sill. By contrast, near the top of this igneous body, where the last melt crystallized, olivine represents only 1 percent of the rock mass.[2]

At any stage in the evolution of a magma, the solid and liquid components can separate into two chemically distinct units. Further, magmatic differentiation within the secondary melt can generate additional chemically distinct fractions. Consequently, magmatic differentiation and separation of the solid and liquid components at various stages of crystallization can produce several chemically diverse magmas and ultimately a variety of igneous rocks (Figure 4.24).

Assimiliation and Magma Mixing

Bowen successfully demonstrated that through magmatic differentiation, a parent magma can generate several mineralogically different igneous rocks. However, more recent work indicates that magmatic differentiation cannot by itself account for the entire compositional spectrum of igneous rocks.

Once a magma body forms, its composition can change through the incorporation of foreign material. For example, as magma migrates upward, it may incorporate some of the surrounding host rock, a process called **assimilation** (Figure 4.25). This process may operate in a near-surface environment where rocks are brittle. As the magma pushes upward, stress causes numerous cracks in the overlying rock. The force of the injected magma is often sufficient to dislodge blocks of "foreign" rock and incorporate them into the magma body. In deeper environments, the magma may be hot enough to simply melt and assimilate some of the surrounding host rock, which is near its melting temperature.

Another means by which the composition of a magma body can be altered is called **magma mixing.** This process occurs whenever one magma body intrudes another (Figure 4.25). Once combined, convective flow may stir the two

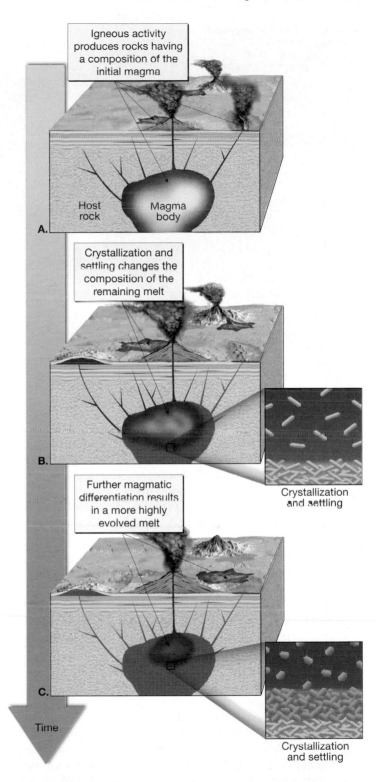

FIGURE 4.24   Illustration of how a magma evolves as the earliest-formed minerals (those richer in iron, magnesium, and calcium) crystallize and settle to the bottom of the magma chamber, leaving the remaining melt richer in sodium, potassium, and silica (SiO_2) **A.** Emplacement of a magma body and associated igneous activity generates rocks having a composition similar to that of the initial magma. **B.** After a period of time, crystallization and settling change the composition of the melt, while generating rocks having a composition quite different than the original magma. **C.** Further magmatic differentiation results in another more highly evolved melt with its associated rock types.

---

[2]Recent studies indicate that this igneous body was produced by multiple injections of magma and represents more than just a simple case of crystal settling.

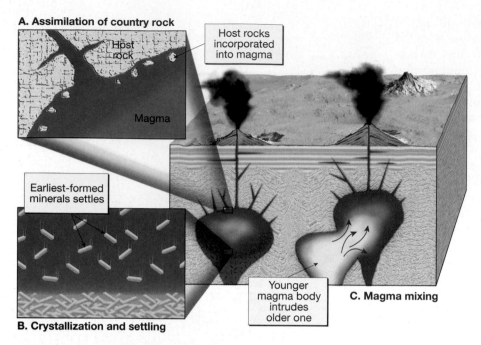

A. Assimilation of country rock

Host rock

Host rocks incorporated into magma

Magma

Earliest-formed minerals settles

B. Crystallization and settling

Younger magma body intrudes older one

C. Magma mixing

FIGURE 4.25 This illustration shows three ways that the composition of a magma body may be altered: magma mixing, assimilation of host rock, and crystallization and settling (magmatic differentiation).

magmas and generate a fluid with an intermediate composition. Magma mixing may occur during the ascent of two chemically distinct magma bodies as the more buoyant mass overtakes the slower moving mass.

In summary, Bowen successfully demonstrated that through magmatic differentiation, a single parent magma can generate several mineralogically different igneous rocks. Thus, this process, in concert with magma mixing and contamination by crustal rocks, accounts in part for the great diversity of magmas and igneous rocks. We will next look at another important process, partial melting, which also generates magmas having varying compositions.

Partial Melting and Magma Composition

Recall that the crystallization of a magma occurs over a temperature range of at least 200°C. As you might expect, melting, the reverse process, spans a similar temperature range. As rock begins to melt, those minerals with the lowest melting temperatures are the first to melt. Should melting continue, minerals with higher melting points begin to melt and the composition of the magma steadily approaches the overall composition of the rock from which it was derived. Most often, however, melting is not complete. The incomplete melting of rocks is known as **partial melting,** a process that produces most, if not all, magma.

Notice in Figure 4.23 that rocks with a granitic composition are composed of minerals with the lowest melting (crystallization) temperatures—namely, quartz and potassium feldspar. Also note that as we move up Bowen's reaction

series, the minerals have progressively higher melting temperatures and that olivine, which is found at the top, has the highest melting point. When a rock undergoes partial melting, it will form a melt that is enriched in ions from minerals with the lowest melting temperatures. The unmelted crystals are those of minerals with higher melting temperatures. Separation of these two fractions would yield a melt with a chemical composition that is richer in silica and nearer to the granitic end of the spectrum than the rock from which it was derived.

Formation of a Basaltic Magma

Most basaltic magmas probably originate from partial melting of the ultramafic rock *peridotite,* the major constituent of the upper mantle. Basaltic magmas that originate from direct melting of mantle rocks are called *primary* magmas because they have not yet evolved. Melting to produce these mantle-derived magmas may be triggered by a reduction in confining pressure (decompression melting). This can occur, for example, where mantle rock ascends as part of slow-moving convective flow at mid-ocean ridges (see Figure 4.20). Recall that basaltic magmas are also generated at subduction zones, where water driven from the descending slab of oceanic crust promotes partial melting of mantle rocks (see Figure 4.21).

Because most basaltic magma forms between about 50 and 250 kilometers (30 and 150 miles) below the surface, we might expect that this material would cool and crystallize at depth. However, as basaltic magma migrates upward, the confining pressure steadily diminishes and reduces its melting temperature. As you will see in the next chapter, environments exist where basaltic magmas ascend rapidly enough that the heat loss to the surrounding environment is offset by the drop in the melting temperature. Consequently, large outpourings of basaltic lavas are common at Earth's surface. In some situations, however, basaltic magmas that are comparatively dense will pond beneath crustal rocks and crystallize at depth.

Formation of Andesitic and Granitic Magmas

If partial melting of mantle rocks generates basaltic magmas, what is the source of the magma that generates andesitic and granitic rocks? Recall that intermediate and felsic magmas are not erupted from volcanoes in the deep-ocean basins; rather, they are found only within, or adjacent to, the continental margins (Figure 4.26). This is strong evidence that interactions between mantle-derived basaltic magmas and more silica-rich components of Earth's crust generate these magmas. For example, as a basaltic magma

FIGURE 4.26 1996 eruption of Mount Ruapehu, Tongariro National Park, New Zealand. Volcanoes that border the Pacific Ocean are fed largely by magmas that have intermediate or felsic compositions. These silica-rich magmas often erupt explosively, generating large plumes of volcanic dust and ash. (Photo by Tui De Roy/Minden Pictures)

migrates upward, it may melt and assimilate some of the crustal rocks through which it ascends. The result is the formation of a more silica-rich magma of andesitic composition (intermediate between basaltic and granitic).

Andesitic magma may also evolve from a basaltic magma by the process of magmatic differentiation. Recall from our discussion of Bowen's reaction series that as basaltic magma solidifies, it is the silica-poor ferromagnesian minerals that crystallize first. If these iron-rich components are separated from the liquid by crystal settling, the remaining melt, now enriched in silica, will have a composition more akin to andesite. These evolved (changed) magmas are termed *secondary magmas*.

Granitic rocks are found in much too large a quantity to be generated solely from the magmatic differentiation of primary basaltic magmas. Most likely they are the end product of the crystallization of an andesitic magma, or the product of partial melting of silica-rich continental rocks. The heat to melt crustal rocks often comes from hot mantle-derived basaltic magmas that formed above a subducting plate and were then emplaced within the crust.

Granitic melts are higher in silica and thus more viscous (thicker) than other magmas. Therefore, in contrast to basaltic magmas that frequently produce vast outpourings of lava, granitic magmas usually lose their mobility before reaching the surface and tend to produce large plutonic structures. On those occasions when silica-rich magmas do reach the surface, explosive pyroclastic eruptions, such as those from Mount St. Helens, are the rule.

In summary, Bowen's reaction series is a useful simplified guide to understanding the partial melting process. In general, the low temperature minerals toward the bottom of Bowen's reaction series melt first and produce magma that is richer in silica (less mafic) than the parent rock. Thus, partial melting of ultramafic rocks in the mantle produces the mafic basalts that form the oceanic crust. In addition, partial melting of basaltic rocks will generate an intermediate (andesitic) magma commonly associated with volcanic arcs.

Summary

- *Igneous rocks* form when *magma cools* and solidifies. *Extrusive*, or *volcanic*, igneous rocks result when *lava* cools at the surface. Magma that solidifies at depth produces *intrusive*, or *plutonic*, igneous rocks.

- As magma cools, the ions that compose it arrange themselves into orderly patterns during a process called *crystallization*. Slow cooling results in the formation of rather large crystals. Conversely, when cooling occurs rapidly,

121

the outcome is a solid mass consisting of tiny intergrown crystals. When molten material is quenched instantly, a mass of unordered atoms, referred to as *glass*, forms.
- Igneous rocks are most often classified by their *texture* and *mineral composition.*
- The texture of an igneous rock refers to the overall appearance of the rock based on the size and arrangement of its interlocking crystals. The most important factor affecting texture is the rate at which magma cools. Common igneous rock textures include *aphanitic,* with grains too small to be distinguished without the aid of a microscope; *phaneritic,* with intergrown crystals that are roughly equal in size and large enough to be identified with the unaided eye; *porphyritic,* which has large crystals (*phenocrysts*) interbedded in a matrix of smaller crystals (*groundmass*); and *glassy.*
- The mineral composition of an igneous rock is the consequence of the chemical makeup of the parent magma and the environment of crystallization. Igneous rocks are divided into broad compositional groups based on the percentage of dark and light silicate minerals they contain. *Felsic rocks* (e.g., granite and rhyolite) are composed mostly of the light-colored silicate minerals potassium feldspar and quartz. Rocks of *intermediate* composition, (e.g., andesite and diorite) contain plagioclase feldspar and amphibole. *Mafic rocks* (e.g., basalt and gabbro) contain abundant olivine, pyroxene, and calcium feldspar. They are high in iron, magnesium, and calcium, low in silicon, and are dark gray to black in color.
- The mineral makeup of an igneous rock is ultimately determined by the chemical composition of the magma from which it crystallizes. N. L. Bowen discovered that as magma cools in the laboratory, those minerals with higher melting points crystallize before minerals with lower melting points. *Bowen's reaction series* illustrates the sequence of mineral formation within magma.
- During the crystallization of magma, if the earlier-formed minerals are denser than the liquid portion, they will settle to the bottom of the magma chamber during a process called *crystal settling*. Owing to the fact that crystal settling removes the earlier-formed minerals, the remaining melt will form a rock with a chemical composition much different from the parent magma. The process of developing more than one magma type from a common magma is called *magmatic differentiation.*
- Once a magma body forms, its composition can change through the incorporation of foreign material, a process termed *assimilation,* or by *magma mixing.*
- Magma originates from essentially solid rock of the crust and mantle. In addition to a rock's composition, its temperature, depth (confining pressure), and water content determine whether it exists as a solid or liquid. Thus, magma can be generated by *raising a rock's temperature,* as occurs when a hot mantle plume "ponds" beneath crustal rocks. A *decrease in pressure* can cause *decompression melting*. Further, the *introduction of volatiles* (water) can lower a rock's melting point sufficiently to generate magma. Because melting is generally not complete, a process called *partial melting* produces a melt made of the lowest-melting-temperature minerals, which are higher in silica than the original rock. Thus, magmas generated by partial melting are nearer to the felsic end of the compositional spectrum than are the rocks from which they formed.

Review Questions

1. What is magma?
2. How does lava differ from magma?
3. How does the rate of cooling influence the crystallization process?
4. In addition to the rate of cooling, what two other factors influence the crystallization process?
5. The classification of igneous rocks is based largely on two criteria. Name these criteria.
6. The statements that follow relate to terms describing igneous rock textures. For each statement, identify the appropriate term.
   a. openings produced by escaping gases
   b. obsidian exhibits this texture
   c. a matrix of fine crystals surrounding phenocrysts
   d. crystals are too small to be seen without a microscope
   e. a texture characterized by two distinctly different crystal sizes
   f. coarse-grained, with crystals of roughly equal size
   g. exceptionally large crystals exceeding 1 centimeter in diameter

7. Why are the crystals in pegmatites so large?
8. What does a porphyritic texture indicate about an igneous rock?
9. How are granite and rhyolite different? In what way are they similar?
10. Compare and contrast each of the following pairs of rocks:
    a. granite and diorite
    b. basalt and gabbro
    c. andesite and rhyolite
11. How do tuff and volcanic breccia differ from other igneous rocks such as granite and basalt?
12. What is the geothermal gradient?
13. Describe the three conditions that are thought to cause rock to melt.
14. What is magmatic differentiation? How might this process lead to the formation of several different igneous rocks from a single magma?
15. Relate the classification of igneous rocks to Bowen's reaction series.

16. What is partial melting?

17. How does the composition of a melt produced by partial melting compare with the composition of the parent rock?

18. How are most basaltic magmas generated?

19. Basaltic magma forms at great depth. Why doesn't most of it crystallize as it rises through the relatively cool crust?

20. Why are rocks of intermediate (andesitic) and felsic (granitic) composition generally *not* found in the ocean basins?

Key Terms

andesitic composition (p. 107)
aphanitic texture (p. 104)
assimilation (p. 119)
basaltic composition (p. 107)
Bowen's reaction series (p. 117)
crystallization (p. 103)
crystal settling (p. 119)
decompression melting (p. 115)
extrusive (p. 102)

felsic (p. 107)
fragmental texture (p. 106)
geothermal gradient (p. 115)
glass (p. 104)
glassy texture (p. 105)
groundmass (p. 105)
granitic composition (p. 107)
igneous rocks (p. 102)
intermediate composition (p. 107)
intrusive (p. 103)

lava (p. 102)
mafic (p. 107)
magma (p. 102)
magma mixing (p. 119)
magmatic differentiation (p. 119)
melt (p. 103)
partial melting (p. 120)
pegmatite (p. 107)
pegmatitic texture (p. 107)
phaneritic texture (p. 105)
phenocryst (p. 105)
plutonic (p. 103)

porphyritic texture (p. 105)
porphyry (p. 105)
pyroclastic texture (p. 106)
texture (p. 104)
ultramafic (p. 108)
vesicular texture (p. 104)
volatiles (p. 103)
volcanic (p. 102)

Web Resources

The *Earth* Website uses the resources and flexibility of the Internet to aid in your study of the topics in this chapter. Written and developed by geology instructors, this site will help improve your understanding of geology. Visit **http://www .prenhall.com/ tarbuck** and click on the cover of *Earth 9e* to find:

- Online review quizzes.
- Critical thinking exercises.
- Links to chapter-specific Web resources.
- Internet-wide key-term searches.

http://www.prenhall.com/tarbuck

GEODe: Earth

GEODe: Earth makes studying faster and more effective by reinforcing key concepts using animation, video, narration, interactive exercises and practice quizzes. A copy is included with every copy of *Earth*.

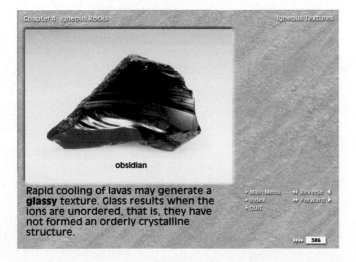

Rapid cooling of lavas may generate a **glassy** texture. Glass results when the ions are unordered, that is, they have not formed an orderly crystalline structure.

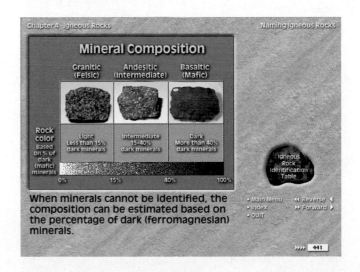

When minerals cannot be identified, the composition can be estimated based on the percentage of dark (ferromagnesian) minerals.

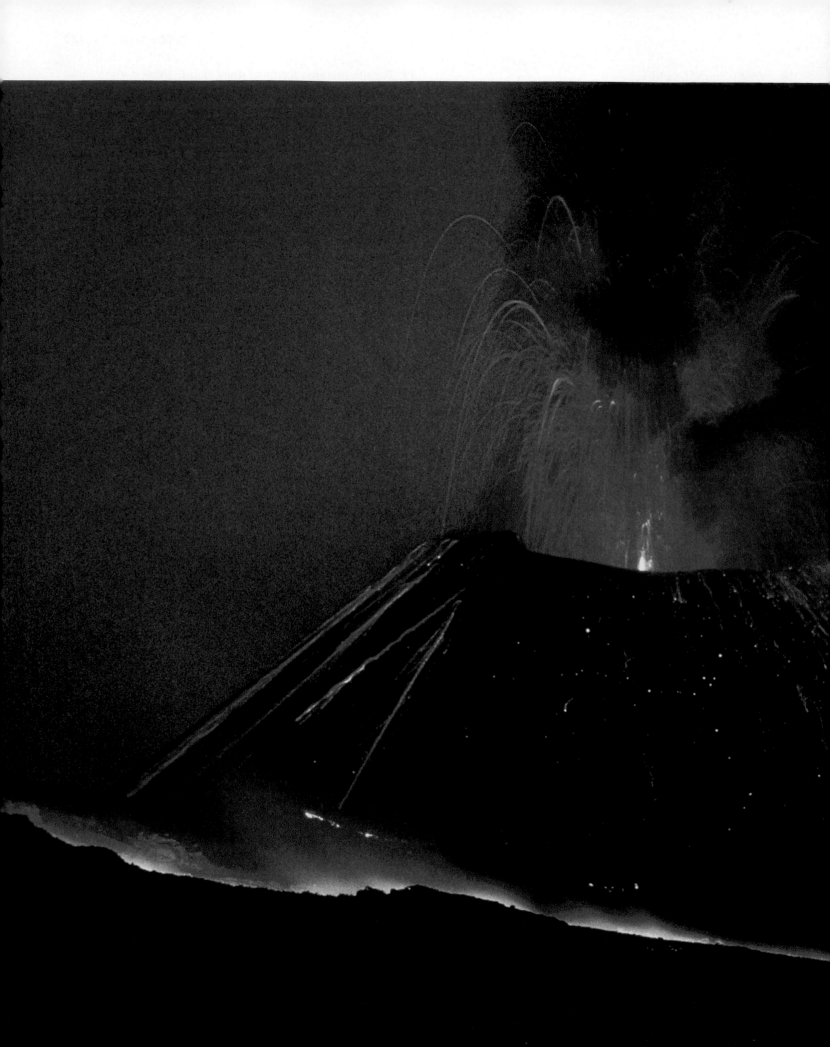

Volcanoes and Other Igneous Activity

A recent eruption of Italy's Mount Etna. (Photo by Art Wolfe)

On Sunday, May 18, 1980, the largest volcanic eruption to occur in North America in historic times transformed a picturesque volcano into a decapitated remnant (Figure 5.1). On this date in southwestern Washington State, Mount St. Helens erupted with tremendous force. The blast blew out the entire north flank of the volcano, leaving a gaping hole. In one brief moment, a prominent volcano whose summit had been more than 2900 meters (9500 feet) above sea level was lowered by more than 400 meters (1350 feet).

The event devastated a wide swath of timber-rich land on the north side of the mountain (Figure 5.2). Trees within a 400-square-kilometer area lay intertwined and flattened, stripped of their branches and appearing from the air like toothpicks strewn about. The accompanying mudflows carried ash, trees, and water-saturated rock debris 29 kilometers (18 miles) down the Toutle River. The eruption claimed 59 lives, some dying from the intense heat and the suffocating cloud of ash and gases, others from being hurled by the blast, and still others from entrapment in the mudflows.

The eruption ejected nearly a cubic kilometer of ash and rock debris. Following the devastating explosion, Mount St. Helens continued to emit great quantities of hot gases and ash. The force of the blast was so strong that some ash was propelled more than 18,000 meters (over 11 miles) into the stratosphere. During the next few days, this very fine-grained material was carried around Earth by strong upper-air winds. Measurable deposits were reported in Oklahoma and Minnesota, with crop damage into central Montana. Meanwhile, ash fallout in the immediate vicinity exceeded 2 meters in depth. The air over Yakima, Washington (130 kilometers to the east), was so filled with ash that residents experienced midnight-like darkness at noon.

Not all volcanic eruptions are as violent as the 1980 Mount St. Helens event. Some volcanoes, such as Hawaii's Kilauea volcano, typically generate relatively quiet outpourings of fluid lavas. These "gentle" eruptions are not without some fiery displays; occasionally fountains of incandescent lava spray hundreds of meters into the air. During Kilauea's most recent active phase, which began in 1983, more than 180 homes and a national park visitor center have been destroyed.

Why do volcanoes like Mount St. Helens erupt explosively, whereas others like Kilauea are relatively quiet? Why do volcanoes occur in chains like the Aleutian Islands or the Cascade Range? Why do some volcanoes form on the ocean floor, while others occur on the continents? This chapter will deal with these and other questions as we explore the nature and movement of magma and lava (Figure 5.3).

The Nature of Volcanic Eruptions

GEODe
EARTH
Volcanoes and Other Igneous Activity
▶ The Nature of Volcanic Eruptions

Volcanic activity is commonly perceived as a process that produces a picturesque, cone-shaped structure that periodically erupts in a violent manner, like Mount St. Helens (Box 5.1). Although some eruptions may be very explosive, many are not. What determines whether a volcano extrudes magma violently or "gently"? The primary factors include the magma's *composition*, its *temperature*, and the amount of *dissolved gases* it contains. To varying degrees, these factors affect the magma's mobility, or **viscosity** (*viscos* = sticky). The more viscous the material, the greater its resistance to flow. (For example, syrup is more viscous than water.) Magma associated with an explosive eruption may be five times more viscous than magma that is extruded in a quiescent manner.

Factors Affecting Viscosity

The effect of temperature on viscosity is easily seen. Just as heating syrup makes it more fluid (less viscous), the

FIGURE 5.1 Before-and-after photographs show the transformation of Mount St. Helens caused by the May 18, 1980, eruption. The dark area in the "after" photo is debris-filled Spirit Lake, partially visible in the "before" photo. (Photos courtesy of U.S. Geological Survey)

FIGURE 5.2   Douglas fir trees were snapped off or uprooted by the lateral blast of Mount St. Helens on May 18, 1980. (Large photo by Lyn Topinka/AP Photo/U.S. Geological Survey; Inset photo by John M. Burnley/Photo Researchers, Inc.)

FIGURE 5.3   A river of basaltic lava from the January 17, 2002, eruption of Mount Nyiragongo destroyed many homes in Goma, Congo. (Photo by Marco Longari/CORBIS)

mobility of lava is strongly influenced by temperature. As lava cools and begins to congeal, its mobility decreases and eventually the flow halts.

A more significant factor influencing volcanic behavior is the chemical composition of the magma. Recall that a major difference among various igneous rocks is their silica (SiO_2) content (Table 5.1). Magmas that produce mafic rocks such as basalt contain about 50 percent silica, whereas magmas that produce felsic rocks (granite and its extrusive equivalent, rhyolite) contain more than 70 percent silica. The intermediate rock types—andesite and diorite—contain about 60 percent silica.

A magma's viscosity is directly related to its silica content. In general, the more silica in magma, the greater its viscosity. The flow of magma is impeded because silica structures link together into long chains, even before crystallization begins. Consequently, because of their high silica content, rhyolitic (felsic) lavas are very viscous and tend to form comparatively short, thick flows. By contrast, basaltic lavas which contain less silica are relatively fluid and have been known to travel 150 kilometers (90 miles) or more before congealing—in other words, crystallizing to become an igneous rock.

The amount of **volatiles** (the gaseous components of magma, mainly water) contained in magma also affects its mobility. Other factors being equal, water dissolved in the magma tends to increase fluidity because it reduces polymerization (formation of long silicate chains) by breaking silicon–oxygen bonds. Therefore, in near-surface environments the loss of gases renders magma (lava) more viscous.

Why Do Volcanoes Erupt?

You learned in Chapter 4 that most magma is generated by partial melting in the upper asthenosphere at depths of about 100 kilometers (60 miles). Once formed, melt which is less dense than the surrounding rock will buoyantly rise toward the surface (Figure 5.4). Because the density of crustal rocks decreases toward the surface, ascending magma may reach a level where the rocks above are less dense. Should this occur, the molten material begins to collect, forming a magma chamber where it may reside for an extended period before it either solidifies, or some of it ascends further. Only a fraction of magma generated at depth ever reaches the surface.

BOX 5.1 ▶ UNDERSTANDING EARTH

Anatomy of an Eruption

The events leading to the May 18, 1980, eruption of Mount St. Helens began about two months earlier as a series of minor Earth tremors centered beneath the awakening mountain (Figure 5.A, part A). The tremors were caused by the upward movement of magma within the mountain. The first volcanic activity took place a week later, when a small amount of ash and steam rose from the summit. Over the next several weeks, sporadic eruptions of varied intensity occurred. Prior to the main eruption, the primary concern had been the potential hazard of mudflows. These moving lobes of saturated soil and rock are created as ice and snow melt from the heat emitted from magma within the volcano.

The only warning of a potential eruption was a bulge on the volcano's north flank (Figure 5.A, part B). Careful monitoring of this dome-shaped structure indicated a very slow but steady growth rate of a few meters per day. If the growth rate of the bulge changed appreciably, an eruption might quickly follow. Unfortunately, no such variation was detected prior to the explosion. In fact, the seismic activity decreased during the two days preceding the huge blast.

Dozens of scientists were monitoring the mountain when it exploded. "Vancouver, Vancouver, this is it!" was the only warning—and last words from one scientist—that preceded the unleashing of tremendous quantities of pent-up gases. The trigger was a medium-sized earthquake. Its vibrations sent the north slope of the cone plummeting into the Toutle River, removing the overburden that had trapped the magma below (Figure 5.A, part C). With the pressure reduced, the water in the magma vaporized and expanded, causing the mountainside to rupture like an overheated steam boiler. Because the eruption originated around the bulge, several hundred meters below the summit, the initial blast was directed laterally rather than ver-

tically. Had the full force of the eruption been upward, far less destruction would have occurred.

Mount St. Helens is one of 15 large volcanoes and innumerable smaller ones that comprise the Cascade Range, which extends from British Columbia to northern

California. Eight of the largest volcanoes have been active in the past few hundred years. Of the remaining seven active volcanoes, the most likely to erupt again are Mount Baker and Mount Rainier in Washington, Mount Shasta and Lassen Peak in California, and Mount Hood in Oregon.

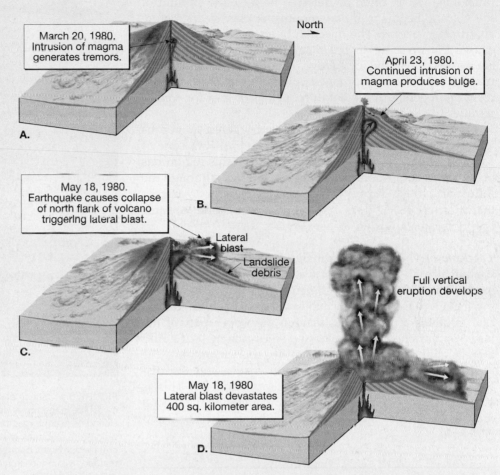

FIGURE 5.A Idealized diagrams showing the events in the May 18, 1980, eruption of Mount St. Helens. **A.** First, a sizable earthquake recorded on Mount St. Helens indicates that renewed volcanic activity is possible. **B.** Alarming growth of a bulge on the north flank suggests increasing magma pressure below. **C.** Triggered by an earthquake, a giant landslide reduced the confining pressure on the magma body and initiated an explosive lateral blast. **D.** Within seconds a large vertical eruption sent a column of volcanic ash to an altitude of about 18 kilometers (11 miles). This phase of the eruption continued for more than 9 hours.

TABLE 5.1	Magmas Have Different Compositions, Which Cause Their Properties to Vary		
Composition	Silica Content	Viscosity	Gas Content
Mafic (Basaltic) Magma	Least (~50%)	Least	Least (1–2%)
Intermediate (Andesitic) magma	Intermediate (~60%)	Intermediate	Intermediate (3–4%)
Felsic (Rhyolitic) magma	Most (~70%)	Greatest	Most (4–6%)

Triggering a Volcanic Eruption   One of the simplest mechanisms for triggering a volcanic eruption is the arrival of a new batch of melt into a near surface magma reservoir. This phenomenon can often be detected because the summit of the volcano begins to inflate months, or even years, before an eruption begins. The injection of a fresh supply of melt causes the pressure in the magma chamber to rise until cracks develop in the rock above. This, in turn, mobilizes the magma, which quickly moves upward along the fractured rock, often generating quiet outpourings of lava for weeks, months, or even years.

This mechanism is thought to have triggered numerous large eruptions of basaltic lava along the East Rift Zone of Kilauea. The most recent eruptive phase in this region has been almost continuous since it began in 1983. Although the influx of a fresh supply of magma is thought to trigger many eruptions, it does not explain why some volcanic eruptions are explosive.

The Role of Volatiles in Explosive Eruptions   All magmas contain a small percentage of water and other volatiles that are held in solution by the immense pressure of the overlying rock. As a magma body evolves, the earlier formed iron-rich minerals tend to sink, leaving the upper part of the reservoir enriched in volatiles and silica-rich melts. Should this volatile-rich magma rise, or the rocks confining the magma fail, a reduction in pressure ensues and the dissolved gases begin to expand and form bubbles. This is analogous to opening a warm soda and allowing the carbon dioxide bubbles to escape. At temperatures of 1000°C and

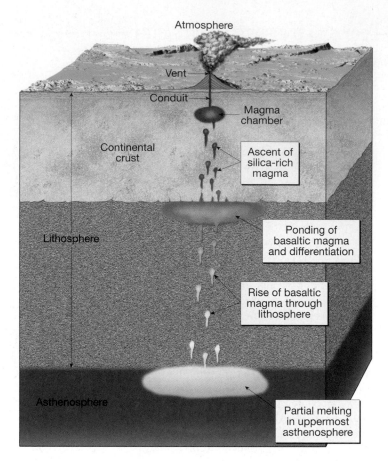

FIGURE 5.4   Schematic drawing showing the movement of magma from its source in the upper asthenosphere through the continental crust. During its ascent, mantle-derived basaltic magma evolves through the process of magmatic differentiation and by incorporating continental crust which has a comparatively low melting temperature. Thus, magmas that feed volcanoes in a continental setting tend to be silica-rich (viscous) and have a high gas content.

low near-surface pressures, these gases can quickly expand to occupy hundreds of times their original volumes.

Very fluid basaltic magmas allow the expanding gases to migrate upward and escape from the vent with relative ease. As they escape, the gases may propel incandescent lava hundreds of meters into the air, producing lava fountains (Figure 5.5). Although spectacular, such fountains are mostly harmless and not generally associated with major explosive events that cause great loss of life and property. Rather, eruptions of fluid basaltic lavas, such as those that occur in Hawaii, are generally quiescent.

At the other extreme, highly viscous, silica-rich magmas may produce explosive clouds of hot ash and gases that evolve into buoyant plumes called **eruption columns** that extend thousands of meters into the atmosphere (Figure 5.6). When viscous magma ascends, the developing gas bubbles do not easily escape but rather cause the molten material to expand resulting in a gradual increase in the internal pressure. Should the pressure exceed the strength of the overlying rock, fracturing occurs. The resulting breaks lead to a further drop in confining pressure which in turn causes even more gas bubbles to form. This chain-reaction often generates an

---

$\mathcal{S}$tudents $\mathcal{S}$ometimes $\mathcal{A}$sk . . .

After all the destruction during the eruption of Mount St. Helens, what does the area look like today?

The area continues to make a slow recovery. Surprisingly, many organisms survived the blast, including animals that live underground and plants (particularly those protected by snow or near streams, where erosion quickly removed the ash). Others have adaptations that allow them to quickly repopulate devastated areas. Fully 20 years after the blast, plants have revegetated the area, first-growth forests are beginning to be established, and many animals have returned. Once the old-growth forest is complete (in a few hundred years), it might be hard to find much evidence of the destruction, other than a thick weathered ash layer in the soil.

The volcano itself is rebuilding, too. A large lava dome is forming inside the summit crater, suggesting that the mountain will build up again. Many volcanoes similar to Mount St. Helens exhibit this behavior: rapid destruction followed by slow rebuilding. If you really want to see what it looks like, go to the Mount St. Helens National Volcanic Home Page at **http://www.fs.fed.us/gpnf/mshnvm/**, where they have a "volcanocam" with real-time images of the mountain.

relative ease as compared to more highly evolved silica-rich magmas. This explains the contrast between "gentle" outflows of fluid basaltic lavas in Hawaii with the explosive and sometimes catastrophic eruptions of viscous lavas from volcanoes such as Mount St. Helens (1980), Mount Pinatubo in the Philippines (1991), and Soufriere Hills on the island of Montserrat (1995).

Materials Extruded during an Eruption

 GEODe Volcanoes and Other Igneous Activity
▶ Materials Extruded during an Eruption

Volcanoes extrude lava, large volumes of gas, and pyroclastic materials (broken rock, lava "bombs," fine ash, and dust). In this section we will examine each of these materials.

Lava Flows

The vast majority of lava on Earth, more than 90 percent of the total volume, is estimated to be basaltic in composition. Andesites and other lavas of intermediate composition account

FIGURE 5.5   Fluid basaltic lava erupting from Kilauea Volcano, Hawaii. (Photo by Douglas Peebles)

explosive eruption, in which magma is literally blown into very fine fragments (ash) that are carried to great heights by the rising hot gases. (As exemplified by the 1980 eruption of Mount St. Helens, a drop in the confining pressure can also be triggered by the collapse of a volcano's flank.)

As magma in the uppermost portion of the magma chamber is forcefully ejected by the escaping gases, pressure on the molten rock directly below drops. Thus, rather than a single "bang," volcanic eruptions are really a series of explosions. This process might logically continue until the magma chamber is emptied, much like a geyser empties itself of water (see Chapter 17). However, this generally does not happen. Only within the upper portion of the magma body is the gas content sufficient to trigger a steam-and-ash explosion.

Most explosive eruptions are followed by the quiet emission of "degassed" lavas. For example, during the six years following the 1980 eruption of Mount St. Helens, episodic extrusions of lava built a large dome on the crater floor. A second dome-building event began in October of 2004. Although these eruptive phases produced some ash plumes, they were benign compared to the May 18, 1980, explosion.

To summarize, the viscosity of magma, plus the quantity of dissolved gases and the ease with which they can escape, determines to a large extent the nature of a volcanic eruption. In general, hot basaltic magmas contain a smaller gaseous component and permit these gases to escape with

FIGURE 5.6   Steam and ash eruption column from Mount Augustine, Cook Inlet, Alaska. (Photo by Steve Kaufman/Peter Arnold, Inc.)

A.

B.

FIGURE 5.7 **A.** Typical slow-moving aa flow **B.** Typical pahoehoe (ropy) lava . Both of these lava flows occurred in association with Hawaii's Kilauea Volcano. (Photo **A** by J. D. Griggs, U.S. Geological Survey and photo **B** by Doug Perrine/DRK)

for most of the rest, while silica-rich rhyolitic flows make up as little as one percent of the total. Recent eruptions of basaltic lavas from two Hawaiian volcanoes, Mauna Loa and Kilauea, had volumes that ranged up to about 0.5 cubic kilometer. One of the largest flows of basaltic lava in historic times came from Iceland's Laki fissure in 1783. The volume of lava produced during this eruption measured 12 cubic kilometers (nearly 3 cubic miles), and some of the lava traveled as far as 88 kilometers (55 miles) from its source. Some prehistoric eruptions, such as those that built the Columbia Plateau in the Pacific Northwest, were even larger. One flow of basaltic lava exceeded 1200 cubic kilometers (almost 300 cubic miles). Such a volume would be sufficient to build three volcanoes the size of Italy's Mount Etna, one of Earth's largest cones.

Because of their lower silica content, hot basaltic lavas are usually very fluid. They flow in thin, broad sheets or streamlike ribbons. On the island of Hawaii, such lavas have been clocked at 30 kilometers (19 miles) per hour down steep slopes. However, flow rates of 10 to 300 meters (30 to 1000 feet) per hour are more common. By contrast, the movement of silica-rich (rhyolitic) lava may be too slow to perceive. Furthermore, most rhyolitic lavas are thick and seldom travel more than a few kilometers from their vents. As you might expect, andesitic lavas, which are intermediate in composition, exhibit characteristics that are between the extremes.

Aa and Pahoehoe Flows Two types of lava flows are known by their Hawaiian names. The most common of these, **aa** (pronounced ah-ah) **flows,** have surfaces of rough jagged blocks with dangerously sharp edges and spiny projections (Figure 5.7A). Crossing an aa flow can be a trying and miserable experience. By contrast, **pahoehoe** (pronounced pah-hoy-hoy) **flows** exhibit smooth surfaces that often resemble the twisted braids of ropes (Figure 5.7B). Pahoehoe means "on which one can walk."

Although aa and pahoehoe lavas can erupt from the same vent, pahoehoe flows have higher temperatures than aa

flows. Consequently, pahoehoe flows are more fluid than lavas that form aa flows. In addition, pahoehoe lavas sometimes transform into aa flows. (Note, however, that aa flows cannot revert to pahoehoe.)

One of the factors thought to facilitate the change from pahoehoe to aa is cooling that occurs as the flow moves from the vent. Cooling promotes bubble formation which, as stated previously, increases viscosity. Escaping gas bubbles produce numerous voids and sharp spines in the surface of the congealing lava. As the molten interior advances, the outer crust is broken further, transforming a rather smooth surface into an advancing mass of rough, clinkery rubble.

Pahoehoe flows are also known to change into aa flows as they move from a gentle to a steep slope. The abrupt change in stress is thought to contribute to the release of gases while breaking the surface to produce fragments that are further disturbed as the flow advances.

The lava associated with the Mexican volcano Parícutin that buried the city of San Juan Parangaricutiro was of the aa type (see Figure 5.17). One of the flows from Parícutin moved an average of only about a meter per day, but it advanced continuously for more than three months.

Pahoehoe lavas have also been known to move at agonizingly slow rates. In 1990, near the Hawaiian village of Kalapana, the villagers had to watch for weeks as a pahoehoe flow crept toward their homes at only a few meters per hour. Although most residents were able to escape injury, they were unable to stop the flow and their houses were eventually incinerated.

Lava Tubes Hardened basaltic flows commonly contain tunnels that once were nearly horizontal conduits carrying lava from the volcanic vent to the flow's leading edge. These openings develop in the interior of a flow where temperatures remain high long after the surface hardens. Under such conditions, the still-molten lava within the conduits continues its forward motion leaving behind cave-like voids called **lava tubes** (Figure 5.8). Lava tubes are important because they allow fluid lavas to advance great distances from their source.

Lava tubes are associated with basaltic volcanoes in most parts of the world and are even present in giant volcanoes on Mars. Some lava tubes exhibit extraordinary dimensions. For example, Kazumura Cave on the southeast slope of Hawaii's Mauna Loa volcano extends for more than 60 kilometers (40 miles). Andesitic and rhyolitic lavas rarely contain lava tubes.

Block Lavas   In contrast to fluid basaltic magmas which typically produce pahoehoe and aa flows, andesitic and rhyolitic magmas tend to generate **block lavas.** Block lava consists largely of detached blocks with slightly curved surfaces that cover unbroken lava in the interior. Although similar to aa flows, these lavas consist of blocks with comparatively smooth surfaces, rather than having rough, clinkery surfaces.

Pillow Lavas   Recall that much of Earth's volcanic output occurs along oceanic ridges (divergent plate boundaries). When outpourings of lava occur on the ocean floor, or when lava enters the ocean, the flow's outer skin quickly congeals. However, the lava is usually able to move forward by breaking through the hardened surface. This process occurs over and over, as molten basalt is extruded—like toothpaste from a tightly squeezed tube. The result is a lava flow composed of elongated structures resembling large bed pillows stacked one atop the other. These structures, called **pillow lavas,** are useful in the reconstruction of Earth history because whenever they are observed, they indicate that the lava flow formed in an underwater environment (see Figure 13.19).

Gases

Magmas contain varying amounts of dissolved gases (*volatiles*) held in the molten rock by confining pressure, just as carbon dioxide is held in soft drinks. As with soft drinks, as soon as the pressure is reduced, the gases begin to escape. Obtaining gas samples from an erupting volcano is difficult and dangerous, so geologists usually estimate the amount of gas originally contained within the magma.

The gaseous portion of most magmas makes up from 1 to 6 percent of the total weight, with most of this in the form of water vapor. Although the percentage may be small, the actual quantity of emitted gas can exceed thousands of tons per day.

The composition of volcanic gases is important because they contribute significantly to the gases that make up our planet's atmosphere. Analyses of samples taken during Hawaiian eruptions indicate that the gases are about 70 percent water vapor, 15 percent carbon dioxide, 5 percent nitrogen, 5 percent sulfur dioxide, and lesser amounts of chlorine, hydrogen, and argon. Sulfur compounds are easily recognized by their pungent odor. Volcanoes are a natural source of air pollution, including sulfur dioxide, which readily combines with water to form sulfuric acid.

In addition to propelling magma from a volcano, gases play an important role in creating the narrow conduit that connects the magma chamber to the surface. First, high temperatures and the buoyant force from the magma body crack the rock above. Then hot blasts of high-pressure gases expand the cracks and develop a passageway to the surface. Once the passageway is completed, the hot gases, armed with rock fragments, erode its walls, producing a larger conduit. Because these erosive forces are concentrated on any protrusion along the pathway, the volcanic pipes that are produced have a circular shape. As the conduit enlarges,

FIGURE 5.8   Lava streams that flow in confined channels often develop a solid crust and become flows within lava tubes. **A.** Thurston Lava Tube, Hawaii Volcanoes National Park. **B.** View of an active lava tube as seen through the collapsed roof. (Photo **A** by Douglas Peebles and Photo **B** by Jeffrey Judd, U.S. Geological Survey)

A.

B.

magma moves upward to produce surface activity. Following an eruptive phase, the volcanic pipe often becomes choked with a mixture of congealed magma and debris that was not thrown clear of the vent. Before the next eruption, a new surge of explosive gases may again clear the conduit.

Occasionally, eruptions emit colossal amounts of volcanic gases that rise high into the atmosphere, where they may reside for several years. Some of these eruptions may have an impact on Earth's climate, a topic we will consider in Chapter 21.

Pyroclastic Materials

When basaltic lava is extruded, dissolved gases escape quite freely and continually. These gases propel incandescent blobs of lava to great heights (see Figure 5.5). Some of this ejected material may land near the vent and build a cone-shaped structure, whereas smaller particles will be carried great distances by the wind. By contrast, viscous (rhyolitic) magmas are highly charged with gases, and upon release they expand a thousandfold as they blow pulverized rock, lava, and glass fragments from the vent. The particles produced in both of these situations are referred to as **pyroclastic materials** (*pyro* = fire, *clast* = fragment). These ejected fragments range in size from very fine dust and sand-sized volcanic ash (less than 2 millimeters) to pieces that weigh several tons.

Ash and *dust* particles are produced from gas-laden viscous magma during an explosive eruption (see Figure 5.6). As magma moves up in the vent, the gases rapidly expand, generating a froth of melt that might resemble froth that flows from a just opened bottle of champagne. As the hot gases expand explosively, the froth is blown into very fine glassy fragments. When the hot ash falls, the glassy shards often fuse to form a rock called *welded tuff*. Sheets of this material, as well as ash deposits that later consolidate, cover vast portions of the western United States.

Also common are pyroclasts that range in size from small beads to walnuts termed *lapilli* ("little stones"). These ejecta are commonly called *cinders* (2–64 millimeters). Particles larger than 64 millimeters (2.5 inches) in diameter are called *blocks* when they are made of hardened lava and *bombs* when they are ejected as incandescent lava. Because bombs are semimolten upon ejection, they often take on a streamlined shape as they hurtle through the air (Figure 5.9). Because of their size, bombs and blocks usually fall on the slopes of a cone; however, they are occasionally propelled far from the volcano. For instance, bombs 6 meters (20 feet) long and weighing about 200 tons were blown 600 meters (2000 feet) from the vent during an eruption of the Japanese volcano Asama.

So far we have distinguished various pyroclastic materials based largely on the size of the fragments. Some materials are also identified by their texture and composition. In particular, **scoria** is the name applied to vesicular (containing voids) ejecta that is a product of basaltic magma (Figure 5.10). These black to reddish-brown fragments are generally found in the size range of lapilli and resemble cinders and clinkers produced by furnaces used to smelt iron. When magma with an intermediate or silica-rich composition generates vesicular ejecta, it is called **pumice** (see Figure 4.15). Pumice is usually lighter in color and less dense than scoria. Furthermore, some pumice fragments have such a preponderance of vesicles that they may float in water for prolonged periods.

Volcanic Structures and Eruptive Styles

 GEODe Volcanoes and Other Igneous Activity
EARTH ▸ Volcanic Structures and Eruptive Styles

The popular image of a volcano is that of a solitary, graceful, snowcapped cone, such as Mount Hood in Oregon or Japan's Fujiyama. These picturesque, conical mountains are produced by volcanic activity that occurred intermittently over thousands, or even hundreds of thousands, of years. However, many volcanoes do not fit this image. Some volcanoes are only 30 meters (100 feet) high and formed during a single eruptive phase that may have lasted only a few days. Further, numerous volcanic landforms are not "volcanoes"

FIGURE 5.9 Volcanic bombs forming during an eruption of Hawaii's Kilauea volcano. Ejected lava masses take on a streamlined shape as they sail through the air. The bomb in the insert is about 10 centimeters long. (Photo by Arthur Roy/National Audubon Society; inset photo by E. J. Tarbuck)

←2 cm→

← 2 cm →

FIGURE 5.10 Scoria is a volcanic rock that exhibits a vesicular texture. Vesicles are small holes left by escaping gas bubbles. (Photo by E. J. Tarbuck)

at all. For example, Alaska's Valley of Ten Thousand Smokes is a flat-topped deposit consisting of 15 cubic kilometers of ash that erupted in less than 60 hours and blanketed a section of river valley to a depth of 200 meters (600 feet).

Volcanic landforms come in a wide variety of shapes and sizes, and each structure has a unique eruptive history. Nevertheless, volcanologists have been able to classify volcanic landforms and determine their eruptive patterns. In this section we will consider the general anatomy of a volcano and look at three major volcanic types: shield volcanoes, cinder cones, and composite cones. This discussion will be followed by an overview of other significant volcanic landforms.

Anatomy of a Volcano

Volcanic activity frequently begins when a fissure (crack) develops in the crust as magma moves forcefully toward the surface. As the gas-rich magma moves up this linear fissure, its path is usually localized into a circular **conduit,** or **pipe,** that terminates at a surface opening called a **vent** (Figure 5.11). Successive eruptions of lava, pyroclastic material, or frequently a combination of both often separated by long periods of inactivity eventually build the structure we call a **volcano.**

Located at the summit of most volcanoes is a somewhat funnel-shaped depression, called a **crater** (*crater* = a bowl). Volcanoes that are built primarily by the ejection of pyroclastic materials typically have craters that

form by gradual accumulation of volcanic debris on the surrounding rim. Other craters form during explosive eruptions as the rapidly ejected particles erode the crater walls. Craters also form when the summit area of a volcano collapses following an eruption (Figure 5.12).

Some volcanoes have very large circular depressions called *calderas.* Whereas most craters have diameters ranging from a few tens to a few hundreds of meters, the diameters of calderas are typically greater than one kilometer and in rare cases can exceed 50 kilometers. We will consider the formation of various types of calderas later in this chapter.

During early stages of growth most volcanic discharges come from a central summit vent. As a volcano matures, material also tends to be emitted from fissures that develop along the flanks, or base, of the volcano. Continued activity from a flank eruption may produce a small **parasitic cone** (*parasitus* = one who eats at the table of another). Mount Etna in Italy, for example, has more than 200 secondary vents, some of which have built cones. Many of these vents, however, emit only gases and are appropriately called **fumaroles** (*fumus* = smoke).

The form of a particular volcano is largely determined by the composition of the contributing magma. As you will see, fluid Hawaiian-type lavas tend to produce broad structures with gentle slopes, whereas more viscous silica-rich lavas (and some gas-rich basaltic lavas) tend to generate cones with moderate to steep slopes.

Shield Volcanoes

Shield volcanoes are produced by the accumulation of fluid basaltic lavas and exhibit the shape of a broad, slightly domed structure that resembles a warrior's shield (Figure 5.13). Most shield volcanoes have grown up from the deep ocean floor to form islands or seamounts. For example, the islands of the Hawaiian chain, Iceland, and the Galápagos are either a single shield volcano or the coalescence of several shields. Some shield volcanoes, however, occur on

FIGURE 5.11 Anatomy of a "typical" composite cone (see also Figures 5.13 and 5.16 for a comparison with a shield and cinder cone, respectively).

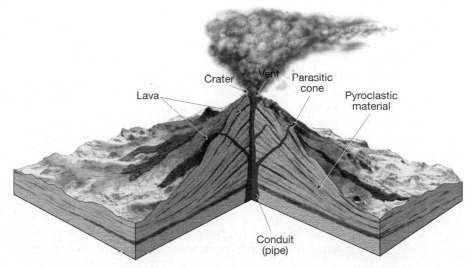

Lava
Crater
Vent
Parasitic cone
Pyroclastic material
Conduit (pipe)

FIGURE 5.12 The crater of Mount Vesuvius, Italy. The city of Naples is located northwest of Vesuvius, while Pompeii, the Roman town that was buried in an eruption in A.D. 79, is located southeast of the volcano. (Courtesy of NASA; inset image by Krafft/Photo Researchers)

continents. Included in this group are rather massive structures located in East Africa, like Suswa in Kenya.

Extensive study of the Hawaiian Islands confirms that each shield was built from a myriad of basaltic lava flows averaging a few meters thick. Also, these islands consist of only about 1 percent pyroclastic ejecta.

Mauna Loa is one of five overlapping shield volcanoes that together comprise the Big Island of Hawaii (Figure 5.13). From its base on the floor of the Pacific Ocean to its summit, Mauna Loa is over 9 kilometers (6 miles) high, exceeding that of Mount Everest. This massive pile of basaltic rock has a volume of 40,000 cubic kilometers that was extruded over a period of nearly a million years. For comparison, the volume of material composing Mauna Loa is roughly 200 times greater than the amount composing a large composite cone such as Mount Rainier (Figure 5.14). Most shields, however, are more modest in size. For example, the classic Icelandic shield, Skjalbreidur, rises to a height of only about 600 meters (2000 feet) and is 10 kilometers (6 miles) across its base.

Despite its enormous size, Mauna Loa is not the largest known volcano in the solar system. Olympus Mons, a huge shield volcano on Mars, is 25 kilometers (15 miles) high and 600 kilometers (375 miles) wide (see Chapter 24).

Young shields, particularly those located in Iceland, emit very fluid lava from a central summit vent and have sides with gentle slopes that vary from 1 to 5 degrees. Mature shields, as exemplified by Mauna Loa, have steeper flanks (about 10 degrees), while their summits are comparatively flat. During the mature stage, lavas are discharged from the summit vents, as well as from rift zones that develop along the slopes. Most lava is discharged as the fluid pahoehoe type, but as these flows cool downslope, many change into a clinkery aa flow. Once an eruption is well established, a large fraction of the lava (perhaps 80 percent) flows through a well-developed system of lava tubes (see Figure 5.8). This greatly increases the distance lava can travel before it solidifies. Thus, lava emitted near the summit often reaches the sea, thereby adding to the width of the cone at the expense of its height.

Another feature common to a mature, active shield volcano is a large, steep-walled caldera that occupies its summit. Mauna Loa's summit caldera measures 2.6 by 4.5 kilometers (1.6 by 2.8 miles) and has a depth that averages about 150 meters (500 feet). Calderas on large shield volcanoes form when the roof above the magma chamber collapses. This can occur through the evacuation of a magma reservoir following a large eruption, or as magma migrates to the flank of a volcano to feed a fissure eruption.

In their final stage of growth, the activity on mature shields is more sporadic and pyroclastic ejections are more prevalent. Further, the lavas increase in viscosity, resulting in thicker, shorter flows. These eruptions tend to steepen the slope of the summit area, which often becomes capped with clusters of cinder cones. This explains why Mauna Kea, a very mature volcano that has not erupted in historic times, has a steeper summit than Mauna Loa, which erupted as recently as 1984. Astronomers are so certain that Mauna Kea is "over the hill" that they have built an elaborate observatory

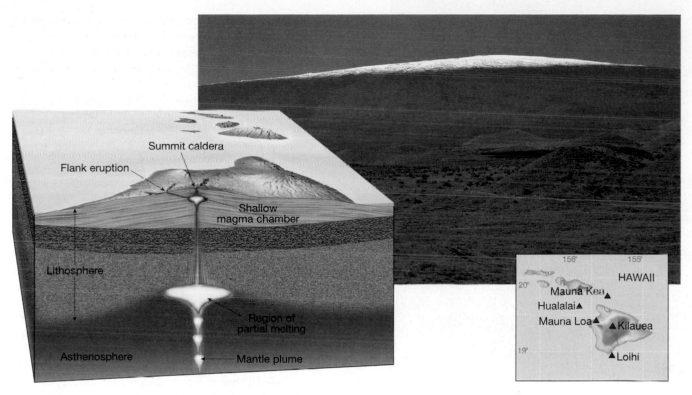

FIGURE 5 13   Mauna Loa is one of five shield volcanoes that together make up the island of Hawaii. Shield volcanoes are built primarily of fluid basaltic lava flows and contain only a small percentage of pyroclastic materials. (Photo by Greg Vaughn)

on its summit, housing some of the world's finest (and most expensive) telescopes.

Kilauea, Hawaii: Eruption of a Shield Volcano   Kilauea, the most active and intensely studied shield volcano in the world, is located on the island of Hawaii in the shadow of Mauna Loa. More than 50 eruptions have been witnessed here since record keeping began in 1823. Several months before each eruptive phase, Kilauea inflates as magma gradually migrates upward and accumulates in a central reservoir located a few kilometers below the summit. For up to 24 hours in advance of an eruption, swarms of small earthquakes warn of the impending activity.

Most of the activity on Kilauea during the past 50 years occurred along the flanks of the volcano in a region called the East Rift Zone. A rift eruption here in 1960 engulfed the coastal village of Kapoho, located nearly 30 kilometers (20 miles) from the source. The longest and largest rift eruption ever recorded on Kilauea began in 1983 and continues to this day, with no signs of abating. The first discharge began along a 6-kilometer (4-mile) fissure where a 100-meter (300-foot) high "curtain of fire" formed as red-hot lava was ejected skyward (Figure 5.15). When the activity became localized, a cinder and spatter cone given the Hawaiian name *Puu Oo* was built. Over the next 3 years the general eruptive pattern consisted of short periods

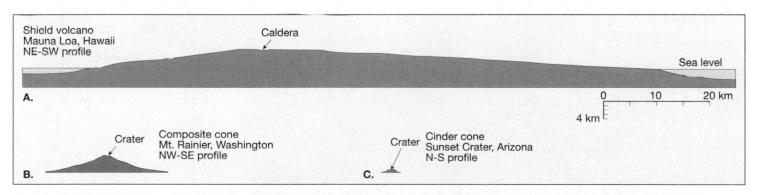

FIGURE 5.14   Profiles illustrating scales of different volcanoes. **A.** Profile of Mauna Loa, Hawaii, the largest shield volcano in the Hawaiian chain. Note size comparison with Mount Rainier, Washington, a large composite cone. **B.** Profile of Mount Rainier, Washington. Note how it dwarfs a typical cinder cone. **C.** Profile of Sunset, Arizona, a typical steep-sided cinder cone.

FIGURE 5.15 Lava extruded along the East Rift Zone, Kilauea, Hawaii. (Photo by Greg Vaughn)

(hours to days) when fountains of gas-rich lava sprayed skyward. Each event was followed by nearly a month of inactivity.

By the summer of 1986 a new vent opened up 3 kilometers downrift. Here smooth-surfaced pahoehoe lava formed

a lava lake. Occasionally the lake overflowed, but more often lava escaped through tunnels to feed pahoehoe flows that moved down the southeastern flank of the volcano toward the sea. These flows destroyed nearly a hundred rural homes, covered a major roadway, and eventually reached the sea. Lava has been intermittently pouring into the ocean ever since, adding new land to the island of Hawaii.

Situated just 32 kilometers (20 miles) off the southern coast of Kilauea, a submarine volcano, Loihi, is also active. It, however, has another 930 meters (3100 feet) to go before it breaks the surface of the Pacific Ocean.

Cinder Cones

As the name suggests, **cinder cones** (also called **scoria cones**) are built from ejected lava fragments that take on the appearance of cinders or clinkers as they begin to harden while in flight (Figure 5.16). These pyroclastic fragments range in size from fine ash to bombs that may exceed a meter in diameter. However, most of the volume of a cinder cone consists of pea- to walnut-sized lapilli that are markedly vesicular (containing voids) and have a black to reddish-brown color. (Recall that these vesicular rock fragments are called *scoria*.) Although cinder cones are composed mostly of loose pyroclastic material, they sometimes extrude lava. On such occasions the discharges come from vents located at or near the base rather than from the summit crater.

Cinder cones are the most abundant of the three major types of volcanoes. They have a very simple, distinctive shape determined by the slope that loose pyroclasic material maintains as it comes to rest (Figure 5.16). Because cinders

Pyroclastic material

Crater

Central vent filled with rock fragments

FIGURE 5.16 SP Crater, a cinder cone north of Flagstaff, Arizona. Cinder cones are built from ejected lava fragments and are usually less than 300 meters (1000 feet) high. (Photo by Michael Collier)

have a high angle of repose (the steepest angle at which material remains stable), young cinder cones are steep-sided, having slopes between 30 and 40 degrees. In addition, cinder cones have large, deep craters in relation to the overall size of the structure. Although relatively symmetrical, many cinder cones are elongated, and higher on the side that was downwind during the eruptions.

Most cinder cones are produced by a single, short-lived eruptive event. One study found that half of all cinder cones examined were constructed in less than one month, and that 95 percent formed in less than one year. However, in some cases, they remain active for several years. Parícutin, shown in Figure 5.17, had an eruptive cycle that spanned 9 years. Once the event ceases, the magma in the "plumbing" connecting the vent to the magma source solidifies, so the volcano never erupts again. As a consequence of this short life span, cinder cones are small, usually between 30 meters (100 feet) and 300 meters (1000 feet). A few rare examples exceed 700 meters (2100 feet) in height (see Figure 5.16).

Cinder cones are found by the thousands all around the globe. Some occur in volcanic fields like the one located near Flagstaff, Arizona, which consists of about 600 cones. Others are parasitic cones of larger volcanoes. Mount Etna, for example, has dozens of cinder cones dotting its flanks (see chapter-opening photo).

Parícutin: Life of a Garden-variety Cinder Cone One of the very few volcanoes studied by geologists from beginning to end is the cinder cone called Parícutin, located about 320 kilometers (200 miles) west of Mexico City. In 1943 its eruptive phase began in a cornfield owned by Dionisio Pulido, who witnessed the event as he prepared the field for planting.

FIGURE 5.17 The village of San Juan Parangaricutiro engulfed by aa lava from Parícutin, shown in the background. Only the church towers remain. (Photo by Tad Nichols)

For two weeks prior to the first eruption, numerous Earth tremors caused apprehension in the nearby village of Parícutin. Then on February 20 sulfurous gases began billowing from a small depression that had been in the cornfield for as long as people could remember. During the night, hot, glowing rock fragments were ejected from the vent, producing a spectacular fireworks display. Explosive discharges continued, throwing hot fragments and ash occasionally as high as 6000 meters (20,000 feet) above the crater rim. Larger fragments fell near the crater, some remaining incandescent as they rolled down the slope. These built an aesthetically pleasing cone, while finer ash fell over a much larger area, burning and eventually covering the village of Parícutin. In the first day the cone grew to 40 meters (130 feet), and by the fifth day it was more than 100 meters (330 feet) high. Within the first year more than 90 percent of the total ejecta had been discharged.

The first lava flow came from a fissure that opened just north of the cone, but after a few months flows began to emerge from the base of the cone itself. In June 1944 a clinkery aa flow 10 meters (30 feet) thick moved over much of the village of San Juan Parangaricutiro, leaving only the church steeple exposed (Figure 5.17). After 9 years of intermittent pyroclastic explosions and nearly continuous discharge of lava from vents at its base, the activity ceased almost as quickly as it had begun. Today, Parícutin is just another one of the scores of cinder cones dotting the landscape in this region of Mexico. Like the others, it will not erupt again.

Composite Cones

Earth's most picturesque yet potentially dangerous volcanoes are **composite cones** or **stratovolcanoes** (Figure 5.18). Most are located in a relatively narrow zone that rims the Pacific Ocean, appropriately called the *Ring of Fire* (see Figure 5.38). This active zone includes a chain of continental volcanoes that are distributed along the west coast of South and North America, including the large cones of the Andes and the Cascade Range of the western United States and Canada. The latter group includes Mount St. Helens, Mount Rainier, and Mount Garibaldi. The most active regions in the Ring of Fire are located along curved belts of volcanic islands situated adjacent to the deep-ocean trenches of the northern and western Pacific. This nearly continuous chain of volcanoes stretches from the Aleutian Islands to Japan and the Philippines and ends on the North Island of New Zealand.

The classic composite cone is a large, nearly symmetrical structure composed of both lava and pyroclastic deposits. Just as shield volcanoes owe their shape to fluid basaltic lavas, composite cones reflect the nature of the erupted material. For the most part, composite cones are the product of gas-rich magma having an andesitic composition. (Composite cones may also emit various amounts of material having a basaltic and/or rhyolitic composition). Relative to shields, the silica-rich magmas typical of composite cones generate thick viscous lavas that travel short distances. In addition, composite cones may generate explosive eruptions that eject huge quantities of pyroclastic material.

FIGURE 5.18 Mount Shasta, California, one of the largest composite cones in the Cascade Range. Shastina is the smaller parasitic cone on the left. (Photo by David Muench)

The growth of a "typical" composite cone begins with both pyroclastic material and lava being emitted from a central vent. As the structure matures, lavas tend to flow from fissures that develop on the lower flanks of the cone. This activity may alternate with explosive eruptions that eject pyroclastic material from the summit crater. Sometimes both activities occur simultaneously.

A conical shape, with a steep summit area and more gradually sloping flanks, is typical of many large composite cones. This classic profile, which adorns calendars and postcards, is partially a consequence of the way viscous lavas and pyroclastic ejecta contribute to the growth of the cone. Coarse fragments ejected from the summit crater tend to accumulate near their source. Because of their high angle of repose, coarse materials contribute to the steep slopes of the summit area. Finer ejecta, on the other hand, are deposited as a thin layer over a large area. This acts to flatten the flank of the cone. In addition, during the early stages of growth, lavas tend to be more abundant and flow greater distances from the vent than do later lavas. This contributes to the cone's broad base. As the volcano matures, the short flows that come from the central vent serve to armor and strengthen the summit area. Consequently, steep slopes exceeding 40 degrees are sometimes possible. Two of the most perfect cones—Mount Mayon in the Philippines and Fujiyama in Japan—exhibit the classic form we expect of a composite cone, with its steep summit and gently sloping flanks.

Despite their symmetrical form, most composite cones have a complex history. Huge mounds of volcanic debris surrounding many cones provide evidence that, in the distant past, a large section of the volcano slid downslope as a massive landslide. Others develop horseshoe-shaped depressions at their summits as a result of explosive eruptions, or as occurred during the 1980 eruption of Mount St. Helens, a combination of a landslide and the eruption of 0.6 cubic kilometer of magma that left a gaping void on the north side of the cone. Often, so much rebuilding has occurred since these eruptions that no trace of the amphitheater-shaped scar remains. Vesuvius in Italy provides us with another example of the complex history of a volcanic region. This young volcano was built on the same site where an older cone was destroyed by an eruption in A.D. 79. In the following section we will look at another aspect of composite cones: their destructive nature.

Living in the Shadow of a Composite Cone

More than 50 volcanoes have erupted in the United States in the past 200 years (Figure 5.19) Fortunately, the most explosive of these eruptions, except for Mount St. Helens in 1980, occurred in sparsely inhabited regions of Alaska. On a global scale numerous destructive eruptions have occurred during the past few thousand years, a few of which may have influenced the course of human civilization (see Box 5.2).

Eruption of Vesuvius A.D. 79

In addition to producing some of the most violent volcanic activity, composite cones can erupt unexpectedly. One of the best documented of these events was the A.D. 79 eruption of

BOX 5.2 ▶ PEOPLE AND THE ENVIRONMENT

The Lost Continent of Atlantis

Anthropologists have proposed that a catastrophic eruption on the island of Santorini (also called Thera) contributed to the collapse of the advanced Minoan civilization that was centered around Crete in the Aegean Sea (Figure 5.B). This event also gave rise to the enduring legend of the lost continent of Atlantis. According to an account by the Greek philosopher Plato, an island empire named Atlantis was swallowed up by the sea in a single day and night. Although the connection between Plato's Atlantis and the Minoan civilization is somewhat tenuous, there is no doubt that a catastrophic eruption took place on Santorini about 1600 B.C.

This eruption generated a tall, billowing *eruption column* composed of huge quantities of pyroclastic materials. Ash and pumice rained from this plume for days, eventually blanketing the surrounding landscape to a maximum depth of 60 meters (200 feet). One nearby Minoan city, now called Akrotiri, was buried and its remains sealed until 1967, when archaeologists began investigating the area. The excavation of beautiful ceramic jars and elaborate wall paintings indicate that Akrotiri was home to a wealthy and sophisticated society.

Following the ejection of this large quantity of material, the summit of Santorini collapsed, producing a caldera 8 kilometers (5 miles) across. This once majestic volcano presently consists of five small islands. The eruption and collapse of Santorini generated large sea waves (*tsunamis*) that caused widespread destruction to the coastal villages of Crete and to nearby islands to the north.

Although some scholars suggest that the eruption of Santorini contributed to the demise of the Minoan civilization, was this eruption the main cause of the dispersal of this great civilization or only one of many contributing factors? Was Santorini the island continent of Atlantis described by Plato? Whatever the answers to these questions, clearly volcanism can dramatically change the course of human events.

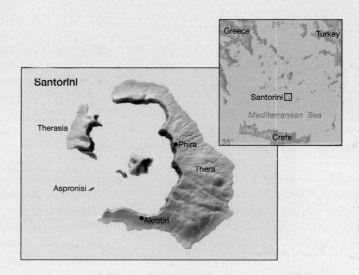

FIGURE 5.B Map showing the remains of the volcanic island of Santorini after the top of the cone collapsed into the emptied magma chamber following an explosive eruption. The location of the recently excavated Minoan town of Akrotiri is shown. Volcanic eruptions over the last 500 years built the central islands. Despite the fact that another destructive eruption is likely, the city of Phira was built on the flanks of the caldera.

---

the Italian volcano we now call Vesuvius. Prior to this eruption, Vesuvius had been dormant for centuries and had vineyards adorning its sunny slopes. On August 24, however, the tranquility ended, and in less than 24 hours the city of Pompeii (near Naples) and more than 2000 of its 20,000 residents perished. Some were entombed beneath a layer of pumice nearly 3 meters (10 feet) thick, while others were encased within a layer of hardened ash (Figure 5.20). They remained this way for nearly 17 centuries, until the city was partially excavated, giving archaeologists a superbly detailed picture of ancient Roman life (Figure 5.21).

By reconciling historical records with detailed scientific studies of the region, volcanologists have pieced together the chronology of the destruction of Pompeii. The eruption most likely began as steam discharges on the morning of August 24. By early afternoon fine ash and pumice fragments formed a tall eruptive cloud emanating from Vesuvius. Shortly thereafter, debris from this cloud began to shower Pompeii, located 9 kilometers (6 miles) downwind of the volcano. Undoubtedly, many people fled during this early phase of the eruption. For the next several hours pumice fragments as large as 5 centimeters (2 inches)

fell on Pompeii. One historical record of this eruption states that people located more distant than Pompeii tied pillows to their heads in order to fend off the flying fragments.

The pumice fall continued for several hours, accumulating at the rate of 12 to 15 centimeters (5 to 6 inches) per hour. Most of the roofs in Pompeii eventually gave way. Despite the accumulation of more than 2 meters of pumice, many of the people that had not evacuated Pompeii were probably still alive the morning of August 25. Then suddenly and unexpectedly a surge of searing hot ash and gas swept rapidly down the flanks of Vesuvius. This blast killed an estimated 2000 people who had somehow managed to survive the pumice fall. Some may have been killed by flying debris, but most died of suffocation as a result of inhaling ash-laden gases. Their remains were quickly buried by the falling ash, which rain cemented into a hard mass before their bodies had time to decay. The subsequent decomposition of the bodies produced cavities in the hardened ash that replicated the form of the entombed bodies, preserving in some cases even facial expressions. Nineteenth-century excavators found these cavities and created casts of the corpses by

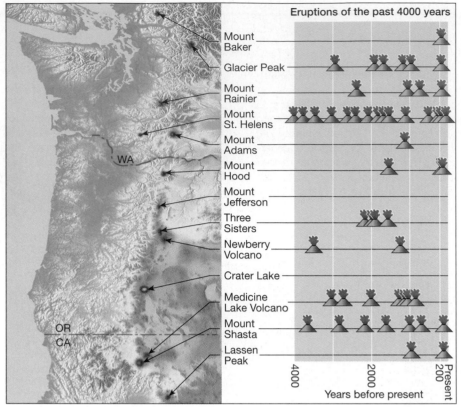

Eruptions of the past 4000 years

Mount Baker
Glacier Peak
Mount Rainier
Mount St. Helens
Mount Adams
Mount Hood
Mount Jefferson
Three Sisters
Newberry Volcano
Crater Lake
Medicine Lake Volcano
Mount Shasta
Lassen Peak

4000   2000   200 Present

Years before present

WA
OR
CA

FIGURE 5.19   Of the 13 potentially active volcanoes in the Cascade Range, 11 have erupted in the past 4000 years and 7 in just the past 200 years. More than 100 eruptions, most of which were explosive, have occurred in the past 4000 years. Mount St. Helens is the most active volcano in the Cascades. Its eruptions have ranged from relatively quiet outflows of lava to explosive events much larger than that of May 18, 1980. Each eruption symbol in the diagram represents from one to several dozen eruptions closely spaced in time. (After U.S. Geological Survey)

pouring plaster of Paris into the voids (Figure 5.20). Some of the plaster casts show victims trying to cover their mouths.

Nuée Ardente: A Deadly Pyroclastic Flow

Although the destruction of Pompeii was catastrophic, **pyroclastic flows** consisting of hot gases infused with incandescent ash and larger rock fragments can be even more devastating. The most destructive of these fiery flows,

called **nuée ardentes** (also referred to as *glowing avalanches*), are capable of racing down steep volcanic slopes at speeds that can approach 200 kilometers (125 miles) per hour (Figure 5.22).

The ground-hugging portion of a glowing avalanche is rich in particulate matter, which is suspended by jets of buoyant gases passing upward through the flow. Some of these gases have escaped from newly erupted volcanic fragments. In addition, air that is overtaken and trapped by an advancing flow may be heated sufficiently to provide buoyancy to the particulate matter of the nuée ardente. Thus, these flows, which can include large rock fragments in addition to ash, travel downslope in a nearly frictionless environment. This helps to explain why some nuée ardente deposits are found more than 100 kilometers (60 miles) from their source.

The pull of gravity is the force that causes these heavier-than-air flows to sweep downslope much like a snow avalanche. Some pyroclastic flows result when a powerful eruption blasts pyroclastic material laterally out the side of a volcano. Probably more often, nuée ardentes form from the collapse of tall eruption columns that form over a volcano

FIGURE 5.20   Plaster casts of several victims of the A.D. 79 eruption of Mount Vesuvius that destroyed the Italian city of Pompeii. (Photo by Leonard von Matt/Photo Researchers, Inc.)

FIGURE 5.21   Ruins of Pompeii, destroyed in 79 AD. Vesuvius stands on the horizon. (Photo by Rodger Ressmeyer/Corbis)

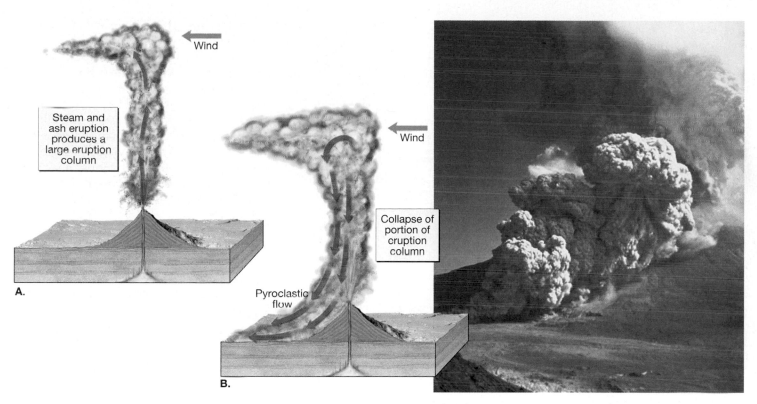

FIGURE 5.22   The sketches illustrate the development of a pyroclastic flow. The photo shows nuée ardente racing down the slope of Mount St. Helens on August 7, 1980, at speeds in excess of 100 kilometers (60 miles) per hour. (Photo by Peter W. Lipman, U.S. Geological Survey)

during an explosive event. Once gravity overcomes the initial upward thrust provided by the escaping gases, the ejecta begin to fall. Massive amounts of incandescent blocks, ash, and pumice fragments that fall onto the summit area begin to cascade downslope under the influence of gravity. The largest fragments have been observed bouncing down the flanks of a cone, while the finer materials travel rapidly as an expanding tongue-shaped cloud.

The Destruction of St. Pierre   In 1902 an infamous nuée ardente from Mount Pelée, a small volcano on the Caribbean island of Martinique, destroyed the port town of St. Pierre. The destruction happened in moments and was so devastating that almost all of St. Pierre's 28,000 inhabitants were killed. Only one person on the outskirts of town—a prisoner protected in a dungeon—and a few people on ships in the harbor were spared (Figure 5.23). Satis N. Coleman, in *Volcanoes, New and Old*, relates a vivid account of this event, which lasted less than five minutes.

I saw St. Pierre destroyed. The city was blotted out by one great flash of fire. . . . Our boat, the *Roraima*, arrived at St. Pierre early Thursday morning. For hours before entering the roadstead, we could see flames and smoke rising from Mt. Pelée. . . . There was a constant muffled roar. It was like the biggest oil refinery in the world burning up on the mountain top. There was a tremendous explosion about 7:45, soon after we got in. The mountain was blown to pieces. There was no warning. The side of the volcano was ripped out and there was hurled straight toward us a solid wall of flame. It sounded like a thousand cannons. . . . The

air grew stifling hot and we were in the thick of it. Wherever the mass of fire struck the sea, the water boiled and sent up vast columns of steam. . . . The blast of fire from the volcano lasted only a few minutes. It shriveled and set fire to everything it touched. Thousands of casks of rum were stored in St. Pierre, and these were exploded by the terrific heat. . . . Before the volcano burst the landings of St. Pierre were covered with people. After the explosion, not one living soul was seen on land.*

Shortly after this calamitous eruption, scientists arrived on the scene. Although St. Pierre was mantled by only a thin layer of volcanic debris, they discovered that masonry walls nearly a meter thick were knocked over like dominoes; large trees were uprooted and cannons were torn from their mounts. A further reminder of the destructive force of this nuée ardente is preserved in the ruins of the mental hospital. One of the immense steel chairs that had been used to confine alcoholic patients can be seen today, contorted, as though it were made of plastic.

Lahars: Mudflows on Active and Inactive Cones

In addition to violent eruptions, large composite cones may generate a type of mudflow referred to by its Indonesian name **lahar.** These destructive flows occur when volcanic

---

*New York: John Day, 1946, pp. 80–8.

FIGURE 5.23   St. Pierre as it appeared shortly after the eruption of Mount Pelée, 1902. (Reproduced from the collection of the Library of Congress)

canic deposits. Thus, lahars can occur even when a volcano is *not* erupting.

When Mount St. Helens erupted in 1980, several lahars formed. These flows and accompanying flood waters raced down the valleys of the north and south forks of the Toutle River at speeds exceeding 30 kilometers per hour. Water levels in the river rose to 4 meters (13 feet) above flood stage, destroying or severely damaging nearly all the homes and bridges along the impacted area. Fortunately, the area was not densely populated.

In 1985 deadly lahars were produced during a small eruption of Nevado del Ruiz, a 5300-meter (17,400-foot) volcano in the Andes Mountains of Colombia. Hot pyroclastic material melted ice and snow that capped the mountain (*nevado* means *snow* in Spanish) and sent torrents of ash and debris down three major river valleys that flank the volcano. Reaching speeds of 100 kilometers (60 miles) per hour, these mudflows tragically took 25,000 lives.

Mount Rainier, Washington, is considered by many to be America's most dangerous volcano because, like Nevado del Ruiz, it has a thick year-round mantle of snow and ice. Adding to the risk is the fact that 100,000 people live in the valleys around Rainier, and many homes are built on lahars that flowed down the volcano hundreds or thousands of years ago. A future eruption, or perhaps just a period of heavy rainfall, may produce lahars that will likely take similar paths.

debris becomes saturated with water and rapidly moves down steep volcanic slopes, generally following gullies and stream valleys. Some lahars may be triggered when large volumes of ice and snow melt during an eruption. Others are generated when heavy rainfall saturates weathered vol-

---

Students Sometimes Ask...

Some of the larger volcanic eruptions, like the eruption of Krakatoa, must have been impressive. What was it like?

On August 27, 1883, in what is now Indonesia, the volcanic island of Krakatoa exploded and was nearly obliterated. The sound of the explosion was heard an incredible 4800 kilometers (3000 miles) away at Rodriguez Island in the western Indian Ocean. Dust from the explosion was propelled into the atmosphere and circled Earth on high-altitude winds. This dust produced unusual and beautiful sunsets for nearly a year.

Not many were killed directly by the explosion, because the island was uninhabited. However, the displacement of water from the energy released during the explosion was enormous. The resulting *seismic sea wave* or *tsunami* exceeded 35 meters (116 feet) in height. It devastated the coastal region of the Sunda Strait between the nearby islands of Sumatra and Java, drowning more than 1000 villages and taking more than 36,000 lives. The energy carried by this wave reached every ocean basin and was detected by tide-recording stations as far away as London and San Francisco.

Other Volcanic Landforms

The most obvious volcanic structure is a cone. But other distinctive and important landforms are also associated with volcanic activity.

Calderas

Calderas (*caldaria* = a cooking pot) are large collapse depressions having a more or less circular form. Their diameters exceed one kilometer, and many are tens of kilometers across. (Those less than a kilometer across are called *collapse pits.*) Most calderas are formed by one of the following processes: (1) the collapse of the summit of a large composite volcano following an explosive eruption of silica-rich pumice and ash fragments (*Crater Lake-type calderas*); (2) the collapse of the top of a shield volcano caused by subterranean drainage from a central magma chamber (*Hawaiian-type calderas*); and (3) the collapse of a large area, caused by the discharge of colossal volumes of silica-rich pumice and ash along ring fractures (*Yellowstone-type calderas*).

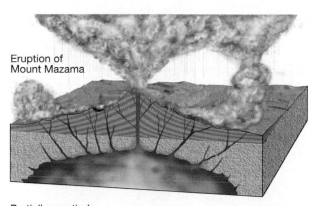

Eruption of
Mount Mazama

Partially emptied
magma chamber

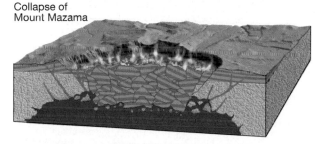

Collapse of
Mount Mazama

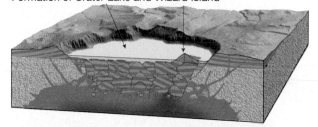

Formation of Crater Lake and Wizard Island

Crater Lake–Type Calderas Crater Lake, Oregon, is located in a caldera that has a maximum diameter of 10 kilometers (6 miles) and is 1175 meters (more than 3800 feet) deep. This caldera formed about 7000 years ago when a composite cone, later named Mount Mazama, violently extruded 50 to 70 cubic kilometers of pyroclastic material (Figure 5.24). With the loss of support, 1500 meters (nearly a mile) of the summit of this once prominent cone collapsed. After the collapse, rainwater filled the caldera (Figure 5.24). Later volcanic activity built a small cinder cone in the lake. Today this cone, called Wizard Island, provides a mute reminder of past activity.

Hawaiian-Type Calderas Although some calderas are produced by *collapse following an explosive eruption,* many are not. For example, Hawaii's active shield volcanoes, Mauna Loa and Kilauea, both have large calderas at their summits. Kilauea's measures 3.3 by 4.4 kilometers (about 2 by 3 miles) and is 150 meters (500 feet) deep. The walls of this caldera are steep-sided, almost vertical, and as a result, it looks like a vast, nearly flat-bottomed pit. Here collapse was not initiated by the eruption of voluminous lavas directly from the caldera, although lava lakes occasionally formed on its floor. Instead, Kilauea's caldera formed by gradual subsidence as magma slowly drained laterally from the underlying magma chamber to the East Rift Zone, leaving the summit unsupported.

Yellowstone-Type Calderas Although the 1980 eruption of Mount St. Helens was spectacular, it pales by comparison to what happened 630,000 years ago in the region now occupied by Yellowstone National Park. Here approximately 1000 cubic kilometers of pyroclastic material erupted, eventually producing a caldera 70 kilometers (43 miles) across. This super-eruption produced showers of ash as far away as the Gulf of Mexico and gave rise to a formation called the Lava Creek Tuff, a hardened ash deposit that is 400 meters (more than 1200 feet) thick in some places. Vestiges of this

FIGURE 5.24 Sequence of events that formed Crater Lake, Oregon. About 7000 years ago a violent eruption partly emptied the magma chamber, causing the summit of former Mount Mazama to collapse. Rainfall and groundwater contributed to form Crater Lake, the deepest lake in the United States. Subsequent eruptions produced the cinder cone called Wizard Island. (After H. Williams, *The Ancient Volcanoes of Oregon.* Photo by Greg Vaughn/Tom Stack and Associates)

event are the many hot springs and geysers in the Yellowstone region.

Based on the extraordinary volume of erupted material, researchers concluded that the magma chambers that produce Yellowstone-type calderas have to be similarly monstrous. As more and more magma accumulates, the pressure within the magma chamber begins to exceed the pressure exerted by the enormous weight of the overlying rocks (Figure 5.25). An eruption ensues when the gas-rich magma raises the overlying strata enough to create vertical fractures that extend to the surface. Magma surges upward along these newly generated cracks forming a ring of eruptions (Figure 5.25C). With a loss of support, the roof of the magma chamber collapses forcing even more gas-rich magma toward the surface.

Caldera-forming eruptions are of colossal proportions, ejecting huge volumes (usually exceeding 100 cubic kilometers) of pyroclastic materials, mainly in the form of ash and pumice fragments. Typically, these materials form pyroclastic flows that spread across the landscape at speeds that may exceed 100 kilometers (60 miles) per hour, destroying most living things in their paths. After coming to rest, the hot fragments of ash and pumice fuse together, forming a welded tuff that closely resembles a solidified lava flow. Despite their size, an entire caldera producing eruption is thought to last a mere 10 hours to perhaps a few days.

A distinctive characteristic of most large caldera eruptions is a slow upheaval, or *resurgence,* of the floor of the caldera that follows the main eruptive phase. As a consequence, these structures consist of large, somewhat circular depressions containing a central elevated region. Most large calderas exhibit a complex history. In the Yellowstone region, for example, three caldera-forming episodes are known to have occurred over the past 2.1 million years. The most recent event was followed by episodic outpourings of rhyolitic and basaltic lavas. Geological evidence suggests that a magma reservoir still exists beneath Yellowstone; thus, another caldera-forming eruption is likely, but not imminent.

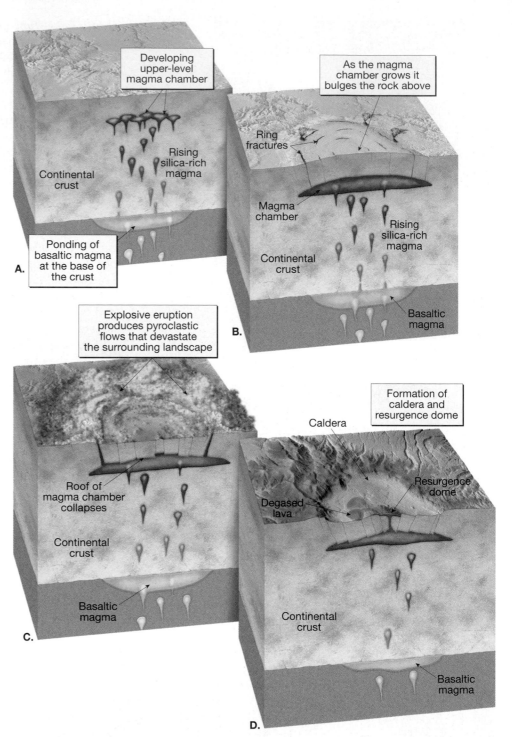

FIGURE 5.25 Formation of a Yellowstone-type caldera. **A.** Partial melting in the upper asthenosphere produces hot, basaltic magma that rises to the base of the continental crust. This magma body melts some of the continental crust, which has a relatively low melting point, generating a highly viscous, silica-rich magma that ascends to form an upper level magma chamber. **B.** As this magma chamber grows it causes the rock above to bulge and crack, producing a set of *ring fractures*. **C.** Once these fractures connect the magma reservoir to the surface, magma surges upward to produce an explosive steam-and-ash eruption. With a loss of support, the roof of the magma chamber collapses, forcing even more gas-rich magma into the atmosphere. The result is a scalding cloud of ash and gases known as pyroclastic flow, which devastates the surrounding landscape, often burying it under tens of meters of ash. **D.** After the main eruption, lava may erupt in the caldera. Most large calderas go through a long period of slow upheaval that produces an elevated region, called a *resurgence dome,* near the center of the caldera.

Calderas produced by Yellowstone-type eruptions are the largest volcanic structures on Earth. Volcanologists have compared their destructive force with that of the impact of a small asteroid. Fortunately, no eruption of this type has occurred in historic times.

Unlike calderas associated with shield volcanoes or composite cones, these depressions are so large and poorly defined that many remained undetected until high-quality aerial, or satellite images became available. Other examples of large calderas located in the United States are California's Long Valley Caldera, LaGarita Caldera, located in the San Juan Mountains of southern Colorado, and the Valles Caldera located west of Los Alamos, New Mexico.

Fissure Eruptions and Basalt Plateaus

We think of volcanic eruptions as building a cone or shield from a central vent. But by far the greatest volume of volcanic material is extruded from fractures in the crust called **fissures** (*fissura* = to split). Rather than building a cone, these long, narrow cracks may emit a low-viscosity basaltic lava, blanketing a wide area.

The extensive Columbia Plateau in the northwestern United States was formed this way (Figure 5.26). Here, numerous **fissure eruptions** extruded very fluid basaltic lava (Figure 5.27). Successive flows, some 50 meters (160 feet) thick, buried the existing landscape as they built a lava plateau nearly a mile thick. The fluid nature of the lava is evident, because some remained molten long enough to flow 150 kilometers (90 miles) from its source. The term **flood basalts** appropriately describes these flows. Massive accumulations of basaltic lava, similar to those of the Co-

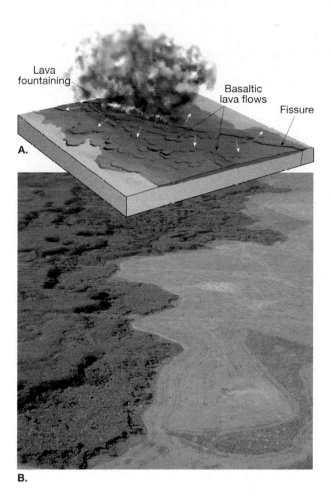

FIGURE 5.27   Basaltic fissure eruption. **A.** Lava fountaining from a fissure and formation of fluid lava flows called flood basalts. **B.** Photo of basalt flows near Idaho Falls. (Photo by John S. Shelton)

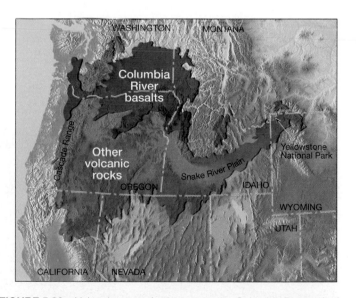

FIGURE 5.26   Volcanic areas that compose the Columbia Plateau in the Pacific Northwest. The Columbia River basalts cover an area of nearly 200,000 square kilometers (80,000 square miles). Activity here began about 17 million years ago as lava began to pour out of large fissures, eventually producing a basalt plateau with an average thickness of more than 1 kilometer. (After U.S. Geological Survey)

lumbia Plateau, occur worldwide. One of the largest is the Deccan Traps, a thick sequence of flat-lying basalt flows covering nearly 500,000 square kilometers (195,000 square miles) of west central India. When the Deccan Traps formed about 66 million years ago, nearly 2 million cubic kilometers of lava were extruded in less than 1 million years. Another huge deposit of flood basalts, called the Ontong Java Plateau, is found on the floor of the Pacific Ocean. A discussion of the origin of large basalt plateaus is provided later in this chapter in the section on "Intraplate Igneous Activity."

Lava Domes

In contrast to mafic lavas, silica-rich lavas near the felsic (rhyolitic) end of the compositional spectrum are so viscous they hardly flow at all. As the thick lava is "squeezed" out of the vent, it may produce a dome-shaped mass of lava called a **lava dome.**

Lava domes come in a variety of shapes that range from pancake-like flows to steep-sided plugs that were pushed upward like pistons. One common type develops over a period of several years following an explosive eruption of gas-rich magma. This is exemplified by the volcanic dome that

If volcanoes are so dangerous, why do people live on or near them?

Realize that many who live near volcanoes did not choose the location; they were simply born there. Their ancestors may have lived in the region for generations. Historically, many have been drawn to volcanic regions because of their fertile soils. Not all volcanoes have explosive eruptions, but all active volcanoes are dangerous. Certainly, choosing to live close to an active composite cone like Mount St. Helens or Soufriére Hills has a high inherent risk. However, the time interval between successive eruptions might be several decades or more—plenty of time for generations of people to forget the last eruption and consider the volcano to be dormant (*dormin* = to sleep) and therefore safe. Other volcanoes, like Mauna Loa or those on Iceland, are continually active, so recent eruptions are fresh in the minds of local populations. Many people that choose to live near an active volcano have the belief that the *relative* risk is no higher than in other hazard-prone places. In essence, they are gambling that they will be able to live out their lives before the next major eruption.

continues to "grow" from the vent that produced the 1980 eruption of Mount St. Helens (Figure 5.28). Although lava domes often form on the summit of a composite cone they can also form structures on the flanks of volcanoes. In addition, some domes form on their own, independent of large volcanic structures. One example is the line of rhyolitic and obsidian domes at Mono Craters, California (Figure 5.29).

FIGURE 5.28 Following the May 1980 eruption of Mount St. Helens, a lava dome began to develop. (Photo by David Falconer/DRK Photo)

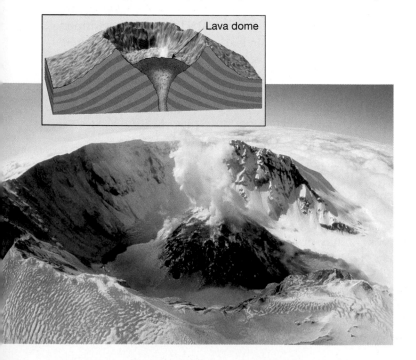

FIGURE 5.29 Mono Craters, California. This chain of rhyolitic and obsidian domes is located near Mono Lake where eruptions ocurred only 200 years ago. (Photo by Jim Sugar/Corbis)

Volcanic Pipes and Necks

Most volcanoes are fed magma through short conduits, called *pipes,* that connect a magma chamber to the surface. In rare circumstances, pipes may extend tubelike to depths exceeding 200 kilometers (125 miles). When this occurs, the ultramafic magmas that migrate up these structures produce rocks that are thought to be samples of the mantle that have undergone very little alteration during their ascent. Geologists consider these unusually deep conduits to be "windows" into Earth, for they allow us to view rock normally found only at great depth.

FIGURE 5.30 This volcanic pipe at Kimberly, South Africa, was excavated because it contains gem-quality diamonds. (Photo by Roger De La Harpe)

FIGURE 5.31   Shiprock, New Mexico, is a volcanic neck. This structure, which stands over 420 meters (1380 feet) high, consists of igneous rock which crystallized in the vent of a volcano that has long since been eroded away. (Photo by Tom Bean)

The best-known volcanic pipes are the diamond-bearing structures of South Africa (Figure 5.30). Here, the rocks filling the pipes originated at depths of at least 150 kilometers (90 miles), where pressure is high enough to generate diamonds and other high-pressure minerals. The task of transporting essentially unaltered magma (along with diamond inclusions) through 150 kilometers of solid rock is exceptional. This fact accounts for the scarcity of natural diamonds.

Volcanoes on land are continually being lowered by weathering and erosion. Cinder cones are easily eroded because they are composed of unconsolidated materials. However, all volcanoes will eventually succumb to relentless erosion over geologic time. As erosion progresses, the rock occupying the volcanic pipe is often more resistant and may remain standing above the surrounding terrain long after most of the cone has vanished. Shiprock, New Mexico, is such a feature and is called a **volcanic neck** (Figure 5.31). This structure, higher than many skyscrapers, is but one of many such landforms that protrude conspicuously from the red desert landscapes of the American Southwest.

Intrusive Igneous Activity

Volcanoes and Other Igneous Activity
▶ **Intrusive Igneous Activity**

Although volcanic eruptions can be among the most violent and spectacular events in nature and therefore worthy of detailed study, most magma is emplaced at depth. Thus, an understanding of intrusive igneous activity is as important to geologists as the study of volcanic events.

The structures that result from the emplacement of igneous material at depth are called **plutons,** named for Pluto, the god of the lower world in classical mythology. Because all plutons form out of view beneath Earth's surface, they can be studied only after uplifting and erosion have exposed them. The challenge lies in reconstructing the events that generated these structures millions or even hundreds of millions of years ago.

For the sake of clarity, we have separated our discussions of volcanism and plutonic activity. Keep in mind, however, that these diverse processes occur simultaneously and involve basically the same materials.

Nature of Plutons

Plutons are known to occur in a great variety of sizes and shapes. Some of the most common types are illustrated in Figure 5.32. Notice that some of these structures have a tabular (tabletop) shape, whereas others are quite massive. Also, observe that some of these bodies cut across existing structures, such as layers of sedimentary rock; others form when magma is injected between sedimentary layers. Because of these differences, intrusive igneous bodies are generally classified according to their shape as either **tabular** (*tabula* = table) or **massive** and by their orientation with respect to the host rock. Plutons are said to be **discordant** (*discordare* = to disagree) if they cut across existing structures and **concordant** (*concordare* = to agree) if they form parallel to features such as sedimentary strata. As you can see

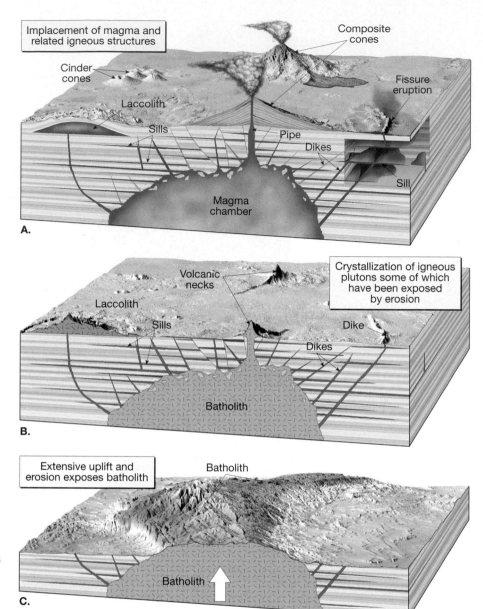

FIGURE 5.32 Illustrations showing basic igneous structures. **A.** This block diagram shows the relationship between volcanism and intrusive igneous activity. **B.** This view illustrates the basic intrusive igneous structures, some of which have been exposed by erosion long after their formation. **C.** After millions of years of uplifting and erosion, a batholith is exposed at the surface.

in Figure 5.32A, plutons are closely associated with volcanic activity. Many of the largest intrusive bodies are the remnants of magma chambers that once fed ancient volcanoes.

Dikes

Dikes are tabular discordant bodies that are produced when magma is injected into fractures. The force exerted by the emplaced magma can be great enough to separate the walls of the fracture further. Once crystallized, these sheetlike structures have thicknesses ranging from less than a centimeter to more than a kilometer. The largest have lengths of hundreds of kilometers. Most dikes, however, are a few meters thick and extend laterally for no more than a few kilometers.

Dikes are often found in groups that once served as vertically oriented pathways followed by molten rock that fed ancient lava flows. The parent pluton is generally not ob-

servable. Some dikes are found radiating, like spokes on a wheel, from an eroded volcanic neck. In these situations the active ascent of magma is thought to have generated fissures in the volcanic cone out of which lava flowed.

Dikes often weather more slowly than the surrounding rock. When exposed by erosion, these dikes have the appearance of a wall, as shown in Figure 5.33.

Sills and Laccoliths

Sills and laccoliths are concordant plutons that form when magma is intruded in a near-surface environment. They differ in shape and usually differ in composition.

Sills **Sills** are tabular plutons formed when magma is injected along sedimentary bedding surfaces (Figure 5.34). Horizontal sills are the most common, although all orientations, even vertical, are known to exist. Because of their

FIGURE 5.33 The vertical structure that looks like a stone wall is a dike, which is more resistant to weathering than the surrounding rock. This dike is located west of Granby, Colorado, near Arapaho National Forest. (Photo by R. Jay Fleisher)

relatively uniform thickness and large areal extent, sills are likely the product of very fluid magmas. Magmas having a low silica content are more fluid, so most sills are composed of the rock basalt.

The emplacement of a sill requires that the overlying sedimentary rock be lifted to a height equal to the thickness of the sill. Although this is a formidable task, in shallow environments it often requires less energy than forcing the magma up the remaining distance to the surface. Consequently, sills form only at shallow depths, where the pressure exerted by the weight of overlying rock layers is low. Although sills are intruded between layers, they can be locally discordant. Large

FIGURE 5.34 Salt River Canyon, Arizona. The dark, essentially horizontal band is a sill of basaltic composition that intruded horizontal layers of sedimentary rock. (Photo by E.J. Tarbuck)

sills frequently cut across sedimentary layers and resume their concordant nature at a higher level.

One of the largest and most studied of all sills in the United States is the Palisades Sill. Exposed for 80 kilometers along the west bank of the Hudson River in southeastern New York and northeastern New Jersey, this sill is about 300 meters thick. Because of its resistant nature, the Palisades Sill forms an imposing cliff that can be seen easily from the opposite side of the Hudson.

In many respects, sills closely resemble buried lava flows. Both are tabular and often exhibit columnar jointing (Figure 5.35). **Columnar joints** form as igneous rocks cool and develop shrinkage fractures that produce elongated, pillarlike columns. Further, because sills generally form in near-surface environments and may be only a few meters thick, the emplaced magma often cools quickly enough to generate an aphanitic texture.

When attempts are made to reconstruct the geologic history of a region, it becomes important to differentiate between sills and buried lava flows. Fortunately, under close examination these two phenomena can be readily distinguished. The upper portion of a buried lava flow usually contains voids produced by escaping gas bubbles. Further, only the rocks beneath a lava flow show evidence of metamorphic alteration. Sills, on the other hand, form when magma has been forcefully intruded between sedimentary layers. Thus, fragments of the overlying rock can be found only in sills. Lava flows, conversely, are extruded before the overlying strata are deposited. Further, "baked" zones in the rock above and below are trademarks of a sill.

Laccoliths **Laccoliths** are similar to sills because they form when magma is intruded between sedimentary layers in a near-surface environment. However, the magma that generates laccoliths is more viscous. This less-fluid magma collects as a lens-shaped mass that arches the overlying strata upward (see Figure 5.32). Consequently, a laccolith can occasionally be detected because of the dome-shaped bulge it creates at the surface.

Most large laccoliths are probably not much wider than a few kilometers. The Henry Mountains in southeastern Utah are largely composed of several laccoliths believed to have been fed by a much larger magma body emplaced nearby.

Batholiths

By far the largest intrusive igneous bodies are **batholiths** (*bathos* = depth, *lithos* = stone). Most often, batholiths occur as linear structures several hundreds of kilometers long and up to 100 kilometers wide, as shown in Figure 5.36. The Idaho batholith, for example, encompasses an area of more than 40,000 square kilometers and consists of many plutons. Indirect evidence gathered from gravitational studies indicates that batholiths are also very thick, possibly extending dozens of kilometers into the crust.

By definition, a plutonic body must have a surface exposure greater than 100 square kilometers (40 square miles) to be considered a batholith. Smaller plutons of this type are

Equally spaced centers of contraction produce six-sided columns

FIGURE 5.35 Columnar jointing in basalt, Giants Causeway National Park, Northern Ireland. These five- to seven-sided columns are produced by contraction and fracturing that results as a lava flow or sill gradually cools. (Photo by Tom Till)

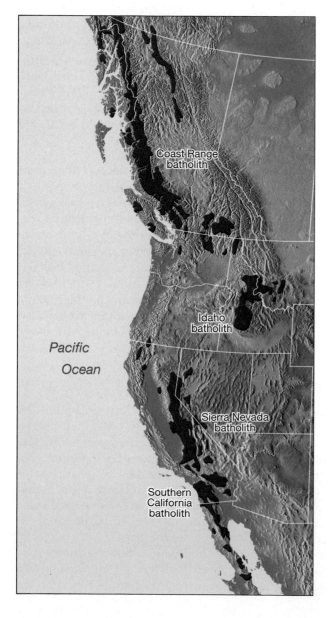

Coast Range batholith

Idaho batholith

Pacific Ocean

Sierra Nevada batholith

Southern California batholith

FIGURE 5.36 Granitic batholiths that occur along the western margin of North America. These gigantic, elongated bodies consist of numerous plutons that were emplaced during the last 150 million years of Earth history.

termed **stocks.** Many stocks appear to be portions of batholiths that are not yet fully exposed.

Batholiths usually consist of rock types having chemical compositions toward the granitic end of the spectrum, although diorite is commonly found. Smaller batholiths can be rather simple structures composed almost entirely of one rock type. However, studies of large batholiths have shown that they consist of large numbers of distinct plutons that were intruded over a period of millions of years. The plutonic activity that created the Sierra Nevada batholith, for example, occurred nearly continuously over a 130-million-year period that ended about 80 million years ago during the Cretaceous period.

Batholiths may compose the core of mountain systems. Here uplifting and erosion have removed the surrounding rock, thereby exposing the resistant igneous body. Some of the highest peaks in the Sierra Nevada, such as Mount Whitney, are carved from such a granitic mass (Figure 5.37).

Large expanses of granitic rock also occur in the stable interiors of the continents, such as the Canadian Shield of North America. These relatively flat exposures are the remains of ancient mountains that have long since been leveled by erosion. Thus, the rocks that make up the batholiths of youthful mountain ranges, such as the Sierra Nevada, were generated near the top of a magma chamber, whereas in shield areas, the roots of former mountains and, thus, the lower portions of batholiths, are exposed. In Chapter 14, we will further consider the role of igneous activity as it relates to mountain building.

FIGURE 5.37 Laurel Mountain, in the Eastern Sierra Nevada of California. This mountain makes up just a tiny portion of the Sierra Nevada batholith, a huge structure that extends for approximately 400 kilometers (Photo by Tim Fitzharris/Minden Pictures)

Emplacement of Batholiths An interesting problem that faced geologists was trying to explain how large granitic batholiths came to reside within only moderately deformed sedimentary and metamorphic rocks. What happened to the rock that was displaced by these igneous masses? How did the magma body make its way through several kilometers of solid rock?

We know that magma rises because it is less dense than the surrounding rock, much like a cork held at the bottom of a container of water will rise when it is released. But Earth's crust is made of solid rock. Nevertheless, at depths of several kilometers, where temperature and pressure are high, even solid rock deforms by flowing. Thus, at great depths a mass of buoyant, rising magma can forcibly make room for itself by pushing aside the overlying rock. As the magma continues to move upward, some of the host rock that was shouldered aside will fill in the space left by the magma body as it passes.*

As a magma body nears the surface, it encounters relatively cool, brittle rock that resists deformation. Further upward movement is accomplished by a process called *stoping*. In this process, fractures that develop in the overlying host rock allow magma to rise and dislodge blocks of rock. Once incorporated into the magma body, these blocks may melt, thereby altering the composition of the magma body. Eventually the magma body will cool sufficiently so that upward movement comes to a halt. Evidence supporting the fact that magma can move through solid rock are inclusions called **xenoliths** (*xenos* = a stranger, *lithos* = stone). These unmelted remnants of the host rock are found in igneous masses that have been exhumed by erosion.

Plate Tectonics and Igneous Activity

Geologists have known for decades that the global distribution of volcanism is not random. Of the more than 800 active volcanoes[†] that have been identified, most are located along the margins of the ocean basins—most notably within the circum-Pacific belt known as the *Ring of Fire* (Figure 5.38). This group of volcanoes consists mainly of composite cones that emit volatile-rich magma having an intermediate (andesitic) composition that occasionally produce awe-inspiring eruptions.

The volcanoes comprising a second group emit very fluid basaltic lavas and are confined to the deep ocean basins, including well-known examples on Hawaii and Iceland. In addition, this group contains many active submarine volcanoes that dot the ocean floor; particularly notable are the innumerable small seamounts that lay along the axis of the mid-ocean ridge. At these depths the pressures are so great that seawater does not boil explosively, even in contact with hot lavas. Thus, firsthand knowledge of these eruptions is limited, coming mainly from deep-diving submersibles.

A third group includes those volcanic structures that are irregularly distributed in the interiors of the continents. None are found in Australia nor in the eastern two-thirds of North and South America. Africa is notable because it has many potentially active volcanoes including Mount Kilimanjaro, the highest point on the continent (5895 meters, 19,454 feet). Volcanism on continents is the most diverse, ranging from eruptions of very fluid basaltic lavas, like those that generated the Columbia Plateau to explosive eruptions of silica-rich rhyolitic magma as occurred in Yellowstone.

Until the late 1960s, geologists had no explanation for the apparently haphazard distribution of continental volcanoes, nor were they able to account for the almost continuous chain of volcanoes that circles the margin of the Pacific basin. With the development of the theory of plate tectonics, the picture was greatly clarified. Recall that most primary (unaltered) magma originates in the upper mantle and that the mantle is essentially solid, *not molten* rock. The basic

---

*An analogous situation occurs when a can of oil-base paint is left in storage. The oily component of the paint is less dense than the pigments used for coloration; thus, oil collects into drops that slowly migrate upward, while the heavier pigments settle to the bottom.

[†]For our purposes, active volcanoes include those with eruptions that have been dated. At least 700 other cones exhibit geologic evidence that they have erupted within the past 10,000 years and are regarded as potentially active. Further innumerable active submarine volcanoes are hidden from view in the depths of the ocean and are not counted in these numbers.

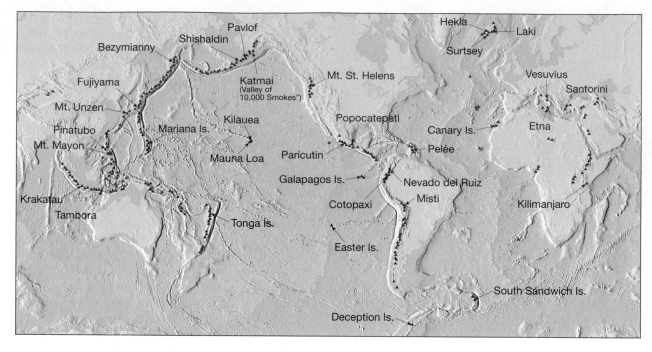

FIGURE 5.38   Locations of some of Earth's major volcanoes.

connection between plate tectonics and volcanism is that *plate motions provide the mechanisms by which mantle rocks melt to generate magma.*

We will examine three zones of igneous activity and their relationship to plate boundaries. These active areas are located (1) along convergent plate boundaries where plates move toward each other and one sinks beneath the other; (2) along divergent plate boundaries, where plates move away from each other and new seafloor is created; and (3) areas within the plates proper that are not associated with any plate boundary. (It should be noted that volcanic activity rarely occurs along transform plate boundaries.) These three volcanic settings are depicted in Figure 5.39. (If you are unclear as to how magma is generated, we suggest that you study the section entitled "Origin of Magma" in Chapter 4 before proceeding.)

Igneous Activity at Convergent Plate Boundaries

Recall that at convergent plate boundaries slabs of oceanic crust are bent as they descend into the mantle, generating an oceanic trench. As a slab sinks deeper into the mantle, the increase in temperature and pressure drives volatiles (mostly) from the oceanic crust. These mobile fluids migrate upward into the wedge-shaped piece of mantle located between the subducting slab and overriding plate (Figure 5.39A). Once the sinking slab reaches a depth of about 100 to 150 kilometers, these water-rich fluids reduce the melting point of hot mantle rock sufficiently to trigger some melting. The partial melting of mantle rock (principally peridotite)

generates magma with a basaltic composition. After a sufficient quantity of magma has accumulated, it slowly migrates upward.

Volcanism at a convergent plate margin results in the development of a linear or slightly curved chain of volcanoes called a *volcanic arc.* These volcanic chains develop roughly parallel to the associated trench—at distances of 200 to 300 kilometers (100 to 200 miles). Volcanic arcs can be constructed on oceanic, or continental, lithosphere. Those that develop within the ocean and grow large enough for their tops to rise above the surface are labeled *island archipelagos* in most atlases. Geologists prefer the more descriptive term **volcanic island arcs,** or simply **island arcs** (Figure 5.39A). Several young volcanic island arcs of this type border the western Pacific basin, including the Aleutians, the Tongas, and the Marianas.

The early stage of island arc volcanism is typically dominated by the eruption of fluid basalts that build numerous shieldlike structures on the ocean floor. Because this activity begins at great depth, volcanic cones must extrude a great deal of lava before their tops rise above the sea to form islands. This cone-building activity, coupled with massive basaltic intrusions as well as magma that is added to the underside of the crust, tend to thicken the arc crust through time. As a result, mature volcanic arcs are underlain by a comparatively thick crust that impedes the upward flow of the mantle-derived basalts. This in turn provides time for magmatic differentiation to occur, in which heavy, iron-rich minerals crystallize and settle out, leaving the melt enriched in silica (see Chapter 4). Consequently, as the arc matures, the magmas that reach the surface tend to erupt silica-rich andesites and even some rhyolites. In addition, magmatic differentiation tends to concentrate the available volatiles (water) into the more silica-rich components of these magmas. Because they emit viscous volatile-rich magma, island arc volcanoes typically have explosive eruptions.

Volcanism associated with convergent plate boundaries may also develop where slabs of oceanic lithosphere are subducted under continental lithosphere to produce a **continental volcanic arc** (Figure 5.39E). The mechanisms that generate these mantle-derived magmas are essentially the same as those operating at island arcs. The major difference is that continental crust is much thicker and is composed of rocks having a higher silica content than oceanic crust. Hence through the assimilation of silica-rich crustal rocks, plus extensive magmatic differentiation, a mantle-derived magma may become highly evolved as it rises through continental crust. Stated another way, the primary magmas generated in the mantle may change from a comparatively dry, fluid basaltic magma to a viscous andesitic or rhyolitic magma having a high concentration of volatiles as it moves up through the continental crust. The volcanic chain of the Andes Mountains along the western edge of South America is perhaps the best example of a mature continental volcanic arc.

Since the Pacific basin is essentially bordered by convergent plate boundaries (and associated subduction zones), it is easy to see why the irregular belt of explosive volcanoes we call the Ring of Fire formed in this region. The volcanoes of the Cascade Range in the northwestern United States, including Mount Hood, Mount Rainier, and Mount Shasta, are included in this group (Figure 5.40).

Igneous Activity at Divergent Plate Boundaries

The greatest volume of magma (perhaps 60 percent of Earth's total yearly output) is produced along the oceanic ridge system in association with seafloor spreading (Figure 5.39B). Here, below the ridge axis where the lithospheric plates are being continually pulled apart, the solid yet mobile mantle responds to the decrease in overburden and

rises upward to fill in the rift. Recall from Chapter 4 that as rock rises, it experiences a decrease in confining pressure and undergoes melting without the addition of heat. This process, called *decompression melting,* is the most common process by which mantle rocks melt.

Partial melting of mantle rock at spreading centers produces basaltic magma. Because this newly formed magma is less dense than the mantle rock from which it was derived, it rises faster than the mantle. Collecting in reservoirs located just beneath the ridge crest, about 10 percent of this melt eventually migrates upward along fissures to erupt as flows on the ocean floor. This activity continuously adds new basaltic rock to the plate margins, temporarily welding them together, only to break again as spreading continues. Along some ridges, outpourings of bulbous pillow lavas build numerous small seamounts. At other locations erupted lavas produce fluid flows that create more subdued topography.

Although most spreading centers are located along the axis of an oceanic ridge, some are not. In particular, the East African Rift is a site where continental lithosphere is being pulled apart, forming a *continental rift* (Figure 5.39F). Here, magma is generated by decompression melting in the same manner it is produced along the oceanic ridge system. Vast outpourings of fluid basaltic lavas are common in this region. The East African Rift zone also contains some large composite cones, as exemplified by Mount Kilimanjaro. Like composite cones that form along convergent plate boundaries, these volcanoes form when mantle-derived basalts evolve into volatile-rich andesitic magma as they migrate up through thick, silica-rich rocks of the continent.

Intraplate Igneous Activity

We know why igneous activity is initiated along plate boundaries, but why do eruptions occur in the interiors of plates? Hawaii's Kilauea is considered the world's most active volcano, yet it is situated thousands of kilometers from the nearest plate boundary in the middle of the vast Pacific plate (Figure 5.39C). Other sites of **intraplate volcanism** (meaning "within the plate") include the Canary Islands, Yellowstone, and several volcanic centers that you may be surprised to learn are located in the Sahara Desert of northern Africa.

We now recognize that most intraplate volcanism occurs where a mass of hotter than normal mantle material called a **mantle plume** ascends toward the surface (Figure 5.41). Although the depth at which (at least some) mantle plumes originate is still hotly debated, many appear to form deep within Earth at the core–mantle boundary. These plumes of solid yet mobile mantle rock rise toward the surface in a manner similar to the blobs that form within a lava lamp. (These are the trendy lamps that contain two immiscible liquids in a glass container. As the base of the lamp is heated, the denser liquid at the bottom becomes buoyant and forms blobs that rise to the top.) Like the blobs in a lava lamp, a mantle plume has a bulbous head that draws out a narrow stalk beneath it as it rises. Once the plume head nears the

---

Students Sometimes Ask . . .

I've heard that explosive volcanic eruptions might cause global air temperatures to drop. Is that possible?

Yes, it is possible. When you view an image such as the one of Mount Augustine in Figure 5.6, it might seem that the cloud of volcanic ash should block the sunlight and cause temperatures to drop. Although this occurs, the effect is very short-lived because volcanic ash quickly settles from the atmosphere. However, the gases emitted by volcanoes, especially sulfur dioxide and carbon dioxide may indeed trigger more enduring episodes of climate change. These possibilities are described at some length in Chapter 21 "Global Climate Change."

A. Convergent plate volcanism
(Island arc)

Volcanic island arc

Trench

Marginal sea

Continental crust

Oceanic crust

Mantle rock melts

Water driven from plate

Subducting oceanic lithosphere

Asthenosphere

Mount Augustine, Alaska (Stephen Kaufman)

C. Intraplate volcanism
(Oceanic)

Oceanic crust

Hot spot

Hawaii

Decompression melting

Rising mantle plume

North America

Hawaii

So... Ame...

Kilauea, Hawaii
(J. D. Griggs/USGS)

Continental volcanic arc

Trench

Oceanic crust

Continental crust

Subducting oceanic lithosphere

Mantle rock melts

Water driven from plate

E. Convergent plate volcanism
(Continental volcanic arc)

FIGURE 5.39 Three zones of volcanism. Two of these zones are plate boundaries, and the third is the interior area of the plates.

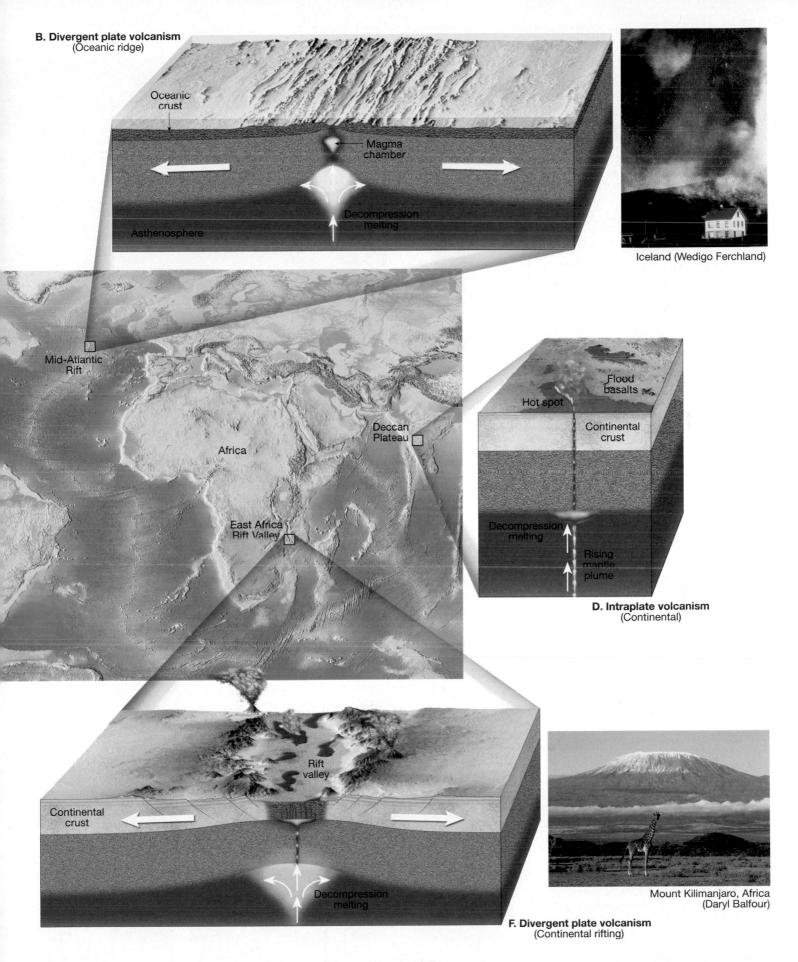

B. Divergent plate volcanism
(Oceanic ridge)

Oceanic crust

Magma chamber

Decompression melting

Asthenosphere

Iceland (Wedigo Ferchland)

Mid-Atlantic Rift

Deccan Plateau

Africa

East Africa Rift Valley

Flood basalts

Hot spot

Continental crust

Decompression melting

Rising mantle plume

D. Intraplate volcanism
(Continental)

Rift valley

Continental crust

Decompression melting

Mount Kilimanjaro, Africa
(Daryl Balfour)

F. Divergent plate volcanism
(Continental rifting)

FIGURE 5.39

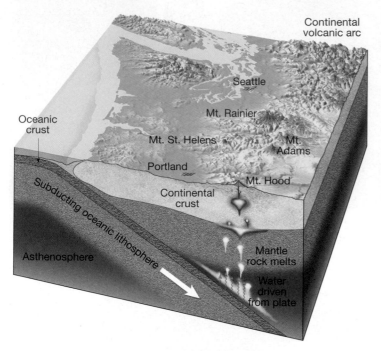

FIGURE 5.42   Lava flowing across road, Kilauea, Hawaii. (Photo by J. D. Griggs/U.S. Geological Survey)

FIGURE 5.40   As an oceanic plate descends into the mantle, water and other volatiles are driven from the subducting crustal rocks. These volatiles lower the melting temperature of mantle rock sufficiently to generate melt.

top of the mantle, decompression melting generates basaltic magma that may eventually trigger volcanism at the surface. The result is a localized volcanic region a few hundred kilometers across called a **hot spot** (Figure 5.41). More than 40 hot spots have been identified, and most have persisted for millions of years. The land surface around hot spots is often elevated, showing that it is buoyed up by a plume of warm low-density material. Furthermore, by measuring the heat flow in these regions, geologists have determined that the mantle beneath hot spots must be 100–150°C hotter than normal.

The volcanic activity on the island of Hawaii, with its outpourings of basaltic lava, is certainly the result of hot-spot volcanism (Figure 5.42). Where a mantle plume has

persisted for long periods of time, a chain of volcanic structures may form as the overlying plate moves over it. In the Hawaiian Islands, hot-spot activity is currently centered on Kilauea. However, over the past 80 million years the same mantle plume generated a chain of volcanic islands (and seamounts) that extend thousands of kilometers from the Big Island in a northwestward direction across the Pacific.

Mantle plumes are also thought to be responsible for the vast outpourings of basaltic lava that create large basalt plateaus such as the Columbia Plateau in the northwestern United States, India's Deccan Plateau, and the Ontong Java Plateau in the western Pacific (Figure 5.43). The most widely accepted explanation for these eruptions, which emit extremely large volumes of basaltic magma over relatively short time intervals, involves a plume with a substantial-sized head. These large structures may have heads that are hundreds of kilometers in diameter connected to a long, narrow tail rising from the core–mantle boundary (Figure

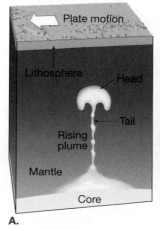
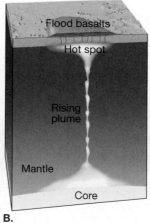
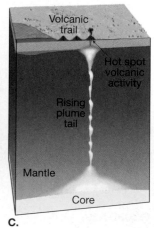

FIGURE 5.41   Model of a mantle plume and associated hot-spot volcanism. **A.** A rising mantle plume with large bulbous head and narrow tail. **B.** Rapid decompression melting of the head of a mantle plume produces vast outpourings of basalt. **C.** Less voluminous activity caused by the plume tail produces a linear volcanic chain on the seafloor.

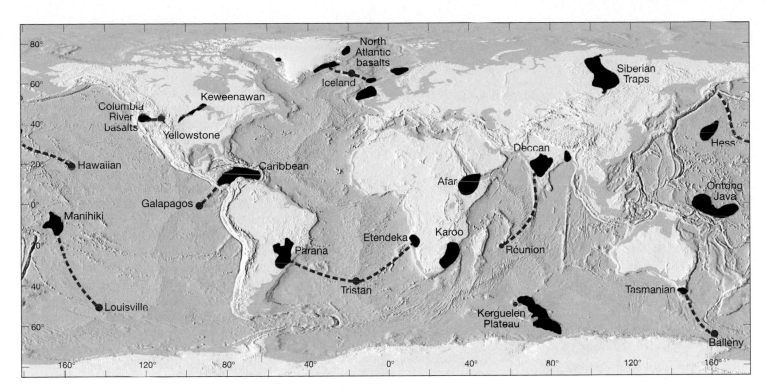

FIGURE 5.43 Global distribution of flood basalt provinces (shown in black) and associated hot-spots (shown as red dots). The red dashed lines are hot-spot tracks, which appear as lines of volcanic structures on the ocean floor. The Keweenawan and Siberian Traps formed in failed continental rifts where the crust had been greatly thinned. Whether there is a connection between the Columbia River basalts and the Yellowstone hot spot is still a matter of ongoing research.

5.41). Upon reaching the base of the lithosphere, the temperature of the material in the plume is estimated to be 200–300°C warmer than the surrounding rock. Thus, as much as 10 to 20 percent of the mantle material making up the plume head rapidly melts. It is this melting that triggers the burst of volcanism that emits voluminous outpourings of lava to form a huge basalt plateau in a matter of a million or so years (Figure 5.41). Substantial evidence supports the idea that massive outpourings of lava associated with a superplume released large quantities of carbon dioxide into the atmosphere, which in turn may have significantly altered the climate of the Cretaceous Period (see Chapter 21). The comparatively short initial eruptive phase is followed by tens of millions of years of less voluminous activity, as the plume tail slowly rises to the surface. Thus, extending away from most large flood basalt provinces is a chain of volcanic structures, similar to the Hawaiian chain, that terminates over an active hot spot marking the current position of the tail of the plume.

Based on the current state of knowledge, it appears that hot-spot volcanism, with its associated mantle plumes, is responsible for most intraplate volcanism. However, there are some widely scattered volcanic regions located far from any plate boundary that are not linked to hot-spot volcanism. Well-known examples are found in the Basin and Range province of the western United States and northwestern Mexico. We will consider the cause of volcanism in this region in Chapter 14.

Living with Volcanoes

About 10 percent of the Earth's population lives in the vicinity of an active volcano. In fact, several major cities including Seattle, Washington; Mexico City, Mexico; Tokyo, Japan; Naples, Italy; and Quito, Ecuador, are located on or near a volcano.

Until recently the dominant view of Western societies was that humans possess the wherewithal to subdue volcanoes and other types of catastrophic natural hazards. Today it is becoming increasingly apparent that volcanoes are not only very destructive but unpredictable as well. With this awareness a new attitude is developing—"How to live with volcanoes."

Volcanic Hazards

As shown in Figure 5.44, volcanoes produce a wide variety of potential hazards that can kill people and wildlife, as well as destroy property. Perhaps the greatest threats to life are pyroclastic flows. These hot mixtures of gas, ash, and pumice that sometimes exceed 800°C, speed down the flanks of volcanoes, giving people little chance of surviving.

Lahars, which can occur even when a volcano is not erupting, are perhaps the next most dangerous volcanic hazard. These mixtures of volcanic debris and water can flow for tens of kilometers down steep volcanic slopes at speeds that may exceed 100 kilometers (60 miles) per hour.

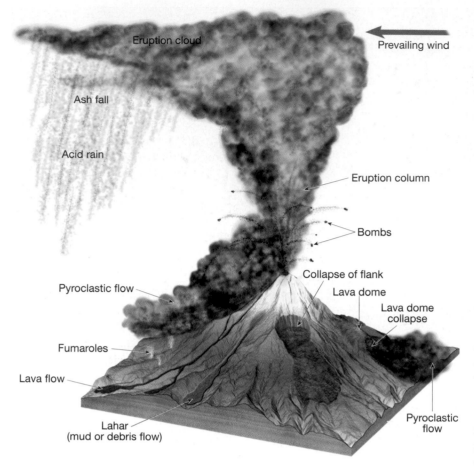

FIGURE 5.44   Simplified drawing showing a wide variety of natural hazards associated with volcanoes. (After U.S. Geological Survey)

Lahars pose a potential threat to many communities downstream from glacier-clad volcanoes such as Mount Rainier. Other potentially destructive mass-wasting events include the rapid collapse of the volcano's summit or flank.

Other obvious hazards include explosive eruptions that can endanger people and property hundreds of miles from a volcano. During the past 15 years at least 80 commercial jets have been damaged by inadvertently flying into clouds of volcanic ash. One of these was a near crash that occurred in 1989 when a Boeing 747, with more than 300 passengers aboard, encountered an ash cloud from Alaska's Redoubt volcano. All four engines stalled after they became clogged with ash. Fortunately, the engines were restarted at the last minute and the aircraft managed to land safely in Anchorage.

Monitoring Volcanic Activity

Today a number of volcano monitoring techniques are employed, with most of them aimed at detecting the movement of magma from a subterranean reservoir (typically several kilometers deep) toward the surface. The four most noticeable changes in a volcanic landscape caused by the migration of magma are: (1) changes in the pattern of vol-

canic earthquakes; (2) expansion of a near-surface magma chamber which leads to inflation of the volcano; (3) changes in the amount and/or composition of the gases that are released from a volcano; and (4) an increase in ground temperature caused by the implacement of new magma.

Almost a third of all volcanoes that have erupted in historic times are now monitored seismically (through the use of instruments that detect earthquake tremors). In general, a sharp increase in seismic unrest followed by a period of relative quiet has shown to be a precursor for many volcanic eruptions. However, some large volcanic structures have exhibited lengthy periods of seismic unrest. For example, Rabaul Caldera in New Guinea recorded a strong increase in seismicity in 1981. This activity lasted 13 years and finally culminated with an eruption in 1994. Occasionally, a large earthquake has triggered a volcanic eruption, or at least disturbed the volcano's plumbing. Kilauea, for example, began to erupt after the Kalapana earthquake of 1975.

The roof of a volcano may rise as new magma accumulates in its interior—a phenomena which precedes many volcanic eruptions. Because the accessibility of many volcanoes is limited, remote sensing devices, including lasers, Doppler radar, and Earth orbiting satellites, are often used to determine whether or not a volcano is swelling. The recent discovery of ground doming at Three Sisters volcanoes in Oregon was first detected using radar images obtained from satellites.

FIGURE 5.45   Photo taken by Jeff Williams from the International Space Station of a steam-and-ash eruption emitted from Cleveland Volcano in the Aleutian Islands. This event was reported to the Alaska Volcano Observatory, which issued a warning to air traffic control. (Photo by NASA)

Volcanologists also frequently monitor the gases that are released from volcanoes in an effort to detect even minor changes in their amount and/or composition. Some volcanoes show an increase in sulfur dioxide (SO_2) emissions months or years prior to an eruption. On the other hand, a few days prior to the 1991 eruption of Mount Pinatubo, emissions of carbon dioxide (CO_2) dropped dramatically.

The development of remote sensing devices has greatly increased our ability to monitor volcanoes. These instruments and techniques are particularly useful for monitoring eruptions in progress. Photographic images and infrared (heat) sensors can detect lava flows and volcanic columns rising from a volcano (Figure 5.45). Furthermore, satellites can detect ground deformation as well as monitor SO_2 emissions (see Figure 21.13A, p. 577).

The overriding goal of all monitoring techniques is to discover precursors that may warn of an imminent eruption. This is accomplished by first diagnosing the current condition of a volcano and then using this baseline data to predict its future behavior. Stated another way, a volcano must be observed over an extended period to recognize significant changes from its "resting state."

Summary

- The primary factors that determine the nature of volcanic eruptions include the magma's *composition*, its *temperature,* and the *amount of dissolved gases* it contains. As lava cools, it begins to congeal and, as *viscosity* increases, its mobility decreases. The *viscosity of magma is directly related to its silica content.* Rhyolitic (felsic) lava, with its high silica content (over 70 percent), is very viscous and forms short, thick flows. Basaltic (mafic) lava, with a lower silica content (about 50 percent), is more fluid and may travel a long distance before congealing. Dissolved gases tend to increase the fluidity of magma and, as they expand, provide the force that propels molten rock from the vent of a volcano.

- The materials associated with a volcanic eruption include (1) *lava flows* (*pahoehoe* flows, which resemble twisted braids; and *aa* flows, consisting of rough, jagged blocks; both form from basaltic lavas); (2) *gases* (primarily *water vapor*); and (3) *pyroclastic material* (pulverized rock and lava fragments blown from the volcano's vent, which include *ash, pumice, lapilli, cinders, blocks,* and *bombs*).

- Successive eruptions of lava from a central vent result in a mountainous accumulation of material known as a *volcano.* Located at the summit of many volcanoes is a steep-walled depression called a *crater. Shield cones* are broad, slightly domed volcanoes built primarily of fluid, basaltic lava. *Cinder cones* have steep slopes composed of pyroclastic material. *Composite cones,* or *stratovolcanoes,* are large, nearly symmetrical structures built of interbedded lavas and pyroclastic deposits. Composite cones produce some of the most violent volcanic activity. Often associated with a violent eruption is a *nuée ardente,* a fiery cloud of hot gases infused with incandescent ash that races down steep volcanic slopes. Large composite cones may also generate a type of mudflow known as a *lahar.*

- Most volcanoes are fed by *conduits* or *pipes.* As erosion progresses, the rock occupying the pipe is often more resistant and may remain standing above the surrounding terrain as a *volcanic neck.* The summits of some volcanoes have large, nearly circular depressions called *calderas* that result from collapse. Calderas also form on shield volcanoes by subterranean drainage from a central magma chamber, and the largest calderas form by the discharge of colossal volumes of silica-rich pumice along ring fractures. Although volcanic eruptions from a central vent are the most familiar, by far the largest amounts of volcanic material are extruded from cracks in the crust called *fissures.* The term *flood basalts* describes the fluid, waterlike, basaltic lava flows that cover an extensive region in the northwestern United States known as the Columbia Plateau. When silica-rich magma is extruded, *pyroclastic flows* consisting largely of ash and pumice fragments usually result.

- Intrusive igneous bodies are classified according to their *shape* and by their *orientation with respect to the host rock,* generally sedimentary rock. The two general shapes are *tabular* (sheetlike) and *massive.* Intrusive igneous bodies that cut across existing sedimentary beds are said to be *discordant;* those that form parallel to existing sedimentary beds are *concordant.*

- *Dikes* are tabular, discordant igneous bodies produced when magma is injected into fractures that cut across rock layers. Tabular, concordant bodies, called *sills,* form when magma is injected along the bedding surfaces of sedimentary rocks. In many respects, sills closely resemble buried lava flows. *Laccoliths* are similar to sills but form from less fluid magma that collects as a lens-shaped mass that arches the overlying strata upward. *Batholiths,* the largest intrusive igneous bodies, frequently make up the cores of mountains, as exemplified by the Sierra Nevada.

- *Most active volcanoes are associated with plate boundaries.* Active areas of volcanism are found along mid-ocean ridges where seafloor spreading is occurring (*divergent plate boundaries*), in the vicinity of ocean trenches where one plate is being subducted beneath another (*convergent plate boundaries*), and in the interiors of plates themselves (intraplate volcanism). Rising plumes of hot mantle rock are the source of most intraplate volcanism.

Review Questions

1. What event triggered the May 18, 1980, eruption of Mount St. Helens? (See Box 5.1.)

2. List three factors that determine the nature of a volcanic eruption. What role does each play?

3. Why is a volcano fed by highly viscous magma likely to be a greater threat to life and property than a volcano supplied with very fluid magma?

4. Describe pahoehoe and aa lava.

5. List the main gases released during a volcanic eruption. Why are gases important in eruptions?

6. How do volcanic bombs differ from blocks of pyroclastic debris?

7. What is scoria? How is scoria different from pumice?

8. Compare a volcanic crater to a caldera.

9. Compare and contrast the three main types of volcanoes (size, composition, shape, and eruptive style).

10. Name a prominent volcano for each of the three types.

11. Briefly compare the eruptions of Kilauea and Parícutin.

12. Contrast the destruction of Pompeii with the destruction of St. Pierre (time frame, volcanic material, and nature of destruction).

13. Describe the formation of Crater Lake. Compare it to the caldera found on shield volcanoes, such as Kilauea.

14. What are the largest volcanic structures on Earth?

15. What is Shiprock, New Mexico, and how did it form?

16. How do the eruptions that created the Columbia Plateau differ from eruptions that create volcanic peaks?

17. Where are fissure eruptions most common?

18. Extensive pyroclastic flow deposits are most often associated with which volcanic structures?

19. Describe each of the four intrusive features discussed in the text (dike, sill, laccolith, and batholith).

20. Why might a laccolith be detected at Earth's surface before being exposed by erosion?

21. What is the largest of all intrusive igneous bodies? Is it tabular or massive? Concordant or discordant?

22. Describe how batholiths are emplaced.

23. Volcanism at divergent plate boundaries is associated with which rock type? What causes rocks to melt in these regions?

24. What is the Ring of Fire?

25. What type of plate boundary is associated with the Ring of Fire?

26. Are volcanoes in the Ring of Fire generally described as quiescent or violent? Name a volcano that would support your answer.

27. Describe the situation that generates magma along convergent plate boundaries.

28. What is the source of magma for intraplate volcanism?

29. What is meant by hot-spot volcanism?

30. How do geologists identify hot spots other than volcanism?

31. The Hawaiian Islands and Yellowstone are associated with which of the three zones of volcanism? Cascade Range? Flood basalt provinces?

32. What are the four changes in a volcanic area that are monitored in order to detect the movement of magma?

Key Terms

aa flow (p. 132)
batholith (p. 151)
block lavas (p. 133)
caldera (p. 144)
cinder cone (p. 138)
columnar joints (p. 151)
composite cone (p. 139)
concordant (p. 149)
conduit (p. 135)
continental volcanic arc (p. 155)
crater (p. 135)
dike (p. 150)

discordant (p. 149)
eruption column (p. 130)
fissure (p. 147)
fissure eruption (p. 147)
flood basalt (p. 147)
fumarole (p. 135)
hot spot (p. 158)
intraplate volcanism (p. 155)
island arc (p. 154)
laccolith (p. 151)
lahar (p. 143)
lava dome (p. 147)

lava tube (p. 133)
mantle plume (p. 155)
massive (p. 149)
nuée ardente (p. 142)
pahoehoe flow (p. 132)
parasitic cone (p. 135)
pillow lava (p. 133)
pipe (p. 135)
pluton (p. 149)
pumice (p. 134)
pyroclastic flow (p. 142)
pyroclastic material (p. 134)
scoria (p. 134)

scoria cone (p. 138)
shield volcano (p. 135)
sill (p. 150)
stock (p. 152)
stratovolcano (p. 139)
tabular (p. 149)
vent (p. 135)
viscosity (p. 126)
volcanic island arc (p. 154)
volcanic neck (p. 149)
volcano (p. 135)
volatiles (p. 128)
xenolith (p. 153)

Web Resources

The *Earth* Website uses the resources and flexibility of the Internet to aid in your study of the topics in this chapter. Written and developed by geology instructors, this site will help improve your understanding of geology. Visit **http://www.prenhall.com/tarbuck** and click on the cover of *Earth 9e* to find:

- Online review quizzes.
- Critical thinking exercises.
- Links to chapter-specific Web resources.
- Internet-wide key-term searches.

http://www.prenhall.com/tarbuck

GEODe: Earth

GEODe: Earth makes studying faster and more effective by reinforcing key concepts using animation, video, narration, interactive exercises and practice quizzes. A copy is included with every copy of *Earth*.

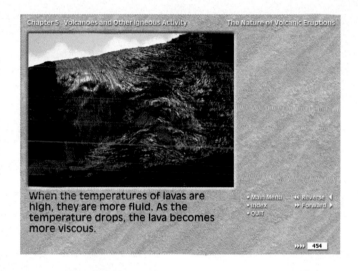

Chapter 5 Volcanoes and Other Igneous Activity   The Nature of Volcanic Eruptions

When the temperatures of lavas are high, they are more fluid. As the temperature drops, the lava becomes more viscous.

• Main Menu   ◀◀ Reverse ◀
• Index   ▶▶ Forward ▶
• QUIT

454

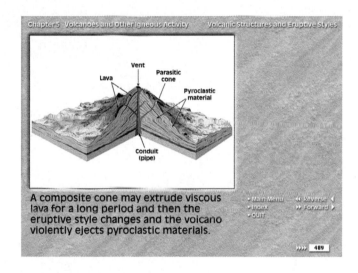

Chapter 5 Volcanoes and Other Igneous Activity   Volcanic Structures and Eruptive Styles

A composite cone may extrude viscous lava for a long period and then the eruptive style changes and the volcano violently ejects pyroclastic materials.

• Main Menu   ◀◀ Reverse ◀
• Index   ▶▶ Forward ▶
• QUIT

489

Weathering and Soil

Differential weathering is responsible for much of the spectacular scenery in Utah's Arches National Park. (Photo by Carr Clifton)

E arth's surface is constantly changing. Rock is disintegrated and decomposed, moved to lower elevations by gravity, and carried away by water, wind, or ice. In this manner Earth's physical landscape is sculptured. This chapter focuses on the first step of this never-ending process—weathering. What causes solid rock to crumble, and why does the type and rate of weathering vary from place to place? Soil, an important product of the weathering process and a vital resource, is also examined.

Earth's External Processes

GEODe
Weathering and Soil
▶ Earth's External Processes

Weathering, mass wasting, and erosion are called **external processes** because they occur at or near Earth's surface and are powered by energy from the Sun. External processes are a basic part of the rock cycle because they are responsible for transforming solid rock into sediment.

To the casual observer, the face of Earth may appear to be without change, unaffected by time. In fact, 200 years ago most people believed that mountains, lakes, and deserts were permanent features of an Earth that was thought to be no more than a few thousand years old. Today we know that Earth is 4.5 billion years old and that mountains eventually succumb to weathering and erosion, lakes fill with sediment or are drained by streams, and deserts come and go with changes in climate.

Earth is a dynamic body. Some parts of Earth's surface are gradually elevated by mountain building and volcanic activity. These **internal processes** derive their energy from Earth's interior. Meanwhile, opposing external processes are continually breaking rock apart and moving the debris to lower elevations. The latter processes include:

1. **Weathering**—the physical breakdown (disintegration) and chemical alteration (decomposition) of rock at or near Earth's surface.
2. **Mass wasting**—the transfer of rock and soil downslope under the influence of gravity.
3. **Erosion**—the physical removal of material by mobile agents such as water, wind, or ice.

In this chapter we will focus on rock weathering and the products generated by this activity. However, weathering cannot be easily separated from mass wasting and erosion because as weathering breaks rocks apart, erosion and mass wasting remove the rock debris. This transport of material by erosion and mass wasting further disintegrates and decomposes the rock.

Weathering

GEODe
Weathering and Soil
▶ Types of Weathering

Weathering goes on all around us, but it seems like such a slow and subtle process that it is easy to underestimate its importance. Yet it is worth remembering that weathering is a basic part of the rock cycle and thus a key process in the Earth system. Weathering is also important to humans—even to those of us who are not studying geology. For example, many of the life-sustaining minerals and elements found in soil, and ultimately in the food we eat, were freed from solid rock by weathering processes. As the chapter-opening photo, Figure 6.1, and many other images in this book illustrate, weathering also contributes to the formation of some of Earth's most spectacular scenery. Of course these same processes are also responsible for causing the deterioration of many of the structures we build (Figure 6.2).

All materials are susceptible to weathering. Consider, for example, the fabricated product concrete, which closely resembles a sedimentary rock called conglomerate. A newly poured concrete sidewalk has a smooth, fresh, unweathered look. However, not many years later the same sidewalk will appear chipped, cracked, and rough, with pebbles exposed at the surface. If a tree is nearby, its roots may heave and buckle the concrete as well. The same natural processes that eventually break apart a concrete sidewalk also act to disintegrate rock.

Weathering occurs when rock is mechanically fragmented (disintegrated) and/or chemically altered (decomposed). **Mechanical weathering** is accomplished by physical forces that break rock into smaller and smaller pieces without changing the rock's mineral composition. **Chemical weathering** involves a chemical transformation of rock into one or more new compounds. These two concepts can be illustrated with a piece of paper. The paper can be disintegrated by tearing it into smaller and smaller pieces, whereas decomposition occurs when the paper is set afire and burned.

Why does rock weather? Simply, weathering is the response of Earth materials to a changing environment. For instance, after millions of years of uplift and erosion, the

pieces from a single large one. Figure 6.3 shows that breaking a rock into smaller pieces increases the surface area available for chemical attack. An analogous situation occurs when sugar is added to a liquid. In this situation, a cube of sugar will dissolve much slower than an equal volume of sugar granules because the cube has much less surface area available for dissolution. Hence, by breaking rocks into smaller pieces, mechanical weathering increases the amount of surface area available for chemical weathering.

In nature, four important physical processes lead to the fragmentation of rock: frost wedging, expansion resulting from unloading, thermal expansion, and biological activity. In addition, although the work of erosional agents such as wind, glacial ice, and running water is usually considered separately from mechanical weathering, it is nevertheless important. As these mobile agents move rock debris, they relentlessly disintegrate these materials.

Frost Wedging

Repeated cycles of freezing and thawing represent an important process of mechanical weathering. Liquid water has the unique property of expanding about 9 percent upon freezing, because water molecules in the regular crystalline structure of ice are farther apart than they are in liquid water near the freezing point. As a result, water freezing in a confined space exerts tremendous outward pressure on the walls of its container. To verify this, consider a tightly sealed glass jar filled with water. As the water freezes, the container cracks.

In nature, water works its way into cracks in rock and, upon freezing, expands and enlarges these openings. After many freeze–thaw cycles, the rock is broken into angular fragments. This process is appropriately called **frost wedging** (Figure 6.4). Frost wedging is most pronounced in mountainous regions where a daily freeze–thaw cycle often exists (see Box 6.1). Here, sections of rock are wedged loose and may tumble into large piles called **talus slopes** that often form at the base of steep rock outcrops (Figure 6.4).

FIGURE 6.1 Bryce Canyon, National Park, Utah. When weathering accentuates differences in rocks, spectacular landforms are sometimes created. As the rock gradually disintegrates and decomposes, mass wasting and erosion remove the products of weathering. (Photo by Tom Bean)

rocks overlying a large, intrusive igneous body may be removed, exposing it at the surface. This mass of crystalline rock—formed deep below ground where temperatures and pressures are high—is now subjected to a very different and comparatively hostile surface environment. In response, this rock mass will gradually change. This transformation of rock is what we call weathering.

In the following sections we will discuss the various modes of mechanical and chemical weathering. Although we will consider these two categories separately, keep in mind that mechanical and chemical weathering processes usually work simultaneously in nature and reinforce each other.

Mechanical Weathering

 GEODe **Weathering and Soil**
EARTH ▶ Mechanical Weathering

When a rock undergoes *mechanical weathering*, it is broken into smaller and smaller pieces, each retaining the characteristics of the original material. The end result is many small

FIGURE 6.2 Even the most "solid" monuments that people erect eventually yield to the day in and day out attack of weathering processes. Temple of Olympian Zeus, Athens, Greece. (Photo by CORBIS)

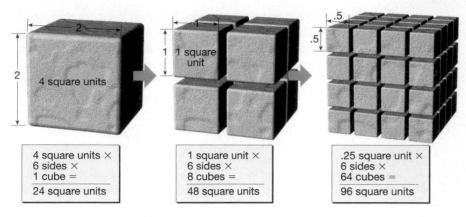

FIGURE 6.3 Chemical weathering can occur only to those portions of a rock that are exposed to the elements. Mechanical weathering breaks rock into smaller and smaller pieces, thereby increasing the surface area available for chemical attack.

Frost wedging also causes great destruction to highways in the northern United States, particularly in the early spring when the freeze–thaw cycle is well established. Roadways acquire numerous potholes and are occasionally heaved and buckled by this destructive force.

Salt Crystal Growth

Another expansive force that can split rocks is created by the growth of salt crystals. Rocky shorelines and arid regions are common settings for this process. It begins when sea spray from breaking waves or salty groundwater penetrates crevices and pore spaces in rock. As this water evaporates, salt crystals form. As these crystals gradually grow larger, they weaken the rock by pushing apart the surrounding grains or enlarging tiny cracks.

This same process can also contribute to crumbling roadways where salt is spread to melt snow and ice in winter. The salt dissolves in water and seeps into cracks that quite likely originated from frost action. When the water evaporates, the growth of salt crystals further breaks the pavement.

Unloading

When large masses of igneous rock, particularly granite, are exposed by erosion, concentric slabs begin to break loose. The process generating these onionlike layers is called

FIGURE 6.4 Frost wedging. As water freezes, it expands, exerting a force great enough to break rock. When frost wedging occurs in a setting such as this, the broken rock fragments fall to the base of the cliff and create a cone-shaped accumulation known as talus. (Photo by Tom & Susan Bean, Inc.)

BOX 6.1 ▶ UNDERSTANDING EARTH

The Old Man of the Mountain

The Old Man of the Mountain, also known as The Great Stone Face or simply The Profile, was one of New Hampshire's (*The Granite State*) best-known and most enduring symbols. Beginning in 1945, it appeared at the center of the official state emblem. It was a natural rock formation sculpted from Conway Red Granite that, when viewed from the proper location, gave the appearance of an old man (Figure 6.A, left). Each year hundreds of thousands of people traveled to view the Old Man, which protruded from high on Cannon Mountain, 360 meters (1200 feet) above Profile Lake in northern New Hampshire's Franconia Notch State Park.

On Saturday morning, May 3, 2003, the people of New Hampshire learned that the famous landmark had succumbed to nature and collapsed (Figure 6.A, right). The collapse ended decades of efforts to protect the state symbol from the same natural processes that created it in the first place. Ultimately, frost wedging and other weathering processes prevailed.

FIGURE 6.A   **(left)** The Old Man of the Mountain, high above Franconia Notch in New Hampshire's White Mountains, as it appeared prior to May 3, 2003. (Associated Press Photo) The inset shows the state emblem of New Hampshire. **(right)** The famous granite outcrop after it collapsed on May 3, 2003. The natural processes that sculpted the Old Man ultimately destroyed it. (Associated Press Photo)

sheeting. It is thought that this occurs, at least in part, because of the great reduction in pressure when the overlying rock is eroded away, a process called *unloading*. Accompanying this unloading, the outer layers expand more than the rock below and thus separate from the rock body (Figure 6.5). Continued weathering eventually causes the slabs to separate and spall off, creating **exfoliation domes** (*ex* = off, *folium* = leaf). Excellent examples of exfoliation domes are Stone Mountain, Georgia, and Half Dome and Liberty Cap in Yosemite National Park (Figure 6.5).

Deep underground mining provides us with another example of how rocks behave once the confining pressure is removed. Large rock slabs have been known to explode off the walls of newly cut mine tunnels because of the abruptly reduced pressure. Evidence of this type, plus the fact that fracturing occurs parallel to the floor of a quarry when large blocks of rock are removed, strongly supports the process of unloading as the cause of sheeting.

Although many fractures are created by expansion, others are produced by contraction during the crystallization of magma (see Figure 5.35 p. 152), and still others by tectonic forces during mountain building. Fractures produced by these activities generally form a definite pattern and are called *joints* (Figure 6.6). Joints are important rock structures that allow water to penetrate to depth and start the process of weathering long before the rock is exposed.

Thermal Expansion

The daily cycle of temperature may weaken rocks, particularly in hot deserts where daily variations may exceed 30°C. Heating a rock causes expansion, and cooling causes contraction. Repeated swelling and shrinking of minerals with different expansion rates should logically exert some stress on the rock's outer shell.

Although this process was once thought to be of major importance in the disintegration of rock, laboratory experiments have not substantiated this. In one test, unweathered rocks were heated to temperatures much higher than those normally experienced on Earth's surface and then cooled. This procedure was repeated many times to simulate hundreds of years of weathering, but the rocks showed little apparent change.

Nevertheless, pebbles in desert areas do show evidence of shattering that may have been caused by temperature changes (Figure 6.7). A proposed solution to this dilemma suggests that rocks must first be weakened by chemical weathering before they can be broken down by thermal activity. Further, this process may be aided by the rapid cooling of a desert rainstorm. Additional data are needed before a definite conclusion can be reached as to the impact of temperature variation on rock disintegration.

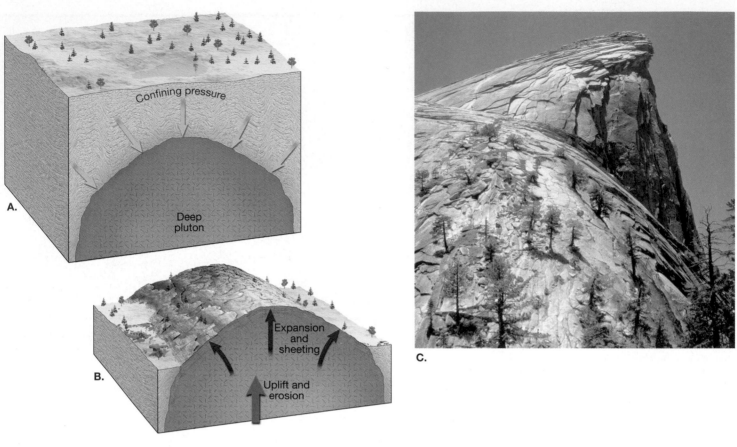

FIGURE 6.5 Sheeting is caused by the expansion of crystalline rock as erosion removes the overlying material. When the deeply buried pluton in **A** is exposed at the surface following uplift and erosion in **B,** the igneous mass fractures into thin slabs. The photo in **C** is of the summit of Half Dome in Yosemite National Park, California. It is an exfoliation dome and illustrates the onionlike layers created by sheeting. (Photo by Breck P. Kent)

FIGURE 6.6 Aerial view of nearly parallel joints near Moab, Utah. (Photo by Michael Collier)

FIGURE 6.7 These stones were once rounded stream gravels; however, long exposure in a hot desert climate disintegrated them. (Photo by C. B. Hunt, U.S. Geological Survey)

Biological Activity

Weathering is also accomplished by the activities of organisms, including plants, burrowing animals, and humans. Plant roots in search of nutrients and water grow into fractures, and as the roots grow, they wedge the rock apart (Figure 6.8). Burrowing animals further break down rock by moving fresh material to the surface, where physical and chemical processes can more effectively attack it. Decaying organisms also produce acids that contribute to chemical weathering. Where rock has been blasted in search of minerals or for road construction, the impact of humans is particularly noticeable.

Chemical Weathering

Weathering and Soil
▸ **Chemical Weathering**

In the preceding discussion of mechanical weathering, you learned that breaking rock into smaller pieces aids chemical weathering by increasing the surface area available for chemical attack. It should also be pointed out that chemical weathering contributes to mechanical weathering. It does so by weakening the outer portions of some rocks, which, in turn, makes them more susceptible to being broken by mechanical weathering processes.

Chemical weathering involves the complex processes that break down rock components and internal structures of minerals. Such processes convert the constituents to new minerals or release them to the surrounding environment. During this transformation, the original rock decomposes into substances that are stable in the surface environment. Consequently, the products of chemical weathering will remain essentially unchanged as long as they remain in an environment similar to the one in which they formed.

Water is by far the most important agent of chemical weathering. Pure water alone is a good solvent, and small amounts of dissolved materials result in increased chemical activity for weathering solutions. The major processes of chemical weathering are dissolution, oxidation, and hydrolysis. Water plays a leading role in each.

Dissolution

Perhaps the easiest type of decomposition to envision is the process of **dissolution.** Just as sugar dissolves in water, so too do certain minerals. One of the most water-soluble minerals is halite (common salt), which as you may recall, is composed of sodium and chloride ions. Halite readily dissolves in water because, although this compound maintains overall electrical neutrality, the individual ions retain their respective charges.

Moreover, the surrounding water molecules are polar—that is, the oxygen end of the molecule has a small residual negative charge; the end with hydrogen has a small positive charge. As the water molecules come in contact with halite, their negative ends approach sodium ions and their positive ends cluster about chloride ions. This disrupts the attractive forces in the halite crystal and releases the ions to the water solution (Figure 6.9).

Although most minerals are, for all practical purposes, insoluble in pure water, the presence of even a small amount

FIGURE 6.8 Root wedging widens fractures in rock and aids the process of mechanical weathering. Harriman State Park, New York. (Photo by Carr Clifton)

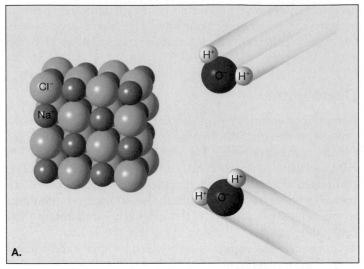

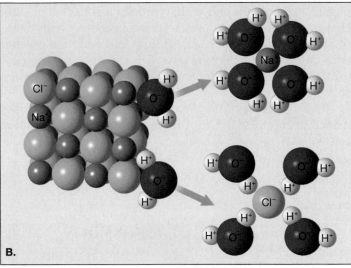

FIGURE 6.9   Illustration of halite dissolving in water. **A.** Sodium and chloride ions are attacked by the polar water molecules. **B.** Once removed, these ions are surrounded and held by a number of water molecules as shown.

of acid dramatically increases the corrosive force of water. (An acidic solution contains the reactive hydrogen ion, H^+.) In nature, acids are produced by a number of processes. For example, carbonic acid is created when carbon dioxide in the atmosphere dissolves in raindrops. As acidic rainwater soaks into the ground, carbon dioxide in the soil may increase the acidity of the weathering solution. Various organic acids are also released into the soil as organisms decay, and sulfuric acid is produced by the weathering of pyrite and other sulfide minerals.

Regardless of the source of the acid, this highly reactive substance readily decomposes most rocks and produces certain products that are water soluble. For example, the mineral calcite, $CaCO_3$, which composes the common building stones marble and limestone, is easily attacked by even a weakly acidic solution.

The overall reaction by which calcite dissolves in water containing carbon dioxide is:

$$CaCO_3 + (H^+ + HCO_3^-) \longrightarrow$$

calcite   carbonic acid

$$Ca^{2+} + 2\,HCO_3^-$$

calcium ion   bicarbonate ion

During this process, the insoluble calcium carbonate is transformed into soluble products. In nature, over periods of thousands of years, large quantities of limestone are dissolved and carried away by underground water. This activity is clearly evidenced by the large number of subsurface caverns found in every one of the contiguous 48 states (Figure 6.10). Monuments and buildings made of limestone or marble are also subjected to the corrosive work of acids, particularly in industrial areas that have smoggy, polluted air (see Box 6.2).

The soluble ions from reactions of this type are retained in our underground water supply. It is these dissolved ions that are responsible for the so-called hard water found in many locales. Simply, hard water is undesirable because the active ions react with soap to produce an insoluble material that renders soap nearly useless in removing dirt. To solve this problem, a water softener can be used to remove these ions, generally by replacing them with others that do not chemically react with soap.

Oxidation

Everyone has seen iron and steel objects that rusted when exposed to water (Figure 6.11). The same thing can happen to iron-rich minerals. The process of rusting occurs when oxygen combines with iron to form iron oxide as follows:

$$4\,Fe \quad + \quad 3\,O_2 \quad \longrightarrow \quad 2\,Fe_2O_3$$

iron   oxygen   iron oxide (hematite)

FIGURE 6.10   The dissolving power of carbonic acid plays an important part in forming limestone caverns. Carlsbad Caverns National Park, New Mexico. (Photo by Hohle Kalkstein/DRK Photo)

FIGURE 6.11 Iron reacts with oxygen to form iron oxide as seen on these rusted barrels. (Photo by Stephen J. Krasemann/DRK Photo)

Hydrolysis

The most common mineral group, the silicates, is decomposed primarily by the process of **hydrolysis** (*hydro* = water, *lysis* = a loosening), which basically is the reaction of any substance with water. Ideally, the hydrolysis of a mineral could take place in pure water as some of the water molecules dissociate to form the very reactive hydrogen (H^+) and hydroxyl (OH^-) ions. It is the hydrogen ion that attacks and replaces other positive ions found in the crystal lattice. With the introduction of hydrogen ions into the crystalline structure, the original orderly arrangement of atoms is destroyed and the mineral decomposes.

In nature, water usually contains other substances that contribute additional hydrogen ions, thereby greatly accelerating hydrolysis. The most common of these substances is carbon dioxide, CO_2, which dissolves in water to form carbonic acid, H_2CO_3. Rain dissolves some carbon dioxide in the atmosphere, and additional amounts, released by decaying organic matter, are acquired as the water percolates through the soil.

In water, carbonic acid ionizes to form hydrogen ions (H^+) and bicarbonate ions (HCO_3^-). To illustrate how a rock undergoes hydrolysis in the presence of carbonic acid, let's examine the chemical weathering of granite, a common continental rock. Recall that granite consists mainly of quartz and potassium feldspar. The weathering of the potassium feldspar component of granite is as follows:

$$2\ KAlSi_3O_8 + 2(H^+ + HCO_3^-) + H_2O \longrightarrow$$

potassium feldspar	carbonic acid	water

$$Al_2Si_2O_5(OH)_4 \quad + \quad 2\,K^+ \quad + \quad 2\,HCO_3^- + \quad 4\,SiO_2$$

kaolinite (residual clay)	potassium ion	bicarbonate ion	silica

in solution

FIGURE 6.12 This water seeping from an abandoned mine in Colorado is an example of *acid mine drainage*. Acid mine drainage is water with a high concentration of sulfuric acid (H_2SO_4) produced by the oxidation of sulfide minerals such as pyrite. When such acid-rich water migrates from its source, it may pollute surface waters and groundwater and cause significant ecological damage. (Photo by Tim Haske/Profiles West/Index Stock Photography, Inc.)

This type of chemical reaction, called **oxidation,*** occurs when electrons are lost from one element during the reaction. In this case, we say that iron was oxidized because it lost electrons to oxygen. Although the oxidation of iron progresses very slowly in a dry environment, the addition of water greatly speeds the reaction.

Oxidation is important in decomposing such ferromagnesian minerals as olivine, pyroxene, and hornblende. Oxygen readily combines with the iron in these minerals to form the reddish-brown iron oxide called *hematite* (Fe_2O_3) or in other cases a yellowish-colored rust called *limonite* [$FeO(OH)$]. These products are responsible for the rusty color on the surfaces of dark igneous rocks, such as basalt, as they begin to weather. However, oxidation can occur only after iron is freed from the silicate structure by another process, called hydrolysis.

Another important oxidation reaction occurs when sulfide minerals such as pyrite decompose. Sulfide minerals are major constituents in many metallic ores, and pyrite is frequently associated with coal deposits as well. In a moist environment, chemical weathering of pyrite (FeS_2) yields sulfuric acid (H_2SO_4) and iron oxide [$FeO(OH)$]. In many mining locales this weathering process creates a serious environmental hazard, particularly in humid areas where abundant rainfall infiltrates spoil banks (waste material left after coal or other minerals are removed). This so-called *acid mine drainage* eventually makes its way to streams, killing aquatic organisms and degrading aquatic habitats (Figure 6.12).

---

*The reader should note that *oxidation* is a term referring to any chemical reaction in which a compound or radical loses electrons. The element oxygen is not necessarily present.

BOX 6.2 ▶ EARTH AS A SYSTEM

Acid Precipitation—A Human Impact on the Earth System

Humans are part of the complex interacting whole we call the Earth system. As such, our actions cause changes to all the other parts of the system. For example, by going about our normal routine, we humans modify the composition of the atmosphere. These atmospheric modifications in turn cause unintended and unwanted changes to occur in the hydrosphere, biosphere, and solid Earth. Acid precipitation is one small but significant example.

Decomposed stone monuments and structures are common sights in many cities (Figure 6.B). Although we expect rock to gradually decompose, many of these monuments have succumbed prematurely. An important cause for this accelerated chemical weathering is acid precipitation.

Rain is naturally somewhat acidic (Figure 6.C). When carbon dioxide from the atmosphere dissolves in water, the product is weak carbonic acid. However, the term *acid precipitation* refers to precipitation that is much more acidic than natural, unpolluted rain and snow.

As a consequence of burning large quantities of fossil fuels, like coal and petroleum products, about 40 million tons of sulfur and nitrogen oxides are released into the atmosphere each year in the United States. The major sources of these emissions include power-generating plants,

FIGURE 6.B   Acid rain accelerates the chemical weathering of stone monuments and structures. (Photo by Adam Hart-Davis/Science Photo Library/Photo Researchers, Inc.)

industrial processes such as ore smelting and petroleum refining, and vehicles of all kinds. Through a series of complex chemical reactions, some of these pollutants are converted into acids that then fall to Earth's surface as rain or snow. Another portion is deposited in dry form and subsequently converted into acid after coming in contact with precipitation, dew, or fog.

Northern Europe and eastern North America have experienced widespread acid rain for some time. Studies have also shown that acid rain occurs in many other regions, including western North America, Japan, China, Russia, and South America. In addition to local pollution sources, a portion of the acidity found in the northeastern United States and eastern Canada originates hundreds of kilometers away in industrialized regions to the south and southwest. This situation occurs because many pollutants remain in the atmosphere as long as five days, during which time they may be transported great distances.

The damaging environmental effects of acid rain are thought to be considerable in some areas and imminent in others (Figure 6.D). The best-known effect is an increased acidity in thousands of lakes in Scandinavia and eastern North America. Accompanying this have been substantial increases in

---

In this reaction, the hydrogen ions (H^+) attack and replace potassium ions (K^+) in the feldspar structure, thereby disrupting the crystalline network. Once removed, the potassium is available as a nutrient for plants or becomes the soluble salt potassium bicarbonate ($KHCO_3$), which may be incorporated into other minerals or carried to the ocean.

The most abundant product of the chemical breakdown of potassium feldspar is the clay mineral kaolinite. Clay minerals are the end products of weathering and are very stable under surface conditions. Consequently, clay minerals make up a high percentage of the inorganic material in soils. Moreover, the most abundant sedimentary rock, shale, contains a high proportion of clay minerals.

In addition to the formation of clay minerals during the weathering of potassium feldspar, some silica is removed from the feldspar structure and carried away by groundwater. This dissolved silica will eventually precipitate, producing nodules of chert or flint, or it will fill in the pore spaces between grains of sediment, or it will be carried to the

ocean, where microscopic animals will remove it from the water to build hard silica shells.

To summarize, the weathering of potassium feldspar generates a residual clay mineral, a soluble salt (potassium bicarbonate), and some silica, which enters into solution.

Quartz, the other main component of granite, is very resistant to chemical weathering; it remains substantially unaltered when attacked by weak acidic solutions. As a result, when granite weathers, the feldspar crystals dull and slowly turn to clay, releasing the once interlocked quartz grains, which still retain their fresh, glassy appearance. Although some of the quartz remains in the soil, much is eventually transported to the sea or to other sites of deposition, where it becomes the main constituent of such features as sandy beaches and sand dunes. In time these quartz grains may be lithified to form the sedimentary rock sandstone.

Table 6.1 lists the weathered products of some of the most common silicate minerals. Remember that silicate minerals make up most of Earth's crust and that these minerals are

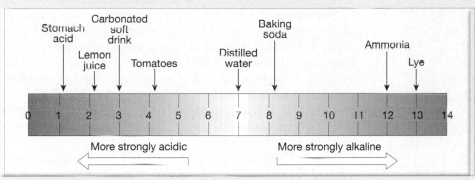

FIGURE 6.C   The *pH scale* is a common measure of the degree of acidity or alkalinity of a solution. The scale ranges from 0 to 14, with a value of 7 denoting a solution that is neutral. Values below 7 indicate greater acidity, whereas numbers above 7 indicate greater alkalinity. The pH values of some familiar substances are shown on the diagram. Although distilled water is neutral (pH 7), rainwater is naturally acidic. It is important to note that the pH scale is logarithmic; that is, each whole number increment indicated a tenfold difference. Thus, pH 4 is 10 times more acidic than pH 5 and 100 times (10×10) more acidic than pH 6.

FIGURE 6.D   Damage to forests by acid precipitation is well documented in Europe and eastern North America. These trees in the Great Smoky Mountains have been injured by acid-laden clouds. (Photo by Doug Locke/Dembinsky Photo Associates)

dissolved aluminum leached from the soil by the acidic water, which is toxic to fish. Consequently, some lakes are virtually devoid of fish, and others are approaching this condition. Ecosystems are characterized by many interactions at many levels of organization, which means that evaluating the effects of acid precipitation on these complex systems is difficult and expensive and far from complete.

In addition to the many lakes that can no longer support fish, research indicates that acid precipitation may also reduce agricultural crop yields and impair the productivity of forests. Acid rain not only harms the foliage but also damages roots and leaches nutrient minerals from the soil. Finally, acid precipitation promotes the corrosion of metals and contributes to the destruction of stone structures.

essentially composed of only eight elements. When chemically weathered, the silicate minerals yield sodium, calcium, potassium, and magnesium ions that form soluble products, which may be removed from groundwater. The element iron combines with oxygen, producing relatively insoluble iron oxides, most notably hematite and limonite, which give soil a reddish-brown or yellowish color. Under most conditions the three remaining elements— aluminum, silicon, and oxygen—join with water to produce residual clay minerals. However, even the highly insoluble clay minerals are very slowly removed by subsurface water.

Alterations Caused by Chemical Weathering

As noted earlier, the most significant result of chemical weathering is the decomposition of unstable minerals and the generation or retention of those materials that are stable at Earth's surface. This accounts for the predominance of certain minerals in the surface material we call soil.

In addition to altering the internal structure of minerals, chemical weathering causes physical changes as well. For instance, when angular rock masses are chemically weathered as water enters along joints, the boulders take on a spherical shape. The gradual rounding of the corners and edges of angular blocks is illustrated in Figure 6.13. The corners are

TABLE 6.1	Products of Weathering	
Mineral	Residual Products	Material in Solution
Quartz	Quartz grains	Silica
Feldspars	Clay minerals	Silica, K^+, Na^+, Ca^{2+}
Amphibole	Clay minerals	Silica
(hornblende)	Limonite	Ca^{2+}, Mg^{2+}
	Hematite	
Olivine	Limonite	Silica
	Hematite	Mg^{2+}

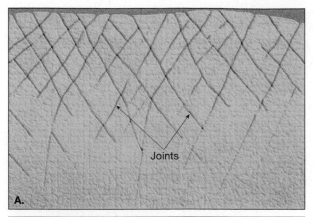

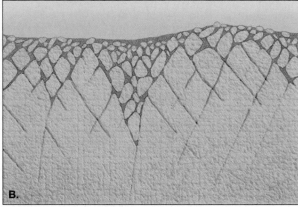

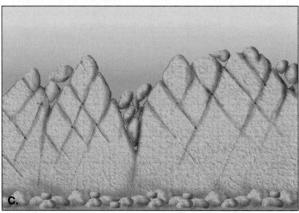

FIGURE 6.13   Spheroidal weathering of extensively jointed rock. Chemical weathering associated with water moving through the joints enlarges them. Because the rocks are attacked more on the corners and edges, they take on a spherical shape. The photo shows spheroidal weathering in Joshua Tree National Park, California. (Photo by E. J. Tarbuck)

attacked most readily because of the greater surface area for their volume as compared to the edges and faces. This process, called **spheroidal weathering,** gives the weathered rock a more rounded or spherical shape (Figure 6.13D).

Sometimes during the formation of spheroidal boulders, successive shells separate from the rock's main body (Figure 6.14). Eventually the outer shells break off, allowing the chemical weathering activity to penetrate deeper into the boulder. This spherical scaling results because, as the minerals in the rock weather to clay, they increase in size through the addition of water to their structure. This increased bulk exerts an outward force that causes concentric layers of rock to break loose and fall off.

Hence, chemical weathering does produce forces great enough to cause mechanical weathering. This type of spheroidal weathering in which shells spall off should not be confused with the phenomenon of sheeting discussed earlier. In sheeting, the fracturing occurs as a result of unloading, and the rock layers that separate from the main body are largely unaltered at the time of separation.

Students Sometimes Ask . . .

Is the clay created by chemical weathering the same clay that's used in making ceramics?

Yes. Kaolinite, the clay described in the section on hydrolysis, is called *china clay* and is used for high-quality porcelain. However, far greater quantities of this clay are used as a coating in the manufacture of high-quality paper, such as that used in this book.

Weathering actually creates many different clay minerals that have many different uses. Clay minerals are used in making bricks, tiles, sewer pipes, and cement. Clays are used as lubricants in the bore holes of oil-drilling rigs and are a common ingredient in paint. Products as varied as your car's catalytic converter and filters used in beer- and wine-making rely on clay minerals.

FIGURE 6.14   Successive shells are loosened as the weathering process continues to penetrate ever deeper into the rock. (Photo by Martin Schmidt, Jr.)

Rates of Weathering

GEODe   Weathering and Soil

▶ Rates of Weathering

Several factors influence the type and rate of rock weathering. We have already seen how mechanical weathering affects the rate of weathering. By breaking rock into smaller pieces, the amount of surface area exposed to chemical weathering increases. Other important factors examined here include the roles of rock characteristics and climate.

Rock Characteristics

Rock characteristics encompass all of the chemical traits of rocks, including mineral composition and solubility. In addition, any physical features, such as joints (cracks), can be important because they influence the ability of water to penetrate rock.

The variations in weathering rates due to the mineral constituents can be demonstrated by comparing old headstones made from different rock types. Headstones of granite, which is composed of silicate minerals, are relatively resistant to chemical weathering. We can see this by examining the inscriptions on the headstones shown in Figure 6.15. In contrast, the marble headstone shows signs of extensive chemical alteration over a relatively short period. Marble is composed of calcite (calcium carbonate), which readily dissolves even in a weakly acidic solution.

The most abundant mineral group, the silicates, weathers in the order shown in Figure 6.16. This arrangement of minerals is identical to Bowen's reaction series. The order in which the silicate minerals weather is essentially the same as their order of crystallization. The explanation for this is related to the crystalline structure of silicate minerals. The strength of silicon–oxygen bonds is great. Because quartz is composed entirely of these strong bonds, it is very resistant to weathering. By contrast, olivine has far fewer silicon–oxygen bonds and is not nearly as resistant to chemical weathering.

Climate

Climatic factors, particularly temperature and moisture, are crucial to the rate of rock weathering. One important example from mechanical weathering is that the frequency of freeze–thaw cycles greatly affects the amount of frost wedging. Temperature and moisture also exert a strong influence on rates of chemical weathering and on the kind and amount of vegetation present. Regions with lush vegetation often have a thick mantle of soil rich in decayed organic matter from which chemically active fluids such as carbonic acid and humic acids are derived.

The optimum environment for chemical weathering is a combination of warm temperatures and abundant moisture. In polar regions chemical weathering is ineffective because frigid temperatures keep the available moisture locked up as ice, whereas in arid regions there is insufficient moisture to foster rapid chemical weathering.

Human activities can influence the composition of the atmosphere, which in turn can impact the rate of chemical weathering. Box 6.2 examines one well-known example—acid rain.

Differential Weathering

Masses of rock do not weather uniformly. Take a moment to look back at the photo of a dike in Figure 5.33 (p. 151) The durable igneous mass stands above the surrounding terrain like a stone wall. A glance at Figure 6.1 shows an additional example of this phenomenon, called **differential weathering.** The results vary in scale from the rough, uneven surface of the marble headstone in Figure 6.15 to the boldly sculpted exposures shown in the chapter-opening photo.

Many factors influence the rate of rock weathering. Among the most important are variations in the composition of the rock. More resistant rock protrudes as ridges (see Figure 14.13, p. 390) or pinnacles, or as steeper cliffs on an irregular hillside (see Figure 7.4, p. 197). The number and spacing of joints can also be a significant factor (see Figures 6.6 and 6.13). Differential weathering and subsequent erosion are responsible for creating many unusual and sometimes spectacular rock formations and landforms.

Soil

Soil covers most land surfaces. Along with air and water, it is one of our most indispensable resources (Figure 6.17). Also like air and water, soil is taken for granted by many of us. The following quote helps put this vital layer in perspective.

FIGURE 6.15   An examination of headstones reveals the rate of chemical weathering on diverse rock types. The granite headstone (left) was erected four years before the marble headstone (right). The inscription date of 1872 on the marble monument is nearly illegible. (Photos by E. J. Tarbuck)

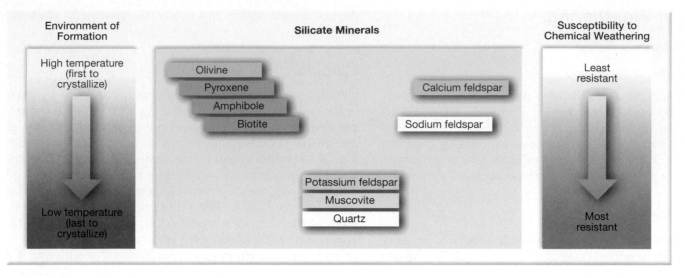

Environment of Formation

High temperature (first to crystallize)

Low temperature (last to crystallize)

Silicate Minerals

Olivine
Pyroxene
Amphibole
Biotite

Calcium feldspar

Sodium feldspar

Potassium feldspar
Muscovite
Quartz

Susceptibility to Chemical Weathering

Least resistant

Most resistant

FIGURE 6.16   The weathering of common silicate minerals. The order in which the silicate minerals chemically weather is essentially the same as their order of crystallization.

Science, in recent years, has focused more and more on the Earth as a planet, one that for all we know is unique—where a thin blanket of air, a thinner film of water, and the thinnest veneer of soil combine to support a web of life of wondrous diversity in continuous change.*

Soil has accurately been called "the bridge between life and the inanimate world." All life—the entire biosphere—owes its existence to a dozen or so elements that must ultimately come from Earth's crust. Once weathering and other processes create soil, plants carry out the intermediary role of assimilating the necessary elements and making them available to animals, including humans.

An Interface in the Earth System

When Earth is viewed as a system, soil is referred to as an *interface*—a common boundary where different parts of a system interact. This is an appropriate designation because soil forms where the geosphere, the atmosphere, the hydrosphere, and the biosphere meet. Soil is a material that develops in response to complex environmental interactions among different parts of the Earth system. Over time, soil gradually evolves to a state of equilibrium or balance with the environment. Soil is dynamic and sensitive to almost every aspect of its surroundings. Thus, when environmental changes occur, such as climate, vegetative cover, and animal (including human) activity, the soil responds. Any such change produces a gradual alteration of soil characteristics until a new balance is reached. Although thinly distributed over the land surface, soil functions as a fundamental interface, providing an excellent example of the integration among many parts of the Earth system.

What Is Soil?

With few exceptions, Earth's land surface is covered by **regolith** (*rhegos* = blanket, *lithos* = stone), the layer of rock and mineral fragments produced by weathering. Some would call this material soil, but soil is more than an accumulation of weathered debris. **Soil** is a combination of mineral and organic matter, water, and air—that portion of the regolith that supports the growth of plants. Although the proportions of the major components in soil vary, the same four components always are present to some extent (Figure 6.17). About one-half

25% air

45% mineral matter

25% water

5% organic matter

FIGURE 6.17   Soil is an essential resource that we often take for granted. Soil is not a living entity, but it contains a great deal of life. Moreover, this complex medium supports nearly all plant life, which in turn supports animal life. The pie chart shows the composition (by volume) of a soil in good condition for plant growth. Although the percentages vary, each soil is composed of mineral and organic matter, water, and air. (Photo by Colin Molyneux/Getty Images)

*Jack Eddy, "A Fragile Seam of Dark Blue Light," in *Proceedings of the Global Change Research Forum.* U.S. Geological Survey Circular 1086, 1993, p. 15.

of the total volume of a good-quality surface soil is a mixture of disintegrated and decomposed rock (mineral matter) and **humus,** the decayed remains of animal and plant life (organic matter). The remaining half consists of pore spaces among the solid particles where air and water circulate.

Although the mineral portion of the soil is usually much greater than the organic portion, humus is an essential component. In addition to being an important source of plant nutrients, humus enhances the soil's ability to retain water. Because plants require air and water to live and grow, the portion of the soil consisting of pore spaces that allow for the circulation of these fluids is as vital as the solid soil constituents.

Soil water is far from "pure" water; instead, it is a complex solution containing many soluble nutrients. Soil water not only provides the necessary moisture for the chemical reactions that sustain life, it also supplies plants with nutrients in a form they can use. The pore spaces not filled with water contain air. This air is the source of necessary oxygen and carbon dioxide for most microorganisms and plants that live in the soil.

Controls of Soil Formation

Soil is the product of the complex interplay of several factors, including parent material, time, climate, plants and animals, and topography. Although all these factors are interdependent, their roles will be examined separately.

Parent Material

The source of the weathered mineral matter from which soils develop is called the **parent material** and is a major factor influencing a newly forming soil. Gradually it undergoes physical and chemical changes as the processes of soil formation progress. Parent material can either be the underlying bedrock or a layer of unconsolidated deposits. When the parent material is bedrock, the soils are termed *residual soils.* By contrast, those developed on unconsolidated sediment are called *transported soils* (Figure 6.18). It should

No soil development because of very steep slope

Transported soil is developed on unconsolidated deposits

Residual soil is developed on bedrock

Thick soil

Bedrock

Thinner soil on steep slope because of erosion

Unconsolidated deposits

FIGURE 6.18 The parent material for residual soils is the underlying bedrock, whereas transported soils form on unconsolidated deposits. Also note that as slopes become steeper, soil becomes thinner. (Left and center photos by E. J. Tarbuck; right photo by Grilly Bernard/Getty Images, Inc./Stone Allstock)

be pointed out that transported soils form *in place* on parent materials that have been carried from elsewhere and deposited by gravity, water, wind, or ice.

The nature of the parent material influences soils in two ways. First, the type of parent material will affect the rate of weathering and thus the rate of soil formation. Also, because unconsolidated deposits are already partly weathered, soil development on such material will likely progress more rapidly than when bedrock is the parent material. Second, the chemical makeup of the parent material will affect the soil's fertility. This influences the character of the natural vegetation the soil can support.

At one time the parent material was thought to be the primary factor causing differences among soils. However, soil scientists came to understand that other factors, especially climate, are more important. In fact, it was found that similar soils often develop from different parent materials and that dissimilar soils have developed from the same parent material. Such discoveries reinforce the importance of other soil-forming factors.

Time

Time is an important component of *every* geological process, and soil formation is no exception. The nature of soil is strongly influenced by the length of time that processes have been operating. If weathering has been going on for a comparatively short time, the character of the parent material strongly influences the characteristics of the soil. As weathering processes continue, the influence of parent material on soil is overshadowed by the other soil-forming factors, especially climate. The amount of time required for various soils to evolve cannot be listed because the soil-forming processes act at varying rates under different circumstances. However, as a rule, the longer a soil has been forming, the thicker it becomes and the less it resembles the parent material.

Climate

Climate is considered to be the most influential control of soil formation. Temperature and precipitation are the elements that exert the strongest impact on soil formation. Variations in temperature and precipitation determine whether chemical or mechanical weathering will predominate and also greatly influence the rate and depth of weathering. For instance, a hot, wet climate may produce a thick layer of chemically weathered soil in the same amount of time that a cold, dry climate produces a thin mantle of mechanically weathered debris. Also, the amount of precipitation influences the degree to which various materials are removed from the soil by percolating waters (a process called *leaching*), thereby affecting soil fertility. Finally, climatic conditions are an important control on the type of plant and animal life present.

Plants and Animals

Plants and animals play a vital role in soil formation. The types and abundance of organisms present have a strong influence on the physical and chemical properties of a soil (Figure 6.19). In fact, for well-developed soils in many regions, the significance of natural vegetation in influencing soil type is frequently implied in the description used by soil scientists. Such phrases as *prairie soil, forest soil,* and *tundra soil* are common.

Plants and animals furnish organic matter to the soil. Certain bog soils are composed almost entirely of organic matter, whereas desert soils might contain as little as a small fraction of 1 percent. Although the quantity of organic matter varies substantially among soils, it is the rare soil that completely lacks it.

The primary source of organic matter in soil is plants, although animals and an infinite number of microorganisms also contribute. When organic matter is decomposed, important nutrients are supplied to plants, as well as to animals and microorganisms living in the soil. Consequently, soil fertility is in part related to the amount of organic matter present. Furthermore, the decay of plant and animal remains causes the formation of various organic acids. These complex acids hasten the weathering process. Organic matter also has a high water-holding ability and thus aids water retention in a soil.

Microorganisms, including fungi, bacteria, and single-celled protozoa, play an active role in the decay of plant and animal remains. The end product is humus, a material that no longer resembles the plants and animals from which it is formed. In addition, certain microorganisms aid soil fertility because they have the ability to convert atmospheric nitrogen into soil nitrogen.

Earthworms and other burrowing animals act to mix the mineral and organic portions of a soil. Earthworms, for example, feed on organic matter and thoroughly mix soils in which they live, often moving and enriching many tons per acre each year. Burrows and holes also aid the passage of water and air through the soil.

Topography

The lay of the land can vary greatly over short distances. Such variations in topography can lead to the development of a variety of localized soil types. Many of the differences exist because the length and steepness of slopes have a significant impact on the amount of erosion and the water content of soil.

On steep slopes, soils are often poorly developed. In such situations the quantity of water soaking in is slight; as a result, the moisture content of the soil may not be sufficient for vigorous plant growth. Further, because of accelerated erosion on steep slopes, the soils are thin or in some cases nonexistent (Figure 6.18).

In contrast, poorly drained and waterlogged soils found in bottomlands have a much different character. Such soils are usually thick and dark. The dark color results from the large quantity of organic matter that accumulates because saturated conditions retard the decay of vegetation. The optimum terrain for soil development is a flat-to-undulating upland surface. Here we find good drainage, minimum erosion, and sufficient infiltration of water into the soil.

FIGURE 6.19 The northern coniferous forest in Alaska's Denali National Park. The type of vegetation strongly influences soil formation. The organic litter received by the soil from the conifers is high in acid resins, which contributes to an accumulation of acid in the soil. As a result, intensive acid leaching is an important soil-forming process here. (Photo by Carr Clifton)

Slope orientation, or the direction the slope is facing, is another consideration. In the midlatitudes of the Northern Hemisphere, a south-facing slope will receive a great deal more sunlight than a north-facing slope. In fact, a steep north-facing slope may receive no direct sunlight at all. The difference in the amount of solar radiation received will cause differences in soil temperature and moisture, which in turn influence the nature of the vegetation and the character of the soil.

Although this section dealt separately with each of the soil-forming factors, remember that all of them work together to form soil. No single factor is responsible for a soil's character; rather, it is the combined influence of parent material, time, climate, plants and animals, and topography that determines this character.

The Soil Profile

Because soil-forming processes operate from the surface downward, variations in composition, texture, structure, and color gradually evolve at varying depths. These vertical differences, which usually become more pronounced as time passes, divide the soil into zones or layers known as **horizons.** If you were to dig a trench in soil, you would see that its walls are layered. Such a vertical section through all of the soil horizons constitutes the **soil profile** (Figure 6.20).

FIGURE 6.20 A soil profile is a vertical cross section from the surface through all of the soil's horizons and into the parent material. **A.** This profile shows a well-developed soil in southeastern South Dakota. (Photo by E. J. Tarbuck) **B.** The boundaries between horizons in this soil in Puerto Rico are indistinct, giving it a relatively uniform appearance. (Photo courtesy of Soil Science Society of America)

A.

B.

Figure 6.21 presents an idealized view of a well-developed soil profile in which five horizons are identified. From the surface downward, they are designated as *O, A, E, B,* and *C.* These five horizons are common to soils in temperate regions. The characteristics and extent of development of horizons vary in different environments. Thus, different localities exhibit soil profiles that can contrast greatly with one another.

The *O* soil horizon consists largely of organic material. This is in contrast to the layers beneath it, which consist mainly of mineral matter. The upper portion of the *O* horizon is primarily plant litter, such as loose leaves and other organic debris that are still recognizable. By contrast, the lower portion of the *O* horizon is made up of partly decomposed organic matter (humus) in which plant structures can no longer be identified. In addition to plants, the *O* horizon is teeming with microscopic life, including bacteria, fungi, algae, and insects. All of these organisms contribute oxygen, carbon dioxide, and organic acids to the developing soil.

Underlying the organic-rich *O* horizon is the *A* horizon. This zone is largely mineral matter, yet biological activity is

high and humus is generally present—up to 30 percent in some instances. Together the *O* and *A* horizons make up what is commonly called the *topsoil.* Below the *A* horizon, the *E* horizon is a light-colored layer that contains little organic material. As water percolates downward through this zone, finer particles are carried away. This washing out of fine soil components is termed **eluviation** (*elu* = get away from, *via* = a way). Water percolating downward also dissolves soluble inorganic soil components and carries them to deeper zones. This depletion of soluble materials from the upper soil is termed **leaching.**

Immediately below the *E* horizon is the *B* horizon, or *subsoil.* Much of the material removed from the *E* horizon by eluviation is deposited in the *B* horizon, which is often referred to as the *zone of accumulation.* The accumulation of the fine clay particles enhances water retention in the subsoil. The *O, A, E,* and *B* horizons together constitute the **solum,** or "true soil." It is in the solum that the soil-forming processes are active and that living roots and other plant and animal life are largely confined.

Below the solum and above the unaltered parent material is the *C* horizon, a layer characterized by partially altered parent material. Whereas the *O, A, E,* and *B* horizons bear little resemblance to the parent material, it is easily identifiable in the *C* horizon. Although this material is undergoing changes that will eventually transform it into soil, it has not yet crossed the threshold that separates regolith from soil.

The characteristics and extent of development can vary greatly among soils in different environments. The boundaries between soil horizons may be sharp, or the horizons may blend gradually from one to another. Consequently, a well-developed soil profile indicates that environmental conditions have been relatively stable over an extended time span and that the soil is *mature.* By contrast, some soils lack horizons altogether.

Such soils are called *immature* because soil building has been going on for only a short time. Immature soils are also

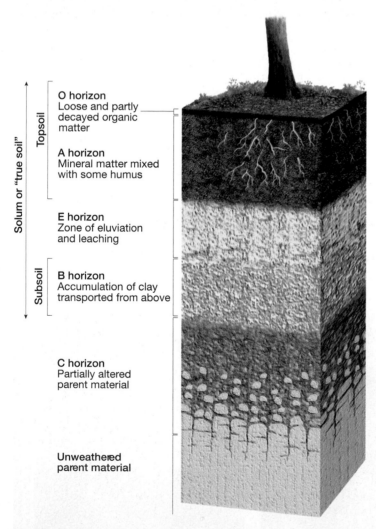

O horizon
Loose and partly decayed organic matter

A horizon
Mineral matter mixed with some humus

E horizon
Zone of eluviation and leaching

B horizon
Accumulation of clay transported from above

C horizon
Partially altered parent material

Unweathered parent material

Topsoil

Subsoil

Solum or "true soil"

FIGURE 6.21 Idealized soil profile from a humid climate in the middle latitudes. The topsoil and subsoil together comprise the solum or "true soil."

characteristic of steep slopes, where erosion continually strips away the soil, preventing full development.

Classifying Soils

There are many variations from place to place and from time to time among the factors that control soil formation. These differences lead to a bewildering variety of soil types. To cope with such variety, it is essential to devise some means of classifying the vast array of data to be studied. By establishing groups consisting of items that have certain important characteristics in common, order and simplicity are introduced. Bringing order to large quantities of information not only aids comprehension and understanding but also facilitates analysis and explanation.

In the United States, soil scientists have devised a system for classifying soils known as the **Soil Taxonomy.** It emphasizes the physical and chemical properties of the soil profile and is organized on the basis of observable soil characteristics. There are six hierarchical categories of classification, ranging from *order*, the broadest category, to *series*, the most specific category. The system recognizes 12 soil orders and more than 19,000 soil series.

The names of the classification units are combinations of syllables, most of which are derived from Latin or Greek. The names are descriptive. For example, soils of the order Aridosol (from the Latin *aridus*, dry, and *solum*, soil) are characteristically dry soils in arid regions. Soils in the order Inceptisols (from

Latin *inceptum*, beginning, and *solum*, soil) are soils with only the beginning or inception of profile development.

Brief descriptions of the 12 basic soil orders are provided in Table 6.2. Figure 6.22 shows the complex worldwide distribution pattern of the Soil Taxonomy's 12 soil orders (see Box 6.3). Like many classification systems, the Soil Taxonomy is not suitable for every purpose. It is especially useful for agricultural and related land-use purposes, but it is not a useful system for engineers who are preparing evaluations of potential construction sites.

Soil Erosion

Soils are just a tiny fraction of all Earth materials, yet they are a vital resource. Because soils are necessary for the growth of rooted plants, they are the very foundation of the human life-support system. Just as human ingenuity can increase the agricultural productivity of soils through fertilization and irrigation, soils can be damaged or destroyed by careless activities. Despite their basic role in providing food, fiber, and other basic materials, soils are among our most abused resources.

Perhaps this neglect and indifference has occurred because a substantial amount of soil seems to remain even where soil erosion is serious. Nevertheless, although the loss of fertile topsoil may not be obvious to the untrained eye, it is a growing problem as human activities expand and disturb more and more of Earth's surface.

TABLE 6.2	World Soil Orders
Alfisols	Moderately weathered soils that form under boreal forests or broadleaf deciduous forests, rich in iron and aluminum. Clay particles accumulate in a subsurface layer in response to leaching in moist environments. Fertile, productive soils, because they are neither too wet nor too dry.
Andisols	Young soils in which the parent material is volcanic ash and cinders, deposited by recent volcanic activity.
Aridosols	Soils that develop in dry places; insufficient water to remove soluble minerals, may have an accumulation of calcium carbonate, gypsum, or salt in subsoil; low organic content.
Entisols	Young soils having limited development and exhibiting properties of the parent material. Productivity ranges from very high for some formed on recent river deposits to very low for those forming on shifting sand or rocky slopes.
Gelisols	Young soils with little profile development that occur in regions with permafrost. Low temperatures and frozen conditions for much of the year; slow soil-forming processes.
Histosols	Organic soils with little or no climatic implications. Can be found in any climate where organic debris can accumulate to form a bog soil. Dark, partially decomposed organic material commonly referred to as *peat.*
Inceptisols	Weakly developed young soils in which the beginning (inception) of profile development is evident. Most common in humid climates, they exist from the Arctic to the tropics. Native vegetation is most often forest.
Mollisols	Dark, soft soils that have developed under grass vegetation, generally found in prairie areas. Humus-rich surface horizon that is rich in calcium and magnesium. Soil fertility is excellent. Also found in hardwood forests with significant earthworm activity. Climatic range is boreal or alpine to tropical. Dry seasons are normal (see Figure 6.20A).
Oxisols	Soils that occur on old land surfaces unless parent materials were strongly weathered before they were deposited. Generally found in the tropics and subtropical regions. Rich in iron and aluminum oxides, oxisols are heavily leached; hence are poor soils for agricultural activity (see Figure 6.20B).
Spodosols	Soils found only in humid regions on sandy material. Common in northern coniferous forests (see Figure 6.19) and cool humid forests. Beneath the dark upper horizon of weathered organic material lies a light-colored horizon of leached material, the distinctive property of this soil.
Ultisols	Soils that represent the products of long periods of weathering. Water percolating through the soil concentrates clay particles in the lower horizons (argillic horizons). Restricted to humid climates in the temperate regions and the tropics, where the growing season is long. Abundant water and a long frost-free period contribute to extensive leaching, hence poorer soil quality.
Vertisols	Soils containing large amounts of clay, which shrink upon drying and swell with the addition of water. Found in subhumid to arid climates, provided that adequate supplies of water are available to saturate the soil after periods of drought. Soil expansion and contraction exert stresses on human structures.

How Soil Is Eroded

Soil erosion is a natural process; it is part of the constant recycling of Earth materials that we call the *rock cycle*. Once soil forms, erosional forces, especially water and wind, move soil components from one place to another. Every time it rains, raindrops strike the land with surprising force (Figure 6.23). Each drop acts like a tiny bomb, blasting movable soil particles out of their positions in the soil mass. Then, water flowing across the surface carries away the dislodged soil particles. Because the soil is moved by thin sheets of water, this process is termed *sheet erosion*.

After flowing as a thin, unconfined sheet for a relatively short distance, threads of current typically develop, and tiny channels called *rills* begin to form. Still deeper cuts in the soil, known as *gullies*, are created as rills enlarge (Figure 6.24). When normal farm cultivation cannot eliminate the channels, we know the rills have grown large enough to be called gullies. Although most dislodged soil particles move only a short distance during each rainfall, substantial quantities eventually leave the fields and make their way downslope to a stream. Once in the stream channel, these soil particles, which can now be called *sediment*, are transported downstream and eventually deposited.

Rates of Erosion

We know that soil erosion is the ultimate fate of practically all soils. In the past, erosion occurred at slower rates than it does today because more of the land surface was covered and protected by trees, shrubs, grasses, and other plants. However, human activities such as farming, logging, and construction, which remove or disrupt the natural vegetation, have greatly accelerated the rate of soil erosion. Without the stabilizing effect of plants, the soil is more easily swept away by the wind or carried downslope by sheet wash.

Natural rates of soil erosion vary greatly from one place to another and depend on soil characteristics as well as such factors as climate, slope, and type of vegetation. Over a broad area, erosion caused by surface runoff may be estimated by determining the sediment loads of the streams that drain the region. When studies of this kind were made on a global scale, they indicated that prior to the appearance of humans, sediment transport by rivers to the ocean amounted to just over 9 billion metric tons per year. By contrast, the amount of material currently transported to the sea by rivers is about 24 billion metric tons per year, or more than two and a half times the earlier rate.

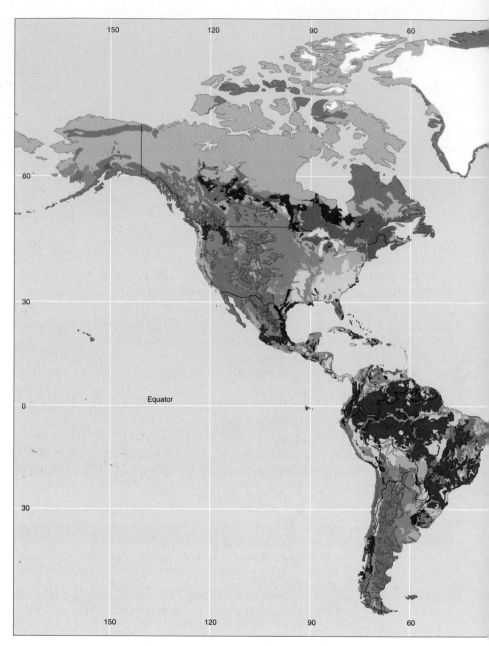

FIGURE 6.22   Global soil regions. Worldwide distribution of the Soil Taxonomy's 12 soil orders. (After U.S. Department of Agriculture, Natural Resources Conservation Service, World Soil Resources Staff)

It is more difficult to measure the loss of soil due to wind erosion. However, the removal of soil by wind is generally much less significant than erosion by flowing water except during periods of prolonged drought. When dry conditions prevail, strong winds can remove large quantities of soil from unprotected fields (Figure 6.25). Such was the case in the 1930s in the portions of the Great Plains that came to be called the Dust Bowl (see Box 6.4).

In many regions the rate of soil erosion is significantly greater than the rate of soil formation. This means that a renewable resource has become nonrenewable in these places. At present, it is estimated that topsoil is eroding faster than

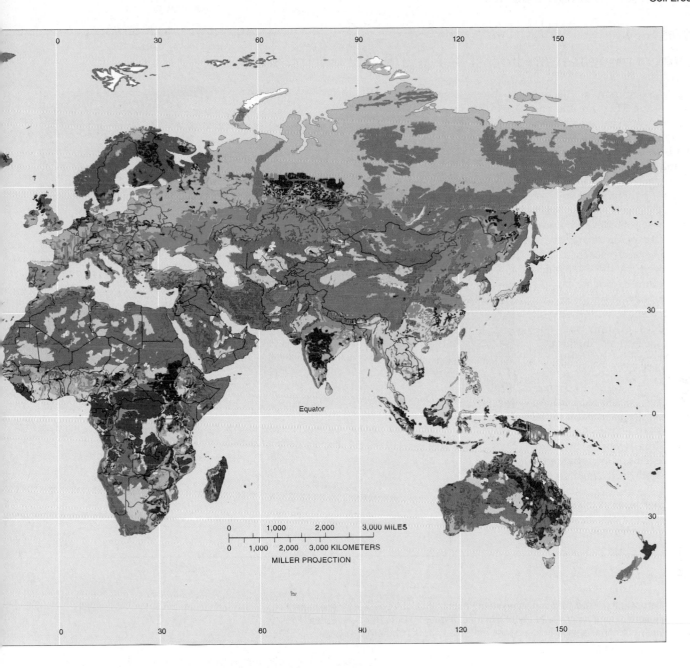

FIGURE 6.23 When it is raining, millions of water drops are falling at velocities approaching 10 meters per second (35 kilometers per hour). When water drops strike an exposed surface, soil particles may splash as high as 1 meter into the air and land more than a meter away from the point of raindrop impact. Soil dislodged by splash erosion is more easily moved by sheet erosion. (Photo courtesy of U.S.D.A./Natural Resources Conservation Service)

BOX 6.3 ▸ PEOPLE AND THE ENVIRONMENT

Clearing the Tropical Rain Forest—The Impact on Its Soils

Thick red soils are common in the wet tropics and subtropics. They are the end product of extreme chemical weathering. Because lush tropical rain forests are associated with these soils, we might assume they are fertile and have great potential for agriculture. However, just the opposite is true—they are among the poorest soils for farming. How can this be?

Because rain forest soils develop under conditions of high temperature and heavy rainfall, they are severely leached. Not only does leaching remove the soluble materials such as calcium carbonate but the great quantities of percolating water also remove much of the silica, with the result that insoluble oxides of iron and aluminum become concentrated in the soil. Iron oxides give the soil its distinctive red color. Because bacterial activity is very high in the tropics, rain forest soils contain practically no humus. Moreover, leaching destroys fertility because most plant nutrients are removed by the large volume of downward-percolating water. Therefore, even though the vegetation may be dense and luxuriant, the soil itself contains few available nutrients.

Most nutrients that support the rain forest are locked up in the trees themselves. As vegetation dies and decomposes, the

FIGURE 6.E   Clearing the tropical rain forest in West Kalimantan (Borneo), Indonesia. The thick soil is highly leached. (Photo by Wayne Lawler/Photo Researchers, Inc.)

FIGURE 6.24   **A.** Soil erosion from this field in northeastern Wisconsin is obvious. Just 1 millimeter of soil lost from a single acre of land amounts to about five tons. (Photo by D. P. Burnside/Photo Researchers, Inc.) **B.** Gully erosion is severe in this poorly protected soil in southern Colombia. (Photo by Carl Purcell/Photo Researchers, Inc.)

A.

B.

FIGURE 6.F This ancient temple at Angkor Wat, Cambodia, was built of bricks made of laterite. (Photo by R. Ian Lloyd/The Stock Market)

roots of the rain forest trees quickly absorb the nutrients before they are leached from the soil. The nutrients are continuously recycled as trees die and decompose.

Therefore, when forests are cleared to provide land for farming or to harvest the timber, most of the nutrients are removed as well (Figure 6.E). What remains is a soil that contains little to nourish planted crops.

The clearing of rain forests not only removes plant nutrients but also accelerates erosion. When vegetation is present, its roots anchor the soil, and its leaves and branches provide a canopy that protects the ground by deflecting the full force of the frequent heavy rains.

The removal of vegetation also exposes the ground to strong direct sunlight. When baked by the Sun, these tropical soils can

harden to a bricklike consistency and become practically impenetrable to water and crop roots. In only a few years, soils in a freshly cleared area may no longer be cultivable.

The term *laterite*, which is often applied to these soils, is derived from the Latin word *latere*, meaning "brick," and was first applied to the use of this material for brickmaking in India and Cambodia. Laborers simply excavated the soil, shaped it, and allowed it to harden in the Sun. Ancient but still well-preserved structures built of laterite remain standing today in the wet tropics (Figure 6.F). Such structures have withstood centuries of weathering because all of the original soluble materials were already removed from the soil by chemical weathering. Laterites are therefore virtually insoluble and very stable.

In summary, we have seen that some rain forest soils are highly leached products of extreme chemical weathering in the warm, wet tropics. Although they may be associated with lush tropical rain forests, these soils are unproductive when vegetation is removed. Moreover, when cleared of plants, these soils are subject to accelerated erosion and can be baked to bricklike hardness by the Sun.

A.

B.

FIGURE 6.25 **A.** Windbreaks protecting wheat fields in North Dakota. These flat expanses are susceptible to wind erosion especially when the fields are bare. The rows of trees slow the wind and deflect it upward, which decreases the loss of fine soil particles. (Photo by Erwin C. Cole/U.S.D.A./Natural Resources Conservation Service) **B.** This windbreak of conifers in Indiana provides year-round protection from wind erosion for this cropland. (Photo by Erwin C. Cole/U.S.D.A./Natural Resources Conservation Service)

BOX 6.4 ▶ PEOPLE AND THE ENVIRONMENT

Dust Bowl—Soil Erosion in the Great Plains

During a span of dry years in the 1930s, large dust storms plagued the Great Plains. Because of the size and severity of these storms, the region came to be called the Dust Bowl, and the time period the Dirty Thirties. The heart of the Dust Bowl was nearly 100 million acres in the panhandles of Texas and Oklahoma and adjacent parts of Colorado, New Mexico, and Kansas (Figure 6.G). To a lesser extent, dust storms were also a problem over much of the Great Plains, from North Dakota to west-central Texas.

At times dust storms were so severe that they were called "black blizzards" and "black rollers" because visibility was reduced to only a few feet. Numerous storms lasted for hours and stripped huge volumes of topsoil from the land.

In the spring of 1934, a windstorm that lasted for a day and a half created a dust cloud 2000 kilometers (1200 miles) long. As the sediment moved east, New York had "muddy rains" and Vermont "black snows." Another storm carried dust more than 3 kilometers (2 miles) into the atmosphere and transported it 3000 kilometers from its source in Colorado to create "midday twilight" in New England and New York.

What caused the Dust Bowl? Clearly, the fact that portions of the Great Plains experienced some of North America's strongest winds is important. However, it was the expansion of agriculture that set the stage for the disastrous period of

FIGURE 6.G An abandoned farmstead shows the disastrous effects of wind erosion and deposition during the Dust Bowl period. This photo of a previously prosperous farm was taken in Oklahoma in 1937. Also see Figure 19.10, p. 000. (Photo courtesy of Soil Conservation Service, U.S. Department of Agriculture)

soil erosion. Mechanization allowed the rapid transformation of the grass-covered prairies of this semiarid region into farms. Between the 1870s and 1930, cultivation expanded nearly tenfold, from about 10 million acres to more than 100 million acres.

As long as precipitation was adequate, the soil remained in place. However, when a prolonged drought struck in the 1930s, the unprotected fields were vulnerable to

the wind. The result was severe soil loss, crop failure, and economic hardship.

Beginning in 1939, a return to rainier conditions brought relief. New farming practices that reduced soil loss by wind were instituted. Although dust storms are less numerous and not as severe as in the Dirty Thirties, soil erosion by strong winds still occurs periodically whenever the combination of drought and unprotected soil exists.

Students Sometimes Ask ...

Is the amount of farmland in the United States and worldwide shrinking?

Yes, indeed. It's been estimated that between 3 and 5 million acres of prime U.S. farmland are lost each year through mismanagement (including soil erosion) and conversion to nonagricultural uses. According to the United Nations, since 1950, more than one-third of the world's farmable land has been lost to soil erosion.

it forms on more than one-third of the world's croplands. The result is lower productivity, poorer crop quality, reduced agricultural income, and an ominous future.

Sedimentation and Chemical Pollution

Another problem related to excessive soil erosion involves the deposition of sediment. Each year in the United States hundreds of millions of tons of eroded soil are deposited in lakes, reservoirs, and streams. The detrimental impact of this process can be significant. For example, as more and more sediment is deposited in a reservoir, the capacity of the reservoir is diminished, limiting its usefulness for flood control, water supply, and/or hydroelectric power generation. In addition, sedimentation in streams and other waterways can restrict navigation and lead to costly dredging operations.

In some cases soil particles are contaminated with pesticides used in farming. When these chemicals are introduced into a lake or reservoir, the quality of the water supply is threatened and aquatic organisms may be endangered. In addition to pesticides, nutrients found naturally in soils as well as those added by agricultural fertilizers make their way into streams and lakes, where they stimulate the growth of plants. Over a period of time, excessive nutrients accelerate the process by which plant growth leads to the depletion of oxygen and an early death of the lake.

The availability of good soils is critical if the world's rapidly growing population is to be fed. On every continent, unnecessary soil loss is occurring because appropriate conservation measures are not being used. Although it is a recognized fact that soil erosion can never be completely eliminated, soil conservation programs can substantially reduce the loss of this basic resource. Windbreaks (rows of trees), terracing, and plowing along the contours of hills are some of the effective measures, as are special tillage practices and crop rotation.

Summary

- External processes include (1) *weathering*—the disintegration and decomposition of rock at or near Earth's surface; (2) *mass wasting*—the transfer of rock material downslope under the influence of gravity; and (3) *erosion*—the removal of material by a mobile agent, usually water, wind, or ice. They are called *external processes* because they occur at or near Earth's surface and are powered by energy from the Sun. By contrast, *internal processes,* such as volcanism and mountain building, derive their energy from Earth's interior.

- *Mechanical weathering* is the physical breaking up of rock into smaller pieces. Rocks can be broken into smaller fragments by *frost wedging* (where water works its way into cracks or voids in rock and, upon freezing, expands and enlarges the openings), *salt crystal growth, unloading* (expansion and breaking due to a great reduction in pressure when the overlying rock is eroded away), *thermal expansion* (weakening of rock as the result of expansion and contraction as it heats and cools), and *biological activity* (by humans, burrowing animals, plant roots, etc.).

- *Chemical weathering* alters a rock's chemistry, changing it into different substances. Water is by far the most important agent of chemical weathering. *Dissolution* occurs when water-soluble minerals such as halite become dissolved in water. Oxygen dissolved in water will *oxidize* iron-rich minerals. When carbon dioxide (CO_2) is dissolved in water, it forms *carbonic acid,* which accelerates the decomposition of silicate minerals by *hydrolysis.* The chemical weathering of silicate minerals frequently produces (1) soluble products containing sodium, calcium, potassium, and magnesium ions, and silica in solution; (2) insoluble iron oxides; and (3) clay minerals.

- The rate at which rock weathers depends on such factors as (1) *particle size*—small pieces generally weather faster than large pieces; (2) *mineral makeup*—calcite readily dissolves in mildly acidic solutions, and silicate minerals that form first from magma are least resistant to chemical weathering; and (3) *climatic factors,* particularly temperature and moisture. Frequently, rocks exposed at Earth's surface do not weather at the same rate. This *differential weathering* of rocks is influenced by such factors as mineral makeup and degree of jointing.

- *Soil* is a combination of mineral and organic matter, water, and air—the portion of the *regolith* (the layer of rock and mineral fragments produced by weathering) that supports the growth of plants. About half of the total volume of a good-quality soil is a mixture of disintegrated and decomposed rock (mineral matter) and *humus* (the decayed remains of animal and plant life); the remaining half consists of pore spaces, where air and water circulate. The most important factors that control soil formation are *parent material, time, climate, plants* and *animals,* and *slope.*

- Soil-forming processes operate from the surface downward and produce zones or layers in the soil that are called *horizons.* From the surface downward, the soil horizons are respectively designated as *O* (largely organic matter), *A* (largely mineral matter), *E* (where the fine soil components and soluble materials have been removed by *eluviation* and *leaching*), *B* (or *subsoil,* often referred to as the *zone of accumulation*), and *C* (partially altered parent material). Together the *O* and *A* horizons make up what is commonly called the *topsoil.*

- In the United States, soils are classified using a system known as the *Soil Taxonomy.* It is based on physical and chemical properties of the soil profile and includes six hierarchical categories. The system is especially useful for agricultural and related land-use purposes.

- Soil erosion is a natural process; it is part of the constant recycling of Earth materials that we call the rock cycle. Once in a stream channel, soil particles are transported downstream and eventually deposited. *Rates of soil erosion* vary from one place to another and depend on the soil's characteristics as well as such factors as climate, slope, and type of vegetation.

Review Questions

1. Describe the role of external processes in the rock cycle.

2. If two identical rocks were weathered, one mechanically and the other chemically, how would the products of weathering for the two rocks differ?

3. In what type of environment is frost wedging most effective?

4. Describe the formation of an exfoliation dome. Give an example of such a feature.

5. How does mechanical weathering add to the effectiveness of chemical weathering?

6. Granite and basalt are exposed at the surface in a hot, wet region.

   a. Which type of weathering will predominate?

   b. Which of these rocks will weather most rapidly? Why?

7. Heat speeds up a chemical reaction. Why then does chemical weathering proceed slowly in a hot desert?

8. How is carbonic acid (H_2CO_3) formed in nature? What results when this acid reacts with potassium feldspar?

9. List some possible environmental effects of acid precipitation (see Box 6.1).

10. What is the difference between soil and regolith?

11. What factors might cause different soils to develop from the same parent material, or similar soils to form from different parent materials?

12. Which of the controls of soil formation is most important? Explain.

13. How can topography influence the development of soil? What is meant by the term *slope orientation?*

14. List the characteristics associated with each of the horizons in a well-developed soil profile. Which of the horizons constitute the solum? Under what circumstances do soils lack horizons?

15. The tropical soils described in Box 6.3 support luxuriant rain forests yet are considered to have low fertility. Explain.

16. List three detrimental effects of soil erosion other than the loss of topsoil from croplands.

17. Briefly describe the conditions that led to the Dust Bowl of the 1930s (see Box 6.3).

Key Terms

chemical weathering (p. 166)
differential weathering (p. 177)
dissolution (p. 171)
eluviation (p. 182)
erosion (p. 166)
exfoliation dome (p. 169)

external process (p. 166)
frost wedging (p. 167)
horizon (p. 181)
humus (p. 179)
hydrolysis (p. 173)
internal process (p. 166)
leaching (p. 182)
mass wasting (p. 166)

mechanical weathering (p. 166)
oxidation (p. 173)
parent material (p. 179)
regolith (p. 178)
sheeting (p. 169)
soil (p. 178)
soil profile (p. 181)

Soil Taxonomy (p. 183)
solum (p. 182)
spheroidal weathering (p. 176)
talus slope (p. 167)
weathering (p. 166)

Web Resources

The *Earth* Website uses the resources and flexibility of the Internet to aid in your study of the topics in this chapter. Written and developed by geology instructors, this site will help improve your understanding of geology. Visit **http://www.prenhall.com/tarbuck** and click on the cover of *Earth 9e* to find:

- Online review quizzes.

- Critical writing exercises.

- Links to chapter-specific Web resources.

- Internet-wide key-term searches.

http://www.prenhall.com/tarbuck

GEODe: Earth

GEODe: Earth makes studying faster and more effective by reinforcing key concepts using animation, video, narration, interactive exercises and practice quizzes. A copy is included with every copy of *Earth*.

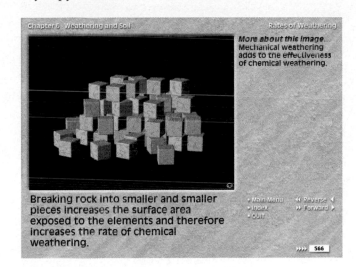

Sedimentary Rocks

The White Chalk Cliffs of Dover. This prominent chalk deposit underlies large portions of southern England as well as parts of northern France. (Photo by Jeremy Woodhouse/ DRK Photo)

Most of the solid Earth consists of igneous and metamorphic rocks. Geologists estimate these two categories represent 90 to 95 percent of the outer 16 kilometers (10 miles) of the crust. Nevertheless, most of Earth's solid surface consists of either sediment or sedimentary rock! About 75 percent of land areas are covered by sediments and sedimentary rocks. Across the ocean floor, which represents about 70 percent of Earth's solid surface, virtually everything is covered by sediment. Igneous rocks are exposed only at the crest of mid-ocean ridges and at some volcanic areas. Thus, while sediment and sedimentary rocks make up only a small percentage of Earth's crust, they are concentrated at or near the surface—the interface among the geosphere, hydrosphere, atmosphere, and biosphere. Because of this unique position, sediments and the rock layers that they eventually form contain evidence of past conditions and events at the surface. Furthermore, it is sedimentary rocks that contain fossils, which are vital tools in the study of the geologic past. Thus, this group of rocks provides geologists with much of the basic information they need to reconstruct the details of Earth history (Figure 7.1).

Such study is not only of interest for its own sake but has practical value as well. Coal, which provides a significant portion of our electrical energy, is classified as a sedimentary rock. Moreover, other major energy sources—oil, natural gas, and uranium—are derived from sedimentary rocks. So are major sources of iron, aluminum, manganese, and phosphate fertilizer, plus numerous materials essential to the construction industry such as cement and aggregate. Sediments and sedimentary rocks are also the primary reservoir of groundwater. Thus, an understanding of this group of rocks and the processes that form and modify them is basic to locating additional supplies of many important resources.

FIGURE 7.1 Sedimentary rocks are exposed at the surface more than igneous and metamorphic rocks. Because they contain fossils and other clues about the geologic past, sedimentary rocks are important in the study of Earth history. The layers shown here south of Soap Creek, Arizona, are known as the Vermillion Cliffs. (Photo by Michael Collier)

Origins of Sedimentary Rock

 GEODe Sedimentary Rocks
▶ Introduction

Figure 7.2 illustrates the portion of the rock cycle that occurs near Earth's surface—the part that pertains to sediments and sedimentary rocks. A brief overview of these processes provides a useful perspective:

- Weathering begins the process. It involves the physical disintegration and chemical decomposition of pre-existing igneous, metamorphic, and sedimentary rocks. Weathering generates a variety of products, including various solid particles and ions in solution. These are the raw materials for sedimentary rocks.
- Soluble constituents are carried away by runoff and groundwater. Solid particles are frequently moved downslope by gravity, a process termed mass wasting, before running water, groundwater, wind, and glacial ice remove them. Transportation moves these materials from the sites where they originated to locations where they accumulate. The transport of sediment is

usually intermittent. For example, during a flood, a rapidly moving river moves large quantities of sand and gravel. As the flood waters recede, particles are temporarily deposited, only to be moved again by a subsequent flood.

- Deposition of solid particles occurs when wind and water currents slow down and as glacial ice melts. The word *sedimentary* actually refers to this process. It is derived from the Latin *sedimentum*, which means "to settle," a reference to solid material settling out of a fluid (water or air). The mud on the floor of a lake, a delta at the mouth of a river, a gravel bar in a stream bed, the particles in a desert sand dune, and even household dust are examples.

- The deposition of material dissolved in water is not related to the strength of wind or water currents. Rather, ions in solution are removed when chemical or temperature changes cause material to crystallize and precipitate or when organisms remove dissolved material to build shells.

- As deposition continues, older sediments are buried beneath younger layers and gradually converted to sedimentary rock (lithified) by compaction and cementation. This and other changes are referred to as *diagenesis* (*dia* = change; *genesis* = origin), a collective term for all of the changes (short of metamorphism) that take place in texture, composition, and other physical properties after sediments are deposited.

Because there are a variety of ways that the products of weathering are transported, deposited, and transformed into solid rock, three categories of sedimentary rocks are recognized. As the overview reminded us, sediment has two principal sources. First, it may be an accumulation of material that originates and is transported as solid particles derived from both mechanical and chemical weathering. Deposits of this type are termed *detrital*, and the sedimentary rocks that they form are called **detrital sedimentary rocks.**

The second major source of sediment is soluble material produced largely by chemical weathering. When these ions in solution are precipitated by either inorganic or biologic processes, the material is known as chemical sediment, and the rocks formed from it are called **chemical sedimentary rocks.**

The third category is **organic sedimentary rocks.** The primary example is coal. This black combustible rock consists of organic carbon from the remains of plants that died and accumulated on the floor of a swamp. The bits and pieces of undecayed plant material that constitute the "sediments" in coal are quite unlike the weathering products that make up detrital and chemical sedimentary rocks.

Detrital Sedimentary Rocks

 GEODe
EARTH

Sedimentary Rocks
▶ Types of Sedimentary Rocks

Though a wide variety of minerals and rock fragments (*clasts*) may be found in detrital rocks, clay minerals and quartz are the chief constituents of most sedimentary rocks in this category. Recall from Chapter 6 that clay minerals are the most abundant product of the chemical weathering of silicate minerals, especially the feldspars. Clays are fine-grained minerals with sheetlike crystalline structures similar to the micas. The other common mineral, quartz, is abundant because it is extremely durable and very resistant to chemical weathering. Thus, when igneous rocks such as granite are attacked by weathering processes, individual quartz grains are freed.

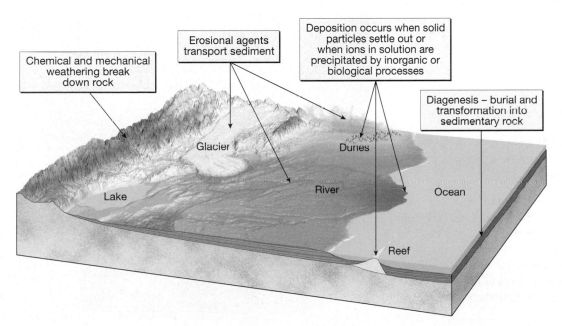

FIGURE 7.2 This diagram outlines the portion of the rock cycle that pertains to the formation of sedimentary rocks. Weathering, transportation, deposition, and diagenesis represent the basic process involved.

TABLE 7.1 Particle Size Classification for Detrital Rocks

Size Range (millimeters)	Particle Name	Common Sediment Name	Detrital Rock
>256	Boulder		
64–256	Cobble	Gravel	Conglomerate or breccia
4–64	Pebble		
2–4	Granule		
1/16–2	Sand	Sand	Sandstone
1/256–1/16	Silt	Mud	Shale, mudstone, or siltstone
<1/256	Clay		

```
        0    10    20    30    40    50    60
        Lшшшшшшшшшшшшшшшшшшшшшшшшшшшшшшш Lшш
                    (Scale in mm)
```

Other common minerals in detrital rocks are feldspars and micas. Because chemical weathering rapidly transforms these minerals into new substances, their presence in sedimentary rocks indicates that erosion and deposition were fast enough to preserve some of the primary minerals from the source rock before they could be decomposed.

Particle size is the primary basis for distinguishing among various detrital sedimentary rocks. Table 7.1 presents the size categories for particles making up detrital rocks. Particle size is not only a convenient method of dividing detrital rocks, the sizes of the component grains also provide useful information about environments of deposition. Currents of water or air sort the particles by size; the stronger the current, the larger the particle size carried. Gravels, for example, are moved by swiftly flowing rivers as well as by landslides and glaciers. Less energy is required to transport sand; thus, it is common to such features as wind-blown dunes and some river deposits and beaches. Very little energy is needed to transport clay, so it settles very slowly. Accumulation of these tiny particles is generally associated with the quiet water of a lake, lagoon, swamp, or certain marine environments.

Common detrital sedimentary rocks, in order of increasing particle size, are shale, sandstone, and conglomerate or breccia. We will now look at each type and how it forms.

Shale

Shale is a sedimentary rock consisting of silt- and clay-size particles (Figure 7.3). These fine-grained detrital rocks account for well over half of all sedimentary rocks. The particles in these rocks are so small that they cannot be readily identified without great magnification and for this reason make shale more difficult to study and analyze than most other sedimentary rocks.

Much of what can be learned is based on particle size. The tiny grains in shale indicate that deposition occurs as the result of gradual settling from relatively quiet, nonturbulent currents. Such environments include lakes, river floodplains, lagoons, and portions of the deep-ocean basins. Even in these "quiet" environments, there is usually enough turbulence to keep clay-size particles suspended almost indefinitely. Consequently, much of the clay is deposited only after the individual particles coalesce to form larger aggregates.

FIGURE 7.3 Shale is a fine-grained detrital rock that is by far the most abundant of all sedimentary rocks. Dark shales containing plant remains are relatively common. (Photo courtesy of E. J. Tarbuck)

Sometimes the chemical composition of the rock provides additional information. One example is black shale, which is black because it contains abundant organic matter (carbon). When such a rock is found, it strongly implies that deposition occurred in an oxygen-poor environment such as a swamp, where organic materials do not readily oxidize and decay.

As silt and clay accumulate, they tend to form thin layers, which are commonly referred to as *laminae* (lamin = a thin sheet). Initially the particles in the laminae are oriented randomly. This disordered arrangement leaves a high percentage of open space (called *pore space*) that is filled with water. However, this situation usually changes with time as additional layers of sediment pile up and compact the sediment below.

During this phase the clay and silt particles take on a more nearly parallel alignment and become tightly packed. This rearrangement of grains reduces the size of the pore spaces and forces out much of the water. Once the grains are pressed closely together, the tiny spaces between particles do not readily permit solutions containing cementing material to circulate. Therefore, shales are often described as being weak because they are poorly cemented and therefore not well lithified.

The inability of water to penetrate its microscopic pore spaces explains why shale often forms barriers to the subsurface movement of water and petroleum. Indeed, rock layers that contain groundwater are commonly underlain by shale beds that block further downward movement. The opposite is true for underground reservoirs of petroleum. They are often capped by shale beds that effectively prevent oil and gas from escaping to the surface.*

It is common to apply the term *shale* to all fine-grained sedimentary rocks, especially in a nontechnical context. However, be aware that there is a more restricted use of the term. In this narrower usage, shale must exhibit the ability to split into thin layers along well-developed closely spaced planes. This property is termed **fissility** (*fissilis* = that which can be cleft or split). If the rock breaks into chunks or blocks, the name *mudstone* is applied. Another fine-grained sedimentary rock that, like mudstone, is often grouped with shale but lacks fissility is *siltstone*. As its name implies, siltstone is composed largely of silt-size particles and contains less clay-size material than shale and mudstone.

Although shale is far more common than other sedimentary rocks, it does not usually attract as much notice as other less abundant members of this group. The reason is that shale does not form prominent outcrops as sandstone and limestone often do. Rather, shale crumbles easily and usually forms a cover of soil that hides the unweathered rock below. This is illustrated nicely in the Grand Canyon, where the gentler slopes of weathered shale are quite inconspicuous and overgrown with vegetation, in sharp contrast with the bold cliffs produced by more durable rocks (Figure 7.4).

Although shale beds may not form striking cliffs and prominent outcrops, some deposits have economic value. Certain shales are quarried to obtain raw material for pottery, brick, tile, and china. Moreover, when mixed with limestone, shale is used to make portland cement. In the future, one type of shale, called oil shale, may become a valuable energy resource. This possibility will be explored in Chapter 23.

Sandstone

Sandstone is the name given rocks in which sand-size grains predominate (Figure 7.5). After shale, sandstone is the most abundant sedimentary rock, accounting for approximately 20 percent of the entire group. Sandstones form in a variety of environments and often contain significant clues about their origin, including sorting, particle shape, and composition.

Sorting and Particle Shape   **Sorting** is the degree of similarity in particle size in a sedimentary rock. For example, if all the grains in a sample of sandstone are about the same size, the sand is considered *well sorted*. Conversely, if the rock contains mixed large and small particles, the sand is said to be *poorly sorted* (Figure 7.6A). By studying the degree of sorting, we can learn much about the depositing current. Deposits of windblown sand are usually better sorted than deposits sorted by wave activity. Particles washed by waves are commonly better sorted than materials deposited by streams. Sediment accumulations that exhibit poor sorting usually result when particles are transported for only a relatively short time and then rapidly deposited. For example, when a turbulent stream reaches the gentler slopes at the base of a steep mountain, its velocity is quickly reduced, and poorly sorted sands and gravels are deposited.

The shapes of sand grains can also help decipher the history of a sandstone (Figure 7.6B). When streams, winds, or

FIGURE 7.4 Sedimentary rock layers exposed in the walls of the Grand Canyon, Arizona. Beds of resistant sandstone and limestone produce bold cliffs. By contrast, weaker, poorly cemented shale crumbles and produces a gentler slope of weathered debris in which some vegetation is growing. (Photo by Tom & Susan Bean, Inc.)

---

*The relationship between impermeable beds and the occurrence and movement of groundwater is examined in Chapter 17. Shale beds as cap rocks in oil traps are discussed in Chapter 21.

A.

B.

FIGURE 7.5 Quartz sandstone. After shale, sandstone is the most abundant sedimentary rock. **A.** This thick layer of sandstone, called the Navajo Sandstone, is exposed in and near Utah's Zion National Park. (Photo by Tom and Susan Bean) **B.** The quartz grains composing the Navajo Sandstone were deposited by the wind as dunes similar to these in Colorado's Great Sand Dunes National Park (Photo by David Muench). **C.** This is a hand sample and close up of the layer in part A. (Photos by E. J. Tarbuck)

waves move sand and other larger sedimentary particles, the grains lose their sharp edges and corners and become more rounded as they collide with other particles during transport. Thus, rounded grains likely have been airborne or waterborne. Further, the degree of rounding indicates the distance or time involved in the transportation of sediment by currents of air or water. Highly rounded grains indicate that a great deal of abrasion and hence a great deal of transport has occurred.

Very angular grains, on the other hand, imply two things: that the materials were transported only a short distance before they were deposited, and that some other medium may have transported them. For example, when glaciers move sediment, the particles are usually made more irregular by the crushing and grinding action of the ice.

In addition to affecting the degree of rounding and the amount of sorting that particles undergo, the length of transport by turbulent air and water currents also influences the mineral composition of a sedimentary deposit. Substantial weathering and long transport lead to the gradual destruction of weaker and less stable minerals, including the feldspars and ferromagnesians. Because quartz is very durable, it is usually the mineral that survives the long trip in a turbulent environment.

The preceding discussion has shown that the origin and history of sandstone can often be deduced by examining the sorting, roundness, and mineral composition of its constituent grains. Knowing this information allows us to infer that a well-sorted quartz-rich sandstone consisting of highly rounded grains must be the result of a great deal of transport. Such a rock, in fact, may represent several cycles of weathering, transport, and deposition. We may also conclude that a sandstone containing significant amounts of feldspar and angular grains of ferromagnesian minerals underwent little chemical weathering and transport and was probably deposited close to the source area of the particles.

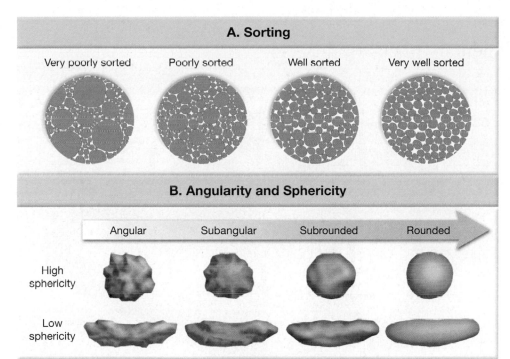

A. Sorting

Very poorly sorted | Poorly sorted | Well sorted | Very well sorted

B. Angularity and Sphericity

Angular | Subangular | Subrounded | Rounded

High sphericity

Low sphericity

FIGURE 7.6 A. Detrital rocks commonly have a variety of different size clasts. *Sorting* refers to the range of sizes present. Rocks with clasts that are nearly all the same size are considered "well-sorted." When sediments are "very poorly sorted," there is a wide range of different sizes. When a rock contains larger clasts surrounded by much smaller ones, the mass of smaller clasts is often referred to as the *matrix*. **B.** Geologists describe a particle's shape in terms of its *angularity* (degree to which the clast's edges and corners are rounded) and *sphericity* (how close the shape of the clast is to a sphere). Transportation reduces the size and angularity of clasts but does not change their general shape.

Composition Owing to its durability, quartz is the predominant mineral in most sandstones. When this is the case, the rock may simply be called *quartz sandstone*. When a sandstone contains appreciable quantities of feldspar (25 percent or more), the rock is called *arkose*. In addition

to feldspar, arkose usually contains quartz and sparkling bits of mica. The mineral composition of arkose indicates that the grains were derived from granitic source rocks. The particles are generally poorly sorted and angular, which suggests short-distance transport, minimal chemical weathering in a relatively dry climate, and rapid deposition and burial.

A third variety of sandstone is known as *graywacke*. Along with quartz and feldspar, this dark-colored rock contains abundant rock fragments and matrix. More than 15 percent of graywacke's volume is matrix. The poor sorting and angular grains characteristic of graywacke suggest that the particles were transported only a relatively short distance from their source area and then rapidly deposited. Before the sediment could be reworked and sorted further, it was buried by additional layers of material. Graywacke is frequently associated with submarine deposits made by dense sediment-choked torrents called turbidity currents.

Conglomerate and Breccia

Conglomerate consists largely of gravels (Figure 7.7). As Table 7.1 indicates, these particles can range in size from large boulders to particles as small as garden peas. The particles are commonly large enough to be identified as distinctive rock types; thus, they can be valuable in identifying the

FIGURE 7.7 Conglomerate is composed primarily of rounded gravel-size particles. (Photo by E. J. Tarbuck)

5 cm

Students Sometimes Ask . . .

Why are many of the sedimentary rocks pictured in this chapter so colorful?

In the western and southwestern United States, steep cliffs and canyon walls made of sedimentary rocks often exhibit a brilliant display of different colors. (For example, see Figures 7.1, 7.4, 7.5, 7.12, and 7.22.) In the walls of Arizona's Grand Canyon we can see layers that are red, orange, purple, gray, brown, and buff. Some of the sedimentary rocks in Utah's Bryce Canyon are a delicate pink color (see Figure 6.1). Sedimentary rocks in more humid places are also colorful, but they are usually covered by soil and vegetation.

The most important "pigments" are iron oxides, and only very small amounts are needed to color a rock. Hematite tints rocks red or pink, whereas limonite produces shades of yellow and brown. When sedimentary rocks contain organic matter, it often colors them black or gray (see Figure 7.3).

BOX 7.1 ▶ EARTH AS A SYSTEM

The Carbon Cycle and Sedimentary Rocks

To illustrate the movement of material and energy in the Earth system, let us take a brief look at the *carbon cycle* (Figure 7.A). Pure carbon is relatively rare in nature. It is found predominantly in two minerals: diamond and graphite. Most carbon is bonded chemically to other elements to form compounds such as carbon dioxide, calcium carbonate, and the hydrocarbons found in coal and petroleum. Carbon is also the basic building block of life as it readily combines with hydrogen and oxygen to form the fundamental organic compounds that compose living things.

In the atmosphere, carbon is found mainly as carbon dioxide (CO_2). Atmospheric carbon dioxide is significant because it is a greenhouse gas, which means it is an efficient absorber of energy emitted by Earth and thus influences the heating of the atmosphere.* Because many of the processes that operate on Earth involve carbon dioxide, this gas is constantly moving into and out of the atmosphere (Figure 7.B). For example, through the process of photosynthesis, plants absorb carbon dioxide from the atmosphere to produce the essential organic compounds needed for growth. Animals that consume these plants (or consume other animals that eat plants) use these organic compounds as a

*For more on this idea, see the discussion of the *greenhouse effect* in Chapter 21.

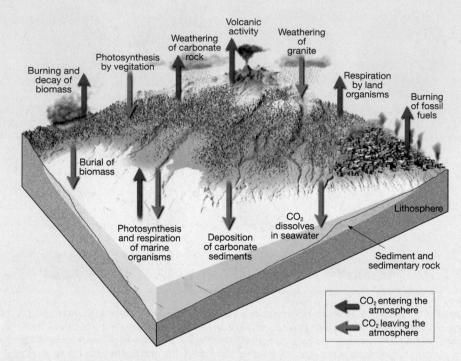

FIGURE 7.A   Simplified diagram of the carbon cycle, with emphasis on the flow of carbon between the atmosphere and the hydrosphere, lithosphere and biosphere. The colored arrows show whether the flow of carbon is into or out of the atmosphere.

source of energy and, through the process of respiration, return carbon dioxide to the atmosphere. (Plants also return some CO_2 to the atmosphere via respiration.) Further, when plants die and decay or are burned,

this biomass is oxidized, and carbon dioxide is returned to the atmosphere.

Not all dead plant material decays immediately back to carbon dioxide. A small percentage is deposited as sediment. Over

source areas of sediments. More often than not, conglomerates are poorly sorted because the opening between the large gravel particles contains sand or mud.

Gravels accumulate in a variety of environments and usually indicate the existence of steep slopes or very turbulent currents. The coarse particles in a conglomerate may reflect the action of energetic mountain streams or result from strong wave activity along a rapidly eroding coast. Some glacial and landslide deposits also contain plentiful gravel.

If the large particles are angular rather than rounded, the rock is called *breccia* (Figure 7.8). Because large particles abrade and become rounded very rapidly during transport, the pebbles and cobbles in a breccia indicate that they did not travel far from their source area before they were deposited. Thus, as with many sedimentary rocks, conglomerates and breccias contain clues to their history. Their particle sizes reveal the strength of the currents that transported them, whereas the degree of rounding indicates how far the particles traveled. The fragments within a sample identify the source rocks that supplied them.

FIGURE 7.8   When the gravel-size particles in a detrital rock are angular, the rock is called breccia. (Photo by E. J. Tarbuck)

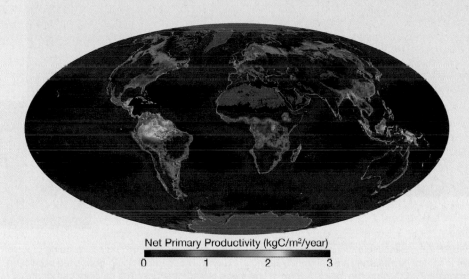

Net Primary Productivity (kgC/m²/year)

0   1   2   3

FIGURE 7.B This map was created using space-based measurements of a range of plant properties and shows the net productivity of vegetation on land and in the oceans in 2002. It is calculated by determining how much CO_2 is taken up by vegetation during photosynthesis minus how much is given off during respiration. Scientists expect this global measure of biological activity to yield new insights into Earth's complex carbon cycle.
(NASA image)

source of much of the carbon dioxide found in the atmosphere. One way that carbon dioxide makes its way back to the hydrosphere and then to the solid Earth is by first combining with water to form carbonic acid (H_2CO_3), which then attacks the rocks that compose the lithosphere. One product of this chemical weathering of solid rock is the soluble bicarbonate ion ($2HCO_3$) which is carried by groundwater and streams to the ocean. Here water-dwelling organisms extract this dissolved material to produce hard parts of calcium carbonate ($CaCO_3$). When the organisms die, these skeletal remains settle to the ocean floor as biochemical sediment and become sedimentary rock. In fact, the lithosphere is by far Earth's largest depository of carbon, where it is a constituent of a variety of rocks, the most abundant being limestone. Eventually the limestone may be exposed at Earth's surface, where chemical weathering will cause the carbon stored in the rock to be released to the atmosphere as CO_2.

In summary, carbon moves among all four of Earth's major spheres. It is essential to every living thing in the biosphere. In the atmosphere carbon dioxide is an important greenhouse gas. In the hydrosphere, carbon dioxide is dissolved in lakes, rivers, and the ocean. In the lithosphere carbon is contained in carbonate sediments and sedimentary rocks and is stored as organic matter dispersed through sedimentary rocks and as deposits of coal and petroleum.

long spans of geologic time, considerable biomass is buried with sediment. Under the right conditions, some of these carbon-rich deposits are converted to fossil fuels—coal, petroleum, or natural gas. Eventually some of the fuels are recovered (mined or pumped from a well) and burned to run factories and fuel our transportation system. One result of fossil-fuel combustion is the release of huge quantities of CO_2 into the atmosphere. Certainly one of the most active parts of the carbon cycle is the movement of CO_2 from the atmosphere to the biosphere and back again.

Carbon also moves from the lithosphere and hydrosphere to the atmosphere and back again. For example, volcanic activity early in Earth's history is thought to be the

Chemical Sedimentary Rocks

 Sedimentary Rocks
▶ **Types of Sedimentary Rocks**

In contrast to detrital rocks, which form from the solid products of weathering, chemical sediments derive from ions that are carried *in solution* to lakes and seas. This material does not remain dissolved in the water indefinitely, however. Some of it precipitates to form chemical sediments. These become rocks such as limestone, chert, and rock salt.

This precipitation of material occurs in two ways. *Inorganic* (*in* = not, *organicus* = life) processes such as evaporation and chemical activity can produce chemical sediments. *Organic* (life) processes of water-dwelling organisms also form chemical sediments, said to be of **biochemical** origin.

One example of a deposit resulting from inorganic chemical processes is the dripstone that decorates many caves (Figure 7.9). Another is the salt left behind as a body of sea-

water evaporates. In contrast, many water-dwelling animals and plants extract dissolved mineral matter to form shells and other hard parts. After the organisms die, their skeletons collect by the millions on the floor of a lake or ocean as biochemical sediment (Figure 7.10).

Limestone

Representing about 10 percent of the total volume of all sedimentary rocks, *limestone* is the most abundant chemical sedimentary rock. It is composed chiefly of the mineral calcite ($CaCO_3$) and forms either by inorganic means or as the result of biochemical processes (see Box 7.1). Regardless of its origin, the mineral composition of all limestone is similar, yet many different types exist. This is true because limestones are produced under a variety of conditions. Those forms having a marine biochemical origin are by far the most common.

Carbonate Reefs Corals are one important example of organisms that are capable of creating large quantities of marine limestone. These relatively simple invertebrate animals secrete a calcareous (calcium carbonate) external skeleton. Although they are small, corals are capable of creating massive structures called *reefs* (Figure 7.11). Reefs consist of coral colonies made up of great numbers of individuals that live side by side on a calcite structure secreted by the animals. In addition, calcium carbonate–secreting algae live with the corals and help cement the entire structure into a solid mass. A wide variety of other organisms also live in and near the reefs.

Certainly the best-known modern reef is Australia's Great Barrier Reef, 2000 kilometers (1240 miles) long, but many lesser reefs also exist. They develop in the shallow, warm waters of the tropics and subtropics equatorward of about 30° latitude. Striking examples exist in the Bahamas and Florida Keys.

FIGURE 7.9 Because many cave deposits are created by the seemingly endless dripping of water over long time spans, they are commonly called *dripstone.* The material being deposited is calcium carbonate ($CaCO_3$) and the rock is a form of limestone called *travertine.* The calcium carbonate is precipitated when some dissolved carbon dioxide escapes from a water drop. (Photo by Guillen Photography; Inset photo by Clifford Stroud/Wind Cave National Park)

Of course, not only modern corals build reefs. Corals have been responsible for producing vast quantities of limestone in the geologic past as well. In addition, other calcium carbonate-secreting organisms are known for being builders of modern and ancient reefs, including certain algae, sponges, and bryozoans. In the United States, reefs of Silurian age are prominent features in Wisconsin, Illinois, and Indiana. In west Texas and adjacent southeastern New Mexico, a massive reef complex formed during the Permian period is strikingly exposed in Guadalupe Mountains National Park (Figure 7.11B)

Coquina and Chalk Although much limestone is the product of biological processes, this origin is not always evident, because shells and skeletons may undergo considerable change before becoming lithified into rock. However, one easily identified biochemical limestone is *coquina,* a coarse rock composed of poorly cemented shells and shell fragments (see Figure 7.10). Another less obvious but nevertheless familiar example is *chalk,* a soft, porous rock made up almost entirely of the hard parts of microscopic marine organisms. Among the most famous chalk deposits are those exposed along the southeast coast of England (see chapter-opening photo).

Inorganic Limestones Limestones having an inorganic origin form when chemical changes or high water temperatures increase the concentration of calcium carbonate to the point that it precipitates. *Travertine,* the type of limestone commonly seen in caves, is an example (Figure 7.9). When travertine is deposited in caves, groundwater is the source of the calcium carbonate. As water droplets become exposed to the air in a cavern, some of the carbon dioxide dissolved in the water escapes, causing calcium carbonate to precipitate.

Another variety of inorganic limestone is *oolitic limestone.* It is a rock composed of small spherical grains called *ooids.* Ooids form in shallow marine waters as tiny "seed" particles (commonly small shell fragments) are moved back and forth by currents. As the grains are rolled about in the warm water, which is supersaturated with calcium carbonate, they become coated with layer upon layer of the chemical precipitate (Figure 7.12).

FIGURE 7.10 This rock, called coquina, consists of shell fragments; therefore, it has a biochemical origin. (Photo by E. J. Tarbuck)

Close up

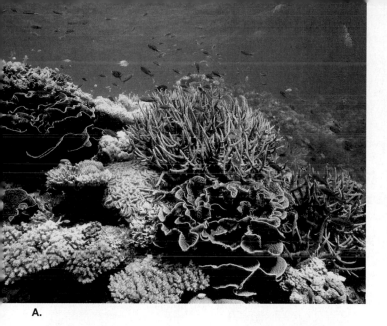
A.

B.

FIGURE 7.11 **A.** This modern coral reef is at Bora Bora in French Polynesia. (Photo by Nancy Sefton/Photo Researchers, Inc.) **B.** View from McKittrick Canyon along the Capitan Reef Escarpment in Guadalupe Mountains National Park, Texas. This image shows just a small portion of a huge Permian-age reef complex that once formed a 600-kilometer (nearly 375-mile) loop around the margin of the Delaware Basin. The reef consisted of a rich community of sponges, bryozoans, crinoids, gastropods, calcareous algae, and rare corals. (Photo by Peter A. Scholle)

Dolostone

Closely related to limestone is *dolostone*, a rock composed of the calcium-magnesium carbonate mineral dolomite [CaMg $(CO_3)_2$]. Although dolostone and limestone sometimes closely resemble one another, they can be easily distinguished by observing their reaction to dilute hydrochloric acid. When a drop of acid is placed on limestone, the reaction (fizzing) is obvious. However, unless dolostone is powdered, it will not visibly react to the acid.

The origin of dolostone is not altogether clear and remains a subject of discussion among geologists. No marine organisms produce hard parts of dolomite and the chemical precipitation of dolomite from seawater occurs only under conditions of unusual water chemistry in certain near shore sites. Yet dolostone is abundant in many ancient sedimentary rock successions.

It appears that significant quantities of dolostone are produced when magnesium-rich waters circulate through limestone and convert calcite to dolomite by the replacement of some calcium ions with magnesium ions (a process called *dolomitization*). However, other dolostones lack evidence that they formed by such a process and their origin remains uncertain.

Chert

Chert is a name used for a number of very compact and hard rocks made of microcrystalline quartz (SiO_2). One well-known form is *flint*, whose dark color results from the organic matter it contains. *Jasper,* a red variety, gets its bright color from the iron oxide it contains. The banded form is usually referred to as *agate* (Figure 7.13). Like glass, most chert has a conchoidal fracture. Its hardness, ease of chip-

ping, and ability to hold a sharp edge made chert a favorite of Native Americans for fashioning "points" for spears and arrows. Because of chert's durability and extensive use, "arrowheads" are found in many parts of North America.

Chert deposits are commonly found in one of two situations: as layered deposits referred to as *bedded cherts* and as *nodules,* somewhat spherical masses varying in diameter from a few millimeters (pea size) to a few centimeters. Most water-dwelling organisms that produce hard parts make them of calcium carbonate. But some, such as diatoms and radiolarians, produce glasslike silica skeletons. These tiny organisms are able to extract silica even though seawater contains only tiny quantities. It is from their remains that most bedded cherts are believed to originate. Some bedded cherts occur in association with lava flows and layers of volcanic ash. For these occurrences it is probable that the silica was derived from the decomposition of the volcanic ash and not from biochemical sources. Chert nodules are sometimes referred to as *secondary* or *replacement cherts* and most often occur within beds of limestone. They form when silica originally deposited in one place, dissolves, migrates, and then chemically precipitates elsewhere, replacing older material.

Evaporites

Very often evaporation is the mechanism triggering deposition of chemical precipitates. Minerals commonly precipitated in this fashion include halite (sodium chloride, NaCl), the chief component of *rock salt*, and gypsum (hydrous calcium sulfate, $CaSO_4 \cdot 2\,H_2O$), the main ingredient of *rock gypsum.* Both have significant importance. Halite is familiar to everyone as the common salt used in cooking and seasoning foods. Of course, it has many other uses, from melting ice on roads to making hydrochloric acid, and has been considered important enough that people have sought, traded, and fought over it for much of human history. Gypsum is the basic ingredient of plaster of Paris. This material is used most extensively in the construction industry for wallboard and interior plaster.

A. Agate

FIGURE 7.12 Oolitic limestone consists of *ooids*, which are small spherical grains formed by the chemical precipitation of calcium carbonate around a tiny nucleus. The carbonate is added in concentric layers as the spheres are rolled back and forth by currents in a warm shallow marine setting. (Top photo by Larry Davis/Washington State University; bottom photo by Marli Miller/Visuals Unlimited)

B. Flint

C. Jasper

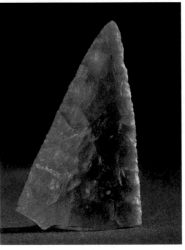

D. Chert arrowhead

FIGURE 7.13 *Chert* is a name used for a number of dense, hard rocks made of microcrystalline quartz. Three examples are shown here. **A.** *Agate* is the banded variety. (Photo by Jeffrey A. Scoville) **B.** The dark color of *flint* results from organic matter. (Photo by E.J. Tarbuck) **C.** The red variety, called *jasper*, gets its color from iron oxide. (Photo by E.J. Tarbuck) **D.** Native Americans frequently made arrowheads and sharp tools from chert. (Photo by LA VENTA/CORBIS SYGMA)

In the geologic past, many areas that are now dry land were basins, submerged under shallow arms of a sea that had only narrow connections to the open ocean. Under these conditions, seawater continually moved into the bay to replace water lost by evaporation. Eventually the waters of the bay became saturated and salt deposition began. Such deposits are called **evaporites.**

FIGURE 7.14 Each year about 30 percent of the world's supply of salt is extracted from seawater. In this process, salt water is held in shallow ponds, while solar energy evaporates the water. The nearly pure salt deposits that eventually form are essentially artificial evaporite deposits. At the southern end of San Francisco Bay, it takes nearly 38,000 liters (10,000 gallons) of water to produce 900 kilograms (1 ton) of salt. (Photo by William E. Townsend Jr./Photo Researchers, Inc.)

When a body of seawater evaporates, the minerals that precipitate do so in a sequence that is determined by their solubility. Less soluble minerals precipitate first, and more soluble minerals precipitate later as salinity increases (Figure 7.14). For example, gypsum precipitates when about 80 percent of the seawater has evaporated, and halite settles out when 90 percent of the water has been removed. During the last stages of this process, potassium and magnesium salts precipitate. One of these last-formed salts, the mineral *sylvite*, is mined as a significant source of potassium ("potash") for fertilizer.

On a smaller scale, evaporite deposits can be seen in such places as Death Valley, California. Here, following rains or periods of snowmelt in the mountains, streams flow from the surrounding mountains into an enclosed basin. As the water evaporates, **salt flats** form when dissolved materials are precipitated as a white crust on the ground (Figure 7.15).

decompose when exposed to the atmosphere or other oxygen-rich environments. One important environment that allows for the buildup of plant material is a swamp (Figure 7.17).

Stagnant swamp water is oxygen-deficient, so complete decay (oxidation) of the plant material is not possible. Instead, the plants are attacked by certain bacteria that partly decompose the organic material and liberate oxygen and hydrogen. As these elements escape, the percentage of carbon gradually increases. The bacteria are not able to finish the job of decomposition because they are destroyed by acids liberated from the plants.

The partial decomposition of plant remains in an oxygen-poor swamp creates a layer of *peat*, a soft brown material in which plant structures are still easily recognized. With shallow burial, peat slowly changes to *lignite*, a soft brown coal.

Coal—An Organic Sedimentary Rock

Coal is quite different from other rocks. Unlike limestone and chert, which are calcite- and silica-rich, coal is made of organic matter. Close examination of coal under a magnifying glass often reveals plant structures such as leaves, bark, and wood that have been chemically altered but are still identifiable. This supports the conclusion that coal is the end product of large amounts of plant material, buried for millions of years (Figure 7.16).

The initial stage in coal formation is the accumulation of large quantities of plant remains. However, special conditions are required for such accumulations, because dead plants readily

FIGURE 7.15 The Bonneville salt flats in Utah are a well-known example of evaporite deposits. (Photo by Tom & Susan Bean, Inc.)

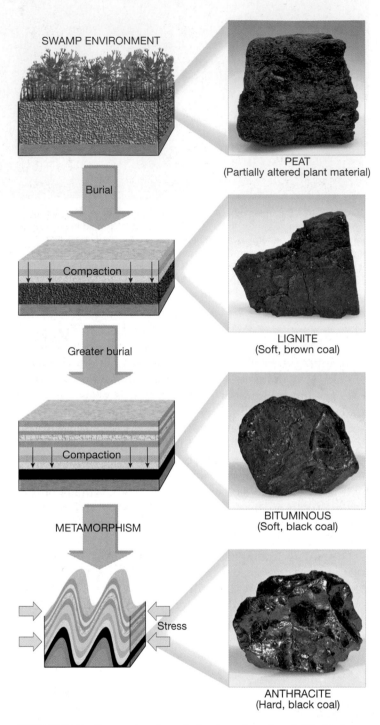

SWAMP ENVIRONMENT

Burial

PEAT
(Partially altered plant material)

Compaction

LIGNITE
(Soft, brown coal)

Greater burial

Compaction

BITUMINOUS
(Soft, black coal)

METAMORPHISM

Stress

ANTHRACITE
(Hard, black coal)

FIGURE 7.16 Successive stages in the formation of coal. (Photos by E. J. Tarbuck)

Burial increases the temperature of sediments as well as the pressure on them.

The higher temperatures bring about chemical reactions within the plant materials and yield water and organic gases (volatiles). As the load increases from more sediment on top of the developing coal, the water and volatiles are pressed out and the proportion of *fixed carbon* (the remaining solid combustible material) increases. The greater the carbon content, the greater the coal's energy ranking as a fuel. During burial, the coal also becomes increasingly com-

pact. For example, deeper burial transforms lignite into a harder, more compacted black rock called *bituminous* coal. Compared to the peat from which it formed, a bed of bituminous coal may be only 1/10 as thick.

Lignite and bituminous coals are sedimentary rocks. However, when sedimentary layers are subjected to the folding and deformation associated with mountain building, the heat and pressure cause a further loss of volatiles and water, thus increasing the concentration of fixed carbon. This metamorphoses bituminous coal into *anthracite*, a very hard, shiny, black *metamorphic* rock. Although anthracite is a clean-burning fuel, only a relatively small amount is mined. Anthracite is not widespread and is more difficult and expensive to extract than the relatively flat-lying layers of bituminous coal.

Coal is a major energy resource. Its role as a fuel and some of the problems associated with burning coal are discussed in Chapter 23.

Turning Sediment into Sedimentary Rock: Diagenesis and Lithification

A great deal of change can occur to sediment from the time it is deposited until it becomes a sedimentary rock and is subsequently subjected to the temperatures and pressures that convert it to metamorphic rock. The term **diagenesis** (*dia* = change, *genesis* = origin) is a collective term for all of the chemical, physical, and biological changes that take place after sediments are deposited and during and after lithification.

Burial promotes diagenesis because as sediments are buried, they are subjected to increasingly higher tempera-

FIGURE 7.17 Because stagnant swamp water does not allow plants to completely decay, large quantities of plant material can accumulate in a swamp. (Copyright © by Carr Clifton. All rights reserved)

tures and pressures. Diagenesis occurs within the upper few kilometers of Earth's crust at temperatures that are generally less than 150° to 200°C. Beyond this somewhat arbitrary threshold, metamorphism is said to occur.

One example of diagenetic change is *recrystallization*, the development of more stable minerals from less stable ones. It is illustrated by the mineral aragonite, the less stable form of calcium carbonate ($CaCO_3$). Aragonite is secreted by many marine organisms to form shells and other hard parts, such as the skeletal structures produced by corals. In some environments, large quantities of these solid materials accumulate as sediment. As burial takes place, aragonite recrystallizes to the more stable form of calcium carbonate, calcite, the main constituent in the sedimentary rock limestone.

Another example of diagenesis was provided in the preceding discussion of coal. It involved the chemical alteration of organic matter in an oxygen-poor environment. Instead of completely decaying, as would occur in the presence of oxygen, the organic matter is slowly transformed to solid carbon.

Diagenesis includes **lithification,** the processes by which unconsolidated sediments are transformed into solid sedimentary rocks (*lithos* = stone, *fic* = making). Basic lithification processes include compaction and cementation.

The most common physical diagenetic change is **compaction.** As sediment accumulates, the weight of overlying material compresses the deeper sediments. The deeper a sediment is buried, the more it is compacted and the firmer it becomes. As the grains are pressed closer and closer, there is considerable reduction in pore space (the open space between particles). For example, when clays are buried beneath several thousand meters of material, the volume of clay may be reduced by as much as 40 percent. As pore space decreases, much of the water that was trapped in the sediments is driven out. Because sands and other coarse sediments are less compressible, compaction is most significant as a lithification process in fine-grained sedimentary rocks.

Cementation is the most important process by which sediments are converted to sedimentary rock. It is a diagenetic change that involves the crystallization of minerals among the individual sediment grains. Groundwater carries ions in solution. Gradually, the crystallization of new minerals from these ions takes place in the pore spaces, cementing the clasts together. Just as the amount of pore space is reduced during compaction, the addition of cement into a sedimentary deposit reduces its porosity as well.

Calcite, silica, and iron oxide are the most common cements. It is often a relatively simple matter to identify the cementing material. Calcite cement will effervesce with dilute hydrochloric acid. Silica is the hardest cement and thus produces the hardest sedimentary rocks. An orange or dark red color in a sedimentary rock means that iron oxide is present.

Most sedimentary rocks are lithified by means of compaction and cementation. However, some initially form as solid masses of intergrown crystals rather than beginning as accumulations of separate particles that later become solid. Other crystalline sedimentary rocks do not begin that way but are transformed into masses of interlocking crystals sometime after the sediment is deposited.

For example, with time and burial, loose sediment consisting of delicate calcareous skeletal debris may be recrystallized into a relatively dense crystalline limestone. Because crystals grow until they fill all the available space, pore spaces are frequently lacking in crystalline sedimentary rocks. Unless the rocks later develop joints and fractures, they will be relatively impermeable to fluids like water and oil.

Classification of Sedimentary Rocks

The classification scheme in Figure 7.18 divides sedimentary rocks into two major groups: detrital and chemical/organic. Further, we can see that the main criterion for subdividing the detrital rocks is particle size, whereas the primary basis for distinguishing among different rocks in the chemical group is their mineral composition.

As is the case with many (perhaps most) classifications of natural phenomena, the categories presented in Figure 7.18 are more rigid than the actual state of nature. In reality, many of the sedimentary rocks classified into the chemical group also contain at least small quantities of detrital sediment. Many limestones, for example, contain varying amounts of mud or sand, giving them a "sandy" or "shaly" quality. Conversely, because practically all detrital rocks are cemented with material that was originally dissolved in water, they too are far from being "pure."

As was the case with the igneous rocks examined in Chapter 4, *texture* is a part of sedimentary rock classification. There are two major textures used in the classification of sedimentary rocks: clastic and nonclastic. The term **clastic** is taken from a Greek word meaning "broken." Rocks that display a clastic texture consist of discrete fragments and particles that are cemented and compacted together. Although cement is present in the spaces between particles, these openings are rarely filled completely. All detrital rocks have a clastic texture. In addition, some chemical sedimentary rocks exhibit this texture. For example, coquina, the limestone composed of shells and shell fragments, is obviously as clastic as a conglomerate or sandstone. The same applies for some varieties of oolitic limestone.

Some chemical sedimentary rocks have a **nonclastic** or **crystalline** texture in which the minerals form a pattern of interlocking crystals. The crystals may be microscopically small or large enough to be visible without magnification. Common examples of rocks with nonclastic textures are those deposited when seawater evaporates (Figure 7.19). The materials that make up many other nonclastic rocks may actually have originated as detrital deposits. In these instances, the particles probably consisted of shell fragments and other hard parts rich in calcium carbonate or silica. The clastic nature of the grains was subsequently obliterated or obscured because the particles recrystallized when they were consolidated into limestone or chert.

Nonclastic rocks consist of intergrown crystals, and some may resemble igneous rocks, which are also crystalline. The two rock types are usually easy to distinguish because the

Detrital Sedimentary Rocks		
ClasticTexture (particle size)	Sediment Name	Rock Name
Coarse (over 2 mm)	Gravel (Rounded particles)	**Conglomerate**
Coarse (over 2 mm)	Gravel (Angular particles)	**Breccia**
Medium (1/16 to 2 mm)	Sand (If abundant feldspar is present the rock is called **Arkose**)	**Sandstone**
Fine (1/16 to 1/256 mm)	Mud	**Siltstone**
Very fine (less than 1/256 mm)	Mud	**Shale or Mudstone**

Chemical and Organic Sedimentary Rocks			
Composition	Texture	Rock Name	
Calcite, CaCO₃	Nonclastic: Fine to coarse crystalline	**Crystalline Limestone**	
Calcite, CaCO₃	Nonclastic: Fine to coarse crystalline	**Travertine**	
Calcite, CaCO₃	Clastic: Visible shells and shell fragments loosely cemented	**Coquina**	**Biochemical Limestone**
Calcite, CaCO₃	Clastic: Various size shells and shell fragments cemented with calcite cement	**Fossiliferous Limestone**	**Biochemical Limestone**
Calcite, CaCO₃	Clastic: Microscopic shells and clay	**Chalk**	**Biochemical Limestone**
Quartz, SiO₂	Nonclastic: Very fine crystalline	**Chert (light colored) Flint (dark colored)**	
Gypsum CaSO₄•2H₂O	Nonclastic: Fine to coarse crystalline	**Rock Gypsum**	
Halite, NaCl	Nonclastic: Fine to coarse crystalline	**Rock Salt**	
Altered plant fragments	Nonclastic: Fine-grained organic matter	**Bituminous Coal**	

FIGURE 7.18 Identification of sedimentary rocks. Sedimentary rocks are divided into three groups, detrital, chemical and organic. The main criterion for naming detrital sedimentary rocks is particle size, whereas the primary basis for distinguishing among chemical sedimentary rocks is their mineral composition.

FIGURE 7.19 Like other evaporites, this sample of rock salt is said to have a nonclastic texture because it is composed of intergrown crystals. (Photos by E. J. Tarbuck)

← 5 cm →

Close up

minerals contained in nonclastic sedimentary rocks are quite unlike those found in most igneous rocks. For example, rock salt, rock gypsum, and some forms of limestone consist of intergrown crystals, but the minerals within these rocks (halite, gypsum, and calcite) are seldom associated with igneous rocks.

Sedimentary Environments

GEODe Sedimentary Rocks
  ▸ Sedimentary Environments

Sedimentary rocks are important in the interpretation of Earth history. By understanding the conditions under which sedimentary rocks form, geologists can often deduce the history of a rock, including information about the origin of its component particles, the method of sediment transport, and the nature of the place where the grains eventually came to rest; that is, the environment of deposition (Figure 7.20).

An **environment of deposition** or **sedimentary environment** is simply a geographic setting where sediment is accumulating. Each site is characterized by a particular combination of geologic processes and environmental conditions. Some sediments, such as the chemical sediments that precipitate in water bodies, are solely the product of their sedimentary environment. That is, their component minerals originated and were deposited in the same place. Other sediments originate far from the site where they accumulate. These materials are transported great distances from their source by some combination of gravity, water, wind, and ice.

At any given time the geographic setting and environmental conditions of a sedimentary environment determine the nature of the sediments that accumulate. Therefore, geologists carefully study the sediments in present-day depositional environments because the features they find can also be observed in ancient sedimentary rocks.

By applying a thorough knowledge of present-day conditions, geologists attempt to reconstruct the ancient environments and geographical relationships of an area at the time a particular set of sedimentary layers were deposited. Such analyses often lead to the creation of maps, which depict the geographic distribution of land and sea, mountains and river valleys, deserts and glaciers, and other environments of deposition. The foregoing description is an excellent example of the application of a fundamental principle of modern geology, namely that "the present is the key to the past."*

Types of Sedimentary Environments

Sedimentary environments are commonly placed into one of three categories: continental, marine, or transitional (shoreline). Each category includes many specific subenvironments. Figure 7.20 is an idealized diagram illustrating a number of important sedimentary environments associated with each category. Realize that this is just a sampling of the great diversity of depositional environments. The remainder of this section provides a brief overview of each category. Later, Chapters 16 through 20 will examine many of these environments in greater detail. Each is an area where sediment accumulates and where organisms live and die. Each produces a characteristic sedimentary rock or assemblage that reflects prevailing conditions.

Continental Environments   Continental environments are dominated by the erosion and deposition associated with streams. In some cold regions, moving masses of glacial ice replace running water as the dominant process. In arid regions (as well as some coastal settings) wind takes on greater importance. Clearly, the nature of the sediments deposited in continental environments is strongly influenced by climate.

Streams are the dominant agent of landscape alteration, eroding more land and transporting and depositing more sediment than any other process. In addition to channel deposits, large quantities of sediment are dropped when flood waters periodically inundate broad, flat valley floors (called *floodplains*). Where rapid streams emerge from a mountainous area onto a flatter surface, a distinctive cone-shaped accumulation of sediment known as an *alluvial fan* forms.

In frigid high-latitude or high-altitude settings, glaciers pick up and transport huge volumes of sediment. Materials deposited directly from ice are typically unsorted mixtures of particles that range in size from clay to boulders. Water from melting glaciers transports and redeposits some of the glacial sediment, creating stratified, sorted accumulations.

The work of wind and its resulting deposits are referred to as *eolian*, after Aeolus, the Greek god of wind. Unlike glacial deposits, eolian sediments are well sorted. Wind can lift fine dust high into the atmosphere and transport it great distances. Where winds are strong and the surface is not anchored by vegetation, sand is transported closer to the ground, where it accumulates in *dunes*. Deserts and coasts are common sites for this type of deposition.

In addition to being areas where dunes sometimes develop, desert basins are the sites where shallow *playa lakes* occasionally form following heavy rains or periods of snowmelt in adjacent mountains. They rapidly dry up, sometimes leaving behind evaporites and other characteristic deposits. In humid regions lakes are more enduring features, and their quiet waters are excellent sediment traps. Small deltas, beaches, and bars form along the lakeshore, with finer sediments coming to rest on the lake floor.

Marine Environments   Marine depositional environments are divided according to depth. The *shallow marine* environment reaches to depths of about 200 meters (nearly 700 feet) and extends from the shore to the outer edge of the continental shelf. The *deep marine* environment lies seaward of the continental shelf in waters deeper than 200 meters.

The shallow marine environment borders all of the world's continents. Its width varies greatly, from practically nonexistent in some places to broad expanses extending as far as 1500 kilometers (more than 900 miles) in other locations. On the average this zone is about 80 kilometers (50 miles) wide. The kind of sediment deposited here depends on several factors, including distance from shore, the elevation of the adjacent land area, water depth, water temperature, and climate.

Due to the ongoing erosion of the adjacent continent, the shallow marine environment receives huge quantities of land-derived sediment. Where the influx of such sediment is small and the seas are relatively warm, carbonate-rich muds may be the predominant sediment. Most of this material consists of the skeletal debris of carbonate-secreting organisms mixed with inorganic precipitates. Coral reefs are also associated with warm, shallow marine environments. In hot regions where the sea occupies a basin with restricted circulation, evaporation triggers the precipitation of soluble materials and the formation of marine evaporite deposits.

Deep marine environments include all the floors of the deep ocean. Far from landmasses, tiny particles from many sources remain adrift for long spans. Gradually these small grains "rain" down on the ocean floor, where they accumulate very slowly. Significant exceptions are thick deposits of relatively coarse sediment that occur at the base of the continental slope. These materials move down from the continental shelf as turbidity currents—dense gravity-driven masses of sediment and water. Box 7.2 takes a closer look at sediments that accumulate in marine environments.

---

*For more on this idea, see "The Birth of Modern Geology" in Chapter 1.

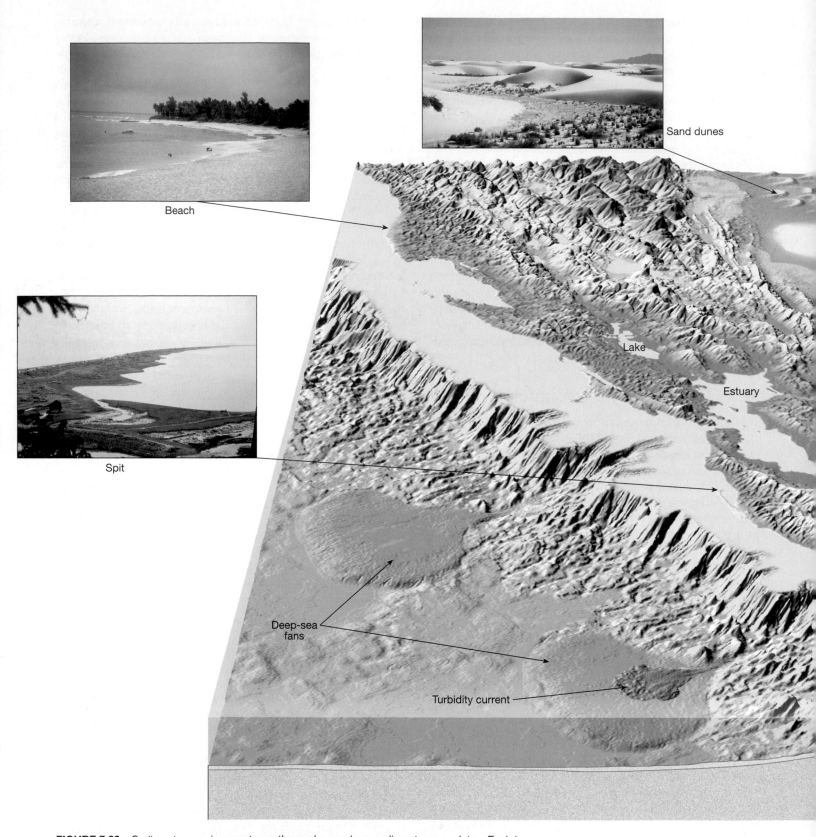

Beach

Sand dunes

Spit

Lake

Estuary

Deep-sea fans

Turbidity current

FIGURE 7.20 Sedimentary environments are those places where sediment accumulates. Each is characterized by certain physical, chemical, and biological conditions. Because each sediment contains clues about the environment in which it was deposited, sedimentary rocks are important in the interpretation of Earth history. A number of important continental, transitional, and marine sedimentary environments are represented in this idealized diagram. (Photos by E. J. Tarbuck, except alluvial fan, by Marli Miller)

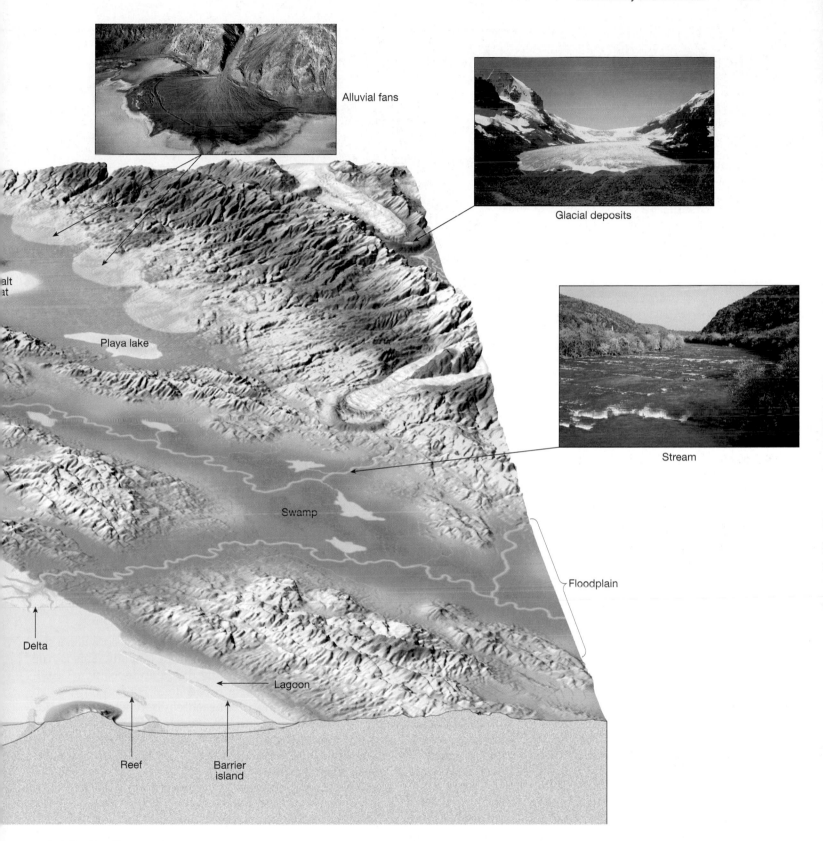

Alluvial fans

Glacial deposits

Stream

Playa lake

Swamp

Floodplain

Delta

Lagoon

Reef

Barrier island

FIGURE 7.20   (Continued)

BOX 7.2 ▶ UNDERSTANDING EARTH

The Nature and Distribution of Seafloor Sediments

Except for steep areas of the continental slope and areas near the crest of the oceanic ridge system, most of the ocean floor is covered with sediment (Figure 7.C). A portion of this material has been deposited by turbidity currents, while much of the rest has slowly settled onto the seafloor from above (see Figure 7.24, page 215).

Seafloor sediments can be classified according to their origin into three broad categories: (1) *terrigenous* (*terra* = land; *generare* = to produce), (2) *biogenous* (*bio* = life; *generare* = to produce), and (3) *hydrogenous* (*hydro* = water; *generare* = to produce). Although each category is discussed separately, seafloor sediments usually come from a variety of sources and are thus mixtures of the various sediment types.

Terrigenous sediment consists primarily of mineral grains that were weathered from continental rocks and transported to the ocean. Larger particles (gravel and sand) usually settle rapidly near shore, whereas finer particles (microscopic clay-size particles) can take years to settle to the ocean floor and may be carried thousands of kilometers by ocean currents. As a consequence, virtually every part of the ocean receives some terrigenous sediment. The rate at which this sediment accumulates on the deep-ocean floor, though, is very slow. To form a 1-centimeter (0.4-inch) *abyssal clay* layer, for example, requires as much as

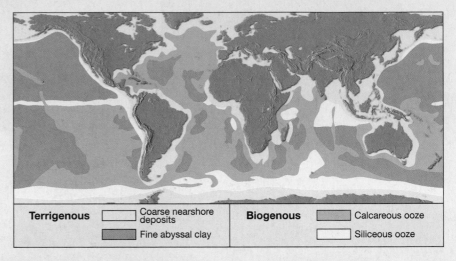

FIGURE 7.C   Distribution of marine sediment. Coarse-grained terrigenous deposits dominate continental margin areas, whereas fine-grained terrigenous material (abyssal clay) is common in deeper areas of the ocean basins. However, deep-ocean deposits are dominated by calcareous oozes, which are found on the shallower portions of deep-ocean areas along the mid-ocean ridge. Siliceous oozes are found beneath areas of unusually high biologic productivity such as the Antarctic and the equatorial Pacific and Indian Oceans. Hydrogenous sediment comprises only a small proportion of deposits in the ocean.

50,000 years. Conversely, on the continental margins near the mouths of large rivers, terrigenous sediment accumulates rapidly and forms thick deposits. In the Gulf of Mexico, for instance, the sediment has reached a depth of many kilometers.

Biogenous sediment consists of shells and skeletons of marine animals and algae. This debris is produced mostly by microscopic organisms living in sunlit waters near the ocean surface. Once these organisms die, their hard *tests* (*testa* = shell)

Transitional Environments   The shoreline is the transition zone between marine and continental environments. Here we find the familiar deposits of sand or gravel called *beaches*. Mud-covered *tidal flats* are alternately covered with shallow sheets of water and then exposed to the air as tides rise and fall. Along and near the shore, the work of waves and currents distributes sand, creating *spits, bars,* and *barrier islands*. Offshore bars and reefs create *lagoons*. The quieter waters in these sheltered areas are another site of deposition in the transition zone.

Deltas are among the most significant deposits associated with transitional environments. The complex accumulations of sediment build outward into the sea when rivers experience an abrupt loss of velocity and drop their load of detrital debris.

Sedimentary Facies

When a series of sedimentary layers is studied, we can see the successive changes in environmental conditions that occurred at a particular place with the passage of time.

Changes in past environments may also be seen when a single unit of sedimentary rock is traced laterally. This is true because at any one time many different depositional environments can exist over a broad area. For example, when sand is accumulating in a beach environment, finer muds are often being deposited in quieter, offshore waters. Still farther out, perhaps in a zone where biological activity is high and land-derived sediments are scarce, the deposits consist largely of the calcareous remains of small organisms. In this example, different sediments are accumulating adjacent to one another at the same time. Each unit possesses a distinctive set of characteristics reflecting the conditions in a particular environment. To describe such sets of sediments, the term **facies** is used. When a sedimentary unit is examined in cross section from one end to the other, each facies grades laterally into another that formed at the same time but which exhibits different characteristics (Figure 7.21). Commonly, the merging of adjacent facies tends to be a gradual transition rather than a sharp boundary, but abrupt changes do sometimes occur.

constantly "rain" down and accumulate on the seafloor.

The most common biogenous sediment is *calcareous* ($CaCO_3$) *ooze*, which, as its name implies, has the consistency of thick mud. This sediment is produced from the tests of organisms such as *coccolithophores* (single-celled algae) and *foraminifers* (small animals) that inhabit warm surface waters. When calcareous tests slowly sink into deeper parts of the ocean, they begin to dissolve. This results because the deeper, cold seawater is rich in carbon dioxide and is thus more acidic than warm water. In seawater deeper than about 4500 meters (15,000 feet), calcareous tests will completely dissolve before they reach bottom. Consequently, calcareous ooze does not accumulate in the deep-ocean basins.

Other biogenous sediments include *siliceous* (SiO_2) *ooze* and phosphate-rich material. The former is composed primarily of tests of *diatoms* (single-celled algae) and *radiolarians* (single-celled animals) that prefer cooler surface waters, whereas the latter is derived from the bones, teeth, and scales of fish and other marine organisms.

Hydrogenous sediment consists of minerals that crystallize directly from seawater through various chemical reactions. Hydrogenous sediments represent a relatively small portion of overall sediment in the ocean. They do, however, have many different compositions and are distributed in diverse environments of deposition.

Some of the most common types of hydrogenous sediment include:

- *Manganese nodules*, which are rounded, hard lumps of manganese, iron, and other metals that precipitate in concentric layers around a central object such as a volcanic pebble or a grain of sand (Figure 7.D).
- *Calcium carbonates*, which form by precipitation directly from seawater in warm climates. If this material is buried and hardened, it forms limestone. Most limestone, however, is composed of biogenous sediment.
- *Metal sulfides*, which are usually precipitated as coatings on rocks near vents associated with the crest of the mid-ocean ridge that spew hot, mineral-rich water. These deposits contain iron, nickel, copper, zinc, silver, and other metals in varying proportions.
- *Evaporites*, which form where evaporation rates are high and there is restricted open-ocean circulation. As water evaporates from such areas, the remaining seawater becomes saturated with dissolved minerals, which then begin to precipitate.

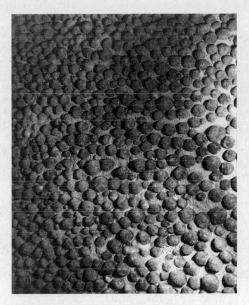

FIGURE 7.D   Manganese nodules photographed at a depth of 2909 fathoms (5323 meters) beneath the *Robert Conrad*, south of Tahiti. (Photo courtesy of Lawrence Sullivan, Lamont-Doherty Earth Observatory/Columbia University)

Sedimentary Structures

In addition to variations in grain size, mineral composition, and texture, sediments exhibit a variety of structures. Some, such as graded beds, are created when sediments are accumulating and are a reflection of the transporting medium. Others, such as *mud cracks*, form after the materials have been deposited and result from processes occurring in the environment. When present, sedimentary structures provide additional information that can be useful in the interpretation of Earth history.

Sedimentary rocks form as layer upon layer of sediment accumulates in various depositional environments. These layers, called **strata,** or **beds,** are probably *the single most common and characteristic feature of sedimentary rocks.* Each stratum is unique. It may be a coarse sandstone, a fossil-rich

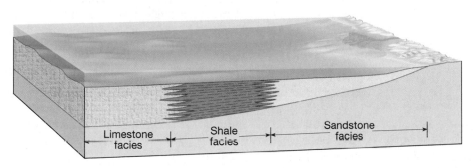

FIGURE 7.21   When a single sedimentary layer is traced laterally, we may find that it is made up of several different rock types. This can occur because many sedimentary environments can exist at the same time over a broad area. The term *facies* is used to describe such sets of sedimentary rocks. Each facies grades laterally into another that formed at the same time but in a different environment.

FIGURE 7.22 This outcrop of sedimentary strata illustrates the characteristic layering of this group of rocks. (Photo by Tom Till/Tom Till Photography)

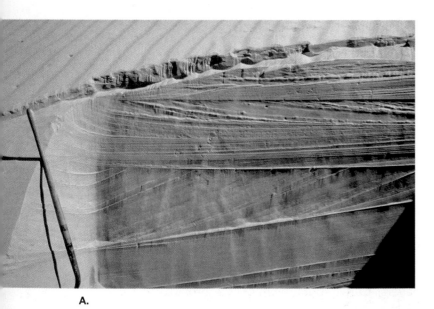

A.

B.

FIGURE 7.23 **A** The cutaway section of this sand dune shows cross-bedding. (Photo by John S. Shelton) **B.** The cross-bedding of this sandstone indicates it was once a sand dune. (Photo by David Muench Photography, Inc.)

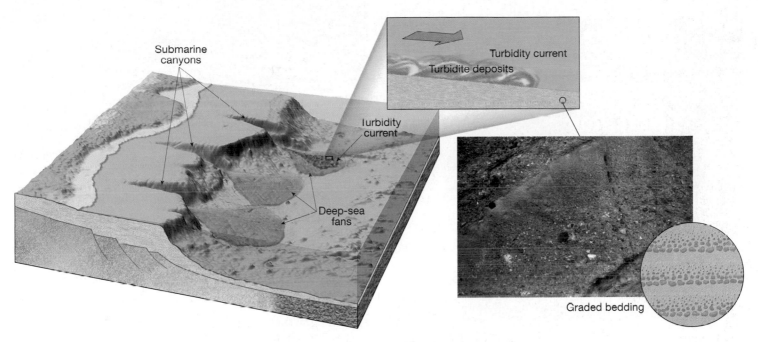

FIGURE 7.24 Turbidity currents are downslope movements of dense, sediment-laden water. They are created when sand and mud on the continental shelf and slope are dislodged and thrown into suspension. Because such mud-choked water is denser than normal seawater, it flows downslope, eroding and accumulating more sediment. Beds deposited by these currents are called *turbidites*. Each event produces a single bed characterized by a decrease in sediment size from bottom to top, a feature known as a *graded bed*. (Photo by Marli Miller)

limestone, a black shale, and so on. When you look at Figure 7.22, or look back through this chapter at Figures 7.1 (p. 194) and 7.4 (p. 198), you will see many such layers, each different from the others. The variations in texture, composition, and thickness reflect the different conditions under which each layer was deposited.

The thickness of beds ranges from microscopically thin to tens of meters thick. Separating the strata are **bedding planes,** flat surfaces along which rocks tend to separate or break. Changes in the grain size or in the composition of the sediment being deposited can create bedding planes. Pauses in deposition can also lead to layering because chances are slight that newly deposited material will be exactly the same as previously deposited sediment. Generally, each bedding plane marks the end of one episode of sedimentation and the beginning of another.

Because sediments usually accumulate as particles that settle from a fluid, most strata are originally deposited as horizontal layers. There are circumstances, however, when sediments do not accumulate in horizontal beds. Sometimes when a bed of sedimentary rock is examined, we see layers within it that are inclined to the horizontal. When this occurs, it is called **cross-bedding** and is most characteristic of sand dunes, river deltas, and certain stream channel deposits (Figure 7.23).

Graded beds represent another special type of bedding. In this case the particles within a single sedimentary layer gradually change from coarse at the bottom to fine at the top. Graded beds are most characteristic of rapid deposition

FIGURE 7.25 In a turbulent stream channel, only large particles settle out. Finer sediments remain suspended and continue their downstream journey. Roaring Fork River near Aspen, Colorado. (Photo by Tom & Susan Bean, Inc.)

A.

B.

FIGURE 7.26   A. Ripple marks preserved in sedimentary rock (Photo by Stephen Trimble). **B.** Ripple marks in the channel of the Rio Negro, a tributary of the Amazon near Manaus, Brazil. (Photo by Galen Rowell/Mountain Light Photography, Inc.)

FIGURE 7.27   Mud cracks form when wet mud or clay dries out and shrinks. (Photo by Gary Yeowell/Stone)

from water containing sediment of varying sizes. When a current experiences a rapid energy loss, the largest particles settle first, followed by successively smaller grains. The deposition of a graded bed is most often associated with a turbidity current, a mass of sediment-choked water that is denser than clear water and that moves downslope along the bottom of a lake or ocean (Figure 7.24).

As geologists examine sedimentary rocks, much can be deduced. A conglomerate, for example, may indicate a high-energy environment, such as a surf zone or rushing stream, where only coarse materials settle out and finer particles are kept suspended (Figure 7.25). If the rock is arkose, it may signify a dry climate, where little chemical alteration of feldspar is possible. Carbonaceous shale is a sign of a low-energy, organic-rich environment, such as a swamp or lagoon.

Other features found in some sedimentary rocks also give clues to past environments. Ripple marks are such a feature. **Ripple marks** are small waves of sand that develop on the surface of a sediment layer by the action of moving water or air (Figure 7.26). The ridges form at right angles to the direction of motion. If the ripple marks were formed by air or water moving in essentially one direction, their form will be asymmetrical. These *current ripple marks* will have steeper sides in the downcurrent direction and more gradual slopes on the upcurrent side. Ripple marks produced by a stream flowing across a sandy channel or by wind blowing over a sand dune are two common examples of current ripples. When present in solid rock, they may be used to determine the direction of movement of ancient wind or water currents. Other ripple marks have a symmetrical form. These features, called *oscillation ripple marks*, result from the back-and-forth movement of surface waves in a shallow nearshore environment.

A.

B.

FIGURE 7.28 Only a tiny fraction of the organisms that lived in the geologic past have been preserved as fossils. The fossil record is slanted toward organisms with hard parts that lived in environments where rapid burial was possible. **A.** Natural cast of a trilobite. These sea-dwelling animals were very abundant during the Paleozoic era. **B.** These dinosaur footprints near Cameron, Arizona were originally made in mud that eventually became a sedimentary rock. Footprints such as these are considered to be *trace fossils* (see Chapter 9). Dinosaurs flourished during the Mesozoic era. (Photo by Tom Bean/DRK Photo)

Mud cracks (Figure 7.27) indicate that the sediment in which they were formed was alternately wet and dry. When exposed to air, wet mud dries out and shrinks, producing cracks. Mud cracks are associated with such environments as tidal flats, shallow lakes, and desert basins.

Fossils, the remains or traces of prehistoric life, are important inclusions in sediment and sedimentary rocks (Figure 7.28). They are important tools for interpreting the geologic past. Knowing the nature of the life forms that ex-

isted at a particular time helps researchers understand past environmental conditions. Further, fossils are important time indicators and play a key role in correlating rocks of similar ages that are from different places.*

---

*The section entitled "Fossils: Evidence of Past Life" in Chapter 9 contains a more detailed discussion of the role of fossils in the interpretation of Earth history.

Summary

- Sedimentary rocks account for about 5 to 10 percent of Earth's outer 16 kilometers (10 miles). Because they are concentrated at Earth's surface, the importance of this group is much greater than this percentage implies. Sedimentary rocks contain much of the basic information needed to reconstruct Earth history. In addition, this group is associated with many important energy and mineral resources.
- *Sedimentary rock* consists of *sediment* that in most cases has been *lithified* into solid rock by the processes of *compaction* and *cementation*. Sediment has two principal sources: (1) as *detrital material,* which originates and is transported as solid particles from both mechanical and chemical weathering, which, when lithified, forms detrital sedimentary rocks; and (2) from soluble material produced largely by chemical weathering, which, when precipitated, forms *chemical sedimentary rocks.* Coal is the

primary example of a third group called *organic sedimentary rocks,* which consist of organic carbon from the remains of partially altered plant material.
- *Particle size* is the primary basis for distinguishing among various detrital sedimentary rocks. The size of the particles in a detrital rock indicates the energy of the medium that transported them. For example, gravels are moved by swiftly flowing rivers, whereas less energy is required to transport sand. Common detrital sedimentary rocks include *shale* (silt- and clay-size particles), *sandstone,* and *conglomerate* (rounded gravel-size particles) or *breccia* (angular gravel-size particles).
- Precipitation of chemical sediments occurs in two ways: (1) by *inorganic processes,* such as evaporation and chemical activity; or by (2) *organic processes* of water-dwelling organisms that produce sediments of *biochemical origin.* *Limestone,* the most abundant chemical sedimentary

rock, consists of the mineral calcite ($CaCO_3$) and forms either by inorganic means or as the result of biochemical processes. Inorganic limestones include *travertine,* which is commonly seen in caves, and *oolitic limestone,* consisting of small spherical grains of calcium carbonate. Other common chemical sedimentary rocks include *dolostone* (composed of the calcium-magnesium carbonate mineral dolomite), *chert* (made of microcrystalline quartz), *evaporites* (such as rock salt and rock gypsum), and *coal* (lignite and bituminous).

- *Diagenesis* refers to all of the physical, chemical, and biological changes that occur after sediments are deposited and during and after the time they are turned into sedimentary rock. Burial promotes diagenesis. Diagenesis includes lithification.

- *Lithification* refers to the processes by which unconsolidated sediments are transformed into solid sedimentary rock. Most sedimentary rocks are lithified by means of *compaction* and/or *cementation.* Compaction occurs when the weight of overlying materials compresses the deeper sediments. Cementation, the most important process by which sediments are converted to sedimentary rocks, occurs when soluble cementing materials, such as *calcite, silica,* and *iron oxide,* are precipitated onto sediment grains, fill open spaces, and join the particles. Although most sedimentary rocks are lithified by compaction or cementation, certain chemical rocks, such as the evaporites, initially form as solid masses of intergrown crystals.

- Sedimentary rocks are divided into three groups: *detrital, chemical* and *organic.* All detrital rocks have a *clastic texture,* which consists of discrete fragments and particles that are cemented and compacted together. The main criterion for subdividing the detrital rocks is particle size. Common detrital rocks include *conglomerate, sandstone,* and *shale.* The primary basis for distinguishing among different rocks in the chemical group is their mineral composition. Some chemical rocks, such as those deposited when seawater evaporates, have a *nonclastic (crystalline) texture* in which the minerals form a pattern of interlocking crystals. However, in reality, many of the sedimentary rocks classified into the chemical group also contain at least small quantities of detrital sediment. Common chemical rocks include *limestone,* chert, and *rock gypsum.* Coal is the primary example of an organic sedimentary rock.

- Sedimentary environments are those places where sediment accumulates. They are grouped into continental, marine, and transitional (shoreline) environments. Each is characterized by certain physical, chemical, and biological conditions. Because sediment contains clues about the environment in which it was deposited, sedimentary rocks are important in the interpretation of Earth's history.

- Sedimentary rocks are particularly important in interpreting Earth's history because, as layer upon layer of sediment accumulates, each records the nature of the environment at the time the sediment was deposited. These layers, called *strata,* or *beds,* are probably the single most characteristic feature of sedimentary rocks. Other features found in some sedimentary rocks, such as *ripple marks, mud cracks, cross-bedding,* and *fossils,* also provide clues to past environments.

Review Questions

1. How does the volume of sedimentary rocks in Earth's crust compare with the volume of igneous rocks in the crust? Are sedimentary rocks evenly distributed throughout the crust?

2. List and briefly distinguish among the three basic sedimentary rock categories.

3. What minerals are most common in detrital sedimentary rocks? Why are these minerals so abundant?

4. What is the primary basis for distinguishing among various detrital sedimentary rocks?

5. Why does shale often crumble easily?

6. How are the degree of sorting and the amount of rounding related to the transportation of sand grains?

7. Distinguish between conglomerate and breccia.

8. Distinguish between the two categories of chemical sedimentary rocks.

9. What are evaporite deposits? Name a rock that is an evaporite.

10. When a body of seawater evaporates, minerals precipitate in a certain order. What determines this order?

11. Each of the following statements describes one or more characteristics of a particular sedimentary rock. For each statement, name the sedimentary rock that is being described.

    a. An evaporite used to make plaster.

    b. A fine-grained detrital rock that exhibits *fissility.*

    c. The primary example of an organic sedimentary rock.

    d. The most abundant chemical sedimentary rock.

    e. A dark-colored hard rock made of microcrystalline quartz.

    f. A variety of limestone composed of small spherical grains.

12. What is the primary basis for distinguishing among different chemical sedimentary rocks?

13. What is diagenesis? Give an example.

14. Compaction is most important as a lithification process with which sediment size?

15. List three common cements for sedimentary rocks. How might each be identified?

16. Distinguish between clastic and nonclastic textures. What type of texture is common to all detrital sedimentary rocks?

17. Some nonclastic sedimentary rocks closely resemble igneous rocks. How might the two be distinguished easily?

18. List three categories of sedimentary environments. Provide one or more examples for each category.

19. Distinguish among the three basic types of seafloor sediment (see Box 7.2, p. 212).

20. What is probably the single most characteristic feature of sedimentary rocks?

21. Distinguish between cross-bedding and graded bedding.

22. How do current ripple marks differ from oscillation ripple marks?

Key Terms

bedding plane (p. 215)
beds (strata) (p. 213)
biochemical (p. 201)
cementation (p. 207)
chemical sedimentary rock
  (p. 195)
clastic texture (p. 207)
compaction (p. 207)
cross-bedding (p. 215)

crystalline texture (p. 207)
detrital sedimentary rock
  (p. 195)
diagenesis (p. 206)
environment of deposition
  (p. 208)
evaporite deposit (p. 204)
facies (p. 212)

fissility (p. 197)
fossil (p. 217)
graded bed (p. 215)
lithification (p. 207)
mud crack (p. 217)
nonclastic texture (p. 207)
organic sedimentary rock
  (p. 195)

ripple mark (p. 216)
salt flat (p. 205)
sedimentary environment
  (p. 208)
sorting (p. 197)
strata (beds) (p. 213)

Web Resources

The *Earth* Website uses the resources and flexibility of the Internet to aid in your study of the topics in this chapter. Written and developed by geology instructors, this site will help improve your understanding of geology. Visit **http://www.prenhall.com/tarbuck** and click on the cover of *Earth 9e* to find:

- Online review quizzes.
- Critical thinking exercises.
- Links to chapter-specific Web resources.
- Internet-wide key-term searches.

http://www.prenhall.com/tarbuck

GEODe: Earth

GEODe: Earth makes studying faster and more effective by reinforcing key concepts using animation, video, narration, interactive exercises and practice quizzes. A copy is included with every copy of *Earth*.

Some sedimentary rocks are not compacted or cemented but form as solid masses of intergrown crystals. Rock salt is one example.

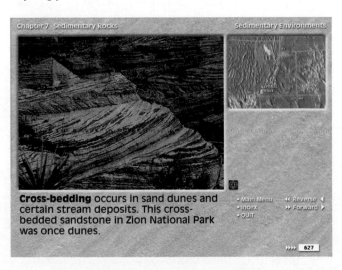

Cross-bedding occurs in sand dunes and certain stream deposits. This cross-bedded sandstone in Zion National Park was once dunes.

Metamorphism and Metamorphic Rocks

Aerial view of metamorphic rocks exposed in the Canadian Shield. (Photo by Robert Hildebrand)

The folded and metamorphosed rocks shown in Figure 8.1 were once flat-lying sedimentary strata. Compressional forces of unimaginable magnitude and temperatures hundreds of degrees above surface conditions prevailed for perhaps thousands or millions of years to produce the deformation displayed by these rocks. Under such extreme conditions, solid rock responds by folding, fracturing, and often by flowing. This chapter looks at the tectonic forces that forge metamorphic rocks and how these rocks change in appearance, mineralogy, and sometimes even in overall chemical composition.

Extensive areas of metamorphic rocks are exposed on every continent in the relatively flat regions known as shields (see chapter opening photo and Figure 8.31). These metamorphic regions are found in eastern Canada, Brazil, Africa, India, Australia, and Greenland. Moreover, metamorphic rocks are an important component of many mountain belts, including the Alps and the Appalachians, where they make up a large portion of a mountain's crystalline core. Even those portions of the stable continental interiors that are covered by sedimentary rocks are underlain by metamorphic basement rocks. In these settings the metamorphic rocks are highly deformed and intruded by large igneous masses. Indeed, significant parts of Earth's continental crust are composed of metamorphic and associated igneous rocks.

Unlike some igneous and sedimentary processes that take place in surface or near-surface environments, metamorphism most often occurs deep within Earth, beyond our direct observation. Notwithstanding this significant obstacle, geologists have developed techniques that have allowed them to learn a great deal about the conditions under which metamorphic rocks form. In turn, the study of metamorphic rocks provides important insights into the tectonic processes that operate within Earth's crust and upper mantle.

FIGURE 8.1   Deformed and folded gneiss, Anza Borrego Desert State Park, California. (Photo by A. P. Trujillo/APT Photos)

What Is Metamorphism?

Metamorphic Rocks
▶ Introduction

Recall from the section on the rock cycle in Chapter 1 that metamorphism is the transformation of one rock type into another. Metamorphic rocks are produced from preexisting igneous, sedimentary, or even from other metamorphic rocks. Thus, every metamorphic rock has a **parent rock**—the rock from which it was formed.

Metamorphism, which means to "change form," is a process that leads to changes in the mineralogy, texture, and sometimes the chemical composition of rocks. Metamorphism takes place where preexisting rock is subjected to a physical or chemical environment that is significantly different from that in which it initially formed. These include changes in temperature, pressure (stress), and the introduction of chemically active fluids. In response to these new conditions, the rock gradually changes until a state of equilibrium with the new environment is reached. Most metamorphic changes occur at the elevated temperatures and pressures that exist in the zone beginning a few kilometers below Earth's surface and extending into the upper mantle.

Metamorphism often progresses incrementally, from slight changes (*low-grade metamorphism*) to substantial changes (*high-grade metamorphism*). For example, in low-grade metamorphism, the common sedimentary rock *shale* becomes the more compact metamorphic rock called *slate*. Hand samples of these rocks are sometimes difficult to distinguish, illustrating that the transition from sedimentary to metamorphic is often gradual and the changes can be subtle.

In more extreme environments, metamorphism causes a transformation so complete that the identity of the parent rock cannot be determined. In high-grade metamorphism, such features as bedding planes, fossils, and vesicles that may have existed in the parent rock are obliterated. Further, when rocks at depth (where temperatures are high) are subjected to directed pressure, they slowly deform to produce a variety of textures as well as large-scale structures such as folds (Figure 8.2).

In the most extreme metamorphic environments, the temperatures approach those at which rocks melt. However, *during metamorphism the rock must remain essentially solid,* for if complete melting occurs, we have entered the realm of igneous activity.

What Drives Metamorphism?

Metamorphic Rocks
▶ Agents of Metamorphism

The agents of metamorphism include *heat, pressure (stress), and chemically active fluids.* During metamorphism, rocks are usually subjected to all three metamorphic agents simultaneously. However, the degree of metamorphism and the contribution of each agent vary greatly from one environment to another.

Heat as a Metamorphic Agent

The most important factor driving metamorphism is *heat* because it provides the energy to drive chemical reactions that result in the recrystallization of existing minerals and/or the formation of new minerals. Recall from the discussion of igneous rocks that an increase in temperature causes the ions within a mineral to vibrate more rapidly. Even in a crystalline *solid,* where ions are strongly bonded, this elevated level of activity allows individual atoms to migrate more freely between sites in the crystalline structure.

Changes Caused by Heat   Heat affects Earth materials, especially those that form in low-temperature environments, in two ways. First, it promotes recrystallization of individual mineral grains. This is particularly true of clays, fine-grained sediments, and some chemical precipitates. Higher temperatures promote recrystallization where fine particles tend to coalesce into larger grains of the same mineralogy.

Second, heat may raise the temperature of a rock to the point where one, or more, of the minerals are no longer chemically stable. In such cases, the constituent ions tend to arrange themselves into crystalline structures that are more stable in the new high-energy environment. Such chemical reactions result in the creation of new minerals with stable configurations that have an overall composition roughly equivalent to that of the original material. (In some environments ions may migrate into or out of a rock unit, thereby changing its overall chemical composition.)

In summary, if we were to traverse a region of metamorphic rocks (now uplifted and exposed) while traveling in the direction of increasing intensity of metamorphism, we would expect to observe two changes largely attributable to increased temperature. The grain size of the rocks would increase and the mineralogy would gradually change.

What Is the Source of Heat?   Earth's internal heat comes mainly from energy that is continually being released by radioactive decay and thermal energy generated during the formation of our planet. Recall that temperatures increase with depth at a

FIGURE 8.2   Deformed metamorphic rocks exposed in a road cut in the Eastern Highland of Connecticut. (Photo by Phil Dombrowski)

rate known as the *geothermal (geo* = Earth, *therm* = heat)*gradient.* In the upper crust, this increase in temperature averages between 20°C and 30°C per kilometer (Figure 8.3). Thus, rocks that formed at Earth's surface will experience a gradual increase in temperature as they are taken to greater depths (Figure 8.3). When buried to a depth of about 8 kilometers (5 miles), where temperatures are about 200°C, clay minerals tend to become unstable and begin to recrystallize into new minerals, such as chlorite and muscovite, that are stable in this environment. (Chlorite is a micalike mineral formed by the metamorphism of dark [mafic] silicate minerals.) However, many silicate minerals, particularly those found in crystalline igneous rocks—quartz and feldspar for example—remain stable at these temperatures. Thus, metamorphic changes in these minerals generally occur at much greater depths.

Environments where rocks may be carried to great depths and heated include convergent plate boundaries where slabs of sediment-laden oceanic crust are being subducted. In addition, rocks may become deeply buried in large basins where gradual subsidence results in very thick accumulations of sediment (Figure 8.3). Such locations, exemplified by the Gulf of Mexico, are known to develop metamorphic conditions near the base of the pile. Furthermore, continental collisions, which result in crustal thickening by folding and faulting, cause rocks to become deeply buried where elevated temperatures may trigger partial melting (see Figure 8.22).

Heat may also be transported from the mantle into even the shallowest layers of the crust by igneous intrusions. Rising mantle plumes, upwelling at mid-ocean ridges, and magma generated by partial melting of mantle rock at subduction zones are three examples (Figure 8.3). Anytime magma forms and buoyantly rises toward the surface, metamorphism occurs. When magma intrudes relatively cool rocks at shallow depths, the host rock is "baked." This process, called *contact metamorphism,* will be considered later in this chapter.

Students Sometimes Ask ...

How hot is it deep in the crust?

The increase in temperature with depth, based on the geothermal gradient, can be expressed as *the deeper one goes, the hotter it gets.* This relationship has been observed by miners in deep mines and in deeply drilled wells. In the deepest mine in the world (the Western Deep Levels mine in South Africa, which is 4 kilometers or 2.5 miles deep), the temperature of the surrounding rock is so hot that it can scorch human skin! In fact, miners often work in groups of two: one to mine the rock, and the other to operate a large fan that keeps the other worker cool.

The temperature is even higher at the bottom of the deepest well in the world, which was completed in the Kola Peninsula of Russia in 1992 and goes down a record 12.3 kilometers (7.7 miles). At this depth it is 245°C (473°F), much greater than the boiling point of water. What keeps water from boiling is the high confining pressure at depth.

Confining Pressure and Differential Stress

Pressure, like temperature, also increases with depth as the thickness of the overlying rock increases. Buried rocks are subjected to **confining pressure,** which is analogous to water pressure, where the forces are applied equally in all directions (Figure 8.4A). The deeper you go in the ocean, the greater the confining pressure. The same is true for rock that is buried. Confining pressure causes the spaces between mineral grains to close, producing a more compact rock having a greater density (Figure 8.4A). Further, at great depths, confining pressure may cause minerals to recrystallize into new minerals that display a more compact crystalline form. Confining pressure does *not,* however, fold and deform rocks like those shown in Figure 8.1.

In addition to confining pressure, rocks may be subjected to directed pressure. This occurs, for example, at convergent plate boundaries where slabs of lithosphere collide. Here the forces that deform rock are unequal in different directions and are referred to as **differential stress.** (A more in-depth discussion of *stress,* which is a force per unit area, is provided in Chapter 10).

Unlike confining pressure, which "squeezes" rock equally in all directions, differential stresses are greater in one direction than in others. As shown in Figure 8.4B, rocks subjected to differential stress are shortened in the direction of greatest stress and elongated, or lengthened, in the direction perpendicular to that stress. As a result, the rocks involved are often *folded* or *flattened* (similar to stepping on a rubber ball).

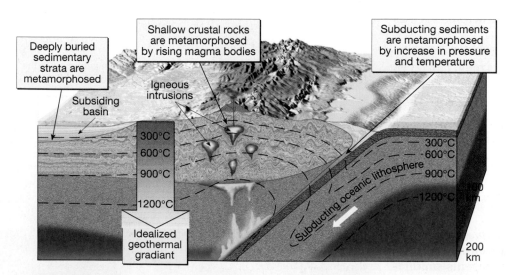

FIGURE 8.3 Illustration of the geothermal gradient and its role in metamorphism. Notice how the geothermal gradient is lowered by the subduction of comparatively cool oceanic lithosphere. By contrast, thermal heating is evident where magma intrudes the upper crust.

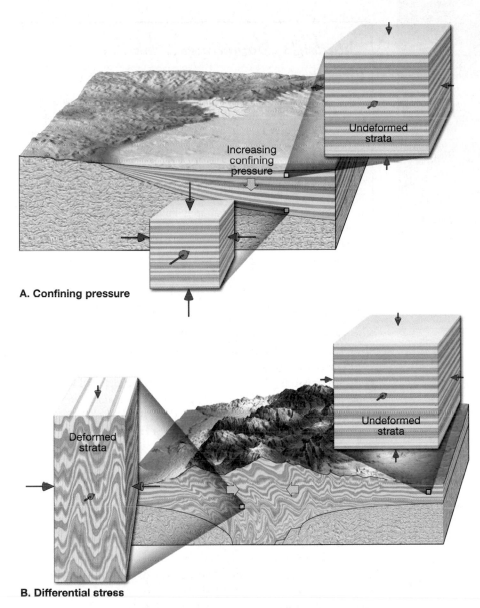

FIGURE 8.4 Confining pressure and differential stress as metamorphic agents. **A.** In a depositional environment, as confining pressure increases, rocks deform by decreasing in volume. **B.** During mountain building, rocks subjected to differential stress are shortened in the direction that pressure is applied, and lengthened in the direction perpendicular to that force.

grains tend to flatten and elongate when subjected to differential stress (Figure 8.5). This accounts for their ability to deform by flowing (rather than fracturing) to generate intricate folds.

Chemically Active Fluids

Fluids composed mainly of water and other volatile components, including carbon dioxide, are believed to play an important role in some types of metamorphism. Fluids that surround mineral grains act as catalysts to promote recrystallization by enhancing ion migration. In progressively hotter environments these ion-rich fluids become correspondingly more reactive. When two mineral grains are squeezed together, the part of their crystalline structures that touch are the most highly stressed. Ions located at these sites are readily dissolved by the hot fluids and migrate along the surface of the grain to the pore spaces located between individual grains. Thus, hydrothermal fluids aid in the recrystallization of mineral grains by dissolving material from regions of high stress and then precipitating (depositing) this material in areas of low stress. As a result, *minerals tend to recrystallize and grow longer in a direction perpendicular to compressional stresses.*

Where hot fluids circulate freely through rocks, ionic exchange may occur between two adjacent rock layers, or ions may migrate great distances before they are finally deposited. The latter situation is particularly common when we consider hot fluids that escape during the crystallization of an igneous pluton. If the rocks that surround the pluton differ markedly in composition from that of the invading fluids, there may be a substantial exchange of ions between the fluids and host rocks. When this occurs, a change in the overall composition of the surrounding rock results. The change in composition by interaction with fluids is called **metasomatism.**

What Is the Source of Chemically Active Fluids? Water is plentiful in the pore spaces of most sedimentary rocks, as well as in fractures in igneous rocks. In addition, many minerals, such as clays, micas, and amphiboles, are *hydrated* (*hydra* = water) and thus contain water in their crystalline structures. Elevated temperatures associated with low to moderate-grade metamorphism cause the dehydration of these minerals. Once expelled, the water moves along the surfaces of individual grains and is available to facilitate ion transport. However, in high-grade metamorphic environments, where temperatures are extreme, these fluids may be

Along convergent plate boundaries the greatest differential stress is directed roughly horizontal in the direction of plate motion, and the least pressure is in the vertical direction. Consequently, in these settings the crust is greatly shortened (horizontally) and thickened (vertically). Although, differential stresses are generally small when compared to confining pressure, they are important in creating the various textures exhibited by metamorphic rocks.

In surface environments where temperatures are comparatively low, rocks are *brittle* and tend to fracture when subjected to differential stress. Continued deformation grinds and pulverizes the mineral grains into small fragments. By contrast, in high-temperature environments rocks are *ductile*. When rocks exhibit ductile behavior, their mineral

FIGURE 8.5 Metaconglomerate, also called stretched pebble conglomerate. These once nearly spherical pebbles have been heated and flattened into elongated structures. (Photo by E. J. Tarbuck)

driven from the rocks. Recall that when oceanic crust is subducted to depths of about 100 kilometers, water expelled from these slabs migrates into the mantle wedge above, where it triggers melting.

The Importance of Parent Rock

Most metamorphic rocks have the same overall chemical composition as the parent rock from which they formed, except for the possible loss or acquisition of volatiles such as water (H_2O) and carbon dioxide (CO_2). For example, the metamorphism of shale results in slate, where clay minerals recrystallize to form micas. (The minute crystals of quartz and feldspar found in shale are not altered in the transformation of shale to slate and thus remain intermixed with the micas.) Although the mineralogy changes in the transformation of shale to slate, the overall chemical composition of slate is comparable to that of the rock from which it was derived. Further, when the parent rock has a mafic composition, such as basalt, the metamorphic product will be rich in minerals containing iron and magnesium unless, of course, there has been a substantial loss of these atoms.

In addition, the mineral makeup of the parent rock determines, to a large extent, the degree to which each metamorphic agent will cause change. For example, when magma forces its way into existing rock, high temperatures and associated hot ion-rich fluids tend to alter the host rock. When the host rock is composed of minerals that are comparatively unreactive, such as the quartz grains found in a clean quartz sandstone, very little alteration may take place. However, if the host rock is a "dirty" limestone that contains abundant silica-rich clay, the calcite ($CaCO_3$) in the limestone may react with silica (SiO_2) in the clays to form wollastonite ($CaSiO_3$) plus carbon dioxide (CO_2). In this situation the zone of metamorphism may extend for several kilometers from the magma body.

Metamorphic Textures

 GEODe Metamorphic Rocks
▶ Textural and Mineralogical Changes

Recall that the term **texture** is used to describe the size, shape, and arrangement of grains within a rock. Most igneous and many sedimentary rocks consist of mineral grains that have a random orientation and thus appear the same when viewed from any direction. By contrast, deformed metamorphic rocks that contain platy minerals (micas) and/or elongated minerals (amphiboles), typically display some kind of *preferred orientation* in which the mineral grains exhibit a parallel to sub-parallel alignment. Like a fistful of pencils, rocks containing elongated minerals that are oriented parallel to each other will appear different when viewed from the side than when viewed head-on. A rock that exhibits a preferred orientation of its minerals is said to possess *foliation*.

Foliation

The term **foliation** (*foliatus* = leaflike) refers to any planar (nearly flat) arrangement of mineral grains or structural features within a rock. Although foliation occurs in some sedimentary and even a few types of igneous rocks, it is a fundamental characteristic of regionally metamorphosed rocks—that is, rock units that have been strongly folded and distorted. In metamorphic environments, foliation is ultimately driven by compressional stresses that shorten rock units, causing mineral grains in preexisting rocks to develop parallel, or nearly parallel, alignments. Examples of foliation include the parallel alignment of platy and/or elongated minerals; the parallel alignment of flattened mineral grains and pebbles; compositional banding where the separation of dark and light minerals generate a layered appearance; and slaty cleavage where rocks can be easily split into thin, tabular slabs along parallel surfaces. These diverse types of foliation can form in many different ways, including:

1. Rotation of platy and/or elongated mineral grains into a new orientation.
2. Recrystallization of minerals to form new grains growing in the direction of preferred orientation.
3. Changing the shape of equidimensional grains into elongated shapes that are aligned in a preferred orientation

The rotation of existing mineral grains is the easiest of these mechanisms to envision. Figure 8.6 illustrates the mechanics by which platy or elongated minerals are rotated. Note that the new alignment is roughly perpendicular to the direction of maximum shortening. Although physical rotation of platy minerals contributes to the development of foliation in low-grade metamorphism, other mechanisms dominate in more extreme environments.

Recall that recrystallization is the creation of new mineral grains out of old ones. During the transformation of shale to slate, minute clay minerals (stable at the surface) recrystallize into microscopic flakes of chlorite and mica (stable at higher temperatures and pressures). In some settings old grains are dissolved and migrate to a different site, where they precipitate to form new mineral grains. The growth of new mineral grains tends to develop on old crystals of similar structure and grow in the same orientation as the older crystals. In this way the new growth "mimics" the growth of old grains and enhances any preexisting preferred orientation. However, recrystallization that accompanies deformation often results in a *new* preferred orientation. As rock units are folded and generally shortened during metamorphism, elongated and platy minerals tend to recrystallize perpendicular to the direction of maximum stress.

Mechanisms that change the shapes of individual grains are especially important for the development of preferred orientations in rocks that contain minerals such as quartz, calcite, and olivine. When these minerals are stressed, they

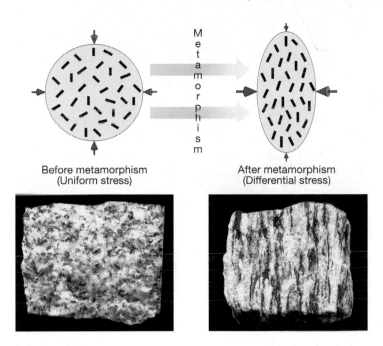

FIGURE 8.6   Mechanical rotation of platy or elongated mineral grains. **A.** Existing mineral grains keep their random orientation if force is uniformly applied. **B.** As differential stress causes rocks to flatten, mineral grains rotate toward the plane of flattening. (Photos by E. J. Tarbuck)

Before metamorphism (Uniform stress)

After metamorphism (Differential stress)

develop elongated grains that align in a direction parallel to maximum flattening (Figure 8.7). This type of deformation occurs in high-temperature environments where ductile deformation is dominant (as opposed to brittle fracturing).

A change in grain shape can occur as units of a mineral's crystalline structure slide relative to one another along discrete planes, thereby distorting the grain as shown in Figure 8.7B. This type of gradual solid-state plastic flow involves slippage that disrupts the crystal lattice as the positions of atoms or ions change. This typically involves the breaking of existing chemical bonds and the formation of new ones. In addition, the shape of a mineral may change as ions move from a point along the margin of the grain that is highly stressed to a less-stressed position on the same grain (Figure 8.7C). This type of deformation occurs by mass transfer of material from one location to another. As you might expect, this mechanism is aided by chemically active fluids and is a type of recrystallization.

Foliated Textures

Various types of foliation exist, depending largely upon the grade of metamorphism and the mineralogy of the parent rock. We will look at three: *rock* or *slaty cleavage, schistosity,* and *gneissic texture.*

Rock or Slaty Cleavage   **Rock cleavage** refers to closely spaced planar surfaces along which rocks split into thin, tabular slabs when hit with a hammer. Rock cleavage is developed in various metamorphic rocks but is best displayed in slates that exhibit an excellent splitting property called **slaty cleavage.**

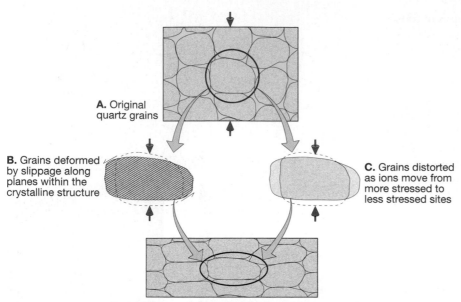

A. Original quartz grains

B. Grains deformed by slippage along planes within the crystalline structure

C. Grains distorted as ions move from more stressed to less stressed sites

D. Flattened rock exhibiting distorted quartz grains

FIGURE 8.7   Development of preferred orientations of minerals such as quartz, calcite, and olivine. **A.** Ductile deformation (flattening) of these roughly equidimensional mineral grains can occur in one of two ways. **B.** The first mechanism is a solid-state plastic flow that involves intracrystalline gliding of individual units within each grain. **C.** The second mechanism involves dissolving material from areas of high stress and depositing that material in locations of low stress. **D.** Both mechanisms change the shape of the grains, but the volume and composition of each grain remains essentially unchanged.

FIGURE 8.8   Development of one type of rock cleavage. As shale is strongly folded (**A, B**) and metamorphosed to form slate, the developing mica flakes are bent into microfolds. **C.** Further metamorphism results in the recrystallization of mica grains along the limbs of these folds to enhance the foliation. **D.** Hand sample of slate illustrates rock cleavage and its orientation to relic bedding surfaces.

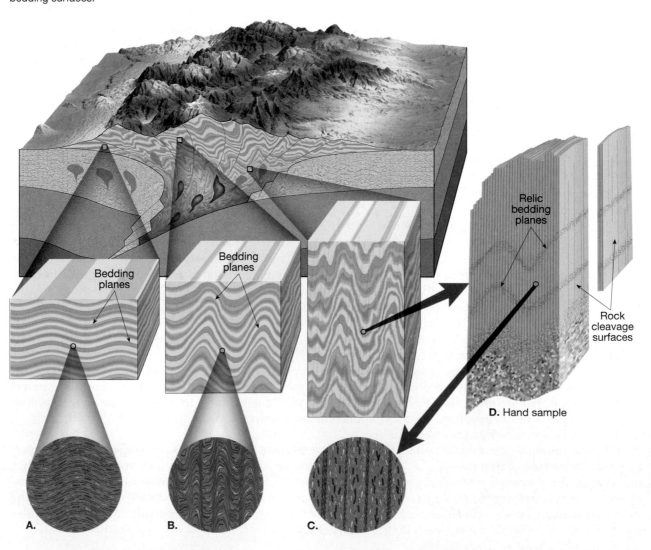

Bedding planes

Bedding planes

Relic bedding planes

Rock cleavage surfaces

D. Hand sample

A.

B.

C.

Depending on the metamorphic environment and the composition of the parent rock, rock cleavage develops in a number of different ways. In a low-grade metamorphic environment, rock cleavage is known to develop where beds of shale (and related sedimentary rocks) are strongly folded and metamorphosed to form slate. The process begins as platy grains are kinked and bent—generating microscopic folds having limbs (sides) that are roughly aligned (Figure 8.8). With further deformation, this new alignment is enhanced as old grains break down and recrystallize preferentially in the direction of the newly developed orientation. In this manner the rock develops narrow parallel zones where mica flakes are concentrated. These planar features alternate with zones containing quartz and other mineral grains that do not exhibit a pronounced linear orientation. It is along these very thin zones, where platy minerals display a parallel alignment, that slate splits (Figure 8.9).

Because slate typically forms during low-grade metamorphism of shale, evidence of the original sedimentary bedding planes are often preserved. However, as shown in Figure 8.8D, the orientation of slate's cleavage usually develops at an oblique angle to the original sedimentary layering. Thus, unlike shale, which splits along bedding planes, slate often splits across them. Other metamorphic rocks, such as schists and gneisses, may also split along planar surfaces and thus exhibit rock cleavage.

Schistosity Under higher temperature-pressure regimes, the minute mica and chlorite grains in slate begin to grow many times larger. When these platy minerals grow large enough to be discernible with the unaided eye and exhibit a planar or layered structure, the rock is said to exhibit a type of foliation called **schistosity.** Rocks having this texture are referred to as *schist.* In addition to platy minerals, schist often contains deformed quartz and feldspar grains that appear as flat, or lens shaped, grains hidden among the mica grains.

Gneissic Texture During high-grade metamorphism, ion migrations can result in the segregation of minerals as

FIGURE 8.10 This rock displays a gneissic texture. Notice that the dark biotite flakes and light silicate minerals are segregated, giving the rock a banded or layered appearance. (Photo by E. J. Tarbuck)

shown in Figure 8.10. Notice that the dark biotite crystals and light silicate minerals (quartz and feldspar) have separated, giving the rock a banded appearance called **gneissic texture.** A metamorphic rock with this texture is called *gneiss* (pronounced "nice"). Although foliated, gneisses will not usually split as easily as slates and some schists. Gneisses that do cleave tend to break parallel to their foliation and expose mica-rich surfaces that resemble schist.

Other Metamorphic Textures

Not all metamorphic rocks exhibit a foliated texture. Those that *do not* are referred to as **nonfoliated.** Nonfoliated metamorphic rocks typically develop in environments where deformation is minimal and the parent rocks are composed of minerals that exhibit equidimensional crystals, such as quartz or calcite. For example, when a fine-grained limestone (made of calcite) is metamorphosed by the intrusion of a hot magma body, the small calcite grains recrystallize to form larger interlocking crystals. The resulting rock, *marble,* displays large, equidimensional grains that are randomly oriented similar to those of a coarse-grained igneous rock.

Another texture common to metamorphic rocks consists of particularly large grains, called *porphyroblasts,* that are surrounded by a fine-grained matrix of other minerals. **Porphyroblastic textures** develop in a wide range of rock types and metamorphic environments when minerals in the parent rock recrystallize to form new minerals. During recrystallization certain metamorphic minerals, including garnet, staurolite, and andalusite, invariably develop *a small number of very large crystals.* By contrast, minerals such as muscovite, biotite, and quartz typically form *a large number of very small grains.* As a result, when metamorphism generates the minerals

FIGURE 8.9 Excellent slaty cleavage is exhibited by the rock in this slate quarry near Alta, Norway. The parallel mineral alignment in this rock allows it to split easily into the flat slabs visible in the photo. (Photo by Fred Bruemmer/DRK Photo)

← 5 cm →

Close up of porphyroblast

FIGURE 8.11 Garnet-mica schist. The dark red garnet crystals (porphyroblasts) are embedded in a matrix of fine-grained micas. (Photo by E. J. Tarbuck)

garnet, biotite, and muscovite in the same setting, the rock will contain large crystals (porphyroblasts) of garnet embedded in a fine-grained matrix consisting of biotite and muscovite (Figure 8.11).

Common Metamorphic Rocks

 Metamorphic Rocks
▸ **Common Metamorphic Rocks**

Recall that metamorphism causes many changes in rocks, including increased density, change in grain size, reorientation of mineral grains into a planar arrangement known as foliation, and the transformation of low-temperature minerals into high-temperature minerals. Moreover, the introduction of ions may generate new minerals, some of which are economically important.

The major characteristics of some common metamorphic rocks are summarized in Figure 8.12. Notice that metamorphic rocks can be broadly classified by the type of foliation exhibited and to a lesser extent on the chemical composition of the parent rock.

Foliated Rocks

Slate Slate is a very fine-grained (less than 0.5 millimeter) foliated rock composed of minute mica flakes that are too small to be visible. Thus, slate generally appears dull and closely resembles shale. A noteworthy characteristic of slate is its excellent rock cleavage, or tendency to break into flat slabs (Figure 8.13).

Slate is most often generated by the low-grade metamorphism of shale, mudstone, or siltstone. Less frequently it is produced when volcanic ash is metamorphosed. Slate's color depends on its mineral constituents. Black (carbonaceous) slate contains organic material, red slate gets its color from iron oxide, and green slate usually contains chlorite.

Phyllite Phyllite represents a gradation in the degree of metamorphism between slate and schist. Its constituent platy minerals are larger than those in slate but not yet large enough to be readily identifiable with the unaided eye. Although phyllite appears similar to slate, it can be easily distinguished from slate by its glossy sheen and its sometimes wavy surface (Figure 8.14). Phyllite usually exhibits rock cleavage and is composed mainly of very fine crystals of either muscovite, chlorite, or both.

Schist Schists are medium- to coarse-grained metamorphic rocks in which platy minerals predominate. These flat components commonly include the micas (muscovite and biotite), which display a planar alignment that gives the rock its foliated texture. In addition, schists contain smaller amounts of other minerals, often quartz and feldspar. Schists composed mostly of dark minerals (amphiboles) are known. Like slate, the parent rock of many schists is shale, which has undergone medium- to high-grade metamorphism during major mountain-building episodes.

The term *schist* describes the texture of a rock and as such it is used to describe rocks having a wide variety of chemical compositions. To indicate the composition, mineral names are used. For example, schists composed primarily of muscovite and biotite are called *mica schist* (Figure 8.15). Depending upon the degree of metamorphism and composition of the parent rock, mica schists often contain *accessory minerals*, some of which are unique to metamorphic rocks. Some common accessory minerals that occur as porphyroblasts include *garnet, staurolite,* and *sillimanite,* in which case the rock is called *garnet-mica schist, staurolite-mica schist,* and so forth (Figure 8.11).

In addition, schists may be composed largely of the minerals chlorite or talc, in which case they are called *chlorite schist* and *talc schist,* respectively. Both chlorite and talc schists can form when rocks with a basaltic composition undergo metamorphism. Others contain the mineral *graphite,* which is used as pencil "lead," graphite fibers (used in fishing rods), and lubricant (commonly for locks).

Gneiss Gneiss is the term applied to medium- to coarse-grained banded metamorphic rocks in which granular and elongated (as opposed to platy) minerals predominate. The most common minerals in gneiss are quartz, potassium feldspar, and sodium-rich plagioclase feldspar. Most gneisses also contain lesser amounts of biotite, muscovite, and amphibole that develop a preferred orientation. Some gneisses will split along the layers of platy minerals, but most break in an irregular fashion.

Rock Name	Texture		Grain Size	Comments	Original Parent Rock
Slate	Foliated		Very fine	Excellent rock cleavage, smooth dull surfaces	Shale, mudstone, or siltstone
Phyllite			Fine	Breaks along wavy surfaces, glossy sheen	Shale, mudstone, or siltstone
Schist			Medium to Coarse	Micaceous minerals dominate, scaly foliation	Shale, mudstone, or siltstone
Gneiss			Medium to Coarse	Compositional banding due to segregation of minerals	Shale, granite, or volcanic rocks
Migmatite			Medium to Coarse	Banded rock with zones of light-colored crystalline minerals	Shale, granite, or volcanic rocks
Mylonite	Weakly Foliated		Fine	When very fine-grained, resembles chert, often breaks into slabs	Any rock type
Metaconglomerate			Coarse-grained	Stretched pebbles with preferred orientation	Quartz-rich conglomerate
Marble	Nonfoliated		Medium to coarse	Interlocking calcite or dolomite grains	Limestone, dolostone
Quartzite			Medium to coarse	Fused quartz grains, massive, very hard	Quartz sandstone
Hornfels			Fine	Usually, dark massive rock with dull luster	Any rock type
Anthracite			Fine	Shiny black rock that may exhibit conchoidal fracture	Bituminous coal
Fault breccia			Medium to very coarse	Broken fragments in a haphazard arrangement	Any rock type

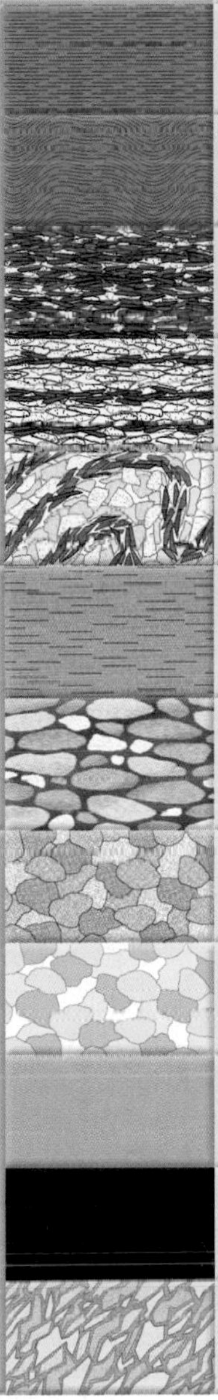

(Increasing Metamorphism indicated alongside Slate through Migmatite)

FIGURE 8.12 Classification of common metamorphic rocks.

FIGURE 8.13 Because slate breaks into flat slabs, it has a number of uses. Here, it is used to roof this house in Switzerland. (Photo by E. J. Tarbuck)

Recall that during high-grade metamorphism the light and dark components separate, giving gneisses their characteristic banded or layered appearance. Thus, most gneisses consist of alternating bands of white or reddish feldspar-rich zones and layers of dark ferromagnesian minerals (see Figure 8.10). These banded gneisses often exhibit evidence of deformation, including folds and sometimes faults (see Figure 8.1).

Most gneisses have a felsic composition and are often derived from granite or its aphanitic equivalent, rhyolite. However, many form from the high-grade metamorphism of shale. In this instance, gneiss represents the last rock in the sequence of shale, slate, phyllite, schist, and gneiss. Like schists, gneisses may also include large crystals of accessory

A. Phyllite   **B.** Slate

FIGURE 8.14 Phyllite (left) can be distinguished from slate (right) by its glossy sheen and wavy surface. (Photo by E. J. Tarbuck)

minerals such as garnet and staurolite. Gneisses made up primarily of dark minerals such as those that compose basalt also occur. For example, an amphibole-rich rock that exhibits a gneissic texture is called *amphibolite*.

Nonfoliated Rocks

Marble   Marble is a coarse, crystalline metamorphic rock whose parent was limestone or dolostone (Figure 8.16). Pure marble is white and composed essentially of the mineral calcite. Because of its relative softness (hardness of 3), marble is easy to cut and shape. White marble is particularly prized as a stone from which to create monuments and statues, such as the famous statue of *David* by Michelangelo (Figure 8.17). Unfortunately, since marble is composed of calcium carbonate, it is readily attacked by acid rain.

The parent rock from which marble forms often contains impurities that tend to color the stone. Thus, marble can be pink, gray, green, or even black and may contain a variety of

FIGURE 8.15 Mica schist. This sample of schist is composed mostly of muscovite and biotite. (Photo by E. J. Tarbuck)

Photomicrograph (6.5x)

FIGURE 8.16 Marble, a crystalline rock formed by the metamorphism of limestone. Photomicrograph shows interlocking calcite crystals using polarized light (Photos by E. J. Tarbuck)

accessory minerals (chlorite, mica, garnet, and wollastonite). When marble forms from limestone interbedded with shales, it will appear banded and exhibit visible foliation. When deformed, these banded marbles may develop highly contorted mica-rich folds that give the rock a rather artistic design. Hence, these decorative marbles have been used as a building stone since prehistoric times.

FIGURE 8.17 Replica of the statue of *David* by Michelangelo. Like the original, this sculpture was created from a large block of very pure, white marble. (Courtesy Getty Images/Photo Disk)

Quartzite *Quartzite* is a very hard metamorphic rock formed from quartz sandstone (Figure 8.18). Under moderate- to high-grade metamorphism, the quartz grains in sandstone fuse together (inset in Figure 8.18). The recrystallization is so complete that when broken, quartzite will split through the quartz grains rather than along their boundaries. In some instances, sedimentary features such as cross-bedding are preserved and give the rock a banded appearance. Pure quartzite is white, but iron oxide may produce reddish or pinkish stains, while dark mineral grains may impart a gray color.

Metamorphic Environments

There are a number of environments in which metamorphism occurs. Most are in the vicinity of plate margins, and many are associated with igneous activity. We will consider the following types of metamorphism: (1) *contact* or *thermal metamorphism;* (2) *hydrothermal metamorphism;* (3) *burial and subduction zone metamorphism;* (4) *regional metamorphism;* (5) *impact metamorphism;* and (6) *metamorphism along faults.* With the exception of impact metamorphism, there is considerable overlap among the other types.

Photomicrograph (26.6x)

FIGURE 8.18 Quartzite is a nonfoliated metamorphic rock formed from quartz sandstone. The photomicrograph shows the interlocking quartz grains typical of quartzite. (Photos by E. J. Tarbuck)

Contact or Thermal Metamorphism

Contact or **thermal metamorphism** occurs when rocks immediately surrounding a molten igneous body are "baked" and therefore altered from their original state. The altered rocks occur in a zone called a metamorphic **aureole** (Figure 8.19). While small intrusions such as dikes and sills typically form aureoles only a few centimeters thick, large intrusions such as batholiths can produce aureoles that extend outward for several kilometers.

In addition to the size of the magma body, the mineral composition of the host rock and the availability of water greatly affect the size of the aureole produced. In chemically active rock such as limestone, the zone of alteration can be 10 kilometers (6 miles) thick. These large aureoles often consist of distinct *zones of metamorphism.* Near the magma body, high-temperature minerals such as garnet may form, whereas farther away low-grade minerals such as chlorite are produced.

Although contact metamorphism is not entirely restricted to shallow crustal depths, it is most easily recognized when it occurs in this setting. Here, the temperature contrast between an intrusion and the surrounding rock is greatest and confining pressure is low. Because contact metamorphism does not involve directed pressure, crystals within the metamorphic aureole are more or less randomly oriented.

During contact metamorphism of mudstones and shales, the clay minerals are baked as if placed in a kiln. The result is a very hard, fine-grained metamorphic rock called *hornfels* (Figure 8.20). Because directed pressure is not a factor in forming these rocks, they are characteristically nonfoliated. Hornfels can form from a variety of materials during contact metamorphism, including volcanic ash and basalt. In some cases, large grains of metamorphic minerals, such as garnet and staurolite, may form, giving the hornfels a porphyroblastic texture (see Figure 8.11).

Two other metamorphic rocks produced in association with hornfels are quartzite and marble (Figure 8.20). Their parent rocks are quartz sandstone and limestone, respectively. Although quartzite and marble are commonly associated with contact metamorphism, unlike hornfels, they also form in other metamorphic environments.

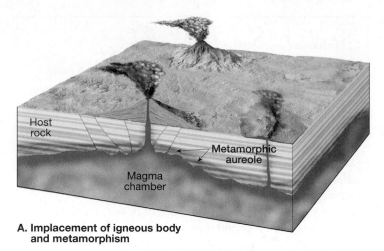

A. Implacement of igneous body and metamorphism

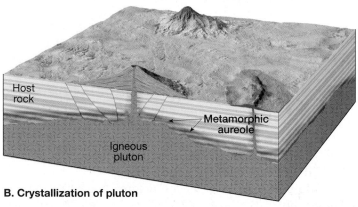

B. Crystallization of pluton

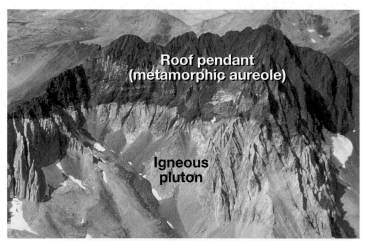

C. Uplift and erosion to expose pluton and metamorphic cap rock

FIGURE 8.19 Contact metamorphism produces a zone of alteration called an *aureole* around an intrusive igneous body. In the photo, the dark layer, called a *roof pendant*, consists of metamorphosed host rock adjacent to the upper part of the light-colored igneous pluton. The term *roof pendant* implies that the rock was once the roof of a magma chamber. Sierra Nevada, near Bishop, California. (Photo by John S. Shelton)

Hydrothermal Metamorphism

When hot, ion-rich fluids circulate through fissures and cracks that develop in rock, a chemical alteration called **hydrothermal metamorphism** occurs. This type of metamor-

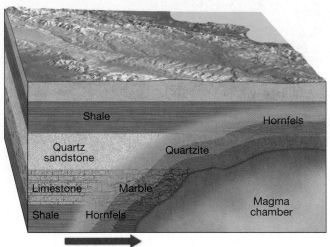

Increasing metamorphic grade

FIGURE 8.20 Contact metamorphism of shale yields hornfels, while contact metamorphism of quartz sandstone and limestone produces quartzite and marble, respectively.

phism is closely associated with igneous activity, since it provides the heat required to circulate these ion-rich solutions. Thus, hydrothermal metamorphism often occurs simultaneously with contact metamorphism in regions where large plutons are emplaced.

As these large magma bodies cool and solidify, the ions that are not incorporated into the crystalline structures of the newly formed silicate minerals, as well as any remaining volatiles (water), are expelled. These ion-rich fluids are called **hydrothermal** (*hydra* = water, *therm* = heat) **solutions** (*solut* = dissolved). In addition to chemically altering the host rocks, the ions contained in hydrothermal solutions sometimes precipitate to form a variety of economically important mineral deposits.

If the host rocks are permeable, as with carbonate rocks such as limestone, these fluids may extend the aureole several kilometers. In addition, these silicate-rich solutions may react with carbonates to produce a variety of calcium-rich silicate minerals that form a rock called *skarn*. Recall that a metamorphic process that alters the overall chemical composition of a rock unit is called metasomatism.

As our understanding of plate tectonics expanded, it became clear that the most widespread occurrence of hydrothermal metamorphism is along the axis of the mid-ocean ridge system. Here, as plates move apart, upwelling magma from the mantle generates new seafloor. As seawater percolates through the young, hot oceanic crust, it is heated and chemically reacts with the newly formed basaltic rocks (Figure 8.21). The result is the conversion of ferromagnesian minerals, such as olivine and pyroxene, into hydrated silicates, including serpentine, chlorite, and talc. In addition, calcium-rich plagioclase feldspars in basalt become increasingly sodium enriched as the salt (NaCl) in seawater exchanges Na ions for Ca ions.

Large amounts of metals, such as iron, cobalt, nickel, silver, gold and copper, are also dissolved from the newly formed crust. These hot, metal-rich fluids eventually rise

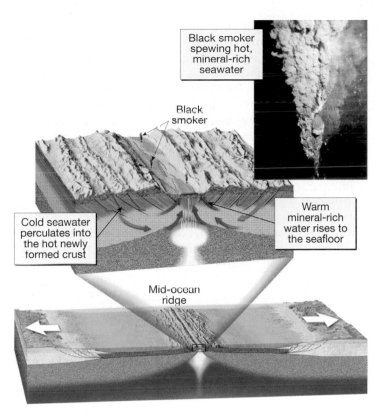

FIGURE 8.21 Hydrothermal metamorphism along a mid-ocean ridge. (Photo by R. Ballard/Woods Hole)

along fractures and gush from the seafloor at temperatures of about 350°C, generating particle-filled clouds called *black smokers*. Upon mixing with the cold seawater, sulfides and carbonate minerals containing these heavy metals precipitate to form metallic deposits some of which are economically valuable. This is believed to be the origin of the copper ores mined today on the island of Cyprus.

Burial and Subduction Zone Metamorphism
Burial metamorphism occurs in association with very thick accumulations of sedimentary strata in a subsiding basin (see Figure 8.3). Here, low-grade metamorphic conditions may be attained within the deepest layers. Confining pressure and geothermal heat drive the recrystallization of the constituent minerals to change the texture and/or mineralogy of the rock without appreciable deformation.

The depth required for burial metamorphism varies from one location to another, depending on the prevailing geothermal gradient. Low-grade metamorphism often begins at depths of about 8 kilometers (5 miles), where temperatures are about 200°C. However, in areas that exhibit large geothermal gradients, such as near the Salton Sea in

California and in northern New Zealand, drilling operations have collected metamorphic minerals from depths of only a few kilometers.

Rocks and sediments can also be carried to great depths along convergent boundaries where oceanic lithosphere is being subducted. This phenomenon, called **subduction zone metamorphism**, differs from burial metamorphism in that differential stresses play a major role in deforming rock as it is metamorphosed. Furthermore, metamorphic rocks that form along subduction zones are often further metamorphosed by the collision of two continental blocks. A discussion of this process is found later in the chapter in the section entitled "Interpreting Metamorphic Environments."

Regional Metamorphism

Most metamorphic rocks are produced during the process of **regional metamorphism** in association with mountain building. During these dynamic events, large segments of Earth's crust are intensely deformed along convergent plate boundaries (Figure 8.22). This activity occurs most often where oceanic lithosphere is subducted to produce continental volcanic arcs and during continental collisions.

Metamorphism associated with continental collisions involves the convergence of an active plate margin with a passive continental margin as shown in Figure 8.22. Such collisions generally result in large segments of Earth's crust being intensely deformed by compressional forces associated with convergent plate motion. Sediments and crustal rocks that form the margins of the colliding continental blocks are folded and faulted, causing them to shorten and thicken like

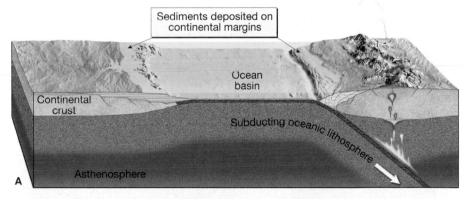

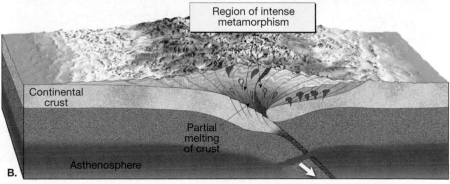

FIGURE 8.22 Regional metamorphism occurs where rocks are squeezed between two converging lithospheric plates during mountain building.

a rumpled carpet (Figure 8.22). This event often involves crystalline continental basement rocks, as well as slices of ocean crust that once floored the intervening ocean basin.

The general thickening of the crust results in buoyant lifting, where deformed rocks are elevated high above sea level to form mountainous terrain. Likewise, crustal thickening results in the deep burial of large quantities of rock as crustal blocks are thrust one beneath another. Here, in the roots of mountains, elevated temperatures caused by deep burial are responsible for the most productive and intense metamorphic activity within a mountain belt. Often, these deeply buried rocks become heated to the point of melting. As a result, magma collects until it forms bodies large enough to buoyantly rise and intrude the overlying metamorphic and sedimentary rocks (Figure 8.22). Consequently, the cores of many mountain ranges consist of folded and faulted metamorphic rocks, often intertwined with igneous bodies. Over time, these deformed rock masses are uplifted, and erosion removes the overlying material to expose the igneous and metamorphic rocks that comprise the central core of the mountain range (see Figure 14.21).

Other Metamorphic Environments

Other types of metamorphism occur that generate comparatively smaller amounts of metamorphic rock in localized concentrations.

Metamorphism along Fault Zones   Near the surface, rock behaves like a brittle solid. Consequently, movement along a fault zone fractures and pulverizes rock (Figure 8.23). The re-

sult is a loosely coherent rock called *fault breccia* that is composed of broken and crushed rock fragments (Figure 8.23). Displacements along California's San Andreas fault have created a zone of fault breccia and related rock types more than 1000 kilometers long and up to 3 kilometers wide.

In some shallow fault zones a soft, uncemented claylike material called *fault gouge* is also produced. Fault gouge is formed by the crushing and grinding of rock material during fault movement. The resulting crushed material is further altered by groundwater that infiltrates the porous fault zone.

Much of the deformation associated with fault zones occurs at great depth and thus at high temperatures. In this environment preexisting minerals deform by ductile flow (Figure 8.23). As large slabs of rock move in opposite directions, the minerals in the fault zone between them tend to form elongated grains that give the rock a foliated or lineated appearance. Rocks formed in these zones of intense ductile deformation are termed *mylonites* (*mylo* = a mill, *ite* = a stone).

Impact Metamorphism   **Impact** (or **shock**) **metamorphism** occurs when high-speed projectiles called *meteorites* (fragments of comets or asteroids) strike Earth's surface. Upon impact the energy of the rapidly moving meteorite is transformed into heat energy and shock waves that pass through the surrounding rocks. The result is pulverized, shattered, and sometimes melted rock. The products of these impacts, called *impactiles,* include mixtures of fused fragmented rock plus glass-rich ejecta that resemble volcanic bombs (see Box 8.1). In some cases, a very dense form of quartz (*coesite*) and

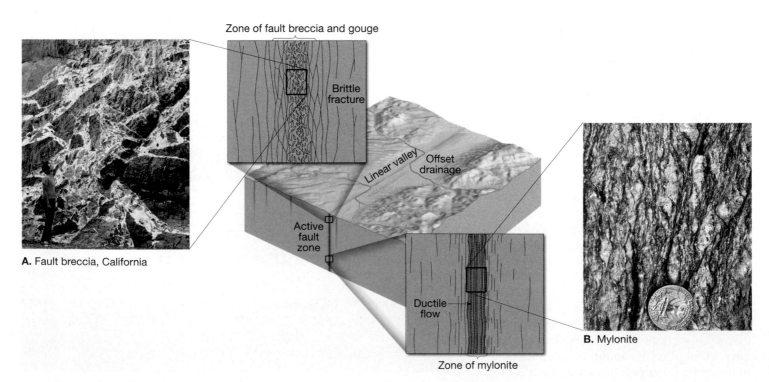

A. Fault breccia, California

B. Mylonite

FIGURE 8.23   Metamorphism along a fault zone. (Photo **A** by A.P. Trujillo, Photo **B** by Ann Bykerk —Kauffman)

BOX 8.1 ▶ UNDERSTANDING EARTH

Impact Metamorphism and Tektites

It is now clear that comets and asteroids have collided with Earth far more frequently than was once assumed. The evidence: More than 100 giant impact structures have been identified to date. Previously many of these structures were thought to have resulted from some poorly understood volcanic process. Most impact structures, like Manicouagan in Quebec, are so old and weathered that they no longer resemble an impact crater (see Figure 24.C). A notable exception is the very fresh-looking Meteor Crater in Arizona (Figure 8.A).

One earmark of impact craters is *shock metamorphism*. When high-velocity projectiles (comets, asteroids) impact Earth's surface, pressures reach millions of atmospheres and temperatures exceed 2000°C momentarily. The result is pulverized, shattered, and melted rock. Where impact craters are relatively fresh, shock-melted ejecta and rock fragments ring the impact site. Although most material is deposited close to its source, some ejecta can travel great distances. One example is *tektites* (*tektos* = molten), beads of silica-rich glass, some of which have been aerodynamically shaped like teardrops during flight (Figure 8.B). Most tektites are no more than a few centimeters across and are jet-black to dark green or yellowish. In Australia, millions of tektites are strewn over an area seven times the size of Texas. Several such tektite groupings have been identified worldwide, one stretching nearly halfway around the globe.

No tektite falls have been observed, so their origin is not known with certainty. Because tektites are much higher in silica than volcanic glass (obsidian), a volcanic origin is unlikely. Most researchers agree that tektites are the result of impacts of large projectiles.

One hypothesis suggests an extraterrestrial origin for tektites. Asteroids may have struck the Moon with such force that ejecta "splashed" outward fast enough to escape the Moon's gravity. Others argue that tektites are terrestrial, but an objection is that some groupings, such as Australia's, lack an identifiable impact crater. However, the object that produced the Australian tek-

FIGURE 8.A   Meteor Crater, located west of Winslow, Arizona. (Photo by Michael Collier)

FIGURE 8.B   Tektites recovered from Nullarbor Plain, Australia. (Photo by Brian Mason/ Smithsonian Institution)

tites might have struck the continental shelf, leaving the remnant crater out of sight below sea level. Evidence supporting a terrestrial origin includes tektite falls in western Africa that appear to be the same age as a crater in the same region.

minute *diamonds* are found. These high-pressure minerals provide convincing evidence that pressures and temperatures at least as great as those existing in the upper mantle must have been attained at least briefly at Earth's surface.

Metamorphic Zones

In areas affected by metamorphism, there usually exist systematic variations in the mineralogy and texture of the rocks that can be observed as we traverse the region. These differences are clearly related to variations in the degree of metamorphism experienced in each metamorphic zone.

Textural Variations

When we begin with a clay-rich sedimentary rock such as mudstone or shale, a gradual increase in metamorphic intensity is accompanied by a general coarsening of the grain size. Thus, we observe shale changing to a fine-grained slate, which then forms phyllite and through continued recrystallization generates a coarse-grained schist (Figure 8.24). Under more intense conditions a gneissic texture that exhibits layers of dark and light minerals may develop. This

systematic transition in metamorphic textures can be observed as we approach the Appalachian Mountains from the west. Beds of shale, which once extended over large areas of the eastern United States, still occur as nearly flat-lying strata in Ohio. However, in the broadly folded Appalachians of central Pennsylvania, the rocks that once formed flat-lying beds are folded and display a preferred orientation of platy mineral grains as exhibited by well-developed slaty cleavage. As we move farther eastward into the intensely deformed crystalline Appalachians, we find large outcrops of schists. The most intense zones of metamorphism are found in Vermont and New Hampshire, where gneissic rocks outcrop.

Index Minerals and Metamorphic Grade

In addition to textural changes, we encounter corresponding changes in mineralogy as we shift from regions of low-grade metamorphism to regions of high-grade metamorphism. An idealized transition in mineralogy that results from the regional metamorphism of shale is shown in Figure 8.25. The first new mineral to form as shale changes to slate is chlorite. At higher temperatures flakes of muscovite and biotite begin to dominate. Under more extreme

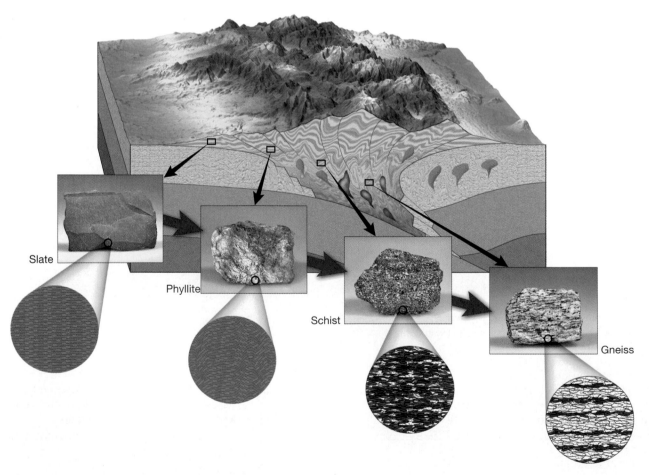

Slate

Phyllite

Schist

Gneiss

FIGURE 8.24 Idealized illustration of progressive regional metamorphism. From left to right, we progress from low-grade metamorphism (slate) to high-grade metamorphism (gneiss). (Photos by E. J. Tarbuck)

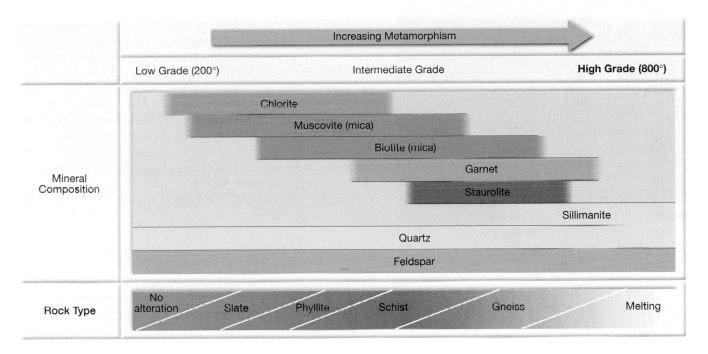

FIGURE 8.25   The typical transition in mineralogy that results from progressive metamorphism of shale.

conditions, metamorphic rocks may contain garnet and staurolite crystals. At temperatures approaching the melting point of rock, sillimanite forms. Sillimanite is a high-temperature metamorphic mineral used to make refractory porcelains such as those used in spark plugs.

Through the study of metamorphic rocks in their natural settings (called *field studies*) and through experimental studies, researchers have learned that certain minerals, such as those in Figure 8.25, are good indicators of the metamorphic environment in which they formed. Using these **index minerals,** geologists distinguish among different zones of regional metamorphism. For example, the mineral chlorite begins to form when temperatures are relatively low, less than 200°C (Figure 8.26). Thus, rocks that contain chlorite (usually slates) are referred to as *low-grade.* By contrast, the mineral sillimanite only forms in very extreme environments where temperatures exceed 600°C, and rocks containing it are considered *high-grade.* By mapping the occurrences of index minerals, geologists are in effect mapping zones of varying metamorphic grade. *Grade* is a term used in a relative sense to refer to the conditions of temperature (or sometimes pressure) to which a rock has been subjected.

Migmatites In the most extreme environments, even the highest-grade metamorphic rocks undergo change. For example, gneissic rocks may be heated sufficiently to cause melting to begin. However, recall from our discussion of igneous rocks that different minerals melt at different temperatures. The light-colored silicates, usually quartz and potassium feldspar, have the lowest melting temperatures and begin to melt first, whereas the mafic silicates, such as amphibole and biotite, remain solid. When this partially melted rock cools, the light bands will be composed of igneous, or igneous-appearing components, while the dark bands will consist of unmelted metamorphic material.

Rocks of this type are called **migmatites** (*migma* = mixture, *ite* = a stone) (Figure 8.27). The light-colored bands in migmatites often form tortuous folds and may contain tabular inclusions of the dark components. Migmatites serve to illustrate the fact that some rocks are transitional and do not clearly belong to any one of the three basic rock groups.

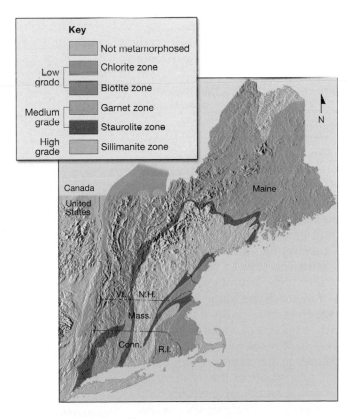

FIGURE 8.26   Zones of metamorphic intensities in New England.

FIGURE 8.27 Migmatite. The lightest-colored layers are igneous rock composed of quartz and feldspar, whereas the darker layers have a metamorphic origin. (Photo by Harlan H. Roepke)

Interpreting Metamorphic Environments

About a century ago geologists came to realize that groups of associated minerals could be used to determine the pressures and temperatures at which rocks undergo metamorphism (see Box 8.2). This discovery let Finnish geologist Pennti Eskola to propose the concept of metamorphic facies. Simply, metamorphic rocks containing the same assemblage of minerals belong to the same **metamorphic facies**—implying that they formed in very similar metamorphic environments. The concept of metamorphic facies is analogous to using a group of plants to define climatic zones—regions that experience similar conditions of precipitation and temperature, exhibit similar plants. For instance, forests consisting of scrawny spruce, fir, larch, and birch trees identify the subarctic or taiga climate zones on Earth.

There are several common metamorphic facies as shown in Figure 8.28. These include the *hornfels, zeolite, greenschist, amphibolite, granulite, blueschist,* and *eclogite facies.* The name for each facies is based on the minerals that define them (Figure 8.28). For example, rocks of the amphibolite facies are characterized by the mineral hornblende (a common amphibole); and the greenschist facies consists of schists in which the green minerals chlorite, epidote, and serpentine are prominent. Similar suites of minerals are found in terranes of all ages and in all parts of the world. Thus, the concept of metamorphic facies has proven useful because rocks that belong to the same metamorphic facies all formed under the same conditions of temperature and pressure, and therefore in similar tectonic settings, regardless of their location or age.

240

It should be noted that the name for each metamorphic facies refers to a metamorphic rock derived specifically from a basaltic parent. This occurs because Pennti Eskola concentrated on the metamorphism of basalts, and his basic terminology, although now slightly modified, remains. The names of Eskola's facies serve as convenient labels for a particular combination of temperatures and pressures no matter what the mineral composition. In other words, even if a nonbasaltic parent rock produces different indicator minerals under a given set of metamorphic conditions, the facies names shown in Figure 8.28 are used for the purpose of denoting the temperature and pressure ranges embodied by that metamorphic rock.

Tectonic Settings and Metamorphic Facies

Figure 8.29 shows how the concept of facies fits into the context of plate tectonics. Near deep-ocean trenches, slabs of relatively cool oceanic lithosphere are subducted. As the lithosphere descends, sediments and crustal rocks are subjected to steadily increasing temperatures and pressures (Figure 8.29). However, temperatures in the slab remain cooler than the surrounding mantle because rock is a poor conductor of heat and therefore warms slowly. The metamorphic facies associated with this type of high-pressure, low-temperature environment is called the *blueschist facies,* because of the presence of the blue-colored variety of amphibole called *glaucophane* (Figure 8.30A). The rocks of the Coast Range of California belong to the blueschist facies. Here, highly deformed rocks that were once deeply buried have been uplifted because of a change in the plate boundary. In some areas subduction carries rocks to even greater depths, producing the *eclogite facies* that is diagnostic of very high temperatures and pressures (Figure 8.30B).

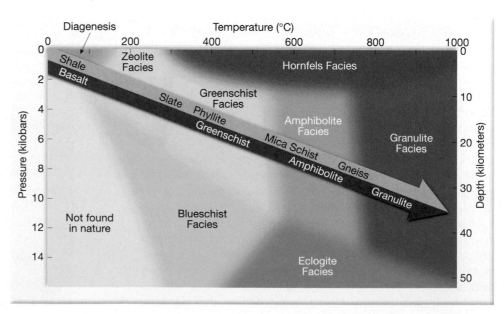

FIGURE 8.28 Metamorphic facies and corresponding temperature and pressure conditions. Note the equivalent metamorphic rocks produced from regional metamorphism of basalt and shale parent rocks.

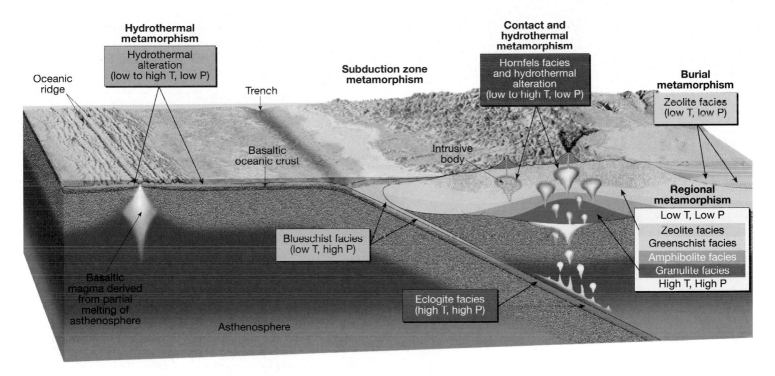

FIGURE 8.29   The association of metamorphic facies with plate tectonic environments.

Along some convergent zones, continental blocks collide to form extensive mountain belts (see Figure 8.22). This activity results in large areas of regional metamorphism that include zones of contact, hydrothermal, and subduction metamorphism. The increasing temperatures and pressures associated with regional metamorphism are recorded by the *greenschist-amphibolite-granulite facies* sequence shown in Figure 8.28.

Ancient Metamorphic Environments

Metamorphic rocks provide valuable and even unique insights into the tectonic evolution of Earth's crust. They contain clues pertaining to the nature of the parent rock—was it deposited at Earth's surface, or does it represent a slab of oceanic crust accreted onto land, or innumerable other possibilities. The mineral compositions and textures of metamorphic rocks record periods of deformation associated with deep burial along subduction zones, or perhaps deformation produced during an episode of mountain building.

In addition to helping unravel the history of relatively recent tectonic activity, metamorphic rocks allow researchers to decipher the history of ancient environments. These include the relatively flat expanses of metamorphic rock and associated igneous plutons known as **shields** that make up the stable interiors of the continents (see Figure 1.18). One such structure, the Canadian Shield, has very little topographic expression and forms the bedrock over much of central Canada, extending from Hudson Bay to northern Minnesota (Figure 8.31). Radiometric dating of the Canadian Shield indicates that it is composed of rocks that range in age from 1.8 billion to 4 billion years. The texture and mineral content of these metamorphic rocks record the ancient mountain-building events that led to the formation of the North American continent, even though these once lofty structures have long since eroded away. We will consider this topic further in Chapter 22.

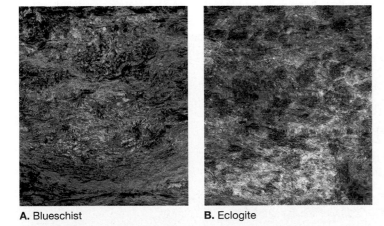

A. Blueschist   **B.** Eclogite

FIGURE 8.30   Rocks produced by subduction zone metamorphism. **A.** Blueschist, representing low temperature, high pressure conditions (note blue-colored amphibole called glaucophane). **B.** Eclogite, representing high temperature, high-pressure conditions of the mantle. Note pink grains of garnet and green grains of pyroxene. (Photos by C. Tsujita)

BOX 8.2 ▶ UNDERSTANDING EARTH

Mineral Stability

In most tectonic environments, such as along subduction zones, rocks experience an increase in *both* pressure and temperature simultaneously. An increase in pressure causes minerals to contract which favors the formation of high density minerals. However, an increased temperature results in expansion, so that mineral phases which occupy greater volume (are less dense) tend to be more stable at high temperatures. Thus, determining the conditions of temperature and pressure at which a mineral is stable (does not change) is not an easy task. To help in this endeavor, researchers have turned to the laboratory. Here materials of various composition are heated and placed under pressures that approximate conditions at various depths within Earth. From these experiments we can determine what minerals are likely to form in various metamorphic environments.

It turns out that some minerals, quartz for example, are stable over a wide range of metamorphic settings. Fortunately, other suites of minerals do provide useful estimates of conditions during metamorphism. One of the most important of these groups includes the minerals *kyanite, andalucite,* and *sillimanite.* All three minerals have identical chemical compositions (Al_2SiO_5) but different crystalline structures, which makes them *polymorphs* (see Chapter 3). Figure 8.C is a *phase diagram* that shows the specific range of pressures and temperatures at which each of these aluminum-rich silicates are stable.

Because shales and mudstones, which are very common, contain the elements found in these minerals, metamorphic products of shale (slate, schist, and gneiss) often contain varying amounts of either kyanite, andalucite, or sillimanite. For example, if shale were buried to a depth of about 35 kilometers (10 kbars), where the temperature was 550°C, the mineral kyanite would form (see the "X" in Figure 8.C).

In general, andalusite is produced by contact metamorphism in near surface environments where temperatures are high but pressures are relatively low. Kyanite is considered the high-pressure polymorph that forms during subduction and deep burial associated with mountain building. Sillimanite, on the other hand, forms only at high temperatures, the result of contact with a very hot magma body and/or very deep burial. Knowing the ranges of temperatures and pressures a rock experienced during metamorphism provides geologists valuable data needed to interpret past tectonic environments.

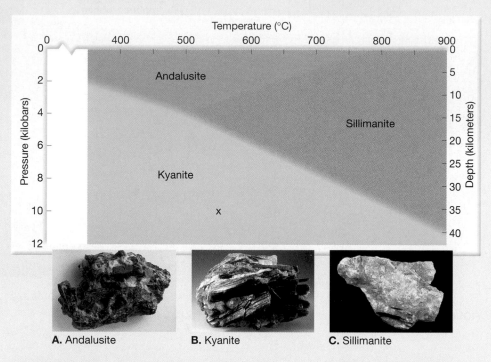

A. Andalusite   **B.** Kyanite   **C.** Sillimanite

FIGURE 8.C Phase diagram illustration of the conditions of pressure and temperature at which the three Al_2SiO_5 minerals are stable. (Photos by A. Harry Taylor/Dorling Kindersley Media Library B. Dennis Tasa, and C. Biophoto Associates/Photo Researchers, Inc.)

FIGURE 8.31 Aerial view showing ancient metamorphic rocks of the Canadian Shield in the Northwest Territories. (Photo by Robert Hildebrand)

Summary

- *Metamorphism* is the transformation of one rock type into another. *Metamorphic rocks* form from preexisting rocks (either igneous, sedimentary, or other metamorphic rocks) that have been altered by the agents of metamorphism, which include *heat, pressure (stress)*, and *chemically active fluids*. During metamorphism the material essentially remains solid. The changes that occur in metamorphosed rocks are textural as well as mineralogical.

- The mineral makeup of the parent rock determines, to a large extent, the degree to which each metamorphic agent will cause change. Heat is the most important agent because it provides the energy to drive chemical reactions that result in the recrystallization of minerals. Pressure, like temperature, also increases with depth. When subjected to *confining pressure*, minerals may recrystallize into more compact forms. During mountain building, rocks are subjected to *differential stress*, which tends to shorten them in the direction pressure is applied and lengthen them in the direction perpendicular to that force. At depth, rocks are warm and *ductile*, which accounts for their ability to deform by flowing when subjected to differential stresses. Chemically active fluids, most commonly water containing ions in solution, also enhance the metamorphic process by dissolving minerals and aiding the migration and precipitation of this material at other sites.

- *The grade of metamorphism is reflected in the texture and mineralogy of metamorphic rocks.* During regional metamorphism, rocks typically display a *preferred orientation* called *foliation* in which their platy and elongated minerals are aligned. Foliation develops as platy or elongated minerals are rotated into parallel alignment, recrystallize to form new grains that exhibit a preferred orientation, or are plastically deformed into flattened grains that exhibit a planar alignment. *Rock cleavage* is a type of foliation in which rocks split cleanly into thin slabs along surfaces where platy minerals are aligned. *Schistosity* is a type of foliation defined by the parallel alignment of medium- to-coarse-grained platy minerals. During high-grade metamorphism, ion migrations can cause minerals to segregate into distinct layers or bands. Metamorphic rocks with a banded texture are called *gneiss*. Metamorphic rocks composed of only one mineral forming equidimensional crystals often appear *nonfoliated*. *Marble* (metamorphosed limestone) is often nonfoliated. Further, metamorphism can cause the transformation of low-temperature minerals into high-temperature minerals and, through the introduction of ions from *hydrothermal solutions*, generate new minerals, some of which form economically important metallic ore deposits.

- Common foliated metamorphic rocks include *slate, phyllite*, various types of *schists* (e.g., garnet-mica schist), and *gneiss*. Nonfoliated rocks include *marble* (parent rock— limestone) and *quartzite* (most often formed from quartz sandstone).

- The four geologic environments in which metamorphism commonly occurs are (1) *contact* or *thermal metamorphism*, (2) *hydrothermal metamorphism*, (3) *burial and subduction zone metamorphism*, and (4) *regional metamorphism*. Contact metamorphism occurs when rocks are in contact with an igneous body, resulting in the formation of zones of alteration around the magma called *aureoles*. Most contact metamorphic rocks are fine-grained, dense, tough rocks of various chemical compositions. Because directional pressure is not a major factor, these rocks are not generally foliated. Hydrothermal metamorphism occurs where hot, ion-rich fluids circulate through rock and cause chemical alteration of the constituent minerals. Most hydrothermal alteration occurs along the mid-ocean ridge system, where seawater migrates through hot oceanic crust and chemically alters newly formed basaltic rocks. Metallic ions that are removed from the crust are eventually carried to the floor of the ocean, where they precipitate from black smokers to form metallic deposits, some of which may be economically important. Regional metamorphism takes place at considerable depths over an extensive area and is associated with the process of mountain building. A gradation in the degree of change usually exists in association with regional metamorphism, in which the intensity of metamorphism (low- to high-grade) is reflected in the texture and mineralogy of the rocks. In the most extreme metamorphic environments, rocks called *migmatites* fall into a transition zone *somewhere between* "true" igneous rocks and "true" metamorphic rocks.

Review Questions

1. What is metamorphism? What are the agents that change rocks?

2. Why is heat considered the most important agent of metamorphism?

3. How is confining pressure different than differential stress?

4. What role do chemically active fluids play in metamorphism?

5. In what two ways can the parent rock affect the metamorphic process?

6. What is foliation? Distinguish between *slaty cleavage, schistosity,* and *gneissic* textures.

7. Briefly describe the three mechanisms by which minerals develop a preferred orientation.

8. List some changes that might occur to a rock in response to metamorphic processes.

9. Slate and phyllite resemble each other. How might you distinguish one from the other?

10. Each of the following statements describes one or more characteristics of a particular metamorphic rock. For each statement, name the metamorphic rock that is being described.

    a. calcite-rich and often nonfoliated

    b. loosely coherent rock composed of broken fragments that formed along a fault zone

    c. represents a grade of metamorphism between slate and schist

    d. very fine-grained and foliated; excellent rock cleavage

    e. foliated and composed predominately of platy minerals

    f. composed of alternating bands of light and dark silicate minerals

    g. hard, nonfoliated rock resulting from contact metamorphism

11. Distinguish between contact metamorphism and regional metamorphism. Which creates the greatest quantity of metamorphic rock?

12. Where does most hydrothermal metamorphism occur?

13. Describe burial metamorphism.

14. How do geologists use index minerals?

15. Briefly describe the textural changes that occur in the transformation of slate to phyllite to schist and then to gneiss.

16. How are gneisses and migmatites related?

17. With which type of plate boundary is regional metamorphism associated?

18. Why do the cores of Earth's major mountain chains contain metamorphic rocks?

19. What are shields? How are these relatively flat areas related to mountains?

20. Briefly describe the tectonic environment that produces each of these metamorphic facies—hornfels, blueschist, and granulite facies.

Key Terms

aureole (p. 233)
burial metamorphism (p. 235)
confining pressure (p. 224)
contact metamorphism (p. 233)
differential stress (p. 224)
foliation (p. 227)
gneissic texture (p. 229)
hydrothermal metamorphism (p. 234)

hydrothermal solution (p. 234)
impact metamorphism (p. 236)
index mineral (p. 239)
metamorphic facies (p. 240)
metamorphism (p. 223)
metasomatism (p. 225)
migmatite (p. 239)

nonfoliated texture (p. 229)
parent rock (p. 223)
porphyroblastic texture (p. 229)
regional metamorphism (p. 235)
rock cleavage (p. 227)
schistosity (p. 229)
shield (p. 241)

shock metamorphism (p. 236)
slaty cleavage (p. 227)
subduction zone metamorphism (p. 235)
texture (p. 226)
thermal metamorphism (p. 233)

Web Resources

The Earth Website uses the resources and flexibility of the Internet to aid in your study of the topics in this chapter. Written and developed by geology instructors, this site will help improve your understanding of geology. Visit **http://www.prenhall.com/tarbuck** and click on the cover of Earth 9e to find:

• Online review quizzes.

• Critical thinking exercises.

• Links to chapter-specific Web resources.

• Internet-wide key-term searches.

http://www.prenhall.com/tarbuck

GEODe: Earth

GEODe: Earth makes studying faster and more effective by reinforcing key concepts using animation, video, narration, interactive exercises and practice quizzes. A copy is included with every copy of *Earth*.

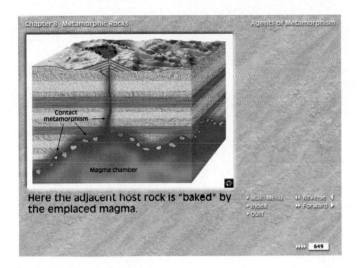

Here the adjacent host rock is "baked" by the emplaced magma.

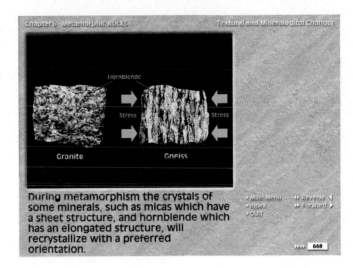

During metamorphism the crystals of some minerals, such as micas which have a sheet structure, and hornblende which has an elongated structure, will recrystallize with a preferred orientation.

Geologic Time

The strata exposed in the Grand Canyon contain clues to millions of years of Earth history. This view is from Yaki Point on the South Rim. (Photo by Tom Bean)

n the late 18th century, James Hutton recognized the immensity of Earth history and the importance of time as a component in all geological processes. In the 19th century, Sir Charles Lyell and others effectively demonstrated that Earth had experienced many episodes of mountain building and erosion, which must have required great spans of geologic time. Although these pioneering scientists understood that Earth was very old, they had no way of knowing its true age. Was it tens of millions, hundreds of millions, or even billions of years old? Rather, a geologic time scale was developed that showed the sequence of events based on relative dating principles. What are these principles? What part do fossils play? With the discovery of radioactivity and radiometric dating techniques, geologists now can assign fairly accurate dates to many of the events in Earth history. What is radioactivity? Why is it a good "clock" for dating the geologic past?

Geology Needs a Time Scale

In 1869 John Wesley Powell, who was later to head the U.S. Geological Survey, led a pioneering expedition down the Colorado River and through the Grand Canyon (Figure 9.1). Writing about the rock layers that were exposed by the downcutting of the river, Powell noted that "the canyons of this region would be a Book of Revelations in the rock-leaved Bible of geology." He was undoubtedly impressed with the millions of years of Earth history exposed along the walls of the Grand Canyon (see chapter-opening photo).

Powell realized that the evidence for an ancient Earth is concealed in its rocks. Like the pages in a long and complicated history book, rocks record the geological events and changing life forms of the past. The book, however, is not complete. Many pages, especially in the early chapters, are missing. Others are tattered, torn, or smudged. Yet enough of the book remains to allow much of the story to be deciphered.

Interpreting Earth history is a prime goal of the science of geology. Like a modern-day sleuth, the geologist must interpret the clues found preserved in the rocks. By studying rocks, especially sedimentary rocks, and the features they contain, geologists can unravel the complexities of the past.

Geological events by themselves, however, have little meaning until they are put into a time perspective. Studying history, whether it be the Civil War or the age of dinosaurs, requires a calendar. Among geology's major contributions to human knowledge are the *geologic time scale* and the discovery that Earth history is exceedingly long.

FIGURE 9.1 **A.** Start of the expedition from Green River station. A drawing from Powell's 1875 book. **B.** Major John Wesley Powell, pioneering geologist and the second director of the U.S. Geological Survey. (Courtesy of the U.S. Geological Survey, Denver)

A.

B.

Relative Dating—Key Principles

GEODe
Geologic Time
▶ Relative Dating—Key Principles

The geologists who developed the geologic time scale revolutionized the way people think about time and perceive our planet. They learned that Earth is much older than anyone had previously imagined and that its surface and interior have been changed over and over again by the same geological processes that operate today.

During the late 1800s and early 1900s, attempts were made to determine Earth's age. Although some of the methods appeared promising at the time, none of these early efforts proved to be reliable. What these scientists were seeking was a **numerical date.** Such dates specify the actual number of years that have passed since an event occurred. Today our understanding of radioactivity allows us to accurately determine numerical dates for rocks that represent important events in Earth's distant past. We will study radioactivity later in this chapter. Prior to the discovery of radioactivity, geologists had no reliable method of numerical dating and had to rely solely on relative dating.

Relative dating means that rocks are placed in their proper *sequence of formation*—which formed first, second, third, and so on. Relative dating cannot tell us how long ago something took place, only that it followed one event and preceded another. The relative dating techniques that were developed are valuable and still widely used. Numerical dating methods did not replace these techniques; they simply supplemented them. To establish a relative time scale, a few basic principles or rules had to be discovered and applied. Although they may seem obvious to us today, they

were major breakthroughs in thinking at the time, and their discovery was an important scientific achievement.

Law of Superposition

Nicolaus Steno, a Danish anatomist, geologist, and priest (1638–1686), is credited with being the first to recognize a sequence of historical events in an outcrop of sedimentary rock layers. Working in the mountains of western Italy, Steno applied a very simple rule that has come to be the

FIGURE 9.2 Applying the law of superposition to these layers exposed in the upper portion of the Grand Canyon, the Supai Group is oldest and the Kaibab Limestone is youngest. (Photo by E. J. Tarbuck)

A.

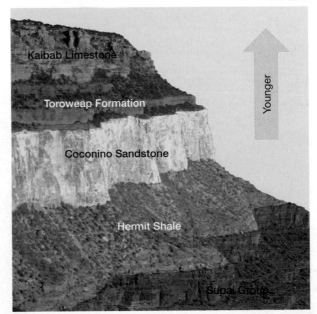

B.

FIGURE 9.3 Most layers of sediment are deposited in a nearly horizontal position. Thus, when we see rock layers that are folded or tilted, we can assume that they must have been moved into that position by crustal disturbances after their deposition. These folds are in Namibia's Lower Ugab Valley. (Michael Fogden/DRK Photo)

most basic principle of relative dating—**the law of superposition** (*super* = above; *positum* = to place). The law simply states that in an undeformed sequence of sedimentary rocks, each bed is older than the one above and younger than the one below. Although it may seem obvious that a rock layer could not be deposited with nothing beneath it for support, it was not until 1669 that Steno clearly stated this principle.

This rule also applies to other surface-deposited materials, such as lava flows and beds of ash from volcanic eruptions. Applying the law of superposition to the beds exposed

in the upper portion of the Grand Canyon (Figure 9.2), we can easily place the layers in their proper order. Among those that are pictured, the sedimentary rocks in the Supai Group are the oldest, followed in order by the Hermit Shale, Coconino Sandstone, Toroweap Formation, and Kaibab Limestone.

Principle of Original Horizontality

Steno is also credited with recognizing the importance of another basic principle, called the **principle of original horizontality.** Simply stated, it means that layers of sediment are generally deposited in a horizontal position. Thus, if we observe rock layers that are flat, it means they have not been disturbed and still have their *original* horizontality. The layers in the Grand Canyon illustrate this in the chapter-opening photo and in Figure 9.2. But if they are folded or inclined at a steep angle, they must have been moved into that position by crustal disturbances sometime *after* their deposition (Figure 9.3).

Principle of Cross-Cutting Relationships

When a fault cuts through other rocks, or when magma intrudes and crystallizes, we can assume that the fault or intrusion is younger than the rocks affected.* For example, in Figure 9.4, the faults and dikes clearly must have occurred after the sedimentary layers were deposited.

This is the **principle of cross-cutting relationships.** By applying the cross-cutting principle, you can see that fault A occurred *after* the sandstone layer was deposited because it "broke" the layer. Likewise, fault A occurred *before* the conglomerate was laid down because that layer is unbroken.

We can also state that dike B and its associated sill are older than dike A because dike A cuts the sill. In the same manner, we know that the batholith was emplaced after movement occurred along fault B but before dike B was formed. This is true because the batholith cuts across fault B, while dike B cuts across the batholith.

Inclusions

Sometimes inclusions can aid the relative dating process. **Inclusions** (*includere* = to enclose) are fragments of one rock unit that have been enclosed within another. The basic principle is logical and straightforward. The rock mass adjacent to the one containing the inclusions must have been there first in order to provide the rock fragments. Therefore, the rock mass

FIGURE 9.4 Cross-cutting relationships represent one principle used in relative dating. An intrusive rock body is younger than the rocks it intrudes. A fault is younger than the rock layers it cuts.

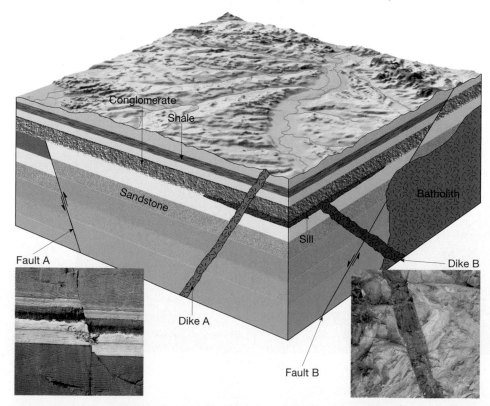

Conglomerate

Shale

Sandstone

Batholith

Sill

Fault A

Dike B

Dike A

Fault B

---

*Faults are fractures in the crust along which appreciable displacement has taken place. Faults are discussed in some detail in Chapter 10.

containing inclusions is the younger of the two. Figure 9.5 provides an example. Here, the inclusions of intrusive igneous rock in the adjacent sedimentary layer indicate that the sedimentary layer was deposited on top of a weathered igneous mass rather than being intruded from below by magma that later crystallized.

Unconformities

When we observe layers of rock that have been deposited essentially without interruption, we call them **conformable.** Particular sites exhibit conformable beds representing certain spans of geologic time. However, no place on Earth has a complete set of conformable strata.

Throughout Earth history, the deposition of sediment has been interrupted over and over again. All such breaks in the rock record are termed unconformities. An **unconformity** represents a long period during which deposition ceased, erosion removed previously formed rocks, and then deposition resumed. In each case, uplift and erosion are followed by subsidence and renewed sedimentation. Unconformities are important features because they represent significant geologic events in Earth history. Moreover, their recognition helps us identify what intervals of time are not represented by strata and thus are missing from the geologic record.

The rocks exposed in the Grand Canyon of the Colorado River represent a tremendous span of geologic history. It is a wonderful place in which to take a trip through time. The canyon's colorful strata record a long history of sedimentation in a variety of environments—advancing seas, rivers and deltas, tidal flats and sand dunes. But the record is not continuous. Unconformities represent vast amounts of time that have not been recorded in the canyon's layers. Figure 9.6 is a geologic cross section of the Grand Canyon. Refer to it as you read about the three basic types of unconformities: angular unconformities, disconformities, and nonconformities.

Angular Unconformity   Perhaps the most easily recognized unconformity is an **angular unconformity.** It consists of tilted or folded sedimentary rocks that are overlain by younger, more flat-lying strata. An angular unconformity indicates that during the pause in deposition, a period of deformation (folding or tilting) and erosion occurred (Figure 9.7).

When James Hutton studied an angular unconformity in Scotland more than 200 years ago, it was clear to him that it represented a major episode of geologic activity.* He and his colleagues also appreciated the immense time span implied by such relationships. When a companion later wrote of their visit to this site, he stated that "the mind seemed to grow giddy by looking so far into the abyss of time."

Disconformity   When contrasted with angular unconformities, **disconformities** are more common but usually far

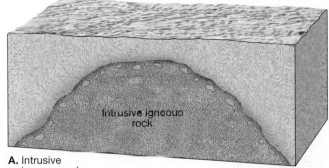

A. Intrusive
igneous rock

B. Exposure and
weathering of intrusive igneous rock

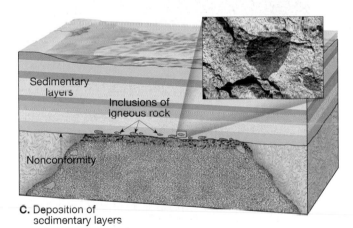

Sedimentary
layers

Inclusions of
igneous rock

Nonconformity

C. Deposition of
sedimentary layers

FIGURE 9.5   These diagrams illustrate two ways that inclusions can form, as well as a type of unconformity termed a nonconformity. In diagram **A,** the inclusions in the igneous mass represent unmelted remnants of the surrounding host rock that were broken off and incorporated at the time the magma was intruded. In diagram **C,** the igneous rock must be older than the overlying sedimentary beds because the sedimentary beds contain inclusions of the igneous rock. When older intrusive igneous rocks are overlain by younger sedimentary layers, a nonconformity is said to exist. The photo shows an inclusion of dark igneous rock in a lighter-colored and younger host rock. (Photo by Tom Bean)

less conspicuous because the strata on either side are essentially parallel. Many disconformities are difficult to identify because the rocks above and below are similar and there is little evidence of erosion. Such a break often resembles an ordinary bedding plane. Other disconformities are easier to identify because the ancient erosion surface is cut deeply into the older rocks below.

---

*This pioneering geologist is discussed in the section on the birth of modern geology in Chapter 1.

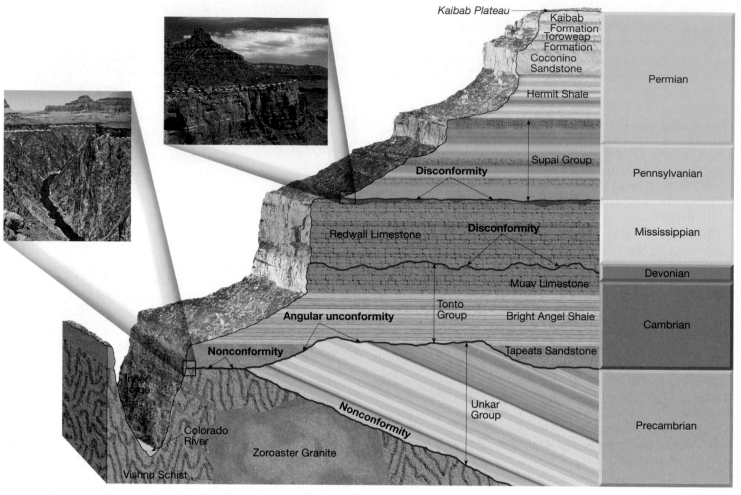

FIGURE 9.6 This cross section through the Grand Canyon illustrates the three basic types of unconformities. An angular unconformity can be seen between the tilted Precambrian Unkar Group and the Cambrian Tapeats Sandstone. Two disconformities are marked, above and below the Redwall Limestone. A nonconformity occurs between the igneous and metamorphic rocks exposed in the inner gorge and the sedimentary strata of the Unkar Group. A nonconformity, highlighted by a photo, also occurs between the rocks of the inner gorge and Tapeats Sandstone.

Nonconformity The third basic type of unconformity is a **nonconformity.** Here the break separates older metamorphic or intrusive igneous rocks from younger sedimentary strata (Figures 9.5 and 9.6). Just as angular unconformities and disconformities imply crustal movements, so too do nonconformities. Intrusive igneous masses and metamorphic rocks originate far below the surface. Thus, for a nonconformity to develop, there must be a period of uplift and the erosion of overlying rocks. Once exposed at the surface, the igneous or metamorphic rocks are subjected to weathering and erosion prior to subsidence and the renewal of sedimentation.

Using Relative Dating Principles

If you apply the principles of relative dating to the hypothetical geologic cross section in Figure 9.8, you can place in proper sequence the rocks and the events they represent. The statements within the figure summarize the logic used to interpret the cross section.

In this example, we establish a relative time scale for the rocks and events in the area of the cross section. Remember that this method gives us no idea of how many years of Earth history are represented, for we have no numerical dates. Nor do we know how this area compares to any other (see Box 9.1).

Correlation of Rock Layers

To develop a geologic time scale that is applicable to the entire Earth, rocks of similar age in different regions must be matched up. Such a task is referred to as **correlation.**

Within a limited area, correlating rocks of one locality with those of another may be done simply by walking along the outcropping edges. However, this may not be possible when the rocks are mostly concealed by soil and vegetation. Correlation over short distances is often achieved by noting the position of a bed in a sequence of strata. Or a layer may be identified in another location if it is composed of distinctive or uncommon minerals.

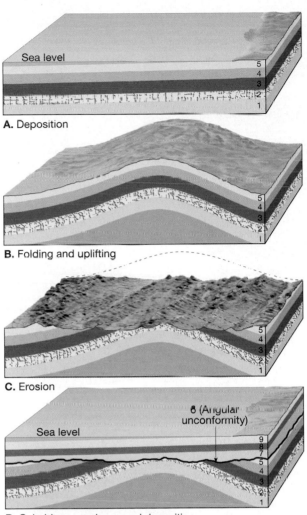

A. Deposition

B. Folding and uplifting

C. Erosion

D. Subsidence and renewed deposition

6 (Angular unconformity)

E.

FIGURE 9.7 Formation of an angular unconformity. An angular unconformity represents an extended period during which deformation and erosion occurred. Part **E** shows an angular unconformity at Siccar Point, Scotland, that was first described by James Hutton more than 200 years ago. (Photo by Edward A. Hay)

By correlating the rocks from one place to another, a more comprehensive view of the geologic history of a region is possible. Figure 9.9, for example, shows the correlation of strata at three sites on the Colorado Plateau in southern Utah and northern Arizona. No single locale exhibits the entire sequence, but correlation reveals a more complete picture of the sedimentary rock record.

Many geologic studies involve relatively small areas. Although they are important in their own right, their full value is realized only when they are correlated with other regions. Although the methods just described are sufficient to trace a rock formation over relatively short distances, they are not adequate for matching up rocks that are separated by great distances. When correlation between widely separated areas or between continents is the objective, geologists must rely on fossils.

Fossils: Evidence of Past Life

Fossils, the remains or traces of prehistoric life, are important inclusions in sediment and sedimentary rocks. They are basic and important tools for interpreting the geologic past. The scientific study of fossils is called **paleontology.** It is an interdisciplinary science that blends geology and biology in an attempt to understand all aspects of the succession of life over the vast expanse of geologic time. Knowing the nature of the life forms that existed at a particular time helps researchers understand past environmental conditions. Further, fossils are important time indicators and play a key role in correlating rocks of similar ages that are from different places.

Types of Fossils

Fossils are of many types. The remains of relatively recent organisms may not have been altered at all. Such objects as teeth, bones, and shells are common examples (Figure 9.10). Far less common are entire animals, flesh included, that have been preserved because of rather unusual circumstances. Remains of prehistoric elephants called mammoths that were frozen in the Arctic tundra of Siberia and Alaska are examples, as are the mummified remains of sloths preserved in a dry cave in Nevada.

Given enough time, the remains of an organism are likely to be modified. Often fossils become *petrified* (literally, "turned into stone"), meaning that the small internal cavities and pores of the original structure are filled with precipitated mineral matter (Figure 9.11A). In other instances *replacement* may occur. Here the cell walls and other solid material are removed and replaced with mineral matter. Sometimes the microscopic details of the replaced structure are faithfully retained.

Molds and casts constitute another common class of fossils. When a shell or other structure is buried in sediment and then dissolved by underground water, a *mold* is created. The mold faithfully reflects only the shape and surface marking of the organism; it does not reveal any information concerning its internal structure. If these hollow spaces are subsequently filled with mineral matter, *casts* are created (Figure 9.11B).

A type of fossilization called *carbonization* is particularly effective in preserving leaves and delicate animal forms. It

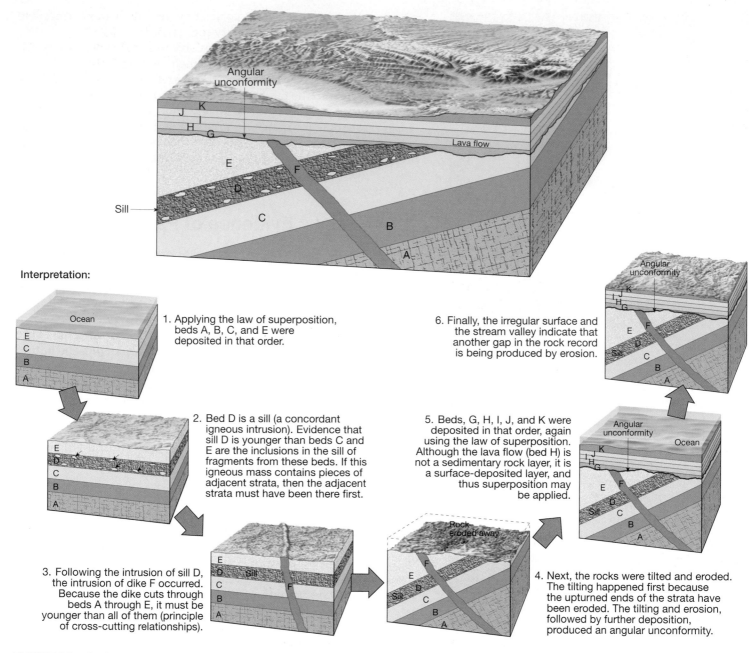

FIGURE 9.8   Geologic cross section of a hypothetical region.

Within the figure:

Angular unconformity

J K I H G

Lava flow

E F D C Sill B A

Interpretation:

Ocean

E C B A

1. Applying the law of superposition, beds A, B, C, and E were deposited in that order.

2. Bed D is a sill (a concordant igneous intrusion). Evidence that sill D is younger than beds C and E are the inclusions in the sill of fragments from these beds. If this igneous mass contains pieces of adjacent strata, then the adjacent strata must have been there first.

E D Sill C B A

3. Following the intrusion of sill D, the intrusion of dike F occurred. Because the dike cuts through beds A through E, it must be younger than all of them (principle of cross-cutting relationships).

E D Sill C F B A

Rock eroded away

F E D Sill C B A

4. Next, the rocks were tilted and eroded. The tilting happened first because the upturned ends of the strata have been eroded. The tilting and erosion, followed by further deposition, produced an angular unconformity.

5. Beds, G, H, I, J, and K were deposited in that order, again using the law of superposition. Although the lava flow (bed H) is not a sedimentary rock layer, it is a surface-deposited layer, and thus superposition may be applied.

Angular unconformity   Ocean

K J I H G E F D Sill C B A

6. Finally, the irregular surface and the stream valley indicate that another gap in the rock record is being produced by erosion.

Angular unconformity

K J I H G E F D Sill C B A

occurs when fine sediment encases the remains of an organism. As time passes, pressure squeezes out the liquid and gaseous components and leaves behind a thin residue of carbon (Figure 9.11C). Black shales deposited as organic-rich mud in oxygen-poor environments often contain abundant carbonized remains. If the film of carbon is lost from a fossil preserved in fine-grained sediment, a replica of the surface, called an *impression*, may still show considerable detail (Figure 9.11D).

Delicate organisms, such as insects, are difficult to preserve, and consequently they are relatively rare in the fossil record. Not only must they be protected from decay but they must not be subjected to any pressure that would crush

them. One way in which some insects have been preserved is in *amber,* the hardened resin of ancient trees. The fly in Figure 9.11E was preserved after being trapped in a drop of sticky resin. Resin sealed off the insect from the atmosphere and protected the remains from damage by water and air. As the resin hardened, a protective pressure-resistant case was formed.

In addition to the fossils already mentioned, there are numerous other types, many of them only traces of prehistoric life. Examples of such indirect evidence include:

1. Tracks—animal footprints made in soft sediment that was later lithified (see Figure 7.28B).

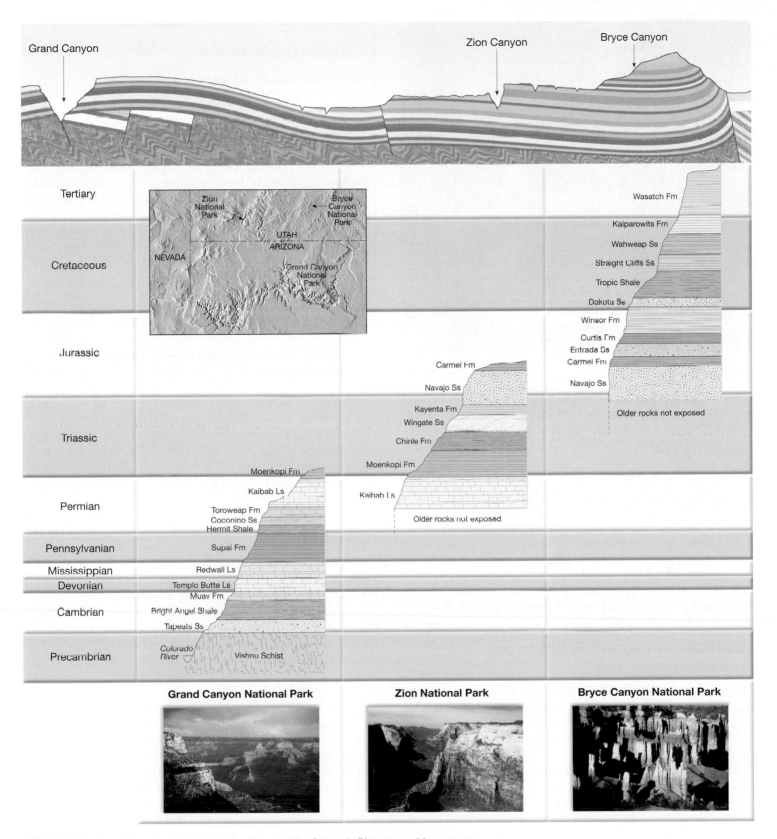

FIGURE 9.9 Correlation of strata at three locations on the Colorado Plateau provides a more complete picture of sedimentary rocks in the region. The diagram at the top is a geologic cross section of the region. (After U.S. Geological Survey; photos by E. J. Tarbuck)

FIGURE 9.10 Fossils of many relatively recent organisms are unaltered remains. Such objects as bones, teeth, and shells are common examples. In this historic 1914 photo of the La Brea tar pits in Los Angeles, bones of an Ice-Age mammal are being excavated. (Photo courtesy of The George C. Page Museum)

2. Burrows—tubes in sediment, wood, or rock made by an animal. These holes may later become filled with mineral matter and preserved. Some of the oldest-known fossils are believed to be worm burrows.
3. Coprolites—fossil dung and stomach contents that can provide useful information pertaining to food habits of organisms (Figure 9.11F).
4. Gastroliths—highly polished stomach stones that were used in the grinding of food by some extinct reptiles.

Conditions Favoring Preservation

Only a tiny fraction of the organisms that have lived during the geologic past have been preserved as fossils. Normally, the remains of an animal or plant are destroyed. Under what circumstances are they preserved? Two special conditions appear to be necessary: rapid burial and the possession of hard parts.

When an organism perishes, its soft parts usually are quickly eaten by scavengers or decomposed by bacteria. Occasionally, however, the remains are buried by sediment. When this occurs, the remains are protected from the environment, where destructive processes operate. Rapid burial therefore is an important condition favoring preservation.

In addition, animals and plants have a much better chance of being preserved as part of the fossil record if they have hard parts. Although traces and imprints of soft-bodied animals such as jellyfish, worms, and insects exist, they are not common. Flesh usually decays so rapidly that preservation is exceedingly unlikely. Hard parts such as shells, bones, and teeth predominate in the record of past life.

Because preservation is contingent on special conditions, the record of life in the geologic past is biased. The fossil record of those organisms with hard parts that lived in areas of sedimentation is quite abundant. However, we get only an occasional glimpse of the vast array of other life forms that did not meet the special conditions favoring preservation.

Fossils and Correlation

The existence of fossils had been known for centuries, yet it was not until the late 1700s and early 1800s that their significance as geologic tools was made evident. During this period an English engineer and canal builder, William Smith, discovered that each rock formation in the canals he worked on contained fossils unlike those in the beds either above or below. Further, he noted that sedimentary strata in widely separated areas could be identified—and correlated—by their distinctive fossil content.

Based on Smith's classic observations and the findings of many geologists who followed, one of the most important and basic principles in historical geology was formulated: *Fossil organisms succeed one another in a definite and determinable order, and therefore any time period can be recognized by its fossil content.* This has come to be known as the **principle of fossil succession.** In other words, when fossils are arranged according to their age, they do not present a random or haphazard picture. To the contrary, fossils document the evolution of life through time.

For example, an Age of Trilobites is recognized quite early in the fossil record. Then, in succession, paleontologists recognize an Age of Fishes, an Age of Coal Swamps, an Age of Reptiles, and an Age of Mammals. These "ages" pertain to groups that were especially plentiful and characteristic

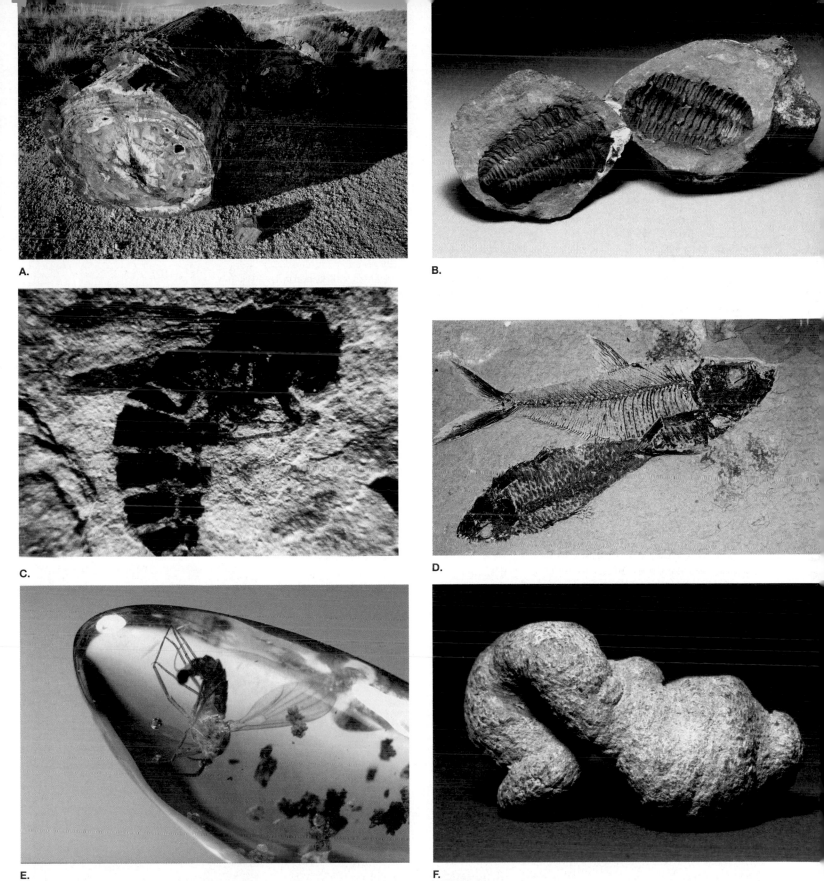

FIGURE 9.11 There are many types of fossilization. Six examples are shown here. **A.** Petrified wood in Petrified Forest National Park, Arizona. **B.** This trilobite photo illustrates mold and cast. **C.** A fossil bee preserved as a thin carbon film. **D.** Impressions are common fossils and often show considerable detail. **E.** Insect in amber. **F.** A coprolite is fossil dung. (Photo A by David Muench; Photos B, D, and F by E. J. Tarbuck; Photo C courtesy of the National Park Service; Photo E by Breck P. Kent)

BOX 9.1 ▶ UNDERSTANDING EARTH

Applying Relative Dating Principles to the Lunar Surface

Just as we use relative dating principles to determine the sequence of geological events on Earth, so too can we apply such principles to the surface of the Moon (and to other planetary bodies as well). For example, the image of the lunar surface in Figure 9.A shows the forward margin of a lava flow "frozen" in place. By applying the law of superposition, we know that this flow is younger than the adjacent layer that disappears beneath it.

Cross-cutting relationships can also be used. In Figure 9.B, when we observe one impact crater that overlaps another, we know that the continuous unbroken crater came after the one that it cuts across.

The most obvious features on the lunar surface are craters. Most were produced by the impact of rapidly moving objects called meteorites. Whereas the Moon has thousands of impact craters, Earth has only a few. This difference can be attributed to Earth's atmosphere. Friction with the air burns up small debris before it reaches the surface. Moreover, evidence for most of the sizable craters that formed in Earth's history has been obliterated by erosion and tectonic processes.

Observations of lunar cratering are used to estimate the relative ages of different locations on the Moon. The principle is straightforward. Older regions have been

FIGURE 9.B   Cross-cutting relationships allow us to say that the smaller, unbroken crater formed after the larger crater. (Photo courtesy of NASA)

exposed to meteorite impact longer and therefore have more craters. Using this technique in conjunction with Figure 9.C, we can infer that the highly cratered highlands are older than the dark areas, called maria. The number of craters per unit area (called *crater density*) is obviously much greater in the highlands. Does this mean that the highlands are *much* older? Although this may seem a logical conclusion, the answer is no. Remember that we are dealing with a principle of *relative* dating. Both the highlands and maria are very old. Radiometric dating of Moon rocks brought back from the *Apollo* missions showed that the age of the highlands is more than 4 billion years, whereas the maria have ages ranging from 3.2 to 3.9 billion years. Thus, the very different crater densities are *not* just the result of different exposure times. Astronomers now realize that the inner solar system experienced a sudden sharp drop in meteoritic bombardment about 3.9 billion years ago. The highlands received most of the craters before that time, and the lava flows that formed the maria solidified afterward.

FIGURE 9.C   Crater density. Younger regions have fewer craters than older regions. The densely cratered highlands are older than the dark areas, called maria. (UCO/Lick Observatory Image)

FIGURE 9.A   By applying the law of superposition, you can determine which lava flow is older. (Photo courtesy of National Space Data Center)

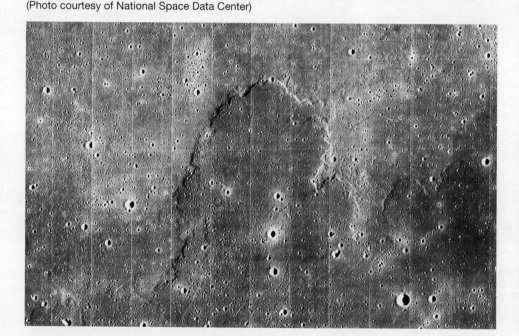

during particular time periods. Within each of the "ages" there are many subdivisions based, for example, on certain species of trilobites and certain types of fish, reptiles, and so on. This same succession of dominant organisms, never out of order, is found on every continent.

When fossils were found to be time indicators, they became the most useful means of correlating rocks of similar age in different regions. Geologists pay particular attention to certain fossils called **index fossils.** These fossils are widespread geographically and are limited to a short span of geologic time, so their presence provides an important method of matching rocks of the same age. Rock formations, however, do not always contain a specific index fossil. In such situations, groups of fossils are used to establish the age of the bed. Figure 9.12 illustrates how an assemblage of fossils may be used to date rocks more precisely than could be accomplished by the use of any one of the fossils.

In addition to being important and often essential tools for correlation, fossils are important environmental indicators. Although much can be deduced about past environments by studying the nature and characteristics of sedimentary rocks, a close examination of the fossils present can usually provide a great deal more information. For example, when the remains of certain clam shells are found in limestone, the geologist quite reasonably assumes that the region was once covered by a shallow sea. Also, by using what we know of living organisms, we can conclude that

fossil animals with thick shells, capable of withstanding pounding and surging waves, inhabited shorelines.

On the other hand, animals with thin, delicate shells probably indicate deep, calm offshore waters. Hence, by looking closely at the types of fossils, the approximate position of an ancient shoreline may be identified. Further, fossils can be used to indicate the former temperature of the water. Certain kinds of present-day corals must live in warm and shallow tropical seas like those around Florida and the Bahamas. When similar types of coral are found in ancient limestones, they indicate the marine environment that must have existed when they were alive. These examples illustrate how fossils can help unravel the complex story of Earth history.

Dating with Radioactivity

 GEODe Geologic Time
EARTH ▸ Dating with Radioactivity

In addition to establishing relative dates by using the principles described in the preceding sections, it is also possible to obtain reliable numerical dates for events in the geologic past. For example, we know that Earth is about 4.5 billion years old and that the dinosaurs became extinct about 65 million years ago. Dates that are expressed in millions and

FIGURE 9.12 Overlapping ranges of fossils help date rocks more exactly than using a single fossil.

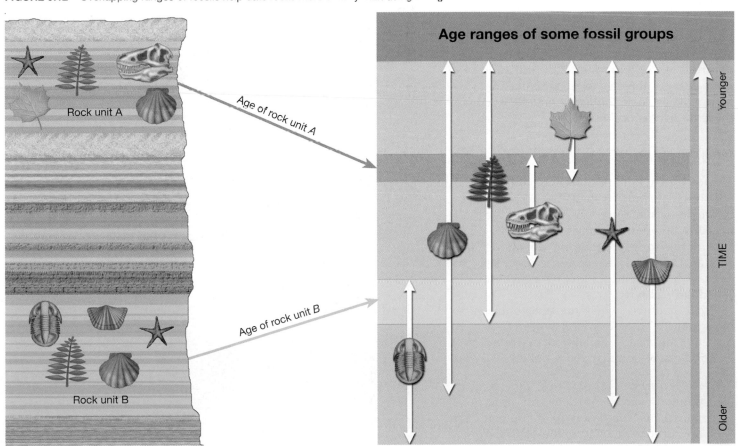

billions of years truly stretch our imagination because our personal calendars involve time measured in hours, weeks, and years. Nevertheless, the vast expanse of geologic time is a reality, and it is radiometric dating that allows us to measure it. In this section you will learn about radioactivity and its application in radiometric dating.

Reviewing Basic Atomic Structure

Recall from Chapter 3 that each atom has a *nucleus* containing protons and neutrons and that the nucleus is orbited by electrons. *Electrons* have a negative electrical charge, and *protons* have a positive charge. A *neutron* is actually a proton and an electron combined, so it has no charge (it is neutral).

The *atomic number* (each element's identifying number) is the number of protons in the nucleus. Every element has a different number of protons and thus a different atomic number (hydrogen = 1, carbon = 6, oxygen = 8, uranium = 92, etc.). Atoms of the same element always have the same number of protons, so the atomic number stays constant.

Practically all of an atom's mass (99.9 percent) is in the nucleus, indicating that electrons have virtually no mass at all. So, by adding the protons and neutrons in an atom's nucleus, we derive the atom's *mass number.* The number of neutrons can vary, and these variants, or *isotopes,* have different mass numbers.

To summarize with an example, uranium's nucleus always has 92 protons, so its atomic number always is 92. But its neutron population varies, so uranium has three isotopes: uranium-234 (protons + neutrons = 234), uranium-235, and uranium-238. All three isotopes are mixed in nature. They look the same and behave the same in chemical reactions.

Radioactivity

The forces that bind protons and neutrons together in the nucleus usually are strong. However, in some isotopes, the nuclei are unstable because the forces binding protons and neutrons together are not strong enough. As a result, the nuclei spontaneously break apart, or decay, a process called **radioactivity.**

What happens when unstable nuclei break apart? Three common types of radioactive decay are illustrated in Figure 9.13 and can be summarized as follows:

1. *Alpha particles* (α particles) may be emitted from the nucleus. An alpha particle is composed of 2 protons and 2 neutrons. Thus, the emission

of an alpha particle means that the mass number of the isotope is reduced by 4 and the atomic number is lowered by 2.

2. When a *beta particle* (β particle), or electron, is given off from a nucleus, the mass number remains unchanged, because electrons have practically no mass. However, because the electron has come from a neutron (remember, a neutron is a combination of a proton and an electron), the nucleus contains one more proton than before. Therefore, the atomic number increases by 1.

3. Sometimes an electron is captured by the nucleus. The electron combines with a proton and forms a neutron. As in the last example, the mass number remains unchanged. However, since the nucleus now contains one less proton, the atomic number decreases by 1.

An unstable radioactive isotope is referred to as the *parent,* and the isotopes resulting from the decay of the parent are termed the *daughter products.* Figure 9.14 provides an example of radioactive decay. Here it can be seen that when the radioactive parent, uranium-238 (atomic number 92, mass number 238) decays, it follows a number of steps,

FIGURE 9.13   Common types of radioactive decay. Notice that in each case the number of protons (atomic number) in the nucleus changes, thus producing a different element.

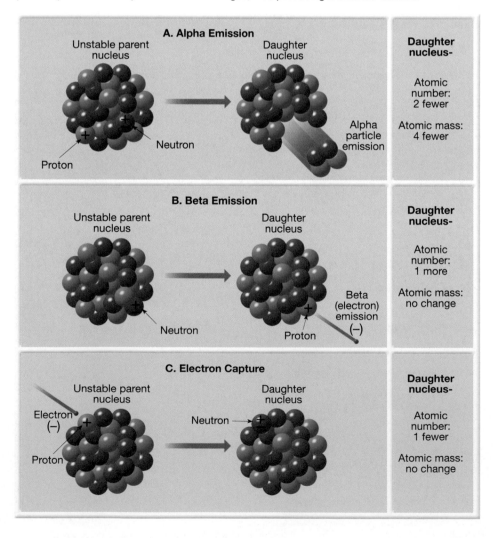

A. Alpha Emission	Daughter nucleus–
Unstable parent nucleus → Daughter nucleus	Atomic number: 2 fewer
Neutron / Proton / Alpha particle emission	Atomic mass: 4 fewer

B. Beta Emission	Daughter nucleus–
Unstable parent nucleus → Daughter nucleus	Atomic number: 1 more
Neutron / Proton / Beta (electron) emission (−)	Atomic mass: no change

C. Electron Capture	Daughter nucleus–
Unstable parent nucleus → Daughter nucleus	Atomic number: 1 fewer
Electron (−) / Proton / Neutron →	Atomic mass: no change

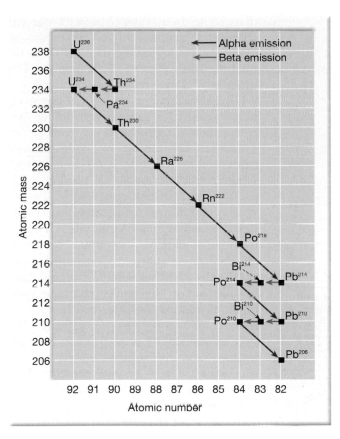

FIGURE 9.14   The most common isotope of uranium (U-238) is an example of a radioactive decay series. Before the stable end product (Pb-206) is reached, many different isotopes are produced as intermediate steps.

emitting 8 alpha particles and 6 beta particles before finally becoming the stable daughter product lead-206 (atomic number 82, mass number 206). One of the unstable daughter products produced during this decay series is radon. Box 9.2 examines the hazards associated with this radioactive gas.

Certainly among the most important results of the discovery of radioactivity is that it provided a reliable means of calculating the ages of rocks and minerals that contain particular radioactive isotopes. The procedure is called **radiometric dating.** Why is radiometric dating reliable? Because the rates of decay for many isotopes have been precisely measured and do not vary under the physical conditions that exist in Earth's outer layers. Therefore, each radioactive isotope used for dating has been decaying at a fixed rate since the formation of the rocks in which it occurs, and the products of decay have been accumulating at a corresponding rate. For example, when uranium is incorporated into a mineral that crystallizes from magma, there is no lead (the stable daughter product) from previous decay. The radiometric "clock" starts at this point. As the uranium in this newly formed mineral disintegrates, atoms of the daughter product are trapped, and measurable amounts of lead eventually accumulate.

Half-Life

The time required for half of the nuclei in a sample to decay is called the **half-life** of the isotope. Half-life is a common way of expressing the rate of radioactive disintegration. Figure 9.15 illustrates what occurs when a radioactive parent decays directly into its stable daughter product. When the quantities of parent and daughter are equal (ratio 1:1), we know that one half-life has transpired. When one-quarter of the original parent atoms remain and three-quarters have decayed to the daughter product, the parent/daughter ratio is 1:3 and we know that two half-lives have passed. After three half-lives, the ratio of parent atoms to daughter atoms is 1:7 (one parent atom for every seven daughter atoms).

If the half-life of a radioactive isotope is known and the parent/daughter ratio can be determined, the age of the sample can be calculated. For example, assume that the half-life of a hypothetical unstable isotope is 1 million years and the parent/daughter ratio in a sample is 1:15. Such a ratio indicates that four half-lives have passed and that the sample must be 4 million years old.

Radiometric Dating

Notice that the *percentage* of radioactive atoms that decay during one half-life is always the same: 50 percent. However, the *actual number* of atoms that decay with the passing of each half-life continually decreases. Thus, as the percentage of radioactive parent atoms declines, the proportion of stable daughter atoms rises, with the increase in daughter atoms just matching the drop in parent atoms. This fact is the key to radiometric dating.

Of the many radioactive isotopes that exist in nature, five have proved particularly useful in providing radiometric ages for ancient rocks (Table 9.1). Rubidium-87, thorium-232,

---

$\mathcal{S}tudents\ \mathcal{S}ometimes\ \mathcal{A}sk\ \ldots$

With radioactive decay, is there ever a time that all of the parent material is converted to the daughter product?

Theoretically, no. During each half-life, half of the parent material is converted to daughter product. Then half again is converted after another half-life, and so on. (Figure 9.15 shows how this logarithmic relationship works—notice that the red line becomes nearly parallel to the horizontal axis after several half-lives.) By converting only half of the remaining parent material to daughter product, there is never a time when all the parent material would be converted. Think about it this way. If you kept cutting a cake in half and eating only half, would you ever eat all of it? (The answer is no, assuming you had a sharp enough knife to slice the cake at an atomic scale!) However, after many half-lives, the parent material can exist in such small amounts that it is essentially undetectable.

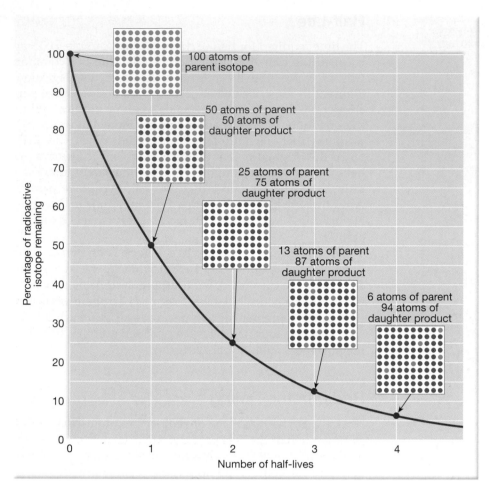

FIGURE 9.15   The radioactive-decay curve shows change that is exponential. Half of the radioactive parent remains after one half-life. After a second half-life one-quarter of the parent remains, and so forth.

does so in two ways. About 11 percent changes to argon-40 (^{40}Ar) by means of electron capture (see Figure 9.13C). The remaining 89 percent of ^{40}K decays to calcium-40 (^{40}Ca) by beta emission (see Figure 9.13B). The decay of ^{40}K to ^{40}Ca, however, is not useful for radiometric dating, because the ^{40}Ca produced by radioactive disintegration cannot be distinguished from calcium that may have been present when the rock formed.

The potassium-argon clock begins when potassium-bearing minerals crystallize from a magma or form within a metamorphic rock. At this point the new minerals will contain ^{40}K but will be free of ^{40}Ar, because this element is an inert gas that does not chemically combine with other elements. As time passes, the ^{40}K steadily decays by electron capture. The ^{40}Ar produced by this process remains trapped within the mineral's crystal lattice. Because no ^{40}Ar was present when the mineral formed, all of the daughter atoms trapped in the mineral must have come from the decay of ^{40}K. To determine a sample's age, the $^{40}K/^{40}Ar$ ratio is measured precisely, and the known half-life for ^{40}K applied.

and the two isotopes of uranium are used only for dating rocks that are millions of years old, but potassium-40 is more versatile.

Potassium-Argon   Although the half-life of potassium-40 is 1.3 billion years, analytical techniques make possible the detection of tiny amounts of its stable daughter product, argon-40, in some rocks that are younger than 100,000 years. Another important reason for its frequent use is that potassium is an abundant constituent of many common minerals, particularly micas and feldspars.

Although potassium (K) has three natural isotopes ^{39}K, ^{40}K, and ^{41}K, only ^{40}K is radioactive. When ^{40}K decays, it

Sources of Error   It is important to realize that an accurate radiometric date can be obtained only if the mineral remained a closed system during the entire period since its formation. A correct date is not possible unless there was neither the addition nor loss of parent or daughter isotopes. This is not always the case. In fact, an important limitation of the potassium-argon method arises from the fact that argon is a gas and it may leak from minerals, throwing off measurements. Indeed, losses can be significant if the rock is subjected to relatively high temperatures.

Of course, a reduction in the amount of ^{40}Ar leads to an underestimation of the rock's age. Sometimes temperatures are high enough for a sufficiently long period that all argon escapes. When this happens, the potassium-argon clock is reset, and dating the sample will give only the time of thermal resetting, not the true age of the rock. For other

TABLE 9.1 Isotopes Frequently Used in Radiometric Dating		
Radioactive Parent	**Stable Daughter Product**	**Currently Accepted Half-life Values**
Uranium-238	Lead-206	4.5 billion years
Uranium-235	Lead-207	713 million years
Thorium-232	Lead-208	14.1 billion years
Rubidium-87	Strontium-87	47.0 billion years
Potassium-40	Argon-40	1.3 billion years

BOX 9.2 ▶ PEOPLE AND THE ENVIRONMENT

Radon

Richard L. Hoffman*

Radioactivity is defined as the spontaneous emission of atomic particles and/or electromagnetic waves from unstable atomic nuclei. For example, in a sample of uranium-238, unstable nuclei decay and produce a variety of radioactive progeny or "daughter" products as well as energetic forms of radiation (Table 9.A). One of its radioactive-decay products is radon—a colorless, odorless, invisible gas.

Radon gained public attention in 1984 when a worker in a Pennsylvania nuclear power plant set off radiation alarms—not when he left work but as he first arrived there. His clothing and hair were contaminated with radon-decay products. Investigation revealed that his basement at home had a radon level 2800 times the average level in indoor air. The home was located along a geological formation known as the Reading Prong—a mass of uranium-bearing rock that runs from near Reading, Pennsylvania, to near Trenton, New Jersey.

*Dr. Hoffman, late Professor of chemistry, Illinois Central College.

Originating in the radio-decay of traces of uranium and thorium found in almost all soils, radon isotopes (Rn-222 and Rn-220) are continually renewed in an ongoing, natural process. Geologists estimate that the top six feet of soil from an average acre of land contains about 50 pounds of uranium (about 2 to 3 parts per million); some types of rock contain more. Radon is continually generated by the gradual decay of this uranium. Because uranium has a half-life of about 4.5 billion years, radon will be with us forever.

Radon itself decays, having a half-life of only about four days. Its decay products (except lead-206) are all radioactive solids that adhere to dust particles, many of which we inhale. During prolonged exposure to a radon-contaminated environment, some decay will occur while the gas is in the lungs, thereby placing the radioactive radon progeny in direct contact with delicate lung tissue. Steadily accumulating evidence indicates radon to be a significant cause of lung cancer second only to smoking.

A house with a radon level of 4.0 picocuries per liter of air has about eight to nine atoms of radon decaying every minute in every liter of air. The EPA suggests that indoor radon levels be kept below this level. EPA risk estimates are conservative—they are based on an assumption that one would spend 75 percent of a 70-year time span (about 52 years) in the contaminated space, which most people would not.

Once radon is produced in the soil, it diffuses throughout the tiny spaces between soil particles. Some radon ultimately reaches the soil surface, where it dissipates into the air. Radon enters buildings and homes through holes and cracks in basement floors and walls. Radon's density is greater than air, so it tends to remain in basements during its short decay cycle.

The source of radon is as enduring as its generation mechanism within Earth; radon will never go away. However, cost-effective mitigation strategies are available to reduce radon to acceptable levels, generally without great expense.

TABLE 9.A   Decay Products of Uranium-238

Some Decay Products of Uranium-238	Decay Particle Produced	Half-Life
Uranium-238	alpha	4.5 billion years
Radium-226	alpha	1600 years
Radon-222	**alpha**	**3.82 days**
Polonium-218	alpha	3.1 minutes
Lead-214	beta	26.8 minutes
Bismuth-214	beta	19.7 minutes
Polonium-214	alpha	1.6×10^{-4} second
Lead-210	beta	20.4 years
Bismuth-210	beta	5.0 days
Polonium-210	alpha	138 days
Lead-206	none	stable

radiometric clocks, a loss of daughter atoms can occur if the rock has been subjected to weathering or leaching. To avoid such a problem, one simple safeguard is to use only fresh, unweathered material and not samples that may have been chemically altered.

Dating with Carbon-14

To date very recent events, carbon-14 is used. Carbon-14 is the radioactive isotope of carbon. The process is often called **radiocarbon dating.** Because the half-life of carbon-14 is only 5730 years, it can be used for dating events from the historic past as well as those from very recent geologic history. In some cases carbon-14 can be used to date events as far back as 70,000 years.

Carbon-14 is continuously produced in the upper atmosphere as a consequence of cosmic-ray bombardment. Cosmic rays (high-energy nuclear particles) shatter the nuclei of gas atoms, releasing neutrons. Some of the neutrons are absorbed by nitrogen atoms (atomic number 7, mass number 14), causing each nucleus to emit a proton. As a result, the atomic number decreases by 1 (to 6), and a different

If parent/daughter ratios are not always reliable, how can meaningful radiometric dates for rocks be obtained?

One common precaution against sources of error is the use of cross checks. Often this simply involves subjecting a sample to two different radiometric methods. If the two dates agree, the likelihood is high that the date is reliable. If, on the other hand, there is an appreciable difference between the two dates, other cross checks must be employed (such as the use of fossils or correlation with other, well-dated marker beds) to determine which date—if either—is correct.

element, carbon-14, is created (Figure 9.16A). This isotope of carbon quickly becomes incorporated into carbon dioxide, which circulates in the atmosphere and is absorbed by living matter. As a result, all organisms contain a small amount of carbon-14, including you.

As long as an organism is alive, the decaying radiocarbon is continually replaced, and the proportions of carbon-14 and carbon-12 remain constant. Carbon-12 is the stable and most common isotope of carbon. However, when any plant or animal dies, the amount of carbon-14 gradually decreases as it decays to nitrogen-14 by beta emission (Figure 9.16B). By comparing the proportions of carbon-14 and carbon-12 in a sample, radiocarbon dates can be determined. It is important to emphasize that carbon-14 is only useful in dating organic materials such as wood, charcoal, bones, flesh, and even cloth made of cotton fibers.

FIGURE 9.16 **A.** Production and **B.** decay of carbon-14. These sketches represent the nuclei of the respective atoms.

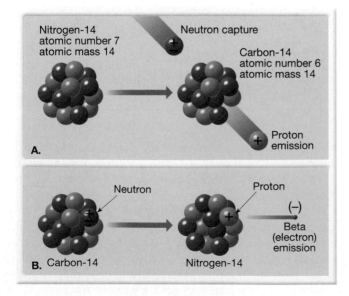

Although carbon-14 is only useful in dating the last small fraction of geologic time, it has become a very valuable tool for anthropologists, archaeologists, and historians, as well as for geologists who study very recent Earth history. In fact, the development of radiocarbon dating was considered so important that the chemist who discovered this application, Willard F. Libby, received a Nobel Prize in 1960.

Importance of Radiometric Dating

Bear in mind that although the basic principle of radiometric dating is simple, the actual procedure is quite complex. The analysis that determines the quantities of parent and daughter must be painstakingly precise. In addition, some radioactive materials do not decay directly into the stable daughter product, as was the case with our hypothetical example, a fact that may further complicate the analysis. In the case of uranium-238, there are 13 intermediate unstable daughter products formed before the 14th and last daughter product, the stable isotope lead-206, is produced (see Figure 9.14).

Radiometric dating methods have produced literally thousands of dates for events in Earth history. Rocks exceeding 3.5 billion years in age are found on all of the continents. Earth's oldest rocks (so far) are gneisses from northern Canada near Great Slave Lake that have been dated at 4.03 billion years (b.y.). Rocks from western Greenland have been dated at 3.7 to 3.8 b.y. and rocks nearly as old are found in the Minnesota River Valley and northern Michigan (3.5 to 3.7 b.y.), in southern Africa (3.4 to 3.5 b.y.) and in western Australia (3.4 to 3.6 b.y.). It is important to point out that these ancient rocks are not from any sort of "primordial crust" but originated as lava flows, igneous intrusions, and sediments deposited in shallow water—an indication that Earth history began *before* these rocks formed. Even older mineral grains have been dated. Tiny crystals of the mineral zircon having radiometric ages as old as 4.3 b.y. have been found in younger sedimentary rocks in western Australia. The source rocks for these tiny durable grains either no longer exist or have not yet been found.

Radiometric dating has vindicated the ideas of Hutton, Darwin, and others, who more than 150 years ago inferred that geologic time must be immense. Indeed, modern dating methods have proved that there has been enough time for the processes we observe to have accomplished tremendous tasks.

The Geologic Time Scale

 Geologic Time
▶ **The Geologic Time Scale**

Geologists have divided the whole of geologic history into units of varying magnitude. Together, they comprise the **geologic time scale** of Earth history (Figure 9.17). The major units of the time scale were delineated during the 19th

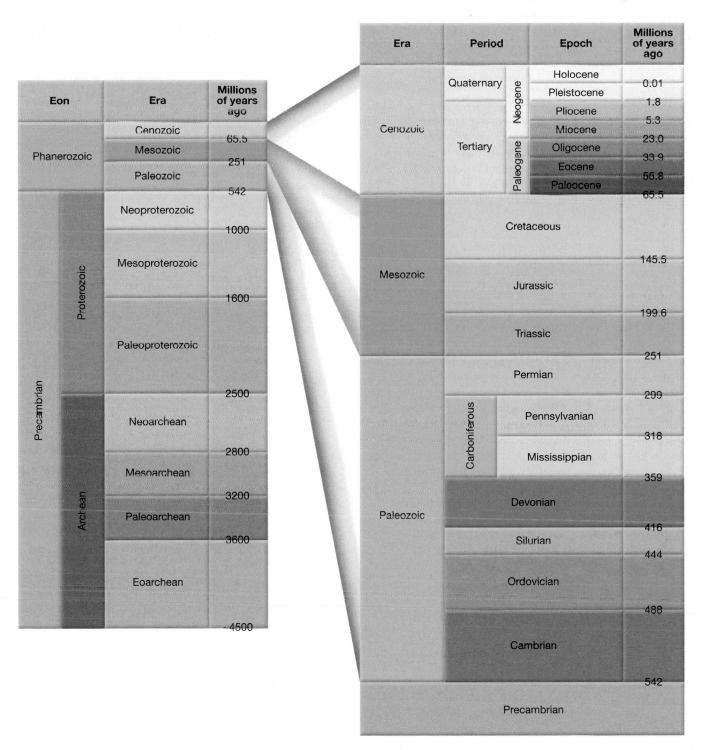

FIGURE 9.17 The geologic time scale. The numerical dates were added long after the time scale had been established using relative dating techniques.

century, principally by workers in Western Europe and Great Britain. Because radiometric dating was unavailable at that time, the entire time scale was created using methods of relative dating. It was only in the 20th century that radiometric methods permitted numerical dates to be added.

Structure of the Time Scale

The geologic time scale subdivides the 4.5-billion-year history of Earth into many different units and provides a meaningful time frame within which the events of the geologic

Students Sometimes Ask ...

Was there ever a time in Earth history when dinosaurs and humans coexisted?

Although some old movies and cartoons have depicted people and dinosaurs living side by side, this was never the case. Dinosaurs flourished during the Mesozoic era and became extinct about 65 million years ago. By contrast, humans and their close ancestors did not appear on the scene until very late in the Cenozoic era, more than 60 million years *after* the demise of the dinosaurs.

TABLE 9.2	Major Divisions of Geologic Time	
Cenozoic Era (Age of Recent Life)	Quaternary period	The several geologic eras were originally named Primary, Secondary, Tertiary, and Quaternary. The first two names are no longer used; Tertiary and Quaternary have been retained but used as period designations.
	Tertiary period	
Mesozoic Era (Age of Middle Life)	Cretaceous period	Derived from Latin word for chalk (creta) and first applied to extensive deposits that form white cliffs along the English Channel (see Chapter 7 opening photo p. 192–193).
	Jurassic period	Named for the Jura Mountains, located between France and Switzerland, where rocks of this age were first studied.
	Triassic period	Taken from the word "trias" in recognition of the threefold character of these rocks in Europe.
Paleozoic Era (Age of Ancient Life)	Permian period	Named after the province of Perm, Russia, where these rocks were first studied.
	Pennsylvanian period*	Named for the state of Pennsylvania where these rocks have produced much coal.
	Mississippian period*	Named for the Mississippi River Valley where these rocks are well exposed.
	Devonian period	Named after Devonshire County, England, where these rocks were first studied.
	Silurian period Ordovician period	Named after Celtic tribes, the Silures and the Ordovices, who lived in Wales during the Roman Conquest.
	Cambrian period	Taken from Roman name for Wales (Cambria), where rocks containing the earliest evidence of complex forms of life were first studied.
Precambrian		The time between the birth of the planet and the appearance of complex forms of life. About 88 percent of Earth's estimated 4.5 billion years fall into this span.

Source: U.S. Geological Survey.
* Outside of North America, the Mississippian and Pennsylvanian periods are combined into the Carboniferous period.

past are arranged. As shown in Figure 9.17, **eons** represent the greatest expanses of time. The eon that began about 542 million years ago is the **Phanerozoic,** a term derived from Greek words meaning *visible life*. It is an appropriate description because the rocks and deposits of the Phanerozoic eon contain abundant fossils that document major evolutionary trends.

Another glance at the time scale reveals that eons are divided into **eras.** The three eras within the Phanerozoic are the **Paleozoic** (*paleo* = ancient, *zoe* = life), the **Mesozoic** (*meso* = middle, *zoe* = life), and the **Cenozoic** (*ceno* = recent, *zoe* = life). As the names imply, these eras are bounded by profound worldwide changes in life forms.*

Each era of the Phaneroic eon is subdivided into time units known as **periods.** The Paleozoic has seven, the Mesozoic three, and the Cenozoic two. Each of these dozen periods is characterized by a somewhat less profound change in life forms as compared with the eras. The eras and periods of the Phanerozoic, with brief explanations of each, are shown in Table 9.2.

Each of the 12 periods is divided into still smaller units called **epochs.** As you can see in Figure 9.17, seven epochs have been named for the periods of the Cenozoic. The epochs of other periods usually are simply termed *early, middle,* and *late.*

Precambrian Time

Notice that the detail of the geologic time scale does not begin until about 542 million years ago, the date for the beginning of the Cambrian period. The nearly 4 billion years prior to the Cambrian are divided into two eons, the **Archean** (*archaios* = ancient), and the **Proterozoic** (*proteros* = before, *zoe* = life).It is also common for this vast expanse of time to simply be referred to as the **Precambrian.** Although it repre-

sents about 88 percent of Earth history, the Precambrian is not divided into nearly as many smaller time units as the Phanerozoic eon.

Why is the huge expanse of Precambrian time not divided into numerous eras, periods, and epochs? The reason is that Precambrian history is not known in great enough detail. The quantity of information that geologists have deciphered about Earth's past is somewhat analogous to the detail of human history. The further back we go, the less that is known. Certainly more data and information exist about

---

*Major changes in life forms are discussed in Chapter 22 "Earth's Evolution Through Geologic Time."

the past 10 years than for the first decade of the 20th century; the events of the 19th century have been documented much better than the events of the 1st century A.D.; and so on. So it is with Earth history. The more recent past has the freshest, least disturbed, and more observable record. The further back in time the geologist goes, the more fragmented the record and clues become. There are other reasons to explain our lack of a detailed time scale for this vast segment of Earth history:

1. The first abundant fossil evidence does not appear in the geologic record until the beginning of the Cambrian period. Prior to the Cambrian, simple life forms such as algae, bacteria, fungi, and worms predominated. All of these organisms lack hard parts, an important condition favoring preservation. For this reason, there is only a meager Precambrian fossil record. Many exposures of Precambrian rocks have been studied in some detail, but correlation is often difficult when fossils are lacking.
2. Because Precambrian rocks are very old, most have been subjected to a great many changes. Much of the Precambrian rock record is composed of highly distorted metamorphic rocks. This makes the interpretation of past environments difficult, because many of the clues present in the original sedimentary rocks have been destroyed.

Radiometric dating has provided a partial solution to the troublesome task of dating and correlating Precambrian rocks. But untangling the complex Precambrian record still remains a daunting task.

accurately determined because the grains composing the rock are not the same age as the rock in which they occur. Rather, the sediments have been weathered from rocks of diverse ages.

Radiometric dates obtained from metamorphic rocks may also be difficult to interpret, because the age of a particular mineral in a metamorphic rock does not necessarily represent the time when the rock initially formed. Instead, the date might indicate any one of a number of subsequent metamorphic phases.

If samples of sedimentary rocks rarely yield reliable radiometric ages, how can numerical dates be assigned to sedimentary layers? Usually the geologist must relate the strata to datable igneous masses, as in Figure 9.18. In this example, radiometric dating has determined the ages of the volcanic ash bed within the Morrison Formation and the dike cutting the Mancos Shale and Mesaverde Formation. The sedimentary beds below the ash are obviously older than the ash, and all the layers above the ash are younger. The dike is younger than the Mancos Shale and the Mesaverde Formation but older than the Wasatch Formation because the dike does not intrude the Tertiary rocks.

From this kind of evidence, geologists estimate that a part of the Morrison Formation was deposited about 160 million years ago, as indicated by the ash bed. Further, they conclude that the Tertiary period began after the intrusion of the dike, 66 million years ago. This is one example of literally thousands that illustrate how datable materials are used to bracket the various episodes in Earth history within specific time periods. It shows the necessity of combining laboratory dating methods with field observations of rocks.

Difficulties in Dating the Geologic Time Scale

Although reasonably accurate numerical dates have been worked out for the periods of the geologic time scale (Figure 9.17), the task is not without difficulty. The primary difficulty in assigning numerical dates to units of time is the fact that not all rocks can be dated by radiometric methods. Recall that for a radiometric date to be useful, all the minerals in the rock must have formed at about the same time. For this reason, radioactive isotopes can be used to determine when minerals in an igneous rock crystallized and when pressure and heat created new minerals in a metamorphic rock.

However, samples of sedimentary rock can only rarely be dated directly by radiometric means. Although a detrital sedimentary rock may include particles that contain radioactive isotopes, the rock's age cannot be

FIGURE 9.18   Numerical dates for sedimentary layers are usually determined by examining their relationship to igneous rocks. (After U.S. Geological Survey)

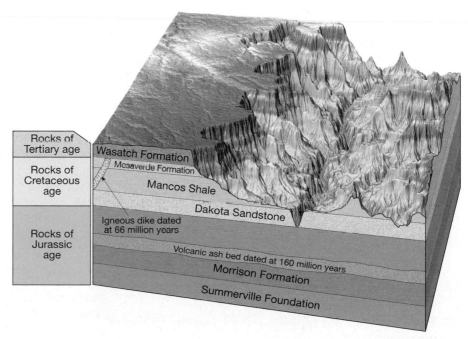

Summary

- The two types of dates used by geologists to interpret Earth history are (1) *relative dates,* which put events in their *proper sequence of formation,* and (2) *numerical dates,* which pinpoint the *time in years* when an event occurred.
- Relative dates can be established using the *law of superposition* (in an underformed sequence of sedimentary rocks or surface-deposited igneous rocks, each bed is older than the one above, and younger than the one below), *principle of original horizontality* (most layers are deposited in a horizontal position), *principle of cross-cutting relationships* (when a fault or intrusion cuts through another rock, the fault or intrusion is younger than the rocks cut through), and *inclusions* (the rock mass containing the inclusion is younger than the rock that provided the inclusion).
- *Unconformities* are gaps in the rock record. Each represents a long period during which deposition ceased, erosion removed previously formed rocks, and then deposition resumed. The three basic types of unconformities are *angular unconformities* (tilted or folded sedimentary rocks that are overlain by younger, more flat-lying strata), *disconformities* (the strata on either side of the unconformity are essentially parallel), and *nonconformities* (where a break separates older metamorphic or intrusive igneous rocks from younger sedimentary strata).
- *Correlation,* the matching up of two or more geologic phenomena of similar age in different areas, is used to develop a geologic time scale that applies to the whole Earth.
- Fossils are the remains or traces of prehistoric life. The special conditions that favor preservation are *rapid burial* and the possession of *hard parts,* such as shells, bones, or teeth.
- Fossils are used to *correlate* sedimentary rocks that are from different regions by using the rocks' distinctive fossil content and applying the *principle of fossil succession.* It is based on the work of *William Smith* in the late 1700s and states that fossil organisms succeed one another in a definite and determinable order, and therefore any time period can be recognized by its fossil content. The use of *index fossils,* those that are widespread geographically and are limited to a short span of geologic time, provides

an important method for matching rocks of the same age.
- Each atom has a nucleus containing *protons* (positively charged particles) and *neutrons* (neutral particles). Orbiting the nucleus are negatively charged *electrons.* The *atomic number* of an atom is the number of protons in the nucleus. The *mass number* is the number of protons plus the number of neutrons in an atom's nucleus. *Isotopes* are variants of the same atom, but with a different number of neutrons and hence a different mass number.
- *Radioactivity* is the spontaneous breaking apart (decay) of certain unstable atomic nuclei. Three common types of radioactive decay are (1) emission of *alpha particles* from the nucleus, (2) emission of *beta particles* from the nucleus, and (3) *capture of electrons* by the nucleus.
- An unstable *radioactive isotope,* called the *parent,* will decay and form stable *daughter products.* The length of time for half of the nuclei of a radioactive isotope to decay is called the *half-life* of the isotope. If the half-life of the isotope is known, and the parent/daughter ratio can be measured, the age of a sample can be calculated. An accurate radiometric date can be obtained only if the mineral containing the radioactive isotope has remained in a closed system during the entire period since its formation.
- The *geologic time scale* divides Earth's history into units of varying magnitude. It is commonly presented in chart form, with the oldest time and event at the bottom and the youngest at the top. The principle subdivisions of the geologic time scale, called *eons,* include the *Archean, Proterozoic* (together, these two eons are commonly referred to as the *Precambrian*), and, beginning about 542 million years ago, the *Phanerozoic.* The Phanerozoic (meaning "visible life") eon is divided into the following *eras: Paleozoic* ("ancient life"), *Mesozoic* ("middle life"), and *Cenozoic* ("recent life").
- A significant problem in assigning numerical dates is that *not all rocks can be radiometrically dated.* A sedimentary rock may contain particles of many ages that have been weathered from different rocks that formed at various times. One way geologists assign numerical dates to sedimentary rocks is to relate them to datable igneous masses, such as volcanic ash beds.

Review Questions

1. Distinguish between numerical and relative dating.
2. What is the law of superposition? How are cross-cutting relationships used in relative dating?
3. Refer to Figure 9.4 (p. 250) and answer the following questions:

   a. Is fault A older or younger than the sandstone layer?

   b. Is dike A older or younger than the sandstone layer?

   c. Was the conglomerate deposited before or after fault A?

   d. Was the conglomerate deposited before or after fault B?

e. Which fault is older, A or B?

f. Is dike A older or younger than the batholith?

4. When you observe an outcrop of steeply inclined sedimentary layers, what principle allows you to assume that the beds were tilted after they were deposited?

5. A mass of granite is in contact with a layer of sandstone. Using a principle described in this chapter, explain how you might determine whether the sandstone was deposited on top of the granite or whether the granite was intruded from below after the sandstone was deposited.

6. Distinguish among angular unconformity, disconformity, and nonconformity.

7. What is meant by the term *correlation?*

8. Describe William Smith's important contribution to the science of geology.

9. List and briefly describe at least five different types of fossils.

10. List two conditions that improve an organism's chances of being preserved as a fossil.

11. Why are fossils such useful tools in correlation?

12. Figure 9.19 is a block diagram of a hypothetical area in the American Southwest. Place the lettered features in the proper sequence, from oldest to youngest. Identify an angular unconformity and a nonconformity.

13. If a radioactive isotope of thorium (atomic number 90, mass number 232) emits 6 alpha particles and 4 beta particles during the course of radioactive decay, what are the atomic number and mass number of the stable daughter product?

14. Why is radiometric dating the most reliable method of dating the geologic past?

15. A hypothetical radioactive isotope has a half-life of 10,000 years. If the ratio of radioactive parent to stable daughter product is 1:3, how old is the rock containing the radioactive material?

16. To provide a reliable radiometric date, a mineral must remain a closed system from the time of its formation until the present. Why is this true?

17. What precautions are taken to ensure reliable radiometric dates?

FIGURE 9.19 Use this block diagram in conjunction with Review Question 12.

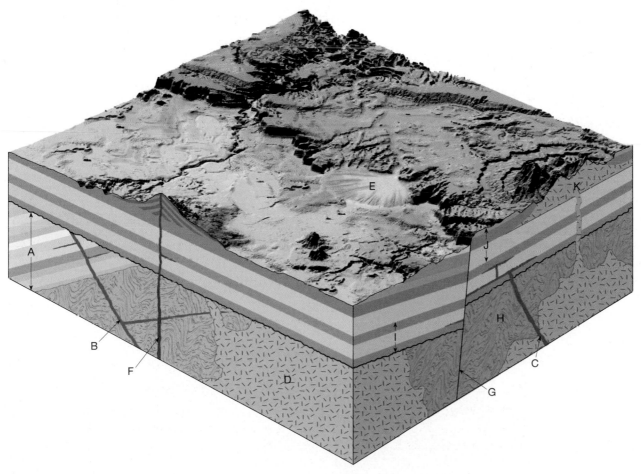

18. To make calculations easier, let us round the age of Earth to 5 billion years.

    a. What fraction of geologic time is represented by recorded history (assume 5000 years for the length of recorded history)?

    b. The first abundant fossil evidence does not appear until the beginning of the Cambrian period (540 million years ago). What percent of geologic time is represented by abundant fossil evidence?

20. What subdivisions make up the geologic time scale?

21. Explain the lack of a detailed time scale for the vast span known as the Precambrian.

22. Briefly describe the difficulties in assigning numerical dates to layers of sedimentary rock.

Key Terms

angular unconformity (p. 251)
Archean eon (p. 266)
Cenozoic era (p. 266)
conformable (p. 251)
correlation (p. 252)
cross-cutting relationships, principle of (p. 250)
disconformity (p. 251)
eon (p. 266)

epoch (p. 266)
era (p. 266)
fossil (p. 253)
fossil succession, principle of (p. 256)
geologic time scale (p. 264)
half-life (p. 261)
inclusions (p. 250)
index fossil (p. 259)
Mesozoic era (p. 266)

nonconformity (p. 252)
numerical date (p. 249)
original horizontality, principle of (p. 250)
paleontology (p. 253)
Paleozoic era (p. 266)
period (p. 266).
Phanerozoic eon (p. 266)
Precambrian (p. 266)
Proterozoic eon (p. 266)

radioactivity (p. 260)
radiocarbon dating (p. 263)
radiometric dating (p. 261)
relative dating (p. 249)
superposition, law of (p. 250)
unconformity (p. 251)

Web Resources

 The *Earth* Web \site uses the resources and flexibility of the Internet to aid in your study of the topics in this chapter. Written and developed by geology instructors, this site will help improve your understanding of geology. Visit **http://www.prenhall.com/tarbuck** and click on the cover of *Earth 9e* to find:

- Online review quizzes.
- Critical thinking exercises.
- Links to chapter-specific Web resources.
- Internet-wide key-term searches.

http://www.prenhall.com/tarbuck

GEODe: Earth

GEODe: Earth makes studying faster and more effective by reinforcing key concepts using animation, video, narration, interactive exercises and practice quizzes. A copy is included with every copy of *Earth*.

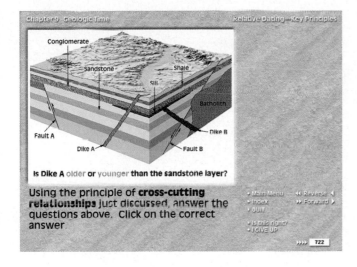

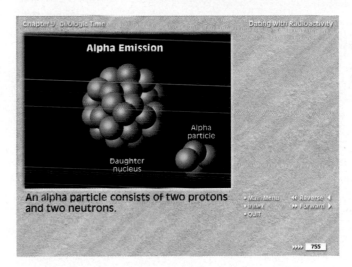

Crustal Deformation

Deformed strata in the Panamint Range, Death Valley National Park, California. (Photo by Michael Collier)

Earth is a dynamic planet. In the preceding chapters, you learned that weathering, mass wasting, and erosion by water, wind, and ice continually sculpture the landscape. In addition, tectonic forces deform rocks in the crust. Evidence demonstrating the operation of enormous forces within Earth includes thousands of kilometers of rock layers that are bent, contorted, overturned, and sometimes riddled with fractures (Figure 10.1). In the Canadian Rockies, for example, some rock units have been thrust for hundreds of kilometers over other layers. On a smaller scale, crustal movements of a few meters occur along faults during major earthquakes. In addition, rifting (spreading) and extension of the crust produce elongated depressions and over long spans of geologic time may even create ocean basins.

Structural Geology: A Study of Earth's Architecture

The results of tectonic activity are strikingly apparent in Earth's major mountain belts (see chapter-opening photo). Here, rocks containing fossils of marine organisms may be found thousands of meters above sea level, and massive rock units show evidence of having been intensely fractured and folded, as though they were made of putty. Even in the stable interiors of the continents, rocks reveal a history of deformation that shows they have been uplifted from much deeper levels in the crust.

Structural geologists study the architecture of Earth's crust and "how it got this way" insofar as it resulted from deformation. By studying the orientations of faults and folds, as well as small-scale features of deformed rocks, structural geologists can often reconstruct the original geologic setting and the nature of the forces that generated these rock structures. In this way, the complex events of Earth's geologic history are unraveled.

An understanding of rock structures is not only important in deciphering Earth history, it is also basic to our economic well-being. For example, most occurrences of oil and natural gas are associated with geologic structures that act to trap these fluids in valuable "reservoirs" (see Chapter 23).

FIGURE 10.1 Uplifted and folded sedimentary strata at Stair Hole, near Lulworth, Dorset, England. These layers of Jurassic-age rock, originally deposited in horizontal beds, have been folded as a result of the collision between the African and European crustal plates. (Photo by Tom & Susan Bean, Inc.)

FIGURE 10.2 Deformation of Earth's crust caused by tectonic forces and associated stresses resulting from the movement of lithospheric plates. **A.** Strata before deformation. **B.** Compressional stresses associated with plate collisions tend to shorten and thicken Earth's crust by folding, flowing, and faulting. **C.** Tensional stresses at divergent plate boundaries tend to lengthen rock bodies by displacement along faults in the upper crust and ductile flow at depth. **D.** Shear stresses at transform plate boundaries tend to produce offsets along fault zones. The right side of each diagram illustrates the deformation (strain) of a cube of rock in response to the differential stresses illustrated in corresponding diagram to the left.

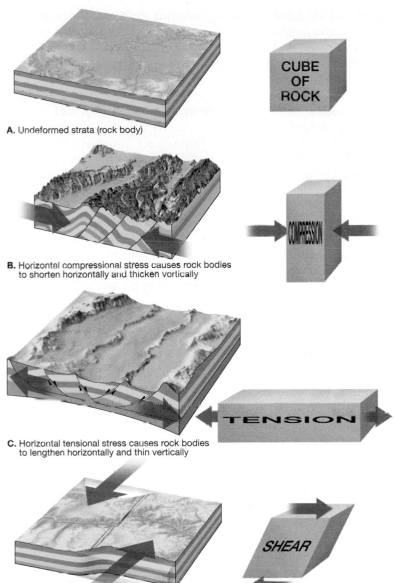

A. Undeformed strata (rock body)

B. Horizontal compressional stress causes rock bodies to shorten horizontally and thicken vertically

C. Horizontal tensional stress causes rock bodies to lengthen horizontally and thin vertically

D. Shear stress causes displacements along fault zones or by ductile flow

Furthermore, rock fractures are sites of hydrothermal mineralization, which means they can be sites of metallic ore deposits. Moreover, the orientation of fracture surfaces, which represent zones of weakness in rocks, must be considered when selecting sites for major construction projects such as bridges, hydroelectric dams, and nuclear power plants. In short, a working knowledge of rock structures is essential to our modern way of life.

In this chapter we will examine the forces that deform rock, as well as the structures that result. The basic geologic structures associated with deformation are folds, faults, joints, and foliation (including rock cleavage). Because rock cleavage and foliation were examined in Chapter 8, this chapter will be devoted to the remaining rock structures and the tectonic forces that produce them.

Deformation

 GEODe Crustal Deformation
▶ Deformation

Every body of rock, no matter how strong, has a point at which it will fracture or flow. **Deformation** (*de* = out, *forma* = form) is a general term that refers to all changes in the size, shape, orientation, or position of a rock mass. Most crustal deformation occurs along plate margins. Plate motions and the interactions along plate boundaries generate the tectonic forces that cause rock units to deform.

Force and Stress

Force is that which tends to put stationary objects in motion or change the motions of moving bodies. From everyday experience you know that if a door is stuck (stationary), you apply force to open it (get it in motion).

To describe the forces that deform rocks, structural geologists use the term **stress**—the amount of force applied to a given area. The magnitude of stress is not simply a function of the amount of force applied but also relates to the area on which the force acts. For example, if you are walking barefoot on a hard surface, the force (weight) of your body is distributed across your entire foot, so the stress acting on any one point of your foot is low. However, if you step on a

small pointed rock (ouch!), the stress concentration at a point on your foot will be high. Thus, you can think of stress as a measure of how concentrated the force is. As you saw in Chapter 8, stress may be applied uniformly in all directions (*confining pressure*) or nonuniformly (*differential stress*).

Types of Stress

When stress is applied unequally in different directions, it is termed **differential stress**. Differential stress that shortens a rock body is known as **compressional** (*com* = together, *premere* = to press) **stress**. Compressional stresses associated with plate collisions tend to shorten and thicken Earth's crust by folding, flowing, and faulting (Figure 10.2B). Recall from our discussion of metamorphic rocks that compressional stress is more concentrated at points where mineral grains are in contact, causing mineral matter to migrate from areas of high stress to areas of low stress (see Figure 8.7). As a result, the mineral grains (and the rock unit) tend to shorten in the direction parallel to the plane of maximum stress and elongate perpendicular to the direction of greatest stress.

When stress tends to elongate or pull apart a rock unit, it is known as **tensional** (*tendere* = to stretch) **stress** (Figure 10.2C). Where plates are being rifted apart (divergent plate boundaries), tensional stresses tend to lengthen those rock bodies located in the upper crust by displacement along faults. At depth, on the other hand, displacement is accomplished by a type of puttylike flow.

Differential stress can also cause rock to **shear** (Figure 10.2D). One type of shearing is similar to the slippage that occurs between individual playing cards when the top of the deck is moved relative to the bottom (Figure 10.3). In near-surface environments, shearing often occurs on closely spaced parallel surfaces of weakness, such as bedding planes, foliation, and microfaults. Further, at transform fault boundaries, shearing stresses produce large-scale offsets along major fault zones. By contrast, at great depths where temperatures and confining pressures are high, shearing is accomplished by solid-state flow.

Strain

Perhaps the easiest type of deformation to visualize occurs along small fault surfaces where differential stress causes rocks to move relative to each other in such a way that their original size and shape are preserved. Stress can also cause an irreversible change in the shape and size of a rock body, referred to as **strain.** Like the circle shown in Figure 10.3B, *strained bodies do not retain their original configuration during deformation.* Figure 10.1 illustrates the strain (deformation) exhibited by rock units near Dorset, England. When studying strained rock units like those shown in Figure 10.1, geologists ask, "What do these deformed structures indicate about the original arrangement of these rocks, and how have they been deformed?"

A. Deck of playing cards

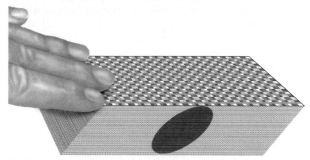

B. Shearing occurs when hand pushes top of deck

FIGURE 10.3 Illustration of shearing and the resulting deformation (strain). **A.** An ordinary deck of playing cards with a circle embossed on its side. **B.** By sliding the top of the deck relative to the bottom, we can illustrate the type of shearing that commonly occurs along closely spaced planes of weakness in rocks. Notice that the circle becomes an ellipse, which can be used to measure the amount and type of strain. Additional displacement (shearing) of the cards would result in further strain and would be indicated by a change in the shape of the ellipse.

tial stress under conditions that simulated those existing at various depths within the crust (Figure 10.5).

Although each rock type deforms somewhat differently, the general characteristics of rock deformation were determined from these experiments. Geologists discovered that when stress is gradually applied, rocks first respond by deforming elastically. Changes that result from *elastic deformation* are recoverable; that is, like a rubber band, the rock will return to nearly its original size and shape when the stress is removed. (As you will see in the next chapter, the energy for most earthquakes comes from stored elastic energy that is released as rock snaps back to its original shape.)

How Rocks Deform

When rocks are subjected to stresses greater than their own strength, they begin to deform, usually by folding, flowing, or fracturing (Figure 10.4). It is easy to visualize how rocks break, because we normally think of them as being brittle. But how can rock units be *bent* into intricate folds without being broken during the process? To answer this question, structural geologists performed laboratory experiments in which rocks were subjected to differen-

FIGURE 10.4 Deformed sedimentary strata exposed in a road cut near Palmdale, California. In addition to the obvious folding, light-colored beds are offset along a fault located on the right side of the photograph. (Photo by E. J. Tarbuck)

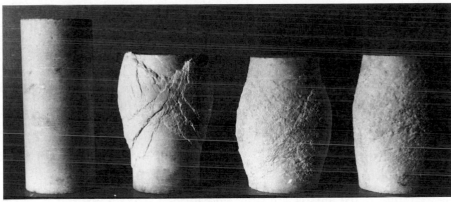

Undeformed   Low confining pressure   Intermediate confining pressure   High confining pressure

FIGURE 10.5 A marble cylinder deformed in the laboratory by applying thousands of pounds of load from above. Each sample was deformed in an environment that duplicated the confining pressure found at various depths. Notice that when the confining pressure was low, the sample deformed by brittle fracture, whereas when the confining pressure was high, the sample deformed plastically. (Photo courtesy of M. S. Patterson, Australian National University)

wax, caramel candy, and most metals. For example, a copper penny placed on a railroad track will be flattened and deformed (without breaking) by the force applied by a passing train. Ductile deformation of a rock—strongly aided by high temperature and high confining pressure—is somewhat similar to the deformation of a penny flattened by a train. One way this type of solid-state flow is accomplished within a rock is by gradual slippage and recrystallization along planes of weakness within the crystal lattice of mineral grains (see Figure 8.7B). This microscopic form of gradual solid-state flow involves slippage that disrupts the crystal lattice and immediate recrystallization that repairs the structure. Rocks that display evidence of ductile flow usually were deformed at great depth and may exhibit contorted folds that give the impression that the strength of the rock was akin to soft putty (Figure 10.6).

Once the elastic limit (strength) of a rock is surpassed, it either flows (*ductile deformation*) or fractures (*brittle deformation*). The factors that influence the strength of a rock and thus how it will deform include temperature, confining pressure, rock type, availability of fluids, and time.

Temperature and Confining Pressure   Rocks near the surface, where temperatures and confining pressures are low, tend to behave like a brittle solid and fracture once their strength is exceeded. This type of deformation is called **brittle** (*bryttian* = to shatter) **failure** or **brittle deformation.** From our everyday experience, we know that glass objects, wooden pencils, china plates, and even our bones exhibit brittle failure once their strength is surpassed. By contrast, at depth, where temperatures and confining pressures are high, rocks exhibit *ductile* behavior. **Ductile deformation** is a type of solid-state flow that produces a change in the size and shape of an object without fracturing, as occurs with brittle failure. Ordinary objects that display ductile behavior include modeling clay, bees-

Rock Type   In addition to the physical environment, the mineral composition and texture of a rock greatly influence how it will deform. For example, crystalline rocks composed of minerals that have strong internal molecular bonds tend to fail by brittle fracture. By contrast, sedimentary rocks that are weakly cemented, or metamorphic rocks that contain zones of weakness, such as foliation, are more susceptible to ductile flow. Rocks that are weak and thus most likely to behave in a ductile manner when subjected to differential stress, include rock salt, gypsum, and shale whereas limestone, schist, and marble are of intermediate

FIGURE 10.6 Rocks in Colorado exhibiting the results of ductile behavior. These rocks were deformed at great depth and were subsequently exposed at the surface. (Photo by Marli Miller)

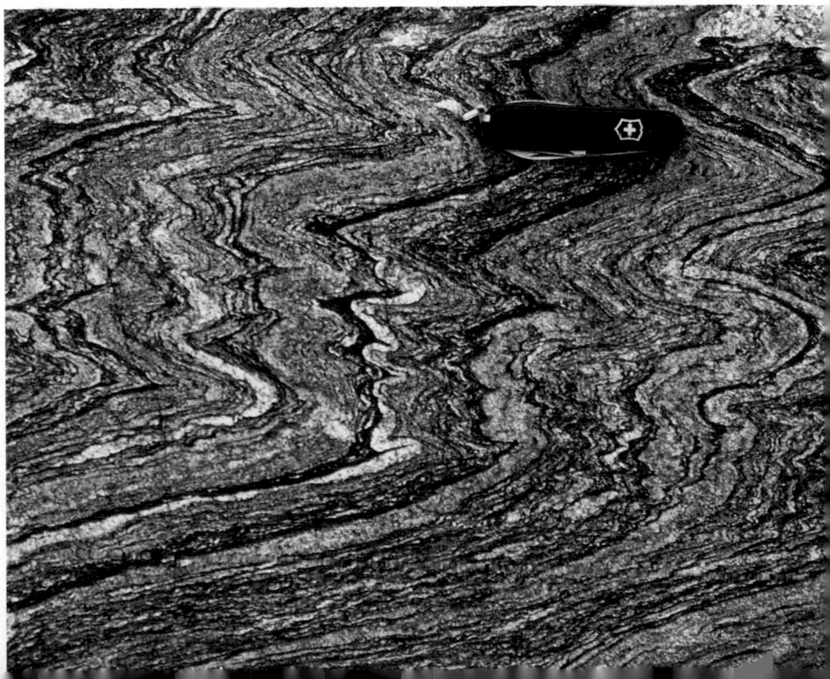

strength. In fact, rock salt is so weak that it deforms under small amounts of differential stress and rises like stone pillars through beds of sediment that lie in and around the Gulf of Mexico. Perhaps the weakest naturally occurring solid to exhibit ductile flow on a large scale is glacial ice. By comparison, granite and basalt are strong and brittle. In a near-surface environment, strong, brittle rocks will fail by fracturing when subjected to stresses that exceed their strength. It is important to note, however, that the presence of even tiny quantities of water in rocks greatly reduces their resistance to ductile deformation.

Time   One key factor that researchers are unable to duplicate in the laboratory is how rocks respond to small amounts of stress applied over long spans of *geologic time.* However, insights into the effects of time on deformation are provided in everyday settings. For example, marble benches have been known to sag under their own weight over a span of a hundred years or so, and wooden bookshelves may bend after being loaded with books for a relatively short period. In nature small stresses applied over long time spans surely play an important role in the deformation of rock. Forces that are unable to deform rock when initially applied may cause rock to flow if the stress is maintained over an extended period of time.

It is important to note that the processes by which rocks deform occur along a continuum that ranges from pure brittle fracture at one end to ductile (viscous) flow at the other. There are no sharp boundaries between different types of deformation. We also need to remind ourselves that the elegant folds and flow patterns we see in deformed rocks are generally achieved by the combined effect of distortion, sliding, and rotation of the individual grains that make up a rock. Further, this distortion and reorganization of mineral grains takes place in rock that is essentially solid.

Mapping Geologic Structures

 GEODe   Crustal Deformation
    ▶ Mapping Geologic Structures

The processes of deformation generate features at many different scales. At one extreme are Earth's major mountain systems. At the other extreme, highly localized stresses create minor fractures in bedrock. All of these phenomena, from the largest folds in the Alps to the smallest fractures in a slab of rock, are referred to as **rock structures.** Before beginning our discussion of rock structures, let us examine the way geologists describe and map them.

When conducting a study of a region, a geologist identifies and describes the dominant structures. A structure often is so large that only a small portion is visible from any particular vantage point. In many situations, most of the bedrock is concealed by vegetation or by recent sedimentation. Consequently, the reconstruction must be done using data gathered from a limited number of *outcrops,* which are sites where bedrock is exposed at the surface (see Box 10.1).

Despite such difficulties, a number of mapping techniques enable geologists to reconstruct the orientation and shape of the existing structures. In recent years this work has been aided by advances in aerial photography, satellite imagery, and the development of the global positioning system (GPS). In addition, seismic reflection profiling (see Chapter 12) and drill holes provide data on the composition and structure of rocks that lie at depth.

Geologic mapping is most easily accomplished where sedimentary strata are exposed. This is because sediments are usually deposited in horizontal layers. If the sedimentary rock layers are still horizontal, this tells geologists that the area is probably undisturbed structurally. But if strata are inclined, bent, or broken, this indicates that a period of deformation occurred following deposition.

Strike and Dip

Geologists use measurements called *strike* (trend) and *dip* (inclination) to help determine the orientation or attitude of a rock layer or fault surface (Figure 10.7). By knowing the strike and dip of rocks at the surface, geologists can predict the nature and structure of rock units and faults that are hidden beneath the surface, beyond their view.

Strike is the compass direction of the line produced by the intersection of an inclined rock layer or fault, with a horizontal plane (Figure 10.7). The strike, or compass bearing, is generally expressed as an angle relative to north. For example, N10° E means the line of strike is 10° to the east of north. The strike of the rock units illustrated in Figure 10.7 is approximately north 75° east (N75° E).

Dip is the angle of inclination of the surface of a rock unit or fault measured from a horizontal plane. Dip includes both an angle of inclination and a direction toward which the rock is inclined. In Figure 10.7 the dip angle of the rock layer is 30°. A good way to visualize dip is to imagine that water will always run down the rock surface parallel to the dip. The direction of dip will always be at a 90° angle to the strike.

FIGURE 10.7   Strike and dip of a rock layer.

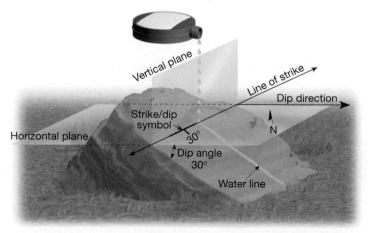

BOX 10.1 ▶ UNDERSTANDING EARTH

Naming Local Rock Units

One of the primary goals of geology is to reconstruct Earth's long and complex history through the systematic study of rocks. In most areas, exposures of rocks are not continuous over great distances. Consequently, the study of rock layers must be done locally and then correlated with data from adjoining areas to produce a larger and more complete picture. The first step in the effort to unravel past geologic events is that of describing and mapping rock units exposed in local outcrops.

Describing anything as complex as a thick sequence of rocks requires subdividing the layers into units of manageable size. The most basic rock division is called a *formation*, which is simply a distinctive series of strata that originated through the same geologic processes. More precisely, a formation is a mappable rock unit that has definite boundaries (or contacts with other units) and certain obvious characteristics (rock type) by which it may be traced from place to place and distinguished from other rock units.

Figure 10.A shows several named formations that are exposed in the walls of the Grand Canyon. Just as these rock strata in the Grand Canyon were subdivided, geologists subdivide rock sequences throughout the world into formations.

Those who have had the opportunity to travel to some of the national parks in the West may already be familiar with the names of certain formations. Well-known formations include the Navajo Sandstone in Zion National Park, the Redwall Limestone in the Grand Canyon, the Entrada Sandstone in Arches National Park, and the Wasatch Formation in Bryce Canyon National Park.

FIGURE 10.A   Grand Canyon with a few of its rock units (formations) named. (Photo by E. J. Tarbuck)

Although formations can consist of igneous or metamorphic rocks, the vast majority of established formations are sedimentary rocks. A formation may be relatively thin and composed of a single rock type, for example, a 1-meter-thick layer of limestone. At the other extreme, formations can be thousands of meters thick and consist of an interbedded sequence of rock types such as sandstones and shales. The most important condition to be met when establishing a formation is that *it constitutes a unit of rock produced by uniform or uniformly alternating conditions.*

In most regions of the world, the name of each formation consists of two parts— for example, the Oswego Sandstone and the Carmel Formation. The first part of the name is generally taken from a geologic structure or a locality where the formation is clearly and completely exposed. For instance, the expansive Morrison Formation is well exposed at Morrison, Colorado. As a result, this particular exposure is known as the *type locality*. Ideally, the second part of the name indicates the dominant rock type as exemplified by such names as the Dakota Sandstone, the Kaibab Limestone, and the Burgess Shale. When no single rock type dominates, the term *formation* is used, such as the well-known Chinle Formation, exposed in Arizona's Petrified Forest National Park.

In summary, describing and naming formations is an important first step in the process of organizing and simplifying the study and analysis of Earth's history.

In the field, geologists measure the strike (trend) and dip (inclination) of sedimentary rocks at as many outcrops as practical. These data are then plotted on a topographic map or an aerial photograph along with a color-coded description of the rock. From the orientation of the strata, an inferred orientation and shape of the structure can be established, as shown in Figure 10.8. Using this information, the geologist can reconstruct the pre-erosional structures and begin to interpret the region's geologic history.

Folds

 Crustal Deformation
▶ Folds

During mountain building, flat-lying sedimentary and volcanic rocks are often bent into a series of wavelike undulations called **folds.** Folds in sedimentary strata are much like those that would form if you were to hold the ends of a

A. Map view

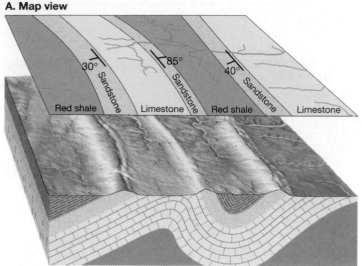

B. Block diagram

FIGURE 10.8 By establishing the strike and dip of outcropping sedimentary beds on a map (A), geologists can infer the orientation of the structure below ground (B).

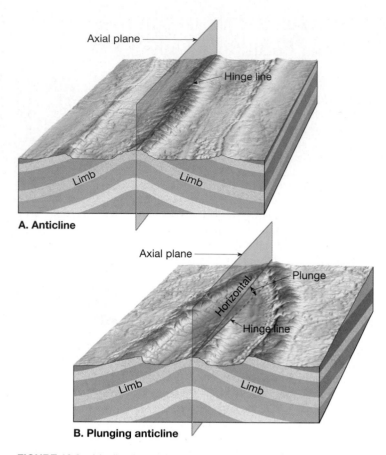

FIGURE 10.9 Idealized sketches illustrating the features associated with symmetrical folds. The hinge line of the fold in **A** is horizontal, whereas the hinge line of the fold in **B** is plunging.

sheet of paper and then push them together. In nature, folds come in a wide variety of sizes and configurations. Some folds are broad flexures in which rock units hundreds of meters thick have been slightly warped. Others are very tight microscopic structures found in metamorphic rocks. Size differences notwithstanding, most folds are the result of compressional stresses that result in the shortening and thickening of the crust. Occasionally, folds are found singly, but most often they occur as a series of undulations.

To aid our understanding of folds and folding, we need to become familiar with the terminology used to name the parts of a fold. As shown in Figure 10.9, the two sides of a fold are called *limbs*. A line drawn along the points of maximum curvature of each layer is termed the hinge line, or simply the *hinge*. In some folds, as Figure 10.9A illustrates, the hinge is horizontal, or parallel to the surface. However, in more complex folding, the hinge is often inclined at an angle known as the *plunge* (Figure 10.9B). Further, the *axial plane* is an imaginary surface that divides a fold as symmetrically as possible.

Types of Folds

The two most common types of folds are called anticlines and synclines (Figure 10.10). An **anticline** is most commonly formed by the upfolding, or arching, of rock layers.* Figure 10.9 is an example of an anticline. Anticlines are sometimes spectacularly displayed where highways have been cut through deformed strata. Often found in associa-

tion with anticlines are downfolds, or troughs, called **synclines.** Notice in Figure 10.11 that the limb of an anticline is also a limb of the adjacent syncline.

Depending on their orientation, these basic folds are described as *symmetrical* when the limbs are mirror images of each other and *asymmetrical* when they are not. An asymmetrical fold is said to be *overturned* if one limb is tilted beyond the vertical (Figure 10.10). An overturned fold can also "lie on its side" so that a plane extending through the axis of the fold actually would be horizontal. These *recumbent* folds are common in mountainous regions such as the Alps (Figure 10.12).

Folds do not continue forever; rather, their ends die out much like the wrinkles in cloth. Some folds *plunge* because the axis of the fold penetrates into the ground (Figure 10.13A, B). As the figure shows, both anticlines and synclines can plunge. Figure 10.13C shows an example of a plunging anticline and the pattern produced when erosion removes the upper layers of the structure and exposes its interior. Note that the outcrop pattern of an anticline points in the direction it is plunging, whereas the opposite is true for a syncline. A good example of the kind of topography that results when erosional forces attack folded sedimentary strata is found in the Valley and Ridge Province of the Appalachians (see Figure 14.13).

---

*By strict definition, an anticline is a structure in which the oldest strata are found in the center. This most typically occurs when strata are upfolded. Further, a syncline is strictly defined as a structure in which the youngest strata are found in the center. This occurs most commonly when strata are downfolded.

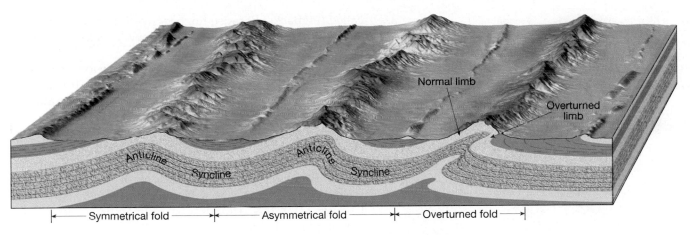

Symmetrical fold ⟶ ⟵ Asymmetrical fold ⟶ ⟵ Overturned fold ⟶

FIGURE 10.10 Block diagram of principal types of folded strata. The upfolded, or arched, structures are anticlines. The downfolds, or troughs, are synclines. Notice that the limb of an anticline is also the limb of the adjacent syncline.

It is important to understand that ridges are not necessarily associated with anticlines, nor are valleys related to synclines. Rather, ridges and valleys result because of differential weathering and erosion. For example, in the Valley and Ridge Province, resistant sandstone beds remain as imposing ridges separated by valleys cut into more easily eroded shale or limestone beds.

Although we have separated our discussion of folds and faults, in the real world folds are generally intimately coupled with faults. Examples of this close association are broad, regional features called *monoclines*. Particularly prominent features of the Colorado Plateau, **monoclines** (*mono* = one, *kleinen* = incline) are large, steplike folds in otherwise horizontal sedimentary strata (Figure 10.14). These folds appear to be the result of the reactivation of steeply dipping fault zones located in basement rocks beneath the plateau. As large blocks of basement rock were displaced upward along ancient faults, the comparatively ductile sedimentary strata above responded by folding. On the Colorado Plateau, monoclines display a narrow zone of steeply inclined beds that flatten out to form the uppermost layers of large elevated areas, including the Zuni Uplift, Echo Cliffs Uplift, and San Rafael Swell (Figure 10.14). Displacement along these reactivated faults often exceeds

1 kilometer (0.6 miles), and the very largest monoclines exhibit displacements that approach 3 kilometers (2 miles).

Domes and Basins

Broad upwarps in basement rock may deform the overlying cover of sedimentary strata and generate large folds. When this upwarping produces a circular or elongated structure,

FIGURE 10.12 Recumbent folds in the Swiss Alps. (Photo by Mike Andrews)

FIGURE 10.11 Syncline (left) and anticline (right) share a common limb. (Photo by E. J. Tarbuck).

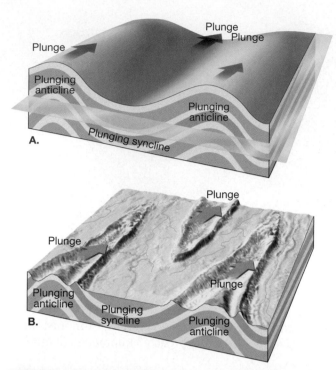

C.

FIGURE 10.13 Plunging folds. **A.** Idealized view of plunging folds in which a horizontal surface has been added. **B.** View of plunging folds as they might appear after extensive erosion. Notice that in a plunging anticline the outcrop pattern "points" in the direction of the plunge, while the opposite is true of plunging synclines. **C.** Sheep Mountain, a doubly plunging anticline. (Photo by John S. Shelton)

the feature is called a **dome** (Figure 10.15A). Downwarped structures having a similar shape are termed **basins** (Figure 10.15B).

The Black Hills of western South Dakota is a large domed structure thought to be generated by upwarping. Here erosion has stripped away the highest portions of the un-warped sedimentary beds, exposing older igneous and metamorphic rocks in the center (Figure 10.16). Remnants of these once continuous sedimentary layers are visible, flank-ing the crystalline core of this mountain range. The more re-sistant strata are easy to identify because differential erosion has left them outcropping as prominent angular ridges

called **hogbacks.** Because hogbacks can form whenever re-sistant strata are steeply inclined, they are also associated with other types of folds.

Domes can also be formed by the intrusion of magma (laccoliths) as shown in Figure 5.32. In addition, the upward migration of salt formations can produce salt domes that are common in the Gulf of Mexico.

Several large basins exist in the United States (Figure 10.17). The basins of Michigan and Illinois have very gently sloping beds similar to saucers. These basins are thought to be the result of large accumulations of sediment, whose weight caused the crust to subside (see section on isostasy in Chapter 14). A few structural basins may have been the re-sult of giant asteroid impacts.

Because large basins contain sedimentary beds sloping at such low angles, they are usually identified by the age of the rocks composing them. The youngest rocks are found near the center, and the oldest rocks are at the flanks. This is just the opposite order of a domed structure, such as the Black Hills, where the oldest rocks form the core.

Faults

GEODe Crustal Deformation
▶ Faults and Fractures

Faults are fractures in the crust along which appreciable dis-placement has taken place. Occasionally, small faults can be recognized in road cuts where sedimentary beds have been offset a few meters, as shown in Figure 10.18. Faults of this scale usually occur as single discrete breaks. By contrast, large faults, like the San Andreas Fault in California, have displacements of hundreds of kilometers and consist of many interconnecting fault surfaces. These *fault zones* can be several kilometers wide and are often easier to identify from high-altitude photographs than at ground level.

Students Sometimes Ask . . .

How do geologists determine which side of a fault has moved?

Surprisingly, for many faults it cannot be definitively estab-lished. For example, in the picture of a fault shown in Figure 10.18, did the left side move down, or did the right side move up? Since the surface (at the top of the photo) has been eroded flat, either side could have moved; or both could have moved, with one side moving more than the other (for instance, they both may have moved up, it's just that the right side moved up more than the left side). That's why geologists talk about *relative* motion across faults. In this case, the left side moved down *relative to* the right side, and the right side moved up *relative to* the left side (note arrows on photo).

FIGURE 10.14 Monocline. **A.** The San Rafael monocline, Utah. (Photo by Stephen Trimble) **B.** Monocline consisting of bent sedimentary beds that were deformed by faulting in the bedrock below. The thrust fault in this diagram is called a *blind thrust* because it does not reach the surface.

Sudden movements along faults are the cause of most earthquakes. However, the vast majority of faults are inactive and thus are remnants of past deformation. The rock along faults often becomes broken and pulverized as crustal blocks on opposite sides of a fault grind past one another. The loosely coherent, clayish material that results from this activity is called *fault gouge*. On some fault surfaces the rocks become highly polished and striated, or grooved, as the crustal blocks slide past one another. These polished and striated surfaces, called *slickensides* (*slicken* = smooth), provide geologists with evidence for the direction of the most recent displacement along the fault. Geologists classify faults by these relative movements, which can be predominantly horizontal, vertical, or oblique.

Dip-Slip Faults

Faults in which the movement is primarily parallel to the dip (or inclination) of the fault surface are called **dip-slip faults**. Vertical displacements along dip-slip faults may produce long, low cliffs called **fault scarps** (*scarpe* = a slope). Fault scarps, such as the one shown in Figure 10.19, are produced by displacements that generate earthquakes.

It has become common practice to call the rock surface that is immediately above the fault the *hanging wall* and to call the rock surface below, the *footwall* (Figure 10.20). This nomenclature arose from prospectors and miners who excavated shafts and tunnels along fault zones because these are frequently sites of ore deposits. In these tunnels, the miners would walk

FIGURE 10.15 Gentle upwarping and downwarping of crustal rocks produce domes (A) and basins (B). Erosion of these structures results in an outcrop pattern that is roughly circular or elongate.

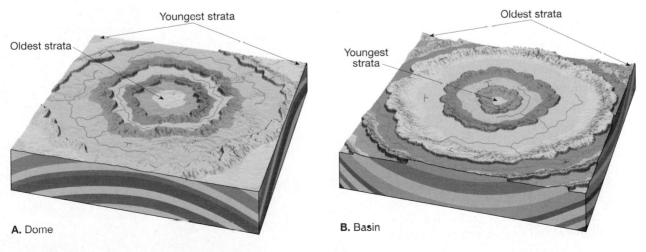

A. Dome

B. Basin

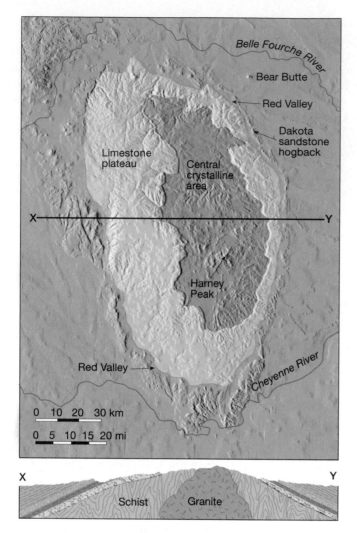

FIGURE 10.16   The Black Hills of South Dakota, a large domal structure with resistant igneous and metamorphic rocks exposed in the core.

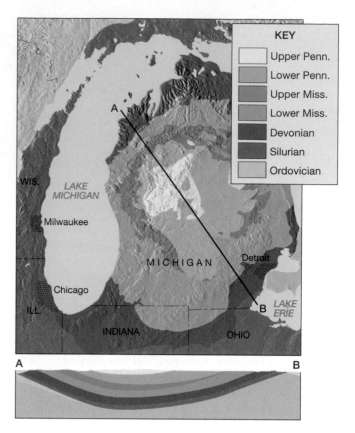

FIGURE 10.17   The bedrock geology of the Michigan Basin. Notice that the youngest rocks are centrally located, while the oldest beds flank this structure.

Examples of fault-block mountains include the Teton Range of Wyoming and the Sierra Nevada of California. Both are faulted along their eastern flanks, which were uplifted as the blocks tilted downward to the west. These precipitous mountain fronts were produced over a period of 5 million to 10 million years by many irregularly spaced

FIGURE 10.18   Faulting caused the vertical displacement of these beds located near Kanab, Utah. Arrows show relative motion of rock units. (Photo by Tom Bean/DRK Photo)

on the rocks below the mineralized fault zone (the footwall) and hang their lanterns on the rocks above (the hanging wall).

Two major types of dip-slip faults are **normal faults** and **reverse faults.** In addition, when a reverse fault has an angle of dip (inclination) less than 45°, it is called a *thrust fault.* We will consider these types of dip-slip faults next.

Normal Faults   Dip-slip faults are classified as normal faults when the hanging wall block moves down relative to the footwall block (Figure 10.21). Most normal faults have steep dips of about 60°, which tend to flatten out with depth. However, some dip-slip faults have much lower dips, with some approaching horizontal. Because of the downward motion of the hanging wall, normal faults accommodate lengthening, or extension, of the crust.

Most normal faults are small, having displacements of only a meter or so, like the one shown in the road cut in Figure 10.18. Others extend for tens of kilometers where they may sinuously trace the boundary of a mountain front. In the western United States, large-scale normal faults like these are associated with structures called **fault-block mountains.**

FIGURE 10.19   A fault scarp located near Joshua Tree National Monument, California. (Photo by A. P. Trujillo/APT Photos)

episodes of faulting. Each event was responsible for just a few meters of displacement.

Other excellent examples of fault-block mountains are found in the Basin and Range Province, a region that encompasses Nevada and portions of the surrounding states (Figure 10.22). Here the crust has been elongated and broken to create more than 200 relatively small mountain ranges. Averaging about 80 kilometers in length, the ranges rise 900 to 1500 meters above the adjacent down-faulted basins.

The topography of the Basin and Range Province has been generated by a system of roughly north to south trending normal faults. Movements along these faults have pro-

duced alternating uplifted fault blocks called **horsts** and down-dropped blocks called **grabens** (*graben* = ditch). Horsts generate elevated ranges, whereas grabens form many of the basins. As Figure 10.22 illustrates, structures called **half-grabens,** which are tilted fault blocks, also contribute to the alternating topographic highs and lows in the Basin and Range Province. The horsts and higher ends of

FIGURE 10.21   Block diagrams illustrating a normal fault. **A.** Rock strata prior to faulting. **B.** The relative movement of displaced blocks. Displacement may continue in a fault-block mountain range over millions of years and consist of many widely spaced episodes of faulting. **C.** How erosion might alter the upfaulted block. **D.** Eventually the period of deformation ends and erosion becomes the dominant geologic process.

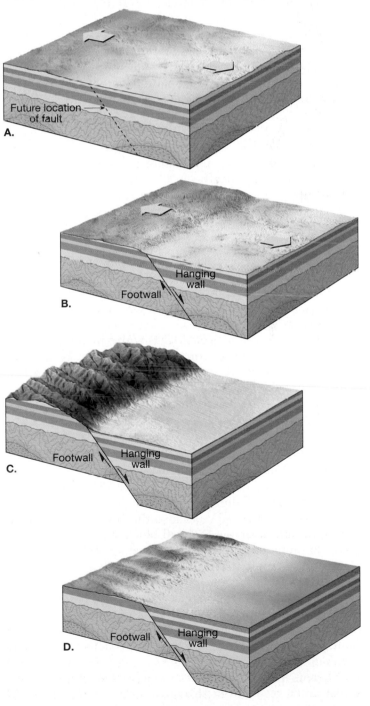

FIGURE 10.20   The rock immediately above a fault surface is the *hanging wall*, and that below is called the *footwall*. These names came from miners who excavated ore along fault zones. The miners hung their lanterns on the rocks above the fault trace (hanging wall) and walked on the rocks below the fault trace (footwall).

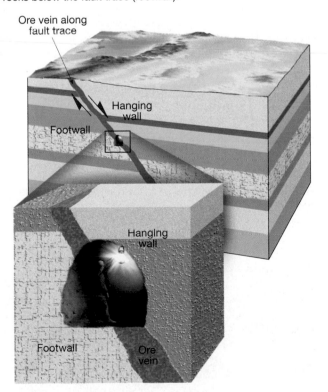

the tilted fault blocks are the source for sediments that have accumulated in the basins that were created by the grabens and lower ends of the tilted blocks.

Notice in Figure 10.22 that the slopes of the normal faults in the Basin and Range Province decrease with depth and eventually join together to form a nearly horizontal fault called a **detachment fault.** These faults may extend for hundreds of kilometers below the surface. Here they form a major boundary between the rocks below, which exhibit ductile deformation, and the rocks above, which demonstrate brittle deformation via faulting.

Normal faulting is also prevalent at spreading centers where plate divergence occurs. Here, a central block (graben) is bounded by normal faults and subsides as the plates separate. These grabens produce an elongated valley bounded by uplifted fault blocks (horsts).

The Rift Valley of East Africa is made up of several large grabens, above which tilted horsts produce a linear mountainous topography. This valley, nearly 6000 kilometers (3600 miles) long, contains the excavation sites of some of the earliest human fossils. Examples of inactive rift valleys include the Rhine Valley in Germany and the Triassic-age grabens of the eastern United States. Even larger inactive normal fault systems are the rifted continental margins, such as along the east coasts of the Americas and the west coasts of Europe and Africa.

FIGURE 10.22   Normal faulting in the Basin and Range Province. Here, tensional stresses have elongated and fractured the crust into numerous blocks. Movement along these fractures has tilted the blocks, producing parallel mountain ranges called fault-block mountains. The downfaulted blocks (grabens) form basins, whereas the upfaulted blocks (horsts) are eroded to form rugged mountainous topography. In addition, numerous tilted blocks (half-grabens) form both basins and mountains. (Photo by Michael Collier)

Fault motion provides geologists with a method of determining the nature of the forces at work within Earth. Normal faults indicate the existence of extensional forces that pull the crust apart. This "pulling apart" can be accomplished either by uplifting that causes the surfaces to stretch and break or by opposing horizontal forces.

Reverse and Thrust Faults   Reverse faults and **thrust faults** are dip-slip faults in which the hanging wall block moves up relative to the footwall block (Figure 10.23). Recall that reverse faults have dips greater than 45° and thrust faults have dips less than 45°. Because the hanging wall block moves up and over the footwall block, reverse and thrust faults accommodate horizontal shortening of the crust.

Most high-angle reverse faults are small and accommodate local displacements in regions dominated by other types of faulting. Thrust faults, on the other hand, exist at all scales. Small thrust faults exhibit displacement on the order of millimeters to a few meters. Some large thrust faults have displacements on the order of tens to hundreds of kilometers.

Whereas normal faults occur in tensional environments, thrust faults result from strong compressional stresses. In these settings, crustal blocks are displaced *toward* one another, with the hanging wall being displaced upward relative to the footwall. Thrust faulting is most pronounced in subduction zones and other convergent boundaries where plates are colliding. Compressional forces generally produce folds as well as faults and result in a thickening and shortening of the material involved.

In mountainous regions such as the Alps, Northern Rockies, Himalayas, and Appalachians, thrust faults have displaced

FIGURE 10.23   Block diagram showing the relative movement along a reverse fault.

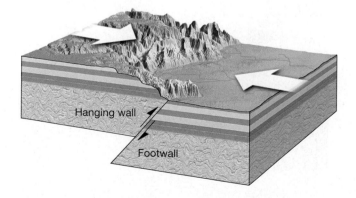

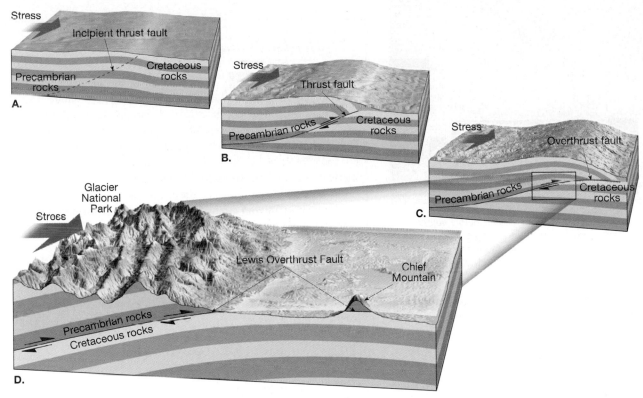

FIGURE 10.24 Idealized development of Lewis Overthrust fault. **A.** Geologic setting prior to deformation. **B., C.** Large-scale movement along a thrust fault displaced Precambrian rock over Cretaceous strata in the region of Glacier National Park. **D.** Erosion by glacial ice and running water sculptured the thrust sheet into a majestic landscape and isolated a remnant of the thrust sheet called Chief Mountain.

strata as far as 50 kilometers over adjacent rock units. The result of this large-scale movement is that older strata end up overlying younger rocks. A classic site of thrust faulting occurs in Glacier National Park (Figure 10.24). Here, mountain peaks that provide the park's majestic scenery have been carved from Precambrian rocks that were displaced over much younger Cretaceous strata. At the eastern edge of Glacier National Park, there is an outlying peak called Chief Mountain. This structure is an isolated remnant of the thrust sheet that was severed by the erosional forces of glacial ice and running water. An isolated block, such as Chief Mountain, is called a **klippe** (*klippe* = cliff) (Figure 10.25).

Strike-Slip Faults

Faults in which the dominant displacement is horizontal and parallel to the strike of the fault surface are called **strike-slip faults.** Because of their large size and linear nature, many strike-slip faults produce a trace that is visible over a great distance (Figure 10.26). Rather than a single fracture along which movement takes place, large strike-slip faults consist of a zone of roughly parallel fractures. The zone may be up to several kilometers in width. The most recent movement, however, is often along a strand only a few meters wide, which may offset features such as stream channels (Figure 10.27). Furthermore,

FIGURE 10.25 Chief Mountain, Glacier National Park, Montana, is a klippe. (Photo by David Muench)

FIGURE 10.26 Aerial view of strike-slip (right-lateral) fault in southern Nevada. Notice that the ridge of white rock in the top right portion of the photo was offset to the right relative to the portion of the same ridge that appears on the bottom left of the image. (Photo by Marli Miller)

Students Sometimes Ask . . .

Has anyone ever seen a fault scarp forming?

Amazingly, yes. There have been several instances where people have been at the fortuitously appropriate place and time to observe the creation of a fault scarp—and have lived to tell about it. In Idaho a large earthquake in 1983 created a 3-meter (10-foot) fault scarp that was witnessed by several people, many of whom were knocked off their feet. More often, though, fault scarps are noticed *after* they form. For example, a 1999 earthquake in Taiwan created a fault scarp that formed a new waterfall and destroyed a nearby bridge.

crushed and broken rocks produced during faulting are more easily eroded, often producing linear valleys or troughs that mark the locations of strike-slip faults.

The earliest scientific records of strike-slip faulting were made following surface ruptures that produced large earthquakes. One of the most noteworthy of these was the great San Francisco earthquake of 1906. During this strong earthquake, structures such as fences that were built across the San Andreas Fault were displaced as much as 4.7 meters (15 feet). Because the movement along the San Andreas causes the crustal block on the opposite side of the fault to move to the right as you face the fault, it is called a *right-lateral* strike-slip fault. The Great Glen fault in Scotland is a well-known example of a *left-lateral* strike-slip fault, which exhibits the opposite sense of displacement. The total displacement along the Great Glen fault is estimated to exceed 100 kilometers (60 miles). Also associated with this fault trace are numerous lakes, including Loch Ness, the home of the legendary monster.

Many major strike-slip faults cut through the lithosphere and accommodate motion between two large crustal plates. Recall that this special kind of strike-slip fault is called a **transform** (*trans* = across, *forma* = form) **fault.** Numerous transform faults cut the oceanic lithosphere and link spreading oceanic ridges. Others accommodate displacement between continental plates that move horizontally with respect to each other. One of the best-known transform faults is California's San Andreas Fault (see Box 10.2). This plate-bounding fault can be traced for about 950 kilometers (600 miles) from the Gulf of California to a point along the Pacific Coast north of San Francisco, where it heads out to sea. Ever since its formation, about 29 million years ago, displacement along the San Andreas Fault has exceeded 560 kilometers. This movement has accommodated the northward displacement of southwestern California and the Baja Peninsula of Mexico in relation to the remainder of North America.

Joints

 Crustal Deformation
▸ Faults and Fractures

Among the most common rock structures are fractures called joints. Unlike faults, **joints** are fractures along which no appreciable displacement has occurred. Although some joints have a random orientation, most occur in roughly parallel groups (see Figure 6.6).

We have already considered two types of joints. Earlier we learned that *columnar joints* form when igneous rocks cool and develop shrinkage fractures that produce elongated, pillarlike columns (Figure 10.28). Also recall that sheeting produces a

FIGURE 10.27 Block diagram illustrating the features associated with strike-slip faults. Note how the stream channels have been offset by fault movement. The faults in this diagram are right-lateral strike-slip faults. (Modified after R. L. Wesson and others)

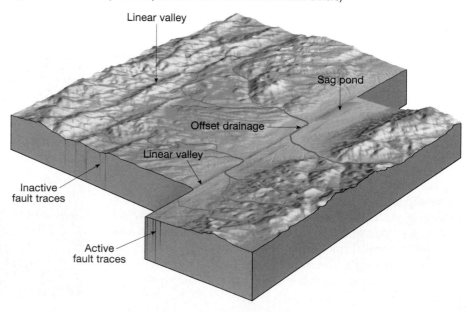

Linear valley

Sag pond

Offset drainage

Linear valley

Inactive fault traces

Active fault traces

Do faults exhibit only strike-slip or dip-slip motion?

No. Strike-slip faults and dip-slip faults are on the opposite ends of a spectrum of fault structures. Faults that exhibit a combination of dip-slip and strike-slip movements are called *oblique-slip faults*. Although most faults could technically be classified as oblique-slip, they predominately exhibit either strike-slip or dip-slip motion.

pattern of gently curved joints that develop more or less parallel to the surface of large exposed igneous bodies such as batholiths. Here the jointing results from the gradual expansion that occurs when erosion removes the overlying load.

In contrast to the situations just described, most joints are produced when rocks in the outermost crust are deformed. Here tensional and shearing stresses associated with crustal movements cause the rock to fail by brittle fracture. For example, when folding occurs, rocks situated at the axes of the folds are elongated and pulled apart to produce tensional

joints. Extensive joint patterns can also develop in response to relatively subtle and often barely perceptible regional upwarping and downwarping of the crust. In many cases, the cause for jointing at a particular locale is not readily apparent.

Many rocks are broken by two or even three sets of intersecting joints that slice the rock into numerous regularly shaped blocks. These joint sets often exert a strong influence on other geologic processes. For example, chemical weathering tends to be concentrated along joints, and in many areas groundwater movement and the resulting dissolution in soluble rocks is controlled by the joint pattern (Figure 10.29). Moreover, a system of joints can influence the direction that stream courses follow. The rectangular drainage pattern described in Chapter 16 is such a case.

Joints may also be significant from an economic standpoint. Some of the world's largest and most important mineral deposits were emplaced along joint systems. Hydrothermal solutions, which are basically mineralized fluids, can migrate into fractured host rocks and precipitate economically important amounts of copper, silver, gold, zinc, lead, and uranium.

Further, highly jointed rocks present a risk to the construction of engineering projects, including highways and dams. On June 5, 1976, 14 lives were lost and nearly 1 billion dollars in property damage occurred when the Teton Dam in Idaho

FIGURE 10.28 Devil's Tower National Monument is in eastern Wyoming. Established in September 1906, it was our nation's first national monument. This nearly vertical monolith rises more than 380 meters (1265 feet) above the surrounding grasslands and pine forests. This imposing structure exhibits columnar joints and the columns that result. Columnar joints form as igneous rocks cool and develop shrinkage fractures that produce five- to seven-sided elongated, pillarlike structures. (Photo by Bob Thomason/Tony Stone Images)

FIGURE 10.29 Chemical weathering is enhanced along joints in granitic rocks near the top of Lembert Dome in Yosemite National Park, California. (Photo by E. J. Tarbuck)

BOX 10.2 ▶ PEOPLE AND THE ENVIRONMENT

The San Andreas Fault System

The San Andreas, the best-known and largest fault system in North America, first attracted wide attention after the great 1906 San Francisco earthquake and fire. Following this devastating event, geologic studies demonstrated that a displacement of as much as 5 meters along the fault had been responsible for the earthquake. It is now known that this dramatic event is just one of many thousands of earthquakes that have resulted from repeated movements along the San Andreas throughout its 29-million-year history.

Where is the San Andreas fault system located? As shown in Figure 10.B, it trends in a northwesterly direction for nearly 1300 kilometers (780 miles) through much of western California. At its southern end, the San Andreas connects with a spreading center located in the Gulf of California. In the north, the fault enters the Pacific Ocean at Point Arena, where it is thought to continue its northwesterly trend, eventually joining the Mendocino fracture zone. In the central section, the San Andreas is relatively simple and straight. However, at its two extremi-

ties, several branches spread from the main trace, so that in some areas the fault zone exceeds 100 kilometers (60 miles) in width.

Over much of its extent, a linear trough reveals the presence of the San Andreas Fault. When the system is viewed from the air, linear scars, offset stream channels, and elongated ponds mark the trace in a striking manner. On the ground, however, surface expressions of the faults are much more difficult to detect. Some of the most distinctive landforms include long, straight escarpments, narrow ridges, and sag ponds

FIGURE 10.B Map showing the extent of the San Andreas Fault system. Inset shows only a few of the many splinter faults that are part of this great fault system.

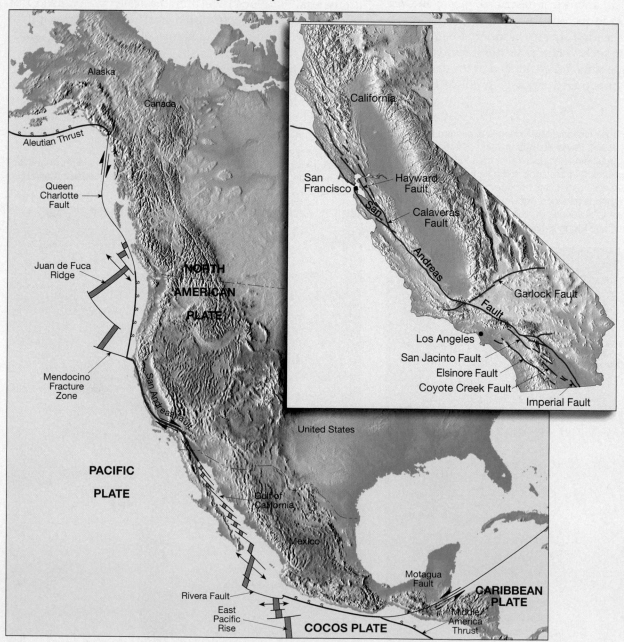

formed by settling of blocks within the fault zone. Further, many stream channels characteristically bend sharply to the right where they cross the fault (Figure 10.C).

With the advent of the theory of plate tectonics, geologists began to realize the significance of this great fault system. The San Andreas Fault is a transform boundary separating two crustal plates that move very slowly. The Pacific plate, located to the west, moves northwestward relative to the North American plate, causing earthquakes along the fault (Table 10.A).

The San Andreas is undoubtedly the most studied of any fault system in the world. Although many questions remain unanswered, geologists have learned that each fault segment exhibits somewhat different behavior. Some portions of the San Andreas exhibit a slow creep with little noticeable seismic activity. Other segments regularly slip, producing small earthquakes, while still other segments may store elastic energy for 200 years or more before rupturing to generate a great earthquake. This knowledge is useful when assigning earthquake hazard potential to a given segment of the fault zone.

Because of the great length and complexity of the San Andreas Fault, it is more appropriately referred to as a "fault sys-

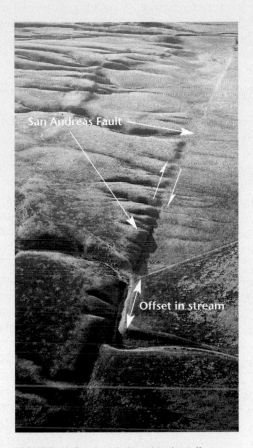

FIGURE 10.C   Aerial view showing offset stream channel across the San Andreas Fault on the Carrizo Plain west of Taft, California. (Photo by Michael Collier/DRK Photo)

tem." This major fault system consists primarily of the San Andreas Fault and several major branches, including the Hayward and Calaveras faults of central California and the San Jacinto and Elsinore faults of southern California (Figure 10.B). These major segments, plus a vast number of smaller faults that include the Imperial Fault, San Fernando Fault, and the Santa Monica Fault, collectively accommodate the relative motion between the North American and Pacific plates.

Blocks on opposite sides of the San Andreas Fault move horizontally in opposite directions, such that if a person stood on one side of the fault, the block on the opposite side would appear to move to the right when slippage occurred. This type of displacement is known as *right-lateral strike-slip* by geologists (Figure 10.C).

Ever since the great San Francisco earthquake of 1906, when as much as 5 meters of displacement occurred, geologists have attempted to establish the cumulative displacement along this fault over its 29-million-year history. By matching rock units across the fault, geologists have determined that the total accumulated displacement from earthquakes and creep exceeds 560 kilometers (340 miles).

TABLE 10.A	Major Earthquakes on the San Andreas Fault System		
Date	Location	Magnitude	Remarks
1812	Wrightwood, CA	7	Church of San Juan Capistrano collapsed, killing 40 worshippers.
1812	Santa Barbara channel	7	Churches and other buildings wrecked in and around Santa Barbara.
1838	San Francisco peninsula	7	At one time thought to have been comparable to the great earthquake of 1906.
1857	Fort Tejon, CA	8.25	One of the greatest U.S. earthquakes. Occurred near Los Angeles, then a city of 4000.
1868	Hayward, CA	7	Rupture of the Hayward fault caused extensive damage in San Francisco Bay area.
1906	San Francisco, CA	8.25	The great San Francisco earthquake. As much as 80 percent of the damage caused by fire.
1940	Imperial Valley	7.1	Displacement on the newly discovered Imperial fault.
1952	Kern County	7.7	Rupture of the White Wolf fault. Largest earthquake in California since 1906. Sixty million dollars in damages and 12 people killed.
1971	San Fernando Valley	6.5	One-half billion dollars in damages and 58 lives claimed.
1989	Santa Cruz Mountains	7.1	Loma Prieta earthquake. Six billion dollars in damages, 62 lives lost, and 3757 people injured.
1994	Northridge (Los Angeles area)	6.9	Over 15 billion dollars in damages, 51 lives lost, and over 5000 injured.

failed. This earthen dam was constructed of very erodible clays and silts and was situated on highly fractured volcanic rocks. Although attempts were made to fill the voids in the jointed rock, water gradually penetrated the subsurface frac-tures and undermined the dam's foundation. Eventually the moving water cut a tunnel into the easily erodible clays and silts. Within minutes the dam failed, sending a 20-meter-high wall of water down the Teton and Snake rivers.

Summary

- *Deformation* refers to changes in the shape and/or volume of a rock body and is most pronounced along plate margins. To describe the forces that deform rocks, geologists use the term *stress*, which is the amount of force applied to a given area. Stress that is uniform in all directions is called *confining pressure*, whereas *differential stresses* are applied unequally in different directions. Differential stresses that shorten a rock body are *compressional stresses*; those that elongate a rock unit are *tensional stresses. Strain* is the change in size and shape of a rock unit caused by stress.

- Rocks deform differently depending on the environment (temperature and confining pressure), the composition and texture of the rock, and the length of time stress is maintained. Rocks first respond by deforming *elastically* and will return to their original shape when the stress is removed. Once their elastic limit (strength) is surpassed, rocks either deform by ductile flow or they fracture. *Ductile deformation* is a solid-state flow that results in a change in size and shape of an object without fracturing. Ductile flow may be accomplished by gradual slippage and recrystallization along planes of weakness within the crystal lattice of mineral grains. Ductile deformation occurs in a high-temperature/high-pressure environment. In a near-surface environment, most rocks deform by *brittle failure*.

- The orientation of rock units or fault surfaces is established with measurements called strike and dip. *Strike* is the compass direction of a line produced by the intersection of an inclined rock layer or fault with a horizontal plane. *Dip* is the angle of inclination of the surface of a rock unit or fault measured from a horizontal plane.

- The most basic geologic structures associated with rock deformation are *folds* (flat-lying sedimentary and volcanic rocks bent into a series of wavelike undulations) and *faults*. The two most common types of folds are *anticlines*, formed by the upfolding, or arching, of rock layers, and *synclines*, which are downfolds. Most folds are the result of horizontal *compressional stresses*. Folds can be *symmetrical, asymmetrical*, or, if one limb has been tilted beyond the vertical, *overturned. Domes* (upwarped structures) and *basins* (downwarped structures) are circular or somewhat elongated folds formed by vertical displacements of strata.

- Faults are fractures in the crust along which appreciable displacement has occurred. Faults in which the movement is primarily vertical are called *dip-slip faults*. Dip-slip faults include both *normal* and *reverse faults*. Low-angle reverse faults are called *thrust faults*. Normal faults indicate *tensional stresses* that pull the crust apart. Along the spreading centers of plates, divergence can cause a central block called a *graben*, bounded by normal faults, to drop as the plates separate.

- Reverse and thrust faulting indicate that *compressional forces* are at work. Large *thrust faults* are found along subduction zones and other convergent boundaries where plates are colliding. In mountainous regions such as the Alps, Northern Rockies, Himalayas, and Appalachians, thrust faults have displaced strata as far as 50 kilometers over adjacent rock units.

- *Strike-slip faults* exhibit mainly horizontal displacement parallel to the strike of the fault surface. Large strike-slip faults, called *transform faults*, accommodate displacement between plate boundaries. Most transform faults cut the oceanic lithosphere and link spreading centers. The San Andreas Fault cuts the continental lithosphere and accommodates the northward displacement of southwestern California.

- *Joints* are fractures along which no appreciable displacement has occurred. Joints generally occur in groups with roughly parallel orientations and are the result of brittle failure of rock units located in the outermost crust.

Review Questions

1. What is rock deformation? How does a rock body change during deformation?

2. List five (5) geologic structures that are associated with deformation.

3. How is *stress* related to *force*?

4. Contrast compressional and tensional stresses.

5. Describe how shearing can deform a rock in a near-surface environment.

6. Compare strain and stress.

7. How is brittle deformation different from ductile deformation?

8. List three factors that determine how rocks will behave when exposed to stresses that exceed their strength. Briefly explain the role of each.

9. What is an outcrop?

10. What two measurements are used to establish the orientation of deformed strata? Distinguish between them.

11. Distinguish between anticlines and synclines, domes and basins, anticlines and domes.

12. How is a monocline different from an anticline?

13. The Black Hills of South Dakota are a good example of what type of structural feature?

14. Contrast the movements that occur along normal and reverse faults. What type of stress is indicated by each fault?

15. Is the fault shown in Figure 10.18 a normal or a reverse fault?

16. Describe a horst and a graben. Explain how a graben valley forms and name one.

17. What type of faults are associated with fault-block mountains?

18. How are reverse faults different from thrust faults? In what way are they the same?

19. The San Andreas Fault is an excellent example of a _____ fault.

20. With which of the three types of plate boundaries does normal faulting predominate? Reverse faulting? Strike-slip faulting?

21. How are joints different from faults?

Key Terms

anticline (p. 280)
basin (p. 282)
brittle failure (p. 277)
brittle deformation (p. 277)
compressional stress (p. 275)
deformation (p. 275)
detachment fault (p. 286)
differential stress (p. 275)
dip (p. 278)
dip-slip fault (p. 293)
dome (p. 292)
ductile deformation (p. 277)
fault (p. 292)
fault-block mountain (p. 294)
fault scarp (p. 293)
fold (p. 288)
force (p. 275)
graben (p. 285)
half-graben (p. 285)
hogback (p. 282)
horst (p. 285)
joint (p. 288)
klippe (p. 287)
monocline (p. 281)
normal fault (p. 284)
reverse fault (p. 284)
rock structure (p. 278)
shear (p. 276)
strain (p. 276)
stress (p. 275)
strike (p. 278)
strike-slip fault (p. 287)
syncline (p. 280)
tensional stress (p. 276)
thrust fault (p. 286)
transform fault (p. 288)

Web Resources

The *Earth* Website uses the resources and flexibility of the Internet to aid in your study of the topics in this chapter. Written and developed by geology instructors, this site will help improve your understanding of geology. Visit **http://www.prenhall.com/tarbuck** and click on the cover of *Earth 9e* to find:

Online review quizzes.
Critical thinking exercises.
Links to chapter-specific Web resources.
Internet-wide key-term searches.
http://www.prenhall.com/tarbuck

GEODe: Earth

GEODe: Earth makes studying faster and more effective by reinforcing key concepts using animation, video, narration, interactive exercises and practice quizzes. A copy is included with every copy of *Earth*.

At greater depth where confining pressures and temperatures are high, rocks become **ductile** and deform by flowing or folding.

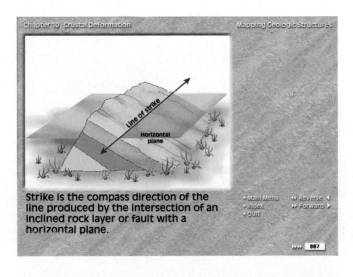

Strike is the compass direction of the line produced by the intersection of an inclined rock layer or fault with a horizontal plane.

Earthquakes

Aftermath of a devastating tsunami at the coastal town of Banda Aceh on the Indonesian island of Sumatra, December 26, 2004. (Photo by Mark Pearson/ Alamy)

295

On October 17, 1989, at 5:04 P.M. Pacific daylight time, millions of television viewers around the world were settling in to watch the third game of the World Series. Instead, they saw their television sets go black as tremors hit San Francisco's Candlestick Park. Although the earthquake was centered in a remote section of the Santa Cruz Mountains, 100 kilometers to the south, major damage occurred in the Marina District of San Francisco.

The most tragic result of the violent shaking was the collapse of some double-decked sections of Interstate 880, also known as the Nimitz Freeway. The ground motions caused the upper deck to sway, shattering the concrete support columns along a mile-long section of the freeway. The upper deck then collapsed onto the lower roadway, flattening cars as if they were aluminum cans. This earthquake, named the Loma Prieta quake for its point of origin, claimed 67 lives.

In mid-January 1994, less than five years after the Loma Prieta earthquake devastated portions of the San Francisco Bay area, a major earthquake struck the Northridge area of Los Angeles. Although not the fabled "Big One," this moderate, 6.7-magnitude earthquake left 51 dead, more than 5000 injured, and tens of thousands of households without water and electricity. The damage exceeded $40 billion and was attributed to an unknown fault that ruptured 18 kilometers (11 miles) beneath Northridge.

The Northridge earthquake began at 4:31 A.M. and lasted roughly 40 seconds. During this brief period the quake terrorized the entire Los Angeles area. In the three-story Northridge Meadows apartment complex, 16 people died when sections of the upper floors collapsed onto the first-floor units. Nearly 300 schools were seriously damaged, and a dozen major roadways buckled. Among these were two of California's major arteries—the Golden State Freeway (Interstate 5), where an overpass collapsed completely and blocked the roadway, and the Santa Monica Freeway. Fortunately, these roadways had practically no traffic at this early morning hour.

In nearby Granada Hills, broken gas lines were set ablaze while the streets flooded from broken water mains. Seventy homes burned in the Sylmar area. A 64-car freight train derailed, including some cars carrying hazardous cargo. But it is remarkable that the destruction was not greater. Unquestionably, the upgrading of structures to meet the requirements of building codes developed for this earthquake-prone area helped minimize what could have been a much greater human tragedy.

What Is an Earthquake?

GEODe Earthquakes
▶ What Is an Earthquake?

An **earthquake** is the vibration of Earth produced by the rapid release of energy (Figure 11.1). Most often, earthquakes are caused by slippage along a fault in Earth's crust. The energy released radiates in all directions from its source, called the **focus** (*foci* = a point) or **hypocenter,** in the form of waves. These waves are analogous to those produced when a stone is dropped into a calm pond (Figure

11.2). Just as the impact of the stone sets water waves in motion, an earthquake generates seismic waves that radiate throughout Earth. Even though the energy dissipates rapidly with increasing distance from the focus, sensitive instruments located around the world record the event.

More than 30,000 earthquakes that are strong enough to be felt occur worldwide annually. Fortunately, most are minor tremors and do very little damage. Generally, only about 75 significant earthquakes take place each year, and many of these occur in remote regions. However, occasionally a large earthquake occurs near a large population center. Under these conditions, an earthquake is among the most destructive natural forces on Earth.

FIGURE 11.1 Destruction caused by a major earthquake that struck northwestern Turkey on August 17, 1999. More than 17,000 people perished. (Photo by Yann Arthus-Bertrand/Peter Arnold, Inc.)

The shaking of the ground, coupled with the liquefaction of some soils, wreaks havoc on buildings and other structures. In addition, when a quake occurs in a populated area, power and gas lines are often ruptured, causing numerous fires. In the famous 1906 San Francisco earthquake, much of the damage was caused by fires (Figure 11.3). They quickly became uncontrollable when broken water mains left firefighters with only trickles of water.

Earthquakes and Faults

The tremendous energy released by atomic explosions or by volcanic eruptions can produce an earthquake, but these events are relatively weak and infrequent. What mechanism produces a destructive earthquake? Ample evidence exists that Earth is not a static planet. We know that Earth's crust has been uplifted at times, because we have found numerous ancient wave-cut benches many meters above the level of the highest tides. Other regions exhibit evidence of extensive subsidence. In addition to these vertical displacements, offsets in fence lines, roads, and other structures indicate that horizontal movement is common (Figure 11.4). These movements are usually associated with large fractures in Earth's crust called **faults.**

Typically, earthquakes occur along preexisting faults that formed in the distant past along zones of weakness in Earth's crust. Some are very large and can generate major earthquakes. One example is the San Andreas Fault, which is a transform fault boundary that separates two great sections of Earth's lithosphere: the North American plate and the Pacific plate. This extensive fault zone trends in a northwesterly direction for nearly 1300 kilometers (780 miles), through much of western California.

FIGURE 11.2 Earthquake focus and epicenter. The focus is the zone within Earth where the initial displacement occurs. The epicenter is the surface location directly above the focus.

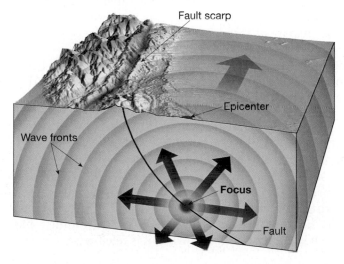

Fault scarp

Epicenter

Wave fronts

Focus

Fault

FIGURE 11.3 San Francisco in flames after the 1906 earthquake. (Reproduced from the collection of the Library of Congress) Inset photo shows fire triggered when a gas line ruptured during the Northridge earthquake in Southern California in 1994. (AFP/Getty Images)

Other faults are small and produce only minor and infrequent earthquakes. However, the vast majority of faults are inactive and do not generate earthquakes at all. Nevertheless, even faults that have been inactive for thousands of years can rupture again if the stresses acting on the region increase sufficiently.

In addition, most faults are not perfectly straight or continuous; instead, they consist of numerous branches and smaller fractures that display kinks and offsets. Such a pattern is displayed in Figure 10.B (p. 290), which shows that the San Andreas Fault is actually a system that consists of several large faults (innumerable small fractures are not shown).

Most of the motion along faults can be satisfactorily explained by the plate tectonics theory, which states that large slabs of Earth's lithosphere are in continual slow motion. These mobile plates interact with neighboring plates, straining and deforming the rocks at their margins. In fact, it is

FIGURE 11.4 Slippage along a fault produced an offset in this orange grove east of Calexico, California. (Photo by John S. Shelton)

along faults associated with plate boundaries that most earthquakes occur. Furthermore, earthquakes are repetitive: As soon as one is over, the continuous motion of the plates adds strain to the rocks until they fail again.

Discovering the Cause of Earthquakes

The actual mechanism of earthquake generation eluded geologists until H. F. Reid of Johns Hopkins University conducted a study following the great 1906 San Francisco earthquake. The earthquake was accompanied by horizontal surface displacements of several meters along the northern portion of the San Andreas Fault. Field investigations determined that during this single earthquake, the Pacific plate lurched as much as 4.7 meters (15 feet) northward past the adjacent North American plate.

The mechanism for earthquake formation that Reid deduced from this information is illustrated in Figure 11.5. In part A of the figure, you see an existing fault, or break in the rock. In part B, tectonic forces ever so slowly deform the crustal rocks on both sides of the fault, as demonstrated by the bent features. Under these conditions, rocks are bending and storing elastic energy, much like a wooden stick does if bent. Eventually, the frictional resistance holding the rocks in place is overcome. As slippage occurs at the weakest point (the focus), displacement will exert stress farther along the fault, where additional slippage will release the built-up strain (Figure 11.5C). This slippage allows the deformed rock to "snap back." The vibrations we know as an earthquake occur as the rock elastically returns to its original shape. The "springing back" of the rock was termed **elastic rebound** by Reid because the rock behaves elastically, much like a stretched rubber band does when it is released.

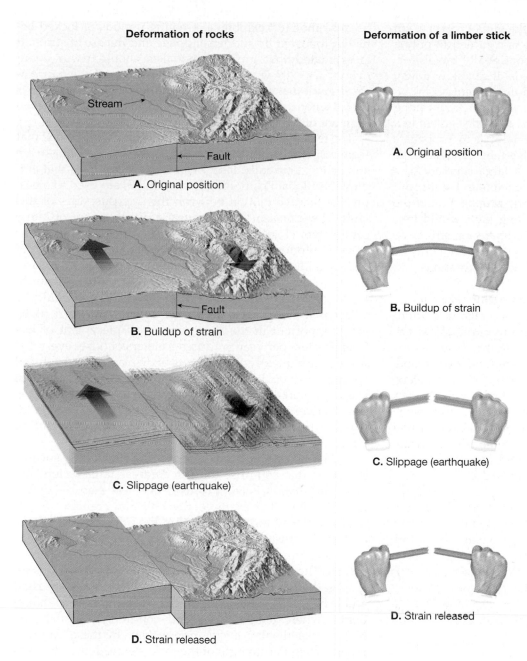

Deformation of rocks

Stream

Fault

A. Original position

Fault

B. Buildup of strain

C. Slippage (earthquake)

D. Strain released

Deformation of a limber stick

A. Original position

B. Buildup of strain

C. Slippage (earthquake)

D. Strain released

FIGURE 11.5   Elastic rebound. As rock is deformed, it bends, storing elastic energy. Once strained beyond its breaking point, the rock cracks, releasing the stored-up energy in the form of earthquake waves.

several days following the main quake. The adjustments that follow a major earthquake often generate smaller earthquakes called **after-shocks**. Although these aftershocks are usually much weaker than the main earthquake, they can sometimes destroy already badly weakened structures. This occurred, for example, during a 1988 earthquake in Armenia. A large aftershock of magnitude 5.8 collapsed many structures that had been weakened by the main tremor.

In addition, small earthquakes called **foreshocks** often precede a major earthquake by days or, in some cases, by as much as several years. Monitoring of these foreshocks has been used as a means of predicting forthcoming major earthquakes, with mixed success. We will consider the topic of earthquake prediction in a later section of this chapter.

Earthquake Rupture and Propagation

We know that the forces (stresses) that cause sudden slippage along faults are ultimately caused by the motions of Earth's plates. It is also clear that most faults are locked, except for brief, abrupt movements that accompany an earthquake rupture. The primary reason most faults are locked is that the confining pressure exerted by the overlying crust is enormous. Because of this, the fractures in the crust are essentially squeezed shut.

In summary, most earthquakes are produced by the rapid release of elastic energy stored in rock that has been subjected to great stress. Once the strength of the rock is exceeded, it suddenly ruptures, causing the vibrations of an earthquake. Earthquakes most often occur along existing faults whenever the frictional forces on the fault surfaces are overcome.

Foreshocks and Aftershocks

The intense vibrations of the 1906 San Francisco earthquake lasted about 40 seconds. Although most of the displacement along the fault occurred in this rather short period, additional movements along this and other nearby faults lasted for

Eventually, the stresses that cause the fault to rupture overcome the frictional resistance to slippage. What actually triggers the initial rupture is still not completely understood. Nevertheless, this event marks the beginning of an earthquake.

Recall that an earthquake begins at a location (at depth) along a fault plane called the focus. Although earthquakes begin at a single point, they involve the slippage along an extended fault surface. Stated another way, the initial rupture begins at the focus and propagates (travels) away from the source, sometimes in both horizontal directions along the fault, but often only in one direction. According to one model, the slippage at any one location along a fault is achieved almost instantaneously, "in the blink of an eye." In addition, at any given time, slippage is confined to only a narrow zone along the fault, which continually travels forward. As this zone of rupture proceeds, it can slow down, speed up, or even jump to a nearby fault segment.

During small earthquakes, the total slippage occurs along a comparatively small fault surface, or a small segment of a larger fault. Thus, the rupture zone is able to propagate quickly, and the earthquake is short-lived. By contrast, large earthquakes involve slippage along a large segment of a fault, occasionally a few hundred kilometers in length, and thus last much longer. For example, the propagation of the rupture zone along a 300-kilometer-long fault would take about 1.5 minutes. Therefore, the accompanying strong vibrations produced by a large earthquake would not only be stronger but would also last longer than the vibrations produced by a small earthquake.

An analogy to the propagation of an earthquake rupture can be made with the development of a crack in a windshield. Imagine a rock hits one corner of your auto's windshield, and a crack develops that rapidly travels across the windshield (a distance of 2 meters) in one-tenth of a second. Now visualize a windshield 300 kilometers (300,000 meters) in width that represents a large segment of a fault. A crack propagating from end to end of that windshield and traveling at the same rate as the crack in your windshield would be en route for roughly four hours. Obviously, an earthquake's propagation is much more rapid and its scale considerably greater than a crack in a windshield.

Having reviewed how earthquake ruptures propagate, the question that follows is, "Why do earthquakes stop rather than continuing along the entire fault?" Evidence suggests that slippage most often stops when the rupture reaches a section of the fault where the rocks have not been sufficiently strained to overcome frictional resistance. This could occur in a section of the fault that has recently experienced an earthquake. The rupture may also stop if it encounters a sufficiently large kink, or an offset along the fault plane.

San Andreas Fault: An Active Earthquake Zone

The San Andreas is undoubtedly the most studied fault system in the world. Over the years, investigations have shown that displacement occurs along discrete segments that are 100 to 200 kilometers long. Further, each fault segment behaves somewhat differently from the others. A few sections of the San Andreas exhibit a slow, gradual displacement known as **fault creep,** which occurs relatively smoothly and therefore with little noticeable seismic activity. Other segments slip at somewhat regular intervals, producing small- to moderate-sized earthquakes.

Still other segments remain locked and store elastic energy for a few hundred years before rupturing in great earthquakes. The latter process is described as *stick-slip* motion,

because the fault exhibits alternating periods of locked behavior followed by sudden slippage and release of strain. It is estimated that great earthquakes should occur about every 50 to 200 years along those sections of the San Andreas Fault that exhibit stick-slip motion. This knowledge is useful when assigning a potential earthquake risk to a given segment of the fault zone.

The tectonic forces along the San Andreas Fault zone that were responsible for the 1906 San Francisco earthquake are still active. Currently, laser beams and techniques that employ the Global Positioning System (GPS) are used to measure the relative motion between the opposite sides of this fault. These measurements reveal a displacement of 2 to 5 centimeters (1 to 2 inches) per year. Although this seems slow, it produces substantial movement over millions of years. To illustrate, in 30 million years this rate of displacement would slide the western portion of California northward so that Los Angeles, on the Pacific plate, would be adjacent to San Francisco on the North American plate! More important in the short term, a displacement of just 2 centimeters per year produces 2 meters of offset every 100 years. Consequently, the 4 meters of displacement produced during the 1906 San Francisco earthquake should occur at least every 200 years along this segment of the fault zone. This fact lies behind California's concern for making buildings earthquake-resistant in anticipation of the inevitable "Big One."

Earthquakes that occur along strike-slip faults, such as the faults that comprise the San Andreas fault system, are generally shallow, having focus depths less than 20 kilometers (12 miles). For example, the 1906 San Francisco earthquake involved movement within the upper 15 kilometers of Earth's crust, and even the comparatively deep 1989 Loma Prieta earthquake had a focus depth of only 19 kilometers. The main reason for the shallow activity in this region is that earthquakes occur only where rocks are rigid and exhibit elastic behavior. Recall from Chapter 10 that at depth, where temperatures and confining pressures are high, rocks display *ductile deformation*. In these environments, when the strength of the rock is exceeded, it deforms by various flow mechanisms that produce slow gradual slippage without storing elastic strain. Thus, rocks at depth are generally not capable of generating an earthquake. The major exception occurs at convergent plate boundaries, where cool lithosphere is being subducted. (For more information on the San Andreas Fault, refer to Box 10.2.)

Seismology: The Study of Earthquake Waves

 GEODe Earthquakes
▶ Seismology

The study of earthquake waves, **seismology** (*seismos* = shake, *ology* = the study of) dates back to attempts made by the Chinese almost 2000 years ago to determine the

FIGURE 11.6 Ancient Chinese seismograph. During an Earth tremor, the dragons located in the direction of the main vibrations would drop a ball into the mouths of the frogs below.

direction from which these waves originated. The seismic instrument used by the Chinese was a large hollow jar that probably contained a mass suspended from the top (Figure 11.6). This suspended mass (similar to a clock pendulum) was connected in some fashion to the jaws of several large dragon figurines that encircled the container. The jaws of each dragon held a metal ball. When earthquake waves reached the instrument, the relative motion between the suspended mass and the jar would dislodge some of the metal balls into the waiting mouths of frogs directly below.

The Chinese were probably aware that the first strong ground motion from an earthquake is directional, and when it is strong enough, all poorly supported items will topple over in the same direction. Apparently the Chinese used this fact, plus the position of the dislodged balls, to detect the direction to an earthquake's source. However, the complex motion of seismic waves makes it unlikely that the actual direction to an earthquake was determined with any regularity.

In principle, at least, modern **seismographs** (*seismos* = shake, *graph* = write), instruments that record seismic waves, are not unlike the device used by the early Chinese. Seismographs have a mass freely suspended from a

support that is attached to the ground (Figure 11.7). When the vibration from a distant earthquake reaches the instrument, the **inertia*** (*iners* = idle) of the mass keeps it relatively stationary while Earth and support move. The movement of Earth in relation to the stationary mass is recorded on a rotating drum or magnetic tape.

Earthquakes cause both vertical and horizontal ground motion; therefore, more than one type of seismograph is needed. The instrument shown in Figure 11.7 is designed so that the mass is permitted to swing from side to side and thus detects horizontal ground motion. Usually two horizontal seismographs are employed, one oriented north–south and the other placed with an east–west orientation. Vertical ground motion can be detected if the mass is suspended from a spring, as shown in Figure 11.8.

To detect very weak earthquakes, or a great earthquake that occurred in another part of the world, seismic instruments are typically designed to magnify ground motion. Conversely, some instruments are designed to withstand the violent shaking that occurs very near the earthquake source.

---

*Inertia: Simply stated, objects at rest tend to stay at rest, and objects in motion tend to remain in motion unless either is acted upon by an outside force. You probably have experienced this phenomenon when you tried to stop your automobile quickly and your body continued to move forward.

FIGURE 11.7 Principle of the seismograph. **A.** The inertia of the suspended mass tends to keep it motionless, while the recording drum, which is anchored to bedrock, vibrates in response to seismic waves. Thus, the stationary mass provides a reference point from which to measure the amount of displacement occurring as the seismic wave passes through the ground. **B.** Seismograph recording earthquake tremors. (Photo courtesy of Zephyr/Photo Researchers, Inc.)

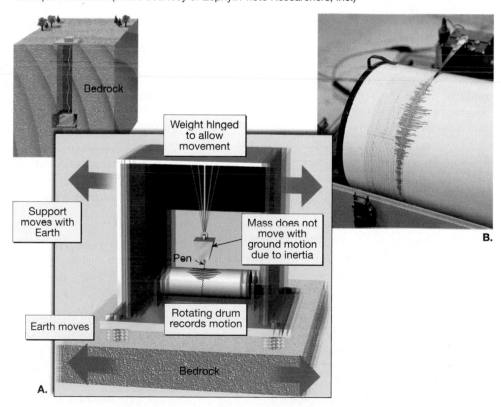

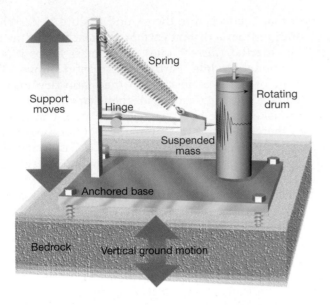

FIGURE 11.8 Seismograph designed to record vertical ground motion.

The records obtained from seismographs, called **seismograms** (*seismos* = shake, *gramma* = what is written), provide a great deal of information concerning the behavior of seismic waves. Simply stated, seismic waves are elastic energy that radiates out in all directions from the focus. The propagation (transmission) of this energy can be compared to the shaking of gelatin in a bowl, which results as some is spooned out. Whereas the gelatin will have one mode of vibration, seismograms reveal that two main groups of seismic waves are generated by the slippage of a rock mass. One of these wave types travels along the outer part of Earth. These are called **surface waves.** Others travel through Earth's interior and are called **body waves.** Body waves are further divided into two types, called **primary,** or **P, waves** and **secondary** or **S, waves.**

Body waves are divided into P and S waves by their mode of travel through intervening materials. P waves are "push-pull" waves—they push (compress) and pull (expand) rocks in the direction the wave is traveling (Figure 11.9A). Imagine holding someone by the shoulders and shaking that person. This push-pull movement is how P waves move through Earth. This wave motion is analogous to that generated by human vocal cords as they move air to create sound. Solids, liquids, and gases resist a change in volume when compressed and will elastically spring back once the force is removed. Therefore, P waves, which are compressional waves, can travel through all these materials.

On the other hand, S waves "shake" the particles at right angles to their direction of travel. This can be illustrated by fastening one end of a rope and shaking the other end, as shown in Figure 11.9B. Unlike P waves, which temporarily change the *volume* of intervening material by alternately compressing and expanding it, S waves temporarily change the *shape* of the material that transmits them. Because fluids (gases and liquids) do not respond elastically to changes in shape, they will not transmit S waves.

The motion of surface waves is somewhat more complex. As surface waves travel along the ground, they cause the ground and anything resting upon it to move, much like ocean swells toss a ship. In addition to their up-and-down motion, surface waves have a side-to-side motion similar to an S wave oriented in a horizontal plane. This latter motion is particularly damaging to the foundations of structures.

By observing a "typical" seismic record, as shown in Figure 11.10, you can see a major difference among these seismic waves: P waves arrive at the recording station first, then S waves, and then surface waves. This is a consequence of their speeds. To illustrate, the velocity of P waves through granite within the crust is about 6 kilometers per second. S waves under the same conditions travel at 3.6 kilometers per second. Differences in density and elastic properties of the rock greatly influence the velocities of these waves. Generally, in any solid material, P waves travel about 1.7 times faster than S waves, and surface waves can be expected to travel at 90 percent of the velocity of the S waves.

In addition to velocity differences, also notice in Figure 11.10 that the height or, more correctly, the amplitude of these wave types varies. The S waves have a slightly greater amplitude than do the P waves, while the surface waves, which cause the greatest destruction, exhibit an even greater amplitude. Because surface waves are confined to a narrow region near the surface and are not spread throughout Earth as P and S waves are, they retain their maximum amplitude longer. Surface waves also have longer periods (time interval between crests); therefore, they are often referred to as **long waves,** or **L waves.**

As we shall see, seismic waves are useful in determining the location and magnitude of earthquakes. In addition, seismic waves provide a tool for probing Earth's interior.

Locating the Source of an Earthquake

 GEODe **Earthquakes**
  ▸ **Locating the Source of an Earthquake**

Recall that the *focus* is the place within Earth where earthquake waves originate. The **epicenter** (*epi* = upon, *centr* = a point) is the location on the surface directly above the focus (see Figure 11.2).

The difference in velocities of P and S waves provides a method for locating the epicenter. The principle used is

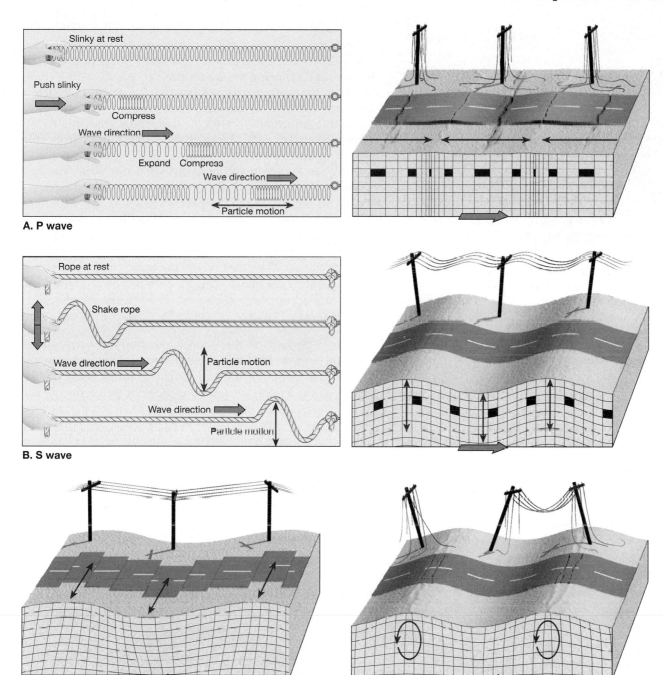

FIGURE 11.9 Types of seismic waves and their characteristic motion. (Note that during a strong earthquake, ground shaking consists of a combination of various kinds of seismic waves.) **A.** As illustrated by a slinky, P waves are compressional waves that alternately compress and expand the material through which they pass. The back-and-forth motion produced as compressional waves travel along the surface can cause the ground to buckle and fracture, and may cause power lines to break. **B.** S waves cause material to oscillate at right angles to the direction of wave motion. Because S waves can travel in any plane, they produce up-and-down and sideways shaking of the ground. **C.** One type of surface wave is essentially the same as that of an S wave that exhibits only horizontal motion. This kind of surface wave moves the ground from side to side and can be particularly damaging to the foundations of buildings. **D.** Another type of surface wave travels along Earth's surface much like rolling ocean waves. The arrows show the elliptical movement of rock as the wave passes.

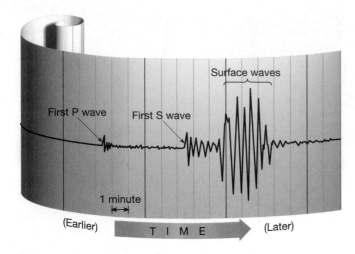

FIGURE 11.10 Typical seismogram. Note the time interval (about 5 minutes) between the arrival of the first P wave and the arrival of the first S wave.

analogous to a race between two autos, one faster than the other. The P wave always wins the race, arriving ahead of the S wave. But the greater the length of the race, the greater will be the difference in the arrival times at the finish line (the seismic station). Therefore, the greater the interval measured on a seismogram between the arrival of the first P wave and the first S wave, the greater the distance to the earthquake source.

A system for locating earthquake epicenters was developed by using seismograms from earthquakes whose epicenters could be easily pinpointed from physical evidence. From these seismograms, travel-time graphs were constructed (Figure 11.11). The first travel-time graphs were greatly improved when seismograms became available from nuclear explosions, because the precise location and time of detonation were known.

Using the sample seismogram in Figure 11.10 and the travel-time curve in Figure 11.11, we can determine the dis-

FIGURE 11.11 A travel-time graph is used to determine the distance to the epicenter. The difference in arrival times of the first P and S waves in the example is 5 minutes. Thus, the epicenter is roughly 3400 kilometers (2100 miles) away.

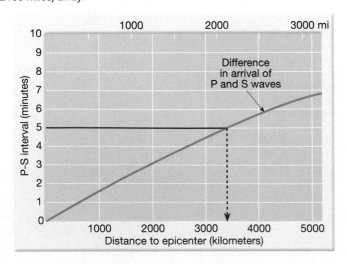

tance separating the recording station from the earthquake in two steps: (1) Using the seismogram, determine the time interval between the arrival of the first P wave and the first S wave, and (2) using the travel-time graph, find the P–S interval on the vertical axis and use that information to determine the distance to the epicenter on the horizontal axis. From this information, we can determine that this earthquake occurred 3400 kilometers (2100 miles) from the recording instrument.

Now we know the *distance*, but what *direction?* The epicenter could be in any direction from the seismic station. As shown in Figure 11.12, the precise location can be found when the distance is known from three or more different seismic stations. On a globe, we draw a circle around each seismic station. Each circle represents the epicenter distance for each station. The point where the three circles intersect is the epicenter of the quake. This method is called *triangulation.*

Earthquake Belts

About 95 percent of the energy released by earthquakes originates in a few relatively narrow zones that wind around the globe (Figure 11.13). The greatest energy is released along a path around the outer edge of the Pacific Ocean known as the *circum-Pacific belt.* Included in this zone are regions of great seismic activity such as Japan, the Philippines, Chile, and numerous volcanic island chains, as exemplified by the Aleutian Islands.

Another major concentration of strong seismic activity runs through the mountainous regions that flank the Mediterranean Sea and continues through Iran and on past the Himalayan complex. Figure 11.13 indicates that yet another continuous belt extends for thousands of kilometers

FIGURE 11.12 An earthquake epicenter is located using the distances obtained from three or more seismic stations.

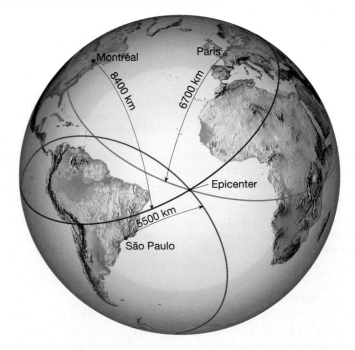

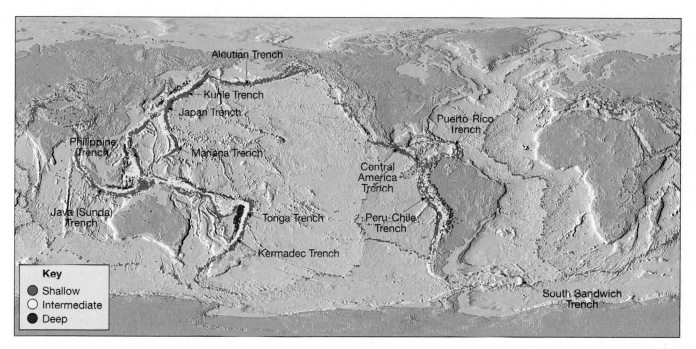

FIGURE 11.13   Distribution of shallow-, intermediate-, and deep-focus earthquakes. Note that deep-focus earthquakes occur only in association with convergent plate boundaries and subduction zones. (Data from NOAA)

through the world's oceans. This zone coincides with the oceanic ridge system, which is an area of frequent but low-intensity seismic activity.

The areas of the United States included in the circum-Pacific belt lie adjacent to California's San Andreas Fault and along the western coastal regions of Alaska, including the Aleutian Islands. In addition to these high-risk areas, other sections of the United States are regarded as regions where strong earthquakes are likely to occur (see Box 11.1).

Earthquake Depths

Evidence from seismic records reveals that earthquakes originate at depths ranging from 5 to about 700 kilometers. In a somewhat arbitrary fashion, earthquake foci have been classified by their depth of occurrence. Those with points of origin within 70 kilometers of the surface are referred to as *shallow*, while those generated between 70 and 300 km are considered *intermediate*, and those with a focus greater than 300 km are classified as *deep*. About 90 percent of all earthquakes occur at depths of less than 100 km, and nearly all very damaging earthquakes appear to originate at shallow depths.

Recall that the 1906 San Francisco earthquake involved movement within the upper 15 km of Earth's crust, whereas the 1964 Alaskan earthquake had a focal depth of 33 km. Seismic data reveal that while shallow-focus earthquakes have been recorded with Richter magnitudes of 8.6, the strongest intermediate-depth quakes have had values below 7.5, and deep-focus earthquakes have not exceeded 6.9 in magnitude.

When earthquake data were plotted according to geographic location and depth, several interesting observations

were noted. Rather than a random mixture of shallow and deep earthquakes, some very definite patterns emerged (Figure 11.13). Earthquakes generated along the oceanic ridge system always have a shallow focus, and none are very strong. Further, it was noted that almost all deep-focus earthquakes occurred in the circum-Pacific belt, particularly in regions situated landward of deep-ocean trenches.

Studies conducted in the Pacific basin established the fact that foci depths increase with increasing distance from deep-ocean trenches. Notice in Figure 11.13 that in South America the foci depths increase landward of the Peru–Chile trench. These seismic regions, called **Wadati-Benioff zones** after the two scientists who were the first to extensively study them, dip at an average angle of about 45 degrees to the surface. Why should earthquakes be oriented along a narrow zone that plunges almost 700 kilometers into Earth's interior? We will consider this question later in the chapter.

Measuring the Size of Earthquakes

Historically, seismologists have employed a variety of methods to obtain two fundamentally different measures that describe the size of an earthquake—intensity and magnitude. The first of these to be used was **intensity**—a measure of the degree of earthquake shaking at a given locale based on the amount of damage. With the development of seismographs, it became clear that a quantitative measure of an earthquake based on seismic records rather than uncertain personal estimates of damage was desirable. The measurement that was developed, called **magnitude,** relies on calculations that use data provided by seismic records (and other

BOX 11.1 ▶ PEOPLE AND THE ENVIRONMENT

Earthquakes East of the Rockies

The majority of earthquakes occur near plate boundaries, as exemplified by California and Japan. However, areas distant from plate boundaries are not necessarily immune. A team of seismologists recently estimated that the probability of a damaging earthquake east of the Rocky Mountains during the next 30 years is roughly two-thirds as likely as an earthquake of comparable damage in California. Like all earthquake-risk assessments, this prediction is based in part on the geographic distribution and average rate of earthquake occurrences in these regions.

At least six major earthquakes have occurred in the central and eastern United States since colonial times. Three of them, having estimated Richter magnitudes of 7.5, 7.3, and 7.8, were centered near the Mississippi River Valley in southeastern Missouri. Occurring over a three-month period in December 1811, January 1812, and February 1812, these earthquakes and numerous smaller tremors destroyed the town of New Madrid, Missouri. They also triggered massive landslides, damaged a six-state area, altered the course of the Mississippi River, and enlarged Tennessee's Reelfoot Lake.

The distance over which these earthquakes were felt is truly remarkable. Chimneys were downed in Cincinnati and Richmond, and even Boston residents— 1770 kilometers (1100 miles) to the northeast—felt the tremor. Although destruction from the New Madrid earthquakes was slight compared to the Loma Prieta earthquake of 1989, the Midwest in the early 1800s was sparsely populated. Memphis, near the epicenter, had not yet been established, and St. Louis was a small frontier town. Other damaging earthquakes—Aurora, Illinois (1909), and Valentine, Texas (1931)—remind us that the central United States is vulnerable.

The greatest historical earthquake in the eastern states occurred in Charleston, South Carolina, in 1886. This one-minute event caused 60 deaths, numerous injuries,

and great economic loss within 200 kilometers (120 miles) of Charleston. Within 8 minutes strong vibrations shook the upper floors of buildings in Chicago and St. Louis, causing people to rush outdoors. In Charleston alone more than a hundred buildings were destroyed, and 90 percent of the remaining structures were damaged. It was difficult to find a chimney that was still standing (Figure 11.A).

New England and adjacent areas have experienced sizable shocks since colonial times, including the 1683 quake in Plymouth and the 1755 quake in Cambridge, Massachusetts. Since records have been kept, New York State has experienced more than 300 earthquakes large enough to be felt by humans.

These eastern and central earthquakes occur far less frequently than do those in California. Yet the shocks east of the Rockies have generally produced structural damage over a larger area than tremors of similar magnitude in California. The reason is that the underlying bedrock in the central and eastern United States is older and more rigid. As a result, seismic waves travel greater distances with less attenuation than in the western United States. For similar earthquakes, the region of maximum ground motion in the East may be up to 10 times larger than in the West. Consequently, the higher rate of earthquakes in the West is partly balanced by more widespread damage in the East.

Despite recent geologic history, Memphis, the largest population center in the area of the New Madrid earthquake, lacks adequate provision for earthquakes in its building code. Worse, Memphis rests on unconsolidated floodplain deposits, so its buildings are more susceptible to damage. A 1985 federal study concluded that a 7.6-magnitude earthquake in this area could cause an estimated 2500 deaths, collapse 3000 structures, cause $25 billion in damages, and displace a quarter of a million people in Memphis alone.

FIGURE 11.A Damage to Charleston, South Carolina, caused by the August 31, 1886, earthquake. Damage ranged from toppled chimneys and broken plaster to total collapse. (Photo courtesy of U.S. Geological Survey)

techniques) to estimate the amount of energy released at the source of the earthquake.

At it turns out, both intensity and magnitude provide useful, although quite different, information about earthquake strength. Consequently, both measures are still used to describe the relative sizes of earthquakes.

Intensity Scales

Until a little more than a century ago, historical records provided the only accounts of the severity of earthquake shaking and destruction. Using these descriptions—which were compiled without any established standards for reporting—

made accurate comparisons of earthquake sizes difficult, at best.

Perhaps the first attempt to "scientifically" describe the aftermath of an earthquake came following the great Italian earthquake in 1857. By systematically mapping effects of the earthquake, a measure of the strength and distribution of ground motion was established. The map generated by this study employed lines to connect places of equal damage and hence equal intensity (Figure 11.14). Using this technique, zones of intensity were identified, with the zone of highest intensity located near the center of maximum ground shaking and often (but not always) the source of seismic waves.

In order to standardize the study of earthquake severity, workers developed various intensity scales that considered damage done to buildings, as well as individual descriptions of the event, and secondary effects—landslides and the extent of ground rupture. By 1902, Giuseppe Mercalli had developed a relatively reliable intensity scale, which in a modified form is still used today (Figure 11.14). The **Modified Mercalli Intensity Scale,** shown in Table 11.1, was developed using California buildings as its standard, but it is appropriate for use throughout most of the United States and Canada to estimate the strength of an earthquake. For example, if some well-built wood structures and most masonry buildings are destroyed by an earthquake, a region would be assigned an intensity of X on the Mercalli scale (Table 11.1).

Despite their usefulness in providing seismologists with a tool to compare earthquake severity, particularly in regions where there are no seismographs, intensity scales

TABLE 11.1	Modified Mercalli Intensity Scale
I	Not felt except by a very few under especially favorable circumstances.
II	Felt only by a few persons at rest, especially on upper floors of buildings.
III	Felt quite noticeably indoors, especially on upper floors of buildings, but many people do not recognize it as an earthquake.
IV	During the day, felt indoors by many, outdoors by few. Sensation like heavy truck striking building.
V	Felt by nearly everyone; many awakened. Disturbances of trees, poles, and other tall objects sometimes noticed.
VI	Felt by all; many frightened and run outdoors. Some heavy furniture moved; few instances of fallen plaster or damaged chimneys. Damage slight.
VII	Everybody runs outdoors. Damage negligible in buildings of good design and construction; slight to moderate in well-built ordinary structures; considerable in poorly built or badly designed structures.
VIII	Damage slight in specially designed structures; considerable in ordinary substantial buildings with partial collapse; great in poorly built structures (fall of chimneys, factory stacks, columns, monuments, walls).
IX	Damage considerable in specially designed structures. Buildings shifted off foundations. Ground cracked conspicuously.
X	Some well-built wooden structures destroyed. Most masonry and frame structures destroyed. Ground badly cracked.
XI	Few, if any (masonry) structures remain standing. Bridges destroyed. Broad fissures in ground.
XII	Damage total. Waves seen on ground surfaces. Objects thrown upward into air.

have severe drawbacks. In particular, intensity scales are based on effects (largely destruction) of earthquakes that depend not only on the severity of ground shaking but also on factors such as population density, building design, and the nature of surface materials. The modest 6.9-magnitude earthquake in Armenia in 1988 was extremely destructive, mainly because of inferior building construction, whereas the 1985 Mexico City quake was deadly because of the soft sediment upon which the city rests. Thus, the destruction wrought by earthquakes may not be a true measure of the earthquake's actual size.

Magnitude Scales

In order to compare earthquakes across the globe, a measure was needed that does not rely on parameters that vary considerably from one part of the world to another, such as types of construction. As a consequence, a number of magnitude scales were developed.

Richter Magnitude   In 1935 Charles Richter of the California Institute of Technology developed the first magnitude scale using seismic records to estimate the relative sizes of earthquakes. As shown in Figure 11.15 (top), the **Richter scale** is based on the amplitude of the largest seismic wave (P, S, or surface wave) recorded on a seismogram. Because seismic waves weaken as the distance between the earthquake focus and the seismograph increases (in a manner

FIGURE 11.14   Zones of destruction associated with the Loma Prieta, California, earthquake of 1989 using the Modified Mercalli Intensity Scale. Roman numerals show the intensity categories. The zone of maximum intensity roughly corresponds to the epicenter. Even higher intensities were experienced at a few sites in San Francisco and Oakland where local conditions amplified the seismic waves. (Data from Plafker and Galloway)

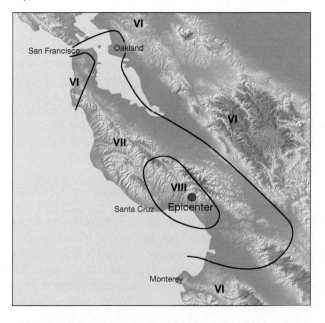

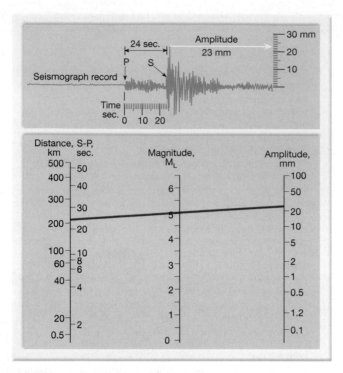

FIGURE 11.15 Illustration showing how the Richter magnitude of an earthquake can be determined graphically using a seismograph record from a Wood-Anderson instrument. First, measure the height (amplitude) of the largest wave on the seismogram (23 mm) and then the distance to the focus using the time interval between S and P waves (24 seconds). Next, draw a line between the distance scale (left) and the wave amplitude scale (right). By doing this, you should obtain the Richter magnitude (M_L) of 5. (Data from California Institute of Technology)

TABLE 11.2	Earthquake Magnitudes and Expected World Incidence	
Richter Magnitudes	Effects Near Epicenter	Estimated Number per Year
<0.2	Generally not felt, but recorded.	600,000
2.0–2.9	Potentially perceptible.	300,000
3.0–3.9	Felt by some.	49,000
4.0–4.9	Felt by most.	6200
5.0–5.9	Damaging shocks.	800
6.0–6.9	Destruction in populous regions.	266
7.0–7.9	Major earthquakes. Inflict serious damage.	18
8.0	Great earthquakes. Cause extensive destruction to communities near epicenter.	1.4

similar to light), Richter developed a method that accounted for the decrease in wave amplitude with increased distance. Theoretically, as long as the same, or equivalent, instruments were used, monitoring stations at various locations would obtain the same Richter magnitude for every recorded earthquake. (Richter selected the Wood-Anderson seismograph as the standard recording device.) In practice, however, different recording stations often obtained slightly different Richter magnitudes for the same earthquake—a consequence of the variations in rock types through which the waves traveled.

Although the Richter scale has no upper limit, the largest magnitude recorded on a Wood-Anderson seismograph was 8.9. These great shocks released approximately 10^{26} ergs of energy—roughly equivalent to the detonation of 1 billion tons of TNT. Conversely, earthquakes with a Richter magnitude of less than 2.0 are not felt by humans. With the development of more sensitive instruments, tremors of a magnitude of minus 2 were recorded. Table 11.2 shows how Richter magnitudes and their effects are related.

Earthquakes vary enormously in strength, and great earthquakes produce wave amplitudes that are thousands of times larger than those generated by weak tremors. To accommodate this wide variation, Richter used a *logarithmic scale* to express magnitude, where a *tenfold* increase in wave amplitude corresponds to an increase of 1 on the magnitude scale. Thus, the amount of ground shaking for a 5-magnitude earthquake is 10 times greater than that produced by an earthquake having a Richter magnitude of 4.

In addition, each unit of Richter magnitude equates to roughly a *32-fold energy increase*. Thus, an earthquake with a magnitude of 6.5 releases 32 times more energy than one with a magnitude of 5.5, and roughly 1000 times more energy than a 4.5-magnitude quake (Table 11.3). A major earthquake with a magnitude of 8.5 releases millions of times more energy than the smallest earthquakes felt by humans.

Other Magnitude Scales Richter's original goal was modest in that he only attempted to rank the earthquakes of southern California (shallow-focus earthquakes) into groups of large, medium, and small magnitude. Hence, Richter magnitude was designed to study nearby (or local) earthquakes and is denoted by the symbol (M_L)—where M is for *magnitude* and L is for *local.*

The convenience of describing the size of an earthquake by a single number that could be calculated quickly from seismograms makes the Richter scale a powerful tool. Further, unlike intensity scales that could only be applied to populated areas of the globe, Richter magnitudes could be assigned to earthquakes in more remote regions and even to events that occurred in the ocean basins. As a result, the method devised by Richter was adapted to a number of different seismographs located throughout the world. In time, seismologists modified Richter's work and developed new magnitude scales.

However, despite their usefulness, none of these "Richter-like" magnitude scales are adequate for describing very large earthquakes. For example, the 1906 San Francisco earthquake and the 1964 Alaskan earthquake had roughly the same Richter magnitudes. However, based on the size of the fault zone and the amount of displacement observed, the Alaskan earthquake released considerably more energy than the San Francisco quake. Thus, the Richter scale (as well as the other related magnitude scales) are said to be

TABLE 11.3	Earthquake Magnitude and Energy Equivalence	
Earthquake Magnitude	Energy Released* (Millions of Ergs)	Approximate Energy Equivalence
0	630,000	1 pound of explosives
1	20,000,000	
2	630,000,000	Energy of lightning bolt
3	20,000,000,000	
4	630,000,000,000	1000 pounds of explosives
5	20,000,000,000,000	
6	630,000,000,000,000	1946 Bikini atomic bomb test 1994 Northridge Earthquake
7	20,000,000,000,000,000	1989 Loma Prieta Earthquake
8	630,000,000,000,000,000	1906 San Francisco Earthquake 1980 Eruption of Mount St. Helens
9	20,000,000,000,000,000,000	1964 Alaskan Earthquake 1960 Chilean Earthquake
10	630,000,000,000,000,000,000	Annual U.S. energy consumption

* For each unit increase in magnitude, the energy released increases about 31.6 times.
Source: U.S. Geological Survey.

had a Richter magnitude of 8.3, would be demoted to 7.9 on the moment magnitude scale, whereas the 1964 Alaskan earthquake with an 8.3 Richter magnitude would be increased to 9.2. The strongest earthquake on record is the 1960 Chilean earthquake, with a moment magnitude of 9.5.

Moment magnitude has gained wide acceptance among seismologists and engineers because: (1) it is the only magnitude scale that estimates adequately the size of very large earthquakes; (2) it is a measure that can be derived mathematically from the size of the rupture surface and the amount of displacement, thus it better reflects the total energy released during an earthquake; and (3) it can be verified by two independent methods—field studies that are based on measurements of fault displacement and by seismographic methods using long-period waves.

Earthquake Destruction

The most violent earthquake ever recorded in North America—the Good Friday Alaskan earthquake—occurred at 5:36 P.M. on March 27, 1964. Felt throughout that state, the earthquake had a moment magnitude (M_W) of 9.2 and reportedly lasted 3 to 4 minutes. This brief event left 131 people dead, thousands homeless, and the economy of the state badly disrupted. Had the schools and business districts been open, the toll surely would have been higher. Within 24 hours of the initial shock, 28 aftershocks were recorded, 10 of which exceeded a magnitude of 6 on the Richter scale. The location of the epicenter and the towns that were hardest hit by the quake are shown in Figure 11.16.

Many factors determine the degree of destruction that will accompany an earthquake. The most obvious is the magnitude of the earthquake and its proximity to a populated area. Fortunately, most earthquakes are small and occur in remote regions of Earth. However, about 20 major earthquakes are reported annually, one or two of which can be catastrophic.

During an earthquake, the region within 20 to 50 kilometers (12.5 to 30 miles) of the epicenter ordinarily will experience roughly the same degree of ground shaking, but beyond this limit the vibration deteriorates rapidly. Occasionally

saturated for large earthquakes because they cannot distinguish between the size of these events.

Moment Magnitude In recent years seismologists have been employing a more precise measure called **moment magnitude** (M_W), which can be calculated using several techniques. In one method the moment magnitude is calculated from field studies using a combination of factors that include the average amount of displacement along the fault, the area of the rupture surface, and the shear strength of the faulted rock—a measure of how much strain energy a rock can store before it suddenly slips and releases this energy in the form of an earthquake (and heat). For example, the energy involved in a 3-meter displacement of a rock body along a rupture a few hundred kilometers long would be much larger than that produced by a 1-meter displacement along a 10-kilometer-long rupture (assuming comparable rupture depths).

The moment magnitude can also be readily calculated from seismograms by examining very long period seismic waves. The values obtained have been calibrated so that small- and moderate-sized earthquakes have moment magnitudes that are roughly equivalent to Richter magnitudes. However, moment magnitudes are much better for describing very large earthquakes. For example, on the moment magnitude scale, the 1906 San Francisco earthquake, which

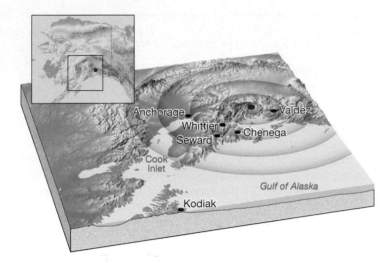

FIGURE 11.16 Region most affected by the Good Friday earthquake of 1964. Note the location of the epicenter (red dot). (After U.S. Geological Survey)

during earthquakes that occur in the stable continental interior, such as the New Madrid earthquake of 1811, the area of influence can be much larger. The epicenter of this earthquake was located directly south of Cairo, Illinois, and the vibrations were felt from the Gulf of Mexico to Canada, and from the Rockies to the Atlantic Seaboard.

Destruction from Seismic Vibrations

The 1964 Alaskan earthquake provided geologists with new insights into the role of ground shaking as a destructive force. As the energy released by an earthquake travels along Earth's surface, it causes the ground to vibrate in a complex manner by moving up and down as well as from side to side. The amount of structural damage attributable to the vibrations depends on several factors, including (1) the intensity and (2) the duration of vibrations, (3) the nature of the material upon which the structure rests, and (4) the design of the structure.

All of the multistory structures in Anchorage were damaged by the vibrations. The more flexible wood-frame residential buildings fared best. However, many homes were destroyed when the ground failed. A striking example of how construction variations affect earthquake damage is shown in Figure 11.17. You can see that the steel-frame building on the left withstood the vibrations, whereas the poorly designed J.C. Penney building was badly damaged. Engineers have learned that unreinforced masonry buildings are the most serious safety threat in earthquakes.

Most large structures in Anchorage were damaged, even though they were built according to the earthquake provisions of the Uniform Building Code. Perhaps some of that destruction can be attributed to the unusually long duration of

this earthquake. Most quakes consist of tremors that last less than a minute. For example, the 1994 Northridge earthquake was felt for about 40 seconds, and the strong vibrations of the 1989 Loma Prieta earthquake lasted less than 15 seconds. But the Alaska quake reverberated for 3 to 4 minutes.

Amplification of Seismic Waves Although the region within 20 to 50 kilometers of the epicenter will experience about the same intensity of ground shaking, the destruction varies considerably within this area (see Box 11.2). This difference is mainly attributable to the nature of the ground on which the structures are built. Soft sediments, for example, generally amplify the vibrations more than solid bedrock. Thus, the buildings located in Anchorage, which were situated on unconsolidated sediments, experienced heavy structural damage. By contrast, most of the town of Whittier, although much nearer the epicenter, rests on a firm foundation of granite and hence suffered much less damage. However, Whittier was damaged by a tsunami (described in the next section).

Liquefaction In areas where unconsolidated materials are saturated with water, earthquake vibrations can generate a phenomenon known as **liquefaction** (*liqueo* = to be fluid, *facio* = to make). Under these conditions, what had been a stable soil turns into a mobile fluid that is not capable of supporting buildings or other structures (Figure 11.18). As a result, underground objects such as storage tanks and sewer lines may literally float toward the surface of their newly liquefied environment. Buildings and other structures may settle and collapse. During the 1989 Loma Prieta earthquake, in San Francisco's Marina District, foundations failed and geysers of sand and water shot from the ground, indicating that liquefaction had occurred (Figure 11.19).

FIGURE 11.17 Damage caused to the five-story J.C. Penney Co. building, Anchorage, Alaska. Very little structural damage was incurred by the adjacent building. (Courtesy of NOAA/Seattle)

BOX 11.2 ▶ UNDERSTANDING EARTH

Wave Amplification and Seismic Risks

Much of the damage and loss of life in the 1985 Mexico City earthquake occurred because downtown buildings were constructed on lake sediment that greatly amplified the ground motion. To understand why this happens, recall that as seismic waves pass through Earth, they cause the intervening material to vibrate much as a tuning fork when it is struck. Although most objects can be "forced" to vibrate over a wide range of frequencies, each has a natural period of vibration that is preferred. Different Earth materials, like different-length tuning forks, also have different natural periods of vibration.*

Ground-motion amplification results when the supporting material has a natural period of vibration (frequency) that matches that of the seismic waves. A common example of this phenomenon occurs when a parent pushes a child on a swing. When the parent periodically pushes the child in rhythm with the frequency of the swing, the child moves back and forth in a greater and greater arc (amplitude). By chance, the column of sediment beneath Mexico City had a natural period of vibration of about 2 seconds, matching that of the strongest seismic waves. Thus, when the seismic waves began shaking the soft sediments, a *resonance* developed, which greatly increased the amplitude of the vibrations. This amplification resulted in vibrations that exhibited 40 centimeters (1.3 feet) of back-and-forth ground motion every 2 seconds for nearly 2 minutes. Such movement was too intense for many poorly designed buildings in the city. In addition, intermediate-height structures (5 to 15 stories) sway back and forth with a period of about 2 seconds. Thus, resonance also developed between these buildings and the ground, with the result that most of the building failures occurred to structures in this height range (Figure 11.B).

Sediment-induced wave amplification is also thought to have contributed significantly to the failure of the two-tiered Cypress section of Interstate 880 during the 1989 Loma Prieta earthquake (Figure 11.C). Studies conducted on the 1.4-kilometer section that did collapse showed that it was built on San Francisco Bay mud. Another section of this interstate that was damaged but did not collapse was constructed on firmer alluvial materials.

FIGURE 11.B During the 1985 Mexican earthquake, multistory buildings swayed back and forth as much as 1 meter. Many, including the hotel shown here, collapsed or were seriously damaged. (Photo by James L. Beck)

FIGURE 11.C The portion of the Cypress Freeway structure in Oakland, California, that stood on soft mud (dashed red line) collapsed during the 1989 Loma Prieta earthquake. Adjacent parts of the structure (solid red) that were built on firmer ground remained standing. Seismograms from an aftershock (upper right) show that shaking is greatly amplified in the soft mud as compared to the firmer materials.

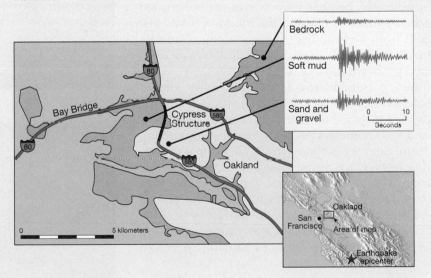

*To demonstrate the natural period of vibration of an object, hold a ruler over the edge of a desk so that most of it is not supported by the desk. Start it vibrating, and notice the noise it makes. By changing the length of the unsupported portion of the ruler, the natural period of vibration will change accordingly.

FIGURE 11.18 Effects of liquefaction. This tilted building rests on unconsolidated sediment that behaved like quicksand during the 1985 Mexican earthquake. (Photo by James L. Beck)

es the shore, it appears as a rapid rise in sea level with a turbulent and chaotic surface. Tsunami can be very destructive (Figure 11.21).

Usually the first warning of an approaching tsunami is a relatively rapid withdrawal of water from beaches. Some residents living near the Pacific basin have learned to heed this warning and move to higher ground, because about 5 to 30 minutes later, the retreat of water is followed by a surge capable of extending hundreds of meters inland. In a successive fashion, each surge is followed by rapid oceanward retreat of the water.

Seiches The effects of great earthquakes may be felt thousands of kilometers from their source. Ground motion may generate *seiches*, the rhythmic sloshing of water in lakes, reservoirs, and enclosed basins such as the Gulf of Mexico. The 1964 Alaskan earthquake, for example, generated 2-meter waves off the coast of Texas, which damaged small craft, while much smaller waves were noticed in swimming pools in both Texas and Louisiana.

Seiches can be particularly dangerous when they occur in reservoirs retained by earthen dams. These waves have been known to slosh over reservoir walls and weaken the structure, thereby endangering the lives of those downstream.

What Is a Tsunami?

Large undersea earthquakes occationally set in motion massive waves of water called **seismic sea waves,** or **tsunami.***
(*tsu* = harbor, *nami* = waves). These destructive waves often are called "tidal waves" by the media. However, this name is inappropriate, because these waves are not created nor influenced by the tidal effect of the Moon or Sun. Most tsunami result from vertical displacement along a fault located on the ocean floor, or from a large underwater landslide triggered by an earthquake (Figure 11.20).

Once formed, a tsunami resembles the ripples formed when a pebble is dropped into a pond. In contrast to ripples, tsunami advance across the ocean at amazing speeds between 500 and 950 kilometers per hour. Despite this striking characteristic, a tsunami in the open ocean can pass undetected because its height is usually less than 1 meter and the distance between wave crests is great, ranging from 100 to 700 kilometers. However, upon entering shallower coastal waters, these destructive waves are slowed down and the water begins to pile up to heights that occasionally exceed 30 meters (Figure 11.20). As the crest of a tsunami approach-

FIGURE 11.19 Liquefaction. **A.** These "mud volcanoes" were produced by the Loma Prieta earthquake of 1989. They formed when geysers of sand and water shot from the ground, an indication that liquefaction occurred. (Photo by Richard Hilton, courtesy of Dennis Fox) **B.** Students experiencing the nature of liquefaction. (Photo by Marli Miller)

A.

B.

---

*Seismic sea waves were given the name *tsunami* by the Japanese, who have suffered a great deal from them. The term *tsunami* is now used worldwide.

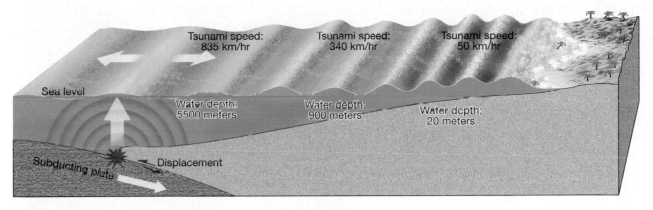

FIGURE 11.20 Schematic drawing of a tsunami generated by displacement of the ocean floor. The speed of a wave correlates with ocean depth. As shown, waves moving in deep water advance at speeds in excess of 800 kilometers per hour. Speed gradually slows to 50 kilometers per hour at depths of 20 meters. Decreasing depth slows the movement of the wave. As waves slow in shallow water, they grow in height until they topple and rush onto shore with tremendous force. The size and spacing of these swells are not to scale.

FIGURE 11.21 A massive earthquake of magnitude 9.0 off the Indonesian island of Sumatra sent a tsunami racing across the Indian Ocean and Bay of Bengal on December 26, 2004. **A.** Unsuspecting foreign tourists, who at first walked on the sand after the water receded, now rush toward shore as the first of six tsunami start to roll toward Hat Rai Lay Beach near Krabi in southern Thailand. (AFP/Getty Images Inc.) **B.** The Indonesian town of Banda Aceh, ten days after the tsunami. (Photo by Derbar Halin/Sipa Press)

A.

B.

The tsunami generated in the 1964 Alaskan earthquake inflicted heavy damage to the communities in the vicinity of the Gulf of Alaska, completely destroying the town of Chenega. Kodiak was also heavily damaged and most of its fishing fleet destroyed when a seismic sea wave carried many vessels into the business district. The deaths of 107 persons have been attributed to this tsunami. By contrast, only nine people died in Anchorage as a direct result of the vibrations.

Tsunami damage following the Alaskan earthquake extended along much of the west coast of North America, and despite a one-hour warning, 12 people perished in Crescent City, California, where all of the deaths and most of the destruction were caused by the fifth wave. The first wave crested about 4 meters (13 feet) above low tide and was followed by three progressively smaller waves. Believing that the tsunami had ceased, people returned to the shore, only to be met by the fifth and most devastating wave, which superimposed upon high tide crested about 6 meters higher than the level of low tide.

Tsunami Damage from the 2004 Indonesian Earthquake A massive undersea earthquake of moment magnitude 9.0 occurred near the island of Sumatra on December 26, 2004, and sent waves of water racing across the Indian Ocean and Bay of Bengal (Figure 11.21A). This tsunami was one of the deadliest natural disasters of any kind in modern times, claiming more than 230,000 lives. As water surged several kilometers inland, cars and trucks were flung around like toys in a bathtub, and fishing boats were rammed into homes. In some locations, the backwash of water dragged bodies and huge amounts of debris out to sea.

The destruction was indiscriminate, destroying luxury resorts and poor fishing hamlets on the Indian Ocean coast (Figure 11.21B). Devastation was most severe along the southeast coast of Sri Lanka, in the Indonesian province of Aceh, in the Indian state of Tamil Nada, and on Thailand's resort island of Phuket. Damages were reported as far away as the Somalia coast of Africa, 4100 kilometers (2500 miles) west of the earthquake epicenter.

The killer waves generated by this massive quake achieved heights as great as 10 meters (33 feet) and struck many unprepared areas within three hours of the event. Although the Pacific basin contains deep-sea buoys and tide gauges that can spot tsunami waves at sea, the Indian Ocean does not. (The deep-sea buoys have pressure sensors that detect changes in pressure as the earthquake's energy travels through the ocean, and tide gauges measure the rise and fall in sea level.) The rarity of tsunami in the Indian Ocean also contributed to the lack of preparedness for such an event. It should come as no surprise that the countries of India, Indonesia, and Thailand have announced plans to establish a tsunami warning system for the Indian Ocean.

Tsunami Warning System In 1946, a large tsunami struck the Hawaiian Islands without warning. A wave more than 15 meters (50 feet) high left several coastal villages in shambles. This destruction motivated the U.S. Coast and Geodetic Survey to establish a tsunami warning system for coastal areas of the Pacific. From seismic observatories throughout the region, large earthquakes are reported to the Tsunami Warning Center in Honolulu. Scientists at the Center use tidal gauges to determine whether a tsunami has formed. Within an hour a warning is issued. Although tsunami travel very rapidly, there is sufficient time to evacuate all but the region nearest the epicenter. For example, a tsunami generated near the Aleutian Islands would take five hours to reach Hawaii, and one generated near the coast of Chile would travel 15 hours before reaching Hawaii (Figure 11.22).

Landslides and Ground Subsidence

In the 1964 Alaskan earthquake, the greatest damage to structures was from landslides and ground subsidence triggered by the vibrations. At Valdez and Seward, the violent shaking caused deltaic materials to experience liquefaction; the subsequent slumping carried both waterfronts away. Because of the threat of recurrence, the entire town of Valdez was relocated about 7 kilometers away on more stable ground. In Valdez, 31 people on a dock died when it slid into the sea.

Most of the damage in the city of Anchorage was also attributed to landslides. Many homes were destroyed in Turnagain Heights when a layer of clay lost its strength and over 200 acres of land slid toward the ocean (Figure 11.23). A

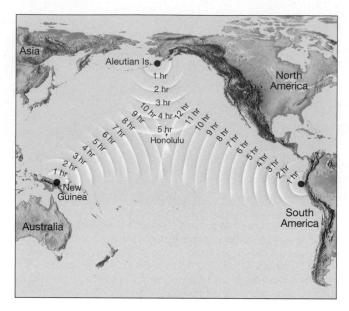

FIGURE 11.22   Tsunami travel times to Honolulu, Hawaii, from selected locations throughout the Pacific. (Data from NOAA)

FIGURE 11.23   Turnagain Heights slide caused by the 1964 Alaskan earthquake. **A.** Vibrations from the earthquake caused cracks to appear near the edge of the bluff. **B.** Within seconds blocks of land began to slide toward the sea on a weak layer of clay. In less than 5 minutes, as much as 200 meters of the Turnagain Heights bluff area had been destroyed. **C.** Photo of a small portion of the Turnagain Heights slide. (Photo courtesy of U.S. Geological Survey)

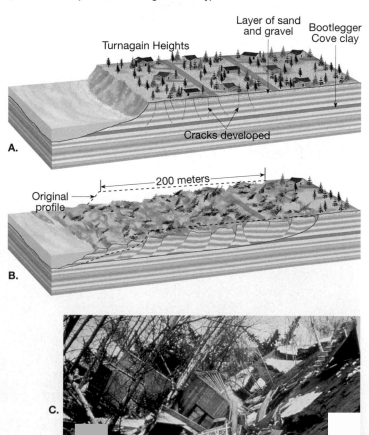

portion of this spectacular landslide was left in its natural condition as a reminder of this destructive event. The site was appropriately named "Earthquake Park." Downtown Anchorage was also disrupted as sections of the main business district dropped by as much as 3 meters (10 feet).

Fire

The 1906 earthquake in San Francisco reminds us of the formidable threat of fire. The central city contained mostly large, older wooden structures and brick buildings. Although many of the unreinforced brick buildings were extensively damaged by vibrations, the greatest destruction was caused by fires, which started when gas and electrical lines were severed. The fires raged out of control for three days and devastated over 500 blocks of the city (see Figure 11.3). The problem was compounded by the initial ground shaking, which broke the city's water lines into hundreds of unconnected pieces.

The fire was finally contained when buildings were dynamited along a wide boulevard to provide a fire break, the same strategy used in fighting a forest fire. Although only a few deaths were attributed to the San Francisco fire, such is not always the case. A 1923 earthquake in Japan triggered an estimated 250 fires, which devastated the city of Yokohama and destroyed more than half the homes in Tokyo. More than 100,000 deaths were attributed to the fires, which were driven by unusually high winds.

Can Earthquakes Be Predicted?

The vibrations that shook Northridge, California, in 1994 inflicted 57 deaths and about $40 billion in damage (Figure 11.24). This was from a brief earthquake (about 40 seconds) of moderate rating (M_W 6.7). Seismologists warn that earthquakes of comparable or greater strength will occur along the San Andreas Fault, which cuts a 1300-kilometer (800-mile) path through the state. The obvious question is: Can earthquakes be predicted?

Short-Range Predictions

The goal of short-range earthquake prediction is to provide a warning of the location and magnitude of a large earthquake within a narrow time frame. Substantial efforts to achieve this objective are being put forth in Japan, the United States, China, and Russia—countries where earthquake risks are high (Table 11.4). This research has concentrated on monitoring possible *precursors*—phenomena that precede and thus provide a warning of a forthcoming earthquake. In California, for example, seismologists are measuring uplift, subsidence, and strain in the rocks near active faults. Some Japanese scientists are studying anomalous animal behavior that may precede a quake. Other researchers are monitoring changes in groundwater levels, while still others are trying to predict earthquakes based on changes in the electrical conductivity of rocks.

FIGURE 11.24   Damage to Interstate 5 caused by the January 17, 1994, Northridge earthquake. (Photo by Tom McHugh/Photo Researchers, Inc.)

Students Sometimes Ask . . .

I've heard that the safest place to be in a house during an earthquake is in a doorframe. Is that really the best place to be while an earthquake is occurring?

It depends. If you're on the road, stay away from tunnels, underpasses, and overpasses. Stop in a safe area and stay in your vehicle until the shaking stops. If you happen to be outside during an earthquake, stand away from buildings, trees, and telephone and electric lines. If you're inside, remember to *duck, cover, and hold.* When you feel an earthquake, *duck* under a desk or sturdy table. Stay away from windows, bookcases, file cabinets, heavy mirrors, hanging plants, and other heavy objects that could fall. Stay under *cover* until the shaking stops. And, *hold* on to the desk or table: If it moves, move with it.

An enduring earthquake image of California is a collapsed adobe home with the doorframe as the only standing part. From this came the belief that a doorway is the safest place to be during an earthquake. This is true only if you live in an old, unreinforced adobe house. In modern homes, doorways are no stronger than any other part of the house and usually have doors that will swing and can injure you. You'd be safer under a table.

Among the most ambitious earthquake experiments is one being conducted along a segment of the San Andreas Fault near the town of Parkfield in central California. Here, earthquakes of moderate intensity have occurred on a regular basis about once every 22 years since 1857. The most recent rupture was a 5.6-magnitude quake that occurred in 1966. With the next event already significantly "overdue," the U.S. Geological Survey has established an elaborate monitoring network. Included are creepmeters, tiltmeters, and bore-hole strain meters that are used to measure the accumulation and release of strain. Moreover, 70 seismographs of various designs have been installed to record foreshocks as well as the main event. Finally, a network of distance-measuring devices that employ lasers measures movement across the fault (Figure 11.25). The object is to identify ground movements that may precede a sizable rupture.

One claim of a successful short-range prediction was made by Chinese seismologists after the February 4, 1975, earthquake in Liaoning Province. According to reports, very few people were killed, although more than 1 million lived near the epicenter, because the earthquake was predicted and the population was evacuated. Recently, some Western seismologists have questioned this claim and suggest

TABLE 11.4	Some Notable Earthquakes			
Year	Location	Deaths (est.)	Magnitude†	Comments
1556	Shensi, China	830,000		Possibly the greatest natural disaster.
1755	Lisbon, Portugal	70,000		Tsunami damage extensive.
*1811–1812	New Madrid, Missouri	Few	7.9	Three major earthquakes.
*1886	Charleston, South Carolina	60		Greatest historical earthquake in the eastern United States.
*1906	San Francisco, California	1500	7.8	Fires caused extensive damage.
1908	Messina, Italy	120,000		
1923	Tokyo, Japan	143,000	7.9	Fire caused extensive destruction.
1960	Southern Chile	5700	9.5	Possibly the largest-magnitude earthquake ever recorded.
*1964	Alaska	131	9.2	Greatest North American earthquake.
1970	Peru	66,000	7.8	Great rockslide.
*1971	San Fernando, California	65	6.5	Damage exceeded $1 billion.
1975	Liaoning Province, China	1328	7.5	First major earthquake to be predicted.
1976	Tangshan, China	240,000	7.6	Not predicted.
1985	Mexico City	9500	8.1	Major damage occurred 400 km from epicenter.
1988	Armenia	25,000	6.9	Poor construction practices.
*1989	Loma Prieta, California	62	6.9	Damages exceeded $6 billion.
1990	Iran	50,000	7.3	Landslides and poor construction practices caused great damage.
1993	Latur, India	10,000	6.4	Located in stable continental interior.
*1994	Northridge, California	57	6.7	Damages in excess of $40 billion.
1995	Kobe, Japan	5472	6.9	Damage estimated to exceed $100 billion.
1999	Izmit, Turkey	17,127	7.4	Nearly 44,000 injured and more than 250,000 displaced
1999	Chi-Chi, Taiwan	2300	7.6	Severe destruction; 8700 injuries.
2001	El Salvador	1000	7.6	Triggered many landslides.
2001	Bhuj, India	20,000	7.9	1 million or more homeless.
2003	Bam, Iran	41,000	6.6	Ancient city with poor construction.
2004	Indian Ocean	230,000	9.0	Devastating tsunami damage.
2005	Pakistan/Kashmir	86,000	7.6	69,000 injured, 4 million homeless, many landslides.

* U.S. earthquakes.
† Widely differing magnitudes have been estimated for some of these earthquakes. When available, moment magnitudes are used.
Source: U.S. Geological Survey

FIGURE 11.25 Lasers used to measure movement along the San Andreas Fault. (Photo by John K. Nakata/U.S. Geological Survey)

instead that an intense swarm of foreshocks, which began 24 hours before the main earthquake, may have caused many people to evacuate spontaneously. Further, an official Chinese government report issued 10 years later stated that 1328 people died and 16,980 injuries resulted from this earthquake.

One year after the Liaoning earthquake, at least 240,000 people died in the Tangshan, China, earthquake, which was not predicted. The Chinese have also issued false alarms. In a province near Hong Kong, people reportedly left their dwellings for over a month, but no earthquake followed. Clearly, whatever method the Chinese employ for short-range predictions, it is *not* reliable.

In order for a prediction scheme to warrant general acceptance, it must be both accurate and reliable. Thus, *it must have a small range of uncertainty as regards to location and timing, and it must produce few failures or false alarms.* Can you imagine the debate that would precede an order to evacuate a large city in the United States, such as Los Angeles or San Francisco? The cost of evacuating millions of people, arranging for living accommodations, and providing for their lost work time and wages would be staggering.

Currently, *no reliable method exists* for making short-range earthquake predictions. In fact, except for a brief period of optimism during the 1970s, the leading seismologists of the past 100 years have generally concluded that short-range earthquake prediction is *not* feasible. To quote Charles Richter, developer of the well-known magnitude scale, "Prediction provides a happy hunting ground for amateurs, cranks, and outright publicity-seeking fakers." This statement was validated in 1990 when Iben Browning, a self-proclaimed expert, predicted that a major earthquake on the New Madrid fault would devastate an area around southeast Missouri on December 2 or 3. Many people in Missouri, Tennessee, and Illinois rushed out to buy earthquake insurance. Some schools and factories closed, while people as far away as northern Illinois stayed home rather than risk traveling to work. The designated date passed without even the slightest tremor.

Long-Range Forecasts

In contrast to short-range predictions, which aim to predict earthquakes within a time frame of hours or at most days, long-range forecasts give the probability of a certain magnitude earthquake occurring on a time scale of 30 to 100 years or more. Stated another way, these forecasts give statistical estimates of the expected intensity of ground motion for a given area over a specified time frame. Although long range forecasts may not be as informative as we might like, these data are important for updating the Uniform Building Code, which contains nationwide standards for designing earthquake-resistant structures.

Long-range forecasts are based on the premise that earthquakes are repetitive or cyclical, like the weather. In other words, as soon as one earthquake is over, the continuing motions of Earth's plates begin to build strain in the rocks again until they fail once more. This has led seismologists to study historical records of earthquakes to see if there are any discernible patterns so that the probability of recurrence might be established.

With this concept in mind, a group of seismologists plotted the distribution of rupture zones associated with great earthquakes that have occurred in the seismically active regions of the Pacific Basin. The maps revealed that individual rupture zones tended to occur adjacent to one another without appreciable overlap, thereby tracing out a plate boundary. Recall that most earthquakes are generated along plate boundaries by the relative motion of large crustal blocks. Because plates are in constant motion, the researchers predicted that over a span of one or two centuries, major earthquakes would occur along each segment of the Pacific plate boundary.

When the researchers studied historical records, they discovered that some zones had not produced a great earthquake in more than a century. These quiet zones, called **seismic gaps,** were identified as probable sites for major earthquakes in the next few decades (Figure 11.26). In the 25 years since the original studies were conducted, some of these gaps have ruptured (see Box 11.3). Included in this group is the zone that produced the earthquake that devastated portions of Mexico City in September 1985.

Another method of long-term forecasting, known as *paleoseismology* (*palaois* = an- cient, *seismos* = shake, *ology* = the study of), has been implemented. One technique involves the study of layered deposits that were offset by prehistoric seismic disturbances. To date, the most complete investigation that employed this method focused on a segment of the San Andreas Fault about 50 kilometers (30 miles) northeast of Los Angeles. Here the drainage of Pallet Creek has been repeatedly disturbed by successive ruptures along the fault zone.

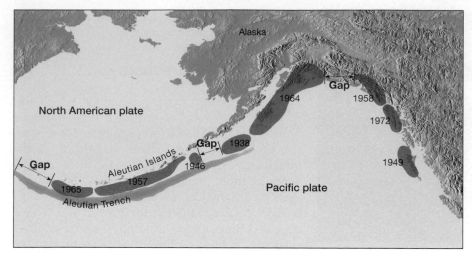

FIGURE 11.26 The distribution of rupture areas of large, shallow earthquakes from 1930 to 1979 along the southwestern coast of Alaska and the Aleutian Islands. The three seismic gaps denote the most likely locations for the next large earthquakes along this plate boundary. (After J. C. Savage et al., U.S. Geological Survey)

Ditches excavated across the creek bed have exposed sediments that had apparently been displaced by nine large earthquakes over a period of 1400 years. From these data it was determined that a great earthquake occurs here an average of once every 140 to 150 years. The last major event occurred along this segment of the San Andreas Fault in 1857. Thus, roughly 140 years have elapsed. If earthquakes are truly cyclic, a major event in southern California seems imminent. Such information led the U.S. Geological Survey to predict that there is a 50 percent probability that an earthquake of magnitude 8.3 will occur along the southern San Andreas Fault within the next 30 years.

Using other paleoseismology techniques, researchers recently discovered strong evidence that very powerful earthquakes (magnitude of 8 or larger) have repeatedly struck the Pacific Northwest over the past several thousand years. The most recent event occurred about 300 years ago. As a result of these findings, public officials have taken steps to strengthen some of the region's existing dams, bridges, and water systems. Even the private sector responded. The U.S. Bancorp building in Portland, Oregon, was strengthened at a cost of $8 million and now exceeds the standards of the Uniform Building Code.

Another U.S. Geological Survey study gives the probability of a rupture occurring along various segments of the San Andreas Fault for the 30 years between 1988 and 2018 (Figure 11.27). From this investigation, the Santa Cruz Mountains region was given a 30 percent probability of producing a 6.5-magnitude earthquake during this time period. In fact, the region experienced the Loma Prieta quake in 1989, of 6.9 magnitude.

The region along the San Andreas Fault given the highest probability (90 percent) of generating a quake is the Parkfield section. This area has been called the "Old Faithful" of earthquake zones because activity here has been very regular since record-keeping began in 1857. In late September 2004, a magnitude 6.0 earthquake again struck this area. Although the event was more than a decade overdue, it did demonstrate the potential usefulness of long-range forecasts. Another section between Parkfield and the Santa Cruz Mountains is given a very low probability of generating an earthquake. This area has experienced very little seismic activity in historical times; rather, it exhibits a slow, continual movement known as *fault creep*. Such movement is beneficial because it prevents strain from building to high levels in the rocks.

In summary, it appears that the best prospects for making useful earthquake predictions involve forecasting magnitudes and locations on time scales of years or perhaps even decades. These forecasts are important because they provide information used to develop the Uniform Building Code and to assist in land-use planning.

FIGURE 11.27 Probabilities of a major earthquake between 1988 and 2018 along the San Andreas Fault.

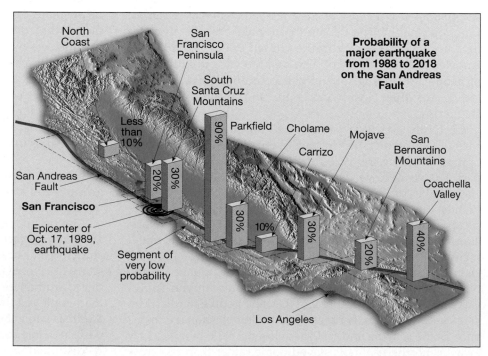

Earthquakes: Evidence for Plate Tectonics

 Earthquakes
▸ **Evidence at Plate Boundaries**

No sooner had the basic outline of the plate tectonics theory been formulated than researchers from various branches of the geosciences began to test its validity. One of the first efforts was undertaken by a group of seismologists, who were able to demonstrate a good fit between the newly developed plate tectonics model and the global distribution of earthquakes shown in Figure 11.13. In particular, these scientists were able to account for the close association between deep-focus earthquakes and subduction zones.

Based on our understanding of the mechanism that generates most earthquakes, one could predict that earthquakes should occur only in Earth's cool, rigid, outermost layer. Recall that as these rocks are deformed, they bend and store elastic energy—like a stretched rubber band. Once the rock is strained sufficiently, it ruptures, releasing the stored energy as the vibrations of an earthquake. By contrast, the hot mobile rocks of the asthenosphere are not capable of storing elastic energy and, therefore, should not generate earthquakes. Yet earthquakes having depths of nearly 700 kilometers (435 miles) are known.

The unique connection between deep-focus earthquakes and oceanic trenches was established through studies conducted in the Tonga Islands. When the depth of earthquake foci and their location within the Tonga arc are plotted, the pattern shown in Figure 11.28 emerges. Most shallow-focus earthquakes occur within or adjacent to the trench, whereas intermediate- and deep-focus earthquakes occur toward the Tonga Islands.

In the plate tectonics model, deep-ocean trenches form where dense slabs of oceanic lithosphere plunge into the mantle (Figure 11.28). Shallow-focus earthquakes are produced in response to the bending and fracturing of the lithosphere as it begins its descent, or as the subducting slab interacts with the overriding plate. As the slab descends farther into the asthenosphere, deeper-focus earthquakes are generated by other mechanisms. Much of the available evidence suggests that the earthquakes occur in the relatively cool subducting slab rather than in the ductile rocks of the mantle. Very few earthquakes have been recorded below 700 kilometers, possibly because the subducting slab has been heated sufficiently to lose its rigidity.

Additional evidence supporting the plate tectonics model came from observations that *only* shallow-focus earthquakes occur along divergent and transform fault boundaries. Recall that along the San Andreas Fault most earthquakes occur in the upper 20 kilometers (12 miles) of the crust. Because oceanic trenches are the only places where cold slabs of oceanic crust plunge to great depths, these should be the only sites of deep focus earthquakes. Indeed, the absence of deep-focus earthquakes along oceanic ridges and transform faults supports the theory of plate tectonics.

FIGURE 11.28   Idealized distribution of earthquake foci in the vicinity of the Tonga Trench. Note that the intermediate- and deep-focus earthquakes occur only within the sinking lithosphere. (Modified after B. Isacks, J. Oliver, and L. R. Sykes)

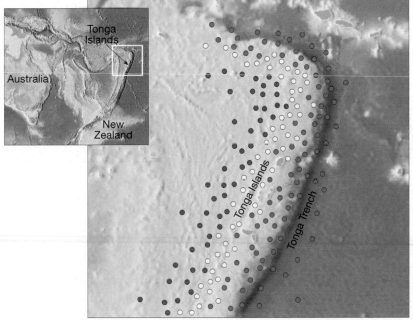
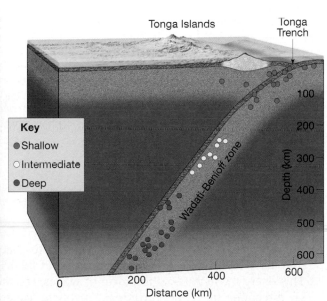

BOX 11.3 ▶ UNDERSTANDING EARTH

A Major Earthquake in Turkey

On August 17, 1999, at 3:02 A.M. local time, northwestern Turkey was shaken by a moment magnitude (M_W) 7.4 earthquake, catching most people in their sleep. The epicenter was 10 kilometers southeast of Ismit in a region that is the industrial center and most densely populated part of the country (see chapter opening photo). Istanbul and its 13 million people are just 70 kilometers (43 miles) to the west.

According to official government estimates, the earthquake killed more than 17,000 people and injured nearly 44,000 (Figure 11.D). More than 250,000 people were forced out of their damaged homes and were sheltered in 120 makeshift "tent cities." Estimates of property losses by the World Bank approached $7 billion. Liquefaction and ground shaking were the dominant causes of damage, but surface faulting and landslides were also responsible for substantial death and destruction. It was the most devastating earthquake to strike Turkey in 60 years.

Turkey is a geologically active region that frequently experiences large earthquakes. Most of the country is part of a small block of continental lithosphere known as the Turkish microplate. This small plate is caught between the northward-moving Arabian and African plates and the relatively stable Eurasian plate (Figure 11.E). The August 1999 earthquake occurred along the western end of the 1500-kilometer- (930-mile-) long North Anatolian fault system. This fault has much in common with California's San Andreas Fault. Both are right-lateral strike-slip faults having similar lengths and similar long-term rates of movement.* Also like its North American

*Recall that if a person is looking across a right-lateral strike-slip fault during an earthquake, that person would see the opposite side move to the right.

FIGURE 11.D Earthquake damage near Ismit, Turkey, 1999. (Photo courtesy of CORBIS/SYGMA)

counterpart, the North Anatolian Fault is a transform plate boundary.

The fact that a large earthquake occurred along this portion of the North Anatolian Fault did not come as a complete surprise. Based on historical records, the region of the epicenter had been identified as a *seismic gap*, a "quiet zone" along the fault where strain had been building for perhaps 300 years. Moreover, during the preceding 60 years an interesting pattern of seismic activity had developed. Beginning in 1939 with a (M_W) 7.9 quake that produced about 350 kilometers of ground rupture, seven earthquakes had broken the fault progressively from east to west as shown in Figure 11.F.

Researchers now understand that as each earthquake occurred, it loaded the zone to the west with additional stress. That is, as a quake released stress on the section of the fault it broke, it transferred stress to adjacent segments. The next segment in line to break is west of Ismit, near Istanbul. It could happen relatively soon. In the sequence since 1939, no earthquake waited longer than 22 years, and some came within a year of the one before.

The 1999 earthquake near Ismit, Turkey, demonstrated the awesome power of a large earthquake and the immense human suffering that can occur when an earthquake strikes an urban area. Although no one knows for sure where or when the next major quake will occur in the region, it appears that the 1999 earthquake near Ismit increased the risk for those living near Istanbul.

Summary

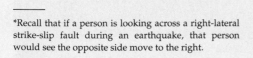

- *Earthquakes* are vibrations of Earth produced by the rapid release of energy from rocks that rupture because they have been subjected to stresses beyond their limit. This energy, which takes the form of waves, radiates in all directions from the earthquake's source, called the *focus*. The movements that produce most earthquakes occur along large fractures, called *faults*, that are usually associated with plate boundaries.

- Along a fault, rocks store energy as they are bent. As slippage occurs at the weakest point (the focus), displacement will exert stress farther along a fault, where additional slippage will occur until most of the built-up

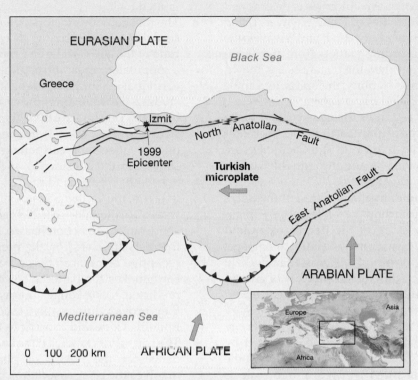

FIGURE 11.E   Earthquakes in Turkey are caused by the northward movement of the Arabian and African plates against the Eurasian plate, squeezing the small Turkish microplate westward. Movement is accommodated along two major strike-slip faults—the North Anatolian Fault and the East Anatolian Fault.

FIGURE 11.F   This map depicts the sequential westward progression of large earthquakes along the North Anatolian Fault between 1939 and 1999. The epicenter and magnitude of each is noted. The length of each colored segment shows the extent of surface rupture along the fault for each event.

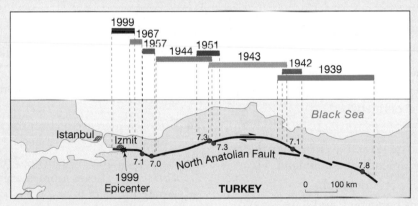

strain is released. An earthquake occurs as the rock elastically returns to its original shape. The "springing back" of the rock is termed *elastic rebound*. Small earthquakes, called *foreshocks*, often precede a major earthquake. The adjustments that follow a major earthquake often generate smaller earthquakes called *aftershocks*.

- Two main types of *seismic waves* are generated during an earthquake: (1) *surface waves*, which travel along the outer layer of Earth, and (2) *body waves*, which travel through Earth's interior. Body waves are further divided into *primary*, or *P, waves*, which push (compress) and pull (expand) rocks in the direction the wave is traveling, and *secondary*, or *S, waves*, which "shake" the particles in rock at right angles to their direction of travel. P waves can travel through solids, liquids, and gases. Fluids (gases and liquids) will not transmit S waves. In any solid material, P waves travel about 1.7 times faster than do S waves.

- The location on Earth's surface directly above the focus of an earthquake is the *epicenter*. An epicenter is determined using the difference in velocities of P and S waves. Using the difference in arrival times between P and S waves, the distance separating a recording station from the earthquake can be determined. When the distances are known from three or more seismic stations, the epicenter can be located using a method called *triangulation*.

- *A close correlation exists between earthquake epicenters and plate boundaries.* The principal earthquake epicenter zones are along the outer margin of the Pacific Ocean, known as the *circum-Pacific belt*, and through the world's oceans along the *oceanic ridge system*.

- Seismologists use two fundamentally different measures to describe the size of an earthquake—intensity and magnitude. *Intensity* is a measure of the degree of ground shaking at a given locale based on the amount of damage. The *Modified Mercalli Intensity Scale* uses damages to buildings in California to estimate the intensity of ground shaking for a local earthquake. *Magnitude* is calculated from seismic records and estimates the amount of energy released at the source of an earthquake. Using the *Richter scale*, the magnitude of an earthquake is estimated by measuring the *amplitude* (maximum displacement) of the largest seismic wave recorded. A logarithmic scale is used to express magnitude, in which a tenfold increase in ground shaking corresponds to an increase of 1 on the magnitude scale. *Moment magnitude* is currently used to estimate the size of moderate and large earthquakes. It is calculated using the average displacement of the fault, the area of the fault surface, and the sheer strength of the faulted rock.

- The most obvious factors determining the amount of destruction accompanying an earthquake are the magnitude of the earthquake and the proximity of the quake to a populated area. Structural damage attributable to earthquake vibrations depends on several factors, including (1) wave amplitudes, (2) the duration of the vibrations, (3) the nature of the material upon which the structure rests, and (4) the design of the structure. Secondary effects of earthquakes include *tsunami*, landslides, ground subsidence, and fire.

- Substantial research to predict earthquakes is under way in Japan, the United States, China, and Russia—countries where earthquake risk is high. No reliable method of short-range prediction has yet been devised. Long-range forecasts are based on the premise that earthquakes are repetitive or cyclical. Seismologists study the history of earthquakes for patterns so their occurrences might be predicted. Long-range forecasts are important because they provide information used to develop the Uniform Building Code and to assist in land-use planning.

- The distribution of earthquakes provides strong evidence for the theory of plate tectonics. One aspect involves the close association between deep-focus earthquakes and subduction zones. Additional evidence involves the fact that only shallow-focus earthquakes occur at divergent and transform fault boundaries.

Review Questions

1. What is an earthquake? Under what circumstances do earthquakes occur?

2. How are faults, foci, and epicenters related?

3. Who was first to explain the actual mechanism by which earthquakes are generated?

4. Explain what is meant by *elastic rebound.*

5. Faults that are experiencing no active creep may be considered "safe." Rebut or defend this statement.

6. Describe the principle of a seismograph.

7. List the major differences between P and S waves.

8. P waves move through solids, liquids, and gases, whereas S waves move only through solids. Explain.

9. Which type of seismic wave causes the greatest destruction to buildings?

10. Using Figure 11.11, determine the distance between an earthquake and a seismic station if the first S wave arrives 3 minutes after the first P wave.

11. Most strong earthquakes occur in a zone on the globe known as the ____.

12. Deep-focus earthquakes occur several hundred kilometers below what prominent features on the deep-ocean floor?

13. Distinguish between the Mercalli scale and the Richter scale.

14. For each increase of 1 on the Richter scale, wave amplitude increases ____ times.

15. An earthquake measuring 7 on the Richter scale releases about ____ times more energy than an earthquake with a magnitude of 6.

16. List three reasons that the moment magnitude scale has gained popularity among seismologists.

17. List four factors that affect the amount of destruction caused by seismic vibrations.

18. What factor contributed most to the extensive damage that occurred in the central portion of Mexico City during the 1985 earthquake? (see Box 11.2.)

19. The 1988 Armenian earthquake had a Richter magnitude of 6.9, less than the 1994 Northridge California

earthquake. Nevertheless, the loss of life was far greater in the Armenian event. Why?

20. In addition to the destruction created directly by seismic vibrations, list three other types of destruction associated with earthquakes.

21. What is a tsunami? How is one generated?

22. Cite some reasons why an earthquake with a moderate magnitude might cause more extensive damage than a quake with a high magnitude.

23. Can earthquakes be predicted?

24. What is the value of long-range earthquake forecasts?

25. Briefly describe how earthquakes can be used as evidence for the theory of plate tectonics.

Key Terms

aftershock (p. 299)
body wave (p. 302)
earthquake (p. 296)
elastic rebound (p. 298)
epicenter (p. 302)
fault (p. 287)
fault creep (p. 300)
focus (p. 296)
foreshock (p. 297)

hypocenter (p. 296)
inertia (p. 301)
intensity (p. 305)
liquefaction (p. 310)
long (L) waves (p. 302)
magnitude (p. 305)
Modified Mercalli Intensity
    Scale (p. 307)
moment magnitude (p. 309)

primary (P) waves (p. 302)
Richter scale (p. 307)
secondary (S) waves
    (p. 302)
seismic gaps (p. 317)
seismic sea wave (p. 312)
seismogram (p. 302)
seismograph (p. 301)
seismology (p. 333)

surface wave (p. 302)
tsunami (p. 312)
Wadati–Benioff zones
    (p. 305)

Web Resources

The *Earth* Website uses the resources and flexibility of the Internet to aid in your study of the topics in this chapter. Written and developed by geology instructors, this site will help improve your understanding of geology. Visit **http://www.prenhall.com/tarbuck** and click on the cover of *Earth 9e* to find:

- Online review quizzes.
- Critical thinking exercises.
- Links to chapter-specific Web resources.
- Internet-wide key-term searches.

http://www.prenhall.com/tarbuck

GEODe: Earth

GEODe: Earth makes studying faster and more effective by reinforcing key concepts using animation, video, narration, interactive exercises and practice quizzes. A copy is included with every copy of *Earth*.

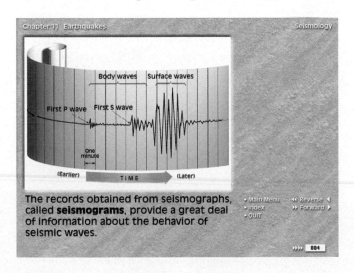

The records obtained from seismographs, called **seismograms**, provide a great deal of information about the behavior of seismic waves.

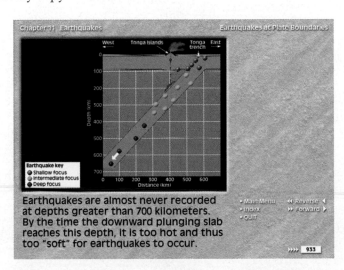

Earthquakes are almost never recorded at depths greater than 700 kilometers. By the time the downward plunging slab reaches this depth, it is too hot and thus too "soft" for earthquakes to occur.

Earth's Interior*

Volcanic eruptions, such as this one at Italy's Mount Etna, provide data about the nature of Earth's interior (Photo by Marco Fulle).

*This chapter was prepared by Professor Michael Wysession, Washington University.

f you could slice any planet in half, the first thing you would notice is that it would be divided into distinct layers. The heaviest materials (metals) would be at the bottom. Lighter solids (rocks) would be in the middle. Liquids and gases would be at the top. Within Earth, we know these layers as the iron core, the rocky mantle and crust, the liquid ocean, and the gaseous atmosphere. More than 95 percent of the variations in composition and temperature within Earth are due to layering. However, this is not the end of the story. If it were, Earth would be a dead, lifeless cinder floating in space.

There are also small horizontal variations in composition and temperature at depth that indicate the interior of our planet is very active. The rocks of the mantle and crust are in constant motion, not only moving about through plate tectonics, but also continuously re-cycling between the surface and the deep interior. It is also from Earth's deep interior that the water and air of our oceans and atmosphere are replenished, allowing life to exist at the surface.

Discovering and identifying the patterns of Earth's deep motions have not been easy. Light does not travel through rock, so we must find other ways to "see" into our planet. The seismic waves associated with earthquakes are one means used to investigate Earth's interior. Other techniques include mineral physics experiments that can recreate the conditions of extreme temperature and pressure inside planets and gravity measurements that show where there are internal variations in the distribution of mass. Examining Earth's magnetic field gives clues to the patterns of flow of liquid iron in the core. Taken together, all these different fields of study give us a picture of Earth as a churning, varied, complex planet that continues to change and evolve over time.

Gravity and Layered Planets

If a bottle were filled with clay, iron filings, water, and air and then shaken, it would appear to have a single, muddy composition. If that bottle were allowed to sit, however, the different materials would settle out into layers. The iron fil-ings, which are the densest, would sink to the bottom. Above the iron would be the clay, then water, then air. This is what happens inside planets. At their birth, planets form from an accumulation of nebular debris but quickly begin to form layers. The iron sinks to form the core, rock forms the mantle and crust, and gases form the atmosphere. All large bodies in the solar system have iron cores and rocky man-tles, even Jupiter, Saturn, and the Sun. A profile of Earth's layered structure is shown in Figure 12.1.

For both the bottle of mud and Earth it is the force of gravity that is responsible for the layering. Figure 12.1 shows another interesting effect of gravity. Not only does the density change between layers, but it changes within layers. This is because materials compress when you squeeze them. Rock having the composition of the upper mantle has a density of about 3.3 g/cm^3 at Earth's surface. But, take that rock to the base of the mantle and its density increases to 5.6 g/cm^3, nearly twice as much. The intense pressure of the overlying rock causes rock at the base of the mantle to be compressed into nearly half its volume!

This increase in density occurs partly because the intense pressure causes atoms to shrink in size. However, atoms do not all compress at the same rate. It is easier to compress negative ions than positive ions. Negative ions have more electrons than protons, and tend to be "fluffier" than posi-tive ions. When rocks are squeezed, the negative ions (such as O^{-2}) compress more easily than the positive ions (such as Si^{+4} and Mg^{+2}), so the ratios of the ionic sizes change. When these ratios change enough, the structure of a mineral is no longer stable, and the atoms rearrange into a more stable and denser structure. This is called a *mineral phase change*, as discussed in Chapter 3. The increase in density of mantle rocks is due both to the compression of existing minerals and to the transition to new "high-pressure" minerals.

Probing Earth's Interior: "Seeing" Seismic Waves

The best way to learn about Earth's interior is to dig or drill a hole and examine it directly. Unfortunately, this is only possible at shallow depths. The deepest a drill has ever pen-etrated is only 12.3 kilometers, which is about 1/500 of the way to Earth's center! Even this was an extraordinary ac-complishment because temperature and pressure increase so rapidly with depth.

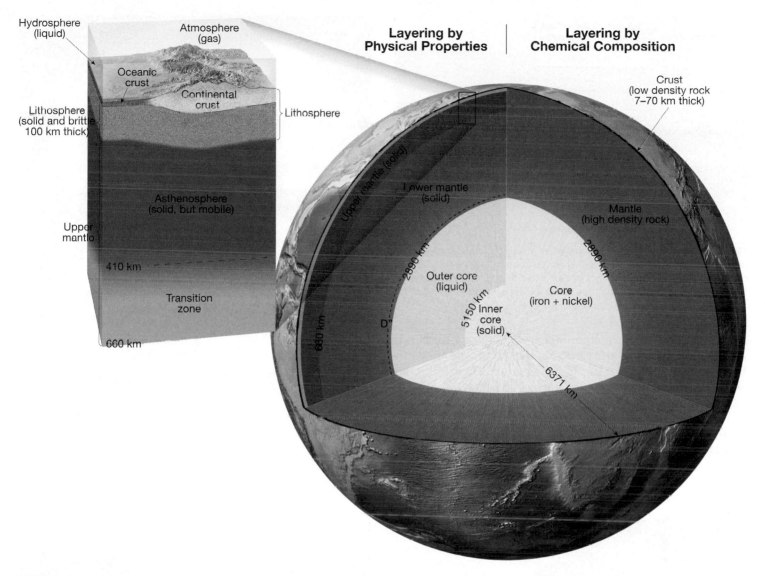

Layering by Physical Properties   **Layering by Chemical Composition**

FIGURE 12.1 The layers of Earth, shown both in terms of physical properties and chemical composition. The physical properties of Earth's layers (left) include the physical state of the material (solid, liquid, or gas) as well as how stiff the material is (for example, the distinction between the lithosphere and asthenosphere). The chemical layers are mainly determined by density, with the heaviest materials at the center and the lightest ones at the surface.

Fortunately, many earthquakes are large enough that their seismic waves travel all the way through Earth and can be recorded on the other side (Figure 12.2). This means that the seismic waves act like medical X rays used to take images of a person's insides. There are about 100 to 200 earthquakes each year that are large enough (about Mw > 6) to be well-recorded by seismographs all around the globe. These large earthquakes provide the means to "see" into our planet. Consequently, they have been the source of most of the data that allowed us to figure out the nature of Earth's interior.

Interpreting the waves recorded on seismograms in order to identify Earth structure is difficult. This is because seismic waves usually do not travel along straight paths. Instead, seismic waves are reflected, refracted, and diffracted as they pass through our planet. They reflect off of

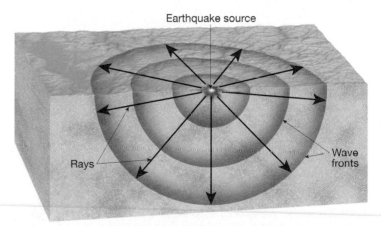

FIGURE 12.2 When traveling through a medium with uniform properties, seismic waves spread out from an earthquake source (focus) as spherically shaped structures called *wave fronts*. It is common practice however, to consider the paths taken by these waves as *seismic rays*, lines drawn perpendicular to the wave front as shown in this diagram.

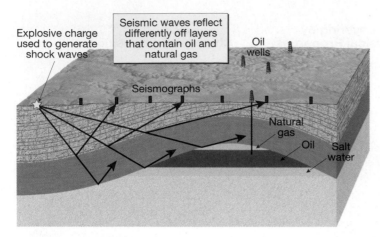

FIGURE 12.3 Reflected seismic waves are used to search for oil and natural gas underground. The seismic waves from explosions reflect differently from layers of rock that contain liquid oil and natural gas, and thus are used to map petroleum reservoirs in Earth's crust.

boundaries between different layers, they refract (or bend) when passing from one layer to another layer, and they diffract around any obstacles they encounter. These different wave behaviors have been used to identify the boundaries that exist within Earth.

As Figure 12.3 shows, changes in the composition or structure of rock cause seismic waves to reflect off of boundaries between different materials. This is especially

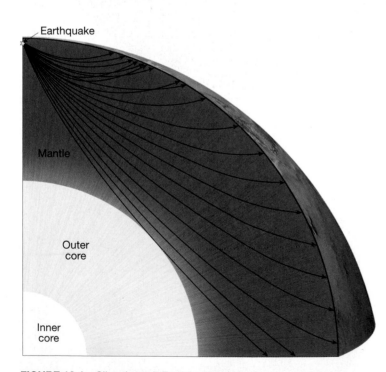

FIGURE 12.4 Slice through Earth's mantle showing some of the ray paths that seismic waves from an earthquake would take. The rays follow curved (refracting) paths rather than straight paths because the seismic velocity of rocks increases with depth in the mantle, a result of increasing pressure with depth. Notice the complicated ray paths in the upper mantle, with some even crossing each other. This is due to the sudden seismic velocity increases that result from mineral phase changes at increasing pressures.

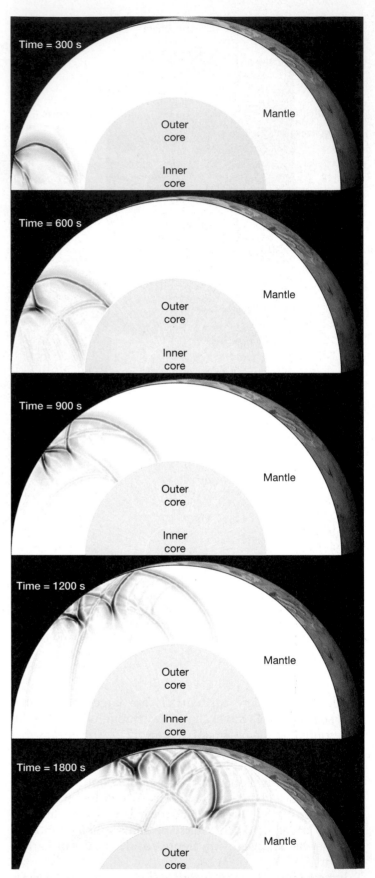

FIGURE 12.5 Five snapshots in time showing the locations of S waves within Earth's mantle following an earthquake. In addition to refracting and diffracting, S waves reflect from boundaries such as the core-mantle boundary. Note that S waves do not penetrate the outer core because S waves do not travel through liquids.

BOX 12.1 ▶ UNDERSTANDING EARTH

Re-creating the Deep Earth

Seismology alone cannot determine what Earth is made of. Additional information must be obtained by some other means so that the seismic velocities can be interpreted in terms of rock type. This is done using *mineral physics* experiments performed in laboratories. By squeezing and heating minerals and rocks, physical properties like stiffness, compressibility, and density (and therefore seismic velocities) can be directly measured. This means that the conditions of the mantle and core can be simulated, and the results compared to seismic modeling.

Most mineral physics experiments are done using giant presses involving very hard carbonized steel. The highest pressures, however, are obtained using diamond-anvil presses like the one shown in Figure 12.A. These take advantage of two important characteristics of diamonds—their hardness and transparency. The tips of two diamonds are cut off, and a small sample of mineral or rock is placed in between. Pressures as great as those in the interior of Jupiter have been obtained by squeezing the two diamonds together. High temperatures are achieved by firing a laser beam through the diamond and into the mineral sample.

Besides measuring seismic velocities at the conditions of different depths within Earth, there are other important mineral

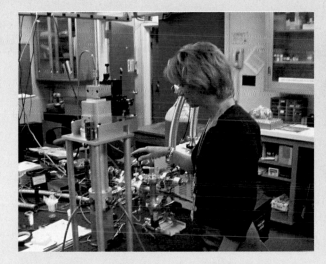

FIGURE 12.A   High-pressure experiments inside a diamond anvil cell (left photo) can recreate the conditions at the center of a planet. The whole apparatus is small, and can fit on top of a table. High pressures are generated by cutting the tips off of high-quality diamonds (right photo), putting a small sample of rock between, squeezing the diamonds together and heating the sample with a laser. (Left photo courtesy of Lawrence Livermore National Laboratory; right photo by Douglass L. Peck Photography)

physics experiments. One experiment determines the temperature at which minerals will begin to melt under various pressures. Another experiment determines (at different temperatures) the pressures at which one mineral phase will become unstable and convert into a new "high-pressure" phase. Yet another involves making these same tests for slightly different mineral compositions. All of these experiments are needed because, as will be discussed later, there are three-dimensional changes in composition and temperature within Earth.

---

important in the exploration for oil and natural gas where artificially generated seismic waves are used to probe the crust. Petroleum tends to get trapped in certain kinds of geological structures, and these structures are identified by mapping out the layering of the upper crust. The price of gasoline would be much more expensive without the existence of seismic imaging because a huge number of drill wells would have to be randomly deployed to find oil. Using seismic waves, companies can just drill in the places that are most likely to have petroleum. Seismic waves also reflect off of boundaries between the crust, mantle, outer core, inner core, and off of other boundaries within the mantle as well.

One of the most noticeable behaviors of seismic waves is that they follow strongly curved paths (Figure 12.4). This occurs because the velocity of seismic waves generally increases with depth. In addition, seismic waves travel faster when rock is stiffer or less compressible. These properties of stiffness and compressibility are then used to interpret the composition and temperature of the rock. For instance, when rock is hotter it becomes less stiff (imagine taking a frozen chocolate bar and then heating it up!), and waves

travel more slowly. Waves also travel at different speeds through rocks of different compositions. Thus the speed that seismic waves travel can help determine both the kind of rock that is inside Earth and how hot it is.

Within Earth's mantle, where there are both sharp boundaries and gradual seismic velocity changes, the pattern of seismic waves becomes quite complex. Figure 12.5 shows what S waves from a deep earthquake look like as they travel through the mantle. Note how the single wave from the shock is soon broken up into many different waves, which appear on the seismograms as separate signals.

Earth's Layers

GEODe   Earth's Interior
▶ Earth's Layered Structure

Combining the data obtained from seismological studies and mineral physics has given us a layer-by-layer understanding of the composition of Earth (see Box 12.1). Seismic velocities,

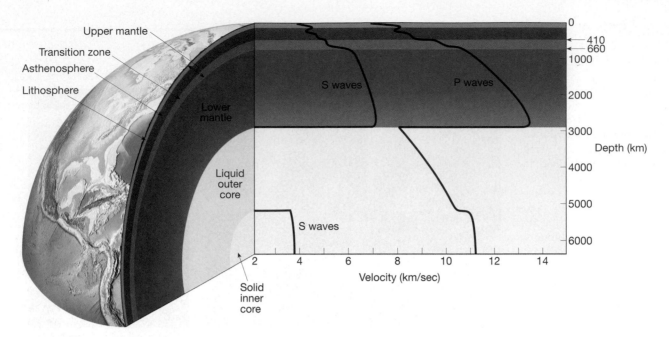

FIGURE 12.6 Cutaway of Earth showing its different layers and the average velocities of P and S waves at each depth. S waves are an indication of how rigid the material is—the inner core is less rigid than the mantle, and the liquid outer core has no rigidity.

as a function of depth, are shown in Figure 12.6. By examining the behavior of a variety of rocks at the pressures corresponding to these depths, geologists have been able to figure out the compositions of Earth's crust, mantle, and core.

Earth's Crust

Earth's **crust** is of two different types—continental crust and oceanic crust. Both share the word "crust," but the similarity ends there. Continental and oceanic crusts have very different compositions, histories, ages, and styles of formation. In fact, the ocean crust is much more similar to rock of the mantle than to rock of the continental crust.

Oceanic Crust Seismic imaging has shown that the ocean crust is usually about 7 kilometers (4.5 miles) thick. All ocean crust forms at mid-ocean ridges, which separate two diverging tectonic plates. Ocean crust has P wave velocities of about 5–7 km/s and a density of about 3.0 g/cm^3, which agrees with experimental values for the rocks basalt and gabbro. The composition and formation of ocean crust is discussed further in Chapter 13.

Continental Crust While oceanic crust is fairly uniform throughout the oceans, no two continental regions have the same structure or composition. Continental crust averages about 40 kilometers (25 miles) in thickness, but can be more than 70 kilometers (45 miles) thick in certain mountainous regions like the Himalayas and Andes. The thinnest crust in North America is beneath the Basin and Range region of the western United States, where the crust can be as thin as 20 kilometers (12 miles). The thickest North American crust, beneath the Rockies, is more than 50 kilometers (30 miles) thick.

Seismic velocities within continents are quite variable, suggesting that the composition of continental crust must also vary greatly. This agrees with what has been learned about the different ways continents can form, discussed in Chapter 22. In general, however, continents have a density of about 2.7 gm/cm^3, which is much lower than both oceanic crust and mantle rock. This lower density explains why continents are buoyant—acting like giant rafts, floating atop tectonic plates, and why they cannot be subducted into the mantle.

Discovering Boundaries: The Moho The boundary between the crust and mantle, called the **Moho,** was one of the first features of Earth's interior discovered using seismic waves. Croatian seismologist Andrija Mohorovičić discovered this boundary in 1909, which is named in his honor. At the base of the continents P waves travel about 6 km/s but abruptly increase to 8 km/s at a slightly greater depth.

Andrija Mohorovičić cleverly used this large jump in seismic velocity to discover the Moho. He noticed that there were two different sets of seismic waves that were recorded at seismographs located within a few hundred kilometers of an earthquake. One set of waves moved across the ground at about 6 km/s, and the other set of waves moved across the ground at about 8 km/s. From these two waves, Mohorovičić correctly determined that the different waves were coming from two different layers, as shown in Figure 12.7.

When a shallow earthquake occurs, there is a *direct wave* that moves straight through the crust and is recorded at nearby seismographs. (In Figure 12.7, the slope of the line gives the velocity of direct waves through the crust.) Seismic waves will also follow a path down through the crust and along the top of the mantle. These are called *refracted waves*

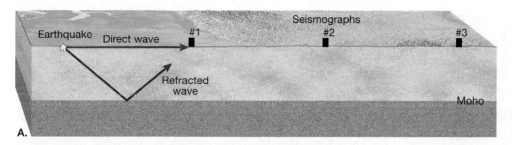

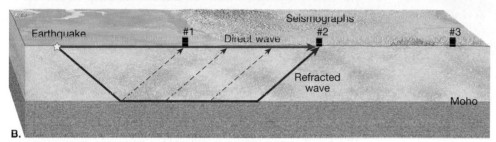

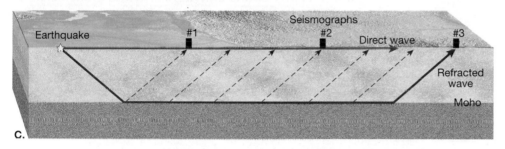

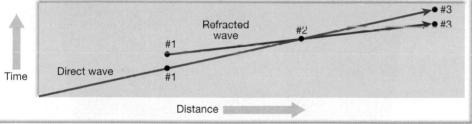

FIGURE 12.7 Diagram showing seismic waves from an earthquake arriving at three different seismographs. Over a short distance, such as at seismograph #1, the direct wave arrives first. For greater distances, such as at seismograph #3, the refracted wave arrives first. At the cross-over point, which in this diagram occurs at seismograph #2, both waves arrive at the same time. The distance to the cross-over point increases with the depth of the Moho, and therefore can be used to determine the thickness of the crust.

er time occurs if you first drive to the interstate and travel along it. The cross-over point, where both routes take the same amount of time, is directly related to how far you are from a major highway. (Or, when determining the depth of the Moho, how far the mantle [fast layer] is from the surface.)

Earth's Mantle

More than 82 percent of Earth's volume is contained within the **mantle,** a nearly 2900-kilometer thick shell extending from the base of the crust (Moho) to the liquid outer core. Because S waves readily travel through the mantle, we know that it is a solid rocky layer composed of silicate minerals that are enriched in iron and magnesium. However, despite its solid nature rock in the mantle is quite hot and capable of flow, albeit at very slow velocities.

The Upper Mantle The upper mantle extends from the Moho down to a depth of about 660 km. The upper mantle can be divided into three different parts. The top layer of the upper mantle is part of the stiff **lithosphere,** and beneath that is the weaker **asthenosphere.** These layers are a result of the temperature structure of Earth, and so are discussed later in this chapter. The bottom part of the upper mantle is called the *transition zone.*

We have a good sense of what the upper mantle is made of because mantle rocks are brought to the surface by several different geological processes. The seismic velocities we observe for the mantle are consistent with a rock called peridotite. Mantle *peridotite* is an ultramafic rock mostly composed of the minerals olivine and pyroxene. It is richer in the metals magnesium and iron than the minerals found in either the continental or oceanic crust.

The olivine crystals in peridotite display a very important property called **seismic anisotropy,** which means that seismic waves travel at different speeds along different paths through the crystals. With olivine, the crystals tend to line up with their fast directions pointing in the same direction that the rock is flowing. This is very fortunate for geologists, because if the fastest seismic wave direction through regions of the upper mantle can be found, then the direction the olivine is moving is also found. Seismology therefore not only provides a snapshot of the structure of Earth's

because they are bent, or refracted, as they enter the mantle. These refracted P waves will travel across the ground at the speed of the waves in the mantle (8 km/s). At nearby distances the direct wave arrives first. However, at greater distances the refracted wave is the first to arrive. The point at which both waves arrive at the same time, called the *crossover,* can be used to determine the depth of the Moho. Thus, using just these two waves and an array of seismographs, you can determine the thickness of the crust for any location.

The difference between direct and refracted waves is analogous to driving along local roads or taking the interstate highway. For short distances you will arrive faster if you just take the local roads. For greater distances, the short-

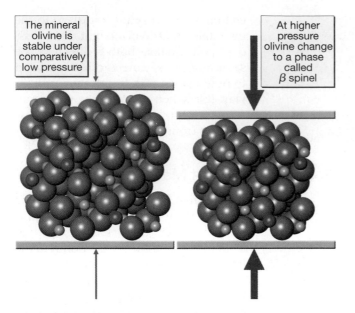

The mineral olivine is stable under comparatively low pressure

At higher pressure olivine change to a phase called β spinel

FIGURE 12.8   Demonstration of the effect of pressure on the structure of minerals. The mineral olivine, which is stable in the upper mantle, is no longer stable at the pressures in the transition zone, and converts to denser phases. In the top of the transition zone olivine converts to a mineral phase called β-spinel. The atoms are the same, but they are compressed into a more compact crystalline structure.

0.1 percent of its weight as water. Because the transition zone is 10 percent of the volume of Earth, it could potentially hold up to five times the volume of Earth's oceans. Water cycles slowly through the planet, brought down into the mantle with subducting oceanic lithosphere and carried upward by rising plumes of mantle rock. How much water is actually contained within the transition zone is not known.

The Lower Mantle   From 660 kilometers deep to the top of the core, at a depth of 2891 kilometers, is the **lower mantle.** Beneath the 660-kilometer discontinuity, both olivine and pyroxene take the form of the mineral *perovskite* (Fe, Mg) SiO_3. The lower mantle is by far the largest layer of Earth, occupying 56 percent of the volume of the planet. This means that perovskite is the single most abundant material within Earth.

The D″ Layer   In the bottom few hundred kilometers of the mantle, just above the core, is a highly variable and unusual layer called the **D″ layer** (pronounced "dee double-prime"). This is a boundary layer between the rocky mantle and the liquid iron outer core (Figure 12.9). The D″ layer is a

interior at this very moment but also a sense of where the rock within Earth will move to in the future.

Transition Zone   From about 410 km to about 660 km in depth is the part of the upper mantle called the **transition zone.** The top of the transition zone is identified by a sudden increase in density from about 3.5 to 3.7 g/cm³. Like the Moho, this boundary reflects seismic waves. Unlike the Moho, this boundary is not due to a change in chemical composition but to a change in mineral phase. The chemical composition above and below the 410-kilometer discontinuity is the same. However, the mineral olivine, which is stable in the upper mantle, is no longer stable at the pressures in the transition zone and converts to denser phases (Figure 12.8). In the top half of the transition zone olivine converts to a phase called β-spinel, and in the bottom half β-spinel converts into a true spinel structure called ringwoodite.

The most unusual thing about the transition zone is that it is capable of holding a great deal of water, up to 2 percent by weight. This is much more than for the peridotite of the upper mantle, which can only hold about

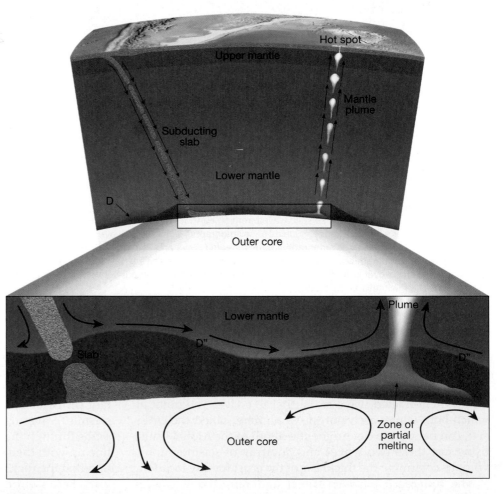

FIGURE 12.9   Schematic of the variable and unusual D″ layer at the base of the mantle. Like the lithosphere at the top of the mantle, the D″ layer contains large horizontal variations in both temperature and composition. Many scientists believe that D″ is the graveyard of some subducted ocean lithosphere and the birthplace of some mantle plumes.

lot like the lithosphere, which is the boundary layer between the mantle and the ocean/atmosphere layer. Both the lithosphere and D" layer have large variations in composition and temperature. The difference in lithospheric temperature between hot mid-ocean ridges and cold abyssal seafloor is more than 1000°C. The horizontal changes in temperature within the D" layer are similar. The composition of the lithosphere varies greatly, with either continental or ocean crust embedded in it. There also seem to be large slabs of differing rock types embedded within D".

The very base of D", the part of the mantle directly in contact with the hot liquid iron core, is like Earth's surface in that there are "upside-down mountains" of rock that protrude into the core. Furthermore, in some regions of the core–mantle boundary, the base of D" seems to be hot enough to be partially molten. This may be the cause of narrow zones at the very base of the mantle where P-wave velocities decrease by 10 percent and S-wave velocities decrease by 30 percent.

Discovering Boundaries: The Core–Mantle Boundary   Evidence that Earth has a distinct central core was uncovered in 1906 by a British geologist, Richard Dixon Oldham. (In 1914 Beno Gutenberg calculated the depth to the core boundary as 2900 kilometers, a value that has stood the test of time.) Oldham observed that at distances of more than about 100° from a large earthquake P and S waves were absent, or very weak. In other words, the central core produced a "*shadow zone*" for seismic waves as shown in Figure 12.10.

As Oldham predicted, Earth's core exhibits markedly different elastic properties from the mantle above, which causes considerable refraction of P waves—similar to how light is refracted (bent) as it passes from air to water. In addition, because the outer core is liquid iron, it blocks the transmission of S waves (recall S waves do not travel through liquids).

Figure 12.10 shows the locations of the P and S wave shadow zones and how the paths of the waves are affected by the core. While there are still P and S waves that arrive in the shadow zone, they are all very different than would be expected for a planet without a core.

Earth's Core

The Outer Core   The boundary between the mantle and the outer core, called the *core–mantle boundary*, is the most significant within Earth in terms of changes in material properties. P waves drop from 13.7 to 8.1 km/s at the core–mantle

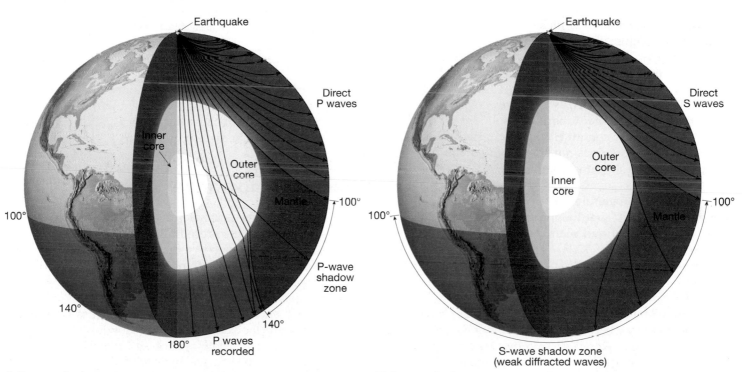

A. P-wave shadow zone   **B. S-wave shadow zone**

FIGURE 12.10   Two views of Earth's interior showing the effects of the outer and inner cores on the ray paths of P and S waves. **A.** When P waves interact with the slow-velocity liquid iron of the outer core, their rays are refracted downward. This creates a shadow zone where no direct P waves are recorded (although diffracted P waves travel there). The P waves that travel through the core are called PKP waves. The "K" represents the path through the core, and comes from the German word for core, which is *kern*. The increase in seismic velocity at the top of the inner core can refract waves sharply so that some arrive within the shadow zone, which is shown here as a single ray. **B.** The core is an obstacle to S waves, because they cannot pass through liquids. Therefore, a large shadow zone exists for S waves. However, some S waves diffract around the core and are recorded on the other side of the planet.

boundary, and S waves drop dramatically from 7.3 km/s to zero. Because S waves do not pass through liquids, the lack of any S waves in the **outer core** means that it is liquid. The change in density, from 5.6 to 9.9 g/cm³, is even larger than the rock–air difference observed at Earth's surface.

Based on our knowledge of the composition of meteorites and the Sun, geologists expect Earth to contain a great deal of iron. However, this iron is mostly missing from the crust and mantle. This fact, plus the great density of the core, tells us that it is mostly made of iron and some nickel, which has a similar density as iron.

The **core** is only about 1/6 of Earth's volume, but because iron is so dense, the core accounts for 1/3 of Earth's mass, and iron is Earth's most abundant element when measured by mass. The outer core is not pure iron, however. Its density and seismic velocities suggest that the outer core contains about 15 percent of other, lighter elements. These are likely to include sulfur, oxygen, silicon, and hydrogen. Based on mineral physics experiments, this is not surprising. For instance, pure iron melts at a very high temperature, but an iron-sulfur mixture melts at a much lower temperature. When Earth was forming and heating up, the iron that sank to form the core melted more easily in the presence of the sulfur, pulling it down into the core as well.

The Inner Core At the center of the core is a solid sphere of iron with lesser amounts of nickel called the **inner core.** In drawings like Figure 12.1 the inner core looks much larger than it really is. The inner core is actually very small, only 1/142 (less than one percent) of the volume of Earth. The inner core did not exist early in Earth's history, when the planet was hotter. However, as the planet cooled, iron began to crystallize at the center to form the solid inner core. Even today, the inner core continues to grow in size as the planet cools. The inner core does not contain the quantity of light elements found in the outer core.

The inner core is separated from the mantle by the liquid outer core, and is therefore free to move independently. Recent studies suggest that the inner core is actually rotating faster than the crust and mantle, lapping them every few hundred years (Figure 12.11). The inner core's small size and great distance from the surface make it the most difficult region within Earth to examine.

Discovering Boundaries: The Inner Core–Outer Core Boundary The boundary between the solid inner core and liquid outer core was discovered in 1936 by Danish seismologist Inge Lehman. She could not tell whether the inner core was actually solid or not, but using basic trigonometry reasoned that some P waves were being strongly refracted by a sudden increase in seismic velocities at the inner core–outer core boundary. This is the opposite situation as what occurs to produce the P-wave shadow zone. When seismic velocities suddenly decrease, such as at the mantle–outer core boundary, waves get bent so that there is a shadow zone where no direct waves arrive. When seismic waves suddenly increase, as they do at the outer core–inner core boundary, waves get bent so that several P waves can sometimes arrive

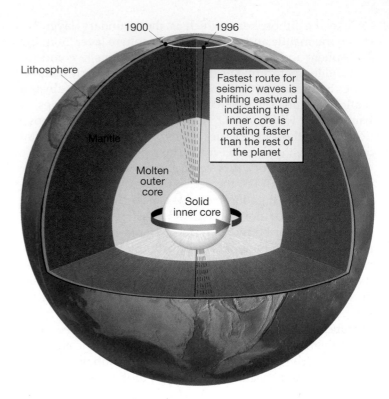

FIGURE 12.11 The solid inner core is separated from the mantle by the liquid outer core, and moves independently. Slight variations in the travel times of seismic waves through the core, measured over many decades, suggest that the inner core actually rotates faster than the mantle. The reason for this is not yet understood.

at a single location. In the case of the inner core, these waves can even be refracted enough to arrive within the P-wave shadow zone. Both of these occurrences, shown in Figure 12.10A, are proof of a distinct inner core.

Earth's Temperature

One way to describe a planet is by the composition of its layers, as was done in the preceding discussions. Another way is to examine the change in temperature with depth. This is very important for understanding the movements of rock within a planet. As you are probably aware, heat flows from hotter regions toward colder regions. Earth is about 5500°C at its center and 0°C at its surface, so heat is continually flowing toward the surface. It is this flow of heat that produces the convective flow of rock and metal in the mantle and core, and, in the process, plate tectonics.

We can measure the rate at which Earth is cooling by measuring the rate at which heat is escaping at Earth's surface. Measurements around the planet have shown that the average flow of heat at the surface is 87 milliwatts per square meter. This is not a lot, as it would take the energy emitted from about 690 square meters, roughly the size of a baseball diamond, to power one 60-watt light bulb. However, because Earth's surface is so large, heat leaves at a rate of 44 terrawatts per year, which is about three times the total world rate of energy consumption.

As Figure 12.12 shows, heat does not leave Earth's surface at the same rate in all locations. The rate of heat flow is highest near mid-ocean ridges, where hot magma is consistently rising toward the surface. Energy flow is also high in many continental regions because of particularly high levels of radioactive isotopes there. Heat flow is lowest in the areas of old, cold, ocean seafloor abyssal plains.

How Did Earth Get So Hot?

Earth, like all planets in our solar system, has had two thermal stages of existence. The first stage occurred during Earth's formation and involved a very rapid increase in internal temperature. The second stage has been the very slow process of cooling down. The first stage was very brief, taking only about 50 million years. The second stage has taken the remaining 4.5 billion years of Earth history and will continue for about another 4.5 billion years, until the Sun becomes a red giant star and Earth is destroyed.

As discussed in Chapter 1, Earth formed through a very violent process involving the collisions of countless planetesimals ("baby planets") during the birth of our solar system. With each collision, the kinetic energy of motion was converted into heat. As the early Earth grew in size, it rapidly got hotter. Several factors contributed to the early increase in temperature. The planet contained many relatively short-lived radioactive isotopes, such as Aluminum-26 and Calcium-41. As these isotopes decayed to stable isotopes, they released a great deal of energy. Further, as Earth's mass increased, so did its gravitational force of attraction, and this force caused the entire Earth to be compressed. This compression led to an increase in Earth's temperature, much the same way that compressing air in a bicycle pump causes the whole pump to get hot.

Two other events caused Earth's temperature to rise suddenly. The first was the collapse of the iron core. At some point during Earth's growth the temperature got high enough that iron started to melt. Droplets of liquid iron began to form and sink toward Earth's center to form the core. The sinking of these iron droplets released additional heat, which caused more iron to melt and sink, which released more heat and so on. The core likely formed quickly by this runaway process.

The second significant event that heated our planet was the collision of a Mars-sized object with Earth that led to the formation of the Moon. At this time the entire core was molten, and most, if not all, of the mantle as well. From that point, about 4.5 billion years ago, to the present, Earth has slowly and steadily cooled down.

If Earth's only source of heat were from its early formation, our planet would have cooled to a frozen cinder billions of years ago. However, Earth's mantle and crust contain enough long-lived radioactive isotopes to keep Earth warm. There are four main radioactive isotopes that keep our planet cooking as if on a slow burner: uranium-235, uranium-238, thorium-232 and potassium-40. As was shown in Table 9.1 (page 262), the half-lives of these four isotopes are on the order of billions of years. As a result, large quantities of these isotopes remain. Radioactivity therefore plays two vital roles in geology. It provides the means for determining the ages of rocks, as discussed in Chapter 9. Even more importantly, however, it has kept mantle convection and plate tectonics active for billions of years.

Heat Flow

Heat travels by three different mechanisms: *radiation, conduction,* and *convection.* Within a planet all three are active, but are more or less significant within different layers. The motions of rock and metal in the interior of a planet are entirely dependent upon how heat is able to move from one layer to the next. The regions where radiation, convection, and conduction are important in controlling the flow of heat out of Earth are shown in Figure 12.13. As you can see, only two of these processes (*convection* and *conduction*) operate within Earth's interior, and these will be considered next.

Convection The transfer of heat by moving material in a fluid-like manner and carrying the heat with it is called **convection.** It is the primary means by which heat is transferred within Earth. You are familiar with convection if you

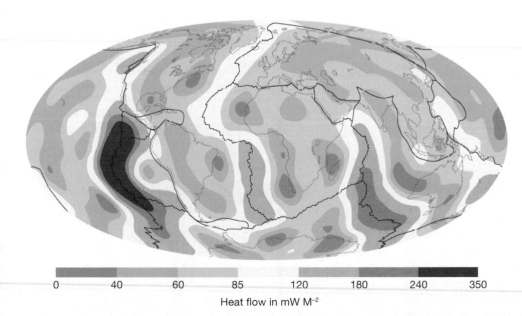

Heat flow in mW M⁻²

FIGURE 12.12 A map of the rate of heat flow out of Earth as it gradually cools over time, measured in milliwatts per square meter. Earth loses most of its heat near mid-ocean ridges, where magma rises toward the surface to fill the cracks formed when tectonic plates pull apart. Continents lose heat faster than old ocean seafloor because they contain higher amounts of heat-producing radioactive isotopes.

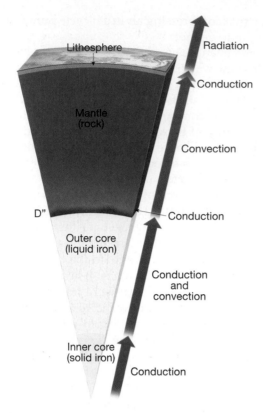

FIGURE 12.13 Diagram showing the dominant style of heat transfer at different depths within Earth as the planet cools down. Earth ultimately loses its heat to space through radiation. However, heat travels from Earth's interior to the surface, through the processes of convection and conduction.

have ever watched a pot of boiling water. The water seems to be rolling—rising up in the middle of the pot, and then down along the sides (Figure 12.14). This pattern is called a *convection cycle,* and occurs within Earth's mantle and outer core, and possibly within the inner core as well.

Convection occurs because of several factors—thermal expansion, gravity, and fluidity. When the water at the bottom of the pot is heated, it expands. The colder and heavier water at the top of the pot sinks and replaces the hot water at the bottom, which then rises to the top. The driving force for convection is the force of gravity, pulling down on the water. If you tried to boil water while floating in outer space, with no strong gravity present, you would find that your pot of boiling water would not convect.

Last, the material has to be fluid enough to be able to flow. Scientists usually measure a material's fluidity in terms of its *resistance* to flow, called its **viscosity.** Water flows easily and has a low viscosity. The liquid iron of Earth's outer core likely has a viscosity close to water, and it also convects very easily. Materials that are very viscous do not flow easily but can still convect. Catsup is 50,000 times more viscous than water, but it still flows. Rock in the lower mantle is 10 trillion trillion (10^{25}) times more viscous than water, but it too flows.

The temperatures at the top and bottom of a convection cycle determine how vigorous the convection is. Earth's surface is very cold, compared to the interior, so newly formed

ocean lithosphere cools rapidly. This causes the ocean lithosphere to contract and become denser and heavier, and in time it sinks back into the mantle at subduction zones. These cold sinking slabs eventually descend to the base of the mantle absorbing heat along the way. When rock in the lower mantle becomes warm enough, it rises back toward the surface, some of it eventually making its way to mid-ocean ridges to become new ocean lithosphere (Figure 12.15). Oceanic lithosphere can therefore be thought of as the top part of the mantle convection cycle. In a similar manner, plate tectonics can be viewed as the surface expression of mantle convection, which is Earth's primary mechanism for cooling down.

Convection can sometimes occur in a way that is not driven by heat flow. This is called *chemical convection,* and it occurs when changes in density result through chemical and not thermal means. Chemical convection is an important mechanism within the outer core. As iron crystallizes and sinks to form the solid inner core, it leaves behind a melt that contains a higher percentage of lighter elements. Because this liquid is more buoyant than the surrounding material it rises upward, creating convection.

Conduction   The flow of heat through a material is called **conduction.** Heat conducts in two ways: (1) through the collisions of atoms and (2) through the flow of electrons. In rocks, atoms are locked in place but are constantly oscillating. If one side of a rock is heated up, its atoms will oscillate more energetically. This will increase the intensity of the collisions with their neighboring atoms, and like a domino ef-

FIGURE 12.14 A simple example of convection, which is heat transfer that involves the actual movement of a substance. Here the flame warms the water in the bottom of the beaker. This heated water expands, becomes less dense (more buoyant), and rises. Simultaneously, the cooler, denser water near the top sinks.

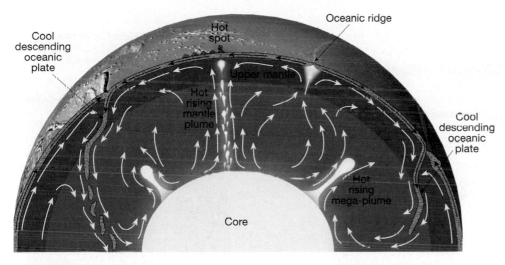

FIGURE 12.15　Diagram showing convection within Earth's mantle. The entire mantle is in motion, driven by the sinking of cold oceanic lithosphere back into the deep mantle. This is like stirring a pot of stew with downward strokes of a spoon. The upward flow of rock likely occurs through a combination of mantle plumes and a broad return flow of rock to replace the ocean lithosphere that leaves the surface at subduction zones.

fect, the energy will slowly propagate all the way through the rock. Conduction occurs much more quickly in metals. Though the atoms of metals are also locked in place, some of their electrons are free to move through the material, and these electrons can carry heat quickly from one side of a metal object to another.

There is a huge variation in the rate of conduction for different materials. For example, heat conducts about 40,000 times more easily through a diamond than through air. Most rocks are poor conductors of heat. Conduction is therefore not an efficient way to move heat through most of Earth. However, there are places where conduction is important; these include the lithosphere, the D″ layer, and the core.

Conduction, through the flow of electrons, is likely very important in both the solid iron inner core and liquid iron outer core. Once heat conducts from the inner core into the outer core, convection may play a significant role in carrying heat to the top of the core. However, heat can only pass from the core to the mantle through conduction, not convection. This is because iron is much too dense to intrude into the lighter mantle floating on top. For heat to leave the core, it must conduct across the core–mantle boundary and thus through the D″ layer. Once the heat reaches the lower mantle it is carried toward the surface through mantle convection.

Next, heat from the mantle moves to Earth's surface through the rocky lithosphere. There are a few places where convection carries heat directly to the surface, in the form of volcanic lava. Everywhere else the heat must make its final journey to the surface by conducting slowly across the stiff, rigid lithosphere.

Earth's Temperature Profile

The profile of Earth's average temperature at each depth is called the **geothermal gradient** or **geotherm,** (Figure 12.16A). Earth's temperature increases from about 0°C at the surface to more than 5000°C at Earth's center. Within Earth's crust, temperature increases rapidly—as great as 30°C per kilometer of depth. You can experience this in deep mines. The deepest diamond mines in South Africa go to depths of more than 3 kilometers, where the temperature is more than 50°C (120°F). The temperature doesn't continue to increase at such a rapid rate, however, or the whole planet would be molten below a depth of 100 kilometers.

At the base of the lithosphere, about 100 kilometers down, the temperature is roughly 1400°C. However, you would need to go to almost the bottom of the mantle before the temperature doubled to 2800°C. For most of the mantle, the temperature increases very slowly—about 0.3°C per kilometer. However, the D″ layer acts as a thermal boundary layer, and the temperature there increases by more than 1000°C from the top to the bottom. Finally, temperatures increase only gradually across the outer and inner cores.

Determining temperatures inside Earth is difficult, and there are large uncertainties. In fact, the temperature at Earth's center may be as high as 8000°C. You may be wondering how geoscientists measure Earth's deep temperatures. The best way is to use mineral physics experiments that measure the temperatures and pressures at which materials change. For example, the basis for the geotherm for the upper mantle (shown in Figure 12.16A) comes from experiments that establish the temperatures at which the mineral olivine makes the phase changes that cause the 410- and 660-kilometer discontinuities. Similar experiments are used to determine the temperature at which liquid and solid iron would coexist at the boundary between the inner and outer core.

Also plotted in Figure 12.16A is the curve for the average melting point of material at each depth. How close the geotherm is to the melting point of a material not only determines whether a material is molten or not, but how stiff it is. Figure 12.16B shows the viscosity of material in the crust and mantle. High-viscosity regions, like the lithosphere, are very stiff. Low-viscosity regions, like the asthenosphere or D″, are much softer. Notice how viscosity is directly related to how close the geotherm and melting point curves are in Figure 12.16A. When rock gets close to its melting point, it begins to weaken and get soft.

Both the geotherm and melting point curves generally increase gradually with depth. This is a result of the continual increase in pressure. Squeezing a material raises its temperature by causing the atoms to collide with each other more often, so the geotherm increases. This is the reason for the gradual increase in temperature seen across the middle of the mantle and across the core. However, squeezing a material

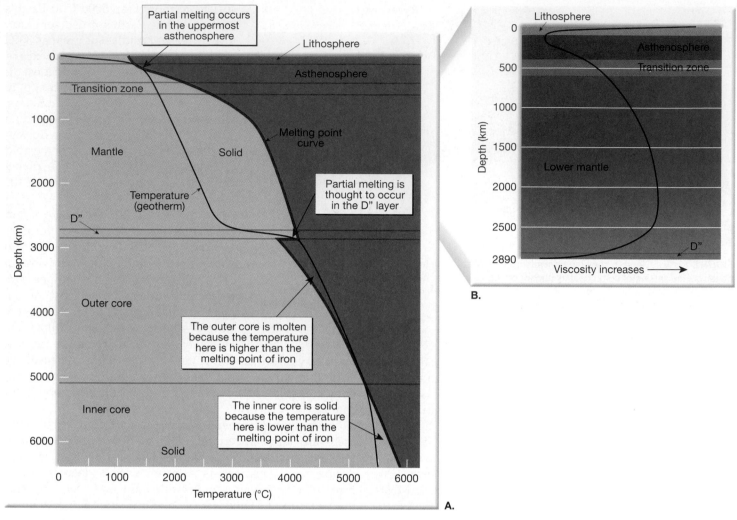

FIGURE 12.16 These graphs show how the viscosity of Earth materials at various depths is related to Earth's geotherm and the melting point of these materials. **A.** Earth's temperature profile with depth, or *geotherm*. Note that Earth's temperature increases gradually in most places. Within Earth's two major thermal boundary layers, the lithosphere and the D″ layer at the core-mantle boundary, the temperature increases rapidly over short distances. Also shown is the melting point curve for the materials (rock or metal) found at various depths. Where the geotherm crosses above (to the right of) the melting point curve, as in the outer core, the material is molten. **B.** This graph shows how viscosity (resistance to flow) changes with depth from Earth's surface to the bottom of the mantle. High viscosities, as in the crust and lithosphere, show rock that is stiffer and flows less easily. If you compare these two figures, you can see that rocks are weakest and flow more easily at depths where the temperature of rocks are close to melting (the asthenosphere and D″ layer).

also makes it harder to melt because liquids usually take up more volume than solids. Higher pressures result in less room for rock to expand into, so materials under pressure tend towards being solid. This causes the melting point curve to also increase with depth. Generally the melting point curve increases more rapidly with depth than the geotherm. However, in two layers, the uppermost asthenosphere and D″ layer, Earth's temperature is high enough that some rock begins to melt.

We can now understand why different layers of Earth behave the way they do. The lithosphere is stiff because its temperature is much colder than its melting temperature. The asthenosphere is weaker and softer because it is very close to its melting temperature, and partial melting likely occurs in some places. The existence of the weak asthenosphere is critical to the existence of plate tectonics on Earth—it allows the stiff sheets of lithosphere to slide across it. Without an asthenosphere, Earth's mantle would still convect, but it would not have tectonic plates.

Most of the lower mantle is very stiff, and rock moves more sluggishly there. It is thought that convective flow occurs several times slower in the lower mantle than in the upper mantle. However, at the very base of the mantle, where Earth's temperature again approaches the melting point of rock, the D″ layer is relatively weak, and rock there flows more easily.

In the core, the temperature increases much more slowly than does the pressure. From the core–mantle boundary to Earth's center, the temperature may only increase by about 40 percent, or from 4000° to 5500°C. However, over the same depth the pressure nearly triples, going from 1.36 to 3.64 megabars. Even though iron in the outer core is cooler than iron in the inner core it is under much less pressure and remains a liquid. Stated another way, iron in the inner core, although very hot, is under such great pressure that the inner core is solid.

Earth's Three-Dimensional Structure

As you have seen, Earth is not perfectly layered. At the surface there are large horizontal differences: oceans, continents, mountains, valleys, trenches, mid-ocean ridges, and so on. Geophysical observations show that horizontal variations are not limited to the surface—they also occur within Earth, and are directly related to the process of mantle convection and plate tectonics. Three-dimensional structure within Earth is identified primarily with a kind of seismic imaging called seismic tomography. It is also studied by examining variations in Earth's gravitational and magnetic fields.

Earth's Gravity

The most significant cause for changes in the force of gravity, at the surface, is due to Earth's rotation. Because Earth rotates around its axis, once every day, the acceleration due to gravity* is less at the equator (9.78 m/s^2) than at the poles (9.83 m/s^2). This happens for two reasons. Earth's rotation causes a centrifugal force that is in proportion to the distance away from the axis of rotation. (This is similar to the force that appears to try to throw you off of a moving merry-go-round.) Centrifugal force acts to throw objects upward at the equator, where the force is greatest.

In addition, Earth's rotation has caused Earth's shape to be slightly flattened, with the equator further from Earth's center (6378 km) than the poles (6357 km). This weakens gravity at the equator because gravitational force is smaller when objects are further apart. Your weight will actually be less by 0.5% at the equator than at the poles (Figure 12.17). Earth is therefore not a perfect sphere, but has the shape of an *oblate ellipsoid.* The amount of flattening is 1 part in 298. This shape was one of the first clues geologists had that Earth's mantle, although essentially solid, behaves like a fluid and is able to flow over very long time scales.

Gravity measurements show that there are more variations in Earth's shape than just its slightly elliptical nature. The density of rock within Earth is different in different locations. This is obvious for the crust, where geologists find rocks with different compositions and densities at different places at the surface. For instance, the igneous rocks that make up extensive lava flows in the northwestern United States are denser than the sedimentary rocks that outcrop in the Midwest. Density differences for rocks of different compositions extend throughout the crust and also into the

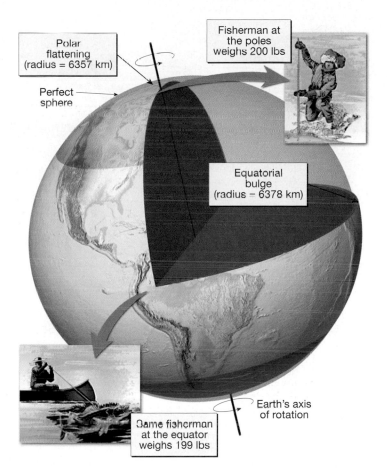

FIGURE 12.17   Drawing of Earth demonstrating the bulge at the equator and flattening of the poles that occur because of Earth's rotation. The combination of Earth's elliptical shape and its daily rotation actually cause the force of gravity to be weaker at the equator than at the poles. This difference is large enough to be measured on a bathroom scale. Imagine two fishermen of equal mass both standing at sea level. If the one at the North Pole weighs 200 pounds, the one at the equator would weigh only 199 pounds.

mantle. If there is denser rock underground, the resulting increased mass will cause a greater gravitational force. Because metals and metal ores tend to be much denser than silicate rocks, gravity anomalies (differences from the expected) have long been used to help prospect for mineral deposits. A map of gravity anomalies for the United States is shown in Figure 12.18. The majority of the variations shown are caused by density differences that result from changes in composition. Another significant cause of gravity variations at the surface is due to topography on land and bathymetry (the topography of the ocean floor). For example, over the oceans, where there are seamounts, water is being replaced by rock, and this increases the gravitational pull in that area. These gravitational anomalies change the level of the sea surface. Contrary to what you might think, the sea surface is elevated above seamounts, ridges, and underwater plateaus, not depressed. Though the downward force of gravity acting on the overlying water is increased, this increase in gravity also causes the surrounding water to be pulled toward these elevated features, causing the sea surface to rise. In fact, the topography of the seafloor is actually

---

*The force of gravity causes objects, such as an apple to accelerate as it falls to the ground, hence the expression "acceleration due to gravity."

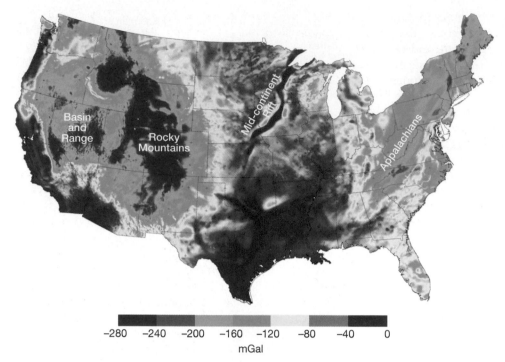

FIGURE 12.18   A map of gravity anomalies beneath the continental United States. Changing elevation changes the strength of Earth's gravity, so values are calculated for what would be measured if you were at sea level at each location. This allows the gravity anomalies to be compared across the map. The negative anomalies (blue) beneath the Rockies and Appalachians show us that the crust has deep roots beneath the mountains there. The negative anomaly (blue) in the Basin and Range Province is the result of hotter, tectonically active crust (rifting and volcanoes). The narrow positive anomaly (red) that runs in a line down the middle of the country is the mid-continent rift, where denser volcanic rocks entered the crust more than a billion years ago.

Seismic Tomography

The three-dimensional changes in composition and density that are detected with gravity measurements can actually be viewed using seismology. In a technique called **seismic tomography,** a very large number of seismic observations are combined to make three-dimensional models of Earth's interior. These models typically involve collecting signals from many different earthquakes recorded at many seismograph stations, in order to "see" all parts of Earth. Seismic tomography is very similar to medical tomography, in which doctors use techniques like CT scans to make three-dimensional images of a person's body.

Seismic tomography usually involves identifying regions where P or S waves travel faster or slower than average for that depth. These seismic velocity "anomalies" are then interpreted as variations in material properties such as temperature, composition, mineral phase, or water content. For instance, increasing the temperature of rock about 100°C can decrease S velocities

measured globally with satellites that use radar to measure the elevation of the sea surface (see Figure 13.5).

Changes in density deep beneath the surface also cause variations in the shape of Earth's surface. The height of the surface of the oceans actually changes vertically by about 200 meters due to very large-scale density variations within the mantle. The shape of this surface, measured from the perfect ellipsoid due to rotation, is called the **geoid.** A map of global geoid variations is shown in Figure 12.19. The width of the geoid anomalies can be an indication of the depth of the density anomalies that cause them. When underground density anomalies are near the surface, the geoid variations are narrow. When density anomalies are very deep, the geoid anomalies are very broad, sometimes thousands of kilometers across. These large-scale geoid anomalies are a result of the large upwellings and downwellings of mantle convection.

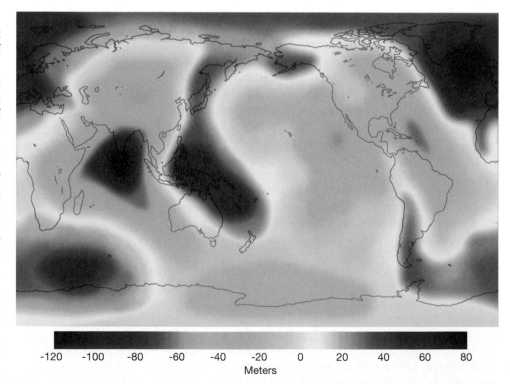

FIGURE 12.19   Map of the large-scale geoid, made from satellite measurements. The geoid is the shape of Earth that differs from what is expected from just Earth's rotation. The geoid is a result of broad differences in density within Earth's interior. The effect is that the sea surface varies in height by more than 200 meters due to movements within Earth's deep interior.

by about 1 percent, so images from seismic tomography are often interpreted in terms of temperature variations.

Figure 12.20 shows an example of S-wave velocity tomography for the mantle beneath North America. *Red colors* show regions where waves travel slower than average and *blue colors* show regions where waves travel faster than average. This colored diagram shows some very significant patterns. Continental lithosphere has fast seismic velocities, when compared to oceanic lithosphere, because it is older and thus has been cooling at the surface for a long time. Seismic imaging also shows that continental lithosphere (deep blue areas) can be very deep, extending more than 300 kilometers into the mantle. This deep continental lithosphere is called the **tectosphere.** Beneath some of the oldest portions of continents the tectosphere connects continuously with the lower mantle, without a strong presence of an asthenosphere. The opposite situation occurs at oceanic ridges, which exhibit slow seismic velocities (bright red areas) because they are very hot (Figure 12.20).

In the mid-mantle beneath North America, you can see a tongue of fast seismic velocities (light blue) that represents a sheet of ancient Pacific Ocean lithosphere known as the Farallon plate. This plate used to subduct beneath North America all along its western edge. Remnants of the plate still subduct beneath Oregon and Washington in the form of the Juan de Fuca plate and beneath Mexico as the Cocos plate. The segment of this former ocean sea floor seen in Figure 12.20 used to descend beneath California, but is now completely subducted. The Farallon slab is now sinking down through the lower mantle toward the core–mantle bound-

ary, where it is slowly heating up and will eventually become mixed back into the mantle. Over time, this slab will become hot enough to begin to rise back to the surface. This kind of upward return flow may be what is observed as the reddish orange areas at the right and left sides of the figure.

The large region of slow seismic velocities at the base of the mantle beneath Africa (the large reddish orange region at the lower right of Figure 12.20) is called the African superplume—a region of upward flow in the mantle. These slow velocities are likely due to both unusually high temperatures and rock that is highly enriched in iron. The rising rock cannot easily break through the African tectosphere, so it seems to be deflected to both sides of Africa, perhaps supplying new magma to both the Mid-Atlantic and Indian Ocean spreading centers.

Images from seismic tomography, like the one shown in Figure 12.20, reveal the whole-mantle cycle of convection. Sheets of cold ancient ocean sea floor sink to the base of the mantle, where they warm, expand, and rise back toward the surface again.

Earth's Magnetic Field

Convection of liquid iron in the outer core is vigorous and gives rise to Earth's magnetic field. Because material in the outer core flows so easily, horizontal temperature variations there are very small—likely less than 1°C. Such small temperature differences create indistinguishable differences in seismic velocities, so the outer core appears uniform at each depth when viewed with seismic waves. However, the patterns of flow in the outer core create variations in Earth's magnetic field, and these can be observed at Earth's surface.

Flow in the outer core is thought to occur for three main reasons:

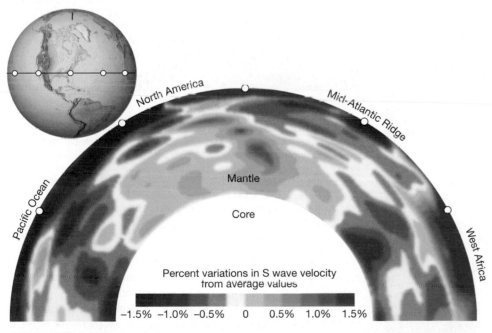

FIGURE 12.20 A seismic tomographic slice through Earth showing mantle structure. Colors show variations in the speed of S waves from their average values. Older portions of continents such as eastern North America and Africa are cold and stiff, so their blue colors show fast S wave speeds. The western United States is hotter and tectonically active, making that portion of the continent warmer and weaker, which slows S waves. The large blue structure extending far below North America is a sheet of cold, dense ancient Pacific seafloor that is sinking toward the base of the mantle. The large orange structures beneath western Africa and the Pacific Ocean are thought to be megaplumes of warm material that are rising toward the surface.

1. As heat conducts out of the core into the surrounding mantle, the outermost core fluid cools, becomes denser, and sinks. This is a form of thermally driven convection.

2. Crystallization of solid iron at the bottom of the outer core, to form the inner core, releases fluid that is depleted in iron and therefore relatively buoyant. As this fluid rises up and away from the inner core boundary it helps drives convection. This is a form of chemically driven convection.

3. There may be radioactive isotopes like potassium-40 within the core that could provide additional heat to drive thermal convection.

The relative importance of these three mechanisms is still uncertain.

The Geodynamo   As the core fluid rises, its path becomes twisted through a phenomenon called the *Coriolis effect*, which is a result of Earth's rotation. The fluid ends up moving in spiraling columns as shown in Figure 12.21. Because the fluid is electrically charged, it generates a magnetic field through a process called a geodynamo that is similar to an electromagnet. If a wire is wrapped around an iron nail and an electric current passed through it, the nail will generate a magnetic field that looks a lot like the field from a bar magnet (Figure 12.22A, B). This is called a dipolar field—a type of magnetic field that has two poles (a north and south magnetic pole). As Figure 12.22C shows, the magnetic field that emanates from Earth's outer core has the same dipolar form.

However, the convection in the outer core is not quite so simple. More than 90% of Earth's magnetic field takes the form of a dipolar field, but the remainder of the field is the result of other more complicated patterns of convection in the core. In addition, some of the features of Earth's magnetic field change over time. For centuries sailors have used compasses to determine direction. Consequently, a great deal of attention has been paid to keeping track of the direction that compass needles point. One observed change in the magnetic field is a gradual "westward drift" of the non-dipole part of the magnetic field. In order to explain this, we first need to look at how the magnetic field is measured.

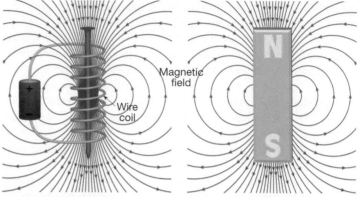

A. Electromagnet (Dipolar field)   **B. Bar magnet (Dipolar field)**

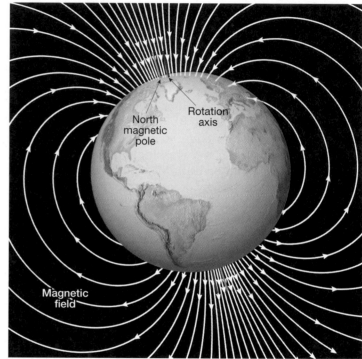

C. Earth's magnetic field (Dipolar field)

FIGURE 12.22   Demonstration of the similarity of Earth's magnetic field to that of an electromagnet (**A**), which consists of an electrical current passed through a coil of wire, or bar magnet (**B**). While it was once thought that Earth's core acts like a large bar magnet, scientists now think that Earth's magnetic field (**C**) is more like an electromagnet, and that the cylinders of spiraling liquid iron shown in Figure 12.21 behave like the coil of current passing through the wires of an electromagnet.

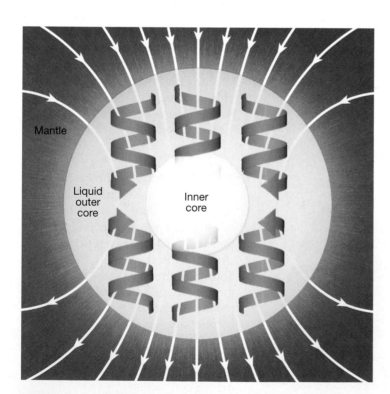

FIGURE 12.21   Illustration of the kind of convection patterns within Earth's liquid iron outer core that could give rise to the magnetic field we measure at the surface. It is thought that convection takes the form of cylindrical gyres of rotating molten iron that are aligned in the direction of Earth's axis of rotation.

At any point on Earth's surface, the direction that the magnetic field is pointing is measured with two angles, called *declination* and *inclination*. The declination measures the direction to the magnetic north pole with respect to the direction to the geographic North Pole (Earth's axis of rotation). The inclination measures the downward tilt of the magnetic lines of force at any location. It is what your compass would read if you could tilt it on its side. At the magnetic north pole the field points directly downward. At the equator it is horizontal (Figure 12.23). In North America it tilts downward at an intermediate angle.

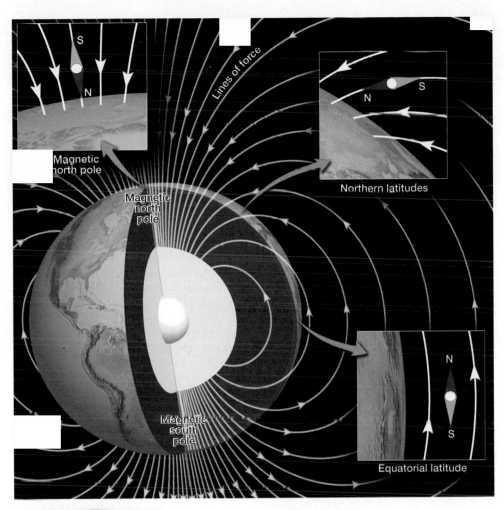

FIGURE 12.23   Drawing that shows the direction of the magnetic field at different locations along Earth's surface. Though a compass measures only the horizontal direction of the magnetic field (the declination), at most locations the field also dips in or out of the surface at a variable angle (inclination).

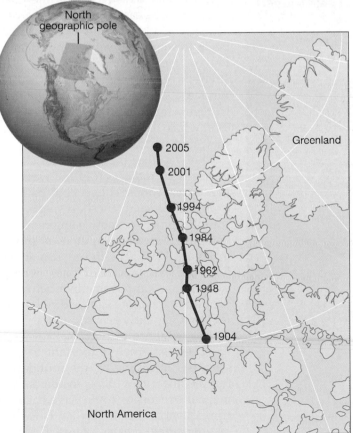

FIGURE 12.24   Map showing the change in measured locations of the north magnetic pole over time. The patterns of convection within the outer core change fast enough that we can see the magnetic field change significantly over our lifetimes.

The location of the magnetic north pole actually moves significant distances during our lifetimes. Earth's magnetic north pole had been located in Canada, but over the past decade moved northward into the Arctic Ocean and is currently moving rapidly toward Siberia at a rate of about 20 kilometers per year (Figure 12.24). The process is not symmetric. Though the magnetic north pole has been moving towards the geographic North Pole, the magnetic south pole has been moving *away* from the South Pole, passing from Antarctica to the Pacific Ocean. This means that core convection changes significantly on a time scale of decades (see Box 12.2).

Magnetic Reversals   Although core convection changes over time, causing the magnetic poles to move, the locations of the magnetic poles averaged over thousands of years are

BOX 12.2 ▶ EARTH AS A SYSTEM

Global Dynamic Connections

The layers of planet Earth are not isolated from one another, but rather their motions are thermally connected. Furthermore, these connected motions don't always occur steadily—sometimes they occur episodically or in pulses. One example shows the possible connection between magnetic reversals, hot-spot volcanoes, and the break-up of the supercontinent of Pangaea.

Pangaea began to break up about 200 million years ago. Plate motions increased and there was a greater amount of subduction of ocean lithosphere. About 80 million years later the core reversal process shut down and Earth's magnetic field did not reverse for 35 million years. In the tens of millions of years that followed, there were several enormous outpourings of lava that have been linked to the arrival of new hot-spot mantle plumes at the surface.

In one hypothesis, these three events are closely connected. The large amount of subducted lithosphere that followed the breakup of Pangaea could have plunged to the base of the mantle. This would have displaced hot rock at the base of the mantle, causing much of it to rise to the surface and erupt as flood basalts such as the Deccan Traps in India. The sudden placement of cold ocean lithosphere next to the hot core at the core–mantle boundary would have chilled the uppermost core, causing more vigorous core convection that would have prevented the field from weakening and reversing. This hypothesis, if correct, is an important reminder that Earth is a complex, churning, pulsing planet that is very active in a variety of geological ways.

A. Normal orientation of magnetic field

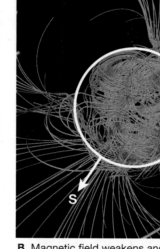

B. Magnetic field weakens and poles begin to wander

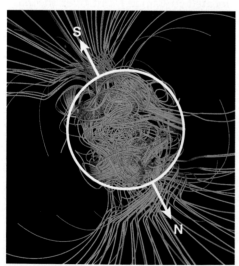

C. Poles wander across the equator

D. Reversal complete with north pole pointing south

FIGURE 12.25 Computer simulations showing how Earth's magnetic field could reverse direction. The white circle represents the core–mantle boundary and the arrows point to the north (N) and south (S) magnetic poles respectively. During a reversal the strength of the magnetic field weakens and the poles begin to wander greatly, going so far as to cross the equator. When the strength of the field returns to normal levels, the field is regenerated with reverse polarity.

the same as Earth's axis of rotation (geographic poles). There is one major exception to this and that is during periods of magnetic field reversals. At apparently random times, Earth's magnetic field reverses polarity so that the *north* needle on your compass would point to the *south*. (The importance of these reversals in the study of paleomagnetism was described already in Chapter 2). What happens during a reversal is that the strength of the magnetic field decreases to about 10 percent of normal and the locations of the poles begin to wander greatly, going so far as to cross the equator (Figure 12.25). The strength of the magnetic field then returns to its normal levels, and the field is regenerated with reverse polarity. The whole process takes only a few thousand years.

The way that the magnetic field reverses is an indication of the way that the outer core's convection patterns change over relatively short spans. This complex process is now being modeled using high-speed computers as shown in Figure 12.25. The figure also shows how the magnetic field lines become twisted in a complex manner before returning to the more simple dipolar pattern that regularly exists.

The existence of magnetic reversals has been extremely important to geoscientists in providing the foundation for the theory of plate tectonics, but magnetic reversals could have harmful consequences for life on land. Earth's magnetic field creates a large magnetic layer in space around the planet known as the *magnetosphere*. Along with the atmosphere, the magnetosphere protects Earth's surface from ionized particles emitted by the Sun. These ionized particles form what is called the *solar wind*. If the strength of the magnetic field decreases greatly during a reversal, the increased amounts of solar wind reaching Earth's surface could cause health hazards for humans and other land-based life forms.

Summary

- Earth is layered with the densest materials at the center and lightest materials forming the outer layer. This layering is a result of gravity, and is similar for all planets. Earth's layers consist of the *inner core* (solid iron), *outer core* (liquid iron), *mantle* (dense rock), *crust* (low-density rock), *ocean* (water), and *atmosphere* (gas). Within layers, the density of materials increases with depth due to compression resulting from the increasing pressure. Within the mantle, increases in density also occur because of mineral phase changes.

- Because it is impossible to drill deep into Earth, *seismic waves* are used to probe Earth's interior. The patterns of seismic waves are complicated because their behavior is influenced by different structures inside the planet before returning to the surface. Seismic waves travel faster through cold rock and slower through hot rock. Seismic waves reflect off of layers composed of different materials. The results of seismic imaging of Earth's interior can be interpreted through comparison with mineral physics experiments. These experiments recreate the temperature and pressure conditions within Earth, and allow scientists to see what rocks and metals are like at various depths.

- *Oceanic crust* and *continental crust* are very different. Oceanic crust is created at mid-ocean ridges, and is fairly similar in composition and thickness everywhere. Continental crust is highly variable, has many different compositions, and is formed in many different ways. Oceanic crust is nowhere older than 200 million years, whereas continental crust can be older than 4 billion years. Oceanic crust is usually about 7 km thick, but continents can be thicker than 70 km. The boundary between the crust and mantle is called the *Moho*.

- The *mantle* comprises most (82%) of Earth's volume. The *upper mantle* extends from the Moho to a depth of 660 km, on average. The upper mantle contains part of the stiff lithosphere, the weak asthenosphere, and the transition zone that may contain significant amounts of water. The lower mantle extends from 660 km down to the core–mantle boundary, 2891 km beneath the surface. At the base of the lower mantle is the variable D″ layer.

- The core is mostly made of iron and nickel, although it contains about 15 percent lighter elements. Because iron is very dense, the core makes up one-third of Earth's mass, and iron is Earth's most abundant element (by mass). The solid inner core grows over time as Earth cools.

- Earth's temperature increases from about 0°C at the surface to about 5500°C at the center of the core (though the exact temperature is very hard to determine). Heat flows unevenly from Earth's interior, with most of the heat loss occurring along the oceanic ridge system. Earth became very hot early in its history (it may have become entirely molten), largely due to the impacts of planetesimals and heat released by radioactive decay (radiogenic heat). Since then, Earth has slowly cooled. Earth is still geologically active because of the radiogenic heat supplied by long-lived radioactive isotopes, including uranium-238, uranium-235, thorium-232, and potassium-40.

- Heat flows from Earth's hot interior to its surface, and does so primarily by convection and conduction. *Convection* transfers heat through the movement of material. *Conduction* transfers heat by collisions between atoms or the motions of electrons. Convection is very important within Earth's mantle and outer core, whereas conduction is most important in the inner core, lithosphere, and D″. Within the asthenosphere and the base of D″ the temperature is close enough to the melting point that some partial melting might occur and the rock is weak enough to flow more easily than elsewhere in the mantle. The weak asthenosphere is very important for plate tectonics because it allows the stiff plates (lithosphere) to move easily across the top of it.

- Rotation causes Earth's shape to take the form of an *oblate ellipsoid*, meaning its equator bulges slightly. The combination of Earth's rotation and its ellipsoidal shape cause gravity to vary significantly—from 9.78 m/s^2 at the equator to 9.83 m/s^2 at the poles. Gravity also varies around Earth's surface due to the presence of rocks of different densities. These density differences actually deform Earth's surface, including the ocean surface, by more than 200 meters. This surface is called the *geoid*.

- Three-dimensional images of structure variations within the mantle are made from large numbers of seismic waves using *seismic tomography*. These images show that continental lithosphere can extend several hundred kilometers into the mantle. They also show cold subducted oceanic lithosphere sinking to the base of the mantle and large superplumes of hot rock rising from the core–mantle boundary. This suggests that convection occurs throughout the mantle.

- Convection of the liquid iron in the outer core causes a magnetic *geodynamo* that is responsible for Earth's magnetic field. The convection takes the form of spiraling cylinders that are a result of the Coriolis effect. This field is primarily dipolar; that is, it resembles the field from a bar magnet or electromagnet. The patterns of convection in the outer core change rapidly enough that the magnetic field varies noticeably over our lifetimes.

- The magnetic field randomly reverses, with the north and south poles swapping positions. A reversal takes only a few thousand years, and involves a significant decrease in the strength of the dipolar field. This is important, because the magnetic field creates a magnetosphere around Earth that protects our planet from much of the Sun's solar wind that would otherwise bombard it. If the magnetosphere weakens, life on land would be adversely affected.

Review Questions

1. What role does gravity play in the layering of planets?

2. What are the two major reasons for the increase in density with depth within Earth's mantle?

3. Why is seismology responsible for gasoline prices being much more affordable than they otherwise might be (without seismology)?

4. Explain three different ways that oceanic crust and continental crust are different. Where would you go to find very thick crust? Very thin crust?

5. What is the importance of determining the *cross-over* distance? (See Figure 12.7.)

6. How do S waves allow us to determine that the mantle is solid?

7. If there were a lot of water in Earth's mantle, in what layer would it most likely reside?

8. What mineral phase changes occur at the top and bottom of the transition zone?

9. What layer of Earth has the greatest volume?

10. How is the D″ layer similar to the lithosphere?

11. True or False: No seismic waves arrive in the *shadow zone*? Explain.

12. Why is Earth's core one-sixth of Earth's volume but one-third of its mass?

13. Why is Earth's inner core growing in size?

14. Why is heat flow from Earth's surface not evenly distributed?

15. What were the sources of heat that caused Earth to get so hot early in its history?

16. What prevents Earth from being a cold, motionless sphere of totally solid rock and metal?

17. Explain the difference between conduction and convection.

18. Why is convection a less efficient means of heat transfer in materials with high viscosity?

19. Why is conduction more important than convection within Earth's crust?

20. What happens to rock as the geotherm approaches the melting point temperature?

21. What happens to rock in regions where the geotherm crosses above the melting point curve?

22. Why would tectonic plates have a hard time moving if it were not for the existence of the asthenosphere?

23. Why is the lithosphere stiffer than the asthenosphere?

24. Earth once rotated much faster than it currently does. How would Earth's shape have been different in the past?

25. Would you expect to find a large layer of iron ore underground in a region with a positive or negative gravity anomaly? Explain.

26. Why does the mid-Atlantic ridge appear as a slow seismic velocity anomaly in Figure 12.20?

27. What are the three sources of convection in the outer core?

28. What happens during a magnetic reversal?

29. Why might a magnetic reversal be dangerous to humans?

30. If plate tectonics got suddenly much faster and a large volume of subducted oceanic lithosphere sank to the bottom of the mantle, do you think magnetic reversals would occur more or less frequently? Explain.

Key Terms

asthenosphere (p. 331)
conduction (p. 336)
convection (p. 335)
core (p. 334)
crust (p. 330)
D″ layer (p. 332)

geothermal gradient,
    or geotherm (p. 337)
geoid (p. 346)
inner core (p. 334)
lithosphere (p. 331)
lower mantle (p. 332)

mantle (p. 331)
Moho (p. 330)
outer core (p. 334)
seismic anisotropy (p. 331)
seismic tomography
    (p. 340)

tectosphere (p. 341)
transition zone (p. 332)
viscosity (p. 336)

Web Resources

The *Earth* Website uses the resources and flexibility of the Internet to aid in your study of the topics in this chapter. Written and developed by geology instructors, this site will help improve your understanding of geology. Visit **http://www.prenhall.com/tarbuck** and click on the cover of *Earth 9e* to find:

- Online review quizzes.
- Critical thinking exercises.
- Links to chapter-specific Web resources.
- Internet-wide key-term searches.

http://www.prenhall.com/tarbuck

GEODe: Earth

GEODe: Earth makes studying faster and more effective by reinforcing key concepts using animation, video, narration, interactive exercises and practice quizzes. A copy is included with every copy of *Earth.*

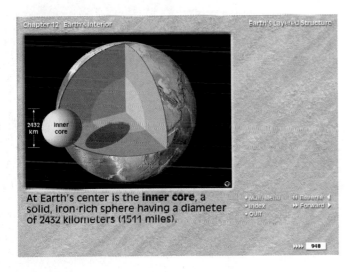

At Earth's center is the **inner core**, a solid, iron-rich sphere having a diameter of 2432 kilometers (1511 miles).

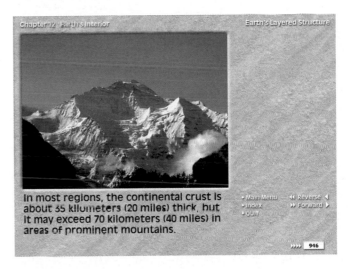

In most regions, the continental crust is about 35 kilometers (20 miles) thick, but it may exceed 70 kilometers (40 miles) in areas of prominent mountains.

Divergent Boundaries:
Origin and Evolution of the Ocean Floor

The University of Hawaii's research vessel, Kilo Moana, arriving at Kodiak, Alaska. (Photo by Marshalena Delaney/AP Photo/US Coast Guard)

The ocean is the largest feature on Earth, covering more than 70 percent of our planet's surface. One of the main reasons that Wegener's continental drift hypothesis was not widely accepted when first proposed was that so little was known about the ocean floor. Until the 20th century, investigators used weighted lines to measure water depth. In deep water these depth measurements, or soundings, took hours to perform and could be wildly inaccurate.

With the development of new marine tools following World War II, our knowledge of the diverse topography of the ocean floor grew rapidly. One of the most interesting discoveries was the global oceanic ridge system. This broad elevated landform, which stands 2 to 3 kilometers above the adjacent deep-ocean basins, is the longest topographic feature on Earth.

Today we know that oceanic ridges mark divergent plate margins where new oceanic lithosphere originates. We also know that deep-ocean trenches represent convergent plate boundaries, where oceanic lithosphere is subducted into the mantle. Because the process of plate tectonics is creating oceanic crust at mid-ocean ridges and consuming it at subduction zones, the oceanic crust is continually being renewed and recycled.

In this chapter, we will examine the topography of the ocean floor and look at the processes that produced its varied features. You will also learn about the composition, structure, and origin of oceanic crust. In addition, you will examine those processes that recycle oceanic lithosphere and consider how this activity causes Earth's landmasses to migrate about the face of the globe.

An Emerging Picture of the Ocean Floor

GEODe Divergent Boundaries
▶ Mapping the Ocean Floor

If all water were drained from the ocean basins, a great variety of features would be seen, including broad volcanic peaks, deep trenches, extensive plains, linear mountain chains, and large plateaus. In fact, the scenery would be nearly as diverse as that on the continents.

An understanding of seafloor features came with the development of techniques that measure the depth of the oceans. **Bathymetry** (*bathos* = depth, *metry* = measurement) is the measurement of ocean depths and the charting of the shape or topography of the ocean floor.

Mapping the Seafloor

The first understanding of the ocean floor's varied topography did not unfold until the historic three-and-a-half-year voyage of the HMS *Challenger* (Figure 13.1). From December 1872 to May 1876, the *Challenger* expedition made the first—and perhaps still most comprehensive—study of the global ocean ever attempted by one agency. The 127,500-kilometer (79,200-mile) trip took the ship and its crew of scientists to every ocean except the Arctic. Throughout the voyage, they sampled a multitude of ocean properties, including water depth, which was accomplished by laboriously lowering a long weighted line overboard. Not many years later, the knowledge gained by the *Challenger* of the ocean's great depth and varied topography was further expanded with the laying of transatlantic communication cables, especially in the North Atlantic Ocean. However, as long as a weighted line was the only way to measure ocean depths, knowledge of seafloor features remained extremely limited.

Bathymetric Techniques Today sound energy is used to measure water depths. The basic approach employs some type of **sonar**, an acronym for *so*und *na*vigation and *r*anging. The first devices that used sound to measure water depth, called **echo sounders**, were developed early in the twentieth century. Echo sounders work by transmitting a sound wave (called a *ping*) into the water in order to produce an echo when it bounces off any object, such as a large marine organism or the ocean floor (Figure 13.2A). A sensitive receiver intercepts the echo reflected from the bottom, and a clock precisely measures the travel time to fractions of a second. By knowing the speed of sound waves in water—about 1500 meters (4900 feet) per second—and the time required for the energy pulse to reach the ocean floor and

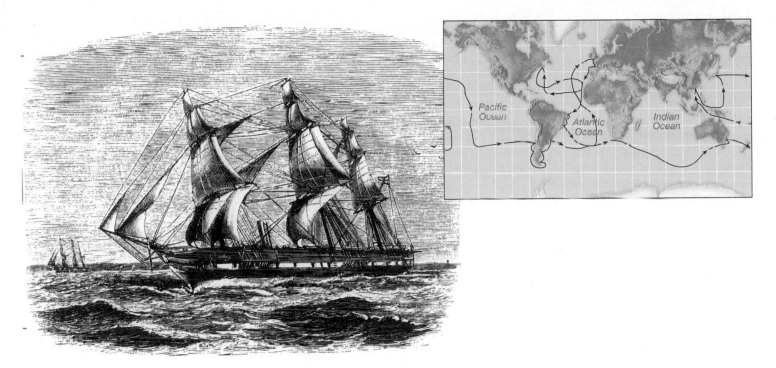

FIGURE 13.1   The first systematic bathymetric measurements of the ocean were made aboard the H.M.S. *Challenger* during its historic three-and-a-half-year voyage. Inset shows route of the H.M.S. *Challenger*, which departed England in December of 1872 and returned in May 1876. (From C.W. Thompson and Sir John Murray, *Report on the Scientific Results of the Voyage of the H.M.S. Challenger*, Vol. 1, Great Britain: Challenger Office, 1895, Plate 1. Library of Congress)

return, depth can be calculated. The depths determined from continuous monitoring of these echoes are plotted so a profile of the ocean floor is obtained. By laboriously combining profiles from several adjacent traverses, a chart of the seafloor can be produced.

Following World War II, the U.S. Navy developed *sidescan sonar* to look for mines and other explosive devices. These torpedo-shaped instruments can be towed behind a ship where they send out a fan of sound extending to either side of the ship's track. By combining swaths of sidescan

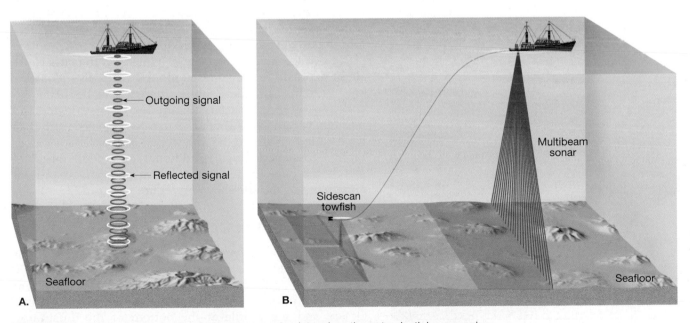

FIGURE 13.2   Various types of sonar. **A.** An echo sounder determines the water depth by measuring the time interval required for an acoustic wave to travel from a ship to the seafloor and back. The speed of sound in water is 1500 m/sec. Therefore, depth = $\frac{1}{2}$(1500 m/sec × echo travel time). **B.** Modern multibeam sonar and sidescan sonar obtain an "image" of a narrow swath of seafloor every few seconds.

sonar data, researchers produced the first photograph-like images of the seafloor. Although sidescan sonar provides valuable views of the seafloor, it does not provide bathymetric (water depth) data.

This problem is not present in the *high-resolution multibeam* instruments developed during the 1990s. These systems use hull-mounted sound sources that send out a fan of sound, then record reflections from the seafloor through a set of narrowly focused receivers aimed at different angles. Thus, rather than obtaining the depth of a single point every few seconds, this technique makes it possible for a survey ship to map the features of the ocean floor along a strip tens of kilometers wide (Figure 13.3). When a ship uses multibeam sonar to make a map of a section of seafloor, it travels through the area in a regularly spaced back-and-forth pattern known as "mowing the lawn." Furthermore, these systems can collect bathymetric data of such high resolution that they can distinguish depths that differ by less than a meter.

Despite their greater efficiency and enhanced detail, research vessels equipped with multibeam sonar travel at a mere 10 to 20 kilometers (6 to 12 miles) per hour. It would take at least 100 vessels outfitted with this equipment hundreds of years to map the entire seafloor. This explains why only about 5 percent of the seafloor has been mapped in detail—and why large areas of the seafloor have not yet been mapped with sonar at all.

Seismic Reflection Profiles   Marine geologists are also interested in viewing the rock structure beneath the sediments that blanket much of the seafloor. This can be accomplished by making a **seismic reflection profile.** To construct such a profile, strong low-frequency sounds are produced by explosions (depth charges) or air guns. These sound waves penetrate beneath the seafloor and reflect off the contacts between rock layers and fault zones, just like sonar reflects

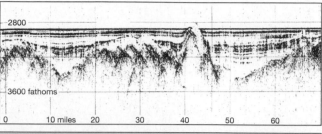

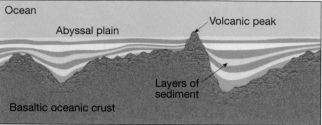

FIGURE 13.4   Seismic cross section and matching sketch across a portion of the Madeira abyssal plain in the eastern Atlantic Ocean, showing the irregular oceanic crust buried by sediments. (Image courtesy of Charles Hollister, Woods Hole Oceanographic Institution)

off the bottom of the sea. Figure 13.4 shows a seismic profile of a portion of the Madeira abyssal plain in the eastern Atlantic. Although the seafloor is flat, notice the irregular ocean crust buried by a thick accumulation of sediments.

Viewing the Ocean Floor from Space

Another technological breakthrough that has led to an enhanced understanding of the seafloor involves measuring the shape of the surface of the global ocean from space. After compensating for waves, tides, currents, and atmospheric effects, it was discovered that the water's surface is not perfectly "flat." This is because gravity attracts water toward regions where massive seafloor features occur. Therefore, mountains and ridges produce elevated areas on the ocean surface, and, conversely, canyons and trenches cause slight depressions. Satellites equipped with *radar altimeters* are able to measure these subtle differences by bouncing microwaves off the sea surface (Figure 13.5). These devices can measure variations as small as 3 to 6 centimeters. Such data have added greatly to the knowledge of ocean-floor topography. Cross-checked with traditional sonar depth measurements, the data are used to produce detailed ocean-floor maps, such as the one shown in Figure 1.17 (p. 22).

Provinces of the Ocean Floor

Oceanographers studying the topography of the ocean floor have delineated three major units: *continental margins, deep-ocean basins,* and *oceanic (mid-ocean) ridges.* The map in Figure 13.6 outlines these provinces for the North Atlantic Ocean, and the profile at the bottom of the illustration shows the varied topography. Such profiles usually have their vertical dimension exaggerated many times—40 times in this case—to make topographic features more conspicuous. Vertical exaggeration, however, makes slopes shown in

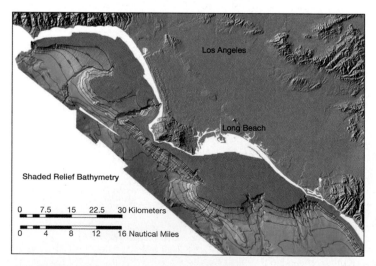

FIGURE 13.3   Color-enhanced perspective map of the seafloor and coastal landforms in the Los Angeles area of California. The ocean floor portion of this map (shown in color) was constructed from data collected using a high-resolution mapping system. (U.S. Geological Survey)

seafloor profiles appear to be *much* steeper than they actually are.

Continental Margins

Two main types of **continental margins** have been identified—*passive* and *active*. Passive margins are found along most of the coastal areas that surround the Atlantic and Indian oceans, including the east coasts of North and South America, as well as the coastal areas of Europe and Africa. Passive margins are *not* situated along an active plate boundary and therefore experience very little volcanism and few earthquakes. Here, weathered materials eroded from the adjacent landmass accumulate to form a thick, broad wedge of relatively undisturbed sediments.

By contrast, active continental margins occur where oceanic lithosphere is being subducted beneath the edge of a continent. The result is a relatively narrow margin, consisting of highly deformed sediments that were scraped from the descending lithospheric slab. Active continental margins are common around the Pacific Rim, where they parallel deep-ocean trenches (see Box 13.1).

Passive Continental Margins

The features comprising a **passive continental margin** include the continental shelf, the continental slope, and the continental rise (Figure 13.7).

Continental Shelf   The **continental shelf** is a gently sloping submerged surface extending from the shoreline toward the

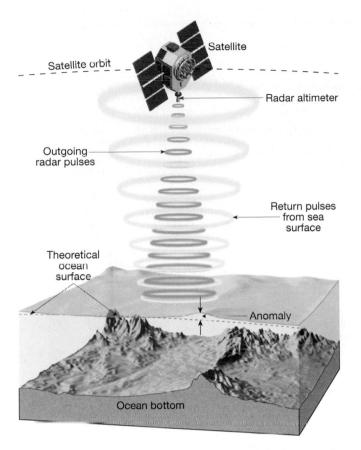

FIGURE 13.5   A satellite altimeter measures the variation in sea surface elevation, which is caused by gravitational attraction and mimics the shape of the seafloor. The sea surface anomaly is the difference between the measured and theoretical ocean surface.

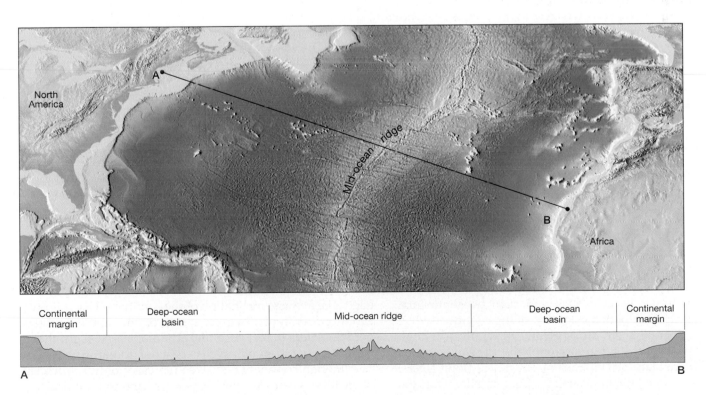

FIGURE 13.6   Major topographic divisions of the North Atlantic and a profile from New England to the coast of North Africa.

BOX 13.1 ▶ UNDERSTANDING EARTH

Susan DeBari—A Career in Geology

I discovered geology the summer I worked doing trail maintenance in the North Cascade mountains of Washington State. I had just finished my freshman year in college and had never before studied earth science. But a coworker (now my best friend) began to describe the geological features of the mountains that we were hiking in—the classic cone shape of Mount Baker volcano, the U-shaped glacial valleys, the advance of active glaciers, and other wonders. I was hooked and went back to college that fall with a geology passion that hasn't abated. As an undergraduate, I worked as a field assistant to a graduate student and did a senior thesis project on rocks from the Aleutian island arc. From that initial spark, island arcs have remained my top research interest, on through Ph.D. research at Stanford University, postdoctoral work at the University of Hawaii, and as a faculty member at San Jose State University and Western Washington University. Of most interest was the deep crust of arcs, the material that lies close to the Mohorovičiĉ discontinuity (fondly known as the Moho).

What kinds of processes are occurring down there at the base of the crust in island arcs? What is the source of magmas that make their way to the surface—the mantle, or the deep crust itself? How do these magmas interact with the crust as they make their way upward? What do these early magmas look like chemically? Are they very different from what is erupted at the surface?

Obviously, geologists cannot go down to the base of the crust (typically 20 to 40 kilometers beneath Earth's surface). So what they do is play a bit of a detective game. They must use rocks that are *now exposed at the surface* that were originally formed in the deep crust of an island arc. The rocks must have been brought to the surface rapidly along fault zones to preserve their original features. Thus, I can walk on rocks of the deep crust without

FIGURE 13.A   Susan DeBari photographed with the Japanese submersible, *Shinkai 6500*, which she used to collect rock samples from the Izu Bonin trench. (Photo courtesy of Susan DeBari)

really leaving the Earth's surface! There are a few places around the world where these rare rocks are exposed. Some of the places that I have worked include the Chugach Mountains of Alaska, the Sierras Pampeanas of Argentina, the Karakorum range in Pakistan, Vancouver Island's west coast, and the North Cascades of Washington. Fieldwork has most commonly involved hiking on foot, along with extensive use of mules and trucks.

I also went looking for exposed pieces of the deep crust of island arcs in a less obvious place, in one of the deepest oceanic trenches of the world, the Izu Bonin trench (Figure 13.A). Here I dove into the ocean in a submersible called the *Shinkai 6500* (pictured to my right in the background). The *Shinkai 6500* is a Japanese submersible that has the capability to dive to 6500 meters below the surface of the ocean (approximately 4 miles). My plan was to take rock

samples from the wall of the trench at its deepest levels using the submersible's mechanical arm. Because preliminary data suggested that vast amounts of rock were exposed for several kilometers in a vertical sense, this could be a great way to sample the deep arc basement. I dove in the submersible three times, reaching a maximum depth of 6497 meters. Each dive lasted nine hours, spent in a space no bigger than the front seat of a Honda, shared with two of the Japanese pilots that controlled the submersible's movements. It was an exhilarating experience!

I am now on the faculty at Western Washington University, where I continue to do research on the deep roots of volcanic arcs, and get students involved as well. I am also involved in science education training for K–12 teachers, hoping to get young people motivated to ask questions about the fascinating world that surrounds them!

---

deep-ocean basin. Because it is underlain by continental crust, it is clearly a flooded extension of the continents.

The continental shelf varies greatly in width. Although almost nonexistent along some continents, the shelf extends seaward more than 1500 kilometers (930 miles) along others. On average, the continental shelf is about 80 kilometers (50 miles) wide and 130 meters (425 feet) deep at its seaward

edge. The average inclination of the continental shelf is only about one-tenth of 1 degree, a drop of only about 2 meters per kilometer (10 feet per mile). The slope is so slight that it would appear to an observer to be a horizontal surface.

Although continental shelves represent only 7.5 percent of the total ocean area, they have economic and political significance because they contain important mineral deposits,

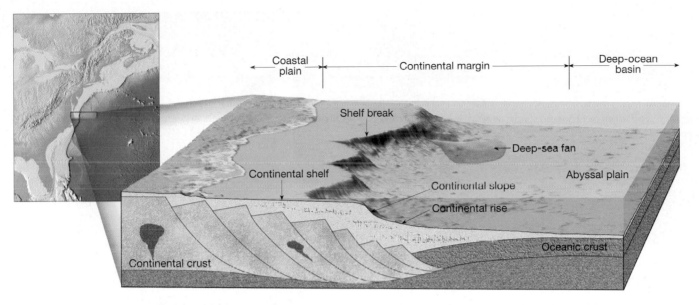

FIGURE 13.7 Schematic view showing the major features of a passive continental margin. Note that the slopes shown for the continental shelf and continental slope are greatly exaggerated. The continental shelf has an average slope of one-tenth of 1 degree, while the continental slope has an average slope of about 5 degrees.

including large reservoirs of oil and natural gas, as well as huge sand and gravel deposits. The waters of the continental shelf also contain many important fishing grounds, which are significant sources of food.

Even though the continental shelf is relatively featureless, some areas are mantled by extensive glacial deposits and thus are quite rugged. In addition, some continental shelves are dissected by large valleys running from the coastline into deeper waters. Many of these *shelf valleys* are the seaward extensions of river valleys on the adjacent landmass. Such valleys appear to have been excavated during the Pleistocene epoch (Ice Age). During this time great quantities of water were stored in vast ice sheets on the continents. This caused sea level to drop by 100 meters (330 feet) or more, exposing large areas of the continental shelves. Because of this drop in sea level, rivers extended their courses, and land-dwelling plants and animals inhabited the newly exposed portions of the continents. Dredging off the coast of North America has retrieved the ancient remains of numerous land dwellers, including mammoths, mastodons, and horses, adding to the evidence that portions of the continental shelves were once above sea level.

Most passive continental shelves, such as those along the East Coast of the United States, consist of shallow-water deposits that can reach several kilometers in thickness. Such deposits have led researchers to conclude that these thick accumulations of sediment are produced along a gradually subsiding continental margin.

Continental Slope   Marking the seaward edge of the continental shelf is the **continental slope,** a relatively steep structure (as compared with the shelf) that marks the boundary between continental crust and oceanic crust (Figure 13.7). Although the inclination of the continental slope varies greatly from place to place, it averages about 5 degrees and

in places may exceed 25 degrees. Further, the continental slope is a relatively narrow feature, averaging only about 20 kilometers (12 miles) in width.

Continental Rise   In regions where trenches do not exist, the steep continental slope merges into a more gradual incline known as the **continental rise.** Here the slope drops to about one-third a degree, or about 6 meters per kilometer (32 feet per mile). Whereas the width of the continental slope averages about 20 kilometers (12 miles), the continental rise may extend for hundreds of kilometers into the deep-ocean basin.

The continental rise consists of a thick accumulation of sediment that moved downslope from the continental shelf to the deep-ocean floor. The sediments are delivered to the base of the continental slope by *turbidity currents* that periodically flow down submarine canyons. When these muddy currents emerge from the mouth of a canyon onto the relatively flat ocean floor, they deposit sediment that forms a **deep-sea fan** (Figure 13.7). As fans from adjacent submarine canyons grow, they merge laterally with one another to produce a continuous covering of sediment at the base of the continental slope forming the continental rise.

Active Continental Margins

Along some coasts the continental slope descends abruptly into a deep-ocean trench. In this situation, the landward wall of the trench and the continental slope are essentially the same feature. In such locations, the continental shelf is very narrow, if it exists at all.

Active continental margins are located primarily around the Pacific Ocean in areas where oceanic lithosphere is being subducted beneath the leading edge of a

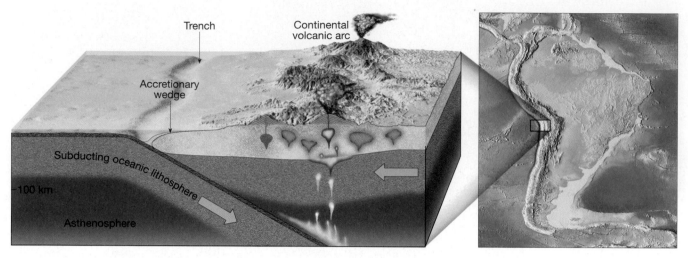

FIGURE 13.8 Active continental margin. Here sediments from the ocean floor are scraped from the descending plate and added to the continental crust as an accretionary wedge.

continent (Figure 13.8). Here sediments from the ocean floor and pieces of oceanic crust are scraped from the descending oceanic plate and plastered against the edge of the overriding continent. This chaotic accumulation of deformed sediment and scraps of oceanic crust is called an **accretionary** (*ad* = toward, *crescere* = to grow) **wedge.** Prolonged plate subduction, along with the accretion of sediments on the landward side of the trench, can produce a large accumulation of sediments along a continental margin. A large accretionary wedge, for example, is found along the northern coast of Japan's Honshu Island.

Some subduction zones have little or no accumulation of sediments, indicating that ocean sediments are being carried into the mantle with the subducting plate. These tend to be regions where old oceanic lithosphere is subducting nearly vertically into the mantle. In these locations the continental margin is very narrow, as the trench may lie a mere 50 kilometers (31 miles) offshore.

Features of Deep-Ocean Basins

Between the continental margin and the oceanic ridge lies the **deep-ocean basin** (see Figure 13.6). The size of this region—almost 30 percent of Earth's surface—is roughly comparable to the percentage of land above sea level. This region includes remarkably flat areas known as *abyssal plains;* tall volcanic peaks called *seamounts* and *guyots; deep-ocean trenches*, which are extremely deep linear depressions in the ocean floor; and large flood basalt provinces called *oceanic plateaus.*

Deep-Ocean Trenches

Deep-ocean trenches are long, relatively narrow creases in the seafloor that form the deepest parts of the ocean (Table 13.1). Most trenches are located along the margins of the Pacific Ocean (Figure 13.9), where many exceed 10,000 meters

(33,000 feet) in depth. A portion of one trench—the Challenger Deep in the Mariana Trench—has been measured at 11,022 meters (36,163 feet) below sea level, making it the deepest known part of the world ocean. Only two trenches are located in the Atlantic—the Puerto Rico Trench adjacent to the Lesser Antilles arc and the South Sandwich Trench.

Although deep-ocean trenches represent only a small portion of the area of the ocean floor, they are nevertheless significant geologic features. Trenches are sites of plate convergence where lithospheric plates subduct and plunge back into the mantle. In addition to earthquakes being created as one plate "scrapes" against another, volcanic activity is also associated with these regions. Thus, trenches are often paralleled by an arc-shaped row of active volcanoes called a *volcanic island arc.* Furthermore, *continental volcanic arcs*, such as those making up portions of the Andes and Cascades, are located parallel to trenches that lie adjacent to continental margins. The large number of trenches and associated volcanic activity along the margins of the Pacific Ocean explains why the region is known as the *Ring of Fire.*

TABLE 13.1	Dimensions of Some Deep-Ocean Trenches		
Trench	Depth (kilometers)	Average Width (kilometers)	Length (kilometers)
Aleutian	7.7	50	3700
Japan	8.4	100	800
Java	7.5	80	4500
Kurile–Kamchatka	10.5	120	2200
Mariana	11.0	70	2550
Central America	6.7	40	2800
Peru–Chile	8.1	100	5900
Philippine	10.5	60	1400
Puerto Rico	8.4	120	1550
South Sandwich	8.4	90	1450
Tonga	10.8	55	1400

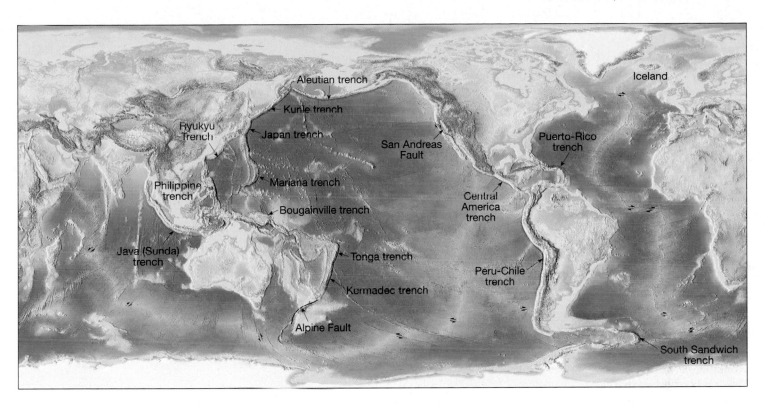

FIGURE 13.9   Distribution of the world's deep-ocean trenches.

Abyssal Plains

Abyssal (*a* = without, *byssus* = bottom) **plains** are deep, incredibly flat features; in fact, these regions are likely the most level places on Earth. The abyssal plain found off the coast of Argentina, for example, has less than 3 meters (10 feet) of relief over a distance exceeding 1300 kilometers (800 miles). The monotonous topography of abyssal plains is occasionally interrupted by the protruding summit of a partially buried volcanic peak.

Using *seismic profilers* (instruments that generate signals designed to penetrate far below the ocean floor), researchers have determined that abyssal plains owe their relatively featureless topography to thick accumulations of sediment that have buried an otherwise rugged ocean floor (see Figure 13.4). The nature of the sediment indicates that these plains consist primarily of fine sediments transported far out to sea by turbidity currents, deposits that have precipitated out of seawater, and shells and skeletons of microscopic marine organisms.

Abyssal plains are found in all oceans. However, the Atlantic Ocean has the most extensive abyssal plains because it has few trenches to act as traps for sediment carried down the continental slope.

Seamounts, Guyots, and Oceanic Plateaus

Dotting the ocean floor are submarine volcanoes called **seamounts,** which may rise hundreds of meters above the surrounding topography. It is estimated that more than a million of these features exist. Some grow large enough to become oceanic islands, but these are rare. Most do not have a sufficiently long eruptive history to build a structure above sea level. Although these conical peaks are found on the floors of all the oceans, the greatest number have been identified in the Pacific. Furthermore, seamounts often form linear chains, or in some cases a more continuous volcanic ridge, not to be confused with mid-ocean ridges.

Students Sometimes Ask . . .

Have humans ever explored the deepest ocean trenches? Could anything live there?

Humans have indeed visited the deepest part of the oceans—where there is crushing high pressure, complete darkness, and near-freezing water temperatures—more than 45 years ago! In January 1960, U.S. Navy Lt. Don Walsh and explorer Jacques Piccard descended to the bottom of the Challenger Deep region of the Mariana Trench in the deep-diving bathyscaphe *Trieste*. At 9906 meters (32,500 feet), the men heard a loud cracking sound that shook the cabin. They were unable to see that a 7.6-centimeter (3-inch) Plexiglas viewing port had cracked (miraculously, it held for the rest of the dive). More than five hours after leaving the surface, they reached the bottom at 10,912 meters (35,800 feet)—a record depth of human descent that has not been broken since. They did see some life forms that are adapted to life in the deep: a small flatfish, a shrimp, and some jellyfish.

BOX 13.2 ▶ UNDERSTANDING EARTH

Explaining Coral Atolls—Darwin's Hypothesis

Coral *atolls* are ring-shaped structures that often extend several thousand meters below sea level (Figure 13.B). What causes atolls to form, and how do they attain such a great thickness?

Corals are colonial animals about the size of an ant that feed with stinging tentacles and are related to jellyfish. Most corals protect themselves by creating a hard external skeleton made of calcium carbonate. Where corals reproduce and grow over many centuries their skeletons fuse into large structures called *coral reefs*. Other corals—as well as sponges and algae—begin to attach to the reef, enlarging it further. Eventually fishes, sea slugs, octopus, and other organisms are attracted to these diverse and productive habitats.

Corals require specific environmental conditions to grow. For example, reef-building corals grow best in waters with an average annual temperature of about 24°C (75°F). They cannot survive prolonged exposure to temperatures below 18°C (64°F) or above 30°C (86°F). In addition, reef-builders require an attachment site (usually other corals) and clear, sunlit water. Consequently, the limiting depth of most active reef growth is only about 45 meters (150 feet).

The restricted environmental conditions required for coral growth create an interesting paradox: How can corals—

FIGURE 13.B An aerial view of Tetiaroa Atoll in the Pacific. The light blue waters of the relatively shallow lagoon contrast with the dark blue color of the deep ocean surrounding the atoll. (Photo by Douglas Peebles Photography)

which require warm, shallow, sunlit water no deeper than a few dozen meters to live—create thick structures such as coral atolls that extend into deep water?

The naturalist Charles Darwin was one of the first to formulate a hypothesis on the origin of atolls. From 1831 to 1836 he sailed aboard the British ship HMS *Beagle* during its famous circumnavigation of the globe. In various places that Darwin visited, he noticed a progression of stages in coral reef development from (1) a *fringing reef* along the margins of a volcano to (2) a *barrier reef* with a volcano in the middle to (3) an *atoll*, which consists of a continuous or broken ring of coral reef surrounded by a central lagoon (Figure 13.C). The essence of Darwin's hypothesis was that as a volcanic island slowly sinks, the corals continue to build the reef complex upward.

Some, like the Hawaiian Island–Emperor Seamount chain in the Pacific, which stretches from the Hawaiian Islands to the Aleutian trench, form over a volcanic hot spot in association with a mantle plume (see Figure 2.28, p. 63). Others are born near oceanic ridges. If the volcano grows large enough before being carried from the magma source by plate movement, the structure may emerge as an island. Examples in the Atlantic include the Azores, Ascension, Tristan da Cunha, and St. Helena.

During the time they exist as islands, some of these volcanic structures are lowered to near sea level by the forces of weathering and erosion. In addition, islands will gradually sink and disappear below the water surface as the moving plate slowly carries them away from the elevated oceanic ridge or hot spot where they originated (see Box 13.2). Submerged, flat-topped seamounts which formed in this manner are called **guyots** or **tablemounts.***

---

*The term *guyot* is named after Princeton University's first geology professor. It is pronounced "GEE-oh" with a hard *g* as in "give."

Mantle plumes have also generated several large **oceanic plateaus,** which resemble the flood basalt provinces found on the continents. Examples of these extensive volcanic structures include the Ontong Java and Caribbean plateaus, which formed from vast outpourings of fluid basaltic lavas onto the ocean floor (Figure 13.10). Hence, oceanic plateaus are composed mostly of basalts and ultramafic rocks that in some cases exceed 30 kilometers in thickness.

Anatomy of the Oceanic Ridge

 GEODe Divergent Boundaries
▶ Oceanic Ridges and Seafloor Spreading

Along well-developed divergent plate boundaries, the seafloor is elevated, forming a broad linear swell called the **oceanic ridge,** or **mid-ocean ridge.** Our knowledge of the oceanic ridge system comes from soundings taken of the ocean floor, core samples obtained from deep-sea drilling, visual inspection using deep-diving submersibles

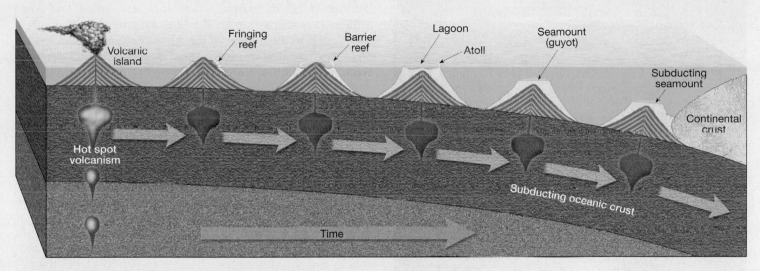

FIGURE 13.C Formation of a coral atoll due to the gradual sinking of oceanic crust and upward growth of the coral reef. A fringing coral reef forms around an active volcanic island. As the volcanic island moves away from the region of hotspot activity it sinks, and the fringing reef gradually becomes a barrier reef. Eventually, the volcano is completely submerged and an atoll remains.

Darwin's hypothesis explained how coral reefs, which are restricted to shallow water, can build structures that now exist in much deeper water. During Darwin's time, however, there was no plausible mechanism to account for how an island might sink.

Today, plate tectonics helps explain how a volcanic island can become extinct and sink to great depths over long periods of time. Volcanic islands often form over a relatively stationary mantle plume, which causes the lithosphere to be buoyantly uplifted. Over a span of millions of years, these volcanic islands become inactive and gradually sink as the moving plate carries them away from the region of hotspot volcanism (Figure 13.C).

Furthermore, drilling through atolls has revealed that volcanic rock does indeed underlie the oldest (and deepest) coral reef structures, confirming Darwin's hypothesis. Thus, atolls owe their existence to the gradual sinking of volcanic islands containing coral reefs that build upward through time.

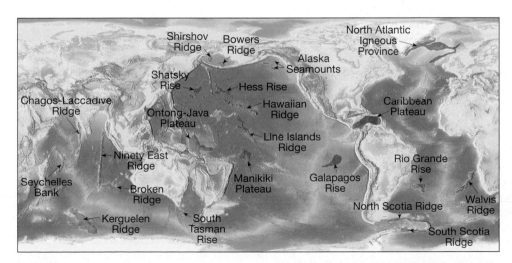

FIGURE 13.10 Distribution of oceanic plateaus, hot spot tracks, and other submerged crustal fragments.

(Figure 13.11), and even firsthand inspection of slices of ocean floor that have been displaced onto dry land along convergent plate boundaries. An elevated position, extensive faulting and associated earthquakes, high heat flow, and numerous volcanic structures characterize the oceanic ridge.

The interconnected oceanic ridge system is the longest topographic feature on Earth's surface, exceeding 70,000 kilometers (43,000 miles) in length. Representing more than 20 percent of Earth's surface, the oceanic ridge winds through all major oceans in a manner similar to the seam on a baseball (Figure 13.12). The crest of this linear structure typically stands 2 to 3 kilometers above the adjacent deep-ocean basins and marks the plate margins where new oceanic crust is created.

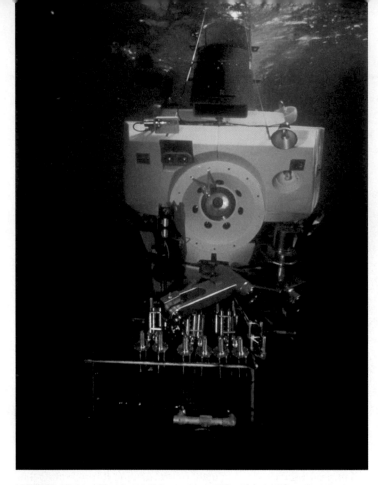

FIGURE 13.11 The deep-diving submersible *Alvin* is 7.6 meters long, weighs 16 tons, has a cruising speed of 1 knot, and can reach depths as great as 4000 meters. A pilot and two scientific observers are along during a normal 6- to 10-hour dive. (Courtesy of Rod Catanach/Woods Hole Oceanographic Institution)

Notice in Figure 13.12 that large sections of the oceanic ridge system have been named based on their locations within the various ocean basins. Some ridges run along the middle of ocean basins, where they are called *mid-ocean* ridges. This holds true for the Mid-Atlantic Ridge, which is positioned in the middle of the Atlantic, roughly paralleling the margins of the continents on either side. This is also true for the Mid-Indian Ridge. However, the East Pacific Rise is not a "mid-ocean" feature. Rather, as its name implies, it is located in the eastern Pacific, far from the center of the ocean. At its northern end there are two branches—one that points toward Central America and another that curves toward South America (Figure 13.12).

The term *ridge* may be misleading, because these features are not narrow and steep as the term implies, but have widths of from 1000 to 4000 kilometers and the appearance of a broad, elongated swell that exhibits various degrees of ruggedness. Furthermore, the ridge system is broken into segments that range from a few tens to hundreds of kilometers in length. Although each segment is offset from the adjacent segment, they are generally connected, one to the next, by a transform fault.

Oceanic ridges are as high as some mountains found on the continents; however, the similarity ends there. Whereas most continental mountains form when compressional forces fold and metamorphose thick sequences of sedimentary rocks along convergent plate boundaries, oceanic ridges form where tensional forces fracture and pull the ocean crust apart. The oceanic ridge consists of layers and piles of newly formed mafic rocks that have

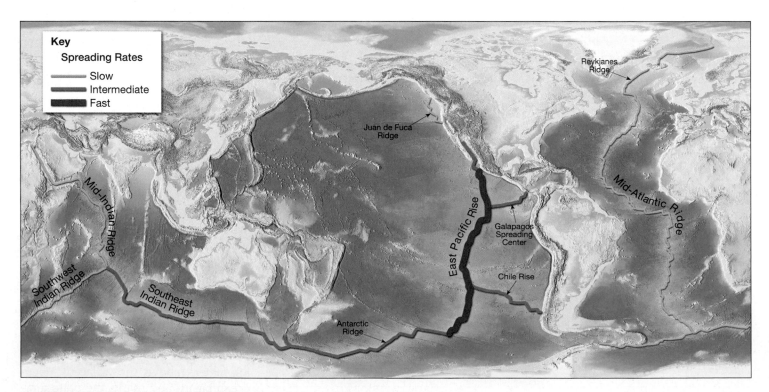

FIGURE 13.12 Distribution of the oceanic ridge system. The map shows ridge segments that exhibit slow, intermediate, and fast spreading rates.

been faulted into elongated blocks that are buoyantly up-lifted.

Along the axis of some segments of the oceanic ridge system are deep, down-faulted structures called **rift valleys** (Figure 13.13). The name *rift valley* has been applied to these features because they are strikingly similar to the continental rift valleys that comprise the East African Rift (see Figure 13.20). Some rift valleys, including many along the Mid-Atlantic Ridge, exceed 30 kilometers in width and have walls that tower 2000 meters above the valley floor. This makes them comparable to the deepest and widest part of Arizona's Grand Canyon.

Oceanic Ridges and Seafloor Spreading

GEODe
Divergent Boundaries
▸ **Oceanic Ridges and Seafloor Spreading**

The greatest volume of magma (more than 60 percent of Earth's total yearly output) is produced along the oceanic ridge system in association with seafloor spreading. As the plates diverge, fractures are created in the oceanic crust that fill with molten rock that wells up from the hot asthenosphere below. This molten material slowly cools to solid rock, producing new slivers of seafloor. This process occurs again and again, generating new lithosphere that moves from the ridge crest in a conveyor belt fashion.

Seafloor Spreading

Harry Hess of Princeton University formulated the concept of seafloor spreading in the early 1960s. Later, geologists were able to verify Hess's view that seafloor spreading occurs along relatively narrow areas, located at the crests of oceanic ridges. Here, below the ridge axis where the lithospheric plates separate, solid hot mantle rocks rise upward to replace the material that has shifted horizontally. Recall from Chapter 4 that as rock rises, it experiences a decrease in confining pressure and may undergo melting without the addition of heat. This process, called *decompression melting*, is how magma is generated along the ridge axis.

Partial melting of mantle rock produces basaltic magma having a composition that is surprisingly consistent along the entire length of the ridge system. This newly formed magma separates from the mantle rock from which it was derived, and rises toward the surface in the form of teardrop-shaped blobs. Although most of this magma is thought to collect in elongated reservoirs (magma chambers) located just beneath the ridge crest, about 10 percent eventually migrates upward along fissures to erupt as lava flows on the ocean floor (Figure 13.13). This activity continuously adds new basaltic rock to the plate margins, temporarily welding them together, only to be broken as spreading continues. Along some ridges, outpourings of bulbous lavas build submerged shield volcanoes (seamounts) as well as elongated lava ridges. At other locations, more voluminous lava flows create a relatively subdued topography.

During seafloor spreading, the magma that is injected into newly developed fractures forms dikes that cool from their outer borders inward toward their centers. Because the warm interiors of these newly formed dikes are weak, continued spreading produces new fractures that tend to split these young rocks roughly in half. As a result, new material is added equally to the two diverging plates. Consequently, new ocean floor grows symmetrically on each side of the centrally located ridge crest. Indeed, the ridge systems of the Atlantic and Indian oceans are located near the middle of these water bodies. However, the East Pacific Rise is situated far from the center of the Pacific Ocean. Despite uniform spreading along the East Pacific Rise, much of the Pacific Basin that once lay east of this divergent boundary has been overridden by the westward migration of the American plates.

When Harry Hess first proposed the concept of seafloor spreading, upwelling in the mantle was thought to be one of the driving forces for plate

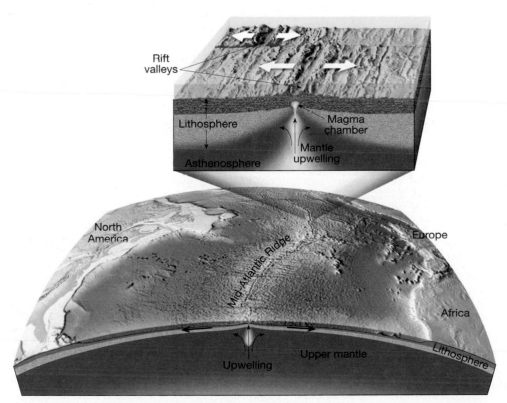

FIGURE 13.13 The axis of some segments of the oceanic ridge system contains deep downfaulted structures called *rift valleys*. Some may exceed 30 kilometers in width and 2000 meters in depth.

motions. Geologists have since discovered that upwelling along the oceanic ridge is a *passive process*. Stated another way, mantle upwelling occurs because "space" is created as oceanic lithosphere moves horizontally away from the ridge axis.

Why Are Oceanic Ridges Elevated?

The primary reason for the elevated position of a ridge system is the fact that newly created oceanic lithosphere is hot, occupies more volume, and is therefore less dense than cooler rocks of the deep-ocean basin. As the newly formed basaltic crust travels away from the ridge crest, it is cooled from above as seawater circulates through the pore spaces and fractures in the rock. It also cools because it is moving away from the zone of upwelling, which is the main source of heat. As a result, the lithosphere gradually cools, contracts, and becomes more dense. This thermal contraction accounts for the greater ocean depths that exist away from the ridge. It takes almost 80 million years before cooling and contraction cease completely. By this time, rock that was once part of an elevated ocean-ridge system is located in the deep-ocean basin, where it may be covered by relatively thick accumulations of sediment.

As the lithosphere is displaced away from the ridge crest, cooling also causes a gradual increase in lithospheric thickness. This occurs because the boundary between the lithosphere and asthenosphere is temperature dependent. Recall that the lithosphere is Earth's cool, rigid outer layer, whereas the asthenosphere is a comparatively hot and weak zone. As material in the upper mantle ages (cools), it becomes rigid. Thus, the upper portion of the asthenosphere is converted to lithosphere simply by cooling. Newly formed oceanic lithosphere will continue to thicken for about 80 million years. Thereafter its thickness remains relatively constant until it is subducted.

Spreading Rates and Ridge Topography

When various segments of the oceanic ridge system were studied in detail, some topographic differences came to light. Many of these differences appear to be controlled by spreading rates. One of the main factors controlled by spreading rates is the amount of magma generated at a rift zone. At fast spreading centers, divergence occurs at a greater rate than at slow spreading centers, resulting in more magma welling up from the mantle. As a result, the magma chambers located below fast spreading centers tend to be larger and more permanent features than those associated with slower spreading centers. Further, spreading along fast spreading centers appears to be a relatively continuous process where rifting and upwelling is occurring along the entire length of the ridge axis. By contrast, rifting at slow spreading centers appears to be more episodic, where segments of the ridge may remain dormant for extended intervals.

At slow spreading rates of 1 to 5 centimeters per year, such as occur at the Mid-Atlantic and Mid-Indian ridges, a prominent rift valley is present along much of the ridge crest and the topography is quite rugged (Figure 13.14A). Recall that these rift valleys can exceed 30 kilometers in width and 2000 meters in depth. Here the upward displacement of large, buoyant slabs of oceanic crust along nearly vertical faults produces the steep walls of the rift valley. In addition, at slow spreading centers volcanism produces numerous volcanic cones within the rift valley which enhance the rugged topography of the ridge crest.

Along the Galápagos ridge an intermediate spreading rate of 5 to 9 centimeters per year is the norm. In settings such as this the rift valleys that develop are shallow, often less than 200 meters deep. In addition, their topography tends to be subdued compared to ridges that exhibit slower spreading rates.

At faster spreading rates (greater than 9 centimeters per year), such as occur along much of the East Pacific Rise, rift valleys are generally absent (Figure 13.14B). Instead, the ridge axis is higher than the ocean floor on either side. Such areas consist of *swells*—volcanic extrusions that tend to overlap or even produce relatively narrow structures (Figure 13.15). In other places along fast spreading centers, sheets of fluid lava have produced areas of relatively subdued topography. Because the depth of the ocean depends on the age of the seafloor, ridge segments that exhibit faster

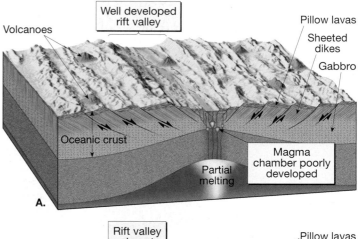

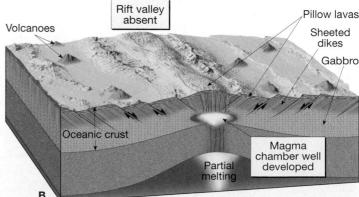

FIGURE 13.14 Topography of the crest of an oceanic ridge. **A.** At slow spreading rates a prominent rift valley develops along the ridge crest, and the topography is typically rugged. **B.** Along fast spreading centers no median rift valleys develop, and the topography is comparatively smooth.

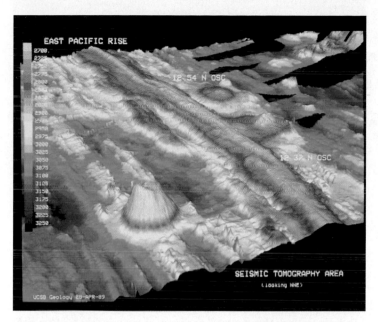

FIGURE 13.15 False-color sonar image of a segment of the East Pacific Rise. The linear pink area is the swell that formed above the ridge axis. Note also the large volcanic cone in the lower left portion of the image. (Courtesy of S. P. Miller)

spreading rates tend to have more gradual profiles than ridges that have slower spreading rates (Figure 13.16). Because of these differences in topography, the gently sloping, less rugged portions of the oceanic ridges are called *rises*.

The Nature of Oceanic Crust

One of the most interesting aspects of the oceanic crust is that its thickness and structure is remarkably consistent throughout the entire ocean basin. Seismic soundings indicate that it

averages only about 7 kilometers (5 miles) in thickness. Furthermore, it is composed almost entirely of mafic (basaltic) rocks that are underlain by a layer of the ultramafic rock peridotite, which forms the lithospheric mantle.

Although most oceanic crust forms out of view, far below sea level, geologists have been able to examine the structure of the ocean floor firsthand. In such locations as Newfoundland, Cyprus, Oman, and California, slivers of oceanic crust have been thrust high above sea level. From these exposures, researchers conclude that the ocean crust consists of four distinct layers (Figure 13.17):

- Layer 1: The upper layer is comprised of a sequence of unconsolidated sediments. Sediments are very thin near the axes of oceanic ridges, but may be several kilometers thick next to continents.
- Layer 2: Below the layer of sediments is a rock unit composed mainly of basaltic lavas that contain abundant pillowlike structures called *pillow basalts.*
- Layer 3: The middle rocky layer is made up of numerous interconnected dikes having a nearly vertical orientation, called the *sheeted dike complex.* These dikes are former pathways where magma rose to feed lava flows on the ocean floor.
- Layer 4: The lower unit is made up mainly of gabbro, the coarse-grained equivalent of basalt, which crystallized in a magma chamber below the ridge axis.

This sequence of rocks composing the oceanic crust is called an **ophiolite complex** (Figure 13.17). From studies of various ophiolite complexes around the globe and related data, geologists have pieced together a scenario for the formation of the ocean floor.

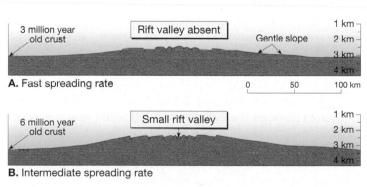

A. Fast spreading rate

3 million year old crust — Rift valley absent — Gentle slope

B. Intermediate spreading rate

6 million year old crust — Small rift valley

C. Slow spreading rate

12 million year old crust — Well developed rift valley — Steep slope

FIGURE 13.16 Schematic of ridge segments that exhibit fast, intermediate, and slow spreading rates. Fast spreading centers have gentle slopes and lack a rift valley. By contrast, ridges that have slow spreading rates have well-developed rift valleys and steep flanks. (The slopes of all these profiles are greatly exaggerated.)

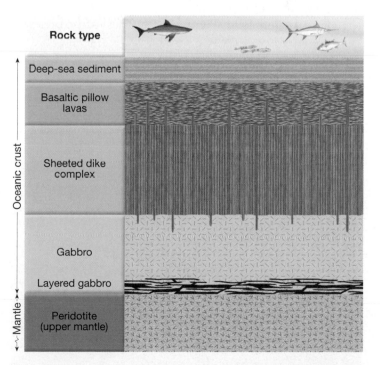

Rock type

Deep-sea sediment

Basaltic pillow lavas

Sheeted dike complex

Gabbro

Layered gabbro

Peridotite (upper mantle)

Oceanic crust

Mantle

FIGURE 13.17 Rock types associated with a typical section of oceanic crust based on data obtained from ophiolite complexes and seismic studies.

How Does Oceanic Crust Form?

Recall that the basaltic magma needed to create new ocean crust originates from partially melted mantle rock (peridotite). Being partially molten and less dense than the surrounding solid rock, the magma gradually moves upward, where it enters a magma chamber that is thought to be less than 10 kilometers wide and located only 1 or 2 kilometers below the ridge crest. Seismic studies conducted along the East Pacific Rise have identified magma chambers along 60 percent of the ridge. Hence, these structures appear to be relatively permanent features, at least along fast spreading centers. However, along slow spreading ridges, where the rate of magma production is less, magma chambers are small and appear to form intermittently.

As seafloor spreading proceeds, numerous vertical fractures develop in the ocean crust that lies above these magma chambers. Molten rock is injected into these fissures, where some of it cools and solidifies to form dikes. New dikes intrude older dikes, which are still warm and weak, to form the **sheeted dike complex.** This portion of the oceanic crust is usually 1 to 2 kilometers thick.

Roughly 10 percent of the magma entering the reservoirs eventually erupts on the ocean floor. Because the surface of a submarine lava flow is chilled quickly by seawater, it rarely travels more than a few kilometers before completely solidifying. The forward motion occurs as lava accumulates behind the congealed margin and then breaks through. This process occurs over and over, as molten basalt is extruded—like toothpaste out of a tightly squeezed tube (Figure 13.18). The result is tube-shaped protuberances resembling large bed pillows stacked one atop the other, hence the name **pillow basalts** (Figure 13.19). In some settings, pillow lavas may build into volcano-size mounds that resemble shield volcanoes, whereas in others they form elongated ridges tens of kilometers long (Figure 13.18). These structures will

FIGURE 13.19 Pillow lava exposed along a sea cliff, Cape Wanbrow, New Zealand. Notice that each "pillow" shows an outer, rapidly cooled, dark glassy layer enclosing a dark gray basalt interior. (Photo by G. R. Roberts)

eventually be cut off from their supply of magma as they are carried away from the ridge crest by seafloor spreading.

The lowest unit of the ocean crust develops from crystallization within the central magma chamber itself. The first minerals to crystallize are olivine, pyroxene, and occasionally chromite (chromium oxide), which fall through the magma to form a layered zone near the floor of the reservoir. The remaining magma tends to cool along the walls of the chamber and forms massive amounts of coarse-grained gabbro. This unit makes up the bulk of the oceanic crust, where it may account for as much as 5 of its 7-kilometer total thickness.

In this manner, the processes at work along the ridge system generate the entire sequence of rocks found in an ophiolite complex. Since the magma chambers are periodically replenished with fresh magma rising from the asthenosphere, the oceanic crust is continuously being generated.

Interactions between Seawater and Oceanic Crust

In addition to serving as a mechanism for the dissipation of Earth's internal heat, the interaction between seawater and the newly formed basaltic crust alters both the seawater and the crust. Because submarine lava flows are very permeable and the upper oceanic crust is highly fractured, seawater can penetrate to a depth of 2 to 3 kilometers. As seawater circulates through the hot crust, it is heated and alters the basaltic rock by a process called *hydrothermal* (hot water) *metamorphism* (see Chapter 8). This alteration causes the dark silicates (olivine and pyroxene) found in basalt to form new minerals such as chlorite and serpentine.

In addition to the basaltic crust being altered, so is the seawater. As the hot seawater circulates through the newly formed rock, it dissolves ions of silica, iron, copper, and sometimes silver and gold from the hot basalts. Once the water is heated to several hundred degrees Celsius, it buoyantly rises along fractures and eventually spews out at the surface (see Box 13.3). Studies conducted by submersibles along the Juan de Fuca Ridge have photographed these metallic-rich solutions as they gush from the seafloor to form particle-filled clouds called **black smokers.** As the

FIGURE 13.18 A photograph taken from the *Alvin* during Project FAMOUS shows lava extrusions in the rift valley of the Mid-Atlantic Ridge. Large toothpastelike extrusions such as this were common features. A mechanical arm is sampling an adjacent blisterlike extrusion. (Photo courtesy of Woods Hole Oceanographic Institution)

BOX 13.3 ▶ EARTH AS A SYSTEM

Deep-Sea Hydrothermal Vents*

Sitting in a computer-filled, darkened room, a group of geologists, biologists, and chemists peer intently at video monitors showing astonishing imagery of giant, smoke-billowing, chimney-like rock formations and an abundance of bizarre animals. Is this a scene from Hollywood's latest sci-fi blockbuster? No, it's a typical scene from an actual research vessel located 250 kilometers southwest of Vancouver Island. The images are being relayed to the ship by the Canadian remotely controlled vehicle ROPOS (Remotely Operated Platform for Ocean Science), which is busy working more than 2 kilometers below the surface along the Juan de Fuca Ridge. This seafloor mountain chain is actively being created by the rifting and pulling apart of the Pacific and Juan de Fuca tectonic plates and the production of new oceanic crust by upwelling magma.

Along ridges like the Juan de Fuca, cold seawater moves several hundreds of meters into the highly fractured basaltic crust, where it is then heated at depth by magmatic sources. Along the way, the heated water strips metals and elements such as sulfur from the surrounding rocks. This heated fluid tends to rise due to convection, following conduits and fractures that are concentrated along the ridges. When it reaches the surface of the crust, the fluid can be over 400°C, but it does not boil because of the extremely high pressures

caused by the water column above the vents. When this hydrothermal fluid comes into contact with the much colder seawater, minerals rapidly precipitate and form a shimmering smoke-like cloud, which gives "black smokers" their name (Figure 13.D). Some minerals immediately solidify and contribute to the formation of spectacular chimney-like structures, which can be as tall as a 15-story building, and are appropriately given names like Godzilla and Inferno. In some cases, these chimneys and related deposits contain concentrated amounts of iron, copper, zinc, lead, silver, and even gold.

The Juan de Fuca vents are also remarkable for the biology that they support. In these environments completely devoid of sunlight, microorganisms take advantage of the mineral-rich hydrothermal fluid to perform what is known as chemosynthesis. The microbial communities, in turn, support larger, more complex animals such as fish, crabs, worms, mussels, and clams. Some species at these vents are not found anywhere else on Earth. The most famous of these, and perhaps the most unique, is the tubeworm (Figure 13.E). With their white chitinous tubes and bright red plumes, these conspicuous creatures rely entirely on bacteria growing in their trophosome, an internal organ designed for harvesting bacteria. These symbiotic bacteria rely on the tubeworm to provide them

FIGURE 13.E   Tube worms up to 3 meters (10 feet) in length are among the organisms found in the extreme environment of hydrothermal vents along the crest of the oceanic ridge, where sunlight is nonexistent. These organisms obtain their food from internal microscopic bacteria-like organisms, which acquire their nourishment and energy through the processes of chemosynthesis. (Photo by Al Giddings Images, Inc.)

with a suitable habitat and, in return, they provide carbon-based building blocks to the tubeworms.

Concern over the possibility of damaging these unique ecosystems by sampling activities of scientists, increasing ecotourism, or the potential exploitation of biological and mineral resources, has recently led the Canadian government to label part of the Juan de Fuca hydrothermal vents as Canada's first Marine Protected Area (MPA). This designation puts in place enforceable regulations to preserve and protect the area and its marine organisms, while encouraging continued scientific study of this unique and remarkable ecosystem located in Canada's backyard.

FIGURE 13.D   A black smoker spewing hot, mineral-rich water along the East Pacific Rise. As heated solutions meet cold seawater, sulfides of copper, iron, and zinc precipitate immediately, forming mounds of minerals around these vents. (Photo by Dudley Foster, Woods Hole Oceanographic Institution)

*This box was prepared by Richard Leveille, a research scientist at the University of Quebec, Montreal, Canada.

hot liquid (about 350°C) mixes with the cold, mineral-laden seawater, the dissolved minerals precipitate to form massive, metallic sulfide deposits, some of which are economically important. Occasionally these deposits grow upward to form large, chimneylike structures as tall as skyscrapers.

Continental Rifting: The Birth of a New Ocean Basin

 GEODe Divergent Boundaries
EARTH ▸ The Formation of Ocean Basins

Why the supercontinent of Pangaea began to split apart nearly 200 million years ago is not known with certainty. Nevertheless, this event serves to illustrate that perhaps most ocean basins get their start when a continent begins to break apart. This clearly is the case for the Atlantic Ocean, which formed as the Americas drifted from Europe and Africa. It is also true for the Indian Ocean, which developed as Africa rifted from Antarctica and India.

Evolution of an Ocean Basin

The development of a new ocean basin begins with the formation of a **continental rift,** an elongated depression in which the entire thickness of the lithosphere has been deformed. Examples of continental rifts include the East African Rift, the Baikal Rift (south central Siberia), the Rhine Valley (northwestern Europe), the Rio Grande Rift, and the Basin and Range province in the western United States. It appears that continental rifts form in a variety of tectonic settings and may result in the breakup of a landmass.

In those settings where rifting continues, the rift system will evolve into a young, narrow ocean basin, exemplified by the present-day Red Sea. Eventually, seafloor spreading results in the formation of a mature ocean basin bordered by rifted continental margins. The Atlantic Ocean is such a feature. What follows is a look at this model of ocean basin evolution using modern examples to represent the various stages of rifting.

East African Rift An example of an active continental rift is the East African Rift, which extends through eastern Africa for approximately 3000 kilometers (2000 miles). Rather than being a single rift, the East African Rift consists of several somewhat interconnected rift valleys that split into an eastern and western section around Lake Victoria (Figure 13.20). Whether this rift will develop into a spreading center, where the Somali subplate separates from the continent of Africa, is still being debated. Nevertheless, the East African Rift is thought to characterize the initial stage in the breakup of a continent.

The most recent period of rifting began about 20 million years ago as upwelling in the mantle forcefully intruded the base of the lithosphere (Figure 13.21A). Buoyant uplifting of the heated lithosphere led to doming of the crust. As a consequence, the upper crust was broken along steep-angle normal faults, producing downfaulted blocks, or *grabens,* while the lower crust deformed by ductile stretching (Figure 13.21B). Thus, this continental rift system closely resembles rifts found along slow spreading centers.

In its early stage of formation, magma generated by decompression melting of the rising mantle plume intrudes the crust. Some of the magma migrates along fractures and erupts at the surface. This activity produces extensive basaltic flows within the rift as well as volcanic cones—some forming more than 100 kilometers from the rift axis. Examples include Mount Kilimanjaro, which is the highest point in Africa, rising almost 6000 meters (20,000 feet) above the Serengeti Plain, and Mount Kenya.

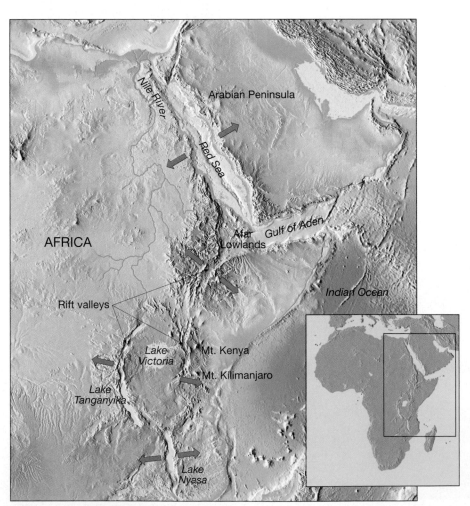

FIGURE 13.20 East African rift valleys and associated features.

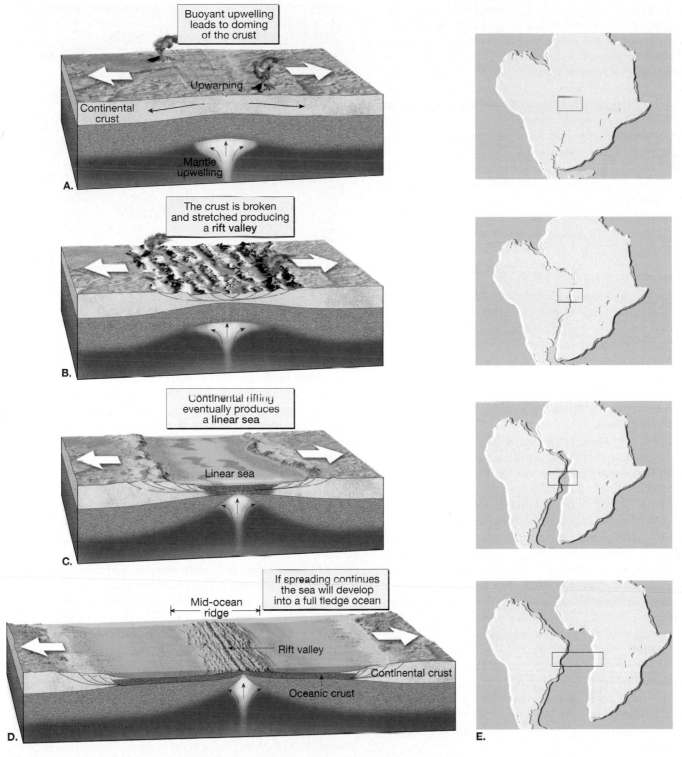

FIGURE 13.21 Formation of an ocean basin. **A.** Tensional forces and buoyant uplifting of the heated lithosphere cause the upper crust to be broken along normal faults, while the lower crust deforms by ductile stretching. **B.** As the crust is pulled apart, large slabs of rock sink, generating a rift zone. **C.** Further spreading generates a narrow sea. **D.** Eventually, an expansive ocean basin and ridge system are created. **E.** Illustration of the separation of South America and Africa to form the South Atlantic.

Red Sea Research suggests that if tensional forces are maintained, a rift valley will lengthen and deepen, eventually extending out to the margin of the continent, thereby splitting it in two (Figure 13.21C). At this point, the continental rift becomes a narrow linear sea with an outlet to the ocean, similar to the Red Sea.

The Red Sea formed when the Arabian Peninsula rifted from Africa beginning about 30 million years ago. Steep fault scarps that rise as much as 3 kilometers above sea level flank the margins of this water body. Thus, the escarpments surrounding the Red Sea are similar to the cliffs that border the East African Rift. Although the Red Sea only reaches oceanic depths (up to 5 kilometers) in a few locations, symmetrical magnetic stripes indicate that typical seafloor spreading here has been taking place for the past 5 million years.

Atlantic Ocean If spreading continues, the Red Sea will grow wider and develop an elevated oceanic ridge similar to the Mid-Atlantic Ridge (Figure 13.21D). As new oceanic crust is added to the diverging plates, the rifted continental margins move ever so slowly away from one another. As a result, the rifted continental margins that were once situated above the region of upwelling are displaced toward the interior of the growing plates. Consequently, as the continental lithosphere moves away from the source of heat, it cools, contracts, and subsides.

In time these continental margins will subside below sea level. Simultaneously, material eroded from the adjacent landmass will be deposited atop the faulted topography of the submerged continental margin. Eventually, this material will accumulate to form a thick, broad wedge of relatively undisturbed sediment and sedimentary rock. Recall that continental margins of this type are called *passive continental margins*. Because passive margins are not associated with plate boundaries, they experience little volcanism and few earthquakes. Recall, however, that this was not the case when these lithospheric blocks made up the flanks of a continental rift.

Not all continental rift valleys develop into full-fledged spreading centers. Running through the central United States is a failed rift that extends from Lake Superior into central Kansas (Figure 13.22). This once active rift valley is filled with volcanic rock that was extruded onto the crust more than a billion years ago. Why one rift valley develops into an active spreading center while others are abandoned is not yet known.

Mechanisms for Continental Rifting

It seems likely that supercontinents existed sporadically during the geologic past. Pangaea, which was the most recent of these, was assembled into a supercontinent between 450 and 230 million years ago, only to break up again shortly thereafter. Thus, geologists have concluded that the formation of a supercontinent followed by continental splitting must be an integral part of plate tectonics. Furthermore, this phenomenon must involve a major change in the direction

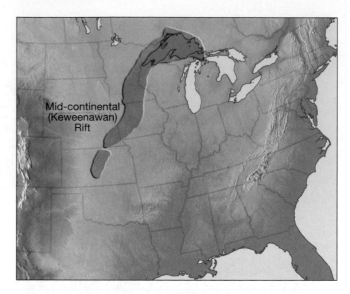

FIGURE 13.22 Map showing location of a failed rift extending from Lake Superior to Kansas.

and nature of the forces that drive plate motion. Stated another way, over long periods of geologic time, the forces that drive plate motions tend to organize crustal fragments into a single supercontinent, only to change directions and disperse them again. Two mechanisms have been proposed for continental rifting—plumes of hot mobile rock rising from deep in the mantle, and forces that arise from plate motions.

Mantle Plumes and Gravity Sliding Recall that a *mantle plume* consists of hotter than normal mantle rock that has a large mushroom-shaped head hundreds of kilometers in diameter attached to a long, but narrow, trailing tail. As the plume head nears the base of the cool lithosphere, it spreads laterally. Decompression melting within the plume head generates huge volumes of basaltic magma that rises and triggers volcanism at the surface. The result is a volcanic region, called a *hot spot,* that can be as much as 2000 kilometers across.

Research suggests that mantle plumes would tend to concentrate beneath a supercontinent, because once assembled, a large landmass forms an insulating "blanket" that traps heat in the mantle. The resulting temperature increase would lead to the formation of mantle plumes that serve as heat dissipation mechanisms.

Evidence for the role that mantle plumes play in continental rifting is available from passive continental margins, the former sites of rifting. In several regions on both sides of the Atlantic, continental rifting was preceded by crustal uplift and massive outpourings of basaltic lava. Examples include the Etendeka flood basalts of southwest Africa and the Paraná basalt province of South America (Figure 13.23A).

About 130 million years ago, when South America and Africa were joined as a single landmass, vast outpourings of lava produced a large continental basalt plateau (Figure 13.23B). Shortly after this event, the South Atlantic began to open, splitting the basalt province into what is now the Etendeka and Paraná basalt plateaus. As the ocean basin grew, the tail of the plume produced a string of seamounts

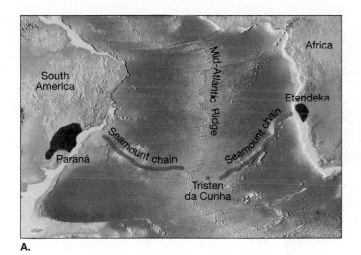

A.

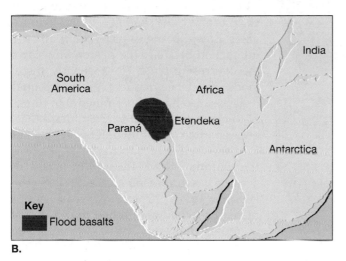

B.

FIGURE 13.23 Evidence for the role that mantle plumes might play in continental rifting. **A.** Relationship of the Paraná and Etendeka basalt plateaus to the Tristan da Cunha hotspot. **B.** Location of these basalt plateaus 130 million years ago, just before the South Atlantic began to open.

on each side of the newly formed ridge (Figure 13.23A). The modern area of hot-spot activity is centered around the volcanic island of Tristan da Cunha, which is located on the Mid-Atlantic Ridge.

About 60 million years ago, yet another mantle plume is thought to have initiated the rifting of Greenland from Northern Europe. Volcanic rocks associated with this activity extend from eastern Greenland to Scotland. The hot spot associated with this event is presently situated beneath Iceland.

From these studies, geologists have concluded that mantle plumes have played a role in the development of at least some continental rifts. In these regions, rifting began when a hot mantle plume reached the base of the lithosphere and caused the overlying crust to dome and weaken. As the crust was buoyantly uplifted, it stretched and developed rifts similar to those in East Africa. Simultaneously, decompression melting of the plume head led to vast outpourings of basaltic lavas. Following these episodes of igneous activity, an ocean basin began to open. The proposed mechanism for rifting is gravity sliding off the uplift caused by plume buoyancy.

It is important to note that not all hot-spot volcanism leads to rifting. For example, vast outpourings of basaltic lava that constitute the Columbia River basalts in the Pacific Northwest, as well as Russia's Siberian Traps, are not associated with the fragmentation of a continent.

Slab Pull and Slab Suction It is generally agreed that tensional forces, which tend to elongate or pull apart a rock unit, are needed in order for a continent to be fragmented. But how do these forces originate?

Recall that old oceanic lithosphere subducts because it is denser than the underlying asthenosphere. In situations where a continent is attached to a subducting slab of oceanic lithosphere, it will be pulled toward the trench. However, continents overlie thick sections of lithospheric mantle. As a result, they tend to resist being towed, which creates tensional stresses that stretch and thin the crust. Whether slab pull can tear a continent apart is still being studied. Perhaps other factors, including the presence of hot spots, or an inherent weakness in the crust, such as a major fault zone, may contribute to rifting.

Investigators have suggested that during the breakup of Pangaea, the Americas were rifted from Europe and Africa as a result of another force—*slab suction*. Recall that when a cold oceanic slab sinks, it causes the trench to retreat, or roll back. This creates flow in the asthenosphere that pulls the overriding plate *toward* the retreating trench (Figure 13.24).

During the breakup of Pangaea a subduction zone extended along the entire western margin of North and South America. As this subduction zone developed, the trench slowly retreated westward toward a spreading center located in the Pacific. Modern remnants of this subduction zone include the Peru–Chile Trench, Central American Trench, and Cascadia subduction zone (see Figure 13.9, p. 357). Slab suction along the entire western margin of the Americas may have provided the tensional forces that rifted Pangaea.

In summary, continental rifting occurs when a landmass is under tension, which tends to elongate and thin the lithosphere. This mechanism may be aided by a series of hotspots that weaken and elevate the crust.

Destruction of Oceanic Lithosphere

Although new lithosphere is continually being produced at divergent plate boundaries, Earth's surface area is not growing larger. In order to balance the amount of newly created lithosphere, there must be a process whereby plates are destroyed. Recall that this occurs along *convergent boundaries*, also called *subduction zones*.

Why Oceanic Lithosphere Subducts

The process of plate subduction is complex, and the ultimate fate of oceanic lithosphere is still being debated. What is known with some certainty is that a slab of oceanic lithosphere subducts because its overall density is greater than that of the underlying mantle. Recall that when ocean

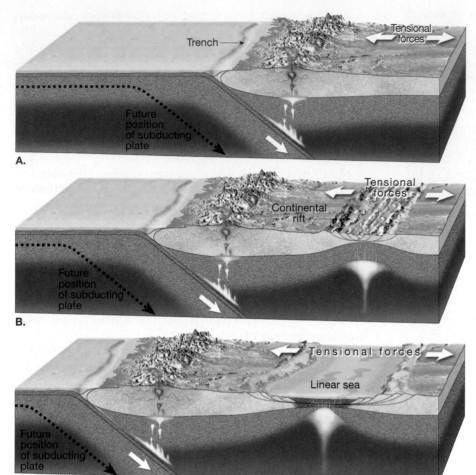

A.

B.

C.

FIGURE 13.24 Illustration of how trench retreat, or roll back, produces slab-suction forces that are thought to contribute to the breakup of a continent.

It is important to note that it is the *lithospheric mantle,* located beneath the oceanic crust, that drives subduction. Even when the oceanic crust is quite old, its density is 3.0 g/cm^3, which is less than the underlying asthenosphere with a density of about 3.2 g/cm^3. Only because the cold lithospheric mantle is denser than the warmer asthenosphere that supports it does subduction occur.

At some locations, the oceanic crust is unusually thick because it contains a chain of seamounts. Here the lithosphere may have sufficient crustal material, and hence buoyancy, to prevent or at least modify subduction. This appears to be the situation in two areas along the Peru–Chile Trench, where the angle of descent is quite shallow—about 10 to 15 degrees. Low dip angles often result in a strong interaction between the descending slab and the overriding plate. Consequently, these regions experience frequent, great earthquakes.

It has also been determined that unusually thick units of oceanic crust, those that are greater than 30 kilometers in thickness, probably will not subduct. An example is the Ontong Java Plateau, which is a thick oceanic basalt plateau located in the western Pacific. About 20 million years ago this plateau reached the trench that formed the boundary between the subducting Pacific plate and the overriding Australian–Indian plate. Apparently too buoyant to subduct, the Ontong Java Plateau clogged the trench and shut down subduction at this site. We will consider what eventually happens to these crustal fragments that are too buoyant to subduct in the next chapter.

Subducting Plates: The Demise of an Ocean Basin

Using magnetic stripes and fracture zones on the ocean floor, geologists began reconstructing the movement of plates over the past 200 million years. From this work they discovered that parts, or even entire ocean basins, have been destroyed along subduction zones. For example, during the breakup of Pangaea shown in Figure 2.A (p. 42), notice that the African plate rotates and moves northward. Eventually the northern margin of Africa collides with Eurasia. During this event, the floor of the intervening Tethys Ocean was almost entirely consumed into the mantle, leaving behind only a small remnant—the Mediterranean Sea.

Reconstructions of the breakup of Pangaea also helped investigators understand the demise of the Farallon plate—a large oceanic plate that once occupied much of the eastern Pacific basin. Prior to the breakup, the Farallon plate, plus one or two smaller plates, were situated opposite the Pacific plate on the eastern side of a spreading center located near

crust forms along a ridge, it is warm and buoyant, a fact that results in the ridge being elevated above the deep-ocean basins. However, as oceanic lithosphere moves away from the site of warm upwelling, it cools and thickens. After about 15 million years, an oceanic slab tends to be denser than the supporting asthenosphere. In parts of the western Pacific some oceanic lithosphere is nearly 180 million years old. This is the thickest and most dense in today's oceans. The subducting slabs in this region typically descend into the mantle at angles approaching 90 degrees (Figure 13.25A). Sites where plates subduct at such steep angles are found in association with the Tonga, Mariana, and Kurile trenches.

When a spreading center is located near a subduction zone, the oceanic lithosphere is still young and therefore warm and buoyant. Hence, the angle of descent for these slabs is small (Figure 13.25B). It is even possible that oceanic lithosphere may be overridden by a continental landmass before it has cooled sufficiently to readily subduct. In this situation, the slab may be so buoyant that rather than plunging into the mantle, it moves horizontally beneath a block of continental lithosphere. This phenomenon is called **buoyant subduction.** Buoyant slabs are thought to eventually sink when they cool sufficiently and their density increases.

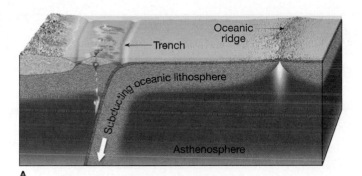

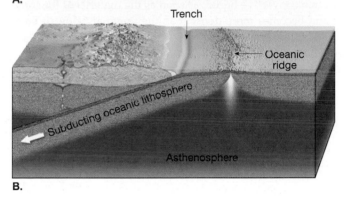

A.

B.

FIGURE 13.25 The angle at which oceanic lithosphere descends into the asthenosphere depends on its density. **A.** In parts of the Pacific, some oceanic lithosphere is older than 160 million years and typically descends into the mantle at angles approaching 90 degrees. **B.** Young oceanic lithosphere is warm and buoyant, hence it tends to subduct at a low angle.

the center of the Pacific basin. A modern remnant of this spreading center, which generated both the Farallon and Pacific plates, is the East Pacific Rise.

Beginning about 180 million years ago, the Americas were propelled westward by seafloor spreading in the Atlantic. Hence, the convergent plate boundaries that formed

along the west coasts of North and South America gradually migrated westward relative to the spreading center located in the Pacific. The Farallon plate, which was subducting beneath the Americas faster than it was being generated, got smaller and smaller (Figure 13.26). As its surface area decreased, it broke into smaller pieces, some of which subducted entirely. The remaining fragments of the once mighty Farallon plate are the Juan de Fuca, Cocos, and Nazca plates.

As the Farallon plate shrank, the Pacific plate grew larger, encroaching on the American plates. About 30 million years ago, a section of the East Pacific Rise collided with the subduction zone that once lay off the coast of California (Figure 13.26B). As this spreading center subducted into the California trench, these structures were mutually destroyed and replaced by a newly generated transform fault system that accommodates the differential motion between the North American and Pacific plates. As more of the ridge was subducted, the transform fault system, which we now call the San Andreas Fault, propagated through western California (Figure 13.26). Farther north, a similar event generated the Queen Charlotte transform fault.

Consequently, much of the present boundary between the Pacific and North American plates lies along transform faults located within the continent. In the United States (outside of Alaska), the only remaining part of the extensive convergent boundary that once ran along the entire West Coast is the Cascadia subduction zone. Here the subduction of the Juan de Fuca plate has generated the volcanoes of the Cascade Range.

Today, the southern end of the San Andreas Fault connects to a young spreading center (an extension of the East Pacific Rise) that generated the Gulf of California (Figure 13.27). Because of this change in plate geometry, the Pacific plate has captured a sliver of North America (the Baja Peninsula) and is carrying it northwestward toward Alaska at a rate of about 6 centimeters per year.

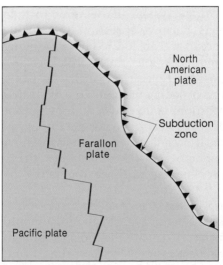
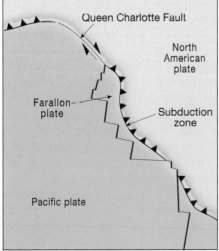
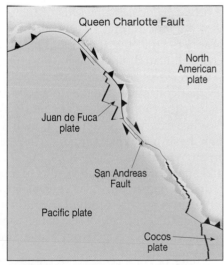

A. 56 million years ago   **B. 37 million years ago**   **C. Today**

FIGURE 13.26 Simplified illustration of the demise of the Farallon plate, which once ran along the western margin of the Americas. Because the Farallon plate was subducting faster than it was being generated, it got smaller and smaller. The remaining fragments of the once mighty Farallon plate are the Juan de Fuca, Cocos, and Nazca plates.

FIGURE 13.27 Satellite image showing the separation between the Baja Penninsula and North America. (Image courtesy of NASA)

Summary

- *Ocean bathymetry is determined using echo sounders and multibeam sonars,* which bounce sonic signals off the ocean floor. Ship-based receivers record the reflected echoes and accurately measure the time interval of the signals. With this information, ocean depths are calculated and plotted to produce maps of ocean-floor topography. Recently, *satellite measurements* of the ocean surface have added data for mapping ocean-floor features.

- Oceanographers studying the topography of the ocean basins have delineated three major units: *continental margins, deep-ocean basins,* and *oceanic (mid-ocean) ridges.*

- The zones that collectively make up a *passive continental margin* include the *continental shelf* (a gently sloping, submerged surface extending from the shoreline toward the deep-ocean basin); *continental slope* (the true edge of the continent, which has a steep slope that leads from the continental shelf into deep water); and in regions where trenches do not exist, the relatively steep continental slope merges into a more gradual incline known as the *continental rise.* The continental rise consists of sediments that have moved downslope from the continental shelf to the deep-ocean floor.

- *Active continental margins* are located primarily around the Pacific Ocean in areas where the leading edge of a continent is overrunning oceanic lithosphere. Here sediment scraped from the descending oceanic plate is plastered against the continent to form a collection of sediments called an *accretionary wedge.* An active continental margin generally has a narrow continental shelf, which grades into a deep-ocean trench.

- The deep-ocean basin lies between the continental margin and the oceanic ridge system. Its features include *deep-ocean trenches* (long, narrow depressions that are the deepest parts of the ocean and are located where moving crustal plates descend back into the mantle); *abyssal plains* (among the most level places on Earth, consisting of thick accumulations of sediments that were deposited atop the low, rough portions of the ocean floor by turbidity currents); *seamounts* (volcanic peaks on the ocean floor that originate near oceanic ridges or in association with volcanic hot spots); and *oceanic plateaus* (large flood basalt provinces similar to those found on the continents).

- *Oceanic (mid-ocean) ridges,* the sites of seafloor spreading, are found in all major oceans and represent more than 20 percent of Earth's surface. They are the most prominent features in the oceans, because they form an almost continuous swell that rises 2 to 3 kilometers above the adjacent ocean basin floor. Ridges are characterized by an *elevated position, extensive faulting,* and *volcanic structures* that have developed on newly formed oceanic crust. Most of the geologic activity associated with ridges occurs along a narrow region on the ridge crest, called the *rift zone,* where magma from the asthenosphere moves upward to create new slivers of oceanic crust. The topography of the various segments of the oceanic ridge is controlled by the rate of seafloor spreading.

- New oceanic crust is formed in a continuous manner by the process of seafloor spreading. The upper crust is composed of *pillow lavas* of basaltic composition. Below this layer are numerous interconnected dikes (*sheeted dike complex*) that are underlain by a thick layer of gabbro. This entire sequence is called an *ophiolite complex*.

- The development of a new ocean basin begins with the formation of a *continental rift* similar to the East African Rift. In those settings where rifting continues, a young, narrow ocean basin develops, exemplified by the Red Sea. Eventually, seafloor spreading creates an ocean basin bordered by rifted continental margins similar to the present-day Atlantic Ocean. Two mechanisms have been proposed for continental rifting—plumes of hot mobile rock rising from deep in the mantle and forces that arise from plate motions.

- Oceanic lithosphere subducts because its overall density is greater than the underlying asthenosphere. The subduction of oceanic lithosphere may result in the destruction of parts—or even entire—ocean basins. A classic example is the Farallon plate, most of subducted beneath the American plates as they were displaced westward by seafloor spreading in the Atlantic.

Review Questions

1. Assuming that the average speed of sound waves in water is 1500 meters per second, determine the water depth if the signal sent out by an echo sounder requires 6 seconds to strike bottom and return to the recorder (see Figure 13.2).

2. Describe how satellites orbiting Earth can determine features on the seafloor without being able to directly observe them beneath several kilometers of seawater.

3. What are the three major topographic provinces of the ocean floor?

4. List the three major features that comprise a passive continental margin. Which of these features is considered a flooded extension of the continent? Which one has the steepest slope?

5. Describe the differences between active and passive continental margins. Be sure to include how various features relate to plate tectonics, and give a geographic example of each type of margin.

6. Why are abyssal plains more extensive on the floor of the Atlantic than on the floor of the Pacific?

7. How does a flat-topped *seamount,* or *guyot,* form?

8. Briefly describe the oceanic ridge system.

9. Although oceanic ridges can be as tall as some mountains found on the continents, how are these features different?

10. What is the source of magma for seafloor spreading?

11. What is the primary reason for the elevated position of the oceanic ridge system?

12. How does hydrothermal metamorphism alter the basaltic rocks that make up the seafloor? How is seawater changed during this process?

13. What is a black smoker?

14. Compare and contrast a slow spreading center such as the Mid-Atlantic Ridge with one that exhibits a faster spreading rate, such as the East Pacific Rise.

15. Briefly describe the four layers of the ocean crust.

16. How does the *sheeted dike complex* form? What about the lower unit?

17. Name a place that exemplifies a continental rift.

18. What role are mantle plumes thought to play in the rifting of a continent?

19. What evidence suggests that hotspot volcanism does not always lead to the breakup of a continent?

20. Explain why oceanic lithosphere subducts even though the oceanic crust is less dense than the underlying asthenosphere.

21. Why does the lithosphere thicken as it moves away from the ridge as a result of seafloor spreading?

22. What happened to the Farallon plate? Name the remaining parts.

Key Terms

abyssal plain (p. 357)	continental rift (p. 366)	guyot (p. 358)	rift valley (p. 361)
accretionary wedge (p. 356)	continental rise (p. 355)	mid-ocean ridge (p. 358)	seamount (p. 357)
active continental margin	continental shelf (p. 353)	oceanic plateau (p. 358)	seismic reflection profile
(p. 355)	continental slope (p. 355)	oceanic ridge (p. 358)	(p. 352)
bathymetry (p. 350)	deep-ocean basin (p. 356)	ophiolite complex (p. 363)	sheeted dike complex
black smokers (p. 364)	deep-ocean trench (p. 356)	passive continental margin	(p. 364)
buoyant subduction (p. 370)	deep-sea fan (p. 355)	(p. 353)	sonar (p. 350)
continental margin (p. 353)	echo sounder (p. 350)	pillow basalts (p. 364)	tablemount (p. 358)

Web Resources

The *Earth* Website uses the resources and flexibility of the Internet to aid in your study of the topics in this chapter. Written and developed by geology instructors, this site will help improve your understanding of geology. Visit **http://www. prenhall.com/tarbuck** and click on the cover of *Earth 9e* to find:

- Online review quizzes.

- Critical thinking exercises.

- Links to chapter-specific Web resources.

- Internet-wide key-term searches.

http://www. prenhall.com/tarbuck

GEODe: Earth

GEODe: Earth makes studying faster and more effective by reinforcing key concepts using animation, video, narration, interactive exercises and practice quizzes. A copy is included with every copy of *Earth*.

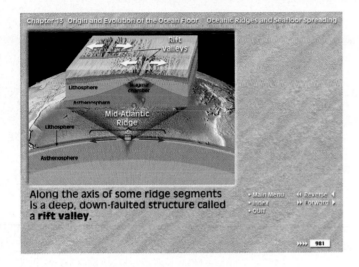

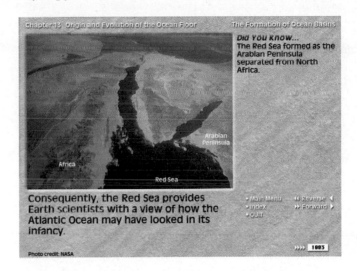

Convergent Boundaries: Origin of Mountains

Elk Mountains, west of Snowmass, Colorado. (Photo by Michael Collier)

Mountains are often spectacular features that rise abruptly above the surrounding terrain (Figure 14.1). Some occur as isolated masses; the volcanic cone Kilimanjaro, for example, stands almost 6000 meters (20,000 feet) above sea level, overlooking the expansive grasslands of East Africa. Other peaks are parts of extensive mountain belts, such as the American Cordillera, which runs almost continuously from the tip of South America through Alaska. Chains such as the Himalayas consist of youthful, towering peaks that are still rising, whereas others, including the Appalachian Mountains in the eastern United States, are much older and have been eroded far below their original lofty heights.

Most major mountain belts show evidence of enormous horizontal forces that have folded, faulted, and generally deformed large sections of Earth's crust. Although folded and faulted strata contribute to the majestic appearance of mountains, much of the credit for their beauty must be given to weathering and mass-wasting processes and to the erosional work of running water and glacial ice, which sculpt these uplifted masses in an unending effort to lower them to sea level. In this chapter, we will examine the nature of mountains and the mechanisms that generate them.

FIGURE 14.1 Uplift along high-angle faults created Mount Sneffels in the Colorado Rockies. (Photo by Art Wolfe, Inc.)

Mountain Building

Convergent Boundaries
▸ Introduction

Mountain building has occurred during the recent geologic past in several locations around the world. Young mountain belts include the American Cordillera, which runs along the western margin of the Americas from Cape Horn to Alaska and includes the Andes and Rocky Mountains; the Alpine–Himalaya chain, that extends from the Mediterranean through Iran to northern India and into Indochina; and the mountainous terrains of the western Pacific, which include volcanic island arcs such as Japan, the Philippines, and Sumatra. Most of these young mountain belts have come into existence within the last 100 million years. Some, including the Himalayas, began their growth as recently as 45 million years ago.

In addition to these young mountain belts, several chains of Paleozoic- and Precambrian-age mountains exist on Earth as well. Although these older structures are deeply eroded and topographically less prominent, they clearly possess the same structural features found in younger mountains. The Appalachians in the eastern United States and the Urals in Russia are classic examples of this older group of mountain belts.

Over the last few decades, geologists have learned a great deal about the tectonic processes that generate mountains. The term for the processes that collectively produce a mountain belt is **orogenesis,** (*oros* = mountain, *genesis* = to come into being). Some mountain belts, including the Andes, are constructed predominantly of lavas and volcanic debris that erupted on the surface, as well as massive amounts of intrusive igneous rocks that have solidified at depth. However, most major mountain belts display striking visual evidence of great tectonic forces that have shortened and thickened the crust. These *compressional mountains* tend to contain large quantities of preexisting sedimentary rocks and crystalline crustal fragments that have been contorted into a series of folds (Figure 14.2). Although folding and thrust faulting are often the most conspicuous signs of orogenesis, metamorphism and igneous activity are always present in varying degrees.

Over the years, several hypotheses have been put forward regarding the formation of Earth's major mountain belts (Figure 14.3). One early proposal suggested that mountains are simply wrinkles in Earth's crust, produced as the planet cooled from its original semimolten state. As Earth lost heat, it contracted and shrank. In response to this process, the crust was deformed similar to how the peel of an orange wrinkles as the fruit dries out. However, neither this nor any other early hypothesis was able to withstand careful scrutiny.

With the development of the theory of plate tectonics, a model for orogenesis with excellent explanatory power has emerged. According to this model, most mountain building occurs at convergent plate boundaries. Here, the subduction of oceanic lithosphere triggers partial melting of mantle rock, providing a source of magma that intrudes the crustal rocks that form the margin of the overlying plate. In addition, colliding plates provide the tectonic forces that fold, fault, and metamorphose the thick accumulations of sediments that have been deposited along the flanks of landmasses. Together, these processes thicken and shorten the continental crust, thereby elevating rocks that may have formed near the ocean floor, to lofty heights.

To unravel the events that produce mountains, researchers examine ancient mountain structures as well as sites where orogenesis is currently active. Of particular

Students Sometimes Ask . . .

You mentioned that most mountains are the result of crustal deformation. Are there areas that exhibit mountainous topography but have been produced without crustal deformation?

Yes. Plateaus—areas of high-standing rocks that are essentially horizontal—are one example of a feature that can be deeply dissected by erosional forces into rugged, mountainlike landscapes. Although these highlands resemble mountains topographically, they lack the structures associated with orogenesis. The opposite situation also exists. For instance, the Piedmont section of the eastern Appalachians exhibits topography that is nearly as subdued as that seen in the Great Plains. Yet because this region is composed of deformed metamorphic rocks, it is clearly part of the Appalachian Mountains.

FIGURE 14.2   Highly deformed sedimentary strata exposed on the face of Alberta's Mount Kidd. These sedimentary rocks are continental shelf deposits that were displaced toward the interior of Canada by low-angle thrust faults. (Photo by Peter French/DRK Photo)

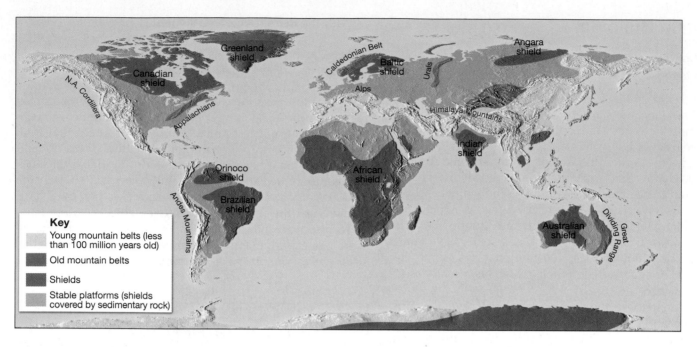

FIGURE 14.3   Earth's major mountain belts.

interest are active subduction zones, where lithospheric plates are converging. Here the subduction of oceanic lithosphere generates Earth's strongest earthquakes and most explosive volcanic eruptions, as well as playing a pivotal role in generating many of Earth's mountain belts.

Convergence and Subducting Plates

As discussed in Chapter 13, upwelling of partially melted mantle rock along divergent plate boundaries results in the formation of new oceanic lithosphere. By contrast, subduction zones located along convergent boundaries are the sites of plate destruction—places where slabs of oceanic lithosphere bend and plunge back into the mantle. As oceanic lithosphere slowly sinks, higher temperatures and pressures gradually alter these rigid slabs until they are fully assimilated into the mantle.

Major Features of Subduction Zones

Subduction zones can be roughly divided into four regions that include: (1) a *deep-ocean trench,* which forms where a subducting slab of oceanic lithosphere bends and descends into the asthenosphere; (2) a *volcanic arc,* which is built upon the overlying plate; (3) a region located between the trench and the volcanic arc (*forearc region*) and; (4) a region on the side of the volcanic arc opposite the trench (*backarc region*). Although all subduction zones exhibit these features, a great deal of variation exists—along the length of an individual subduction zone, as well as among different subduction zones (Figure 14.4).

Subduction zones can also be placed into one of two categories—those in which oceanic lithosphere is subducted beneath another oceanic slab and those in which oceanic lithosphere descends beneath a continental block. (An exception is the Aleutian subduction zone, where the western part is an oceanic–oceanic subduction zone, while subduction along the eastern section occurs under the Alaskan mainland.)

Volcanic Arcs   Perhaps the most obvious structure generated by subduction is a *volcanic arc,* which is built upon the overlying plate. Where two oceanic slabs converge, one is subducted beneath the other, initiating partial melting of the mantle wedge located above the subducting plate. This eventually leads to the growth of a **volcanic island arc,** or simply an **island arc,** on the ocean floor. Examples of active island arcs include the Mariana, New Hebrides, Tonga, and Aleutian arcs (Figure 14.5).

In locations where oceanic lithosphere is subducted beneath a continental block, a **continental volcanic arc** results. Here the volcanic arc builds upon the higher topography of older continental rocks, resulting in volcanic peaks that may reach 6000 meters (nearly 20,000 feet) above sea level.

Deep-Ocean Trenches   Another major feature associated with subduction is a deep-ocean trench. Trench depth appears to be strongly related to the age and hence the temperature of the subducting oceanic slab. In the western Pacific, where oceanic lithosphere is cold, relatively dense oceanic slabs descend into the mantle and produce deep trenches. A well-known example is the Mariana Trench, where the deepest area is more than 11,000 meters (36,000 feet) below sea level. By contrast, the Cascadia subduction zone lacks a well-defined trench. Here the warm, buoyant Juan de Fuca plate is actively subducting at a very low angle beneath southwestern Canada and the northwestern United States. The Peru–Chile subduction zone, on the other hand, displays

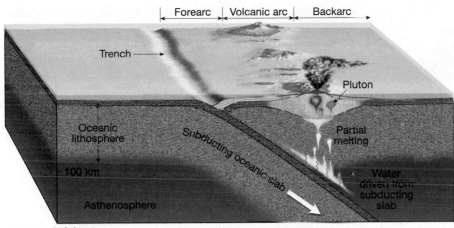

| Forearc | Volcanic arc | Backarc |

Trench

Pluton

Oceanic
lithosphere

Subducting oceanic slab

100 km

Partial
melting

Asthenosphere

Water
driven from
subducting
slab

A. Volcanic island arc

Continental volcanic arc

Trench

Accretionary
wedge

Forearc
basin

Magma
chambers

Oceanic
lithosphere

100 km

Subducting oceanic slab

Water
driven from
subducting
slab

Partial
melting

Asthenosphere

B. Andean-type plate margin

FIGURE 14.4 Diagrammatic comparison of a volcanic island arc and an Andean-type plate margin.

trench depths between these extremes. Much of this trench is 2 to 3 kilometers shallower than those in the western Pacific, averaging between 7 and 8 kilometers deep. One exception occurs in central Chile, where the plate boundary has a very shallow dip, making the trench virtually nonexistent.

Forearc and Backarc Regions   Located between developing volcanic arcs and deep-ocean trenches are the *forearc* regions (Figure 14.4). Here pyroclastic material from the volcanic arc as well as sediments eroded from the adjacent landmass accumulate. In addition, ocean-floor sediments are carried to the forearc region by the subducting plate.

Another site where sediments and volcanic debris may accumulate is the *backarc* region, which is located on the side of the volcanic arc opposite the trench. In these regions, tensional forces often dominate, causing the crust to be stretched and thinned.

Dynamics at Subduction Zones

Because subduction zones form where two plates are converging, it is natural to assume that large compressional forces are at work to deform the plate margins. Indeed, this

is the case along many convergent plate boundaries. However, convergent margins are not always regions dominated by compressional forces.

Extension and Backarc Spreading   Along some convergent plate margins, the overlying plates are under tension, which causes stretching and thinning of the crust. But how do extensional processes operate where two plates are moving together?

The age of the subducting oceanic slab is thought to play a significant role in determining the dominant forces acting on the overriding plate. Recall that when a relatively cold, dense slab subducts, it does *not* follow a fixed path into the asthenosphere. Rather, it sinks vertically as it descends, causing the trench to retreat, or "roll back," as shown in Figure 14.6. As the subducting plate sinks, it creates a flow (*slab suction*) in the asthenosphere that "pulls" the upper plate toward the retreating trench. (Visualize what would happen if you were sitting in a lifeboat near the *Titanic* as it sank!) As a result, the overriding plate is under tension and may be elongated and thinned. If tension is maintained long enough, a **backarc basin** may form.

Recall from Chapter 13 that thinning and rifting of the lithosphere results in upwelling of hot mantle rock and accompanying decompression melting. Continued extension initiates a type of seafloor spreading that generates new ocean crust, thereby increasing the size of a developing backarc basin.

Active backarc basins are found behind the Mariana and Tonga islands, whereas inactive basins contain the South

FIGURE 14.5   Three of many volcanic islands that comprise the Aleutian arc. This narrow band of volcanism results from the subduction of the Pacific plate. In the distance is the Great Sitkin volcano (1772 meters), which the Aleuts call the "Great Emptier of Bowels" because of its frequent activity. (Photo by Bruce D. Marsh)

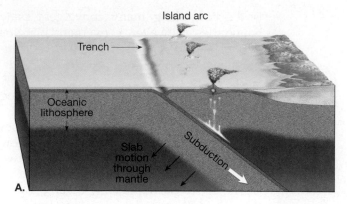

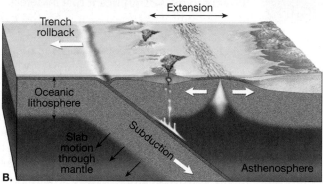

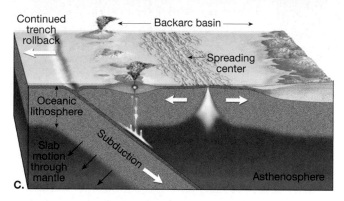

FIGURE 14.6 Model showing the formation of a backarc basin. Subduction and "roll back" of an oceanic slab creates flow in the mantle that "pulls" the upper plate toward the retreating trench.

China Sea and the Sea of Japan. Backarc spreading that formed the Sea of Japan is thought to have rifted a small piece of continental crust from Asia. Gradually, this crustal fragment migrated seaward along with the retreating trench. Seafloor spreading, in turn, created the oceanic crust that floors the Sea of Japan.

Compressional Regimes At some subduction zones compressional forces are dominant (see Box 14.1). This appears to be the case in the central Andes, where an episode of deformation began about 30 million years ago. During this span, the western margin of South America has been actively overrunning the subducting Nazca plate at a rate of about 3 centimeters per year. Put another way, the South American plate has been advancing toward the Peru–Chile trench at a rate faster than the trench has been retreating, because of slab sinking. Thus, in the case of the Andes, the descending

slab of oceanic lithosphere serves as a "wall" that resists the westward motion of the South American plate. The resulting tectonic forces have shortened and thickened the western margin of South America. (It is important to note that continental crust is generally weaker than oceanic crust, hence most deformation occurs in the continental blocks.) In this region, the Andean crustal block has its greatest thickening, about 70 kilometers (40 miles), and a mountainous topography that occasionally exceeds 6000 meters (20,000 feet) in elevation.

Subduction and Mountain Building

As noted earlier, the subduction of oceanic lithosphere gives rise to two different types of mountain belts. Where *oceanic lithosphere* subducts beneath an *oceanic plate,* an island arc and related tectonic features develop. Subduction beneath a *continental block,* on the other hand, results in the formation of a volcanic arc along the margin of a continent. Plate boundaries that generate continental volcanic arcs are often referred to as **Andean-type plate margins.**

Island Arcs

Island arcs represent what are perhaps the simplest mountain belts. These structures result from the steady subduction of oceanic lithosphere, which may last for 100 million years or more. Somewhat sporadic volcanic activity, the emplacement of plutonic bodies at depth, and the accumulation of sediment that is scraped from the subducting plate gradually increase the volume of crustal material capping the upper plate. Some mature volcanic island arcs, such as Japan, appear to have been built upon a preexisting fragment of continental crust.

The continued development of a mature volcanic island arc can result in the formation of mountainous topography consisting of belts of igneous and metamorphic rocks. This activity, however, is viewed as just one phase in the development of a major mountain belt. As you will see later, some volcanic arcs are carried by a subducting plate to the margin of a large continental block, where they become involved in a major mountain-building episode.

Mountain Building along Andean-type Margins

The first stage in the development of an Andean-type mountain belt occurs prior to the formation of the subduction zone. At this time, the continental margin is a **passive margin**; that is, it is not a plate boundary but a part of the same plate as the adjoining oceanic crust. The East Coast of the United States provides a present-day example of a passive continental margin. In such settings, deposition of sediment on the continental shelf produces a thick platform of shallow-water sandstones, limestones, and shales (Figure 14.7A). Beyond the continental shelf, turbidity currents deposit sediments on the floor of the deep-ocean basin (see

BOX 14.1 ▶ UNDERSTANDING EARTH

Earthquakes in the Pacific Northwest

Seismic studies have shown that the Cascadia subduction zone exhibits less earthquake activity than any other subduction zone along the margin of the Pacific basin. Does this mean that earthquakes pose no major threat to the population centers of the Pacific Northwest (Figure 14.A)? For some time, that was the conventional wisdom. However, that view changed with the discovery of buried marshes and coastal forests that are best explained by the rapid subsidence that accompanies a large earthquake.

The Cascadia subduction zone is very similar to the convergent margin in central Chile, where the oceanic slab descends at a shallow angle of about 10–15 degrees. In Chile, the effects of large compressional forces are being felt regularly in the form of strong earthquakes. The strongest earthquake ever recorded occurred there in 1960, Mw 9.5. Research predicts that subduction at shallow dip angles results in an environment that is conducive to great earthquakes (Mw 8.0 or greater). A partial explanation lies in the fact that in such settings, a large area of contact exists between the upper plate and the subducting slab.

Like the central Chilean subduction zone, the Cascadia boundary has a gently dipping plate and lacks a trench. This suggests that the Cascadia subduction zone is capable of great earthquakes. Evidence for past events of great magnitude include buried peat deposits found in some bay areas. These discoveries are consistent with episodes of rapid subsidence similar to what occurred during the 1964 Alaskan

FIGURE 14.A   Scene outside a historic Pioneer Square building following the February 28, 2001, Seattle earthquake. (Photo by Tim Crosby/Newsmakers/Liaison Agency, Inc.)

Earthquake (see Chapter 11). In addition, a fault near Seattle apparently ruptured about 1100 years ago, producing a large tsunami.

However, evidence has also been uncovered suggesting that a great earthquake is not very likely, at least over the short term. Geodetic studies conducted along coastal areas of the Pacific Northwest over the past few decades indicate that elastic strain is not accumulating to any great extent.

Which view is correct? Is a major earthquake in the Pacific Northwest imminent or unlikely? Hopefully further research will resolve this question. In the meantime, those living in the region bordering the Cascadia subduction zone should become aware of the precautions that can be taken to mitigate the effects of a great earthquake.

Chapter 13). In this environment, three distinct structural elements of a developing mountain belt gradually take form: volcanic arcs, accretionary wedges, and forearc basins (Figure 14.7).

Building a Volcanic Arc   Recall that as oceanic lithosphere descends into the mantle, increasing temperatures and pressures drive volatiles (mostly water) from the crustal rocks. These mobile fluids migrate upward into the wedge-shaped piece of mantle located between the subducting slab and upper plate. Once the sinking slab reaches a depth of about 100 kilometers (60 miles), these water-rich fluids reduce the melting point of hot mantle rock sufficiently to trigger some melting (Figure 14.7B). Partial melting of mantle rock (principally peridotite) generates *primary magmas,* with basaltic

compositions. Because they are less dense than the rocks from which they originated, these newly formed basaltic magmas will buoyantly rise. Upon reaching the base of the continental crust, which consists of low-density, rocky components, these basaltic magmas typically collect, or pond. However, recent volcanism at modern arcs— the eruption of Mount Etna, for example—indicates that some magma must reach the surface.

In order to continue to ascend, magma bodies must remain buoyant relative to the crust. In subduction zones, this is generally achieved through magmatic differentiation, in which heavy iron-rich minerals crystallize and settle out, leaving the remaining melt enriched in silica and other "light" components (see Chapter 4). Hence, through magmatic differentiation, a comparatively dense basaltic magma

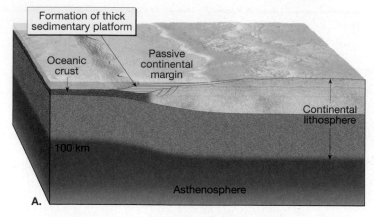

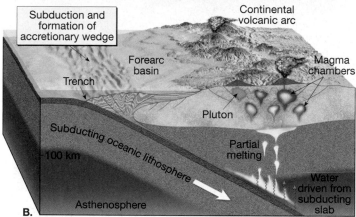

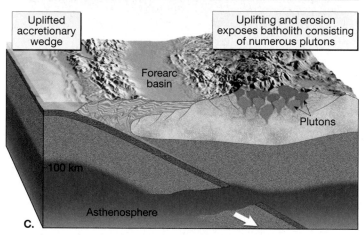

FIGURE 14.7 Orogenesis along an Andean-type subduction zone.
A. Passive continental margin with an extensive platform of sediments.
B. Plate convergence generates a subduction zone, and partial melting produces a volcanic arc. Continued convergence and igneous activity further deform and thicken the crust, elevating the mountain belt, while an accretionary wedge develops. **C.** Subduction ends and is followed by a period of uplift and erosion.

can generate low-density, buoyant melts that have an andesitic (intermediate) or even a rhyolitic (felsic) composition.

Volcanism along continental arcs is dominated by the eruption of lavas and pyroclastic materials of andesitic composition, whereas lesser amounts of basaltic and rhyolitic rocks may be generated. Because water driven from the subducting plate is necessary for melting, these mantle-derived magmas are enriched in water and other volatiles (the

gaseous component of magma). It is these gas-laden andesitic magmas that produce the explosive eruptions that characterize continental volcanic arcs and mature island arcs.

Emplacement of Plutons   Thick continental crust greatly impedes the ascent of magma. Consequently, a high percentage of the magma that intrudes the crust never reaches the surface—instead, it crystallizes at depth to form plutons. The emplacement of these massive igneous bodies will metamorphose the host rock by the process called contact metamorphism (see Chapter 8).

Eventually, uplifting and erosion exhume these igneous bodies and associated metamorphic rocks. Once they are exposed at the surface, these massive structures are called *batholiths* (Figure 14.7C). Composed of numerous plutons, batholiths form the core of the Sierra Nevada in California and are prevalent in the Peruvian Andes. Most batholiths are composed of intrusive igneous rocks with an intermediate to felsic composition, such as diorite and granodiorite, although granites have been observed. (Granite is sparse in the batholiths along the western margin of North America, but significant amounts occur in the core of the Appalachian Mountains.)

Development of an Accretionary Wedge   During the development of volcanic arcs, sediments that are carried on the subducting plate, as well as fragments of oceanic crust, may be scraped off and plastered against the edge of the overriding plate. The resulting chaotic accumulation of deformed and thrust-faulted sediments and scraps of ocean crust is called an **accretionary wedge** (Figure 14.7B). The processes that deform these sediments have been likened to what happens to a wedge of soil as it is scraped and pushed in front of an advancing bulldozer.

Some of the sediments that comprise an accretionary wedge are muds that accumulated on the ocean floor and were subsequently carried to the subduction zone by plate motion. Other materials are derived from the adjacent volcanic arc and consist of volcanic ash and other pyroclastic materials, as well as sediments eroded from these elevated landforms.

Some subduction zones have minimal accretionary wedges or lack them all together. The Mariana Trench, for example, lacks an accretionary wedge, partly because of the distance to a significant source region. (Another proposed explanation for the lack of an accretionary wedge is that much of the available sediment is subducted.) By contrast, the Cascadia subduction zone has a large accretionary wedge. Here, the Juan de Fuca plate has a 3-kilometer- (2-mile-) thick mantle of sediments, contributed mainly by the Columbia River.

Prolonged subduction, in regions where sediment is plentiful, may thicken an accretionary wedge enough so it protrudes above sea level. This has occurred along the southern end of the Puerto Rico Trench, where the Orinoco River Basin of Venezuela is a major source region. The resulting wedge emerges to form the island of Barbados.

Not all the available sediment becomes part of the accretionary wedge; rather, some is subducted to great depths. As these sediments descend, pressure steadily increases, but the temperatures within the sediments remain relatively low, because they are in contact with the cool, plunging plate. This activity generates a suite of high-pressure, low-temperature metamorphic minerals. Because of their low density, some of the subducted sediments and associated metamorphic components will buoyantly rise toward the surface. This "backflow" tends to mix and churn the sediment within the accretionary wedge. Thus, an accretionary wedge evolves into a complex structure consisting of faulted and folded sedimentary rocks and scraps of oceanic crust that may be intermixed with metamorphic rocks formed during the subduction process. The unique structure of accretionary wedges has greatly aided geologists in their efforts to piece together the events that have generated our modern continents.

Forearc Basins As the accretionary wedge grows upward, it tends to act as a barrier to the movement of sediment from the volcanic arc to the trench. As a result, sediments begin to collect between the accretionary wedge and volcanic arc. This region, which is composed of relatively undeformed layers of sediment and sedimentary rocks, is called a **forearc basin** (Figure 14.7B). Subsidence, and continued sedimentation in forearc basins, can generate a sequence of horizontal, sedimentary strata that is several kilometers thick.

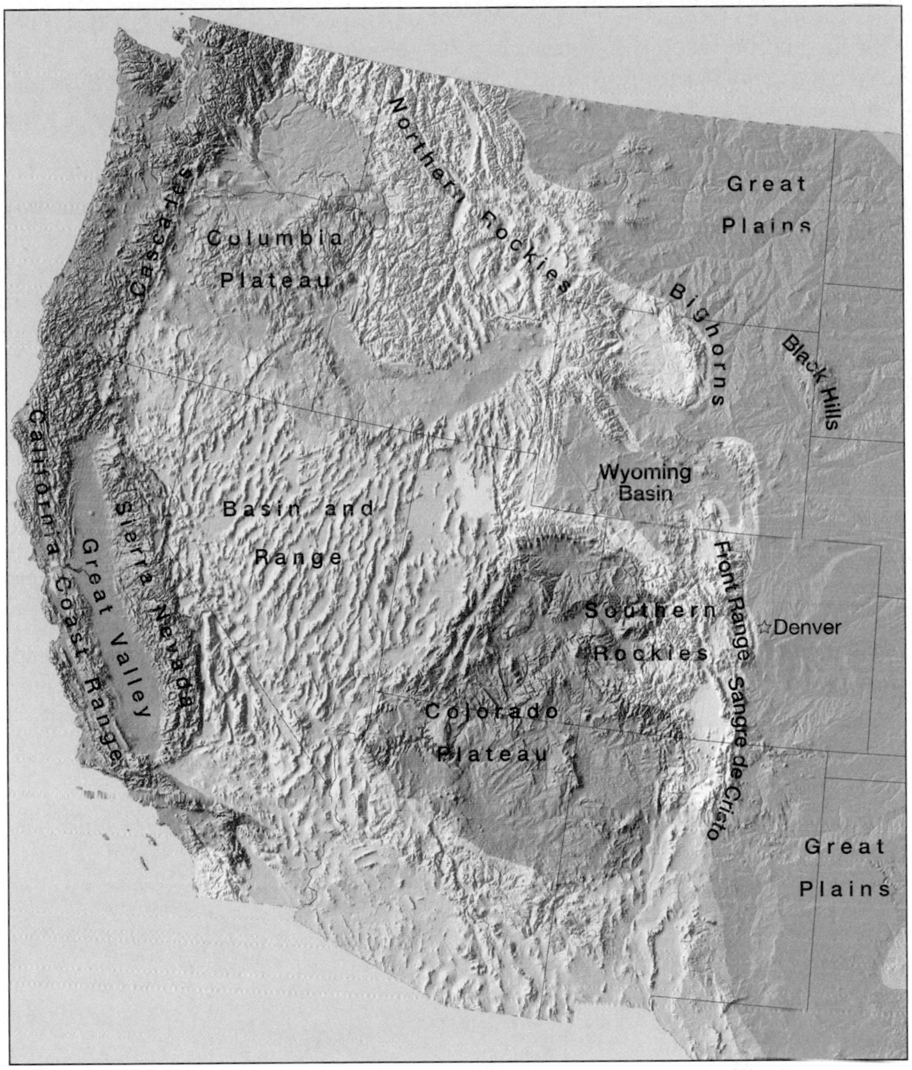

FIGURE 14.8 Map of mountains and landform regions in the western United States. (After Thelin and Pike, U.S. Geological Survey)

Sierra Nevada and Coast Ranges

During the Jurassic period, when the North Atlantic began to open, a subduction zone formed along the western margin of the North American plate. Evidence for this episode of subduction is found in a nearly continuous belt of igneous plutons that include Mexico's Baja batholith, the Sierra Nevada and Idaho batholiths found in the western United States, and the Coast Range batholith in Canada (see Figure 5.36).

Part of what formed this convergent plate boundary is now an excellent example of an inactive Andean-type orogenic belt. It includes the Sierra Nevada and the Coast Ranges in California (Figure 14.8). These parallel mountain belts were produced by the subduction of a portion of the Pacific basin (Farallon plate) under the western margin of California.

The Sierra Nevada batholith is a remnant of the continental volcanic arc that was produced by many surges of magma over tens of millions of years. The Coast Ranges represent an accretionary wedge that formed when sediments scraped from the subducting plate and provided by the eroding continental volcanic arc were intensely folded and faulted. (Portions of the Coast Ranges are composed of a chaotic mixture of sedimentary and metamorphic rocks, plus fragments of oceanic crust called the Franciscan Formation.)

Beginning about 30 million years ago, subduction gradually ceased along much of the margin of North America as the spreading center that produced the Farallon plate entered the California trench (see Figure 13.26). Both the spreading center and subduction zone were subsequently destroyed. Uplifting and erosion that followed this event have removed most of the evidence of past volcanic activity and exposed a core of crystalline, igneous, and associated metamorphic rocks that make up the Sierra Nevada. The Coast Ranges were uplifted only recently, as evidenced by the young, unconsolidated sediments that still mantle portions of these highlands.

California's Great Valley is a remnant of the forearc basin that formed between the developing Sierra Nevada and the Coast Ranges. Throughout much of its history, portions of the Great Valley lay below sea level. This sediment-laden basin contains thick marine deposits and debris eroded from the continental volcanic arc.

From this example, we can see that Andean-type mountain belts are composed of two roughly parallel zones of deformation. A continental volcanic arc, which forms along the continental margins, consists of volcanoes and large intrusive igneous bodies and associated metamorphosed rocks. Seaward of the continental volcanic arc, where subducting plates descend beneath the continent, an accretionary wedge is generated. This feature consists mainly of sediments and volcanic debris that have been folded, faulted, and in some places metamorphosed (Figure 14.7). Between these regions of deformation lies a forearc basin, composed mostly of horizontal marine strata.

In summary, the growth of mountain belts at subduction zones is a response to crustal thickening caused by the addition of mantle-derived igneous rocks. In addition, crustal shortening and thickening may occur along the continental margins as a result of convergence.

Continental Collisions

GEODe
Convergent Boundaries
▶ Continental Collisions

As you have seen, when a slab of oceanic lithosphere subducts beneath a continental margin, an Andean-type mountain belt develops. If the subducting plate also contains a continent, continued subduction eventually carries the continental block to the trench. Although oceanic lithosphere is relatively dense and readily subducts, continental crust contains significant amounts of low-density materials and is too buoyant to undergo appreciable subduction. Consequently, the arrival of continental lithosphere at the trench results in a collision with the margin of the overlying continental block and an end to subduction (Figure 14.9).

Continental collisions result in the development of mountains that are characterized by shortened and thickened crust. Thicknesses of 50 kilometers (30 miles) are common, and some regions have crustal thicknesses in excess of 70 kilometers (40 miles). In these settings, crustal thickening is typically achieved through folding and faulting.

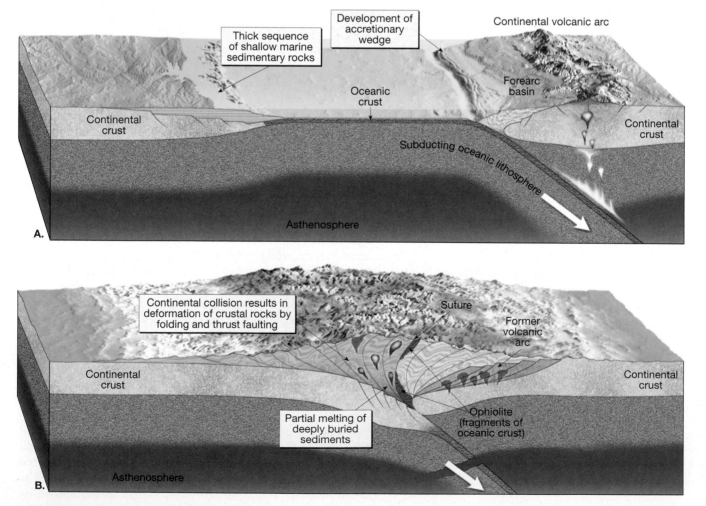

FIGURE 14.9   Illustration showing the formation of the major features of a compressional mountain belt, including the fold-and-thrust belt.

Noteworthy features of most mountain ranges that result from continental collisions are **fold-and-thrust belts.** These mountainous terrains often result from deformation of thick sequences of shallow marine sedimentary rocks similar to those that make up the passive continental margins of the Atlantic. During a continental collision, these sedimentary rocks are pushed inland, away from the core of the developing mountain belt and over the stable continental interior. In essence, crustal shortening is achieved by displacement along thrust faults (low angle reverse faults), where once relatively flat-lying strata are stacked one upon another as illustrated in Figure 14.9. During this displacement, material caught between the thrust faults is often folded, thereby forming the other major structure of a fold-and-thrust belt. Excellent examples of fold-and-thrust belts are found in the Appalachian Valley and Ridge province, the Canadian Rockies, the Lesser (southern) Himalayas, and the northern Alps.

The zone where two continents collide is called the **suture.** This portion of the mountain belt often preserves slivers of the oceanic lithosphere that were entrapped between the colliding plates. As a result of their unique ophiolite structure (see Chapter 13), these pieces of oceanic lithosphere help identify the location of the collision boundary. It is along suture zones that continents are described as being "welded" together.

We will take a closer look at two examples of collision mountains—the Himalayas and the Appalachians. The Himalayas are the youngest collision mountains on Earth and are still rising. The Appalachians are a much older mountain belt, in which active mountain building ceased about 250 million years ago.

The Himalayas

The mountain-building episode that created the Himalayas began roughly 45 million years ago when India began to collide with Asia. Prior to the breakup of Pangaea, India was a part of Gondwana in the Southern Hemisphere (see Figure 2.A, p. 42). Upon splitting from that continent, India moved rapidly, geologically speaking, a few thousand kilometers in a northward direction (see Figure 2.A).

The subduction zone that facilitated India's northward migration was located near the southern margin of Asia. Continuing subduction along Asia's margin created an Andean-type plate margin that contained a well-developed volcanic arc and accretionary wedge. India's northern margin, on the other hand, was a passive continental margin consisting of a thick platform of shallow-water sediments and sedimentary rocks.

Although the details remain somewhat sketchy, one, or perhaps more, small continental fragments were positioned on the subducting plate somewhere between India and Asia. During the closing of the intervening ocean basin, a relatively small crustal fragment,

which now forms southern Tibet, reached the trench. This event was followed by the docking of India itself. The tectonic forces involved in the collision of India with Asia were immense and caused the more deformable materials located on the seaward edges of these landmasses to be highly folded and faulted. The shortening and thickening of the crust elevated great quantities of crustal material, thereby generating the spectacular Himalayan mountains (Figure 14.10).

In addition to uplift, crustal shortening produced a thick mass of material in which the lower layers experienced elevated temperatures and pressures (Figure 14.9). Partial melting within the deepest and most deformed region of the developing mountain belt produced plutons that intruded and further deformed the overlying rocks. It is in such environments that the metamorphic and igneous core of compressional mountains are generated.

The formation of the Himalayas was followed by a period of uplift that raised the Tibetan Plateau. Evidence from seismic studies suggests that a portion of the Indian subcontinent was thrust beneath Tibet a distance of perhaps 400 kilometers. If so, the added crustal thickness would account for the lofty landscape of southern Tibet, which has an average elevation higher than Mount Whitney, the highest point in the contiguous United States. Other researchers disagree with this scenario. Instead, they suggest that extensive thrust faulting and folding within the upper crust, as well as uniform ductile deformation of the lower crust and underlying lithospheric mantle, produced the great crustal thickness that accounts for this extremely high plateau. Further research is necessary to resolve this issue.

The collision with Asia slowed but did not stop the northward migration of India, which has since penetrated at least 2000 kilometers (1200 miles) into the mainland of Asia. Some of this motion can be accounted for by crustal shortening. Much of the remaining penetration into Asia is thought to have resulted in the lateral displacement of large

FIGURE 14.10 Himalaya Mountains with Mount Everest in the background (left center) and Nuptse in the foreground (right center). (Photo by David Woodfall/DRK Photo)

blocks of the Asian crust by a mechanism described as *continental escape*. As shown in Figure 14.11, as India collided with Asia, parts of Asia were "squeezed" eastward out of the collision zone. These displaced crustal blocks included much of present-day Indochina and sections of mainland China.

Why has the interior of Asia deformed to such a large degree while India proper has remained essentially undisturbed? The answer lies in the nature of these diverse crustal blocks. Much of India is a shield composed mainly of crystalline Precambrian rocks (see Figure 14.3). This thick, cold slab of crustal material has been intact for more than 2 billion years. By contrast, Southeast Asia was assembled more recently from several smaller crustal fragments—during and even after the formation of Pangaea. Consequently, it is still relatively "warm and weak" from recent periods of mountain building. The deformation of Asia has been re-created in the laboratory with a rigid block representing India pushed into a mass of deformable modeling clay as shown in Figure 14.11. India continues to be thrust into Asia at an estimated rate of a few centimeters each year.

The Appalachians

The Appalachian Mountains provide great scenic beauty near the eastern margin of North America from Alabama to Newfoundland. In addition, mountains that formed contemporaneously with the Appalachians are found in the British Isles, Scandinavia, northwestern Africa, and Greenland (see Figure 2.6, p. 40). The orogeny that generated this extensive mountain system lasted a few hundred million years and was one of the stages in assembling the supercontinent of Pangaea. Detailed studies in the central and southern Appalachians indicate that the formation of this mountain belt was more complex than once thought. Rather than forming during a single continental collision, the Appalachians resulted from three distinct episodes of mountain building.

This oversimplified scenario begins roughly 750 million years ago with the breakup of a pre-Pangaea supercontinent (Rodinia), which rifted North America from Europe and Africa. This episode of continental rifting and seafloor spreading generated the ancestral North Atlantic. Located within this developing ocean basin was a fragment of continental crust that had been rifted from North America (Figure 14.12A).

Then, about 600 million years ago, plate motion dramatically changed and the ancestral North Atlantic began to close. Two subduction zones probably formed. One of these was located seaward of the coast of Africa and gave rise to a volcanic arc similar to those that presently rim the western Pacific. The other developed on the continental fragment that lay off the coast of North America, as shown in Figure 14.12.

Between 450 and 500 million years ago, the marginal sea located between this crustal fragment and North America began to close. The ensuing collision deformed the continental shelf and sutured the crustal fragment to the North American plate. The metamorphosed remnants of the continental fragment are recognized today as the crystalline rocks of the Blue Ridge and western Piedmont regions of the Appalachians (Figure 14.12B). In addition to the pervasive regional metamorphism, igneous activity placed numerous plutonic bodies along the entire continental margin, particularly in New England.

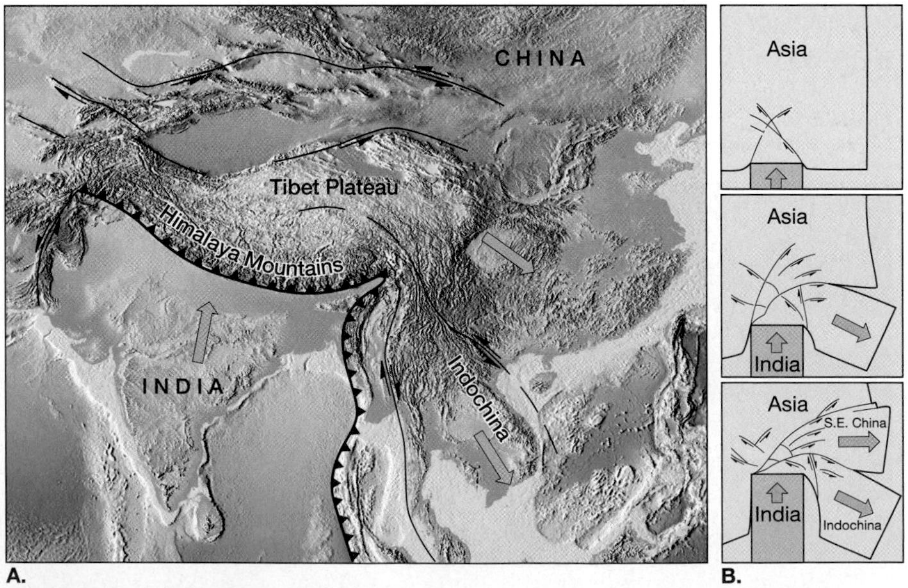

A second episode of mountain building occurred about 400 million years ago. In the southern Appalachians, the continued closing of the ancestral North Atlantic resulted in the collision of the developing volcanic arc with North America (Figure 14.12C). Evidence for this event is visible in the Carolina Slate Belt of the eastern Piedmont, which contains metamorphosed sedimentary and volcanic rocks characteristic of an island arc.

The final orogeny occurred somewhere between 250 and 300 million years ago, when Africa collided with North America. At some locations the total landward displacement of the Blue Ridge and Piedmont provinces may have exceeded 250 kilometers (155 miles). This event displaced and further deformed the shelf sediments and sedimentary rocks that had once flanked the eastern margin of North America (Figure 14.12D). Today these folded and thrust-faulted sandstones, limestones, and shales make up the largely unmetamorphosed rocks of the Valley and Ridge Province. Outcrops of the fold-

FIGURE 14.11 The collision between India and Asia that generated the Himalayas and Tibetan Plateau also severely deformed much of Southeast Asia. **A.** Map view of some of the major structural features of Southeast Asia thought to be related to this episode of mountain building. **B.** Re-creation of the deformation of Asia, with a rigid block representing India pushed into a mass of deformable modeling clay.

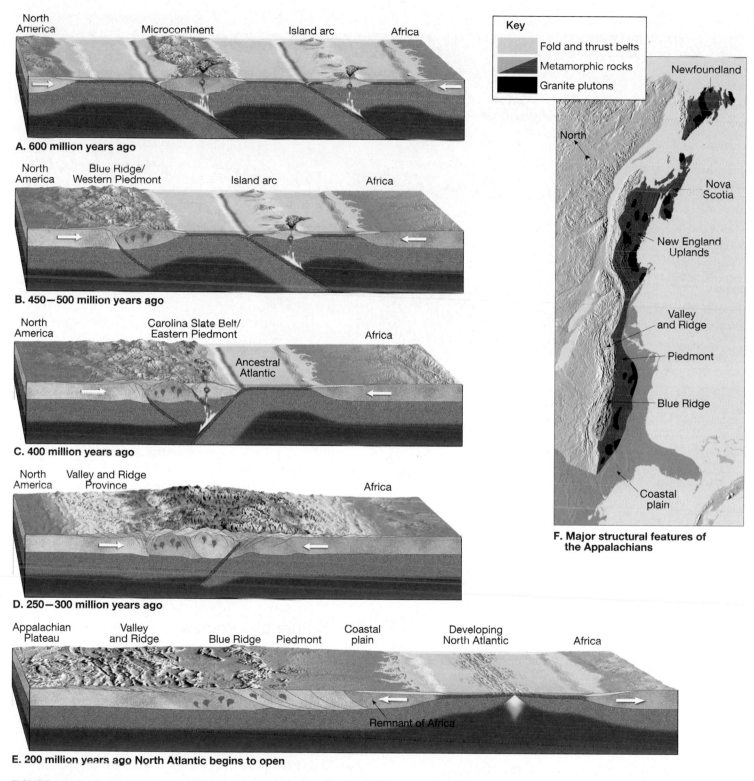

A. 600 million years ago

North America   Microcontinent   Island arc   Africa

B. 450—500 million years ago

North America   Blue Ridge/ Western Piedmont   Island arc   Africa

C. 400 million years ago

North America   Carolina Slate Belt/ Eastern Piedmont   Ancestral Atlantic   Africa

D. 250—300 million years ago

North America   Valley and Ridge Province   Africa

E. 200 million years ago North Atlantic begins to open

Appalachian Plateau   Valley and Ridge   Blue Ridge   Piedmont   Coastal plain   Developing North Atlantic   Africa

Remnant of Africa

Key

Fold and thrust belts
Metamorphic rocks
Granite plutons

Newfoundland
North
Nova Scotia
New England Uplands
Valley and Ridge
Piedmont
Blue Ridge
Coastal plain

F. Major structural features of the Appalachians

FIGURE 14.12   These simplified diagrams depict the development of the southern Appalachians as the ancient North Atlantic was closed during the formation of Pangaea. Three separate stages of mountain-building activity spanned more than 300 million years. (After Zve Ben-Avraham, Jack Oliver, Larry Brown, and Frederick Cook)

ed thrust-faulted structures that characterize collision mountains are found as far inland as central Pennsylvania and western Virginia (Figure 14.13).

Geologically speaking, shortly after the formation of the Appalachian Mountains, the newly formed supercontinent of Pangaea began to break into smaller fragments. Because this new zone of rifting occurred east of the suture that formed between Africa and North America, a remnant of Africa remains "welded" to the North American plate (Figure 14.12E).

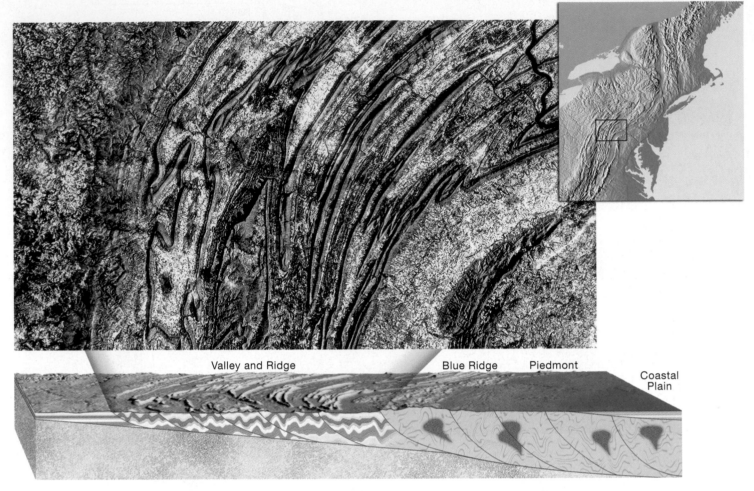

FIGURE 14.13   The Valley and Ridge Province. This portion of the Appalachian Mountains consists of folded and faulted sedimentary strata that were displaced landward with the closing of the proto-Atlantic. (LANDSAT image, courtesy of Phillips Petroleum Company, Exploration Projects Section)

Other mountain ranges that exhibit evidence of continental collisions include the Alps and the Urals. The Alps are thought to have formed as a result of a collision between Africa and Europe during the closing of the Tethys Sea. The Urals, on the other hand, formed during the assembly of Pangaea when Baltica (northern Europe) and Siberia (northern Asia) collided.

Terranes and Mountain Building

 Convergent Boundaries
▶ Crustal Fragments and Mountain Building

Mountain belts can also develop as a result of the collision and merger of an island arc, or some other small crustal fragment, to a continental block. The process of collision and accretion (joining together) of comparatively small crustal fragments to a continental margin has generated many of the mountainous regions rimming the Pacific.

The Nature of Terranes

Geologists refer to these accreted crustal blocks as terranes. Simply, the term **terrane** refers to any crustal fragment that has a geologic history distinct from that of the adjoining terranes. Terranes come in varied shapes and sizes.

What is the nature of these crustal fragments, and from where do they originate? Research suggests that prior to their accretion to a continental block, some of the fragments may have been **microcontinents** similar to the present-day island of Madagascar, located east of Africa in the Indian Ocean. Many others were island arcs similar to Japan, the Philippines, and the Aleutian Islands. Still others may have been submerged crustal fragments, such as those occurring on the floor of the western Pacific (see Figure 13.10). More than 100 of these comparatively small crustal fragments are presently known to exist. Their origins vary. Some are submerged fragments consisting mainly of continental crust, whereas others are extinct volcanic islands, such as the Hawaiian Island–Emperor Seamount chain. Still others are

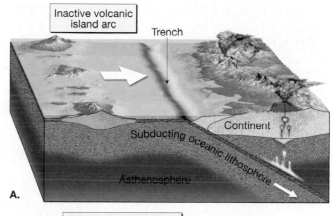

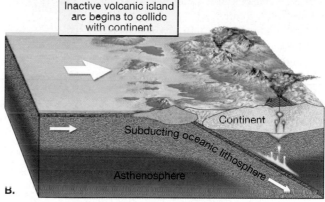

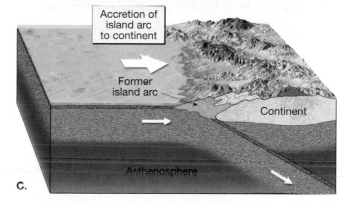

FIGURE 14.14   Sequence of events showing the collision and accretion of an island arc to a continental margin.

submerged oceanic plateaus created by massive outpourings of basaltic lavas associated with hot-spot activity.

Accretion and Orogenesis

The widely accepted view is that as oceanic plates move, they carry embedded oceanic plateaus, volcanic island arcs, and microcontinents to an Andean-type subduction zone. When an oceanic plate contains a chain of small seamounts, these structures are generally subducted along with the descending oceanic slab. However, very thick units of oceanic crust, such as the Ontong Java Plateau, or a mature island arc composed of abundant "light" igneous rocks produced by magmatic differentiation, may render the oceanic lithosphere too buoyant to subduct. In these situations, a collision between the crustal fragment and the continent occurs.

The sequence of events that occurs when a mature island arc reaches an Andean-type margin is shown in Figure 14.14. Because of its buoyancy, a mature island arc will not subduct beneath the continental plate. Instead, the upper portions of these thickened zones are peeled from the descending plate and thrust in relatively thin sheets upon the adjacent continental block. In some settings continued subduction may carry another crustal fragment to the continental margin. When this fragment collides with the continental margin, it displaces the accreted island arc further inland, adding to the zone of deformation and to the thickness and lateral extent of the continental margin.

The North American Cordillera   The idea that mountain building occurs in association with the accretion of crustal fragments to a continental mass arose principally from studies conducted in the North American Cordillera (Figure 14.15). Here it was determined that some mountainous areas, principally those in the orogenic belts of Alaska and British Columbia, contain fossil and paleomagnetic evidence indicating that these strata once lay nearer the equator.

It is now assumed that many of the other terranes found in the North American Cordillera were once scattered throughout the eastern Pacific, much as we find island arcs and oceanic plateaus distributed in the western Pacific today (see Figure 13.10). Since before the breakup of Pangaea, the eastern portion of the Pacific basin (Farallon plate)

has been subducting under the western margin of North America. Apparently, this activity resulted in the piecemeal addition of crustal fragments to the entire Pacific margin of the continent—from Mexico's Baja Peninsula to northern Alaska (Figure 14.15). In a like manner, many modern microcontinents will eventually be accreted to active continental margins, producing new orogenic belts.

Fault-Block Mountains

Most mountain belts, including the Alps, Himalayas, and Appalachians, form in compressional environments, as evidenced by the predominance of large thrust faults and folded strata. However, other tectonic processes, such as

BOX 14.2 ▶ UNDERSTANDING EARTH

The Southern Rockies

The portion of the Rocky Mountains that extends from southern Montana to New Mexico was produced by a period of deformation known as the *Laramide Orogeny*. This event, which created some of the most picturesque scenery in the United States, peaked about 60 million years ago (Figure 14.B). The mountain ranges generated during the Laramide Orogeny include the Front Range of Colorado, the Sangre de Cristo of New Mexico and Colorado, and the Bighorns of Wyoming.

These mountains are structurally much different from the northern Rockies, which include the Canadian Rockies and those portions of the Rockies found in Idaho, western Wyoming, and western Montana. The northern Rockies are compressional mountains, composed of thick sequences of sedimentary rocks that were deformed by folding and low-angle thrust faulting. Most investigators agree that the collision of one or more microcontinents with the western margin of North America generated the driving force behind the formation of the northern Rockies.

The southern Rockies, on the other hand, formed when deeply buried crystalline rock was lifted nearly vertically along steeply dipping faults, upwarping the overlying layers of younger sedimentary rocks. The resulting mountainous topography consists of large blocks of ancient basement rocks that are separated by sediment-filled basins. Since their formation, much of the sedimentary cover has been eroded from the highest portions of the uplifted blocks, exposing their igneous and metamorphic cores. Examples include a number of granitic outcrops that project as steep summits, such as Pike's Peak and Long's Peak in Colorado's Front Range. In many areas, remnants of the sedimentary strata that once covered this region are visible as prominent angular ridges, called *hogbacks*, flanking the crystalline cores of the mountains (Figure 14.C).

It was once assumed that like other regions of mountainous topography, the southern Rockies stood tall because the crust had been thickened by past tectonic events. However, seismic studies conducted across the American Southwest revealed a crustal thickness no greater than that found below Denver. These data ruled out crustal buoyancy as the cause for the abrupt 2-kilometer (1.2-mile) jump in elevation that occurs where the Great Plains meet the Rockies.

FIGURE 14.B The spectacular Maroon Bells are part of the Colorado Rockies. (Photo by Peter Saloutos/The Stock Market)

Although the southern Rockies have been extensively studied for more than a century, there is still a good deal of debate regarding the mechanisms that led to uplift. One hypothesis proposes that this period of uplift started with the nearly horizontal subduction of the Farallon plate eastward beneath North America as far inland as the Black Hills of South Dakota. As the subducted slab scraped beneath the continent, compressional forces initiated a period of tectonic activity. As the comparatively cool Farallon plate sank, it was replaced by hot rock that upwelled from the mantle. Thus, according to this scenario, the hot mantle provided the buoyancy to raise the southern Rockies, as well as the Colorado Plateau and the mountains of the Basin and Range.

Others disagree, maintaining that there is no need to invoke the process of buoyant subduction. Rather, they suggest that plate convergence and the collision of one or more microcontinents to the western margin of North America generated the driving force behind the Laramide Orogeny (see the section entitled "Terranes and Mountain Building").

It should be pointed out that neither of these proposals has gained widespread support. As one geologist familiar with this region put it, "We just don't know."

FIGURE 14.C Hogback ridges in the Rocky Mountains of Colorado. Shown is a view looking south along the east flank of the Front Range. These upturned sedimentary rocks are remnants of strata that once covered the Precambrian igneous and metamorphic core of the mountains to the west (right). (Photo by Tom Till/DRK Photo)

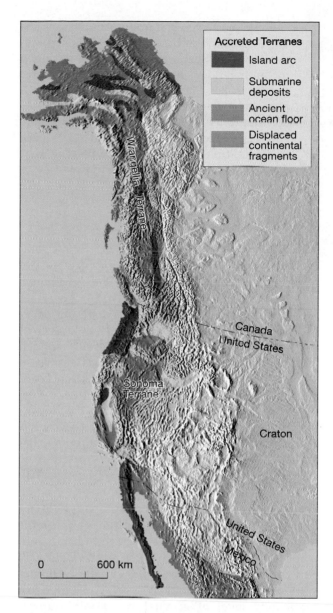

FIGURE 14.15 Map showing terranes that have been added to western North America during the past 200 million years. Paleomagnetic studies and fossil evidence indicate that some of these terranes originated thousands of kilometers to the south of their present location. (After D. R. Hutchinson and others)

Legend — Accreted Terranes:
- Island arc
- Submarine deposits
- Ancient ocean floor
- Displaced continental fragments

Map labels: Wrangellia Terrane, Canada / United States, Sonoma Terrane, Craton, United States / Mexico. Scale: 0 — 600 km

Students Sometimes Ask ...

What are some modern-day examples of material that may wind up as terranes in the future?

The southwestern Pacific Ocean is a good place to find pieces and fragments of land that may one day become terranes. Here there are many island arcs, oceanic plateaus, and microcontinents that will likely become accreted to the edges of a continent. One of the best-known terranes is the area west of the San Andreas Fault, which includes southwest California and the Baja California peninsula of Mexico. Already named the "California Terrane," it is moving to the northwest and will probably detach from North America in about 50 million years. Continued movement toward the northwest will bring it to southern Alaska, where it will become the next in a long procession of terranes that have rafted toward and "docked" against Alaska during the past 200 million years.

continental rifting, can also produce uplift and the formation of topographic mountains. The mountains that form in these settings, termed **fault-block mountains,** are bounded by high-angle normal faults that gradually flatten with depth. Most fault-block mountains form in response to broad uplifting, which causes elongation and faulting. Such a situation is exemplified by the fault blocks that rise high above the rift valleys of East Africa.

Mountains in the United States in which faulting and gradual uplift have contributed to their lofty stature include the Sierra Nevada of California and the Grand Tetons of Wyoming. Both are faulted along their eastern flanks, which were uplifted as the blocks tilted downward to the west. Looking west from Owens Valley, California, and Jackson Hole, Wyoming, the eastern fronts of these ranges (the Sier-

ra Nevada and the Tetons, respectively) rise more than 2 kilometers, making them two of the most imposing mountain fronts in the United States (Figure 14.16).

Basin and Range Province

Located between the Sierra Nevada and Rocky Mountains is one of Earth's largest regions of fault-block mountains—the Basin and Range Province. This region extends in a roughly north to south direction for nearly 3000 kilometers (2000 miles) and encompasses all of Nevada and portions of the surrounding states, as well as parts of southern Canada and western Mexico. Here, the brittle upper crust has literally been broken into hundreds of fault blocks. Tilting of these faulted structures (half-grabens) gave rise to nearly parallel mountain ranges, averaging about 80 kilometers in length, which rise above adjacent sediment-laden basins (see Figure 10.22).

Extension in the Basin and Range Province began about 20 million years ago and appears to have "stretched" the crust as much as twice its original width. Figure 14.17 shows a rough outline of the boundaries of the western states before and after this period of extension. High heat flow in the region, three times average, and several episodes of volcanism provide strong evidence that mantle upwelling caused doming of the crust, which in turn contributed to extension in the region.

It has also been suggested that the change in the nature of the plate boundary along the western margin of California may have contributed to the formation of the Basin and Range. About 40 million years ago the dominant forces acting on the western margin of North America were compressional, caused by the buoyant subduction of a segment of the Pacific basin (Figure 14.18). Subduction gradually ceased along the coast of California as the convergent boundary separating the Pacific and North American plates turned into the transform boundary we call the San

FIGURE 14.16 The Grand Tetons of Wyoming are an example of fault-block mountains. (Photo by Art Wolfe, Inc.)

Andreas Fault. Approximately 20 million years ago, a warm, rising mantle plume began to uplift and fault the crust between the Sierra Nevada and Rocky Mountains. According to one model, these elevated crustal blocks began to gravitationally slide off their lofty perches to generate the fault-block topography of the Basin and Range Province (Figure 14.18).

Vertical Movements of the Crust

In addition to the large crustal displacements driven mainly by plate tectonics, gradual up-and-down motions of the continental crust are observed at many locations around the globe. Although much of this vertical movement occurs along plate margins and is associated with active mountain building, some of it is not.

FIGURE 14-17 Extension in the Basin and Range Province has "stretched" the crust in some locations by as much as twice its original width. Shown here is a rough outline of the western states before (left) and after (right) extension.

Evidence for crustal uplift occurs along the West Coast of the United States. When the elevation of a coastal area remains unchanged for an extended period, a wavecut platform develops (see Figure 20.11). In parts of California, ancient wave-cut platforms can now be found as terraces hundreds of meters above sea level (Figure 14.19). Such evidence of crustal uplift is easy to find; unfortunately, the reason for uplift is not always as easy to determine.

Isostasy

Early workers discovered that Earth's less-dense crust floats on top of the denser and deformable rocks of the mantle. The concept of a floating crust in gravitational balance is called **isostasy** (*iso* = equal, *stasis* = standing). Perhaps the easiest way to grasp the concept of isostasy is to envision a series of wooden blocks of different heights floating in water, as shown in Figure 14.20. Note that the thicker wooden blocks float higher than the thinner blocks.

Similarly, many mountain belts stand high above the surrounding terrain because of crustal thickening. These compressional mountains have buoyant crustal "roots" that extend deep into the supporting material below, just like the thicker wooden blocks shown in Figure 14.20 (see Box 14.3).

Isostatic Adjustment Visualize what would happen if another small block of wood were placed atop one of the blocks in Figure 14.20. The combined block would sink until a new isostatic (gravitational) balance was reached. However, the top of the combined block would actually be higher than before, and the bottom would be lower. This process of establishing a new level of gravitational equilibrium is called **isostatic adjustment.**

Applying the concept of isostatic adjustment, we should expect that when weight is added to the crust, it will respond by subsiding, and when weight is removed, the crust will rebound. (Visualize what happens to a ship as cargo is being loaded and unloaded.) Evidence for crustal subsidence followed by crustal rebound is provided by Ice Age glaciers. When continental ice sheets occupied portions of North America during the Pleistocene epoch, the added weight of 3-kilometer-thick masses of ice caused downwarping of Earth's crust by hundreds of meters. In the 8000 years since the last ice sheet melted, uplifting of as much as 330 meters (1000 feet) has occurred in Canada's Hudson Bay region, where the thickest ice had accumulated (see Figure 18.26).

One of the consequences of isostatic adjustment is that as erosion lowers the summits of mountains, the crust will rise in response to the reduced load (Figure 14.21). However, each episode of isostatic uplift is somewhat less than the elevation loss due to erosion. The processes of uplifting and erosion will continue until the mountain block reaches

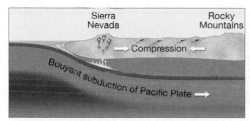

A. 40 million years ago (compressional forces dominate)

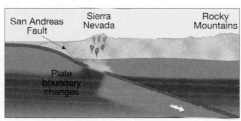

B. 30 million years ago (change from a convergent to transform boundary)

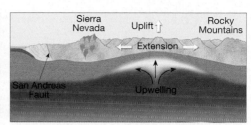

C. 20 million years ago (uplift and extension of the crust)

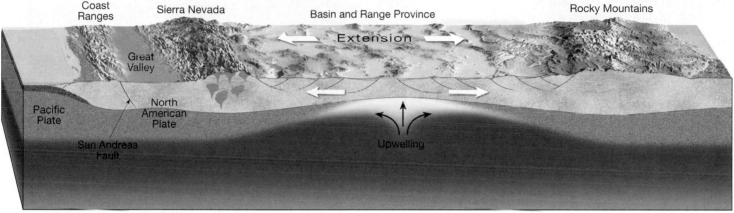

D. 20 million years ago to present (uplift and gravitational sliding creates the topography of the Basin and Range)

FIGURE 14.18 The Basin and Range Province consists of numerous fault-block mountains that were generated during the last 20 million years of Earth history. Upwelling of hot mantle rock and perhaps gravitational collapse (crustal sliding) contributed to considerable stretching and thinning of the crust.

"normal" crustal thickness. When this occurs, the mountains will be eroded to near sea level, and the once deeply buried interior of the mountain will be exposed at the surface. In addition, as mountains are worn down, the eroded sediment is deposited on adjacent landscapes, causing these areas to subside (Figure 14.21).

How High Is Too High? Where compressional forces are great, such as those driving India into Asia, mountains such as the Himalayas result. But is there a limit on how high a mountain can rise? As mountaintops are elevated, gravity-driven processes such as erosion and mass wasting accelerate, carving the deformed strata into rugged landscapes. Just as impor-

FIGURE 14.19 Former wave-cut platforms now exist as a series of elevated terraces on the west side of San Clemente Island off the southern California coast. Once at sea level, the highest terraces have risen about 400 meters. (Photo by John S. Shelton)

tant, however, is the fact that gravity also acts on the rocks within these mountainous masses. The higher the mountain, the greater the downward force on the rocks near the base. (Visualize a group of cheerleaders at a sporting event building a human pyramid.) At some point the rocks deep within the developing mountain, which are comparatively warm and weak, will begin to flow laterally, as shown in Figure 14.22. This is analogous to what happens when a ladle of very thick pancake batter is poured on a hot griddle. As a result, the mountain will experience a **gravitational collapse,** which involves normal faulting and subsidence in the upper, brittle portion of the crust and ductile spreading at depth.

You then might ask, What keeps the Himalayas standing? Simply, the horizontal compressional forces that are driving India into Asia are greater than the vertical force of gravity. However, once India's northward trek ends, the downward pull of gravity will become the dominant force acting on this mountainous region.

Mantle Convection: A Cause of Vertical Crustal Movement

Based on studies of Earth's gravitational field, it became clear that up-and-down convective flow in the mantle also affects the elevation of Earth's major landforms. The buoyancy of hot rising material accounts for broad upwarping in the overlying lithosphere, while downward flow causes downwarping.

BOX 14.3 ▶ UNDERSTANDING EARTH

Do Mountains Have Roots?

One of the major advances in determining the structure of mountains occurred in the 1840s when Sir George Everest (after whom Mount Everest is named) conducted the first topographical survey in India. During this survey the distance between the towns of Kalianpur and Kaliana, located south of the Himalayan range, was measured using two different methods. One method employed the conventional surveying technique of triangulation, and the other method determined the distance astronomically. Although the two techniques should have given similar results, the astronomical calculations placed these towns nearly 150 meters closer to each other than did the triangulation survey.

The discrepancy was attributed to the gravitational attraction exerted by the massive Himalayas on the plumb bob used for leveling the astronomical instrument. (A plumb bob is a metal weight suspended by a cord, used to determine a vertical orientation.) It was suggested that the deflection of the plumb bob would be greater at Kaliana than at Kalianpur because it is closer to the mountains (Figure 14.D).

A few years later, J. H. Pratt estimated the mass of the Himalayas and calculated the error that should have been caused by the gravitational influence of the mountains. To his surprise, Pratt discovered that the mountains should have produced an error three times larger than was actually observed. Simply stated, the mountains were not "pulling their weight." It was as if they had a hollow central core.

A hypothesis to explain the apparent "missing" mass was developed by George

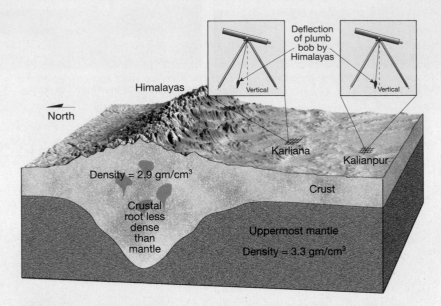

FIGURE 14.D During the first survey of India, an error in measurement occurred because the plumb bob on an instrument was deflected by the massive Himalayas. Later work by George Airy predicted that the mountains have roots of light crustal rocks. Airy's model explained why the plumb bob was deflected much less than expected.

Airy. Airy suggested that Earth's lighter crustal rocks float on the denser, more easily deformed mantle. Further, he correctly argued that the crust must be thicker under mountains than beneath the adjacent lowlands. In other words, mountainous terrains are supported by light crustal material that extends as "roots" into the denser mantle (Figure 14.D). This phenomenon is exhibited by icebergs, which are buoyed up by the weight of the displaced water. If the Himalayas do have roots of light crustal rocks

that extend far beneath them, then these mountains would exert less gravitational attraction, as Pratt had calculated. Hence, Airy's model explained why the plumb bob was deflected much less than expected.

Seismological and gravitational studies have confirmed the existence of crustal roots under some mountain ranges. The thickness of continental crust is normally about 35 kilometers, but crustal thicknesses exceeding 70 kilometers have been determined for some mountain belts.

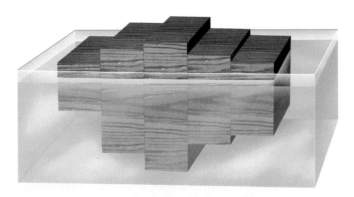

FIGURE 14.20 This drawing illustrates how wooden blocks of different thicknesses float in water. In a similar manner, thick sections of crustal material float higher than do thinner crustal slabs.

Uplifting Whole Continents Southern Africa is one region where large-scale vertical motion is evident. Here much of the region exists as an expansive plateau having an average elevation of nearly 1500 meters (5000 feet). Geologic studies have shown that southern Africa and the surrounding seafloor have been slowly rising for the past 100 million years, even though it has not experienced a plate collision for nearly 400 million years.

Evidence from seismic tomography (see Figure 12.20, p. 341) indicates that a large, mushroom-shaped mass of hot mantle rock is centered below the southern tip of Africa. This *superplume* extends upward about 2900 kilometers (1800 miles) from the mantle–core boundary and spreads out over several thousand kilometers. Researchers have concluded that the upward flow of this huge mantle plume is sufficient to elevate southern Africa.

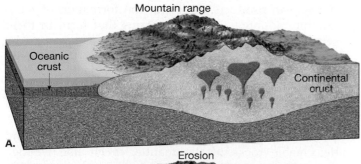

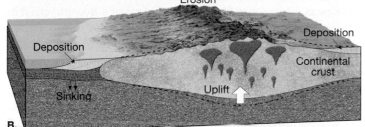

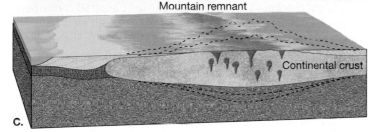

FIGURE 14.21   This sequence illustrates how the combined effects of erosion and isostatic adjustment result in a thinning of the crust in mountainous regions. **A.** When mountains are young, the continental crust is thickest. **B.** As erosion lowers the mountains, the crust rises in response to the reduced load. **C.** Erosion and uplift continue until the mountains reach "normal" crustal thickness.

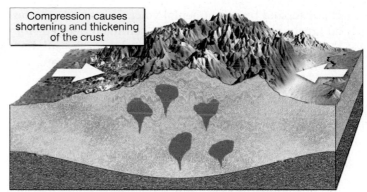

A. Horizontal compressional forces dominate

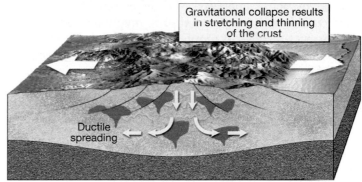

B. Gravitational forces dominate

FIGURE 14.22   Block diagram of a mountain belt that is collapsing under its own "weight." Gravitational collapse involves normal faulting in the upper, brittle portion of the crust and ductile spreading at depth.

Crustal Subsidence   Extensive areas of downwarping have also been discovered. For example, large, nearly circular basins are found in the interiors of some continents. Studies indicate that many major episodes of crustal downwarping are not caused by the weight of accumulating sediments. Rather, they show that the formation of basins promoted the accumulation of vast quantities of sediments. Several of these downwarped structures exist in the United States, including the large basins of Michigan and Illinois.

Similar episodes of large-scale downwarping are known on other continents, including Australia. The cause of these downward movements followed by rebound may be linked to the subduction of slabs of oceanic lithosphere. One proposal suggests that when subduction ceases along a continental margin, the subducting slab detaches from the trailing lithosphere and continues its descent into the mantle. As this detached lithospheric slab sinks, it creates a downward flow in its wake that tugs on the base of the overriding continent. In some situations the crust is apparently pulled down sufficiently to allow the ocean to extend inland. As the oceanic slab sinks deeper into the mantle, the pull of the trailing wake weakens and the continent "floats" back into isostatic balance.

Summary

- The name for the processes that collectively produce a *compressional mountain belt* is *orogenesis*. Most compressional mountains consist of folded and faulted sedimentary and volcanic rocks, portions of which have been strongly metamorphosed and intruded by younger igneous bodies.
- Plate convergence can result in a subduction zone consisting of four regions: (1) a *deep-ocean trench* that forms where a subducting slab of oceanic lithosphere bends and descends into the asthenosphere; (2) a *volcanic arc,* which is built upon the overlying plate; (3) a region located between the trench and the volcanic arc (*forearc region*); and (4) a region on the side of the volcanic arc opposite the trench (*backarc region*). Along some subduction zones backarc spreading results in the formation of *backarc basins,* such as those that floor the Sea of Japan and China Sea.
- Subduction of oceanic lithosphere under a continental block gives rise to an *Andean-type plate margin* that is

characterized by a continental volcanic arc and associated igneous plutons. In addition, sediment derived from the land, as well as material scraped from the subducting plate, becomes plastered against the landward side of the trench, forming an *accretionary wedge*. An excellent example of an inactive Andean-type mountain belt is found in the western United States and includes the Sierra Nevada and the Coast Range in California.

- Continued subduction of oceanic lithosphere beneath an Andean-type continental margin will eventually close an ocean basin. The result will be a *continental collision* and the development of compressional mountains that are characterized by shortened and thickened crust as exhibited by the Himalayas. The development of a major mountain belt is often complex, involving two or more distinct episodes of mountain building. A common feature of compressional mountains are *fold-and-thrust belts*. Continental collisions have generated many mountain belts, including the Alps, Urals, and Appalachians.

- Mountain belts can develop as a result of the collision and merger of an island arc, oceanic plateau, or some other small crustal fragment to a continental block. Many of the mountain belts of the North American Cordillera, principally those in Alaska and British Columbia, were generated in this manner.

- Although most mountains form along convergent plate boundaries, other tectonic processes, such as continental rifting, can produce uplift and the formation of topographic mountains. The mountains that form in these settings, termed *fault-block mountains,* are bounded by high-angle normal faults that gradually flatten with depth. The Basin and Range Province in the western United States consists of hundreds of faulted blocks that give rise to nearly parallel mountain ranges that stand above sediment-laden basins.

- Earth's less dense crust floats on top of the denser and deformable rocks of the mantle, much like wooden blocks floating in water. The concept of a floating crust in gravitational balance is called *isostasy.* Most mountainous topography is located where the crust has been shortened and thickened. Therefore, mountains have deep crustal roots that isostatically support them. As erosion lowers the peaks, *isostatic adjustment* gradually raises the mountains in response. The processes of uplifting and erosion will continue until the mountain block reaches "normal" crustal thickness. Gravity also causes elevated mountainous structures to collapse under their own "weight."

- Convective flow in the mantle contributes to the up-and-down bobbing of the crust. The upward flow of a large superplume located beneath southern Africa is thought to have elevated this region during the last 100 million years. Crustal subsidence has produced large basins and may have allowed the ocean to invade the continents several times in the geologic past.

Review Questions

1. In the plate tectonics model, which type of plate boundary is most directly associated with mountain building?

2. List the four main structures of a subduction zone, and describe where each is located relative to the others.

3. Briefly describe how backarc basins form.

4. Describe the process that generates most basaltic magma at subduction zones.

5. How are magmas that exhibit an intermediate to felsic composition thought to be produced from mantle-derived basaltic magmas at Andean-type plate margins?

6. What is a batholith? In what modern tectonic setting are batholiths being generated?

7. In what ways are the Sierra Nevada and the Andes similar? How are they different?

8. What is an accretionary wedge? Briefly describe its formation.

9. What is a passive margin? Give an example. Give an example of an active continental margin.

10. The formation of mountainous topography at a volcanic island arc, such as Japan, is considered just one phase in the development of a major mountain belt. Explain.

11. What tectonic structure is exhibited by the Coast Ranges of California?

12. Suture zones are often described as the place where continents are "welded" together. Why might that statement be misleading?

13. During the formation of the Himalayas, the continental crust of Asia was deformed more than India proper. Why do we think that happened?

14. Where might magma be generated in a newly formed collision mountain?

15. Suppose a sliver of oceanic crust was discovered in the interior of a continent. Would this support or refute the theory of plate tectonics? Explain.

16. How can the Appalachian Mountains be considered a collision-type mountain range when the nearest continent is 5000 kilometers (3000 miles) away?

17. How does the plate tectonics theory help explain the existence of fossil marine life in rocks atop compressional mountains?

18. In your own words, briefly describe the stages in the formation of a major mountain belt according to the plate tectonics model.

19. Define the term *terrane*. How is it different from the term *terrain?*

20. In addition to microcontinents, what other structures are thought to be carried by the oceanic lithosphere and eventually accreted to a continent?

21. Briefly describe the major differences between the evolution of the Appalachian Mountains and the North American Cordillera.

22. Compare the processes that generate fault-block mountains to those associated with most other major mountain belts.

23. Give one example of evidence that supports the concept of crustal uplift.

24. What happens to a floating object when weight is added? Subtracted? How does this principal apply to changes in the elevations of mountains? What term is applied to the adjustment that causes crustal uplift of this type?

25. How do some researchers explain the elevated position of southern Africa?

Key Terms

accretionary wedge (p. 384)
Andean-type plate margins (p. 382)
backarc basin (p. 381)
continental volcanic arc (p. 380)

fault-block mountains (p. 393)
fold-and-thrust belts (p. 387)
forearc basin (p. 385)

gravitational collapse (p. 395)
island arcs (p. 380)
isostasy (p. 394)
isostatic adjustment (p. 394)
microcontinent (p. 390)

orogenesis (p. 379)
passive margin (p. 382)
suture (p. 387)
terrane (p. 390)
volcanic island arc (p. 380)

Web Resources

The *Earth* Website uses the resources and flexibility of the Internet to aid in your study of the topics in this chapter. Written and developed by geology instructors, this site will help improve your understanding of geology. Visit **http://www.prenhall.com/tarbuck** and click on the cover of *Earth 9e* to find:

• Online review quizzes.

• Critical thinking exercises.

• Links to chapter-specific Web resources.

• Internet-wide key-term searches.

http://www.prenhall.com/tarbuck

GEODe: Earth

GEODe: Earth makes studying faster and more effective by reinforcing key concepts using animation, video, narration, interactive exercises and practice quizzes. A copy is included with every copy of *Earth*.

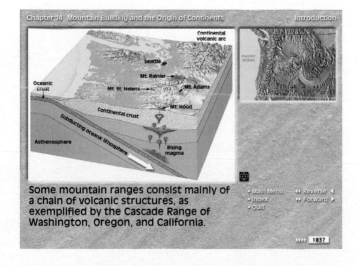

Some mountain ranges consist mainly of a chain of volcanic structures, as exemplified by the Cascade Range of Washington, Oregon, and California.

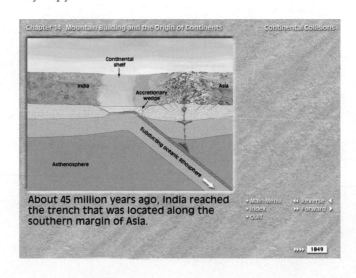

About 45 million years ago, India reached the trench that was located along the southern margin of Asia.

Mass Wasting: The Work of Gravity

CHAPTER

15

This debris flow at La Conchita, California, in January 2005, was triggered by extraordinary rains. Read a case study in Box 15.1. (Photo by David McNew/ Getty Images)

401

arth's surface is never perfectly flat, but instead consists of slopes of many different varieties. Some are steep and precipitous; others are moderate or gentle. Some are long and gradual; others are short and abrupt. Slopes can be mantled with soil and covered by vegetation or consist of barren rock and rubble. Taken together, slopes are the most common elements in our physical landscape. Although most slopes may appear to be stable and unchanging, the force of gravity causes material to move downslope. At one extreme, the movement may be gradual and practically imperceptible. At the other extreme, it may consist of a roaring debris flow or a thundering rock avalanche. Landslides are a worldwide natural hazard. When these hazardous processes lead to loss of life and property, they become natural disasters.

Landslides as Natural Disasters

Even in areas with steep slopes, catastrophic landslides are relatively rare occurrences. As a result people living in susceptible areas often do not appreciate the risk of living where they do. However, media reports remind us that such events occur with some regularity around the world (Figure 15.1). The three examples described here occurred during a span of just four months.

On October 8, 2005, a magnitude 7.6 earthquake struck the Kashmir region between India and Pakistan. Compounding the tragic effects caused directly by the severe ground shaking were hundreds of landslides triggered by the quake and its many aftershocks. Rockfalls and debris slides thundered down steep mountain slopes and were focused into narrow valleys where many people made their homes. The landslides also blocked roads and trails, delaying attempts to reach those in need.

Just three days earlier, on October 5, 2005, torrential rains from Hurricane Stan triggered mudflows in Guatemala. A slurry of mud 1 kilometer wide and up to 12 meters (40 feet) deep buried the village of Panabaj. Death toll estimates for the area approached 1400 people. Flows such as this can travel at speeds of 50 kilometers (30 miles) per hour down rugged mountain slopes.

On February 17, 2006, only a few months after the tragedy in Central America, a lethal mudflow triggered by extraordinary rains buried a small town on the Philippine island of Leyte. A mass of mud engulfed this remote coastal area to depths as great as 10 meters (30 feet). Although an accurate count of fatalities was difficult, as many as 1800 people perished. This region is prone to such events due in part to the fact that deforestation has denuded the nearby mountain slopes. In the pages that follow you will take a closer look at rapid mass-wasting events in an attempt to better understand their causes and effects.

Mass Wasting and Landform Development

Landslides are spectacular examples of a basic geologic process called mass wasting. **Mass wasting** refers to the downslope movement of rock, regolith, and soil under the

FIGURE 15.1 **A.** On October 8, 2005, a major earthquake in Kashmir triggered hundreds of landslides including the one shown here. (AP Photo/Burhan Ozblici) **B.** In February 2006, heavy rains triggered this mudflow that buried a small town on the Philippine island of Leyte. (AP Photo/ Pat Roque)

A.

B.

direct influence of gravity. It is distinct from the erosional processes that are examined in subsequent chapters because mass wasting does not require a transporting medium such as water, wind, or glacial ice.

The Role of Mass Wasting

In the evolution of most landforms, mass wasting is the step that follows weathering. By itself, weathering does not produce significant landforms. Rather, landforms develop as products of weathering are removed from the places where they originate. Once weathering weakens and breaks rock apart, mass wasting transfers the debris downslope, where a stream, acting as a conveyor belt, usually carries it away. Although there may be many intermediate stops along the way, the sediment is eventually transported to its ultimate destination: the sea.

The combined effects of mass wasting and running water produce stream valleys, which are the most common and conspicuous of Earth's landforms. If streams alone were responsible for creating the valleys in which they flow, the valleys would be very narrow features. However, the fact that most river valleys are much wider than they are deep is a strong indication of the significance of mass-wasting processes in supplying material to streams. This is illustrated by the Grand Canyon (Figure 15.2). The walls of the canyon extend far from the Colorado River, owing to the transfer of weathered debris downslope to the river and its tributaries by mass-wasting processes. In this manner, streams and mass wasting combine to modify and sculpt the surface. Of course, glaciers, groundwater, waves, and wind are also important agents in shaping landforms and developing landscapes.

Students Sometimes Ask . . .

It seems as though you have used the term "landslide" to refer to several different things—from mudflows to rock avalanches. What exactly is the definition of "landslide"?

Although many people, including geologists, frequently use the word *landslide*, the term has no specific definition in geology. Rather, it is a popular nontechnical term used to describe any or all relatively rapid forms of mass wasting.

FIGURE 15.2   The walls of the Grand Canyon extend far from the channel of the Colorado River. This results primarily from the transfer of weathered debris downslope to the river and its tributaries by mass wasting processes. (Photo by Tom and Susan Bean, Inc.)

Slopes Change through Time

It is clear that if mass wasting is to occur, there must be slopes that rock, soil, and regolith can move down. It is Earth's mountain building and volcanic processes that produce these slopes through sporadic changes in the elevations of landmasses and the ocean floor. If dynamic internal processes did not continually produce regions having higher elevations, the system that moves debris to lower elevations would gradually slow and eventually cease.

Most rapid and spectacular mass-wasting events occur in areas of rugged, geologically young mountains. Newly formed mountains are rapidly eroded by rivers and glaciers into regions characterized by steep and unstable slopes. It is in such settings that massive destructive landslides, such as the Yungay disaster, occur. As mountain building subsides, mass wasting and erosional processes lower the land. Through time, steep and rugged mountain slopes give way to gentler, more subdued terrain. Thus, as a landscape ages, massive and rapid mass-wasting processes give way to smaller, less dramatic downslope movements.

Controls and Triggers of Mass Wasting

GEODe **Mass Wasting**
▶ Controls and Triggers of Mass Wasting

Gravity is the controlling force of mass wasting, but several factors play an important role in overcoming inertia and creating downslope movements. Long before a landslide occurs, various processes work to weaken slope material, gradually making it more and more susceptible to the pull of gravity. During this span, the slope remains stable but gets closer and closer to being unstable. Eventually, the strength of the slope is weakened to the point that something causes it to cross the threshold from stability to instability. Such an event that initiates downslope movement is called a *trigger.* Remember that the trigger is not the sole cause of the mass-wasting event but just the last of many causes. Among the common factors that trigger mass-wasting processes are saturation of material with water, oversteepening of slopes, removal of anchoring vegetation, and ground vibrations from earthquakes.

The Role of Water

Mass wasting is sometimes triggered when heavy rains or periods of snowmelt saturate surface materials. This was the case in October 1998 when torrential downpours associated with Hurricane Mitch triggered devastating mudflows in Central America (Figure 15.3). Box 15.1 presents a case study of another event that occurred at La Conchita, California, in January 2005.

When the pores in sediment become filled with water, the cohesion among particles is destroyed, allowing them to slide past one another with relative ease. For example, when sand is slightly moist, it sticks together quite well. However, if enough water is added to fill the openings between the grains, the sand will ooze out in all directions (Figure 15.4). Thus, saturation reduces the internal resistance of materials, which are then easily set in motion by the force of gravity. When clay is wetted, it becomes very slick— another example of the "lubricating" effect of water. Water also adds considerable weight to a mass of material. The added weight in itself may be enough to cause the material to slide or flow downslope.

Oversteepened Slopes

Oversteepening of slopes is another trigger of many mass movements. There are many situations in nature where oversteepening takes place. A stream undercutting a valley wall and waves pounding against the base of a cliff are but two familiar examples. Furthermore, through their activities, people often create oversteepened and unstable slopes that become prime sites for mass wasting.

FIGURE 15.3 Heavy rains from Hurricane Mitch in the fall of 1998 triggered devastating mudflows in Central America. Water plays an important role in many mass-wasting processes. (Associated Press Photo)

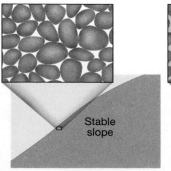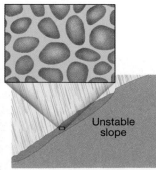

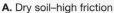

A. Dry soil–high friction   **B.** Saturated soil

FIGURE 15.4 The effect of water on mass wasting can be great. **A.** When little or no water is present, friction among the closely packed soil particles on the slope holds them in place. **B.** When the soil is saturated, the grains are forced apart and friction is reduced, allowing the soil to move downslope.

FIGURE 15.5 The angle of repose for this granular material is about 30°. (Photo by G. Leavens/Photo Researchers)

Unconsolidated, granular particles (sand-size or coarser) assume a stable slope called the **angle of repose** (*reposen* = to be at rest). This is the steepest angle at which material remains stable (Figure 15.5). Depending on the size and shape of the particles, the angle varies from 25 to 40 degrees. The larger, more angular particles maintain the steepest slopes. If the angle is increased, the rock debris will adjust by moving downslope.

Oversteepening is not just important because it triggers movements of unconsolidated granular materials. Oversteepening also produces unstable slopes and mass movements in cohesive soils, regolith, and bedrock. The response will not be immediate, as with loose, granular material, but sooner or later, one or more mass-wasting processes will eliminate the oversteepening and restore stability to the slope.

Removal of Vegetation

Plants protect against erosion and contribute to the stability of slopes because their root systems bind soil and regolith together. In addition, plants shield the soil surface from the erosional effects of raindrop impact (see Figure 6.23, p. 185). Where plants are lacking, mass wasting is enhanced, especially if slopes are steep and water is plentiful. When anchoring vegetation is removed by forest fires or by people (for timber, farming, or development), surface materials frequently move downslope.

An unusual example illustrating the anchoring effect of plants occurred several decades ago on steep slopes near Menton, France. Farmers replaced olive trees, which have deep roots, with a more profitable but shallow-rooted crop: carnations. When the less stable slope failed, the landslide took 11 lives.

In July 1994 a severe wildfire swept Storm King Mountain west of Glenwood Springs, Colorado, denuding the slopes of vegetation. Two months later heavy rains resulted in numerous debris flows, one of which blocked Interstate 70 and threatened to dam the Col-

orado River. A 5-kilometer (3-mile) length of the highway was inundated with tons of rock, mud, and burned trees. The closure of Interstate 70 imposed costly delays on this major highway. Following extensive wildfires that occurred in the summer of 2000, similar types of mass wasting threaten highways and other development near fire-ravaged hillsides throughout the West (Figure 15.6).

In addition to eliminating plants that anchor the soil, fire can promote mass wasting in other ways. Following a wildfire, the upper part of the soil may become dry and loose. As a result, even in dry weather, the soil tends to move down steep slopes. Moreover, fire can also "bake" the ground, creating a water-repellant layer at a shallow depth. This nearly impermeable barrier prevents or slows the infiltration of water, resulting in increased surface runoff during rains. The consequence can be dangerous torrents of viscous mud and rock debris.

Earthquakes as Triggers

Conditions that favor mass wasting may exist in an area for a long time without movement occurring. An additional factor is sometimes necessary to trigger the movement. Among the more important and dramatic triggers are earthquakes. An earthquake and its aftershocks can dislodge enormous volumes of rock and unconsolidated material. The event in the Kashmir region described near the beginning of the chapter is one tragic example.

Landslides Triggered by the Northridge Earthquake In January 1994 an earthquake struck the Los Angeles region of southern California. Named for its epicenter in the town of Northridge, the 6.7-magnitude event produced estimated losses of $20 billion. Some of these losses were the result of more than 11,000 landslides in an area of about 10,000 square kilometers (3900 square miles) that were set in

FIGURE 15.6 During summer, wildfires are common occurrences in many parts of the West. Millions of acres are burned each year. The loss of anchoring vegetation sets the stage for accelerated mass wasting. (Photo by Raymond Gehman)

BOX 15.1 ▶ PEOPLE AND THE ENVIRONMENT

Landslide Hazards at La Conchita, California*

Southern California lies astride a major plate boundary defined by the San Andreas Fault and numerous other related faults that are spread across the region. It is a dynamic environment characterized by rugged mountains and steep-walled canyons. Unfortunately this scenic landscape presents serious geologic hazards. Just as tectonic forces are steadily pushing the landscape upward, gravity is relentlessly pulling it downward. When gravity prevails, landslides occur.

As you might expect, some of the region's landslides are triggered by earthquakes. Many others, however, are related to periods of prolonged and intense rainfall. A tragic example of the latter situation occurred on January 10, 2005, when a massive debris flow (popularly called a *mudslide*) swept through La Conchita, California, a small town located about 80 kilometers (50 miles) northwest of Los Angeles (Figure 15.A).

Although the rapid torrent of mud took many of the town's inhabitants by surprise, such an event should not have been unexpected. Let's briefly examine the factors that contributed to the deadly debris flow at La Conchita.

The town is situated on a narrow coastal strip about 250 meters (800 feet) wide between the shoreline and a steep 180-meter (600-foot) bluff (Figure 15.B). The bluff consists of poorly sorted marine sediments and weakly cemented layers of shale, siltstone, and sandstone.

The deadly 2005 debris flow involved little or no newly failed material, but rather consisted of the remobilization of a portion of a large landslide that destroyed several homes in 1995. In fact, historical accounts

FIGURE 15.A View of the La Conchita debris flow taken four days after the January 2005 event. The light-colored exposed rock in the upper part of the photo is the main scarp of a slide that occurred 10 years earlier in 1995. The January 2005 event (arrow) was a remobilization of a portion of the 1995 landslide. (Photo by Randall Jibson/U.S. Geological Survey)

dating back to 1865 indicate that landslides in the immediate area have been a regular occurrence. Further, geologic evidence shows that landsliding of a variety of types and scales has probably been occurring at La Conchita for thousands of years.

The most significant contributing factor to the tragic 2005 debris flow was prolonged and intense rain. The event occurred at the end of a span that produced near record amounts of rainfall in Southern California. Wintertime rainfall at near-

by Ventura totaled 49.3 centimeters (19.4 inches) as compared to an average value of just 12.2 centimeters (4.8 inches). As Figure 15.C indicates, much of that total fell during the two weeks immediately preceding the debris flow.

This was not the first destructive landslide to strike La Conchita, nor is it likely to

---

*Based in part on material prepared by the U.S. Geological Survey.

motion by the quake (Figure 15.7). Most were shallow rock falls and slides, but some were much larger and filled canyon bottoms with jumbles of soil, rock, and plant debris. The debris in canyon bottoms created a secondary threat because it can mobilize during rainstorms, producing debris flows. Such flows are common and often disastrous in southern California.

The mass-wasting processes triggered by the Northridge earthquake destroyed dozens of homes and caused extensive damage to roads, pipelines, and well machinery in oil

fields. In some places more than 75 percent of slope areas were denuded by landslides, making them vulnerable to subsequent mass wasting triggered by heavy rains.

Liquefaction Intense ground shaking during earthquakes can cause water-saturated surface materials to lose their strength and behave as fluidlike masses that flow. This process, called *liquefaction,* was a major cause of property damage in Anchorage, Alaska, during the massive 1964 Good Friday earthquake described in Chapter 11.

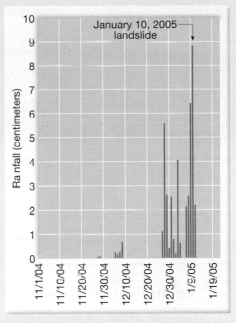

FIGURE 15.B   The larger image is a view down the length of the 2005 La Conchita debris flow. It also depicts the setting of the small town between the ocean and a steep cliff. The arrow on the larger photo is pointing to the house shown on the inset. The flow was quite viscous and moved houses in its path rather than flowing around them. As you can see, the left side of the house was detached and moved. (Photo by Randall Jibson/U.S. Geological Survey)

FIGURE 15.C   Daily rainfall at the nearby town of Ventura during the weeks leading up to the January 2005 La Conchita event. Each line on the bar graph shows rain for a particular day. The 2005 debris flow occurred at the culmination of the heaviest rainfall of the season. About 80 percent of the season's exceptional total fell in this short span. (After National Weather Service)

be the last. The town's geologic setting and history of rapid mass-wasting events clearly support this notion. When the amount and intensity of rainfall is sufficient, debris flows are to be expected. The concluding paragraph from a U.S. Geological Survey report puts it this way:

> The La Conchita area has experienced, and will likely continue to experience, a rather bewildering variety of landslide hazards. Different landslide scenarios are more or less likely to occur as a result of different specific rainfall conditions, and no part of the community can be considered safe from landslides. Unfortunately, we currently lack the understanding to accurately forecast what might happen in each possible rainfall scenario. Prudence would certainly dictate, however, that we anticipate renewed landslide activity during or after future periods of prolonged and/or intense rainfall. Future earthquakes, of course, also could trigger landsliding in the area.*

*Jibsen, Randall W. "Landslide Hazards at La Conchita, California," *U.S. Geological Survey Open-File Report 2005-1067*, p. 11.

Landslides without Triggers?

Do rapid mass-wasting events always require some sort of trigger such as heavy rains or an earthquake? The answer is no; such events sometimes occur without being triggered. For example, on the afternoon of May 9, 1999, a landslide killed 10 hikers and injured many others at Sacred Falls State Park near Hauula on the north shore of Oahu, Hawaii. The tragic event occurred when a mass of rock from a canyon wall plunged 150 meters (500 feet) down a nearly vertical slope to the valley floor. Because of safety concerns, the park was closed so that landslide specialists from the U.S. Geological Survey could investigate the site. Their study concluded that the landslide occurred *without triggering* from any discernible external conditions.

Many rapid mass-wasting events occur without a discernible trigger. Slope materials gradually weaken over time under the influence of long-term weathering, infiltration of water, and other physical processes. Eventually, if the strength falls below what is necessary to maintain slope stability, a landslide will occur. The timing of such events is

FIGURE 15.7 Various forms of mass wasting can be triggered by earthquakes. This home in Pacific Palisades, California, was destroyed by a landslide triggered by the January 1994 Northridge earthquake. In some cases, damages from earthquake-induced mass wasting are greater than damages caused directly by an earthquake's ground vibrations. (Photo by Chromo Sohm/Corbis/The Stock Market)

Students Sometimes Ask . . .

How many deaths result from landslides each year?

The U.S. Geological Survey estimates that between 25 and 50 people are killed by landslides annually in the United States. The worldwide death toll, of course, is much higher.

random, and thus accurate prediction is impossible (see Box 15.2).

Classification of Mass-Wasting Processes

There is a broad array of different processes that geologists call mass wasting. Generally, the different types are classified based on the type of material involved, the kind of motion displayed, and the velocity of the movement.

Type of Material

The classification of mass-wasting processes on the basis of the material involved in the movement depends upon whether the descending mass began as unconsolidated material or as bedrock. If soil and regolith dominate, terms such as debris, mud, or earth are used in the description. In contrast, when a mass of bedrock breaks loose and moves downslope, the term rock may be part of the description.

Type of Motion

In addition to characterizing the type of material involved in a mass-wasting event, the way in which the material moves may also be important. Generally, the kind of motion is described as either a fall, a slide, or a flow.

Fall When the movement involves the freefall of detached individual pieces of any size, it is termed a **fall.** Fall is a common form of movement on slopes that are so steep that loose material cannot remain on the surface. The rock may fall directly to the base of the slope or move in a series of leaps and bounds over other rocks along the way. Rockfalls are the primary way in which *talus slopes* are built and maintained (Figure 15.8). Many falls result when freeze and thaw cycles and/or the action of plant roots loosen rock to the point that gravity takes over. Although signs along bedrock cuts on highways warn of falling rock, few of us have actually witnessed such an event. However, as the image in Figure 15.9 illustrates, they do indeed occur.

When large masses of rock plunge from great heights they hit the ground with enormous force and often trigger additional mass-wasting events. One deadly example occurred in Peru. In May 1970, an earthquake caused a huge mass of rock and ice to break free from the precipitous north face of Nevados Huascaran, the loftiest peak in the Peruvian Andes. The material plunged nearly a kilometer and pulverized on impact. The rock avalanche that followed rushed down the mountainside made fluid by trapped air and ice. Along the way it ripped loose millions of tons of additional debris that ultimately and tragically buried more than 20,000 people in the towns of Yungay and Ranrahirca.

A different effect triggered by a rockfall occurred in Yosemite National Park less than 6 months before the event pictured in Figure 15.9. On July 10, 1996, two large rock masses broke loose from steep cliffs and fell about 500 meters (1640 feet) to the floor of Yosemite Valley. The impacts were great enough to be recorded at seismic stations 200 kilometers (125

FIGURE 15.8 Talus is a slope built of angular rock fragments. Mechanical weathering, especially frost wedging, loosens the pieces of bedrock, which then fall to the base of the cliff. With time, a series of steep, cone-shaped accumulations build up at the base of the vertical slope. These talus cones are in Banff National Park, Alberta, Canada. (Photo by Marli Miller)

BOX 15.2 ▶ PEOPLE AND THE ENVIRONMENT

The Vaiont Dam Disaster

Most often landslides are triggered by a natural event, such as heavy rains or an earthquake. However, sometimes the rapid downslope movement of surface material is caused by the actions of people. Such is the case with the example discussed in this box. Unfortunately it had disastrous consequences.

In 1960 a large dam, almost 265 meters (870 feet) tall, was built across Vaiont Canyon in the Italian Alps. It was engineered without good geological input, and the result was a disaster only three years later.

The bedrock in Vaiont Canyon slanted steeply downward toward the lake impounded behind the dam. The bedrock was weak, highly fractured limestone strata with beds of clay and numerous solution cavities. As the reservoir filled behind the completed dam, rocks became saturated and the clays became swollen and more plastic. The rising water reduced the internal friction that had kept the rock in place.

Measurements made shortly after the reservoir was filled hinted at the problem, because they indicated that a portion of the mountain was slowly creeping downhill at the rate of 1 centimeter per week. In September 1963 the rate increased to 1 centimeter per day, then 10–20 centimeters per day, and eventually as much as 80 centimeters on the day of the disaster.

Finally, the mountainside let loose. In just an instant, 240 million cubic meters of

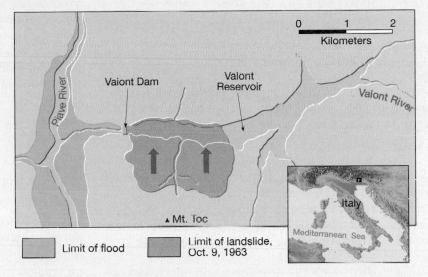

FIGURE 15.D Sketch map of the Vaiont River area showing the limits of the landslide, the portion of the reservoir that was filled with debris, and the extent of flooding downstream. (After G. A. Kiersch, "Vaiont Reservoir Disaster," *Civil Engineering* 34 (1964): 32–39.)

rock and rubble slid down the mountainside and filled nearly 2 kilometers of the gorge to heights of 150 meters above the reservoir level (Figure 15.D). This pushed the water completely over the dam in a wave more than 90 meters high. More than 1.5 kilometers downstream, the wall of water was still 70 meters high, destroying everything in its path.

The entire event lasted less than seven minutes, yet it claimed an estimated 2600 lives. Although this is known as the worst dam disaster in history, the Vaiont Dam itself remained intact. And while the catastrophe was triggered by human interference with the Vaiont River, the slide eventually would have occurred on its own; however, the effects would not have been nearly as tragic.

FIGURE 15.9 In January 1997, this rockfall blocked Highway 140 near the Arch Rock entrance to Yosemite National Park, California. (Photo by Roger J. Wyan/AP/Wide World Photos)

miles) from the site. When the dislodged rock masses struck the ground, they generated atmospheric pressure waves that were comparable in velocity to a tornado or hurricane. The force of the airblasts uprooted and snapped more than a thousand trees, including some that were 40 meters tall.

Slide Many mass-wasting processes are described as **slides.** The term refers to mass movements in which there is a distinct zone of weakness separating the slide material from the more stable underlying material. Two basic types of slides are recognized. *Rotational slides* are those in which the surface of rupture is a concave-upward curve that resembles the shape of a spoon and the descending material exhibits a downward and outward rotation. By contrast, a *translational slide* is one in which a mass of material moves along a relatively flat surface such as a joint, fault, or bedding plane. Such slides exhibit little rotation or backward tilting.

Flow The third type of movement common to mass-wasting processes is termed **flow.** Flow occurs when material moves downslope as a viscous fluid. Most flows are saturated with water and typically move as lobes or tongues.

Rate of Movement

Some of the events that have been described so far in this chapter involved very rapid rates of movement. For example, it is estimated that the debris that rushed down the slopes of Peru's Nevados Huascaran moved at speeds in excess of 200 kilometers (125 miles) per hour. This most rapid type of mass movement is termed a **rock avalanche** (*avaler* = to descend). Many researchers believe that rock avalanches, such as the one that produced the scene in Figure 15.10, must literally "float on air" as they move downslope. That is, high velocities result when air becomes trapped and compressed beneath the falling mass of debris, allowing it to move as a buoyant, flexible sheet across the surface.

Most mass movements, however, do not move with the speed of a rock avalanche. In fact, a great deal of mass wasting is imper-

ceptibly slow. One process that we will examine later, termed *creep*, results in particle movements that are usually measured in millimeters or centimeters per year. Thus, as you can see, rates of movement can be spectacularly sudden or exceptionally gradual. Although various types of mass wasting are often classified as either rapid or slow, such a distinction is highly subjective because a wide range of rates exists between the two extremes. Even the velocity of a single process at a particular site can vary considerably.

Slump

GEODe Mass Wasting
▸ Types of Mass Wasting

Slump refers to the downward sliding of a mass of rock or unconsolidated material moving as a unit along a curved surface (Figure 15.11). Usually the slumped material does

FIGURE 15.10 This 4-kilometer-long tongue of rubble was deposited atop Alaska's Sherman Glacier by a rock avalanche. The event was triggered by a tremendous earthquake in March 1964. (Photo by Austin Post, U.S. Geological Survey)

Rock avalanche debris

FIGURE 15.11   **A.** Slump occurs when material slips downslope en masse along a curved surface of rupture. It is an example of a rotational slide. Earthflows frequently form at the base of the slump. **B.** Slump at Point Fermin, California. Slump is often triggered when slopes become oversteepened by erosional processes such as wave action. (Photo by John S. Shelton)

not travel spectacularly fast nor very far. This is a common form of mass wasting, especially in thick accumulations of cohesive materials such as clay. The ruptured surface is characteristically spoon-shaped and concave upward or outward. As the movement occurs, a crescent-shaped scarp is created at the head, and the block's upper surface is sometimes tilted backward. Although slump may involve a single mass, it often consists of multiple blocks. Sometimes water collects between the base of the scarp and the top of the tilted block. As this water percolates downward along the surface of rupture, it may promote further instability and additional movement.

Slump commonly occurs because a slope has been oversteepened. The material on the upper portion of a slope is held in place by the material at the bottom of the slope. As this anchoring material at the base is removed, the material above is made unstable and reacts to the pull of gravity. One relatively common example is a valley wall that becomes oversteepened by a meandering river. The photo in Figure 15.11 provides another example in which a coastal cliff has been undercut by wave action at its base. Slumping may also occur when a slope is overloaded, causing internal

stress on the material below. This type of slump often occurs where weak, clay-rich material underlies layers of stronger, more resistant rock such as sandstone. The seepage of water through the upper layers reduces the strength of the clay below, resulting in slope failure.

Rockslide

 GEODe   Mass Wasting
EARTH   ▶ Types of Mass Wasting

Rockslides occur when blocks of bedrock break loose and slide down a slope (Figure 15.12). If the material involved is largely unconsolidated, the term **debris slide** is used instead. Such events are among the fastest and most destructive mass movements. Usually rockslides take place in a geologic setting where the rock strata are inclined, or where joints and fractures exist parallel to the slope. When such a rock unit is undercut at the base of the slope, it loses support and the rock eventually gives way. Sometimes the rockslide is triggered when rain or melting snow lubricates the

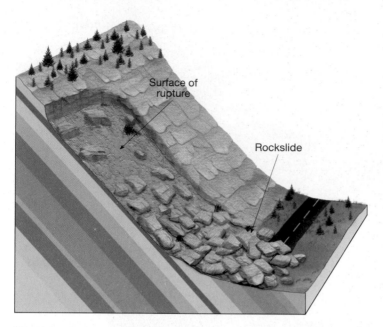

FIGURE 15.12 Rockslides and debris slides are rapid movements that are classified as translational slides in which the material moves along a relatively flat surface with little or no rotation or backward tilting.

underlying surface to the point that friction is no longer sufficient to hold the rock unit in place. As a result, rockslides tend to be more common during the spring, when heavy rains and melting snow are most prevalent.

As was mentioned earlier, earthquakes can trigger rockslides and other mass movements. There are many well-known examples. The 1811 earthquake at New Madrid, Missouri, caused slides in an area of more than 13,000 square kilometers (5000 square miles) along the Mississippi River valley. On August 17, 1959, a severe earthquake west of Yellowstone National Park triggered a massive slide in the canyon of the Madison River in southwestern Montana. In a matter of moments an estimated 27 million cubic meters of rock, soil, and trees slid into the canyon. The debris dammed the river and buried a campground and highway. More than 20 unsuspecting campers perished.

Not far from the site of the Madison Canyon slide, the legendary Gros Ventre rockslide occurred 34 years earlier. The Gros Ventre River flows west from the northernmost part of the Wind River Range in northwestern Wyoming, through Grand Teton National Park, and eventually empties into the Snake River. On June 23, 1925, a massive rockslide took place in its valley, just east of the small town of Kelly. In the span of just a few minutes a great mass of sandstone, shale, and soil crashed down the south side of the valley, carrying with it a dense pine forest. The volume of debris, estimated at 38 million cubic meters (50 million cubic yards), created a 70-meter-high dam on the Gros Ventre River. Because the river was completely blocked, a lake was formed. It filled so quickly that a house that had been 18 meters (60 feet) above the river was floated off its foundation 18 hours after the slide. In 1927 the lake overflowed the dam,

partially draining the lake and resulting in a devastating flood downstream.

Why did the Gros Ventre rockslide take place? Figure 15.13 shows a diagrammatic cross-sectional view of the geology of the valley. You will notice that (1) the sedimentary strata in this area dip (tilt) 15–21 degrees; (2) underlying the bed of sandstone is a relatively thin layer of clay; and (3) at the bottom of the valley the river had cut through much of the sandstone layer. During the spring of 1925, water from heavy rains and melting snow seeped through the sandstone, saturating the clay below. Because much of the sandstone layer had been cut through by the Gros Ventre River, the layer had virtually no support at the bottom of the slope. Eventually the sandstone could no longer hold its position on the wetted clay, and gravity pulled the mass down the side of the valley. The circumstances at this location were such that the event was inevitable.

Debris Flow

 GEODe   Mass Wasting
EARTH   ▸ Types of Mass Wasting

Debris flow is a relatively rapid type of mass wasting that involves a flow of soil and regolith containing a large amount of water. Debris flows, which are also called **mudflows,** are most characteristic of semiarid mountainous regions and are also common on the slopes of some volcanoes. Because of their fluid properties, debris flows frequently follow canyons and stream channels. In populated areas, debris flows can pose a significant hazard to life and property (Figure 15.14).

Debris Flows in Semiarid Regions

When a cloudburst or rapidly melting mountain snows create a sudden flood in a semiarid region, large quantities of soil and regolith are washed into nearby stream channels because there is usually little vegetation to anchor the surface material. The end product is a flowing tongue of well-mixed mud, soil, rock, and water. Its consistency may range from that of wet concrete to a soupy mixture not much thicker than muddy water. The rate of flow therefore depends not only on the slope but also on the water content. When dense, debris flows are capable of carrying or pushing large boulders, trees, and even houses with relative ease.

Debris flows pose a serious hazard to development in relatively dry mountainous areas such as southern California. The construction of homes on canyon hillsides and the removal of native vegetation by brush fires and other means have increased the frequency of these destructive events. Moreover, when a debris flow reaches the end of a steep, narrow canyon, it spreads out, covering the area beyond the mouth of the canyon with a mixture of wet debris. This material contributes to the buildup of fanlike deposits at canyon mouths. The fans are relatively easy to build on, often have nice views, and are close to the mountains; in

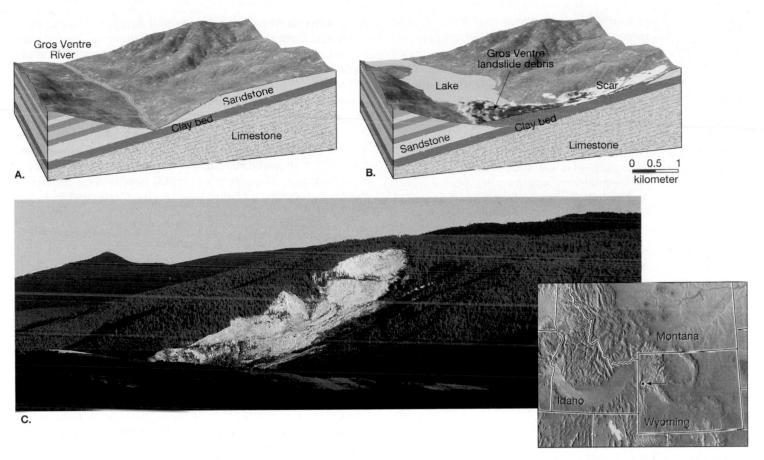

FIGURE 15.13   Parts A and B show a before-and-after cross-sectional view of the Gros Ventre rockslide. The slide occurred when the tilted and undercut sandstone bed could no longer maintain its position atop the saturated bed of clay. As the photo in part C illustrates, even though the Gros Ventre rockslide occurred in 1925, the scar left on the side of Sheep Mountain is still a prominent feature. (Parts A and B after W. C. Alden, "Landslide and Flood at Gros Ventre, Wyoming," Transactions (AIME) 76 (1928): 348: part C photo by Stephen Trimble.)

fact, like the nearby canyons, many have become preferred sites for development. Because debris flows occur only sporadically, the public is often unaware of the potential hazard of such sites (see Box 15.3).

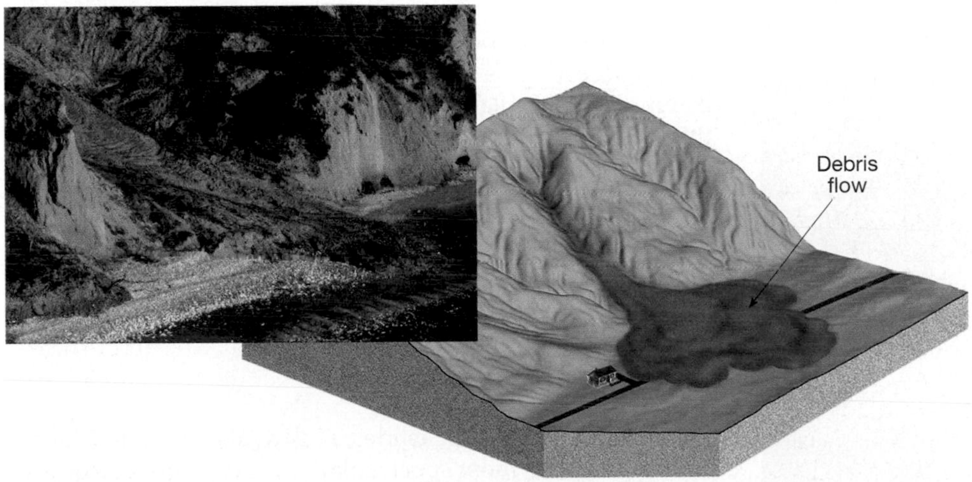

FIGURE 15.14   Debris flow is a moving tongue of well-mixed mud, soil, rock, and water. Its consistency may range from that of wet concrete to a soupy mixture not much thicker than muddy water. (Photo by Tony Waltham)

Lahars

Debris flows composed mostly of volcanic materials on the flanks of volcanoes are called **lahars**. The word originated in Indonesia, a volcanic region that has experienced many of these often destructive events. Historically, lahars have been one of the deadliest volcano hazards. They can occur either during an eruption or when a volcano is quiet. They take place when highly unstable layers of ash and debris become saturated with water and flow down steep volcanic slopes, generally following existing stream channels. Heavy rainfalls often trigger these flows. Others are initiated when large volumes of ice and snow are melted by heat flowing to the surface from within the volcano or by the hot gases and near-molten debris emitted during a violent eruption.

When Mount St. Helens erupted in May 1980, several lahars were created. The flows and accompanying floods raced down the valleys of the north and south forks of the Toutle River at speeds that were often in excess of 30 kilometers (20 miles) per hour. Fortunately, the affected area was not densely settled. Nevertheless, more than 200 homes were destroyed or severely damaged (Figure 15.15). Most bridges met a similar fate. According to the U.S. Geological Survey:

> Even after traveling many tens of miles from the volcano and mixing with cold waters, the mudflows maintained temperatures in the range of about 84° to 91° C; they undoubtedly had higher temperatures closer to the eruption source. . . . Locally the mudflows surged up the valley walls as much as 360 feet and over hills as high as 250 feet. From the evidence left by the "bathtub-ring" mudlines, the larger mudflows at their peak averaged from 33 to 66 feet deep.*

Eventually the lahars in the Toutle River drainage area carried more than 50 million cubic meters of material to the lower Cowlitz and Columbia rivers. The deposits temporarily reduced the water-carrying capacity of the Cowlitz River by 85 percent, and the depth of the Columbia River navigational channel was decreased from 12 meters to less than 4 meters.

In November 1985, lahars were produced during the eruption of Nevado del Ruiz, a 5300-meter (17,400-foot) volcano in the Andes Mountains of Colombia. The eruption melted much of the snow and ice that capped the uppermost 600 meters (2000 feet) of the peak, producing torrents of hot viscous mud, ash, and debris. The lahars moved outward from the volcano, following the valleys of three rainswollen rivers that radiate from the peak. The flow that moved down the valley of the Lagunilla River was the most

---

*Robert I. Tilling, *Eruptions of Mount St. Helens: Past, Present, and Future.* Washington, DC: U.S. Government Printing Office, 1987.

FIGURE 15.15 A house damaged by a lahar along the Toutle River, west-northwest of Mount St. Helens. The end section of the house was torn free and lodged against trees. (Photo by D. R. Crandell, U.S. Geological Survey)

destructive. It devastated the town of Armero, 48 kilometers (30 miles) from the mountain. Most of the more than 25,000 deaths caused by the event occurred in this once thriving agricultural community.

Death and property damage due to the lahars also occurred in 13 other villages within the 180-square-kilometer (70-square-mile) disaster area. Although a great deal of pyroclastic material was explosively ejected from Nevado del Ruiz, it was the lahars triggered by this eruption that made this such a devastating natural disaster. In fact, it was the worst volcanic disaster since 28,000 people died following the 1902 eruption of Mount Pelée on the Caribbean island of Martinique.**

Earthflow

 Mass Wasting
 ▸ **Types of Mass Wasting**

We have seen that debris flows are frequently confined to channels in semiarid regions. In contrast, **earthflows** most often form on hillsides in humid areas during times of heavy precipitation or snowmelt. When water saturates the soil and regolith on a hillside, the material may break away, leaving a scar on the slope and forming a tongue- or teardrop-shaped mass that flows downslope (Figure 15.16).

The materials most commonly involved are rich in clay and silt and contain only small proportions of sand and coarser particles. Earthflows range in size from bodies a few meters long, a few meters wide, and less than a meter deep to masses more than 1 kilometer long, several hundred meters wide, and more than 10 meters deep. Because earthflows are quite viscous, they generally move at slower rates than the more fluid debris flows described in the preceding section. They are characterized by a slow and persistent movement and may remain active for periods ranging from days to years. Depending on the steepness of the slope and the material's consistency, measured velocities range from less than a millimeter a day up to several meters a day. Over the time span that earthflows are active, movement is typically faster during wet periods than during drier times. In addition to occurring as isolated hillside phenomena, earthflows commonly take place in association with large slumps. In this situation, they may be seen as tonguelike flows at the base of the slump block (see Figure 15.11, p. 411).

Slow Movements

 Mass Wasting
 ▸ **Types of Mass Wasting**

Movements such as rockslides, rock avalanches, and lahars are certainly the most spectacular and catastrophic forms of mass wasting. As these events have been known to kill

---

**A discussion of the Mount Pelée eruption can be found in Chapter 5.

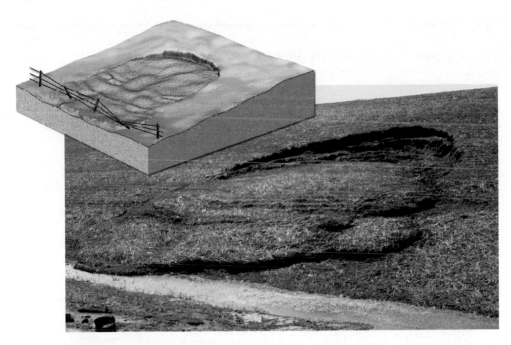

FIGURE 15.16 This small, tongue-shaped earthflow occurred on a newly formed slope along a recently constructed highway. It formed in clay-rich material following a period of heavy rain. Notice the small slump at the head of the earthflow. (Photo by E. J. Tarbuck)

Solifluction

When soil is saturated with water, the soggy mass may flow downslope at a rate of a few millimeters or a few centimeters per day or per year. Such a process is called **solifluction** (literally, "soil flow"). It is a type of mass wasting that is common wherever water cannot escape from the saturated surface layer by infiltrating to deeper levels. A dense clay hardpan in soil or an impermeable bedrock layer can promote solifluction.

Solifluction is also common in regions underlain by *permafrost*. Permafrost refers to the permanently frozen ground that occurs in association with Earth's harsh tundra and ice-cap climates. (There is more about permafrost in the next section.) Solifluction occurs in a zone above the permafrost called the *active layer*, which thaws to a depth of about a meter during the brief high-latitude summer and then refreezes in winter. During the summer season, water is unable to percolate into the impervious permafrost layer below. As a result, the active layer becomes saturated and slowly flows. The process can occur on slopes as gentle as 2 to 3 degrees. Where there is a well-developed mat of vegetation, a solifluction sheet may move in a series of well-defined lobes or as a series of partially overriding folds (Figure 15.19).

The Sensitive Permafrost Landscape

Many of the mass-wasting disasters described in this chapter had sudden and disastrous impacts on people. When the activities of people cause ice contained in permanently frozen ground to melt, the impact is more gradual and less

thousands, they deserve intensive study so that through more effective prediction, timely warnings, and better controls can help save lives. However, because of their large size and spectacular nature, they give us a false impression of their importance as a mass-wasting process. Indeed, sudden movements are responsible for moving less material than the slower and far more subtle action of creep. Whereas rapid types of mass wasting are characteristic of mountains and steep hillsides, creep takes place on both steep and gentle slopes and is thus much more widespread.

Creep

Creep is a type of mass wasting that involves the gradual downhill movement of soil and regolith. One factor that contributes to creep is the alternate expansion and contraction of surface material caused by freezing and thawing or wetting and drying. As shown in Figure 15.17, freezing or wetting lifts particles at right angles to the slope, and thawing or drying allows the particles to fall back to a slightly lower level. Each cycle therefore moves the material a tiny distance downslope. Creep is aided by anything that disturbs the soil, such as raindrop impact and disturbance by plant roots and burrowing animals. Creep is also promoted when the ground becomes saturated with water. Following a heavy rain or snowmelt, a waterlogged soil may lose its internal cohesion, allowing gravity to pull the material downslope. Because creep is imperceptibly slow, the process cannot be observed in action. What can be observed, however, are the effects of creep. Creep causes fences and utility poles to tilt and retaining walls to be displaced (Figure 15.18).

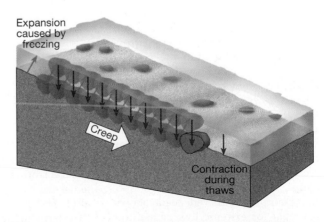

FIGURE 15.17 The repeated expansion and contraction of the surface material causes a net downslope migration of rock particles—a process called *creep.*

BOX 15.3 ▶ PEOPLE AND THE ENVIRONMENT

Debris Flows on Alluvial Fans: A Case Study from Venezuela*

FIGURE 15.E Area of Venezuela affected by disastrous debris flows and flash floods in 1999.

FIGURE 15.F Aerial view of the highly developed alluvial fan at Caraballeda, Venezuela, covered by material from a massive debris flow. (Kimberly White/REUTERS/Corbis/Bettmann)

In December 1999, heavy rains triggered thousands of landslides along the coast of Venezuela (Figure 15.E). Debris flows and flash floods caused severe property damage and the tragic loss of an estimated 19,000 lives. The sites of most of the death and destruction were *alluvial fans.* These landforms are gently sloping, cone- to fan-shaped accumulations of sediment that are commonly found where high-gradient streams leave narrow valleys in mountainous areas and abruptly meet flat terrain.**

Several hundred thousand people live in the narrow coastal zone north of Caracas, Venezuela. They occupy alluvial fans located at the base of steep mountains that rise to elevations of more than 2000 meters (6600 feet) because these sites are the only areas that are not too steep to build on (Figure 15.F). Such settings are highly vulnerable to rainfall-induced landslides.

An unusually wet period in December 1999 included rains of 20 centimeters (8 inches) on December 2 and 3, followed by an additional 91 centimeters (36 inches) between December 14 and 16. The heavy rains triggered thousands of debris flows and other types of mass wasting. Once created, these moving masses of mud and rock coalesced to form giant debris flows that moved rapidly through steep, narrow canyons before exiting onto the alluvial fans.

On virtually every alluvial fan in the area, debris flows and flash floods brought massive amounts of sediment, including boulders as large as 10 meters (33 feet) in diameter. Hundreds of houses and other structures were damaged or destroyed (Figure 15.G). Total damages approached $2 billion.

This example from Venezuela shows the potential for extreme loss of life and property damage where large numbers of people occupy alluvial fans. The possibili-

FIGURE 15.G Debris-flow damage. Huge boulders (in excess of 300 tons) were transported by some flows. (AP/Wide World Photo)

ty for similar events of comparable magnitude exists in other parts of the world.

Building communities on alluvial fans can transform natural processes into major lethal events. Kofi Annan, Secretary General of the United Nations, put it this way: "The term 'natural disaster' has become an increasingly anachronistic misnomer. In reality, human behavior transforms natural

hazards into what should really be called unnatural disasters."***

---

*Based on material prepared by the U.S. Geological Survey.

**For more about alluvial fans, see p. 443 in Chapter 16 and p. 523 in Chapter 19.

***Matthew C. Larsen, et al., *Natural Hazards on Alluvial Fans: The Venezuela Debris Flow and Flash Flood Disaster,* U.S. Geological Survey Fact Sheet FS 103, p. 4.

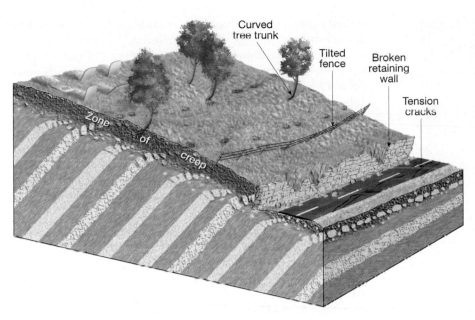

FIGURE 15.18 Although creep is an imperceptibly slow movement, its effects are often visible.

deadly. Nevertheless, because permafrost regions are sensitive and fragile landscapes, the scars resulting from poorly planned actions can remain for generations.

Permanently frozen ground, known as **permafrost**, occurs where summers are too cool to melt more than a shallow surface layer. Deeper ground remains frozen year-round. Strictly speaking, permafrost is defined only on the basis of temperature; that is, it is ground with temperatures that have remained below 0°C (32°F) continuously for two years or more. The degree to which ice is present in the ground strongly affects the behavior of the surface material. Knowing how much ice is present and where it is located is very important when it comes to constructing roads, buildings, and other projects in areas underlain by permafrost.

Permafrost is extensive in the lands surrounding the Arctic Ocean. It covers more than 80 percent of Alaska, about 50 percent of Canada, and a substantial portion of northern Siberia (Figure 15.20). Near the southern margins of the region, the permafrost consists of relatively thin, isolated masses. Farther north, the area and thickness gradually increase to the point where the permafrost is essentially continuous and its thickness may approach or even exceed 500 meters. In the discontinuous zone, land-use planning is frequently more difficult than in the continuous zone farther north because the occurrences of permafrost are patchy and difficult to predict.

When people disturb the surface, such as by removing the insulating vegetation mat or by building roads and buildings, the delicate thermal balance is disturbed, and the permafrost can thaw (Figure 15.21A). Thawing produces unstable ground that may slide, slump, subside, and undergo severe frost heaving. When a heated structure is built directly on permafrost that contains a high proportion of ice, thawing creates soggy material into which a building can sink. One solution is to place buildings and other structures on piles, like stilts. Such piles allow subfreezing air to circulate between the floor of the building and the soil and thereby keep the ground frozen.

When oil was discovered on Alaska's North Slope, many people were concerned about the building of a pipeline linking the oil fields at Prudhoe Bay to the ice-free port of Valdez 1300 kilometers to the south. There was serious concern that such a massive project might damage the sensitive permafrost environment. Many also worried about possible oil spills.

Because oil must be heated to about 60°C to flow properly, special engineering procedures had to be developed to isolate this heat from the permafrost. Methods included insulating the pipe, elevating portions of the pipeline above ground level, and even placing cooling devices in the ground to keep it frozen (Figure 15.21B). The Alaskan pipeline is clearly one of the most complex and costly projects ever built in the Arctic tundra. Detailed studies and careful engineering helped minimize adverse effects resulting from the disturbance of frozen ground.

Submarine Landslides

As you might imagine, mass-wasting processes are not confined to land. The development of high-quality instruments that perform ocean-floor imaging has allowed us to determine that submarine mass wasting is a common and widespread phenomenon. For example, studies reveal enormous submarine landslides on the flanks of the Hawaiian chain as well as along the continental shelf and slope of the U.S. mainland. In

FIGURE 15.19 Solifluction lobes northeast of Fairbanks, Alaska. Solifluction occurs in permafrost regions when the active layer thaws in summer. (Photo by James E. Patterson)

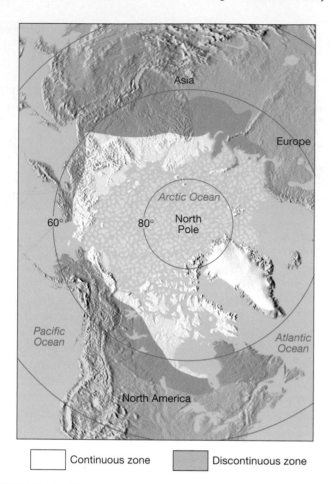

Continuous zone   Discontinuous zone

FIGURE 15.20 Distribution of permafrost in the Northern Hemisphere. More than 80 percent of Alaska and about 50 percent of Canada are underlain by permafrost. Two zones are recognized. In the continuous zone, the only ice-free areas are beneath deep lakes or rivers. In the higher-latitude portions of the discontinuous zone, there are only scattered islands of thawed ground. Moving southward, the percentage of unfrozen ground increases until all the ground is unfrozen. (After the U.S. Geological Survey)

fact, many submarine landslides, mostly in the form of slumps and debris avalanches, appear to be much larger than any similar mass-wasting events that occur on land.

Among the most spectacular underwater landslides are those that occur on the flanks of submarine volcanoes

(called *seamounts*) and volcanic islands such as Hawaii. On the submerged flanks of the Hawaiian Islands dozens of major landslides more than 20 kilometers (13 miles) long have been identified. Some have truly spectacular dimensions. One of the largest yet mapped, known as the Nuuanu debris avalanche, is on the northeastern side of Oahu. It extends for nearly 25 kilometers (15 miles) across the ocean floor, then rises up a 300-meter (nearly 1000-foot) slope at its terminus, indicating that it must have had great power and momentum. Giant blocks many kilometers across were transported by this giant landslide. It is probable that when such large and rapid events occur, they produce giant sea waves called tsunami that race across the Pacific.*

The massive submarine slides discovered on the flanks of the Hawaiian Islands are almost certainly related to the movement of magma while a volcano is active. As huge quantities of lava are added to the seaward margin of a volcano, the buildup of material eventually triggers a great landslide. In the Hawaiian chain, it appears that this process of growth and collapse is repeated at intervals of from 100,000 to 200,000 years while the volcano is active.

Along the continental margins of the U.S. mainland, large slumps and debris-flow scars mark the continental slope. These processes are triggered by the rapid buildup of unstable sediments or by such forces as storm waves and earthquakes. Submarine mass wasting is especially active near deltas, which are massive deposits of sediment at the mouths of rivers. Here, as great loads of water-saturated clay and organic-rich sediments accumulate, they become unstable and readily flow down even gentle slopes. Some of these movements have been forceful enough to damage large offshore drilling platforms.

Mass wasting appears to be an integral part of the growth of passive continental margins. Sediments supplied to the continental shelf by rivers move across the shelf to the upper continental slope. From here slumps, slides, and debris flows move sediment down to the continental rise and sometimes beyond.

_____

*For more on these destructive waves, see the section on tsunamis in Chapter 11.

FIGURE 15.21 **A.** When a rail line was built across this permafrost landscape in Alaska, the ground subsided. (Photo by Lynn A. Yehle, U.S. Geological Survey) **B.** In places in Alaska, a pipeline is suspended above ground to prevent melting of delicate permafrost. (Tom & Pat Lesson/Photo Researchers)

A.

B.

Summary

- *Mass wasting* refers to the downslope movement of rock, regolith, and soil under the direct influence of gravity. In the evolution of most landforms, mass wasting is the step that follows weathering. The combined effects of mass wasting and erosion by running water produce stream valleys.

- *Gravity is the controlling force of mass wasting.* Other factors that influence or trigger downslope movements are saturation of the material with water, oversteepening of slopes beyond the *angle of repose*, removal of vegetation, and ground shaking by earthquakes.

- The various processes included under the name of mass wasting are divided and described on the basis of (1) the type of material involved (debris, mud, earth, or rock); (2) the type of motion (fall, slide, or flow); and (3) the rate of movement (rapid or slow).

- The more rapid forms of mass wasting include *slump*, the downward sliding of a mass of rock or unconsolidated material moving as a unit along a curved surface; *rockslide*, blocks of bedrock breaking loose and sliding downslope; *debris flow*, a relatively rapid flow of soil and regolith containing a large amount of water; and *earthflow*, an unconfined flow of saturated clay-rich soil that most often occurs on a hillside in a humid area following heavy precipitation or snowmelt.

- The slowest forms of mass wasting include *creep*, the gradual downhill movement of soil and regolith; and *solifluction*, the gradual flow of a saturated surface layer that is underlain by an impermeable zone. Common sites for solifluction are regions underlain by *permafrost* (permanently frozen ground associated with tundra and ice-cap climates).

- *Permafrost*, permanently frozen ground, covers large portions of North America and Siberia. Thawing produces unstable ground that may slide, slump, subside, and undergo severe frost heaving.

- Mass wasting is not confined to land; it also occurs underwater. Many *submarine landslides*, mostly slumps and debris avalanches, are much larger than those that occur on land.

Review Questions

1. Describe how mass-wasting processes contribute to the development of stream valleys.

2. What is the controlling force of mass wasting?

3. How does water affect mass-wasting processes?

4. Describe the significance of the angle of repose.

5. How might the removal of vegetation by fire or logging promote mass wasting?

6. How are earthquakes linked to landslides?

7. How did the building of a dam contribute to the Vaiont Canyon disaster? Was the disaster avoidable? (See Box 15.2, p. 409).

8. Distinguish among fall, slide, and flow.

9. Why can rock avalanches move at such great speeds?

10. Both slump and rockslide move by sliding. In what ways do these processes differ?

11. What factors led to the massive rockslide at Gros Ventre, Wyoming?

12. Explain why building a home on an alluvial fan might not be a good idea (see Box 15.3, p. 416).

13. Compare and contrast mudflow and earthflow.

14. Describe the mass wasting that occurred at Mount St. Helens during its active period in 1980 and at Nevado del Ruiz in 1985.

15. Because creep is an imperceptibly slow process, what evidence might indicate that this phenomenon is affecting a slope?

16. What is permafrost? What portion of Earth's land surface is affected?

17. During what season does solifluction occur in permafrost regions?

Key Terms

angle of repose (p. 404)
creep (p. 415)
debris flow (p. 412)
debris slide (p. 411)

earthflow (p. 414)
fall (p. 408)
flow (p. 410)
lahar (p. 413)

mass wasting (p. 402)
mudflow (p. 412)
permafrost (p. 417)
rock avalanche (p. 410)

rockslide (p. 411)
slide (p. 410)
slump (p. 410)
solifluction (p. 415)

Web Resources

 The *Earth* Website uses the resources and flexibility of the Internet to aid in your study of the topics in this chapter. Written and developed by geology instructors, this site will help improve your understanding of geology. Visit **http://www.prenhall.com/tarbuck** and click on the cover of *Earth 9e* to find:

• Online review quizzes.
• Critical thinking exercises.
• Links to chapter-specific Web resources.
• Internet-wide key-term searches.

http://www.prenhall.com/tarbuck

GEODe: Earth

GEODe: Earth makes studying faster and more effective by reinforcing key concepts using animation, video, narration, interactive exercises and practice quizzes. A copy is included with every copy of *Earth.*

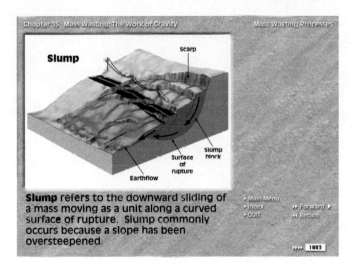

Running Water

Rainbow Falls on the Middle Fork of the San Joaquin River, Devil's Postpile National Monument, California. (Photo by Carr Clifton)

R ivers are very important to people. We use them as highways for moving goods, as sources of water for irrigation, and as an energy source. Their fertile floodplains have been cultivated since the dawn of civilization. When viewed as part of the Earth system, rivers and streams represent a basic link in the constant cycling of the planet's water. Moreover, running water is the dominant agent of landscape alteration, eroding more terrain and transporting more sediment than any other process. Because so many people live near rivers, floods are among the most destructive of all geologic hazards. Despite huge investments in levees and dams, rivers cannot always be controlled (Figure 16.1).

Earth As a System: The Hydrologic Cycle

 Running Water
▸ Hydrologic Cycle

All the rivers run into the sea;
yet the sea is not full;
unto the place from whence the rivers come, thither they
return again.   (Ecclesiastes 1:7)

As the perceptive writer of Ecclesiastes indicated, water is continually on the move, from the ocean to the land and back again in an endless cycle. Water is just about everywhere on Earth—in the oceans, glaciers, rivers, lakes, the air, soil, and in living tissue. All of these "reservoirs" constitute Earth's hydrosphere. In all, the water content of the hydrosphere is about 1.36 billion cubic kilometers (326 million cubic miles).

The vast bulk of it, about 97.2 percent, is stored in the global oceans (Figure 16.2). Ice sheets and glaciers account for another 2.15 percent, leaving only 0.65 percent to be

FIGURE 16.1 Aerial view of a portion of Portland, Pennsylvania, flooded by the Delaware River in late June 2006. Extremely heavy rainfall during June 24-28 caused record or near-record flood crests along many streams throughout the Delaware River basin. (AP Photo/*Pocono Record*/David Kidwell) (Associated Press Photo)

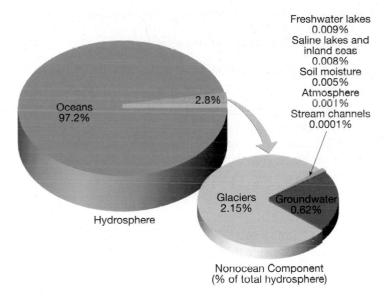

Freshwater lakes
0.009%
Saline lakes and
inland seas
0.008%
Soil moisture
0.005%
Atmosphere
0.001%
Stream channels
0.0001%

Oceans
97.2%

2.8%

Hydrosphere

Glaciers
2.15%

Groundwater
0.62%

Nonocean Component
(% of total hydrosphere)

FIGURE 16.2   Distribution of Earth's water.

The water found in each of the reservoirs depicted in Figure 16.2 does not remain in these places indefinitely. Water can readily change from one state of matter (solid, liquid, or gas) to another at the temperatures and pressures that occur at Earth's surface. Therefore, water is constantly moving among the hydrosphere, the atmosphere, the geosphere, and the biosphere. This unending circulation of Earth's water supply is called the **hydrologic cycle**. The cycle shows us many critical interrelationships among different parts of the Earth system.

The hydrologic cycle is a gigantic worldwide system powered by energy from the Sun in which the atmosphere provides the vital link between the oceans and continents (Figure 16.3). Water evaporates into the atmosphere from the ocean and to a much lesser extent from the continents. Winds transport this moisture-laden air often great distances until conditions cause the moisture to condense into clouds and precipitation to fall. The precipitation that falls into the ocean has completed its cycle and is ready to begin another. The water that falls on land, however, must make its way back to the ocean.

What happens to precipitation once it has fallen on the land? A portion of the water soaks into the ground (called **infiltration**) moving downward, then laterally, and finally seeping into lakes, streams, or directly into the ocean. When

divided among lakes, streams, subsurface water, and the atmosphere (Figure 16.2). Although the percentages of Earth's water found in each of the latter sources is just a small fraction of the total inventory, the absolute quantities are great.

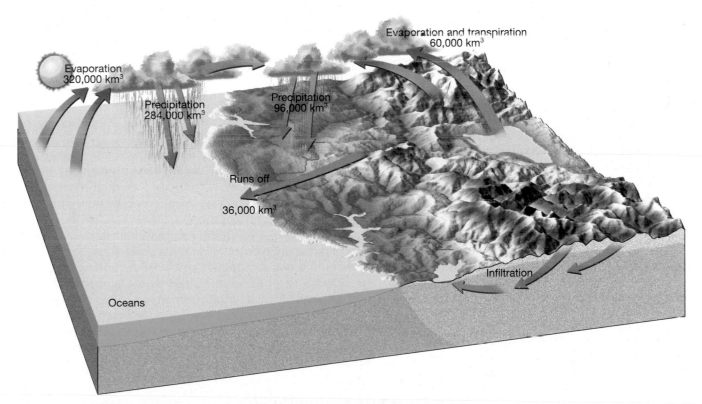

Evaporation and transpiration
60,000 km³

Evaporation
320,000 km³

Precipitation
284,000 km³

Precipitation
96,000 km³

Runs off

36,000 km³

Infiltration

Oceans

FIGURE 16.3   Earth's water balance. Each year, solar energy evaporates about 320,000 cubic kilometers of water from the oceans, while evaporation from the land (including lakes and streams) contributes 60,000 cubic kilometers of water. Of this total of 380,000 cubic kilometers of water, about 284,000 cubic kilometers fall back to the ocean, and the remaining 96,000 cubic kilometers fall on the land surface. Of that 96,000 cubic kilometers, only 60,000 cubic kilometers of water return to the atmosphere by evaporation and transpiration, leaving 36,000 cubic kilometers of water to erode the land during the journey back to the oceans.

the rate of rainfall is greater than the land's ability to absorb it, the additional water flows over the surface into lakes and streams, a process called **runoff.** Much of the water that infiltrates or runs off eventually finds its way back to the atmosphere because of evaporation from the soil, lakes, and streams. Also, some of the water that infiltrates the ground is absorbed by plants, which later release it into the atmosphere. This process is called **transpiration** (*trans* = across, *spiro* = to breathe). The term **evapotranspiration** is used to refer to the loss of water from land areas through the combined effect of evaporation and transpiration.

When precipitation falls in very cold areas—at high elevations or high latitudes—the water may not immediately soak in, run off, or evaporate. Instead, it may become part of a snowfield or a glacier. In this way, glaciers store large quantities of water on land. If present-day glaciers were to melt and release their stored water, sea level would rise by several tens of meters worldwide and submerge many heavily populated coastal areas. As you will see in Chapter 18, over the past 2 million years, huge ice sheets have formed and melted on several occasions, each time changing the balance of the hydrologic cycle.

Figure 16.3 also shows Earth's overall *water balance,* or the volume of water that passes through each part of the cycle annually. The amount of water vapor in the air is just a tiny fraction of Earth's total water supply. But the absolute quantities that are cycled through the atmosphere over a one-year period are immense—some 380,000 cubic kilometers. Estimates show that over North America almost six times more water is carried by moving currents of air than is transported by all the continent's rivers.

It is important to know that the hydrologic cycle is balanced. Because the total water vapor in the atmosphere remains about the same, average annual precipitation over Earth must be equal to the quantity of water evaporated. However, for all of the continents taken together, precipitation exceeds evaporation. Conversely, over the oceans, evaporation exceeds precipitation. Since the level of the world ocean is not dropping, the system must be in balance.

The erosional work accomplished by the 36,000 cubic kilometers of water that flows annually from the land to the ocean is enormous. Arthur Bloom effectively described it as follows:

> The average continental height is 823 meters above sea level. . . . If we assume that the 36,000 cubic kilometers of annual runoff flow downhill an average of 823 meters, the potential mechanical power of the system can be calculated. Potentially, the runoff from all lands would continuously generate almost 9×10^9 kW. If all this power were used to erode the land, it would be comparable to having . . . one horse-drawn scraper or scoop at work on each 3-acre piece of land, day and night, year round. Of course, a large part of the potential energy of runoff is wasted as frictional heat by turbulent flow and splashing of water.*

Geomorphology: A Systematic Analysis of Late Cenozoic Landforms (Englewood Cliffs, N.J.: Prentice Hall, 1978), p. 97.

Although only a small percentage of the energy of running water is used to erode the surface, running water nevertheless is *the single most important agent sculpturing Earth's land surface.*

To summarize, the hydrologic cycle represents the continuous movement of water from the oceans to the atmosphere, from the atmosphere to the land, and from the land back to the sea. The wearing down of Earth's land surface is largely attributable to the last of these steps and is the primary focus of the remainder of this chapter.

Running Water

 GEODe EARTH **Running Water**
▸ **Stream Characteristics**

Although we have always depended on running water, its source eluded us for centuries. Not until the 1500s did we realize that streams were supplied by surface runoff and underground water, which ultimately had their sources as rain and snow.

Runoff initially flows in broad, thin sheets across the ground, appropriately termed **sheet flow.** The amount of water that runs off in this manner rather than sinking into the ground depends upon the **infiltration capacity** of the soil. Infiltration capacity is controlled by many factors, including: (1) the intensity and duration of the rainfall, (2) the prior wetted condition of the soil, (3) the soil texture, (4) the slope of the land, and (5) the nature of the vegetative cover. When the soil becomes saturated, sheet flow begins as a layer only a few millimeters thick. After flowing as a thin, unconfined sheet for only a short distance, threads of current typically develop and tiny channels called **rills** begin to form and carry the water to a stream. At first the streams are small, but as one intersects another, larger and larger ones form. Eventually rivers develop that carry water from a broad region to the ocean.

Drainage Basins

The land area that contributes water to a river system is called a **drainage basin.** The drainage basin of one stream is separated from another by an imaginary line called a **divide** (Figure 16.4). Divides range in scale from a ridge separating two small gullies on a hillside to a *continental divide,* which

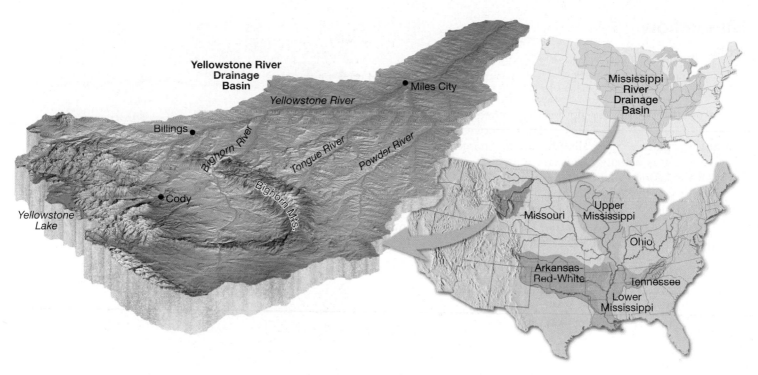

FIGURE 16.4 A *drainage basin* is the land area drained by a stream and its tributaries. *Divides* are the boundaries separating drainage basins. Drainage basins and divides exist for all streams, regardless of size. The drainage basin of the Yellowstone River is one of many that contribute water to the Missouri River, which in turn, is one of many that make up the drainage basin of the Mississippi River. The drainage basin of the Mississippi River, North America's largest, covers about 3 million square kilometers.

splits entire continents into enormous drainage basins. The Mississippi River has the largest drainage basin in North America. Extending between the Rocky Mountains in the West and the Appalachian Mountains in the East, the Mississippi River and its tributaries collect water from more than 3.2 million square kilometers (1.2 million square miles) of the continent.

River Systems

Rivers and streams can be simply defined as water flowing in a channel. They have three important roles in the formation of a landscape: They erode the channels in which they flow, they transport materials provided by weathering and slope processes, and they produce a wide variety of erosional and depositional landforms. In fact, in most areas, including many arid regions, river systems have shaped the varied landscapes that we humans inhabit.

A river system consists of three main parts in which different processes dominate: a zone of erosion, a zone of sediment transport, and a zone of sediment deposition. It is important to realize that some erosion, transport, and deposition occur in all three zones; however, within each zone, one of these processes is usually dominant. In addition, the parts of a river system are interdependent, so that the processes occurring in one part influence the others.

In large river systems erosion is the dominant process in the upstream area, which generally consists of mountainous or hilly topography. Here small tributary streams erode the channels in which they flow and carry material provided by weathering and mass wasting.

The region within a river system that is dominated by deposition is usually located where the stream enters a large body of water. Here sediments are moved offshore by currents, accumulate to form a delta, or are reworked by wave action to form a variety of coastal features. Between the zones of erosion and deposition is the *trunk stream* that serves to transport sediments. Taken together, erosion, transportation, and deposition are the processes by which rivers move Earth's surface materials and sculpt landscapes.

Students Sometimes Ask . . .

What's the difference between a stream and a river?

In common usage, these terms imply relative size (a river is larger than a stream, both of which are larger than a creek or a brook). However, in geology this is not the case: The word stream is used to denote channelized flow of any size, from a small creek to the mightiest river. It is important to note that although the terms river and stream are sometimes used interchangeably, the term river is often preferred when describing a main stream into which several tributaries flow.

Streamflow

Running Water
▶ Stream Characteristics

Water may flow in one of two ways, either as **laminar flow** or **turbulent flow.** In very slow moving streams the flow is often laminar and the water particles move in roughly straight-line paths that parallel the stream channel. However, streamflow is usually turbulent, with the water moving in an erratic fashion that can be characterized as a swirling motion. Strong, turbulent flow may exhibit whirlpools and eddies, as well as rolling whitewater rapids. Even streams that appear smooth on the surface often exhibit turbulent flow near the bottom and sides of the channel. Turbulence contributes to the stream's ability to erode its channel because it acts to lift sediment from the streambed.

Water makes its way to the sea under the influence of gravity. The time required for the journey depends on the velocity of the stream. Velocity is the distance that water travels in a unit of time.

Some sluggish streams flow at less than 1 kilometer per hour, whereas a few rapid ones may exceed 30 kilometers per hour. Velocities are determined at gaging stations where measurements are taken at several locations across the stream channel and then averaged. This is done because the rate of water movement is not uniform within a stream channel. When the channel is straight, the highest velocities occur in the center just below the surface (Figure 16.5). It is here that friction is least. Minimum velocities occur along the sides and bottom (bed) of the channel where friction is always greatest. When a stream channel is crooked or curved, the fastest flow is not in the center. Rather, the zone of maximum velocity shifts toward the outside of each bend. As we shall see later, this shift plays an important part in eroding the stream's channel on that side.

The ability of a stream to erode and transport material is directly related to its velocity. Even slight variations in velocity can lead to significant changes in the load of sediment that water can transport. Several factors determine the velocity of a stream and therefore control the amount of erosional work a stream may accomplish. These factors include (1) the gradient; (2) the shape, size, and roughness of the channel; and (3) the discharge.

Gradient and Channel Characteristics

The slope of a stream channel expressed as the vertical drop of a stream over a specified distance is **gradient.** Portions of the lower Mississippi River, for example, have very low gradients of 10 centimeters per kilometer or less. By contrast, some mountain stream channels decrease in elevation at a rate of more than 40 meters per kilometer, or a gradient 400 times steeper than the lower Mississippi (Figure 16.6). Gradient varies not only among different streams but also over a particular stream's length. The steeper the gradient the more energy available for streamflow. If two streams

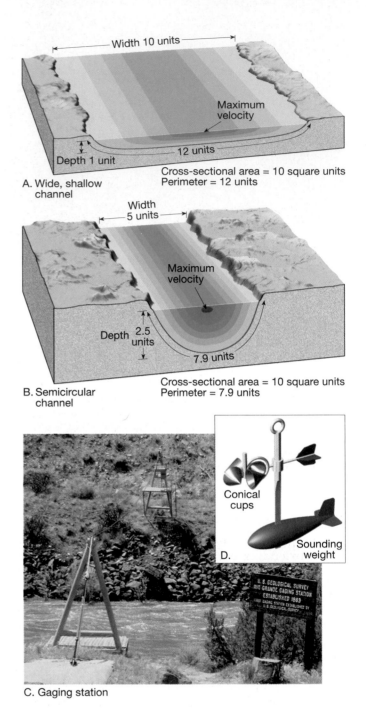

FIGURE 16.5 Influence of channel shape on velocity. **A.** The stream in this wide, shallow channel moves more slowly than does water in the semicircular channel because of greater frictional drag. **B.** The cross-sectional area of this semicircular channel is the same as the one in part A, but it has less water in contact with its channel and therefore less frictional drag. Thus, water will flow more rapidly in channel B, all other factors being equal. **C.** Continuous records of stage and discharge are collected by the U.S. Geological Survey at more than 7000 gaging stations in the United States. Average velocities are determined by using measurements from several spots across the stream. This station is on the Rio Grande south of Taos, New Mexico. (Photo by E. J. Tarbuck) **D.** Current meter used to measure stream velocity at a gaging station.

roughness is obvious. A smooth channel promotes a more uniform flow, whereas an irregular channel filled with boulders creates enough turbulence to significantly retard the stream's forward motion.

Discharge

The **discharge** of a stream is the volume of water flowing past a certain point in a given unit of time. This is usually measured in cubic meters per second or cubic feet per second. Discharge is determined by multiplying a stream's cross-sectional area by its velocity:

$$\text{discharge}(\text{m}^3/\text{second}) = \\ \text{channel width (meters)} \\ \times \text{channel depth (meters)} \\ \times \text{velocity (meters/second)}$$

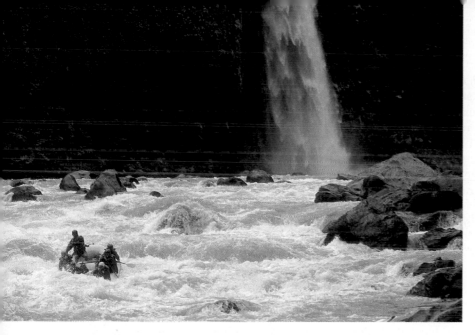

FIGURE 16.6   Rapids are common in mountain streams where the gradient is steep and the channel is rough and irregular. Although most streamflow is turbulent, it is usually not as rough as that experienced by these river runners at Lost Yak Rapids on Chile's Rio Bio Bio. (Photo © by Carr Clifton)

were identical in every respect except gradient, the stream with the higher gradient would obviously have the greater velocity.

A stream's channel is a conduit that guides the flow of water, but the water encounters friction as it flows. The *cross-sectional shape* of a channel determines the amount of water in contact with the channel and hence affects the frictional drag. The most efficient channel is one with the least perimeter for its cross-sectional area. Figure 16.5 compares two channel shapes. Although the cross-sectional area of both is identical, the semicircular shape has less water in contact with the channel and therefore less frictional drag. As a result, if all other factors are equal, the water will flow more rapidly in the semicircular channel.

The size and roughness of the channel also affect the amount of friction. An increase in the size of a channel reduces the ratio of perimeter to cross-sectional area and therefore increases the efficiency of flow. The effect of

Table 16.1 lists the world's largest rivers in terms of discharge. The largest river in North America, the Mississippi, discharges an average 17,300 cubic meters per second. Although this is a huge quantity of water, it is nevertheless dwarfed by the mighty Amazon, the world's largest river. Draining an area that is nearly three-quarters the size of the conterminous United States and that averages about 200 centimeters of rain per year, the Amazon discharges 12 times more water than the Mississippi. In fact, it has been estimated that the flow of the Amazon accounts for about 15 percent of all the fresh water discharged into the ocean by all of the world's rivers. Just one day's discharge would supply the water needs of New York City for nine years!

The discharges of most rivers are far from constant. This is true because of such variables as rainfall and snowmelt. When discharge changes, the factors noted earlier must also change. When discharge increases, the width or depth of the channel must increase or the water must flow faster, or some

			Drainage Area		Average Discharge	
Rank	River	Country	Square kilometers	Square miles	Cubic meters	Cubic feet
1	Amazon	Brazil	5,778,000	2,231,000	212,400	7,500,000
2	Congo	Zaire	4,014,500	1,550,000	39,650	1,400,000
3	Yangtze	China	1,942,500	750,000	21,800	770,000
4	Brahmaputra	Bangladesh	935,000	361,000	19,800	700,000
5	Ganges	India	1,059,300	409,000	18,700	660,000
6	Yenisei	Russia	2,590,000	1,000,000	17,400	614,000
7	Mississippi	United States	3,222,000	1,244,000	17,300	611,000
8	Orinoco	Venezuela	880,600	340,000	17,000	600,000
9	Lena	Russia	2,424,000	936,000	15,500	547,000
10	Parana	Argentina	2,305,000	890,000	14,900	526,000

TABLE 16.1   World's Largest Rivers Ranked by Discharge

combination of these factors must change. Indeed, measurements show that when the amount of water in a stream increases, the width, depth, and velocity all increase in an orderly fashion (Figure 16.7). To handle the additional water, the stream will increase the size of its channel by widening and deepening it. As we saw earlier, when the size of the channel increases, proportionally less of the water is in contact with the bed and banks of the channel. This means that friction, which acts to retard the flow, is reduced. The less friction, the more swiftly the water will flow.

Changes from Upstream to Downstream

GEODe Running Water
EARTH ▸ Stream Characteristics

One useful way of studying a stream is to examine its **longitudinal profile.** Such a profile is simply a cross-sectional view of a stream from its source area (called the **head** or **headwaters**) to its **mouth,** the point downstream where the

river empties into another water body. By examining Figure 16.8, you can see that the most obvious feature of a typical longitudinal profile is a constantly decreasing gradient from the head to the mouth. Although many local irregularities may exist, the overall profile is a relatively smooth concave-upward curve.

The longitudinal profile shows that the gradient decreases downstream. To see how other factors change in a downstream direction, observations and measurements must be made. When data are collected from successive gaging stations along a river, they show that in humid regions discharge increases toward the mouth. This should come as no surprise because, as we move downstream, more and more tributaries contribute water to the main channel. In the case of the Amazon, for example, about 1000 tributaries join the main river along its 6500-kilometer course across South America.

Furthermore, in most humid regions additional water is added from the groundwater supply. Because this is the case, the width, depth, and velocity all must change in response to the increased volume of water carried by the stream. Indeed, the downstream changes in these variables have been shown to vary in a manner similar to what occurs when discharge increases at one place; that is, width, depth, and velocity all increase.

The observed increase in average velocity that occurs downstream contradicts our intuitive impressions concerning wild, turbulent mountain streams and wide, placid rivers in the lowlands. The mountain stream has much higher instantaneous, turbulent velocities, but the water moves vertically, laterally, and actually upstream in some cases. Thus, the average flow velocity may be lower than in a wide, placid river just "rollin' along" very efficiently with far less turbulence.

In the headwaters region where the gradient is steepest, the water must flow in a relatively small and often boulder-strewn channel. The small channel and rough bed create great drag and inhibit movement by sending water in all

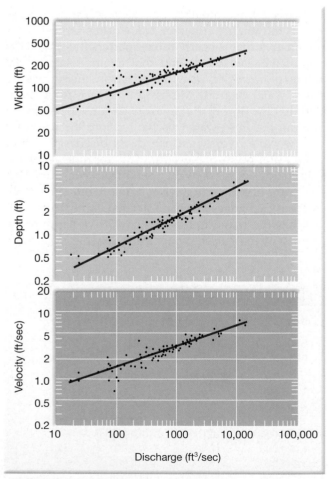

FIGURE 16.7 Relationship of width, depth, and velocity to discharge of the Powder River at Locate, Montana. As discharge increases, width, depth, and velocity all increase in an orderly fashion. (After L. B. Leopold and Thomas Maddock, Jr., U.S. Geological Survey Professional Paper 252, 1953.)

Students Sometimes Ask . . .

The weather report where I live usually includes information about the stage of the river that runs through the area. What exactly is the "stage"?

It is one of the basic measurements made at each of the more than 7000 river gaging stations in the United States. Stage is simply the height of the water surface relative to an arbitrarily fixed reference point. The measurement is commonly made in a gage house. This structure consists of a well dug along the riverbank with a surrounding shelter that protects the equipment inside. Water enters or leaves through one or more pipes that allow the water in the well to rise or fall to the same level as the river. Recording equipment in the gage house records the water level in the well (the river stage). Recorded data can then be accessed by telephone or transmitted via satellite. Data are used to issue flood warnings, among other things.

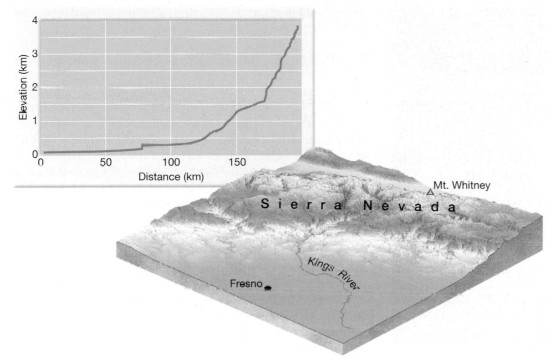

FIGURE 16.8 Longitudinal profile of California's Kings River. It originates in the Sierra Nevada and flows westward into the San Joaquin Valley. A longitudinal profile is a cross-section along the length of a stream. Note the concave-upward curve of the profile, with a steeper gradient upstream and a gentler gradient downstream.

When the ground is saturated, rainwater cannot infiltrate, so it flows downslope, transporting some of the material it has dislodged. On barren slopes the flow of muddy water (sheet flow) often produces small channels (rills), which in time often evolve into larger gullies (see Figure 6.24 on p. 186).

Once the surface flow reaches a stream, its ability to erode is greatly enhanced by the increase in water volume. When the flow of water is sufficiently strong, it can dislodge particles from the channel and lift them into the moving water. In this manner, the force of running water swiftly erodes poorly consolidated materials on the bed and sides of a stream channel. On occasion, the banks of the channel may be undercut, dumping even more loose debris into the water to be carried downstream.

Observing a muddy stream demonstrates that currents of water can lift and carry debris. However, it is not as obvious that a stream is capable of eroding solid rock in a manner similar to sandpapering. Just as the particles of grit on sandpaper can wear down a piece of wood, so too can the sand and gravel carried by a stream abrade a bedrock channel. Many steep-sided gorges that cut through solid rock by the ceaseless bombardment of particles against the bed and banks of a channel serve as testimony to this erosional strength (Figure 16.9). In addition, the individual sediment grains are also abraded by their many impacts with the channel and with one another. Thus, by scraping, rubbing, and bumping, abrasion erodes a bedrock channel and simultaneously smoothes and rounds the abrading particles.

Common features on some river beds are rounded depressions known as **potholes** (Figure 16.10). They are created by the abrasive action of particles swirling in fast-moving eddies. The rotational motion of the sand and pebbles acts like a drill to bore the holes. As the particles wear down to nothing, they are replaced by new ones that continue to drill the stream bed. Eventually, smooth depressions several meters across and just as deep may result.

directions with almost as much backward motion as forward motion. However, as one progresses downstream, the material on the bed of the stream becomes much smaller, offering less resistance to flow, and the width and depth of the channel increase to accommodate the greater discharge. These factors, especially the wider and deeper channel, permit the water to flow more freely and hence more rapidly.

In summary, we have seen that there is an inverse relationship between gradient and discharge. Where the gradient is high, the discharge is small, and where the discharge is great, the gradient is small. Stated another way, a stream can maintain a higher velocity near its mouth even though it has a lower gradient than upstream because of the greater discharge, larger channel, and smoother bed.

The Work of Running Water

Streams are Earth's most important erosional agent. Not only do they have the ability to downcut and widen their channels, but streams also have the capacity to transport the enormous quantities of sediment that are delivered to them by sheet flow, mass wasting, and groundwater. Eventually much of this material is deposited to create a variety of features.

Stream Erosion

A stream's ability to accumulate and transport soil and weathered rock is aided by the work of raindrops, which knock sediment particles loose (see Figure 6.23 on p. 185).

Transport of Sediment by Streams

All streams, regardless of size, transport some rock material. Streams also sort the solid sediment they transport because finer, lighter material is carried more readily than larger heavier particles. Streams transport their load of sediment in three ways: (1) in solution (**dissolved load**), (2) in suspension

FIGURE 16.9 This steep-sided gorge in Zion National Park, Utah, has been cut through solid rock by stream erosion. Canyons such as this are narrow due to rapid downcutting by the stream and very slow weathering of the canyon walls. (Photo by Zandria Muench Beraldo/CORBIS)

FIGURE 16.10 Potholes in the bed of a small stream in Cataract Falls State Park, Indiana. The rotational motion of swirling pebbles acts like a drill to create potholes. (Photo by Tom Till Photography)

(**suspended load**), and (3) sliding or rolling along the bottom (**bed load**).

Dissolved Load   Most of the dissolved load is brought to a stream by groundwater and is dispersed throughout the flow. As water percolates through the ground, it first acquires soluble soil compounds. As the water seeps deeper through cracks and pores in the bedrock below, it may dissolve additional mineral matter. Eventually much of this mineral-rich water finds its way into streams.

The velocity of streamflow has essentially no effect on a stream's ability to carry its dissolved load. Once material is in solution, it goes wherever the stream goes, regardless of velocity. Precipitation occurs only when the chemistry of the water changes.

The quantity of material carried in solution is highly variable and depends on such factors as climate and the geological setting. Usually the dissolved load is expressed as parts of dissolved material per million parts of water (parts per million, or ppm). Although some rivers may have a dissolved load of 1000 ppm or more, the average figure for the world's rivers is estimated at between 115 and 120 ppm. Almost 4 billion metric tons of dissolved mineral matter are supplied to the oceans each year by streams.

Suspended Load   Most streams (but not all) carry the largest part of their load in *suspension*. Indeed, the visible cloud of sediment suspended in the water is the most obvious portion of a stream's load. Usually only fine sand-, silt-, and clay-size particles can be carried this way, but during flood stage, larger particles are carried as well. Also during flood stage, the total quantity of material carried in suspension increases dramatically, as can be verified by people whose homes have become sites for the deposition of this material. During flood stage, the Hwang Ho (Yellow River) of China is reported to carry an amount of sediment equal in weight to the water that carries it. Rivers like this are appropriately described as "too thick to drink but too thin to cultivate."

The type and amount of material carried in suspension are controlled by two factors: the velocity of the water and the settling velocity of each sediment grain. **Settling velocity** is defined as the speed at which a particle falls through a still fluid. The larger the particle, the more rapidly it settles toward the stream bed. In addition to size, the shape and specific gravity of particles also influence settling velocity. Flat grains sink through water more slowly than do spherical grains, and dense particles fall toward the bottom more rapidly than do less dense particles. The slower the settling velocity and the stronger the turbulence, the longer a sediment particle will stay in suspension, and the farther it will be carried downstream with the flowing water.

Bed Load   A portion of a stream's load of solid material consists of sediment that is too large to be carried in suspension. These coarser particles move along the bottom (bed) of the stream and constitute the *bed load* (Figure 16.11). In terms of the erosional work accomplished by a downcutting stream, the grinding action of the bed load is of great importance.

The particles that make up the bed load move along the bottom by rolling, sliding, and saltation. Sediment moving by **saltation** (*saltare* = to leap) appears to jump or skip along the stream bed. This occurs as particles are propelled upward by collisions or lifted by the current and then carried downstream a short distance until gravity pulls them back to the bed of the stream. Particles that are too large or heavy to move by saltation either roll or slide along the bottom, depending on their shapes.

The bed load usually does not exceed 10 percent of a stream's total load, although it may constitute up to 50 percent of the total load of a few streams. For example, consider the distribution of the 750 million tons of material carried to the Gulf of Mexico by the Mississippi River each year. Of this total, it is estimated that approximately 67 percent is carried in suspension, 26 percent in solution, and the remaining 7 percent as bed load. Estimates of a stream's bed load should be viewed cautiously, however, because the transport of sediment as bed load or as suspended load changes frequently. The proportions of each depend upon the characteristics of streamflow at any time and these may fluctuate over short intervals. With an increase in velocity, parts of the bed load are thrown into suspension. Conversely, when velocity decreases, a portion of the bed load is deposited and some sediment that had been suspended changes to moving as bed load.

Capacity and Competence   A stream's ability to carry solid particles is typically described using two criteria. First, the maximum load of solid particles that a stream can transport is termed its **capacity.** The greater the amount of water flowing in a stream (discharge), the greater the stream's capacity for hauling sediment. Second, the **competence** of a stream indicates the maximum particle size that a stream can transport. The stream's velocity determines its competence; the stronger the flow, the larger the particles it can carry in suspension and as bed load.

It is a general rule that the competence of a stream increases as the square of its velocity. Thus, if the velocity of a stream doubles, the impact force of the water increases four times; if the velocity triples, the force increases nine times, and so forth. Hence, the large boulders that are often visible during a low-water stage and that seem immovable can, in fact, be transported during flood stage because of the stream's increased competence (Figure 16.11).

By now it should be clear why the greatest erosion and transportation of sediment occur during floods (Figure 16.12). The increase in discharge results in greater capacity;

FIGURE 16.11   Although the bed load of many rivers consists of sand, the bed load of this stream is made up of boulders and is easily seen during periods of low water. During floods, the seemingly immovable rocks in this channel are rolled along the bed of the stream. The maximum-size particle a stream can move is determined by the velocity of the water. (Photo by E. J. Tarbuck)

FIGURE 16.12   During floods both capacity and competency increase. Therefore, the greatest erosion and sediment transport occur during these high-water periods. Here we see the sediment-filled floodwaters of Kenya's Mara River. (Photo by Tim Davis/Stone)

the increased velocity produces greater competency. With rising velocity the water becomes more turbulent, and larger and larger particles are set in motion. In the course of just a few days, or perhaps just a few hours, a stream in flood stage can erode and transport more sediment than it does during months of normal flow.

Deposition of Sediment by Streams

Whenever a stream slows down, the situation reverses. As its velocity decreases, its competence is reduced and sediment begins to drop out, largest particles first. Each particle size has a *critical settling velocity.* As streamflow drops below the critical setting velocity of a certain particle size, sediment in that category begins to settle out. Thus, stream transport provides a mechanism by which solid particles of various sizes are separated. This process, called **sorting,** explains why particles of similar size are deposited together.

The material deposited by a stream is called **alluvium,** the general term for any stream-deposited sediment. Many different depositional features are composed of alluvium. Some occur within stream channels, some occur on the valley floor adjacent to the channel, and some exist at the mouth of the stream. We will consider the nature of these features later.

Stream Channels

A basic characteristic of streamflow that distinguishes it from overland flow is that it is usually confined to a channel. A stream channel can be thought of as an open conduit that consists of the streambed and banks that act to confine the flow except during floods.

Although somewhat oversimplified, we can divide stream channels into two types. *Bedrock channels* are those in which the streams are actively cutting into solid rock. In contrast, when the bed and banks are composed mainly of unconsolidated sediment, the channel is called an *alluvial channel.*

Bedrock Channels

In their headwaters, where the gradient is steep, many rivers cut into bedrock. Such streams typically transport coarse particles that actively abrade the bedrock channel. Potholes are often visible evidence of the erosional forces at work.

Bedrock channels typically alternate between relatively gently sloping segments where alluvium tends to accumulate, and steeper segments where bedrock is exposed. These steeper areas may contain rapids or occasionally a waterfall. The channel pattern exhibited by streams cutting into

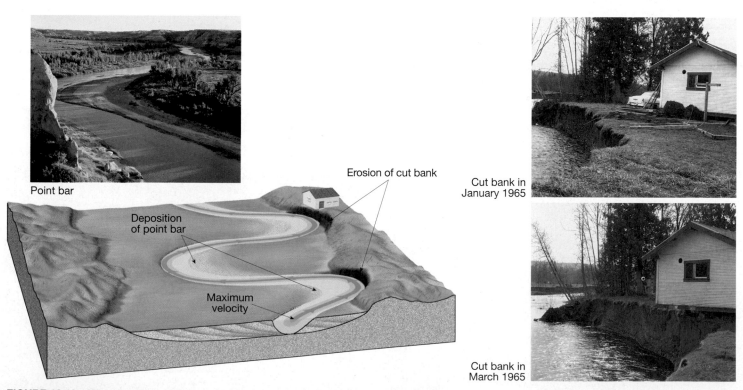

FIGURE 16.13 When a stream meanders, its zone of maximum speed shifts toward the outer bank. A point bar is deposited where the water on the inside of a meander slows. The point bar shown here is on the Missouri River in North Dakota. The black and white photos show erosion of a cut bank along the Newaukum River in Washington State. By eroding its outer bank and depositing material on the inside of the bend, a stream is able to shift its channel. (Point bar photo by Carr Clifton; cut bank photos by P. A. Glancy, U.S. Geological Survey)

bedrock is controlled by the underlying geologic structure. Even when flowing over rather uniform bedrock, streams tend to exhibit a winding or irregular pattern rather than flowing in a straight channel. Anyone who has gone on a white-water rafting trip has observed the steep, winding nature of a stream flowing in a bedrock channel.

Alluvial Channels

Many stream channels are composed of loosely consolidated sediment (alluvium) and therefore can undergo major changes in shape because the sediments are continually being eroded, transported, and redeposited. The major factors affecting the shapes of these channels is the average size of the sediment being transported, the channel gradient, and the discharge.

Alluvial channel patterns reflect a stream's ability to transport its load at a uniform rate, while expending the least amount of energy. Thus, the size and type of sediment being carried help determine the nature of the stream channel. Two common types of alluvial channels are *meandering channels* and *braided channels*.

Meandering Streams   Streams that transport much of their load in suspension generally move in sweeping bends called **meanders.** These streams flow in relatively deep, smooth channels and transport mainly mud (silt and clay). The lower Mississippi River exhibits a channel of this type.

Because of the cohesiveness of consolidated mud, the banks of stream channels carrying fine particles tend to resist erosion. As a consequence, most of the erosion in such channels occurs on the outside of the meander, where velocity and turbulence are greatest. In time, the outside bank is undermined, especially during periods of high water. Because the outside of a meander is a zone of active erosion, it is often referred to as the **cut bank** (Figure 16.13). The debris acquired by the stream at the cut bank moves downstream with the coarser material generally being deposited as **point bars** in zones of decreased velocity on the insides of meanders. In this manner, meanders migrate laterally by eroding the outside of the bends and depositing on the inside.

In addition to migrating laterally, the bends in a channel also migrate down the valley. This occurs because erosion is more effective on the downstream (downslope) side of the meander. Sometimes the downstream migration of a meander is slowed when it reaches a more resistant material. This allows the next meander upstream to "catch up" and overtake it as shown in Figure 16.14. Gradually the neck of land between the meanders is narrowed. Eventually the river may erode through the narrow neck of land to the next loop (Figure 16.14). The new, shorter channel segment is called **a cutoff** and, because of its shape, the abandoned bend is called an **oxbow lake** (Figure 16.15).

Braided Streams   Some streams consist of a complex network of converging and diverging channels that thread

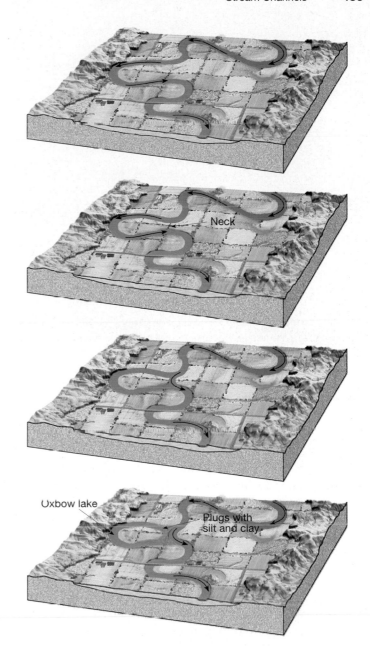

FIGURE 16.14   Formation of a cutoff and oxbow lake.

their way among numerous islands or gravel bars (Figure 16.16). Because these channels have an interwoven appearance, these streams are said to be **braided.** Braided channels form where a large proportion of the stream's load consists of coarse material (sand and gravel) and the stream has a highly variable discharge. Because the bank material is readily erodable, braided channels are wide and shallow.

One circumstance in which braided streams form is at the end of a glacier where there is a large seasonal variation in discharge. Here, large amounts of ice-eroded sediment are dumped into the meltwater streams flowing away from the glacier. When flow is sluggish, the stream is unable to move

FIGURE 16.15 Oxbow lakes occupy abandoned meanders. As they fill with sediment, oxbow lakes gradually become swampy meander scars. Aerial view of oxbow lake created by the meandering Green River near Bronx, Wyoming. (Photo by Michael Collier)

all of the sediment and therefore deposits the coarsest material as bars that force the flow to split and follow several paths. Usually the laterally shifting channels completely rework most of the surface sediments each year, thereby transforming the entire streambed. In some braided streams, however, the bars have built up to form islands that are anchored by vegetation.

In summary, meandering channels develop where the load consists largely of fine-grained particles that are transported as suspended load in a deep, smooth channel. By contrast, wide, shallow braided channels develop where coarse-grained alluvium is transported as bedload.

Base Level and Graded Streams

In 1875 John Wesley Powell, the pioneering geologist who first explored the Grand Canyon and later headed the U.S. Geological Survey, introduced the concept that there is a downward limit to stream erosion, which he called **base level.** Although the idea is relatively straightforward, it is nevertheless a key concept in the study of stream activity. Base level is defined as the lowest elevation to which a stream can erode its channel. Essentially this is the level at which the mouth of a stream enters the ocean, a lake, or another stream. Base level accounts for the fact that most stream profiles have low gradients near their mouths, because the streams are approaching the elevation below which they cannot erode their beds. Powell recognized that two types of base level exist: We may consider the level of the sea to be a grand base level, below which the dry lands

cannot be eroded; but we may also have, for local and temporary purposes, other base levels of erosion.*

Sea level, which Powell called "grand base level," is now referred to as **ultimate base level. Local** or **temporary base levels** include lakes, resistant layers of rock, and main streams that act as base levels for their tributaries. All have the capacity to limit a stream at a certain level.

For example, when a stream enters a lake, its velocity quickly approaches zero and its ability to erode ceases. Thus, the lake prevents the stream from eroding below its level at any point upstream from the lake. However, because the outlet of the lake can cut downward and drain the lake, the lake is only a temporary hindrance to the stream's ability to downcut its channel. In a similar manner the layer of resistant rock at the lip of the waterfall in Figure 16.17 acts as a temporary base level. Until the ledge of hard rock is eliminated, it will limit the amount of downcutting upstream.

Any change in base level will cause a corresponding readjustment of stream activities. When a dam is built along a stream course, the reservoir that forms behind it raises the base level of the stream (Figure 16.18). Upstream from the reservoir the stream gradient is reduced, lowering its velocity and, hence, its sediment-transporting ability. The stream, now unable to transport all of its load, will deposit material, thereby building up its channel. This process continues until the stream again has a gradient sufficient to carry its load. The profile of the new channel would be similar to the old, except that it would be somewhat higher.

If, on the other hand, the base level should be lowered, either by uplifting of the land or by a drop in sea level, the stream would again readjust. The stream, now above base level, would have excess energy and downcut its channel to establish a balance with its new base level. Erosion would first progress near the mouth, then work upstream until the stream profile was adjusted along its full length.

Exploration of the Colorado River of the West (Washington, D.C.: Smithsonian Institution, 1875), p. 203.

FIGURE 16.16 Braided stream choked with sediment near the edge of a melting glacier. (Photo by Bradford Washburn)

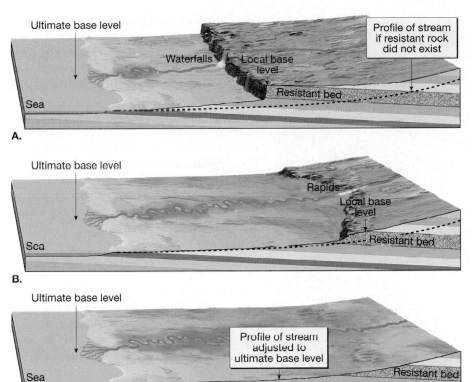

A.

B.

C.

FIGURE 16.17 A resistant layer of rock can act as a local (temporary) base level. Because the durable layer is eroded more slowly, it limits the amount of downcutting upstream.

The observation that streams adjust their profiles to changes in base level led to the concept of a graded stream. A **graded stream** has the correct slope and other channel characteristics necessary to maintain just the velocity required to transport the material supplied to it. On the average, a graded system is neither eroding nor depositing material but is simply transporting it. Once a stream has reached this state of equilibrium, it becomes a self-regulating system in which a change in one characteristic causes an adjustment in the others to counteract the effect. Referring again to our example of a stream adjusting to a lowering of its base level, the stream would not be graded while it was cutting its new channel but would achieve this state after downcutting had ceased.

Shaping Stream Valleys

 Running Water
▸ **Reviewing Valleys and Stream-related Features**

Valleys are the most common landforms on Earth's surface. In fact, they exist in such large numbers that they have never been counted except in limited areas used for study. Prior to the turn of the 19th century, it was generally believed that valleys were created by catastrophic events that pulled the crust apart and created troughs in which streams

could flow. Today, however, we know that with a few exceptions, streams create their own valleys.

One of the first clear statements of this fact was made by English geologist John Playfair in 1802. In his well-known work, *Illustrations of the Huttonian Theory of the Earth*, Playfair stated the principle that has come to be called *Playfair's law:*

Every river appears to consist of a main trunk, fed from a variety of branches, each running in a valley proportioned to its size, and all of them together forming a system of valleys, communicating with one another, and having such a nice adjustment of their declivities, that none of them join the principal valley, either on too high or too low a level; a circumstance that would be indefinitely improbable, if each of these valleys were not the work of the stream that flows in it.[*]

Not only were Playfair's observations essentially correct but they were written in a style that is seldom achieved in scientific prose.

Streams, with the aid of weathering and mass wasting, shape the landscape through which they flow. As a result, streams continuously modify the valleys that they occupy. A **stream valley** consists not only of the channel but also the surrounding terrain that directly contributes water to the stream. Thus it includes the valley bottom, which is the lower, flatter area that is partially or totally occupied by the stream channel, and the sloping valley walls that rise above the valley bottom on both sides. Most stream valleys are much broader at the top than is the width of their channel at the bottom. This would not be the case if the only agent responsible for eroding valleys were the streams flowing through them. The sides of most valleys are shaped by a combination of weathering, overland flow, and mass wasting (see Figure 15.2, p. 403). In some arid regions, where weathering is slow and where rock is particularly resistant, narrow valleys having nearly vertical walls are common (see Figure 16.9, p. 432).

Stream valleys can be divided into two general types—narrow V-shaped valleys and wide valleys with flat floors—with many gradations between.

Valley Deepening

When a stream's gradient is steep and the channel is well above base level, downcutting is the dominant activity. Abrasion caused by bed load sliding and rolling along the bottom, and the hydraulic power of fast-moving water, slowly lowers the streambed. The result is usually a

*Playfair, John. *Illustrations of the Huttonian Theory of the Earth,* New York: Dover Publications, p. 102 (facsimile reprint, 1964).

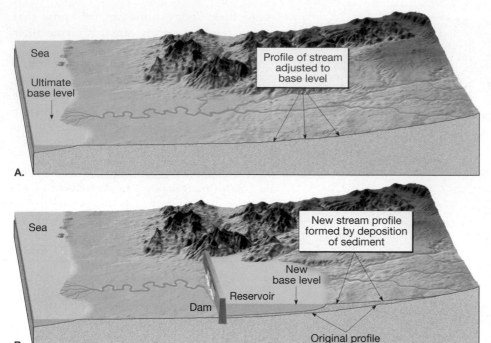

FIGURE 16.18   When a dam is built and a reservoir forms, the stream's base level is raised. This reduces the stream's velocity and leads to deposition and a reduction of the gradient upstream from the reservoir.

takes on a meandering pattern, and more of the stream's energy is directed from side to side. The result is a widening of the valley as the river cuts away first at one bank and then at the other (Figure 16.21). The continuous lateral erosion caused by shifting of the stream's meanders gradually produces a broad, flat valley floor covered with alluvium. This feature, called a **floodplain,** is appropriately named because when a river overflows its banks during flood stage, it inundates the floodplain. Over time the floodplain will widen to the point that the stream is only actively eroding the valley walls in a few places. In the case of the lower Mississippi River, the distance from one valley wall to another sometimes exceeds 150 kilometers.

When a river erodes laterally and creates a floodplain as just described, it is called an *erosional floodplain.* However, floodplains can be depositional as well.

V-shaped valley with steep sides. A classic example of a V-shaped valley is the section of the Yellowstone River shown in Figure 16.19.

The most prominent features of a V-shaped valley are *rapids* and *waterfalls.* Both occur where the stream's gradient increases significantly, a situation usually caused by variations in the erodability of the bedrock into which a stream channel is cutting. Resistant beds create rapids by acting as a temporary base level upstream while allowing downcutting to continue downstream. In time erosion usually eliminates the resistant rock.

Waterfalls are places where the stream makes a vertical drop. One type of waterfall is exemplified by Niagara Falls (Figure 16.20). Here, the falls are supported by a resistant bed of dolostone that is underlain by a less resistant shale. As the water plunges over the lip of the falls, it erodes the less resistant shale, undermining a section of dolostone, which eventually breaks off. In this manner the waterfall retains its vertical cliff while slowly but continually retreating upstream. Over the past 12,000 years, Niagara Falls has retreated more than 11 kilometers (7 miles) upstream.

Valley Widening

Once a stream has cut its channel closer to base level, it approaches a graded condition, and downward erosion becomes less dominant. At this point the stream's channel

FIGURE 16.19   V-shaped valley of the Yellowstone River. The rapids and waterfalls indicate that the river is vigorously downcutting. (Photo by Art Wolfe, Inc.)

FIGURE 16.20 The smaller American Falls at Niagara Falls. The river plunges over the falls and erodes the shale beneath the more resistant Lockport Dolostone. As a section of dolostone is undercut, it loses support and breaks off. (Photo by David Ball/Stone)

Depositional floodplains are produced by a major fluctuation in conditions, such as a change in base level. The floodplain in California's Yosemite Valley is one such feature; it was produced when a glacier gouged the former stream valley about 300 meters (1000 feet) deeper than it had been. After the glacial ice melted, the stream readjusted itself to its former base level by refilling the valley with alluvium.

Incised Meanders and Stream Terraces

We usually expect a stream with a highly meandering course to be on a floodplain in a wide valley. However, certain rivers exhibit meandering channels that flow in steep, narrow valleys. Such meanders are called **incised** (*incisum* = to cut into) **meanders** (Figure 16.22). How do such features form?

Originally the meanders probably developed on the floodplain of a stream that was relatively near base level. Then, a change in base level caused the stream to begin downcutting. One of two events could have occurred. Either base level dropped or the land upon which the river was flowing was uplifted.

An example of the first circumstance happened during the Ice Age when large quantities of water were withdrawn from the ocean and locked up in glaciers on land. The result was that sea level (ultimate base level) dropped, causing rivers flowing into the ocean to begin to downcut. Of course, this activity ceased at the close of the Ice Age when the glaciers melted and the ocean rose to its former level.

Regional uplift of the land, the second cause for incised meanders, is exemplified by the Colorado Plateau in the

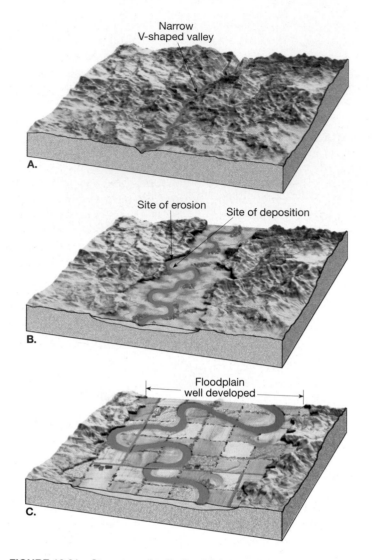

FIGURE 16.21 Stream eroding its floodplain.

southwestern United States. Here, as the plateau was gradually uplifted, numerous meandering rivers adjusted to being higher above base level by downcutting (Figure 16.22).

After a river has adjusted to a relative drop in base level by downcutting, it may once again produce a floodplain at a level below the old one. The remnants of a former floodplain are sometimes present in the form of flat surfaces called **terraces** (Figure 16.23).

Depositional Landforms

As indicated earlier, whenever a stream's velocity slows, it begins to deposit some of the sediment it is carrying. Also recall that streams continually pick up sediment in one part of their channel and redeposit it downstream. These channel deposits are most often composed of sand and gravel and are commonly referred to as **bars.** Such features, however, are only temporary, for the material will be picked up again and eventually carried to the ocean. In addition to sand and gravel bars, streams also create other depositional features that have a somewhat longer life span. These include deltas, natural levees, and alluvial fans.

Deltas

A **delta** forms when a stream enters an ocean or a lake. Figure 16.24A depicts the structure of a simple delta that might form in the relatively quiet waters of a lake. As the stream's forward motion is checked upon entering the lake, the dying current deposits its load of sediments. These deposits occur in three types of beds. *Foreset beds* are composed of coarser particles that drop almost immediately upon

FIGURE 16.22 **A.** This high-altitude image shows incised meanders of the Delores River in western Colorado. (Courtesy of USDA-ASCS) **B.** A close-up view of incised meanders of the Colorado River in Canyonlands National Park, Utah. (Photo by Michael Collier) In both places, meandering streams began downcutting because of the uplift of the Colorado Plateau.

A.

B.

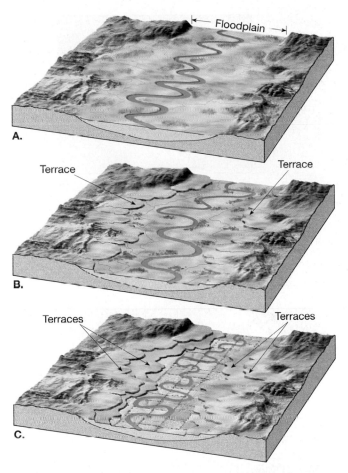

FIGURE 16.23 Parts **A, B,** and **C** show the development of multiple stream terraces. Terraces can form when a stream downcuts through previously deposited alluvium. This may occur in response to a lowering of base level or as a result of regional uplift. The photo in **D** shows terraces of the Snake River in Wyoming. (Photo by Lester Lefkowitz)

entering the lake to form layers that slope downcurrent from the delta front. The foreset beds are usually covered by thin, horizontal *topset beds* that are deposited during flood stage. The finer silts and clays settle out some distance from the mouth in nearly horizontal layers called *bottomset beds*.

As the delta grows outward, the stream's gradient continually lessens. This circumstance eventually causes the channel to become choked with sediment from the slowing

water. As a consequence, the river seeks a shorter, higher-gradient route to base level, as illustrated in Figure 16.24B. This illustration shows the main channel dividing into several smaller ones, called **distributaries.** Most deltas are characterized by these shifting channels that act in an opposite way to that of tributaries. Rather than carrying water into the main channel, distributaries carry water away from the main channel in varying paths to base level. After numerous shifts of the channel, the simple delta may grow into the triangular shape of the Greek letter delta (Δ), for which it was named (Figure 16.25). Note, however, that many deltas do not exhibit this shape. Differences in the configurations of shorelines, and variations in the nature and strength of wave activity, result in many different shapes.

Although deltas that form in the ocean generally exhibit the same basic form as the simple lake-deposited feature just described, most large marine deltas are far more complex and have foreset beds that are inclined at a much lower angle than those depicted in Figure 16.24A. Indeed, many of the world's great rivers have created massive deltas, each with its own peculiarities and none as simple as the one illustrated in Figure 16.24B.

The Mississippi Delta

Many large rivers have deltas that extend over thousands of square kilometers. The delta of the Mississippi River is one such feature. It resulted from the accumulation of huge quantities of sediment derived from the vast region drained by the river and its tributaries. Today New Orleans rests where there was ocean less than 5000 years ago. Figure 16.26 shows that portion of the Mississippi delta that has been built over the past 5000 to 6000 years. As the figure illustrates, the delta is actually a series of seven coalescing subdeltas. Each was formed when the river left its existing channel to find a shorter, more direct path to the Gulf of Mexico. The individual subdeltas interfinger and partially cover one another to produce a very complex structure. It is also apparent from Figure 16.26 that after each portion was abandoned, coastal erosion modified the features. The present subdelta, called a *bird-foot* delta because of the configuration of its distributaries, has been built by the Mississippi in the last 500 years.

At present, this active bird-foot delta has extended about as far as natural forces will allow. In fact, for many years the river has been struggling to cut through a narrow neck of land and shift its course to that of the Atchafalaya River (see inset in Figure 16.26). If this were to happen, the Mississippi would abandon its lowermost 500-kilometer path in favor of the Atchafalaya's much shorter 225-kilometer route. From the early 1940s until the 1950s, an increasing portion of the Mississippi's discharge was diverted to this new path, indicating that the river was ready to shift and begin building a new subdelta.

To prevent such an event and to keep the Mississippi following its present course, a damlike structure was erected at the site where the channel was trying to break through. A flood in 1973 weakened the control structure, and the river

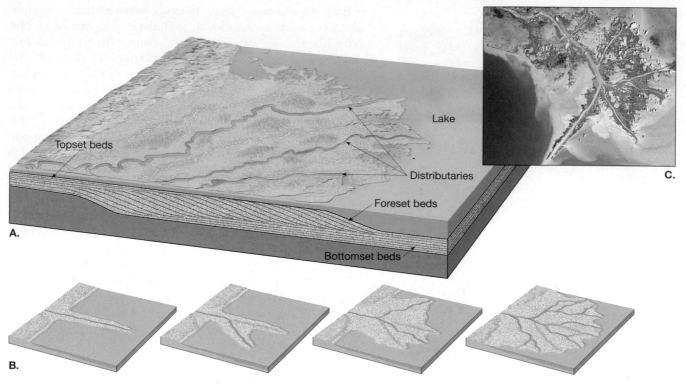

FIGURE 16.24 **A.** Structure of a simple delta that forms in the relatively quiet waters of a lake. **B.** Growth of a simple delta. As a stream extends its channel, the gradient is reduced. Frequently, during flood stage the river is diverted to a higher-gradient route, forming a new distributary. Old, abandoned distributaries are gradually invaded by aquatic vegetation and filled with sediment. (After Ward's Natural Science Establishment, Inc., Rochester, N.Y.) **C.** Satellite image of a portion of the delta of the Mississippi River in May 2001. For the past 500 years or so, the main flow of the river has been along its present course, extending southeast from New Orleans. During that span, the delta advanced into the Gulf of Mexico at a rate of about 10 kilometers (6 miles) per century. Notice the numerous distributaries. (Image courtesy of JPL/Cal Tech/NASA)

again threatened to shift until a massive auxiliary dam was completed in the mid-1980s. For the time being, at least, the inevitable has been avoided, and the Mississippi River will continue to flow past Baton Rouge and New Orleans on its way to the Gulf of Mexico (see Box 16.1).

Students Sometimes Ask . . .

I know that all rivers carry sediment. Do all rivers have deltas?

Surprisingly, no. Streams that transport large loads of sediment may lack deltas at their mouths because ocean waves and powerful currents quickly redistribute the material as soon as it is deposited (the Colombia River in the Pacific Northwest is one such example). In other cases, rivers do not carry sufficient quantities of sediment to build a delta. The St. Lawrence River, for example, has little opportunity to pick up much sediment between Lake Ontario and its mouth in the Gulf of St. Lawrence.

Natural Levees

Some rivers occupy valleys with broad floodplains and build **natural levees** (*lever* = to raise) that parallel their channels on both banks (Figure 16.27). Natural levees are built by successive floods over many years. When a stream overflows its banks onto the floodplain, the water flows over the surface as a broad sheet. Because such a flow pattern significantly reduces the water's velocity and turbulence, the coarser portion of the suspended load is deposited in strips bordering the channel. As the water spreads out over the floodplain, a lesser amount of finer sediment is laid down over the valley floor. This uneven distribution of material produces the gentle, almost imperceptible slope of the natural levee.

The natural levees of the lower Mississippi River rise 6 meters above the lower portions of the valley floor. The area behind the levee is characteristically poorly drained for the obvious reason that water cannot flow up over the levee and into the river. Marshes called **back swamps** often result. When a tributary stream enters a valley having substantial natural levees, it may not be able to make its way into the main channel. As a consequence, the tributary may flow through the back swamp parallel to the main river for many kilometers before crossing the natural levee and joining the main river. Such streams are called **yazoo tributaries**, after the Yazoo River, which parallels the lower Mississippi River for more than 300 kilometers.

Alluvial Fans

As Figure 16.28 illustrates, **alluvial fans** are aptly named. These fan-shaped deposits are built from the accumulation of sediments deposited at the mouth of a valley that emerges from a mountainous or upland area onto a relatively flat lowland or larger mainstream valley. As these streams encounter reduced slopes, they deposit a portion of their sediment load. By doing this they maintain an adequate gradient. The fan shape is produced by the stream swinging back and forth from the fixed apex of the fan. Although alluvial fans are more obvious in desert areas, they are also found in humid regions.

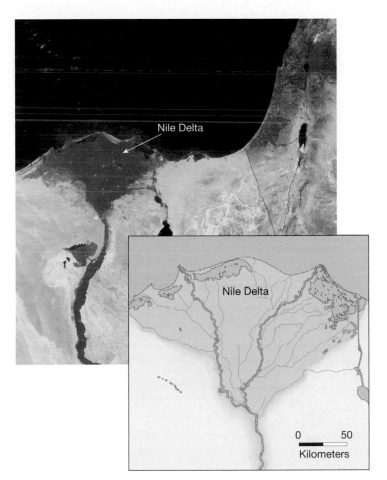

FIGURE 16.25   The shapes of deltas vary and depend on such factors as a river's sediment load and the strength and nature of shoreline processes. The triangular shape of the Nile delta was the basis for naming this feature. In this satellite image, the delta and the well watered areas adjacent to the Nile River stand out in sharp contrast to the surrounding Sahara Desert. (NASA/GSFC/METI/ERSDAC/JAROS, and U.S./Japan ASTER Science Team)

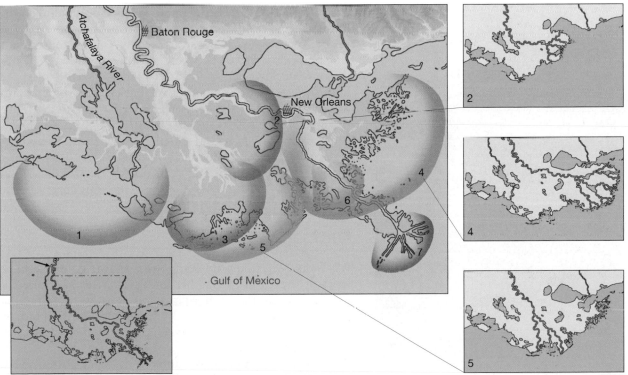

FIGURE 16.26   During the past 5000 to 6000 years, the Mississippi River has built a series of seven coalescing subdeltas. The numbers indicate the order in which the subdeltas were deposited. The present bird-foot delta (number 7) represents the activity of the past 500 years. Without ongoing human efforts, the present course will shift and follow the path of the Atchafalaya River. The inset on left shows the point where the Mississippi may someday break through (arrow) and the shorter path it would take to the Gulf of Mexico. (After C. R. Kolb and J. R. Van Lopik)

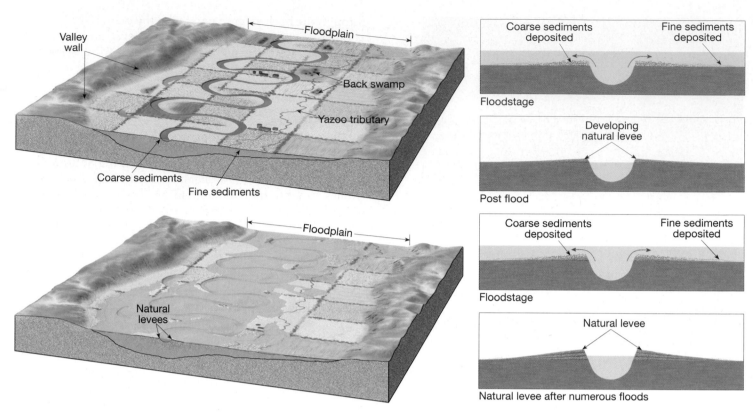

FIGURE 16.27 Natural levees are gently sloping structures that are created by repeated floods. The diagrams on the right show the sequence of development. Because the ground next to the stream channel is higher than the adjacent floodplain, backswamps and yazoo tributaries may develop.

FIGURE 16.28 Alluvial fans are built from sediments deposited at the mouth of a valley that emerges from a mountainous or upland area onto a relatively flat lowland. The surface of the fan slopes outward in a broad arc from the apex at the mouth of the stream valley. Usually, coarse material is dropped near the apex of the fan, while finer materials are carried toward the base of the deposit. California's Death Valley has many large alluvial fans. (Photo by Michael Collier)

When a stream emerges from its valley onto an alluvial fan, it usually splits into many distributary channels. Because sediments composing fans are often coarse sands and gravels, water quickly soaks in. Between rainy periods in deserts, little or no water flows across an alluvial fan. An examination of a fan's surface in such places shows that it is crossed by many dry channels. Thus fans in dry regions grow intermittently, receiving considerable water and sediment only during rainy periods. As you learned in Chapter 15, steep canyons in dry regions are prime locations for debris flows. Therefore, it should be expected that many alluvial fans in arid areas have debris-flow deposits interbedded with the alluvium.

Drainage Patterns

All drainage systems are made up of an interconnected network of streams that together form particular patterns. The nature of a drainage pattern can vary greatly from one type of terrain to another, primarily in response to the kinds of rock on which the streams developed or the structural pattern of faults and folds.

The most commonly encountered drainage pattern is the **dendritic pattern** (Figure 16.29A). This pattern is characterized by irregular branching of tributary streams that resembles the branching pattern of a deciduous tree. In fact, the word *dendritic* means "treelike." The dendritic pattern forms where underlying bedrock is relatively uniform, such as flat-lying sedimentary strata or massive igneous rocks. Because

BOX 16.1 ▶ PEOPLE AND THE ENVIRONMENT

Coastal Wetlands Are Vanishing on the Mississippi Delta

Coastal wetlands form in sheltered environments that include swamps, tidal flats, coastal marshes, and bayous. They are rich in wildlife and provide nesting grounds and important stopovers for waterfowl and migratory birds, as well as spawning areas and valuable habitats for fish.

The delta of the Mississippi River in Louisiana contains about 40 percent of all coastal wetlands in the lower 48 states. Louisiana's wetlands are sheltered from the wave action of hurricanes and winter storms by low-lying offshore barrier islands. Both the wetlands and the barrier islands have formed as a result of the shifting of the Mississippi River during the past 7000 years.

The dependence of Louisiana's coastal wetlands and offshore islands on the Mississippi River and its distributaries as a direct source of sediment leaves them vulnerable to changes in the river system. Moreover, the reliance on barrier islands for protection from storm waves leaves coastal wetlands vulnerable when these narrow offshore islands are eroded.

Today, the coastal wetlands of Louisiana are disappearing at an alarming rate. Although Louisiana contains 40 percent of the wetlands in the lower 48 states, it accounts for 80 percent of the wetland loss. According to the U.S. Geological Survey, Louisiana lost nearly 5000 square kilometers (1900 square miles) of coastal land between 1932 and 2000. The state continues to lose between 65 and 91 square kilometers (25 to 35 square miles) each year. At this rate another 1800 to 4500 square kilometers (700 to 1750 square miles) will vanish under the Gulf of Mexico by the year 2050.* Global climate change could increase the severity of the problem because rising sea level and stronger tropical storms accelerate rates of coastal erosion.** Unfortunately, this was observed firsthand during the extraordinary 2005 hurricane season when hurricanes Katrina and Rita devastated portions of the Gulf Coast.

By nature, the delta, its wetlands, and the adjacent barrier islands are dynamic features. Over the millennia, as sediment accumulated and built the delta in one area, erosion and subsidence caused losses elsewhere (Figure 16.A). Whenever the river shifted, the zones of delta growth and destruction also shifted. However, with the arrival of people, this relative balance between formation and destruction changed—the rate at which the delta and its wetlands were destroyed accelerated and now greatly exceeds the rate of formation. Why are Louisiana's wetlands shrinking?

Before Europeans settled the delta, the Mississippi River regularly overflowed its banks in seasonal floods. The huge quantities of sediment that were deposited renewed the soil and kept the delta from sinking below sea level. However, with settlement came flood-control efforts and the desire to maintain and improve navigation on the river. Artificial levees were constructed to contain the rising river during flood stage. Over time the levees were extended all the way to the mouth of the Mississippi to keep the channel open for navigation.

The effects have been straightforward. The levees prevent sediment and fresh water from being dispersed into the wetlands. Instead, the river is forced to carry its load to the deep waters at the mouth. Meanwhile, the processes of compaction, subsidence, and wave erosion continue. Because not enough sediment is added to offset these forces, the size of the delta and the extent of its wetlands gradually shrink.

The problem has been aggravated by a decline in the sediment transported by the Mississippi, decreasing by approximately 50 percent over the past 100 years. A substantial portion of the reduction results from trapping of sediment in large reservoirs created by dams built on tributaries to the Mississippi.

Another factor contributing to wetland decline is the fact that the delta is laced with 13,000 kilometers (8000 miles) of navigation channels and canals. These artificial openings to the sea allow salty Gulf waters to flow far inland. The invasion of saltwater and tidal action causes massive "brownouts" or marsh die-offs.

Understanding and modifying the impact of people is a necessary basis for any plan to reduce the loss of wetlands in the Mississippi delta. The U.S. Geological Survey estimates that restoring Louisiana's coasts will require about $14 billion over the next 40 years. What if nothing is done? State and federal officials estimate that costs of inaction could exceed $100 billion.

*See "Louisiana's Vanishing Wetlands: Going, Going . . ." in *Science*, Vol. 289, 15 September 2000, pp. 1860–63. Also see Elizabeth Kolbert, "Watermark—Can Southern Louisiana be Saved?" *The New Yorker*, February 27, 2006, pp. 46–57.

**For more on this possibility, see Box 20.3, "Coastal Vulnerability to Sea-Level Rise."

FIGURE 16.A   Dead cypress trees, known as ghost forests, killed by encroaching salt water in Terrebonne Parish, Louisiana. (Photo by Robert Caputo/Aurora Photos)

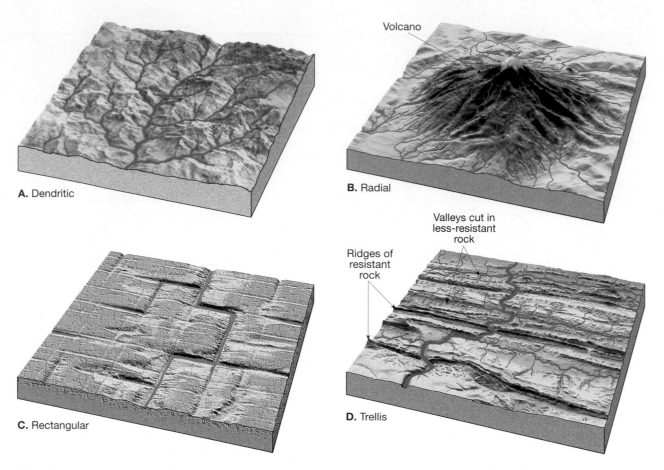

A. Dendritic

Volcano

B. Radial

C. Rectangular

Ridges of resistant rock

Valleys cut in less-resistant rock

D. Trellis

FIGURE 16.29 Drainage patterns. **A.** Dendritic. **B.** Radial. **C.** Rectangular. **D.** Trellis.

the underlying material is essentially uniform in its resistance to erosion, it does not control the pattern of streamflow. Rather, the pattern is determined chiefly by the direction of the slope of land.

When streams diverge from a central area, like spokes from the hub of a wheel, the pattern is said to be **radial** (Figure 16.29B). This pattern typically develops on isolated volcanic cones and domal uplifts.

Figure 16.29C illustrates a **rectangular pattern,** with many right-angle bends. This pattern develops when the bedrock is crisscrossed by a series of joints and faults. Because these structures are eroded more easily than unbroken rock, their geometric pattern guides the directions of streams as they carve their valleys.

Figure 16.29D illustrates a **trellis drainage pattern,** a rectangular pattern in which tributary streams are nearly parallel to one another and have the appearance of a garden trellis. This pattern forms in areas underlain by alternating bands of resistant and less resistant rock and is particularly well displayed in the folded Appalachian Mountains, where both weak and strong strata outcrop in nearly parallel belts.

Headward Erosion and Stream Piracy

We have seen that a stream can lengthen its course by building a delta at its mouth. A stream also lengthens its course by **headward erosion,** that is, by extending the head of its

valley upslope. As sheet flow converges and becomes concentrated at the head of a stream channel, its velocity, and hence its power to erode, increases. The result can be vigorous erosion at the head of the valley. Thus, through headward erosion, the valley becomes extended into previously undissected terrain (Figure 16.30).

Headward erosion by streams plays a major role in the dissection of upland areas. In addition, an understanding of this process helps explain changes that take place in drainage patterns. One cause for changes that occur in the pattern of streams is **stream piracy,** the diversion of the drainage of one stream because of the headward erosion of another stream. Piracy can occur, for example, if a stream on one side of a divide has a steeper gradient than a stream on the other side. Because the stream with the steeper gradient has more energy, it can extend its valley headward, eventually breaking down the divide and capturing part or all of the drainage of the slower stream. In Figure 16.31 the flow of stream *A* was captured when a more swiftly flowing tributary of stream *B* breached the divide at its head and diverted stream *A*.

Stream piracy also explains the existence of narrow, steep-sided gorges that have no active streams running through them. These abandoned water gaps (called *wind gaps*) form when the stream that cut the notch has its course changed by a pirate stream. In Figure 16.31, one water gap that had been created by stream *A* became a wind gap as a result of stream piracy.

FIGURE 16.30 By headward erosion, valleys extend into previously undissected terrain. The San Rafael River is shown above its confluence with the Green River in Utah. (Photo by Michael Collier)

valley was deepened and the structure was encountered, the river would continue to cut its valley into it. The folded Appalachians provide some good examples. Here a number of major rivers, such as the Potomac and the Susquehanna, cut across the folded strata on their way to the Atlantic.

Floods and Flood Control

When the discharge of a stream becomes so great that it exceeds the capacity of its channel, it overflows its banks as a flood. **Floods** are among the most deadly and most destructive of all geologic hazards. They are, nevertheless, simply part of the *natural* behavior of streams.

Most floods have a meteorological origin caused by atmospheric processes that can vary greatly in both time and space. Just an hour or less of intense thunderstorm rainfall can trigger floods in small valleys. By contrast, major floods in large river valleys are often the result

Formation of a Water Gap

Sometimes to understand fully the pattern of streams in an area, we must understand the history of the streams. For example, in many places a river valley can be seen cutting through a ridge or mountain that lies across its path. The steep-walled notch followed by the river through the structure is called a **water gap** (Figure 16.32).

Why does a stream cut across such a structure and not flow around it? One possibility is that the stream existed before the ridge or mountain was formed. In this situation, the stream, called an **antecedent stream,** would have to keep pace downcutting while the uplift progressed. That is, the stream would maintain its course as folding or faulting raised an area of the crust across the path of the stream.

A second possibility is that the stream was **superposed,** or let down, upon the structure (Figure 16.32). This can occur when a ridge or mountain is buried beneath layers of relatively flat-lying sediments or sedimentary strata. Streams originating on this cover would establish their courses without regard to the structures below. Then, as the

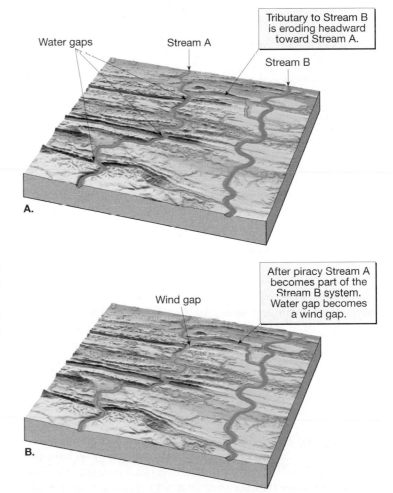

FIGURE 16.31 Stream piracy and the formation of wind gaps. A tributary of stream **B** erodes headward until it eventually captures and diverts stream **A.** A water gap through which stream **A** flowed is abandoned because of the piracy. As a result, this feature is now a wind gap. In this valley and ridge-type setting, the softer rocks in the valleys are eroded more easily than the resistant ridges. Consequently, as the valleys are lowered, the ridges and wind gaps become elevated relative to the valleys.

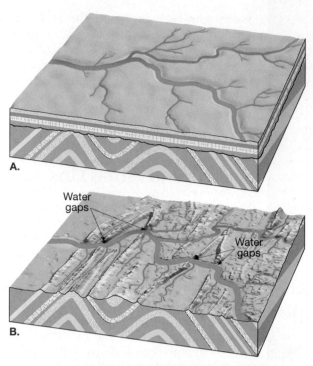

FIGURE 16.32   Development of a superposed stream. **A.** The river establishes its course on relatively uniform strata. **B.** It then encounters and cuts through the underlying structure. **C.** Harpers Ferry gap at the confluence of the Shenandoah and Potomac rivers near the West Virginia–Maryland border. Water gaps such as this one are common in parts of the Appalachians. (Photo by John S. Shelton)

of an extraordinary series of precipitation events over a broad region for an extended time span.

Land use planning in river basins requires an understanding of the frequency and magnitude of floods. Probably the greatest immediate practical use of the data collected at stream-gaging stations is estimating the probability of various flood magnitudes.

Floods are often described in terms of **recurrence interval** or **return period.** This is the case when you hear about a *100-year flood* (or a 30-year flood or a 50-year flood). What does this mean? The flood discharge that has a 1 percent (1 in 100) probability of being exceeded in any one year is called a 100-year flood. This phrase is misleading because it leads people

to believe that only one such flood will occur in a 100-year span or that such floods occur regularly every 100 years. Neither is accurate. The fact is that uncommonly big floods occur at irregular intervals and can happen in *any* year.

Many flood designations are reevaluated and changed over time as more data are collected or when a river basin is altered in a way that affects the flow of water. Dams and urban development are examples of some human influences in a basin that can affect recurrence intervals.

Causes and Types of Floods

Floods can be the result of several naturally occurring and human-induced factors. Among the common types of floods are regional floods, flash floods, ice-jam floods, and dam-failure floods.

Regional Floods   Some regional floods are seasonal. Rapid melting of snow in spring and/or heavy spring rains often overwhelm a river. The extensive 1997 flood along the Red River of the North is a notable example of an event triggered by rapid snowmelt. The flood was preceded by an especially snowy winter. As April began, the snow was melting and flooding seemed imminent, but on April 5 and 6, a blizzard rebuilt the shrinking snowdrifts to heights of 6 meters (20 feet) in some places. Then rapidly rising temperatures melted the snow in a matter of days, causing a record-breaking 500-year flood. Roughly 4.5 million acres were underwater, and the losses in the Grand Forks, North Dakota region exceeded $3.5 billion. Another factor contributing to this extraordinary event is the fact that the Red River flows northward. Therefore, in early spring, it flows into areas in which the ground is still frozen solid. Such conditions reduce infiltration into the soil, thereby increasing runoff.*

Extended wet periods any time of the year can create saturated soils, after which any additional rain runs off into

_____
*Ice jams also contribute to floods on the Red River of the North. See the section on "Ice-Jam Floods" on p. 450.

streams until capacities are exceeded. Regional floods are often caused by slow-moving storm systems, including decaying hurricanes. The extensive and costly floods in eastern North Carolina in September 1999 were the result of torrential rains on already waterlogged soils from decaying Hurricane Floyd. Persistent wet weather patterns led to the exceptional rains and devastating floods in the upper Mississippi River Valley during the summer of 1993 (Figure 16.33).

Flash Floods A flash flood can occur with little warning and can be deadly because it produces a rapid rise in water levels and can have a devastating flow velocity (see Box 16.2). Several factors influence flash flooding. Among them are rainfall intensity and duration, surface conditions, and

FIGURE 16.33 Satellite views of the Missouri River flowing into the Mississippi River. St. Louis is just south of their confluence. The upper image shows the rivers during a drought that occurred in summer 1988. The lower image depicts the peak of the record-breaking 1993 flood. Exceptional rains produced the wettest spring and early summer of the twentieth century in the upper Mississippi River basin. In all, nearly 14 million acres were inundated, displacing at least 50,000 people. (Courtesy of Spaceimaging.com)

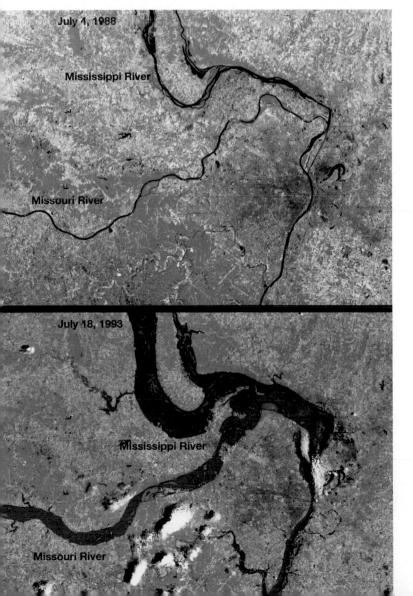

the topography. Mountainous areas are especially susceptible because steep slopes can funnel runoff into narrow canyons with disastrous consequences. This is illustrated by the Big Thompson River flood of July 31, 1976, in Colorado. (see Figure 16.C in Box 16.2). During a four-hour period, more than 30 centimeters (12 inches) of rain fell on a portion of the river's small drainage basin. The flash flood in the narrow canyon lasted only a few hours but took 139 lives and caused tens of millions of dollars in damages.

Urban areas are susceptible to flash floods because a high percentage of the surface area is composed of impervious roofs, streets, and parking lots, where runoff is very rapid. In fact, a recent study determined that the area of impervious surfaces in the United States (excluding Alaska and Hawaii) amounts to more than 112,600 square kilometers (nearly 44,000 square miles), which is slightly less than the area of the state of Ohio.*

To better understand the effect of urbanization on streamflow, examine Figure 16.34. Part A of the figure is a hypothetical

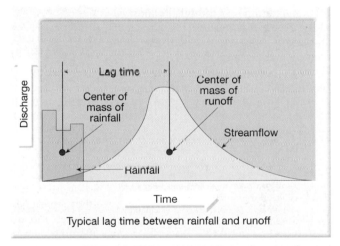

Typical lag time between rainfall and runoff

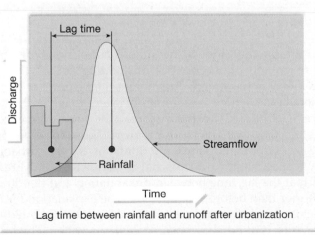

Lag time between rainfall and runoff after urbanization

FIGURE 16.34 When an area changes from rural to urban, the lag time between rainfall and flood peak is shortened. The flood peak is also higher following urbanization. (After L. B. Leopold, U.S. Geological Survey)

*C. D. Elvidge, et al. "U. S. Constructed Area Approaches the Size of Ohio," in *EOS, Transactions, American Geophysical Union*, vol. 85, no. 24, (June 15, 2004) p. 233.

BOX 16.2 ▶ PEOPLE AND THE ENVIRONMENT

Flash Floods

Tornadoes and hurricanes are nature's most awesome storms. Because of this status, they are logically the focus of much well-deserved attention. Yet, surprisingly, in most years these dreaded events are not responsible for the greatest number of storm-related deaths. That distinction is reserved for flash floods.* For the 10-year period 1995–2004, the number of storm-related deaths in the United States from flooding averaged 84 per year. By contrast, tornado fatalities averaged 65 annually and hurricanes, 15.

Flash floods are local floods of great volume and short duration (Figure 16.B). The rapidly rising surge of water usually occurs with little advance warning and can destroy roads, bridges, homes, and other substantial structures. Discharges quickly reach a maximum and diminish almost as rapidly. Flood flows often contain large quantities of sediment and debris as they sweep channels clean.

Frequently, flash floods result from the torrential rains associated with a slow-moving severe thunderstorm or take place when a series of thunderstorms repeatedly

FIGURE 16.B   Flash flooding in Las Vegas, Nevada, in August 2003. Parts of the city received nearly half the average annual rainfall in a matter of hours. Here firefighters are rescued from a firetruck that was caught in a torrent of water. (Photo by John Locher/*Las Vegas Review-Journal*)

*The year 2005 was certainly an exception. Hurricanes Dennis (12), Katrina (1300), Rita (119), and Wilma (35) were responsible for nearly 1500 deaths in the United States. If that number is factored into our calculations, the annual average for hurricanes is about 65 (30 years) and 169 (10 years).

pass over the same location. Sometimes they are triggered by heavy rains from hurricanes and tropical storms. Occasionally, floating debris or ice can accumulate at a

natural or artificial obstruction and restrict the flow of water. When such temporary dams fail, torrents of water can be released as a flash flood.

---

hydrograph that shows the time relationship between a rainstorm and the occurrence of flooding. Notice that the water level in the stream does not rise at the onset of precipitation, because time is required for water to move from the place where it fell to the stream. This time difference is called the *lag time*. The hydrograph in Figure 16.34B depicts the same hypothetical area and rainfall event *after* urbanization. Notice that the peak discharge during a flood is greater and that the lag time between precipitation and flood peak is shorter than before urbanization. The explanation for this effect is straightforward. Streets, parking lots, and buildings cover the ground that once soaked up water. Thus, less water infiltrates and the rate and amount of runoff increase. Further, because much less water soaks into the ground, the low-water (dry-season) flow in many urban streams, which is maintained by the movement of groundwater into the channel, is greatly reduced. As one might expect, the magnitude of these effects is a function of the percentage of land that is covered by impermeable surfaces.

Ice-Jam Floods   Frozen rivers are susceptible to ice-jam floods. As the level of a stream rises, it will break up the ice and create ice flows that can pile up on channel obstructions. Such an ice jam creates a dam across the channel. Water upstream from the ice dam can rise rapidly and overflow the channel banks. When the ice dam fails, the water stored behind the dam is released, causing a flash flood downstream.

Dam-Failure Floods   Human interference with a stream system can cause floods. A prime example is the failure of a dam or an artificial levee. Dams and artificial levees are built for flood protection. They are designed to contain floods of a certain magnitude. If a larger flood occurs, the dam or levee is overtopped. If the dam or levee fails or is washed out, the water behind it is released to become a flash flood. The bursting of a dam in 1889 on the Little Conemaugh River caused the devastating Johnstown, Pennsylvania, flood that took over 2200 lives (Figure 16.35).

Flash floods can take place in almost any area of the country. They are particularly common in mountainous terrain, where steep slopes can quickly channel runoff into narrow valleys. The hazard is most acute when the soil is already nearly saturated from earlier rains or consists of impermeable materials. A disaster in Shadydale, Ohio, demonstrates what can happen when even moderately heavy rains fall on saturated ground with steep slopes.

On the evening of 14 June 1990, 26 people lost their lives as rains estimated to be in the range of 3 to 5 inches fell on saturated soil, which generated flood waves in streams that reached tens of feet in height, destroying near-bank residences and businesses. Preceding months of above-normal rainfall had generated soil moisture contents of near saturation. As a result, moderate amounts of rainfall caused large amounts of surface and near-surface runoff. Steep valleys with practically vertical walls channeled the floods, creating very fast, high, and steep wave crests.*

Why do so many people perish in flash floods? Aside from the factor of surprise

*"Prediction and Mitigation of Flash Floods: A Policy Statement of the American Meteorological Society," *Bulletin of the American Meteorological Society*, vol. 74, no. 8 (1993), p. 1586.

FIGURE 16.C   The disastrous nature of flash floods is illustrated by the Big Thompson River flood of July 31, 1976, in Colorado. During a four-hour span more than 30 centimeters (12 inches) of rain fell on portions of the river's small drainage basin. This amounted to nearly three-quarters of the average yearly total. The flash flood in the narrow canyon lasted only a few hours but cost 139 people their lives. Damages were estimated at $00 million. (U.S. Geological Survey, Denver)

(many are caught sleeping), people do not appreciate the power of moving water. A glance at Figure 16.C helps illustrate the force of a flood wave. Just 15 centimeters (6 inches) of fast-moving floodwater can knock a person down. Most automobiles will float and be swept away in only 0.6 meter (2 feet) of water. *More than half of all U.S. flash-flood fatalities are auto related!* Clearly, people should never attempt to drive over a flooded road. The depth of water is not always obvious. Also, the road bed may have been washed out underwater. Present-day flash floods are calamities with potential for very high death tolls and huge property losses. Although efforts are being made to improve observations and warnings, flash floods remain elusive natural killers.

Flood Control

Several strategies have been devised to eliminate or lessen the catastrophic effects of floods. Engineering efforts include the construction of artificial levees, the building of flood-control dams, and river channelization.

Artificial Levees   Artificial levees are earthen mounds built on the banks of a river to increase the volume of water the channel can hold. These most common of stream-containment structures have been used since ancient times and continue to be used today. Artificial levees are usually easy to distinguish from natural levees because their slopes are much steeper. In some locations, especially urban areas, concrete floodwalls are sometimes constructed that serve the same purpose as artificial levees.

Such structures do not always provide the flood protection that was intended. Many artificial levees were not built to withstand periods of extreme flooding. For example,

levee failures were numerous in the Midwest during the summer of 1993, when the upper Mississippi and many of its tributaries experienced record floods (Figure 16.36). During that same event, floodwalls at St. Louis, Missouri, created a bottleneck for the river that led to increased flooding upstream of the city.

Flood-Control Dams   Flood-control dams are built to store floodwater and then let it out slowly. This lowers the flood crest by spreading it out over a longer time span. Since the 1920s, thousands of dams have been built on nearly every major river in the United States. Many dams have significant nonflood-related functions such as providing water for irrigated agriculture and for hydroelectric power generation. Many reservoirs are also major regional recreational facilities.

Although dams may reduce flooding and provide other benefits, building these structures also has significant costs and consequences. For example, reservoirs created by dams

FIGURE 16.35 Aftermath of the historic Johnstown, Pennsylvania, flood. The flood occurred on May 31, 1889, when the South Fork Dam on the Conemaugh River collapsed. The flood destroyed 1600 homes and killed 2209 people. (Photo by CORBIS)

may cover fertile farmland, useful forests, historic sites, and scenic valleys. Of course, dams trap sediment. Therefore, deltas and floodplains downstream erode because they are no longer replenished with silt during floods. Large dams can also cause significant ecological damage to river environments that took thousands of years to establish.

Building a dam is not a permanent solution to flooding. Sedimentation behind a dam means that the volume of its

FIGURE 16.36 Water rushes through a break in an artificial levee in Monroe County, Illinois. During the record-breaking 1993 Midwest floods, many artificial levees could not withstand the force of the floodwaters. Sections of many weakened structures were overtopped or simply collapsed. (Photo by James A. Finley/AP/Wide World Photos)

reservoir will gradually diminish, reducing the effectiveness of this flood-control measure.

Channelization Channelization involves altering a stream channel in order to speed the flow of water to prevent it from reaching flood height. This may simply involve clearing a channel of obstructions or dredging a channel to make it wider and deeper.

Another alteration involves straightening a channel by creating *artificial cutoffs*. The idea is that by shortening the stream, the gradient and hence the velocity are increased. By increasing velocity, the larger discharge associated with flooding can be dispersed more rapidly.

Since the early 1930s, the Army Corps of Engineers has created many artificial cutoffs on the Mississippi for the purpose of increasing the efficiency of the channel and reducing the threat of flooding. In all, the river has been shortened more than 240 kilometers (150 miles). The program has been somewhat successful in reducing the height of the river in flood stage. However, because the river's tendency toward meandering still exists, preventing the river from returning to its previous course has been difficult.

Artificial cutoffs increase a stream's velocity and may also accelerate erosion of the bed and banks of the channel. A case in point is the Blackwater River in Missouri, whose meandering course was shortened in 1910. Among the many effects of this project was a dramatic increase in the width of the channel caused by the increased velocity of the stream. One particular bridge over the river collapsed because of bank erosion in 1930. Over the next 17 years the same bridge was replaced on three more occasions, each time with a longer span.

A Nonstructural Approach All of the flood-control measures described so far have involved structural solutions aimed at "controlling" a river. These solutions are expensive and often give people residing on the floodplain a false sense of security.

Today many scientists and engineers advocate a nonstructural approach to flood control. They suggest that an alternative to artificial levees, dams, and channelization is sound floodplain management. By identifying high-risk areas, appropriate zoning regulations can be implemented to minimize development and promote more appropriate land use.

Summary

- The *hydrologic cycle* describes the continuous interchange of water among the oceans, atmosphere, and continents. Powered by energy from the Sun, it is a global system in which the atmosphere provides the link between the oceans and continents. The processes involved in the hydrologic cycle include *precipitation, evaporation, infiltration* (the movement of water into rocks or soil through cracks and pore spaces), *runoff* (water that flows over the land), and *transpiration* (the release of water vapor to the atmosphere by plants). *Running water is the single most important agent sculpturing Earth's land surface.*

- The amount of water running off the land rather than sinking into the ground depends upon the *infiltration capacity* of the soil. Initially, runoff flows as broad, thin sheets across the ground, appropriately termed *sheet flow*. After a short distance, threads of current typically develop, and tiny channels called *rills* form.

- The land area that contributes water to a stream is its *drainage basin.* Drainage basins are separated by imaginary lines called *divides*.

- River systems consist of three main parts: the zones of erosion, transportation, and deposition.

- The factors that determine a stream's *velocity* are *gradient* (slope of the stream channel), *shape, size,* and *roughness* of the channel and the stream's *discharge* (amount of water passing a given point per unit of time frequently measured in cubic feet per second). Most often, the gradient and roughness of a stream decrease downstream, while width, depth, discharge, and velocity increase.

- Streams transport their load of sediment in solution (*dissolved load*), in suspension (*suspended load*), and along the bottom of the channel (*bed load*). Much of the dissolved load is contributed by groundwater. Most streams carry the greatest part of their load in suspension.

- A stream's ability to transport solid particles is described using two criteria: *capacity* (the maximum load of solid particles a stream can carry) and *competence* (the maximum particle size a stream can transport). Competence increases as the square of stream velocity, so if velocity doubles, water's force increases fourfold.

- Streams deposit sediment when velocity slows and competence is reduced. This results in *sorting,* the process by which like-sized particles are deposited together. Stream deposits are called *alluvium* and may occur as channel deposits called *bars,* as floodplain deposits, which include *natural levees,* and as *deltas* or *alluvial fans* at the mouths of streams.

- Stream channels are of two basic types: *bedrock channels* and *alluvial channels.* Bedrock channels are most common in headwaters regions where gradients are steep. Rapids and waterfalls are common features. Two types of alluvial channels are *meandering channels* and *braided channels.*

- The two general types of *base level* (the lowest point to which a stream may erode its channel) are (1) *ultimate base level* and (2) *temporary,* or *local base level.* Any change in base level will cause a stream to adjust and establish a new balance. Lowering base level will cause a stream to downcut, whereas raising base level results in deposition of material in the channel.

- Although many gradations exist, the two general types of stream valleys are (1) *narrow V-shaped valleys* and (2) *wide valleys with flat floors.* Because the dominant activity is downcutting toward base level, narrow valleys often contain *waterfalls* and *rapids.*

- When a stream has cut its channel closer to base level, its energy is directed from side to side, and erosion produces a flat valley floor, or *floodplain.* Streams that flow upon floodplains often move in sweeping bends called *meanders.* Widespread meandering may result in shorter channel segments, called *cutoffs,* and/or abandoned bends, called *oxbow lakes.*

- Common *drainage patterns* (the form of a network of streams) produced by a main channel and its tributaries include (1) *dendritic,* (2) *radial,* (3) *rectangular,* and (4) *trellis.*

- *Headward erosion* lengthens a stream course by extending the head of its valley upslope. This process can lead to *stream piracy* (the diversion of the drainage of one stream by another). Former water gaps called *wind gaps* can result from stream piracy.

- *Floods* are triggered by heavy rains and/or snowmelt. Sometimes human interference can worsen or even cause floods. Flood-control measures include the building of *artificial levees* and dams, as well as *channelization,* which could involve creating *artificial cutoffs.* Many scientists and engineers advocate a nonstructural approach to flood control that involves more appropriate land use.

Review Questions

1. Describe the movement of water through the hydrologic cycle. Once precipitation has fallen on land, what paths might the water take?

2. Over the oceans, evaporation exceeds precipitation, yet sea level does not drop. Can you explain why?

3. List several factors that influence infiltration capacity.

4. What are the three main parts (zones) of a river system?

5. A stream originates at 2000 meters above sea level and travels 250 kilometers to the ocean. What is its average gradient in meters per kilometer?

6. Suppose the stream mentioned in Question 5 developed extensive meanders so that its course was lengthened to 500 kilometers. Calculate the new gradient. How does meandering affect gradient?

7. When the discharge of a river increases, what happens to the river's velocity?

8. What typically happens to channel width, channel depth, velocity, and discharge from the point where a stream begins to the point where it ends? Briefly explain why these changes take place.

9. In what three ways does a stream transport its load?

10. If you collect a jar of water from a stream, what part of its load will settle to the bottom of the jar? What portion will remain in the water indefinitely? What part of the stream's load would probably not be present in your sample?

11. Distinguish between capacity and competency.

12. What is settling velocity? What factors influence settling velocity?

13. Are bedrock channels more likely to be found near the head or near the mouth of a stream?

14. Describe a situation that might cause a stream channel to become braided.

15. Define *base level*. Name the main river in your area. For what streams does it act as base level? What is base level for the Mississippi River? The Missouri River?

16. Describe two situations that would trigger the formation of incised meanders.

17. Briefly describe the formation of a natural levee. How is this feature related to back swamps and yazoo tributaries?

18. List two major depositional features, other than natural levees, that are associated with streams. Under what circumstances does each form?

19. How has the construction of artificial levees and dams on the Mississippi River and its tributaries contributed to a shrinking of the Mississippi's delta and extensive wetlands (see Box 16.1)?

20. Each of the following statements refers to a particular drainage pattern. Identify the pattern.
    a. Streams diverging from a central high area, such as a dome.
    b. Branching treelike pattern.
    c. A pattern that develops when bedrock is crisscrossed by joints and faults.

21. Describe how a water gap might form.

22. Contrast regional floods and flash floods. Which type is deadliest?

23. List and briefly describe three basic flood-control strategies. What are some drawbacks of each?

Key Terms

alluvial fan (p. 443)
alluvium (p. 434)
antecedent stream (p. 447)
back swamp (p. 442)
bar (p. 440)
base level (p. 436)
bed load (p. 432)
braided stream (p. 435)
capacity (p. 433)
competence (p. 433)
cut bank (p. 435)
cutoff (p. 435)
delta (p. 440)
dendritic pattern (p. 444)
discharge (p. 429)
dissolved load (p. 434)
distributary (p. 441)

divide (p. 426)
drainage basin (p. 426)
evapotranspiration (p. 426)
flood (p. 447)
floodplain (p. 438)
graded stream (p. 437)
gradient (p. 428)
head (headwaters) (p. 430)
headward erosion (p. 446)
hydrologic cycle (p. 425)
incised meander (p. 439)
infiltration (p. 425)
infiltration capacity (p. 426)
laminar flow (p. 428)
local (temporary) base level
   (p. 436)
longitudinal profile (p. 430)

meander (p. 435)
mouth (p. 430)
natural levee (p. 444)
oxbow lake (p. 435)
point bar (p. 435)
pothole (p. 431)
radial pattern (p. 446)
rectangular pattern (p. 446)
recurrence interval (p. 448)
return period (p. 448)
rills (p. 426)
runoff (p. 426)
saltation (p. 433)
settling velocity (p. 432)
sheet flow (p. 426)
sorting (p. 434)
stream piracy (p. 446)

stream valley (p. 437)
superposed stream (p. 447)
suspended load (p. 432)
temporary (local) base level
   (p. 446)
terrace (p. 440)
transpiration (p. 426)
trellis drainage pattern
   (p. 446)
turbulent flow (p. 428)
ultimate base level (p. 436)
water gap (p. 447)
yazoo tributary (p. 443)

Web Resources

The *Earth* Website uses the resources and flexibility of the Internet to aid in your study of the topics in this chapter. Written and developed by geology instructors, this site will help improve your understanding of geology. Visit **http://www .prenhall.com/tarbuck** and click on the cover of *Earth 9e* to find:

• Online review quizzes.
• Critical thinking exercises.
• Links to chapter-specific Web resources.
• Internet-wide key-term searches.

http://www.prenhall.com/tarbuck

GEODe: Earth

GEODe: Earth makes studying faster and more effective by reinforcing key concepts using animation, video, narration, interactive exercises and practice quizzes. A copy is included with every copy of *Earth*.

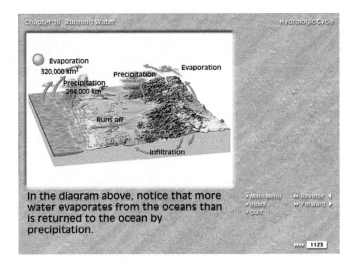

In the diagram above, notice that more water evaporates from the oceans than is returned to the ocean by precipitation.

North America's largest river, the Mississippi, has a huge drainage basin covering about 3 million square kilometers (nearly 1.2 million square miles).

Groundwater

Groundwater provides over 190 billion liters (50 billion gallons) per day in support of the agricultural economy of the United States. (Photo by Michael Collier)

Worldwide, wells and springs provide water for cities, crops, livestock, and industry. In the United States, groundwater is the source of about 40 percent of the water used for all purposes (except hydroelectric power generation and power-plant cooling). Groundwater is the drinking water for more than 50 percent of the population, is 40 percent of the water used for irrigation, and provides more than 25 percent of industry's needs. In some areas, however, overuse of this basic resource has resulted in water shortage, streamflow depletion, land subsidence, contamination by saltwater, increased pumping cost, and groundwater pollution.

Importance of Groundwater

 GEODe Groundwater
▶ Importance and Distribution of Groundwater

Groundwater is one of our most important and widely available resources, yet people's perceptions of the subsurface environment from which it comes are often unclear and incorrect. The reason is that the groundwater environment is largely hidden from view except in caves and mines, and the impressions people gain from these subsurface openings are misleading. Observations on the land surface give an impression that Earth is "solid." This view remains when we enter a cave and see water flowing in a channel that appears to have been cut into solid rock.

Because of such observations, many people believe that groundwater occurs only in underground "rivers." In reality, most of the subsurface environment is not "solid" at all. It includes countless tiny *pore spaces* between grains of soil and sediment, plus narrow joints and fractures in bedrock. Together, these spaces add up to an immense volume. It is in these small openings that groundwater collects and moves.

Considering the entire hydrosphere, or all of Earth's water, only about six-tenths of 1 percent occurs underground. Nevertheless, this small percentage, stored in the rocks and sediments beneath Earth's surface, is a vast quan-

tity. When the oceans are excluded and only sources of fresh water are considered, the significance of groundwater becomes more apparent.

Table 17.1 contains estimates of the distribution of fresh water in the hydrosphere. Clearly the largest volume occurs as glacial ice. Second in rank is groundwater, with slightly more than 14 percent of the total. However, when ice is excluded and just liquid water is considered, more than 94 percent of all fresh water is groundwater. Without question, *groundwater represents the largest reservoir of fresh water that is readily available to humans.* Its value in terms of economics and human well-being is incalculable.

Geologically, groundwater is important as an erosional agent. The dissolving action of groundwater slowly removes soluble rock such as limestone, allowing surface depressions known as *sinkholes* to form as well as creating subterranean caverns (Figure 17.1). Groundwater is also an equalizer of streamflow. Much of the water that flows in rivers is not direct runoff from rain and snowmelt. Rather, a large percentage of precipitation soaks in and then moves slowly underground to stream channels. Groundwater is thus a form of storage that sustains streams during periods when rain does not fall. Therefore, when we see water flowing in a river during a dry period, it is rain that fell at some earlier time and was stored underground.

TABLE 17.1 Fresh Water of the Hydrosphere		
Parts of the Hydrosphere	Volume of Fresh Water (km³)	Share of Total Volume of Fresh Water (percent)
Ice sheets and glaciers	24,000,000	84.945
Groundwater	4,000,000	14.158
Lakes and reservoirs	155,000	0.549
Soil moisture	83,000	0.294
Water vapor in the atmosphere	14,000	0.049
River water	1,200	0.004
Total	28,253,200	100.00

Source: U.S. Geological Survey Water Supply Paper 2220, 1987.

Distribution of Groundwater

 Groundwater
▸ Importance and Distribution of Groundwater

When rain falls, some of the water runs off, some returns to the atmosphere by evaporation and transpiration, and the remainder soaks into the ground. This last path is the primary source of practically all subsurface water. The amount of water that takes each of these paths, however, varies greatly both in time and space. Influential factors include

FIGURE 17.1   **A.** A view of the interior of Three Fingers Cave, Lincoln County, New Mexico. The dissolving action of groundwater created the cavern. Later, groundwater deposited the limestone decorations. (Photo by Harris Photographic/Tom Stack and Associates) **B.** Groundwater was responsible for creating these sinkholes in a limestone plateau north of Jajce, Bosnia and Herzegovina. (Photo by Jerome Wyckoff)

A.

B.

steepness of slope, nature of surface material, intensity of rainfall, and type and amount of vegetation. Heavy rains falling on steep slopes underlain by impervious materials will obviously result in a high percentage of the water running off. Conversely, if rain falls steadily and gently upon more gradual slopes composed of materials that are easily penetrated by the water, a much larger percentage of water soaks into the ground.

Some of the water that soaks in does not travel far, because it is held by molecular attraction as a surface film on soil particles. This near-surface zone is called the **zone of soil moisture.** It is crisscrossed by roots, voids left by decayed roots, and animal and worm burrows that enhance the infiltration of rainwater into the soil. Soil water is used by plants in life functions and transpiration. Some water also evaporates directly back into the atmosphere.

Water that is not held as soil moisture percolates downward until it reaches a zone where all of the open spaces in sediment and rock are completely filled with water (Figure 17.2). This is the **zone of saturation** (also called the *phreatic zone*). Water within it is called **groundwater.** The upper limit of this zone is known as the **water table.** Extending upward from the water table is the **capillary fringe** (*capillus* = hair). Here groundwater is held by surface tension in tiny passages between grains of soil or sediment. The area above the water table that includes the capillary fringe and the zone of soil moisture is called the **unsaturated zone** (also known as the *vadose zone*). Although a considerable amount of water can be present in the unsaturated zone this water cannot be pumped by wells because it clings too tightly to rock and soil particles. By contrast, below the water table the water pressure is great enough to allow water to enter wells, thus permitting groundwater to be withdrawn for use. We will examine wells more closely later in the chapter.

The Water Table

 Groundwater
▸ Importance and Distribution of Groundwater

The water table, the upper limit of the zone of saturation, is a very significant feature of the groundwater system. The water table level is important in predicting the productivity of wells, explaining the changes in the flow of springs and streams, and accounting for fluctuations in the levels of lakes.

Variations in the Water Table

The depth of the water table is highly variable and can range from zero, when it is at the surface, to hundreds of meters in some places. An important characteristic of the water table is that its configuration varies seasonally and from year to year because the addition of water to the groundwater system is closely related to the quantity, distribution, and timing of precipitation. Except where the water table is at the surface, we cannot observe it directly. Nevertheless, its elevation can

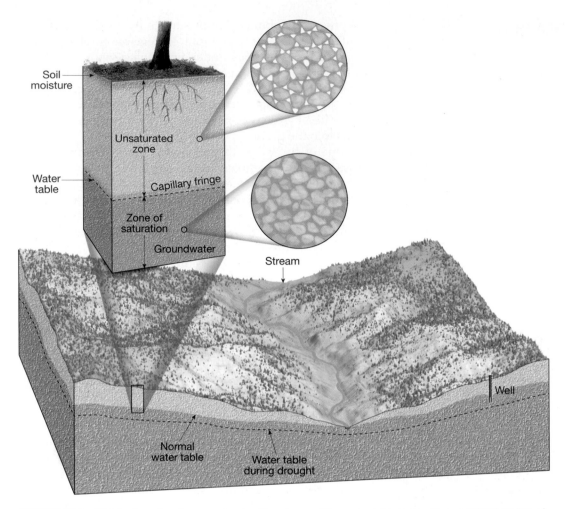

FIGURE 17.2 Distribution of underground water. The shape of the water table is usually a subdued replica of the surface topography. During periods of drought, the water table falls, reducing streamflow and drying up some wells.

be mapped and studied in detail where wells are numerous because the water level in wells coincides with the water table (Figure 17.3). Such maps reveal that the water table is rarely level, as we might expect a table to be. Instead, its shape is usually a subdued replica of the surface topography, reaching its highest elevations beneath hills and then descending toward valleys (Figure 17.2). Where a wetland (swamp) is encountered, the water table is right at the surface. Lakes and streams generally occupy areas low enough that the water table is above the land surface.

Several factors contribute to the irregular surface of the water table. One important influence is the fact that groundwater moves very slowly and at varying rates under different conditions. Because of this, water tends to "pile up" beneath high areas between stream valleys. If rainfall were to cease completely, these water-table "hills" would slowly subside and gradually approach the level of the valleys. However, new supplies of rainwater are usually added frequently enough to prevent this. Nevertheless, in times of extended drought (see Box 17.1), the water table may drop enough to dry up shallow wells (Figure 17.2). Other causes for the uneven water table are variations in rainfall and permeability from place to place.

Interaction between Groundwater and Streams

The interaction between the groundwater system and streams is a basic link in the hydrologic cycle. It can take place in one of three ways. Streams may gain water from the inflow of groundwater through the streambed. Such streams are called **gaining streams** (Figure 17.4A). For this to occur, the elevation of the water table must be higher than the level of the surface of the stream. Streams may lose water to the groundwater system by outflow through the streambed. The term **losing stream** is applied to this situation (Figure 17.4B, C). When this happens, the elevation of the water table must be lower than the surface of the stream. The third possibility is a combination of the first two—a stream gains in some sections and loses in others.

Losing streams can be connected to the groundwater system by a continuous saturated zone or they can be disconnected from the groundwater system by an unsaturated zone. Compare parts B and C in Figure 17.4. When the stream is disconnected, the water table may have a discernible bulge beneath the stream if the rate of water movement through the streambed and zone of aeration is greater

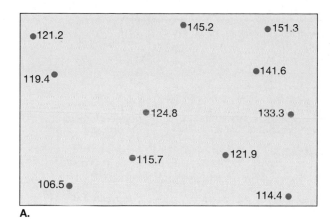

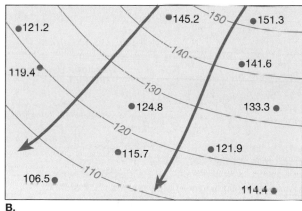

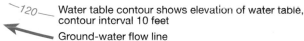

EXPLANATION

- Location of well and elevation of water table above sea level, in feet

~120— Water table contour shows elevation of water table, contour interval 10 feet

⟵ Ground-water flow line

FIGURE 17.3 Preparing a map of the water table. The water level in wells coincides with the water table. **A.** First, the locations of wells and the elevation of the water table above sea level are plotted on a map. **B.** These data points are used to guide the drawing of water-table contour lines at regular intervals. On this sample map the interval is 10 feet. Groundwater flow lines can be added to show water movement in the upper portion of the zone of saturation. Groundwater tends to move approximately perpendicular to the contours and down the slope of the water table. (After U.S Geological Survey)

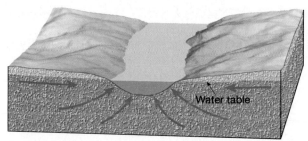

A. Gaining stream

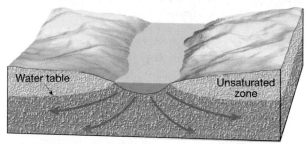

B. Losing stream (connected)

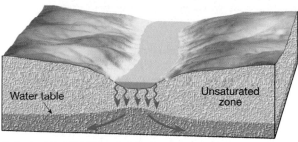

C. Losing stream (disconnected)

FIGURE 17.4 Interaction between the groundwater system and streams. **A.** Gaining streams receive water from the groundwater system. **B.** Losing streams lose water to the groundwater system. **C.** When losing streams are separated from the groundwater system by the unsaturated zone, a bulge may form in the water table. (After U.S. Geological Survey)

than the rate of groundwater movement away from the bulge.

In some settings, a stream might always be a gaining stream or always be a losing stream. However, in many situations flow direction can vary a great deal along a stream; some sections receive groundwater and other sections lose water to the groundwater system. Moreover, the direction of flow can change over a short time span as the result of storms adding water near the streambank or when temporary flood peaks move down the channel.

Groundwater contributes to streams in most geologic and climatic settings. Even where streams are primarily losing water to the groundwater system, certain sections may receive groundwater inflow during some seasons. In one study of 54 streams in all parts of the United States, the analysis indicated that 52 percent of the streamflow was

contributed by groundwater. The groundwater contribution ranged from a low of 14 percent to a maximum of 90 percent. Groundwater is also a major source of water for lakes and wetlands.

Factors Influencing the Storage and Movement of Groundwater

The nature of subsurface materials strongly influences the rate of groundwater movement and the amount of groundwater that can be stored. Two factors are especially important—porosity and permeability.

Porosity

Water soaks into the ground because bedrock, sediment, and soil contain countless voids or openings. These openings are similar to those of a sponge and are often called *pore*

| BOX 17.1 ▶ EARTH AS A SYSTEM |

Drought Impacts the Hydrologic System*

Drought is a period of abnormally dry weather that persists long enough to produce a significant hydrologic imbalance such as crop damage or water supply shortages. Drought severity depends upon the degree of moisture deficiency, its duration, and the size of the affected area.

Although natural disasters such as floods and hurricanes usually generate more attention, droughts can be just as devastating and carry a bigger price tag. On the average, droughts cost the United States $6 to $8 billion annually compared to $2.4 billion for floods and $1.2 to $4.8 billion for hurricanes. Direct economic losses from a severe drought in 1988 were estimated at $61 billion (2002 dollars).

Drought is different from other natural hazards in several ways. First, it occurs in a gradual, "creeping" way, making its onset and end difficult to determine. The effects of drought accumulate slowly over an extended time span and sometimes linger for years after the drought has ended. Second, there is not a precise and universally accepted definition of drought. This adds to the confusion about whether or not a drought is actually occurring and if it is, its severity. Third, drought seldom produces structural damages, so its social and economic effects are less obvious than damages from other natural disasters.

Definitions reflect four basic approaches to measuring drought: meteorological, agricultural, hydrological, and socioeconomic. *Meteorological drought* deals with the degree of dryness based on the departure of precipitation from normal values and the duration of the dry period. *Agricultural drought* is usually linked to a deficit of soil moisture. A plant's need for water depends on prevailing weather conditions, biological characteristics of the particular plant, its stage of growth, and various soil properties. *Hydrological drought* refers to deficiencies in surface and subsurface water supplies. It is measured as streamflow and as lake, reservoir, and groundwater levels. There is a time lag between the onset of dry conditions and a drop in streamflow, or the lowering of lakes, reservoirs, or groundwater levels. So hydrological measurements are not the earliest indicators of drought. *Socioeconomic drought* is a reflection of what hap-

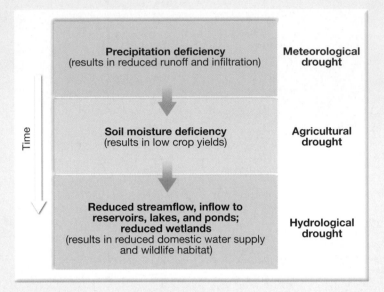

FIGURE 17.A   Sequence of drought impacts. After the onset of meteorological drought, agriculture is affected first, followed by reductions in streamflow and water levels in lakes, reservoirs, and underground. When meteorological drought ends, agricultural drought ends as soil moisture is replenished. It takes a considerably longer time span for hydrological drought to end.

pens when a physical water shortage affects people. Socioeconomic drought occurs when the demand for an economic good exceeds supply as a result of a shortfall in water supply. For example, drought can result in significantly reduced hydroelectric power production, which in turn may require conversion to more expensive fossil fuels and/or significant energy shortfalls.

There is a sequence of impacts associated with meteorological, agricultural, and hydrological drought (Figure 17.A). When meteorological drought begins, the agricultural sector is usually the first to be affected because of its heavy dependence on soil moisture. Soil moisture is rapidly depleted during extended dry periods. If precipitation deficiencies continue, those dependent on rivers, reservoirs, lakes, and groundwater may be affected.

When precipitation returns to normal, meteorological drought comes to an end. Soil moisture is replenished first, followed by streamflow, reservoirs and lakes, and finally groundwater. Thus, drought impacts may diminish rapidly in the agricultural

sector because of its reliance on soil moisture but linger for months or years in other sectors that depend on stored surface or subsurface water supplies. Groundwater users, who are often the last to be affected following the onset of meteorological drought, may also be the last to experience a return to normal water levels. The length of the recovery period depends upon the intensity of the meteorological drought, its duration, and the quantity of precipitation received when the drought ends.

The impacts suffered because of drought are the product of both the meteorological event and the vulnerability of society to periods of precipitation deficiency. As demand for water increases as a result of population growth and regional population shifts, future droughts can be expected to produce greater impacts whether or not there is any increase in the frequency or intensity of meteorological drought.

*Based in part on material prepared by the National Drought Mitigation Center (http:// drought.unl.edu).

spaces. The quantity of groundwater that can be stored depends on the **porosity** of the material, which is the percentage of the total volume of rock or sediment that consists of pore spaces. Voids most often are spaces between sedimentary particles, but also common are joints, faults, cavities formed by the dissolving of soluble rock such as limestone, and vesicles (voids left by gases escaping from lava).

Variations in porosity can be great. Sediment is commonly quite porous, and open spaces may occupy 10 to 50 percent of the sediment's total volume. Pore space depends on the size and shape of the grains, how they are packed together, the degree of sorting, and in sedimentary rocks, the amount of cementing material. For example, clay may have a porosity as high as 50 percent, whereas some gravels may have only 20 percent voids.

Where sediments are poorly sorted, the porosity is reduced because the finer particles tend to fill the openings among the larger grains (see Figure 7.6 on p. 199). Most igneous and metamorphic rocks, as well as some sedimentary rocks, are composed of tightly interlocking crystals so the voids between the grains may be negligible. In these rocks, fractures must provide the voids.

Permeability, Aquitards, and Aquifers

Porosity alone cannot measure a material's capacity to yield groundwater. Rock or sediment may be very porous yet still not allow water to move through it. The pores must be *connected* to allow water flow, and they must be *large enough* to allow flow. Thus, the **permeability** (*permeare* = to penetrate) of a material—its ability to *transmit* a fluid—is also very important.

Groundwater moves by twisting and turning through small interconnected openings. The smaller the pore spaces, the slower the water moves. This idea is clearly illustrated by examining the information about the water-yielding potential of different materials in Table 17.2. Here groundwater is divided into two categories: (1) the portion that will drain under the influence of gravity (called *specific yield*) and (2) the part that is retained as a film on particle and rock surfaces and in tiny openings (called *specific retention*). Specific yield indicates how much water is actually available for use,

whereas specific retention indicates how much water remains bound in the material. For example, clay's ability to store water is great, owing to its high porosity, but its pore spaces are so small that water is unable to move through it. Thus, clay's porosity is high but because its permeability is poor, clay has a very low specific yield.

Impermeable layers that hinder or prevent water movement are termed **aquitards** (*aqua* = water, *tard* = slow). Clay is a good example. On the other hand, larger particles, such as sand or gravel, have larger pore spaces. Therefore, the water moves with relative ease. Permeable rock strata or sediment that transmit groundwater freely are called **aquifers** (*aqua* = water, *fer* = carry). Sands and gravels are common examples.

In summary, you have seen that porosity is not always a reliable guide to the amount of groundwater that can be produced, and permeability is significant in determining the rate of groundwater movement and the quantity of water that might be pumped from a well.

Movement of Groundwater

The movement of water in the atmosphere and on the land surface is relatively easy to visualize, but the movement of groundwater is not. Near the beginning of the chapter we mentioned the common misconception that groundwater occurs in underground rivers that resemble surface streams. Although subsurface streams do exist, they are *not* common. Rather, as you learned in the preceding sections, groundwater exists in the pore spaces and fractures in rock and sediment. Thus, contrary to any impressions of rapid flow that an underground river might evoke, the movement of most groundwater is exceedingly slow, from pore to pore. By exceedingly slow, we mean typical rates of a few centimeters each day.

Figure 17.5 depicts a simple example of a *groundwater flow system*—a three-dimensional body of Earth material saturated with moving groundwater. It shows groundwater moving along flow paths from areas of recharge to a zone of discharge along a stream. Discharge also occurs at springs, lakes, or wetlands, and in coastal areas, as groundwater seeps into bays or the ocean. Transpiration by plants whose

TABLE 17.2 Selected Values of Porosity, Specific Yield, and Specific Retention*			
Material	Porosity	Specific Yield	Specific Retention
Clay	50	2	48
Sand	25	22	3
Gravel	20	19	1
Limestone	20	18	2
Sandstone (semiconsolidated)	11	6	5
Granite	0.1	0.09	0.01
Basalt (fresh)	11	8	3

*Values in percent by volume.
Source: U.S. Geological Survey Water Supply Paper 2220, 1987.

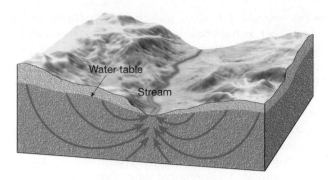

FIGURE 17.5 Arrows indicate groundwater movement through uniformly permeable material. The looping curves may be thought of as a compromise between the downward pull of gravity and the tendency of water to move toward areas of reduced pressure.

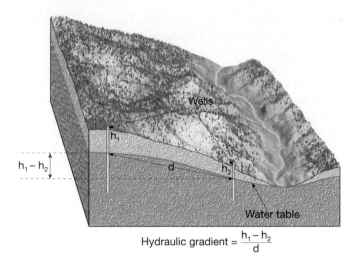

Hydraulic gradient $= \dfrac{h_1 - h_2}{d}$

FIGURE 17.6 The hydraulic gradient is determined by measuring the difference in elevation between two points on the water table ($h_1 - h_2$) divided by the distance between them, d. Wells are used to determine the height of the water table.

roots extend to near the water table represents another form of groundwater discharge.

The energy that makes groundwater move is provided by the force of gravity. In response to gravity, water moves from areas where the water table is high to zones where the water table is lower. Although some water takes the most direct path down the slope of the water table, much of the water follows long, curving paths.

Figure 17.5 shows water percolating into a stream from all possible directions. Some paths clearly turn upward, apparently against the force of gravity, and enter through the bottom of the channel. This is easily explained: The deeper you go into the zone of saturation, the greater is the water pressure. Thus, the looping curves followed by water in the saturated zone may be thought of as a compromise between the downward pull of gravity and the tendency of water to move toward areas of reduced pressure. As a result, water at any given height is under greater pressure beneath a hill than beneath a stream channel, and the water tends to migrate toward points of lower pressure.

Darcy's Law

The foundations of our modern understanding of groundwater movement began in the mid-19th century with the work of the French engineer Henri Darcy. During this time, Darcy made measurements and conducted experiments in an attempt to determine whether the water needs of the city of Dijon in east–central France could be met by tapping local supplies of groundwater. Among the experiments carried out by Darcy was one that showed that the velocity of groundwater flow is proportional to the slope of the water table—the steeper the slope, the faster the water moves (because the steeper the slope, the greater the pressure difference between two points). The water-table slope is known as the **hydraulic gradient** and can be expressed as follows:

$$\text{hydraulic gradient} = \frac{h_1 - h_2}{d}$$

where h_1 is the elevation of one point on the water table, h_2 the elevation of a second point, and d is the horizontal distance between the two points (Figure 17.6).

Darcy also experimented with different materials such as coarse sand and fine sand by measuring the rate of flow through sediment-filled tubes that were tilted at varying angles. He found that the flow velocity varied with the permeability of the sediment—groundwater flows more rapidly through sediments having greater permeability than through materials having lower permeability. This factor is known as **hydraulic conductivity** and is a coefficient that takes into account the permeability of the aquifer and the viscosity of the fluid.

To determine discharge (Q), that is, the actual volume of water that flows through an aquifer in a specified time, the following equation is used:

$$Q = \frac{KA(h_1 - h_2)}{d}$$

where $\dfrac{h_1 - h_2}{d}$ is the hydraulic gradient, K is the coefficient that represents hydraulic conductivity, and A is the cross-sectional area of the aquifer. This expression has come to be called **Darcy's law** in honor of the pioneering French scientist-engineer.

Different Scales of Movement

The extent of groundwater flow systems varies from a few square kilometers or less to tens of thousands of square kilometers. The length of flow paths ranges from a few meters to tens and sometimes hundreds of kilometers. Figure 17.7 is a cross section of a hypothetical region in which a deep

into a more distant one. Finally, the black arrows show groundwater movement in a deep regional system that lies beneath the more shallow ones and is connected to them. The horizontal scale of the figure could range from tens to hundreds of kilometers.

Springs

 Groundwater
▸ **Springs and Wells**

Springs have aroused the curiosity and wonder of people for thousands of years. The fact that springs were, and to some people still are, rather mysterious phenomena is not difficult to understand, for here is water flowing freely from the ground in all kinds of weather in seemingly inexhaustible supply, but with no obvious source.

Not until the middle of the 17th century did the French physicist Pierre Perrault invalidate the age-old assumption that precipitation could not adequately account for the amount of water emanating from springs and flowing in rivers. Over several years Perrault computed the quantity of water that fell on France's Seine River basin. He then calculated the mean annual runoff by measuring the river's discharge. After allowing for the loss of water by evaporation, he showed that there was sufficient water remaining to feed the springs. Thanks to Perrault's pioneering efforts and the measurements by many afterward, we now know that the

How does the rate of groundwater movement compare to the velocity of streams?

Velocities of groundwater flow are generally orders of magnitude less than velocities of streamflow. A velocity of 1 foot per day or greater is a high rate of movement for groundwater, and groundwater velocities can be as low as 1 foot per year or 1 foot per decade. In contrast, rates of streamflow are generally measured in feet per second. A velocity of 1 foot per second equals about 16 miles per day.

groundwater flow system is overlain by and connected to several, more shallow local flow systems. The subsurface geology exhibits a complicated arrangement of high hydraulic-conductivity aquifer units and low hydraulic-conductivity aquitard units.

Starting near the top of Figure 17.7, the blue arrows represent water movement in several local groundwater systems that occur in the upper water table aquifer. They are separated by groundwater divides at the center of the hills and discharge into the nearest surface water body. Beneath these most shallow systems, red arrows show water movement in a somewhat deeper system in which groundwater does not discharge into the nearest surface water body, but

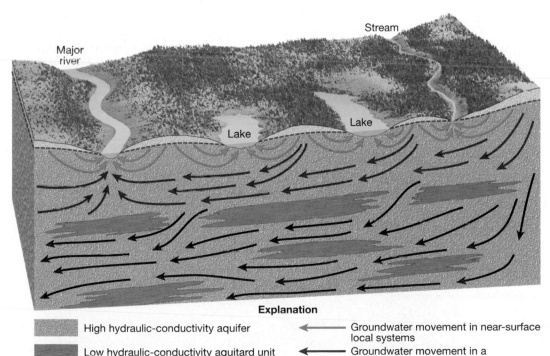

FIGURE 17.7 A hypothetical groundwater flow system that includes subsystems at different scales. Variations in surface topography and subsurface geology can produce a complex situation. The horizontal scale of the figure could range from tens to hundreds of kilometers. (After U. S. Geological Survey)

Explanation

- High hydraulic-conductivity aquifer
- Low hydraulic-conductivity aquitard unit
- ------- Water table
- ← Groundwater movement in near-surface local systems
- ← Groundwater movement in a subregional system
- ← Groundwater movement in a deep regional system

FIGURE 17.8 Spring in Arizona's Marble Canyon. (Photo by Michael Collier)

source of springs is water from the zone of saturation and that the ultimate source of this water is precipitation.

Whenever the water table intersects Earth's surface, a natural outflow of groundwater results, which we call a **spring** (Figure 17.8). Springs such as the one pictured in Figure 17.8 form when an aquitard blocks the downward movement of groundwater and forces it to move laterally. Where the permeable bed outcrops, a spring results. Another situation leading to the formation of a spring is illustrated in Figure 17.9. Here an aquitard is situated above the main water table. As water percolates downward, a portion of it is intercepted by the aquitard, thereby creating a localized zone of saturation and a **perched water table.**

Springs, however, are not confined to places where a perched water table creates a flow at the surface. Many geological situations lead to the formation of springs because subsurface conditions vary greatly from place to place. Even in areas underlain

by impermeable crystalline rocks, permeable zones may exist in the form of fractures or solution channels. If these openings fill with water and intersect the ground surface along a slope, a spring will result.

Hot Springs and Geysers

By definition, the water in **hot springs** is 6–9°C (10–15°F) warmer than the mean annual air temperature for the localities where they occur. In the United States alone, there are more than 1000 such springs (Figure 17.10).

Temperatures in deep mines and oil wells usually rise with increasing depth, an average of about 2°C per 100 meters (1°F per 100 feet). Therefore, when groundwater circulates at great depths, it becomes heated. If it rises to the surface, the water may emerge as a hot spring. The water of some hot springs in the eastern United States is heated in this manner. However, the great majority (more than 95 percent) of the hot springs (and geysers) in the United States are found in the West (Figure 17.10). The reason for such a distribution is that the source of heat for most hot springs is cooling igneous rock, and it is in the West that igneous activity has occurred more recently.

Geysers are intermittent hot springs or fountains in which columns of water are ejected with great force at various intervals, often rising 30–60 meters (100–200 feet) into the air. After the jet of water ceases, a column of steam rushes out, usually with a thunderous roar. Perhaps the most famous geyser in the world is Old Faithful in Yellowstone National Park, which erupts about once each hour (Figure 17.11). The great abundance, diversity, and spectacular nature of Yellowstone's geysers and other thermal features undoubtedly was the primary reason for its becoming the first national park in the United States. Geysers are also found in other parts of the world, notably New Zealand and Iceland.

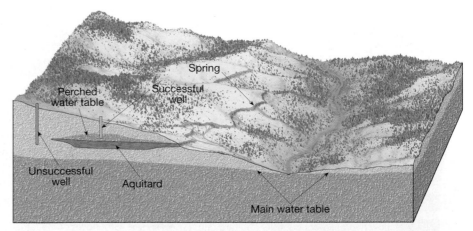

FIGURE 17.9 When an aquitard is situated above the main water table, a localized zone of saturation may result. Where the perched water table intersects the side of the valley, a spring flows. The perched water table also caused the well on the right to be successful, whereas the well on the left will be unsuccessful unless it is drilled to a greater depth.

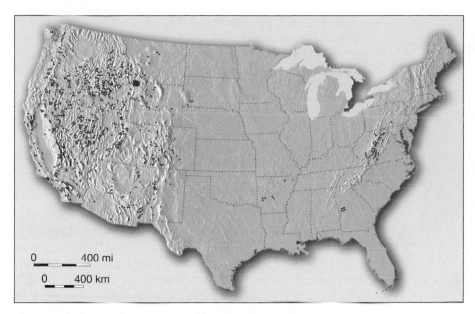

FIGURE 17.10   Distribution of hot springs and geysers in the United States. Note the concentration in the West, where igneous activity has been most recent. (After G. A. Waring, U.S. Geological Survey Professional Paper 492, 1965)

In fact, the Icelandic word *geysa*, to gush, gives us the name *geyser*.

Geysers occur where extensive underground chambers exist within hot igneous rocks. How they operate is shown

FIGURE 17.11   A wintertime eruption of Old Faithful, one of the world's most famous geysers. It emits as much as 45,000 liters (almost 12,000 gallons) of hot water and steam about once each hour. (Photo by Marc Muench/David Muench Photography, Inc.)

in Figure 17.12. As relatively cool groundwater enters the chambers, it is heated by the surrounding rock. At the bottom of the chambers, the water is under great pressure because of the weight of the overlying water. This great pressure prevents the water from boiling at the normal surface temperature of 100°C (212°F). For example, water at the bottom of a 300-meter (1000-foot) water-filled chamber must attain nearly 230°C before it will boil. The heating causes the water to expand, with the result that some is forced out at the surface. This loss of water reduces the pressure on the remaining water in the chamber, which lowers the boiling point. A portion of the water deep within the chamber quickly turns to steam, and the geyser erupts (Figure 17.12). Following eruption, cool groundwater again seeps into the chamber and the cycle begins anew.

When groundwater from hot springs and geysers flows out at the surface, material in solution is often precipitated, producing an accumulation of chemical sedimentary rock. The material deposited at any given place commonly reflects the chemical makeup of the rock through which the water circulated. When the water contains dissolved silica, a material called *siliceous sinter* or *geyserite* is deposited around the spring. When the water contains dissolved calcium carbonate, a form of limestone called *travertine* or *calcareous tufa* is deposited. The latter term is used if the material is spongy and porous.

The deposits at Mammoth Hot Springs in Yellowstone National Park are more spectacular than most (Figure 17.13). As the hot water flows upward through a series of channels and then out at the surface, the reduced pressure allows carbon dioxide to separate and escape from the

Students Sometimes Ask . . .

I know Old Faithful in Yellowstone National Park is the most famous geyser. Is it the largest?

No. It appears as though that distinction goes to Yellowstone's Steamboat Geyser, at least if we use the word "large" to mean "tall." During a major eruption, Steamboat Geyser can spew plumes of water 90 meters (300 feet) high for as long as 40 minutes. Following this water phase, the steam phase features powerful bursts of hot mist that extend 150 meters (500 feet) into the sky. Like most of Yellowstone's geysers, Steamboat Geyser is not reliable like Old Faithful. Intervals between eruptions can vary from three days to 50 years. The geyser, which was completely dormant from 1911 to 1961, has erupted fewer than 10 times since 1989.

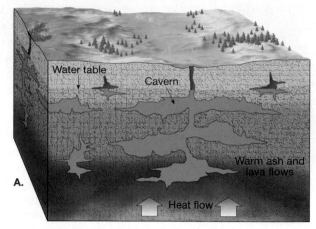

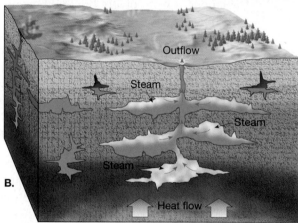

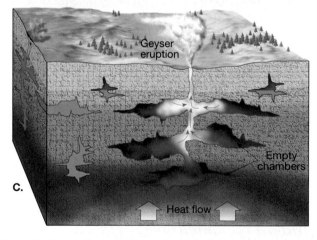

FIGURE 17.12 Idealized diagrams of a geyser. A geyser can form if the heat is not distributed by convection. **A.** In this figure, the water near the bottom is heated to near its boiling point. The boiling point is higher there than at the surface because the weight of the water above increases the pressure. **B.** The water higher in the geyser system is also heated; therefore, it expands and flows out at the top, reducing the pressure on the water at the bottom. **C.** At the reduced pressure on the bottom, boiling occurs. Some of the bottom water flashes into steam, and the expanding steam causes an eruption.

water. The loss of carbon dioxide causes the water to become supersaturated with calcium carbonate, which then precipitates. In addition to containing dissolved silica and calcium carbonate, some hot springs contain sulfur, which gives water a poor taste and unpleasant odor. Undoubtedly, Rotten Egg Spring, Nevada, is such a situation.

Wells

 GEODe Groundwater
▶ Springs and Wells

The most common method for removing groundwater is the **well,** a hole bored into the zone of saturation. Wells serve as small reservoirs into which groundwater migrates and from which it can be pumped to the surface. The use of wells dates back many centuries and continues to be an important method of obtaining water today. By far the single greatest use of this water in the United States is irrigation for agriculture. More than 65 percent of the groundwater used each year is for this purpose. Industrial uses rank a distant second, followed by the amount used in city water systems and rural homes.

The water-table level may fluctuate considerably during the course of a year, dropping during the dry seasons and rising following periods of rain. Therefore, to ensure a continuous supply of water, a well must penetrate below the water table. Whenever water is withdrawn from a well, the water table around the well is lowered. This effect, termed **drawdown,** decreases with increasing distance from the well. The result is a depression in the water table, roughly conical in shape, known as a **cone of depression** (Figure 17.14). Because the cone of depression increases the hydraulic gradient near the well, groundwater will flow more rapidly toward the opening. For most smaller domestic wells, the cone of depression is negligible. However, when wells are heavily pumped for irrigation or for industrial

---

Students Sometimes Ask . . .

I have heard people say that supplies of groundwater can be located using a forked stick. Can this actually be done?

What you describe is a practice called "water dowsing." In the classic method, a person holding a forked stick walks back and forth over an area. When water is detected, the bottom of the "Y" is supposed to be attracted downward.

Geologists and engineers are dubious, to say the least. Case histories and demonstrations may seem convincing, but when dowsing is exposed to scientific scrutiny, it fails. Most "successful" examples of water dowsing occur in places where water would be hard to miss. In a region of adequate rainfall and favorable geology, it is difficult to drill and not find water!

FIGURE 17.13 Mammoth Hot Springs at Yellowstone National Park. Although most of the deposits associated with geysers and hot springs in Yellowstone Park are silica-rich geyserite, the deposits at Mammoth Hot Springs consist of a form of limestone called travertine. (Photo by Stephen Trimble)

purposes, the withdrawal of water can be great enough to create a very wide and steep cone of depression. This may substantially lower the water table in an area and cause nearby shallow wells to become dry. Figure 17.14 illustrates this situation.

Digging a successful well is a familiar problem for people in areas where groundwater is the primary source of supply. One well may be successful at a depth of 10 meters (33 feet), whereas a neighbor may have to go twice as deep to find an adequate supply. Still others may be forced to go deeper or try a different site altogether. When subsurface materials are heterogeneous, the amount of water a well is capable of providing may vary a great deal over short distances. For example, when two nearby wells are drilled to the same level and only one is successful, it may be caused by the presence of a perched water table beneath one of them. Such a case is shown in Figure 17.9. Massive igneous and metamorphic rocks provide a second example. These crystalline rocks are usually not very permeable except where they are cut by many intersecting joints and fractures. Therefore, when a well drilled into such rock does not intersect an adequate network of fractures, it is likely to be unproductive.

Artesian Wells

 GEODe Groundwater
▸ Springs and Wells

In most wells, water cannot rise on its own. If water is first encountered at 30 meters depth, it remains at that level, fluctuating perhaps a meter or two with seasonal wet and dry

periods. However, in some wells, water rises, sometimes overflowing at the surface. Such wells are abundant in the *Artois* region of northern France, and so we call these self-rising wells *artesian.*

To many people the term *artesian* is applied to any well drilled to great depths. This use of the term is incorrect. Others believe that an artesian well must flow freely at the surface (Figure 17.15). Although this is a more correct notion than the first, it represents too narrow a definition. The term **artesian** is applied to *any* situation in which groundwater under pressure rises above the level of the aquifer. As we shall see, this does not always mean a free-flowing surface discharge.

For an artesian system to exist, two conditions usually are met (Figure 17.15): (1) water is confined to an aquifer that is inclined so that one end can receive water; and (2) aquitards, both above and below the aquifer, must be present to prevent the water from escaping. (Such an aquifer is called a *confined aquifer.*) When such a layer is tapped, the pressure created by the weight of the water above will force the water to rise. If there were no friction, the water in the well would rise to the level of the water at the top of the aquifer. However, friction reduces the height of the pressure surface. The greater the distance from the recharge area (where water enters the inclined aquifer), the greater the friction and the less the rise of water.

In Figure 17.16, well 1 is a **nonflowing artesian well,** because at this location the pressure surface is below ground level. When the pressure surface is above the ground and a well is drilled into the aquifer, a **flowing artesian well** is created (well 2, Figure 17.16). Not all artesian systems are wells. *Artesian springs* also exist. Here groundwater may reach the surface by rising along a natural fracture such as a fault rather than through an artificially produced hole. In deserts, artesian springs are sometimes responsible for creating an oasis.

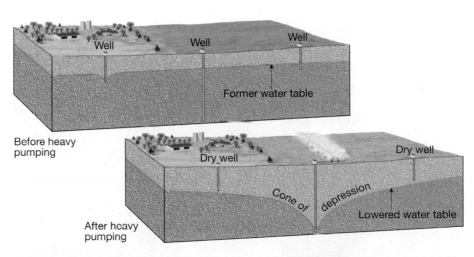

FIGURE 17.14 A cone of depression in the water table often forms around a pumping well. If heavy pumping lowers the water table, the shallow wells may be left dry.

FIGURE 17.15 Sometimes water flows freely at the surface when an artesian well is developed. However, for most artesian wells, the water must be pumped to the surface. (Photo by James E. Patterson)

Artesian systems act as conduits, often transmitting water great distances from remote areas of recharge to points of discharge. A well-known artesian system in South Dakota is a good example of this. In the western part of the state, the edges of a series of sedimentary layers have been bent up to the surface along the flanks of the Black Hills. One of these beds, the permeable Dakota Sandstone, is sandwiched between impermeable strata and gradually dips into the ground toward the east. When the aquifer was first tapped, water poured from the ground surface, creating fountains many meters high (Figure 17.17). In some places the force of the water was sufficient to power waterwheels. Scenes such as the one pictured in Figure 17.17, however, can no longer occur, because thousands of additional wells now tap the same aquifer. This depleted the reservoir, and the water table in the recharge area was lowered. As a consequence, the pressure dropped to the point where many wells stopped flowing altogether and had to be pumped.

On a different scale, city water systems can be considered examples of artificial artesian systems (Figure 17.18). The water tower, into which water is pumped, would represent the area of recharge, the pipes the confined aquifer, and the faucets in homes the flowing artesian wells.

Problems Associated with Groundwater Withdrawal

As with many of our valuable natural resources, groundwater is being exploited at an increasing rate. In some areas, overuse threatens the groundwater supply. In other places, groundwater withdrawal has caused the ground and everything resting on it to sink. Still other localities are concerned with the possible contamination of the groundwater supply.

Treating Groundwater as a Nonrenewable Resource

Many natural systems tend to establish a condition of equilibrium. The groundwater system is no exception. The water table's height reflects a balance between the rate of infiltration and the rate of discharge and withdrawal. Any imbalance will either raise or lower the water table. Long-term imbalances can lead to a significant drop in the water table if there is either a decrease in recharge due to prolonged drought, or an increase in groundwater discharge or withdrawal.

For many, groundwater appears to be an endlessly renewable resource because it is continually replenished by rainfall and melting snow. But in some regions, groundwater has been and continues to be treated as a *nonrenewable* resource. Where this occurs, the water available to recharge the aquifer falls significantly short of the amount being withdrawn.

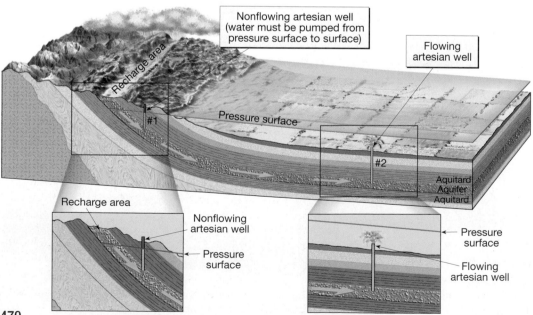

FIGURE 17.16 Artesian systems occur when an inclined aquifer is surrounded by impermeable beds.

FIGURE 17.17 A "gusherlike" flowing artesian well in South Dakota in the early part of the twentieth century. Thousands of additional wells now tap the same confined aquifer; thus, the pressure has dropped to the point that many wells stopped flowing altogether and have to be pumped. (Photo by N. H. Darton, U.S. Geological Survey)

The High Plains provides one example. Here an extensive agricultural economy is largely dependent on irrigation (Figure 17.19). In some parts of the region, where intense irrigation has been practiced for an extended period, depletion of groundwater has been severe. Under these circumstances, it can be said that the groundwater is literally being "mined." Even if pumping were to cease immediately, it could take hundreds or thousands of years for the groundwater to be fully replenished.

Subsidence

As you will see later in this chapter, surface subsidence can result from natural processes related to groundwater. However, the ground may also sink when water is pumped from

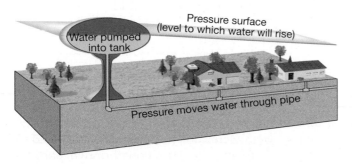

FIGURE 17.18 City water systems can be considered to be artificial artesian systems.

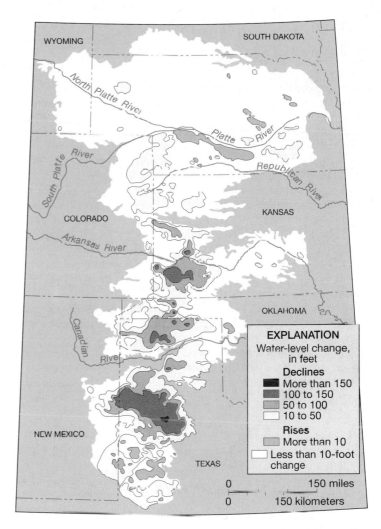

FIGURE 17.19 Changes in groundwater levels in the High Plains aquifer from predevelopment to 1997. Extensive pumping for irrigation has led to water level declines in excess of 100 feet in parts of Kansas, Oklahoma, Texas, and New Mexico. Water level rises have occurred where surface water is used for irrigation, such as along the Platte River in Nebraska. (After U.S. Geological Survey)

wells faster than natural recharge processes can replace it. This effect is particularly pronounced in areas underlain by thick layers of unconsolidated sediments. As the water is withdrawn, the water pressure drops and the weight of the overburden is transferred to the sediment. The greater pressure packs the sediment grains tightly together, and the ground subsides. The size of the area affected by such subsidence is significant. In the contiguous United States, it amounts to an estimated 26,000 square kilometers—an area about the same size as the state of Massachusetts!

Many areas may be used to illustrate land subsidence caused by the excessive pumping of groundwater from relatively loose sediment. A classic example in the United States occurred in the San Joaquin Valley of California and is discussed in Box 17.2. Many other cases of land subsidence due to groundwater pumping exist in the United States, including Las Vegas, Nevada; New Orleans and Baton Rouge, Louisiana; and the Houston–Galveston area of Texas. In the

BOX 17.2 ▶ PEOPLE AND THE ENVIRONMENT

Land Subsidence in the San Joaquin Valley

The San Joaquin Valley is a broad structural basin that contains a thick fill of sediments. The size of Maryland, it constitutes the southern two-thirds of California's Central Valley, a flatland separating two mountain ranges—the Coast Ranges to the west and the Sierra Nevada to the east (Figure 17.B). The valley's aquifer system is a mixture of alluvial materials derived from the surrounding mountains. Sediment thicknesses average about 870 meters (about half a mile). The valley's climate is arid to semiarid, with average annual precipitation ranging from 12 to 35 centimeters (5 to 14 inches).

The San Joaquin Valley has a strong agricultural economy that requires large quantities of water for irrigation. For many years up to 50 percent of this need was met by groundwater. In addition, nearly every city in the region uses groundwater as its principal source for homes and industry.

FIGURE 17.B   The shaded area shows California's San Joaquin Valley.

Although development of the valley's groundwater for irrigation began in the late 1800s, land subsidence did not begin until the mid-1920s when withdrawals were substantially increased. By the early 1970s, water levels had declined up to 120 meters (400 feet). The resulting ground subsidence exceeded 8.5 meters (29 feet) at one place in the region (Figure 17.C). At that time, areas within the valley were subsiding faster than 0.3 meter (1 foot) per year.

Then, because surface water was being imported and groundwater pumping was being decreased, water levels in the aquifer recovered and subsidence ceased. However, during a drought in 1976–1977, heavy groundwater pumping led to renewed subsidence. This time water levels dropped much faster because of the reduced storage capacity caused by earlier compaction of sediments. In all, half the entire valley was affected by subsidence. According to the U.S. Geological Survey:

> Subsidence in the San Joaquin Valley probably represents one of the greatest single manmade alterations in the configuration of the Earth's surface. . . . It has caused serious and costly problems in construction and maintenance of water-transport structures, highways, and highway structures; also many millions of dollars have been spent on the repair or replacement of deep-water wells. Subsidence, besides changing the gradient and course of valley creeks and streams, has caused unexpected flooding, costing farmers many hundreds of thousands of dollars in recurrent land leveling.*

Similar effects have been documented in the San Jose area of the Santa Clara Valley, California, where between 1916 and 1966 subsidence approached 4 meters. Flooding of lands bordering the southern part of San Francisco Bay was one of the results. As was the case in the San Joaquin Valley, the subsidence stopped when imports of surface

FIGURE 17.C   The marks on this utility pole indicate the level of the surrounding land in preceding years. Between 1925 and 1975 this part of the San Joaquin Valley subsided almost 9 meters because of the withdrawal of groundwater and the resulting compaction of sediments. (Photo courtesy of U.S. Geological Survey)

water were increased, allowing groundwater withdrawal rates to be decreased.

---

*Ireland R. L., J. F. Poland, and F. S. Riley, *Land Subsidence in the San Joaquin Valley, California, as of 1980*, U.S. Geological Survey Professional Paper 437-I (Washington, D.C.: U.S. Government Printing Office, 1984), p. 11.

low-lying coastal area between Houston and Galveston, land subsidence ranges from 1.5 to 3 meters (5 to 9 feet). The result is that about 78 square kilometers (30 square miles) are permanently flooded.

Outside the United States, one of the most spectacular examples of subsidence occurred in Mexico City, which is built on a former lake bed. In the first half of the 20th century, thousands of wells were sunk into the water-saturated sediments beneath the city. As water was withdrawn, portions of the city subsided by as much as 6 to 7 meters. In some places buildings have sunk to such a point that access to them from the street is located at what used to be the second-floor level!

Saltwater Contamination

In many coastal areas the groundwater resource is being threatened by the encroachment of saltwater. To understand this problem, let us examine the relationship between fresh groundwater and salt groundwater. Figure 17.20A is a diagrammatic cross section that illustrates this relationship in a coastal area underlain by permeable homogeneous materials. Fresh water is less dense than saltwater, so it floats on the saltwater and forms a large lens-shaped body that may extend to considerable depths below sea level. In such a situation, if the water table is 1 meter above sea level, the base of the freshwater body will extend to a depth of about 40 meters below sea level. Stated another way, the depth of the fresh water below sea level is about 40 times greater than the elevation of the water table above sea level. Thus, when excessive pumping lowers the water table by a certain amount, the bottom of the freshwater zone will rise by 40 times that amount. Therefore, if groundwater withdrawal continues to exceed recharge, there will come a time when the elevation of the saltwater will be sufficiently high to be drawn into wells, thus contaminating the freshwater supply (Figure 17.20B). Deep wells and wells near the shore are usually the first to be affected.

In urbanized coastal areas, the problems created by excessive pumping are compounded by a decrease in the rate of natural recharge. As more and more of the surface is covered by streets, parking lots, and buildings, infiltration into the soil is diminished.

In an attempt to correct the problem of saltwater contamination of groundwater resources, a network of recharge wells may be used. These wells allow wastewater to be pumped back into the groundwater system. A second method of correction is accomplished by building large basins. These basins collect surface drainage and allow it to seep into the ground. On New York's Long Island, where the problem of saltwater contamination was recognized more than 40 years ago, both of these methods have been employed with considerable success.

Contamination of freshwater aquifers by saltwater is primarily a problem in coastal areas, but it can also threaten noncoastal locations. Many ancient sedimentary rocks of marine origin were deposited when the ocean covered places that are now far inland. In some instances significant quantities of seawater were trapped and still remain in the rock. These strata sometimes contain quantities of fresh water and may be pumped for use by people. However, if fresh water is removed more rapidly than it is replenished, saline water may encroach and render the wells unusable. Such a situation threatened users of a deep (Cambrian age) sandstone aquifer in the Chicago area. To counteract this, water from Lake Michigan was allocated to the affected communities to offset the rate of withdrawal from the aquifer.

Groundwater Contamination

The pollution of groundwater is a serious matter, particularly in areas where aquifers provide a large part of the water supply. One common source of groundwater pollution is sewage. Its sources include an ever increasing number of septic tanks, as well as inadequate or broken sewer systems and farm wastes.

If sewage water that is contaminated with bacteria enters the groundwater system, it may become purified through natural processes. The harmful bacteria may be mechanically filtered by the sediment through which the water percolates, destroyed by chemical oxidation, and/or assimilated by other organisms. For purification to occur, however, the aquifer must be of the correct composition. For example, extremely permeable aquifers (such as highly fractured crystalline rock, coarse gravel, or cavernous limestone) have such large openings that contaminated groundwater may travel long distances without being cleansed. In this case, the water flows too rapidly and is not in contact with the surrounding material long enough for purification to occur. This is the problem at well 1 in Figure 17.21A.

On the other hand, when the aquifer is composed of sand or permeable sandstone, it can sometimes be purified after traveling only a few dozen meters through it. The openings between sand grains are large enough to permit water movement, yet the movement of the water is slow enough to allow ample time for its purification (well 2, Figure 17.21B).

Sometimes sinking a well can lead to groundwater pollution problems. If the well pumps a sufficient quantity of water, the cone of depression will locally increase the slope

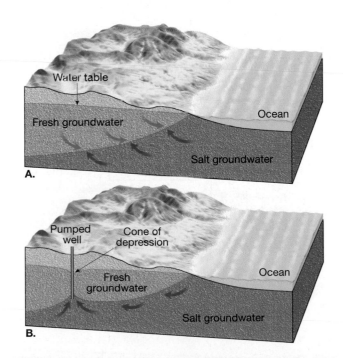

FIGURE 17.20 **A.** Because fresh water is less dense than saltwater, it floats on the saltwater and forms a lens-shaped body that may extend to considerable depths below sea level. **B.** When excessive pumping lowers the water table, the base of the freshwater zone will rise by 40 times that amount. The result may be saltwater contamination of wells.

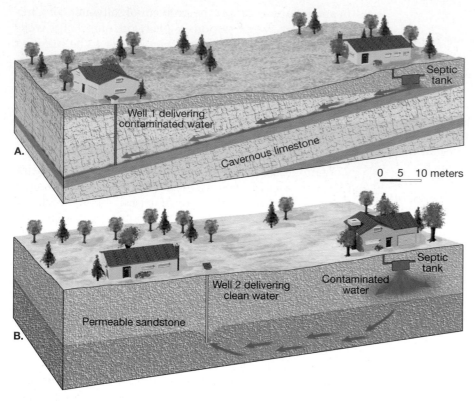

FIGURE 17.21 **A.** Although the contaminated water has traveled more than 100 meters before reaching well 1, the water moves too rapidly through the cavernous limestone to be purified. **B.** As the discharge from the septic tank percolates through the permeable sandstone, it is purified in a relatively short distance.

contamination is sometimes discovered only after drinking water has been affected and people become ill. By this time, the volume of polluted water may be very large, and even if the source of contamination is removed immediately, the problem is not solved. Although the sources of groundwater contamination are numerous, there are relatively few solutions.

Once the source of the problem has been identified and eliminated, the most common practice is simply to abandon the water supply and allow the pollutants to be flushed away gradually. This is the least costly and easiest solution, but the aquifer must remain unused for many years. To accelerate this process, polluted water is sometimes pumped out and treated. Following removal of the tainted water, the aquifer is allowed to recharge naturally, or in some cases the treated water or other fresh water is pumped back in. This process is costly, time-consuming, and it may be risky because there is no way to be certain that all of the contamination has been removed. Clearly, the most effective solution to groundwater contamination is prevention.

of the water table. In some instances the original slope may even be reversed. This could lead to the contamination of wells that yielded unpolluted water before heavy pumping began (Figure 17.22). Also recall that the rate of groundwater movement increases as the slope of the water table steepens. This could produce problems because a faster rate of movement allows less time for the water to be purified in the aquifer before it is pumped to the surface.

Other sources and types of contamination also threaten groundwater supplies (Figure 17.23). These include widely used substances such as highway salt, fertilizers that are spread across the land surface, and pesticides. In addition, a wide array of chemicals and industrial materials may leak from pipelines, storage tanks, landfills, and holding ponds. Some of these pollutants are classified as *hazardous*, meaning that they are either flammable, corrosive, explosive, or toxic. In land disposal, potential contaminants are heaped onto mounds or spread directly over the ground. As rainwater oozes through the refuse, it may dissolve a variety of organic and inorganic materials. If the leached material reaches the water table, it will mix with the groundwater and contaminate the supply. Similar problems may result from leakage of shallow excavations called holding ponds into which a variety of liquid wastes are disposed.

Because groundwater movement is usually slow, polluted water can go undetected for a long time. In fact,

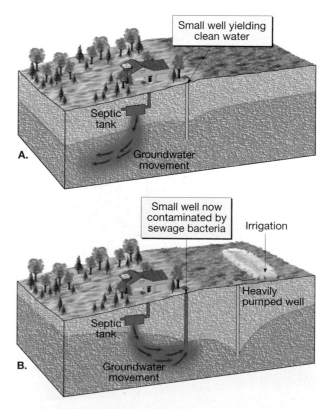

FIGURE 17.22 **A.** Originally the outflow from the septic tank moved away from the small well. **B.** The heavily pumped well changed the slope of the water table, causing contaminated groundwater to flow toward the small well.

A.

B.

FIGURE 17.23 Sometimes agricultural chemicals **A.** and materials leached from landfills **B.** find their way into the groundwater. These are two of the potential sources of groundwater contamination. (Photo **A** by Roy Morsch/Corbis/The Stock Market; Photo **B** by F. Rossotto/Corbis/The Stock Market)

The Geologic Work of Groundwater

Groundwater dissolves rock. This fact is key to understanding how caverns and sinkholes form. Because soluble rocks, especially limestone, underlie millions of square kilometers of Earth's surface, it is here that the groundwater carries on its important role as an erosional agent. Limestone is nearly insoluble in pure water but is quite easily dissolved by water containing small quantities of carbonic acid, and most groundwater contains this acid. It forms because rainwater readily dissolves carbon dioxide from the air and from decaying plants. Therefore, when groundwater comes in contact with limestone, the carbonic acid reacts with the calcite (calcium carbonate) in the rocks to form calcium bicarbonate, a soluble material that is then carried away in solution.

Students Sometimes Ask . . .

Is carbonic acid the only acid that creates limestone caverns?

No. It appears as though sulfuric acid (H_2SO_4) creates some caves. One example is Lechuquilla Cave in the Guadalupe Mountains near Carlsbad, New Mexico. Here solutions under pressure containing hydrogen sulfide (H_2S) derived from deep petroleum-rich sediments migrated upward through rock fractures. When these solutions mixed with groundwater containing oxygen, they formed sulfuric acid that then dissolved the limestone. Lechuquilla Cave is one of the deepest-known caves in the United States, with a vertical range of 478 meters (1568 feet), and is also one of the largest in the country, with 170 kilometers (105 miles) of passages.

Caverns

The most spectacular results of groundwater's erosional handiwork are limestone **caverns.** In the United States alone about 17,000 caves have been discovered and new ones are being found every year. Although most are relatively small, some have spectacular dimensions. Mammoth Cave in Kentucky and Carlsbad Caverns in southeastern New Mexico are famous examples. The Mammoth Cave system is the most extensive in the world, with more than 540 kilometers of interconnected passages. The dimensions at Carlsbad Caverns are impressive in a different way. Here we find the largest and perhaps most spectacular single chamber. The Big Room at Carlsbad Caverns has an area equivalent to 14 football fields and enough height to accommodate the U.S. Capitol building.

Most caverns are created at or just below the water table in the zone of saturation. Here acidic groundwater follows lines of weakness in the rock, such as joints and bedding planes. As time passes, the dissolving process slowly creates cavities and gradually enlarges them into caverns. Material that is dissolved by the groundwater is eventually discharged into streams and carried to the ocean.

In many caves, development has occurred at several levels, with the current cavern-forming activity occurring at the lowest elevation. This situation reflects the close relationship between the formation of major subterranean passages and the river valleys into which they drain. As streams cut their valleys deeper, the water table drops as the elevation of the river drops. Consequently, during periods when surface streams are rapidly downcutting, surrounding groundwater levels drop rapidly and cave passages are abandoned by the water while the passages are still relatively small in cross-sectional area. Conversely, when the entrenchment of streams is slow or negligible, there is time for large cave passages to form.

Certainly the features that arouse the greatest curiosity for most cavern visitors are the stone formations that give some caverns a wonderland appearance. These are not erosional features, like the cavern itself, but depositional features created by the seemingly endless dripping of water over great spans of time. The calcium carbonate that is left behind produces the limestone we call travertine. These cave deposits, however, are also commonly called *dripstone,* an obvious reference to their mode of origin. Although the formation of caverns takes place in the zone of saturation, the deposition of dripstone is not possible until the caverns are above the water table in the unsaturated zone. As soon as the chamber is filled with air, the stage is set for the decoration phase of cavern building to begin.

The various dripstone features found in caverns are collectively called **speleothems** (*spelaion* = cave, *them* = put), no two of which are exactly alike (Figure 17.24). Perhaps the most familiar speleothems are **stalactites** (*stalaktos* = trickling). These icicle-like pendants hang from the ceiling of the cavern and form where water seeps through cracks above. When the water reaches air in the cave, some of the dissolved carbon dioxide escapes from the drop, and calcite precipitates. Deposition occurs as a ring around the edge of the water drop. As drop after drop follows, each leaves an infinitesimal trace of calcite behind, and a hollow limestone tube is created. Water then moves through the tube, remains suspended momentarily at the end, contributes a tiny ring of calcite, and falls to the cavern floor. The stalactite just described is appropriately called a *soda straw* (Figure 17.25).

Often the hollow tube of the soda straw becomes plugged or its supply of water increases. In either case, the water is forced to flow and hence deposit along the outside of the tube. As deposition continues, the stalactite takes on the more common conical shape.

Speleothems that form on the floor of a cavern and reach upward toward the ceiling are called **stalagmites** (*stalagmos* = dropping). The water supplying the calcite for stalagmite growth falls from the ceiling and splatters over the surface. As a result, stalagmites do not have a central tube and are usually more massive in appearance and rounded on their upper ends than stalactites. Given enough time, a downward-growing stalactite and an upward-growing stalagmite may join to form a *column.*

Karst Topography

Many areas of the world have landscapes that to a large extent have been shaped by the dissolving power of groundwater. Such areas are said to exhibit **karst topography,** named for the Kras Plateau in Slovenia located along the northeastern shore of the Adriatic Sea where such topography is strikingly developed. In the United States, karst landscapes occur in many areas that are underlain by limestone, including portions of Kentucky, Tennessee, Alabama, southern Indiana, and central and northern Florida. Generally, arid and semiarid areas are too dry to develop karst topography. When solution features exist in such regions, they are likely to be remnants of a time when rainier conditions prevailed.

Karst areas typically have irregular terrain punctuated with many depressions, called **sinkholes** or **sinks.** In the limestone areas of Florida, Kentucky, and southern Indiana, there are literally tens of thousands of these depressions varying in depth from just a meter or two to a maximum of more than 50 meters (Figure 17.26).

Sinkholes commonly form in two ways. Some develop gradually over many years without any physical disturbance to the rock. In these situations the limestone immediately below the soil is dissolved by downward-sweeping rainwater that is freshly charged with carbon dioxide. With time the bedrock surface is lowered and the fractures into which the water seeps are enlarged. As the fractures grow in size, soil subsides into the widening voids, from which it is removed by groundwater flowing in the passages below. These depressions are usually shallow and have gentle slopes.

By contrast, sinkholes can also form abruptly and without warning when the roof of a cavern collapses under its own weight. Typically, the depressions created in this manner are steep-sided and deep. When they form in populous areas, they may represent a

FIGURE 17.24 Speleothems are of many types, including stalactites, stalagmites, and columns. Big Room, Carlsbad Caverns National Park. The large stalagmite is called the Totem Pole. (Photo by David Muench/David Muench Photography)

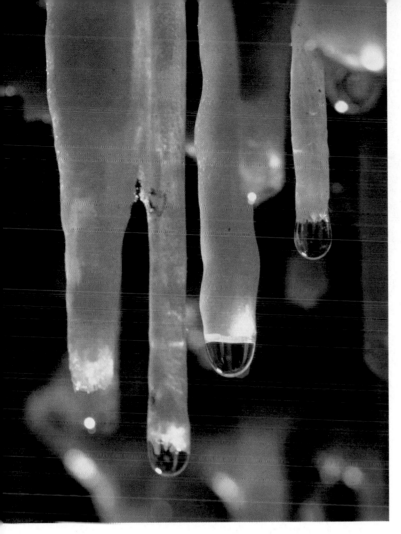

FIGURE 17.25 "Live" soda straw stalactites. Lehman Caves, Great Basin National Park, Nevada. (Photo by Tom & Susan Bean, Inc.)

serious geologic hazard. Such a situation is clearly the case in Figure 17.26B and Box 17.3.

In addition to a surface pockmarked by sinkholes, karst regions characteristically show a striking lack of surface drainage (streams). Following a rainfall, the runoff is quickly funneled below ground through sinks. It then flows through caverns until it finally reaches the water table. Where streams do exist at the surface, their paths are usually short. The names of such streams often give a clue to their fate. In the Mammoth Cave area of Kentucky, for example, there is Sinking Creek, Little Sinking Creek, and Sinking Branch. Some sinkholes become plugged with clay and debris, creating small lakes or ponds. The development of a karst landscape is depicted in Figure 17.26.

Some regions of karst development exhibit landscapes that look very different from the sinkhole-studded terrain depicted in Figure 17.27. One striking example is an extensive region in southern China that is described as exhibiting *tower karst*. As Figure 17.28 shows, the term *tower* is appropriate because the landscape consists of a maze of isolated steep-sided hills that rise abruptly from the ground. Each is riddled with interconnected caves and passageways. This type of karst topography forms in wet tropical and subtropical regions having thick beds of highly jointed limestone. Here groundwater has dissolved large volumes of limestone, leaving only these residual towers. Karst development is more rapid in tropical climates due to the abundant rainfall and the greater availability of carbon dioxide from the decay of lush tropical vegetation. The extra carbon dioxide in the soil means there is more carbonic acid for dissolving limestone. Other tropical areas of advanced karst development include portions of Puerto Rico, western Cuba, and northern Vietnam.

FIGURE 17.26 **A.** This high-altitude infrared image shows an area of karst topography in central Florida. The numerous lakes occupy sinkholes. (Courtesy of USDA–ASCS) **B.** This small sinkhole formed suddenly in 1991 when the roof of a cavern collapsed, destroying this home in Frostproof, Florida. (Photo by *St. Petersburg Times*/Liaison Agency, Inc.)

A.

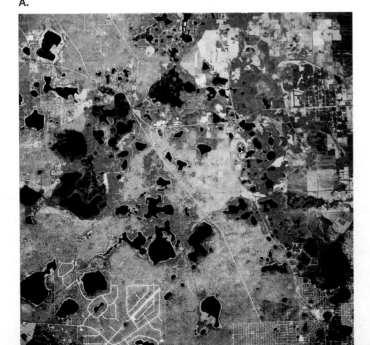

B.

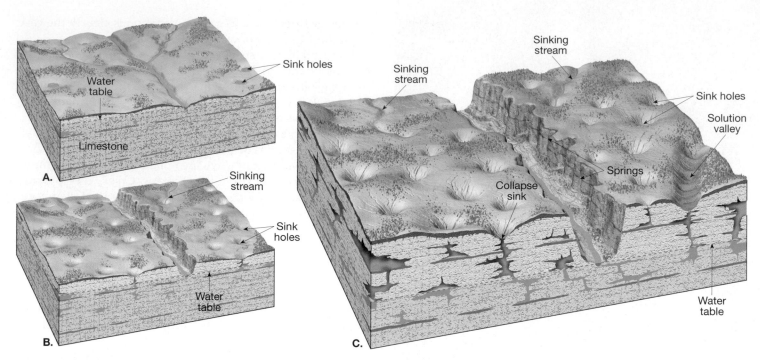

FIGURE 17.27   Development of a karst landscape. **A.** During early stages, groundwater percolates through limestone along joints and bedding planes. Solution activity creates and enlarges caverns at and below the water table. **B.** In this view, sinkholes are well developed and surface streams are funneled below ground. **C.** With the passage of time, caverns grow larger and the number and size of sinkholes increase. Collapse of caverns and coalescence of sinkholes form larger, flat-floored depressions. Eventually, solution activity may remove most of the limestone from the area, leaving only isolated remnants as in Figure 17.28 below.

A.

FIGURE 17.28   A. Painting of Chinese tower karst, "Peach Garden Land of Immortals," by Qiu Ying. (Asian Art and Archaeology, Inc./CORBIS–NY) **B.** One of the best-known and most distinctive regions of tower karst development is the Guilin District of southeastern China. (Photo by A.C. Waltham/Robert Harding World Imagery)

B.

BOX 17.3 ▶ PEOPLE AND THE ENVIRONMENT

The Case of the Disappearing Lake

Lake Chesterfield was a pleasant 9.3-hectare (23-acre) manmade lake in a quiet suburb of St. Louis, where people living along its shore could fish from their small paddleboats—until it disappeared! In June 2004 residents witnessed the entire lake drain in less than three days (Figure 17.D).

"It was like someone pulled the plug," said Donna Ripp, who lives across the street from the lake, which is now a giant mud hole. Ripp said she began to notice the water level sinking and that by the second day, the lake was half empty. A day later it was completely gone.

What happened? The culprit is clear. At the north end of the lake there is a gaping sinkhole estimated to be about 20 meters (70 feet) in diameter. What geologists are now investigating is what the larger subterranean network looks like. This part of Missouri has many caves, including many that are large enough for humans to explore.

Geologist David Taylor, who inspected the lake shortly after the water drained into the ground, said the sinkhole itself "is really not that big." But it doesn't take a very large sinkhole to knock out an entire lake. Taylor said a hole 0.3 meter (1 foot) in diameter can drain at least 3800 liters (about 1000 gallons) a minute. Taylor is the head of a St. Charles–based company called Strata Services, Inc., that specializes in repairing lakes that are draining into subterranean cavities. "In my business I have fixed hundreds of leaky lakes," he said.

But before Taylor can consider repairing Lake Chesterfield, he and his colleagues

FIGURE 17.D Residents examine emptied Lake Chesterfield, a 23-acre reservoir that drained in three days when a sinkhole opened beneath it. (Photo by Hillary Levin/*St. Louis Post Dispatch*)

first must get a sense of the network of cavities under the lake—a task that he said is exceedingly difficult. "There's all kinds of crazy stuff going on down there," he said. "This is all subsurface work. It's very unpredictable and very difficult."

Taylor found that the subsurface cavity responsible for the sinkhole under Lake Chesterfield runs laterally underground for several kilometers. A tracing dye placed near the sinkhole reemerges in a spring about 5.5 kilometers (3.5 miles) from the lake. In order to develop a better picture of the subterranean cavities, Taylor drilled five test holes at 12-meter (40-foot) intervals, finding that two revealed empty

cavities below. But he estimated that it would require 600 holes in a 12-meter grid to even begin to understand the region completely.

Once that picture emerges, Taylor's company then fills the cavities with a cementlike substance so that other sinkholes don't open and create a similar problem. "If we just put a Band-Aid over the hole and fill the lake back up, the same thing will happen again," he said.

In the meantime, nearby residents are getting a crash course on karst topography. "I didn't even know there were underground caves here until all this happened." Donna Ripp said.

Students Sometimes Ask . . .

Is limestone the only rock type that develops karst features?

No. For example, karst development can occur in other carbonate rocks such as marble and dolostone. In addition, evaporites such as gypsum and salt (halite) are highly soluble and are readily dissolved to form karst features including sinkholes, caves, and disappearing streams. This latter situation is termed *evaporite karst.*

Summary

- As a resource, *groundwater* represents the largest reservoir of fresh water that is readily available to humans. Geologically, the dissolving action of groundwater produces *caves* and *sinkholes*. Groundwater is also an equalizer of streamflow.

- Groundwater is water that completely fills the pore spaces in sediment and rock in the subsurface *zone of saturation*. The upper limit of this zone is the *water table*. The *zone of aeration* is above the water table where the soil, sediment, and rock are not saturated.

- The interaction between streams and groundwater takes place in one of three ways: streams gain water from the inflow of groundwater (*gaining stream*); they lose water through the streambed to the groundwater system (*losing stream*); or they do both, gaining in some sections and losing in others.

- Materials with very small pore spaces (such as clay) hinder or prevent groundwater movement and are called *aquitards*. *Aquifers* consist of materials with larger pore spaces (such as sand) that are permeable and transmit groundwater freely.

- Groundwater moves in looping curves that are a compromise between the downward pull of gravity and the tendency of water to move toward areas of reduced pressure.

- The primary factors influencing the velocity of groundwater flow are the slope of the water table (*hydraulic gradient*) and the permeability of the aquifer (*hydraulic conductivity*).

- *Springs* occur whenever the water table intersects the land surface and a natural flow of groundwater results. *Wells,* openings bored into the zone of saturation, withdraw groundwater and create roughly conical depressions in the water table known as *cones of depression*. *Artesian wells* occur when water rises above the level at which it was initially encountered.

- When groundwater circulates at great depths, it becomes heated. If it rises, the water may emerge as a *hot spring*. *Geysers* occur when groundwater is heated in underground chambers, expands, and some water quickly changes to steam, causing the geyser to erupt. The source of heat for most hot springs and geysers is hot igneous rock.

- Some of the current environmental problems involving groundwater include (1) *overuse* by intense irrigation, (2) *and subsidence* caused by groundwater withdrawal, (3) *saltwater contamination,* and (4) contamination by pollutants.

- Most *caverns* form in limestone at or below the water table when acidic groundwater dissolves rock along lines of weakness, such as joints and bedding planes. The various *dripstone* features found in caverns are collectively called *speleothems*. Landscapes that to a large extent have been shaped by the dissolving power of groundwater exhibit *karst topography*, an irregular terrain punctuated with many depressions, called *sinkholes* or *sinks*.

Review Questions

1. What percentage of fresh water is groundwater? If glacial ice is excluded and only liquid fresh water is considered, about what percentage is groundwater?

2. Geologically, groundwater is important as an erosional agent. Name another significant geological role for groundwater.

3. Compare and contrast the zones of aeration and saturation. Which of these zones contains groundwater?

4. Although we usually think of tables as being flat, the water table generally is not. Explain.

5. Although meteorological drought may have ended, hydrological drought may still continue. Explain. (See Box 17.1.)

6. Contrast a gaining stream and a losing stream.

7. Distinguish between porosity and permeability.

8. What is the difference between an aquitard and an aquifer?

9. Under what circumstances can a material have a high porosity but not be a good aquifer?

10. As illustrated in Figure 17.5, groundwater moves in looping curves. What factors cause the water to follow such paths?

11. Briefly describe the important contribution to our understanding of groundwater movement made by Henri Darcy.

12. When an aquitard is situated above the main water table, a localized saturated zone may be created. What term is applied to such a situation?

13. What is the source of heat for most hot springs and geysers? How is this reflected in the distribution of these features?

14. Two neighbors each dig a well. Although both wells penetrate to the same depth, one neighbor is successful and the other is not. Describe a circumstance that might explain what happened.

15. What is meant by the term *artesian?*

16. In order for artesian wells to exist, two conditions must be present. List these conditions.

17. When the Dakota Sandstone was first tapped, water poured freely from many artesian wells. Today these wells must be pumped. Explain.

18. What problem is associated with the pumping of groundwater for irrigation in the southern part of the High Plains?

19. Briefly explain what happened in the San Joaquin Valley as the result of excessive groundwater withdrawal (see Box 17.2).

20. In a particular coastal area, the water table is 4 meters above sea level. Approximately how far below sea level does the fresh water reach?

21. Why does the rate of natural groundwater recharge decrease as urban areas develop?

22. Which aquifer would be most effective in purifying polluted groundwater: coarse gravel, sand, or cavernous limestone?

23. What is meant when a groundwater pollutant is classified as hazardous?

24. Name two common speleothems and distinguish between them.

25. Areas whose landscapes largely reflect the erosional work of groundwater are said to exhibit what kind of topography?

26. Describe two ways in which sinkholes are created.

Key Terms

aquifer (p. 463)
aquitard (p. 463)
artesian (p. 469)
capillary fringe (p. 459)
cavern (p. 475)
cone of depression (p. 468)
Darcy's law (p. 464)
drawdown (p. 468)
flowing artesian well (p. 469)

gaining stream (p. 460)
geyser (p. 466)
groundwater (p. 459)
hot spring (p. 466)
hydraulic conductivity (p. 464)
hydraulic gradient (p. 464)
karst topography (p. 476)
losing stream (p. 460)

nonflowing artesian well (p. 469)
perched water table (p. 466)
permeability (p. 463)
porosity (p. 463)
sinkhole (sink) (p. 476)
speleothem (p. 476)
spring (p. 466)
stalactite (p. 476)

stalagmite (p. 476)
unsaturated zone (p. 459)
water table (p. 459)
well (p. 468)
zone of saturation (p. 459)
zone of soil moisture (p. 459)

Web Resources

The *Earth* Website uses the resources and flexibility of the Internet to aid in your study of the topics in this chapter. Written and developed by geology instructors, this site will help improve your understanding of geology. Visit **http://www.prenhall.com/tarbuck** and click on the cover of *Earth 9e* to find:

- Online review quizzes.
- Critical thinking exercises.
- Links to chapter-specific Web resources.
- Internet-wide key-term searches.

http://www.prenhall.com/tarbuck

GEODe: Earth

GEODe: Earth makes studying faster and more effective by reinforcing key concepts using animation, video, narration, interactive exercises and practice quizzes. A copy is included with every copy of *Earth*.

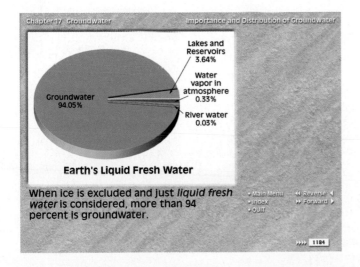

When ice is excluded and just *liquid fresh water* is considered, more than 94 percent is groundwater.

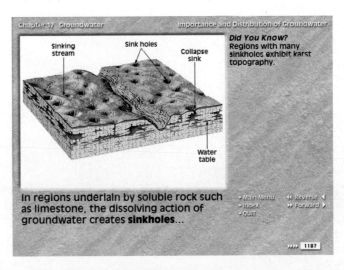

In regions underlain by soluble rock such as limestone, the dissolving action of groundwater creates **sinkholes**...

Glaciers and Glaciation

A small boat nears the seaward margin of an Antarctic glacier. (Photo by Sergio Pitamitz/ CORBIS)

Climate has a strong influence on the nature and intensity of Earth's external processes. This fact is dramatically illustrated in this chapter because the existence and extent of glaciers is largely controlled by Earth's changing climate.

Like the running water and groundwater that were the focus of the preceding two chapters, glaciers represent a significant erosional process. These moving masses of ice are responsible for creating many unique landforms and are part of an important link in the rock cycle in which the products of weathering are transported and deposited as sediment.

Today glaciers cover nearly 10 percent of Earth's land surface; however, in the recent geologic past, ice sheets were three times more extensive, covering vast areas with ice thousands of meters thick. Many regions still bear the mark of these glaciers (Figure 18.1). The basic character of such diverse places as the Alps, Cape Cod, and Yosemite Valley was fashioned by now vanished masses of glacial ice. Moreover, Long Island, the Great Lakes, and the fiords of Norway and Alaska all owe their existence to glaciers. Glaciers, of course, are not just a phenomenon of the geologic past. As you will see, they are still sculpting and depositing debris in many regions today.

FIGURE 18.1 Glacier National Park, Montana, presents a landscape carved by glaciers. St. Mary Lake occupies a glacially eroded valley (glacial trough) and exists because of glacial deposits that act as a dam. The sharp ridges on the left (arêtes) and the pointed peak in the background (a horn) were also sculptured by glacial ice. (Copyright © by Carr Clifton. All rights reserved.)

Glaciers: A Part of Two Basic Cycles

GEODe Glaciers and Glaciation
▶ Introduction

Glaciers are a part of two fundamental cycles in the Earth system—the hydrologic cycle and the rock cycle. Earlier you learned that the water of the hydrosphere is constantly cycled through the atmosphere, biosphere, and geosphere. Time and time again, water evaporates from the oceans into the atmosphere, precipitates upon the land, and flows in rivers and underground back to the sea. However, when precipitation falls at high elevations or high latitudes, the water may not immediately make its way toward the sea. Instead, it may become part of a glacier. Although the ice will eventually melt, allowing the water to continue its path to the sea, water can be stored as glacial ice for many tens, hundreds, or even thousands of years.

A **glacier** is a thick ice mass that forms over hundreds or thousands of years. It originates on land from the accumulation, compaction, and recrystallization of snow. A glacier appears to be motionless, but it is not—glaciers move very slowly. Like running water, groundwater, wind, and waves, glaciers are dynamic erosional agents that accumulate, transport, and deposit sediment. As such, glaciers are among the processes that perform a basic function in the rock cycle. Although glaciers are found in many parts of the world today, most are located in remote areas, either near Earth's poles or in high mountains.

Valley (Alpine) Glaciers

Literally thousands of relatively small glaciers exist in lofty mountain areas, where they usually follow valleys that were originally occupied by streams. Unlike the rivers that previously flowed in these valleys, the glaciers advance slowly, perhaps only a few centimeters per day. Because of their setting, these moving ice masses are termed **valley glaciers** or **alpine glaciers** (Figure 18.2). Each glacier actually is a stream of ice, bounded by precipitous rock walls, that flows downvalley from an accumulation center near its head. Like rivers, valley glaciers can be long or short, wide or narrow, single or with branching tributaries. Generally, the widths of alpine glaciers are small compared to their lengths. Some extend for just a fraction of a kilometer, whereas others go on for many tens of kilometers. The west branch of the Hubbard Glacier, for example, runs through 112 kilometers (nearly 70 miles) of mountainous terrain in Alaska and the Yukon Territory.

Ice Sheets

In contrast to valley glaciers, **ice sheets** exist on a much larger scale. The low total annual solar radiation reaching the poles makes these regions eligible for great ice accumulations. Although many ice sheets have existed in the past, just two achieve this status at present (Figure 18.3). In the area of the North Pole, Greenland is covered by an imposing ice sheet that occupies 1.7 million square kilometers, or about 80 percent of this large island. Averaging nearly 1500 meters (5000 feet) thick, the ice extends 3000 meters (10,000 feet) above the island's bedrock floor in some places.

In the Southern Hemisphere, the huge Antarctic Ice Sheet attains a maximum thickness of about 4300 meters (14,000 feet) and covers nearly the entire continent, an area of more than 13.9 million square kilometers (5.4 million square miles). Because of the proportions of these huge features, they often are called *continental ice sheets*. Indeed, the combined areas of present-day continental ice sheets represent almost 10 percent of Earth's land area.

These enormous masses flow out in all directions from one or more snow-accumulation centers and completely obscure all but the highest areas of underlying terrain. Even sharp variations in the topography beneath the glacier usually appear as relatively subdued undulations on the surface of the ice. Such topographic differences, however, do affect the behavior of the ice sheets, especially near their margins, by guiding flow in certain directions and creating zones of faster and slower movement.

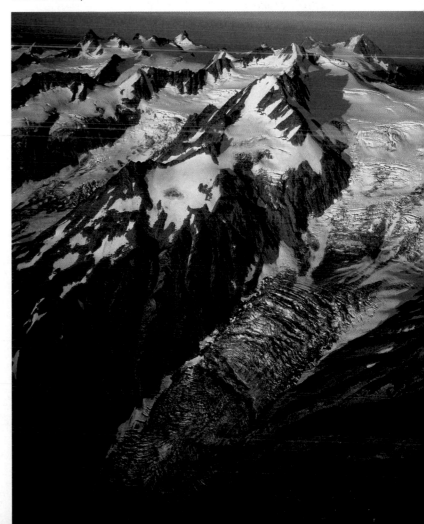

FIGURE 18.2   Aerial view of Chigmit Mountains in Lake Clark National Park, Alaska. Valley glaciers continue to modify this landscape. (Photo by Michael Collier)

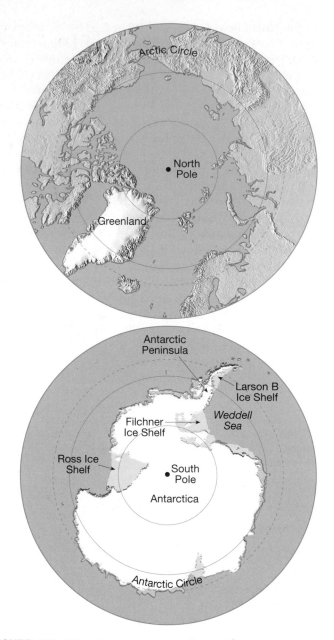

FIGURE 18.3 The only present-day continental ice sheets are those covering Greenland and Antarctica. Their combined areas represent almost 10 percent of Earth's land area. Greenland's ice sheet occupies 1.7 million square kilometers, or about 80 percent of the island. The area of the Antarctic Ice Sheet is almost 14 million square kilometers. Ice shelves occupy an additional 1.4 million square kilometers adjacent to the Antarctic Ice Sheet.

Along portions of the Antarctic coast, glacial ice flows into the adjacent ocean, creating features called **ice shelves.** They are large, relatively flat masses of floating ice that extend seaward from the coast but remain attached to the land along one or more sides. The shelves are thickest on their landward sides, and they become thinner seaward. They are sustained by ice from the adjacent ice sheet as well as being nourished by snowfall and the freezing of seawater to their bases. Antarctica's ice shelves extend over approximately 1.4 million square kilometers (0.6 million square miles). The

Ross and Filchner ice shelves are the largest, with the Ross Ice Shelf alone covering an area approximately the size of Texas (Figure 18.3). In recent years, satellite monitoring has shown that some ice shelves are unstable and breaking apart. Box 18.1 explores this topic.

Other Types of Glaciers

In addition to valley glaciers and ice sheets, other types of glaciers are also identified. Covering some uplands and plateaus are masses of glacial ice called **ice caps.** Like ice sheets, ice caps completely bury the underlying landscape but are much smaller than the continental-scale features. Ice caps occur in many places, including Iceland and several of the large islands in the Arctic Ocean (Figure 18.4).

Often ice caps and ice sheets feed **outlet glaciers.** These tongues of ice flow down valleys extending outward from the margins of these larger ice masses. The tongues are essentially valley glaciers that are avenues for ice movement from an ice cap or ice sheet through mountainous terrain to the sea. Where they encounter the ocean, some outlet glaciers spread out as floating ice shelves. Often large numbers of icebergs are produced.

Piedmont glaciers occupy broad lowlands at the bases of steep mountains and form when one or more alpine glaciers emerge from the confining walls of mountain valleys. Here the advancing ice spreads out to form a broad sheet. The size of individual piedmont glaciers varies greatly. Among the largest is the broad Malaspina Glacier along the coast of southern Alaska. It covers thousands of square kilometers of the flat coastal plain at the foot of the lofty St. Elias range (Figure 18.5).

FIGURE 18.4 The ice cap in this satellite image is the Vantnajükull in southeastern Iceland (*jükull* means "ice cap" in Danish). In 1996 the Grimsvötn Volcano erupted beneath the ice cap, producing large quantities of melted glacial water that created floods. (*Landsat* image from NASA)

FIGURE 18.5   Malaspina glacier in southeastern Alaska is considered a classic example of a piedmont glacier. Piedmont glaciers occur where valley glaciers exit a mountain range onto broad lowlands, are no longer laterally confined, and spread to become wide lobes. Malaspina Glacier is actually a compound glacier, formed by the merger of several valley glaciers, the most prominent of which seen here are Agassiz Glacier (left) and Seward Glacier (right). In total, Malaspina Glacier is up to 65 kilometers (40 miles) wide and extends up to 45 kilometers (28 miles) from the mountain front nearly to the sea. This perspective view looking north covers an area about 55 kilometers × 55 kilometers (34 × 34 miles). It was created from a *Landsat* satellite image and an elevation model generated by the Shuttle Radar Topography Mission (SRTM). Such images are excellent tools for mapping the geographic extent of glaciers and for determining whether such glaciers are thinning or thickening. (Image from NASA/JPL)

melts during the summer. Glaciers develop in the high latitude polar realm because, even though annual snowfall totals are modest, temperatures are so low that little of the snow melts. Glaciers can form in mountains because temperatures drop with an increase in altitude. So even near the equator, glaciers may form at elevations above about 5000 meters (16,400 feet). For example, Tanzania's Mount Kilimanjaro, which is located practically astride the equator at an altitude of 5895 meters (19,336 feet), has glaciers at its summit (Figure 18.7). The elevation above which snow remains throughout the year is called the **snowline**. As you would expect the elevation of the snowline varies with latitude. Near the equator, it occurs high in the mountains, whereas near the 60th parallel it is at sea level. Before a glacier is created, snow must be converted into glacial ice.

What if the Ice Melted?

How much water is stored as glacial ice? Estimates by the U.S. Geological Survey indicate that only slightly more than 2 percent of the world's water is tied up in glaciers. But even 2 percent of a vast quantity is still large. The total volume of just valley glaciers is about 210,000 cubic kilometers, comparable to the combined volume of the world's largest saline and freshwater lakes.

As for ice sheets, Antarctica's represents 80 percent of the world's ice and an estimated 65 percent of Earth's fresh water. It covers an area almost one and a half times the area of the entire United States. If this ice melted, sea level would rise an estimated 60 to 70 meters and the ocean would inundate many densely populated coastal areas (Figure 18.6).

The hydrologic importance of the Antarctic ice can be illustrated in another way. If the ice sheet were melted at a uniform rate, it could feed (1) the Mississippi River for more than 50,000 years, (2) all the rivers in the United States for about 17,000 years, (3) the Amazon River for approximately 5000 years, or (4) all the rivers of the world for about 750 years.

Formation and Movement of Glacial Ice

GEODe   Glaciers and Glaciation
▶ Budget of a Glacier

Snow is the raw material from which glacial ice originates; therefore, glaciers form in areas where more snow falls in winter than

Glacial Ice Formation

When temperatures remain below freezing following a snowfall, the fluffy accumulation of delicate hexagonal crystals soon changes. As air infiltrates the spaces between the crystals, the extremities of the crystals evaporate and the water vapor condenses near the centers of the crystals. In this manner, snowflakes become smaller, thicker, and more spherical, and the large pore spaces disappear. By this process air is forced out and what was once light, fluffy snow is recrystallized into a much denser mass of small grains having the consistency of coarse sand. This granular

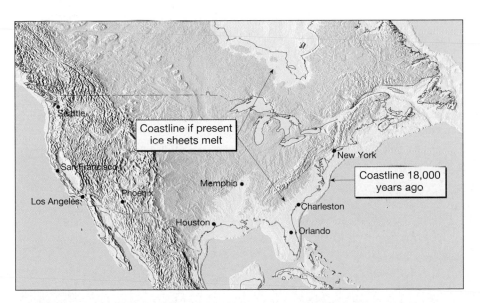

FIGURE 18.6   This map of a portion of North America shows the present-day coastline compared to the coastline that existed during the last ice age (18,000 years ago) and the coastline that would exist if present ice sheets on Greenland and Antarctica melted. (After R. H. Dott, Jr., and R. L. Battan, *Evolution of the Earth*, New York: McGraw Hill, 1971. Reprinted by permission of the publisher.)

BOX 18.1 ▶ UNDERSTANDING EARTH

The Collapse of Antarctic Ice Shelves

Studies using recent satellite imagery show that portions of some ice shelves are breaking apart. For example, during a 35-day span in February and March 2002 an ice shelf on the eastern side of the Antarctic Peninsula, known as the Larsen B ice shelf, broke apart and separated from the continent (Figure 18.A). The event sent thousands of icebergs adrift in the adjacent Weddell Sea (see Figure 18.3). In all, about 3250 square kilometers (1250 square miles)

FIGURE 18.A This satellite image shows the Larsen B ice shelf during its collapse in early 2002. (Image courtesy of NASA)

of the ice shelf broke apart. (For reference, the entire state of Rhode Island covers 2717 square kilometers.) This was not an isolated event but part of a trend. Over a five-year span, the Larsen B ice shelf shrank by 5700 square kilometers (more than 2200 square miles). Moreover, since 1974 the extent of seven ice shelves surrounding the Antarctic Peninsula declined by about 13,500 square kilometers (nearly 5300 square miles).

Why did these masses of floating ice break apart? What if the trend continues? Could there be any serious consequences?

Scientists attribute the breakup of the ice shelves to a strong regional climate warming. Since about 1950, Antarctic temperatures have risen by 2.5°C (4.5°F). The rate of warming has been about 0.5°C (nearly 1°F) per decade. If temperatures continue to rise, an ice shelf adjacent to Larsen B may start to recede in coming decades. Moreover, regional warming of just a few degrees Celsius may be sufficient to cause portions of the huge Ross Ice Shelf to become unstable and begin to break apart (see Figure 18.3).

What might the consequences be? Scientists at the National Snow and Ice Data Center (NSIDC) suggest the following:

> While the breakup of the ice shelves in the Peninsula has little consequence for sea level rise, the breakup of other

shelves in the Antarctic could have a major effect on the rate of ice flow off the continent. Ice shelves act as a buttress, or braking system, for glaciers. Further, the shelves keep warmer marine air at a distance from the glaciers; therefore, they moderate the amount of melting that occurs on the glaciers' surfaces. Once their ice shelves are removed, the glaciers increase in speed due to meltwater percolation and/or a reduction of braking forces, and they may begin to dump more ice into the ocean. Glacier ice speed increases are already observed in Peninsula areas where ice shelves disintegrated in prior years.[*]

The addition of large quantities of glacial ice to the ocean could indeed cause a significant rise in sea level.

Remember that what is being suggested here is still speculative because our knowledge of the dynamics of Antarctica's ice shelves and glaciers is incomplete. Additional satellite monitoring and field studies will be necessary if we are to more accurately predict potential rises in global sea level triggered by the mechanism described here.

---

[*]National Snow and Ice Data Center, "Antarctic Ice-Shelf Collapses," 21 March 2002, http://nsidc.org/iceshelves/larsenb2002.

FIGURE 18.7 Mount Kilimanjaro, the highest peak in Africa, is capped by small alpine glaciers. (Photo by Ulrich Doering/Alamy)

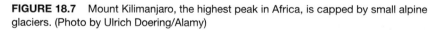

recrystallized snow is called **firn** and is commonly found making up old snowbanks near the end of winter. As more snow is added, the pressure on the lower layers gradually increases, compacting the ice grains at depth. Once the thickness of ice and snow exceeds 50 meters (160 feet), the weight is sufficient to fuse firn into a solid mass of interlocking ice crystals. Glacial ice has now been formed.

The rate at which this transformation occurs varies. In regions where the annual snow accumulation is great, burial is relatively rapid and snow may turn to glacial ice in a matter of a decade or less. Where the yearly addition of snow is less abundant, burial is slow and the transformation of snow to glacial ice may take hundreds of years.

In chapter 8 you said that glacial ice is a metamorphic rock which makes it part of the geosphere. Glacial ice is also a part of the hydrologic cycle, so is it considered part of the hydrosphere, too?

Yes, it's okay to place ice in both "spheres." Moreover, scientists also place ice in its own "sphere"—the cryosphere. The word comes from the Greek word *kryos,* meaning "frost or icy cold." Cryosphere refers to that portion of Earth's surface where water is in a solid form. This includes snow, glaciers, sea ice, freshwater ice, and frozen ground (permafrost).

Movement of a Glacier

The movement of glacial ice is generally referred to as *flow.* The fact that glacial movement is described in this way seems paradoxical—how can a solid flow? The way in which ice flows is complex and is of two basic types. The first of these, **plastic flow,** involves movement *within* the ice. Ice behaves as a brittle solid until the pressure upon it is equivalent to the weight of about 50 meters (165 feet) of ice. Once that load is surpassed, ice behaves as a plastic material and flow begins. Such flow occurs because of the molecular structure of ice. Glacial ice consists of layers of molecules stacked one upon the other. The bonds between layers are weaker than those within each layer. Therefore, when a stress exceeds the strength of the bonds between the layers, the layers remain intact and slide over one another.

A second and often equally important mechanism of glacial movement consists of the entire ice mass slipping along the ground. With the exception of some glaciers located in polar regions where the ice is probably frozen to the solid bedrock floor, most glaciers are thought to move by this sliding process, called **basal slip.** In this process, meltwater probably acts as a hydraulic jack and perhaps as a lubricant helping the ice over the rock. The source of the liquid water is related in part to the fact that the melting point of ice decreases as pressure increases. Therefore, deep within a glacier the ice may be at the melting point even though its temperature is below 0°C.

In addition, other factors may contribute to the presence of meltwater deep within the glacier. Temperatures may be increased by plastic flow (an effect similar to frictional heating), by heat added from the Earth below, and by the refreezing of meltwater that has seeped down from above. This last process relies on the property that as water changes state from liquid to solid, heat (termed latent heat of fusion) is released.

Figure 18.8 illustrates the effects of these two basic types of glacial motion. This vertical profile through a glacier also shows that not all the ice flows forward at the same rate. Frictional drag with the bedrock floor causes the lower portions of the glacier to move more slowly.

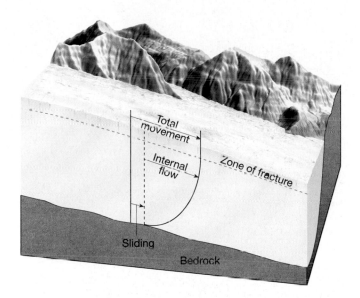

FIGURE 18.8 Vertical cross section through a glacier to illustrate ice movement. Glacial movement is divided into two components. Below about 50 meters (160 feet), ice behaves plastically and flows. In addition, the entire mass of ice may slide along the ground. The ice in the zone of fracture is carried along "piggyback" style. Notice that the rate of movement is slowest at the base of the glacier, where frictional drag is greatest.

In contrast to the lower portion of the glacier, the upper 50 meters or so are not under sufficient pressure to exhibit plastic flow. Rather, the ice in this uppermost zone is brittle and is appropriately referred to as the **zone of fracture.** The ice in the zone of fracture is carried along "piggyback" style by the ice below. When the glacier moves over irregular terrain, the zone of fracture is subjected to tension, resulting in cracks called **crevasses** (Figure 18.9). These gaping cracks can make travel across glaciers dangerous and may extend to depths of 50 meters. Below this depth, plastic flow seals them off.

I've heard that icebergs might be used as a source of water in deserts. Is that possible?

It is true that people in arid lands have seriously studied the possibility of towing icebergs from Antarctica to serve as a source of fresh water. There are certainly ample supplies. Each year, in the waters surrounding Antarctica, about 1000 cubic kilometers (240 cubic miles) of glacial ice breaks off and creates icebergs. However, there are significant technological problems that are not likely to be overcome soon. For instance, vessels capable of towing huge icebergs (1 to 2 kilometers across) have not yet been developed. In addition, there would be a substantial loss of ice to melting and evaporation that would take place as an iceberg is slowly rafted (for up to one year) through warm ocean waters.

FIGURE 18.9 Crevasses form in the brittle ice of the zone of fracture. They can extend to depths of 50 meters and can obviously make travel across glaciers dangerous. Gasherbrum II expedition, Pakistan. (Photo by Galen Rowell/Mountain Light Photography, Inc.)

Rates of Glacial Movement

Unlike streamflow, glacial movement is not obvious. If we could watch an alpine glacier move, we would see that like the water in a river, all of the ice in the valley does not move at an equal rate. Just as friction with the valley floor slows the movement of the ice at the bottom of the glacier, the drag created by the valley walls causes the rate of flow to be greatest in the center of the glacier. This was first demonstrated by experiments during the nineteenth century, in which markers were carefully placed in a straight line across the top of a valley glacier. Periodically, the positions of the stakes were recorded, revealing the type of movement just described. More about these experiments may be found in Box 1.2 on page 13.

How rapidly does glacial ice move? Average velocities vary considerably from one glacier to another. Some move so slowly that trees and other vegetation may become well established in the debris that has accumulated on the glacier's surface, whereas others may move at rates of up to several meters per day. For example, Byrd Glacier, an outlet glacier in Antarctica that was the subject of a 10-year study using satellite images, moved at an average rate of 750 to 800 meters per year (about 2 meters per day). Other glaciers in the study advanced at one-fourth that rate.

The advance of some glaciers is characterized by periods of extremely rapid movements called **surges.** Glaciers that exhibit such movement may flow along in an apparently normal manner, then speed up for a relatively short time before returning to the normal rate again. The flow rates during surges are as much as 100 times the normal rate. Evidence indicates that many glaciers may be of the surging type.

It is not yet clear whether the mechanism that triggers these rapid movements is the same for each surging-type glacier. However, researchers studying the Variegated Glacier shown in Figure 18.10 have determined that the surges of this ice mass take the form of a rapid increase in basal sliding that is caused by increases in water pressure beneath the ice. The increased water pressure at the base of the glacier acts to reduce friction between the underlying bedrock and the moving ice. The pressure buildup, in turn, is related to changes in the system of passageways that conduct water along the glacier's bed and deliver it as outflow to the terminus.

Budget of a Glacier

Snow is the raw material from which glacial ice originates; therefore, glaciers form in areas where more snow falls in winter than melts during the summer. Glaciers are constantly gaining and losing ice. Snow accumulation and ice formation occur in the **zone of accumulation.** Its outer limits are defined by the snowline. As noted earlier, the elevation of the snowline varies greatly, from sea level in polar regions to altitudes approaching 5000 meters near the equator. Above the snowline, in the zone of accumulation, the addition of snow thickens the glacier and promotes movement. Beyond the snowline is the **zone of wastage.** Here there is a net loss to the glacier as all of the snow from the previous winter melts, as does some of the glacial ice (Figure 18.11).

In addition to melting, glaciers also waste as large pieces of ice break off the front of the glacier in a process called **calving.** Calving creates *icebergs* in places where the glacier has reached the sea or a lake (Figure 18.12). Because icebergs are just slightly less dense than seawater, they float very low in the water, with more than 80 percent of their mass submerged. Along the margins of Antarctica's ice shelves, calving is the primary means by which these masses lose ice. The relatively flat icebergs produced here can be several kilometers across and 600 meters thick. By comparison, thousands of irregularly shaped icebergs are produced by outlet glaciers flowing from the margins of the Greenland Ice Sheet. Many drift southward and find their way into the North Atlantic, where they can pose a hazard to navigation.

August 1964

August 1965

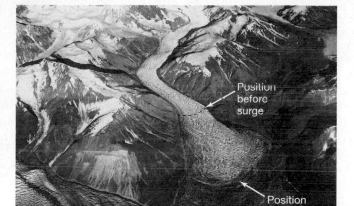

FIGURE 18.10   The surge of Variegated Glacier, a valley glacier near Yakutat, Alaska, northwest of Juneau is captured in these two aerial photographs taken one year apart. During a surge, ice velocities in Variegated Glacier are 20 to 50 times greater than during a quiescent phase. (Photos by Austin Post, U.S. Geological Survey)

Whether the margin of a glacier is advancing, retreating, or remaining stationary depends on the budget of the glacier. The **glacial budget** is the balance, or lack of balance, between accumulation at the upper end of the glacier and loss at the lower end. This loss is termed **ablation.** If ice accumulation exceeds ablation, the glacial front advances until the two factors balance. When this happens, the terminus of the glacier becomes stationary.

If a warming trend increases ablation and/or if a drop in snowfall decreases accumulation, the ice front will retreat.

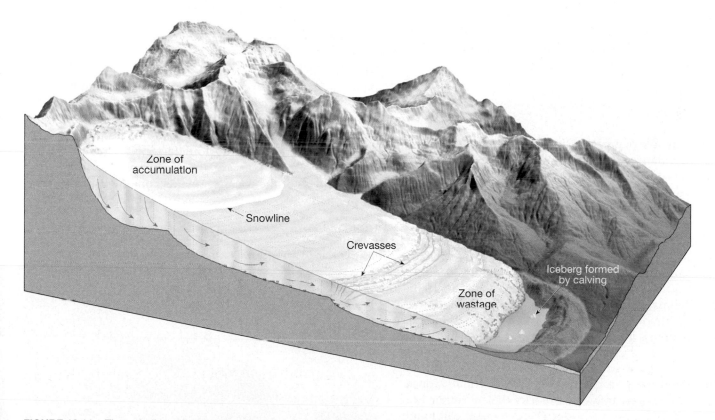

FIGURE 18.11   The snowline separates the zone of accumulation and the zone of wastage. Above the snowline, more snow falls each winter than melts each summer. Below the snowline, the snow from the previous winter completely melts as does some of the underlying ice. Whether the margin of a glacier advances, retreats, or remains stationary depends on the balance between accumulation and wastage (ablation). When a glacier moves across irregular terrain, *crevasses* form in the brittle portion.

FIGURE 18.12 Icebergs are created when large pieces calve from the front of a glacier after it reaches a water body. Here ice is calving from the terminus of Alaska's Hubbard Glacier in Wrangell–St. Elias National Park, Alaska. Only about 20 percent of an iceberg protrudes above the water line. (Photo by Tom & Susan Bean, Inc.)

As the terminus of the glacier retreats, the extent of the zone of wastage diminishes. Therefore, in time a new balance will be reached between accumulation and wastage, and the ice front will again become stationary.

Whether the margin of a glacier is advancing, retreating, or stationary, the ice within the glacier continues to flow forward. In the case of a receding glacier, the ice still flows forward, but not rapidly enough to offset ablation. This point is illustrated well in Figure 1.C on page 13. As the line of stakes within Rhône Glacier continued to move downvalley, the terminus of the glacier slowly retreated upvalley.

Glacial Erosion

Glaciers are capable of great erosion. For anyone who has observed the terminus of an alpine glacier, the evidence of its erosive force is clear. You can witness firsthand the release of rock material of various sizes from the ice as it melts. All signs lead to the conclusion that the ice has scraped, scoured, and torn rock from the floor and walls of the valley and carried it downvalley. It should be pointed out, however, that in mountainous regions mass-wasting processes also make substantial contributions to the sediment load of a glacier. A glance back at Figure 15.10 on page 410 provides a striking example. The image shows a 4-kilometer-long tongue of rock avalanche debris deposited atop Alaska's Sherman Glacier.

Once rock debris is acquired by the glacier, the enormous competency of ice will not allow the debris to settle out like the load carried by a stream or by the wind. Indeed, as a medium of sediment transport, ice has no equal. Consequently, glaciers can carry huge blocks that no other erosional agent could possibly budge (Figure 18.13). Although today's glaciers are of limited importance as erosional agents, many landscapes that were modified by the widespread glaciers of the most recent Ice Age still reflect to a high degree the work of ice.

Glaciers erode the land primarily in two ways—plucking and abrasion. First, as a glacier flows over a fractured bedrock surface, it loosens and lifts blocks of rock and incorporates them into the ice. This process, known as **plucking,** occurs when meltwater penetrates the cracks and joints of bedrock beneath a glacier and freezes. As the water expands, it exerts tremendous leverage that pries the rock loose. In this manner, sediment of all sizes becomes part of the glacier's load.

The second major erosional process is **abrasion** (Figure 18.14). As the ice and its load of rock fragments slide over bedrock, they function like sandpaper to smooth and polish the surface below. The pulverized rock produced by the glacial "grist mill" is appropriately called **rock flour.** So much rock flour may be produced that meltwater streams flowing out of a glacier often have the grayish appearance of skim milk and offer visible evidence of the grinding power of ice.

When the ice at the bottom of a glacier contains large rock fragments, long scratches and grooves called **glacial striations** may even be gouged into the bedrock (Figure 18.14A). These linear grooves provide clues to the direction of ice flow. By mapping the striations over large areas, patterns of glacial flow can often be reconstructed. On the other hand, not all abrasive action produces striations. The rock surfaces over which the glacier moves may also become highly polished by the ice and its load of finer particles. The broad expanses of smoothly polished granite in Yosemite National Park provide an excellent example (Figure 18.14B).

As is the case with other agents of erosion, the rate of glacial erosion is highly variable. This differential erosion by ice is largely controlled by four factors: (1) rate of glacial movement; (2) thickness of the ice; (3) shape, abundance, and hardness of the rock fragments contained in the ice at

FIGURE 18.13 A large, glacially transported boulder in central Wyoming. Such boulders are called glacial erratics. (Photo by Yva Momatiuk and John Eastcott/Photo Researchers, Inc.)

the base of the glacier; and (4) the erodibility of the surface beneath the glacier. Variations in any or all of these factors from time to time and/or from place to place mean that the features, effects, and degree of landscape modification in glaciated regions can vary greatly.

Landforms Created by Glacial Erosion

The erosional effects of valley glaciers and ice sheets are quite different. A visitor to a glaciated mountain region is likely to see a sharp and angular topography (see Figure 18.1, p. 484). The reason is that as alpine glaciers move downvalley, they tend to accentuate the irregularities of the mountain landscape by creating steeper canyon walls and making bold peaks even more jagged. By contrast, continental ice sheets generally override the terrain and hence subdue rather than accentuate the irregularities they encounter. Although the erosional potential of ice sheets is enormous, landforms carved by these huge ice masses usually do not inspire the same wonderment and awe as do the erosional features created by valley glaciers. Much of the rugged mountain scenery so celebrated for its majestic beauty is the product of erosion by alpine glaciers. Figure 18.15 shows a hypothetical mountain area before, during, and after glaciation.

Glaciated Valleys

A hike up a glaciated valley reveals a number of striking ice-created features. The valley itself is often a dramatic sight. Unlike streams, which create their own valleys, glaciers take the path of least resistance by following the course of existing stream valleys. Prior to glaciation, mountain valleys are characteristically narrow and V-shaped because streams are well above base level and are therefore downcutting. However, during glaciation these narrow valleys undergo a transformation as the glacier widens and deepens them, creating a U-shaped **glacial trough** (Figure 18.15). In addition to producing a broader and deeper valley, the glacier also straightens the valley. As ice flows around sharp curves, its great erosional force removes the spurs of land that extend into the valley. The results of this activity are triangular-shaped cliffs called **truncated spurs** (Figure 18.15).

The amount of glacial erosion that takes place in different valleys in a mountainous area varies. Prior to glaciation, the mouths of tributary streams join the main (trunk) valley at the elevation of the stream in that valley. During glaciation, the amount of ice flowing through the main valley can be much greater than the amount advancing down each tributary. Consequently, the valley containing the trunk glacier is eroded deeper than the smaller valleys that feed it. Thus, when the glaciers eventually recede, the valleys of tributary glaciers are left standing above the main glacial trough and are termed **hanging valleys** (Figure 18.15). Rivers flowing

A. **B.**

FIGURE 18.14 **A.** Glacial abrasion created the scratches and grooves in this bedrock. Glacier Bay National Park, Alaska. (Photo copyright © Carr Clifton. All rights reserved.) **B.** Glacially polished granite in California's Yosemite National Park. (Photo by E. J. Tarbuck)

through hanging valleys may produce spectacular waterfalls, such as those in Yosemite National Park (Figure 18.15).

As hikers walk up a glacial trough, they may pass a series of bedrock depressions on the valley floor that were probably created by plucking and then scoured by the abrasive force of the ice. If these depressions are filled with water, they are called **pater noster lakes** (Figure 18.15). The Latin name means "our Father" and is a reference to a string of rosary beads.

At the head of a glacial valley is a very characteristic and often imposing feature called a **cirque.** As the photo in Figure 18.15 illustrates, these bowl-shaped depressions have precipitous walls on three sides but are open on the downvalley side. The cirque is the focal point of the glacier's growth because it is the area of snow accumulation and ice formation. Cirques begin as irregularities in the mountainside that are subsequently enlarged by the frost wedging and plucking that occur along the sides and bottom of the

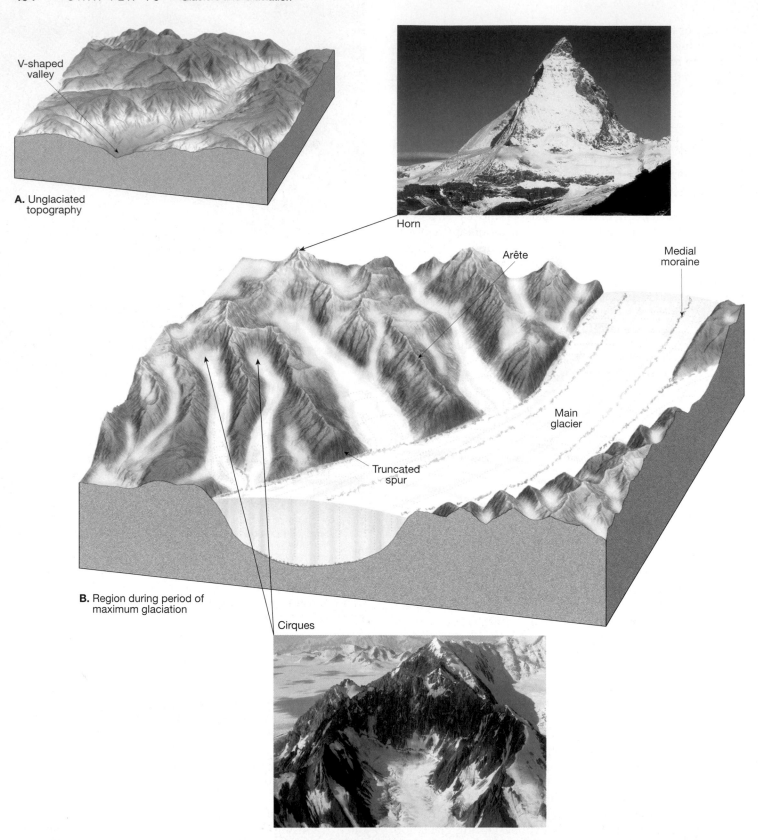

FIGURE 18.15 Prior to glaciation, a mountain valley is typically narrow and V-shaped. During glaciation, an alpine glacier widens, deepens, and straightens the valley, creating a U-shaped glacial trough. These diagrams of a hypothetical area show the development of erosional landforms created by alpine glaciers. The unglaciated landscape in part **A** is modified by valley glaciers in part **B**. After the ice recedes, in part **C,** the terrain looks very different than before glaciation. (Cirque photo by Marli Miller, horn photo by Gavriel Jecan, glacial trough photo by John Montagne, hanging valley photo by Marc Muench, tarn photo by Stephen J. Krasemann, arête photo by James E. Patterson)

Tarn

Horn

Arête

Cirques

Pater noster lakes

C. Glaciated topography

Hanging valley

Glacial trough

FIGURE 18.15   (Continued)

glacier. After the glacier has melted away, the cirque basin is often occupied by a small lake called a **tarn** (Figure 18.15).

Sometimes when two glaciers exist on opposite sides of a divide, each flowing away from the other, the dividing ridge between their cirques is largely eliminated as plucking and frost action enlarge each one. When this occurs, the two glacial troughs come to intersect, creating a gap or pass from one valley into the other. Such a feature is termed a **col.** Some important and well-known mountain passes that are cols include St. Gotthard Pass in the Swiss Alps, Tioga Pass in California's Sierra Nevada, and Berthoud Pass in the Colorado Rockies.

Before leaving the topic of glacial troughs and their associated features, one more rather well-known feature should be discussed: fiords. **Fiords** are deep, often spectacular steep-sided inlets of the sea that are present at high latitudes where mountains are adjacent to the ocean (Figure 18.16). They are drowned glacial troughs that became submerged as the ice left the valley and sea level rose following the Ice Age. The depths of fiords may exceed 1000 meters (3300 feet). However, the great depths of these flooded troughs are only partly explained by the post–Ice Age rise in sea level. Unlike the situation governing the downward erosional work of rivers, sea level does not act as base level for glaciers. As a consequence, glaciers are capable of eroding their beds far below the surface of the sea. For example, a 300-meter-thick (1000-foot-thick) glacier can carve its valley floor more than 250 meters (820 feet) below sea level before downward erosion ceases and the ice begins to float. Norway, British Columbia, Greenland, New Zealand, Chile, and Alaska all have coastlines characterized by fiords.

Arêtes and Horns

A visit to the Alps and the Northern Rockies, or many other scenic mountain landscapes carved by valley glaciers, reveals not only glacial troughs, cirques, pater noster lakes, and the other related features just discussed. You also are likely to see sinuous sharp-edged ridges called **arêtes** (French for *knife-edge*) and sharp pyramid-like peaks called **horns** projecting above the surroundings. Both features can originate from the same basic process, the enlargement of cirques produced by plucking and frost action (Figure 18.15). In the case of the spires of rock called horns, groups of cirques around a single high mountain are responsible. As the cirques enlarge and converge, an isolated horn is produced. The most famous example is the Matterhorn in the Swiss Alps (Figure 18.15).

Arêtes can be formed in a similar manner, except that the cirques are not clustered around a point but rather exist on opposite sides of a divide. As the cirques grow, the divide separating them is reduced to a narrow knifelike partition. An arête, however, may also be created in another way. When two glaciers occupy parallel valleys, an arête can form when the divide separating the moving tongues of ice is progressively narrowed as the glaciers scour and widen their adjacent valleys.

Roches Moutonnées

In many glaciated landscapes, but most frequently where continental ice sheets have modified the terrain, the ice carves small streamlined hills from protruding bedrock

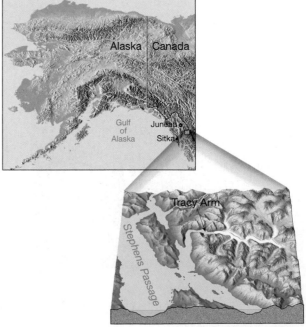

FIGURE 18.16 Like other fiords, this one at Tracy Arm, Alaska, is a drowned glacial trough. (Photo by Tom & Susan Bean, Inc.)

FIGURE 18.17 Roche Moutonnée, in Yosemite National Park, California. The gentle slope was abraded and the steep side was plucked. The ice moved from right to left. (Photo by E. J. Tarbuck)

nized that it remained in the basic glacial vocabulary. Today **glacial drift** is an all-embracing term for sediments of glacial origin, no matter how, where, or in what shape they were deposited.

One of the features distinguishing drift from sediments laid down by other erosional agents is that glacial deposits consist primarily of mechanically weathered rock debris that underwent little or no chemical weathering prior to deposition. Thus, minerals that are notably prone to chemical decomposition, such as hornblende and the plagioclase feldspars, are often conspicuous components in glacial sediments.

Glacial drift is divided by geologists into two distinct types: (1) materials deposited directly by the glacier, which are known as **till,** and (2) sediments laid down by glacial meltwater, called **stratified drift.** We will now look at landforms made of each type.

knobs. Such an asymmetrical knob of bedrock is called a **roche moutonnée** (French for *sheep rock*). They are formed when glacial abrasion smoothes the gentle slope facing the oncoming ice sheet and plucking steepens the opposite side as the ice rides over the knob (Figure 18.17). Roches moutonnées indicate the direction of glacial flow because the gentler slope is generally on the side from which the ice advanced.

Glacial Deposits

GEODe
Glaciers and Glaciation
▸ Reviewing Glacial Features

Glaciers pick up and transport a huge load of debris as they slowly advance across the land. Ultimately these materials are deposited when the ice melts. In regions where glacial sediment is deposited, it can play a truly significant role in forming the physical landscape. For example, in many areas once covered by the continental ice sheets of the recent Ice Age, the bedrock is rarely exposed because glacial deposits that are tens or even hundreds of meters thick completely mantle the terrain. The general effect of these deposits is to reduce the local relief and thus level the topography. Indeed, rural country scenes that are familiar to many of us—rocky pastures in New England, wheat fields in the Dakotas, rolling farmland in the Midwest—result directly from glacial deposition.

Long before the theory of an extensive Ice Age was ever proposed, much of the soil and rock debris covering portions of Europe was recognized as coming from somewhere else. At the time, these "foreign" materials were believed to have been "drifted" into their present positions by floating ice during an ancient flood. As a consequence, the term *drift* was applied to this sediment. Although rooted in an incorrect concept, this term was so well established by the time the true glacial origin of the debris became widely recog-

Landforms Made of Till

Till is deposited as glacial ice melts and drops its load of rock fragments. Unlike moving water and wind, ice cannot sort the sediment it carries; therefore, deposits of till are characteristically unsorted mixtures of many particle sizes (Figure 18.18). A close examination of this sediment shows that many of the pieces are scratched and polished as the result of being dragged along by the glacier. Such pieces help distinguish till from other deposits that are a mixture of different sediment sizes, such as material from a debris flow or a rockslide.

Boulders found in the till or lying free on the surface are called **glacial erratics** if they are different from the bedrock below (see Figure 18.13, p. 492). Of course, this means that they must have been derived from a source outside the area where they are found. Although the locality of origin for most erratics is unknown, the origin of some can be determined. In many cases, boulders were transported as far as 500 kilometers (300 miles) from their source area and, in a few instances, more than 1000 kilometers (600 miles). Therefore, by studying glacial erratics as well as the mineral composition of the remaining till, geologists are sometimes able to trace the path of a lobe of ice.

In portions of New England and other areas, erratics dot pastures and farm fields. In fact, in some places these large rocks were cleared from fields and piled to make fences and walls (Figure 18.19). Keeping the fields clear, however, is an ongoing chore because each spring newly exposed erratics appear. Wintertime frost heaving has lifted them to the surface.

FIGURE 18.19 Land cleared of glacial erratics, which were then piled atop one another to build this stone wall near West Bend, Wisconsin. (Photo by Tom Bean)

Alps. Today, however, moraine has a broader meaning, because it is applied to a number of landforms, all of which are composed primarily of till.

Alpine glaciers produce two types of moraines that occur exclusively in mountain valleys. The first of these is called a **lateral moraine** (Figure 18.20). As we learned earlier, when an alpine glacier moves downvalley, the ice erodes the sides of the valley with great efficiency. In addition, large quantities of debris are added to the glacier's surface as rubble falls or slides from higher up on the valley walls and collects on the edges of the moving ice. When the ice eventually melts, this accumulation of debris is dropped next to the valley walls. These ridges of till paralleling the sides of the valley constitute the lateral moraines.

FIGURE 18.20 Well-developed lateral moraines deposited by the shrinking Athabaska Glacier in Canada's Jasper National Park. (Photo by David Barnes/The Stock Market)

Close up of cobble

FIGURE 18.18 Glacial till is an unsorted mixture of many different sediment sizes. A close examination often reveals cobbles that have been scratched as they were dragged along by the glacier. (Photos by E. J. Tarbuck)

Lateral and Medial Moraines

The most common term for landforms made of glacial deposits is *moraine*. Originally this term was used by French peasants when referring to the ridges and embankments of rock debris found near the margins of glaciers in the French

FIGURE 18.21   Medial moraines form when the lateral moraines of merging valley glaciers join. Saint Elias National Park, Alaska. (Photo by Tom Bean)

The second type of moraine that is unique to alpine glaciers is the **medial moraine** (Figure 18.21). Medial moraines are created when two alpine glaciers coalesce to form a single ice stream. The till that was once carried along the sides of each glacier joins to form a single dark stripe of debris within the newly enlarged glacier. The creation of these dark stripes within the ice stream is one obvious proof that glacial ice moves, because the moraine could not form if the ice did not flow downvalley. It is quite common to see several medial moraines within a single large alpine glacier, because a streak will form whenever a tributary glacier joins the main valley.

End and Ground Moraines

An **end moraine** is a ridge of till that forms at the terminus of a glacier. These relatively common landforms are deposited when a state of equilibrium is attained between ablation and ice accumulation. That is, the end moraine forms when the ice is melting and evaporating near the end of the glacier at a rate equal to the forward advance of the glacier from its region of nourishment. Although the terminus of the glacier is now stationary, the ice continues to flow forward, delivering a continuous supply of sediment in the same manner a conveyor belt delivers goods to the end of a production line. As the ice melts, the till is dropped and the end moraine grows. The longer the ice front remains stable, the larger the ridge of till will become.

Eventually the time comes when ablation exceeds nourishment. At this point, the front of the glacier begins to recede in the direction from which it originally advanced. However, as the ice front retreats, the conveyor-belt action of the glacier continues to provide fresh supplies of sediment to the terminus. In this manner a large quantity of till is deposited as the ice melts away, creating a rock-strewn, undulating plain. This gently rolling layer of till deposited as the ice front recedes is termed **ground moraine.** Ground moraine has a leveling effect, filling in low spots and clogging old stream channels, often leading to a derangement of the existing drainage system. In areas where this layer of till is still relatively fresh, such as the northern Great Lakes region, poorly drained swampy lands are quite common.

Periodically, a glacier will retreat to a point where ablation and nourishment once again balance. When this happens, the ice front stabilizes and a new end moraine forms.

The pattern of end moraine formation and ground moraine deposition may be repeated many times before the glacier has completely vanished. Such a pattern is illustrated by Figure 18.22. It should be pointed out that the outermost end moraine marks the limit of the glacial advance. Because of its special status, this end moraine is also called the **terminal moraine.** On the other hand, the end moraines that were created as the ice front occasionally stabilized during retreat are termed **recessional moraines.** Note that both terminal and recessional moraines are essentially alike; the only difference between them is their relative positions.

End moraines deposited by the most recent major stage of Ice Age glaciation are prominent features in many parts of the Midwest and Northeast. In Wisconsin the wooded, hilly

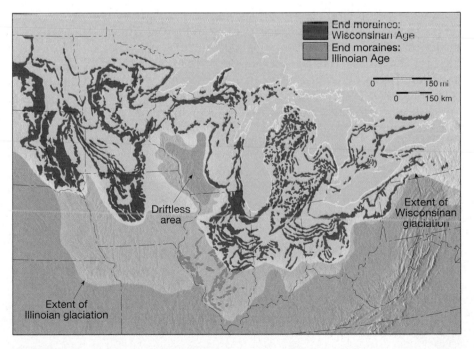

FIGURE 18.22   End moraines of the Great Lakes region. Those deposited during the most recent (Wisconsinan) stage are most prominent.

terrain of the Kettle Moraine near Milwaukee is a particularly picturesque example. A well-known example in the Northeast is Long Island. This linear strip of glacial sediment that extends northeastward from New York City is part of an end moraine complex that stretches from eastern Pennsylvania to Cape Cod, Massachusetts (Figure 18.23). The end moraines that make up Long Island represent materials that were deposited by a continental ice sheet in the relatively shallow waters off the coast and built up many meters above sea level. Long Island Sound, the narrow body of water separating the island and the mainland, was not built up as much by glacial deposition and was therefore subsequently flooded by the rising sea following the Ice Age.

Figure 18.24 represents a hypothetical area during glaciation and following the retreat of ice sheets. It shows the moraines that were described in this section, as well as the depositional features that are discussed in the sections that follow. This figure depicts landscape features similar to what might be encountered if you were traveling in the upper Midwest or New England. As you read upcoming sections dealing with other glacial deposits, you will be referred to this figure several times.

Drumlins

Moraines are not the only landforms deposited by glaciers. In some areas that were once covered by continental ice sheets, a special variety of glacial landscape exists—one characterized by smooth, elongate, parallel hills called **drumlins** (Figure 18.24). Certainly one of the best-known drumlins is Bunker Hill in Boston, the site of the famous Revolutionary War battle in 1775.

An examination of Bunker Hill or other less famous drumlins would reveal that drumlins are streamlined asymmetrical hills composed largely of till. They range in height from about 15 to 50 meters and may be up to 1 kilometer long. The steep side of the hill faces the direction from which the ice advanced, whereas the gentler, longer slope points in the direction the ice moved. Drumlins are not found as isolated landforms but rather occur in clusters called *drumlin fields* (Figure 18.25). One such cluster, east of Rochester, New York, is estimated to contain about 10,000 drumlins. Although drumlin formation is not fully understood, their

streamlined shape indicates that they were molded in the zone of plastic flow within an active glacier. It is believed that many drumlins originate when glaciers advance over previously deposited drift and reshape the material.

Landforms Made of Stratified Drift

As the name implies, stratified drift is sorted according to the size and weight of the particles. Because ice is not capable of such sorting activity, these materials are not deposited directly by the glacier as till is, but instead reflect the sorting action of glacial meltwater. Accumulations of stratified drift often consist largely of sand and gravel—that is, bed-load material—because the finer rock flour remains suspended and therefore is commonly carried far from the glacier by the meltwater streams.

Outwash Plains and Valley Trains

At the same time that an end moraine is forming, water from the melting glacier cascades over the till, sweeping some of it out in front of the growing ridge of unsorted debris. Meltwater generally emerges from the ice in rapidly moving streams that are often choked with suspended material and carry a substantial bed load as well. As the water leaves the glacier, it moves onto the relatively flat surface beyond and rapidly

Students Sometimes Ask . . .

Are any glacial deposits valuable?

Yes. In glaciated regions, landforms made of stratified drift, such as eskers, are often excellent sources of sand and gravel. Although the value per ton is low, huge quantities of these materials are used in the construction industry. In addition, glacial sands and gravels are valuable because they make excellent aquifers and thus are significant sources of groundwater in some areas. Clays from former glacial lakes have been used in the manufacture of bricks.

FIGURE 18.23 End moraines make up substantial parts of Long Island, Cape Cod, Martha's Vineyard, and Nantucket. Although portions are submerged, the Ronkonkoma moraine (a terminal moraine) extends through central Long Island, Martha's Vineyard, and Nantucket. It was deposited about 20,000 years ago. The recessional Harbor Hill moraine, which formed about 14,000 years ago, extends along the north shore of Long Island, through southern Rhode Island and Cape Cod.

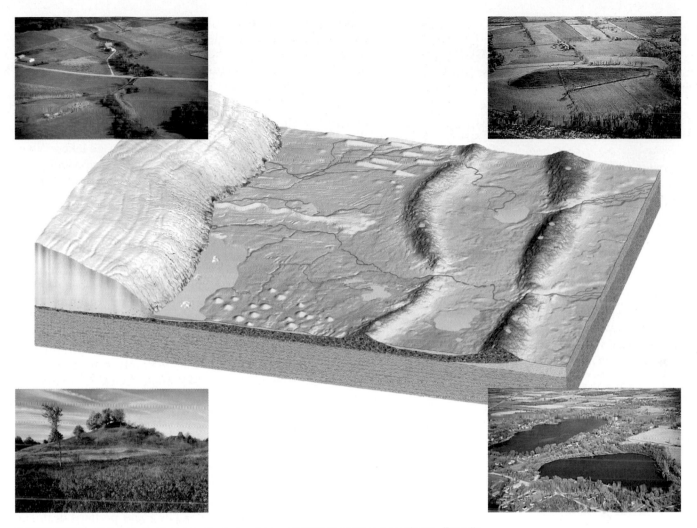

FIGURE 18.24   This hypothetical area illustrates many common depositional landforms. (Drumlin photo courtesy of Ward's Natural Science Establishment; kame, esker, and kettle photos by Richard R. Jacobs/JLM Visuals)

loses velocity. As a consequence, much of its bed load is dropped and the meltwater begins weaving a complex pattern of braided channels (Figure 18.24). In this way, a broad, ramplike surface composed of stratified drift is built adjacent to the downstream edge of most end moraines. When the feature is formed in association with an ice sheet, it is termed an **outwash plain,** and when largely confined to a mountain valley, it is usually referred to as a **valley train.**

Outwash plains and valley trains often are pockmarked with basins or depressions known as **kettles** (Figure 18.24). Kettles also occur in deposits of till. Kettles are formed when blocks of stagnant ice become wholly or partly buried in drift and eventually melt, leaving pits in the glacial sediment. Although most kettles do not exceed 2 kilometers in diameter, some with diameters exceeding 10 kilometers occur in Minnesota. Likewise, the typical depth of most kettles is less than 10 meters, although the vertical dimensions of some approach 50 meters. In many cases water eventually fills the depression and forms a pond or lake. One well-known example is Walden Pond near Concord, Massachusetts. It is here that Henry David Thoreau lived alone for two years in the 1840s and about which he wrote his famous book *Walden,* or *Life in the Woods.*

Ice-Contact Deposits

When the melting terminus of a glacier shrinks to a critical point, flow virtually stops and the ice becomes stagnant. Meltwater that flows over, within, and at the base of the motionless ice lays down deposits of stratified drift. Then, as the supporting ice melts away, the stratified sediment is left behind in the form of hills, terraces, and ridges. Such accumulations are collectively termed **ice-contact deposits** and are classified according to their shapes.

When the ice-contact stratified drift is in the form of a mound or steep-sided hill, it is called a **kame** (Figure 18.24). Some kames represent bodies of sediment deposited by meltwater in openings within, or depressions on top of, the ice. Others originate as deltas or fans built outward from the ice by meltwater streams. Later, when the stagnant ice melts away, these various accumulations of sediment collapse to form isolated, irregular mounds.

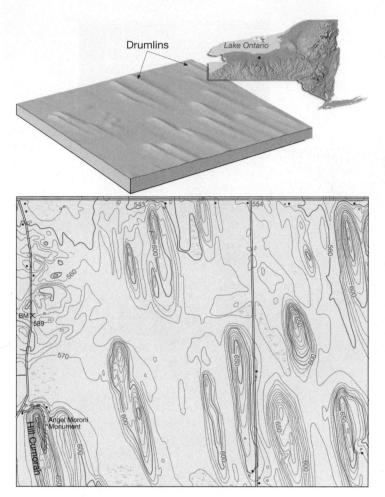

FIGURE 18.25 Portion of a drumlin field shown on the Palmyra, New York, 7.5-minute topographic map. North is at the top. The drumlins are steepest on the north side, indicating that the ice advanced from this direction.

When glacial ice occupies a valley, **kame terraces** may be built along the sides of the valley. These features commonly are narrow masses of stratified drift laid down between the glacier and the side of the valley by streams that drop debris along the margins of the shrinking ice mass.

A third type of ice-contact deposit is a long, narrow, sinuous ridge composed largely of sand and gravel. Some are more than 100 meters high with lengths in excess of 100 kilometers. The dimensions of many others are far less spectacular. Known as **eskers,** these ridges are deposited by meltwater rivers flowing within, on top of, and beneath a mass of motionless, stagnant glacial ice (Figure 18.24). Many sediment sizes are carried by the torrents of meltwater in the ice-banked channels, but only the coarser material can settle out of the turbulent stream.

Other Effects of Ice-Age Glaciers

In addition to the massive erosional and depositional work carried on by Pleistocene glaciers, the ice sheets had other effects, sometimes profound, on the landscape. For example, as the ice advanced and retreated, animals and plants were forced to migrate. This led to stresses that some organisms could not tolerate. Hence, a number of plants and animals became extinct. Other effects of Ice-Age glaciers that are described in this section involve adjustments in Earth's crust due to the addition and removal of ice and sea-level changes associated with the formation and melting of ice sheets. The advance and retreat of ice sheets also led to significant changes in the routes taken by rivers. In some regions, glaciers acted as dams that created large lakes. When these ice dams failed, the effects on the landscape were profound. In areas that today are deserts, lakes of another type, called pluvial lakes, formed during the Pleistocene.

Crustal Subsidence and Rebound

In areas that were centers of ice accumulation, such as Scandinavia and the Canadian Shield, the land has been slowly rising over the past several thousand years. Uplifting of almost 300 meters (1000 feet) has occurred in the Hudson Bay region. This, too, is the result of the continental ice sheets. But how can glacial ice cause such vertical crustal movement? We now understand that the land is rising because the added weight of the 3-kilometer-thick (2-mile-thick) mass of ice caused downwarping of Earth's crust. Following the removal of this immense load, the crust has been adjusting by gradually rebounding upward ever since (Figure 18.26).*

Sea-Level Changes

Certainly, one of the most interesting and perhaps dramatic effects of the Ice Age was the fall and rise of sea level that accompanied the advance and retreat of the glaciers. Earlier in the chapter it was pointed out that sea level would rise by an estimated 60 or 70 meters if the water locked up in the Antarctic Ice Sheet were to melt completely. Such an occurrence would flood many densely populated coastal areas.

Although the total volume of glacial ice today is great, exceeding 25 million cubic kilometers, during the Ice Age the volume of glacial ice amounted to about 70 million cubic kilometers, or 45 million cubic kilometers more than at present. Because we know that the snow from which glaciers are made ultimately comes from the evaporation of ocean water, the growth of ice sheets must have caused a worldwide drop in sea level (Figure 18.27). Indeed, estimates suggest that sea level was as much as 100 meters lower than today. Thus, land that is presently flooded by the oceans was dry. The Atlantic Coast of the United States lay more than 100 kilometers to the east of New York City; France and Britain were joined where the famous English Channel is today; Alaska and Siberia were connected across the Bering Strait; and Southeast Asia was tied by dry land to the islands of Indonesia.

---

*For a more complete discussion of this concept, termed *isostatic adjustment,* see the section on Isostatic Adjustment in Chapter 14.

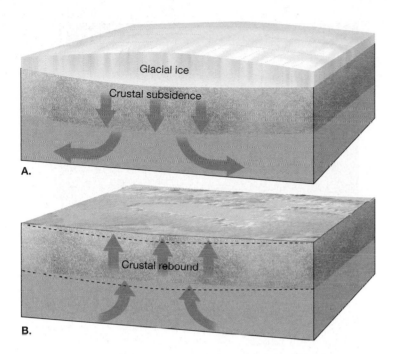

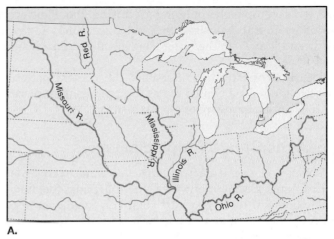

A.

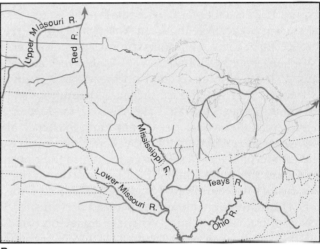

B.

FIGURE 18.26   Simplified illustration showing crustal subsidence and rebound resulting from the addition and removal of continental ice sheets. **A.** In northern Canada and Scandinavia, where the greatest accumulation of glacial ice occurred, the added weight caused downwarping of the crust. **B.** Ever since the ice melted, there has been gradual uplift or rebound of the crust.

Rivers before and after the Ice Age

Figure 18.28A shows the familiar present-day pattern of rivers in the central United States, with the Missouri, Ohio, and Illinois rivers as major tributaries to the Mississippi. Figure 18.28B depicts drainage systems in this region prior

FIGURE 18.28   **A.** This map shows the Great Lakes and the familiar present-day pattern of rivers in the central United States. Pleistocene ice sheets played a major role in creating this pattern. **B.** Reconstruction of drainage systems in the central United States prior to the Ice Age. The pattern was very different from today, and there were no Great Lakes.

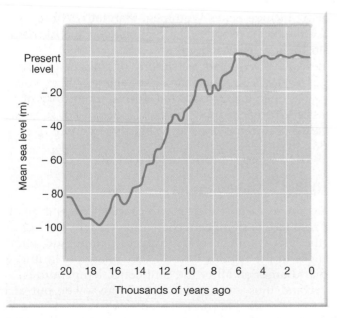

FIGURE 18.27   Changing sea level during the past 20,000 years. The lowest level shown on the graph represents the time about 18,000 years ago when the most recent ice advance was at a maximum.

to the Ice Age. The pattern is *very* different from the present. This remarkable transformation of river systems resulted from the advance and retreat of the ice sheets.

Notice that prior to the Ice Age, a significant part of the Missouri River drained north toward Hudson Bay. Moreover, the Mississippi River did not follow the present Iowa–Illinois boundary but rather flowed across west-central Illinois, where the lower Illinois River flows today. The preglacial Ohio River barely reached to the present-day state of Ohio, and the rivers that today feed the Ohio in western Pennsylvania flowed north and drained into the North Atlantic. The Great Lakes were created by glacial erosion during the Ice Age. Prior to the Pleistocene, the basins occupied by these huge lakes were lowlands with rivers that ran eastward to the Gulf of St. Lawrence.

The large Teays River was a significant feature prior to the Ice Age (Figure 18.28B). It flowed from West Virginia across Ohio, Indiana, and Illinois, where it discharged into the Mississippi River not far from present-day Peoria. This river valley, which would have rivaled the Mississippi in size, was completely obliterated during the Pleistocene, buried by glacial deposits hundreds of feet thick. Today the sands and gravels in the buried Teays valley make it an important aquifer.

Clearly, if we are to understand the present pattern of rivers in the central United States (and many other places as well), we must be aware of glacial history.

Ice Dams Create Proglacial Lakes

Ice sheets and alpine glaciers can act as dams to create lakes by trapping glacial meltwater and blocking the flow of rivers. Some of these lakes are relatively small, short-lived impoundments. Others can be large and exist for hundreds or thousands of years.

Figure 18.29 is a map of Lake Agassiz—the largest lake to form during the Ice Age in North America. With the retreat of the ice sheet came enormous volumes of meltwater. The Great Plains generally slope upward to the west. As the terminus of the ice sheet receded northeastward, meltwater was trapped between the ice on one side and the sloping land on the other, causing Lake Agassiz to deepen and spread across the landscape. It came into existence about 12,000 years ago and lasted for about 4500 years. Such water bodies are termed *proglacial lakes,* referring to their position just beyond the outer limits of a glacier or ice sheet. The lake's history is complicated by the dynamics of the ice sheet, which at various times readvanced and affected lake levels and drainage systems. Where drainage occurred depended upon the water level of the lake and the position of the ice sheet.

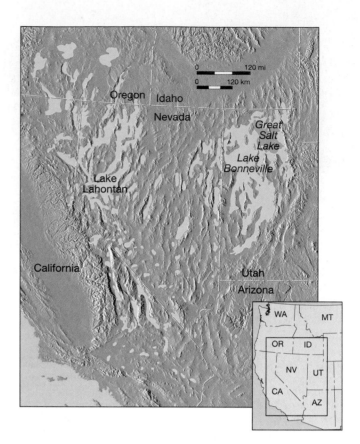

FIGURE 18.30 Pluvial lakes of the western United States. (After R. F. Flint, *Glacial and Quaternary Geology,* New York: John Wiley & Sons)

Lake Agassiz left marks over a broad region. Former beaches, many kilometers from any water, mark former shorelines. Several modern river valleys, including the Red River and the Minnesota River were originally cut by water entering or leaving the lake. Present-day remnants of Lake Agassiz include Lakes Winnipeg, Manitoba, Winnipegosis, and Lake of the Woods. The sediments of the former lake basin are now fertile agricultural land.

Research shows that the shifting of glaciers and the failure of ice dams can cause the rapid release of huge volumes of water. Such events occurred during the history of Lake Agassiz. One of the most dramatic examples of such *glacial outbursts* occurred in the Pacific Northwest and is described in Box 18.2.

Pluvial Lakes

While the formation and growth of ice sheets was an obvious response to significant changes in climate, the existence of the glaciers themselves triggered important climatic changes in the regions beyond their margins. In arid and semiarid areas on all of the continents, temperatures were lower and thus evaporation rates were lower, but at the same time moderate precipitation totals were experienced. This cooler, wetter climate formed many **pluvial lakes** (from the Latin term *pluvia,* meaning *rain*). In North America the greatest concentration of pluvial lakes occurred in the

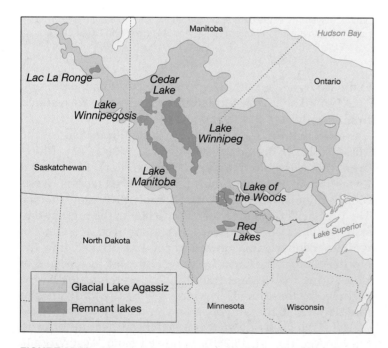

FIGURE 18.29 Map showing the extent of glacial Lake Agassiz. It was an immense feature—bigger than all of the present-day Great Lakes combined. The modern-day remnants of this proglacial water body are still major landscape features.

vast Basin and Range region of Nevada and Utah (Figure 18.30). By far the largest of the lakes in this region was Lake Bonneville. With maximum depths exceeding 300 meters and an area of 50,000 square kilometers, Lake Bonneville was nearly the same size as present-day Lake Michigan. As the ice sheets waned, the climate again grew more arid, and the lake levels lowered in response. Although most of the lakes completely disappeared, a few small remnants of Lake Bonneville remain, the Great Salt Lake being the largest and best known.

The Glacial Theory and the Ice Age

In the preceding pages, we mentioned the Ice Age, a time when ice sheets and alpine glaciers were far more extensive than they are today. As noted, there was a time when the most popular explanation for what we now know to be glacial deposits was that the materials had been drifted in by means of icebergs or perhaps simply swept across the landscape by a catastrophic flood. What convinced geologists that an extensive ice age was responsible for these deposits and many other glacial features?

In 1821 a Swiss engineer, Ignaz Venetz, presented a paper suggesting that glacial landscape features occurred at considerable distances from the existing glaciers in the Alps. This implied that the glaciers had once been larger, and occupied positions farther downvalley. Another Swiss scientist, Louis Agassiz, doubted the proposal of widespread glacial activity put forth by Venetz. He set out to prove that the idea was not valid. Ironically, his 1836 fieldwork in the Alps convinced him of the merits of his colleague's hypothesis. In fact, a year later Agassiz hypothesized a great ice age that had extensive and far-reaching effects—an idea that was to give Agassiz widespread fame.

The proof of the glacial theory proposed by Agassiz and others constitutes a classic example of applying the principle of uniformitarianism. Realizing that certain features are produced by no other known process but glacial action, they were able to begin reconstructing the extent of now vanished ice sheets based on the presence of features and deposits found far beyond the margins of present-day glaciers. In this manner, the development and verification of the glacial theory continued during the 19th century, and through the efforts of many scientists, a knowledge of the nature and extent of former ice sheets became clear.

By the beginning of the 20th century, geologists had largely determined the areal extent of the Ice Age glaciation. Further, during the course of their investigations, they had discovered that many glaciated regions had not one layer of drift but several. Moreover, close examination of these older deposits showed well-developed zones of chemical weathering and soil formations as well as the remains of plants that require warm temperatures. The evidence was clear; there had been not just one glacial advance but many, each separated by extended periods when climates were as warm or warmer than the present. The Ice Age had not simply been a time when the ice advanced over the land, lingered for a while, and then receded. Rather, the period was a very complex event, characterized by a number of advances and withdrawals of glacial ice.

By the early 20th century a fourfold division of the Ice Age had been established for both North America and Europe. The divisions were based largely on studies of glacial deposits. In North America each of the four major stages was named for the midwestern state where deposits of that stage were well exposed and/or were first studied. These are, in order of occurrence, the Nebraskan, Kansan, Illinoian, and Wisconsinan. These traditional divisions remained in place until relatively recently, when it was learned that sediment cores from the ocean floor contain a much more complete record of climate change during the Ice Age. Unlike the glacial record on land, which is punctuated by many unconformities, seafloor sediments provide an uninterrupted record of climatic cycles for this period. Studies of these seafloor sediments showed that glacial/interglacial cycles had occurred about every 100,000 years. About 20 such cycles of cooling and warming were identified for the span we call the Ice Age.

During the glacial age, ice left its imprint on almost 30 percent of Earth's land area, including about 10 million square kilometers of North America, 5 million square kilometers of Europe, and 4 million square kilometers of Siberia (Figure 18.31). The amount of glacial ice in the Northern Hemisphere was roughly twice that of the Southern Hemisphere. The primary reason is that the southern polar ice could not spread far beyond the margins of Antarctica. By contrast, North America and Eurasia provided great expanses of land for the spread of ice sheets.

Today we know that the Ice Age began between 2 million and 3 million years ago. This means that most of the major

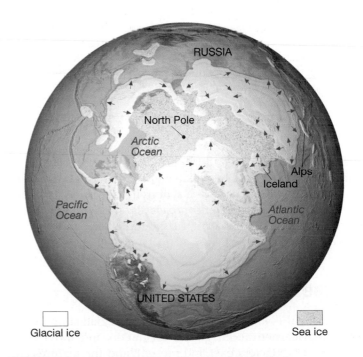

FIGURE 18.31   Maximum extent of ice sheets in the Northern Hemisphere during the Ice Age.

BOX 18.2 ▶ UNDERSTANDING EARTH

Glacial Lake Missoula, Megafloods, and the Channeled Scablands

Lake Missoula was a prehistoric proglacial lake in western Montana that existed as the Pleistocene ice age was drawing to a close between about 15,000 and 13,000 years ago. It was a time when the climate was gradually warming and the Cordilleran ice sheet, which covered much of western Canada and portions of the Pacific Northwest, was melting and retreating.

The lake was the result of an ice dam that formed when a southward protruding mass of glacial ice called the Purcell Lobe blocked the Clark Fork River (Figure 18.B). As the water rose behind the 600 meter- (2000 foot-) high dam, it flooded the valleys of western Montana. At its greatest extent, Lake Missoula extended eastward for more than 300 kilometers (200 miles). Its volume exceeded 2500 cubic kilometers (600 cubic miles)—greater than present day Lake Ontario.

Eventually, the lake became so deep that the ice dam began to float. That is, the rising waters behind the dam lifted the buoyant ice so that it no longer functioned as a dam. The result was a catastrophic flood as Lake Missoula's waters suddenly poured out under the failed dam. The outburst rushed across the lava plains of eastern Washington and down the Columbia River to the Pacific Ocean. Based on the sizes of boulders moved during the event, the velocity of the torrent approached or exceeded 70 kilometers (nearly 45 miles) per hour. The entire lake was emptied in a matter of a few days. Due to the temporary impounding of water behind narrow gaps along the flood's path, flooding probably continued through the devastated region for a few weeks.

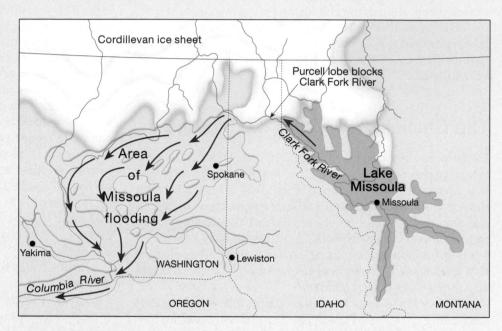

FIGURE 18.B   Lake Missoula was a proglacial lake created when the Purcell Lobe of the Cordilleran ice sheet formed an ice dam on the Clark Fork River. Periodically, the ice dam would fail sending a huge torrent of water flooding across the landscape of eastern Washington.

The erosional and depositional results of such a megaflood were dramatic. The towering mass of rushing water stripped away thick layers of sediment and soil and cut deep canyons (*coulees*) into the underlying basalt (Figure 18.C). Today the region is called the *Channeled Scablands*—a landscape consisting of a bizarre assemblage of landforms. Perhaps the most striking features are the table-like lava mesas called *scabs* left between braided interlocking channels (Figure 18.D).

It was not a single flood from glacial Lake Missoula that created this extraordinary landscape. What makes Lake Missoula remarkable is that it alternately filled and emptied in a cycle that was repeated more than 40 times over a span of 1500 years. After each flood, the glacial lobe would again block the valley and create a new ice dam, and the cycle of lake growth, dam failure, and megaflood would follow at intervals of 20 to 60 years.

glacial stages occurred during a division of the geologic time scale called the **Pleistocene epoch.** Although the Pleistocene is commonly used as a synonym for the Ice Age, note that this epoch does not encompass all of the last glacial period. The Antarctic ice sheet, for example, probably formed at least 30 million years ago.

Causes of Glaciation

A great deal is known about glaciers and glaciation. Much has been learned about glacier formation and movement, the extent of glaciers past and present, and the features created by glaciers, both erosional and depositional. However,

a widely accepted theory for the causes of glacial ages has not yet been established. Although more than 160 years have elapsed since Louis Agassiz proposed his theory of a great Ice Age, no complete agreement exists as to the causes of such events.

Although widespread glaciation has been rare in Earth's history, the Ice Age that encompassed the Pleistocene epoch is not the only glacial period for which a record exists. Earlier glaciations are indicated by deposits called **tillite,** a sedimentary rock formed when glacial till becomes lithified. Such deposits, found in strata of several different ages, usually contain striated rock fragments, and some overlie grooved and polished bedrock surfaces or are associated with sandstones and conglomerates that show features of

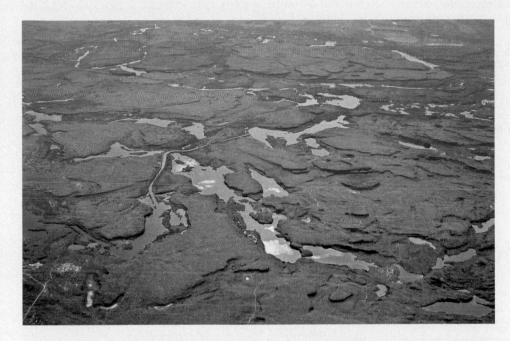

FIGURE 18.C The landscape of the Channeled Scablands was sculpted by the megafloods associated with glacial Lake Missoula. (Photo by Jim Wark/AirPhotoNA)

FIGURE 18.D These table-like mesas carved from basalt are called *scabs* and were produced by floods from Lake Missoula near the end of the Pleistocene. (Photo by John S. Shelton)

outwash deposits. Two Precambrian glacial episodes have been identified in the geologic record, the first approximately 2 billion years ago and the second about 600 million years ago. Further, a well-documented record of an earlier glacial age is found in late Paleozoic rocks that are about 250 million years old, which exist on several landmasses.

Any theory that attempts to explain the causes of glacial ages must successfully answer two basic questions. (1) *What causes the onset of glacial conditions?* For continental ice sheets to have formed, average temperature must have been somewhat lower than at present and perhaps substantially lower than throughout much of geologic time. Thus, a successful theory would have to account for the cooling that finally leads to glacial conditions. (2) *What caused the alternation of*

glacial and interglacial stages that have been documented for the Pleistocene epoch? The first question deals with long-term trends in temperature on a scale of millions of years, but this second question relates to much shorter-term changes.

Although the literature of science contains a vast array of hypotheses relating to the possible causes of glacial periods, we will discuss only a few major ideas to summarize current thought.

Plate Tectonics

Probably the most attractive proposal for explaining the fact that extensive glaciations have occurred only a few times in the geologic past comes from the theory of plate tectonics.

Because glaciers can form only on land, we know that land-masses must exist somewhere in the higher latitudes before an ice age can commence. Many scientists suggest that ice ages have occurred only when Earth's shifting crustal plates have carried the continents from tropical latitudes to more poleward positions.

Glacial features in present-day Africa, Australia, South America, and India indicate that these regions, which are now tropical or subtropical, experienced an ice age near the end of the Paleozic era, about 250 million years ago. However, there is no evidence that ice sheets existed during this same period in what are today the higher latitudes of North America and Eurasia. For many years this puzzled scientists. Was the climate in these relatively tropical latitudes once like it is today in Greenland and Antarctica? Why did glaciers not form in North America and Eurasia? Until the plate tectonics theory was formulated, there had been no reasonable explanation.

Today scientists understand that the areas containing these ancient glacial features were joined together as a single supercontinent located at latitudes far to the south of their present positions. Later this landmass broke apart, and its pieces, each moving on a different plate, migrated toward their present locations (Figure 18.32). Now we know that during the geologic past, plate movements accounted for many dramatic climate changes as landmasses shifted in relation to one another and moved to different latitudinal positions. Changes in oceanic circulation also must have occurred, altering the transport of heat and moisture and consequently the climate as well. Because the rate of plate movement is very slow—a few centimeters annually—appreciable changes in the positions of the continents occur only over great spans of geologic time. Thus, climate changes triggered by shifting plates are extremely gradual and happen on a scale of millions of years.

Variations in Earth's Orbit

Because climatic changes brought about by moving plates are extremely gradual, the plate tectonics theory cannot be used to explain the alternation between glacial and interglacial climates that occurred during the Pleistocene epoch. Therefore, we must look to some other triggering mechanism that might cause climate change on a scale of thousands rather than millions of years. Many scientists today believe, or strongly suspect, that the climate oscillations that characterized the Pleistocene may be linked to variations in Earth's orbit. This hypothesis was first developed and strongly advocated by the Serbian astrophysicist Milutin Milankovitch and is based on the premise that variations in incoming solar radiation are a principal factor in controlling Earth's climate.

Milankovitch formulated a comprehensive mathematical model based on the following elements (Figure 18.33):

1. Variations in the shape (*eccentricity*) of Earth's orbit about the Sun;
2. Changes in *obliquity*; that is, changes in the angle that the axis makes with the plane of Earth's orbit;
3. The wobbling of Earth's axis, called *precession*.

Using these factors, Milankovitch calculated variations in the receipt of solar energy and the corresponding surface temperature of Earth back into time in an attempt to correlate these changes with the climate fluctuations of the Pleistocene. In explaining climate changes that result from these three variables, note that they cause little or no variation in the total solar energy reaching the ground. Instead, their impact is felt because they change the degree of contrast between the seasons. Somewhat milder winters in the middle to high latitudes mean greater snowfall totals, whereas cooler summers would bring a reduction in snowmelt.

Among the studies that have added credibility to the astronomical theory of Milankovitch is one in which deep-sea sediments containing certain climatically sensitive microorganisms were analyzed to establish a chronology of temperature changes going back nearly one-half million years.* This time scale of climate change was then compared to astronomical calculations of eccentricity, obliquity, and precession to determine if a correlation did indeed exist. Although the study was very involved and mathematically complex, the conclusions were straightforward. The researchers found that major variations in climate over the past several hundred thousand years were closely associated with changes in the geometry of Earth's orbit; that is, cycles of climate

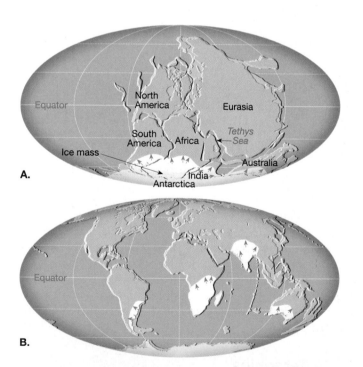

FIGURE 18.32   **A.** The supercontinent Pangaea showing the area covered by glacial ice 300 million years ago. **B.** The continents as they are today. The white areas indicate where evidence of the former ice sheets exists.

*J. D. Hays, John Imbrie, and N. J. Shackelton, "Variations in the Earth's Orbit: Pacemaker of the Ice Ages," *Science* 194 (1976): 1121–32.

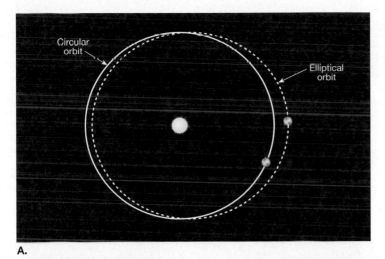

A.

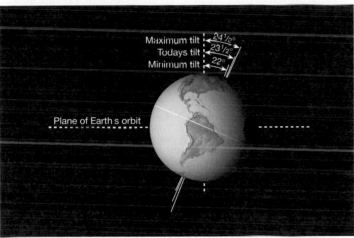

B.

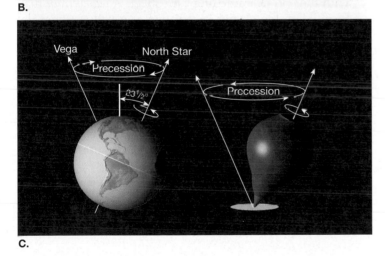

C.

FIGURE 18.33 Orbital variations. **A.** The shape of Earth's orbit changes during a cycle that spans about 100,000 years. It gradually changes from nearly circular to one that is more elliptical and then back again. This diagram greatly exaggerates the amount of change. **B.** Today the axis of rotation is tilted about 23.5° to the plane of Earth's orbit. During a cycle of 41,000 years, this angle varies from 21.5° to 24.5°. **C.** Precession. Earth's axis wobbles like that of a spinning top. Consequently, the axis points to different spots in the sky during a cycle of about 26,000 years.

change were shown to correspond closely with the periods of obliquity, precession, and orbital eccentricity. More specifically, the authors stated: "It is concluded that changes in the earth's orbital geometry are the fundamental cause of the succession of Quaternary ice ages."*

Let us briefly summarize the ideas that were just described. The theory of plate tectonics provides us with an explanation for the widely spaced and nonperiodic onset of glacial conditions at various times in the geologic past, whereas the theory proposed by Milankovitch and supported by the work of J. D. Hays and his colleagues furnishes an explanation for the alternating glacial and interglacial episodes of the Pleistocene.

Other Factors

Variations in Earth's orbit correlate closely with the timing of glacial–interglacial cycles. However, the variations in solar energy reaching Earth's surface caused by these orbital changes do not adequately explain the magnitude of the temperature changes that occurred during the most recent ice age. Other factors must also have contributed. One factor involves variations in the composition of the atmosphere. Other influences involve changes in the reflectivity of Earth's surface and in ocean circulation. Let's take a brief look at these factors.

Chemical analyses of air bubbles that become trapped in glacial ice at the time of ice formation indicate that the ice-age atmosphere contained less of the gases carbon dioxide and methane than the post ice-age atmosphere. Carbon dioxide and methane are important "greenhouse" gases, which means that they trap radiation emitted by Earth and contribute to the heating of the atmosphere. When the amount of carbon dioxide and methane in the atmosphere increase, global temperatures rise, and when there is a reduction in these gases, as occurred during the Ice Age, temperatures fall. Therefore, reductions in the concentrations of greenhouse gases help explain the magnitude of the temperature drop that occurred during glacial times. Although scientists know that concentrations of carbon dioxide and methane dropped, they do not know what caused the drop. As often occurs in science, observations gathered during one investigation yield information and raise questions that require further analysis and explanation.

Obviously, whenever Earth enters an ice age, extensive areas of land that were once ice free are covered with ice and snow. In addition, a colder climate causes the area covered by sea ice (frozen surface sea water) to expand as well. Ice and snow reflect a large portion of incoming solar energy back to space. Thus, energy that would have warmed Earth's surface and the air above is lost and global cooling is reinforced.**

---

*J. D. Hays et al., p. 1131. The term *Quaternary* refers to the period on the geologic time scale that encompasses the last 1.8 million years.
**Recall from Chapter 1 that something that reinforces (adds to) the initial change is called a *positive feedback mechanism*. To review this idea, see the discussion on feedback mechanisms in the section on "Earth As a System" in Chapter 1.

Yet another factor that influences climate during glacial times relates to ocean currents. Research has shown that ocean circulation changes during ice ages. For example, studies suggest that the warm current that transports large amounts of heat from the tropics toward higher latitudes in the North Atlantic was significantly weaker during the Ice Age. This would lead to a colder climate in Europe, amplifying the cooling attributable to orbital variations.

In conclusion, we emphasize that the ideas just discussed do not represent the only possible explanations for glacial ages. Although interesting and attractive, these proposals are certainly not without critics, nor are they the only possibilities currently under study. Other factors may be, and probably are, involved.

Summary

- A *glacier* is a thick mass of ice originating on land as a result of the compaction and recrystallization of snow, and it shows evidence of past or present flow. Today *valley* or *alpine glaciers* are found in mountain areas where they usually follow valleys that were originally occupied by streams. *Ice sheets* exist on a much larger scale, covering most of Greenland and Antarctica.

- Near the surface of a glacier, in the *zone of fracture*, ice is brittle. However, below about 50 meters, pressure is great, causing ice to *flow* like *plastic material*. A second important mechanism of glacial movement consists of the entire ice mass *slipping* along the ground.

- The average velocity of glacial movement is generally quite slow but varies considerably from one glacier to another. The advance of some glaciers is characterized by periods of extremely rapid movements called *surges*.

- Glaciers form in areas where more snow falls in winter than melts during summer. Snow accumulation and ice formation occur in the *zone of accumulation*. Its outer limits are defined by the *snowline*. Beyond the snowline is the *zone of wastage*, where there is a net loss to the glacier. The *glacial budget* is the balance, or lack of balance, between accumulation at the upper end of the glacier, and loss, called *ablation*, at the lower end.

- Glaciers erode land by *plucking* (lifting pieces of bedrock out of place) and *abrasion* (grinding and scraping of a rock surface). Erosional features produced by valley glaciers include *glacial troughs, hanging valleys, pater noster lakes, fiords, cirques, arêtes, horns,* and *roches moutonnées*.

- Any sediment of glacial origin is called *drift*. The two distinct types of glacial drift are (1) *till*, which is unsorted sediment deposited directly by the ice; and (2) *stratified drift*, which is relatively well-sorted sediment laid down by glacial meltwater.

- The most widespread features created by glacial deposition are layers or ridges of till, called *moraines*. Associated with valley glaciers are *lateral moraines*, formed along the sides of the valley, and *medial moraines*, formed between two valley glaciers that have joined. *End moraines*, which mark the former position of the front of a glacier, and *ground moraines*, undulating layers of till deposited as the ice front retreats, are common to both valley glaciers and ice sheets. An *outwash plain* is often associated with the end moraine of an ice sheet. A *valley train* may form when the glacier is confined to a valley. Other depositional features include *drumlins* (streamlined asymmetrical hills composed of till), *eskers* (sinuous ridges composed largely of sand and gravel deposited by streams flowing in tunnels beneath the ice, near the terminus of a glacier), and *kames* (steep-sided hills consisting of sand and gravel).

- The *Ice Age*, which began about 2 million years ago, was a very complex period characterized by a number of advances and withdrawals of glacial ice. Most of the major glacial episodes occurred during a division of the geologic time scale called the *Pleistocene epoch*. Perhaps the most convincing evidence for the occurrence of several glacial advances during the Ice Age is the widespread existence of *multiple layers of drift* and an uninterrupted record of climate cycles preserved in *seafloor sediments.*

- In addition to massive erosional and depositional work, other effects of Ice Age glaciers included the *forced migration of organisms, changes in stream courses, formation of large proglacial lakes, adjustment of the crust* by rebounding after the removal of the immense load of ice, and *climate changes* caused by the existence of the glaciers themselves. In the sea, the most far-reaching effect of the Ice Age was the *worldwide change* in *sea level* that accompanied each advance and retreat of the ice sheets.

- Any theory that attempts to explain the causes of glacial ages must answer two basic questions: (1) What causes the onset of glacial conditions? and (2) What caused the alternating glacial and interglacial stages that have been documented for the Pleistocene epoch? Two of the many hypotheses for the cause of glacial ages involve (1) plate tectonics and (2) variations in Earth's orbit. Other factors that are related to climate change during glacial ages include: changes in atmospheric composition, variations in the amount of sunlight reflected by Earth's surface, and changes in ocean circulation.

Review Questions

1. Where are glaciers found today? What percentage of Earth's land area do they cover? How does this compare to the area covered by glaciers during the Pleistocene?

2. Describe how glaciers fit into the hydrologic cycle. What role do they play in the rock cycle?

3. Each statement below refers to a particular type of glacier. Name the type of glacier.

    a. The term *continental* is often used to describe this type of glacier.

    b. This type of glacier is also called an *alpine glacier.*

    c. This is a stream of ice leading from the margin of an ice sheet through the mountains to the sea.

    d. This is a glacier formed when one or more valley glaciers spreads out at the base of a steep mountain front.

    e. Greenland is the only example in the Northern Hemisphere.

4. Describe the two components of glacial flow. At what rates do glaciers move? In a valley glacier, does all of the ice move at the same rate? Explain.

5. Why do crevasses form in the upper portion of a glacier but not below 50 meters (160 feet)?

6. Under what circumstances will the front of a glacier advance? Retreat? Remain stationary?

7. Describe the processes of glacial erosion.

8. How does a glaciated mountain valley differ in appearance from a mountain valley that was not glaciated?

9. List and describe the erosional features you might expect to see in an area where valley glaciers exist or have recently existed.

10. What is glacial drift? What is the difference between till and stratified drift? What general effect do glacial deposits have on the landscape?

11. List the four basic moraine types. What do all moraines have in common? What is the significance of terminal and recessional moraines?

12. Why are medial moraines proof that valley glaciers must move?

13. How do kettles form?

14. What direction was the ice sheet moving that affected the area shown in Figure 18.25? Explain how you were able to determine this.

15. What are ice-contact deposits? Distinguish between kames and eskers.

16. Describe at least four effects of Ice Age glaciers aside from the formation of major erosional and depositional features.

17. The development of the glacial theory is a good example of applying the principle of uniformitarianism. Explain briefly.

18. During the Pleistocene epoch the amount of glacial ice in the Northern Hemisphere was about twice as great as in the Southern Hemisphere. Briefly explain why this was the case.

19. How might plate tectonics help explain the cause of ice ages? Can plate tectonics explain the alternation between glacial and interglacial climates during the Pleistocene?

Key Terms

ablation (p. 491)
abrasion (p. 492)
alpine glacier (p. 485)
arête (p. 496)
basal slip (p. 489)
calving (p. 490)
cirque (p. 493)
col (p. 496)
crevasse (p. 489)
drumlin (p. 500)
end moraine (p. 499)
esker (p. 502)
fiord (p. 496)
firn (p. 488)
glacial budget (p. 491)

glacial drift (p. 497)
glacial erratic (p. 497)
glacial striations (p. 492)
glacial trough (p. 493)
glacier (p. 485)
ground moraine (p. 499)
hanging valley (p. 493)
horn (p. 496)
ice cap (p. 486)
ice-contact deposit (p. 501)
ice sheet (p. 485)
ice shelf (p. 486)
kame (p. 501)
kame terrace (p. 502)
kettle (p. 501)

lateral moraine (p. 498)
medial moraine (p. 499)
outlet glacier (p. 486)
outwash plain (p. 501)
pater noster lakes (p. 493)
piedmont glacier (p. 486)
plastic flow (p. 489)
Pleistocene epoch (p. 506)
plucking (p. 492)
pluvial lake (p. 504)
recessional moraine (p. 499)
roche moutonnée (p. 497)
rock flour (p. 492)
snowline (p. 487)
stratified drift (p. 497)

surge (p. 490)
tarn (p. 496)
terminal moraine (p. 499)
till (p. 497)
tillite (p. 506)
truncated spur (p. 493)
valley glacier (p. 486)
valley train (p. 501)
zone of accumulation
   (p. 490)
zone of fracture (p. 489)
zone of wastage (p. 490)

Web Resources

The *Earth* Website uses the resources and flexibility of the Internet to aid in your study of the
topics in this chapter. Written and developed
by geology instructors, this site will help improve your understanding of geology. Visit **http://www**
.prenhall.com/tarbuck and click on the cover of *Earth 9e* to
find:

- Online review quizzes.

- Critical thinking exercises.

- Links to chapter-specific Web resources.

- Internet-wide key-term searches.

http://www.prenhall.com/tarbuck

GEODe: Earth

GEODe: Earth makes studying faster and more effective by reinforcing key concepts using animation, video, narration, interactive exercises and practice quizzes. A copy is included with every copy of *Earth*.

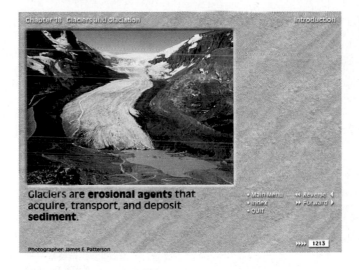

Glaciers are **erosional agents** that acquire, transport, and deposit **sediment**.

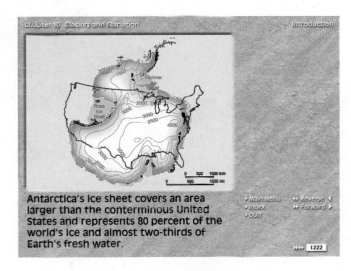

Antarctica's ice sheet covers an area larger than the conterminous United States and represents 80 percent of the world's ice and almost two-thirds of Earth's fresh water.

Deserts and Winds

*Arizona's Organ
Pipe Cactus
National
Monument is
in the Sonoran
Desert. Ajo
Mountains in
the background.
(Photo by Jeff
Lepore/Photo
Researchers, Inc.)*

515

Climate has a strong influence on the nature and intensity of Earth's external processes. This was clearly demonstrated in the preceding chapter on glaciers. Another excellent example of the strong link between climate and geology is seen when we examine the development of arid landscapes. The word *desert* literally means deserted or unoccupied. For many dry regions this is a very appropriate description, although where water is available in deserts, plants and animals thrive. Nevertheless, the world's dry regions are probably the least familiar land areas on Earth outside of the polar realm.

Desert landscapes frequently appear stark. Their profiles are not softened by a carpet of soil and abundant plant life. Instead, barren rocky outcrops with steep, angular slopes are common. At some places the rocks are tinted orange and red. At others they are gray and brown and streaked with black. For many visitors, desert scenery exhibits a striking beauty; to others, the terrain seems bleak. No matter which feeling is elicited, it is clear that deserts are very different from the more humid places where most people live.

As you will see, arid regions are not dominated by a single geologic process. Rather, the effects of tectonic forces, running water, and wind are all apparent. Because these processes combine in different ways from place to place, the appearance of desert landscapes varies a great deal as well (Figure 19.1).

FIGURE 19.1 A scene in Southern Utah near the San Juan River. The appearance of desert landscapes varies a great deal from place to place. (Photo © by Carr Clifton. All rights reserved.)

Distribution and Causes of Dry Lands

 GEODe
Deserts and Winds
▶ **Distribution and Causes of Dry Lands**

The dry regions of the world encompass about 42 million square kilometers, a surprising 30 percent of Earth's land surface. No other climatic group covers so large a land area. Within these water-deficient regions, two climatic types are commonly recognized: **desert,** or arid, and **steppe,** or semi-arid. The two share many features; their differences are primarily a matter of degree (see Box 19.1). The steppe is a marginal and more humid variant of the desert and is a transition zone that surrounds the desert and separates it from bordering humid climates. The world map showing the distribution of desert and steppe regions reveals that dry lands are concentrated in the subtropics and in the middle latitudes (Figure 19.2).

Low-Latitude Deserts

The heart of the low-latitude dry climates lies in the vicinities of the Tropics of Cancer and Capricorn. Figure 19.2 shows a virtually unbroken desert environment stretching for more than 9300 kilometers (5800 miles) from the Atlantic coast of North Africa to the dry lands of northwestern India. In addition to this single great expanse, the Northern Hemisphere contains another, much smaller area of tropical desert and steppe in northern Mexico and the southwestern United States.

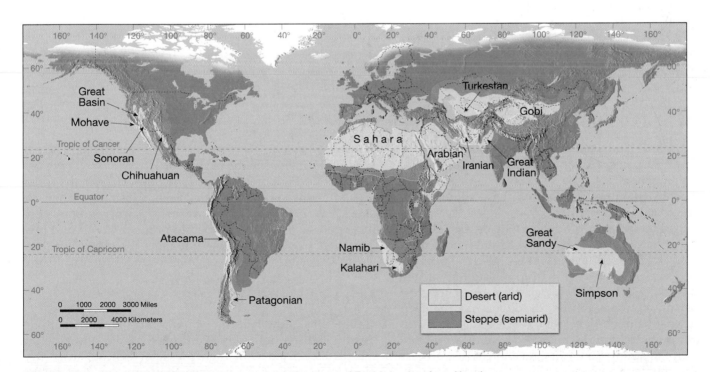

FIGURE 19.2 Arid and semiarid climates cover about 30 percent of Earth's land surface. No other climate group covers so large an area.

BOX 19.1 ▶ UNDERSTANDING EARTH

What Is Meant by "Dry"?

Albuquerque, New Mexico, in the southwestern United States, receives an average of 20.7 centimeters (8.07 inches) of rainfall annually. As you might expect, because Albuquerque's precipitation total is modest, the station is classified as a desert when the commonly used Köppen climate classification is applied. The Russian city of Verkhoyansk is a remote station located near the Arctic Circle in Siberia. The yearly precipitation total there averages 15.5 centimeters (6.05 inches), about 5 centimeters less than Albuquerque's. Although Verkhoyansk receives less precipitation than Albuquerque, its classification is that of a humid climate. How can this occur?

We all recognize that deserts are dry places, but just what is meant by the term *dry*? That is, how much rain defines the boundary between humid and dry regions? Sometimes it is arbitrarily defined by a single rainfall figure, for example, 25 centimeters (10 inches) per year of precipitation. However, the concept of dryness is a relative one that refers to any situation in which a water deficiency exists. Hence, climatologists define *dry climate* as one in which yearly precipitation is not as great as the potential loss of water by evaporation. Dryness, then, not only is related to annual rainfall totals but is also a function of evaporation, which in turn is closely dependent upon temperature.

As temperatures climb, potential evaporation also increases. Fifteen to 25 centi-

meters of precipitation may be sufficient to support coniferous forests in northern Scandinavia or Siberia, where evaporation into the cool, humid air is slight and a surplus of water remains in the soil. However, the same amount of rain falling on New Mexico or Iran supports only a sparse vegetative cover because evaporation into the hot, dry air is great. So, clearly, no specific amount of precipitation can serve as a universal boundary for dry climates.

To establish the boundary between dry and humid climates, the widely used Köppen classification system uses formulas that involve three variables: average annual precipitation, average annual temperature, and seasonal distribution of precipitation. The use of average annual temperature reflects its importance as an index of evaporation. The amount of rainfall defining the humid–

dry boundary will be larger where mean annual temperatures are high, and smaller where temperatures are low. The use of seasonal precipitation distribution as a variable is also related to this idea. If rain is concentrated in the warmest months, loss to evaporation is greater than if the precipitation is concentrated in the cooler months.

Table 19.A summarizes the precipitation amounts that divide dry and humid climates. Notice that a station with an annual mean of 20°C (68°F) and a summer rainfall maximum of 68 centimeters (26.5 inches) is classified as dry. If the rain falls primarily in winter, however, the station must receive only 40 centimeters (15.6 inches) or more to be considered humid. If the precipitation is more evenly distributed, the figure defining the humid–dry boundary is between the other two.

TABLE 19.A Average Annual Precipitation Defining the Boundary between Dry and Humid Climates

Average Rainfall Annual Temp. (C°)	Winter Rainfall Maximum (centimeters)	Even Distribution (centimeters)	Summer Maximum (centimeters)
5	10	24	38
10	20	34	48
15	30	44	58
20	40	54	68
25	50	64	78
30	60	74	88

In the Southern Hemisphere, dry climates dominate Australia. Almost 40 percent of the continent is desert, and much of the remainder is steppe. In addition, arid and semiarid areas occur in southern Africa and make a limited appearance in coastal Chile and Peru.

What causes these bands of low-latitude desert? The answer is the global distribution of air pressure and winds. Figure 19.3A, an idealized diagram of Earth's general circulation, helps visualize the relationship. Heated air in the pressure belt known as the *equatorial low* rises to great heights (usually between 15 and 20 kilometers) and then spreads out. As the upper-level flow reaches 20° to 30° latitude, north or south, it sinks toward the surface. Air that rises through the atmosphere expands and cools, a process that leads to the development of clouds and precipitation. For this reason, the areas under the influence of the equato-

rial low are among the rainiest on Earth. Just the opposite is true for the regions in the vicinity of 30° north and south latitude, where high pressure predominates. Here, in the zones known as the *subtropical highs*, air is subsiding. When air sinks, it is compressed and warmed. Such conditions are just opposite of what is needed to produce clouds and precipitation. Consequently, these regions are known for their clear skies, sunshine, and ongoing drought (Figure 19.3B).

Middle-Latitude Deserts

Unlike their low-latitude counterparts, middle-latitude deserts and steppes are not controlled by the subsiding air masses associated with high pressure. Instead, these dry lands exist principally because they are sheltered in the deep interiors of large landmasses. They are far removed

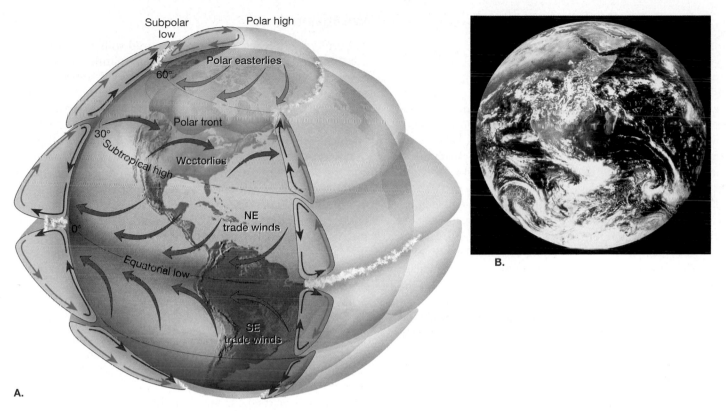

FIGURE 19.3   A. Idealized diagram of Earth's general circulation. The deserts and steppes that are centered in the latitude belt between 20° and 30° north and south coincide with the subtropical high-pressure belts. Here dry, subsiding air inhibits cloud formation and precipitation. By contrast, the pressure belt known as the equatorial low is associated with areas that are among the rainiest on Earth. **B.** In this view of Earth from space, North Africa's Sahara Desert, the adjacent Arabian Desert, and the Kalahari and Namib deserts in southern Africa are clearly visible as tan-colored, cloud free zones. The band of clouds that extends across central Africa and the adjacent oceans coincides with the equatorial low-pressure belt. (Photo courtesy of NASA/Science Source/Photo Researchers, Inc.)

from the ocean, which is the ultimate source of moisture for cloud formation and precipitation. One well-known example is the Gobi Desert of central Asia, shown on the map north of India.

The presence of high mountains across the paths of prevailing winds further separates these areas from water-bearing, maritime air masses; plus, the mountains force the air to lose much of its water. The mechanism is simple: As prevailing winds meet mountain barriers, the air is forced to ascend. When air rises, it expands and cools, a process that can produce clouds and precipitation. The windward sides of mountains, therefore, often have high precipitation. By contrast, the leeward sides of mountains are usually much drier (Figure 19.4). This situation exists because air reaching the leeward side has lost much of its moisture, and if the air descends, it is compressed and warmed, making cloud formation even less likely. The dry region

that results is often referred to as a **rainshadow desert.** Because many middle-latitude deserts occupy sites on the leeward sides of mountains, they can also be classified as rainshadow deserts. In North America, the foremost

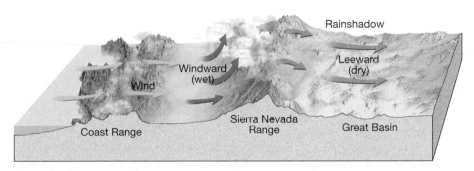

FIGURE 19.4   Many deserts in the middle latitudes are rainshadow deserts. As moving air meets a mountain barrier, it is forced to rise. Clouds and precipitation on the windward side often result. Air descending the leeward side is much drier. The mountains effectively cut the leeward side off from the sources of moisture, producing a rainshadow desert. The Great Basin desert is a rainshadow desert that covers nearly all of Nevada and portions of adjacent states.

Students Sometimes Ask . . .

Are all deserts hot?

No, but many deserts do experience some very high temperatures. For instance, the highest authentically recorded temperature in the United States—as well as the entire Western Hemisphere—is 57°C (134°F), measured at Death Valley, California, on July 10, 1913. The world-record high temperature of nearly 59°C (137°F) was recorded in Azizia, Libya, in North Africa's Sahara Desert on September 13, 1922.

Despite these remarkably high figures, cold temperatures are also experienced in desert regions. For example, the average daily minimum temperature in January in Phoenix, Arizona, is 1.7°C (35°F), just barely above freezing. At Ulan Bator in Mongolia's Gobi Desert the average *high* temperature on January days is only −19°C (−2°F). Dry climates are found from the tropics poleward to the high middle latitudes. Although tropical deserts lack a cold season, deserts in the middle latitudes do experience seasonal temperature changes, which cause some to get quite cold.

mountain barriers to moisture from the Pacific are the Coast Ranges, Sierra Nevada, and Cascades. (Figure 19.4). In Asia, the great Himalayan chain prevents the summertime monsoon flow of moist Indian Ocean air from reaching the interior (see Box 19.2).

Because the Southern Hemisphere lacks extensive land areas in the middle latitudes, only a small area of desert and steppe occurs in this latitude range, existing primarily near the southern tip of South America in the rainshadow of the towering Andes.

The middle-latitude deserts provide an example of how tectonic processes affect climate. Rainshadow deserts exist by virtue of the mountains produced when plates collide. Without such mountain-building episodes, wetter climates would prevail where many dry regions exist today.

Geologic Processes in Arid Climates

 GEODe Deserts and Winds
▶ Common Misconceptions About Deserts

The angular hills, the sheer canyon walls, and the desert surface of pebbles or sand contrast sharply with the rounded hills and curving slopes of more humid places. Indeed, to a visitor from a humid region, a desert landscape may seem to have been shaped by forces different from those operating in well-watered areas. However, although the contrasts may be striking, they do not reflect different processes. They merely disclose the differing effects of the same processes that operate under contrasting climatic conditions.

Weathering

In humid regions, relatively fine-textured soils support an almost continuous cover of vegetation that mantles the surface. Here the slopes and rock edges are rounded, reflecting the strong influence of chemical weathering in a humid climate. By contrast, much of the weathered debris in deserts consists of unaltered rock and mineral fragments—the result of mechanical weathering processes. In dry lands, rock weathering of any type is greatly reduced because of the lack of moisture and the scarcity of organic acids from decaying plants. However, chemical weathering is not completely lacking in deserts. Over long spans of time, clays and thin soils do form, and many iron-bearing silicate minerals oxidize, producing the rust-colored stain that tints some desert landscapes.

The Role of Water

Permanent streams are normal in humid regions, but practically all desert streambeds are dry most of the time (Figure 19.5A). Deserts have **ephemeral** (*ephemero* = short-lived) **streams,** which means they carry water only in response to specific episodes of rainfall. A typical ephemeral stream might flow only a few days or perhaps just a few hours during the year. In some years the channel might carry no water at all.

This fact is obvious even to the casual traveler who notices numerous bridges with no streams beneath them or numerous dips in the road where dry channels cross. However, when the rare heavy showers do come, so much rain falls in such a short time that all of it cannot soak in (Figure 19.6). Because desert vegetative cover is sparse, runoff is largely unhindered and consequently rapid, often creating flash floods along valley floors (Figure 19.5B). These floods are quite unlike floods in humid regions. A flood on a river like the Mississippi may take several days to reach its crest and then subside. But desert floods arrive suddenly and subside quickly. Because much surface material in a desert is not anchored by vegetation, the amount of erosional work that occurs during a single short-lived rain event is impressive.

In the dry western United States, different names are used for ephemeral streams, including *wash* and *arroyo*. In other parts of the world, a dry desert stream may be a *wadi* (Arabia and North Africa), a *donga* (South America), or a *nullah* (India).

Humid regions are notable for their integrated drainage systems. But in arid regions, streams usually lack an extensive system of tributaries. In fact, a basic characteristic of desert streams is that they are small and die out before reaching the sea. Because the water table is usually far below the surface, few desert streams can draw upon it as streams do in humid regions (see Figure 17.4, p. 461). Without a steady supply of water, the combination of evaporation and infiltration soon depletes the stream.

The few permanent streams that do cross arid regions, such as the Colorado and Nile rivers, originate *outside* the desert, often in well-watered mountains. Here the water supply must be great to compensate for the losses occurring as the stream crosses the desert. For example, after the Nile

BOX 19.2 ▶ PEOPLE AND THE ENVIRONMENT

The Disappearing Aral Sea

The Aral Sea lies on the border between Uzbekistan and Kazakhstan in central Asia (Figure 19.A). The setting is the Turkestan desert, a middle-latitude desert in the rainshadow of Afghanistan's high mountains. In this region of interior drainage, two large rivers, the Amu Darya and the Syr Darya, carry water from the mountains of northern Afghanistan across the desert to the Aral Sea. Water leaves the sea by evaporation. Thus, the size of the water body depends upon the balance between river inflow and evaporation.

In 1960 the Aral Sea was one of the world's largest inland water bodies, with an area of about 67,000 square kilometers (26,000 square miles). Only the Caspian Sea, Lake Superior, and Lake Victoria were larger. By the year 2000 the sea had shrunk by 75 percent and split into 2 parts joined by a narrow passage. The water level had dropped 22 meters (72 feet) and the sea had lost 90 percent of its volume. The shrinking of this water body is depicted in Figure 19.B. By about 2010 all that will likely remain will be three shallow remnants.

What caused the Aral Sea to dry up over the past 40 years? The answer is that the flow of water from the mountains that supplied the sea was significantly reduced and then all but eliminated. As recently as 1965, the Aral Sea received about 50 cubic kilometers (12 cubic miles) of fresh water per year. By the early 1980s this number fell to nearly zero. The reason was that the waters of the Amu Darya and Syr Darya were diverted to supply a major expansion of irrigated agriculture in this dry realm.

The intensive irrigation greatly increased agricultural productivity, but not without significant costs. The deltas of the two major rivers have lost their wetlands, and wildlife has disappeared. The once thriving fishing industry is dead, and the 24 species of fish that once lived in the Aral Sea are no longer there. The shoreline is now tens of kilometers from the towns that were once fishing centers (Figure 19.C).

The shrinking sea has exposed millions of acres of former seabed to sun and wind. The surface is encrusted with salt and with agrochemicals brought by the rivers. Strong winds routinely pick up and deposit thousands of tons of newly exposed material every year. This process has not only contributed to a significant reduction in air quality for people living in the region but has also appreciably affected crop yields due to the deposition of salt-rich sediments on arable land.

The shrinking Aral Sea has had a noticeable impact on the region's climate. Without the moderating effect of a large water body, there are greater extremes of temperature, a shorter growing season, and reduced local precipitation. These changes have caused many farms to switch from growing cotton to growing rice, which demands even more diverted water.

Environmental experts agree that the current situation cannot be sustained. Could this crisis be reversed if enough fresh water were to once again flow into the Aral Sea?

Prospects appear grim. Experts estimate that restoring the Aral Sea to about twice its present size would require stopping all irrigation from the two major rivers for 50 years. This could not be done without ruining the economies of the countries that rely on that water.*

The decline of the Aral Sea is a major environmental disaster that sadly is of human making.

*For more on this, see "Coming to Grips with the Aral Sea's Grim Legacy," in *Science*, vol. 284, April 2, 1999, pp. 30–31. and "To Save a Vanishing Sea" in *Science* vol. 307, February 18, 2005, pp. 1032–33.

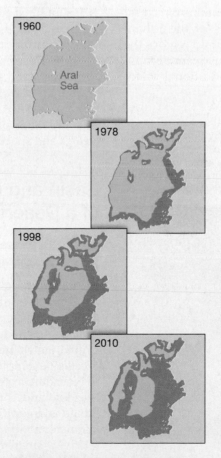

FIGURE 19.A The Aral Sea lies east of the Caspian Sea in the Turkestan Desert. Two rivers, the Amu Darya and Syr Darya, bring water from the mountains to the south.

FIGURE 19.B The shrinking Aral Sea. By the year 2010 all that will remain are three small remnants.

FIGURE 19.C In the town of Jamboul, Kazakhstan, boats now lie in the sand because the Aral Sea has dried up. (Photo by Ergun Cagatay/Liaison Agency, Inc.)

A.   B.

FIGURE 19.5   **A.** Most of the time, desert stream channels are dry. **B.** An ephemeral stream shortly after a heavy shower. Although such floods are short-lived, large amounts of erosion occur. (Photos by E. J. Tarbuck)

leaves its headwaters in the lakes and mountains of central Africa, it traverses almost 3000 kilometers of the Sahara without a single tributary. By contrast, in humid regions the discharge of a river grows as it flows downstream because tributaries and groundwater contribute additional water along the way.

FIGURE 19.6   Desert thunderstorm over Tucson, Arizona. There are often many weeks, months, or occasionally even years separating periods of rain in the desert. When rains do occur, they are often heavy and of relatively short duration. Because the rainfall intensity is high, all of the water cannot soak in, and rapid runoff results. (Photo by Warren Faidley/DRK Photo)

It should be emphasized that *running water, although infrequent, nevertheless does most of the erosional work in deserts.* This is contrary to the common belief that wind is the most important erosional agent sculpturing desert landscapes. Although wind erosion is indeed more significant in dry areas than elsewhere, most desert landforms are carved by running water. As you will see shortly, the main role of wind is in the transportation and deposition of sediment, which creates and shapes the ridges and mounds we call dunes.

Basin and Range: The Evolution of a Desert Landscape

 GEODe   Deserts and Winds
EARTH   ▸ Reviewing Landforms and Landscapes

Because arid regions typically lack permanent streams, they are characterized as having **interior drainage.** This means that they have a discontinuous pattern of intermittent streams that do not flow out of the desert to the ocean. In the United States, the dry Basin and Range region provides an excellent example. The region includes southern Oregon, all of Nevada, western Utah, southeastern California, southern Arizona, and southern New Mexico. The name Basin and Range is an apt description for this almost 800,000-square-kilometer region because it is characterized by more than 200 relatively small mountain ranges that rise 900 to 1500 meters (3000 to 5000 feet) above the basins that separate them.

In this region, as in others like it around the world, erosion mostly occurs without reference to the ocean (ultimate base level) because the interior drainage never reaches the sea.

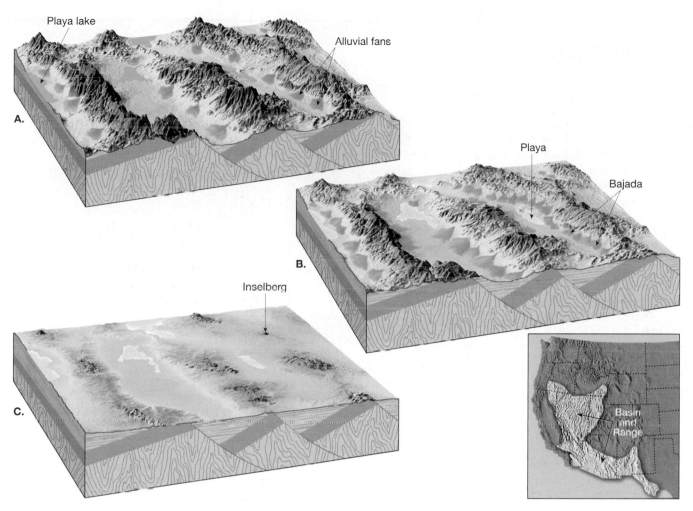

FIGURE 19.7 Stages of landscape evolution in a mountainous desert such as the Basin and Range region of the West. As erosion of the mountains and deposition in the basins continue, relief diminishes. **A.** Early stage. **B.** Middle stage. **C.** Late stage.

FIGURE 19.8 Aerial view of alluvial fans in Death Valley, California. The size of the fan depends on the size of the drainage basin. As the fans grow, they eventually coalesce to form a bajada. (Photo by Michael Collier)

Even where permanent streams flow to the ocean, few tributaries exist, and thus only a narrow strip of land adjacent to the stream has sea level as its ultimate level of land reduction.

The block models in Figure 19.7 depict how the landscape has evolved in the Basin and Range region. During and following the uplift of the mountains, running water begins carving the elevated mass and depositing large quantities of debris in the basin. During this early stage, relief is greatest, because as erosion lowers the mountains and sediment fills the basins, elevation differences gradually diminish.

When the occasional torrents of water produced by sporadic rains move down the mountain canyons, they are heavily loaded with sediment. Emerging from the confines of the canyon, the runoff spreads over the gentler slopes at the base of the mountains and quickly loses velocity. Consequently, most of its load is dumped within a short distance. The result is a cone of debris at the mouth of a canyon known as an alluvial fan (Figure 19.8). Because the coarsest material is dropped first, the head of the fan is steepest, having a slope of perhaps 10 to 15 degrees. Moving down the fan, the size of the sediment and the steepness of the slope decrease and merge imperceptibly with the basin

BOX 19.3 ▶ UNDERSTANDING EARTH

Australia's Mount Uluru

When travelers contemplating a trip to Australia consult brochures and other tourist literature, they are bound to see a photograph or read a description of Mount Uluru (formerly Ayers Rock). As Figure 19.D illustrates, this well-known attraction is a massive feature that rises steeply from the surrounding plain. Located in Uluru–Kata Tjuta National Park, southwest of Alice Springs in the dry center of the continent, the roughly circular monolith is more than 350 meters (1200 feet) high, and its base is more than 9.5 kilometers (6 miles) in circumference. Its summit is flattened, its sides furrowed. The rock type is sandstone, and the hues of red and orange change with the light of day. In addition to being a striking geological attraction, Mount Uluru is of interest because it is a sacred place for the aboriginal tribes of the region.

Mount Uluru is a spectacular example of a feature known as an inselberg. *Inselberg* is a German word meaning "island mountain" and seems appropriate because these masses clearly resemble rocky islands standing above the surface of a broad sea. Similar features are scattered throughout many other arid and semiarid regions of the world. Mount Uluru is a special type of inselberg that consists of a very resistant rock mass exhibiting a rounded or domed form. Such masses are termed *bornhardts* for the 19th-century German explorer Wilhelm Bornhardt, who described similar features in parts of Africa.

Bornhardts form in regions where massive or resistant rock such as granite or sandstone is surrounded by rock that is more susceptible to weathering. The greater susceptibility of the adjacent rock is often the result of its being more highly jointed. Joints allow water and therefore weathering processes to penetrate to greater depths. When the adjacent, deeply weathered rock is stripped away by erosion, the far less weathered rock mass remains standing high. After a bornhardt forms, it tends to shed water. By contrast, the surrounding a bornhardt helps to perpetuate its existence by reinforcing the processes that created it. In fact, masses such as Mount Uluru can remain a part of the landscape for tens of millions of years.

Bornhardts are more common in the lower latitudes because the weathering that is responsible for their formation proceeds more rapidly in warmer climates. In regions that are now arid or semiarid, bornhardts may reflect times when the climate was wetter than it is today.

FIGURE 19.D   Mount Uluru (formerly Ayers Rock) rises conspicuously above the dry plains of central Australia. It is a type of inselberg known as a *bornhardt*. As erosion gradually lowers the surface, the less weathered massive rock remains standing high above the more jointed and more easily weathered rock that surrounds it. (Photo by Art Wolfe, Inc.)

floor. An examination of the fan's surface would likely reveal a braided channel pattern because of the water shifting its course as successive channels became choked with sediment. Over the years a fan enlarges, eventually coalescing with fans from adjacent canyons to produce an apron of sediment called a **bajada** along the mountain front.

On the rare occasions of abundant rainfall, streams may flow across the bajada to the center of the basin, converting the basin floor into a shallow **playa lake.** Playa lakes are temporary features that last only a few days or at best a few weeks before evaporation and infiltration remove the water. The dry, flat lake bed that remains is called a **playa.** Playas are typically composed of fine silts and clays and are occasionally encrusted with salts precipitated during evaporation (see Figure 7.15, p. 205). These precipitated salts may be unusual. A case in point is the sodium borate (better known as borax) mined from ancient playa lake deposits in Death Valley, California.

With the ongoing erosion of the mountain mass and the accompanying sedimentation, the local relief continues to diminish. Eventually nearly the entire mountain mass is gone. Thus, by the late stages of erosion, the mountain areas are reduced to a few large bedrock knobs projecting above the surrounding sediment-filled basin. These isolated erosional remnants on a late-stage desert landscape are called **inselbergs,** a German word meaning "island mountains" (see Box 19.3).

Each of the stages of landscape evolution in an arid climate depicted in Figure 19.7 can be observed in the Basin and Range region. Recently uplifted mountains in an early stage of erosion are found in southern Oregon and northern Nevada. Death Valley, California, and southern Nevada fit into the more advanced middle stage, whereas the late stage, with its inselbergs, can be seen in southern Arizona.

Transportation of Sediment by Wind

Moving air, like moving water, is turbulent and able to pick up loose debris and transport it to other locations. Just as in a stream, the velocity of wind increases with height above the surface. Also like a stream, wind transports fine particles in suspension while heavier ones are carried as bed load. However, the transport of sediment by wind differs from that of running water in two significant ways. First, wind's lower density compared to water renders it less capable of picking up and transporting coarse materials. Second, because wind is not confined to channels, it can spread sediment over large areas, as well as high into the atmosphere.

Bed Load

The **bed load** carried by wind consists of sand grains. Observations in the field and experiments using wind tunnels indicate that windblown sand moves by skipping and bouncing along the surface—a process termed **saltation.** The term is not a reference to salt, but instead derives from the Latin word meaning "to jump."

The movement of sand grains begins when wind reaches a velocity sufficient to overcome the inertia of the resting particles. At first the sand rolls along the surface. When a moving sand grain strikes another grain, one or both of them may jump into the air. Once in the air, the grains are carried forward by the wind until gravity pulls them back toward the surface. When the sand hits the surface, it either bounces back into the air or dislodges other grains, which then jump upward. In this manner, a chain reaction is established, filling the air near the ground with saltating sand grains in a short period of time (Figure 19.9).

Bouncing sand grains never travel far from the surface. Even when winds are very strong, the height of the saltating sand seldom exceeds a meter and usually is no greater than a half meter. Some sand grains are too large to be thrown into the air by impact from other particles. When this is the case, the energy provided by the impact of the smaller saltating grains drives the larger grains forward. Estimates indicate that between 20 and 25 percent of the sand transported in a sandstorm is moved in this way.

Suspended Load

Unlike sand, finer particles of dust can be swept high into the atmosphere by the wind. Because dust is often composed of rather flat particles that have large surface areas compared to their weight, it is relatively easy for turbulent air to counterbalance the pull of gravity and keep these fine particles airborne for hours or even days. Although both silt and clay can be carried in suspension, silt commonly makes up the bulk of the **suspended load** because the reduced level of chemical weathering in deserts provides only small amounts of clay.

Fine particles are easily carried by the wind, but they are not so easily picked up to begin with. The reason is that the wind velocity is practically zero within a very thin layer close to the ground. Thus, the wind cannot lift the sediment by itself. Instead, the dust must be ejected or spattered into the moving air by bouncing sand grains or other disturbances. This idea is illustrated nicely by a dry, unpaved country road on a windy day. Left undisturbed, little dust is raised by the wind. However, as a car or truck moves over the road, the layer of silt is kicked up, creating a thick cloud of dust.

Although the suspended load is usually deposited relatively near its source, high winds are capable of carrying large quantities of dust great distances (Figure 19.10). In the 1930s, silt that was picked up in Kansas was transported to New England and beyond into the North Atlantic. Similarly, dust blown from the Sahara has been traced as far as the West Indies (Figure 19.11).

FIGURE 19.9 A cloud of saltating sand grains moving up the gentle slope of a dune. (Photo by Stephen Trimble)

FIGURE 19.10   Dust blackens the sky on May 21, 1937, near Elkhart, Kansas. It was because of storms like this that portions of the Great Plains were called the "Dust Bowl" in the 1930s. (Photo reproduced from the collection of the Library of Congress)

Wind Erosion

GEODe Deserts and Winds
EARTH ▶ Common Misconceptions About Deserts

Compared to running water and glaciers, wind is a relatively insignificant erosional agent. Recall that even in deserts, most erosion is performed by intermittent running water, not by the wind. Wind erosion is more effective in arid lands than in humid areas because in humid places moisture binds particles together and vegetation anchors the soil. For wind to be an effective erosional force, dryness and scanty vegetation are important prerequisites (see Box 19.4). When such circumstances exist, wind may pick up, transport, and deposit great quantities of fine sediment. During the 1930s, parts of the Great Plains experienced vast dust storms. The plowing

FIGURE 19.11   This satellite image shows thick plumes of dust from the Sahara Desert blowing across the Mediterranean Sea toward Italy on July 16, 2003. Such dust storms are common in arid North Africa. In fact, this region is the largest dust source in the world. Satellites are an excellent tool for studying the transport of dust on a global scale. They show that dust storms can cover huge areas and that dust can be transported great distances. (Image courtesy of NASA)

under of the natural vegetative cover for farming, followed by severe drought, exposed the land to wind erosion and led to the area's being labeled the Dust Bowl.*

Deflation and Blowouts

One way that wind erodes is by **deflation** (*de* = out, *flat* = blow), the lifting and removal of loose material. Deflation sometimes is difficult to notice because the entire surface is being lowered at the same time, but it can be significant. In portions of the 1930s Dust Bowl, vast areas of land were lowered by as much as a meter in only a few years.

The most noticeable results of deflation in some places are shallow depressions appropriately called **blowouts** (Figure 19.12). In the Great Plains region, from Texas north to Montana, thousands of blowouts are visible on the landscape. They range from small dimples less than a meter deep and 3 meters wide to depressions that approach 50 meters in depth and several kilometers across. The factor that controls the depths of these basins (that is, acts as base level) is the local water table. When blowouts are lowered to the water table, damp ground and vegetation prevent further deflation.

Desert Pavement

In portions of many deserts, the surface consists of a closely packed layer of coarse particles. This veneer of pebbles and cobbles, called **desert pavement,** is only one or two stones thick (Figure 19.13). Beneath is a layer containing a significant proportion of silt and sand. When desert pavement is present, it is an important control on wind erosion because pavement stones are too large for deflation to remove. When this armor is disturbed, wind can easily erode the exposed fine silt.

For many years, the most common explanation for the formation of desert pavement was that it develops when wind removes sand and silt within poorly sorted surface deposits. As Figure 19.14A illustrates, the concentration of larger particles at the surface gradually increases as the finer particles are blown away. Eventually the surface is completely covered with pebbles and cobbles too large to be moved by the wind.

Studies have shown that the process depicted in Figure 19.14A is not an adequate explanation for all environments in which desert pavement exists. For example, in many places, desert pavement is underlain by a relatively thick layer of silt that contains few if any pebbles and cobbles. In such a setting, deflation of fine sediment could not leave behind a layer of coarse particles. Studies also showed that in some areas the pebbles and cobbles composing desert pavement have all been exposed at the surface for about the same length of time. This would not be the case for the process shown in Figure 19.14A. Here, the coarse particles that make up the pavement reach the surface over an extended time span as deflation gradually removes the fine material.

---

*For more information, see Box 6.4 (p. 188), "Dust Bowl—Soil Erosion in the Great Plains."

BOX 19.4 ▶ PEOPLE AND THE ENVIRONMENT

Deserts Are Expanding

The transition zones surrounding deserts have very fragile, delicately balanced ecosystems. In these marginal areas, human activities may stress the ecosystem beyond its tolerance limit, resulting in degradation of the land. If such degradation is severe, it is referred to as *desertification*. Desertification means the expansion of desertlike conditions into nondesert areas. Although such a transformation can also result from natural processes that act over decades, centuries, or even millennia, in recent years, desertification has come to mean the rapid alteration of land to desertlike conditions as the result of human activities.

The United Nations has recognized desertification as one of the most serious environmental challenges of the 21st century. According to the U.N.'s International Fund for Agricultural Development, each year desertification claims another 10 million acres of agricultural drylands. In response, more than 190 countries (including the United States) have ratified a treaty known as the Convention to Combat Desertification.

The advancement of desertlike conditions into areas that were previously useful for agriculture is not a uniform, clear-cut shifting of desert borders. Rather, degeneration into desert usually occurs as a patchy transformation of dry but habitable land into dry, uninhabitable land. It results primarily from inappropriate land use and is aided and accelerated by drought. Unfortunately, an area undergoing desertification comes to our attention only after the process is well under way.

Desertification begins when land near the desert's edge is used for growing crops or for grazing livestock. Either way, the natural vegetation is removed by plowing or grazing.

If crops are planted and drought occurs, the unprotected soil is exposed to the forces of erosion. Gullying of slopes and accumulations of sediment in stream channels are visible signs on the landscape, as are the clouds of dust created as topsoil is removed by the wind.

Where livestock are raised, the land is also degraded. Although the modest natural vegetation on marginal lands can maintain local wildlife, it cannot support the intensive grazing of large domesticated herds. Overgrazing reduces or eliminates plant cover. When the vegetative cover is destroyed beyond the minimum required to hold the soil against erosion, the destruction becomes irreversible. Moreover, by pounding the ground with their hooves, livestock compact the soil, which reduces the amount of water the land can soak up when it does rain. The pounding hooves also pulverize the soil, increasing the proportion of fine material, which is then more easily removed when winds are strong.

Desertification first received worldwide attention when drought struck a region in Africa called the *Sahel* in the late 1960s (Figure 19.E). During that period, and many subsequent episodes, the people in this vast expanse south of the Sahara Desert have suffered malnutrition and death by starvation. Livestock herds have been decimated, and the loss of productive land has been great (Figure 19.F). Hundreds of thousands of people have been forced to migrate. As agricultural lands shrink, people must rely on smaller areas for food production. This, in turn, stresses the environment and accelerates the desertification process.

Although human suffering from desertification is most serious in the Sahel, the problem is by no means confined to that region. Desertification exists in other parts of Africa and on every other continent except Antarctica. Recurrent droughts may seem to be the obvious reason for desertification, but the chief cause is stress placed by people on a tenuous environment with fragile soils.

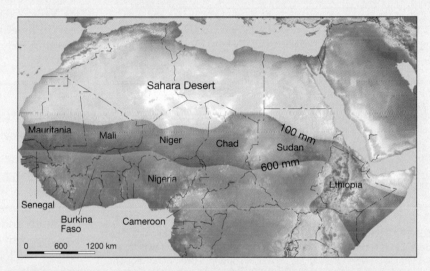

FIGURE 19.E Desertification is most serious in the southern margin of the Sahara in a region known as the Sahel. The lines defining the approximate boundaries of the Sahel represent average annual rainfall in millimeters.

FIGURE 19.F Overgrazing of marginal lands in Africa south of the Sahara has contributed to desertification. (Photo by Sean Sprague/Peter Arnold, Inc.)

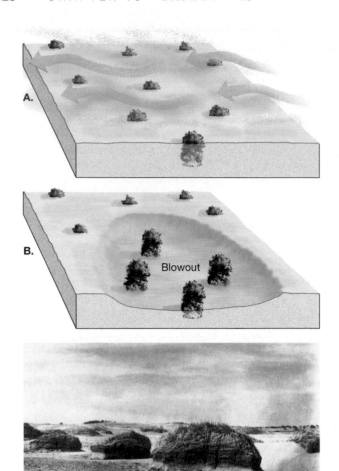

FIGURE 19.12 Formation of a blowout. **A.** Area prior to deflation. **B.** Area after deflation has created a shallow depression. **C.** This photo was taken north of Granville, North Dakota, in July 1936, during a prolonged drought. Strong winds removed the soil that was not anchored by vegetation. The mounds are 1.2 meters (4 feet) high and show the level of the land prior to deflation. (Photo courtesy of the State Historical Society of North Dakota, col 278-1)

As a result, an alternate explanation for desert pavement was formulated (Figure 19.14B). This hypothesis suggests that pavement develops on a surface that initially consists of coarse particles. Over time, protruding cobbles trap fine, windblown grains that settle and sift downward through the spaces between the larger surface stones. The process is aided by infiltrating rainwater. In this model, the cobbles composing the pavement were never buried. Moreover, it successfully explains the lack of coarse particles beneath the desert pavement.

Ventifacts and Yardangs

Like glaciers and streams, wind also erodes by **abrasion** (*ab* = away, *radere* = to scrape). In dry regions as well as along some beaches, windblown sand cuts and polishes exposed rock surfaces. Abrasion sometimes creates interestingly shaped stones called **ventifacts** (Figure 19.15A). The side of the stone exposed to the prevailing wind is abraded,

leaving it polished, pitted, and with sharp edges. If the wind is not consistently from one direction, or if the pebble becomes reoriented, it may have several faceted surfaces.

Unfortunately, abrasion is often given credit for accomplishments beyond its capabilities. Such features as balanced rocks that stand high atop narrow pedestals, and intricate detailing on tall pinnacles, are not the results of abrasion. Sand seldom travels more than a meter above the surface, so the wind's sandblasting effect is obviously limited in vertical extent.

In addition to ventifacts, wind erosion is responsible for creating much larger features, called yardangs (from the Turkistani word *yar*, meaning "steep bank"). A **yardang** is a streamlined, wind-sculpted landform that is oriented parallel to the prevailing wind (Figure 19.15B). Individual yardangs are generally small features that stand less than 5 meters (16 feet) high and no more than about 10 meters (32 feet) long. Because the sandblasting effect of wind is greatest near the ground, these abraded bedrock remnants are usually narrower at their base. Sometimes yardangs are large features. Peru's Ica Valley contains yardangs that approach 100 meters (330 feet) in height and several kilometers in length. Some in the desert of Iran reach 150 meters (nearly 500 feet) in height.

Wind Deposits

GEODe Deserts and Winds
▶ Reviewing Landforms and Landscapes

Although wind is relatively unimportant in producing *erosional* landforms, significant *depositional* landforms are created by the wind in some regions. Accumulations of windblown sediment are particularly conspicuous in the world's dry lands and along many sandy coasts. Wind deposits are of two distinctive types: (1) mounds and ridges of sand from the wind's bed load, which we call dunes, and (2) extensive blankets of silt, called loess, that once were carried in suspension.

FIGURE 19.13 Desert pavement consists of a closely packed veneer of pebbles and cobbles that is only one or two stones thick. Beneath the pavement is material containing a significant proportion of finer particles. If left undisturbed, desert pavement will protect the surface from deflation. (Photo by Bobbé Christopherson)

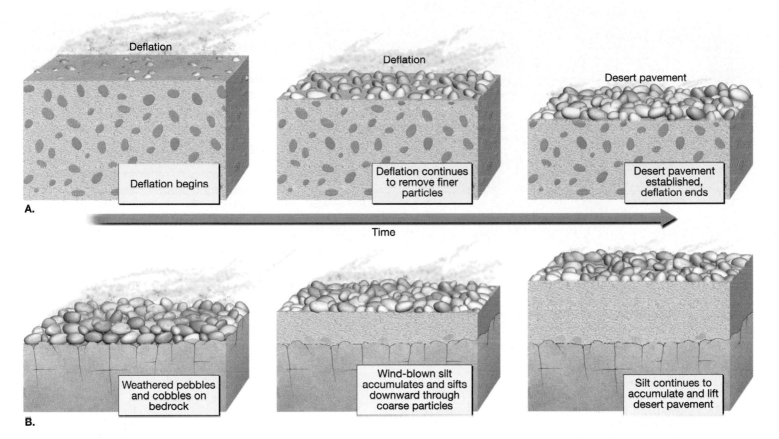

FIGURE 19.14 **A.** This model portrays an area with poorly sorted surface deposits. Coarse particles gradually become concentrated into a tightly packed layer as deflation lowers the surface by removing sand and silt. Here desert pavement is the result of wind erosion. **B.** This model shows the formation of desert pavement on a surface initially covered with coarse pebbles and cobbles. Windblown dust accumulates at the surface and gradually sifts downward through spaces between coarse particles. Infiltrating rainwater aids the process. This depositional process raises the surface and produces a layer of coarse pebbles and cobbles underlain by a substantial layer of fine sediment.

Sand Deposits

As is the case with running water, wind drops its load of sediment when velocity falls and the energy available for transport diminishes. Thus, sand begins to accumulate wherever an obstruction across the path of the wind slows its movement. Unlike many deposits of silt, which form blanketlike layers over large areas, winds commonly deposit sand in mounds or ridges called **dunes** (Figure 19.16).

As moving air encounters an object, such as a clump of vegetation or a rock, the wind sweeps around and over it, leaving a shadow of slower-moving air behind the obstacle, as well as a smaller zone of quieter air just in front of the obstacle. Some of the saltating sand grains moving with the wind come to rest

FIGURE 19.15 **A.** Ventifacts are rocks that are polished and shaped by sandblasting (Photo by Stephen Trimble). **B.** Yardangs are usually small, wind-sculpted landforms that are aligned parallel with the wind. (Photo by David Love, New Mexico Bureau of Geology and Mineral Resources)

A.

B.

FIGURE 19.16 Sand sliding down the steep slipface of a dune, in White Sands National Monument, New Mexico. (Photo by Michael Collier)

Nevada, open to traffic, sand must be taken away about three times a year. Each time, between 1500 and 4000 cubic meters of sand are removed. Attempts at stabilizing the dunes by planting different varieties of grasses have been unsuccessful because the meager rainfall cannot support the plants.

Types of Sand Dunes

Dunes are not just random heaps of wind-blown sediment. Rather, they are accumulations that usually assume patterns that are surprisingly consistent (Figure 19.18). Addressing this point, a leading early investigator of dunes, the British engineer R. A. Bagnold, observed: "Instead of finding chaos and disorder, the observer never fails to be amazed at a

in these wind shadows. As the accumulation of sand continues, it becomes a more imposing barrier to the wind and thus a more efficient trap for even more sand. If there is a sufficient supply of sand and the wind blows steadily for a long enough time, the mound of sand grows into a dune.

Many dunes have an asymmetrical profile, with the leeward (sheltered) slope being steep and the windward slope more gently inclined (Figure 19.17). Sand moves up the gentler slope on the windward side by saltation. Just beyond the crest of the dune, where the wind velocity is reduced, the sand accumulates. As more sand collects, the slope steepens and eventually some of it slides under the pull of gravity (Figure 19.16). In this way, the leeward slope of the dune, called the **slipface,** maintains an angle of about 34 degrees, the angle of repose for loose dry sand. (Recall from Chapter 15 that the angle of repose is the steepest angle at which loose material remains stable.) Continued sand accumulation, coupled with periodic slides down the slipface, results in the slow migration of the dune in the direction of air movement.

As sand is deposited on the slipface, layers form that are inclined in the direction the wind is blowing. These sloping layers are called **cross beds** (Figure 19.17). When the dunes are eventually buried under other layers of sediment and become part of the sedimentary rock record, their asymmetrical shape is destroyed, but the cross beds remain as testimony to their origin. Nowhere is cross-bedding more prominent than in the sandstone walls of Zion Canyon in southern Utah (Figure 19.17).

For some areas, moving sand is troublesome. In Figure 19.18 dunes are advancing across irrigated fields in Egypt. In portions of the Middle East, valuable oil rigs must be protected from encroaching dunes. In some cases, fences are built sufficiently upwind of the dunes to stop their migration. As sand continues to collect, however, the fences must be built higher. In Kuwait protective fences extend for almost 10 kilometers around one important oil field. Migrating dunes can also pose a problem to the construction and maintenance of highways and railroads that cross sandy desert regions. For example, to keep a portion of Highway 95 near Winnemucca,

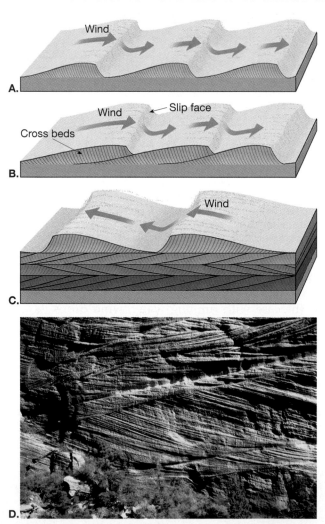

FIGURE 19.17 As parts **A** and **B** illustrate, dunes commonly have an asymmetrical shape. The steeper leeward side is called the *slipface.* Sand grains deposited on the slipface at the angle of repose create the cross-bedding of the dunes. **C.** A complex pattern develops in response to changes in wind direction. Also notice that when dunes are buried and become part of the sedimentary record, the cross-bedded structure is preserved. **D.** Cross beds are an obvious characteristic of the Navajo Sandstone in Zion National Park, Utah. (Photo by Marli Miller)

FIGURE 19.18 These desert dunes (called *barchans*) in Egypt are advancing from right to left across irrigated fields. (Photo by Georg Gerster/Photo Researchers, Inc.)

simplicity of form, an exactitude of repetition, and a geometric order. . . ." A broad assortment of dune forms exists, generally simplified to a few major types for discussion.

Of course, gradations exist among different forms as well as irregularly shaped dunes that do not fit easily into any category. Several factors influence the form and size that dunes ultimately assume. These include wind direction and velocity, availability of sand, and the amount of vegetation present. Six basic dune types are shown in Figure 19.19, with arrows indicating wind directions.

Barchan Dunes Solitary sand dunes shaped like crescents and with their tips pointing downwind are called **barchan dunes** (Figures 19.18 and 19.19A). These dunes form where

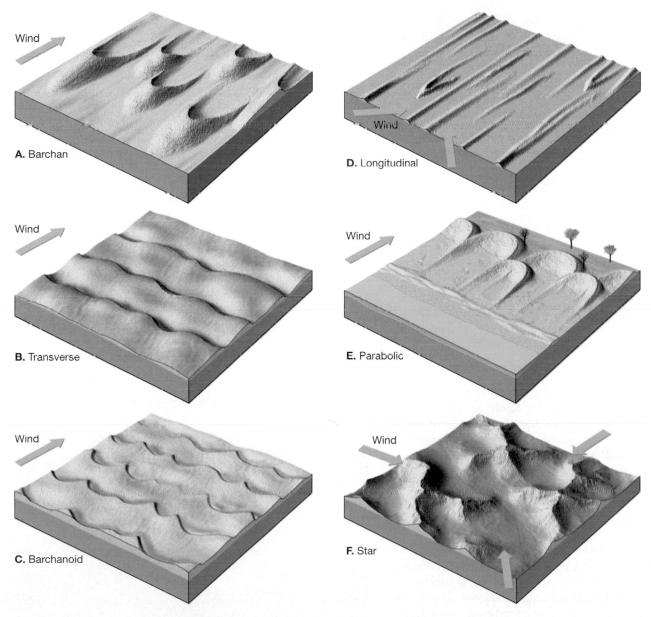

FIGURE 19.19 Sand dune types. **A.** Barchan dunes. **B.** Transverse dunes. **C.** Barchanoid dunes. **D.** Longitudinal dunes. **E.** Parabolic dunes. **F.** Star dunes.

FIGURE 19.20 Barchanoid dunes represent a type that is intermediate between isolated barchans on the one hand and extensive transverse dunes on the other. The gypsum dunes at White Sands National Monument, New Mexico, are an example. (Photo by Michael Collier)

Students Sometimes Ask . . .

Aren't deserts mostly covered with sand dunes?

A common misconception about deserts is that they consist of mile after mile of drifting sand dunes. It is true that sand accumulations do exist in some areas and may be striking features. But, perhaps surprisingly, sand accumulations worldwide represent only a small percentage of the total desert area. For example, in the Sahara—the world's largest desert—accumulations of sand cover only *one-tenth* of its area. The sandiest of all deserts is the Arabian, one-third of which consists of sand.

supplies of sand are limited and the surface is relatively flat, hard, and lacking vegetation. They migrate slowly with the wind at a rate of up to 15 meters (50 feet) per year. Their size is usually modest, with the largest barchans reaching heights of about 30 meters (100 feet) while the maximum spread between their horns approaches 300 meters (1000 feet). When the wind direction is nearly constant, the crescent form of these dunes is nearly symmetrical. However, when the wind direction is not perfectly fixed, one tip becomes larger than the other.

Transverse Dunes   In regions where the prevailing winds are steady, sand is plentiful, and vegetation is sparse or absent, the dunes form a series of long ridges that are separated by troughs and oriented at right angles to the prevailing wind. Because of this orientation, they are termed **transverse dunes** (Figure 19.19B). Typically, many coastal dunes are of this

type. In addition, transverse dunes are common in many arid regions where the extensive surface of wavy sand is sometimes called a *sand sea.* In some parts of the Sahara and Arabian deserts, transverse dunes reach heights of 200 meters, are 1 to 3 kilometers across, and can extend for distances of 100 kilometers or more.

There is a relatively common dune form that is intermediate between isolated barchans and extensive waves of transverse dunes. Such dunes, called **barchanoid dunes,** form scalloped rows of sand oriented at right angles to the wind (Figure 19.19C). The rows resemble a series of barchans that have been positioned side by side. Visitors exploring the gypsum dunes at White Sands National Monument, New Mexico, will recognize this form (Figure 19.20).

Longitudinal Dunes   **Longitudinal dunes** are long ridges of sand that form more or less parallel to the prevailing wind and where sand supplies are moderate (Figure 19.19D). Apparently the prevailing wind direction must vary somewhat but still remain in the same quadrant of the compass. Although the smaller types are only 3 or 4 meters high and several dozens of meters long, in some large deserts longitudinal dunes can reach great size. For example, in portions of North Africa, Arabia, and central Australia, these dunes may approach a height of 100 meters (300 feet) and extend for distances of more than 100 kilometers (62 miles).

Parabolic Dunes   Unlike the other dunes that have been described thus far, **parabolic dunes** form where vegetation partially covers the sand. The shape of these dunes resembles the shape of barchans except that their tips point into the wind rather than downwind (Figure 19.19E). Parabolic dunes often form along coasts where there are strong onshore winds and abundant sand. If the sand's sparse vegetative cover is disturbed at some spot, deflation creates a blowout. Sand is then transported out of the depression and deposited as a curved rim, which grows higher as deflation enlarges the blowout.

Star Dunes   Confined largely to parts of the Sahara and Arabian deserts, **star dunes** are isolated hills of sand that exhibit a complex form (Figure 19.19F). Their name is derived from

FIGURE 19.21   Star dune in the Namib Desert in southwestern Africa. (Photo by Comstock)

the fact that the bases of these dunes resemble multipointed stars. Usually three or four sharp-crested ridges diverge from a central high point that in some cases may approach a height of 90 meters (Figure 19.21). As their form suggests, star dunes develop where wind directions are variable.

Loess (Silt) Deposits

In some parts of the world the surface topography is mantled with deposits of windblown silt, called **loess**. Over periods of perhaps thousands of years, dust storms deposited this material. When loess is breached by streams or road cuts, it tends to maintain vertical cliffs and lacks any visible layers, as you can see in Figure 19.22.

The distribution of loess worldwide indicates that there are two primary sources for this sediment: deserts and glacial outwash deposits. The thickest and most extensive deposits of loess on Earth occur in western and northern China. They were blown here from the extensive desert basins of central Asia. Accumulations of 30 meters (100 feet) are common, and thicknesses of more than 100 meters have been measured. It is this fine, buff-colored sediment that gives the Yellow River (Huang Ho) its name.

In the United States, deposits of loess are significant in many areas, including South Dakota, Nebraska, Iowa, Missouri, and Illinois, as well as portions of the Columbia Plateau in the Pacific Northwest. The correlation between the distribution of loess and important farming regions in the Midwest and eastern Washington State is not just a coincidence because soils derived from this wind-deposited sediment are among the most fertile in the world.

Unlike the deposits in China, which originated in deserts, the loess in the United States (and Europe) is an indirect product of glaciation. Its source is deposits of stratified drift. During the retreat of the ice sheets, many river valleys were choked with sediment deposited by meltwater. Strong westerly winds sweeping across the barren floodplains picked up the finer sediment and dropped it as a blanket on the eastern sides of the valleys. Such an origin is confirmed by the fact that loess deposits are thickest and coarsest on the lee side of such major glacial drainage outlets as the Mississippi and Illinois rivers and rapidly thin with increasing distance from the valleys. Furthermore, the angular, mechanically weathered particles composing the loess are essentially the same as the rock flour produced by the grinding action of glaciers.

A.

B.

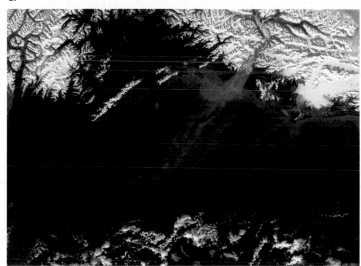
C.

FIGURE 19.22 **A.** This vertical loess bluff near the Mississippi River in southern Illinois is about 3 meters high. (Photo by James E. Patterson) **B.** In parts of China loess has sufficient structural strength to permit the excavation of dwellings. (Photo by Betty Crowell) **C.** This satellite image from March 13, 2003, shows streamers of windblown dust moving southward into the Gulf of Alaska. It illustrates a process similar to the one that created many loess deposits in the American Midwest during the Ice Age. Fine silt is produced by the grinding action of glaciers, then transported beyond the margin of the ice by running water and deposited. Later, the fine silt is picked up by strong winds and deposited as loess. (NASA image)

Students Sometimes Ask . . .

Where are the largest sand dunes found, and how big are they?

The highest dunes in the world are located along the southwest coast of Africa in the Namib Desert. In places, these huge dunes reach heights of 300 to 350 meters (1000 to 1167 feet). The dunes at Great Sand Dunes National Park in southern Colorado are the highest in North America, rising over 210 meters (700 feet) above the surrounding terrain.

Summary

- The *concept of dryness is relative;* it refers to any situation in which a water deficiency exists. Dry regions encompass about 30 percent of Earth's land surface. Two climatic types are commonly recognized: *desert,* which is arid, and *steppe* (a marginal and more humid variant of desert), which is semiarid. *Low-latitude deserts* coincide with the zones of subtropical highs in lower latitudes. On the other hand, *middle-latitude deserts* exist principally because of their positions in the deep interiors of large landmasses far removed from the ocean.

- The same geologic processes that operate in humid regions also operate in deserts, but under contrasting climatic conditions. In dry lands *rock weathering of any type is greatly reduced* because of the lack of moisture and the scarcity of organic acids from decaying plants. Much of the weathered debris in deserts is the result of *mechanical weathering.* Practically all desert streams are dry most of the time and are said to be *ephemeral.* Stream courses in deserts are seldom well integrated and lack an extensive system of tributaries. Nevertheless, *running water is responsible for most of the erosional work in a desert.* Although wind erosion is more significant in dry areas than elsewhere, the main role of wind in a desert is in the transportation and deposition of sediment.

- Because arid regions typically lack permanent streams, they are characterized as having *interior drainage.* Many of the landscapes of the Basin and Range region of the western and southwestern United States are the result of streams eroding uplifted mountain blocks and depositing the sediment in interior basins. *Alluvial fans, playas,* and *playa lakes* are features often associated with these landscapes. In the late stages of erosion, the mountain areas are reduced to a few large bedrock knobs, called *inselbergs,* projecting above sediment-filled basins.

- The transport of sediment by wind differs from that by running water in two ways. First, wind has a low density compared to water; thus, it is not capable of picking up and transporting coarse materials. Second, because wind is not confined to channels, it can spread sediment over large areas. The *bed load* of wind consists of sand grains skipping and bouncing along the surface in a process termed *saltation.* Fine dust particles are capable of being carried by the wind great distances as *suspended load.*

- Compared to running water and glaciers, wind is a less significant erosional agent. *Deflation,* the lifting and removal of loose material, often produces shallow depressions called *blowouts.* Wind also erodes by *abrasion,* often creating interestingly shaped stones called *ventifacts. Yardangs* are narrow, streamlined, wind-sculpted landforms.

- *Desert pavement* is a thin layer of coarse pebbles and cobbles that covers some desert surfaces. Once established, it protects the surface from further deflation. Depending upon circumstances, it may develop as a result of deflation or deposition of fine particles.

- Wind deposits are of two distinct types: (1) *mounds and ridges of sand,* called *dunes,* which are formed from sediment that is carried as part of the wind's bed load; and (2) extensive *blankets of silt,* called *loess,* that once were carried by wind in *suspension.* The profile of a dune shows an asymmetrical shape with the leeward (sheltered) slope being steep and the windward slope more gently inclined. The *types of sand dunes* include (1) *barchan dunes;* (2) *transverse dunes;* (3) *barchanoid dunes;* (4) *longitudinal dunes;* (5) *parabolic dunes;* and (6) *star dunes.* The thickest and most extensive deposits of loess occur in western and northern China. Unlike the deposits in China, which originated in deserts, the loess in the United States and Europe is an indirect product of glaciation.

Review Questions

1. How extensive are the desert and steppe regions of Earth?

2. What is the primary cause of subtropical deserts? Of middle-latitude deserts?

3. In which hemisphere (Northern or Southern) are middle-latitude deserts most common?

4. Why is the amount of precipitation that is used to determine whether a place has a dry climate or a humid climate a variable figure? (See Box 19.1, p. 518)

5. *Deserts are hot, lifeless, sand-covered landscapes shaped largely by the force of wind.* The preceding statement summarizes the image of arid regions that many people hold, especially those living in more humid places. Is it an accurate view?

6. Why is rock weathering reduced in deserts?

7. As a permanent stream such as the Nile River crosses a desert, does discharge increase or decrease? How does this compare to a river in a humid region?

8. What is the most important erosional agent in deserts?

9. Why is sea level (ultimate base level) not a significant factor influencing erosion in desert regions?

10. Why is the Aral Sea shrinking (see Box 19.2, p. 521)?

11. Describe the features and characteristics associated with each of the stages in the evolution of a mountainous desert. Where in the United States can these stages be observed?

12. Describe the way in which wind transports sand. During very strong winds, how high above the surface can sand be carried?

13. Why is wind erosion relatively more important in arid regions than in humid areas?

14. In what ways do human activities contribute to desertification (see Box 19.4, p. 527)?

15. What factor limits the depths of blowouts?

16. Briefly describe two hypotheses used to explain the formation of desert pavement.

17. How do sand dunes migrate?

18. List three factors that influence the form and size of a sand dune.

19. Six major dune types are recognized. Indicate which type of dune is associated with each of the following statements.

   a. dunes whose tips point into the wind

b. long sand ridges oriented at right angles to the wind

c. dunes that often form along coasts where strong winds create a blowout

d. solitary dunes whose tips point downwind

e. long sand ridges that are oriented more or less parallel to the prevailing wind

f. an isolated dune consisting of three or four sharp-crested ridges diverging from a central high point

g. scalloped rows of sand oriented at right angles to the wind

20. Although sand dunes are the best-known wind deposits, accumulations of loess are very significant in some parts of the world. What is loess? Where are such deposits found? What are the origins of this sediment?

Key Terms

abrasion (p. 528)
alluvial fan (p. 523)
bajada (p. 524)
barchan dune (p. 531)
barchanoid dune (p. 531)
bed load (p. 525)
blowout (p. 526)
cross beds (p. 530)

deflation (p. 526)
desert (p. 517)
desert pavement (p. 526)
dune (p. 529)
ephemeral stream (p. 520)
inselberg (p. 524)
interior drainage (p. 522)
loess (p. 532)

longitudinal dune (p. 531)
parabolic dune (p. 531)
playa (p. 524)
playa lake (p. 524)
rainshadow desert (p. 519)
saltation (p. 525)
slipface (p. 530)
star dune (p. 531)

steppe (p. 517)
suspended load (p. 525)
transverse dune (p. 531)
ventifact (p. 528)
yardang (p. 528)

Web Resources

The *Earth* Website uses the resources and flexibility of the Internet to aid in your study of the topics in this chapter. Written and developed by geology instructors, this site will help improve your understanding of geology. Visit **http://www.prenhall.com/tarbuck** and click on the cover of *Earth 9e* to find:

- Online review quizzes.
- Critical thinking exercises.
- Links to chapter-specific Web resources.
- Internet-wide key-term searches.

http://www.prenhall.com/tarbuck

GEODe: Earth

GEODe: Earth makes studying faster and more effective by reinforcing key concepts using animation, video, narration, interactive exercises and practice quizzes. A copy is included with every copy of *Earth*.

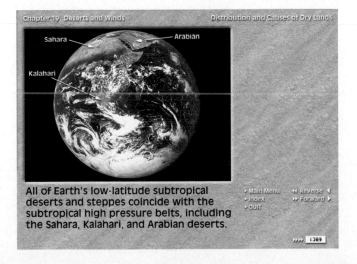

All of Earth's low-latitude subtropical deserts and steppes coincide with the subtropical high pressure belts, including the Sahara, Kalahari, and Arabian deserts.

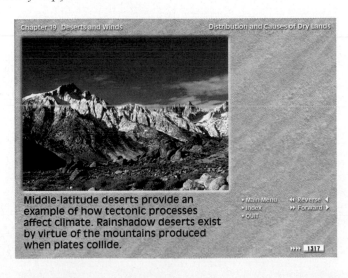

Middle-latitude deserts provide an example of how tectonic processes affect climate. Rainshadow deserts exist by virtue of the mountains produced when plates collide.

Shorelines

The shore of Kaho'olawe, a small volcanic island south of Maui (in background) in the Hawaiian islands. (Photo by David Muench)

537

The restless waters of the ocean are constantly in motion. Winds generate surface currents, the gravity of the Moon and Sun produces tides, and density differences create deep-ocean circulation. Further, waves carry the energy from storms to distant shores, where their impact erodes the land.

Shorelines are dynamic environments. Their topography, geologic makeup, and climate vary greatly from place to place. Continental and oceanic processes converge along coasts to create landscapes that frequently undergo rapid change. When it comes to the deposition of sediment, they are transition zones between marine and continental environments.

The Shoreline: A Dynamic Interface

Nowhere is the restless nature of the ocean's water more noticeable than along the shore—the dynamic interface among air, land, and sea. An *interface* is a common boundary where different parts of a system interact. This is certainly an appropriate designation for the coastal zone. Here we can see the rhythmic rise and fall of tides and observe waves constantly rolling in and breaking. Sometimes the waves are low and gentle. At other times they pound the shore with awesome fury.

Although it may not be obvious, the shoreline is constantly being modified by waves. For example, along Cape Cod, Massachusetts, wave activity is eroding cliffs of poorly consolidated glacial sediment so aggressively that the cliffs are retreating inland up to 1 meter per year (Figure 20.1A). By contrast, at Point Reyes, California, the far more durable bedrock cliffs are less susceptible to wave attack and

FIGURE 20.1   A. This satellite image includes the familiar outline of Cape Cod. Boston is to the upper left. The two large islands off the south shore of Cape Cod are Martha's Vineyard (left) and Nantucket (right). Although the work of waves constantly modifies this coastal landscape, shoreline processes are not primarily responsible for creating it. Rather, the present size and shape of Cape Cod result from the positioning of moraines and other glacial materials deposited during the Pleistocene epoch. (Satellite image courtesy of Earth Satellite Corporation/Science Photo Library/Photo Researchers, Inc.) **B.** High-altitude image of the Point Reyes area north of San Francisco, California. The 5.5-kilometer-long south-facing cliffs at Point Reyes (bottom of photo) are exposed to the full force of the waves from the Pacific Ocean. Nevertheless, this promontory retreats slowly because the bedrock from which it formed is very resistant. (Image courtesy of USDA-ASCS)

A.

B.

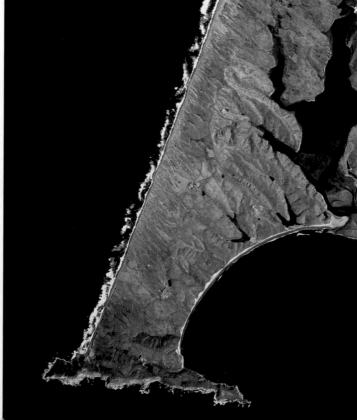

therefore are retreating much more slowly (Figure 20.1B). Along both coasts, wave activity is moving sediment along the shore and building narrow sandbars that protrude into and across some bays.

The nature of present-day shorelines is not just the result of the relentless attack of the land by the sea. Indeed, the shore has a complex character that results from multiple geologic processes. For example, practically all coastal areas were affected by the worldwide rise in sea level that accompanied the melting of glaciers at the close of the Pleistocene epoch. As the sea encroached landward, the shoreline retreated, becoming superimposed upon existing landscapes that had resulted from such diverse processes as stream erosion, glaciation, volcanic activity, and the forces of mountain building.

Today the coastal zone is experiencing intensive human activity. Unfortunately, people often treat the shoreline as if it were a stable platform on which structures can safely be built. This attitude inevitably leads to conflicts between people and nature. As you will see, many coastal landforms, especially beaches and barrier islands, are relatively fragile, short-lived features that are inappropriate sites for development.

The Coastal Zone

In general conversation a number of terms are used when referring to the boundary between land and sea. In the preceding section, the terms *shore, shoreline, coastal zone,* and *coast* were all used. Moreover, when many think of the land–sea interface, the word *beach* comes to mind. Let's take a moment to clarify these terms and introduce some other terminology used by those who study the land–sea boundary zone. You will find it helpful to refer to Figure 20.2, which is an idealized profile of the coastal zone.

The **shoreline** is the line that marks the contact between land and sea. Each day, as tides rise and fall, the position of the shoreline migrates. Over longer time spans, the average position of the shoreline gradually shifts as sea level rises or falls.

The **shore** is the area that extends between the lowest tide level and the highest elevation on land that is affected by storm waves. By contrast, the **coast** extends inland from the shore as far as ocean-related features can be found. The **coastline** marks the coast's seaward edge, whereas the inland boundary is not always obvious or easy to determine.

As Figure 20.2 illustrates, the shore is divided into the *foreshore* and the *backshore*. The **foreshore** is the area exposed when the tide is out (low tide) and submerged when the tide is in (high tide). The **backshore** is landward of the high-tide shoreline. It is usually dry, being affected by waves only during storms. Two other zones are commonly identified. The **nearshore zone** lies between the low-tide shoreline and the line where waves break at low tide. Seaward of the nearshore zone is the **offshore zone.**

For many, a beach is the sandy area where people lie in the sun and walk along the water's edge. Technically, a **beach** is an accumulation of sediment found along the landward margin of the ocean or a lake. Along straight coasts, beaches may extend for tens or hundreds of kilometers. Where coasts are irregular, beach formation may be confined to the relatively quiet waters of bays.

Beaches consist of one or more **berms,** which are relatively flat platforms often composed of sand that are adjacent to coastal dunes or cliffs and marked by a change in slope at the seaward edge. Another part of the beach is the **beach face,** which is the wet, sloping surface that extends from the berm to the shoreline. Where beaches are sandy, sunbathers usually prefer the berm, whereas joggers prefer the wet, hard-packed sand of the beach face.

Beaches are composed of whatever material is locally abundant. The sediment for some beaches is derived from the erosion of adjacent cliffs or nearby coastal mountains. Other beaches are built from sediment delivered to the coast by rivers.

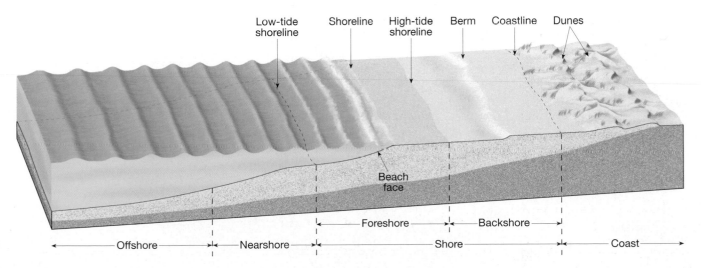

FIGURE 20.2 The coastal zone consists of several parts. The beach is an accumulation of sediment on the landward margin of the ocean or a lake. It can be thought of as material in transit along the shore.

Although the mineral makeup of many beaches is dominated by durable quartz grains, other minerals may be dominant. For example, in areas such as southern Florida, where there are no mountains or other sources of rock-forming minerals nearby, most beaches are composed of shell fragments and the remains of organisms that live in coastal waters. Some beaches on volcanic islands in the open ocean are composed of weathered grains of the basaltic lava that comprise the islands, or of coarse debris eroded from coral reefs that develop around islands in low latitudes.

Regardless of the composition, the material that comprises the beach does not stay in one place. Instead, crashing waves are constantly moving it. Thus, beaches can be thought of as material in transit along the shore.

Waves

GEODe Shorelines
▸ Waves and Beaches

Ocean waves are energy traveling along the interface between ocean and atmosphere, often transferring energy from a storm far out at sea over distances of several thousand kilometers. That's why even on calm days the ocean still has waves that travel across its surface. When observing waves, always remember that you are watching *energy* travel through a medium (water). If you make waves by tossing a pebble into a pond, or by splashing in a pool, or by blowing across the surface of a cup of coffee, you are imparting *energy* to the water, and the waves you see are just the visible evidence of the energy passing through.

Wind-generated waves provide most of the energy that shapes and modifies shorelines. Where the land and sea meet, waves that may have traveled unimpeded for hundreds or thousands of kilometers suddenly encounter a barrier that will not allow them to advance farther and must absorb their energy. Stated another way, the shore is the location where a practically irresistible force confronts an almost immovable object. The conflict that results is never-ending and sometimes dramatic.

Wave Characteristics

Most ocean waves derive their energy and motion from the wind. When a breeze is less than 3 kilometers (2 miles) per hour, only small wavelets appear. At greater wind speeds, more stable waves gradually form and advance with the wind.

Characteristics of ocean waves are illustrated in Figure 20.3, which shows a simple, nonbreaking waveform. The tops of the waves are the *crests,* which are separated by *troughs.* Halfway between the crests and troughs is the *still water level,* which is the level the water would occupy if there were no waves. The vertical distance between trough and crest is called the **wave height,** and the horizontal distance between successive crests (or troughs) is the **wavelength.** The time it takes one full wave—one wavelength—to pass a fixed position is the **wave period.**

The height, length, and period that are eventually achieved by a wave depend on three factors: (1) the wind speed, (2) the length of time the wind has blown, and (3) the **fetch,** or distance that the wind has traveled across open water. As the quantity of energy transferred from the wind to the water increases, the height and steepness of the waves increase as well. Eventually a critical point is reached where waves grow so tall that they topple over, forming ocean breakers called *whitecaps.*

For a particular wind speed, there is a maximum fetch and duration of wind beyond which waves will no longer increase in size. When the maximum fetch and duration are reached for a given wind velocity, the waves are said to be "fully developed." The reason that waves can grow no further is that they are losing as much energy through the breaking of whitecaps as they are receiving from the wind.

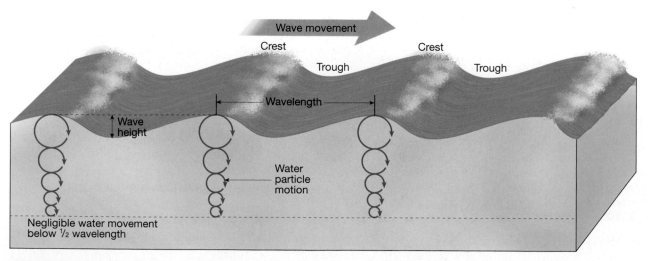

FIGURE 20.3 Diagrammatic view of an idealized nonbreaking ocean wave showing the basic parts of a wave as well as the movement of water particles at depth. Negligible water movement occurs below a depth equal to one half the wavelength (lower dashed line).

When wind stops or changes direction, or if waves leave the stormy area where they were created, they continue on without relation to local winds. The waves also undergo a gradual change to *swells*, which are lower and longer and may carry a storm's energy to distant shores. Because many independent wave systems exist at the same time, the sea surface acquires a complex, irregular pattern. Hence, the sea waves we watch from the shore are often a mixture of swells from faraway storms and waves created by local winds.

Circular Orbital Motion

Waves can travel great distances across ocean basins. In one study, waves generated near Antarctica were tracked as they traveled through the Pacific Ocean basin. After more than 10,000 kilometers (over 6000 miles), the waves finally expended their energy a week later along the shoreline of the Aleutian Islands of Alaska. The water itself doesn't travel the entire distance, but the waveform does. As the wave travels, the water passes the energy along by moving in a circle. This movement is called *circular orbital motion.*

Observation of an object floating in waves reveals that it moves not only up and down but also slightly forward and backward with each successive wave. Figure 20.4 shows that a floating object moves up and backward as the crest approaches, up and forward as the crest passes, down and forward after the crest, down and backward as the trough approaches, and rises and moves backward again as the next crest advances. When the movement of the toy boat shown in Figure 20.4 is traced as a wave passes, it can be seen that the boat moves in a circle and it returns to essentially the same place. Circular orbital motion allows a waveform (the wave's shape) to move forward *through the water* while the individual water particles that transmit the wave move in a circle. Wind moving across a field of wheat causes a similar phenomenon: The wheat itself doesn't travel across the field, but the waves do.

The energy contributed by the wind to the water is transmitted not only along the surface of the sea but also downward. However, beneath the surface the circular motion rapidly diminishes until, at a depth equal to one half the wavelength measured from still water level, the movement of water particles becomes negligible. This depth is known as the *wave base.* The dramatic decrease of wave energy with depth is shown by the rapidly diminishing diameters of water-particle orbits in Figure 20.3.

Waves in the Surf Zone

As long as a wave is in deep water, it is unaffected by water depth (Figure 20.5, *left*). However, when a wave approaches the shore, the water becomes shallower and influences wave behavior. The wave begins to "feel bottom" at a water depth equal to its wave base. Such depths interfere with water movement at the base of the wave and slow its advance (Figure 20.5, *center*).

As a wave advances toward the shore, the slightly faster waves farther out to sea catch up, decreasing the wave-

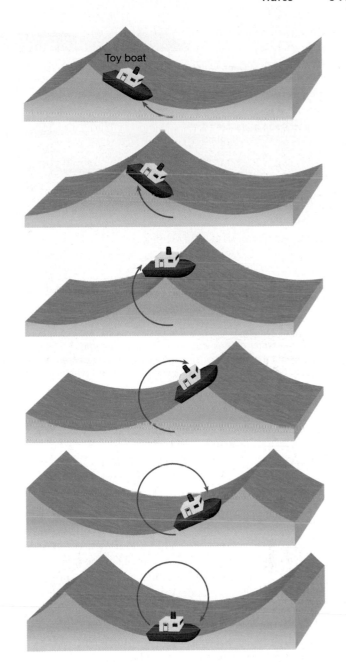

FIGURE 20.4 The movements of the toy boat show that the wave form advances, but the water does not advance appreciably from the original position. In this sequence, the wave moves from left to right as the boat (and the water in which it is floating) rotates in an imaginary circle.

length. As the speed and length of the wave diminish, the wave steadily grows higher. Finally, a critical point is reached when the wave is too steep to support itself and the wave front collapses, or *breaks* (Figure 20.5, *right*), causing water to advance up the shore.

The turbulent water created by breaking waves is called **surf.** On the landward margin of the surf zone the turbulent sheet of water from collapsing breakers, called *swash*, moves up the slope of the beach. When the energy of the swash has been expended, the water flows back down the beach toward the surf zone as *backwash.*

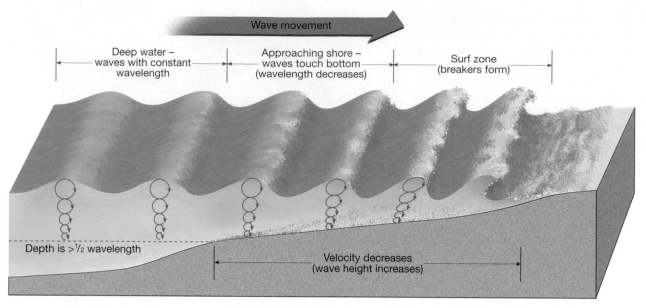

FIGURE 20.5   Changes that occur when a wave moves onto shore. The waves touch bottom as they encounter water depths less than half a wavelength. The wave speed decreases, and the waves stack up against the shore, causing the wavelength to decrease. This results in an increase in wave height to the point where the waves pitch forward and break in the surf zone.

Students Sometimes Ask ...

What are tidal waves?

Tidal waves, more accurately known as *tsunami* (*tsu* = harbor, *nami* = wave), have nothing to do with the tides! They are long-wavelength, fast-moving, often large, and sometimes destructive waves that originate from sudden changes in the topography of the seafloor. They are caused by underwater fault slippage, underwater avalanches, or underwater volcanic eruptions. Since the mechanisms that trigger tsunami are frequently seismic events, tsunami are appropriately termed *seismic sea waves*. For more information about characteristics of tsunami and their destructive effects, see Chapter 11, "Earthquakes."

average nearly 10,000 kilograms per square meter (more than 2000 pounds per square foot). The force during storms is even greater. During one such storm a 1350-ton portion of a steel-and-concrete breakwater was ripped from the rest of the structure and moved to a useless position toward the shore at Wick Bay, Scotland. Five years later the 2600-ton unit that replaced the first met a similar fate.

There are many such stories that demonstrate the great force of breaking waves. It is no wonder that cracks and crevices are quickly opened in cliffs, seawalls, breakwaters,

FIGURE 20.6   When waves break against the shore, the force of the water can be powerful and the erosional work that is accomplished can be great. These storm waves are breaking along the coast of Wales. (The Photolibrary Wales/Alamy)

Wave Erosion

 Shorelines
▸ Wave Erosion

During calm weather, wave action is minimal. However, just as streams do most of their work during floods, so too do waves accomplish most of their work during storms. The impact of high, storm-induced waves against the shore can be awesome in its violence (Figure 20.6). Each breaking wave may hurl thousands of tons of water against the land, sometimes causing the ground to literally tremble. The pressures exerted by Atlantic waves in wintertime, for example,

and anything else that is subjected to these enormous shocks. Water is forced into every opening, causing air in the cracks to become highly compressed by the thrust of crashing waves. When the wave subsides, the air expands rapidly, dislodging rock fragments and enlarging and extending fractures.

In addition to the erosion caused by wave impact and pressure, **abrasion**—the sawing and grinding action of the water armed with rock fragments—is also important. In fact, abrasion is probably more intense in the surf zone than in any other environment. Smooth, rounded stones and pebbles along the shore are obvious reminders of the relentless grinding action of rock against rock in the surf zone (Figure 20.7A). Further, such fragments are used as "tools" by the waves as they cut horizontally into the land (Figure 20.7B).

Along shorelines composed of unconsolidated material rather than hard rock, the rate of erosion by breaking waves can be extraordinary. In parts of Britain, where waves have the easy task of eroding glacial deposits of sand, gravel, and clay, the coast has been worn back 3 to 5 kilometers since Roman times (2000 years ago), sweeping away many villages and ancient landmarks.

Sand Movement on the Beach

Beaches are sometimes called "rivers of sand." The reason is that the energy from breaking waves often causes large quantities of sand to move along the beach face and in the surf zone roughly parallel to the shoreline. Wave energy also causes sand to move perpendicular to (toward and away from) the shoreline.

Movement Perpendicular to the Shoreline

If you stand ankle deep in water at the beach, you will see that swash and backwash move sand toward and away from the shoreline. Whether there is a net loss or addition of sand depends on the level of wave activity. When wave activity is relatively light (less energetic waves), much of the swash soaks into the beach, which reduces the backwash. Consequently, the swash dominates and causes a net movement of sand up the beach face toward the berm.

When high-energy waves prevail, the beach is saturated from previous waves, so much less of the swash soaks in. As a result, the berm erodes because backwash is strong and causes a net movement of sand down the beach face.

Along many beaches, light wave activity is the rule during the summer. Therefore, a wide sand berm gradually develops. During winter, when storms are frequent and more powerful, strong wave activity erodes and narrows the

Students Sometimes Ask . . .

During heavy wave activity, where does the sand from the berm go?

The orbital motion of waves is too shallow to move sand very far offshore. Consequently, the sand accumulates just beyond where the surf zone ends and forms one or more offshore sandbars called longshore bars.

FIGURE 20.7   **A.** Abrasion can be intense in the surf zone. Smooth rounded rocks along the shore are an obvious reminder of this fact. Garrapata State Park, California. (Photo by Carr Clifton) **B.** Sandstone cliff undercut by wave erosion at Gabriola Island, British Columbia, Canada. (Photo by Fletcher and Baylis/Photo Researchers, Inc.)

A.

B.

FIGURE 20.8   Wave bending around the end of a beach at Stinson Beach, California. (Photo by James E. Patterson)

berm. A wide berm that may have taken months to build can be dramatically narrowed in just a few hours by the high-energy waves created by a strong winter storm.

Wave Refraction

The bending of waves, called **wave refraction,** plays an important part in shoreline processes (Figure 20.8). It affects the distribution of energy along the shore and thus strongly influences where and to what degree erosion, sediment transport, and deposition will take place.

Waves seldom approach the shore straight on. Rather, most waves move toward the shore at an angle. However, when they reach the shallow water of a smoothly sloping bottom, they are bent and tend to become parallel to the shore. Such bending occurs because the part of the wave nearest the shore reaches shallow water and slows first, whereas the end that is still in deep water continues forward at its full speed. The net result is a wave front that may approach nearly parallel to the shore regardless of the original direction of the wave.

Because of refraction, wave impact is concentrated against the sides and ends of headlands that project into the water, whereas wave attack is weakened in bays. This differential wave attack along irregular coastlines is illustrated in Figure 20.9. As the waves reach the shallow water in front of the headland sooner than they do in adjacent bays, they are bent more nearly parallel to the protruding land and strike it from all three sides. By contrast, refraction in

the bays causes waves to diverge and expend less energy. In these zones of weakened wave activity, sediments can accumulate and form sandy beaches. Over a long period, erosion of the headlands and deposition in the bays will straighten an irregular shoreline.

Beach Drift and Longshore Currents

Although waves are refracted, most still reach the shore at some angle, however slight. Consequently, the uprush of water from each breaking wave (the swash) is at an oblique angle to the shoreline. However, the backwash is straight down the slope of the beach. The effect of this pattern of water movement is to transport sediment in a zigzag pattern along the beach face (Figure 20.10). This movement is called **beach drift,** and it can transport sand and pebbles hundreds or even thousands of meters each day. However, a more typical rate is 5 to 10 meters per day.

Oblique waves also produce currents within the surf zone that flow parallel to the shore and move substantially more sediment than beach drift. Because the water here is turbulent, these **longshore currents** easily move the fine suspended sand and roll larger sand and gravel along the bottom. When the sediment transported by longshore currents is added to the quantity moved by beach drift, the total amount can be very large. At Sandy Hook, New Jersey, for example, the quantity of sand transported along the shore over a 48-year period averaged almost 750,000 tons annually. For a 10-year period in Oxnard, California, more than 1.5 million tons of sediment moved along the shore each year.

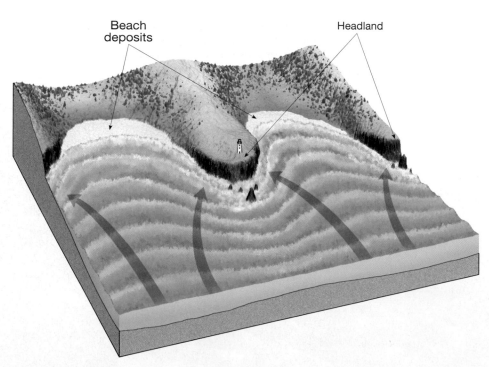

FIGURE 20.9   Wave refraction along an irregular coastline. As waves first touch bottom in the shallows off the headlands, they are slowed, causing the waves to refract and align nearly parallel to the shoreline. This causes wave energy to be concentrated at headlands (resulting in erosion) and dispersed in bays (resulting in deposition).

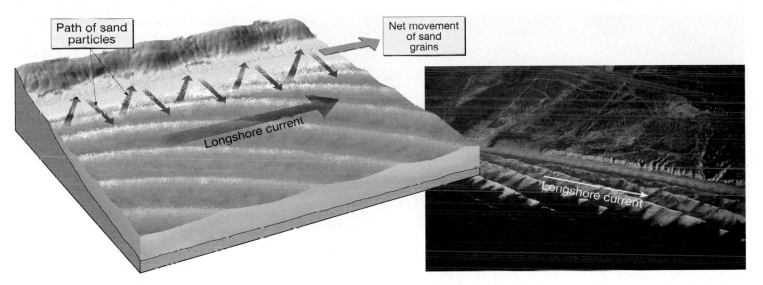

FIGURE 20.10   Beach drift and longshore currents are created by obliquely breaking waves. Beach drift occurs as incoming waves carry sand obliquely up the beach, while the water from spent waves carries it directly down the slope of the beach. Similar movements occur offshore in the surf zone to create the longshore current. These processes transport large quantities of material along the beach and in the surf zone. In the photo, waves approaching the beach at a slight angle near Oceanside, California, produce a longshore current moving from left to right. (Photo by John S. Shelton)

Both rivers and coastal zones move water and sediment from one area (*upstream*) to another (*downstream*). As a result, the beach has often been characterized as a "river of sand." Beach drift and longshore currents, however, move in a zigzag pattern, whereas rivers flow mostly in a turbulent, swirling fashion. Additionally, the direction of flow of longshore currents along a shoreline can change, whereas rivers flow in the same direction (downhill). Longshore currents change direction because the direction that waves approach the beach changes seasonally. Nevertheless, longshore currents generally flow southward along both the Atlantic and Pacific shores of the United States.

Students Sometimes Ask ...

Are rip currents the same thing as longshore currents?

No. Longshore currents occur in the surf zone and move roughly parallel to the shore. By contrast, rip currents occur perpendicular to the coast and move in the opposite direction of breaking waves. Most of the backwash from spent waves finds its way back to the open ocean as an unconfined flow across the ocean bottom called *sheet flow*. However, a portion of the returning water moves seaward in more concentrated surface *rip currents*. Rip currents do not travel far beyond the surf zone before breaking up, and can be recognized by the way they interfere with incoming waves or by the sediment that is often held in suspension within the rip current. They can also be a hazard to swimmers, who, if caught in them, can be carried out away from the shore.

Shoreline Features

A fascinating assortment of shoreline features can be observed along the world's coastal regions. These shoreline features vary depending on the type of rocks exposed along the shore, the intensity of wave activity, the nature of coastal currents, and whether the coast is stable, sinking, or rising. Features that owe their origin primarily to the work of erosion are called *erosional features*, whereas accumulations of sediment produce *depositional features*.

Erosional Features

Many coastal landforms owe their origin to erosional processes. Such erosional features are common along the rugged and irregular New England coast and along the steep shorelines of the West Coast of the United States.

Wave-Cut Cliffs, Wave-Cut Platforms, and Marine Terraces
Wave-cut cliffs, as the name implies, originate by the cutting action of the surf against the base of coastal land. As erosion progresses, rocks overhanging the notch at the base of the cliff crumble into the surf and the cliff retreats. A relatively flat, benchlike surface, called a **wave-cut platform,** is left behind by the receding cliff (Figure 20.11 *left*). The platform broadens as wave attack continues. Some debris produced by the breaking waves remains along the water's edge as sediment on the beach, while the remainder is transported farther seaward. If a wave-cut platform is uplifted above sea level by tectonic forces, it becomes a **marine terrace** (Figure 20.11, *right*). Marine terraces are easily recognized by their gentle seaward-sloping shape and are often desirable sites for coastal roads, buildings, or agriculture.

FIGURE 20.11 Wave-cut platform and marine terrace. A wave-cut platform is exposed at low tide along the California coast at Bolinas Point near San Francisco. A wave-cut platform has been uplifted to create the marine terrace. (Photo by John S. Shelton)

Sea Arches and Sea Stacks Headlands that extend into the sea are vigorously attacked by waves because of refraction. The surf erodes the rock selectively, wearing away the softer or more highly fractured rock at the fastest rate. At first, sea caves may form. When two caves on opposite sides of a headland unite, a **sea arch** results (Figure 20.12). Eventually the arch falls in, leaving an isolated remnant, or **sea stack,** on the wave-cut platform (Figure 20.12). In time, it too will be consumed by the action of the waves.

Depositional Features

Sediment eroded from the beach is transported along the shore and deposited in areas where wave energy is low. Such processes produce a variety of depositional features.

FIGURE 20.12 Sea arch and sea stack at the tip of Mexico's Baja Peninsula. (Photo by Mark A. Johnson/ The Stock Market)

Spits, Bars, and Tombolos Where beach drift and longshore currents are active, several features related to the movement of sediment along the shore may develop. A **spit** (*spit* = spine) is an elongated ridge of sand that projects from the land into the mouth of an adjacent bay. Often the end in the water hooks landward in response to the dominant direction of the longshore current (Figure 20.13). The term **baymouth bar** is applied to a sandbar that completely crosses a bay, sealing it off from the open ocean (Figure 20.13B). Such a feature tends to form across bays where currents are weak, allowing a spit to extend to the other side. A **tombolo** (*tombolo* = mound), a ridge of sand that connects an island to the mainland or to another island, forms in much the same manner as a spit.

Barrier Islands The Atlantic and Gulf Coastal Plains are relatively flat and slope gently seaward. The shore zone is characterized by **barrier islands.** These low ridges of land parallel the coast at distances from 3 to 30 kilometers offshore. From Cape Cod, Massachusetts, to Padre Island, Texas, nearly 300 barrier islands rim the coast (Figure 20.14).

Most barrier islands are from 1 to 5 kilometers wide and between 15 and 30 kilometers long. The tallest features are sand dunes, which usually reach heights of 5 to 10 meters; in a few areas, unvegetated dunes are more than 30 meters high. The lagoons separating these narrow islands from the shore represent zones of relatively quiet water that allow small craft traveling between New York and northern Florida to avoid the rough waters of the North Atlantic.

Barrier islands probably form in several ways. Some originate as spits that were subsequently severed from the mainland by wave erosion or by the general rise in sea level following the last episode of glaciation. Others are created when turbulent waters in the line of breakers heap up sand scoured from the bottom. Because these sand barriers rise above normal sea level, the piling of sand likely is the result of the work of storm waves at high tide. Finally, some barrier islands may be former sand-dune ridges that originated along the shore during the last glacial period, when sea level was lower. When the ice sheets melted, sea level rose and flooded the area behind the beach-dune complex.

A.

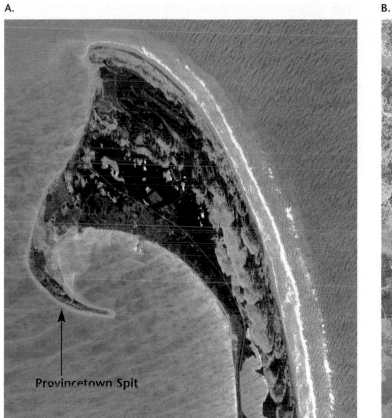

FIGURE 20.13   A. This photograph, taken from the International Space Station, shows Province-town Spit located at the tip of Cape Cod. Can you pick this feature out in the satellite image in Figure 20.1A? (NASA image) **B.** High-altitude image of a well-developed spit and baymouth bar along the coast of Martha's Vineyard, Massachusetts. (Image courtesy of USDA-ASCS)

FIGURE 20.14   Nearly 300 barrier islands rim the Gulf and Atlantic coasts. The islands along the south Texas coast and along the coast of North Carolina are excellent examples.

The Evolving Shore

A shoreline continually undergoes modification regardless of its initial configuration. At first most coastlines are irregular, although the degree of and reason for the irregularity may vary considerably from place to place. Along a coastline that is characterized by varied geology, the pounding surf may at first increase its irregularity because the waves will erode the weaker rocks more easily than the stronger ones. However, if a shoreline remains stable, marine erosion and deposition will eventually produce a straighter, more regular coast. Figure 20.15 illustrates the evolution of an initially irregular coast. As waves erode the headlands, creating cliffs and a wave-cut platform, sediment is carried along the shore. Some material is deposited in the bays, while other debris is formed into spits and baymouth bars. At the same time, rivers fill the bays with sediment. Ultimately a generally straight, smooth coast results.

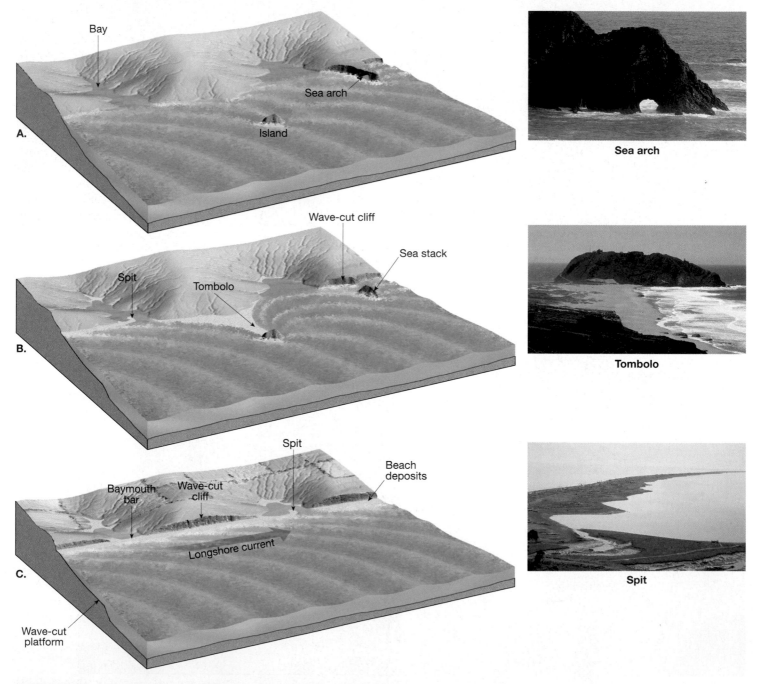

FIGURE 20.15 These diagrams illustrate the changes that can take place through time along an initially irregular coastline that remains relatively stable. The coastline shown in part **A** gradually evolves to **B,** and then **C.** The diagrams also serve to illustrate many of the features described in the section on shoreline features. (Photos by E. J. Tarbuck)

Stabilizing the Shore

GEODe Shorelines
  ▸ Waves and Beaches

Today the coastal zone teems with human activity. Unfortunately, people often treat the shoreline as if it were a stable platform on which structures can be built safely. This approach jeopardizes both people and the shoreline because many coastal landforms are relatively fragile, short-lived features that are easily damaged by development. And as anyone who has endured a tropical storm knows, the shoreline is not always a safe place to live. We will examine this latter idea more closely in the section on "Hurricanes–The Ultimate Coastal Hazard."

Compared with natural hazards such as earthquakes, volcanic eruptions, and landslides, shoreline erosion is often perceived to be a more continuous and predictable process that appears to cause relatively modest damage to limited areas. In reality, the shoreline is a dynamic place that can change rapidly in response to natural forces. Exceptional storms are capable of eroding beaches and cliffs at rates that are far in excess of the long-term average. Such bursts of accelerated erosion not only have a significant impact on the natural evolution of a coast but can also have a profound impact on people who reside in the coastal zone (Figure 20.16). Erosion along our coasts causes significant property damage. Huge sums are spent annually not only to repair damage but also to prevent or control erosion. Already a problem at many sites, shoreline erosion is certain to become an increasingly serious problem as extensive coastal development continues.

Although the same processes cause change along every coast, not all coasts respond in the same way. Interactions among different processes and the relative importance of each process depend on local factors. The factors include (1) the proximity of a coast to sediment-laden rivers, (2) the degree of tectonic activity, (3) the topography and composition of the land, (4) prevailing winds and weather patterns, and (5) the configuration of the coastline and nearshore areas.

During the past 100 years, growing affluence and increasing demands for recreation have brought unprecedented development to many coastal areas. As both the number and the value of buildings have increased, so too have efforts to protect property from storm waves by stabilizing the shore. Also, controlling the natural migration of sand is an ongoing struggle in many coastal areas. Such interference can result in unwanted changes that are difficult and expensive to correct.

Hard Stabilization

Structures built to protect a coast from erosion or to prevent the movement of sand along a beach are known as **hard stabilization**. Hard stabilization can take many forms and often results in predictable yet unwanted outcomes. Hard stabilization includes jetties, groins, breakwaters, and seawalls.

Jetties From relatively early in America's history a principal goal in coastal areas was the development and maintenance of harbors. In many cases, this involved the construction of jetty systems. **Jetties** are usually built in pairs and extend into the ocean at the entrances to rivers and harbors. With the flow of water confined to a narrow zone, the ebb and flow caused by the rise and fall of the tides keep the sand in motion and prevent deposition in the channel. However, as illustrated in Figure 20.17, the jetty may act as a dam against which the longshore current and beach drift deposit sand. At the same time, wave activity removes sand on the other side. Because the other side is not receiving any new sand, there is soon no beach at all.

Groins To maintain or widen beaches that are losing sand, groins are sometimes constructed. A **groin** (*groin* = ground) is a barrier built at a right angle to the beach to trap sand that is moving parallel to the shore. Groins are usually constructed of large rocks but may also be composed of wood. These structures often do their job so effectively that the longshore current beyond the groin becomes sand starved. As a result, the current erodes sand from the beach on the downstream side of the groin.

To offset this effect, property owners downstream from the structure may erect a groin on their property. In this manner, the number of groins multiplies, resulting in a *groin field* (Figure 20.18). An example of such proliferation is the shoreline of New Jersey, where hundreds of these structures have been built. Because it has been shown that groins often do not provide a

FIGURE 20.16 Wave erosion caused by strong storms forced abandonment of this highly developed shoreline area in Long Island, New York. Coastal areas are dynamic places that can change rapidly in response to natural forces. (Photo by Mark Wexler/Woodfin Camp & Associates)

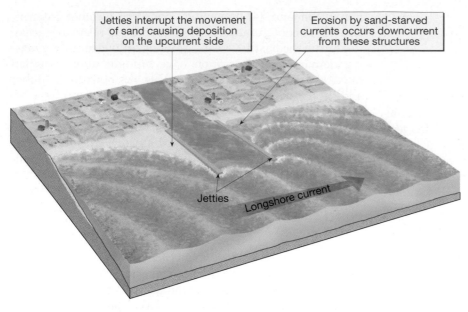

FIGURE 20.17 Jetties are built at the entrances to rivers and harbors and are intended to prevent deposition in the navigation channel. Jetties interrupt the movement of sand by beach drift and longshore currents. Beach erosion often occurs downcurrent from the site of the structure.

satisfactory solution, they are no longer the preferred method of keeping beach erosion in check.

Breakwaters and Seawalls Hard stabilization can also be built parallel to the shoreline. One such structure is a **breakwater,** the purpose of which is to protect boats from the force of large breaking waves by creating a quiet water zone near the shoreline. However, when this is done, the reduced wave activity along the shore behind the structure may allow sand to accumulate. If this happens, the marina will eventually fill with sand while the downstream beach erodes and retreats. At Santa Monica, California, where the building of a breakwater created such a problem, the city uses a dredge to remove sand from the protected quiet water zone and deposit it downstream where longshore currents and beach drift continue to move the sand down the coast (Figure 20.19).

Another type of hard stabilization built parallel to the shoreline is a **seawall,** which is designed to armor the coast and defend property from the force of breaking waves. Waves expend much of their energy as they move across an open beach. Seawalls cut this process short by reflecting the force of unspent waves seaward. As a consequence, the beach to the seaward side of the seawall experiences significant erosion and may in some instances be eliminated entirely (Figure 20.20). Once the width of the beach is reduced, the seawall is subjected to even greater pounding by the waves. Eventually this battering will cause the wall to fail, and a larger, more expensive wall must be built to take its place.

The wisdom of building temporary protective structures along shorelines is increasingly questioned. The opinions of many coastal scientists and engineers are expressed in the following excerpt from a position paper that grew out of a conference on America's Eroding Shoreline:

> It is now clear that halting the receding shoreline with protective structures benefits only a few and seriously degrades or destroys the natural beach and the value it holds for the majority. Protective structures divert the ocean's energy temporarily from private properties, but usually refocus that energy on the adjacent natural beaches. Many interrupt the natural sand flow in coastal currents, robbing many beaches of vital sand replacement.*

_____

*"Strategy for Beach Preservation Proposed," *Geotimes* 30 (No. 12, December 1985): p. 15.

FIGURE 20.18 Groins along the New Jersey shore at Cape May. (Photo by John S. Shelton)

FIGURE 20.19 Aerial view of a breakwater at Santa Monica, California. The breakwater appears as a faint line in the water behind which many boats are anchored. The construction of the breakwater disrupted longshore transport and caused the seaward growth of the beach. (Photo by John S. Shelton)

FIGURE 20.20   Seabright in northern New Jersey once had a broad, sandy beach. A seawall 5 to 6 meters (16 to 20 feet) high and 8 kilometers (5 miles) long was built to protect the town and the railroad that brought tourists to the beach. As you can see, after the wall was built, the beach narrowed dramatically. (Photo by Rafael Macia/Photo Researchers, Inc.)

Alternatives to Hard Stabilization

Armoring the coast with hard stabilization has several potential drawbacks, including the cost of the structure and the loss of sand on the beach. Alternatives to hard stabilization include beach nourishment and relocation.

Beach Nourishment   **Beach nourishment** represents an approach to stabilizing shoreline sands without hard stabilization. As the term implies, this practice involves the addition of large quantities of sand to the beach system (Figure 20.21). By building the beaches seaward, beach quality and storm protection are both improved. Beach nourishment, however, is not a permanent solution to the problem of shrinking beaches. The same processes that removed the sand in the first place will eventually remove the replacement sand as well. In addition, beach nourishment is very expensive because huge volumes of sand must be transported to the beach from offshore areas, nearby rivers, or other source areas. Orrin Pilkey, a respected coastal scientist, described the situation this way:

> Nourishment has been carried out on beaches on both sides of the continent, but by far the greatest effort in dollars and sand volume has been expended on the East Coast barrier islands between the southern shore of Long Island, N.Y., and southern Florida. Over this shoreline reach, communities have added a total of 500 million cubic yards of sand on 195 beaches in 680 separate instances. Some beaches. . .have been renourished more than 20 times each since 1965. Virginia Beach has been renourished more than 50 times. Nourished beaches typically cost between $2 million and $10 million per mile.*

*"Beaches Awash with Politics," in *Geotimes*, July 2005, pp. 38–39.

A.

B.

FIGURE 20.21   Miami Beach. **A.** Before beach nourishment and **B.** After beach nourishment. (Courtesy of the U.S. Army Corps of Engineers, Vicksburg District)

In some instances, beach nourishment can lead to unwanted environmental effects. For example, beach replenishment at Waikiki Beach, Hawaii, involved replacing coarse calcareous sand with softer, muddier calcareous sand. Destruction of the soft beach sand by breaking waves increased the water's turbidity and killed offshore coral reefs. At Miami Beach increased turbidity also damaged local coral communities.

Beach nourishment appears to be an economically viable long-range solution to the beach preservation problem only in areas where there exists dense development, large supplies of sand, relatively low wave energy, and reconcilable environmental issues. Unfortunately, few areas possess all these attributes.

Relocation   Instead of building structures such as groins and seawalls to hold the beach in place, or adding sand to replenish eroding beaches, another option is available. Many coastal scientists and planners are calling for a policy shift from defending and rebuilding beaches and coastal property in high hazard areas to *relocating* storm-damaged buildings in those places and letting nature reclaim the beach (see Box 20.1). This approach is similar to that adopted by the federal government for river floodplains following the devastating 1993 Mississippi River floods in which vulnerable structures are abandoned and relocated on higher, safer ground.

Such proposals, of course, are controversial. People with significant nearshore investments shudder at the thought of not rebuilding and defending coastal developments from the erosional wrath of the sea. Others, however, argue that with sea level rising, the impact of coastal storms will only get worse in the decades to come. This group advocates that oft-damaged structures be abandoned or relocated to improve personal safety and to reduce costs. Such ideas will no doubt be the focus of much study and debate as states and communities evaluate and revise coastal land-use policies.

Erosion Problems along U.S. Coasts

The shoreline along the Pacific Coast of the United States is strikingly different from that characterizing the Atlantic and Gulf coast regions. Some of the differences are related to plate tectonics. The West Coast represents the leading edge of the North American plate, and because of this it experiences active uplift and deformation. By contrast, the East Coast is a tectonically quiet region that is far from any active plate margin. Because of this basic geological difference, the nature of shoreline erosion problems along America's opposite coasts is different.

Atlantic and Gulf Coasts

Much of the coastal development along the Atlantic and Gulf coasts has occurred on barrier islands. Typically, barrier islands, also termed *barrier beaches* or *coastal barriers*, consist of a wide beach that is backed by dunes and separated from the mainland by marshy lagoons. The broad expanses of sand and exposure to the ocean have made barrier islands exceedingly attractive sites for development. Unfortunately, development has taken place more rapidly than our understanding of barrier island dynamics.

Because barrier islands face the open ocean, they receive the full force of major storms that strike the coast. When a storm occurs, the barriers absorb the energy of the waves primarily through the movement of sand. This process and the dilemma that results have been described as follows:

> Waves may move sand from the beach to offshore areas or, conversely, into the dunes; they may erode the dunes, depositing sand onto the beach or carrying it out to sea; or they may carry sand from the beach and the dunes into the marshes behind the barrier, a process known as overwash. The common factor is movement. Just as a flexible reed may survive a wind that destroys an oak tree, so the barriers survive hurricanes and nor'easters not through unyielding strength but by giving before the storm.
>
> This picture changes when a barrier is developed for homes or as a resort. Storm waves that previously rushed harmlessly through gaps between the dunes now encounter buildings and roadways. Moreover, since the dynamic nature of the barriers is readily perceived only during storms, homeowners tend to attribute damage to a particular storm, rather than to the basic mobility of coastal barriers. With their homes or investments at stake, local residents are more likely to seek to hold the sand in place and the waves at bay than to admit that development was improperly placed to begin with.*

Pacific Coast

In contrast to the broad, gently sloping coastal plains of the Atlantic and Gulf Coasts, much of the Pacific Coast is characterized by relatively narrow beaches that are backed by steep cliffs and mountain ranges. Recall that America's western margin is a more rugged and tectonically active region than the eastern margin. Because uplift continues, a rise in sea level in the West is not so readily apparent. Nevertheless, like the shoreline erosion problems facing the East's barrier islands, West Coast difficulties also stem largely from the alteration of a natural system by people.

A major problem facing the Pacific shoreline, and especially portions of southern California, is a significant narrowing of many beaches. The bulk of the sand on many of these beaches is supplied by rivers that transport it from the mountains to the coast. Over the years this natural flow of material to the coast has been interrupted by dams built for irrigation and flood control. The reservoirs effectively trap the sand that would otherwise nourish the beach environment. When the beaches were wider, they served to protect the cliffs behind them from the force of storm waves. Now, however, the waves move across the narrowed beaches

---

*Frank Lowenstein, "Beaches or Bedrooms—The Choice as Sea Level Rises," *Oceanus* 28 (No. 3, Fall 1985): p. 22.

BOX 20.1 ▶ PEOPLE AND THE ENVIRONMENT

The Move of the Century—Relocating the Cape Hatteras Lighthouse*

In spite of efforts to protect structures that are too close to the shore, they can still be in danger of being destroyed by receding shorelines and the destructive power of waves. Such was the case for one of the nation's most prominent landmarks, the candy-striped lighthouse at Cape Hatteras, North Carolina, which is 21 stories tall—the nation's tallest lighthouse.

The lighthouse was built in 1870 on the Cape Hatteras barrier island 457 meters (1500 feet) from the shoreline to guide mariners through the dangerous offshore shoals known as the "Graveyard of the Atlantic." As the barrier island began migrating toward the mainland, its beach narrowed. When the waves began to lap just 37 meters (120 feet) from its brick and granite base, there was concern that even a moderate-strength hurricane could trigger beach erosion sufficient to topple the lighthouse.

In 1970 the U.S. Navy built three groins in front of the lighthouse in an effort to protect the beach from further erosion. The groins initially slowed erosion but disrupted sand flow in the surf zone, which caused the flattening of nearby dunes and the formation of a bay south of the lighthouse. Attempts to increase the width of the beach in front of the lighthouse included beach nourishment and artificial offshore beds of seaweed, both of which failed to widen the beach substantially. In the 1980s the Army Corps of Engineers proposed building a massive stone seawall around the lighthouse but decided the eroding coast would eventually move out from under the structure, leaving it stranded at sea on its own island. In 1988 the National Academy of Sciences determined that the shoreline in front of the lighthouse would retreat so far as to destroy the lighthouse and recommended relocation of the tower as had been done with smaller lighthouses. In 1999 the National Park Service, which owns the lighthouse, finally authorized moving the structure to a safer location (Figure 20.A).

Moving the lighthouse, which weighs 4395 metric tons (4830 short tons), was accomplished by severing it from its foundation and carefully hoisting it onto a platform of steel beams fitted with roller dollies. Once on the platform, it was slowly rolled along a specially designed steel track using a series of hydraulic jacks. A strip of vegetation was cleared to make a runway along which the lighthouse traveled 1.5 meters (5 feet) at a time, with the track picked up from behind and reconstructed in front of the tower as it moved. In less than a month, the lighthouse was gingerly transported 884 meters (2900 feet) from its original location, making it one of the largest structures ever successfully moved.

After its $12 million move, the lighthouse now resides in a scrub oak and pine woodland (Figure 20.B). Although it now stands farther inland, the light's slightly higher elevation makes it visible just as far out to sea, where it continues to warn mariners of the hazardous shoals. At the current rate of shoreline retreat, the lighthouse should be safe from the threat of waves for at least another century.

---

*This box was prepared by Professor Alan P. Trujillo, Palomar College.

FIGURE 20.A When North Carolina's historic Cape Hatteras Lighthouse was threatened by coastal erosion, it was moved in the summer of 1999. Here you see the lighthouse before it was moved when it was only about 100 meters (330 feet) from the water. (Photo by Don Smetzer/Getty Images Inc.—Stone Allstock)

Former location of lighthouse

FIGURE 20.B Cape Hatteras Lighthouse after being moved 488 meters (1600 feet) from the shoreline. At its new location it should be safe for 50 years or more. (Photo by Reuters/Stringer/Getty Images, Inc.—Hulton Archive Photos)

without losing much energy and cause more rapid erosion of the sea cliffs.

Although the retreat of the cliffs provides material to replace some of the sand impounded behind dams, it also endangers homes and roads built on the bluffs. In addition, development atop the cliffs aggravates the problem. Urbanization increases runoff, which if not carefully controlled can result in serious bluff erosion. Watering lawns and gardens adds significant quantities of water to the slope. This water percolates downward toward the base of the cliff, where it may emerge in small seeps. This action reduces the slope's stability and facilitates mass wasting.

Shoreline erosion along the Pacific Coast varies considerably from one year to the next, largely because of the sporadic occurrence of storms. As a consequence, when the infrequent but serious episodes of erosion occur, the damage is often blamed on the unusual storms and not on coastal development or the sediment-trapping dams that may be great distances away. If, as predicted, sea level rises at an increasing rate in the years to come, increased shoreline erosion and sea-cliff retreat should be expected along many parts of the Pacific Coast. Coastal vulnerability to sea-level rise is examined in some detail as part of a discussion of the possible consequences of global warming in Chapter 21 on "Global Climate Change."

Hurricanes—The Ultimate Coastal Hazard

The whirling tropical cyclones that occasionally have wind speeds exceeding 300 kilometers (185 miles) per hour are known in the United States as *hurricanes*—the greatest storms on Earth (Figure 20.22). In the western Pacific they are called *typhoons*, and in the Indian Ocean they are simply called *cyclones*. No matter which name is used, these storms are among the most destructive of natural disasters. When a hurricane reaches land, it is capable of annihilating coastal areas and killing tens of thousands of people.

The vast majority of hurricane-related deaths and damage are caused by relatively infrequent yet powerful storms. The deadliest and most costly storm in more than a century occurred in August 2005, when Hurricane Katrina devastated the Gulf Coast of Louisiana, Mississippi, and Alabama (see Box 20.2). Although hundreds of thousands fled before the storm made landfall, thousands of others were caught by the storm. In addition to the human suffering and tragic loss of life that were left in the wake of Hurricane Katrina, the financial losses caused by the storm are practically incalculable.

Our coasts are vulnerable. People are flocking to live near the ocean. The proportion of the U.S. population residing within 75 kilometers (45 miles) of a coast in 2010 is projected to be well in excess of 50 percent. The concentration of such large numbers of people near the shoreline means that hurricanes place millions at risk. Moreover, the potential costs of property damage are incredible.

The amount of damage caused by a hurricane depends on several factors, including the size and population density of the area affected and the shape of the ocean bottom near the shore. The most significant factor, of course, is the strength of the storm itself. By studying past storms, a scale has been established to rank the relative intensities of hurricanes. As Table 20.1 indicates, a *category 5* storm is the worst possible, whereas a *category 1* hurricane is least severe.

During the hurricane season it is common to hear scientists and reporters alike use the numbers from the *Saffir–Simpson Hurricane Scale.* When Hurricane Katrina made landfall, sustained winds were 225 kilometers (140 miles) per hour, making it a strong category 4 storm. Storms that fall into category 5 are rare. Hurricane Camille, a 1969 storm that caused catastrophic damage along the coast of Mississippi, is one well-known example (Figure 20.23).

FIGURE 20.22   A. Satellite image of Hurricane Katrina in late August 2005 shortly before it devastated the Gulf Coast. (NASA image) **B.** An estimated 80 percent of New Orleans was flooded after several levees failed in the wake of Katrina. The storm surge caused the level of Lake Pontchartrain to rise, straining the levee system protecting the city. (AP Photo/David J. Phillip)

A.

B.

TABLE 20.1	Saffir–Simpson Hurricane Scale					
Scale Number (Category)	Central Pressure (Millibars)	Wind Speed (KPH)	Wind Speed (MPH)	Storm Surge (Meters)	Storm Surge (Feet)	Damage
1	≥980	119–153	74–95	1.2–1.5	4–5	*Minimal*
2	965–979	154–177	96–110	1.6–2.4	6–8	*Moderate*
3	945–964	178–209	111–130	2.5–3.6	9–12	*Extensive*
4	920–944	210–250	131–155	3.7–5.4	13–18	*Extreme*
5	<920	>250	>155	>5.4	>18	*Catastrophic*

Damage caused by hurricanes can be divided into three categories: (1) storm surge, (2) wind damage, and (3) inland flooding.

Storm Surge

Without question, the most devastating damage in the coastal zone is caused by the storm surge. It not only accounts for a large share of coastal property losses but is also responsible for a high percentage of all hurricane-caused deaths. A **storm surge** is a dome of water 65 to 80 kilometers

A.

B.

FIGURE 20.23 In 1969 Hurricane Camille hit the coast of Mississippi. It was a rare category 5 storm. These classic photos document the devastating force of the hurricane's 7.5-meter (25-foot) storm surge at Pass Christian. **A.** The Richelieu Apartments before the hurricane. This substantial-looking, three-story building was directly across the highway from the beach. **B.** The same apartments after the hurricane. (Estate of Chauncey T. Hinman)

(40 to 50 miles) wide that sweeps across the coast near the point where the eye makes landfall. If all wave activity were smoothed out, the storm surge would be the height of the water above normal tide level. In addition, tremendous wave activity is superimposed on the surge. We can easily imagine the damage that this surge of water could inflict on low-lying coastal areas (Figure 20.23). The worst surges occur in places like the Gulf of Mexico, where the continental shelf is very shallow and gently sloping. In addition, local features such as bays and rivers can cause the surge height to double and increase in speed.

As a hurricane advances toward the coast in the Northern Hemisphere, storm surge is always most intense on the right side of the eye where winds are blowing *toward* the shore. In addition, on this side of the storm, the forward movement of the hurricane also contributes to the storm surge. In Figure 20.24, assume a hurricane with peak winds of 175 kilometers (109 miles) per hour is moving toward the shore at 50 kilometers (31 miles) per hour. In this case, the net wind speed on the right side of the advancing storm is 225 kilometers (140 miles) per hour. On the left side, the hurricane's winds are blowing opposite the direction of storm movement, so the net winds are *away* from the coast at 125 kilometers (78 miles) per hour. Along the shore facing the left side of the oncoming hurricane, the water level may actually decrease as the storm makes landfall.

Wind Damage

Destruction caused by wind is perhaps the most obvious of the classes of hurricane damage. Debris such as signs, roofing materials, and small items left outside become dangerous

| BOX 20.2 ▶ UNDERSTANDING EARTH |

Examining Hurricane Katrina from Space

Satellites allow us to track the formation, movement, and growth of hurricanes. In addition, their specialized instruments provide data that can be transformed into images that allow scientists to analyze the internal structure and workings of these huge storms. The images and captions in this box provide a unique perspective of Hurricane Katrina, the most devastating

storm to strike the United States in more than a century. Figure 20.22A and Figure 20.C are from NASA's *Terra* satellite and are relatively "traditional" images that show Katrina in different stages of development. Figure 20.D is a color-enhanced infrared image from the *GOES-East* satellite. The coldest cloud tops (red) are associated with the most intense storms and are easily seen

in this image of Hurricane Katrina taken a few hours before landfall. Color-enhanced imagery is a method scientists use to aid them with satellite interpretation. The colors enable them to easily and quickly see features that are of special interest.

Figure 20.E from NASA's *QuikSCAT* satellite is very different in appearance. It provides a detailed look at Katrina's

FIGURE 20.C Image of Tropical Storm Katrina on August 24, 2005, shortly after it became the 11th named storm of the 2005 Atlantic hurricane season. When this image was taken, Katrina had winds of 64 kilometers (40 miles) per hour, and was just beginning to take on the recognizable swirling shape of a hurricane. (NASA image)

FIGURE 20.D Color-enhanced infrared image from the *GOES-East* satellite of Hurricane Katrina several hours before making landfall on August 29, 2005. The most intense activity is associated with red and orange. (NOAA)

surface winds shortly before the storm made landfall. This image depicts relative wind speeds rather than actual values. The satellite sends out high-frequency radio waves, some of which bounce off the ocean and return to the satellite. Rough, storm-tossed seas create a strong signal, whereas a smooth surface returns a weaker signal. In order to match wind speeds with the type of signal that returns to the satellite, scientists compare wind measurements taken by data buoys in the ocean to the strength of the signal received by the satellite. When there are too few data buoy measurements to compare to the satellite data, exact wind speeds cannot be determined. Instead, the image provides a clear picture of relative wind speeds.

Finally, Figure 20.F shows the *Multi-satellite Precipitation Analysis (MPA)* of the storm. This image, which also depicts the track of storm, shows the overall pattern of rainfall. It was constructed from data collected over several days by the *Tropical Rainfall Measuring Mission (TRMM)* satellite and other satellites.

FIGURE 20.E NASA's *QuikSCAT* satellite was the source of data for this image of Hurricane Katrina on August 28, 2005. The image depicts relative wind speeds. The strongest winds, shown in shades of purple, circle a well-defined eye. The barbs show wind direction. (NASA image)

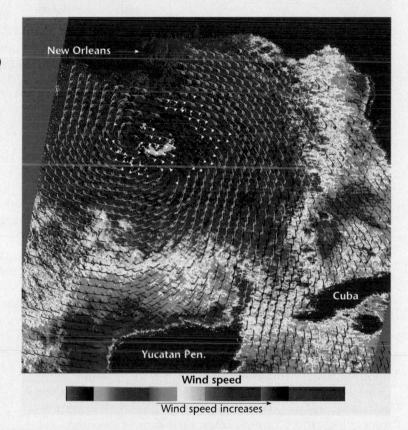

FIGURE 20.F Storm track and rainfall values for Hurricane Katrina for the period August 23 to 31, 2005. Rainfall amounts are derived from satellite data. The highest totals (dark red) exceeded 30 centimeters (12 inches) over north-western Cuba and the Florida Keys. Amounts over southern Florida (green to yellow) were 12–20 centimeters (5–8 inches). Rainfall along the Mississippi coast (yellow to orange) was between 15–23 centimeters (6–9 inches). After coming ashore the storm moved rapidly, rainfall totals (green to blue) were generally less than 13 centimeters (5 inches). (NASA image)

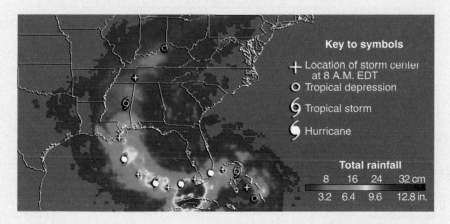

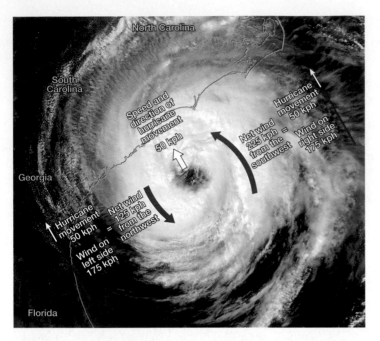

FIGURE 20.24 Winds associated with a Northern Hemisphere hurricane advancing toward the coast. This hypothetical storm, with peak winds of 175 kilometers (109 miles) per hour, is moving toward the coast at 50 kilometers (31 miles) per hour. On the right side of the advancing storm, the 175-kilometer-per-hour winds are in the same direction as the movement of the storm (50 kilometers per hour). Therefore, the *net* wind speed on the right side of the storm is 225 kilometers (140 miles) per hour. On the left side, the hurricane's winds are blowing opposite the direction of storm movement, so the *net* winds of 125 kilometers (78 miles) per hour are away from the coast. Storm surge will be greatest along that part of the coast hit by the right side of the advancing hurricane.

flying missiles in hurricanes. For some structures, the force of the wind is sufficient to cause total ruin. Mobile homes are particularly vulnerable. High-rise buildings are also susceptible to hurricane-force winds. Upper floors are most vulnerable because wind speeds usually increase with height. Recent research suggests that people should stay below the 10th floor but remain above any floors at risk for flooding. In regions with good building codes, wind damage is usually not as catastrophic as storm-surge damage. However, hurricane-force winds affect a much larger area than storm surge and can cause huge economic losses. For example, in 1992 it was

largely the winds associated with Hurricane Andrew that produced more than $25 billion of damage in southern Florida and Louisiana.

Hurricanes sometimes produce tornadoes that contribute to the storm's destructive power. Studies have shown that more than half of the hurricanes that make landfall produce at least one tornado. In 2004 the number of tornadoes associated with tropical storms and hurricanes was extraordinary. Tropical Storm Bonnie and five landfalling hurricanes—Charley, Frances, Gaston, Ivan, and Jeanne—produced nearly 300 tornadoes that affected the southeast and mid-Atlantic states.

Inland Flooding

The torrential rains that accompany most hurricanes represent a third significant threat: flooding. Whereas the effects of storm surge and strong winds are concentrated in coastal areas, heavy rains may affect places hundreds of kilometers from the coast for up to several days after the storm has lost its hurricane-force winds.

In September 1999 Hurricane Floyd brought flooding rains, high winds, and rough seas to a large portion of the Atlantic seaboard. More than 2.5 million people evacuated their homes from Florida north to the Carolinas and beyond. It was the largest peacetime evacuation in U.S. history up to that time. Torrential rains falling on already saturated ground created devastating inland flooding. Altogether Floyd dumped more than 48 centimeters (19 inches) of rain on Wilmington, North Carolina, 33.98 cm (13.38 inches) in a single 24-hour span.

To summarize, extensive damage and loss of life in the coastal zone can result from storm surge, strong winds, and torrential rains. When loss of life occurs, it is commonly caused by the storm surge, which can devastate entire barrier islands or zones within a few blocks of the coast. Although wind damage is usually not as catastrophic as the storm surge, it affects a much larger area. Where building codes are inadequate, economic losses can be especially severe. Because hurricanes weaken as they move inland, most wind damage occurs within 200 kilometers (125 miles) of the coast. Far from the coast, a weakening storm can produce extensive flooding long after the winds have diminished below hurricane levels. Sometimes the damage from inland flooding exceeds storm-surge destruction.

Coastal Classification

The great variety of shorelines demonstrates their complexity. Indeed, to understand any particular coastal area, many factors must be considered, including rock types, size and direction of waves, frequency of storms, tidal range, and offshore topography. In addition, practically all coastal areas were affected by the worldwide rise in sea level that accompanied the melting of Ice Age glaciers at the close of the Pleistocene epoch. Finally, tectonic events that elevate or drop the land or change the volume of ocean basins must be taken into account. The large number of factors that influence coastal areas make shoreline classification difficult.

Students Sometimes Ask . . .

What's the difference between a hurricane and a tropical storm?

Both are tropical cyclones—circular zones of low pressure with strong inward-spiraling winds. The difference relates to intensity. When sustained winds are between 61–119 kilometers (37–74 miles) per hour, the cyclone is called a tropical storm. It is during this phase that a name is given (Andrew, Fran, Rita, and so forth). When the cyclone's sustained winds exceed 119 kilometers (74 miles) per hour, it has reached hurricane status.

Many geologists classify coasts based on the changes that have occurred with respect to sea level. This commonly used classification divides coasts into two very general categories: emergent and submergent. **Emergent coasts** develop either because an area experiences uplift or as a result of a drop in sea level. Conversely, **submergent coasts** are created when sea level rises or the land adjacent to the sea subsides.

Emergent Coasts

In some areas the coast is clearly emergent because rising land or a falling water level exposes wave-cut cliffs and platforms above sea level. Excellent examples include portions of coastal California, where uplift has occurred in the recent geological past (see Figure 14.19 p. 395). The marine terrace shown in Figure 20.11 also illustrates this situation. In the case of the Palos Verdes Hills, south of Los Angeles, seven different terrace levels exist, indicating seven episodes of uplift. The ever persistent sea is now cutting a new platform at the base of the cliff. If uplift follows, it too will become an elevated marine terrace.

Other examples of emergent coasts include regions that were once buried beneath great ice sheets. When glaciers were present, their weight depressed the crust, and when the ice melted, the crust began gradually to spring back. Consequently, prehistoric shoreline features may now be found high above sea level. The Hudson Bay region of Canada is such an area, portions of which are still rising at a rate of more than a centimeter annually.

Submergent Coasts

In contrast to the preceding examples, other coastal areas show definite signs of submergence. Shorelines that have been submerged in the relatively recent past are often highly irregular because the sea typically floods the lower reaches of river valleys flowing into the ocean. The ridges separating the valleys, however, remain above sea level and project into the sea as headlands. These drowned river mouths, which are called **estuaries** (*aestus* = tide) characterize many coasts today. Along the Atlantic coastline, the Chesapeake and Delaware bays are examples of large estuaries created by submergence (Figure 20.25). The picturesque coast of Maine, particularly in the vicinity of Acadia National Park, is another excellent example of an area that was flooded by the post-glacial rise in sea level and transformed into a highly irregular coastline.

Keep in mind that most coasts have complicated geologic histories. With respect to sea level, many have at various times emerged and then submerged. Each time they may retain some of the features created during the previous situation.

Tides

Tides are daily changes in the elevation of the ocean surface. Their rhythmic rise and fall along coastlines have been known since antiquity. Other than waves, they are the easiest ocean movements to observe (Figure 20.26).

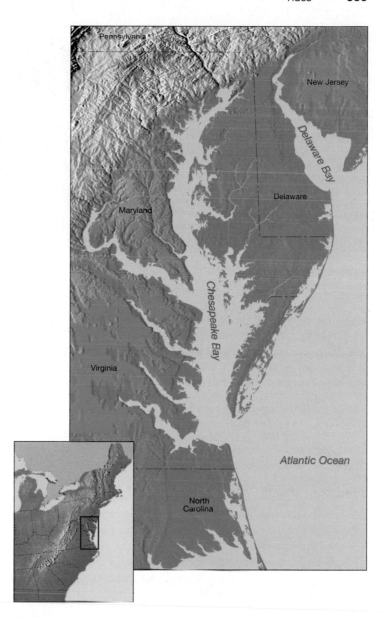

FIGURE 20.25   Major estuaries along the East Coast of the United States. The lower portions of many river valleys were submerged by the rise in sea level that followed the end of the Ice Age, creating large estuaries such as Chesapeake and Delaware bays.

Although known for centuries, tides were not explained satisfactorily until Sir Isaac Newton applied the law of gravitation to them. Newton showed that there is a mutual attractive force between two bodies and that since oceans are free to move, they are deformed by this force. Hence, ocean tides result from the gravitational attraction exerted upon Earth by the Moon and, to a lesser extent, by the Sun.

Causes of Tides

It is easy to see how the Moon's gravitational force can cause the water to bulge on the side of Earth nearest the Moon. In addition, however, an equally large tidal bulge is produced on the side of Earth directly opposite the Moon (Figure 20.27).

FIGURE 20.26 High tide and low tide on Nova Scotia's Minas Basin in the Bay of Fundy. The areas exposed during low tide and flooded during high tide are called *tidal flats*. Tidal flats here are extensive. (Courtesy of Nova Scotia Department of Tourism and Culture)

Students Sometimes Ask . . .

Where are the world's largest tides?

The world's largest *tidal range* (the difference between successive high and low tides) is found in the northern end of Nova Scotia's 258-kilometer- (160-mile-) long Bay of Fundy. During maximum spring tide conditions, the tidal range at the mouth of the bay (where it opens to the ocean) is only about 2 meters (6.6 feet). However, the tidal range progressively increases from the mouth of the bay northward because the natural geometry of the bay concentrates tidal energy. In the northern end of Minas Basin, the maximum spring tidal range is about 17 meters (56 feet). This extreme tidal range leaves boats high and dry during low tide (see Figure 20.26).

Both tidal bulges are caused, as Newton discovered, by the pull of gravity. Gravity is inversely proportional to the square of the distance between two objects, meaning simply that it quickly weakens with distance. In this case, the two objects are the Moon and Earth. Because the force of gravity decreases with distance, the Moon's gravitational pull on Earth is slightly greater on the near side of Earth than on the far side. The result of this differential pulling is to stretch (elongate) the "solid" Earth very slightly. In contrast, the world ocean, which is mobile, is deformed quite dramatically by this effect to produce the two opposing tidal bulges.

Because the position of the Moon changes only moderately in a single day, the tidal bulges remain in place while Earth rotates "through" them. For this reason, if you stand on the seashore for 24 hours, Earth will rotate you through alternating areas of deeper and shallower water. As you are carried into each tidal bulge, the tide rises, and as you are carried into the intervening troughs between the tidal bulges, the tide falls. Therefore, most places on Earth experience two high tides and two low tides each day.

Further, the tidal bulges migrate as the Moon revolves around Earth about every 29 days. As a result, the tides, like the time of moonrise, shift about 50 minutes later each day. After 29 days the cycle is complete and a new one begins.

In many locations, there may be an inequality between the high tides during a given day. Depending on the position of the Moon, the tidal bulges may be inclined to the equator as in Figure 20.27. This figure illustrates that one high tide experienced by an observer in the Northern Hemisphere is considerably higher than the high tide half a day

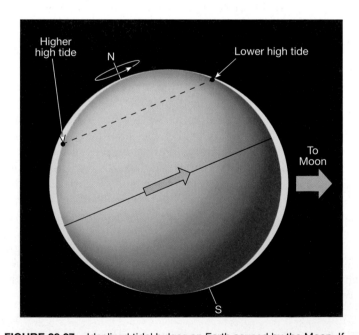

FIGURE 20.27 Idealized tidal bulges on Earth caused by the Moon. If Earth were covered to a uniform depth with water, there would be two tidal bulges: one on the side of Earth facing the Moon (right), and the other on the opposite side of Earth (left). Depending on the Moon's position, tidal bulges may be inclined relative to Earth's equator. In this situation, Earth's rotation causes an observer to experience two unequal high tides during a day.

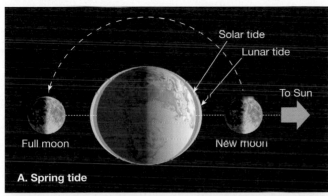

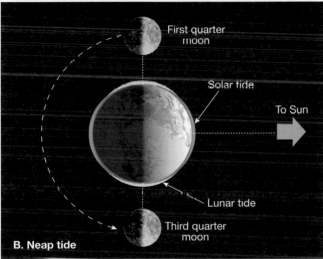

FIGURE 20.28 Earth–Moon–Sun positions and the tides. **A.** When the Moon is in the full or new position, the tidal bulges created by the Sun and Moon are aligned, there is a large tidal range on Earth, and *spring tides* are experienced. **B.** When the Moon is in the first- or third-quarter position, the tidal bulges produced by the Moon are at right angles to the bulges created by the Sun. Tidal ranges are smaller, and *neap tides* are experienced.

later. In contrast, a Southern Hemisphere observer would experience the opposite effect.

Monthly Tidal Cycle

The primary body that influences the tides is the Moon, which makes one complete revolution around Earth every 29 and a half days. The Sun, however, also influences the tides. It is far larger than the Moon, but because it is much farther away, its effect is considerably less. In fact, the Sun's tide-generating effect is only about 46 percent that of the Moon's.

Near the times of new and full moons, the Sun and Moon are aligned and their forces are added together (Figure 20.28A). Accordingly, the combined gravity of these two tide-producing bodies causes larger tidal bulges (higher high tides) and deeper tidal troughs (lower low tides), producing a large tidal range. These are called the **spring** (*springen* = to rise up) **tides,** which have no connection with the spring season but occur twice a month during the

time when the Earth–Moon–Sun system is aligned. Conversely, at about the time of the first and third quarters of the Moon, the gravitational forces of the Moon and Sun act on Earth at right angles, and each partially offsets the influence of the other (Figure 20.28B). As a result, the daily tidal range is less. These are called **neap** (*nep* = scarcely or barely touching) **tides,** which also occur twice each month. Each month, then, there are two spring tides and two neap tides, each about one week apart.

Tidal Patterns

So far, the basic causes and types of tides have been explained. Keep in mind, however, that these theoretical considerations cannot be used to predict either the height or the time of actual tides at a particular place. This is because many factors—including the shape of the coastline, the configuration of ocean basins, and water depth—greatly influence the tides. Consequently, tides at various locations respond differently to the tide-producing forces. This being the case, the nature of the tide at any coastal location can be determined most accurately by actual observation. The predictions in tidal tables and tidal data on nautical charts are based on such observations.

Three main tidal patterns exist worldwide. A **diurnal** (*diurnal* = dailiy) **tidal pattern** is characterized by a single high tide and a single low tide each tidal day (Figure 20.29). Tides of this type occur along the northern shore of the Gulf of Mexico, among other locations. A **semidiurnal** (*semi* = twice, *diurnal* = daily) **tidal pattern** exhibits two high tides and two low tides each tidal day, with the two highs about the same height and the two lows about the same height (Figure 20.29). This type of tidal pattern is common along the Atlantic Coast of the United States. A **mixed tidal pattern** is similar to a semidiurnal pattern except that it is characterized by a large inequality in high water heights, low water heights, or both (Figure 20.29). In this case, there are usually two high and two low tides each day, with high tides of different heights and low tides of different heights. Such tides are prevalent along the Pacific Coast of the United States and in many other parts of the world.

Tidal Currents

Tidal current is the term used to describe the *horizontal* flow of water accompanying the rise and fall of the tide. These water movements induced by tidal forces can be important in some coastal areas. Tidal currents flow in one direction during a portion of the tidal cycle and reverse their flow during the remainder. Tidal currents that advance into the coastal zone as the tide rises are called **flood currents.** As the tide falls, seaward-moving water generates **ebb currents.** Periods of little or no current, called *slack water,* separate flood and ebb. The areas affected by these alternating tidal currents are **tidal flats** (see Figure 20.26). Depending on the nature of the coastal zone, tidal flats vary from narrow strips seaward of the beach to extensive zones that may extend for several kilometers.

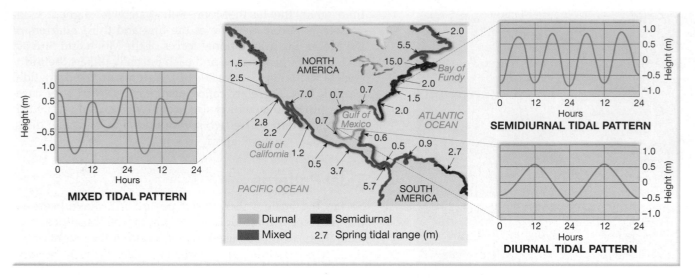

FIGURE 20.29 Tidal patterns and their occurrence along portions of the coastlines of North and South America. A diurnal tidal pattern (lower right) shows one high and one low tide each tidal day. A semidiurnal pattern (upper right) shows two highs and two lows of approximately equal heights during each tidal day. A mixed tidal pattern (left) shows two highs and two lows of unequal heights during each tidal day.

Although tidal currents are not important in the open sea, they can be rapid in bays, river estuaries, straits, and other narrow places. Off the coast of Brittany in France, for example, tidal currents that accompany a high tide of 12 meters (40 feet) may attain a speed of 20 kilometers (12 miles) per hour. While tidal currents are not generally major agents of erosion and sediment transport, notable exceptions occur where tides move through narrow inlets. Here they constantly scour the small entrances to many good harbors that would otherwise be blocked.

Sometimes deposits called **tidal deltas** are created by tidal currents (Figure 20.30). They may develop either as *flood deltas* landward of an inlet or as *ebb deltas* on the seaward side of an inlet. Because wave activity and longshore currents are reduced on the sheltered landward side, flood deltas are more common and more prominent (see Figure 20.13B). They form after the tidal current moves rapidly through an inlet. As the current emerges from the narrow passage into more open waters, it slows and deposits its load of sediment.

Tides and Earth's Rotation

The tides by friction against the floor of the ocean basins act as weak brakes that are steadily slowing Earth's rotation. The rate of slowing, however, is not great. Astronomers, who have precisely measured the length of the day over the past 300 years, have found that the day is increasing by 0.002 second per century. Although this may seem inconsequential, over millions of years this small effect will become very large.

If Earth's rotation is slowing, the length of each day must have been shorter and the number of days per year must have been greater in the geologic past. One method used to investigate this phenomenon involves the microscopic examination of shells of certain invertebrates. Clams and corals, as well as other organisms, grow a microscopically thin layer of new shell material each day. By studying the daily growth rings of some well-preserved fossil specimens, we can determine the number of days in a year. Studies using this ingenious technique indicate that early in the Cambrian period, about 540 million years ago, the length of the day was only 21 hours. Because the length of the year, which is determined by Earth's revolution about the Sun, does not change, the Cambrian year contained 424 21-hour days. By late Devonian time, about 365 million years ago, a year consisted of about 410 days, and as the Permian period opened, about 290 million years ago, there were 390 days in a year.

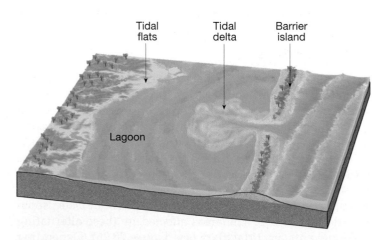

FIGURE 20.30 Because this tidal delta is forming in the relatively quiet waters on the landward side of a barrier island, it is termed a flood delta. As a rapidly moving tidal current emerges from the inlet, it slows and deposits sediment. The shapes of tidal deltas are variable.

Summary

- The *shore* is the area extending between the lowest tide level and the highest elevation on land that is affected by storm waves. The *coast* extends inland from the shore as far as ocean-related features can be found. The shore is divided into the *foreshore* and *backshore*. Seaward of the foreshore are the *nearshore* and *offshore* zones.

- A *beach* is an accumulation of sediment found along the landward margin of the ocean or a lake. Among its parts are one or more *berms* and the *beach face*. Beaches are composed of whatever material is locally abundant and should be thought of as material in transit along the shore.

- *Waves are moving energy* and *most ocean waves are initiated by the wind*. The three factors that influence the *height, length,* and *period* of a wave are (1) *wind speed,* (2) *length of time the wind has blown,* and (3) *fetch,* the distance the wind has traveled across open water. Once waves leave a storm area, they are termed *swells,* which are symmetrical, longer-wavelength waves.

- As waves travel, *water particles transmit energy by circular orbital motion,* which extends to a depth equal to one half the wavelength. When a wave travels into shallow water, it experiences physical changes that can cause the wave to collapse, or *break,* and form *surf.*

- Wave erosion is caused by *wave impact pressure* and *abrasion* (the sawing and grinding action of water armed with rock fragments). The bending of waves is called *wave refraction.* Owing to refraction, wave impact is concentrated against the sides and ends of headlands.

- Most waves reach the shore at an angle. The uprush (swash) and backwash of water from each breaking wave moves the sediment in a zigzag pattern along the beach. This movement, called *beach drift,* can transport sand hundreds or even thousands of meters each day. Oblique waves also produce *longshore currents* within the surf zone that flow parallel to the shore.

- Features produced by *shoreline erosion* include *wave-cut cliffs* (which originate from the cutting action of the surf against the base of coastal land), *wave-cut platforms* (relatively flat, benchlike surfaces left behind by receding cliffs), *sea arches* (formed when a headland is eroded and two caves from opposite sides unite), and *sea stacks* (formed when the roof of a sea arch collapses).

- Some of the depositional features that form when sediment is moved by beach drift and longshore currents are *spits* (elongated ridges of sand that project from the land into the mouth of an adjacent bay), *baymouth bars* (sandbars that completely cross a bay), and *tombolos* (ridges of sand that connect an island to the mainland or to another island). Along the Atlantic and Gulf Coastal Plains, the shore zone is characterized by *barrier islands,* low ridges of sand that parallel the coast at distances from 3 to 30 kilometers offshore.

- Local factors that influence shoreline erosion are (1) the proximity of a coast to sediment-laden rivers, (2) the degree of tectonic activity, (3) the topography and composition of the land, (4) prevailing winds and weather patterns, and (5) the configuration of the coastline and nearshore areas.

- *Hard stabilization* involves building hard, massive structures in an attempt to protect a coast from erosion or prevent the movement of sand along the beach. Hard stabilization includes *groins* (short walls constructed at a right angle to the shore to trap moving sand), *breakwaters* (structures built parallel to the shore to protect it from the force of large breaking waves), and *seawalls* (armoring the coast to prevent waves from reaching the area behind the wall). *Alternatives to hard stabilization* include *beach nourishment,* which involves the addition of sand to replenish eroding beaches, and *relocation* of damaged or threatened buildings.

- Because of basic geological differences, the *nature of shoreline erosion problems along America's Pacific and Atlantic coasts is very different.* Much of the development along the Atlantic and Gulf coasts has occurred on barrier islands, which receive the full force of major storms. Much of the Pacific Coast is characterized by narrow beaches backed by steep cliffs and mountain ranges. A major problem facing the Pacific shoreline is a narrowing of beaches caused by the natural flow of materials to the coast being interrupted by dams built for irrigation and flood control.

- Although damages caused by a hurricane depend on several factors, including the size and population density of the area affected and the nearshore bottom configuration, the most significant factor is the strength of the storm itself. The *Saffir–Simpson* scale ranks the relative intensities of hurricanes. A category 5 storm is most severe and a category 1 storm is least severe. Damage caused by hurricanes can be divided into three categories: (1) *storm surge,* which is most intense on the right side of the eye where winds are blowing toward the shore, occurs when a dome of water sweeps across the coast near the point where the eye makes landfall; (2) *wind damage;* and (3) *inland flooding,* which is caused by torrential rains that accompany most hurricanes.

- One commonly used classification of coasts is based upon changes that have occurred with respect to sea level. *Emergent coasts,* often with wave-cut cliffs and wave-cut platforms above sea level, develop either because an area experiences uplift or as a result of a drop in sea level. Conversely, *submergent coasts,* with their drowned river mouths, called *estuaries,* are created when sea level rises or the land adjacent to the sea subsides.

- *Tides,* the daily rise and fall in the elevation of the ocean surface at a specific location, are caused by the *gravitational attraction* of the Moon and, to a lesser extent, the Sun. The Moon and the Sun each produce a pair of *tidal bulges* on Earth. These tidal bulges remain in fixed positions relative to the generating bodies as Earth rotates through them, resulting in alternating high and low tides. *Spring tides* occur near the times of new and full moons when the Sun and Moon are aligned and their bulges are added together to produce especially high and low tides (a *large daily tidal range*). Conversely, *neap tides* occur at about the times of the first and third quarters of the Moon when the bulges of the Moon and Sun are at right angles, producing a *smaller daily tidal range*.

- *Three main tidal patterns exist worldwide. A diurnal tidal pattern* exhibits one high and low tide daily; a *semidiurnal tidal pattern* exhibits two high and low tides daily of

about the same height; and a *mixed tidal pattern* usually has two high and low tides daily of different heights.

- *Tidal currents* are horizontal movements of water that accompany the rise and fall of tides. *Tidal flats* are the areas that are affected by the advancing and retreating tidal currents. When tidal currents slow after emerging from narrow inlets, they deposit sediment that may eventually create *tidal deltas*.

Review Questions

1. Distinguish among shore, shoreline, coast, and coastline.

2. What is a beach? Briefly distinguish between beach face and berm. What are the sources of beach sediment?

3. List three factors that determine the height, length, and period of a wave.

4. Describe the motion of a floating object as a wave passes (see Figure 20.4).

5. Describe the physical changes that occur to a wave's speed, wavelength, and height as it moves into shallow water and breaks.

6. Describe two ways in which waves cause erosion.

7. What is wave refraction? What is the effect of this process along irregular coastlines (see Figure 20.9)?

8. Why are beaches often called "rivers of sand"?

9. Describe the formation of the following features: wave-cut cliff, wave-cut platform, marine terrace, sea stack, spit, baymouth bar, and tombolo.

10. List three ways that barrier islands may form.

11. In what direction is beach drift and longshore currents moving sand in Figure 20.18 p. 550? Is it moving toward the top or toward the bottom of the photo?

12. List some examples of hard stabilization and describe what each is intended to do. What effect does each one have on the distribution of sand on the beach?

13. List two alternatives to hard stabilization, indicating potential problems with each one.

14. Relate the damming of rivers to the shrinking of beaches at many locations along the West Coast of the United States. Why do narrower beaches lead to accelerated sea-cliff retreat?

15. Hurricane damage can be divided into three broad categories. Name them. Which category is responsible for the greatest number of hurricane-related deaths?

16. What observable features would lead you to classify a coastal area as emergent?

17. Are estuaries associated with submergent or emergent coasts? Explain.

18. Discuss the origin of ocean tides. Explain why the Sun's influence on Earth's tides is only about half that of the Moon's, even though the Sun is so much more massive than the Moon.

19. Explain why an observer can experience two unequal high tides during one day (see Figure 20.27).

20. How do diurnal, semidiurnal, and mixed tidal patterns differ?

21. Distinguish between flood current and ebb current.

22. How have tides affected Earth's rotation? How did geologists substantiate this idea?

Key Terms

abrasion (p. 543)
backshore (p. 549)
barrier island (p. 546)
baymouth bar (p. 546)
beach (p. 539)
beach drift (p. 544)
beach face (p. 539)
beach nourishment (p. 551)

berm (p. 539)
breakwater (p. 550)
coast (p. 539)
coastline (p. 539)
diurnal tidal pattern
  (p. 561)
ebb current (p. 561)
emergent coast (p. 559)

estuary (p. 559)
fetch (p. 540)
flood current (p. 561)
foreshore (p. 539)
groin (p. 549)
hard stabilization (p. 549)
jetty (p. 549)
longshore current (p. 544)

marine terrace (p. 545)
mixed tidal pattern (p. 561)
neap tide (p. 561)
nearshore (p. 539)
offshore zone (p. 539)
sea arch (p. 546)
sea stack (p. 546)
seawall (p. 550)

Web Resources

The *Earth* Website uses the resources and flexibility of the Internet to aid in your study of the topics in this chapter. Written and developed by geology instructors, this site will help improve your understanding of geology. Visit **http://www.prenhall.com/tarbuck** and click on the cover of *Earth 9e* to find:

- Online review quizzes.
- Critical thinking exercises.
- Links to chapter-specific Web resources.
- Internet-wide key-term searches.

http://www.prenhall.com/tarbuck

GEODe: Earth

GEODe: Earth makes studying faster and more effective by reinforcing key concepts using animation, video, narration, interactive exercises and practice quizzes. A copy is included with every copy of *Earth*.

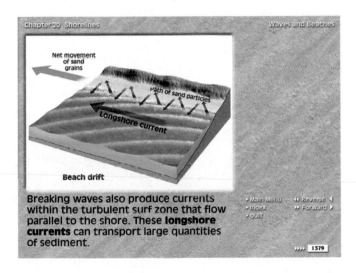

Breaking waves also produce currents within the turbulent surf zone that flow parallel to the shore. These **longshore currents** can transport large quantities of sediment.

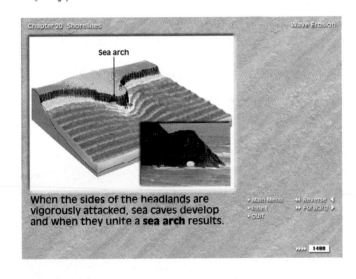

When the sides of the headlands are vigorously attacked, sea caves develop and when they unite a **sea arch** results.

Global Climate Change

Ancient bristlecone
pines in
California's
White Mountains.
The study of
tree-growth rings
is one way that
scientists
reconstruct past
climates. Some
of these trees are
more than 4000
years old. (Photo
by Dennis Flaherty/
Photo Researchers,
Inc.)

Climate, the behavior of Earth's atmosphere over extended time spans, has a profound effect on many geologic processes. Thus, if climate were to change, these geologic processes would be affected as well. A glance back at the rock cycle (Figure 1.23, p. 29) reminds us about many of the connections. Of course, rock weathering has an obvious climate connection, as do the processes that operate in arid and glacial landscapes. Events such as debris flows and floods are often triggered by atmospheric happenings such as periods of heavy rainfall. Clearly the hydrologic cycle has a strong link to the atmosphere (Figure 21.1). Other climate–geology connections involve the impact of internal processes on the atmosphere. For example, the particles and gases emitted by a volcano can change the composition of the atmosphere, and mountain building can have a significant impact on regional temperature, precipitation, and wind patterns.

The geologic record has shown us that climate is variable. The study of sediments, sedimentary rocks, and fossils clearly demonstrates that, through the ages, practically every place on our planet has experienced wide swings in climate, such as from ice ages to conditions associated with subtropical coal swamps or desert dunes. Chapter 22, *"Earth's Evolution through Geologic Time,"* reinforces this fact. Time scales for climate change vary from decades to millions of years.

What factors have been responsible for climate variations during Earth history? One of the topics treated in this chapter involves natural causes of climate change.

Today global climate change is more than just a topic that is of "academic interest" to a group of scientists who are curious about Earth history. Rather, the subject is making headlines. Many members of the general public are not only curious but concerned about the possibilities. Why is climate change newsworthy? The reason is that research focused on human activities and their impact on the environment has demonstrated that people are inadvertently changing the climate. Unlike changes in the past, modern climate change is dominated by human influences that are sufficiently large that they exceed the bounds of natural variability. Moreover, these changes are likely to continue for many centuries. The effects of this venture into the unknown with climate could be very disruptive not only to humans but to many other life forms as well.

How are the details of past climates determined? How is this information useful to us now? In what ways are humans changing global climate? What are the possible consequences?

FIGURE 21.1 The atmosphere is a basic link in the hydrologic cycle. Heavy rains such as those shown here in southern Utah can trigger floods and debris flows. (Photo by Michael Collier.)

The Climate System

Throughout this book you have been reminded that Earth is a multidimensional system that consists of many interacting parts. A change in any one part can produce changes in any or all of the other parts—often in ways that are neither obvious nor immediately apparent. This fact is certainly true when it comes to the study of climate and climate change.

To understand and appreciate climate, it is important to realize that climate involves more than just the atmosphere:

The atmosphere is the central component of the complex, connected, and interactive global environmental system upon which all life depends. Climate may be broadly defined as the long-term behavior of this environmental system. To understand fully and to predict changes in the atmospheric component of the climate system, one must understand the sun, oceans, ice sheets, solid earth, and all forms of life.*

Indeed, we must recognize that there is a **climate system** that includes the atmosphere, hydrosphere, geosphere, biosphere, and cryosphere. (The *cryosphere* refers to the ice and snow that exist at Earth's surface.) The climate system *involves the exchanges of energy and moisture that occur among the five spheres.* These exchanges link the atmosphere to the other spheres so that the whole functions as an extremely complex interactive unit. Changes to the climate system do not occur in isolation. Rather, when one part of it changes, the other components also react. The major components of the climate system are shown in Figure 21.2.

---

*The American Meteorological Society and the University Corporation for Atmospheric Research, "Weather and the Nation's Well-Being," *Bulletin of the American Meteorological Society*, 73, No. 12 (December 2001), p. 2038.

Students Sometimes Ask ...

What's the difference between weather and climate?

The term weather refers to the state of the atmosphere at a given time and place. Changes in the weather are frequent and sometimes seemingly erratic. Climate is a description of aggregate weather conditions based on observations over many decades. Climate is often defined simply as "average weather," but this definition is inadequate because variations and extremes are also part of a climate description.

How Is Climate Change Detected?

High-technology and precision instrumentation are now available to study the composition and dynamics of the atmosphere. But such tools are recent inventions and therefore have been providing data for only a short time span. To understand fully the behavior of the atmosphere and to anticipate future climate change, we must somehow discover how climate has changed over broad expanses of time.

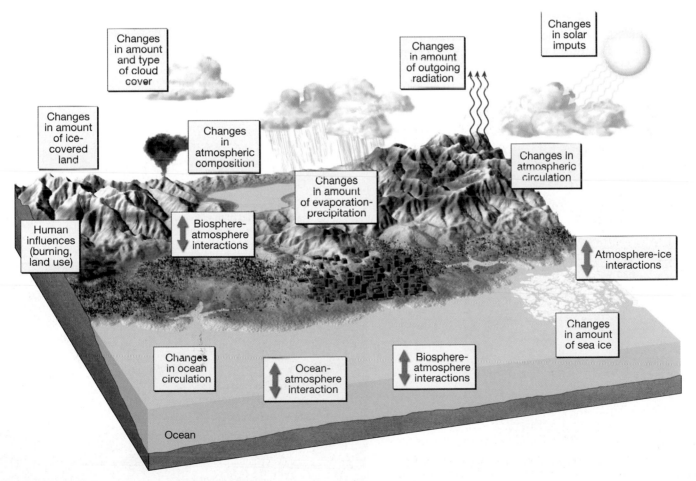

FIGURE 21.2   Schematic view showing several components of Earth's climate system. Many interactions occur among the various components on a wide range of space and time scales, making the system extremely complex.

Instrumental records go back only a couple of centuries at best, and the further back we go, the less complete and more unreliable the data become. To overcome this lack of direct measurements, scientists must decipher and reconstruct past climates by using indirect evidence. Such **proxy data** comes from natural recorders of climate variability, such as seafloor sediments, glacial ice, fossil pollen, and tree-growth rings, as well as from historical documents. Scientists who analyze proxy data and reconstruct past climates are engaged in the study of **paleoclimatology.** The main goal of such work is to understand the climate of the past in order to assess the current and potential future climate in the context of natural climate variability. In the following discussion, we will briefly examine some of the important sources of proxy data.

Seafloor Sediment—A Storehouse of Climate Data

We know that the parts of the Earth system are linked so that a change in one part can produce changes in any or all of the other parts. In this example you will see how changes in atmospheric and oceanic temperatures are reflected in the nature of life in the sea.

Most seafloor sediments contain the remains of organisms that once lived near the sea surface (the ocean–atmosphere interface). When such near-surface organisms die, their shells slowly settle to the floor of the ocean, where they become part of the sedimentary record (Figure 21.3). These seafloor sediments are useful recorders of worldwide climate change because the numbers and types of organisms living near the sea surface change with the climate:

> We would expect that in any area of the ocean/atmosphere interface the average annual temperature of the surface water of the ocean would approximate that of the contiguous atmosphere. The temperature equilibrium established between surface seawater and the air above it should mean that . . . changes in climate should be reflected in changes in organisms living near the surface of the deep sea. . . . When we recall that the seafloor sediments in vast areas of the ocean consist mainly of shells of pelagic foraminifers, and that these animals are sensitive to variations in water temperature, the connection between such sediments and climate change becomes obvious.*

Thus, in seeking to understand climate change as well as other environmental transformations, scientists are tapping the huge reservoir of data in seafloor sediments. The sediment cores gathered by drilling ships and other research vessels have provided invaluable data that have significantly expanded our knowledge and understanding of past climates (Figure 21.4).

One notable example of the importance of seafloor sediments to our understanding of climate change relates to unraveling the fluctuating atmospheric conditions of the Ice Age. The records of temperature changes contained in cores of sediment from the ocean floor have proven critical to our present understanding of this recent span of Earth history.**

---

*Richard F. Flint, *Glacial and Quaternary Geology* (New York: John Wiley & Sons, 1971), p. 718.
**For more information on this topic, see "Causes of Glaciation" in Chapter 18.

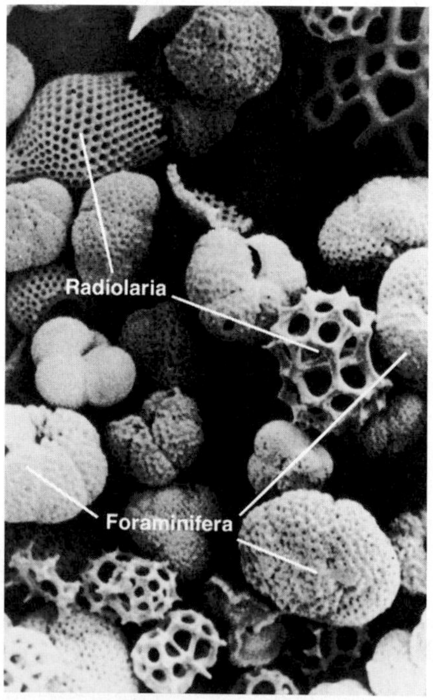

FIGURE 21.3 Microscopic hard parts of radiolarians and foraminifera. This photomicrograph has been enlarged hundreds of times. These organisms are sensitive to even small fluctuations in environmental conditions. (Photo courtesy of Deep Sea Drilling Project, Scripps Institution of Oceanography, University of California, San Diego)

FIGURE 21.4 Scientists examine a sediment core aboard the *JOIDES Resolution*, the drilling ship of the Ocean Drilling Program. The seafloor represents a huge reservoir of data relating to global environmental change. (Photo courtesy of Ocean Drilling Program.)

Oxygen Isotope Analysis

Oxygen isotope analysis is based on precise measurement of the ratio between two isotopes of oxygen: ^{16}O, which is the most common, and the heavier ^{18}O. A molecule of H_2O can form from either ^{16}O or ^{18}O. But the lighter isotope, ^{16}O, evaporates more readily from the oceans. Because of this, precipitation (and hence the glacial ice that it may form) is enriched in ^{16}O. This leaves a greater concentration of the heavier isotope, ^{18}O, in the ocean water. Thus, during periods when glaciers are extensive, more of the lighter ^{16}O is tied up in ice, so the concentration of ^{18}O in seawater increases. Conversely, during warmer interglacial periods when the amount of glacial ice decreases dramatically, more ^{16}O is returned to the sea, so the proportion of ^{18}O relative to ^{16}O in ocean water also drops. Now, if we had some ancient recording of the changes of the $^{18}O/^{16}O$ ratio, we could determine when there were glacial periods and therefore when the climate grew cooler.

Fortunately, we do have such a recording. As certain microorganisms secrete their shells of calcium carbonate ($CaCO_3$), the prevailing $^{18}O/^{16}O$ ratio is reflected in the composition of these hard parts. When the organisms die, their hard parts settle to the ocean floor, becoming part of the sediment layers there. Consequently, periods of glacial activity can be determined from variations in the oxygen isotope ratio found in shells of certain microorganisms buried in deep-sea sediments.

The $^{18}O/^{16}O$ ratio also varies with temperature. Thus, more ^{18}O is evaporated from the oceans when temperatures are high, and less is evaporated when temperatures are low. Therefore, the heavy isotope is more abundant in the precipitation of warm eras and less abundant during colder periods. Using this principle, scientists studying the layers of ice and snow in glaciers have been able to produce a record of past temperature changes.

Climate Change Recorded in Glacial Ice

Ice cores are an indispensable source of data for reconstructing past climates. Research based on vertical cores taken from the Greenland and Antarctic ice sheets has changed our basic understanding of how the climate system works.

Scientists collect samples with a drilling rig, like a small version of an oil drill. A hollow shaft follows the drill head into the ice, and an ice core is extracted. In this way, cores that sometimes exceed 2000 meters (6500 feet) in length and may represent more than 200,000 years of climate history are acquired for study (Figure 21.5A).

The ice provides a detailed record of changing air temperatures and snowfall. Air bubbles trapped in the ice record variations in atmospheric composition. Changes in carbon dioxide and methane are linked to fluctuating temperatures. The cores also include atmospheric fallout such as wind-blown dust, volcanic ash, pollen, and modern-day pollution.

Past temperatures are determined by *oxygen isotope analysis*. Using this technique, scientists are able to produce a record of past temperature changes. A portion of such a record is shown in Figure 21.5B.

Tree Rings—Archives of Environmental History

If you look at the end of a log, you will see that it is composed of a series of concentric rings. Each of these *tree rings* becomes larger in diameter outward from the center (Figure 21.6). Every year in temperate regions trees add a layer of new

A.

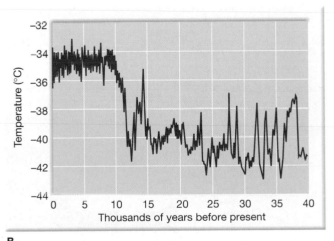

B.

FIGURE 21.5 A. The National Ice Core Laboratory is a physical plant for storing and studying cores of ice taken from glaciers around the world. These cores represent a long-term record of material deposited from the atmosphere. The lab provides scientists the capability to conduct examinations of ice cores, and it preserves the integrity of these samples in a repository for the study of global climate change and past environmental conditions. (Photo by USGS/National Ice Core Laboratory)
B. This graph showing temperature variations over the past 40,000 years is derived from oxygen isotope analysis of ice cores recovered from the Greenland ice sheet. (After U.S. Geological Survey)

FIGURE 21.6 Each year a growing tree produces a layer of new cells beneath the bark. If the tree is felled and the trunk examined (or if a core is taken, to avoid cutting the tree), each year's growth can be seen as a ring. Because the amount of growth (thickness of a ring) depends upon precipitation and temperature, tree rings are useful records of past climates. The dating and study of tree rings is called *dendrochronology*. (Photo by Stephen J. Krasemann/DRK Photo)

wood under the bark. Characteristics of each tree ring, such as size and density, reflect the environmental conditions (especially climate) that prevailed during the year when the ring formed. Favorable growth conditions produce a wide ring; unfavorable ones produce a narrow ring. Trees growing at the same time in the same region show similar tree-ring patterns.

Because a single growth ring is usually added each year, the age of the tree when it was cut can be determined by counting the rings. If the year of cutting is known, the age of the tree and the year in which each ring formed can be determined by counting back from the outside ring. Scientists are not limited to working with trees that have been cut down. Small, nondestructive core samples can be taken from living trees.

To make the most effective use of tree rings, extended patterns known as *ring chronologies* are established. They are produced by comparing the patterns of rings among trees in an area. If the same pattern can be identified in two samples, one of which has been dated, the second sample can be dated from the first by matching the ring patterns common to both. This technique, called *cross dating*, is illustrated in Figure 21.7. Tree-ring chronologies extending back for thousands of years have been established for some regions. To date a timber sample of unknown age, its ring pattern is matched against the reference chronology.

Tree-ring chronologies are unique archives of environmental history and have important applications in such disciplines as climate, geology, ecology, and archaeology. For example, tree rings are used to reconstruct climate variations within a region for spans of thousands of years prior to human historical records. Knowledge of such long-term variations is of great value in making judgments regarding the recent record of climate change.

Other Types of Proxy Data

Other sources of proxy data that are used to gain insight into past climates include fossil pollen, corals, and historical documents.

Fossil Pollen   Climate is a major factor influencing the distribution of vegetation, so knowing the nature of the plant community occupying an area is a reflection of the climate. Pollen and spores are parts of the life cycles of many plants, and because they have very resistant walls, they are often the most abundant, easily identifiable, and best-preserved plant remains in sediments. By analyzing pollen from accurately dated sediments, it is possible to obtain high-resolution records of vegetational changes in an area. From such information, past climates can be reconstructed.

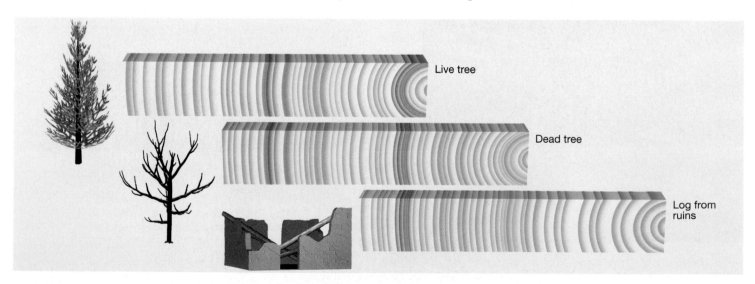

FIGURE 21.7 Cross dating is a basic principle in dendrochronology. Here it was used to date an archaeological site by correlating tree-ring patterns for wood from trees of three different ages. First, a tree-ring chronology for the area is established using cores extracted from living trees. This chronology is extended further back in time by matching overlapping patterns from older, dead trees. Finally, cores taken from beams inside the ruin are dated using the chronology established from the other two sites.

FIGURE 21.8   Coral colonies thrive in warm, shallow tropical waters. The tiny invertebrates extract calcium carbonate from seawater to build hard parts. They live atop the solid foundation left by past corals. Chemical analysis of the changing composition of coral reefs with depth can provide useful data on past near-surface water temperatures. Moreover, because corals are associated with shallow depths, relic reefs can sometimes provide clues to changes in sea level. (Photo courtesy of Jeff Hunter/Photographer's Choice/Getty)

Corals   Coral reefs consist of colonies of corals that live in warm shallow waters and form atop the hard material left behind by past corals (Figure 21.8). Corals build their hard skeletons from calcium carbonate ($CaCO_3$) extracted from seawater. The carbonate contains isotopes of oxygen that can be used to determine the temperature of the water in which the coral grew. Useful information on past climate conditions are determined by analyzing the changing chemistry of coral reefs with depth.

Historical Data   Historical documents sometimes contain helpful information. Although it might seem that such records should readily lend themselves to climate analysis, such is not the case. Most manuscripts were written for purposes other than climate description. Furthermore, writers understandably neglected periods of relatively stable atmospheric conditions and mention only droughts, severe storms, memorable blizzards, and other extremes. Nevertheless, records of crops, floods, and the migration of people have furnished useful evidence of the possible influences of changing climate.

Some Atmospheric Basics

In order to better understand climate change, it is helpful to possess some knowledge about the composition of the atmosphere and the process by which it is heated—the *greenhouse effect.*

Composition of the Atmosphere

Air is *not* a unique element or compound. Rather, air is a *mixture* of many discrete gases, each with its own physical properties, in which varying quantities of tiny solid and liquid particles are suspended.

As you can see in Figure 21.9, clean, dry air is composed almost entirely of two gases—78 percent nitrogen and 21 percent oxygen. Although these gases are the most plentiful components of air and are of great significance to life on Earth, they are of little or no importance in affecting weather phenomena. The remaining 1 percent of dry air is mostly the inert gas argon (0.93 percent) plus tiny quantities of a number of other gases. Carbon dioxide, although present in only minute amounts (0.038 percent), is nevertheless an important constituent of air, because it has the ability to absorb heat energy radiated by Earth and thus influences the heating of the atmosphere.

Air includes many gases and particles that vary significantly from time to time and from place to place. Important examples include water vapor and tiny solid and liquid particles.

The amount of water vapor in the air varies considerably, from practically none at all up to about 4 percent by volume. Why is such a small fraction of the atmosphere so significant? Certainly the fact that water vapor is the source of all clouds and precipitation would be enough to explain its importance. However, water vapor has other roles. Like carbon dioxide, it has the ability to absorb heat energy given off by Earth as well as some solar energy. It is therefore important when we examine the heating of the atmosphere.

The movements of the atmosphere are sufficient to keep a large quantity of solid and liquid particles suspended within it. Although visible dust sometimes clouds the sky, these relatively large particles are too heavy to stay in the air for very long. Still, many particles are microscopic and remain

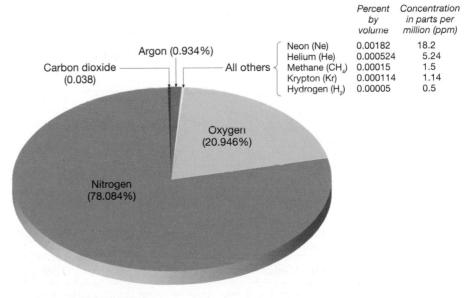

	Percent by volume	Concentration in parts per million (ppm)
Neon (Ne)	0.00182	18.2
Helium (He)	0.000524	5.24
Methane (CH_4)	0.00015	1.5
Krypton (Kr)	0.000114	1.14
Hydrogen (H_2)	0.00005	0.5

Argon (0.934%)
Carbon dioxide (0.038)
All others
Oxygen (20.946%)
Nitrogen (78.084%)

FIGURE 21.9   Proportional volume of gases composing dry air. Nitrogen and oxygen obviously dominate.

---

suspended for considerable periods of time. They may originate from many sources, both natural and human made, and include sea salts from breaking waves, fine soil blown into the air, smoke and soot from fires, pollen and microorganisms lifted by the wind, ash and dust from volcanic eruptions, and more. Collectively, these tiny solid and liquid particles are called **aerosols.**

From a meteorological standpoint, these tiny, often invisible particles can be significant. First, many act as surfaces on which water vapor can condense, an important function in the formation of clouds and fog. Second, aerosols can absorb or reflect incoming solar radiation. Thus, when an air pollution episode is occurring, or when ash fills the sky following a volcanic eruption, the amount of sunlight reaching Earth's surface can be measurably reduced.

Energy from the Sun

Nearly all of the energy that drives Earth's variable weather and climate comes from the Sun.

From our everyday experience, we know that the Sun emits light and heat as well as the ultraviolet rays that cause suntan. Although these forms of energy comprise a major portion of the total energy that radiates from the Sun, they are only part of a large array of energy called *radiation* or *electromagnetic radiation.* This array, or spectrum, of electromagnetic energy is shown in Figure 21.10. All radiation—whether X rays, microwaves, or radio waves—transmits energy through the vacuum of space at 300,000 kilometers (186,000 miles) per second and only slightly slower through our atmosphere. When an object absorbs any form of radiant energy, the result is an increase in molecular motion, which causes a corresponding increase in temperature.

To better understand how the atmosphere is heated, it is useful to have a general understanding of the basic laws governing radiation.

1. *All objects, at whatever temperature, emit radiant energy.* Hence, not only hot objects like the Sun but also Earth, including its polar ice caps, continually emit energy.
2. *Hotter objects radiate more total energy per unit area than do colder objects.*
3. *The hotter the radiating body, the shorter the wavelength of maximum radiation.* The Sun, with a surface temperature of about 5700°C, radiates maximum energy at 0.5 micrometer, which is in the visible range. The maximum radiation for Earth occurs at a wavelength of 10 micrometers, well within the infrared (heat) range. Because the maximum Earth radiation is roughly 20 times longer than the maximum solar radiation, Earth radiation is often called *longwave radiation,* and solar radiation is called *shortwave radiation.*
4. *Objects that are good absorbers of radiation are good emitters as well.* Earth's surface and the Sun approach being perfect radiators because they absorb and radiate with nearly 100 percent efficiency for their respective temperatures. On the other hand, *gases are selective absorbers and emitters of radiation.* For some wavelengths the atmosphere is nearly transparent (little radiation absorbed). For others, however, it is nearly opaque (a good absorber). Experience tells us that the atmosphere is transparent to visible light; hence, these wavelengths readily reach Earth's surface. This is not the case for the longer wavelength radiation emitted by Earth.

The Fate of Incoming Solar Energy

Figure 21.11 shows the fate of incoming solar radiation averaged for the entire globe. Notice that the atmosphere is quite transparent to incoming solar radiation. On average, about 50 percent of the solar energy reaching the top of the atmosphere is absorbed at Earth's surface. Another 30 percent is

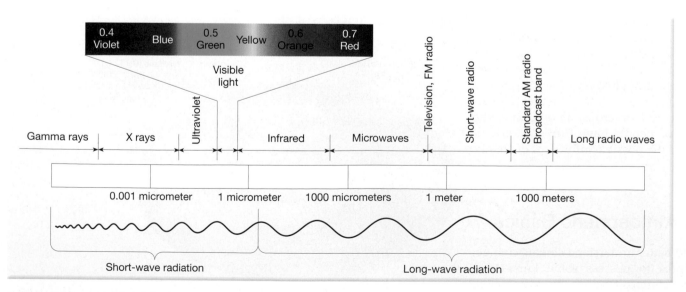

FIGURE 21.10 The electromagnetic spectrum, illustrating the wavelengths and names of various types of radiation.

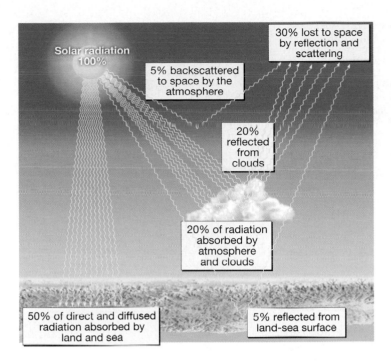

Solar radiation
100%

30% lost to space
by reflection and
scattering

5% backscattered
to space by the
atmosphere

20%
reflected
from
clouds

20% of radiation
absorbed by
atmosphere
and clouds

50% of direct and diffused
radiation absorbed by
land and sea

5% reflected from
land-sea surface

FIGURE 21.11 Average distribution of incoming solar radiation by percentage. More solar energy is absorbed by Earth's surface than by the atmosphere. Consequently, the air is not heated directly by the Sun, but is heated indirectly from Earth's surface. These percentages can vary. For example, if cloud cover increases or the surface is brighter, the percentage of light reflected increases. This situation leaves less solar energy to take the other two paths.

reflected back to space by the atmosphere, clouds, and reflective surfaces such as snow and water. The remaining 20 percent is absorbed directly by clouds and the atmosphere's gases. What determines whether solar radiation will be transmitted to the surface, scattered, or reflected outward? It depends greatly on the wavelength of the energy being transmitted, as well as on the nature of the intervening material.

The numbers shown in Figure 21.11 represent global averages. The actual percentages can vary greatly. An important reason for much of this variation has to do with changes in the percentage of light reflected and scattered back to space. For example, if the sky is overcast, a higher percentage of light is reflected back to space than when the sky is clear.

Heating the Atmosphere:
The Greenhouse Effect

Approximately 50 percent of the solar energy that strikes the top of the atmosphere reaches Earth's surface and is absorbed. Most of this energy is then reradiated skyward. Because Earth has a much lower surface temperature than the Sun, the radiation that it emits has longer wavelengths than solar radiation.

The atmosphere as a whole is an efficient absorber of the longer wavelengths emitted by Earth (*terrestrial radiation*). Water vapor and carbon dioxide are the principal absorbing gases. Water vapor absorbs roughly five times more terres-

trial radiation than do all other gases combined and accounts for the warm temperature found in the lower atmosphere (a layer known as the *troposphere*).* Because the atmosphere is quite transparent to shorter-wavelength solar radiation and more readily absorbs longer-wavelength terrestrial radiation, the atmosphere is heated from the ground up rather than vice versa. This explains the general drop in temperature with increasing altitude experienced in the troposphere. The farther from the "radiator," the colder it becomes.

When the gases in the atmosphere absorb terrestrial radiation, they warm, but they eventually radiate this energy away. Some travels skyward, where it may be reabsorbed by other gas molecules, a possibility that is less likely with increasing height because the concentration of water vapor decreases with altitude. The remainder travels Earthward and is again absorbed by Earth. For this reason, Earth's surface is continually being supplied with heat from the atmosphere as well as from the Sun. Without these absorptive gases in our atmosphere, Earth would not be a suitable habitat for humans and numerous other life forms.

This very important phenomenon has been termed the **greenhouse effect** because it was once thought that greenhouses were heated in a similar manner (Figure 21.12). The gases of our atmosphere, especially water vapor and carbon dioxide, act very much like the glass in the greenhouse. They allow shorter-wavelength solar radiation to enter, where it is absorbed by the objects inside. These objects in turn radiate energy, but at longer wavelengths, to which glass is nearly opaque. The heat therefore is "trapped" in the greenhouse. However, a more important factor in keeping a greenhouse warm is the fact that the greenhouse itself prevents mixing of air inside with cooler air outside. Nevertheless, the term *greenhouse effect* is still used.

Natural Causes of Climate Change

A great variety of hypotheses have been proposed to explain climate change. Several have gained wide support, only to lose it and then sometimes to regain it again. Some explanations are controversial. This is to be expected, because planetary atmospheric processes are so large-scale and complex that they cannot be reproduced physically in laboratory experiments. Rather, climate and its changes must be simulated mathematically (modeled) using powerful computers.

In Chapter 18, the section on "Causes of Glaciation" (pp. 506–510) described two "natural" mechanisms of climate change. Recall that the movement of lithospheric plates gradually moves Earth's continents closer or farther from the equator. Although these shifts in latitude are slow, they can have a dramatic impact on climate over spans of millions of years. Moving landmasses can also lead to significant shifts

---

*The bottom layer of the atmosphere is called the *troposphere*. On average, it is characterized by a decrease in temperature with an increase in height of about 6.5°C per km (3.5°F per 1000 ft). The thickness of this layer is variable, but averages about 12 kilometers (7.4 miles).

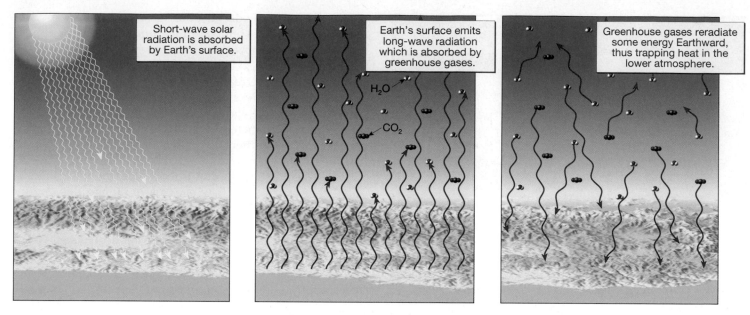

FIGURE 21.12   The heating of the atmosphere. Most of the short-wavelength radiation from the Sun passes through the atmosphere and is absorbed by Earth s land-sea surface. This energy is then emitted from the surface as longer-wavelength radiation, much of which is absorbed by certain gases in the atmosphere. Some of the energy absorbed by the atmosphere will be reradiated Earthward. This so-called greenhouse effect is responsible for keeping Earth's surface much warmer than it would be otherwise.

in ocean circulation, which influences heat transport around the globe.*

A second natural mechanism of climate change discussed in Chapter 18 involved variations in Earth's orbit. Changes in the shape of the orbit (*eccentricity*), variations in the angle that Earth's axis makes with the plane of its orbit (*obliquity*), and the wobbling of the axis (precession) cause fluctuations in the seasonal and latitudinal distribution of solar radiation. These variations, in turn, contributed to the alternating glacial-interglacial episodes of the Ice Age.

In this section, we describe two additional hypotheses that have earned serious consideration from the scientific community. One involves the role of volcanic activity. How do the gases and particles emitted by volcanoes impact climate? A second natural cause of climate change discussed in this section involves solar variability. Does the Sun vary in its radiation output? Do sunspots affect the output?

After this look at natural factors, we will examine human-made climate changes, including the effect of rising carbon dioxide levels caused primarily by our burning of fossil fuels.

As you read this section, you will find that more than one hypothesis may explain the same climatic change. In fact, several mechanisms may interact to shift climate. Also, no single hypothesis can explain climate change on all time scales. A proposal that explains variations over millions of years generally cannot explain fluctuations over hundreds of years. If our atmosphere and its changes ever become fully understood, we will probably see that climate change

*For more on this, see the section titled "Climates and Supercontinents" in chapter 22.

is caused by many of the mechanisms discussed here, plus new ones yet to be proposed.

Volcanic Activity and Climate Change

The idea that explosive volcanic eruptions might alter Earth's climate was first proposed many years ago. It is still regarded as a plausible explanation for some aspects of climatic variability. Explosive eruptions emit huge quantities of gases and fine-grained debris into the atmosphere (Figure 21.13). The greatest eruptions are sufficiently powerful to inject material high into the atmosphere, where it spreads around the globe and remains for many months or even years.

The Basic Premise   The basic premise is that this suspended volcanic material will filter out a portion of the incoming solar radiation, which in turn will lower temperatures in the troposphere. More than 200 years ago Benjamin Franklin used this idea to argue that material from the eruption of a large Icelandic volcano could have reflected sunlight back to space and therefore might have been responsible for the unusually cold winter of 1783–1784.

Perhaps the most notable cool period linked to a volcanic event is the "year without a summer" that followed the 1815 eruption of Mount Tambora in Indonesia. The eruption of Tambora is the largest of modern times. During April 7–12, 1815, this nearly 4000-meter-high (13,000-foot) volcano violently expelled more than 100 cubic kilometers (24 cubic miles) of volcanic debris. The impact of the volcanic aerosols

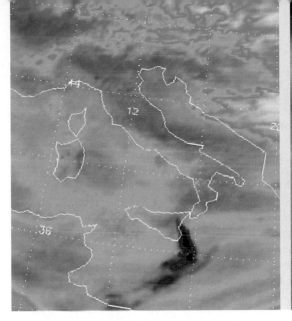

FIGURE 21.13 Mount Etna, a volcano on the island of Sicily, erupting in late October 2002. Mount Etna is Europe's largest and most active volcano. **A.** This image from the Atmospheric Infrared Sounder on NASA's *Aqua* satellite shows the sulfur dioxide (SO_2) plume in shades of purple and black. Climate may be affected when large quantities of SO_2 are injected into the atmosphere. **B.** This photo of Mount Etna looking southeast was taken by a member of the International Space Station. It shows a plume of volcanic ash streaming southeastward from the volcano. (Images courtesy of NASA)

on climate is believed to have been widespread in the Northern Hemisphere. From May through September 1816 an unprecedented series of cold spells affected the northeastern United States and adjacent portions of Canada. There was heavy snow in June and frost in July and August. Abnormal cold was also experienced in much of Western Europe. Similar, although apparently less dramatic, effects were associated with other great explosive volcanoes, including Indonesia's Krakatoa in 1883.

Three major volcanic events have provided considerable data and insight regarding the impact of volcanoes on global temperatures. The eruptions of Washington State's Mount St. Helens in 1980, the Mexican volcano El Chichón in 1982, and the Philippines' Mount Pinatubo in 1991 have given scientists an opportunity to study the atmospheric effects of volcanic eruptions with the aid of more sophisticated technology than had been available in the past. Satellite images and remote-sensing instruments allowed scientists to monitor closely the effects of the clouds of gases and ash that these volcanoes emitted.

Mount St. Helens When Mount St. Helens erupted, there was immediate speculation about the possible effects on our climate. Could such an eruption cause our climate to change? There is no doubt that the large quantity of volcanic ash emitted by the explosive eruption had significant local and regional effects for a short period. Still, studies indicated that any longer-term lowering of hemispheric temperatures was negligible. The cooling was so slight, probably less than 0.1°C (0.2°F), that it could not be distinguished from other natural temperature fluctuations.

El Chichón Two years of monitoring and studies following the 1982 El Chichón eruption indicated that its cooling effect on global mean temperature was greater than that of Mount St. Helens, on the order of 0.3 to 0.5°C (0.5 to 0.9°F). The eruption of El Chichón was *less explosive* than the Mount St. Helens blast, so why did it have a greater impact on global temperatures? The reason is that the material emitted by Mount St. Helens was largely fine ash that settled out in a relatively short time. El Chichón, on the other hand, emitted far greater quantities of sulfur dioxide gas (an estimated 40 times more) than Mount St. Helens. This gas combines with water vapor in the stratosphere to produce a dense cloud of tiny sulfuric-acid particles (Figure 21.14).* The particles, called *aerosols*, take several years to settle out completely. They lower the troposphere's mean temperature because they reflect solar radiation back into space.

We now understand that volcanic clouds that remain in the stratosphere for a year or more are composed largely of sulfuric-acid droplets and not of dust, as was once thought. Thus, the volume of fine debris emitted during an explosive event is not an accurate criterion for predicting the global atmospheric effects of an eruption.

Mount Pinatubo The Philippines volcano, Mount Pinatubo, erupted explosively in June 1991, injecting 25 million to 30 million tons of sulfur dioxide into the stratosphere. The event provided scientists with an opportunity to study the climatic impact of a major explosive volcanic eruption using NASA's spaceborne Earth Radiation Budget Experiment. During the next year the haze of tiny aerosols increased reflectivity and lowered global temperatures by 0.5°C (0.9°F).

It may be true that the impact on global temperature of eruptions like El Chichón and Mount Pinatubo is relatively minor, but many scientists agree that the cooling produced

---

*The stratosphere is the atmospheric layer directly above the troposphere. It extends from a height of about 12 kilometers to a height of about 50 kilometers.

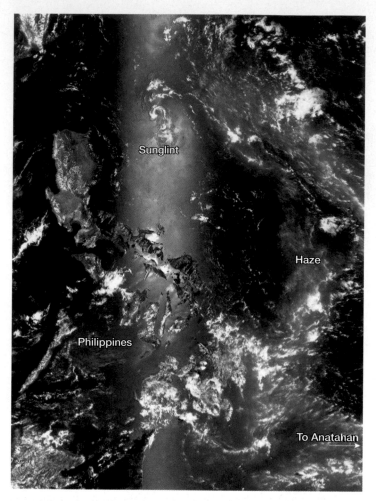

FIGURE 21.14 This satellite image shows a plume of white haze from Anatahan Volcano, blanketing a portion of the Philippine Sea following a large eruption in April 2005. The haze is *not* volcanic ash. Rather, it consists of tiny droplets of sulfuric acid formed when sulfur dioxide from the volcano combined with water in the atmosphere. The haze is bright and reflects sunlight back to space. To the left of the haze, the Sun is reflecting from the smooth surface of the ocean. The effect is a silvery mirror called *sunglint* that extends down a narrow strip of the image where the Sun's angle was just right to reflect light directly into the sensor. (NASA image)

could alter the general pattern of atmospheric circulation for a limited period. Such a change, in turn, could influence the weather in some regions. Predicting or even identifying specific regional effects still presents a considerable challenge to atmospheric scientists.

The preceding examples illustrate that the impact on climate of a single volcanic eruption, no matter how great, is relatively small and short-lived. Therefore, if volcanism is to have a pronounced impact over an extended period, many great eruptions, closely spaced in time, need to occur. If this happens, the stratosphere would be loaded with enough gases and volcanic dust to seriously diminish the amount of solar radiation reaching the surface. Because no such period of explosive volcanism is known to have occurred in historic times, it is most often mentioned as a possible contributor to prehistoric climatic shifts. Another way in which volcanism may influence climate is described in Box 21.1.

Students Sometimes Ask . . .

Could a meteorite colliding with Earth cause the climate to change?

Yes, it is possible. For example, the most strongly supported hypothesis for the extinction of dinosaurs (about 65 million years ago) is related to such an event. When a large (about 10 kilometers in diameter) meteorite struck Earth, huge quantities of debris were blasted high into the atmosphere. For months the encircling dust cloud greatly restricted the amount of light reaching Earth's surface. Without sufficient sunlight for photosynthesis, delicate food chains collapsed. When the sunlight returned, more than half of the species on Earth, including the dinosaurs and many marine organisms, had become extinct. There is more about this in Chapter 22.

Solar Variability and Climate

Among the most persistent hypotheses of climate change have been those based on the idea that the Sun is a variable star and that its output of energy varies through time. The effect of such changes would seem direct and easily understood: Increases in solar output would cause the atmosphere to warm, and reductions would result in cooling. This notion is appealing because it can be used to explain climate change of any length or intensity. However, no major *long-term* variations in the total intensity of solar radiation have yet been measured outside the atmosphere. Such measurements were not even possible until satellite technology became available. Now that it is possible, we will need many years of records before we begin to sense how variable (or invariable) energy from the Sun really is.

Several proposals for climate change, based on a variable Sun, relate to sunspot cycles. The most conspicuous and best-known features on the surface of the Sun are the dark blemishes called **sunspots** (Figure 21.15). Sunspots are huge magnetic storms that extend from the Sun's surface deep into the interior. Moreover, these spots are associated with the Sun's ejection of huge masses of particles that, on reaching Earth's upper atmosphere, interact with gases there to produce auroral displays such as the Aurora Borealis or Northern Lights in the Northern Hemisphere.

Along with other solar activity, the numbers of sunspots seem to increase and decrease in a regular way, creating a cycle of about 11 years. The graph in Figure 21.16 shows the annual number of sunspots, beginning in the early 1700s. However, this pattern does not always occur. There have been periods when the Sun was essentially free of sunspots. In addition to the well-known 11-year cycle, there is also a 22-year cycle. This longer cycle is based on the fact that the magnetic polarities of sunspot clusters reverse every successive 11 years.

Interest in possible Sun–climate effects has been sustained by an almost continuous effort to find correlations on time scales ranging from days to tens of thousands of years. Two widely debated examples are briefly described here.

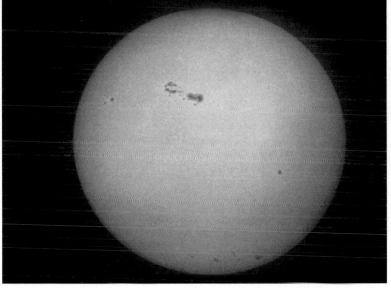

A.

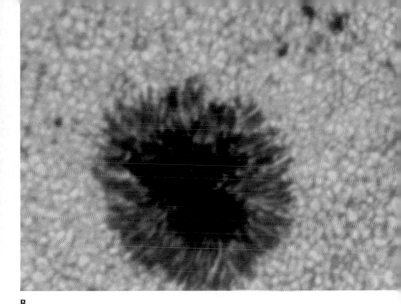

B.

FIGURE 21.15 **A.** Large sunspot group on the solar disk. (Celestron 8 photo courtesy of Celestron International) **B.** Sunspots having visible umbra (dark central area) and penumbra (lighter area surrounding umbra). (Courtesy of National Optical Astronomy Observatories)

A Possible Link Between Volcanism and Climate Change in the Geologic Past

The Cretaceous Period is the last period of the Mesozoic Era, the era of *middle life* that is often called the "age of dinosaurs." It began about 145.5 million years ago and ended about 65.5 million years ago with the extinction of the dinosaurs (and many other life forms as well).*

The Cretaceous climate was among the warmest in Earth's long history. Dinosaurs, which are associated with mild temperatures, ranged north of the Arctic Circle. Tropical forests existed in Greenland and Antarctica, and coral reefs grew as much as 15 degrees latitude closer to the poles than at present. Deposits of peat that would eventually form widespread coal beds accumulated at high latitudes. Sea level was as much as 200 meters (650 feet) higher than today, indicating that there were no polar ice sheets.

What was the cause of the unusually warm climates of the Cretaceous Period? Among the significant factors that may have contributed was an enhancement of the greenhouse effect due to an increase in the amount of carbon dioxide in the atmosphere.

Where did the additional CO_2 come from that contributed to the Cretaceous warming? Many geologists suggest that the probable source was volcanic activity. Carbon dioxide is one of the gases emitted during volcanism, and there is now considerable geologic evidence that the Middle Cretaceous was a time when there was an unusually high rate of volcanic activity.

Several huge oceanic lava plateaus were produced on the floor of the western Pacific during this span. These vast features were associated with hot spots that may have been the product of large mantle plumes (see Figure 5.41). Massive outpourings of lava over millions of years would have been accompanied by the release of huge quantities of CO_2, which in turn would have enhanced the atmospheric greenhouse effect. *Thus, the warmth that characterized the Cretaceous may have had its origins deep in Earth's mantle.*

There were other probable consequences of this extraordinarily warm period that are linked to volcanic activity. For example, the high global temperatures and enriched atmospheric CO_2 in the Cretaceous led to increases in the quantity and types of phytoplankton (tiny, mostly microscopic plants such as algae) and other life forms in the ocean. This expansion in marine life is reflected in the widespread chalk deposits associated with the Cretaceous Period (Figure 21.A). Chalk consists of the calcite-rich hard parts of microscopic marine organisms. Oil and gas originate from the alteration of biological remains (chiefly phytoplankton). Some of the world's most important oil and gas fields occur in marine sediments of the Cretaceous age, a consequence of the greater abundance of marine life during this warm time.

This list of possible consequences linked to the extraordinary period of volcanism during the Cretaceous Period is far from complete, yet it serves to illustrate the interrelationships among parts of the Earth system. Materials and processes that at first might seem to be completely unrelated turn out to be linked. Here you have seen how processes originating deep in Earth's interior are connected directly or indirectly to the atmosphere, the oceans, and the biosphere.

*For more on the end of the Cretaceous, see Chapter 22.

FIGURE 21.A The famous chalk deposits, known as the White Cliffs of Dover, are associated with the expansion of marine life that occurred during the exceptional warmth of the Cretaceous Period. (Photo by Laguna Photo/Getty Images, Inc.—Liaison)

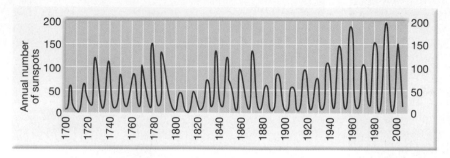

FIGURE 21.16 Mean annual sunspot numbers.

Sunspots and Temperature Studies indicate prolonged periods when sunspots have been absent or nearly so. Moreover, these events correspond closely with cold periods in Europe and North America. Conversely, periods characterized by plentiful sunspots have correlated well with warmer times in these regions.

Referring to these matches, some scientists have suggested that such correlations make it appear that changes on the Sun are an important cause of climate change. But other scientists seriously question this notion. Their hesitation stems in part from subsequent investigations using different climate records from around the world that failed to find a significant correlation between solar activity and climate. Even more troubling is that no testable physical mechanism exists to explain the purported effect.

Sunspots and Drought A second possible Sun–climate connection, on a time scale different from the preceding example, relates to variations in precipitation rather than temperature. An extensive study of tree rings revealed a recurrent period of about 22 years in the pattern of droughts in the western United States. This periodicity coincides with the 22-year magnetic cycle of the Sun mentioned earlier.

Commenting on this possible connection, a panel of the National Research Council pointed out:

> No convincing mechanism that might connect so subtle a feature of the sun to drought patterns in limited regions has yet appeared. Moreover, the cyclic pattern of droughts found in tree rings is itself a subtle feature that shifts from place to place within the broad region of the study.*

Possible connections between solar variability and climate would be much easier to determine if researchers could identify physical linkages between the Sun and the lower atmosphere. But despite much research, no connection between solar variations and weather has yet been well established. Apparent correlations have almost always faltered when put to critical statistical examination or when tested with different data sets. As a result, the subject has been characterized by ongoing controversy and debate.

Solar Variability, Weather, and Climate (Washington, D.C.: National Academy Press, 1982), p. 7.

Human Impact on Global Climate

So far we have examined four potential causes of climate change, each of them natural. In this section we discuss how humans may contribute to global climate change (see Box 21.2). One impact largely results from the addition of carbon dioxide and other greenhouse gases to the atmosphere. A second impact is related to the addition of human-generated aerosols to the atmosphere.

Many people assume that human influence on regional and global climate began with the onset of the modern industrial period, but this probably is not so. There is good evidence that people have been modifying the environment over extensive areas for thousands of years. The use of fire and the overgrazing of marginal lands by domesticated animals have both reduced the abundance and distribution of vegetation. By altering ground cover, humans have modified such important climatological factors as surface reflectivity, evaporation rates, and surface winds. Commenting on this aspect of human-induced climate modification, the late astronomer Carl Sagan noted: "In contrast to the prevailing view that only modern humans are able to alter climate, we believe it is more likely that the human species has made a substantial and continuing impact on climate since the invention of fire."*

Recently, Carl Sagan's ideas were reinforced and expanded upon by a study that used data collected from Antarctic ice cores. This research suggested that humans may have started to have a significant impact on atmospheric composition and global temperatures thousands of years ago.

> Humans started slowly ratcheting up the thermostat as early as 8000 years ago, when they began clearing forests for agriculture, and 5000 years ago with the arrival of wet-rice cultivation. The greenhouse gases carbon dioxide and methane given off by these changes would have warmed the world. . . .**

Carbon Dioxide, Trace Gases, and Climate Change

Earlier you learned that carbon dioxide (CO_2) represents only about 0.038 percent of the gases that make up clean, dry air. Nevertheless, it is a very significant component meteorologically. Carbon dioxide is influential because it is transparent to incoming short-wavelength solar radiation, but it is not transparent to some of the longer-wavelength outgoing Earth radiation. A portion of the energy leaving

---

*Carl Sagan et al., "Anthropogenic Albedo Changes and the Earth's Climate," *Science*, 206, No. 4425 (1980), p. 367.
**An Early Start for Greenhouse Warming?", *Science*, Vol. 303, 16 January 2004. This article is a report on a paper given by paleoclimatologist William Ruddiman at a meeting of the American Geophysical Union in December 2003.

BOX 21.2 ▶ UNDERSTANDING EARTH

Computer Models of Climate: Important Yet Imperfect Tools

Earth's climate system is amazingly complex. Comprehensive state-of-the-science climate simulation models are among the basic tools used to develop possible climate-change scenarios. Called *General Circulation Models*, or *GCMs*, they are based on fundamental laws of physics and chemistry and incorporate human and biological interactions. GCMs are used to simulate many variables, including temperature, rainfall, snow cover, soil moisture, winds, clouds, sea ice, and ocean circulation over the entire globe through the seasons and over spans of decades.

In many other fields of study, hypotheses can be tested by direct experimentation in the laboratory or by observations and measurements in the field. However, this is often not possible in the study of climate. Rather, scientists must construct computer models of how our planet's climate system works. If we understand the climate system correctly and construct the model appropriately, then the behavior of the model climate system should mimic the behavior of Earth's climate system.

What factors influence the accuracy of climate models? Clearly, mathematical models are *simplified* versions of the real Earth and cannot capture its full complexity, especially at smaller geographic scales. Moreover, when computer models are used to simulate future climate change, many assumptions have to be made that significantly influence the outcome. They must consider a wide range of possibilities for future changes in population, economic growth, consumption of fossil fuels, technological development, improvements in energy efficiency, and more.

Despite many obstacles, our ability to use supercomputers to simulate climate continues to improve. Although today's models are far from infallible, they are powerful tools for understanding what Earth's future climate might be like.

---

the ground is absorbed by atmospheric CO_2. This energy is subsequently reemitted, part of it back toward the surface, thereby keeping the air near the ground warmer than it would be without CO_2.

Thus, along with water vapor, carbon dioxide is largely responsible for the *greenhouse effect* of the atmosphere. Carbon dioxide is an important heat absorber, and it follows logically that any change in the air's CO_2 content could alter temperatures in the lower atmosphere.

CO_2 Levels Are Rising

Earth's tremendous industrialization of the past two centuries has been fueled—and still is fueled—by burning fossil fuels: coal, natural gas, and petroleum (see Figure 23.4, p. 630). Combustion of these fuels has added great quantities of carbon dioxide to the atmosphere.

The use of coal and other fuels is the most prominent means by which humans add CO_2 to the atmosphere, but it is not the only way. The clearing of forests also contributes substantially because CO_2 is released as vegetation is burned or decays. Deforestation is particularly pronounced in the tropics, where vast tracts are cleared for ranching and agriculture or subjected to inefficient commercial logging operations. According to U.N. estimates, the destruction of tropical forests exceeded 15 million hectares (38 million acres) per year during the 1990s.

Although some of the excess CO_2 is taken up by plants or is dissolved in the ocean, it is estimated that 45 to 50 percent remains in the atmosphere. Figure 21.17A shows CO_2

concentrations over the past thousand years based on ice-core records and (since 1958) measurements taken at Mauna Loa Observatory, Hawaii. The rapid increase in the

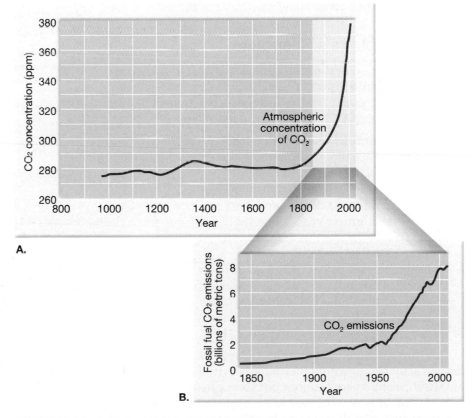

A.

B.

FIGURE 21.17 **A.** Carbon dioxide (CO_2) concentrations over the past 1000 years. Most of the record is based on data obtained from Antarctic ice cores. Bubbles of air trapped in the glacial ice provide samples of past atmospheres. The record since 1958 comes from direct measurements of atmospheric CO_2 taken at Mauna Loa Observatory, Hawaii. **B.** Fossil-fuel CO_2 emissions. The rapid increase in CO_2 concentration since the onset of industrialization has followed closely the rise in CO_2 emissions from fossil fuels.

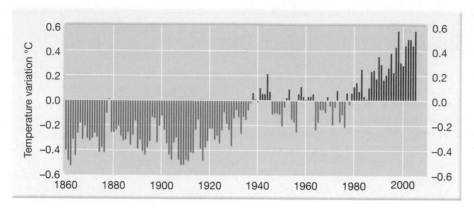

FIGURE 21.18 Annual average global temperature variations for the period 1860–2005. The basis for comparison is the average for the 1961–90 period (the 0.0 line on the graph). Each narrow bar on the graph represents the departure of the global mean temperature from the 1961–90 average for one year. For example, the global mean temperature for 1862 was more than 0.5°C (°F) *below* the 1961–90 average, whereas the global mean for 1998 was more than 0.5°C above. (Specifically, 1998 was 0.56°C warmer.) The bar graph clearly indicates that there can be *significant variations from year to year.* But the graph also shows a trend. Estimated global mean temperatures have been above the 1961–90 average every year since 1978. Globally the 1990s was the warmest decade—and the years 1998, 2005, 2002, 2003, 2001, and 2004 the warmest years—since 1861. (Modified and updated after G. Bell, et al. "Climate Assessment for 1998, *Bulletin of the American Meteorological Society,* Vol. 80, No. 5, May 1999, p. 54.)

CO_2 concentration since the onset of industrialization is obvious and has closely followed the increase in CO_2 emissions from burning fossil fuels (Figure 21.17B).

The Atmosphere's Response

Given the increase in the atmosphere's carbon dioxide content, have global temperatures actually increased? The answer is yes. A report by the Intergovernmental Panel on Climate Change (IPCC)* indicates the following:

- During the 20th century, the global average surface temperature increased by about 0.6°C (1°F).
- Globally it is very likely that the 1990s was the warmest decade and 1998 and 2005 the warmest years since 1861. The next four warmest years all occurred after 2000 (Figure 21.18).
- New analyses of data for the Northern Hemisphere indicate that the increase in temperature in the 20th century is likely to have been the largest in any century during the past 1000 years (Figure 21.19).

Are these temperature trends caused by human activities, or would they have occurred anyway? Scientists are cautious but seem convinced that human activities have played a significant role. An IPCC report in 1996 stated that "the balance of evidence suggests a discernible human influence on global climate.* Five years later the IPCC stated that "there is new and stronger evidence that most of the warming observed over the last 50 years is attributable to human activities."** What about the future? By the year 2100, models project atmospheric CO_2 concentrations of 540 to 970 ppm. With such an increase, how will global temperatures change? Here is some of what the 2001 IPCC report has to say:***

- The globally averaged surface temperature is projected to increase by 1.4 to 5.8°C by the year 2100.
- The projected rate of warming is much larger than the observed changes during the 20th century and is very likely to be without precedent during at least the last 10,000 years.
- It is very likely that nearly all land areas will warm more rapidly than the global average, particularly those at northern high latitudes in the cold season.

The Role of Trace Gases

Carbon dioxide is not the only gas contributing to a possible global increase in temperature. In recent years atmospheric scientists have come to realize that the industrial and

*Intergovernmental Panel on Climate Change, *Climate Change 1995: The Science of Climate Change,* New York: Cambridge University Press, 1996, p. 4.

**IPCC, *Climate Change 2001: The Scientific Basis,* p. 10.

***IPCC, *Climate Change 2001: The Scientific Basis,* p. 13.

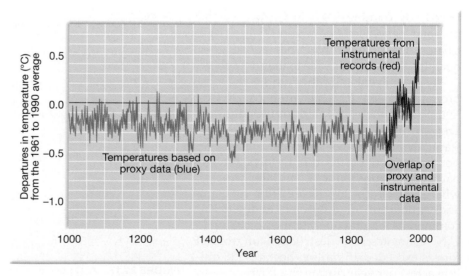

FIGURE 21.19 Year-by-year variations in average surface temperatures for the Northern Hemisphere for the past 1000 years reconstructed from tree rings, ice cores, corals, and historical records (blue portion of the line) and air temperatures directly measured (red portion of line). The warming during the 20 century was much greater than in any of the previous nine centuries. (After U.S. Global Change Research Program and IPCC)

*Intergovernmental Panel on Climate Change, *Climate Change 2001: The Scientific Basis.* Cambridge, UK: Cambridge University Press, 2001, p. 2.

Students Sometimes Ask ...

What is the Intergovernmental Panel on Climate Change?

Recognizing the problem of potential global climate change, the World Meteorological Organization and the United Nation's Environment Program established the *Intergovernmental Panel on Climate Change* (IPCC, for short). The IPCC assesses the scientific, technical, and socioeconomic information that is relevant to an understanding of human-induced climate change. This authoritative group provides advice to the world community through periodic reports that assess the state of knowledge of causes of climate change.

agricultural activities of people are causing a buildup of several trace gases that may also play a significant role. The substances are called *trace gases* because their concentrations are so much smaller than that of carbon dioxide. The trace gases that appear to be most important are methane (CH_4), nitrous oxide (N_2O), and chlorofluorocarbons (CFCs). These gases absorb wavelengths of outgoing radiation from Earth that would otherwise escape into space. Although individually their impact is modest, taken together the effects of these trace gases may be as great as CO_2 in warming the troposphere.

Methane is present in much smaller amounts than CO_2, but its significance is greater than its relatively small concentration of about 1.7 ppm (parts per million) would indicate. The reason is that methane is 20 to 30 times more effective than CO_2 at absorbing infrared radiation emitted by Earth.

Methane is produced by *anaerobic* bacteria in wet places where oxygen is scarce (anaerobic means "without air," specifically oxygen). Such places include swamps, bogs, wetlands, and the guts of termites and grazing animals like cattle and sheep. Methane is also generated in flooded paddy fields ("artificial swamps") used for growing rice. Mining of coal and drilling for oil and natural gas are other sources because methane is a product of their formation (Figure 21.20).

The concentration of methane in the atmosphere is believed to have about doubled since 1800, an increase that has been in step with the growth in human population. This relationship reflects the close link between methane formation and agriculture. As population has risen, so have the number of cattle and rice paddies.

Nitrous oxide, sometimes called "laughing gas," is also building in the atmosphere, although not as rapidly as methane. The increase is believed to result primarily from agricultural activity. When farmers use nitrogen fertilizers to boost crop yield, some of the nitrogen enters the air as nitrous oxide. This gas is also produced by high-temperature combustion of fossil fuels. Although the annual release into the atmosphere is small, the lifetime of a nitrous oxide molecule is about 150 years! If the use of nitrogen fertilizers and fossil fuels grows at projected rates, nitrous oxide may make

a contribution to greenhouse warming that approaches half that of methane.

Unlike methane and nitrous oxide, chlorofluorocarbons (CFCs) are not naturally present in the atmosphere. CFCs are manufactured chemicals with many uses that have gained notoriety because they are responsible for ozone depletion in the stratosphere. The role of CFCs in global warming is less well known. CFCs are very effective greenhouse gases. They were not developed until the 1920s and were not used in great quantities until the 1950s, but they already contribute to the greenhouse effect at a level equal to methane. Although corrective action has been taken, CFC levels will *not* drop rapidly. CFCs remain in the atmosphere for decades, so even if all CFC emissions were to stop immediately, the atmosphere would not be free of them for many years.

Carbon dioxide is clearly the most important single cause for the projected global greenhouse warming. However, it is not the only contributor. When the effects of all human-generated greenhouse gases other than CO_2 are added together and projected into the future, their collective impact significantly increases the impact of CO_2 alone.

Sophisticated computer models show that the warming of the lower atmosphere caused by CO_2 and trace gases will not be the same everywhere. Rather, the temperature response in polar regions could be two to three times greater than the global average. One reason is that the polar troposphere is very stable, which suppresses vertical mixing and thus limits the amount of surface heat that is transferred

FIGURE 21.20 Methane is produced by anaerobic bacteria in wet places, where oxygen is scarce (anaerobic means "without air," specifically oxygen). Such places include swamps, bogs, wetlands, and the guts of termites and grazing animals, like cattle and sheep. Methane is also generated in flooded paddy fields ("artificial swamps") used for growing rice. These paddies are in India's Ganges lowlands. Mining of coal and drilling for oil and natural gas are other sources because methane is a product of their formation. (Photo by George Holton/Photo Researchers, Inc.)

Students Sometimes Ask ...

If Earth's atmosphere had no greenhouse gases, what would surface-air temperatures be like?

Cold! Earth's average surface temperature would be a chilly −18°C (−0.4°F) instead of the relatively comfortable 14.5°C (58°F) that it is today.

upward. In addition, an expected reduction in sea ice would also contribute to the greater temperature increase. This topic will be explored more fully in the next section.

Climate-Feedback Mechanisms

Climate is a very complex interactive physical system. Thus, when any component of the climate system is altered, scientists must consider many possible outcomes. These possible outcomes are called **climate-feedback mechanisms.** They complicate climate-modeling efforts and add greater uncertainty to climate predictions.

What climate-feedback mechanisms are related to carbon dioxide and other greenhouse gases? One important mechanism is that warmer surface temperatures increase evaporation rates. This in turn increases the water vapor in the atmosphere. Remember that water vapor is an even more powerful absorber of radiation emitted by Earth than is carbon dioxide. Therefore, with more water vapor in the air, the temperature increase caused by carbon dioxide and the trace gases is reinforced.

Recall that the temperature increase at high latitudes may be two to three times greater than the global average. This assumption is based in part on the likelihood that the area covered by sea ice will decrease as surface temperatures rise. Because ice reflects a much larger percentage of incoming solar radiation than does open water, the melting of the sea ice would replace a highly reflecting surface with a relatively dark surface (Figure 21.21). The result would be a substantial increase in the solar energy absorbed at the surface. This in turn would feed back to the atmosphere and magnify the initial temperature increase created by higher levels of greenhouse gases.

So far the climate-feedback mechanisms discussed have magnified the temperature rise caused by the buildup of carbon dioxide. Because these effects reinforce the initial change, they are called **positive-feedback mechanisms.** However, other effects must be classified as **negative-feedback mechanisms** because they produce results that are just the opposite of the initial change and tend to offset it.

One probable result of a global temperature rise would be an accompanying increase in cloud cover due to the higher moisture content of the atmosphere. Most clouds are good reflectors of solar radiation. At the same time, however, they are also good absorbers and emitters of radiation emitted by Earth. Consequently, clouds produce two opposite effects. They are a negative-feedback mechanism because they increase the reflection of solar radiation and thus diminish the amount of solar energy available to heat the atmosphere. On the other hand, clouds act as a positive-feedback mechanism by absorbing and emitting radiation that would otherwise be lost from the troposphere.

Which effect, if either, is stronger? Atmospheric modeling shows that the negative effect of a higher reflectivity is dominant. Therefore, the net result of an increase in cloudiness

FIGURE 21.21 This satellite image shows the springtime breakup of sea ice near Antarctica. The inset shows a likely feedback loop. A reduction in sea ice acts as a positive-feedback mechanism because surface reflectivity would decrease and the amount of energy absorbed at the surface would increase. (Photo by George Holton/Photo Researchers Inc.)

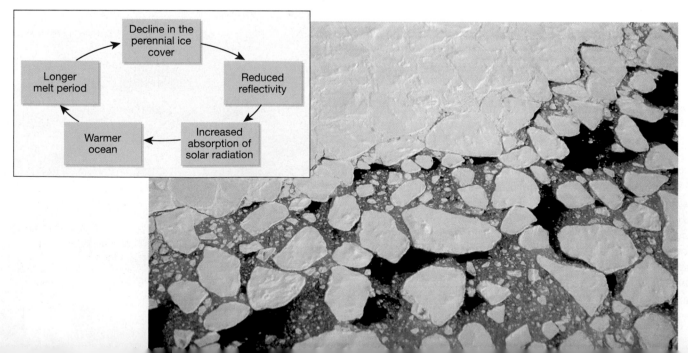

should be a decrease in air temperature. The magnitude of this negative feedback, however, is not believed to be as great as the positive feedback caused by added moisture and decreased sea ice. Thus, although increases in cloud cover may partly offset a global temperature increase, climate models show that the ultimate effect of the projected increase in CO_2 and trace gases will still be a temperature increase.

The problem of global warming caused by human-induced changes in atmospheric composition continues to be one of the most studied aspects of climate change. Although no models yet incorporate the full range of potential factors and feedbacks, the scientific consensus is that the increasing levels of atmospheric carbon dioxide and trace gases will lead to a warmer planet with a different distribution of climate regimes.

How Aerosols Influence Climate

Increasing the levels of carbon dioxide and other greenhouse gases in the atmosphere is the most direct human influence on global climate. But it is not the only impact. Global climate is also affected by human activities that contribute to the atmosphere's aerosol content. Recall that *aerosols* are the tiny, often microscopic, liquid and solid particles that are suspended in the air. Unlike cloud droplets, aerosols are present even in relatively dry air. Atmospheric aerosols are composed of many different materials, including soil, smoke, sea salt, and sulfuric acid. Natural sources are numerous and include such phenomena as dust storms and volcanoes.

Presently the human contribution of aerosols to the atmosphere *equals* the quantity emitted by natural sources. Most human-generated aerosols come from the sulfur dioxide emitted during the combustion of fossil fuels and as a consequence of burning vegetation to clear agricultural land. Chemical reactions in the atmosphere convert the sulfur dioxide into sulfate aerosols, the same material that produces acid precipitation.

How do aerosols affect climate? Aerosols act directly by reflecting sunlight back to space and indirectly by making clouds "brighter" reflectors. The second effect relates to the fact that many aerosols (such as those composed of salt or sulfuric acid) attract water and thus are especially effective as cloud condensation nuclei. The large quantity of aerosols produced by human activities (especially industrial emissions) trigger an increase in the number of cloud droplets that form within a cloud. A greater number of small droplets increases the cloud's brightness—that is, more sunlight is reflected back to space.

By reducing the amount of solar energy available to the climate system, aerosols have a net cooling effect. Studies indicate that the cooling effect of human-generated aerosols could offset a portion of the global warming caused by the growing quantities of greenhouse gases in the atmosphere. Unfortunately, the magnitude and extent of the cooling effect of aerosols is highly uncertain. This uncertainty is a significant hurdle in advancing our understanding of how humans alter Earth's climate.

It is important to point out some significant differences between global warming by greenhouse gases and aerosol cooling. After being emitted, greenhouse gases such as carbon dioxide remain in the atmosphere for many decades. By contrast, aerosols released into the troposphere remain there for only a few days or, at most, a few weeks before they are "washed out" by precipitation. Because of their short lifetime in the troposphere, aerosols are distributed unevenly over the globe. As expected, human-generated aerosols are concentrated near the areas that produce them, namely industrialized regions that burn fossil fuels and land areas where vegetation is burned (Figure 21.22).

Because their lifetime in the atmosphere is short, the effect of aerosols on today's climate is determined by the amount emitted during the preceding couple of weeks. By contrast, the carbon dioxide and trace gases released into the atmosphere remain for much longer spans and thus influence climate for many decades.

Some Possible Consequences of Global Warming

What consequences can be expected if the carbon dioxide content of the atmosphere reaches a level that is twice what it was early in the 20th century? Because the climate system is so complex, predicting the distribution of particular regional changes is very speculative. It is not yet possible to pinpoint specifics, such as where or when it will become drier or wetter. Nevertheless, plausible scenarios can be given for larger scales of space and time.

As noted, the magnitude of the temperature increase will not be the same everywhere. The temperature rise will probably be smallest in the tropics and increase toward the poles. As for precipitation, the models indicate that some

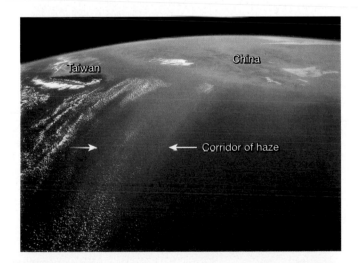

FIGURE 21.22 Human-generated aerosols are concentrated near the areas that produce them. Because aerosols reduce the amount of solar energy available to the climate system, they have a net cooling effect. This satellite image shows a dense blanket of pollution moving away from the coast of China. The plume is about 200 kilometers wide and more than 600 kilometers long. (NASA Image)

regions will experience significantly more precipitation and runoff. However, others will experience a decrease in run-off (due to reduced precipitation or greater evaporation caused by higher temperatures).

Table 21.1 summarizes some of the more likely effects and their possible consequences. The table also provides the

IPCC's estimate of the probability of each effect. Levels of confidence for these projections vary from "*likely*" (67 to 90 percent probability) to "*very likely*" (90 to 99 percent probability). Box 21.3 looks at key findings regarding the possible consequences of climate change for the United States in the 21st century.

Students Sometimes Ask . . .

What are scenarios, and why are they used?

A scenario is an example of what might happen under a particular set of assumptions. Scenarios are a way of examining questions about an uncertain future. For example, future trends in fossil-fuel use and other human activities are uncertain. Therefore, scientists have developed a set of scenarios for how climate may change based on a wide range of possibilities for these variables.

Water Resources and Agriculture

Such changes could profoundly alter the distribution of the world's water resources and hence affect the productivity of agricultural regions that depend on rivers for irrigation water. For example, a 2°C (3.6°F) warming and 10 percent precipitation decrease in the region drained by the Colorado River could diminish the river's flow by 50 percent or more. Because the present flow of the river barely meets current demand for irrigation agriculture, the negative effect would be serious. Many other rivers are the basis for extensive irrigated agriculture, and the projected reduction of their flow could have equally grave consequences. In contrast, large

TABLE 21.1 Projected Changes and Effects of Global Warming in the 21st Century

Projected changes and estimated probability*	Examples of projected impacts
Higher maximum temperatures; more hot days and heat waves over nearly all land areas (*very likely*).	• Increased incidence of death and serious illness in older age groups and urban poor. • Increased heat stress in livestock and wildlife. • Shift in tourist destinations. • Increased risk of damage to a number of crops. • Increased electric cooling demand and reduced energy-supply reliability.
Higher minimum temperatures; fewer cold days, frost days, and cold waves over nearly all land areas (*very likely*).	• Decreased cold-related human morbidity and mortality. • Decreased risk of damage to a number of crops, and increased risk to others. • Extended range and activity of some pest and disease vectors. • Reduced heating energy demand.
More intense precipitation events (*very likely* over many areas).	• Increased flood, landslide, avalanche, and mudslide damage. • Increased soil erosion. • Increased flood runoff could increase recharge of some floodplain aquifers. • Increased pressure on government and private flood insurance systems and disaster relief.
Increased summer drying over most mid-latitude continental interiors and associated risk of drought (*likely*).	• Decreased crop yields. • Increased damage to building foundations caused by ground shrinkage. • Decreased water-resource quantity and quality. • Increased risk of forest fire.
Increase in tropical cyclone peak wind intensities, mean and peak precipitation intensities (*likely* over some areas).	• Increased risks to human life, risk of infectious-disease epidemics, and many other risks. • Increased coastal erosion and damage to coastal buildings and infrastructure. • Increased damage to coastal ecosystems, such as coral reefs and mangroves.
Intensified droughts and floods associated with El Niño events in many different regions (*likely*).	• Decreased agricultural and rangeland productivity in drought-and flood-prone regions. • Decreased hydropower potential in drought-prone regions.
Increased Asian summer monsoon precipitation variability (*likely*).	• Increased flood and drought magnitude and damages in temperate and tropical Asia.
Increased intensity of mid-latitude storms (*uncertain*).	• Increased risks to human life and health. • Increased property and infrastructure losses. • Increased damage to coastal ecosystems.

Very likely indicates a probability of 90–99 percent. *Likely* indicates a probability of 67–90 percent.
Source: IPCC, 2001.

BOX 21.3 ▶ PEOPLE AND THE ENVIRONMENT

Possible Consequences of Climate Change on the United States

What might people living in the United States expect as a consequence of climate change during the 21st century? This box summarizes the key findings from *Climate Change Impacts on the United States.** It is a report prepared by the National Assessment Synthesis Team (NAST) as part of the United States Global Change Research Program (USGCRP). Its purpose was to "synthesize, evaluate, and report on what we presently know about the potential consequences of climate variability and change for the U.S. in the 21st century." The report's key findings are as follows:

1. *Increased warming.* Assuming continued growth in world greenhouse-gas emissions, the primary climate models used in this assessment project that temperatures in the United States will rise 5–9°F (3–5°C) on average in the next 100 years. A wider range of outcomes is possible.
2. *Differing regional impacts.* Climate change will vary widely across the United States. Temperature increases will vary somewhat from one region to the next. Heavy and extreme precipitation events are likely to become more frequent, yet some regions will get drier. The potential impacts of climate change will also vary widely across the nation.
3. *Vulnerable ecosystems.* Many ecosystems are highly vulnerable to the projected rate and magnitude of climate change. A few, such as alpine meadows in the Rocky Mountains and some barrier islands, are likely to disappear entirely in some areas. Others, such as forests of the Southeast, are likely to experience major species shifts or break up into a mosaic of grasslands, woodlands, and forests. The goods and services lost

through the disappearance or fragmentation of certain ecosystems are likely to be costly or impossible to replace.

4. *Widespread water concerns.* Water is an issue in every region, but the nature of the vulnerabilities varies. Drought is an important concern in every region. Floods and water quality are concerns in many regions. Snowpack changes are especially important in the West, Pacific Northwest, and Alaska.
5. *Secure food supply.* At the national level the agriculture sector is likely to be able to adapt to climate change. Overall, U.S. crop productivity is very likely to increase over the next few decades, but the gains will not be uniform across the nation. Falling prices and competitive pressures are very likely to stress some farmers while benefiting consumers.
6. *Near-term increase in forest growth.* Forest productivity is likely to increase over the next several decades in some areas as trees respond to higher carbon dioxide levels. Over the longer term, changes in larger-scale processes, such as fire, insects, droughts, and disease, will possibly decrease forest productivity. In addition, climate change is likely to cause long-term shifts in forest species, such as sugar maples moving north out of the United States.
7. *Increased damage in coastal and permafrost areas.* Climate change and the resulting rise in sea level are likely to exacerbate threats to buildings, roads, powerlines, and other infrastructure in climatically sensitive places. For example, infrastructure damage is related to permafrost melting in Alaska and to sea-level rise and storm surge in low-lying coastal areas.

8. *Adaptation determines health outcomes.* A range of negative health impacts is possible from climate change, but adaptation is likely to help protect much of the U.S. population. Maintaining our nation's public health and community infrastructure, from water-treatment systems to emergency shelters, will be important for minimizing the impacts of water-borne diseases, heat stress, air pollution, extreme weather events, and diseases transmitted by insects, ticks, and rodents.
9. *Other stresses magnified by climate change.* Climate change will very likely magnify the cumulative impacts of other stresses, such as air and water pollution and habitat destruction due to human-development patterns. For some systems, such as coral reefs, the combined effects of climate change and other stresses are very likely to exceed a critical threshold, bringing large, possibly irreversible, impacts.
10. *Uncertainties remain and surprises are expected.* Significant uncertainties remain in the science underlying regional climate changes and their impacts. Further research would improve understanding and our ability to project societal and ecosystem impacts and provide the public with additional useful information about options for adaptation. However, it is likely that some aspects and impacts of climate change will be totally unanticipated as complex systems respond to ongoing climate change in unforeseeable ways.

---

*National Assessment Synthesis Team, *Climate Change Impacts on the United States: The Potential Consequences of Climate Variability and Change*, Washington, DC: US Global Change Research Program, 2000, p. 19.

---

precipitation increases in other areas would increase the flow of some rivers and bring more frequent destructive floods.

Harder to estimate is the effect on nonirrigated crops that depend on direct rainfall and snowfall for moisture. Some places will no doubt experience productivity loss due to a decrease in rainfall or increase in evaporation. Still, these losses may be offset by gains elsewhere. Warming in the higher latitudes could lengthen the growing season, for instance. This in turn could allow expansion of agriculture into areas presently unsuited to crop production.

Sea-Level Rise

Another impact of a human-induced global warming is a probable rise in sea level. How is a warmer atmosphere related to a global rise in sea level? The most obvious connection, the melting of glaciers, is important, but *not* the most significant. Far more significant is that a warmer atmosphere causes an increase in ocean volume due to thermal expansion. Higher air temperatures warm the adjacent upper layers of the ocean, which in turn causes the water to expand and the sea level to rise.

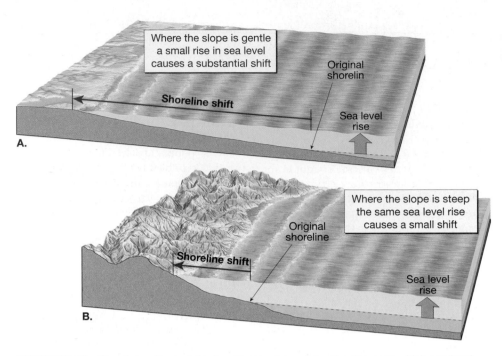

FIGURE 21.23 The slope of a shoreline is critical to determining the degree to which sea-level changes will affect it. **A.** When the slope is gentle, small changes in the sea level cause a substantial shift. **B.** The same sea-level rise along a steep coast results in only a small shoreline shift. **C.** As sea level gradually rises, the shoreline retreats, and structures that were once thought to be safe from wave attack are exposed to the force of the sea. (Photo by Kenneth Hasson)

Research indicates that sea level has risen between 10 and 25 centimeters (4 and 8 inches) over the past century and that the trend will continue at an accelerated rate. Some models indicate that the rise may approach or even exceed 50 centimeters (20 inches) by the end of the 21st century. Such a change may seem modest, but scientists realize that any rise in sea level along a *gently* sloping shoreline, such as the Atlantic and Gulf coasts of the United States, will lead to significant erosion and severe, permanent inland flooding (Figure 21.23). If this happens, many beaches and wetlands will be eliminated and coastal civilization will be severely disrupted.

Because rising sea level is a gradual phenomenon, it may be overlooked by coastal residents as an important contributor to shoreline erosion problems. Rather, the blame may be assigned to other forces, especially storm activity. Although a given storm may be the immediate cause, the magnitude of its destruction may result from the relatively small sea-level rise that allowed the storm's power to cross a much greater land area (Figure 21.23C).

As mentioned, a warmer climate will cause glaciers to melt. In fact, a portion of the 10- to 25-centimeter (4- to 8-inch) rise in sea level over the past century is attributed to the melting of mountain glaciers (Figure 21.24). This contribution is projected to continue through the 21st century. Of course, if the Greenland and Antarctic ice sheets were to experience a significant increase in melting, it would lead to a much greater rise in sea level and a major encroachment by the sea in coastal zones.

The Changing Arctic

A recent study of climate change in the Arctic began with the following statement:

For nearly 30 years, Arctic sea ice extent and thickness have been falling dramatically. Permafrost temperatures are rising and coverage is decreasing. Mountain glaciers and the Greenland ice sheet are shrinking. Evidence suggests we are witnessing the early stage of an anthropogenically induced global warming superimposed on natural cycles, reinforced by reductions in Arctic ice.*

---

*J. T. Overpeck, et al. "Arctic System on Trajectory to New, Seasonally Ice-Free States," *EOS, Transactions, American Geophysical Union,* Vol. 86, No. 34, 23 August 2005, p. 309.

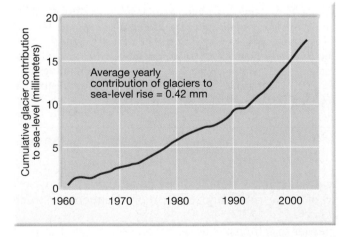

FIGURE 21.24 The contribution of melting glacial ice to the rise in sea level 1961–2003. Mountain glaciers cover an area of about 785,000 square kilometers—about 4 percent of the total land area covered by glacial ice. It is estimated that these glaciers contributed approximately 20 to 30 percent of the total sea-level rise over the past 100 years. Since 1990, the relative contribution of melting glaciers has increased. As much as 40 percent of the overall sea-level rise during that span may have come from melting glacial ice. (National Snow and Ice Data Center)

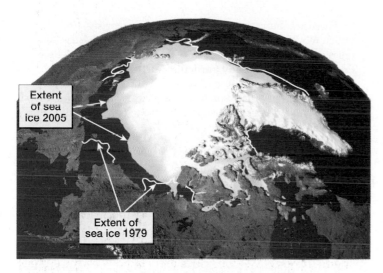

FIGURE 21.25   A comparison of the extent of sea ice at the end of the summer melting period—1979 vs. 2005. The decrease in the area covered by sea ice exceeds 20 percent. (NASA)

Arctic Sea Ice   Climate models are in general agreement that one of the strongest signals of global warming should be a loss of sea ice in the Arctic. This is indeed occurring. The map in Figure 21.25 compares the extent of sea ice in September, at the end of the summer melt period, for the years 1979 and 2005. Over this span, September sea ice extent has decreased more than 20 percent. The trend is also clear when you examine the graph in Figure 21.26. Is it possible that this trend may be part of a natural cycle? Yes, but it is more likely that the sea-ice decline represents a combination of natural variability and human-induced global warming, with the latter being increasingly evident in coming decades. As was noted in the section on "Climate Feedback Mechanisms," a reduction in sea ice represents a positive feedback mechanism that reinforces global warming.

Permafrost   During the past decade, evidence has mounted to indicate that the extent of permafrost in the Northern Hemisphere has decreased, as would be expected under long-term warming conditions. Figure 21.27 presents one example that such a decline is occurring.

In the Arctic, short summers thaw only the top layer of frozen ground. The permafrost beneath this *active layer* is like the cement bottom of a swimming pool. In summer, water cannot percolate downward, so it saturates the soil above the permafrost and collects on the surface in thousands of lakes. However, as Arctic temperatures climb, the bottom of the "pool" seems to be "cracking." Satellite imagery shows that over a 20-year span, a significant number of lakes have shrunk or disappeared altogether. As the permafrost thaws, lake water drains deeper into the ground.

Thawing permafrost represents a potentially significant positive feedback mechanism that may reinforce global warming. When vegetation dies in the Arctic, cold temperatures inhibit its total decomposition. As a consequence, over thousands of years a great deal of organic matter has become stored in the permafrost. When the permafrost thaws,

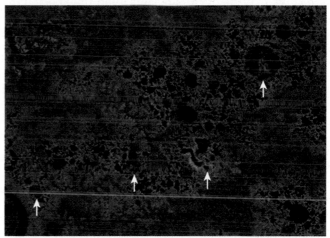

A. June 27, 1973

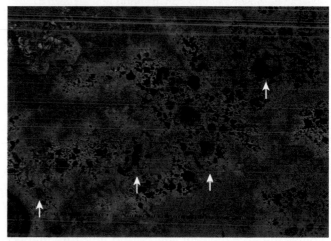

B. July 2, 2002

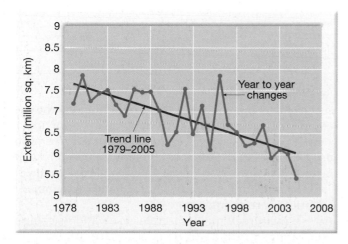

FIGURE 21.26   This graph depicts the decline in Arctic sea ice from 1979 to 2005. The rate of decline exceeds 8 percent per decade. The year 2005 exhibited the smallest area of Arctic sea ice yet measured in the satellite record up to that time. (National Snow and Ice Data Center)

FIGURE 21.27   This image pair shows lakes dotting the tundra in northern Siberia in 1973 and 2002. The tundra vegetation is colored a faded red, whereas lakes appear blue or blue-green. Many lakes have clearly disappeared or shrunk considerably between 1973 and 2002. Compare the areas highlighted by the white arrowheads in each image. After studying satellite imagery of about 10,000 large lakes in a 500,000-square-kilometer area in northern Siberia, scientists documented an 11 percent decline in the number of lakes, with at least 125 disappearing completely.

organic matter that may have been frozen for millennia comes out of "cold storage" and decomposes. The result is the release of carbon dioxide and methane—greenhouse gases the contribute to global warming.

The Potential for "Surprises"

In summary, you have seen that climate in the 21st century, unlike the preceding thousand years, is not expected to be stable. Rather, a constant state of change is very likely. Many of the changes will probably be gradual environmental shifts, imperceptible from year to year. Nevertheless, the effects, accumulated over decades, will have powerful economic, social, and political consequences.

Despite our best efforts to understand future climate shifts, there is also the potential for "surprises." This simply means that due to the complexity of Earth's climate system, we might experience relatively sudden, unexpected changes or see some aspects of climate shift in an unexpected manner. The report on *Climate Change Impacts on the United States* describes the situation like this:

> Surprises challenge humans' ability to adapt, because of how quickly and unexpectedly they occur. For example, what if the Pacific Ocean warms in such a way that El Niño events become much more extreme? This could reduce the frequency, but perhaps not the strength, of hurricanes along the East Coast, while on the West Coast, more severe winter storms, extreme precipitation events, and damaging winds could become common. What if large quantities of methane, a potent greenhouse gas currently frozen in icy Arctic tundra and sediments, began to be released to the atmosphere by warming, potentially creating an amplifying "feedback loop" that would cause even more warming? We simply do not know how far the climate system or other systems it affects can be pushed before they respond in unexpected ways.

There are many examples of potential surprises, each of which would have large consequences. Most of these potential outcomes are rarely reported, in this study or elsewhere. Even if the chance of any particular surprise happening is small, the chance that at least one such surprise will occur is much greater. In other words, while we can't know which of these events will occur, it is likely that one or more will eventually occur.*

Clearly, the impact on climate of an increase in atmospheric CO_2 and trace gases is obscured by many unknowns and uncertainties. Policy makers are confronted with responding to the risks posed by emissions of greenhouse gases in the face of significant scientific uncertainties. However, they are also faced with the fact that climate-induced environmental changes cannot be reversed quickly, if at all, owing to the lengthy time scales associated with the climate system. Addressing this issue, the Intergovernmental Panel on Climate Change states:

> Uncertainty does not mean that a nation or the world community cannot position itself better to cope with the broad range of possible climate changes or protect against potentially costly future outcomes. Delaying such measures may leave a nation or the world poorly prepared to deal with adverse changes and may increase the possibility of irreversible or very costly consequences. Options for adapting to change or mitigating change that can be justified for other reasons today (e.g., abatement of air and water pollution) and make society more flexible or resilient to anticipated adverse effects of climate change appear particularly desirable.**

---

*National Assessment Synthesis Team, *Climate Change Impacts on the United States: The Potential Consequences of Climate Variability and Change.* Washington, D.C.: U.S. Global Research Program, 2000, p. 19.
**Intergovernmental Panel on Climate Change. *Climate Change 1995: Impacts, Adaptations and Mitigation of Climate Change: Scientific-Technical Analysis.* New York: Cambridge University Press, 1996, p. 23.

Summary

- The *climate system* includes the atmosphere, hydrosphere, geosphere, biosphere, and cryosphere (the ice and snow that exists at Earth's surface). The system involves the exchanges of energy and moisture that occur among the five spheres.
- Techniques for analyzing Earth's climate history on a scale of hundreds to thousands of years include evidence from *seafloor sediments* and *oxygen isotope analysis.* Seafloor sediments are useful recorders of worldwide climate change because the numbers and types of organic remains included in the sediment are indicative of past sea-surface temperatures. Using oxygen isotope analysis, scientists can use the $^{18}O/^{16}O$ ratio found in the shells of microorganisms in sediment and layers of ice and snow to detect past temperatures. Other sources of data used for the study of past climates (called *proxy data*) include the growth rings of trees, pollen contained in sediments, coral reefs, and information contained in historical documents.

- Air is a mixture of many discrete gases, and its composition varies from time to time and place to place. After water vapor, dust, and other variable components are removed, two gases, *nitrogen* and *oxygen,* make up 99 percent of the volume of the remaining clean, dry air. *Carbon dioxide*, although present in only minute amounts (0.038 percent or 380 parts per million), is an efficient absorber of energy emitted by Earth and thus influences the heating of the atmosphere.

- Two important variable components of air are water vapor and aerosols. Like carbon dioxide, water vapor can absorb heat given off by Earth. *Aerosols* (tiny solid and liquid particles) are important because these often invisible particles act as surfaces on which water vapor can condense and are also good absorbers and reflectors (depending on the particles) of incoming solar radiation.

- *Electromagnetic radiation* is energy emitted in the form of rays, or waves, called electromagnetic waves. All radiation is capable of transmitting energy through the vacuum of space. One of the most important differences among electromagnetic waves is their *wavelengths*, which range from very long *radio waves* to very short *gamma rays*. *Visible light* is the only portion of the electromagnetic spectrum we can see. Some of the basic laws that govern radiation as it heats the atmosphere are (1) all objects emit radiant energy, (2) hotter objects radiate more total energy than do colder objects, (3) the hotter the radiating body, the shorter the wavelengths of maximum radiation, and (4) objects that are good absorbers of radiation are good emitters as well. Gases are selective absorbers, meaning that they absorb and emit certain wavelengths but not others.

- Approximately 50 percent of the solar energy that strikes the top of the atmosphere reaches Earth's surface. About 30 percent is reflected back to space. The remaining 20 percent of the energy is absorbed by clouds and the atmosphere's gases. The wavelengths of the energy being transmitted, as well as the size and nature of the absorbing or reflecting substance, determine whether solar radiation will be scattered and reflected back to space, or absorbed.

- Radiant energy that is absorbed heats Earth and eventually is reradiated skyward. Because Earth has a much lower surface temperature than the Sun, its radiation is in the form of long-wave infrared radiation. Because the atmospheric gases, primarily water vapor and carbon dioxide, are more efficient absorbers of terrestrial (long-wave) radiation, the atmosphere is heated from the ground up. The transmission of shortwave solar radiation by the atmosphere, coupled with the selective absorption of Earth radiation by atmospheric gases, results in the warming of the atmosphere and is referred to as the *greenhouse effect*.

- Several explanations have been formulated to explain climate change. Current hypotheses for the "natural" mechanisms (causes unrelated to human activities) of climate change include (1) plate tectonics, rearranging Earth's continents closer or farther from the equator, (2) variations in Earth's orbit, involving changes in the shape of the orbit (*eccentricity*), angle that Earth's axis makes with the plane of its orbit (*obliquity*), and/or the wobbling of the axis (*precession*), (3) volcanic activity, reducing the solar radiation that reaches the surface, and (4) changes in the Sun's output associated with *sunspots*.

- Humans have been modifying the environment for thousands of years. By altering ground cover with the use of fire and the overgrazing of land, people have modified such important climatological factors as surface reflectivity, evaporation rates, and surface winds.

- By adding carbon dioxide and other trace gases (methane, nitrous oxide, and chlorofluorocarbons) to the atmosphere, modern humans are likely contributing to global climate change in a significant way.

- When any component of the climate system is altered, scientists must consider the many possible outcomes, called *climate-feedback mechanisms*. Changes that reinforce the initial change are called *positive-feedback mechanisms*. For example, warmer surface temperatures cause an increase in evaporation, which further increases temperature as the additional water vapor absorbs more radiation emitted by Earth. On the other hand, *negative-feedback mechanisms* produce results that are the opposite of the initial change and tend to offset it. An example would be the negative effect that increased cloud cover has on the amount of solar energy available to heat the atmosphere.

- Global climate is also affected by human activities that contribute to the atmosphere's *aerosol* (tiny, often microscopic, liquid and solid particles that are suspended in air) content. By reflecting sunlight back to space, aerosols have a net cooling effect. The effect of aerosols on today's climate is determined by the amount emitted during the preceding couple of weeks, while carbon dioxide remains for much longer spans and influences climate for many decades.

- Because the climate system is so complex, predicting specific regional changes that may occur as the result of increased levels of carbon dioxide in the atmosphere is speculative. However, some possible consequences of greenhouse warming include (1) altering the distribution of the world's water resources (2) a probable rise in sea level, (3) a greater intensity of tropical cyclones, and (4) changes in the extent of Arctic sea ice and permafrost.

- Due to the complexity of the climate system, not all future shifts can be foreseen. Thus, "surprises" (relatively sudden unexpected changes in climate) are possible.

Review Questions

1. List the five parts of the climate system.

2. What are *proxy data?* List several examples. Why are such data necessary in the study of climate change?

3. Why are seafloor sediments useful in the study of past climates?

4. Briefly describe why tree rings are helpful in studying the geologic past.

5. What are the major components of clean, dry air? List two significant variable components of the atmosphere.

6. Compared to Earth, does the Sun emit most of its energy as longer or shorter wavelengths of electromagnetic radiation?

7. What are the three basic paths taken by incoming solar radiation? On average, what percentage takes each path? What might cause their percentages to vary?

8. Explain why the atmosphere is heated chiefly by radiation from Earth's surface rather than by direct solar radiation.

9. Which gases are the primary heat absorbers in the lower atmosphere?

10. Describe or outline the greenhouse effect.

11. The volcanic eruptions of El Chichón in Mexico and Mount Pinatubo in the Philippines had measurable short-term effects on global temperatures. Describe and briefly explain these effects.

12. List two examples of possible climate change linked to solar variability. Are these Sun–climate connections widely accepted?

13. Why has the carbon dioxide level of the atmosphere been rising for more then 150 years?

14. How are temperatures in the lower atmosphere likely to change as carbon dioxide levels continue to increase?

15. Aside from carbon dioxide, what other trace gases are contributing to a future global temperature change?

16. What are climate-feedback mechanisms? Give some examples.

17. What are the main sources of human-generated aerosols? What effect do these aerosols have on temperatures in the troposphere? How long do aerosols remain in the lower atmosphere before they are removed?

18. List four potential consequences of global warming.

Key Terms

aerosols (p. 574)
climate-feedback
    mechanism (p. 584)
climate system (p. 569)
greenhouse effect (p. 575)

negative-feedback
    mechanism (p. 584)
oxygen isotope analysis
    (p. 571)
paleoclimatology (p. 570)

positive-feedback
    mechanism (p. 584)
proxy data (p. 570)
sunspots (p. 578)

Web Resources

 The *Earth* Website uses the resources and flexibility of the Internet to aid in your study of the topics in this chapter. Written and developed by geology instructors, this site will help improve your understanding of geology. Visit **http://www. prenhall.com/tarbuck** and click on the cover of *Earth 9e* to find:

- Online review quizzes.
- Critical thinking exercises.
- Links to chapter-specific Web resources.
- Internet-wide key-term searches.

http://www.prenhall.com/tarbuck

Earth's Evolution through Geologic Time

Maroon Bells in Autumn, Colorado Rockies. (Photo by Tim Fitzharris/ Minden Pictures)

Earth has a long and complex history. Time and again, the splitting and colliding of continents has resulted in the formation of new ocean basins and the creation of great mountain ranges. Furthermore, the nature of life on our planet has experienced dramatic changes through time.

Many of the changes on planet Earth occur at a "snail's pace," generally too slow for people to perceive. Thus, human awareness of evolutionary change is fairly recent. Evolution is not confined to life forms, for all Earth's "spheres" have evolved together: the atmosphere, hydrosphere, geosphere, and biosphere (Figure 22.1). These changes can be observed in the air we breathe, the composition of the world's oceans, the ponderous movements of crustal plates that give rise to mountains, and the evolution of a vast array of life-forms. As each of Earth's spheres has evolved, it has powerfully influenced the others.

FIGURE 22.1 Earth's spheres have evolved together through the long expanse of geologic time. (Photos by A. Momatiuk Eastcott, B. and C. Michael Collier, D. Carr Clifton/Minden Pictures)

A. Atmosphere

B. Hydrosphere

C. Geosphere

D. Biosphere

Is Earth Unique?

There is only one place in the entire universe, as far as we know, that can support life—a modest-sized planet called Earth that orbits an average-sized star, the Sun. Life on Earth is ubiquitous; it is found in boiling mudpots and hot springs, in the deep abyss of the ocean, and even under the Antarctic ice sheet. However, living space on our planet is greatly limited when we consider the needs of individual organisms, particularly humans. The global ocean covers 71 percent of Earth's surface, but only a few hundred meters below the water's surface pressures are so great that our lungs would begin to collapse. Further, many continental areas are too steep, too high, or too cold for us to inhabit (Figure 22.2). Nevertheless, based on what we know about other bodies in the solar system—and the 80 or so planets recently discovered orbiting around other stars—Earth is still, by far, the most accommodating.

What fortuitous events produced a planet so hospitable to living organisms like us? Earth was not always as we find it today. During its formative years, our planet became hot enough to support a magma ocean. It also went through a several-hundred-million-year period of extreme bombardment, to which the heavily cratered lunar surface testifies. Even the oxygen rich atmosphere that makes higher life forms possible is a relatively recent event, geologically speaking. Nevertheless, Earth seems to be the right planet, in the right location, at the right time.

The Right Planet

What are some of the characteristics that make Earth unique among the planets? Consider the following:

1. If Earth were considerably larger (more massive) the force of gravity would be proportionately greater. Like the giant planets, Earth would have retained a thick, hostile atmosphere consisting of ammonia and methane, and possibly even hydrogen and helium.
2. If Earth were much smaller, oxygen, water vapor and other volatiles would escape into space and be lost forever. Thus, like the Moon and Mercury, which lack an atmosphere, Earth would be void of life.
3. If Earth did not have a rigid lithosphere overlaying a weak asthenosphere, plate tectonics would not operate. Our continental crust (Earth's "highlands") would not have formed without the recycling of plates. Consequently, the entire planet would likely be covered by an ocean a few kilometers deep. As the author Bill Bryson so aptly stated, "There might be life in that lonesome ocean, but there certainly wouldn't be baseball."
4. Most surprisingly, perhaps, is the fact that if our planet did not have a molten metallic core, most of the life forms on Earth would not exist. Although this may seem like a stretch of the imagination, without the flow of iron in the core, Earth could not support a magnetic field. It is the magnetic field which prevents lethal cosmic rays (the solar wind) from showering Earth's surface.

The Right Location

One of the primary factors that determines whether or not a planet is suitable for higher life-forms is its location in the solar system. Earth is in a great location.

1. If Earth were about 10 percent closer to the Sun, like Venus, our atmosphere would consist mainly of the greenhouse gas, carbon dioxide. As a result, Earth's surface temperature would be too hot to support higher life-forms.
2. If Earth were about 10 percent farther from the Sun, the problem would be the opposite—too cold rather than too hot. The oceans would freeze over and Earth's active water cycle would not exist. Without liquid water most life-forms would perish.
3. Earth is located near a star of modest size. Stars like the Sun have a life span of roughly 10 billion years. During most of this time radiant energy is emitted at a fairly constant level. Giant stars on the other hand consume their nuclear fuel at very high rates and thus "burn out" in a few hundred million years. This is simply not enough time for the evolution of humans, which first appeared on this planet only a few million years ago.

The Right Time

The last, but certainly not the least fortuitous factor is timing. The first life-forms to inhabit Earth were extremely primitive and came into existence

FIGURE 22.2 Climbers near the top of Mount Everest. At this altitude the level of oxygen is only one-third the amount available at sea level. (Photo courtesy of Woodfin Camp and Associates)

roughly 3.8 billion years ago. From this point in Earth's history innumerable changes occurred—life-forms came and went along with changes in the physical environment of our planet. Two of many timely, Earth-altering events include:

1. The development of our modern atmosphere. Earth's primitive atmosphere is thought to have been composed mostly of water vapor and carbon dioxide, with small amounts of other gases, but no free oxygen. Fortunately, microorganisms evolved that produced oxygen by the process of *photosynthesis*. By about 2.2 billion years ago an atmosphere with free oxygen came into existence. The result was the evolution of the forbearers of life-forms that occupy Earth today.

2. About 65 million years ago our planet was struck by an asteroid 10 kilometers in diameter. This impact caused a mass extinction during which nearly three-quarters of all plant and animal species died out—including the dinosaurs (Figure 22.3). Although this may not seem like a fortuitous event, the extinction of the dinosaurs opened new habitats for the small mammals that survived the impact. These habitats, along with evolutionary forces, led to the development of the many large mammals that occupy our modern world. Without this event, mammals might still be small rodentlike creatures that live in burrows.

As various observers have noted, Earth developed under "just right" conditions to support higher life-forms. Astronomers like to refer to this as the *Goldilocks scenario*. As in the classic Goldilocks and the Three Bears fable, Venus is too hot (the papa bear's porridge), Mars is too cold (the mama bear's porridge), but Earth is just right (the baby bear's porridge). Did these "just right" conditions come about purely by chance as some researchers suggest, or as others have argued, might Earth's hospitable environment have developed for the evolution and survival of higher life-forms?

The remainder of this chapter will focus on the origin and evolution of planet Earth—the one place in the Universe we know fosters life. As you learned in Chapter 9, researchers

FIGURE 22.3 The excavation of a dinosaur fossil from Dry Mesa Quarry, Colorado. (Photo by Michael Collier)

utilize many tools to interpret the clues about Earth's past. Using these tools, and clues that are contained in the rock record, scientists have been able to unravel many of the complex events of the geologic past. The goal of this chapter is to provide a brief overview of the history of our planet and its life-forms. The journey takes us back about 4.5 billion years to the formation of Earth and its atmosphere. Next we will consider how our physical world assumed its present form and how Earth's inhabitants changed through time. We suggest that you reacquaint yourself with the *geologic time scale* presented in Figure 22.4 and refer to it as needed throughout the chapter.

Birth of a Planet

According to the Big Bang theory, the formation of our home planet began about 13.7 billion years ago with a cataclysmic explosion that created all matter and space almost instantaneously (Figure 22.5). Initially atomic particles (protons, neutrons, and electrons) formed, then later as this debris cooled, atoms of hydrogen and helium, the two lightest elements, began to form. Within a few hundred million years, clouds of these gases condensed into stars that compose the galactic systems we now observe fleeing from their birthplace.

As these gases contracted to form the first stars, heating triggered the process of *nuclear fusion*. Within stars' interiors, hydrogen atoms are converted to helium atoms, releasing enormous amounts of energy in the form of radiation (heat, light, cosmic rays). Astronomers have determined that in stars more massive than our Sun, other thermonuclear reactions occur that generate all the elements on the periodic table up to number 26, iron. The heaviest elements (beyond number 26) are only created at extreme temperatures during the explosive death of a star perhaps 10 to 20 times more massive than the Sun. During one of these cataclysmic **supernova** events, an exploding star produces all of the elements heavier than iron and spews them into interstellar space. It is from such debris that our Sun and solar system formed. According to the Big Bang scenario, the atoms in your body were produced billions of years ago in the hot interior of now defunct stars, and the gold in your jewelry was formed during a supernova explosion that occurred trillions of miles away.

From Planetesimals to Protoplanets

Recall that Earth, along with the rest of the solar system, formed about 4.5 billion years ago from the **solar nebula,** a large rotating cloud of interstellar dust and gas (see Chapter 1). As the solar nebula contracted, most of the matter collected in the center to form the hot *protosun*, while the remainder became a flattened spinning disk. Within this spinning disk, matter gradually formed clumps that collided and stuck together to form asteroid-size objects called

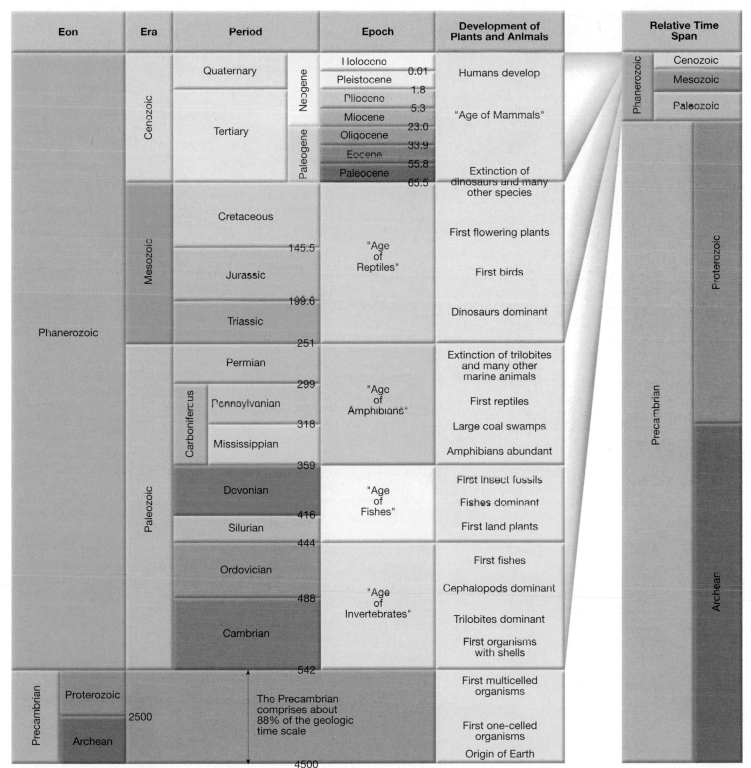

FIGURE 22.4   The geologic time scale. Numbers represent time in millions of years before the present. These dates were added long after the time scale had been established using relative dating techniques. The Precambrian accounts for about 88 percent of geologic time.

planetesimals. The composition of each planetesimal was governed largely by its distance from the hot protosun.

Near the present day orbit of Mercury only metallic grains condensed from the solar nebula. Further out, near Earth's orbit, metallic as well as rocky substances con-

densed, and beyond Mars, ices of water, carbon dioxide, methane, and ammonia formed. It was from these clumps of matter that the planetesimals formed and through repeated collisions and accretion (sticking together) grew into eight **protoplanets** and their moons (Figure 22.5).

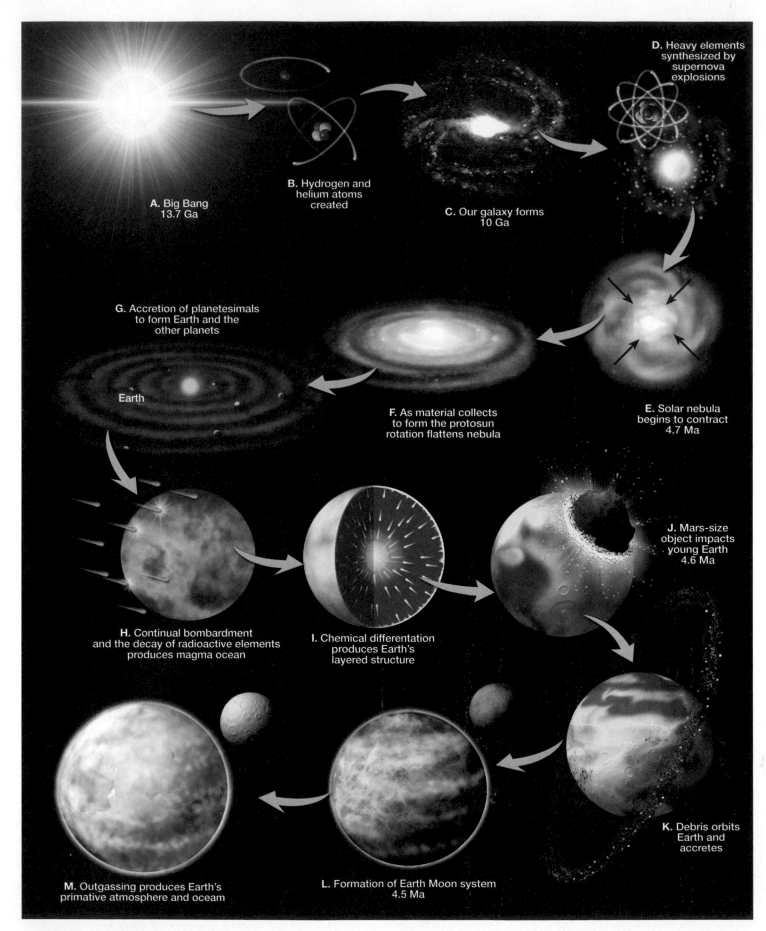

FIGURE 22.5 Major events that led to the formation of early Earth.

At some point in Earth's evolution a giant impact occurred between a Mars-sized planetesimal and a young, semi-molten Earth. This collision ejected huge amounts of debris into space, some of which coalesced (joined together) to form the Moon (see Figure 24.9, p. 663).

Earth's Early Evolution

As material continued to accumulate, the high-velocity impact of interplanetary debris (planetesimals) and the decay of radioactive elements caused the temperature of our planet to steadily increase. During this period of intense heating, Earth became hot enough that iron and nickel began to melt. Melting produced liquid blobs of heavy metal that sank under their own weight. This process occurred rapidly on the scale of geologic time and produced Earth's dense iron-rich core. The formation of a molten iron core was only the first stage of chemical differentiation, in which Earth was converted from a homogeneous body, with roughly the same stuff at all depths, to a layered planet with material sorted by density (Figure 22.5).

This early period of heating also resulted in a magma ocean, perhaps several hundred kilometers deep. Within the magma ocean buoyant masses of molten rock rose toward the surface, where they eventually solidified to produce a thin, primitive crust. Earth's first crust was probably basaltic in composition, not unlike modern oceanic crust. Whether or not plate tectonics was active at this time is not known. However, vigorous, fluidlike motions in the hot, upper mantle must have continually recycled the crust over and over again.

This period of chemical differentiation established the three major divisions of Earth's interior—the iron-rich *core*, the thin *primitive crust*, and Earth's largest layer, the *mantle*, which is located between the core and the crust. In addition, the lightest material, including water vapor, carbon dioxide, and other gases, escaped to form a primitive atmosphere and, shortly thereafter, the oceans (Figure 22.6).

Origin of the Atmosphere and Oceans

Thank goodness for our atmosphere; without it Earth would be nearly 60 degrees Fahrenheit colder. Although not as cold as the surface of Mars, most water bodies on Earth would be frozen over and the hydrological cycle, where water leaves the ocean as a vapor and returns as a liquid, would be meager at best. Recall that the warming effect of certain gases in the atmosphere, mainly carbon dioxide and water vapor, is called the greenhouse effect.

Today, the air we breathe is a stable mixture of 78 percent nitrogen, 21 percent oxygen, about 1 percent argon (an inert gas), and small amounts of gases such as carbon dioxide and water vapor (see Figure 21.9, p. 573). But our planet's original atmosphere 4.5 billion years ago was very different.

Earth's Primitive Atmosphere

When Earth formed, any atmosphere it might have had would have consisted of the gases most common in the early solar system—hydrogen, helium, methane, ammonia, carbon dioxide, and water vapor (see Chapter 24). The lightest of these gases, hydrogen and helium, would have escaped into space as Earth's gravity is too weak to hold them. Most of the remaining gases were probably blown off by strong *solar winds* (a vast stream of particles) from a young, active Sun. (All stars, including the Sun, apparently experience a highly active stage early in their evolution known as the *T-Tauri phase*, during which their solar winds are very intense.)

Earth's first enduring atmosphere formed by a process called **outgassing**, through which gases trapped in the planet's interior are released. Outgassing continues today from hundreds of active volcanoes worldwide (Figure 22.7). However, early in Earth's history, when massive heating and fluidlike motion occurred in the mantle, the gas output must have been immense. The composition of the gases emitted then were probably roughly equivalent to those released during volcanism today. Depending on the chemical makeup of the magma, the gaseous components of modern

FIGURE 22.6 Artistic depiction of Earth over 4 billion years ago. This was a time of intense volcanic activity that produced Earth's primitive atmosphere and oceans, while early life forms produced mound-like structures called stromatolites.

FIGURE 22.7 Earth's first enduring atmosphere formed by a process called outgassing, which continues today from hundreds of active volcanoes worldwide. (Photo by Marco Fulle/www.stromboli.net)

Once much of the available iron had precipitated and the numbers of oxygen-generating organisms increased, oxygen began to accumulate in the atmosphere. Chemical analysis of rocks suggest that a significant amount of oxygen appeared in the atmosphere as early as 2.2 billion years ago and that the amount increased steadily until it reached a stable level about 1.5 billion years ago. The availability of free oxygen had a major impact on the development of life.

Another significant benefit of the "oxygen explosion" is that when oxygen (O_2) molecules in the atmosphere are bombarded by ultraviolet radiation they rearrange themselves to form *ozone* (O_3). Today ozone is concentrated above the surface in a layer called the *stratosphere* where it absorbs much of the ultraviolet radiation that strikes the atmosphere. For the first time, Earth's surface was protected from this form of solar radiation, which is particularly harmful to DNA. Marine organisms had

eruptions consist of between 35–90 percent water vapor, 5–30 percent carbon dioxide, 2–30 percent sulfur dioxide, and lesser amounts of nitrogen, chorine, hydrogen, and argon. Thus, Earth's primitive atmosphere probably consisted of mostly water vapor, carbon dioxide, and sulfur dioxide with minor amounts of other gases, but no free oxygen and little nitrogen.

Oxygen in the Atmosphere

As Earth cooled, water vapor condensed to form clouds, and torrential rains began to fill low-lying areas forming the oceans. It was in the oceans nearly 3.5 billion years ago that photosynthesizing bacteria began to release oxygen into the water. During *photosynthesis,* the Sun's energy is used by organisms to produce organic material (energetic molecules of sugar containing hydrogen and carbon) from carbon dioxide (CO_2) and water (H_2O). The first bacteria probably used hydrogen sulfide (H_2S), rather than water, as the source of hydrogen. Nevertheless, one of the earliest types of bacteria, the *cyanobacteria* began to produce oxygen as a by-product of photosynthesis.

Initially, the newly liberated oxygen was readily consumed by chemical reactions with other atoms and molecules in the ocean, especially iron. The source of most iron appears to be submarine volcanism and associated hydrothermal vents (black smokers). Iron has tremendous affinity for oxygen, and these two elements joined to form iron oxide (also known as rust), which accumulated on the seafloor as sediment. These early iron oxide deposits consist of alternating layers of iron-rich rocks and chert, and are called **banded iron formations** (Figure 22.8). Most banded iron deposits were laid down in the Precambrian between 3.5 and 2 billion years ago, and represent the world's most important reservoir of iron ore.

FIGURE 22.8 These layered iron-rich rocks, called banded iron formations, were deposited during the Precambrian. Much of the oxygen generated as a by-product of photosynthesis was readily consumed by chemical reactions with iron to produce these rocks. (Courtesy Spencer R. Titley)

always been shielded from ultraviolet radiation by the oceans, but with the development of the protective ozone layer the continents became a more hospitable place for life to develop.

Evolution of the Oceans

About 4 billion years ago, as much as 90 percent of the current volume of seawater was contained in the ocean basins. Because the primitive atmosphere was rich in carbon dioxide as well as sulfur dioxide and hydrogen sulfide, the earliest rain water was highly acidic—even more so than the acid rain that recently damaged lakes and streams in eastern North America. Consequently, weathering of Earth's rocky surface occurred at an accelerated rate. The products released by chemical weathering included atoms and molecules of various substances, including sodium, calcium, potassium, and silica, that were carried into the newly formed oceans. Some of these dissolved substances precipitated to form chemical sediment that mantled the ocean floor. Others gradually built up, increasing the salinity of seawater. Today seawater contains an average of 3.5 percent dissolved salts, most of which is common table salt (sodium chloride). Research suggests that the salinity of the oceans increased rapidly at first, but has not changed dramatically in the last few billion years.

Earth's oceans also served as a depository for tremendous volumes of carbon dioxide, a major constituent in the primitive atmosphere—and they still do today. This is significant because carbon dioxide is a greenhouse gas that strongly influences the heating of the atmosphere. Venus, which was once thought to very similar to Earth, has an atmosphere composed of 97 percent carbon dioxide that produced a "runaway" greenhouse effect. The surface of Venus has a temperature of 475°C (900°F)—hot enough to melt lead.

Carbon dioxide is readily soluble in seawater where it often joins with other atoms or molecules to produce various chemical precipitates. By far the most common compound generated by this process is calcium carbonate ($CaCO_3$), which makes up the most abundant chemical sedimentary rock, *limestone*. Later in Earth's history, marine organisms began to remove calcium carbonate from seawater to make their shells and other hard parts. Included were trillions of tiny marine organisms such as foraminifera, that died and were deposited on the seafloor. Today some of these deposits make up the chalk beds exposed along the White Cliffs of Dover, England shown in Figure 22.9. By locking up carbon dioxide, these limestone deposits prevent this greenhouse gas from easily re-entering the atmosphere.*

---

*For more on this, see Box 7.1, "The Carbon Cycle and Sedimentary Rocks," on p. 200.

Precambrian History: The Formation of Earth's Continents

The first 4 billion years of Earth's history are encompassed in the span of time called the *Precambrian*. Representing nearly 90 percent of Earth's history, the Precambrian is divided into the *Archean eon* ("ancient age") and the succeeding *Proterozoic eon* ("early life"). To get a visual sense of the proportion of time represented by the Precambrian, look at the right side of Figure 22.4, which shows relative time spans for the Precambrian and the eras of the Phanerozoic eon.

Our knowledge of this ancient time is sketchy, for much of the early rock record has been obscured by the very Earth processes you have been studying, especially plate tectonics, erosion, and deposition. Most Precambrian rocks are devoid of fossils, which hinders correlation of rock units. In addition, rocks of this great age are metamorphosed and deformed, extensively eroded, and sometimes obscured by overlying strata of younger age. Indeed, Precambrian history is written in scattered, speculative episodes, like a long book with many missing chapters.

We are, however, relatively certain that during the early Archean, Earth was covered by a magma ocean. It is from this material that Earth's atmosphere, oceans and first continents arose.

Earth's First Continents

More than 95 percent of Earth's population lives on the continents—not included are people living on volcanic islands such as the Hawaiian Islands and Iceland. These islanders

FIGURE 22.9 This prominent chalk deposit, the White Cliffs of Dover, is found in southern England. Similar deposits are also found in northern France. (Photo by Jon Arnold/Getty Images/Taxi)

inhabit unusually thick pieces of oceanic crust, thick enough to rise above sea level.

What differentiates continental crust from oceanic crust? Recall that oceanic crust is a comparatively dense (3.0g/ cm^3), homogeneous layer of basaltic rocks derived from partial melting of the rocky, upper mantle. Furthermore, oceanic crust is thin, averaging only 7 kilometers in thickness. Unusually thick blocks of oceanic crust, such as ocean plateaus, tend to form over mantle plumes (hot spot volcanism). Continental crust, on the other hand, is composed of a variety of rock types, has an average thickness of nearly 40 kilometers, and contains a large percentage of low-density (2.7 g/cm^3) silica-rich rocks such as granite.

These are very important differences. Oceanic crust, because it is relatively thin and dense, occurs several kilometers below sea level—unless, of course it has been shoved up onto a landmass by tectonic forces. Continental crust, because of its great thickness and lower density, extends well above sea level. Also, recall that oceanic crust of normal thickness will readily subduct, whereas thick, buoyant blocks of continental crust resist being recycled into the mantle.

Making Continental Crust Earth's first crust was probably basalt, like that generated at modern oceanic ridges. But we do not know for sure because none has ever been found. The hot turbulent mantle that existed during the Archean eon probably recycled most of this material back into the mantle. In fact, it may have been recycled over and over again, much like the "crust" that forms on a lava lake is continually being replaced with fresh lava from below (Figure 22.10).

The oldest preserved continental rocks (greater than 3.5 billion years old) occur as small, highly deformed terranes, which are incorporated within somewhat younger blocks of continental crust (Figure 22.11). The oldest of these is the 4 billion-year-old Acasta gneiss located in the Slave Province of Canada's Northwest Territories. (A few tiny crystals of zircon, found in the Jack Hills area of Australia have radiometric dates between 3.8 and 4.4 billion years.)

The formation of continental crust is simply a continuation of the gravitational segregation of Earth materials that began during the final stage in the accretion of our planet. After the metallic core and rocky mantle formed, low density, silica-rich minerals were gradually extracted from the mantle to form continental crust. This occurs through a multi-stage process during which partial melting of ultramafic mantle rocks (peridotite) generates basaltic rocks and remelting of basalts produces magmas that crystallize to form felsic, quartz-bearing rocks (see Chapter 4). However, little is known about the details of the mechanisms that operated during the Archean to generate these silica-rich rocks.

Many geologists, but certainly not all, conclude that some type of plate-like motion that included subduction operated early in Earth's history. In addition, hot spot volcanism likely played a role as well. However, because the mantle was hotter in the Archean than it is today, both of these phenomena would have progressed at higher rates than their modern counterparts. Hot spot volcanism is thought to have created immense shield volcanoes as well as oceanic plateaus. At the same time, subduction of oceanic crust generated volcanic island arcs. These relatively small, thin crustal fragments represent the first phase in creating stable, continental-size landmasses.

From Continental Crust to Continents According to one model, the growth of large continental masses was accomplished through the collision and accretion of various types of terranes as illustrated in Figure 22.12. This type of collision tectonics deformed and metamorphosed the sediments caught between the converging crustal fragments, shortening and thickening the developing crust. Within the deepest regions of these collision zones, partial melting of the thickened crust generated silica-rich magmas that ascended and intruded the rocks above. The result was the formation of large crustal provinces that, in turn, accreted with others to form even larger crustal blocks called **cratons.** (That portion of a modern craton that is exposed at the surface is referred to as a *shield.*) The assembly of a large craton involved several major mountain building episodes such as occurred when the Indian subcontinent collided with Asia. Figure 22.13 shows the extent of crustal material that was produced during the Archean and Proterozoic eons. This was accomplished by the collision and accretion of many thin and highly mobile terranes into nearly recognizable continental masses.

The Precambrian was a time when much of Earth's continental crust was generated. However, a substantial amount of crustal ma-

FIGURE 22.10 Rift pattern on lava lake. The crust covering this lava lake is continually being replaced with fresh lava from below, much like Earth's crust was recycled early in its history. (Photo by Juerg Alean/www.stromboli.net)

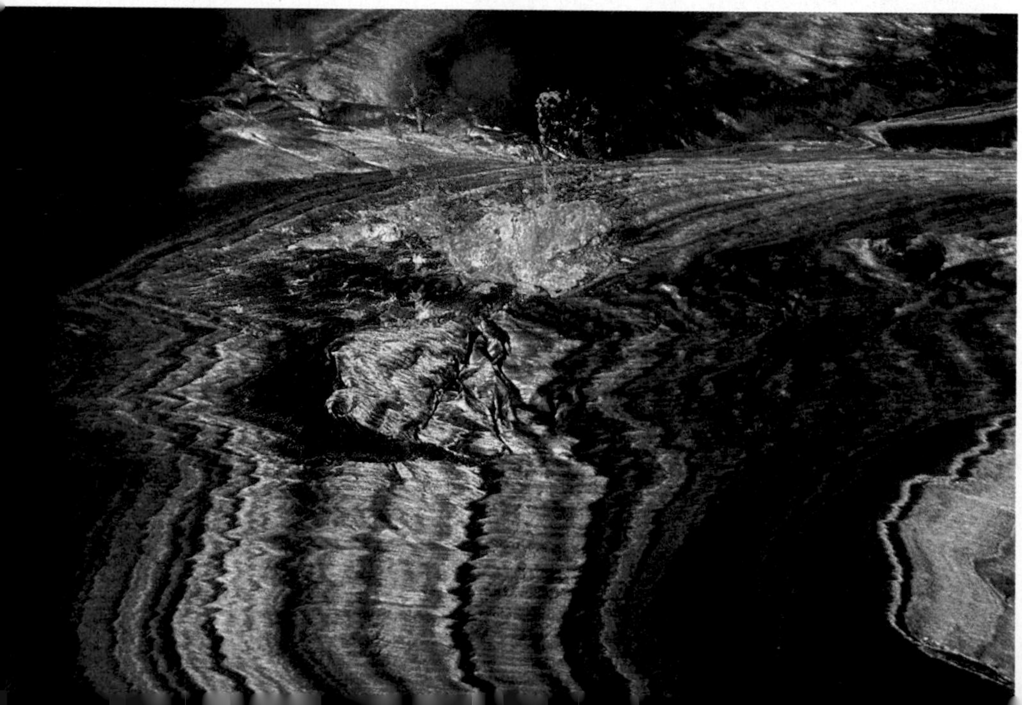

FIGURE 22.11 These rocks at Isua, Greenland, some of the world's oldest, have been dated at 3.8 billion years. (Photo courtesy of Corbis/Bettmann)

the end of the Precambrian most of the modern continental crust had formed—perhaps 85 percent.

In summary, terranes are the basic building blocks of continents and terrane collisions are the major means by which continents grow.

The Making of North America

North America provides an excellent example of the development of continental crust and its piecemeal assembly into a continent. Notice in Figure 22.14 that very little continental crust older that 3.5 billion years still remains. In the late Archean, between 3–2.5 billion years ago, there was a period of major crustal growth. During this time span, the accretion of numerous island arcs and other crustal fragments generated several large crustal provinces. North America contains some of these crustal units, including the Superior and Hearne/Rae cratons shown in Figure 22.14. The locations of these ancient continental blocks during their formation are not known.

About 1.9 billion years ago these crustal provinces collided to produce the Trans-Hudson mountain belt (Figure 22.14). (This mountain-building episode was not restricted to North America, because ancient deformed strata of similar age are also found on other continents.) This event built the North America craton, around which several large and numerous small crustal fragments were later added. Examples of these late arrivals include the Blue Ridge and Piedmont

terial was destroyed as well. Crust can be lost in two ways, by weathering and erosion or by direct reincorporation into the mantle through subduction. Evidence suggests that during much of the Archean, thin slabs of continental crust were destroyed mainly by subduction into the mantle. However, by about 3 billion years ago, cratons grew sufficiently large and thick to resist direct reincorporation into the mantle. From this point onward, weathering and erosion took over as the primary processes of crustal destruction. By

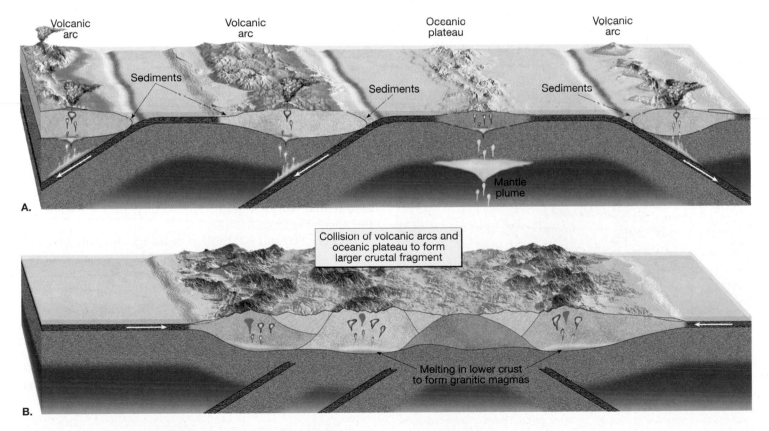

FIGURE 22.12 According to one model, the growth of large continental masses was accomplished through the collision and accretion of various types of terranes.

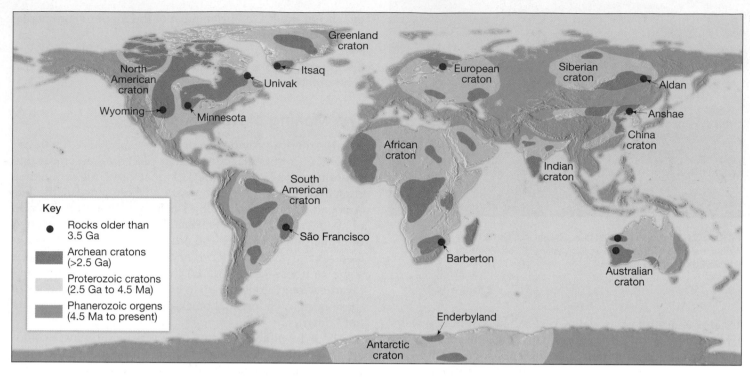

FIGURE 22.13 Illustration showing the extent of crustal material remaining from the Archean and Proterozoic eons.

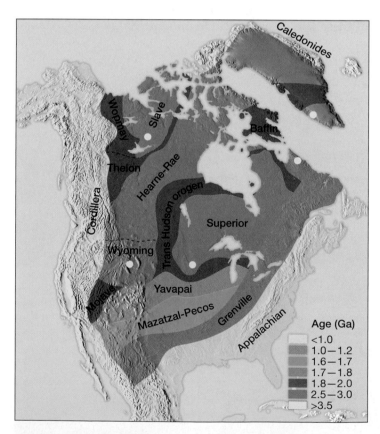

FIGURE 22.14 Map showing the major geological provinces of North America and their ages in billions of years (Ga). It appears that North America was assembled from crustal blocks that were joined by processes very similar to modern plate tectonics. These ancient collisions produced mountainous belts that include remnant volcanic island arcs trapped by the colliding continental fragments.

provinces of the Appalachians and several terranes that were added to the western margin of North America during the Mesozoic and Cenozoic eras to generate the mountainous North American Cordillera.

Supercontinents of the Precambrian

Supercontinents are large landmasses that contain all, or nearly all, of the existing continents. Pangaea was the most recent but certainly not the only supercontinent to exist in the geologic past. The earliest well-documented supercontinent, Rodinia, formed during the Proterozoic eon about 1.1 billion years ago. Although its reconstruction is still being researched, it is clear that Rodinia had a much different configuration than Pangaea (Figure 22.15). One obvious difference is that North America was located near the center of this ancient landmass.

Between 800 and 600 million years ago, Rodinia gradually split apart and the pieces dispersed. By the close of the Precambrian, many of the fragments had reassembled to produce a large landmass located in the Southern Hemisphere called *Gondwana*. Sometimes considered a supercontinent in its own right, Gondwana was comprised mainly of present-day South America, Africa, India, Australia, and Antarctica (Figure 22.16). Other continental fragments also formed—North America, Siberia, and northern Europe. We will consider the fate of these Precambrian landmasses later in the chapter.

Supercontinent Cycle The idea that rifting and dispersal of one supercontinent is followed by a long period during

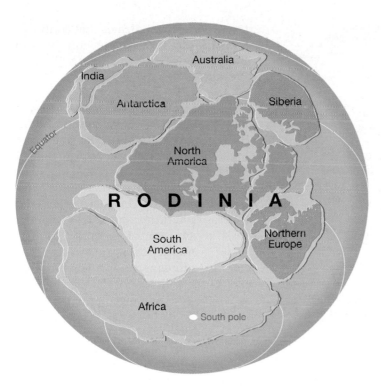

A. Continent of Gondwana

FIGURE 22.15   Simplified drawing showing one of several possible configurations of the supercontinent Rodinia. For clarity, the continents are drawn with somewhat modern shapes, which was not the case 1 billion years ago. (After P. Hoffman, J. Rogers, and others)

which the fragments are gradually reassembled into a new supercontinent having a different configuration is called the **supercontinent cycle.** As indicated earlier the assembly and dispersal of supercontinents had a profound impact on the evolution of Earth's continents. In addition, this phenomenon has greatly influenced global climates as well as contributing to periodic episodes of rising and falling sea level.

Climate and Supercontinents   Moving continents change the patterns of ocean currents and affect global wind patterns, resulting in a change in the distribution of temperature and precipitation. One relatively recent example of how the dispersal of a supercontinent influenced climate relates to the formation of the Antarctic ice sheet. Although eastern Antarctica remained over the South Pole for more than 100 million years, it was not glaciated until about 25 million years ago. Prior to this time South America was connected to the Antarctic Peninsula. This arrangement of landmasses helped maintain a circulation pattern in which warm ocean currents reached the coast of Antarctica as shown in Figure 22.17A. This is similar to how the modern Gulf Stream keeps Iceland mostly ice free—despite its name. However, as South America separated from Antarctica, it moved northward, permitting ocean circulation to flow from west to east around the entire continent of Antarctica (Figure 22.17B). This current, called the West Wind Drift, effectively cut off the entire Antarctic coast from the warm, poleward-directed currents in the southern oceans. This led to eventual covering of almost the entire Antarctic landmass with glacial ice.

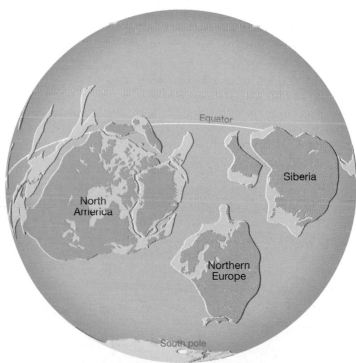

B. Continents not a part of Gondwana

FIGURE 22.16   Reconstruction of Earth as it may have appeared in late Precambrian time. The southern continents were joined into a single landmass called Gondwana. Other landmasses that were not part of Gondwana include North America, northwestern Europe and northern Asia. (After P. Hoffman, J. Rogers, and others)

Local and regional climates have also been impacted by large mountain systems that formed through the collision of large cratons. Because of their high elevations, mountains exhibit markedly lower average temperatures than the surrounding lowlands. In addition, when air is forced to rise

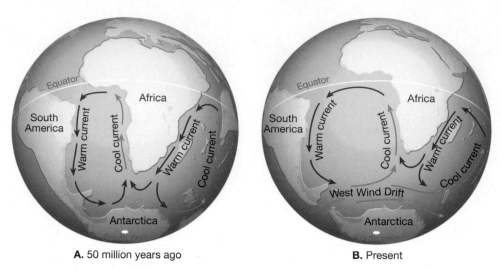

A. 50 million years ago

B. Present

FIGURE 22.17 Comparison of the oceanic circulation pattern 50 million years ago with that of the present. When South America separated from Antarctica the West Wind Drift developed, which effectively isolated the entire Antarctic coast from the warm, poleward-directed currents in the southern oceans. This led to the eventual covering of much of Antarctica with glacial ice.

the rock record indicates that continental glaciation occurred several times in the geologic past, including the late Precambrian.

Sea Level Changes and Supercontinents
Significant sea level changes have been documented numerous times in geologic history and many appear related to the assembly and dispersal of supercontinents. If sea level rises, or the average elevation of a landmass is lowered by erosional or tectonic forces, shallow seas advance onto the continents. The result is the deposition of widespread marine sediments, often a few hundred meters thick. Evidence for such periods when the seas advanced onto the continents include thick sequences of ancient sedimentary rocks that blanket large areas of modern landmasses.

Sea level tends to rise during a period of "global warming" that results in melting of glacial ice. (This appears to be happening today, see chapter 21.) Naturally, during periods of cooling, glacial ice will accumulate, sea level will drop, and shallow inland seas will retreat, thereby exposing large areas of the continental margins.

The supercontinent cycle and sea level changes are directly related to the rates of *seafloor spreading*. When the rate of spreading is rapid, as it is along the East Pacific Rise today, the production of warm oceanic crust is also high. Because warm oceanic crust is less dense (takes up more space) than cold crust, fast spreading ridges occupy more

over these lofty structures, lifting "squeezes" moisture from the air, leaving the region downwind relatively dry. A modern analogy is the wet, heavily forested western slopes of the Sierra Nevada and the dry climate of the Great Basin desert that lies directly to the east (see Figure 19.4, p. 519). Furthermore, large mountain systems, depending on their elevation and latitude, may support extensive valley glaciation, as the Himalayas do today.

Because early Precambrian life was very primitive (mostly bacteria) and left few remains, little is known about Earth's climate during this period. However, evidence from

A.

B.

FIGURE 22.18 Fossils of common Paleozoic life forms. **A.** Natural cast of a trilobite. Trilobites dominated the early Paleozoic ocean, scavenging food from the bottom. **B.** Extinct coiled cephalopods. Like their modern descendants, these were highly developed marine organisms. (Photos courtesy of E.J. Tarbuck)

volume in the ocean basins than slow spreading centers. (Think of getting into a bathtub full of water.) As a result, when the rates of seafloor spreading increase, sea level rises. This, in turn, causes shallow seas to advance onto the low-lying portions of the continents.

Phanerozoic History: The Formation of Earth's Modern Continents

The time span since the close of the Precambrian, called the *Phanerozoic eon*, encompasses 542 million years and is divided into three eras: Paleozoic, Mesozoic, and Cenozoic. The beginning of the Phanerozoic is marked by the appearance of the first life forms with hard parts such as shells, scales, bones, or teeth that greatly enhance the chance of an organism being preserved in the fossil record (Figure 22.18).* Consequently, the study of Phanerozoic crustal history was aided by the availability of fossils, which facilitated much more refined methods for dating geologic events. Moreover, because every organism is associated with its own particular niche, the greatly improved fossil record provided invaluable information for deciphering ancient environments.

Paleozoic History

As the Paleozoic era opened, North America was a land with no living things, either plant or animal. There were no Appalachian or Rocky Mountains; the continent was a largely barren lowland. Several times during the early Paleozoic, shallow seas moved inland and then receded from the interior of the continent. Deposits of clean sandstones mark the shorelines of these shallow seas in the mid-continent. One deposit, the St. Peter sandstone, is mined extensively in Missouri and Illinois for the manufacture of glass, filters, abrasives, and for "tract sand" used in oil and natural gas drilling.

Formation of Pangaea   One of the major events of the Paleozoic was the formation of the supercontinent of Pangaea. It began with a series of collisions that gradually joined North America, Europe, Siberia, and other smaller crustal fragments (Figure 22.19). These events eventually generated a large northern continent called *Laurasia*. This landmass was

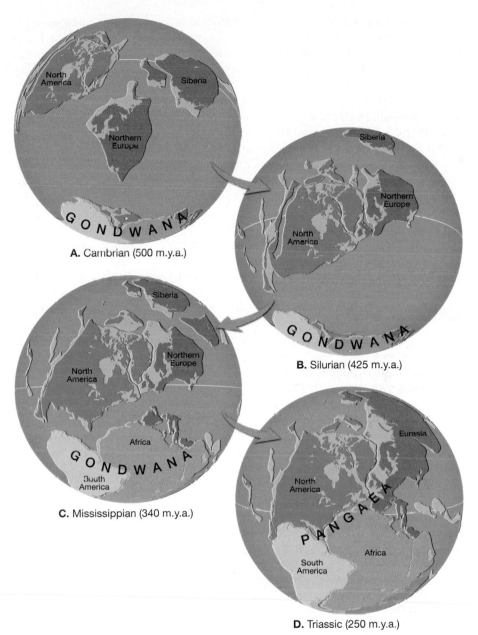

A. Cambrian (500 m.y.a.)

B. Silurian (425 m.y.a.)

C. Mississippian (340 m.y.a.)

D. Triassic (250 m.y.a.)

FIGURE 22.19   During the late Paleozoic, plate movements were joining together the major landmasses to produce the supercontinent of Pangaea. (After P. Hoffman, J. Rogers, and others)

located in the tropics where warm wet conditions led to the formation of vast swamps that ultimately became the coal which fueled the Industrial Revolution of the 1800s and that we still use in large quantities today. During the early Paleozoic, the vast southern continent of Gondwana encompassed five continents—South America, Africa, Australia, Antarctica, India, and perhaps portions of China. Evidence of an extensive continental glaciation places this landmass near the South Pole! By the end of the Paleozoic, Gondwana had migrated northward to collide with Laurasia, culminating in the formation of the supercontinent of Pangaea.

The accretion of Pangaea spans more than 200 million years and resulted in the formation of several mountain belts. This time period saw the collision of northern Europe

---

*For more about this see the discussion on "Conditions Favoring Preservation" in Chapter 9, p. 256.

(mainly Norway) with Greenland to produce the Caledonian Mountains. At roughly the same time at least two microcontinents collided with and deformed the sediments that had accumulated along the eastern margin of North America. This event was an early phase in the formation of the Appalachian Mountains.

By the late Paleozoic, the joining of northern Asia (Siberia) and Europe created the Ural Mountains. Northern China is also thought to have accreted to Asia by the end of the Paleozoic, whereas southern China may not have become part of Asia until after Pangaea had begun to rift apart. (Recall that India did not accrete to Asia until about 45 million years ago.)

Pangaea reached its maximum size about 250 million years ago as Africa collided with North America (Figure 22.19D). This event marked the final episode of growth in the long history of the Appalachian Mountains (see Chapter 14).

Mesozoic History

Spanning about 186 million years, the Mesozoic era is divided into three periods: the Triassic, Jurassic, and Cretaceous. Major geologic events of the Mesozoic include the breakup of Pangaea and the evolution of our modern ocean basins.

The Mesozoic era began with much of the world's land above sea level. In fact, in North America no period exhibits a more meager sedimentary record than the Triassic period. Of the exposed Triassic strata, most are red sandstones and mudstones that lack marine fossils, features that indicate a terrestrial environment. (The red color in sandstone comes from the oxidation of iron.)

As the Jurassic period opened, the sea invaded western North America. Adjacent to this shallow sea, extensive continental sediments were deposited on what is now the Colorado Plateau. The most prominent is the Navajo Sandstone, windblown, white quartz sandstone that, in

places, approaches a thickness of 300 meters (1000 feet). These remnants of massive dunes indicate that a major desert occupied much of the American Southwest during early Jurassic times (Figure 22.20). Another well-known Jurassic deposit is the Morrison Formation—the world's richest storehouse of dinosaur fossils. Included are the fossilized bones of huge dinosaurs such as Apatosaurus (formerly Brontosaurus), Brachiosaurus, and Stegosaurus.

As the Jurassic period gave way to the Cretaceous, shallow seas once again invaded much of western North America, as well as the Atlantic and Gulf coastal regions. This led to the formation of great swamps similar to those of the Paleozoic era. Today the Cretaceous coal deposits in the western United States and Canada are very important economically. For example, on the Crow Native American reservation in Montana, there are nearly 20 billion tons of high-quality coal of Cretaceous age.

Another major event of the Mesozoic era was the breakup of Pangaea (Figure 22.21). About 165 million years ago a rift developed between what is now North America and western Africa, marking the birth of the Atlantic Ocean. As Pangaea gradually broke apart, the westward-moving North American plate began to override the Pacific basin (see Figure 13.26, p. 371). This tectonic event marked the beginning of a continuous wave of deformation that moved inland along the entire western margin of North America. By Jurassic times, subduction of the Farallon plate had begun to produce the chaotic mixture of rocks that exist today in the Coast Ranges of California. Further inland, igneous activity was widespread, and for nearly 60 million years volcanism was rampant as huge masses of magma rose to within a few miles of the surface. The remnants of this activity include the granitic plutons of the Sierra Nevada as well as the Idaho batholith, and British Columbia's Coast Range batholith.

FIGURE 22.20 These massive, cross-bedded sandstone cliffs in Zion National Park are the remnants of ancient sand dunes. (Photo by Michael Collier)

Tectonic activity that began in the Jurassic continued throughout the Cretaceous. Compressional forces moved huge rock units in a shinglelike fashion toward the east. Across much of North America's western margin, older rocks were thrust eastward over younger strata, for distances exceeding 150 kilometers (90 miles). This ultimately formed the vast Northern Rockies that extend from Wyoming to Alaska.

As the Mesozoic came to an end, the southern ranges of the Rocky Mountains formed. This mountain-building event, called the Laramide Orogeny, occurred when large blocks of deeply buried Precambrian rocks were lifted nearly vertically along steeply dipping faults, upwarping the overlying younger sedimentary strata. The mountain ranges produced by the Laramide Orogeny include the Front Range of Colorado, the Sangre de Cristo of New Mexico and Colorado, and the Bighorns of Wyoming (see Box 14.2, p. 392).

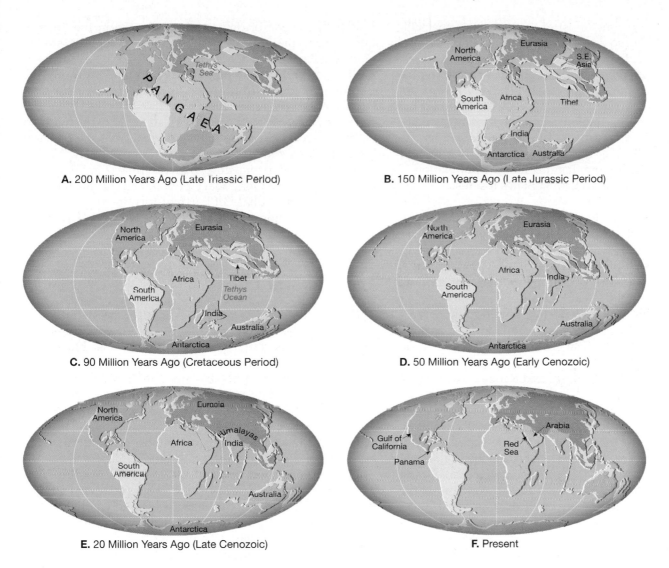

A. 200 Million Years Ago (Late Triassic Period)

B. 150 Million Years Ago (Late Jurassic Period)

C. 90 Million Years Ago (Cretaceous Period)

D. 50 Million Years Ago (Early Cenozoic)

E. 20 Million Years Ago (Late Cenozoic)

F. Present

FIGURE 22.21   Several views of the breakup of Pangaea over a period of 200 million years.

Cenozoic History

The Cenozoic era, or "era of recent life," encompasses the last 65.5 million years of Earth history. It is during this span that the physical landscapes and life-forms of our modern world came into being. The Cenozoic era represents a much smaller fraction of geologic time than either the Paleozoic or the Mesozoic. Although shorter, it nevertheless possesses a rich history because the completeness of the geologic record improves as time approaches the present. The rock formations of this time span are more widespread and less disturbed than those of any preceding time period.

The Cenozoic era is divided into two periods of very unequal duration—the Tertiary and the Quaternary. The Tertiary period includes five epochs and embraces about 63 million years, practically all of the Cenozoic era. The Quaternary period consists of two epochs that represent only the last 2 million years of geologic time.

Most of North America was above sea level throughout the Cenozoic era. However, the eastern and western margins of the continent experienced markedly contrasting events because of their different relationships with plate boundaries. The Atlantic and Gulf coastal regions, far removed from an active plate boundary, were tectonically stable. By contrast, western North America was the leading edge of the North American plate. As a result, plate interactions during the Cenozoic gave rise to many events of mountain building, volcanism, and earthquakes.

Eastern North America   The stable continental margin of eastern North America was the site of abundant marine sedimentation. The most extensive deposition surrounded the Gulf of Mexico, from the Yucatán Peninsula to Florida. Here, the great buildup of sediment caused the crust to downwrap and produced numerous faults. In many instances, the faults created structures in which oil and natural gas accumulated. Today, these and other petroleum traps are the most economically important resource of the Gulf Coast, as evidenced by the numerous offshore drilling platforms.

By early Cenozoic time, most of the original Appalachians had been eroded to a low plain. Later, isostatic adjustments raised the region once again, rejuvenating its rivers. Streams eroded with renewed vigor, gradually sculpting the surface into its present-day topography. The sediments from this erosion were deposited along the eastern margin of the continent, where they attained a thickness of many kilometers. Today, portions of the strata deposited during the Cenozoic are exposed as the gently sloping Atlantic and Gulf coastal plains. It is here that much of the population of the eastern and southeastern United States resides.

Western North America In the West, the Laramide Orogeny that built the southern Rocky Mountains was coming to an end (Figure 22.22). As erosion lowered the mountains, the basins between uplifted ranges filled with sediments. Eastward, a great wedge of sediment from the eroding Rockies was creating the Great Plains.

Beginning in the Miocene epoch about 20 million years ago, a broad region from northern Nevada into Mexico experienced crustal extension that created more than 150 fault-block mountain ranges. Today, they rise abruptly above the adjacent basins, creating the Basin and Range Province (see Chapter 14).

As the Basin and Range Province was forming, the entire western interior of the continent was gradually uplifted. This event re-elevated the Rockies and rejuvenated many of the West's major rivers. As the rivers became incised, many spectacular gorges were formed, including the Grand Canyon of the Colorado River, the Grand Canyon of the Snake River, and the Black Canyon of the Gunnison River.

Volcanic activity was also common in the West during much of the Cenozoic. Beginning in the Miocene epoch, great volumes of fluid basaltic lava flowed from fissures in portions of present-day Washington, Oregon, and Idaho. These eruptions built the extensive (1.3 million square miles) Columbia Plateau. Immediately west of the Columbia Plateau, volcanic activity was different in character. Here, more viscous magmas with higher silica contents erupted explosively, creating the Cascades, a chain of stratovolcanoes extending from northern California into Canada. Some of these volcanoes are still classified as active.

A final episode of deformation occurred in the West in late Tertiary time, creating the Coast Ranges that stretch along the Pacific Coast. Meanwhile, the Sierra Nevada were faulted and uplifted along their eastern flank, forming the imposing mountain front we know today.

As the Tertiary period drew to a close, the effect of mountain building, volcanic activity, isostatic adjustments, and extensive erosion and sedimentation had created a physical landscape very similar to the configuration of today. All that remained of Cenozoic time was the final 2 million year episode called the Quaternary period. During this most recent (and current) phase of Earth history, in which humans evolved, the action of the glacial ice, wind, and running water added the finishing touches.

Earth's First Life

The oldest fossils show that life on Earth was established at least 3.5 billion years ago. Microscopic fossils similar to modern cyanobacteria (formerly known as blue-green algae) have been found in silica-rich chert deposits in locations worldwide. Two notable areas are in southern Africa, where the rocks date to more than 3.1 billion years, and in the Gunflint Chert (named for its use in flintlock rifles) of Lake Superior. Chemical traces of organic matter in even older rocks have led paleontologists to conclude that life may have existed 3.8 billion years ago.

How did life begin? A requirement for life, in addition to a hospitable environment, is the chemical raw materials needed to form life's critical molecules, DNA, RNA, and proteins. One of the building blocks of these substances are organic compounds called *amino acids*. The first amino acids may have been synthesized from methane and ammonia which were plentiful in Earth's primitive atmosphere. The question remains whether these gases could have been easily reorganized into useful organic molecules by ultraviolet light. Or lightning may have been the impetus, as the well-known experiments conducted by Stanley Miller and Harold Urey attempted to demonstrate.

Other researchers suggest that amino acids arrived ready-made, delivered by asteroids or comets that collided

FIGURE 22.22 San Juan Mountains near Telluride, Colorado, are one of several ranges that make up the Rocky Mountains. (Photo by Jim Steinberg/Photo Researchers, Inc.)

with a young Earth. A group of meteorites (debris from asteroids and comets that strike Earth) called *carbonaceous chrondrites* are known to contain amino acidlike organic compounds. Maybe early life had an extraterrestrial beginning.

Yet another hypothesis proposes that the organic material needed for life came from the methane and hydrogen sulfide that spews from deep-sea hydrothermal vents. The study of modern bacteria and other "hyperthermophiles" that live around hydrothermal vents (black smokers) suggests that life may have formed in this extreme environment, where temperatures exceed the boiling point of water.

Is it possible that life originated near a hydrothermal vent deep on the ocean floor, or within a hot spring similar to those in Yellowstone National Park (Figure 22.23)? Some origin-of-life researchers think that this scenario is highly improbable as the scalding temperatures would have destroyed any early types of self-replicating molecules. They argue that life's first home would have been along sheltered stretches of ancient beaches, where waves and tides would have brought together various organic materials formed in the Precambrian oceans.

Regardless of where life originated, change was inevitable (Figure 22.24). The first known organisms were single-cell bacteria that belong to the group called **prokaryotes** which means their genetic material (DNA) is not separated from the rest of the cell by a nucleus. Because oxygen was absent from Earth's early atmosphere and oceans, the first organisms employed anaerobic (without oxygen) metabolism to extract energy from "food." Their food source was likely organic molecules in their surroundings, but the supply of this material was limited. Then a type of bacteria evolved that used solar energy to synthesize organic compounds (sugars). This event was an important turning point in evolution—for the first time organisms had the capability of producing food for themselves as well as for other organisms.

Recall that photosynthesis by ancient cyanobacteria, a type of prokaryote, contributed to the gradual rise in the level of oxygen, first in the ocean and then in the atmosphere. Thus, these early organisms radically transformed our planet. Fossil evidence for the existence of these microscopic bacteria includes distinctively layered mounds of calcium carbonate, called **stromatolites** (Figure 22.25A). Stromatolites are not actually the remains of organisms, but limestone mats built up by lime-accreting bacteria. Strong evidence for the origin of these ancient fossils is the close similarity they have to modern stromatolites found in Shark Bay, Australia (Figure 22.25B).

The oldest fossils of more advanced organisms, called **eukaryotes,** are about 2.1 billion years old. Like prokaryotes, the first eukaryotes were microscopic, water-dwelling organisms. Their cellular structure did, however, contain nuclei. It is these primitive organisms that gave rise to essentially all of the multicelled organisms that now inhabit our planet—trees, birds, fishes, reptiles, and even humans.

During much of the Precambrian, life consisted exclusively of single-celled organisms. It wasn't until perhaps 1.5 billion years ago that multicelled eukaryotes evolved. Green algae, one of the first multicelled organisms, contained chloroplasts (used in photosynthesis) and were the forbears of modern plants. The first primitive marine animals did not appear until somewhat later, perhaps 600 million years ago; we just do not know for sure.

Fossil evidence suggests that organic evolution progressed at an excruciatingly slow pace until the end of the Precambrian. At this time, Earth's continents were barren, and the oceans were populated primarily by organisms too small to be seen with the naked eye. Nevertheless, the stage was set for the evolution of larger and more complex plants and animals at the dawn of the Paleozoic.

Paleozoic Era: Life Explodes

The Cambrian period marks the beginning of the Paleozoic era, about 542 million years ago. This time span saw the emergence of new animal forms, the likes of which have never been seen, before or since. All major invertebrate (animals lacking backbones) groups made their appearance, including jellyfish, sponges, worms, mollusks (clams), and arthropods (insects, crabs). This huge expansion in biodiversity is often referred to as the *Cambrian explosion* (see Box 22.1).

But did it happen? Evidence suggests that these life forms may have gradually diversified late in the Precambrian, but were not preserved in the rock record. After all, the Cambrian period marks the first time organisms developed hard parts. Is it possible that the Cambrian event was an explosion of animal forms that grew in size and became "hard" enough to be fossilized?

Paleontologists may never definitively answer that question. They know, however, that hard parts clearly served many useful purposes and aided adaptations to new lifestyles. Sponges, for example, developed a network of fine interwoven silica spicules that allowed them to grow larger and more erect, and thus capable of extending above the seafloor in search of food. Mollusks (clams and snails)

FIGURE 22.23 The Grand Prismatic Pool, Yellowstone National Park, Wyoming. This hot-water pool gets its blue color from several species of heat-tolerant cynobacteria. (Photo by Jim Brandenburg/Minden Pictures)

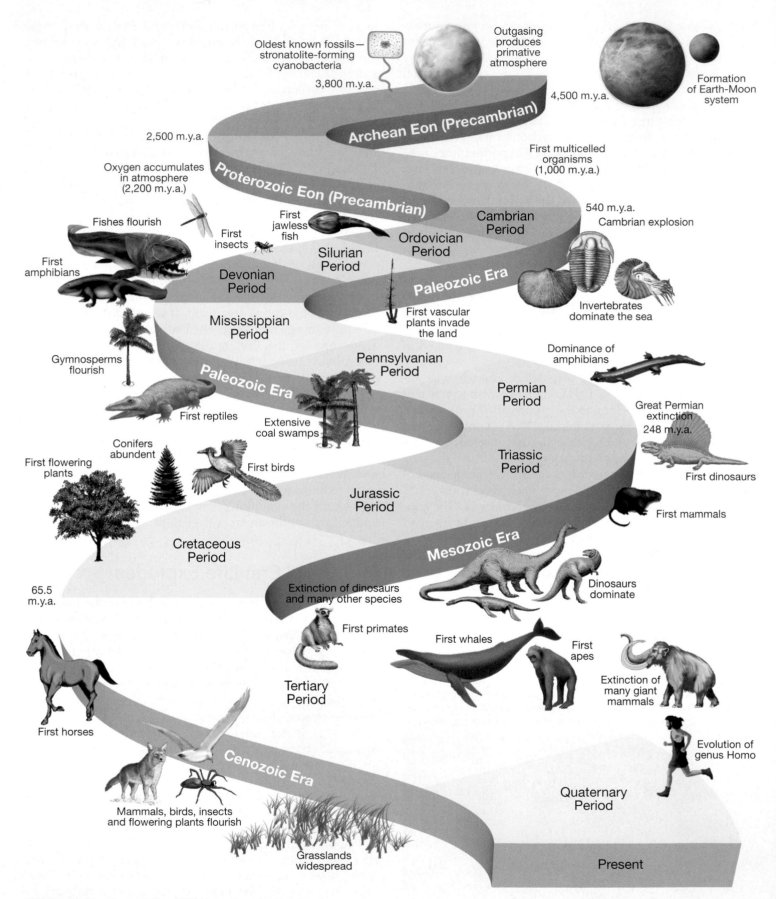

FIGURE 22.24 The evolution of life through geologic time.

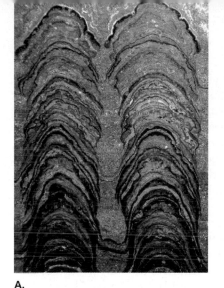

A.   **B.**

FIGURE 22.25 Stromatolites are among the most common Precambrian fossils. **A.** Precambrian fossil stromatolites composed of calcium carbonate deposited by algae in the Helena Formation, Glacier National Park. (Photo by Ken M. Johns/Photo Researchers, Inc.) **B.** Modern stromatolites growing in shallow saline seas, western Australia. (Photo by Bill Bachman/Photo Researchers, Inc.)

BOX 22.1 ▶ UNDERSTANDING EARTH

The Burgess Shale

The possession of hard parts greatly enhances the likelihood of organisms being preserved in the fossil record. Nevertheless, there have been rare occasions in geologic history when large numbers of soft-bodied organisms have been preserved. The Burgess Shale is one well-known example. Located in the Canadian Rockies near the town of Field in southeastern British Columbia, the site was discovered in 1909 by Charles D. Walcott of the Smithsonian Institution.

The Burgess Shale is a site of exceptional fossil preservation and records a diversity of animals found nowhere else (Figure 22.A). The animals of the Burgess Shale lived shortly after the *Cambrian explosion*, a time when there had been a huge expansion of marine biodiversity. Its beautifully preserved fossils represent our most complete and authoritative snapshot of Cambrian life, far better than deposits containing only fossils of organisms with hard parts. To date, more than 100,000 unique fossils have been found.

The animals preserved in the Burgess Shale inhabited a warm, shallow sea adjacent to a large reef that was part of the continental margin of North America. During the Cambrian, the North American continent was in the tropics astride the equator. Life was restricted to the ocean, and the land was barren and uninhabited.

What were the circumstances that led to the preservation of the many life forms found in the Burgess Shale? The animals lived in and on underwater mudbanks that formed as sediment accumulated on the

outer margins of a reef adjacent to a steep escarpment (cliff). Periodically the accumulation of muds became unstable and the slumping and sliding sediments moved down the escarpment as turbidity currents. These flows transported the animals in a turbulent cloud of sediment to the base of the reef where they were buried. Here, in an environment lacking oxygen, the buried carcasses were protected from scavengers and decomposing bacteria. This process occurred again and again, building a thick

sequence of fossil-rich sedimentary layers. Beginning about 175 million years ago, mountain-building forces elevated these strata from the seafloor and moved them many kilometers eastward along huge faults to their present location in the Canadian Rockies.

The Burgess Shale is one of the most important fossil discoveries of the 20th century. Its layers preserve for us an intriguing glimpse of early animal life that is more than a half billion years old.

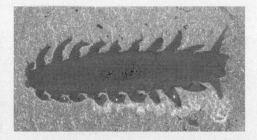

FIGURE 22.A Two examples of Burgess Shale fossils. *Thumatilon walcotti* (left) was a relatively large (up to 20 centimeters, or 8 inches, long) leaflike animal. (Photo by National Museum of Natural History) *Aysheaia pedunculata* (right) was an ancient relative of modern velvet worms and may have clung to soft sponges with tiny hooks on its feet. (Photo by Royal Tyrrell Museum)

secreted external shells of calcium carbonate that provided protection and allowed body organs to function in a more controlled environment. The successful trilobites developed an exoskeleton of a protein called chitin (similar to a human fingernail), which permitted them to search for food by burrowing through soft sediment (Figure 22.18A).

Early Paleozoic Life-Forms

The Cambrian period was the golden age of *trilobites*. More than 600 genera of these mud-burrowing scavengers flourished worldwide. The Ordovician marked the appearance of abundant cephalopods—mobile, highly developed mollusks that became the major predators of their time (Figure 22.26). The descendants of these cephalopods include the squid, octopus, and chambered nautilus that inhabit our modern oceans. Cephalopods were the first truly large organisms on Earth, one species reaching a length of nearly 10 meters (30 feet).

The early diversification of animals was partly driven by the emergence of predatory lifestyles. The larger mobile cephalopods preyed on trilobites that were mostly smaller than a child's hand. The evolution of efficient movement was often associated with the evolution of greater sensory capabilities and more complex nervous systems. These animals developed sensory devices for detecting light, smells, and touch.

Approximately 400 million years ago, green algae that had adapted to survive at the water's edge, gave rise to the first multicellular land plants. The primary difficulty of sustaining plant life on land was obtaining water and staying upright despite gravity and winds. These earliest land plants were leafless, vertical spikes about the size of your index finger. However, by the end of the Devonian period, 40 million years later, the fossil record indicates the existence of forests with trees tens of meters tall.

In the oceans, fishes perfected a new form of support for the body, an internal skeleton, and were the first creatures to have jaws. Armor-plated fishes that had evolved during the Ordovician continued to adapt (Figure 22.27). Their armor plates thinned to lightweight scales that permitted increased speed and mobility. Other fishes evolved during the Devonian, including primitive sharks that had skeletons made of cartilage and bony fishes, the groups to which many modern fishes belong. Fishes, the first large vertebrates, proved to be faster swimmers than invertebrates and possessed more acute senses and larger brains. Hence, they became the dominant predators of the sea. Because of this, the Devonian period is often referred to as the "Age of the Fishes."

Vertebrates Move to Land

During the Devonian, a group of fishes called the *lobe-finned fish* began to adapt to terrestrial environments (Figure 22.28). Like their modern relative, these fishes had sacks that could be filled with air to supplement their "breathing" through gills. The first lobe-finned fish probably occupied freshwater tidal flats or small ponds near the ocean. Some began to use their fins to move from one pond to another in search of food, or to evacuate a pond that was drying up. This favored the evolution of a group of animals that could stay out of water longer and move about on land more efficiently. By the late Devonian, lobe-finned fish had evolved into air-breathing amphibians. Although they had developed strong legs, they retained a fishlike head and tail.

FIGURE 22.26 During the Ordovician period (490–443 million years ago), the shallow waters of an inland sea over central North America contained an abundance of marine invertebrates. Shown in this reconstruction are straight-shelled cephalopods, trilobites, brachiopods, snails, and corals. (© The Field Museum, Neg. # GEO80820c, Chicago)

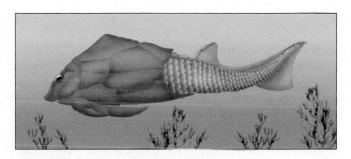

FIGURE 22.27   These placoderms, or "plate-skinned" fish were abundant during the Devonian (417–354 million years ago). (Drawing after A.S. Romer)

Modern amphibians, such as frogs, toads, and salamanders, are small and occupy limited biological niches. But conditions during the late Paleozoic were ideal for these newcomers to the land. Large tropical swamps extended across North America, Europe and Siberia that were teeming with large insects and millipedes (Figure 22.29). With no predators to speak of, amphibians diversified rapidly. Some groups took on lifestyles and forms similar to modern reptiles, such as crocodiles.

Despite their success, the early amphibians were not fully adapted to life out of the water. In fact, amphibian means "double life," because these creatures need both the watery world from which they came and the land onto which they moved. Amphibians are born in the water, as exemplified by tadpoles, complete with gills and tails. In time, these features disappear and an air breathing adult with legs emerges.

Near the end of the Paleozoic, Earth's major landmasses were joined to form the supercontinent of Pangaea (see Figure 22.19, p. 609). This redistribution of land and water along with changes in the elevations of landmasses brought pronounced changes in world climates. Broad areas of the northern continents became elevated above sea level, and the climate grew drier. These changes apparently resulted in the decline of the amphibians (see Box 22.2).

Mesozoic Era: Age of the Dinosaurs

As the Mesozoic era dawned, its life forms were the survivors of the great Permian extinction. These organisms diversified in many ways to fill the biological voids created at the close of the Paleozoic. On land, conditions favored those that could adapt to drier climates. Among plants, the gymnosperms were one such group. Unlike the first plants to invade the land, the seed-bearing gymnosperms did not depend on freestanding water for fertilization. Consequently, those plants were not restricted to a life near water's edge.

The gymnosperms quickly became the dominant trees of the Mesozoic. They included the following: cycads that

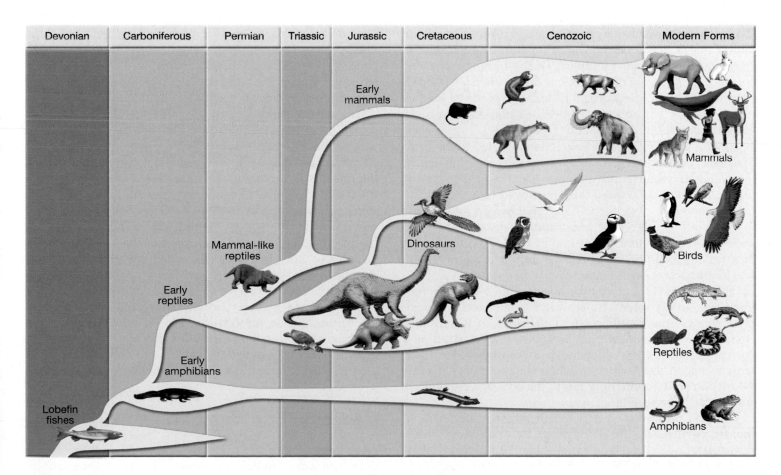

FIGURE 22.28   Relationships of various vertebrates and their evolution from a fish-like ancestor.

FIGURE 22.29 Restoration of a Pennsylvanian-age coal swamp (323 million to 290 million years ago). Shown are scale trees (left), seed ferns (lower left), and scouring rushes (right). Also note the large dragonfly. (© The Field Museum, Neg. # GEO85637c, Chicago. Photographer John Weinstein.)

resembled a large pineapple plant; ginkgoes that had fan-shaped leaves, much like their modern relatives; and the largest plants, the conifers, whose modern descendants include the pines, firs, and junipers. The best-known fossil occurrence of these ancient trees is in northern Arizona's Petrified Forest National Park. Here, huge petrified logs lie exposed at the surface, having been weathered from rocks of the Triassic Chinle Formation (Figure 22.30).

Among the animals, reptiles readily adapted to the drier Mesozoic environment, thereby relegating amphibians to the wetlands where most remain today. Reptiles were the first true terrestrial animals with improved lungs for an active lifestyle and "waterproof" skin that helped prevent the loss of body fluids. Most importantly, reptiles developed shell-covered eggs that can be laid on land. The elimination of a water-dwelling stage (like the tadpole stage in frogs) was an important evolutionary step.

Of interest is the fact that the watery fluid within the reptilian egg closely resembles seawater in chemical composition. Because the reptile embryo develops in this watery environment, the shelled egg has been characterized as a "private aquarium" in which the embryos of these land vertebrates spend their water-dwelling stage of life. With this "sturdy egg," the remaining ties to the oceans were broken, and reptiles moved inland.

The first reptiles were small, but larger forms evolved rapidly, particularly the dinosaurs. One of the largest was *Apatosaurus,* which weighed more than 30 tons and measured over 25 meters (80 feet) from head to tail. For nearly 160 million years, dinosaurs reigned supreme.

Some of the largest dinosaurs were carnivorous (*Tyrannosaurus*), whereas others were herbivorous (like ponderous *Apatosaurus*). The extremely long neck of *Apatosaurus* may have been an adaptation for feeding on tall conifer trees. However, not all dinosaurs were large. Some small forms closely resembled modern, fleet-footed lizards.

The reptiles made one of the most spectacular adaptive radiations in all of Earth history. One group, the pterosaurs, took to the air. These "dragons of the sky" possessed huge membranous wings that allowed them rudimentary flight (Figure 22.31). Another group of reptiles, exemplified by the fossil *Archaeopteryx*, led to more successful flyers: the birds. Whereas some reptiles took to the skies, others returned to the sea, including the fish-eating plesiosaurs and ichthyosaurs (Figure 22.32). These reptiles became proficient swimmers, but they retained their reptilian teeth and breathed by means of lungs.

At the close of the Mesozoic, many reptile groups became extinct. Only a few types survived to recent times, including the turtles, snakes, crocodiles, and lizards (Figure 22.33). The huge, land-dwelling dinosaurs, the marine plesiosaurs, and the flying pterosaurs are known only through the fossil record. What caused this great extinction? (See Box 22.3.)

Cenozoic Era: Age of Mammals

During the Cenozoic, mammals replaced reptiles as the dominant land animals. At nearly the same time, angiosperms (flowering plants with covered seeds) replaced

The Great Permian Extinction

By the close of the Permian period, a mass extinction destroyed 70 percent of all vertebrate species on land, and perhaps as much as 90 percent of all marine organisms. The late Permian extinction was the greatest of at least five mass extinctions to occur over the past 500 million years. Each extinction wreaked havoc with the existing biosphere, wiping out large numbers of species. In each case, however, the survivors formed new biological communities that were eventually more diverse than their predecessors. Thus, mass extinctions actually invigorated life on Earth, as the few hardy survivors eventually filled more niches than the ones left behind by the victims.

Several mechanisms have been proposed to explain these ancient mass extinctions. Initially, paleontologists believed they were gradual events caused by a combination of climate change and biological forces, such as predation and competition. Then, in the 1980s, a research team proposed that the mass extinction that happened 65 million years ago occurred swiftly as a result of an explosive impact by an asteroid about 10 kilometers in diameter. This event, which caused the extinction of the dinosaurs, is described in Box 22.3.

Was the Permian extinction also caused by a giant impact, like the now-famous dinosaur extinction? For many years, researchers thought so. However, scant evidence could be found of debris that would have been generated by an impact large enough to destroy many of Earth's lifeforms.

Another possible mechanism for the Permian extinction was the voluminous eruptions of basaltic lavas which began about 251 million years ago and are known to have covered thousands of square kilometers of the land. (This period of volcanism produced the Siberian Traps located in northern Russia.) The release of carbon dioxide would certainly have enhanced greenhouse warming, and the emissions of sulfur dioxide probably resulted in copious amounts of acid rain.

A recent hypothesis begins with this period of volcanism and the ensuing period of global warming but adds a new twist. These researchers agree that the additional carbon dioxide released into the atmosphere would cause rapid greenhouse warming. This alone, however, would not destroy most plants because they tend to be heat tolerant and consume CO_2 in photosynthesis. They contend, instead, that the trouble begins in the ocean rather than on land.

Most organisms on Earth use oxygen to metabolize food, as do humans. However, some forms of bacteria employ *anaerobic*

(without oxygen) metabolism. Under normal conditions, oxygen from the atmosphere is readily dissolved in seawater, and is then evenly distributed to all depths by deep-water currents. This oxygen "rich" water relegates "oxygen-hating" anaerobic bacteria to anoxic (oxygen free) environments found in deep-water sediments.

The greenhouse warming associated with the vast outpouring of volcanic debris, however, would have warmed the ocean surface, thereby significantly reducing the amount of oxygen that seawater would absorb (Figure 22.B). This condition favors deep-sea anaerobic bacteria, which generate toxic hydrogen sulfide as a waste gas. As these organisms proliferated, the amount of hydrogen sulfide dissolved in seawater would have steadily increased. Eventually, the concentration of hydrogen sulfide reached a threshold and great bubbles of this toxin exploded into the atmosphere (Figure 22.B). On land, hydrogen sulfide was lethal to both plants and animals, but oxygen-breathing marine life would have been hit hardest.

How plausible is this scenario? Remember that the ideas that were just described represent a hypothesis, a tentative explanation regarding a set of observations. Additional research about this and other hypotheses that relate to the Permian extinction continues.

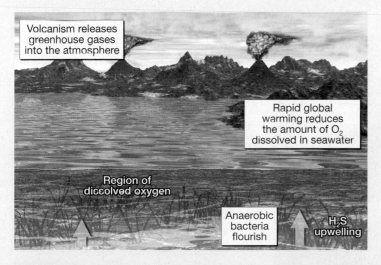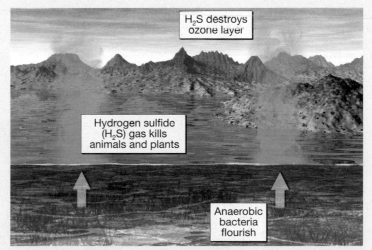

FIGURE 22.B Model for the "Great Permian Extinction". Extensive volcanism released greenhouse gases which resulted in extreme global warming. This condition reduced the amount of oxygen dissolved by seawater. This, in turn, favored "oxygen-hating" anaerobic bacteria, which generated toxic hydrogen sulfide as a waste gas. Eventually, the concentration of hydrogen sulfide reached a threshold and great bubbles of this toxin exploded into the atmosphere, wreaking havoc on organisms on land, but oxygen-breathing marine life was hit the hardest.

BOX 22.3 ▶ UNDERSTANDING EARTH

Demise of the Dinosaurs

The boundaries between divisions on the geologic time scale represent times of significant geological and/or biological charge. Of special interest is the boundary between the Mesozoic era ("middle life") and Cenozoic era ("recent life"), about 65 million years ago. Around this time, about three-quarters of all plant and animal species died out in a *mass extinction*. This boundary marks the end of the era in which dinosaurs and other reptiles dominated the landscape and the beginning of the era when mammals become very important (Figure 22.C). Because the last period of the Mesozoic is the Cretaceous (abbreviated K to avoid confusion with other "C" periods), and the first period of the Cenozoic is the Tertiary (abbreviated T), the time of this mass extinction is called the *Cretaceous–Tertiary* or *KT boundary*.

The extinction of the dinosaurs is generally attributed to this group's inability to adapt to some radical change in environmental conditions. What event could have triggered the rapid extinction of the dinosaurs—one of the most successful groups of land animals ever to have lived?

The most strongly supported hypothesis proposes that about 65 million years ago our planet was struck by a large carbonaceous meteorite, a relic from the formation of the solar system. The errant mass of rock was approximately 10 kilometers in diameter and was traveling at about 90,000 kilometers per hour at impact. It collided with the southern portion of North America in what is now Mexico's Yucatán Peninsula but at the time was a shallow tropical sea (Figure 22.D). The energy released by the impact is estimated to have been equivalent to 100 million megatons (*mega* = million) of high explosives.

For a year or two after the impact, suspended dust greatly reduced the sunlight reaching Earth's surface. This caused global cooling ("impact winter") and inhibited photosynthesis, greatly disrupting food production. Long after the dust settled, carbon dioxide, water vapor, and sulfur oxides that had been added to the atmosphere by the blast remained. If significant quantities of sulfate aerosols formed, their high reflectivity would have helped to perpetuate the cooler surface temperatures for a few more years. Eventually sulfate aerosols leave the atmosphere as acid precipitation. By contrast, carbon dioxide has a much longer residence time in the atmosphere. Carbon dioxide is a *greenhouse gas*, a gas that traps a portion of the radiation emitted by Earth's surface. With the aerosols gone, the enhanced greenhouse effect caused by the carbon dioxide would have led to a long-term rise in average global temperatures. The likely result was that some of the plant and animal life that had survived the initial environmental assault finally fell victim to stresses associated with global cooling, followed by acid precipitation and global warming.

The extinction of the dinosaurs opened up habitats for the small mammals that survived. These new habitats, along with evolutionary forces, led to the development of the large mammals that occupy our modern world.

What evidence points to such a catastrophic collision 65 million years ago? First, a thin layer of sediment nearly 1 centimeter thick has been discovered at the KT boundary, worldwide. This sediment contains a high level of the element *iridium*, rare in Earth's crust but found in high proportions in stony meteorites. Could this layer be the scattered remains of the meteorite that was responsible for the environmental changes that led to the demise of many reptile groups?

Despite growing support, some scientists disagree with the impact hypothesis. They suggest instead that huge volcanic eruptions may have led to a breakdown in the food chain. To support this hypothesis, they cite enormous outpourings of lavas in the Deccan Plateau of northern India about 65 million years ago.

Whatever caused the KT extinction, we now have a greater appreciation of the role of catastrophic events in shaping the history of our planet and the life that occupies it. Could a catastrophic event having similar results occur today? This possibility may explain why an event that occurred 65 million years ago has captured the interest of so many.

FIGURE 22.C Dinosaurs dominated the Mesozoic landscape until their extinction at the close of the Cretaceous period. This skeleton of *Tyrannosaurus* stands on display in New York's Museum of Natural History. (Photo by Gail Mooney/CORBIS Photo)

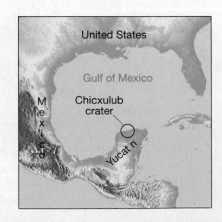

FIGURE 22.D Chicxulub crater is a giant impact crater that formed about 65 million years ago and has since been filled with sediments. About 180 kilometers (110 miles) in diameter, Chicxulub crater is regarded by some researchers to be the impact site that resulted in the demise of the dinosaurs.

FIGURE 22.30   Petrified logs of Triassic age in Arizona's Petrified Forest National Park. (Photo by David Muench)

FIGURE 22.32   Marine reptiles such as this *Ichthyosaur* were the most spectacular of sea animals. (Photo by Chip Clark)

gymnosperms as the dominant plants. The Cenozoic is often called the "Age of the Mammals," but could also appropriately be called the "Age of Flowering Plants," for the angiosperms enjoy a similar status in the plant world.

The development of the flowering plants strongly influenced the evolution of both birds and mammals that feed on seeds and fruits. During the middle Tertiary, grasses (angiosperms) developed rapidly and spread over the plains. This fostered the emergence of herbivorous (plant-eating) mammals which, in turn, established the setting for the evolution of the large, predatory mammals.

During the Cenozoic the ocean was teeming with modern fish such as tuna, swordfish and barracuda. In addition,

some mammals, including seals, whales, and walruses returned to the sea.

From Reptiles to Mammals

The earliest mammals coexisted with dinosaurs for nearly 100 million years but were small rodentlike creatures that gathered food at night when the dinosaurs were less active. Then, about 65 million years ago, fate intervened when a large asteroid collided with Earth and dealt a crashing blow to the reign of the dinosaurs. This transition, during which one dominant group is replaced by another, is clearly visible in the fossil record.

Mammals are distinct from reptiles in that they give birth to live young (which they suckle on milk) and they are warm-blooded. This latter adaptation allowed mammals to

FIGURE 22.33   Fossil skull of a huge crocodile of Mesozoic age. (Photo courtesy of Project Exploration)

FIGURE 22.31   Fossils of the great flying *Pteranodon* have been recovered from Cretaceous chalk deposits located in Kansas. *Pteranodon* had a wingspan of 7 meters (22 feet), but flying reptiles with twice this wingspan have been discovered in west Texas.

FIGURE 22.34 After the breakup of Pangaea, the Australian marsupials evolved differently than their relatives in the Americas. (Photo by Martin Harvey)

their diet, and (4) specialization of limbs to better equip the animal for a particular lifestyle or environment.

Marsupial and Placental Mammals Two groups of mammals, the marsupials and the placentals, evolved and diversified during the Cenozoic. The groups differ principally in their modes of reproduction. Young marsupials are born live but at a very early stage of development. At birth, the tiny and immature young enter the mother's pouch to suckle and complete their development. Today, marsupials are found primarily in Australia, where they underwent a separate evolutionary expansion largely isolated from placental mammals. Modern marsupials include kangaroos, opossums, and koalas (Figure 22.34).

Placental mammals, conversely, develop within the mother's body for a much longer period, so that birth occurs after the young are comparatively mature. Most modern mammals are placental, including humans.

In South America, primitive marsupials and placentals coexisted in isolation for about 40 million years after the breakup of Pangaea. Evolution and specialization of both groups continued undisturbed until about 3 million years ago when the Panamanian land-bridge connected the two American continents. This event permitted the exchange of fauna between the two continents. Monkeys, armadillos, sloths, and opossums arrived in North America, while various types of horses, bears, rhinos, camels, and wolves migrated southward. Many animals that had been unique to South America disappeared completely after this event, including hoofed mammals, rhino-sized rodents, and a number of carnivorous marsupials. Because this period of extinction coincided with the formation of the Panamanian land-bridge, it was thought that the advanced carnivores from North America were responsible. Recent research, however, suggests that other factors, including climatic changes, may have played a significant role.

lead more active lives and to occupy more diverse habitats than reptiles because they could survive in cold regions. (Most modern reptiles are dormant during cold weather.) Other mammalian adaptations included the development of insulating body hair and more efficient heart and lungs.

With the demise of the large Mesozoic reptiles, Cenozoic mammals diversified rapidly. The many forms that exist today evolved from small primitive mammals that were characterized by short legs, flat five-toed feet, and small brains. Their development and specialization took four principal directions: (1) increase in size, (2) increase in brain capacity, (3) specialization of teeth to better accommodate

Large Mammals and Extinction

As you have seen, mammals diversified rapidly during the Cenozoic era. One tendency was for some groups to become very large. For example, by the Oligocene epoch, a hornless rhinoceros that stood nearly 5 meters (16 feet) high had evolved. It is the largest land mammal known to have existed. As time passed, many other mammals evolved to larger sizes—more, in fact, than now exist. Many of these large forms were common as recently as 11,000 years ago. However, a wave of late Pleistocene extinctions rapidly eliminated these animals from the landscape.

In North America, the mastodon and mammoth, both huge relatives of the elephant, became extinct. In addition, saber-toothed cats, giant beavers, large ground sloths, horses, camels, giant bison, and others died out (Figure 22.35). In Europe, late Pleistocene extinctions included woolly rhinos,

FIGURE 22.35 Artist's depiction of extinct Cenozoic mammals. (Photo by Chase Studio/Photo Researchers, Inc.)

large cave bears, and the Irish elk. The reason for this recent wave of extinctions that targeted large animals puzzles scientists. These animals had survived several major glacial advances and interglacial periods, so it is difficult to ascribe these extinctions to climate change. Some scientists hypothesize that early humans hastened the decline of these mammals by selectively hunting large forms.

Summary

- The history of Earth began about 13.7 billion years ago when the first elements were created during the *Big Bang*. It was from this material, plus other elements ejected into interstellar space by now defunct stars, that Earth along with the rest of the solar system formed. As material collected, high velocity impacts of chunks of matter called *planetesimals* and the decay of radioactive elements caused the temperature of our planet to steadily increase. Iron and nickel melted and sank to form the metallic core, while rocky material rose to form the mantle and Earth's initial crust.

- Earth's primitive atmosphere, which consisted mostly of water vapor and carbon dioxide, formed by a process called *outgassing*, which resembles the steam eruptions of modern volcanoes. About 3.5 billions years ago, photosynthesizing bacteria began to release oxygen, first into the oceans and then into the atmosphere. This began the evolution of our modern atmosphere. The oceans, formed early in Earth's history, as water vapor condensed to form clouds, and torrential rains filled low-lying areas. The salinity in seawater came from volcanic outgassing and from elements weathered and eroded from Earth's primitive crust.

- The Precambrian, which is divided into the Archean and Proterozoic eons, spans nearly 90 percent of Earth's history, beginning with the formation of Earth about 4.5 billion years ago and ending approximately 542 million years ago. During this time, much of Earth's stable continental crust was created through a multi-stage process. First, partial melting of the mantle generated magma that rose to form volcanic island arcs and oceanic plateaus. These thin crustal fragments collided and accreted to form larger crustal provinces, which, in turn assembled into larger blocks called *cratons*. Cratons, which form the core of modern continents, were created mainly during the Precambrian.

- Supercontinents are large landmasses that consist of all, or nearly all, existing continents. *Pangaea* was the most recent supercontinent, but a massive southern continent called *Gondwana*, and perhaps an even larger one, *Rodinia*, preceded it. The splitting and reassembling of supercontinents have generated most of Earth's major mountain belts. In addition, the movement of these crustal blocks have profoundly affected Earth's climate, and have caused sea level to rise and fall.

- The time span following the close of the Precambrian, called the *Phanerozoic eon*, encompasses 542 million years and is divided into three eras: *Paleozoic*, *Mesozoic*, and *Cenozoic*. The Paleozoic era was dominated by continental collisions as the supercontinent of Pangaea assembled—forming the Caledonian, Appalachian, and Ural Mountains. Early in the Mesozoic, much of the land was above sea level. However, by the middle Mesozoic, seas invaded western North America. As Pangaea began to break up, the westward-moving North American plate began to override the Pacific plate, causing crustal deformation along the entire western margin of North America. Most of North America was above sea level throughout the Cenozoic. Owing to their different relations with plate boundaries, the eastern and western margins of the continent experienced contrasting events. The stable eastern margin was the site of abundant sedimentation as isostatic adjustment raised the Appalachians, causing streams to erode with renewed vigor and deposit their sediment along the continental margin. In the West, building of the Rocky Mountains (the *Laramide Orogeny*) was coming to an end, the Basin and Range Province was forming, and volcanic activity was extensive.

- The first known organisms were single-celled bacteria, *prokaryotes*, which lack a nucleus. One group of these organisms, called cyanobacteria, that used solar energy to synthesize organic compounds (sugars) evolved. For the first time, organisms had the ability to produce their own food. Fossil evidence for the existence of these bacteria includes layered mounds of calcium carbonate called *stromatolites*.

- The beginning of the Paleozoic is marked by the *appearance of the first life-forms with hard parts* such as shells. Therefore, abundant Paleozoic fossils occur, and a far more detailed record of Paleozoic events can be constructed. Life in the early Paleozoic was restricted to the seas and consisted of several invertebrate groups, including trilobites, cephalopods, sponges and corals. During the Paleozoic, organisms diversified dramatically. Insects and plants moved onto land, and lobe-finned fishes that adapted to land became the first amphibians. By the Pennsylvanian period, large tropical swamps, which became the major coal deposits of today, extended across North America, Europe, and Siberia. At the close of the Paleozoic, a mass extinction destroyed 70 percent of all vertebrate species on land and 90 percent of all marine organisms.

- The Mesozoic era, literally the era of middle life, is often called the "*Age of Reptiles.*" Organisms that survived the extinction at the end of the Paleozoic began to diversity in spectacular ways. *Gymnosperms* (cycads, conifers, and ginkgoes) became the dominant trees of the Mesozoic because they could adapt to the drier climates. Reptiles became the dominant land animals. The most awesome of the Mesozoic reptiles were the *dinosaurs*. At the close of the Mesozoic, many large reptiles, including the dinosaurs, became extinct.

- The Cenozoic is often called the *"Age of Mammals"* because these animals replaced the reptiles as the dominant vertebrate life forms on land. Two groups of mammals, the marsupials and the placentals, evolved

and expanded during this era. One tendency was for some mammal groups to become very large. However, a wave of late *Pleistocene* extinctions rapidly eliminated these animals from the landscape. Some scientists suggest that early humans hastened their decline by selectively hunting the larger animals. The Cenozoic could also be called the *"Age of Flowering Plants."* As a source of food, flowering plants (angiosperms) strongly influenced the evolution of both birds and herbivorous (plant-eating) mammals throughout the Cenozoic era.

Review Questions

1. Why is Earth's molten, metallic core important to humans living today?

2. What two elements made up most of the very early universe?

3. What is the cataclysmic event called in which an exploding star produces all of the elements heavier than iron?

4. Briefly describe the formation of the planets from the solar nebula.

5. What is meant by outgassing and what modern phenomenon serves that role today?

6. Outgassing produced Earth's early atmosphere, which was rich in what two gases?

7. Why is the evolution of a type of bacteria that employed photosynthesis to produce food important to most modern organisms?

8. What was the source of water for the first oceans?

9. How does the ocean remove carbon dioxide from the atmosphere? What role do tiny marine organisms, such as foraminifera, play?

10. Explain why Precambrian history is more difficult to decipher than more recent geological history.

11. Briefly describe how cratons come into being.

12. What is the supercontinent cycle?

13. How can the movement of continents trigger climate change?

14. Match the following words and phrases to the most appropriate time span. Select from the following: *Precambrian, Paleozoic, Mesozoic, Cenozoic.*

    a. Pangaea came into existence.

    b. First trace fossils.

    c. The era that encompasses the least amount of time.

d. Earth's major cratons formed.

e. "Age of Dinosaurs."

f. Formation of the Rocky Mountains.

g. Formation of the Appalachian Mountains.

h. Coal swamps extended across North America, Europe, and Siberia.

i. Gulf Coast oil deposits formed.

j. Formation of most of the world's major iron-ore deposits.

k. Massive sand dunes covered a large portion of the Colorado Plateau region.

l. The "Age of the Fishes" occurred during this span.

m. Pangaea began to break apart and disperse.

n. "Age of Mammals."

o. Animals with hard parts first appeared in abundance.

p. Gymnosperms were the dominant trees.

q. Stromatolites were abundant.

r. Fault-block mountains formed in the Basin and Range region.

15. Contrast the eastern and western margins of North America during the Cenozoic era in terms of their relationships to plate boundaries.

16. What did plants have to overcome to move onto land?

17. What group of animals is thought to have left the ocean to become the first amphibians?

18. Why are amphibians not considered "true" land animals?

19. What major development allowed reptiles to move inland?

20. What event is thought to have ended the reign of the dinosaurs?

Key Terms

banded iron formations
    (p. 602)
cratons (p. 604)
eukaryotes (p. 613)

outgassing (p. 601)
planetesimals (p. 598)
prokaryotes (p. 613)
protoplanets (p. 599)

solar nebula (p. 598)
stromatolites (p. 613)
supercontinent (p. 606)

supercontinent cycle
    (p. 607)
supernova (p. 598)

Web Resources

The *Earth* Website uses the resources and flexibility of the Internet to aid in your study of the topics in this chapter. Written and developed by geology instructors, this site will help improve your understanding of geology. Visit **http://www.prenhall.com/tarbuck** and click on the cover of *Earth 9e* to find:

- Online review quizzes.
- Critical thinking exercises.
- Links to chapter-specific Web resources.
- Internet-wide key-term searches.

http://www.prenhall.com/tarbuck

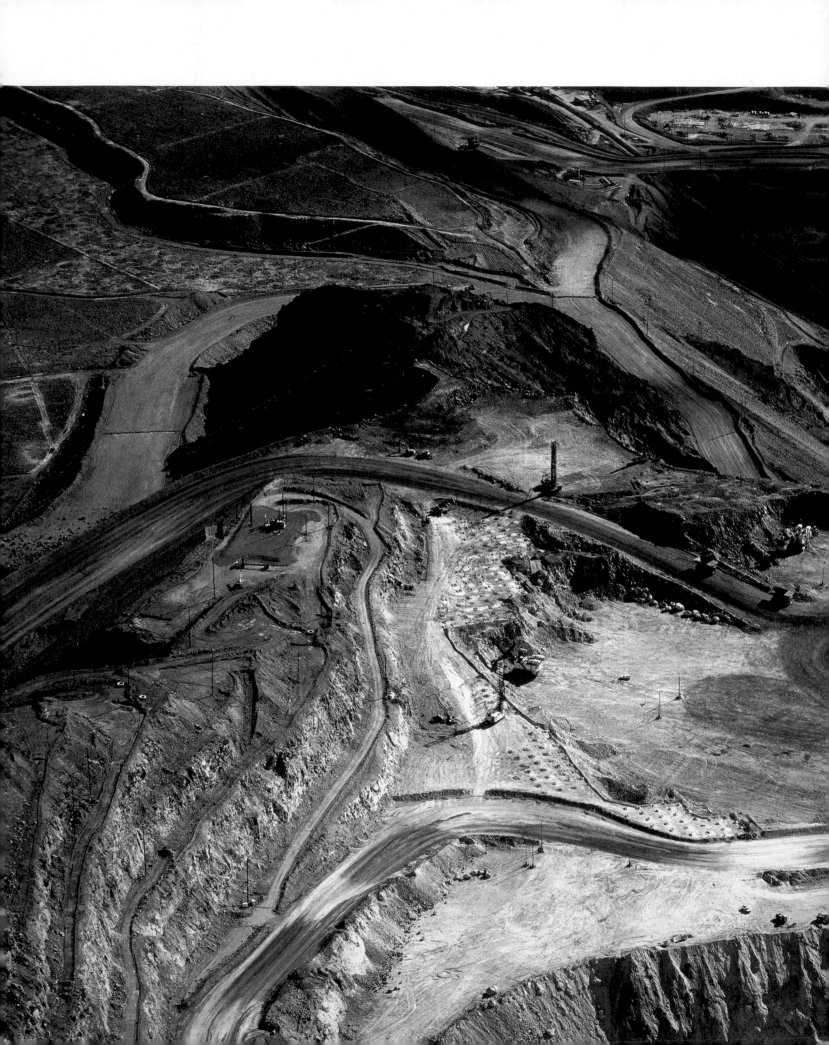

Energy and Mineral Resources

A large open pit copper mine at Morenci, Arizona. Copper mining at this site has been going on since the 1870s. (Photo by Michael Collier)

627

M aterials that we extract from Earth are the basis of modern civilization (Figure 23.1). Mineral and energy resources from the crust are the raw materials from which the products used by society are made. Like most people who live in highly industrialized nations, you may not realize the quantity of resources needed to maintain your present standard of living. Figure 23.2 shows the annual per capita consumption of several important metallic and nonmetallic mineral resources for the United States. This is each person's prorated share of the materials required by industry to provide the vast array of homes, cars, electronics, cosmetics, packaging, and so on that modern society demands. Figures for other highly industrialized countries, such as Canada, Australia, and several nations in Western Europe are comparable.

The number of different mineral resources required by modern industries is large. Although some countries, including the United States, have substantial deposits of many important minerals, no nation is entirely self-sufficient. This reflects the fact that important deposits are limited in number and localized in occurrence. All countries must rely on international trade to fulfill at least some of their needs.

Renewable and Nonrenewable Resources

Resources are commonly divided into two broad categories—renewable and nonrenewable. **Renewable resources** can be replenished over relatively short time spans such as months, years, or decades. Common examples are plants and animals for food, natural fibers for clothing, and trees for lumber and paper. Energy from flowing water, wind, and the Sun are also considered renewable.

By contrast, **nonrenewable resources** continue to be formed in Earth, but the processes that create them are so slow that significant deposits take millions of years to

FIGURE 23.1   As this scene of Houston, Texas, reminds us, mineral and energy resources are the basis of modern civilization. (Photo by H. R. Bramaz/Peter Arnold, Inc.)

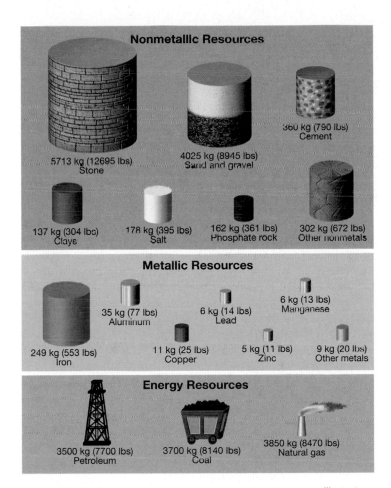

Nonmetallic Resources

5713 kg (12695 lbs)
Stone

4025 kg (8945 lbs)
Sand and gravel

360 kg (790 lbs)
Cement

137 kg (304 lbs)
Clays

178 kg (395 lbs)
Salt

162 kg (361 lbs)
Phosphate rock

302 kg (672 lbs)
Other nonmetals

Metallic Resources

35 kg (77 lbs)
Aluminum

6 kg (14 lbs)
Lead

6 kg (13 lbs)
Manganese

249 kg (553 lbs)
Iron

11 kg (25 lbs)
Copper

5 kg (11 lbs)
Zinc

9 kg (20 lbs)
Other metals

Energy Resources

3500 kg (7700 lbs)
Petroleum

3700 kg (8140 lbs)
Coal

3850 kg (8470 lbs)
Natural gas

FIGURE 23.2 The annual per capita consumption of nonmetallic and metallic mineral resources for the United States is about 11,000 kilograms (12 tons)! About 97 percent of the materials used are nonmetallic. The per capita use of oil, coal, and natural gas exceeds 11,000 kilograms. (After U.S. Geological Survey)

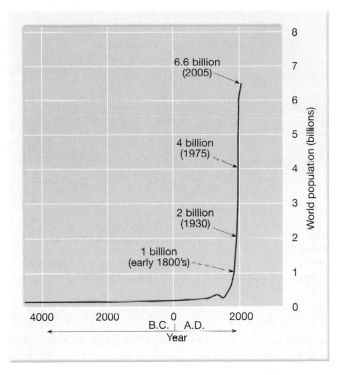

FIGURE 23.3 Growth of world population. It took until 1800 for the number to reach 1 billion. By the year 2015, more than 7 billion people may inhabit the planet. The demand for basic resources is growing faster than the rate of population increase. (Data from the Population Reference Bureau)

accumulate. For human purposes, Earth contains fixed quantities of these substances. When the present supplies are mined or pumped from the ground, there will be no more. Examples are fuels (coal, oil, natural gas) and many important metals (iron, copper, uranium, gold). Some of these nonrenewable resources, such as aluminum, can be used over and over again; others, such as oil, cannot be recycled.

Occasionally some resources can be placed in either category, depending on how they are used. Groundwater is one such example. Where it is pumped from the ground at a rate that can be replenished, groundwater can be classified as a renewable resource. However, in places where groundwater is withdrawn faster than it is replenished, the water table drops steadily. In this case, the groundwater is being "mined" just like other nonrenewable resources.*

Figure 23.3 shows that the population of our planet is growing rapidly. Although the number did not reach 1 billion until the beginning of the 19th century, just 130 years later the population doubled to 2 billion. Between 1930 and 1975 the figure doubled again to 4 billion, and by 2015 more

than 7 billion people may inhabit the planet. Clearly, as population grows, the demand for resources expands as well. However, the rate of mineral-and-energy resource usage has climbed faster than population growth. This results from an increasing standard of living. In the United States only 6 percent of the world's population uses approximately 30 percent of the world's annual production of mineral and energy resources!

How long can our remaining resources sustain the rising standard of living in today's industrialized countries and still provide for the growing needs of developing regions? How much environmental deterioration are we willing to accept in pursuit of resources? Can alternatives be found? If we are to cope with an increasing per capita demand and a growing world population, we must understand our resources and their limits.

Energy Resources

Coal, petroleum, and natural gas are the primary fuels of our modern industrial economy (Figure 23.4). About 86 percent of the energy consumed in the United States today comes from these basic fossil fuels. Although major shortages of oil and gas will not occur for many years, proven reserves are declining. Despite new exploration, even in very remote regions and severe environments, new sources are not keeping pace with consumption.

*The problem of declining water-table levels is discussed in Chapter 17.

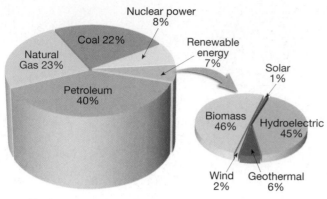

Total = 100 quadrillion btu

FIGURE 23.4 U.S. energy consumption, 2004. The total was 100 quadrillion Btu. A quadrillion, by the way, is 10 raised to the 12th power, or a million million—a quadrillion Btu is a convenient unit for referring to U.S. energy use as a whole. (Source: U.S. Department of Energy, Energy Information Administration)

Unless large, new petroleum reserves are discovered (which is possible, but not likely), a greater share of our future needs will have to come from coal and from alternative energy sources such as nuclear, geothermal, solar, wind, tidal, and hydroelectric power (see Box 23.1). Two fossil-fuel alternatives—oil sands and oil shale—are sometimes mentioned as promising sources of liquid fuels. In the following sections, we will briefly examine the fuels that have traditionally supplied our energy needs, as well as sources that will provide an increasing share of our future requirements.

Students Sometimes Ask . . .

Figure 23.4 shows biomass as a form of renewable energy. What exactly is biomass?

Biomass refers to organic matter that can be burned directly as fuel or converted into a different form and then burned. Biomass is a relatively new name for some of the oldest human fuels. Examples include firewood, charcoal, crop residues, and animal waste. These fuels are especially important in countries with emerging economies. More modern biomass fuels include such products as ethanol (a type of alcohol produced from corn and other crops that is added to gasoline) and biodiesel (a diesel-equivalent fuel derived from natural renewable sources such as vegetable oils).

Coal

Along with oil and natural gas, coal is commonly called a **fossil fuel.** Such a designation is appropriate because each time we burn coal we are using energy from the Sun that was stored by plants many millions of years ago. We are indeed burning a "fossil."

Coal has been an important fuel for centuries. In the 19th and early 20th centuries, cheap and plentiful coal powered the Industrial Revolution. By 1900 coal was providing 90 percent of the energy used in the United States. Although still important, coal currently accounts for about 22 percent of the nation's energy needs (Figure 23.4).

Until the 1950s, coal was an important domestic heating fuel as well as a power source for industry. However, its direct use in the home has been largely replaced by oil, natural gas, and electricity. These fuels are preferred because they are more readily available (delivered via pipes, tanks, or wiring) and cleaner to use.

Nevertheless, coal remains the major fuel used in power plants to generate electricity, and it is therefore indirectly an important source of energy for our homes. More than 70 percent of present-day coal usage is for the generation of electricity. As oil reserves gradually diminish in the years to come, the use of coal may increase. Expanded coal production is possible because the world has enormous reserves and the technology to mine coal efficiently. In the United States, coal fields are widespread and contain supplies that should last for hundreds of years (Figure 23.5).

Although coal is plentiful, its recovery and use present a number of problems. Surface mining can turn the countryside into a scarred wasteland if careful (and costly) reclamation is not carried out to restore the land. (Today all U.S. surface mines must reclaim the land.) Although underground mining does not scar the landscape to the same degree, it has been costly in terms of human life and health.

Moreover, underground mining long ago ceased to be a pick-and-shovel operation and is today a highly mechanized and computerized process (Figure 23.6). Strong federal safety regulations have made U.S. mining quite safe. However, the hazards of collapsing roofs, gas explosions, and working with heavy equipment remain.

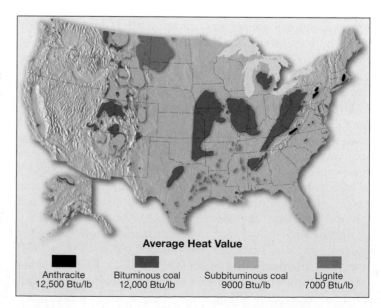

Average Heat Value

Anthracite 12,500 Btu/lb — Bituminous coal 12,000 Btu/lb — Subbituminous coal 9000 Btu/lb — Lignite 7000 Btu/lb

FIGURE 23.5 Coal fields of the United States. (Data from the Bureau of Mines, U.S. Department of the Interior)

Air pollution is a major problem associated with the burning of coal. Much coal contains significant quantities of sulfur. Despite efforts to remove sulfur before the coal is burned, some remains; when the coal is burned, the sulfur is converted into noxious sulfur oxide gases. Through a series of complex chemical reactions in the atmosphere, the sulfur oxides are converted to sulfuric acid, which then falls to Earth's surface as rain or snow. This acid precipitation can have adverse ecological effects over widespread areas (see Box 6.2, p. 174).

As is the case when other fossil fuels are burned, the combustion of coal produces carbon dioxide. This major "greenhouse gas" plays a significant role in the heating of our atmosphere. Chapter 21, "Global Climate Change," examines this issue in some detail.

Oil and Natural Gas

Petroleum and natural gas are found in similar environments and frequently occur together. Both consist of various hydrocarbon compounds (compounds consisting of hydrogen and carbon) mixed together. They may also contain small quantities of other elements, such as sulfur, nitrogen, and oxygen. Like coal, petroleum and natural gas are biological products derived from the remains of organisms. However, the environments in which they form are very different, as are the organisms. Coal is formed mostly from plant material that accumulated in a swampy environment above sea level. Oil and gas are derived from the remains of both plants and animals having a marine origin.

Petroleum Formation

Petroleum formation is complex and not completely understood. Nonetheless, we know that it begins with the accumulation of sediment in ocean areas that are rich in plant and animal remains. These accumulations must occur where biological activity is high, such as in nearshore areas. However, most marine environments are oxygen-rich, which leads to the decay of organic remains before they can be buried by other sediments. Therefore, accumulations of oil and gas are not as widespread as are the marine environments that support abundant biological activity. This limiting factor notwithstanding, large quantities of organic matter are buried and protected from oxidation in many offshore sedimentary basins. With increasing burial over millions of years, chemical reactions gradually transform some of the original organic matter into the liquid and gaseous hydrocarbons we call petroleum and natural gas.

Unlike the organic matter from which they formed, the newly created petroleum and natural gas are mobile. These fluids are gradually squeezed from the compacting, mud-rich layers where they originate into adjacent permeable beds such as sandstone, where openings between sediment grains are larger. Because this occurs under water, the rock layers containing the oil and gas are saturated with water. But oil and gas are less dense than water, so they migrate upward through the water-filled pore spaces of the enclosing rocks. Unless something halts this upward migration, the fluids will eventually reach the surface, at which point the volatile components will evaporate.

FIGURE 23.6   A. Modern underground coal mining is highly mechanized and relatively safe. (Photo by Melvin Grubb/Grubb Photo Services, Inc.) **B.** Strip mining of coal at Black Mesa, Arizona. Surface mining is common when coal seams are near the surface. (Photo by Richard W. Brooks/Photo Researchers, Inc.)

A.

B.

BOX 23.1 ▶ UNDERSTANDING EARTH

Gas Hydrates—A Fuel from Ocean-Floor Sediments

Gas hydrates are unusually compact chemical structures made of water and natural gas. The most common type of natural gas is methane, which produces *methane hydrate.* Gas hydrates occur beneath permafrost areas on land and under the ocean floor at depths below 525 meters (1720 feet).

Most oceanic gas hydrates are created when bacteria break down organic matter trapped in seafloor sediments, producing methane gas with minor amounts of ethane and propane. These gases combine with water in deep-ocean sediments (where pressures are high and temperatures are low) in such a way that the gas is trapped inside a latticelike cage of water molecules.

Vessels that have drilled into gas hydrates have retrieved cores of mud mixed with chunks or layers of gas hydrates (Figure 23.A) that fizzle and evaporate quickly when they are exposed to the relatively warm, low-pressure conditions at the ocean surface. Gas hydrates resemble chunks of ice but ignite when lit by a flame because methane and other flammable gases are released as gas hydrates vaporize (Figure 23.B).

Some estimates indicate that as much as 20 quadrillion cubic meters (700 quadrillion cubic feet) of methane are locked up in sed-

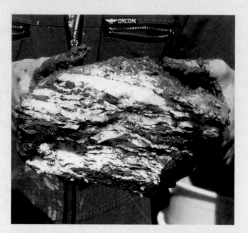

FIGURE 23.A This sample retrieved from the ocean floor shows layers of white icelike gas hydrate mixed with mud. (Photo courtesy of GEOMAR Research Center, Kiel, Germany)

iments containing gas hydrates. This is equivalent to about *twice* as much carbon as Earth's coal, oil, and conventional gas reserves combined, so gas hydrates have great potential. One major drawback in exploiting reserves of gas hydrate is that they rapidly decompose at surface temperatures

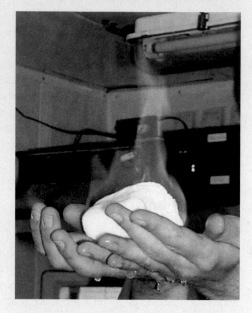

FIGURE 23.B Gas hydrates evaporate when exposed to surface conditions and release natural gas, which can be ignited. (Photo courtesy of GEOMAR Research Center, Kiel, Germany)

and pressures. In the future, however, these vast seafloor reserves of energy may help power modern society.

Oil Traps

Sometimes the upward migration is halted. A geologic environment that allows for economically significant amounts of oil and gas to accumulate underground is termed an **oil trap.** Several geologic structures may act as oil traps, but all have two basic conditions in common: a porous, permeable **reservoir rock** that will yield petroleum and natural gas in sufficient quantities to make drilling worthwhile; and a **cap rock,** such as shale, that is virtually impermeable to oil and gas. The cap rock halts the upwardly mobile oil and gas and keeps the oil and gas from escaping at the surface (Figure 23.7).

Figure 23.8 illustrates some common oil and natural-gas traps. One of the simplest traps is an *anticline,* an uparched series of sedimentary strata (Figure 23.8A). As the strata are bent, the rising oil and gas collect at the apex (top) of the fold. Because of its lower density, the natural gas collects above the oil. Both rest upon the denser water that saturates the reservoir rock. One of the world's largest oil fields, El Nala in Saudi Arabia, is the result of an anticlinal trap, as is the famous Teapot Dome in Wyoming.

Fault traps form when strata are displaced in such a manner as to bring a dipping reservoir rock into position oppo-

site an impermeable bed, as shown in Figure 23.8B. In this case the upward migration of the oil and gas is halted where it encounters the fault.

In the Gulf coastal plain region of the United States, important accumulations of oil occur in association with *salt domes.* Such areas have thick accumulations of sedimentary strata, including layers of rock salt. Salt occurring at great depths has been forced to rise in columns by the pressure of overlying beds. These rising salt columns gradually deform the overlying strata. Because oil and gas migrate to the highest level possible, they accumulate in the upturned sandstone beds adjacent to the salt column (Figure 23.8C).

Yet another important geologic circumstance that may lead to significant accumulations of oil and gas is termed a *stratigraphic trap.* These oil-bearing structures result primarily from the original pattern of sedimentation rather than structural deformation. The stratigraphic trap illustrated in Figure 23.8D exists because a sloping bed of sandstone thins to the point of disappearance.

When the lid created by the cap rock is punctured by drilling, the oil and natural gas, which are under pressure, migrate from the pore spaces of the reservoir rock to the drill hole. On rare occasions, when fluid pressure is great, it

A.

B.

FIGURE 23.7 Oil accumulates in oil traps that consist of porous, permeable *reservoir rock* overlain by an impermeable *cap rock*. **A.** Modern offshore oil production platform in the North Sea. (Photo by Peter Bowater/Photo Researchers, Inc.) **B.** The first successful oil well was completed by Edwin Drake (right) on August 27, 1859, near Titusville, Pennsylvania. The oil-bearing reservoir rock was encountered at a depth of 21 meters (69 feet). (Photo by CORBIS/Bettman)

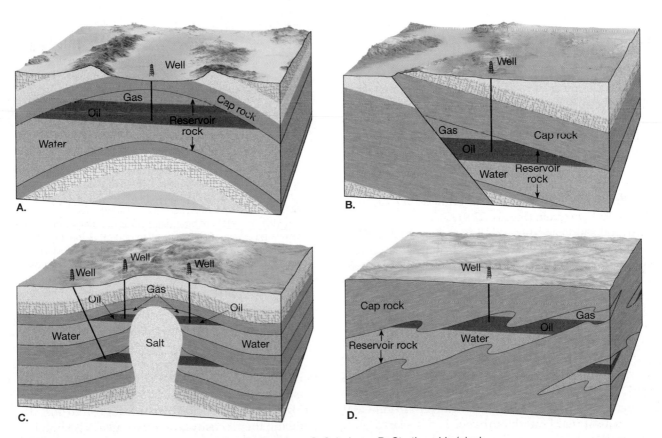

FIGURE 23.8 Common oil traps. **A.** Anticline. **B.** Fault trap. **C.** Salt dome. **D.** Stratigraphic (pinch-out) trap.

may force oil up the drill hole to the surface, causing a "gusher" or oil fountain at the surface. Usually, however, a pump is required to lift the oil out.

A drill hole is not the only means by which oil and gas can escape from a trap. Traps can be broken by natural forces. For example, Earth movements may create fractures that allow the hydrocarbon fluids to escape. Surface erosion may breach a trap with similar results. The older the rock strata, the greater the chance that deformation or erosion has affected a trap. Indeed, not all ages of rock yield oil and gas in the same proportions. The greatest production comes from the youngest rocks, those of Cenozoic age. Older Mesozoic rocks produce considerably less, followed by even smaller yields from the still older Paleozoic strata. There is virtually no oil produced from the most ancient rocks, those of Precambrian age.

Oil Sands and Oil Shale— Petroleum for the Future?

In years to come, world oil supplies will dwindle. When this occurs, a lower grade of hydrocarbon may have to be substituted. Are the fuels derived from oil sands and oil shales good candidates?

Oil Sands

Oil sands (also called *tar sands*) are usually mixtures of clay and sand combined with water and varying amounts of a black, highly viscous tar known as *bitumen*. The use of the term *sand* can be misleading, because not all deposits are associated with sands and sandstones. Some occur in other materials, including shales and limestones. The oil in these deposits is very similar to heavy crude oils pumped from wells. The major difference between conventional oil reservoirs and oil sand deposits is in the viscosity (resistance to flow) of the oil they contain. In oil sands, the oil is much more viscous and cannot simply be pumped out.

Substantial oil sand deposits occur in several locations around the world. The two largest are the Athabasca Oil Sands in the Canadian province of Alberta and the Orinoco River deposit in Venezuela (Figure 23.9).

Currently, oil sands are mined at the surface in a manner similar to the strip mining of coal. The excavated material is then heated with pressurized steam until the bitumen softens and rises. Once collected, the oily material is treated to remove impurities and then hydrogen is added. This last step upgrades the material to a synthetic crude, which can then be refined. Extracting and refining oil sands requires a great deal of energy—nearly half as much as the end product yields! Nevertheless, oil sands from Alberta's vast deposits are the source of about one-third of Canada's oil production.

Obtaining oil from oil sand has significant environmental drawbacks. Substantial land disturbance is associated with mining huge quantities of rock and sediment. Moreover,

FIGURE 23.9 **A.** In North America, the largest oil sand deposits occur in the Canadian province of Alberta. Known as the Athabasca Oil Sands, these deposits cover an area of more than 42,000 square kilometers (16,300 square miles). Albertas major oil sands deposits contain more than 1.7 trillion barrels of bitumen. However, much of the bitumen cannot be recovered at reasonable cost. With current technology, it is estimated that only about 300 billion barrels are recoverable. **B.** Mining of oil sands in Alberta. Notice the "tar" on the hands of the worker. (Photo by Burkard/ Bilderberg/Peter Arnold, Inc.)

large quantities of water are required for processing, and when processing is completed, contaminated water and sediment accumulate in toxic disposal ponds.

About 80 percent of the oil sands in Alberta are buried too deeply for surface mining. Oil from these deep deposits must be recovered by *in situ* techniques. Using drilling technology, steam is injected into the deposit to heat the oil sand, reducing the viscosity of the bitumen. The hot, mobile bitumen migrates towards producing wells, which pump it to the surface, while the sand is left in place ("*in situ*" is Latin for "in place"). *In situ* technology is expensive and requires certain conditions like a nearby water source. Production using *in situ* techniques already rivals open pit mining and in the future may well replace mining as the main source of bitumen production from the oil sands.

Challenges facing *in situ* processes include increasing the efficiency of oil recovery, management of water used to make steam, and reducing the costs of energy required for the process.

Oil Shale

Oil shale contains enormous amounts of untapped oil. Worldwide, the U.S. Geological Survey estimates that there are more than 3000 billion barrels of oil contained in shales that would yield more than 38 liters (10 gallons) of oil per ton of shale. But this figure is misleading because fewer than 200 billion barrels are known to be recoverable with present technology. Still, estimated U.S. resources are about 14 times as great as those of conventionally recoverable oil, and estimates will probably increase as more geological information is gathered.

Roughly half of the world supply is in the Green River Formation in Colorado, Utah, and Wyoming (Figure 23.10). Within this region the oil shales are part of sedimentary layers that accumulated at the bottoms of two vast, shallow

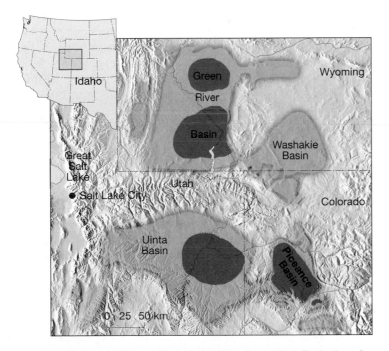

FIGURE 23.10   Distribution of oil shale in the Green River Formation of Colorado, Utah, and Wyoming. The areas shaded with the darker color represent the richest deposits. Government and industry have invested large sums to make these oil shales an economic resource, but costs have always been higher than the price of oil. However, as prices of competing fuels rise, these vast deposits may someday become economically more attractive. (After D. C. Duncan and V. E. Swanson, U.S. Geological Survey Circular 523, 1965)

lakes during the Eocene epoch (57 million to 36 million years ago).

Oil shale has been suggested as a partial solution to dwindling fuel supplies. However, the heat energy in oil shale is about one-eighth that in crude oil, owing to the large proportion of mineral matter in the shales.

This mineral matter adds cost to mining, processing, and waste disposal. Producing oil from oil shale has the same problems as producing oil from oil sands. Surface mining causes widespread land disturbance and presents significant waste-disposal problems. Moreover, processing requires large amounts of water, something that is in short supply in the semiarid region occupied by the Green River Formation.

At present, oil is plentiful and relatively inexpensive on world markets. Therefore, with current technologies, most oil shales are not worth mining. Oil shale research-and-development efforts have been nearly abandoned by industry. Nevertheless, the U.S. Geological Survey suggests that the large amount of oil that potentially can be extracted from oil shales in the United States probably ensures its eventual inclusion in the national energy mix.

Alternate Energy Sources

An examination of Figure 23.4 (see p. 630) clearly shows that we live in the era of fossil fuels. More than 85 percent of United States's energy needs comes from these nonrenewable resources. Present estimates indicate that the amount of recoverable fossil fuels may equal 10 trillion barrels of oil, which is enough to last 170 years at the present rate of consumption. Of course, as world population soars, the rate of consumption will climb. Thus, reserves will eventually be in short supply. In the meantime, the environmental impact of burning huge quantities of fossil fuels will undoubtedly have an adverse effect.

How can a growing demand for energy be met without radically affecting the planet we inhabit? Although no clear answer has yet been formulated, the need to rely more heavily on alternate energy sources must be considered. In this section, we will examine several possible sources, including nuclear, solar, wind, hydroelectric, geothermal, and tidal energy.

Nuclear Energy

Roughly 8 percent of the energy demand of the United States is being met by nuclear power plants. The fuel for these facilities comes from radioactive materials that release energy by the process of **nuclear fission.** Fission is accomplished by bombarding the nuclei of heavy atoms, commonly uranium-235, with neutrons. This causes the uranium nuclei to split into smaller nuclei and to emit neutrons and heat energy. The ejected neutrons, in turn, bombard the nuclei of adjacent uranium atoms, producing a *chain reaction.* If the supply of fissionable material is sufficient and if the reaction is allowed to proceed in an uncontrolled manner, an enormous amount of energy would be released in the form of an atomic explosion.

In a nuclear power plant, the fission reaction is controlled by moving neutron-absorbing rods into or out of the nuclear reactor. The result is a controlled nuclear chain reaction that releases great amounts of heat. The energy produced is transported from the reactor and used to drive steam turbines that turn electrical generators, which is similar to what occurs in most conventional power plants.

Uranium Uranium-235 is the only naturally occurring isotope that is readily fissionable and is therefore the primary fuel used in nuclear power plants.* Although large quantities of uranium ore have been discovered, most contain less than 0.05 percent uranium. Of this small amount, 99.3 percent is the nonfissionable isotope uranium-238 and just 0.7 percent consists of the fissionable isotope uranium-235. Because most nuclear reactors operate with fuels that are at least 3 percent uranium-235, the two isotopes must be separated in order to concentrate the fissionable uranium-235. The process of separating the uranium isotopes is difficult and substantially increases the cost of nuclear power.

Although uranium is a rare element in Earth's crust, it does occur in enriched deposits. Some of the most important occurrences are associated with what are believed to be ancient placer deposits in stream beds.** For example, in Witwatersrand, South Africa, grains of uranium ore (as well as rich gold deposits) were concentrated by virtue of their high density in rocks made largely of quartz pebbles. In the United States the richest uranium deposits are found in Jurassic and Triassic sandstones in the Colorado Plateau and in younger rocks in Wyoming. Most of these deposits have formed through the precipitation of uranium compounds from groundwater. Here precipitation of uranium occurs as a result of a chemical reaction with organic matter, as evidenced by the concentration of uranium in fossil logs and organic-rich black shales.

Obstacles to Development At one time nuclear power was heralded as the clean, cheap source of energy to replace fossil fuels. However, several obstacles have emerged to hinder the development of nuclear power as a major energy source. Not the least of these is the skyrocketing costs of building nuclear facilities that contain numerous safety features. More important, perhaps, is the concern over the possibility of a serious accident at one of the nearly 200 nuclear plants in existence worldwide (Figure 23.11). The 1979 accident at Three Mile Island near Harrisburg, Pennsylvania, helped bring this point home. Here, a malfunction led the plant operators to believe there was too much water in the primary system instead of too little. This confusion allowed the reactor core to lie uncovered for several hours. Although there was little danger to the public, substantial damage to the reactor resulted.

FIGURE 23.11 Diablo Canyon nuclear power plant near San Luis Obispo, California. Reactors are in the dome-shaped buildings, and cooling water is being released to the ocean. The setting of this facility was controversial because of its close proximity to faults capable of producing potentially damaging earthquakes. (Photo by Comstock)

Unfortunately, the 1986 accident at Chernobyl in the former Soviet Union was far more serious. In this incident, the reactor ran out of control and two small explosions lifted the roof off the structure, allowing pieces of uranium to be thrown over the immediate area. During the 10 days that it took to quench the fire that ensued, high levels of radioactive material were carried by the atmosphere and detected as far away as Norway. In addition to the 18 people who died within six weeks of the accident, many thousands more face an increased risk of death from cancers associated with the fallout.

It should be emphasized that the concentrations of fissionable uranium-235 and the design of reactors are such that nuclear power plants cannot explode like an atomic bomb. The dangers arise from the possible escape of radioactive debris during a meltdown of the core or other malfunction. In addition, hazards such as the disposal of nuclear waste and the relationship that exists between nuclear energy programs and the proliferation of nuclear weapons must be considered as we evaluate the pros and cons of employing nuclear power.

Among the "pros" for nuclear energy is the fact that nuclear power plants do not emit carbon dioxide—the greenhouse gas that contributes significantly to global warming (see Chapter 21). By contrast, the generation of electricity from fossil fuels produces large quantities of carbon dioxide. Thus substituting nuclear power for power generated by fossil fuels represents one option for reducing carbon emissions.

Solar Energy

The term *solar energy* generally refers to the direct use of the Sun's rays to supply energy for the needs of people. The simplest and perhaps most widely used *passive solar collectors* are south-facing windows. As sunlight passes through the glass, its energy is absorbed by objects in the room. These objects in turn radiate heat that warms the air. In the

---

*Thorium, although not capable by itself of sustaining a chain reaction, can be used with uranium-235 as a nuclear fuel.

**Placer deposits are discussed later in the chapter.

A.

B.

FIGURE 23.12 **A.** Solar One, a solar installation used to generate electricity in the Mojave Desert near Barstow, California. (Photo by Thomas Braise/Corbis/Stock Market) **B.** Solar cells, or photovoltaics, convert sunlight directly into electricity. Many photovoltaic systems in use today are in remote areas, but this array of solar panels is near Sacramento, California. (Photo by Martin Bond/Science Photo Library/Photo Researchers, Inc.)

over the long term, solar energy is economical in many parts of the United States and will become even more cost effective as the price of other fuels increases.

Research is currently under way to improve the technologies for concentrating sunlight. One method being examined uses mirrors that track the Sun and keep its rays focused on a receiving tower. A solar collection facility with 2000 mirrors has been constructed near Barstow, California (Figure 23.12A). Solar energy focused on a central tower heats water in pressurized panels to more than 500°C. The superheated water is then transferred to turbines, which turn electrical generators.

Another type of collector uses photovoltaic (solar) cells that convert the Sun's energy directly into electricity. An experimental facility that uses photovoltaic cells is located near Sacramento, California (Figure 23.12B).

Recently small rooftop photovoltaic systems have begun being used in rural households of some Third World countries, including the Dominican Republic, Sri Lanka, and Zimbabwe. These units are about the size of an open briefcase and use a battery to store electricity that is generated during the daylight hours. In the tropics these small photovoltaic systems are capable of running a television or radio, plus a few lightbulbs, for three to four hours. Although much cheaper than building conventional electric generators, these units are still too expensive for poor families. Consequently, an estimated 2 billion people in developing countries still lack electricity.

United States we often use south-facing windows, along with better insulated and more airtight construction, to reduce heating costs substantially.

More elaborate systems used for home heating involve an *active solar collector.* These roof-mounted devices are usually large, blackened boxes that are covered with glass. The heat they collect can be transferred to where it is needed by circulating air or fluids through piping. Solar collectors are also used successfully to heat water for domestic and commercial needs. For example, solar collectors provide hot water for more than 80 percent of Israel's homes.

Although solar energy is free, the necessary equipment and its installation are not. The initial costs of setting up a system, including a supplemental heating unit for times when solar energy is diminished (cloudy days and winter) or unavailable (nighttime), can be substantial. Nevertheless,

Students Sometimes Ask . . .

Are electric vehicles better for the environment?

Yes, but probably not as much as you might think. This is because much of the electricity that electric-powered vehicles use comes from power plants that use nonrenewable fossil fuels. Thus, the pollutants are not coming directly from the car; rather, they are coming from the power plant that generated electricity for the car. Nonetheless, modern electric-powered vehicles are engineered to be more fuel efficient than traditional gasoline-powered vehicles, so they generate fewer pollutants per mile.

Wind Energy

Air has mass, and when it moves (that is, when the wind blows), it contains the energy of that motion—kinetic energy. A portion of that energy can be converted into other forms—mechanical force or electricity—that we can use to perform work (Figure 23.13).

Mechanical energy from wind is commonly used for pumping water in rural or remote places. The "farm windmill," still a familiar sight in many rural areas, is an example. Mechanical energy converted from wind can also be used for other purposes, such as sawing logs, grinding grain, and propelling sailboats. By contrast, wind-powered electric turbines generate electricity for homes, businesses, and for sale to utilities.

Approximately 0.25 percent (one quarter of 1 percent) of the solar energy that reaches the lower atmosphere is transformed into wind. Although it is just a minuscule percentage, the absolute amount of energy is enormous. According to one estimate, North Dakota alone is theoretically capable of producing enough wind-generated power to meet more than one-third of U.S. electricity demand. Wind speed is a crucial element in determining whether a place is a suitable site for installing a wind-energy facility. Generally a minimum, annual average wind speed of 21 kilometers (13 miles) per hour is necessary for a utility-scale wind-power plant.

The power available in the wind is proportional to the cube of its speed. Thus, a turbine operating at a site with an average wind speed of 12 mph could in theory generate about 30 percent more electricity than one at an 11-mph site, because the cube of 12 (1728) is 30 percent larger than the cube of 11 (1331). (In the real world, the turbine will not produce quite that much more electricity, but it will still generate much more than the 9 percent difference in wind speed.) The important thing to understand is that what seems like a small difference in wind speed can mean a large difference in available energy and in electricity produced, and therefore a large difference in the cost of the electricity generated. Also, there is little energy to be harvested at very low wind speeds (6-mph winds contain less than one-eighth the energy of 12-mph winds).*

As technology has improved, efficiency has increased and the costs of wind-generated electricity have become more competitive. Between 1983 and 2005 technological advances cut the cost of wind power by more than 85 percent. As a result, the growth of installed capacity has grown dramatically. Worldwide the total amount of installed wind power grew from 7636 megawatts in 1997 to about 59,000 in 2005 (Table 23.1). 59,000 megawatts is enough to supply more than 13 million average American households, or as much as could be generated by 17 large nuclear power plants. By mid-2006, U.S. capacity reached nearly 10,000 megawatts (Figure 23.14).

The U.S. Department of Energy has announced a goal of obtaining 5 percent of U.S. electricity from wind by the year 2020—a goal that seems consistent with the current growth

*American Wind Energy Association, "Wind Energy Basics," http://www.awea.org.

FIGURE 23.13 **A.** Farm windmills are still familiar sights in some areas. Mechanical energy from wind is commonly used to pump water. (Photo by Darren Bennett) **B.** The use of wind turbines for generating electricity is growing rapidly. These turbines are at Crowley Ridge, Alberta, Canada. (Photo by Alan Sirulnikoff/Photo Researchers, Inc.)

A.

B.

TABLE 23.1	World leaders in wind energy capacity (2005)
Country	Capacity (megawatts*)
Germany	18,428
Spain	10,027
United States	9149
India	4430
Denmark	3122
Italy	1717
United Kingdom	1353
China	1260
Netherlands	1219
Japan	1078
Rest of the world	7301
Total	59,084

*1 megawatt is enough electricity to supply about 250–300 average American households.

Source: Global Wind Energy Council.

rate of wind energy nationwide. Thus, wind-generated electricity seems to be shifting from being an "alternative" to being a "mainstream" energy source.

Hydroelectric Power

Falling water has been an energy source for centuries. Through much of human history, the mechanical energy produced by waterwheels was used to power mills and other machinery. Today the power generated by falling water is used to drive turbines that produce electricity, hence the term **hydroelectric power.** In the United States, hydroelectric power plants contribute about 3 percent of the country's demand. Most of this energy is produced at large dams,

which allow for a controlled flow of water (Figure 23.15). The water impounded in a reservoir is a form of stored energy that can be released at any time to produce electricity.

Although water power is considered a renewable resource, the dams built to provide hydroelectricity have finite lifetimes. All rivers carry suspended sediment that is deposited behind the dam as soon as it is built. Eventually the sediment will completely fill the reservoir. This takes 50 to 300 years, depending on the quantity of suspended material transported by the river. An example is Egypt's huge Aswan High Dam, which was completed in the 1960s. It is estimated that half of the reservoir will be filled with sediment from the Nile River by the year 2025.

The availability of appropriate sites is an important limiting factor in the development of large-scale hydroelectric power plants. A good site provides a significant height for the water to fall and a high rate of flow. Hydroelectric dams exist in many parts of the United States, with the greatest concentrations occurring in the Southeast and the Pacific Northwest. Most of the best U.S. sites have already been developed, limiting the future expansion of hydroelectric power. The total power produced by hydroelectric sources might still increase, but the relative share provided by this source will likely decline because other alternate energy sources will increase at a faster rate.

In recent years a different type of hydroelectric power production has come into use. Called a *pumped water-storage system*, it is actually a type of energy management. During times when demand for electricity is low, unneeded power produced by nonhydroelectric sources is used to pump water from a lower reservoir to a storage area at a higher elevation. Then, when demand for electricity is great, the water stored in the higher reservoir is available to drive turbines and produce electricity to supplement the power supply.

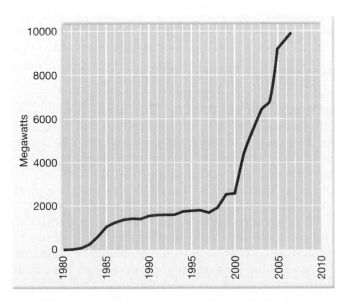

FIGURE 23.14 U.S.-installed wind-power capacity (in megawatts in mid-2006). Growth in recent years has been dramatic. (Data from U.S. Department of Energy and American Wind Energy Association)

Students Sometimes Ask . . .

What is the world's largest hydropower project?

That distinction goes to the Three Gorges Project on China's Yangtze River. Construction began in 1993. When fully operational in 2009, it is expected to generate 85 billion kilowatt-hours of electricity each year, equal to about 6.5 percent of China's electrical needs in 2001. Flood control was presumably the primary motivation for building the controversial dam. Its reservoir will inundate 632 square kilometers (244 square miles) of land stretched over about 660 kilometers (412 miles) of the river.

Geothermal Energy

Geothermal energy is harnessed by tapping natural underground reservoirs of steam and hot water. These occur where subsurface temperatures are high, owing to relatively recent volcanic activity. Geothermal energy is put to use in

FIGURE 23.15 Lake Powell is the reservoir that was created when Glen Canyon Dam was built across the Colorado River. As water in the reservoir is released, it drives turbines and produces electricity. Eventually the reservoir will be filled with sediment deposited by the Colorado River. (Photo by Michael Collier)

two ways: The steam and hot water are used for heating and to generate electricity.

Iceland is a large volcanic island with current volcanic activity (Figure 23.16). In Iceland's capital, Reykjavik, steam and hot water are pumped into buildings throughout the city for space heating. They also warm greenhouses, where fruits and vegetables are grown all year. In the United States, localities in several western states use hot water from geothermal sources for space heating.

As for generating electricity geothermally, the Italians were first to do so in 1904, so the idea is not new. By the turn of the 21st century, more than 250 geothermal power plants in 22 countries were producing more than 8000 megawatts (million watts). These plants provide power to more than 60 million people. The leading producers of geothermal power are listed in Table 23.2.

The first commercial geothermal power plant in the United States was built in 1960 at The Geysers, north of San Francisco (Figure 23.17). The Geysers remains the world's largest geothermal power plant, generating nearly 1000 megawatts of U.S. geothermal power (see Box 23.2). In addition to The Geysers, geothermal development is occurring elsewhere in the western United States, including Nevada, Utah, and the Imperial Valley in southern California. The U.S. geothermal generating capacity of more than 2500 megawatts is enough to power more than 3 million homes. This is an amount of electricity comparable to burning about 60 million barrels of oil each year.

What geologic factors favor a geothermal reservoir of commercial value?

1. *A potent source of heat,* such as a large magma chamber deep enough to ensure adequate pressure and slow cooling, yet not so deep that the natural water circulation is inhibited. Such magma chambers are most likely in regions of recent volcanic activity.
2. *Large and porous reservoirs with channels connected to the heat source,* near which water can circulate and then be stored in the reservoir.
3. *A cap of low-permeability rocks* that inhibits the flow of water and heat to the surface. A deep, well-insulated reservoir contains much more stored energy than a similar but uninsulated reservoir.

As with other alternative methods of power production, geothermal sources are not expected to provide a high percentage of the world's growing energy needs. Nevertheless, in regions where its potential can be developed, its use will no doubt continue to grow.

Tidal Power

Several methods of generating electrical energy from the oceans have been proposed, but the ocean's energy potential remains largely untapped. The development of tidal power is the principal example of energy production from the ocean.

Tides have been used as a source of power for centuries. Beginning in the 12th century, waterwheels driven by the tides were used to power gristmills and sawmills. During the 17th and 18th centuries, much of Boston's flour was produced at a tidal mill. Today far greater energy demands must be satisfied, and more sophisticated ways of using the force created by the perpetual rise and fall of the ocean must be employed.

Tidal power is harnessed by constructing a dam across the mouth of a bay or an estuary in a coastal area having a large tidal range (Figure 23.18A). The narrow opening between the bay and the open ocean magnifies the variations in water level that occur as the tides rise and fall. The strong in-and-out flow that results at such a site is then used to drive turbines and electrical generators.

Tidal energy utilization is exemplified by the tidal power plant at the mouth of the Rance River in France. By far the largest yet constructed, this plant went into operation in 1966 and produces enough power to satisfy the needs of Brittany and also contribute to the demands of other regions. Much smaller experimental facilities have been built near Murmansk in Russia and near Taliang in China, and on the Annapolis River estuary, an arm of the Bay of Fundy in the Canadian province of Nova Scotia (Figure 23.18B).

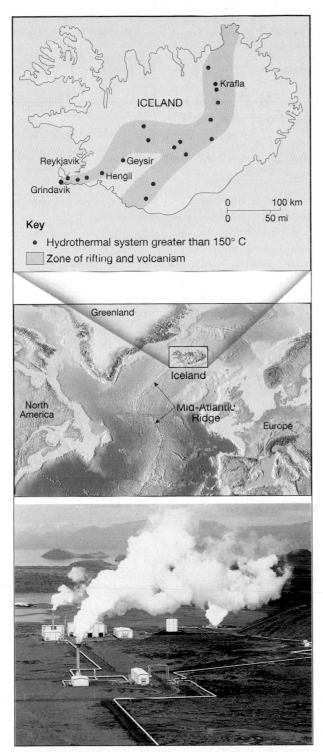

FIGURE 23.16 Iceland straddles the Mid-Atlantic Ridge. This divergent plate boundary is the site of numerous active volcanoes and geothermal systems. Because the entire country consists of geologically young volcanic rocks, warm water can be encountered in holes drilled almost anywhere. More than 45 percent of Iceland's energy comes from geothermal sources. The photo shows a power station in southwestern Iceland. The steam is used to generate electricity. Hot (83°C) water from the plant is sent via an insulated pipeline to Reykjavik for space heating. (Photo by Simon Fraser/Science Photo Library/Photo Researchers, Inc.)

TABLE 23.2   Worldwide Geothermal Power Production, 2005

Producing Country	Megawatts
United States	2564
Philippines	1931
Mexico	953
Indonesia	797
Italy	791
Japan	535
New Zealand	435
Iceland	202
Costa Rica	163
El Salvador	151
All others	411
Total	8933

Source: Geothermal Resources Council.

Along most of the world's coasts it is not possible to harness tidal energy. If the tidal range is less than 8 meters (25 feet), or if narrow, enclosed bays are absent, tidal power development is uneconomical. For this reason the tides will never provide a very high portion of our ever increasing electrical energy requirements. Nevertheless, the development of tidal power may be worth pursuing at feasible sites because electricity produced by the tides consumes no exhaustible fuels and creates no noxious wastes.

FIGURE 23.17 The Geysers, near the city of Santa Rosa in northern California, is the world's largest electricity-generating geothermal development. Most of the steam wells are about 3000 meters deep. (AP Photo/Calpine)

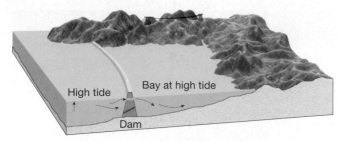

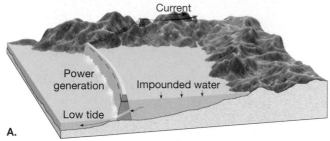

A.

B.

FIGURE 23.18 **A.** Simplified diagram showing the principle of the tidal dam. Electricity is generated only when a sufficient water-height difference exists between the bay and the ocean. **B.** Tidal power plant at Annapolis Royal, Nova Scotia, on the Bay of Fundy. (Photo by James P. Blair)

Students Sometimes Ask . . .

Is power from ocean waves a practical alternative energy source?

Wave energy technology is very young compared to generating electricity from hydro or wind turbines, but methods for harnessing wave energy are being developed. In November 2000 the world's first commercial wave power station began operating on the Scottish island of Islay, providing power to the United Kingdom power grid. The 500-kilowatt power station uses a technology called the oscillating water column, in which the incoming waves push air up and down inside a concrete tube that is partially submerged in the ocean. The air rushing into and out of the top of the tube is used to drive a turbine to produce electricity. If this technology proves successful, it may open the door for wave power to be a significant contributor of renewable energy in appropriate coastal settings.

Mineral Resources

Earth's crust is the source of a wide variety of useful and essential substances. In fact, practically every manufactured product contains substances derived from minerals. Table 23.3 lists some important examples.

Mineral resources are the endowment of useful minerals ultimately available commercially. Resources include already identified deposits from which minerals can be

Maintaining the Flow of Geothermal Energy at The Geysers*

As a result of the rapid development at The Geysers during the 1980s and some subsequent development, there was a decline in the rate of steam production (and electrical generation) due to loss of pressure in production wells. Steam production peaked in 1988 and declined for the next decade.

Most of the geothermal energy of this system remains intact, stored in hot rocks that constitute the hydrothermal reservoir. To mitigate the decline of steam pressure in production wells and thereby extend the useful life of the resource, a team of private industry and governmental agencies devised a clever and effective solution. The solution also provided an effective method for

disposing of large volumes of wastewater from nearby communities. Simply put, the wastewater of treated sewage is injected underground through appropriately positioned wells. As it flows toward the intake zones of production wells, the wastewater is heated by contact with hot rocks. Production wells then tap the natural steam augmented by vaporized wastewater.

In 1997, a 50-kilometer-long pipeline began delivering about 30 million liters of wastewater a day for injection into the southern part of The Geysers geothermal field. This slowed the pressure decline and resulted in the recovery of 75 megawatts of generating output that had been lost to the preinjection pressure decline.

The initial experiment was so successful that a second pipeline was completed in 2003 that delivers another 40 million liters per day to the central part of the field. Together, these two sources of recharge water replace nearly all of the geothermal fluid being lost to electricity production. The injection program is expected to maintain total electrical output from The Geysers at about 1000 megawatts for at least two more decades and possibly much longer.

---

*Based on material appearing in Wendell A. Duffield and John H. Sass, *Geothermal Energy—Clean Power From the Earth's Heat*, U.S. Geological Survey Circular 1249, p. 14.

TABLE 23.3	Occurrence of Metallic Minerals	
Metal	Principal Ores	Geological Occurrences
Aluminum	Bauxite	Residual product of weathering
Chromium	Chromite	Magmatic segregation
Copper	Chalcopyrite Bornite Chalcocite	Hydrothermal deposits; contact metamorphism; enrichment by weathering processes
Gold	Native gold	Hydrothermal deposits; placers
Iron	Hematite Magnetite Limonite	Banded sedimentary formations; magmatic segregation
Lead	Galena	Hydrothermal deposits
Magnesium	Magnesite Dolomite	Hydrothermal deposits
Manganese	Pyrolusite	Residual product of weathering
Mercury	Cinnabar	Hydrothermal deposits
Molybdenum	Molybdenite	Hydrothermal deposits
Nickel	Pentlandite	Magmatic segregation
Platinum	Native platinum	Magmatic segregation; placers
Silver	Native silver Argentite	Hydrothermal deposits; enrichment by weathering processes
Tin	Cassiterite	Hydrothermal deposits; placers
Titanium	Ilmenite Rutile	Magmatic segregation; placers
Tungsten	Wolframite Scheelite	Pegmatites; contact metamorphic deposits; placers
Uranium	Uraninite (pitchblende)	Pegmatites; sedimentary deposits
Zinc	Sphalerite	Hydrothermal deposits

extracted profitably, called **reserves,** as well as known deposits that are not yet economically or technologically recoverable. Deposits inferred to exist but not yet discovered are also considered as mineral resources. In addition, the term **ore** is used to denote those useful metallic minerals that can be mined at a profit. In common usage the term *ore* is also applied to some nonmetallic minerals such as fluorite and sulfur. However, materials used for such purposes as building stone, road aggregate, abrasives, ceramics, and fertilizers are not usually called ores; rather, they are classified as industrial rocks and minerals.

Recall that more than 98 percent of the crust is composed of only eight elements. Except for oxygen and silicon, all other elements make up a relatively small fraction of common crustal rocks (see Figure 3.26, p. 89). Indeed, the natural concentrations of many elements are exceedingly small. A deposit containing the average percentage of a valuable element is worthless if the cost of extracting it exceeds the value of the material recovered. To be considered valuable, an element must be concentrated above the level of its average crustal abundance. Generally, the lower the crustal abundance, the greater the concentration.

For example, copper makes up about 0.0135 percent of the crust. However, for a material to be considered a copper ore, it must contain a concentration that is about 50 times this amount. Aluminum, in contrast, represents 8.13 percent of the crust and must be concentrated to only about four

times its average crustal percentage before it can be extracted profitably.

It is important to realize that a deposit may become profitable to extract or lose its profitability because of economic changes. If demand for a metal increases and prices rise, the status of a previously unprofitable deposit changes, and it becomes an ore. The status of unprofitable deposits may also change if a technological advance allows the useful element to be extracted at a lower cost than before. This occurred at the copper-mining operation located at Bingham Canyon, Utah, one of the largest open-pit mines on Earth (Box 23.3). Mining was halted here in 1985 because outmoded equipment had driven the cost of extracting the copper beyond the selling price. The owners responded by replacing an antiquated 1000-car railroad with conveyor belts and pipelines for transporting the ore and waste. These devices achieved a cost reduction of nearly 30 percent and returned this mining operation to profitability.

Over the years, geologists have been keenly interested in learning how natural processes produce localized concentrations of essential metallic minerals. One well-established fact is that occurrences of valuable mineral resources are closely related to the rock cycle. That is, the mechanisms that generate igneous, sedimentary, and metamorphic rocks, including the processes of weathering and erosion, play a major role in producing concentrated accumulations of useful elements. Moreover, with the development of the plate tectonics theory, geologists added yet another tool for understanding the processes by which one rock is transformed into another.

Mineral Resources and Igneous Processes

Some of the most important accumulations of metals, such as gold, silver, copper, mercury, lead, platinum, and nickel, are produced by igneous processes (see Table 23.3). These mineral resources, like most others, result from processes that concentrate desirable materials to the extent that extraction is economically feasible.

Magmatic Segregation

The igneous processes that generate some of these metal deposits are quite straightforward. For example, as a large magma body cools, the heavy minerals that crystallize early tend to settle to the lower portion of the magma chamber. This type of magmatic segregation is particularly active in large basaltic magmas where chromite (ore of chromium), magnetite, and platinum are occasionally generated. Layers of chromite, interbedded with other heavy minerals, are mined from such deposits in the Stillwater Complex of Montana. Another example is the Bushveld Complex in South Africa, which contains over 70 percent of the world's known reserves of platinum.

Magmatic segregation is also important in the late stages of the magmatic process. This is particularly true of granitic

BOX 23.3 ▶ PEOPLE AND THE ENVIRONMENT

Utah's Bingham Canyon Mine

In the photo, a mountain once stood where there is now a huge pit (Figure 23.C). This is the world's largest open-pit mine, Bingham Canyon copper mine, about 40 kilometers (25 miles) southwest of Salt Lake City, Utah. The rim is nearly 4 kilometers (2.5 miles) across and covers almost 8 square kilometers (3 square miles). Its depth is 900 meters (3000 feet). If a steel tower were erected at the bottom, it would have to be five times taller than the Eiffel Tower to reach the top of the pit!

It began in the late 1800s as an underground mine for veins of silver and lead. Later copper was discovered. Similar deposits occur at several sites in the American Southwest and in a belt from southern Alaska to northern Chile.

As in other places in this belt, the ore at Bingham Canyon is disseminated throughout *porphyritic* igneous rocks; hence, the name *porphyry copper deposits*. The deposit formed after magma was intruded to shallow depths. Following this, shattering created extensive fractures that were penetrated by hydrothermal solutions from which the ore minerals precipitated.

Although the percentage of copper in the rock is small, the total volume of copper is huge. Ever since open-pit operations started in 1906, some 5 billion tons of material have been removed, yielding more than 12 million tons of copper. Significant amounts of gold, silver, and molybdenum have also been recovered.

The ore body is far from exhausted today. Over the next 25 years, plans call for

FIGURE 23.C Aerial view of Bingham Canyon copper mine near Salt Lake City, Utah. This huge open-pit mine is about 4 kilometers across and 900 meters deep. Although the amount of copper in the rock is less than 1 percent, the huge volumes of material removed and processed each day (about 200,000 tons) yield significant quantities of metal. (Photo by Michael Collier)

an additional 3 billion tons of material to be removed and processed. This largest of artificial excavations has generated most of Utah's mineral production for more than 80 years and has been called the "richest hole on Earth."

Like many older mines, the Bingham pit was unregulated during most of its history.

Development occurred prior to the present-day awareness of the environmental impacts of mining and prior to effective environmental legislation. Today problems of groundwater and surface water contamination, air pollution, solid and hazardous wastes, and land reclamation are receiving long overdue attention at Bingham Canyon.

---

magmas in which the residual melt can become enriched in rare elements and heavy metals. Further, because water and other volatile substances do not crystallize along with the bulk of the magma body, these fluids make up a high percentage of the melt during the final phase of solidification. Crystallization in a fluid-rich environment, where ion migration is enhanced, results in the formation of crystals several centimeters, or even a few meters, in length. The resulting rocks, called **pegmatites,** are composed of these unusually large crystals (Figure 4.9, p. 107).

Most pegmatites are granitic in composition and consist of unusually large crystals of quartz, feldspar, and muscovite. Feldspar is used in the production of ceramics, and muscovite is used for electrical insulation and glitter. Further, pegmatites often contain some of the least abundant elements. Thus, in addition to the common silicates, some pegmatites include semiprecious gems such as beryl, topaz,

and tourmaline. Moreover, minerals containing the elements lithium, cesium, uranium, and the rare earths* are occasionally found (Figure 23.19). Most pegmatites are located within large igneous masses or as dikes or veins that cut into the host rock surrounding the magma chamber (Figure 23.20).

Not all late-stage magmas produce pegmatites, nor do all have a granitic composition. Rather, some magmas become enriched in iron or occasionally copper. For example, at Kiruna, Sweden, magma composed of over 60 percent magnetite solidified to produce one of the largest iron deposits in the world.

---

*The rare earths are a group of 15 elements (atomic numbers 57 through 71) that possess similar properties. They are useful catalysts in petroleum refining and are used to improve color retention in television picture tubes.

FIGURE 23.19   This pegmatite in the Black Hills of South Dakota was mined for its large crystals of spodumene, an important source of lithium. Arrows are pointing to impressions left by crystals. Note person in upper center of photo for scale. (Photo by James G. Kirchner)

Diamonds

Another economically important mineral with an igneous origin is diamond. Although best known as gems, diamonds are used extensively as abrasives. Diamonds originate at depths of nearly 200 kilometers, where the confining pressure is great enough to generate this high-pressure form of carbon. Once crystallized, the diamonds are carried upward through pipe-shaped conduits that increase in diameter toward the surface. In diamond-bearing pipes, nearly the entire pipe contains diamond crystals that are disseminated throughout an ultramafic rock called *kimberlite*. The most

productive kimberlite pipes are those found in South Africa. The only equivalent source of diamonds in the United States is located near Murfreesboro, Arkansas, but this deposit is exhausted and today serves merely as a tourist attraction.

Hydrothermal Solutions

Among the best-known and most important ore deposits are those generated from **hydrothermal** (hot-water) **solutions.** Included in this group are the gold deposits of the Homestake mine in South Dakota; the lead, zinc, and silver ores near Coeur d'Alene, Idaho; the silver deposits of the Comstock Lode in Nevada; and the copper ores of the Keweenaw Peninsula in Michigan (Figure 23.21).

The majority of hydrothermal deposits originate from hot, metal-rich fluids that are remnants of late-stage magmatic processes. During solidification, liquids plus various metallic ions accumulate near the top of the magma chamber. Because of their mobility, these ion-rich solutions can migrate great distances through the surrounding rock before they are eventually deposited, usually as sulfides of various metals (Figure 23.20). Some of this fluid moves along openings such as fractures or bedding planes, where it cools and precipitates the metallic ions to produce **vein deposits** (Figure 23.22). Many of the most productive deposits of gold, silver, and mercury occur as hydrothermal vein deposits.

Another important type of accumulation generated by hydrothermal activity is called a **disseminated deposit.** Rather than being concentrated in narrow veins and dikes, these ores are distributed as minute masses throughout the entire rock mass. Much of the world's copper is extracted from disseminated deposits, including those at Chuquicamata, Chile, and the huge Bingham Canyon copper mine in Utah (see Box 23.3). Because these accumulations contain only 0.4 to

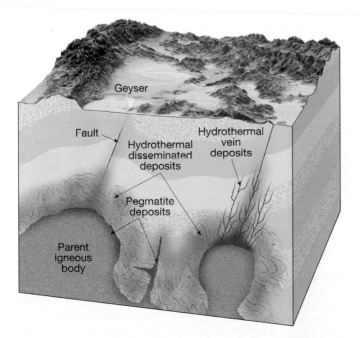

FIGURE 23.20   Illustration of the relationship between a parent igneous body and the associated pegmatite and hydrothermal deposits.

FIGURE 23.21   Native copper from northern Michigan's Keweenaw Peninsula is an excellent example of a hydrothermal deposit. At one time this area was an important source of copper, but it is now largely depleted. (Photo by E. J. Tarbuck)

FIGURE 23.22 Gneiss laced with quartz veins at Diablo Lake Overlook, North Cascades National Park, Washington. (Photo by James E. Patterson)

0.8 percent copper, between 125 and 250 kilograms of ore must be mined for every kilogram of metal recovered. The environmental impact of these large excavations, including the problem of waste disposal, is significant.

Some hydrothermal deposits have been generated by the circulation of ordinary groundwater in regions where magma was emplaced near the surface. The Yellowstone National Park area is a modern example of such a situation (Figure 23.23). When groundwater invades a zone of recent igneous activity, its temperature rises, greatly enhancing its ability to dissolve minerals. These migrating hot waters remove metallic ions from intrusive igneous rocks and carry them upward, where they may be deposited as an ore body. Depending on the conditions, the resulting accumulations may occur as vein deposits, as disseminated deposits, or, where hydrothermal solutions reach the surface in the form of hot springs or geysers, as surface deposits.

FIGURE 23.23 Some hydrothermal deposits are formed by the circulation of heated groundwater in regions where magma is near the surface, as in the Yellowstone National Park area. (Photo by Craig J. Brown/Liaison Agency, Inc.)

With the development of the plate tectonics theory it became clear that some hydrothermal deposits originated along ancient oceanic ridges. A well-known example is found on the island of Cyprus, where copper has been mined for more than 4000 years. Apparently, these deposits represent ores that formed on the seafloor at an ancient oceanic spreading center.

Since the mid-1970s, active hot springs and metal-rich sulfide deposits have been detected at several sites, including study areas along the East Pacific Rise and the Juan de Fuca Ridge. The deposits are forming where heated seawater, rich in dissolved metals and sulfur, gushes from the seafloor as particle-filled clouds called *black smokers*. As shown in Figure 23.24, seawater infiltrates the hot oceanic crust along the flanks of the ridge. As the water moves through the newly formed material, it is heated and chemically interacts with the basalt, extracting and transporting sulfur, iron, copper, and other metals. Near the ridge axis, the hot, metal-rich fluid rises along faults. Upon reaching the seafloor, the spewing liquid mixes with the cold seawater, and the sulfides precipitate to form massive sulfide deposits.

Mineral Resources and Metamorphic Processes

The role of metamorphism in producing mineral deposits is frequently tied to igneous processes. For example, many of the most important metamorphic ore deposits are produced by contact metamorphism. Here the host rock is recrystallized and chemically altered by heat, pressure, and hydrothermal solutions emanating from an intruding igneous body. The extent to which the host rock is altered depends on the nature of the intruding igneous mass as well as the nature of the host rock.

Some resistant materials, such as quartz sandstone, may show very little alteration, whereas others, including limestone, might exhibit the effects of metamorphism for several kilometers from the igneous pluton. As hot, ion-rich fluids move through limestone, chemical reactions take place, which produce useful minerals such as garnet and corundum. Further, these reactions release carbon dioxide, which greatly facilitates the outward migration of metallic ions. Thus, extensive aureoles of metal-rich deposits commonly surround igneous plutons that have invaded limestone strata.

The most common metallic minerals associated with contact metamorphism are sphalerite (zinc), galena (lead), chalcopyrite (copper), magnetite (iron), and bornite (copper). The hydrothermal ore deposits may be disseminated throughout the altered zone or exist as concentrated masses that are located either next to the intrusive body or along the margins of the metamorphic zone.

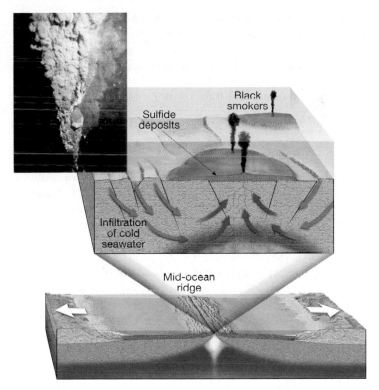

FIGURE 23.24 Massive sulfide deposits can result from the circulation of seawater through the oceanic crust along active spreading centers. As seawater infiltrates the hot basaltic crust, it leaches sulfur, iron, copper, and other metals. The hot, enriched fluid returns to the seafloor near the ridge axis along faults and fractures. Some metal sulfides may be precipitated in these channels as the rising fluid begins to cool. When the hot liquid emerges from the seafloor and mixes with cold seawater, the sulfides precipitate to form massive deposits. Photo shows a close-up view of a black smoker spewing hot, mineral-rich seawater along the East Pacific Rise. (Photo © Robert Ballard, Woods Hole Oceanographic Institution)

found in low concentrations near the surface are removed and carried to lower zones, where they are redeposited and become more concentrated.

Bauxite

The formation of *bauxite,* the principal ore of aluminum, is one important example of an ore created as a result of enrichment by weathering processes (Figure 23.25). Although aluminum is the third most abundant element in Earth's crust, economically valuable concentrations of this important metal are not common, because most aluminum is tied up in silicate minerals from which it is extremely difficult to extract.

Bauxite forms in rainy tropical climates in association with laterites. (In fact, bauxite is sometimes referred to as aluminum laterite.) When aluminum-rich source rocks are subjected to the intense and prolonged chemical weathering of the tropics, most of the common elements, including calcium, sodium, and silicon, are removed by leaching. Because aluminum is extremely insoluble, it becomes concentrated in the soil as bauxite, a hydrated aluminum oxide. Thus, the formation of bauxite depends both on climatic conditions in which chemical weathering and leaching are pronounced and on the presence of an aluminum-rich source rock. Important deposits of nickel and cobalt are also found in laterite soils that develop from igneous rocks rich in ferromagnesian silicate minerals.

Other Deposits

Many copper and silver deposits result when weathering processes concentrate metals that are deposited through a low-grade primary ore. Usually such enrichment occurs in deposits containing pyrite (FeS_2), the most common and

Regional metamorphism can also generate useful mineral deposits. Recall that at convergent plate boundaries the oceanic crust, along with sediments that have accumulated at the continental margins, are carried to great depths. In these high-temperature, high-pressure environments the mineralogy and texture of the subducted materials are altered, producing deposits of nonmetallic minerals such as talc and graphite.

Weathering and Ore Deposits

Weathering creates many important mineral deposits by concentrating minor amounts of metals that are scattered through unweathered rock into economically valuable concentrations. Such a transformation is often termed **secondary enrichment** and takes place in one of two ways. In one situation chemical weathering coupled with downward-percolating water removes undesirable materials from decomposing rock, leaving the desirable elements enriched in the upper zones of the soil. The second way is basically the reverse of the first. That is, the desirable elements that are

FIGURE 23.25 Bauxite is the ore of aluminum and forms as a result of weathering processes under tropical conditions. Its color varies from red or brown to nearly white. (Photo by E. J. Tarbuck)

widespread sulfide mineral. Pyrite is important because when it chemically weathers, sulfuric acid forms, which enables percolating waters to dissolve the ore metals. Once dissolved, the metals gradually migrate downward through the primary ore body until they are precipitated. Deposition takes place because of changes that occur in the chemistry of the solution when it reaches the groundwater zone (the zone beneath the surface where all pore spaces are filled with water). In this manner the small percentage of dispersed metal can be removed from a large volume of rock and redeposited as a higher-grade ore in a smaller volume of rock.

This enrichment process is responsible for the economic success of many copper deposits, including one located in Miami, Arizona. Here the ore was upgraded from less than 1 percent copper in the primary deposit to as much as 5 percent copper in some localized zones of enrichment. When pyrite weathers (oxidizes) near the surface, residues of iron oxide remain. The presence of these rusty-colored caps at the surface indicates the possibility of an enriched ore below, and this represents a visible sign for prospectors.

Placer Deposits

Sorting typically results in like-sized grains being deposited together. However, sorting according to the specific gravity of particles also occurs. This latter type of sorting is responsible for the creation of **placers,** which are deposits formed when heavy minerals are mechanically concentrated by currents. Placers associated with streams are among the most common and best known, but the sorting action of waves can also create placers along the shore. Placer deposits usually involve minerals that are not only heavy but also durable (to withstand physical destruction during transportation) and chemically resistant (to endure weathering processes). Placers form because heavy minerals settle quickly from a current, whereas less dense particles remain suspended and are carried onward. Common sites of accumulation include point bars on the insides of meanders as well as cracks, depressions, and other irregularities on stream beds.

Many economically important placer deposits exist, with accumulations of gold the best known. Indeed, it was the placer deposits discovered in 1848 that led to the famous California gold rush. Years later similar deposits created a gold rush to Alaska as well (Figure 23.26). Panning for gold by washing sand and gravel from a flat pan to concentrate the fine "dust" at the bottom was a common method used by early prospectors to recover the precious metal, and it is a process similar to that which created the placer in the first place.

In addition to gold, other heavy and durable minerals form placers. These include platinum, diamonds, and tin. The Ural Mountains contain placers rich in platinum, and placers are important sources of diamonds in southern Africa. Significant portions of the world's supply of cassiterite, the principal ore of tin, have come from placer deposits in Malaysia and Indonesia. Cassiterite is often widely disseminated through granitic igneous rocks. In this state the mineral is not sufficiently concentrated to be extracted profitably. However, as the enclosing rock dissolves and disintegrates, the heavy and durable cassiterite grains are set free. Eventually the freed particles are washed to a stream, where they are deposited in placers that are significantly more concentrated than the original deposit. Similar circumstances and events are common to many minerals mined from placers.

FIGURE 23.26 **A.** It was placer deposits that led to the 1848 California gold rush. Here, a prospector in 1850 swirls his gold pan, separating sand and mud from flecks of gold. (Photo courtesy of Seaver Center for Western History Research, Los Angeles County Museum of Natural History) **B.** Modern gold mining of a placer deposit near Nome, Alaska. The dredge scoops up thawed gravel. (Photo by Fred Bruemmer/DRK Photo)

A.

B.

In some cases, if the source rock for a placer deposit can be located, it too may become an important ore body. By following placer deposits upstream, one can sometimes locate the original deposit. This is how the gold-bearing veins of the Mother Lode in California's Sierra Nevada batholith were found, as well as the famous Kimberly diamond mines of South Africa. The placers were discovered first, their source at a later time.

Nonmetallic Mineral Resources

Earth materials that are not used as fuels or processed for the metals they contain are referred to as **nonmetallic mineral resources.** Realize that use of the word "mineral" is very broad in this economic context and is quite different from the geologist's strict definition of mineral found in Chapter 3. Nonmetallic mineral resources are extracted and processed either for the nonmetallic elements they contain or for the physical and chemical properties they possess.

People often do not realize the importance of nonmetallic minerals because they see only the products that resulted from their use and not the minerals themselves. That is, many nonmetallics are used up in the process of creating other products. Examples include the fluorite and limestone that are part of the steelmaking process, the abrasives required to make a piece of machinery, and the fertilizers needed to grow a food crop (see Table 23.4).

The quantities of nonmetallic minerals used each year are enormous. A glance at Figure 23.2 (p. 629) reminds us that the per capita consumption of nonfuel resources in the United States totals more than 11 metric tons, of which about 94 percent are nonmetallics. Nonmetallic mineral resources are commonly divided into two broad groups—*building materials* and *industrial minerals.* Because some substances have many different uses, they are found in both categories. Limestone, perhaps the most versatile and widely used rock of all, is the best example. As a building material, it is used not only as crushed rock and building stone but also in the making of cement. Moreover, as an industrial mineral, limestone is an ingredient in the manufacture of steel and is used in agriculture to neutralize soils.

Building Materials

Natural aggregate consists of crushed stone, sand, and gravel. From the standpoint of quantity and value, aggregate is a very important building material. The United States produces nearly 2 billion tons of aggregate per year, which represents about one half of the entire nonenergy mining volume in the country. It is produced commercially in every state and is used in nearly all building construction and in most public works projects (Figure 23.27).

TABLE 23.4	Occurrences and Uses of Nonmetallic Minerals	
Mineral	Uses	Geological Occurrences
Apatite	Phosphorus fertilizers	Sedimentary deposits
Asbestos	Incombustible fibers	Metamorphic alteration (chrysotile)
Calcite	Aggregate; steelmaking; soil conditioning; chemicals; cement; building stone	Sedimentary deposits
Clay minerals	Ceramics; china	Residual product of weathering (kaolinite)
Corundum	Gemstones; abrasives	Metamorphic deposits
Diamond	Gemstones; abrasives	Kimberlite pipes; placers
Fluorite	Steelmaking; aluminum refining; glass; chemicals	Hydrothermal deposits
Garnet	Abrasives; gemstones	Metamorphic deposits
Graphite	Pencil lead; lubricant; refractories	Metamorphic deposits
Gypsum	Plaster of Paris	Evaporite deposits
Halite	Table salt; chemicals; ice control	Evaporite deposits; salt domes
Muscovite	Insulator in electrical applications	Pegmatites
Quartz	Primary ingredient in glass	Igneous intrusions; sedimentary deposits
Sulfur	Chemicals; fertilizer manufacture	Sedimentary deposits; hydrothermal deposits
Sylvite	Potassium fertilizers	Evaporite deposits
Talc	Powder used in paints, cosmetics, etc.	Metamorphic deposits

FIGURE 23.27 Crushed stone and sand and gravel are primarily used for aggregate in the construction industry, especially in cement concrete for residential and commercial buildings, bridges, and airports, and as cement concrete or bituminous concrete (asphalt) for highway construction. A large percentage is used without a binder as road base, for road surfacing, and as railroad ballast. (Photo by Robert Ginn/PhotoEdit)

Besides aggregate, other important building materials include gypsum for plaster and wallboard, clay for tile and bricks, and cement, which is made from limestone and shale. Cement and aggregate go into the making of concrete, a material that is essential to practically all construction. Aggregate gives concrete its strength and volume, and cement binds the mixture into a rock-hard substance. Just 2 kilometers of four-lane highway require more than 85 metric tons of aggregate. On a smaller scale, 90 tons of aggregate are needed just to build an average six-room house.

Because most building materials are widely distributed and present in almost unlimited quantities, they have little intrinsic value. Their economic worth comes only after the materials are removed from the ground and processed. Since their per-ton value compared with metals and industrial minerals is low, mining and quarrying operations are usually undertaken to satisfy local needs. Except for special types of cut stone used for buildings and monuments, transportation costs greatly limit the distance most building materials can be moved.

Industrial Minerals

Many nonmetallic resources are classified as industrial minerals. In some instances these materials are important because they are sources of specific chemical elements or compounds. Such minerals are used in the manufacture of chemicals and the production of fertilizers. In other cases their importance is related to the physical properties they exhibit. Examples include minerals such as corundum and garnet, which are used as abrasives. Although supplies are generally plentiful, most industrial minerals are not nearly as abundant as building materials. Moreover, deposits are far more restricted in distribution and extent. As a result, many of these nonmetallic resources must be transported considerable distances, which of course adds to their cost. Unlike most building materials, which need a minimum of processing before they are ready to use, many industrial minerals require considerable processing to extract the desired substance at the proper degree of purity for its ultimate use.

Fertilizers The growth in world population toward 7 billion requires that the production of basic food crops continues to expand. Therefore, fertilizers—primarily nitrate, phosphate, and potassium compounds—are extremely important to agriculture. The synthetic nitrate industry, which derives nitrogen from the atmosphere, is the source of practically all the world's nitrogen fertilizers. The primary source of phosphorus and potassium, however, remains Earth's crust. The mineral apatite is the primary source of phosphate. In the United States most production comes from marine sedimentary deposits in Florida and North Carolina (Figure 23.28). Although potassium is an abundant element in many minerals, the primary commercial sources are evaporite deposits containing the mineral sylvite. In the United States, deposits near Carlsbad, New Mexico, have been especially important.

Sulfur Because it has many uses, sulfur is an important nonmetallic resource. In fact, the quantity of sulfur used is considered one index of a country's level of industrialization. More than 80 percent is used to produce sulfuric acid. Although its principal use is in the manufacture of phosphate fertilizer, sulfuric acid has a large number of other applications as well. Sources include deposits of native sulfur associated with salt domes and volcanic areas, as well as common iron sulfides such as pyrite. In recent years an increasingly important source has been the sulfur removed from coal, oil, and natural gas in order to make these fuels less polluting.

Salt Common salt, known by the mineral name *halite*, is another important and versatile resource. It is among the more prominent nonmetallic minerals used as a raw material in the chemical industry. In addition, large quantities are used to "soften" water and to keep streets and highways free of ice. Of course, most people are aware that it is also a basic nutrient and a part of many food products.

Salt is a common evaporite, and thick deposits are exploited using conventional underground mining techniques. Subsurface deposits are also tapped, using brine wells in which a pipe is introduced into a salt deposit and water is pumped down the pipe. The salt dissolved by the water is brought to the surface through a second pipe. In addition, seawater continues to serve as a source of salt as it has for centuries. The salt is harvested after the Sun evaporates the water.

FIGURE 23.28   Large open-pit phosphate mine in Florida. The phosphorus-bearing mineral apatite is a calcium phosphate associated with bones and teeth. Fish and other marine organisms extract phosphate from seawater to form apatite. These sedimentary deposits are associated with the floor of a shallow sea. (Photo by C. Davidson/ Comstock)

Summary

- *Renewable resources* can be replenished over relatively short time spans. Examples include natural fibers for clothing and trees for lumber. *Nonrenewable resources* form so slowly that, from a human standpoint, Earth contains fixed supplies. Examples include fuels such as oil and coal, and metals such as copper and gold. A rapidly growing world population and the desire for an improved living standard cause nonrenewable resources to become depleted at an increasing rate.

- *Coal, petroleum,* and *natural gas,* the *fossil fuels* of our modern economy, are all associated with sedimentary rocks. Coal originates from large quantities of plant remains that accumulate in an oxygen-deficient environment, such as a swamp. More than 70 percent of present-day coal usage is for the generation of electricity. Air pollution produced by the sulfur oxide gases that form from burning most types of coal is a significant environmental problem.

- Oil and natural gas, which commonly occur together in the pore spaces of some sedimentary rocks, consist of various *hydrocarbon compounds* (compounds made of hydrogen and carbon) mixed together. Petroleum formation is associated with the accumulation of sediment in ocean areas that are rich in plant and animal remains that become buried and isolated in an oxygen-deficient environment. As the mobile petroleum and natural gas form, they migrate and accumulate in adjacent permeable beds such as sandstone. If the upward migration is halted by an impermeable rock layer, referred to as a *cap rock,* a geologic environment that allows for economically significant amounts of oil and gas to accumulate underground, called an *oil trap,* develops. The two basic conditions common to all oil traps are (1) a porous, permeable *reservoir rock* that will yield petroleum and/or natural gas in sufficient quantities, and (2) a cap rock.

- When conventional petroleum resources are no longer adequate, fuels derived from *oil sands* and *oil shale* may become substitutes. Presently, oil sands from the province of Alberta are the source of about 15 percent of Canada's oil production. Oil from oil shale is presently uneconomical to produce. Oil production from both oil sands and oil shale has significant environmental drawbacks.

- More than 85 percent of our energy is derived from fossil fuels. In the United States the most important alternative energy sources are *nuclear energy* and *hydroelectric power.* Other alternative energy sources are locally important but collectively provide about 1 percent of the U.S. energy demand. These include *solar power, geothermal energy, wind energy,* and *tidal power.*

- *Mineral resources* are the endowment of useful minerals ultimately available commercially. Resources include already identified deposits from which minerals can be extracted profitably, called *reserves,* as well as known deposits that are not yet economically or technologically recoverable. Deposits inferred to exist but not yet discovered are also considered mineral resources. The term *ore* is used to denote those useful metallic minerals that can be mined for a profit, as well as some nonmetallic minerals, such as fluorite and sulfur, that contain useful substances.

- Some of the most important accumulations of metals, such as gold, silver, lead, and copper, are produced by igneous processes. The best-known and most important ore deposits are generated from *hydrothermal* (hot-water) *solutions.* Hydrothermal deposits originate from hot, metal-rich fluids that are remnants of late-stage magmatic processes. These ion-rich solutions move along fractures or bedding planes, cool, and precipitate the metallic ions to produce *vein deposits.* In a *disseminated deposit* (e.g., much of the world's copper deposits) the ores from hydrothermal solutions are distributed as minute masses throughout the entire rock mass.

- Many of the most important metamorphic ore deposits are produced by contact metamorphism. Extensive aureoles of metal-rich deposits commonly surround igneous bodies where ions have invaded limestone strata. The most common metallic minerals associated with contact metamorphism are sphalerite (zinc), galena (lead), chalcopyrite (copper), magnetite (iron), and bornite (copper). Of equal economic importance are the metamorphic rocks themselves. In many regions, slate, marble, and quartzite are quarried for a variety of construction purposes.

- Weathering creates ore deposits by concentrating minor amounts of metals into economically valuable deposits. The process, often called *secondary enrichment,* is accomplished by either (1) removing undesirable materials and leaving the desired elements enriched in the upper zones of the soil, or (2) removing and carrying the desirable elements to lower zones where they are redeposited and become more concentrated. *Bauxite,* the principal ore of aluminum, is one important ore created as a result of enrichment by weathering processes. In addition, many copper and silver deposits result when weathering processes concentrate metals that were formerly dispersed through low-grade primary ore.

- Earth materials that are not used as fuels or processed for the metals they contain are referred to as *nonmetallic resources.* Many are sediments or sedimentary rocks. The two broad groups of nonmetallic resources are *building materials* and *industrial minerals.* Limestone, perhaps the most versatile and widely used rock of all, is found in both groups.

Review Questions

1. Contrast renewable and nonrenewable resources. Give one or more examples of each.
2. What is the estimated world population for the year 2015? How does this compare to the figures for 1930 and 1975? Is demand for resources growing as rapidly as world population?
3. More than 70 percent of present-day coal usage is for what purpose?
4. What is an oil trap? List two conditions common to all oil traps.
5. List two drawbacks associated with the processing of oil sands recovered by surface mining.
6. The United States has huge oil shale deposits but does not produce oil shale commercially. Explain.
7. What is the main fuel for nuclear fission reactors?
8. List two obstacles that have hindered the development of nuclear power as a major energy source. What environmental advantage does nuclear power have compared to fossil fuels?
9. Briefly describe two methods by which solar energy might be used to produce electricity.
10. Explain why dams built to provide hydroelectricity do not last indefinitely.
11. What advantages does tidal power production offer? Is it likely that tides will ever provide a significant proportion of the world's electrical energy requirements?
12. Contrast *resource* and *reserve.*
13. What might cause a mineral deposit that had not been considered an ore to be reclassified as an ore?
14. List two general types of hydrothermal deposits.
15. Metamorphic ore deposits are often related to igneous processes. Provide an example.
16. Name the primary ore of aluminum and describe its formation.
17. A rusty-colored zone of iron oxide at the surface may indicate the presence of a copper deposit at depth. Briefly explain.
18. Briefly describe the way in which minerals accumulate in placers. List four minerals that are mined from such deposits.
19. Which is greater, the per capita consumption of metallic resources or that of nonmetallic mineral resources?
20. Nonmetallic resources are commonly divided into two broad groups. List the two groups and some examples of materials that belong to each. Which group is most widely distributed?

Key Terms

cap rock (p. 632)
disseminated deposit
  (p. 645)
fossil fuel (p. 630)
geothermal energy (p. 639)
hydroelectric power (p. 639)
hydrothermal solution
  (p. 645)

mineral resource (p. 642)
nonmetallic mineral
  resource (p. 649)
nonrenewable resource
  (p. 620)
nuclear fission (p. 625)

oil trap (p. 632)
ore (p. 643)
pegmatite (p. 644)
placer (p. 648)
renewable resource (p. 628)
reserve (p. 643)

reservoir rock (p. 632)
secondary enrichment
  (p. 647)
vein deposit (p. 645)

Web Resources

 The *Earth* Website uses the resources and flexibility of the Internet to aid in your study of the topics in this chapter. Written and developed by geology instructors, this site will help improve your understanding of geology. Visit **http://www.prenhall.com/tarbuck** and click on the cover of *Earth 9e* to find:

- Online review quizzes.
- Critical thinking exercises.
- Links to chapter-specific Web resources.
- Internet-wide key-term searches.

http://www.prenhall.com/tarbuck

Planetary Geology*

Panoramic view of the Martian landscape taken by the Spirit rover in 2004. (NASA/JPL/ Cornell/Peter Arnold, Inc.)

*This chapter was revised by Teresa Tarbuck and Mark Watry, Rocky Mountain College.

655

When people first recognized that the planets resembled Earth more than the stars, excitement grew. Could intelligent life exist on these other planets, or elsewhere in the universe? Space exploration has rekindled this interest. So far, no evidence of extraterrestrial life within our solar system has emerged. Nevertheless, we study the other planets to learn about Earth's formation and early history. Recent space explorations have been organized with this goal in mind.

The Planets: An Overview

The Sun is the hub of a huge rotating system of eight classical planets, their satellites, and numerous smaller asteroids, comets, meteoroids, and dwarf planets. An estimated 99.85 percent of the mass of our solar system is contained within the Sun. The planets collectively make up most of the remaining 0.15 percent. The planets, traveling outward from the Sun, are Mercury, Venus, Earth, Mars, Jupiter, Saturn, Uranus, and Neptune (Figure 24.1). Pluto was recently reclassified as a dwarf planet.

Under the Sun's control, each planet is tethered by gravity in a nearly circular orbit—all of them traveling in the same direction. The nearest planet to the Sun, Mercury, has the fastest orbital motion, 48 kilometers per second, and the shortest period of revolution around the Sun, 88 Earth-days. By contrast, the distant dwarf planet Pluto has an orbital speed of 5 kilometers per second and requires 248 Earth-years to complete one revolution. Furthermore, the orbital planes of seven planets lie within 3 degrees of the plane of the Sun's equator. The other, Mercury, is inclined 7 degrees.

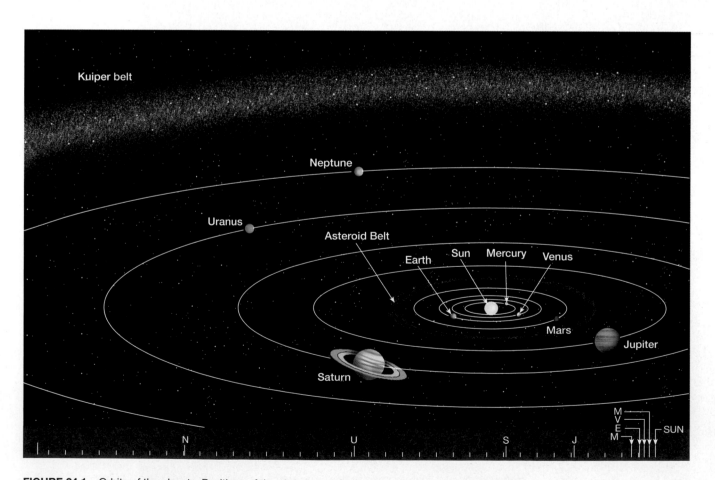

FIGURE 24.1   Orbits of the planets. Positions of the planets are shown to scale along bottom of diagram.

How Did the Planets Form?

The Sun and planets formed at the same time from a large rotating cloud of interstellar dust and gas called the **solar nebula.** As the solar nebula contracted, the vast majority of material collected in the center to form the hot *protosun.* The remainder formed a flattened, spinning disk. Within this spinning disk, matter gradually formed clumps of material that collided, stuck together and grew into asteroid-sized objects called **planetesimals.** The composition of these planetesimals depended largely on their location with respect to the protosun. (Temperatures were greater near the protosun and much lower in the outer reaches of the disk.) This was critical since only those materials that could condense—form solid or liquid clumps—in a particular location would be available to form planetesimals.

Near the present orbit of Mercury only metallic grains condensed—it was simply too hot for anything else to exist. Further out, near Earth's orbit, metallic as well as rocky substances condensed, and beyond Mars, ices of water, carbon dioxide, ammonia, and methane formed. It was from these clumps of matter that the planetesimals formed and through repeated collisions and accretion (sticking together) grew into eight **protoplanets** and their moons. It took roughly a billion years after the protoplanets formed to gravitationally sweep the solar system clear of interplanetary debris. This was a period of intense bombardment that is clearly visible on the Moon and elsewhere in the solar system. Only a small amount of the interplanetary matter escaped capture by a planet or moon and became the asteroids, comets, and meteoroids.

Terrestrial and Jovian Planets

Careful examination of Table 24.1 shows that the planets fall quite nicely into two groups: the **terrestrial** (Earth-like) **planets** (Mercury, Venus, Earth, and Mars), and the **Jovian** (Jupiter-like) **planets** (Jupiter, Saturn, Uranus, and Neptune). Pluto was recently demoted to a *dwarf planet*—a new class of solar system objects that have an orbit around the Sun but share their space with other celestial bodies. We will consider the nature of dwarf planets later in the chapter.

The most obvious difference between the terrestrial and the Jovian planets is their size (Figure 24.2). The largest terrestrial planets (Earth and Venus) have diameters only one-quarter as great as the diameter of the smallest Jovian planet (Neptune). Also, their masses are only 1/17 as great as Neptune's. Hence, the Jovian planets are often called *giants.* Because of their relative locations, the four Jovian planets are also referred to as the **outer planets,** whereas the terrestrial planets are called the **inner planets.** As we shall see, there appears to be a correlation between the location of these planets within the solar system and their sizes.

TABLE 24.1 Planetary Data

Planet	Symbol	Mean Distance from Sun: AU*	Mean Distance from Sun: Millions of Miles	Mean Distance from Sun: Millions of Kilometers	Period of Revolution	Inclination of Orbit	Orbital Velocity: mi/s	Orbital Velocity: km/s
Mercury	☿	0.39	36	58	88^d	7°00′	29.5	47.5
Venus	♀	0.72	67	108	225^d	3°24′	21.8	35.0
Earth	⊕	1.00	93	150	365.25^d	0°00′	18.5	29.8
Mars	♂	1.52	142	228	687^d	1°51′	14.9	24.1
Jupiter	♃	5.20	483	778	12^{yr}	1°18′	8.1	13.1
Saturn	♄	9.54	886	1427	29.5^{yr}	2°29′	6.0	9.6
Uranus	⛢	19.18	1783	2870	84^{yr}	0°46′	4.2	6.8
Neptune	♆	30.06	2794	4497	165^{yr}	1°46′	3.3	5.3

Planet	Period of Rotation	Diameter: Miles	Diameter: Kilometers	Relative Mass (Earth = 1)	Average Density (g/cm³)	Polar Flattening (%)	Eccentricity†	Number of Known Satellites**
Mercury	59^d	3015	4878	0.06	5.4	0.0	0.206	0
Venus	244^d	7526	12,104	0.82	5.2	0.0	0.007	0
Earth	$23^h56^m04^s$	7920	12,756	1.00	5.5	0.3	0.017	1
Mars	$24^h37^m23^s$	4216	6794	0.11	3.9	0.5	0.093	2
Jupiter	9^h50^m	88,700	143,884	317.87	1.3	6.7	0.048	63
Saturn	10^h14^m	75,000	120,536	95.14	0.7	10.4	0.056	56
Uranus	17^h14^m	29,000	51,118	14.56	1.2	2.3	0.047	27
Neptune	16^h03^m	28,900	50,530	17.21	1.7	1.8	0.009	13

* AU = astronomical unit, Earth's mean distance from the Sun.

**Includes all satellites discovered as of August 2006.

† Eccentricity is a measure of the amount an orbit deviates from a circular shape. The larger the number, the less circular the orbit.

FIGURE 24.2 The planets drawn to scale.

Other dimensions in which the terrestrial and the Jovian planets differ include density, chemical makeup, and rate of rotation. The densities of the terrestrial planets average about five times the density of water, whereas the Jovian planets have densities that average only 1.5 times that of water. One of the outer planets, Saturn, has a density only 0.7 times that of water, which means that Saturn would float if placed in a large enough water tank! Differences in the chemical compositions of the planets are largely responsible for these density differences.

The Compositions of the Planets

The substances that make up the planets are divided into three compositional groups: *gases*, *rocks*, and *ices*, based on their melting points.

1. *Gases*, hydrogen and helium, are those with melting points near absolute zero (0 Kelvin). These two gases were the most abundant constituents of the solar nebula.
2. *Rocks* are principally silicate minerals and metallic iron, which have melting points that exceed 700°C.
3. *Ices* include ammonia, methane, carbon dioxide, and water. They have intermediate melting points (for example, water has a melting point of 0°C).

The terrestrial planets are dense, consisting mostly of rocky and metallic substances, with minor amounts of ices. The Jovian planets, on the other hand, contain large amounts of gases (hydrogen and helium) and ices (mostly water, ammonia, and methane). This accounts for their low densities. The Jovian planets also contain substantial amounts of rocky and metallic materials, which are concentrated in their central cores.

The Atmospheres of the Planets

The Jovian planets have very thick atmospheres of hydrogen, helium, methane, and ammonia. By contrast, the terrestrial planets, including Earth, have meager atmospheres at best. There are two reasons for this difference.

The first reason is the location of each planet within the solar nebula during its formation. The outer planets formed where the temperature was low enough to allow water vapor, ammonia, and methane to condense into ices. Hence, the Jovian planets contain large amounts of these volatiles. However, in the inner regions of the developing solar system the environment was too hot for ices to survive. Consequently, one of the long-standing questions for the nebular hypothesis was "How did Earth acquire water and other volatile gases?" The answer seems to be that during its protoplanet stage, Earth was bombarded with icy fragments (planetesimals) that originated beyond the orbit of Mars.

But why do Mercury and our Moon lack an atmosphere? Like the inner planets, they surely would have been bombarded by icy bodies. That brings us to the second reason that the inner planets (and the Moon) lack substantial atmospheres. A planet's ability to retain an atmosphere depends on its mass and its temperature. Simply stated, more massive planets have a better chance of retaining their atmospheres because atoms and molecules need a higher speed to escape. On the Moon the **escape velocity** is only 2.4 km/s compared with over 11 km/s for Earth.

Because of their strong gravitational fields, the Jovian planets have escape velocities that are much higher than

Why are the Jovian planets so much larger than the terrestrial planets?

According to the nebular hypothesis, the planets formed from a rotating disk of dust and gases that surrounded the Sun. The growth of planets began as solid bits of matter began to collide and clump together. In the inner solar system, the temperatures were so high that only metals and silicate minerals could form solid grains. It was too warm for ices of water, carbon dioxide, and methane to form. Thus, the innermost (terrestrial) planets grew mainly from the high melting point substances found in the solar nebula. By contrast, in the frigid outer reaches of the solar system, it was cold enough for clumps of water and other substances to form. Consequently, the outer planets grew not only from accumulations of solid bits of metals and silicate minerals but also from large quantities of ices. Eventually, the outer planets became large enough to gravitationally capture and hold onto even the lightest gases (hydrogen and helium), and thus grow to become "giant" planets.

that of Earth, which is the largest terrestrial planet. Consequently, it is much more difficult for gases to escape from the outer planets. Also, because the molecular motion of a gas is temperature-dependent, at the low temperatures of the Jovian planets even the lightest gases (hydrogen and helium) are unlikely to acquire the speed needed to escape.

By contrast, a comparatively warm body with a small surface gravity, such as Mercury and our Moon, is unable to hold even heavy gases such as carbon dioxide and radon. (Mercury does have trace amounts of gas present.) The slightly larger terrestrial planets of Earth, Venus, and Mars retain some heavy gases such as water vapor, nitrogen, and carbon dioxide, but even their atmospheres make up only a very small portion of their total mass.

Earth's Moon

Today Earth has hundreds of satellites, but only one, the Moon, is natural. Although other planets have moons, our planet-satellite system is unique in the solar system because Earth's Moon is unusually large compared to its parent planet. The diameter of the Moon is 3475 kilometers (2150 miles), about one-fourth of Earth's 12,756 kilometers (7920 miles).

From a calculation of the Moon's mass, its density is 3.3 times that of water. This density is comparable to that of *mantle* rocks on Earth

but is considerably less than Earth's average density, which is 5.5 times that of water. Geologists have suggested that this difference can be accounted for if the Moon's iron core is small.

The gravitational attraction at the lunar surface is one-sixth of that experienced on Earth's surface (a 150-pound person on Earth weighs only 25 pounds on the Moon although they still have the same mass). This difference allows an astronaut to carry a heavy life-support system with relative ease. If not burdened with such a load, an astronaut could jump six times higher than on Earth.

The Lunar Surface

When Galileo first pointed his telescope toward the Moon, he saw two different types of terrain—dark lowlands and brighter, highly cratered highlands (Figure 24.3). Because the dark regions resembled seas on Earth, they were called **maria** (*mar* = sea, singular **mare**). Today we know that the maria are not oceans, but instead are flat plains that resulted from immense outpouring of fluid basaltic lavas. By contrast, the light-colored areas resemble Earth's continents, so the first observers dubbed them **terrae** (Latin for "land"). Today these areas are generally referred to as lunar **highlands**, because

FIGURE 24.3 Telescopic view from Earth of the lunar surface. The major features are the dark "seas" (maria) and the light, highly cratered highlands. (UCO/Lick Observatory Image)

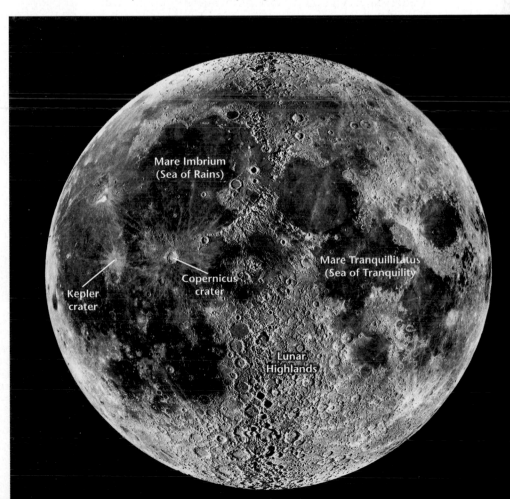

Mare Imbrium
(Sea of Rains)

Mare Tranquillitatus
(Sea of Tranquility)

Copernicus
crater

Kepler
crater

Lunar
Highlands

they are elevated several kilometers above the maria. Together the arrangement of terrae and maria result in the well-known "face of the moon."

Impact Craters   The most obvious features of the lunar surface are craters. They are so profuse that craters-within-craters are the rule! The larger ones in the lower portion of Figure 24.3 are about 250 kilometers (150 miles) in diameter, roughly the width of Indiana. **Impact craters** are produced by the impact of rapidly moving debris (meteoroids, asteroids, and comets), a phenomenon that was considerably more common in the early history of the solar system than it is today.

By contrast, Earth has only about a dozen easily recognized impact craters. This difference can be attributed to Earth's atmosphere, erosion, and tectonic processes. Friction with the air burns up small debris and makes large meteoroids smaller before they reach the ground. In addition, evidence for most of the craters that formed in Earth's history has been obliterated by erosional or tectonic processes.

The formation of an impact crater is illustrated in Figure 24.4. Upon impact, the high-speed meteoroid compresses the material it strikes, then almost instantaneously the compressed rock rebounds, ejecting material from the crater. This process is analogous to the splash that occurs when a rock is dropped into water. Craters excavated by objects several kilometers across often exhibit a central peak, as seen in the large crater in Figure 24.5. Most of the ejected material (*ejecta*) lands in the crater or nearby, where it builds a rim around it. Depending on the size of the meteoroid, the heat generated by the impact may be sufficient to melt some of the impacted rock. Astronauts have brought back samples of glass beads produced in this manner, as well as rock formed when broken fragments and dust were welded together by the heat of an impact.

A meteoroid only 3 meters (10 feet) in diameter can blast out a 150-meter- (500-foot-) wide crater. A few of the large craters, such as Kepler and Copernicus, shown in Figure 24.3, formed from the bombardment of bodies 1 kilometer or more in diameter. These two large craters are thought to be relatively young because of the bright *rays* (splash marks) that radiate outward for hundreds of kilometers.

Highlands and "Seas"   Densely pockmarked highland areas make up most of the lunar surface (Figure 24.6). In fact, most of the back side of the Moon is characterized by such topography. (Only astronauts have directly observed the farside, because the Moon rotates on its axis once with each revolution around Earth, always keeping the same side facing Earth.) As seen in Figure 24.3, highlands consist of an apparently endless sequence of overlapping craters. Indeed, the great number of impact craters is evidence of the Moon's violent early history. This activity crushed and repeatedly mixed at least the upper few kilometers of the lunar crust. As a result, the highlands are very rugged.

The highlands are made of plutonic rocks that contain over 90 percent plagioclase feldspar that rose like "scum" from a magma ocean early in lunar history. Maria, on the

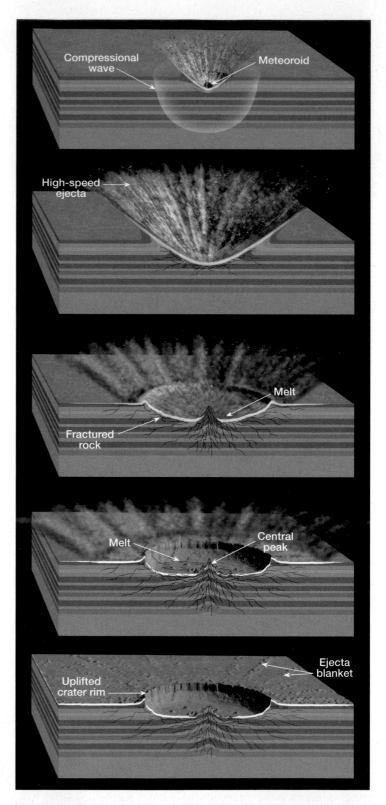

FIGURE 24.4   Formation of an impact crater. The energy of the rapidly moving meteoroid is transformed into heat energy and compressional waves. The rebound of the compressed rock causes debris to be ejected from the crater. Heat melts some material, producing glass beads. Small secondary craters are formed by the material "splashed" from the impact crater. (After E. M. Shoemaker)

FIGURE 24.5   The 20-kilometer-wide lunar crater Euler in the southwestern part of Mare Imbrium. Clearly visible are the bright rays, central peak, secondary craters, and the large accumulation of ejecta near the crater rim. (Courtesy of NASA)

other hand, are composed of volcanic rocks with a mafic composition and, hence, are similar to flood basalts on Earth.

The dark, flat maria make up only about 16 percent of the Moon's landscape and are concentrated on the side of the Moon facing Earth. (Figure 24.3). More than 4 billion years ago, asteroids having diameters as large as Rhode Island excavated several huge craters on the lunar surface. Because the crust was sufficiently fractured, magma began to bleed out. Apparently the craters were flooded with layer upon layer of very fluid lavas resembling those of the Columbia Plateau in the Pacific Northwest (Figure 24.7). In most cases, these lavas welled up long after the craters formed. It seems that volcanism occurred because the crust was thinner and more fractured in these regions rather than because of heat generated by the impact itself.

Weathering and Erosion   The Moon has no atmosphere or flowing water. Therefore, the processes of weathering and erosion that continually modify Earth's surface are virtually lacking on the Moon. In addition, tectonic forces are no longer active on the Moon, so earthquakes and volcanic eruptions do not occur. However, because the Moon is unprotected by an atmosphere, a different kind of erosion occurs: Tiny particles from space (micrometeorites) continually bombard its surface and ever so gradually smooth the landscape.

Both the maria and terrae are mantled with a layer of gray, unconsolidated

debris derived from a few billion years of meteoric bombardment (Figure 24.8). This soil-like layer, properly called **lunar regolith** (*rhegos* = blanket, *lithos* = stone) is composed of igneous rocks, breccia, glass beads, and fine *lunar dust*. In the maria that have been explored by *Apollo* astronauts, the lunar regolith is apparently just over 3 meters (10 feet) thick.

Lunar History

Although the Moon is our nearest planetary neighbor and astronauts have sampled its surface, much is still unknown about its origin. The most widely favored scenario is that during the formative period of the solar system, a glancing collision occurred between a Mars-sized body and a youthful semi-molten Earth (Figure 24.9). (Collisions of this type were probably frequent events in the early solar system.) During such a catastrophic collision most of the ejected debris would have collected in an orbit around Earth. Gradually this debris coalesced to form the Moon. Computer simulations show that most of the ejected material would have come from the mantles of both the impacting object and Earth (Figure 24.9). (Recall that the mantle makes up 82 percent of Earth's volume.) Further, if the impacting object had an iron core, this dense material would eventually have been incorporated into Earth's core. The *impact theory* is consistent with the Moon's internal structure—the Moon has a large mantle and a small core.

Planetary geologists have also worked out the basic details of the Moon's later history. One method of dating topographic features on the lunar surface is to observe variations in crater density (quantity per unit area). The greater the crater density, the longer the feature must have existed. From such evidence, scientists conclude that the Moon evolved in four phases: (1) the formation of the original crust, (2) the lunar highlands, (3) the maria basins, and (4) the rayed craters.

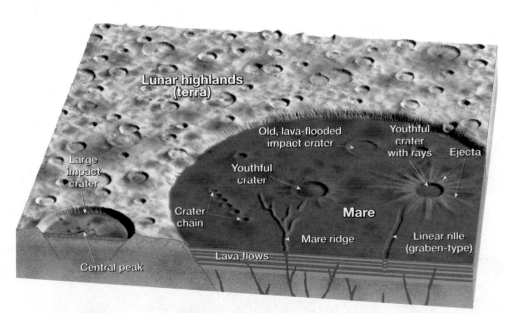

FIGURE 24.6   Block diagram illustrating major topographic features on the lunar surface.

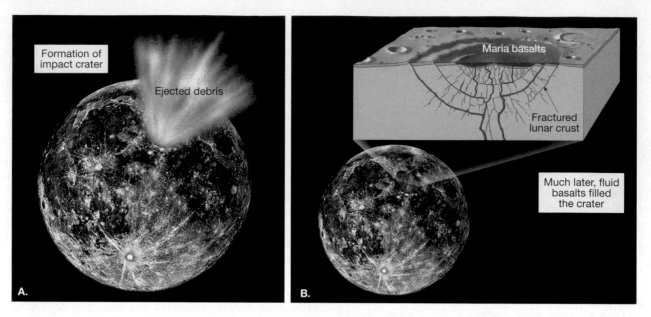

FIGURE 24.7   Formation of lunar maria. **A.** Impact of an asteroid-size mass produced a huge crater hundreds of kilometers in diameter and disturbed the lunar crust far beyond the crater. **B.** Filling of the impact area with fluid basalts, perhaps derived from partial melting deep within the lunar mantle.

During the late stages of its accretion, the Moon's upper mantle was partially or completely melted, resulting in a magma ocean. Then about 4.4 billion years ago, the magma ocean began to cool and underwent magmatic differentiation (see Chapter 4). Most of the dense minerals, olivine and pyroxene, sank while the less dense plagioclase feldspar floated to form the Moon's crust. Once formed, the lunar crust was continually impacted as it swept up debris from the solar nebula. Then about 3.9 billion years ago the Moon, and probably the entire solar system, experienced a sudden drop in the rate of meteoritic bombardment. (Since that time, the rate of cratering has been roughly constant.) Remnants of this original crust are represented by the densely cratered highlands, which have been estimated to be as much as 4.4 billion years old.

The next major event was the formation of the large maria basins (see Figure 24.7). Radiometric dating of the maria basalts puts their age between 3.2 billion and 3.8 billion years, considerably younger than the initial lunar crust. In places, the lava flows overlap the highlands, another testimonial to the younger age of the maria deposits. Evidence suggests that some mare-forming eruptions may have occurred as recently as a billion years ago. However, no volcanism occurs today, perhaps because cooling caused the Moon's crust to become too thick for magma to penetrate.

The last prominent features to form on the lunar surface were the rayed craters, as exemplified by the 90 kilometer-wide Copernicus crater shown in Figure 24.3. Material ejected from these younger depressions is clearly seen blanketing the surfaces of the maria and many older rayless craters. Even a relatively young crater like Copernicus is thought to be about a billion years old. Had it formed on Earth, erosional forces would have long since obliterated it.

If photos of the Moon taken several hundreds of millions of years ago were available, they would reveal that the Moon has changed little in the intervening years. By all measures, the Moon is essentially a geologically dead body wandering through space and time.

FIGURE 24.8   Astronaut Harrison Schmitt sampling the lunar surface. Notice the footprints (inset) in the lunar "soil." (Courtesy of NASA)

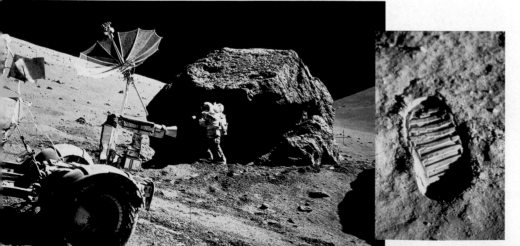

The Planets: A Brief Tour

Mercury: The Innermost Planet

Mercury, the innermost and smallest planet, is hardly larger than Earth's Moon and is smaller than three other moons in the solar system. Mercury revolves quickly but rotates slowly. One full day–night cycle on Earth takes 24 hours, but on Mercury it requires 179 Earth-days. Thus, a night on Mercury lasts for about three months and is followed by three months of daylight. Nighttime temperatures drop as low as −173°C (−280°F) and noontime temperatures exceed 427°C (800°F), hot enough to melt tin and lead. Mercury has the greatest temperature extremes of any planet. The odds of life as we know it existing on Mercury are nil.

Like our own Moon, Mercury absorbs most of the sunlight that strikes it, reflecting only 6 percent into space (Figure 24.10). In contrast, the Earth reflects about 30 percent of the light that strikes it, much of it from clouds. The low reflectivity of sunlight from Mercury is characteristic of terrestrial bodies that have virtually no atmosphere. The minuscule amounts of gas present on Mercury may have originated as ionized gas emitted from the Sun; from the ices of a recent comet impact; from surface rocks subject to solar winds; and/or from outgassing of the planet's interior.

The probe *Messenger*, scheduled to arrive in 2011, will investigate the composition of Mercury's core and the nature of its magnetic field. Mercury is very dense (5.4 g/cm³), which implies that it contains a very large iron core for its size. Mercury has cratered highlands, much like the Moon, and vast smooth terrains that resemble maria. Mercury also has very long scarps that cut across the plains and numerous craters. These scarps are thought to be the result of

FIGURE 24.10   Photomosaic of Mercury. This view of Mercury is remarkably similar to the "far side" of Earth's Moon. (Courtesy of NASA)

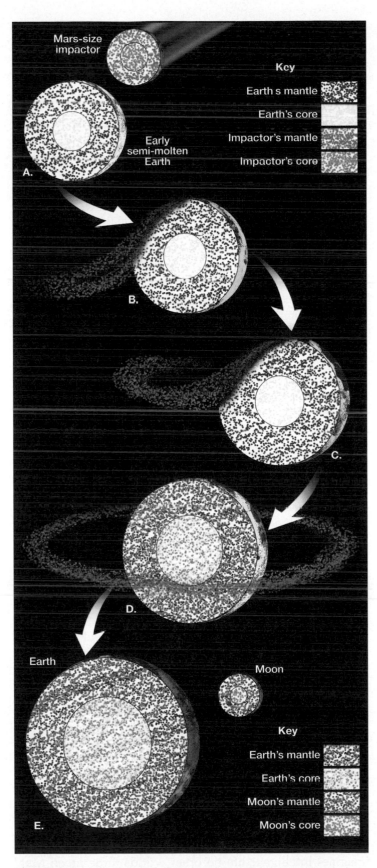

FIGURE 24.9   The formation of the Moon based on a scenario often called the impact theory. According to this model a protoplanet the size of Mars was involved in a glancing collision with a young semi-molten Earth. Most of the pieces of the protoplanet and splattered fragments of Earth coalesced in an orbit to form the Moon. Note that much of the impactor's core became part of Earth, leaving the Moon with a large rocky mantle and a comparatively small metallic core.

crustal shortening as the planet cooled and shrank early in its history. The shrinking caused the crust to fracture.

Venus: The Veiled Planet

Venus, second only to the Moon in brilliance in the night sky, is named for the goddess of love and beauty. It orbits the Sun in a nearly perfect circle once every 255 Earth-days. Venus is similar to Earth in size, density, mass, and location in the solar system. Thus, it has been referred to as "Earth's twin." Because of these similarities, it is hoped that a detailed study of Venus will provide geologists with a better understanding of Earth's evolutionary history.

FIGURE 24.11 This global view of the surface of Venus is computer-generated from two years of *Magellan* project radar mapping. The twisting bright features that cross the globe are highly fractured mountains and canyons of the eastern Aphrodite highland. (Courtesy of NASA/JPL)

Venus is shrouded in thick clouds impenetrable to visible light. The Venusian atmosphere contains an opaque cloud layer about 25 kilometers (15 miles) thick, and has an atmospheric pressure that is 90 times that at Earth's surface. Before the advent of space vehicles, Venus was considered to be a potentially hospitable site for living organisms. However, evidence from space probes indicates otherwise. The surface of Venus reaches temperatures as great as 475°C (900°F), and the Venusian atmosphere is 97 percent carbon dioxide. Only scant water vapor and nitrogen have been detected. This hostile environment makes it unlikely that life as we know it exists on Venus.

The composition of the Venusian core is likened to that of Earth. However, since Venus only has a small induced magnetic field, the internal dynamics must be very different. Mantle convection still operates on Venus, but the processes of plate tectonics, which recycle rigid lithosphere, do not appear to have contributed to the present Venusian topography. Tectonic activity on Venus seems to be restricted to upwelling (mantle plumes) and downwelling of material in the planet's interior, rather than lateral movement and subduction of lithospheric plates.

Radar mapping by the unmanned *Magellan* spacecraft and by instruments on Earth has revealed a varied topography with features somewhat between those of Earth and Mars (Figure 24.11). Radar pulses in the microwave range are sent toward the Venusian surface, and the heights of plateaus and mountains are measured by timing the return of the radar echo. These data have confirmed that basaltic volcanism and tectonic deformation are the dominant processes operating on Venus. Further, based on the low density of impact craters, volcanism and tectonic deformation must have been very active during the recent geologic past.

About 80 percent of the Venusian surface consists of subdued plains that are covered by volcanic flows. Some lava channels extend hundreds of kilometers; one meanders 6800 kilometers across the planet. Thousands of volcanic structures have been identified, mostly small shield volcanoes, although more than 1500 volcanoes greater than 20 kilometers (12 miles) across have been mapped (Figure 24.12). One is Sapas Mons, 400 kilometers (250 miles) across and 1.5 kilometers (0.9 mile) high. Many flows from this volcano erupted from its flanks rather than the summit, in the manner of Hawaiian shield volcanoes. Only 8 percent of the Venusian surface consists of highlands that may be likened to continental areas on Earth.

Mars: The Red Planet

Mars has evoked greater interest than any other planet, both for scientists and for nonscientists (see Box 24.1). When one imagines intelligent life on other worlds, little green Martians may come to mind. Interest in Mars stems mainly from this planet's accessibility to observation. All other planets within telescopic range have their surfaces hidden by clouds, except for Mercury, whose nearness to the Sun makes viewing difficult. Mars is approximately half the size

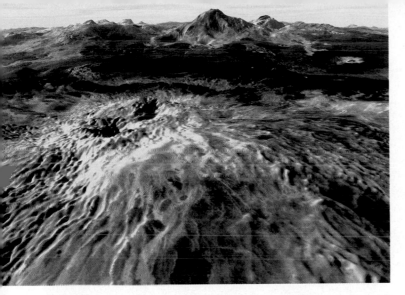

FIGURE 24.12   Computer-generated image of Venus. Near the horizon is Maat Mons, a large volcano. The bright feature below is the summit of Sapas Mons. (Courtesy David P. Anderson/SMU/NASA/Science Photo Library/Photo Researchers, Inc.)

of the Earth and revolves around the Sun in 687 Earth-days. Through the telescope, Mars appears as a reddish ball interrupted by some permanent dark regions that change intensity during the Martian year. The most prominent telescopic features of Mars are its brilliant white polar caps, resembling Earth's.

The Martian Atmosphere   The Martian atmosphere has only 1 percent the density of Earth's, and it is primarily carbon dioxide with tiny amounts of water vapor. Data from Mars probes confirm that the polar caps of Mars are made of water ice, covered by a thin layer of frozen carbon dioxide. As winter nears in either hemisphere, we see the equatorward growth of that hemisphere's ice cap as temperatures drop to $-125°C$ ($-193°F$) and additional carbon dioxide is deposited.

Although the atmosphere of Mars is very thin, extensive dust storms occur and may cause the color changes observed from Earth-based telescopes. Hurricane-force winds up to 270 kilometers (170 miles) per hour can persist for weeks. Images from *Viking 1* and *Viking 2* revealed a Martian landscape remarkably similar to a rocky desert on Earth (see chapter-opening photo), with abundant dunes and impact craters partially filled with dust.

Mars' History   Although the Martian core is believed to be solid because of the lack of a global magnetic field, evidence suggests that some of the planet's oldest rocks formed in the presence of a strong magnetic field. In the past, Mars may have had a hot interior with a molten core.

Most Martian surface features are old by Earth standards. Evidence suggests that weathering accounts for almost all surface changes during the last 3.5 billion years. Denser cratering in the highlands indicates that these areas are very old, whereas lighter cratering in the lowlands suggest volcanic eruptions 3.7–3.8 billion years ago. The highly cratered southern hemisphere is probably similar in age to the lunar highlands (nearly 4.5 billion years old). Even the relatively

fresh-appearing volcanic features of the northern hemisphere are most likely older than several hundred million years. These facts and the absence of Marsquake recordings by *Viking* seismographs point to a tectonically dead planet.

Mars' Dramatic Geological Past   *Mariner 9*, the first spacecraft to orbit another planet, reached Mars in 1971 amid a raging dust storm. When the dust cleared, images of Mars' northern hemisphere revealed numerous large volcanoes. The biggest, Olympus Mons, is the size of Ohio and 23 kilometers (75,000 feet) tall, nearly three times higher than Mount Everest. This gigantic volcano was last active about 100 million years ago and resembles Hawaiian shield volcanoes on Earth (Figure 24.13).

Why are the volcanoes on Mars many times larger than even the biggest volcanic structures on Earth? The largest volcanoes on terrestrial planets tend to form where plumes of hot rock rise from deep within its interior. Earth is tectonically active, with moving plates that keep the crust in constant motion. The Hawaiian Islands, for example, consist of a chain of shield volcanoes that formed as the Pacific plate moved over a relatively stationary mantle plume. On Mars, volcanoes such as Olympus Mons have grown to great size because the crust there remains stationary. Successive eruptions occur from a mantle plume at the same location and add to the bulk of a single volcano rather than producing several smaller structures, as occurs on Earth.

Another surprising find made by *Mariner 9* was the existence of several canyons that dwarf even Earth's Grand Canyon of the Colorado River. One of the largest, Valles

FIGURE 24.13   Image of Olympus Mons, an inactive shield volcano on Mars that covers an area about the size of the state of Ohio. (Courtesy of the U.S. Geological Survey)

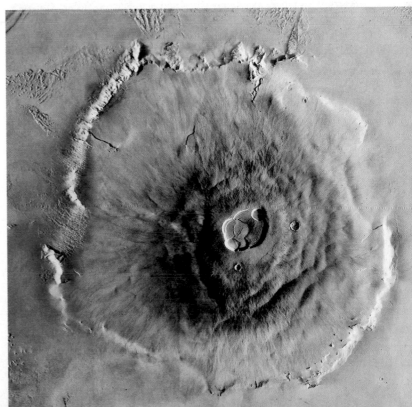

| BOX 24.1 ▶ UNDERSTANDING EARTH

Pathfinder—The First Geologist on Mars

On July 4, 1997, the *Mars Pathfinder* bounced onto the rock-littered surface of Mars and deployed its wheeled companion, *Sojourner*. Over the next three months the lander sent three gigabits of data back to Earth, including 16,000 images and 20 chemical analyses. The landing site was a vast rolling landscape carved by ancient floods. The flood-deposit locale was selected with hope that a variety of rock types would be available for the rover *Sojourner* to examine.

Sojourner carried an alpha photon X-ray spectrometer (APXS) used to determine the chemical composition of rocks and Martian "soil" (regolith) at the landing site (Figure 24.A). In addition, the rover was able to take close-up images of the rocks. From these images, researchers concluded that the rocks were igneous. However, one hard, white, flat object named Scooby Doo was originally thought to be a sedimentary rock, but the APXS data suggest its chemistry is like that of the soil found at the site. Thus, Scooby Doo is probably a well-cemented soil.

During its first week on Mars, *Sojourner's* APXS obtained data for a patch of windblown soil and a medium-sized rock, known affectionately as Barnacle Bill. Preliminary evaluation of the APXS data from Barnacle Bill showed that it contained over 60 percent silica. This could indicate that the volcanic rock andesite is present on Mars. However, researchers had expected that most volcanic rocks on Mars would be basalt, which is lower in silica (less than

FIGURE 24.A   *Pathfinder's* rover *Sojourner* (left) obtaining data on the chemical composition of a Martian rock known as Yogi. (Photo courtesy of NASA)

50 percent). On Earth, andesites are associated with tectonically active regions where oceanic crust is subducted into the mantle. Examples include the volcanoes of South America's Andes Mountains and the Cascades of North America. It is not clear whether the rocks are andesite or basalt with an andesite coating caused by weathering of the basalt.

In all, *Sojourner* analyzed eight rocks and seven soils, and the results are controversial. Because these rocks are covered with a reddish dust that is high in sulfur, the exact composition of the rocks is debated. Some researchers claim they are all of the same composition, and that the differences between the measurements are due to varying thicknesses of dust.

Marineris, is thought to have formed by slippage of material along huge faults in the outer crustal layer. In this respect, it would be comparable to the rift valleys of East Africa (Figure 24.14).

Water on Mars?   Liquid water does not appear to exist anywhere on the Martian surface. However, poleward of about 30 degrees latitude ice can be found within a meter of the surface and in the polar regions it forms small permanent ice caps. In addition, considerable evidence indicates that in the first billion years of the planet's history, liquid water flowed on the surface creating valleys and related features. In particular, some areas of Mars exhibit treelike drainage patterns similar to those created by streams on Earth. When these streamlike channels were first discovered, some ob-

servers speculated that a thick, water-rich atmosphere capable of generating torrential downpours once existed on Mars. Other researchers, however, were skeptical.

Today, most planetary geologists agree that running water was involved in carving at least some of the valleys on Mars. Some examples can be seen on an image from the *Mars Global Surveyor* in Figure 24.15A. Researchers have proposed that melting of subsurface ice caused springlike seeps to emerge along this valley wall, slowly creating the gullies shown.

Other channels have streamlike banks and contain numerous teardrop-shaped islands (24.15B). These valleys appear to have been cut by catastrophic floods that had discharge rates that were more than 1000 times greater than the Missippippi River. Most of these large flood channels emerge from areas of chaotic topography that appear to have formed when the surface collapsed. The most likely source of water for these flood valleys is the melting of subsurface ice. If the meltwater was trapped beneath a thick layer of permafrost, pressure could mount until a catastrophic release of groundwater occurred. As a result, the overlying surface layer would collapse, creating the chaotic terrain.

Not all Martian valleys appear to be the result of water released by the melting of subsurface ice. Some that exhibit branching treelike patterns that closely resemble the drainage networks on Earth were likely part of an active hydrologic cycle.

Recent evidence for surface water comes from the *Spirit* and *Opportunity* rovers. *Opportunity* investigated geologic structures similar to formations created by water on Earth—layered sedimentary rock, playas, and lake beds. Minerals that form only in the presence of water such as hydrated sulfates and micas were detected. Small spheres of hematite, called "blueberries," were also found that probably precipitated from water to form lake sediments. Although these findings are new, water does not appear to have significantly altered the topography of Mars for billions of years with the exception of the polar regions. Near the poles, the *Mars Express Orbiter* and *Mars Odyssey* detected geologically recent glacial and snow deposits. Polygonal patterned ground

FIGURE 24.14 This image shows the entire Valles Marineris canyon system, over 5000 kilometers long and up to 8 kilometers deep. The dark spots on the left edge of the image are huge volcanoes, each about 25 kilometers high. (Courtesy of U.S. Geological Survey)

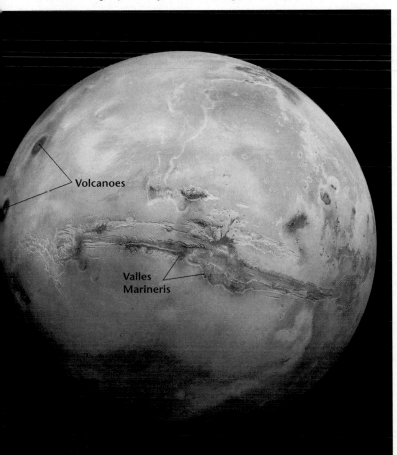

Volcanoes

Valles
Marineris

similar to features associated with permafrost are also commonly seen on Mars (Figure 24.16).

The Martian Satellites   Tiny Phobos and Deimos, the two satellites of Mars, were not discovered until 1877 because they are only 24 and 15 kilometers, in diameter. Phobos is nearer to its parent than any other natural satellite in the solar system—only 5500 kilometers (3400 miles)—and requires just 7 hours and 39 minutes for one revolution. *Mariner 9* revealed that both satellites are irregularly shaped and have numerous impact craters.

It is likely that these moons are asteroids captured by Mars. A most interesting coincidence in astronomy and literature is the close resemblance of Phobos and Deimos to two fictional satellites of Mars described by Jonathan Swift in *Gulliver's Travels*, written about 150 years before these satellites were actually discovered.

Jupiter: Lord of the Heavens

Jupiter, truly a giant among planets, has a mass two and a half times greater than the combined mass of all the remaining planets, satellites, and asteroids. In fact, had Jupiter been about 10 times larger, it would have evolved into a small star. Despite its great size, however, it is only 1/800 as massive as the Sun.

Jupiter revolves around the Sun once every 12 Earth-years, and rotates more rapidly than any other planet, completing one rotation in slightly less than 10 hours. The effect of this fast spin is to make the equatorial region bulge and to make the polar dimension contract (see the Polar Flattening column in Table 24.1).

Atmosphere, Structure, and Composition   When viewed through a telescope or binoculars, Jupiter appears to be covered with alternating bands of multicolored clouds aligned parallel to its equator (Figure 24.17). The most striking feature is the *Great Red Spot* in the southern hemisphere (Figure 24.17). The Great Red Spot has been a prominent feature since it was first seen more than three centuries ago. When *Voyager 2* swept by Jupiter in 1979, the Great Red Spot was the size of two Earth-size circles placed side by side. On occasion, it has grown even larger. Images from *Pioneer 11* as it moved near Jupiter's cloud tops in 1974 indicated that the Great Red Spot is a counterclockwise-rotating storm caught between two jetstream-like bands flowing in opposite directions. This huge, hurricane-like storm makes a complete rotation about once every 12 days.

As on other planets, the winds on Jupiter are the product of differential heating, which generates vertical convective motions in the atmosphere. Jupiter's convective flow produces alternating dark-colored *bands* and light-colored *zones* as shown in Figure 24.18. The light clouds (zones) are regions where warm material is ascending and cooling, whereas cool material is sinking in the dark belts. This convective circulation, along with Jupiter's rapid rotation, generates the high-speed, east–west flow observed between the

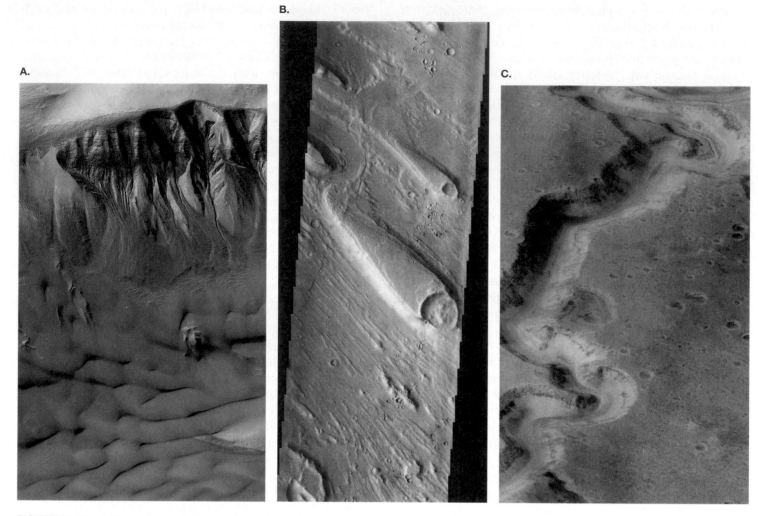

FIGURE 24.15 Evidence for erosion by running water. **A.** Image from *Mars Global Surveyor* showing crater wall with large gullies that may have been cut by liquid water, mixed with soil, rocks, and ice. (NASA/JPL photo courtesy of National Geographic) **B.** Streamlined islands in Ares Valles formed where running water encountered obstacles along its path. (NASA image) **C.** Terraces and a small central channel suggest that Nanedi Vallis may have been carved by running water. (NASA image)

FIGURE 24.16 Polygonal patterned ground, shown here on Mars, is common in areas of permafrost in polar latitudes on Earth. (Courtesy of JPL/NASA)

belts and zones. Unlike winds on Earth, which are driven by solar energy, Jupiter gives off nearly twice as much heat as it receives from the Sun. Thus, it is the heat emanating from Jupiter's interior that produces the huge convection currents observed in its atmosphere.

Jupiter's atmosphere is composed mainly of hydrogen and helium but also contains lesser amounts of methane, ammonia, and water which form clouds composed of liquid droplets or ice crystals. Atmospheric pressure at the top of the clouds is equal to sea-level pressure on Earth. Because of Jupiter's immense gravity, the pressure increases rapidly toward its surface. At 1000 kilometers below the clouds, the pressure is great enough to compress hydrogen gas into a liquid. Consequently, Jupiter's surface is thought to be a gigantic ocean of liquid hydrogen. Less than halfway into Jupiter's interior, extreme pressures cause the liquid hydrogen to turn into *liquid metallic* hydrogen. The fast rotation and liquid metallic core are a possible explanation for the

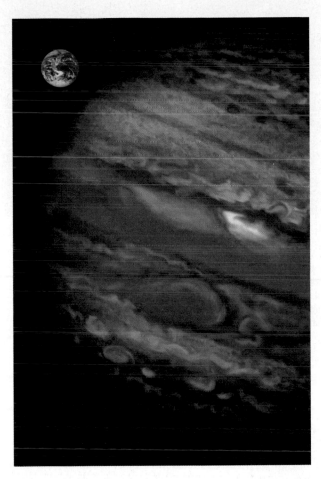

FIGURE 24.17   Artist's view of Jupiter with Great Red Spot visible in its southern hemisphere. Earth for scale.

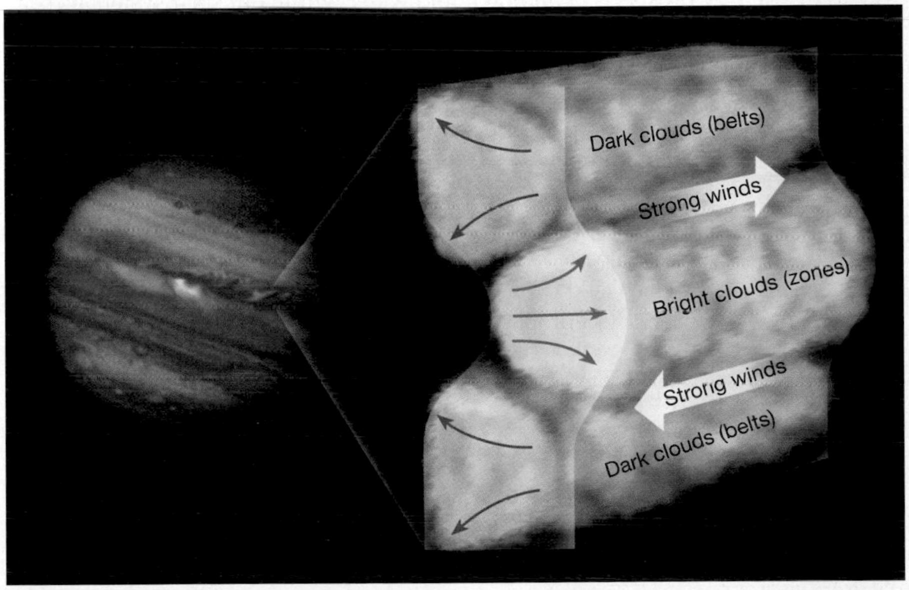

FIGURE 24.18   The structure of Jupiter's atmosphere. The areas of light clouds (zones) are regions where gases are ascending and cooling. Sinking dominates the flow in the darker cloud layers (belts). This convective circulation, along with the rapid rotation of the planet, generates the high-speed winds observed between the belts and zones.

intense magnetic field surrounding Jupiter. Jupiter is also believed to contain as much rocky and metallic material as is found in the terrestrial planets, probably located in a central core.

Jupiter's Moons   Jupiter's satellite system, consisting of 63 moons discovered so far, resembles a miniature solar system. The four largest satellites, discovered by Galileo, travel in nearly circular orbits around the planet, with periods of from 2 to 17 Earth-days (Figure 24.19). The two largest Galilean satellites, Callisto and Ganymede, surpass Mercury in size, whereas the two smaller ones, Europa and Io, are about the size of Earth's Moon. The Galilean moons can be observed with binoculars or a small telescope and are interesting in their own right.

Images from *Voyagers 1* and 2 revealed, to the surprise of almost everyone, that each of the four Galilean satellites is a unique geological world (Figure 24.19). A further surprise from the *Galileo* mission was that the composition of each satellite is strikingly different, which implies a different evolution for each satellite. For example, Ganymede's has a dynamic core that generates a strong magnetic field not observed on the other satellites.

The innermost of the Galilean moons, Io, is perhaps the most volcanically active body in our solar system. In all, more than 80 active sulfurous volcanic centers have been discovered. Umbrella-shaped plumes have been observed rising from the surface of Io to heights approaching 200 kilometers (120 miles) (Figure 24.20A). The heat source for volcanic activity is tidal energy generated by a relentless "tug of war" between Io and Jupiter and the other Galilean satellites. The gravitational field of Jupiter and the other nearby satellites pulls and pushes on Io's tidal bulge as its slightly eccentric orbit takes it alternately closer to and farther from its parent planet. This gravitational flexing of Io is transformed into heat energy (similar to the back-and-forth bending of a paper clip) and results in Io's spectacular sulfurous volcanic eruptions. Moreover, lava thought to be mostly composed of silicate minerals regularly erupts on its surface (Figure 24.20B).

In addition, Jupiter has numerous satellites that are very small (about 20 kilometers in diameter), revolve in a direction that is opposite (*retrograde motion*) of the largest moons, and have orbits that are steeply inclined to the Jovian equator. These satellites appear to be asteroids that passed near enough to be captured gravitationally by Jupiter.

Jupiter's Rings   One of the interesting aspects of the *Voyager 1* mission was a study of Jupiter's ring system. By analyzing how these rings scatter light, researchers determined that the rings are composed of fine, dark particles, similar in size to smoke particles. Further, the faint nature of the rings indicates that these minute particles are

A. Io   B. Europa   C. Ganymede   D. Callisto

FIGURE 24.19 Jupiter's four largest moons (from left to right) are called the Galilean moons because they were discovered by Galileo. **A.** The innermost moon, Io, is one of only three volcanically active bodies in the solar system. **B.** Europa, smallest of the Galilean moons, has an icy surface that is criss-crossed by many linear features. **C.** Ganymede, the largest Jovian satellite, contains cratered areas, smooth regions, and areas covered by numerous parallel grooves. **D.** Callisto, the outermost of the Galilean satellites, is densely cratered, much like Earth's moon. (Courtesy of NASA/NGS Image Collection)

A.

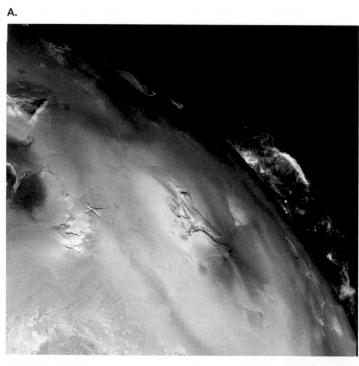

B.

Students Sometimes Ask . . .

Besides Earth, do any other bodies in the solar system have liquid water?

The planets closer to the Sun than Earth are considered too warm to contain liquid water, and those farther from the Sun are generally too cold to have water in the liquid form (although some features on Mars indicate that it probably had abundant liquid water at some point in its history). The best prospects of finding liquid water within our solar system lie beneath the icy surfaces of some of Jupiter's moons. For instance, Europa is suspected to have an ocean of liquid water hidden under its outer covering of ice. Detailed images sent back from the *Galileo* spacecraft have revealed that Europa's icy surface is quite young and exhibits cracks apparently filled with dark fluid from below. This suggests that under its icy shell, Europa must have a warm, mobile interior—and perhaps an ocean. Because the presence of water in the liquid form is a necessity for life as we know it, there has been much interest in sending an orbiter to Europa—and eventually a lander capable of launching a robotic submarine—to determine if it may harbor life.

FIGURE 24.20 A volcanic eruption on Jupiter's Moon Io. **A.** This plume of volcanic gases and debris is rising over 100 kilometers (60 miles) above Io's surface. **B.** The bright area on the left side of the image is newly erupted hot lava. (Courtesy of NASA)

widely dispersed. The main ring is composed of particles believed to be fragments blasted by meteorite impacts from the surfaces of Metis and Adrastea, two small moons of Jupiter. Impacts on Jupiter's moon's Amalthea and Thebe are believed to be the sources of the outer Gossamer ring.

Saturn: The Elegant Planet

Requiring 29.46 Earth-years to make one revolution, Saturn is almost twice as far from the Sun as Jupiter, yet its atmosphere, composition, and internal structure are believed to be remarkably similar to Jupiter's. The most prominent feature of Saturn is its system of rings (Figure 24.21), first seen by Galileo in 1610. With his primitive telescope, the rings appeared as two small bodies adjacent to the planet. Their ring nature was determined 50 years later by the Dutch astronomer Christian Huygens.

Structure of Saturn Saturn's atmosphere is very dynamic, with winds roaring at up to 1500 kilometers (930 miles) per hour. Cyclonic "storms" similar to Jupiter's Great Red Spot occur in Saturn's atmosphere as does intense lightning. Although the atmosphere is nearly 75 percent hydrogen and 25 percent helium, the clouds are composed of ammonia, ammonia hydrosulfide, and water, each segregated by temperature. The central core is composed of rock and ice layered with liquid metallic hydrogen, and then liquid hydrogen. Like Earth, Saturn's magnetic field is believed to be created within the core. In this process, helium condenses in the liquid hydrogen layers, releasing the heat necessary for convection.

In 1980 and 1981, the nuclear-powered *Voyagers* 1 and 2 space probes came within 100,000 kilometers of Saturn. More information was gained in a few days than had been acquired in all the time since Galileo first viewed this elegant planet telescopically. More recently, observations from ground-based telescopes and the *Hubble Space Telescope* have added to our knowledge of Saturn's ring system. In 1995 and 1996, when the positions of Earth and Saturn allowed the rings to be viewed edge-on, reducing the glare from the main rings, Saturn's faintest rings and satellites became visible.

Planetary Ring Systems Until the recent discovery that Jupiter, Uranus, and Neptune also have ring systems, this phenomenon was thought to be unique to Saturn. Although the four known ring systems differ in detail, they share many attributes. They all consist of multiple concentric rings separated by gaps of various widths. In addition, each ring is composed of individual particles—"moonlets" of ice and rock—that circle the planet while regularly impacting one another.

Most rings fall into one of two categories based on particle density. Saturn's main rings (designated A and B in Figure 24.21) and the bright rings of Uranus are tightly packed and contain "moonlets" that range in size from a few centimeters (pebble-size) to several meters (house-size). These particles are thought to collide frequently as they orbit the parent planet. Despite the fact that Saturn's dense rings stretch across several hundred kilometers, they are very thin, perhaps less than 100 meters (300 feet) from top to bottom.

At the other extreme, the faintest rings, such as Jupiter's ring system and Saturn's outermost ring (designated E in Figure 24.21), are composed of very fine (smoke-size) particles that are widely dispersed. In addition to having very low particle densities, these rings tend to be thicker than Saturn's bright rings.

Recent studies have shown that the moons that coexist with the rings play a major role in determining their structure. In particular, the gravitational influence of these moons tends to shepherd the ring particles by altering their orbits. The narrow rings appear to be the work of satellites located on either side that confine the ring by pulling back particles that try to escape.

More importantly, the ring particles are believed to be debris ejected from these moons, including the volcanic "ash" that spews from Jupiter's moon Io. Recent evidence from the *Cassini* mission shows the moon Enceladus has a hot spot with geysers spewing water ice that contributes to Saturn's E ring. It is possible that material is continually being recycled between the rings and the ring moons. The moons gradually sweep up particles, which are subsequently ejected by collisions with large chunks of ring material, or perhaps by energetic collisions with other moons. It seems then, that planetary rings are not the timeless features

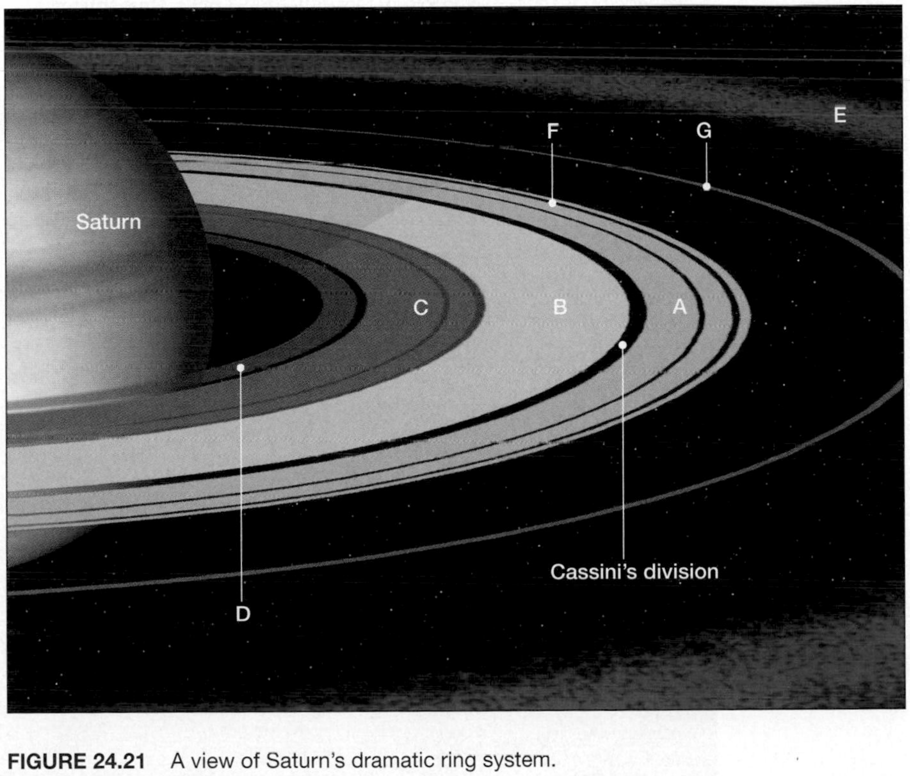

FIGURE 24.21 A view of Saturn's dramatic ring system.

FIGURE 24.22 Montage of the Saturnian satellite system. The moon Dione is in foreground; Tethys and Mimas are at lower right; Enceladus and Rhea are off ring's left; and Titan is upper right. (Photo courtesy of NASA)

that we once thought; rather, they continually reinvent themselves.

The origin of planetary ring systems is still being debated. Perhaps the rings formed out of a flattened cloud of dust and gases that encircled the parent planet. In this scenario, the rings formed at the same time and of the same material as the planets and moons. Perhaps the rings formed later, when a moon or large asteroid was gravitationally pulled apart after straying too close to a planet. Yet another hypothesis suggests that a foreign body blasted apart one of the planet's moons; the fragments of which would tend to jostle one another and form a flat, thin ring. Researchers expect more light to be shed on the origin of planetary rings as the Cassini spacecraft continues its four-year tour of Saturn.

Saturn's Moons The Saturnian satellite system consists of 56 known moons (Figure 24.22). (If you count the "moonlets" that comprise Saturn's rings, this planet has millions of satellites.) The largest, Titan, is bigger than Mercury and is the second-largest satellite in the solar system (after Jupiter's Ganymede). Titan and Neptune's Triton are the only satellites in the solar system known to have a substantial atmosphere. Because of its dense gaseous cover, the atmospheric pressure at Titan's surface is about 1.5 times that at the Earth's surface. The Cassini-Huygens probe determined the atmospheric composition to be about 95 percent nitrogen and 5 percent methane with additional organic compounds—similar to Earth's primitive atmosphere prior to the onset of life. Recent evidence suggests that Titan has Earth-like geological landforms and geological processes, such as dune formation and fluvial erosion caused by methane rain. Another satellite, Phoebe, exhibits retrograde motion. It, like other moons with retrograde orbits, is most likely a captured asteroid or large planetesimal left over from the episode of planetary formation.

Uranus and Neptune: The Twins

While Earth and Venus have many similar traits, Uranus and Neptune are nearly twins, having similar structures and compositions. They are less than 1 percent different in diameter (about four times the size of Earth), and they are both bluish in appearance, which is attributable to the methane in their atmospheres (Figures 24.23 and 24.24). Uranus and Neptune take 84 and 165 Earth-years, respectively, to complete one revolution around the Sun. The core composition is similar to the other gas giants with a rocky silicate and iron core, but with less liquid metallic hydrogen and more ice than Jupiter and Saturn. Neptune, however, is colder, because it is half again as distant from the Sun as is Uranus.

Uranus: The Sideways Planet A unique feature of Uranus is that it rotates "on its side." Its axis of rotation, instead of being generally perpendicular to the plane of its orbit, like the other planets, lies nearly parallel to the plane of its orbit. Its rotational motion, therefore, has the appearance of a rolling ball, rather than the toplike spinning of other planets. This is likely due to a huge impact early in the planet's evolution that literally knocked Uranus on its side. Because the axis of Uranus is inclined more than 90 degrees, the Sun is nearly overhead at one of its poles once each revolution, and then half a revolution later it is overhead at the other pole.

FIGURE 24.23 This image of Uranus was sent back to Earth by *Voyager 2* as it passed by this planet on January 24, 1986. Taken from a distance of nearly 1 million kilometers, little detail of its atmosphere is visible, except a few streaks (clouds) in the northern hemisphere. (Courtesy of NASA)

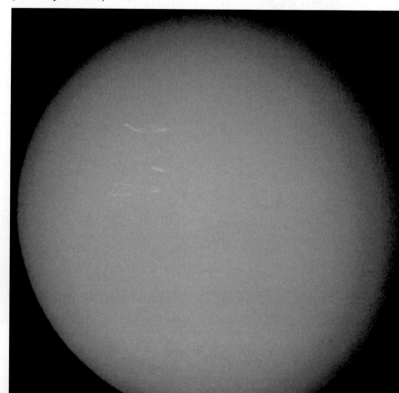

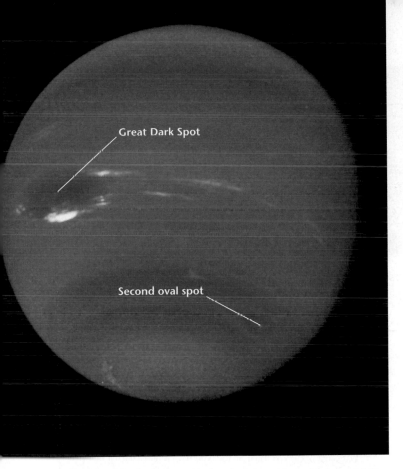

FIGURE 24.24 This image of Neptune shows the Great Dark Spot (left center). Also visible are bright cirruslike clouds that travel at high speed around the planet. A second oval spot is at south latitude on the east side of the planet. (Courtesy of the Jet Propulsion Laboratory)

A surprise discovery in 1977 revealed that Uranus has a ring system. This find occurred as Uranus passed in front of a distant star and blocked its view, a process called *occultation* (*occult* = hidden). Observers saw the star "wink" briefly five times (meaning five rings) before the primary occultation and again five times afterward. Later studies indicate that Uranus has *at least* nine distinct belts of debris orbiting its equatorial region.

Spectacular views from *Voyager 2* of the five largest moons of Uranus show quite varied terrains. Some have long, deep canyons and linear scars, whereas others possess large, smooth areas on otherwise crater-riddled surfaces. The Jet Propulsion Laboratory described Miranda, the innermost of the five largest moons, as having a greater variety of landforms than any body yet examined in the solar system. These features indicate that Miranda was recently geologically active.

Neptune: The Windy Planet Even when the most powerful telescope is focused on Neptune, it appears a bluish fuzzy disk. Until the 1989 *Voyager 2* encounter, astronomers knew very little about this planet. However, the 12-year, nearly 3-billion-mile journey of *Voyager 2* provided investigators with so much new information about Neptune and its satellites that years will still be needed to analyze it all.

Neptune has a dynamic atmosphere, much like those of Jupiter and Saturn (Figure 24.24). Winds exceeding 1000 kilometers (600 miles) per hour encircle the planet, making

it one of the windiest places in the solar system. It also has an Earth-size blemish called the *Great Dark Spot* that is reminiscent of Jupiter's Great Red Spot and is assumed to be a large rotating storm. About five years after the *Voyager 2* encounter, when the *Hubble Space Telescope* viewed Neptune, the spot had vanished, and was replaced by another dark spot in the planet's northern hemisphere.

Perhaps most surprising are white, cirruslike clouds that form a layer about 50 kilometers above the main cloud deck, probably frozen methane. Six new satellites were discovered in the *Voyager* images, bringing Neptune's family to 8, and more recent observations bring the total to 13. All of the newly discovered moons orbit the planet in a direction opposite that of the two larger satellites. *Voyager* images also revealed a ring system around Neptune.

Triton, Neptune's largest moon, is a most interesting object. It is the only large moon in the solar system that exhibits retrograde motion. This indicates that Triton formed independently of Neptune and was gravitationally captured.

Triton, along with other moons of the Jovian planets, exhibits one of the most amazing manifestations of volcanism, the eruption of ices. This type of volcanism is termed **cryovolcanism** (from the Greek *Kryos,* meaning frost) and refers to the eruption of magmas derived from the partial melting of ice rather than silicate rocks. Triton's icy magma originates from a mixture of water-ice, methane, and probably ammonia. When partially melted, this mixture behaves just as molten rock does on Earth. In fact, upon reaching the surface these magmas can generate quiet outpourings of ice lavas, or even explosive eruptions. When volatiles are released instantaneously, an explosive eruptive column will generate the ice equivalent of volcanic ash. In 1989, *Voyager 2* detected active plumes on Triton that rose 8 kilometers above the surface and were blown downwind for more than 100 kilometers. In other environments ice lavas develop that can flow great distances from their source—not unlike the fluid basaltic flows on Hawaii.

Minor Members of the Solar System: Asteroids, Comets, Meteoroids, and Dwarf Planets

In the vast spaces separating the eight planets as well as the expanses that extend to the outer reaches of the solar system are countless small chunks of debris, ranging from several hundred kilometers in diameter down to minute grains of dust. The objects found in this vast interplanetary space include *asteroids, comets, meteoroids,* and *dwarf planets.* Asteroids and meteoroids are fragments of rocky and metallic material with compositions somewhat like the terrestrial planets. They are distinguished from each other on the basis of size—those larger than 100 meters are asteroids, whereas anything smaller is a meteoroid. By contrast, comets are predominantly ices, with only small amounts of rocky material. The newest class of solar-system objects—dwarf planets—

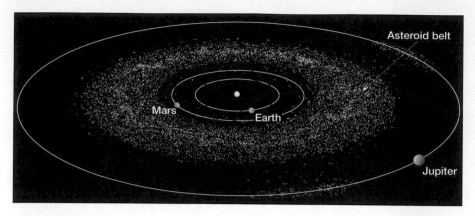

FIGURE 24.25 The orbits of most asteroids lie between Mars and Jupiter. Also shown are the orbits of a few known near-Earth asteroids. Perhaps a thousand or more asteroids have near-Earth orbits. Luckily, only a few dozen are thought to be larger than 1 kilometer in diameter.

includes Ceres, the largest known asteroid, and Pluto, which is thought to have a composition similar to Neptune's icy moon, Triton.

Asteroids: Planetesimals

Asteroids are small fragments (planetesimals) about 4.5 billion years old left over from the formation of the solar system. Most asteroids have a lower density than was first thought. This suggests that asteroids are porous bodies and not solid rocky or metallic objects, but more like a "rubble pile" of fragments bound together by gravity. The largest asteroid, Ceres, is 940 kilometers in diameter, but most of the 100,000 known asteroids are much smaller.

Most asteroids lie roughly midway between the orbits of Mars and Jupiter in the region known as the **asteroid belt** (Figure 24.25). Some travel along eccentric orbits that take them very near the Sun, and a few larger ones regularly pass close to Earth and the Moon. Many of the recent large impact craters on the Moon and Earth were probably the result of collisions with asteroids. About 2000 Earth-crossing asteroids are known, one-third of which are more than one kilometer in diameter. Inevitably, future Earth–asteroid collisions will occur (see Box 24.2).

Because most asteroids have irregular shapes, planetary geologists first speculated that they might be fragments of a broken planet that once orbited between Mars and Jupiter (Figure 24.25). However, the total mass of the asteroids is estimated to be only 1/1000 that of Earth, which itself is not a large planet. Today, most researchers agree that asteroids are the leftover debris from the solar nebula. Because of their location near Jupiter, with its huge gravitational field that continuously disrupted their motion, these planetesimals never accreted into a planet.

In February 2001 an American spacecraft became the first visitor to an asteroid. Although it was not designed for landing, *NEAR Shoemaker* landed successfully and generated information that has planetary geologists intrigued and perplexed. Images obtained as the spacecraft drifted toward the surface of Eros revealed a barren, rocky surface composed

of particles ranging in size from fine dust to boulders up to 10 meters (30 feet) across (Figure 24.26). Researchers unexpectedly discovered that fine debris tends to concentrate in the low areas where it forms flat deposits that resemble ponds. Surrounding the low areas, the landscape is marked by an abundance of large boulders.

One of several hypotheses to explain the boulder-strewn topography is seismic shaking, which causes the boulders to move upward as the finer material sinks. This is analogous to what happens when a can of mixed nuts is shaken—the larger nuts rise to the top while the smaller pieces settle to the bottom.

Indirect evidence from meteorites suggests asteroids might retain much of the heat generated from impact events. Some may even have completely melted, which would cause them to differentiate into a dense iron and nickel core, and a rocky mantle. In November 2005, a Japanese probe, *Hayabusa,* landed on a small near-Earth asteroid named 25143 Itokawa and is scheduled to return samples to Earth by June 2010.

Comets: Dirty Snowballs

Comets, like asteroids, are also left over from the formation of the solar system. Unlike asteroids, comets are composed of ices (water, ammonia, methane, carbon dioxide, and

Close-up of surface

FIGURE 24.26 Image of asteroid Eros obtained by the *NEAR-Shoemaker* probe. Inset shows a close-up of Eros' barren rocky surface. (Courtesy of NASA)

BOX 24.2 ▶ EARTH AS A SYSTEM

Is Earth on a Collision Course?

The solar system is cluttered with meteoroids, asteroids, active comets, and extinct comets. These fragments travel at great speeds and can strike Earth with an explosive force many times greater than a powerful nuclear weapon.

In recent decades, it has become increasingly clear that comets and asteroids collide with Earth far more frequently than was previously known. The evidence is the 100 or so giant impact structures that have been identified (Figure 24.B). (Many impact craters were once mistaken for volcanic structures.) Most impact structures are so old and heavily eroded that they were discovered using satellite photography (Figure 24.C). One notable exception is a very fresh-looking crater near Winslow, Arizona, known as Meteor Crater (Figure 24.31). Note that this crater was produced by a meteorite perhaps only 50 meters in diameter.

About 65 million years ago a large asteroid about 10 kilometers (6 miles) in diameter collided with Earth off the Yucatan peninsula in Mexico. This impact is thought to have caused the demise of the dinosaurs, as well as the extinction of near-

ly 50 percent of all plant and animal species (see more on this in Chapter 22 *Earth's Evolution through Geologic Time*).

More recently, a spectacular explosion has been attributed to the collision of our planet with a comet or asteroid. In 1908, in a remote region of Siberia, a "fireball" that appeared more brilliant than the Sun exploded violently. The shock waves rattled windows and triggered reverberations heard up to 1000 kilometers away. The "Tunguska event," as it is called, scorched, delimbed, and flattened trees up to 30 kilometers from the epicenter. But expeditions to the area found no evidence of an impact crater or any metallic fragments. Evidently, the explosion, which equaled at least a 10-megaton nuclear bomb, occurred several kilometers above the surface. Why it exploded prior to impact is uncertain.

The dangers of living with these small but deadly objects from space again came to public attention in 1989 when an asteroid nearly 1 kilometer across shot past Earth. It was a near miss, about twice the distance to the Moon. Traveling at 70,000 kilometers (44,000 miles) per hour, it could have produced a crater 10 kilometers

FIGURE 24.C Manicouagan, Quebec, is a 200-million-year-old eroded impact structure. The lake outlines the crater remnant, which is 70 kilometers (42 miles) across. Fractures related to this event extend outward for an additional 30 kilometers. (Courtesy of U.S. Geological Survey)

FIGURE 24.B World map of major impact structures. Others are being identified every year. (Data from Griffith Observatory)

(6 miles) in diameter and perhaps 2 kilometers (1.2 miles) deep. As an observer noted, "Sooner or later it will be back." As it was, it crossed our orbit just six hours ahead of Earth. Statistics show that collisions of this magnitude should take place every few hundred million years and could have drastic consequences for life on Earth.

Scientists at NASA are tracking near-Earth objects (NEOS). When comets or asteroids pass close to any of the other planets, their orbits may be altered by the gravitational interaction, which may send them toward Earth. Currently, more than 800 asteroids with Earth-crossing orbits are being tracked.

carbon monoxide) that hold together small pieces of rocky and metallic materials, thus the nickname "dirty snowballs." Comets are among the most interesting and unpredictable bodies in the solar system. Many comets travel in very elongated orbits that carry them far beyond Pluto. These comets take hundreds of thousands of years to complete a single orbit around the Sun. However, a few *short-period comets* (those having orbital periods of less than 200

years), such as Halley's comet, make regular encounters with the inner solar system.

When first observed, a comet appears very small, but as it approaches the Sun, solar energy begins to vaporize the ices, producing a glowing head called the **coma** (Figure 24.27). The size of the coma varies greatly from one comet to another. Extremely rare ones exceed the size of the Sun, but most approximate the size of Jupiter. Within the coma, a small

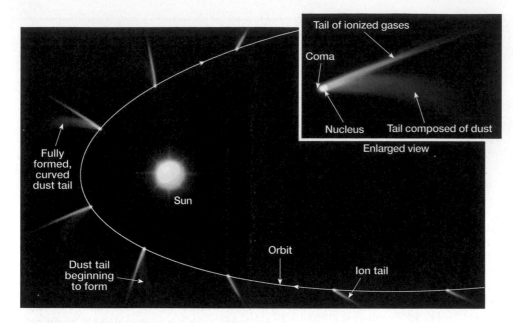

FIGURE 24.27 Orientation of a comet's tail as it orbits the Sun.

glowing nucleus with a diameter of only a few kilometers can sometimes be detected. As comets approach the Sun, some, but not all, develop a tail that extends for millions of kilometers. Despite the enormous size of their tails and comas, comets are relatively small members of the solar system.

The fact that the tail of a comet points away from the Sun in a slightly curved manner (Figure 24.27) led early astronomers to propose that the Sun has a repulsive force that pushes the particles of the coma away, thus forming the tail. Today, two solar forces are known to contribute to this formation. One, *radiation pressure*, pushes dust particles away from the coma. The second, known as *solar wind*, is responsible for moving ionized gases, particularly carbon monoxide. Sometimes a single tail composed of both dust and ionized gases is produced, but often two tails are observed (Figure 24.28).

As a comet moves away from the Sun, the gases forming the coma recondense, the tail disappears, and the comet returns to cold storage. Material that was blown from the

FIGURE 24.28 Comet Hale-Bopp. The two tails seen in the photograph are between 10 million and 15 million miles long. (Peoria Astronomical Society photograph by Eric Clifton and Craig Neaveill)

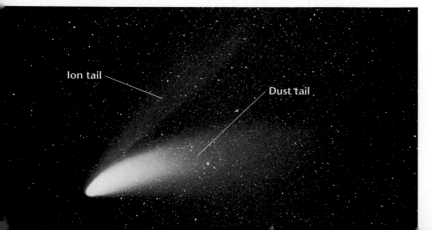

coma to form the tail is lost from the comet forever. Consequently, it is believed that most comets cannot survive more than a few hundred close orbits of the Sun. Once all the gases are expelled, the remaining material—a swarm of tiny metallic and stony particles—continues the orbit without a coma or a tail.

Most comets are found in two regions of the outer solar system. The short period comets are thought to orbit beyond Neptune in a region called the **Kuiper belt,** in honor of astronomer Gerald Kuiper, who had predicted their existence (Figure 24.29). (During the past decade, more than a hundred of these icy bodies have been discovered.) Like the asteroids in the inner solar system, most Kuiper belt comets move in nearly circular orbits that lie roughly in the same plane as the planets. A chance collision between two Kuiper belt comets, or the gravitational influence of one of the Jovian planets, may occasionally alter the orbit of a comet enough to send it to the inner solar system, and into our view.

Unlike Kuiper belt comets, long-period comets have orbits that are *not* confined to the plane of the solar system. These comets appear to be distributed in all directions from the Sun, forming a spherical shell around the solar system, called the **Oort cloud,** after the Dutch astronomer Jan Oort. Millions of comets are believed to orbit the Sun at distances greater than 10,000 times the Earth–Sun distance. The gravitational effect of a distant passing star is believed to send an occasional Oort cloud comet into a highly eccentric orbit that carries it toward the Sun. However, only a tiny fraction of Oort cloud comets have orbits that bring them into the inner solar system.

But why are comets found in both the Kuiper belt and the Oort cloud? It appears that comets, like asteroids, are planetesimal bodies formed by accretion of material in the solar nebula. In other words, they are leftovers from the planetary formation process. In contrast to asteroids that formed in the inner solar system of mostly metallic and rocky materials, comets are icy planetesimals that formed in the region of the giant planets and beyond. The ones that formed beyond the orbit of Neptune and did not accrete into a large planetary body make up the Kuiper belt objects.

The icy planetesimals of the Oort cloud, on the other hand, are thought to have formed in the region of the Jovian planets. While the vast majority of icy objects would have accreted into protoplanets, some survived. Under the strong gravitational fields of the giants, their orbits would have been greatly altered. Some cometary objects would have been hurled toward the inner solar system where they collided with the terrestrial planets. The remainder were "thrown out" into space to form the Oort cloud located at the outermost reaches of the solar system.

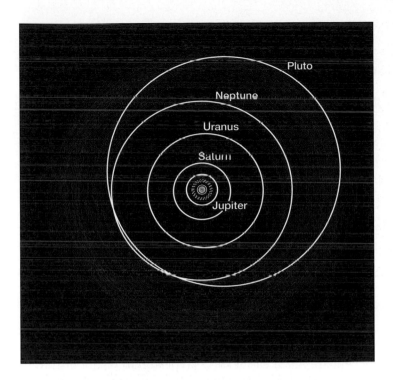

FIGURE 24.29   The orbits of some known Kuiper belt objects.

The most famous short-period comet is Halley's Comet. Its orbital period averages 76 years, and every one of its 29 appearances since 240 B.C. has been recorded by Chinese astronomers. This record is a testimonial to their dedication as astronomical observers and to the endurance of their culture. When seen in 1910, Halley's Comet had developed a tail nearly 1.6 million kilometers (1 million miles) long and was visible during the daytime.

In 1986 the unspectacular showing of Halley's Comet was a disappointment to many people in the Northern Hemisphere. Yet it was during this most recent visit to the inner solar system that a great deal of new information was learned about this most famous of comets. The new data were gathered by space probes sent to rendezvous with the comet. Most notably, the European probe *Giotto* approached to within 600 kilometers of the comet's nucleus and obtained the first close up images of comet structure.

We now know that the nucleus of Halley's Comet is potato-shaped and 16 by 8 kilometers in size. The surface is irregular and full of craterlike pits. Gases and dust that escape from the nucleus to form the coma and tail appear to gush from its surface as bright jets or streams. Only about 10 percent of the comet's total surface was emitting these jets at the time of the rendezvous. The remaining surface area of the comet appeared to be covered with a dark layer that may be organic material.

In 1997 the comet Hale-Bopp made for spectacular viewing around the globe. As comets go, the nucleus of Hale-Bopp was unusually large, about 40 kilometers (25 miles) in diameter. As shown in Figure 24.28, two tails nearly 15 million miles long extended from this comet. The bluish gas-tail is composed of positively charged ions, and it points almost directly away from the Sun. The brighter tail is composed of dust and other rocky debris. Because the rocky material is more massive than the ionized gases, it is less affected by the solar wind and follows a different trajectory away from the comet.

Meteoroids: Visitors to Earth

Nearly everyone has seen a **meteor,** popularly (but inaccurately) called a "shooting star." This streak of light lasts from an eyeblink to a few seconds and occurs when a small solid particle, a **meteoroid,** enters Earth's atmosphere from interplanetary space. Friction between the meteoroid and the air heats both and produces the light we see. Most meteoroids originate from any one of the following three sources: (1) interplanetary debris that was not gravitationally swept up by the planets during the formation of the solar system, (2) material that is continually being lost from the asteroid belt, or (3) the solid remains of comets that once passed through Earth's orbit. A few meteoroids are believed to be fragments of the Moon, or possibly Mars, that were ejected when an asteroid impacted these bodies.

Meteoroids less than about a meter in diameter generally vaporize before reaching Earth's surface. Some, called *micrometeorites,* are so tiny that their rate of fall becomes too slow to cause them to burn up, so they drift down as space dust. Each day, the number of meteoroids that enter Earth's atmosphere must reach into the thousands. After sunset on a clear night, a half dozen or more are bright enough to be seen with the naked eye each hour from anywhere on Earth.

Occasionally, meteor sightings increase dramatically to 60 or more per hour. These displays, called **meteor showers,** result when Earth encounters a swarm of meteoroids traveling in the same direction and at nearly the same speed as Earth. The close association of these swarms to the orbits of some short-term comets strongly suggests that they represent material lost by these comets (Table 24.2). Some swarms not associated with the orbits of known comets are probably the remains of the nucleus of a long-defunct comet. The notable Perseid meteor shower that occurs each year around

TABLE 24.2	Major Meteor Showers	
Shower	Approximate Dates	Associated Comet
Quadrantids	January 4–6	
Lyrids	April 20–23	Comet 1861 I
Eta Aquarids	May 3–5	Halley's comet
Delta Aquarids	July 30	
Perseids	August 12	Comet 1862 III
Draconids	October 7–10	Comet Giacobini-Zinner
Orionids	October 20	Halley's comet
Taurids	November 3–13	Comet Encke
Andromedids	November 14	Comet Biela
Leonids	November 18	Comet 1866 I
Geminids	December 4–16	

FIGURE 24.30 Iron meteorite found near Meteor Crater, Arizona. (Courtesy of Meteor Crater Enterprises, Inc.)

August 12 is believed to be the remains of the Comet 1862 III, which has a period of 110 years.

Meteoroids that are the remains of comets tend to be small and only occasionally reach the ground. Most meteoroids large enough to survive the heated fall are thought to originate among the asteroids, where chance collisions modify their orbits and send them toward Earth. Earth's gravity does the rest.

The remains of meteoroids, when found on Earth, are referred to as **meteorites** (Figure 24.30). A few very large meteoroids have blasted out craters on Earth's surface that strongly resemble those on the lunar surface. The most famous is Meteor Crater in Arizona (Figure 24.31). This huge cavity is about 1.2 kilometers (0.75 miles) across, 170 meters (560 feet) deep, and has an upturned rim that rises 50 meters (165 feet) above the surrounding countryside. Over 30 tons of iron fragments have been found in the immediate area, but attempts to locate the main body have been unsuccessful. Based on the amount of erosion, the impact likely occurred within the last 50,000 years.

Prior to the Moon rocks brought back by lunar explorers, meteorites were the only extraterrestrial materials that could be directly examined (Figure 24.30). Meteorites are classified by their composition: (1) *irons,* mostly iron with 5 percent to 20 percent nickel; (2) *stony* silicate minerals with inclusions of other minerals; and (3) *stony-irons* mixtures. Although stony meteorites are much more common, people tend to find irons. This is understandable because metallic meteorites withstand the impact better, weather more slowly, and are much easier for a layperson to distinguish from terrestrial rocks. Iron meteorites are probably fragments of once-molten cores of large asteroids or small planets.

One type of meteorite, called a *carbonaceous chondrite,* contains simple amino acids and other organic compounds, which are the basic building blocks of life. This discovery confirms similar findings in observational astronomy, which indicate that numerous organic compounds exist in the frigid realm of outer space.

FIGURE 24.31 Meteor Crater, near Winslow, Arizona. This cavity is about 1.2 kilometers (0.75 mile) across and 170 meters (560 feet) deep. The solar system is cluttered with meteoroids and other objects that can strike Earth with explosive force. (Photo by Michael Collier)

FIGURE 24.32   Pluto, a dwarf planet compared to some large Kuiper belt objects. The Earth–Moon system and Neptune's moon, Triton, added for scale. (Courtesy of NASA)

If meteorites represent the makeup of Earth-like planets, as some planetary geologists suggest, then Earth must contain a much larger percentage of iron than is indicated by surface rocks. This is one reason that geologists suggest that Earth's core must be mostly iron and nickel. In addition, radiometric dating of meteorites indicates that our solar system's age certainly exceeds 4.5 billion years. This "old age" has been confirmed by data obtained from lunar samples.

Dwarf Planets

Since Pluto's discovery in 1930, it has been a mystery on the edge of the solar system. At first, Pluto was thought to be about as large as Earth, but as better images were obtained, Pluto's diameter was estimated to be a little less than one half that of Earth. Then, in 1978, astronomers discovered that Pluto has a satellite (Charon), whose brightness combined with its parent made Pluto appear much larger than it really was (Figure 24.32). Recent images obtained by the *Hubble Space Telescope* established the diameter of Pluto at only 2300 kilometers (1425 miles). This is about one-fifth that of Earth and less than half that of Mercury, long considered the runt of the solar system. In fact, seven moons, including Earth's moon, are larger than Pluto.

Even more attention was given to Pluto's status as a planet when in 1992 astronomers discovered another icy body in orbit beyond Neptune. Soon hundreds of these *Kuiper belt objects* were discovered forming a band of small objects, similar to the asteroid belt between Mars and Jupiter. However, these orbiting bodies are made up of dust and ices, like

comets, rather than metallic and rocky substances, like asteroids. Many other planetary objects, some larger than Pluto, are thought to exist in this belt of icy worlds found beyond the orbit of Neptune.

The International Astronomical Union, a group that has the power to determine whether or not Pluto is a planet, voted August 24, 2006, to add a new class of planets called **dwarf planets.** These include celestial bodies that orbit around the Sun, are essentially round due to their self-gravity, but are not the only objects to occupy their area of space. By this definition, Pluto is recognized as a dwarf planet and the prototype of a new category of planetary objects. Other dwarf planets include 2003 UB313, a Kuiper Belt object, and Ceres, the largest known asteroid.

This is not the first time a planet has been demoted. In the mid-1800s astronomy textbooks listed as many as 11 planets in our solar system, including the asteroids Vesta, Juno, Ceres, and Pallas. Soon astronomers discovered dozens of other "planets," which made it clear that these small bodies represent another class of objects separate from the planets. Consequently, the number of major planets fell to eight. (Neptune had just been discovered, but Pluto's discovery did not occur until 1930.)

It is now clear that Pluto was unique among the classical planets, being very different from the four rocky innermost planets, and unlike the four gaseous giants. With the new classification, astronomers believe hundreds of additional dwarf planets might be found. *New Horizons*, the first spacecraft to explore the outer solar system, was launched in January 2006. It is scheduled to fly by Pluto in July 2015 and continue on to explore the Kuiper Belt.

Summary

- The planets can be arranged into two groups: the *terrestrial* (Earth-like) *planets* (Mercury, Venus, Earth, and Mars) and the *Jovian* (Jupiter-like) *planets* (Jupiter, Saturn, Uranus, and Neptune). *When compared to the Jovian planets, the terrestrial planets are smaller, more dense, contain proportionally more rocky material, have slower rates of rotation, and meager atmospheres..*

- The lunar surface exhibits several types of features. *Impact craters* were produced by the collision of rapidly moving interplanetary debris (*meteoroids*). Bright, densely cratered *highlands* make up most of the lunar surface. The dark, fairly smooth lowlands are called *maria.* Maria basins are enormous impact craters that were later flooded with layer upon layer of very fluid basaltic lava. All lunar terrains are mantled with a soil-like layer of gray, unconsolidated debris, called *lunar regolith,* which has been derived from a few billion years of meteoric bombardment. One hypothesis for the Moon's origin suggests that a Mars-sized object collided with Earth to produce the Moon. Scientists conclude that the *lunar surface evolved in four stages:* (1) *the original crust (highlands),* (2) *the highlands,* (3) *maria basins,* and (4) *youthful rayed craters.*

- *Mercury* is a small, dense planet that has virtually no atmosphere and exhibits the greatest temperature extremes of any planet. *Venus,* the brightest planet in the sky, has a thick, heavy atmosphere composed of 97 percent carbon dioxide, a surface of relatively subdued plains and inactive volcanic features, a surface atmospheric pressure 90 times that of Earth's, and surface temperatures of 475°C (900°F). *Mars,* the Red Planet, has a carbon dioxide atmosphere only 1 percent as dense as Earth's, extensive dust storms, numerous inactive volca-noes, many large canyons, and several valleys of debatable origin exhibiting drainage patterns similar to stream valleys on Earth. *Jupiter,* the largest planet, rotates rapidly, has a banded appearance caused by huge convection currents driven by the planet's interior heat, a *Great Red Spot* that varies in size, a thin ring system, and at least 63 moons (one of the moons, *Io,* is perhaps the most volcanically active body in the solar system). *Saturn* is best known for its system of rings. It also has a dynamic atmosphere with winds up to 930 miles per hour and storms similar to Jupiter's Great Red Spot. *Uranus* and *Neptune* are often called "the twins" because of their similar structure and composition. A unique feature of Uranus is the fact that it rotates on its side. Neptune has white, cirruslike clouds above its main cloud deck and an Earth-size *Great Dark Spot,* assumed to be a large rotating storm similar to Jupiter's Great Red Spot.

- The minor members of the solar system include *asteroids, comets, meteoroids,* and *dwarf planets.* Most asteroids lie between the orbits of Mars and Jupiter. Asteroids are leftover rocky and metallic debris from the solar nebula that never accreted into a planet. Comets are made of ices (water, ammonia, methane, carbon dioxide, and carbon monoxide) with small pieces of rocky and metallic material. Many travel in very elongated orbits that carry them beyond Pluto. Meteoroids, small solid particles that travel through interplanetary space, become *meteors* when they enter Earth's atmosphere and vaporize with a flash of light. *Meteor showers* occur when Earth encounters a swarm of meteoroids, probably material lost by a comet. *Meteorites* are the remains of meteoroids found on Earth. Recently, Pluto was placed into a new class of solar system objects called *dwarf planets.*

Review Questions

1. By what criteria are the planets placed into either the Jovian or terrestrial group?

2. What are the three types of materials thought to make up the planets? How are they different? How does their distribution account for the density differences between the terrestrial and Jovian planetary groups?

3. Explain why the terrestrial planets have meager atmospheres, as compared to the Jovian planets.

4. How is crater density used in the relative dating of features on the Moon?

5. Briefly outline the history of the Moon.

6. How are the maria of the Moon thought to be similar to the Columbia Plateau?

7. Venus has been referred to as "Earth's twin." In what ways are these two planets similar? How do they differ?

8. What surface features does Mars have that are also common on Earth?

9. Why are the largest volcanoes on Earth so much smaller than the largest ones on Mars?

10. Why might astrobiologists be intrigued by evidence that groundwater has seeped onto the surface of Mars?

11. The two "moons" of Mars were once suggested to be artificial. What characteristics do they have that would cause such speculation?

12. What is the nature of Jupiter's Great Red Spot?

13. Why are the Galilean satellites of Jupiter so named?

14. What is distinctive about Jupiter's satellite Io?

15. Why are the *outer* satellites of Jupiter thought to have been captured rather than having been formed with the rest of the satellite system?

16. How are Jupiter and Saturn similar?

17. What two roles do ring moons play in the nature of planetary ring systems?

18. How are Saturn's satellite Titan and Neptune's satellite Triton similar?

19. Name three bodies in the solar system that exhibit active volcanism.

20. Where are most asteroids found?

21. What do you think would happen if Earth passed through the tail of a comet?

22. Where are most comets thought to reside? What eventually becomes of comets that orbit close to the Sun?

23. Compare meteoroid, meteor, and meteorite.

24. What are the three main sources of meteoroids?

25. Why are impact craters more common on the Moon than on Earth, even though the Moon is a much smaller target and has a weaker gravitational field?

26. It has been estimated that Halley's Comet has a mass of 100 billion tons. Further, this comet is estimated to lose 100 million tons of material during the few months that its orbit brings it close to the Sun. With an orbital period of 76 years, what is the maximum remaining life span of Halley's Comet?

Key Terms

asteroids (p. 674)	highlands (p. 659)	meteor (p. 677)	protoplanets (p. 657)
asteroid belt (p. 674)	inner planet (p. 657)	meteorite (p. 678)	solar nebula (p. 657)
coma (p. 675)	impact craters (p. 660)	meteoroid (p. 677)	terrae (p. 659)
comet (p. 674)	Jovian planet (p. 657)	meteor shower (p. 677)	terrestrial planet (p. 657)
cryovolcanism (p. 673)	Kuiper belt (p. 676)	Oort cloud (p. 676)	
dwarf planet (p. 679)	lunar regolith (p. 661)	outer planet (p. 657)	
escape velocity (p. 658)	maria (p. 659)	planetesimals (p. 657)	

Web Resources

 The *Earth* Website uses the resources and flexibility of the Internet to aid in your study of the topics in this chapter. Written and developed by geology instructors, this site will help improve your understanding of geology. Visit **http://www.prenhall.com/tarbuck** and click on the cover of *Earth 9e* to find:

- Online review quizzes.
- Critical thinking exercises.
- Links to chapter-specific Web resources.
- Internet-wide key-term searches.

http://www.prenhall.com/tarbuck

Metric and English Units Compared

Units

1 kilometer (km)	=	1000 meters (m)
1 meter (m)	=	100 centimeters (cm)
1 centimeter (cm)	=	0.39 inch (in.)
1 mile (mi)	=	5280 feet (ft)
1 foot (ft)	=	12 inches (in.)
1 inch (in.)	=	2.54 centimeters (cm)
1 square mile (mi^2)	=	640 acres (a)
1 kilogram (kg)	=	1000 grams (g)
1 pound (lb)	=	16 ounces (oz)
1 fathom	=	6 feet (ft)

Conversions

When you want
to convert: multiply by: to find:

Length

inches	2.54	centimeters
centimeters	0.39	inches
feet	0.30	meters
meters	3.28	feet
yards	0.91	meters
meters	1.09	yards
miles	1.61	kilometers
kilometers	0.62	miles

Area

square inches	6.45	square centimeters
square centimeters	0.15	square inches
square feet	0.09	square meters
square meters	10.76	square feet
square miles	2.59	square kilometers
square kilometers	0.39	square miles

Volume

cubic inches	16.38	cubic centimeters
cubic centimeters	0.06	cubic inches
cubic feet	0.028	cubic meters
cubic meters	35.3	cubic feet
cubic miles	4.17	cubic kilometers
cubic kilometers	0.24	cubic miles
liters	1.06	quarts
liters	0.26	gallons
gallons	3.78	liters

Masses and Weights

ounces	28.35	grams
grams	0.035	ounces
pounds	0.45	kilograms
kilograms	2.205	pounds

Temperature

When you want to convert degrees Fahrenheit (°F) to degrees Celsius (°C), subtract 32 degrees and divide by 1.8.

When you want to convert degrees Celsius (°C) to degrees Fahrenheit (°F), multiply by 1.8 and add 32 degrees.

When you want to convert degrees Celsius (°C) to Kelvins (K), delete the degree symbol and add 273. When you want to convert Kelvins (K) to degrees Celsius (°C), add the degree symbol and subtract 273.

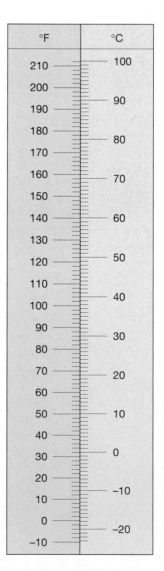

▶ **FIGURE A.1** A comparison of Fahrenheit and Celsius temperature scales.

Landforms of the Conterminous United States

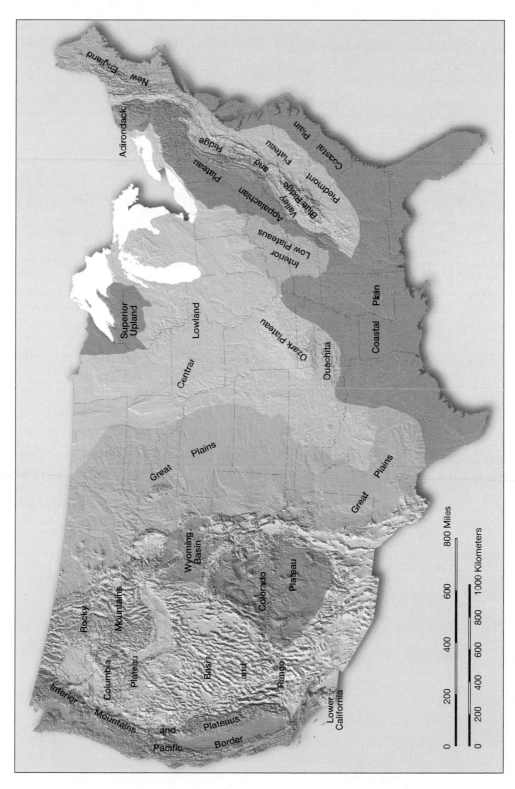

▲ **FIGURE B.1** Outline map showing major landform regions of the United States.

Landforms of the Conterminous United States

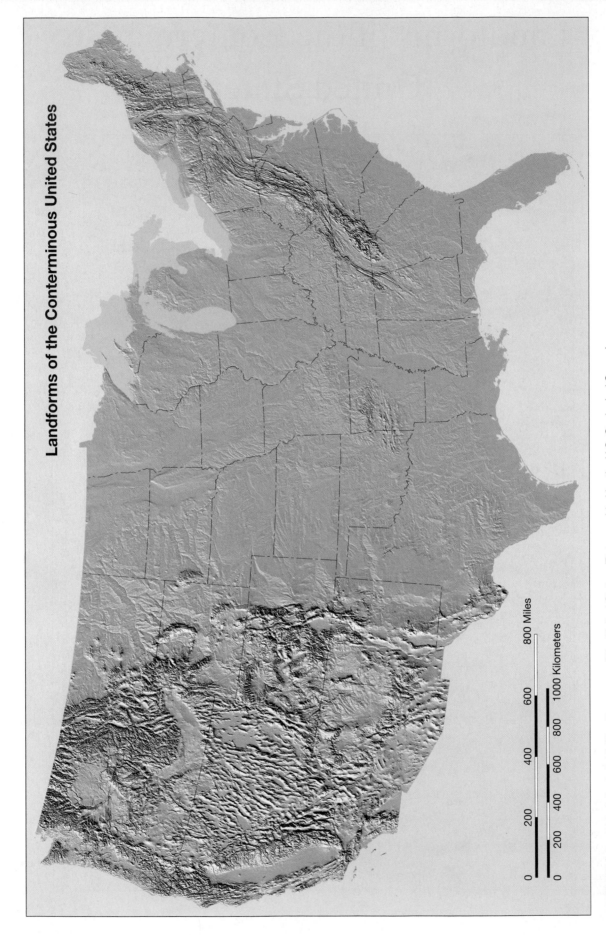

▲ FIGURE B.2 Digital shaded relief landform map of the United States. (Data provided by the U.S. Geological Survey)

Glossary

Aa flow A type of lava flow that has a jagged, blocky surface.

Ablation A general term for the loss of ice and snow from a glacier.

Abrasion The grinding and scraping of a rock surface by the friction and impact of rock particles carried by water, wind, and ice.

Abyssal plain Very level area of the deep-ocean floor, usually lying at the foot of the continental rise.

Accretionary wedge A large wedge-shaped mass of sediment that accumulates in subduction zones. Here sediment is scraped from the subducting oceanic plate and accreted to the overriding crustal block.

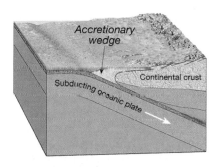

Active continental margin Usually narrow and consisting of highly deformed sediments. Such margins occur where oceanic lithosphere is being subducted beneath the margin of a continent.

Active layer The zone above the permafrost that thaws in summer and refreezes in winter.

Aerosols Tiny solid and liquid particles suspended in the atmosphere.

Aftershock A smaller earthquake that follows the main earthquake.

Alluvial fan A fan-shaped deposit of sediment formed when a stream's slope is abruptly reduced.

Alluvium Unconsolidated sediment deposited by a stream.

Alpine glacier A glacier confined to a mountain valley, which in most instances had previously been a stream valley.

Alluvial fan

Andesitic composition See *Intermediate composition*.

Angle of repose The steepest angle at which loose material remains stationary without sliding downslope.

Angular unconformity An unconformity in which the older strata dip at an angle different from that of the younger beds.

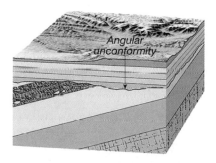

Antecedent stream A stream that continued to downcut and maintain its original course as an area along its course was uplifted by faulting or folding.

Anthracite A hard, metamorphic form of coal that burns cleanly and hot.

Anticline A fold in sedimentary strata that resembles an arch.

Aphanitic texture A texture of igneous rocks in which the crystals are too small for individual minerals to be distinguished without the aid of a microscope.

Aquifer Rock or sediment through which groundwater moves easily.

Aquitard An impermeable bed that hinders or prevents groundwater movement.

Archean eon The first eon of Precambrian time. The eon preceding the Proterozoic. It extends between 4.5 and 2.5 billion years ago.

Arête A narrow, knifelike ridge separating two adjacent glaciated valleys.

Arkose A feldspar-rich sandstone.

Artesian well A well in which the water rises above the level where it was initially encountered.

Assimilation In igneous activity, the process of incorporating country rock into a magma body.

Asteroid One of thousands of small planetlike bodies, ranging in size from a few hundred kilometers to less than one kilometer across. Most asteroids' orbits lie between those of Mars and Jupiter.

Asthenosphere A subdivision of the mantle situated below the lithosphere. This zone of weak material exists below a depth of about 100 kilometers and in some regions extends as deep as 700 kilometers. The rock within this zone is easily deformed.

Atmosphere The gaseous portion of a planet, the planet's envelope of air. One of the traditional subdivisions of Earth's physical environment.

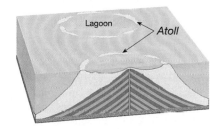

Atoll A coral island consisting of a nearly continuous ring of coral reef surrounding a central lagoon.

Atom The smallest particle that exists as an element.

Atomic mass unit A mass unit equal to exactly one-twelfth the mass of a carbon-12 atom.

Atomic number The number of protons in the nucleus of an atom.

Atomic weight The average of the atomic masses of isotopes for a given element.

Aureole A zone or halo of contact metamorphism found in the country rock surrounding an igneous intrusion.

Backarc basin A basin that forms on the side of a volcanic arc away from the trench.

Backshore The inner portion of the shore, lying landward of the high-tide shoreline. It is usually dry, being affected by waves only during storms.

Backswamp A poorly drained area on a floodplain resulting when natural levees are present.

Bajada An apron of sediment along a mountain front created by the coalescence of alluvial fans.

Banded iron formations A finely layered iron and silica-rich (chert) layer deposited mainly during the Precambrian.

Bar Common term for sand and gravel deposits in a stream channel.

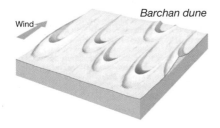
Barchan dune

Wind

Barchan dune A solitary sand dune shaped like a crescent with its tips pointing downwind.

Barrier island

Barchanoid dune Dunes forming scalloped rows of sand oriented at right angles to the wind. This form is intermediate between isolated barchans and extensive waves of transverse dunes.

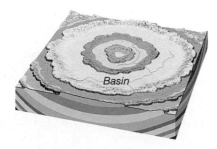
Basin

Barrier island A low, elongate ridge of sand that parallels the coast.

Baymouth bar

Basal slip A mechanism of glacial movement in which the ice mass slides over the surface below.

Basalt A fine-grained igneous rock of mafic composition.

Basaltic composition A compositional group of igneous rocks indicating that the rock contains substantial dark silicate minerals and calcium-rich plagioclase feldspar.

Base level The level below which a stream cannot erode.

Basin A circular downfolded structure.

Batholith A large mass of igneous rock that formed when magma was emplaced at depth, crystallized, and subsequently exposed by erosion.

Bathymetry The measurement of ocean depths and the charting of the topography of the ocean floor.

Baymouth bar A sandbar that completely crosses a bay, sealing it off from the main body of water.

Beach An accumulation of sediment found along the landward margin of the ocean or a lake.

Beach drift The transport of sediment in a zigzag pattern along a beach, caused by the uprush of water from obliquely breaking waves.

Beach face The wet, sloping surface that extends from the berm to the shoreline.

Beach nourishment Process in which large quantities of sand are added to the beach system to offset losses caused by wave erosion. Building beaches seaward improves beach quality and storm protection.

Bed See *Strata.*

Bedding plane A nearly flat surface separating two beds of sedimentary rock. Each bedding plane marks the end of one deposit and the beginning of another having different characteristics.

Bed load Sediment moved along the bottom of a stream by moving water, or particles moved along the ground surface by wind.

Benioff zone See *Wadati–Benioff zone.*

Berm The dry, gently sloping zone on the backshore of a beach at the foot of the coastal cliffs or dunes.

Biochemical A type of chemical sediment that forms when material dissolved in water is precipitated by water-dwelling organisms. Shells are common examples.

Biogenous sediment Seafloor sediments consisting of material of marine-organic origin.

Biosphere The totality of life forms on Earth.

Blowout

Bituminous coal The most common form of coal, often called soft, black coal.

Black smoker A hydrothermal vent on the ocean floor that emits a black cloud of hot, metal-rich water.

Block lava Lava having a surface of angular blocks associated with material having andesitic and rhyolitic compositions.

Blowout A depression excavated by wind in easily eroded materials.

Body wave A seismic wave that travels through Earth's interior.

Bottomset bed A layer of fine sediment deposited beyond the advancing edge of a delta and then buried by continued delta growth.

Bowen's reaction series A concept proposed by N. L. Bowen that illustrates the relationships between magma and the minerals crystallizing from it during the formation of igneous rocks.

Braided stream A stream consisting of numerous intertwining channels.

Breakwater A structure protecting a nearshore area from breaking waves.

Breccia A sedimentary rock composed of angular fragments that were lithified.

Brittle failure The loss of strength by a material usually in the form of sudden fracturing.

Buoyant subduction Subduction in which the angle of descent is small because the oceanic lithosphere is still warm and buoyant. It occurs where a spreading center is near a subduction zone.

Burial metamorphism Low-grade metamorphism that occurs in the lowest layers of very thick accumulations of sedimentary strata.

Caldera A large depression typically caused by collapse or ejection of the summit area of a volcano.

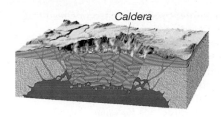

Caldera

Caliche A hard layer, rich in calcium carbonate, that forms beneath the *B* horizon in soils of arid regions.

Calving Wastage of a glacier that occurs when large pieces of ice break into the water.

Capacity The total amount of sediment a stream is able to transport.

Capillary fringe A relatively narrow zone at the base of the zone of aeration. Here water rises from the water table in tiny, threadlike openings between grains of soil or sediment.

Cap rock A necessary part of an oil trap. The cap rock is impermeable and hence keeps upwardly mobile oil and gas from escaping at the surface.

Cassini gap A wide gap in the ring system of Saturn between the *A* ring and the *B* ring.

Catastrophism The concept that Earth was shaped by catastrophic events of a short-term nature.

Cavern A naturally formed underground chamber or series of chambers most commonly produced by solution activity in limestone.

Cementation One way in which sedimentary rocks are lithified. As material precipitates from water that percolates through the sediment, open spaces are filled and particles are joined into a solid mass.

Cenozoic era A time span on the geologic time scale beginning about 65.5 million years ago, following the Mesozoic era.

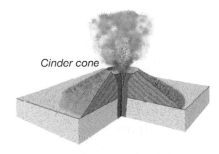

Cinder cone

Chemical sedimentary rock Sedimentary rock consisting of material that was precipitated from water by either inorganic or organic means.

Chemical weathering The processes by which the internal structure of a mineral is altered by the removal and/or addition of elements.

Cinder cone A rather small volcano built primarily of ejected lava fragments that consist mostly of pea- to walnut-size lapilli.

Cirque An amphitheater-shaped basin at the head of a glaciated valley produced by frost wedging and plucking.

Clastic texture A sedimentary rock texture consisting of broken fragments of preexisting rock.

Cleavage The tendency of a mineral to break along planes of weak bonding.

Climate feedback mechanism Because the atmosphere is a complex interactive physical system, several different possible outcomes may result when one of the system's elements is altered. These various possibilities are called *climate-feedback mechanisms*.

Climate system The exchanges of energy and moisture occurring among the atmosphere, hydrosphere, lithosphere, biosphere, and cryosphere.

Closed system A system that is self-contained with regard to matter—that is, no matter enters or leaves.

Coast A strip of land that extends inland from the coastline as far as ocean-related features can be found.

Coastline The coast's seaward edge. The landward limit of the effect of the highest storm waves on the shore.

Col A pass between mountain valleys where the headwalls of two cirques intersect.

Color A phenomenon of light by which otherwise identical objects may be differentiated.

Column A feature found in caves that is formed when a stalactite and stalagmite join.

Columnar joints A pattern of cracks that forms during cooling of molten rock to generate columns.

Coma The fuzzy, gaseous component of a comet's head.

Comet A small body that generally revolves about the Sun in an elongated orbit.

Compaction A type of lithification in which the weight of overlying material compresses more deeply buried sediment. It is most important in the fine-grained sedimentary rocks such as shale.

Competence A measure of the largest particle a stream can transport; a factor dependent on velocity.

Composite cone A volcano composed of both lava flows and pyroclastic material.

Compound A substance formed by the chemical combination of two or more elements in definite proportions and usually having properties different from those of its constituent elements.

Compressional stress Differential stress that shortens a rock body.

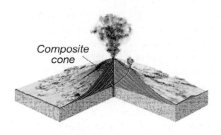

Composite cone

Compressional stress

Concordant A term used to describe intrusive igneous masses that form parallel to the bedding of the surrounding rock.

Conduction The transfer of heat through matter by molecular activity.

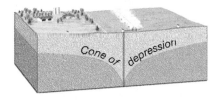

Cone of depression

Conduit A pipelike opening through which magma moves toward Earth's surface. It terminates at a surface opening called a vent.

Cone of depression A cone-shaped depression in the water table immediately surrounding a well.

Confining pressure An equal, all-sided pressure.

Conformable layers Rock layers that were deposited without interruption.

Conglomerate A sedimentary rock composed of rounded, gravel-size particles.

Contact metamorphism Changes in rock caused by the heat from a nearby magma body.

Continental drift A hypothesis, credited largely to Alfred Wegener, that suggested all present continents once existed as a single supercontinent. Further, beginning about 200 million years ago, the supercontinent began breaking into smaller continents, which then "drifted" to their present positions.

Continental margin That portion of the seafloor adjacent to the continents. It may include the continental shelf, continental slope, and continental rise.

Continental rift A linear zone along which continental lithosphere stretches and pulls apart. Its creation may mark the beginning of a new ocean basin.

Continental rise The gently sloping surface at the base of the continental slope.

Continental shelf The gently sloping submerged portion of the continental margin, extending from the shoreline to the continental slope.

Continental slope The steep gradient that leads to the deep-ocean floor and marks the seaward edge of the continental shelf.

Continental volcanic arc Mountains formed in part by igneous activity associated with the subduction of oceanic lithosphere beneath a continent. Examples include the Andes and the Cascades.

Convection The transfer of heat by the mass movement or circulation of a substance.

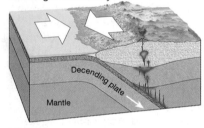

Convergent boundary

Convergent plate boundary A boundary in which two plates move together, resulting in oceanic lithosphere being thrust beneath an overriding plate, eventually to be reabsorbed into the mantle. It can also involve the collision of two continental plates to create a mountain system.

Coral reef Structure formed in a warm, shallow, sunlit ocean environment that consists primarily of the calcite-rich remains of corals as well as the limy secretions of algae and the hard parts of many other small organisms.

Core The innermost layer of Earth based on composition. It is thought to be largely an iron-nickel alloy with minor amounts of oxygen, silicon, and sulfur.

Correlation Establishing the equivalence of rocks of similar age in different areas.

Covalent bond A chemical bond produced by the sharing of electrons.

Crater The depression at the summit of a volcano, or that which is produced by a meteorite impact.

Craton That part of the continental crust that has attained stability; that is, it has not been affected by significant tectonic activity during the Phanerozoic eon. It consists of the shield and stable platform.

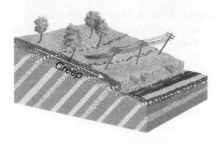

Creep The slow downhill movement of soil and regolith.

Crevasse A deep crack in the brittle surface of a glacier.

Cross-bedding Structure in which relatively thin layers are inclined at an angle to the main bedding. Formed by currents of wind or water.

Cross-cutting A principle of relative dating. A rock or fault is younger than any rock (or fault) through which it cuts.

Crust The very thin outermost layer of Earth.

Crystal Any natural solid with an ordered, repetitive atomic structure.

Crystalline See *Crystal*.

Crystal shape See *Habit*.

Crystal settling During the crystallization of magma, the earlier-formed minerals are denser than the liquid portion and settle to the bottom of the magma chamber.

Crystalline texture See *Nonclastic texture*.

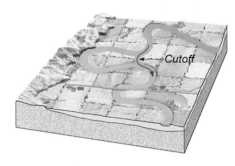

Crystallization The formation and growth of a crystalline solid from a liquid or gas.

Curie point The temperature above which a material loses its magnetization.

Cut bank The area of active erosion on the outside of a meander.

Cutoff A short channel segment created when a river erodes through the narrow neck of land between meanders.

Darcy's law An equation stating that groundwater discharge depends on the hydraulic gradient, hydraulic conductivity, and cross-sectional area of an aquifer.

Dark silicate Silicate minerals containing ions of iron and/or magnesium in their structure. They are dark in color and have a higher specific gravity than nonferromagnesian silicates.

Daughter product An isotope resulting from radioactive decay.

Debris flow A flow of soil and regolith containing a large amount of water. Most common in semiarid mountainous regions and on the slopes of some volcanoes.

Debris slide See *Rockslide*.

Decompression melting Melting that occurs as rock ascends due to a drop in confining pressure.

Deep-ocean basin The portion of seafloor that lies between the continental margin and the oceanic ridge system. This region comprises almost 30 percent of Earth's surface.

Deep-ocean trench A narrow, elongated depression of the seafloor.

Deep-sea fan A cone-shaped deposit at the base of the continental slope. The sediment is transported to the fan by turbidity currents that follow submarine canyons.

Deflation The lifting and removal of loose material by wind.

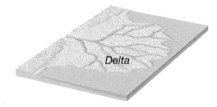

Delta

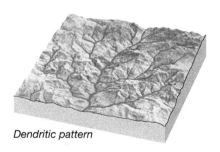

Dendritic pattern

Deformation General term for the processes of folding, faulting, shearing, compression, or extension of rocks as the result of various natural forces.

Delta An accumulation of sediment formed where a stream enters a lake or an ocean.

Dendritic pattern A stream system that resembles the pattern of a branching tree.

Density A property of matter defined as mass per unit volume.

Desalination The removal of salts and other chemicals from seawater.

Desert One of the two types of dry climate; the driest of the dry climates.

Desert pavement A layer of coarse pebbles and gravel created when wind removed the finer material.

Detachment fault A nearly horizontal fault that may extend for hundreds of kilometers below the surface. Such faults represent a boundary between rocks that exhibit ductile deformation and rocks that exhibit brittle deformation.

Detrital sedimentary rocks Rocks that form from the accumulation of materials that originate and are transported as solid particles derived from both mechanical and chemical weathering.

Diagenesis A collective term for all the chemical, physical, and biological changes that take place after sediments are deposited and during and after lithification.

Differential stress Forces that are unequal in different directions.

Differential weathering The variation in the rate and degree of weathering caused by such factors as mineral make-up, degree of jointing, and climate.

Dike A tabular-shaped intrusive igneous feature that cuts through the surrounding rock.

Dip The angle at which a rock layer or fault is inclined from the horizontal. The direction of dip is at a right angle to the strike.

Dip-slip fault A fault in which the movement is parallel to the dip of the fault.

Discharge The quantity of water in a stream that passes a given point in a period of time.

Disconformity A type of unconformity in which the beds above and below are parallel.

Discontinuity A sudden change with depth in one or more of the physical properties of the material making up Earth's interior. The boundary between two dissimilar materials in Earth's interior as determined by the behavior of seismic waves.

Discordant A term used to describe plutons that cut across existing rock structures, such as bedding planes.

Disseminated deposit Any economic mineral deposit in which the desired mineral occurs as scattered particles in the rock but in sufficient quantity to make the deposit an ore.

Dissolution A common form of chemical weathering, it is the process of dissolving into a homogeneous solution, as when an acidic solution dissolves limestone.

Divergent boundary

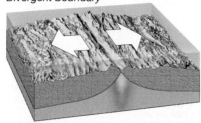

Dissolved load That portion of a stream's load carried in solution.

Distributary A section of a stream that leaves the main flow.

Divergent plate boundary A boundary in which two plates move apart, resulting in upwelling of material from the mantle to create new seafloor.

Divide An imaginary line that separates the drainage of two streams, often found along a ridge.

D″ layer A region in roughly the lowermost 200 kilometers of the mantle where P-waves experience a sharp decrease in velocity.

Dome A roughly circular upfolded structure.

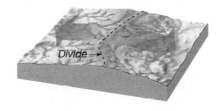

Drainage basin The land area that contributes water to a stream.

Drawdown The difference in height between the bottom of a cone of depression and the original height of the water table.

Drift See *Glacial drift*.

Drumlin A streamlined symmetrical hill composed of glacial till. The steep side of the hill faces the direction from which the ice advanced.

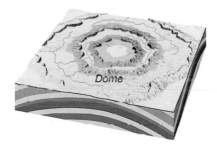

Dry climate A climate in which yearly precipitation is less than the potential loss of water by evaporation.

Ductile deformation A type of solid-state flow that produces a change in the size and shape of a rock body without fracturing. Occurs at depths where temperatures and confining pressures are high.

Dune A hill or ridge of wind-deposited sand.

Earthflow The downslope movement of water-saturated, clay-rich sediment. Most characteristic of humid regions.

Earthquake Vibration of Earth produced by the rapid release of energy.

Ebb current The movement of tidal current away from the shore.

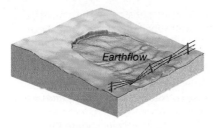

Echo sounder An instrument used to determine the depth of water by measuring the time interval between emission of a sound signal and the return of its echo from the bottom.

Elastic rebound The sudden release of stored strain in rocks that results in movement along a fault.

Electron A negatively charged subatomic particle that has a negligible mass and is found outside an atom's nucleus.

Element A substance that cannot be decomposed into simpler substances by ordinary chemical or physical means.

Eluviation The washing out of fine soil components from the *A* horizon by downward-percolating water.

Emergent coast A coast where land formerly below sea level has been exposed by crustal uplift or a drop in sea level or both.

End moraine A ridge of till marking a former position of the front of a glacier.

Energy levels or shells Spherically shaped, negatively charged zones that surround the nucleus of an atom.

Environment of deposition A geographic setting where sediment accumulates. Each site is characterized by a particular combination of geologic processes and environmental conditions.

Eon The largest time unit on the geologic time scale, next in order of magnitude above era.

Ephemeral stream A stream that is usually dry because it carries water only in response to specific episodes of rainfall. Most desert streams are of this type.

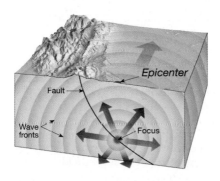

Epicenter The location on Earth's surface that lies directly above the focus of an earthquake.

Epoch A unit of the geologic time scale that is a subdivision of a period.

689

Era A major division on the geologic time scale; eras are divided into shorter units called periods.

Erosion The incorporation and transportation of material by a mobile agent, such as water, wind, or ice.

Eruption column Buoyant plumes of hot, ash-laden gases that can extend thousands of meters into the atmosphere.

Escape velocity The initial velocity an object needs to escape from the surface of a celestial body.

Esker Sinuous ridge composed largely of sand and gravel deposited by a stream flowing in a tunnel beneath a glacier near its terminus.

Estuary

Estuary A funnel-shaped inlet of the sea that formed when a rise in sea level or subsidence of land caused the mouth of a river to be flooded.

Eukaryotes An organism whose genetic material is enclosed in a nucleus; plants, animals, and fungi are eukaryotes.

Evaporite A sedimentary rock formed of material deposited from solution by evaporation of the water.

Evapotranspiration The combined effect of evaporation and transpiration.

Exfoliation dome Large, dome-shaped structure, usually composed of granite, formed by sheeting.

Exotic stream A permanent stream that traverses a desert and has its source in well-watered areas outside the desert.

External process Process such as weathering, mass wasting, or erosion that is powered by the Sun and contributes to the transformation of solid rock into sediment.

Extrusive Igneous activity that occurs at Earth's surface.

Facies A portion of a rock unit that possesses a distinctive set of characteristics that distinguishes it from other parts of the same unit.

Fall A type of movement common to mass-wasting processes that refers to the freefalling of detached individual pieces of any size.

Fault A break in a rock mass along which movement has occurred.

Fault-block mountain A mountain formed by the displacement of rock along a fault.

Fault creep Gradual displacement along a fault. Such activity occurs relatively smoothly and with little noticeable seismic activity.

Fault scarp A cliff created by movement along a fault. It represents the exposed surface of the fault prior to modification by weathering and erosion.

Felsic composition See *Granitic composition.*

Ferromagnesian silicate See *Dark silicate.*

Fetch The distance that the wind has traveled across the open water.

Fiord A steep-sided inlet of the sea formed when a glacial trough was partially submerged.

Firn Granular recrystallized snow. A transitional stage between snow and glacial ice.

Fissility The property of splitting easily into thin layers along closely spaced, parallel surfaces, such as bedding planes in shale.

Fission (nuclear) The splitting of a heavy nucleus into two or more lighter nuclei, caused by the collision with a neutron. During this process a large amount of energy is released.

Fissure A crack in rock along which there is a distinct separation.

Fissure eruption An eruption in which lava is extruded from narrow fractures or cracks in the crust.

Flood The overflow of a stream channel that occurs when discharge exceeds the channel's capacity. The most common and destructive geologic hazard.

Flood basalts Flows of basaltic lava that issue from numerous cracks or fissures and commonly cover extensive areas to thicknesses of hundreds of meters.

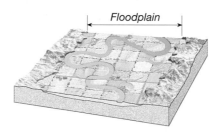
Floodplain

Floodplain The flat, low-lying portion of a stream valley subject to periodic inundation.

Flood current The tidal current associated with the increase in the height of the tide.

Flow A type of movement common to mass-wasting processes in which water-saturated material moves downslope as a viscous fluid.

Flowing artesian well An artesian well in which water flows freely at Earth's surface because the pressure surface is above ground level.

Fluorescence The absorption of ultraviolet light, which is reemitted as visible light.

Focus (earthquake) The zone within Earth where rock displacement produces an earthquake.

Fold A bent layer or series of layers that were originally horizontal and subsequently deformed.

Fold-and-thrust belts Regions within compressional mountain systems where large areas have been shortened and thickened by the processes of folding and thrust faulting, as exemplified by the Valley and Ridge Province of the Appalachains.

Foliated texture A texture of metamorphic rocks that gives the rock a layered appearance.

Foliation A term for a linear arrangement of textural features often exhibited by metamorphic rocks.

Forearc basin The region located between a volcanic arc and an accretionary wedge where shallow-water marine sediments typically accumulate.

Foreset bed An inclined bed deposited along the front of a delta.

Foreshocks Small earthquakes that often precede a major earthquake.

Foreshore That portion of the shore lying between the normal high and low water marks; the intertidal zone.

Force That which tends to put stationary objects in motion or change motions of moving bodies.

Fossil The remains or traces of organisms preserved from the geologic past.

Fossil fuel General term for any hydrocarbon that may be used as a fuel, including coal, oil, natural gas, bitumen from tar sands, and shale oil.

Fossil magnetism See *Paleomagnetism.*

Fossil succession Fossil organisms succeed one another in a definite and determinable order, and any time period can be recognized by its fossil content.

Fracture Any break or rupture in rock along which no appreciable movement has taken place.

Fracture zone Linear zone of irregular topography on the deep-ocean floor that follows transform faults and their inactive extensions.

Fragmental texture See *Pyroclastic texture.*

Frost wedging The mechanical breakup of rock caused by the expansion of freezing water in cracks and crevices. (See art next page)

Fumarole A vent in a volcanic area from which fumes or gases escape.

Gaining stream Streams that gain water from the inflow of groundwater through the streambed.

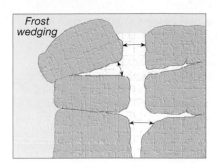

Frost wedging

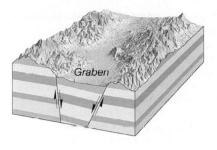

Graben

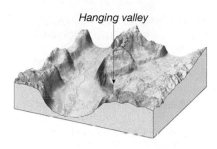

Hanging valley

Geologic time scale The division of Earth history into blocks of time—eons, eras, periods, and epochs. The time scale was created using relative dating principles.

Geology The science that examines Earth, its form and composition, and the changes that it has undergone and is undergoing.

Geosphere The solid Earth; one of Earth's four basic spheres.

Geothermal energy Natural steam used for power generation.

Geothermal gradient The gradual increase in temperature with depth in the crust. The average is 30°C per kilometer in the upper crust.

Geyser A fountain of hot water ejected periodically from the ground.

Glacial budget The balance, or lack of balance, between ice formation at the upper end of a glacier, and ice loss in the zone of wastage.

Glacial drift An all-embracing term for sediments of glacial origin, no matter how, where, or in what shape they were deposited.

Glacial erratic An ice-transported boulder that was not derived from the bedrock near its present site.

Glacial striations Scratches and grooves on bedrock caused by glacial abrasion.

Glacial trough A mountain valley that has been widened, deepened, and straightened by a glacier.

Glacier A thick mass of ice originating on land from the compaction and recrystallization of snow that shows evidence of past or present flow.

Glass (volcanic) Natural glass produced when molten lava cools too rapidly to permit recrystallization. Volcanic glass is a solid composed of unordered atoms.

Glassy A term used to describe the texture of certain igneous rocks, such as obsidian, that contain no crystals.

Gneissic texture A texture of metamorphic rocks in which dark and light silicate minerals are separated, giving the rock a banded appearance.

Gondwanaland The southern portion of Pangaea consisting of South America, Africa, Australia, India, and Antarctica.

Graben A valley formed by the downward displacement of a fault-bounded block.

Graded bed A sediment layer characterized by a decrease in sediment size from bottom to top.

Graded stream A stream that has the correct channel characteristics to maintain exactly the velocity required to transport the material supplied to it.

Gradient The slope of a stream, generally expressed as the vertical drop over a fixed distance.

Granitic composition A compositional group of igneous rocks indicating the rock is composed almost entirely of light-colored silicates.

Gravitational collapse The gradual subsidence of mountains caused by lateral spreading of weak material located deep within these structures.

Greenhouse effect The transmission of shortwave solar radiation by the atmosphere coupled with the selective absorption of longer-wavelength terrestrial radiation, especially by water vapor and carbon dioxide, resulting in warming of the atmosphere.

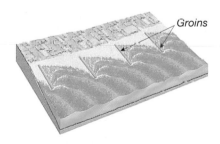

Groins

Groin A short wall built at a right angle to the seashore to trap moving sand.

Groundmass The matrix of smaller crystals within an igneous rock that has porphyritic texture.

Ground moraine An undulating layer of till deposited as the ice front retreats.

Groundwater Water in the zone of saturation.

Guyot A submerged, flat-topped seamount.

Habit Refers to the common or characteristic shape of a crystal, or aggregate of crystals.

Half graben A tilted fault block in which the higher side is associated with mountainous topography and the lower side is a basin that fills with sediment.

Half-life The time required for one-half of the atoms of a radioactive substance to decay.

Hanging valley A tributary valley that enters a glacial trough at a considerable height above the floor of the trough.

Hardness A mineral's resistance to scratching and abrasion.

Head (stream) The beginning or source area for a stream. Also called the headwaters.

Headward erosion The extension upslope of the head of a valley due to erosion.

Historical geology A major division of geology that deals with the origin of Earth and its development through time. Usually involves the study of fossils and their sequence in rock beds.

Hogback A narrow, sharp-crested ridge formed by the upturned edge of a steeply dipping bed of resistant rock.

Horizon A layer in a soil profile.

Horn A pyramid-like peak formed by glacial action in three or more cirques surrounding a mountain summit.

Horst

Horst An elongate, uplifted block of crust bounded by faults.

Hot spot A concentration of heat in the mantle, capable of producing magma that in turn extrudes onto Earth's surface. The intraplate volcanism that produced the Hawaiian Islands is one example.

Hot spot tracks Chain of volcanic structures produced as a lithospheric plate moves over a mantle plume.

Hot spring A spring in which the water is 6–9°C (10–15°F) warmer than the mean annual air temperature of its locality.

Humus Organic matter in soil produced by the decomposition of plants and animals.

Hydraulic conductivity A factor relating to groundwater flow; it is a coefficient that takes into account the permeability of the aquifer and the viscosity of the fluid.

Hydraulic gradient The slope of the water table. It is determined by finding the height difference between two points on the water table and dividing by the horizontal distance between the two points.

Hydroelectric power Electricity generated by falling water that is used to drive turbines.

Hydrogenous sediment Seafloor sediment consisting of minerals that crystallize from seawater. An important example is manganese nodules.

Hydrologic cycle The unending circulation of Earth's water supply. The cycle is powered by energy from the Sun and is characterized by continuous exchanges of water among the oceans, the atmosphere, and the continents.

Hydrolysis A chemical weathering process in which minerals are altered by chemically reacting with water and acids.

Hydrosphere The water portion of our planet; one of the traditional subdivisions of Earth's physical environment.

Hydrothermal metamorphism Chemical alterations that occur as hot, ion-rich water circulates through fractures in rock.

Hydrothermal solution The hot, watery solution that escapes from a mass of magma during the latter stages of crystallization. Such solutions may alter the surrounding country rock and are frequently the source of significant ore deposits.

Hypocenter See *Focus* (earthquake).

Hypothesis A tentative explanation that is then tested to determine if it is valid.

Ice cap A mass of glacial ice covering a high upland or plateau and spreading out radially.

Ice-contact deposit An accumulation of stratified drift deposited in contact with a supporting mass of ice.

Ice sheet A very large, thick mass of glacial ice flowing outward in all directions from one or more accumulation centers.

Ice shelf Forming where glacial ice flows into bays, it is a large, relatively flat mass of floating ice that extends seaward from the coast but remains attached to the land along one or more sides.

Igneous rock Rock formed from the crystallization of magma.

Immature soil A soil lacking horizons.

Impact metamorphism Metamorphism that occurs when meteorites strike Earth's surface.

Incised meander Meandering channel that flows in a steep, narrow valley. These features form either when an area is uplifted or when base level drops.

Inclusion A piece of one rock unit contained within another. Inclusions are used in relative dating. The rock mass adjacent to the one containing the inclusion must have been there first in order to provide the fragment.

Index fossil A fossil that is associated with a particular span of geologic time.

Index mineral A mineral that is a good indicator of the metamorphic environment in which it formed. Used to distinguish different zones of regional metamorphism.

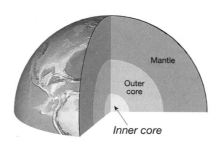

Inertia Objects at rest tend to remain at rest, and objects in motion tend to stay in motion unless either is acted upon by an outside force.

Infiltration The movement of surface water into rock or soil through cracks and pore spaces.

Infiltration capacity The maximum rate at which soil can absorb water.

Inner core The solid innermost layer of Earth, about 1216 kilometers (754 miles) in radius.

Inner planets The innermost planets of our solar system, which include Mercury, Venus, Earth, and Mars. Also known as the terrestrial planets because of their Earth-like internal structure and composition.

Inselberg An isolated mountain remnant characteristic of the late stage of erosion in a mountainous arid region.

Intensity (earthquake) A measure of the degree of earthquake shaking at a given locale based on the amount of damage.

Interface A common boundary where different parts of a system interact.

Interior drainage A discontinuous pattern of intermittent streams that do not flow to the ocean.

Intermediate composition A compositional group of igneous rocks, indicating that the rock contains at least 25 percent dark silicate minerals. The other dominant mineral is plagioclase feldspar.

Internal process A process such as mountain building or volcanism that derives its energy from Earth's interior and elevates Earth's surface.

Intraplate volcanism Igneous activity that occurs within a tectonic plate away from plate boundaries.

Intrusive rock Igneous rock that formed below Earth's surface.

Ion An atom or molecule that possesses an electrical charge.

Ionic bond A chemical bond between two oppositely charged ions formed by the transfer of valence electrons from one atom to the other.

Volcanic island arc

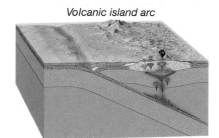

Iron meteorite One of the three main categories of meteorites. This group is composed largely of iron with varying amounts of nickel (5–20 percent). Most meteorite finds are irons.

Island arc See *Volcanic island arc*.

Isostasy The concept that Earth's crust is "floating" in gravitational balance upon the material of the mantle.

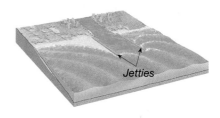

Jetties

Isotatic adjustment Compensation of the lithosphere when weight is added or removed. When weight is added, the lithosphere will respond by subsiding, and when weight is removed, there will be uplift.

Isotopes Varieties of the same element that have different mass numbers; their nuclei contain the same number of protons but different numbers of neutrons.

Jetties A pair of structures extending into the ocean at the entrance to a harbor or river that are built for the purpose of protecting against storm waves and sediment deposition.

Joint A fracture in rock along which there has been no movement.

Jovian planet One of the Jupiter-like planets, Jupiter, Saturn, Uranus, and

Neptune. These planets have relatively low densities.

Kame A steep-sided hill composed of sand and gravel, originating when sediment collected in openings in stagnant glacial ice.

Kame terrace A narrow, terracelike mass of stratified drift deposited between a glacier and an adjacent valley wall.

Karst A type of topography formed on soluble rock (especially limestone) primarily by dissolution. It is characterized by sinkholes, caves, and underground drainage.

Kettle holes Depressions created when blocks of ice become lodged in glacial deposits and subsequently melt.

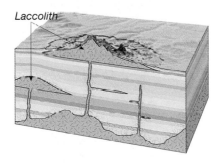

Laccolith

Klippe A remnant or outlier of a thrust sheet that was isolated by erosion.

Kuiper belt A region outside the orbit of Neptune where most short-period comets are thought to originate.

Laccolith A massive igneous body intruded between preexisting strata.

Lag time The amount of time between a rainstorm and the occurrence of flooding.

Lahar Debris flows on the slopes of volcanoes that result when unstable layers of ash and debris become saturated and flow downslope, usually following stream channels.

Laminar flow The movement of water particles in straight-line paths that are parallel to the channel. The water particles move downstream without mixing.

Lateral moraine A ridge of till along the sides of a valley glacier composed primarily of debris that fell to the glacier from the valley walls.

Laterite A red, highly leached soil type found in the tropics that is rich in oxides of iron and aluminum.

Lava dome

Laurasia The northern portion of Pangaea, consisting of North America and Eurasia.

Lava Magma that reaches Earth's surface.

Lava dome A bulbous mass associated with an old-age volcano, produced when thick lava is slowly squeezed from the vent. Lava domes may act as plugs to deflect subsequent gaseous eruptions.

Lava tube Tunnel in hardened lava that acts as a horizontal conduit for lava flowing from a volcanic vent. Lava tubes allow fluid lavas to advance great distances.

Law A formal statement of the regular manner in which a natural phenomenon occurs under given conditions, (e.g., the "law of superposition").

Law of Constancy of Interfacial Angles A law stating that the angle between equivalent faces of the same mineral are always the same.

Law of Superposition In any undeformed sequence of sedimentary rocks, each bed is older than the one above and younger than the one below.

Leaching The depletion of soluble materials from the upper soil by downward-percolating water.

Light silicate Silicate minerals that lack iron and/or magnesium. They are generally lighter in color and have lower specific gravities than dark silicates.

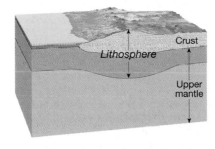

Liquefaction The transformation of a stable soil into a fluid that is often unable to support buildings or other structures.

Lithification The process, generally cementation and/or compaction, of converting sediments to solid rock.

Lithosphere The rigid outer layer of Earth, including the crust and upper mantle.

Lithospheric plate A coherent unit of Earth's rigid outer layer that includes the crust and upper unit.

Local base level See *Temporary base level.*

Loess Deposits of windblown silt, lacking visible layers, generally buff-colored, and capable of maintaining a nearly vertical cliff.

Longitudinal dunes Long ridges of sand oriented parallel to the prevailing wind; these dunes form where sand supplies are limited (see art in next column).

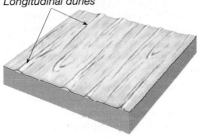

Longitudinal dunes

Longitudinal profile A cross section of a stream channel along its descending course from the head to the mouth.

Longshore current A nearshore current that flows parallel to the shore.

Long (L) waves These earthquake-generated waves travel along the outer layer of Earth and are responsible for most of the surface damage. L waves have longer periods than other seismic waves.

Losing stream Streams that lose water to the groundwater system by outflow through the streambed.

Lower mantle See *Mesosphere.*

Low-velocity zone A subdivision of the mantle located between 100 and 250 kilometers and discernible by a marked decrease in the velocity of seismic waves. This zone does not encircle Earth.

Lunar breccia A lunar rock formed when angular fragments and dust are welded together by the heat generated by the impact of a meteoroid.

Lunar regolith A thin, gray layer on the surface of the Moon, consisting of loosely compacted, fragmented material believed to have been formed by repeated meteoritic impacts.

Luster The appearance or quality of light reflected from the surface of a mineral.

Mafic composition See *Basaltic composition.*

Magma A body of molten rock found at depth, including any dissolved gases and crystals.

Magma mixing The process of altering the composition of a magma through the mixing of material from another magma body.

Magmatic differentiation The process of generating more than one rock type from a single magma.

Magnetometer A sensitive instrument used to measure the intensity of Earth's magnetic field at various points.

Magnitude (earthquake) An estimate of the total amount of energy released during an earthquake, based on seismic records.

Manganese nodules A type of hydrogenous sediment scattered on the ocean floor, consisting mainly of manganese

693

and iron and usually containing small amounts of copper, nickel, and cobalt.

Mantle One of Earth's compositional layers. The solid rocky shell that extends from the base of the crust to a depth of 2900 kilometers.

Mantle plume A mass of hotter-than-normal mantle material that ascends toward the surface, where it may lead to igneous activity. These plumes of solid yet mobile material may originate as deep as the core–mantle boundary.

Maria The smooth areas on our Moon's surface that were incorrectly thought to be seas.

Massive An igneous pluton that is not tabular in shape.

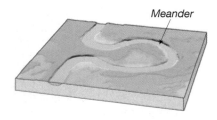
Meander

Mass number The sum of the number of neutrons and protons in the nucleus of an atom.

Mass wasting The downslope movement of rock, regolith, and soil under the direct influence of gravity.

Meander A looplike bend in the course of a stream.

Mechanical weathering The physical disintegration of rock, resulting in smaller fragments.

Medial moraine A ridge of till formed when lateral moraines from two coalescing alpine glaciers join.

Melt The liquid portion of magma excluding the solid crystals.

Mercalli intensity scale See *Modified Mercalli intensity scale.*

Mesosphere The part of the mantle that extends from the core–mantle boundary to a depth of 660 kilometers. Also known as the lower mantle.

Mesozoic era A time span on the geologic time scale between the Paleozoic and Cenozoic eras—from about 248 to 65.5 million years ago.

Metallic bond A chemical bond present in all metals that may be characterized as an extreme type of electron sharing in which the electrons move freely from atom to atom.

Metamorphic facies A group of associated minerals that are used to establish the pressures and temperatures at which rocks undergo metamorphism.

Metamorphic rock Rock formed by the alteration of preexisting rock deep within Earth (but still in the solid state) by heat, pressure, and/or chemically active fluids.

Metamorphism The changes in mineral composition and texture of a rock subjected to high temperatures and pressures within Earth.

Meteor The luminous phenomenon observed when a meteoroid enters Earth's atmosphere and burns up; popularly called a "shooting star."

Meteorite Any portion of a meteoroid that survives its traverse through Earth's atmosphere and strikes the surface.

Meteoroid Any small, solid particle that has an orbit in the solar system.

Meteor shower Numerous meteoroids traveling in the same direction and at nearly the same speed. They are thought to be material lost by comets.

Micrometeorite A very small meteorite that does not create sufficient friction to burn up in the atmosphere but slowly drifts down to Earth.

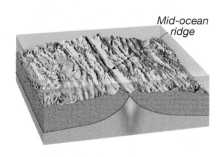
Mid-ocean ridge

Microcontinents Relatively small fragments of continental crust that may lie above sea level, such as the island of Madagascar, or be submerged, as exemplified by the Campbell Plateau located near New Zealand.

Mid-ocean ridge A continuous mountainous ridge on the floor of all the major ocean basins and varying in width from 500 to 5000 kilometers (300 to 3000 miles). The rifts at the crests of these ridges represent divergent plate boundaries.

Migmatite A rock exhibiting both igneous and metamorphic rock characteristics. Such rocks may form when light-colored silicate minerals melt and then crystallize, while the dark silicate minerals remain solid.

Mineral A naturally occurring, inorganic crystalline material with a unique chemical structure.

Mineralogy The study of minerals.

Mineral resource All discovered and undiscovered deposits of a useful mineral that can be extracted now or at some time in the future.

Model A term often used synonymously with hypothesis but is less precise because it is sometimes used to describe a theory as well.

Modified Mercalli intensity scale A 12-point scale developed to evaluate earthquake intensity based on the amount of damage to various structures.

Mohorovičić discontinuity (Moho) The boundary separating the crust and the mantle, discernible by an increase in seismic velocity.

Mohs scale A series of 10 minerals used as a standard in determining hardness.

Moment magnitude A more precise measure of earthquake magnitude than the Richter scale that is derived from the amount of displacement that occurs along a fault zone.

Monocline A one-limbed flexure in strata. The strata are usually flat-lying or very gently dipping on both sides of the monocline.

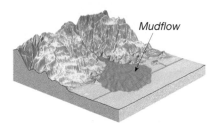
Mudflow

Mouth The point downstream where a river empties into another stream or water body.

Mud crack A feature in some sedimentary rocks that forms when wet mud dries out, shrinks, and cracks.

Mudflow See *Debris flow.*

Natural levee An elevated landform composed of alluvium that parallels some streams and acts to confine their waters, except during floodstage.

Neap tide The lowest tidal range, occurring near the times of the first and third quarters of the Moon.

Nearshore The zone of a beach that extends from the low-tide shoreline seaward to where waves break at low tide.

Nebular hypothesis A model for the origin of the solar system that supposes a rotating nebula of dust and gases that contracted to form the Sun and planets.

Negative feedback mechanism As used in climatic change, any effect that is opposite of the initial change and tends to offset it.

Neutron A subatomic particle found in the nucleus of an atom. The neutron is electrically neutral, with a mass approximately equal to that of a proton.

Nonclastic texture A term for the texture of sedimentary rocks in which the minerals form a pattern of interlocking crystals.

Nonconformity An unconformity in which older metamorphic or intrusive igneous rocks are overlain by younger sedimentary strata.

Nonferromagnesian silicate See *Light silicate.*

Nonflowing artesian well An artesian well in which water does not rise to the surface, because the pressure surface is below ground level.

Nonfoliated Metamorphic rocks that do not exhibit foliation.

Normal fault

Nonmetallic mineral resource A mineral resource that is not a fuel or processed for the metals it contains.

Nonrenewable resource A resource that forms or accumulates over such long time spans that it must be considered as fixed in total quantity.

Normal fault A fault in which the rock above the fault plane has moved down relative to the rock below.

Normal polarity A magnetic field the same as that which presently exists.

Nuclear fission The splitting of atomic nuclei into smaller nuclei, causing neutrons to be emitted and heat energy to be released.

Nucleus The small, heavy core of an atom that contains all of its positive charge and most of its mass.

Nuée ardente Incandescent volcanic debris buoyed up by hot gases that moves downslope in an avalanche fashion.

Numerical date The number of years that have passed since an event occurred.

Occultation The disappearance of light resulting when one object passes behind an apparently larger one. For example, the passage of Uranus in front of a distant star.

Oceanic ridge See *Mid-ocean ridge.*

Oceanic plateau An extensive region on the ocean floor composed of thick accumulations of pillow basalts and other mafic rocks that in some cases exceed 30 kilometers in thickness.

Octet rule Atoms combine in order that each may have the electron arrangement

of a noble gas; that is, the outer energy level contains eight neutrons.

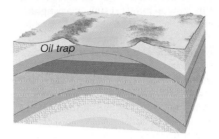

Oil trap

Offshore The relatively flat submerged zone that extends from the breaker line to the edge of the continental shelf.

Oil trap A geologic structure that allows for significant amounts of oil and gas to accumulate.

Oort cloud A spherical shell composed of comets that orbit the Sun at distances generally greater than 10,000 times the Earth–Sun distance.

Open system A system in which both matter and energy flow into and out of the system. Most natural systems are of this type.

Ophiolite complex The sequence of rocks that make up the oceanic crust. The three-layer sequence includes an upper layer of pillow basalts, a middle zone of sheeted dikes, and a lower layer of gabbro.

Ore Usually a useful metallic mineral that can be mined at a profit. The term is also applied to certain nonmetallic minerals such as fluorite and sulfur.

Organic sedimentary rock Sedimentary rock composed of organic carbon from the remains of plants that died and accumulated on the floor of a swamp. Coal is the primary example.

Original horizontality Layers of sediment that are generally deposited in a horizontal or nearly horizontal position.

Orogenesis The processes that collectively result in the formation of mountains.

Outer core A layer beneath the mantle about 2270 kilometers (1410 miles) thick, which has the properties of a liquid.

Outer planets The outermost planets of our solar system, which include Jupiter, Saturn, Uranus, Neptune, and Pluto. Except for Pluto these bodies are known as the Jovian planets.

Outgassing The escape of dissolved gases from molten rocks.

Outlet glacier A tongue of ice normally flowing rapidly outward from an ice cap or ice sheet, usually through mountainous terrain to the sea.

Outwash plain A relatively flat, gently sloping plain consisting of materials

deposited by meltwater streams in front of the margin of an ice sheet.

Oxbow lake A curved lake produced when a stream cuts off a meander.

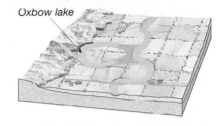

Oxbow lake

Oxidation The removal of one or more electrons from an atom or ion. So named because elements commonly combine with oxygen.

Oxygen isotope analysis A method of deciphering past temperatures based on the precise measurement of the ratio between two isotopes of oxygen, ^{16}O and ^{18}O. Analysis is commonly made of seafloor sediments and cores from ice sheets.

Pahoehoe flow A lava flow with a smooth-to-ropy surface.

Paleoclimatology The study of ancient climates; the study of climate and climate change prior to the period of instrumental records using proxy data.

Paleomagnetism The natural remnant magnetism in rock bodies. The permanent magnetization acquired by rock that can be used to determine the location of the magnetic poles and the latitude of the rock at the time it became magnetized.

Paleontology The systematic study of fossils and the history of life on Earth.

Paleozoic era A time span on the geologic time scale between the Precambrian and Mesozoic eras—from about 542 million to 251 million years ago.

Pangaea The proposed supercontinent that 200 million years ago began to break apart and form the present landmasses.

Parabolic dune A sand dune similar in shape to a barchan dune except that its tips point into the wind. These dunes often form along coasts that have strong onshore winds, abundant sand, and vegetation that partly covers the sand. (See art on next page)

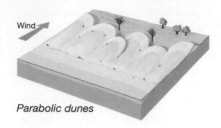

Parabolic dunes

Paradigm Theory that is held with a very high degree of confidence and is comprehensive in scope.

Parasitic cone A volcanic cone that forms on the flank of a larger volcano.

Parent material The material upon which a soil develops.

Parent rock The rock from which a metamorphic rock formed.

Partial melting The process by which most igneous rocks melt. Since individual minerals have different melting points, most igneous rocks melt over a temperature range of a few hundred degrees. If the liquid is squeezed out after some melting has occurred, a melt with a higher silica content results.

Passive continental margin A margin that consists of a continental shelf, continental slope, and continental rise. They are not associated with plate boundaries and therefore experience little volcanism and few earthquakes.

Pater noster lakes A chain of small lakes in a glacial trough that occupies basins created by glacial erosion.

Pegmatite A very coarse-grained igneous rock (typically granite) commonly found as a dike associated with a large mass of plutonic rock that has smaller crystals. Crystallization in a water-rich environment is believed to be responsible for the very large crystals.

Pegmatitic texture A texture of igneous rocks in which the interlocking crystals are all larger than one centimeter in diameter.

Perched water table A localized zone of saturation above the main water table, created by an impermeable layer (aquiclude).

Peridotite An igneous rock of ultramafic composition thought to be abundant in the upper mantle.

Period A basic unit of the geologic time scale that is a subdivision of an era. Periods may be divided into smaller units called epochs.

Periodic table An arrangement of the elements in which atomic number increases from the left to right and elements with similar properties appear in columns called families or groups.

Permafrost Any permanently frozen subsoil. Usually found in the subarctic and arctic regions.

Permeability A measure of a material's ability to transmit water.

Phaneritic texture An igneous rock texture in which the crystals are roughly equal in size and large enough so the individual minerals can be identified without the aid of a microscope.

Phanerozoic eon That part of geologic time represented by rocks containing abundant fossil evidence. The eon extending from the end of the Proterozoic eon (540 million years ago) to the present.

Phenocryst Conspicuously large crystal embedded in a matrix of finer-grained crystals.

Physical geology A major division of geology that examines the materials of Earth and seeks to understand the processes and forces acting beneath and upon Earth's surface.

Piedmont glacier A glacier that forms when one or more alpine glaciers emerge from the confining walls of mountain valleys and spread out to create a broad sheet in the lowlands at the base of the mountains.

Pillow basalts Basaltic lava that solidifies in an underwater environment and develops a structure that resembles a pile of pillows.

Pipe A vertical conduit through which magmatic materials have passed.

Placer Deposit formed when heavy minerals are mechanically concentrated by currents, most commonly streams and waves. Placers are sources of gold, tin, platinum, diamonds, and other valuable minerals.

Planetesimal A solid celestial body that accumulated during the first stages of planetary formation. Planetesimals aggregated into increasingly larger bodies, ultimately forming the planets.

Plastic flow A type of glacial movement that occurs within the glacier, below a depth of approximately 50 meters, in which the ice is not fractured.

Plate See *Lithospheric plate.*

Plate resistance A force that counteracts plate motion as a subducting plate scrapes against an overriding plate.

Plate tectonics The theory that proposes that Earth's outer shell consists of individual plates that interact in various ways and thereby produce earthquakes, volcanoes, mountains, and the crust itself.

Playa The flat central area of an undrained desert basin.

Playa lake A temporary lake in a playa.

Pleistocene epoch An epoch of the Quaternary period beginning about 1.8 million years ago and ending about 10,000 years ago. Best known as a time of extensive continental glaciation.

Plucking The process by which pieces of bedrock are lifted out of place by a glacier.

Pluton A structure that results from the emplacement and crystallization of magma beneath the surface of Earth.

Plutonic rock Igneous rocks that form at depth. After Pluto, the god of the lower world in classical mythology.

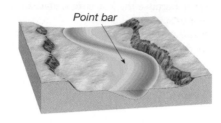

Point bar

Pluvial lake A lake formed during a period of increased rainfall. For example, this occurred in many nonglaciated areas during periods of ice advance elsewhere.

Point bar A crescent-shaped accumulation of sand and gravel deposited on the inside of a meander.

Polymorphs Two or more minerals having the same chemical composition but different crystalline structures. Exemplified by the diamond and graphite forms of carbon.

Porosity The volume of open spaces in rock or soil.

Porphyritic texture An igneous rock texture characterized by two distinctively different crystal sizes. The larger crystals are called phenocrysts, whereas the matrix of smaller crystals is termed the groundmass.

Porphyroblastic texture A texture of metamorphic rocks in which particularly large grains (porphyroblasts) are surrounded by a fine-grained matrix of other minerals.

Porphyry An igneous rock with a porphyritic texture.

Positive feedback mechanism As used in climatic change, any effect that acts to reinforce the initial change.

Pothole A depression formed in a stream channel by the abrasive action of the water's sediment load.

Precambrian All geologic time prior to the Phanerozoic eon. A term encompassing both the Archean and Proterozoic eons.

Primary (P) wave A type of seismic wave that involves alternating compression and expansion of the material through which it passes.

Principal shell The shell or energy level an electron occupies.

Principle of faunal succession Fossil organisms succeed one another in a definite and determinable order, and any time period can be recognized by its fossil content.

Principle of original horizontality Layers of sediment are generally deposited in a horizontal or nearly horizontal position.

Prokaryotes Refers to the cells or organisms such as bacteria whose genetic material is not enclosed in a nucleus.

Proterozoic eon The eon following the Archean and preceding the Phanerozoic. It extends between 2500 and 542 million years ago.

Proton A positively charged subatomic particle found in the nucleus of an atom.

Protoplanets A developing planetary body that grows by the accumulation of planetesimals.

Proxy data Data gathered from natural recorders of climate variability such as tree rings, ice cores, and ocean-floor sediments.

Pumice A light-colored glassy vesicular rock commonly having a granitic composition.

P wave The fastest earthquake wave, which travels by compression and expansion of the medium.

Pyroclastic texture An igneous rock texture resulting from the consolidation of individual rock fragments that are ejected during a violent volcanic eruption.

Pyroclastic flow A highly heated mixture, largely of ash and pumice fragments, traveling down the flanks of a volcano or along the surface of the ground.

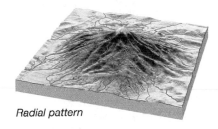

Radial pattern

Pyroclastic material The volcanic rock ejected during an eruption. Pyroclastics include ash, bombs, and blocks.

Radial pattern A system of streams running in all directions away from a central elevated structure, such as a volcano.

Radioactive decay The spontaneous decay of certain unstable atomic nuclei.

Radioactivity See *Radioactive decay*.

Radiocarbon (carbon-14) dating The radioactive isotope of carbon is produced continuously in the atmosphere and used in dating events from the very recent geologic past (the last few tens of thousands of years).

Radiometric dating The procedure of calculating the absolute ages of rocks and minerals that contain certain radioactive isotopes.

Rainshadow desert

Rainshadow desert A dry area on the lee side of a mountain range. Many middle-latitude deserts are of this type.

Rapids A part of a stream channel in which the water suddenly begins flowing more swiftly and turbulently because of an abrupt steepening of the gradient.

Rays Bright streaks that appear to radiate from certain craters on the lunar surface. The rays consist of fine debris ejected from the primary crater.

Recessional moraine An end moraine formed as the ice front stagnated during glacial retreat.

Rectangular pattern A drainage pattern characterized by numerous right angle bends that develops on jointed or fractured bedrock.

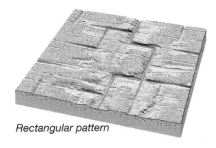

Rectangular pattern

Recurrence interval The average time interval between occurrences of hydrological events such as floods of a given or greater magnitude.

Refraction See *Wave refraction*.

Regional metamorphism Metamorphism associated with large-scale mountain building.

Regolith The layer of rock and mineral fragments that nearly everywhere covers Earth's land surface.

Rejuvenation A change in relation to base level, often caused by regional uplift, which causes the forces of erosion to intensify.

Relative dating Rocks and structures are placed in their proper sequence or order. Only the chronological order of events is determined.

Renewable resource A resource that is virtually inexhaustible or that can be replenished over relatively short time spans.

Reserve Already identified deposits from which minerals can be extracted profitably.

Reservoir rock The porous, permeable portion of an oil trap that yields oil and gas.

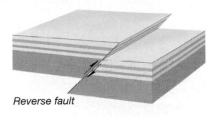

Reverse fault

Residual soil Soil developed directly from the weathering of the bedrock below.

Return period See *Recurrence interval*.

Reverse fault A fault in which the material above the fault plane moves up in relation to the material below.

Reverse polarity A magnetic field opposite to that which presently exists.

Richter scale A scale of earthquake magnitude based on the amplitude of the largest seismic wave.

Ridge push A mechanism that may contribute to plate motion. It involves the oceanic lithosphere sliding down the oceanic ridge under the pull of gravity.

Rift valley A long, narrow trough bounded by normal faults. It represents a region where divergence is taking place.

Rills Tiny channels that develop as unconfined flow begins producing threads of current.

Ripple marks Small waves of sand that develop on the surface of a sediment layer by the action of moving water or air.

Roche moutonnée An asymmetrical knob of bedrock formed when glacial abrasion smoothes the gentle slope facing the advancing ice sheet and plucking steepens the opposite side as the ice overrides the knob.

Rock A consolidated mixture of minerals.

Rock avalanche The very rapid downslope movement of rock and debris. These rapid movements may be aided by a layer of air trapped beneath the debris, and they have been known to reach speeds in excess of 200 kilometers per hour.

Rock cleavage The tendency of rocks to split along parallel, closely spaced surfaces. These surfaces are often highly inclined to the bedding planes in the rock.

Rock cycle A model that illustrates the origin of the three basic rock types and the interrelatedness of Earth materials and processes.

Rock flour Ground-up rock produced by the grinding effect of a glacier.

Rockslide The rapid slide of a mass of rock downslope along planes of weakness.

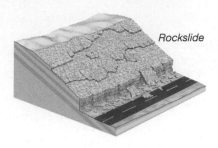

Rockslide

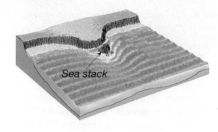

Sea stack

Rock structure All features created by the processes of deformation from minor fractures in bedrock to a major mountain chain.

Runoff Water that flows over the land rather than infiltrating into the ground.

Salinity The proportion of dissolved salts to pure water, usually expressed in parts per thousand (0/000).

Saltation Transportation of sediment through a series of leaps or bounces.

Salt flat A white crust on the ground produced when water evaporates and leaves its dissolved materials behind.

Schistosity A type of foliation characteristic of coarser-grained metamorphic rocks. Such rocks have a parallel arrangement of platy minerals such as the micas.

Scoria Vesicular ejecta that is the product of basaltic magma.

Scoria cone See *Cinder cone.*

Sea arch An arch formed by wave erosion when caves on opposite sides of a headland unite.

Seafloor spreading The hypothesis first proposed in the 1960s by Harry Hess, which suggested that new oceanic crust is produced at the crests of mid-ocean ridges, which are the sites of divergence.

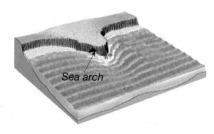

Sea arch

Seamount An isolated volcanic peak that rises at least 1000 meters (3300 feet) above the deep-ocean floor.

Sea stack An isolated mass of rock standing just offshore, produced by wave erosion of a headland.

Seawall A barrier constructed to prevent waves from reaching the area behind the wall. Its purpose is to defend property from the force of breaking waves.

Secondary enrichment The concentration of minor amounts of metals that are scattered through unweathered rock into economically valuable concentrations by weathering processes.

Secondary (S) wave A seismic wave that involves oscillation perpendicular to the direction of propagation.

Sediment Unconsolidated particles created by the weathering and erosion of rock, by chemical precipitation from solution in water, or from the secretions of organisms, and transported by water, wind, or glaciers.

Sedimentary environment See *Environment of deposition.*

Sedimentary rock Rock formed from the weathered products of preexisting rocks that have been transported, deposited, and lithified.

Seismic gap A segment of an active fault zone that has not experienced a major earthquake over a span when most other segments have. Such segments are probable sites for future major earthquakes.

Seismic reflection profile A method of viewing the rock structure beneath a blanket of sediment by using strong, low-frequency sound waves that penetrate the sediments and reflect off the contacts between rock layers and fault zones.

Seismic sea wave A rapidly moving ocean wave generated by earthquake activity, which is capable of inflicting heavy damage in coastal regions.

Seismogram The record made by a seismograph.

Seismograph An instrument that records earthquake waves.

Seismology The study of earthquakes and seismic waves.

Settling velocity The speed at which a particle falls through a still fluid. The size, shape, and specific gravity of particles influence settling velocity.

Shadow zone The zone between 105 and 140 degrees' distance from an earthquake epicenter, which direct waves do not penetrate because of refraction by Earth's core.

Shear Stress that causes two adjacent parts of a body to slide past one another.

Sheeted dike complex A large group of nearly parallel dikes.

Sheet flow Runoff moving in unconfined thin sheets.

Sheeting A mechanical weathering process characterized by the splitting off of slablike sheets of rock.

Shelf break The point at which a rapid steepening of the gradient occurs, marking the outer edge of the continental shelf and the beginning of the continental slope.

Shield A large, relatively flat expanse of ancient igneous and metamorphic rocks within the craton.

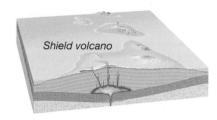

Shield volcano

Shield volcano A broad, gently sloping volcano built from fluid basaltic lavas.

Shore Seaward of the coast, this zone extends from the highest level of wave action during storms to the lowest tide level.

Shoreline The line that marks the contact between land and sea. It migrates up and down as the tide rises and falls.

Silicate mineral Any one of numerous minerals that have the silicon-oxygen tetrahedron as their basic structure.

Silicon-oxygen tetrahedron A structure composed of four oxygen atoms surrounding a silicon atom that constitutes the basic building block of silicate minerals.

Sill A tabular igneous body that was intruded parallel to the layering of preexisting rock.

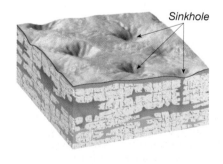

Sinkhole

Sinkhole A depression produced in a region where soluble rock has been removed by groundwater.

Slab-pull A mechanism that contributes to plate motion in which cool, dense oceanic crust sinks into the mantle and "pulls" the trailing lithosphere along.

Slab suction One of the driving forces of plate motion, it arises from the drag of the subducting plate on the adjacent mantle. It is an induced mantle circulation that pulls both the subducting and overriding plates toward the trench.

Slaty cleavage The type of foliation characteristic of slates in which there is a parallel arrangement of fine-grained metamorphic minerals.

Slide A movement common to mass-wasting processes in which the material moving downslope remains fairly coherent and moves along a well-defined surface.

Slip face The steep, leeward surface of a sand dune that maintains a slope of about 34 degrees.

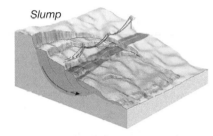

Slump

Slump The downward slipping of a mass of rock or unconsolidated material moving as a unit along a curved surface.

Snowfield An area where snow persists throughout the year.

Snowline The lower limit of perennial snow.

Soil A combination of mineral and organic matter, water, and air; that portion of the regolith that supports plant growth.

Soil horizon A layer of soil that has identifiable characteristics produced by chemical weathering and other soil-forming processes.

Soil profile A vertical section through a soil, showing its succession of horizons and the underlying parent material.

Soil taxonomy A soil classification system consisting of six hierarchical categories based on observable soil characteristics. The system recognizes 12 soil orders.

Solar nebula The cloud of interstellar gas and/or dust from which the bodies of our solar system formed.

Solifluction Slow, downslope flow of water-saturated materials common to permafrost areas.

Solum The *O, A,* and *B* horizons in a soil profile. Living roots and other plant and animal life are largely confined to this zone.

Sonar Instrument that uses acoustic signals (sound energy) to measure water depths. Sonar is an acronym for *so*und *na*vigation and *r*anging.

Sorting The degree of similarity in particle size in sediment or sedimentary rock.

Specific gravity The ratio of a substance's weight to the weight of an equal volume of water.

Speleothem A collective term for the dripstone features found in caverns.

Spheroidal weathering Any weathering process that tends to produce a spherical shape from an initially blocky shape.

Spit An elongate ridge of sand that projects from the land into the mouth of an adjacent bay.

Spreading center See *Divergent plate boundary.*

Spring A flow of groundwater that emerges naturally at the ground surface.

Spring tide The highest tidal range. Occurs near the times of the new and full moons.

Stable platform That part of the craton that is mantled by relatively undeformed sedimentary rocks and underlain by a basement complex of igneous and metamorphic rocks.

Stalactite The icicle-like structure that hangs from the ceiling of a cavern.

Stalagmite The columnlike form that grows upward from the floor of a cavern.

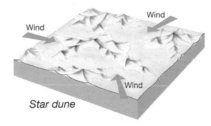

Star dune

Star dune An isolated hill of sand that exhibits a complex form and develops where wind directions are variable.

Steno's Law See *Law of Constancy of Interfacial Angles.*

Steppe One of the two types of dry climate. A marginal and more humid variant of the desert that separates it from bordering humid climates.

Stock A pluton similar to but smaller than a batholith.

Stony-iron meteorite One of the three main categories of meteorites. This group, as the name implies, is a mixture of iron and silicate minerals.

Stony meteorite One of the three main categories of meteorites. Such meteorites are composed largely of silicate minerals with inclusions of other minerals.

Storm surge The abnormal rise of the sea along a shore as a result of strong winds.

Strain An irreversible change in the shape and size of a rock body caused by stress.

Strata Parallel layers of sedimentary rock.

Stratified drift Sediments deposited by glacial meltwater.

Stratovolcano See *Composite cone.*

Streak The color of a mineral in powdered form.

Stream A general term to denote the flow of water within any natural channel. Thus, a small creek and a large river are both streams.

Stream piracy The diversion of the drainage of one stream, resulting from the headward erosion of another stream.

Stream valley The channel, valley floor, and sloping valley walls of a stream.

Stress The force per unit area acting on any surface within a solid.

Striations (glacial) Scratches or grooves in a bedrock surface caused by the grinding action of a glacier and its load of sediment.

Strike The compass direction of the line of intersection created by a dipping bed or fault and a horizontal surface. Strike is always perpendicular to the direction of dip.

Strike-slip fault A fault along which the movement is horizontal.

Stromatolites Distinctively layered mounds of calcium carbonate, which are fossil evidence for the existence of ancient microscopic bacteria.

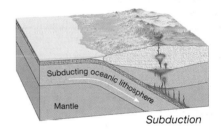

Subduction

Subduction The process by which oceanic lithosphere plunges into the mantle along a convergent zone.

Subduction zone A long, narrow zone where one lithospheric plate descends beneath another.

Subduction zone metamorphism High pressure, low temperature metamorphism that occurs where sediments are carried to great depths by a subducting plate.

Submarine canyon A seaward extension of a valley that was cut on the continental shelf during a time when sea level was lower, or a canyon carved into the outer continental shelf, slope, and rise by turbidity currents.

Submergent coast A coast whose form is largely the result of the partial drowning of a former land surface due to a rise of sea level or subsidence of the crust, or both.

Subsoil A term applied to the *B* horizon of a soil profile.

Sunspot A dark area on the Sun associated with powerful magnetic storms that extend from the Sun's surface deep into the interior.

Supercontinent A large landmass that contains all, or nearly all, of the existing continents.

Supercontinent cycle The idea that the rifting and dispersal of one supercontinent is followed by a long period during which the fragments gradually reassemble into a new supercontinent.

Supernova An exploding star that increases its brightness many thousands of times.

Superposed stream A stream that cuts through a ridge lying across its path. The stream established its course on uniform layers at a higher level without regard to underlying structures and subsequently downcut.

Superposition, law of In any undeformed sequence of sedimentary rocks, each bed is older than the one above and younger than the one below.

Surf A collective term for breakers; also the wave activity in the area between the shoreline and the outer limit of breakers.

Surface waves Seismic waves that travel along the outer layer of Earth.

Surge A period of rapid glacial advance. Surges are typically sporadic and short-lived.

Suspended load The fine sediment carried within the body of flowing water or air.

Suture The zone along which two crustal fragments are jointed together. For example, following a continental collision the two continental blocks are sutured together.

S wave An earthquake wave, slower than a P wave, that travels only in solids.

Swells Wind-generated waves that have moved into an area of weaker winds or calm.

Syncline A linear downfold in sedimentary strata; the opposite of anticline.

System A group of interacting or interdependent parts that form a complex whole.

Tablemount See *Guyot*.

Tabular Describing a feature such as an igneous pluton having two dimensions that are much longer than the third.

Talus An accumulation of rock debris at the base of a cliff.

Tarn A small lake in a cirque.

Tectonic plate See *Lithospheric plate*.

Tectonics The study of the large-scale processes that collectively deform Earth's crust.

Temporary (local) base level The level of a lake, resistant rock layer, or any other base level that stands above sea level.

Tenacity Describes a mineral's toughness or its resistance to breaking or deforming.

Tensional stress The type of stress that tends to pull a body apart.

Terminal moraine The end moraine marking the farthest advance of a glacier.

Terrace A flat, benchlike structure produced by a stream, which was left elevated as the stream cut downward.

Terrane A crustal block bounded by faults, whose geologic history is distinct from the histories of adjoining crustal blocks.

Terrestrial planet One of the Earthlike planets: Mercury, Venus, Earth, and Mars. These planets have similar densities.

Terrigenous sediment Seafloor sediments derived from terrestrial weathering and erosion.

Texture The size, shape, and distribution of the particles that collectively constitute a rock.

Theory A well-tested and widely accepted view that explains certain observable facts.

Thermal metamorphism See *Contact metamorphism*.

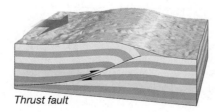
Thrust fault

Thrust fault A low-angle reverse fault.

Tidal current The alternating horizontal movement of water associated with the rise and fall of the tide.

Tidal delta A deltalike feature created when a rapidly moving tidal current emerges from a narrow inlet and slows, depositing its load of sediment.

Tidal flat A marshy or muddy area that is alternately covered and uncovered by the rise and fall of the tide.

Tide Periodic change in the elevation of the ocean surface.

Till Unsorted sediment deposited directly by a glacier.

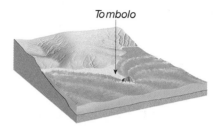

Tillite A rock formed when glacial till is lithified.

Tombolo A ridge of sand that connects an island to the mainland or to another island.

Topset bed An essentially horizontal sedimentary layer deposited on top of a delta during floodstage.

Transform fault A major strike-slip fault that cuts through the lithosphere and accommodates motion between two plates.

Transform fault boundary A boundary in which two plates slide past one another without creating or destroying lithosphere.

Transpiration The release of water vapor to the atmosphere by plants.

Transported soil Soils that form on unconsolidated deposits.

Transverse dunes A series of long ridges oriented at right angles to the prevailing wind; these dunes form where vegetation is sparse and sand is very plentiful.

Trellis drainage

Travertine A form of limestone (CaCO$_3$) that is deposited by hot springs or as a cave deposit.

Trellis drainage pattern A system of streams in which nearly parallel tributaries occupy valleys cut in folded strata.

Trench See *Deep-ocean trench*.

Truncated spurs Triangular-shaped cliffs produced when spurs of land that extend into a valley are removed by the great erosional force of a valley glacier.

Turbidity current

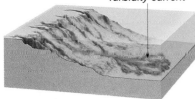

Tsunami The Japanese word for a seismic sea wave.

Turbidite Turbidity current deposit characterized by graded bedding.

Turbidity current A downslope movement of dense, sediment-laden water created when sand and mud on the continental shelf and slope are dislodged and thrown into suspension.

Turbulent flow The movement of water in an erratic fashion often characterized by swirling, whirlpool-like eddies. Most streamflow is of this type.

Ultimate base level Sea level; the lowest level to which stream erosion could lower the land.

Ultramafic composition A compositional group of igneous rocks containing mostly olivine and pyroxene.

Unconformity A surface that represents a break in the rock record, caused by erosion and nondeposition.

Uniformitarianism The concept that the processes that have shaped Earth in the geologic past are essentially the same as those operating today.

Unsaturated zone The area above the water table where openings in soil, sediment, and rock are not saturated but filled mainly with air.

Valence electron The electrons involved in the bonding process; the electrons occupying the highest principal energy level of an atom.

Valley glacier See *Alpine glacier*.

Valley train A relatively narrow body of stratified drift deposited on a valley floor by meltwater streams that issue from the terminus of an alpine glacier.

Vein deposit A mineral filling a fracture or fault in a host rock. Such deposits have a sheetlike, or tabular, form.

Vent The surface opening of a conduit or pipe.

Ventifact A cobble or pebble polished and shaped by the sandblasting effect of wind.

Vesicles Spherical or elongated openings on the outer portion of a lava flow that were created by escaping gases.

Vesicular texture A term applied to aphanitic igneous rocks that contain many small cavities called vesicles.

Viscosity A measure of a fluid's resistance to flow.

Volatiles Gaseous components of magma dissolved in the melt. Volatiles will readily vaporize (form a gas) at surface pressures.

Volcanic Pertaining to the activities, structures, or rock types of a volcano.

Volcanic bomb A streamlined pyroclastic fragment ejected from a volcano while still semimolten.

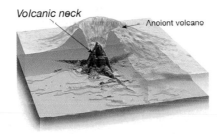

Volcanic neck — *Ancient volcano*

Volcanic island arc A chain of volcanic islands generally located a few hundred kilometers from a trench where there is active subduction of one oceanic plate beneath another.

Volcanic neck An isolated, steep-sided, erosional remnant consisting of lava that once occupied the vent of a volcano.

Volcano A mountain formed from lava and/or pyroclastics.

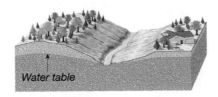

Water table

Wadati-Benioff zone The narrow zone of inclined seismic activity that extends from a trench downward into the asthenosphere.

Waterfall A precipitous drop in a stream channel that causes water to fall to a lower level.

Water gap A pass through a ridge or mountain in which a stream flows.

Water table The upper level of the saturated zone of groundwater.

Wave-cut cliff A seaward-facing cliff along a steep shoreline formed by wave erosion at its base and mass wasting.

Wave-cut platform A bench or shelf along a shore at sea level, cut by wave erosion.

Wave height The vertical distance between the trough and crest of a wave.

Wavelength The horizontal distance separating successive crests or troughs.

Wave of oscillation A water wave in which the wave form advances as the water particles move in circular orbits.

Wave of translation The turbulent advance of water created by breaking waves.

Wave period The time interval between the passage of successive crests at a stationary point.

Wave refraction A change in direction of waves as they enter shallow water. The portion of the wave in shallow water is slowed, which causes the waves to bend and align with the underwater contours.

Weathering The disintegration and decomposition of rock at or near the surface of Earth.

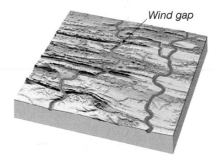

Wind gap

Welded tuff A pyroclastic deposit composed of particles fused together by the combination of heat still contained in the deposit after it has come to rest and the weight of overlying material.

Well An opening bored into the zone of saturation.

Wilson Cycle See *Supercontinent cycle*.

Wind gap An abandoned water gap. These gorges typically result from stream piracy.

Xenolith An inclusion of unmelted country rock in an igneous pluton.

Xerophyte A plant highly tolerant of drought.

701

Yardang A streamlined, wind-sculpted ridge having the appearance of an inverted ship's hull that is oriented parallel to the prevailing wind.

Yazoo tributary A tributary that flows parallel to the main stream because a natural levee is present.

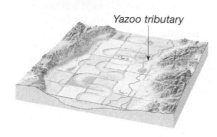

Yazoo tributary

Zone of accumulation The part of a glacier characterized by snow accumulation

and ice formation. The outer limit of this zone is the snowline.

Zone of aeration The area above the water table where openings in soil, sediment, and rock are not saturated but filled mainly with air.

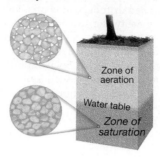

Zone of aeration

Water table

Zone of saturation

Zone of fracture The upper portion of a glacier consisting of brittle ice.

Zone of saturation The zone where all open spaces in sediment and rock are completely filled with water.

Zone of soil moisture A zone in which water is held as a film on the surface of soil particles and may be used by plants or withdrawn by evaporation. The uppermost subdivision of the unsaturated zone.

Zone of wastage The part of a glacier beyond the snowline where annually there is a net loss of ice.

Index